General Anatomy and Musculoskeletal System

THIEME Atlas of Anatomy

Fourth Edition

Thieme

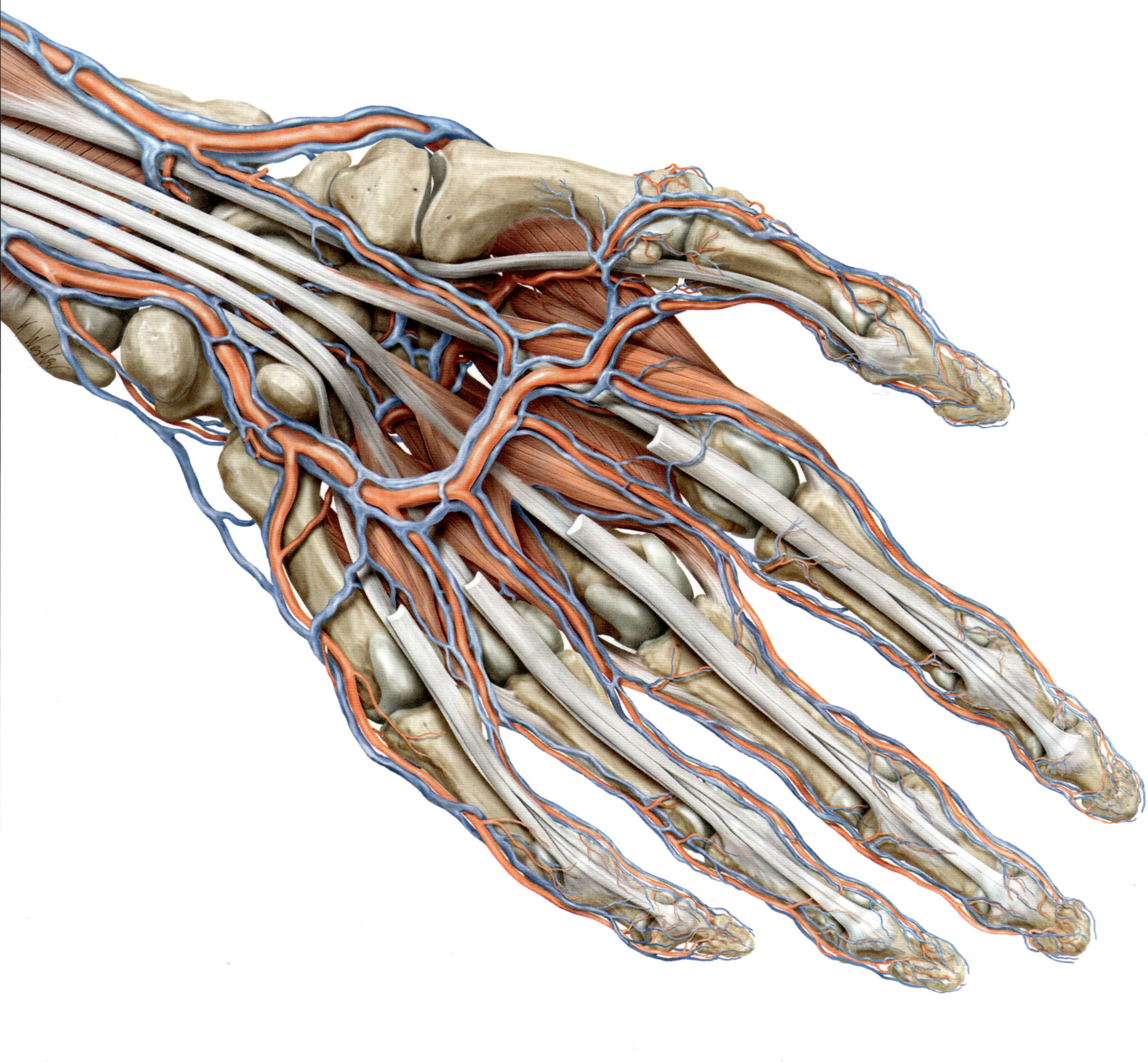

General Anatomy and Musculoskeletal System

THIEME Atlas of Anatomy

Fourth Edition

Authors

Michael Schuenke, MD, PhD
Institute of Anatomy
Christian Albrechts University, Kiel

Erik Schulte, MD
Institute of Functional and Clinical Anatomy
Johannes Gutenberg University, Mainz

Udo Schumacher, MD
FRCPath, CBiol, FIBiol, DSc
Institute of Anatomy and Experimental Morphology University Medical Center, Hamburg-Eppendorf

Consulting Editors

Wayne A. Cass, PhD
Department of Neuroscience
University of Kentucky College of Medicine
Lexington, Kentucky

Nathan F. Johnson, PT, PhD, DPT
Division of Physical Therapy
University of Kentucky College of Health Sciences
Lexington, Kentucky

Illustrations by
Markus Voll
Karl Wesker

2120 Illustrations

Thieme
New York • Stuttgart • Delhi • Rio de Janeiro

Translators: Terry Telger, John Grossman, and Judith Tomat
Illustrators: Markus Voll and Karl Wesker
Compositor: DiTech Process Solutions

Library of Congress Cataloging-in-Publication Data is available from the publisher.

Thieme Medical Publishers, Inc.
Thieme Publishers New York
333 Seventh Avenue, New York, NY 10001 USA
+1 800 782 3488, customerservice@thieme.com

Cover design: © Thieme
Cover art: © Karl Wesker, Berlin

Typesetting: DiTech Process Solutions

Printed in India by Replika Press Pvt. Ltd. 5 4 3 2 1

DOI: 10.1055/b000000889
ISBN: 978-1-68420-589-9
Also available as eBook:
eISBN (PDF) 978-1-68420-598-1
eISBN (epub) 978-1-68420-599-8

MIX
Paper | Supporting responsible forestry
FSC™ C016779

Foreword

Each of the authors of the single volume *Thieme Atlas of Anatomy* was impressed with the extraordinary detail, accuracy, and beauty of the illustrations that were created for the *Thieme* three-volume series of anatomy atlases. We felt these images were one of the most significant additions to anatomic education in the past 50 years. The effective pedagogical approach of this series, with two-page learning units that combined the outstanding illustrations and captions that emphasized the functional and clinical significance of structures, coupled with the numerous tables summarizing key information, was unique. We also felt that the overall organization of each region, with structures presented first systemically—musculoskeletal, vascular, and nervous—and then topographically, supported classroom learning and active dissection in the laboratory.

This series combines the best of a clinically oriented text and an atlas. Its detail and pedagogical presentation make it a complete support for classroom and laboratory instruction and a reference for life in all the medical, dental, and allied health fields. Each of the volumes—*General Anatomy and Musculoskeletal System, Internal Organs,* and *Head, Neck, and Neuroanatomy*—can also be used as a stand-alone text/atlas for an in-depth study of systems often involved in the allied health/medical specialty fields.

We were delighted when *Thieme* asked us to work with them to create a single-volume atlas from this groundbreaking series, and we owe a great debt to the authors and illustrators of this series inasmuch as their materials and vision formed the general framework for the single volume *Thieme Atlas of Anatomy*.

We thank the authors and illustrators for this very special contribution to the teaching of anatomy and recommend it for thorough mastery of anatomy and its clinically functional importance in all fields of health care-related specialties.

Lawrence M. Ross, Brian R. MacPherson, and Anne M. Gilroy

Preface of the Authors and Illustrators

When Thieme started planning the first edition of this atlas, they sought the opinions of students and instructors alike in both the United States and Europe on what constituted an "ideal" atlas of anatomy—ideal to learn from, to master extensive amounts of information while on a busy class schedule, and, in the process, to acquire sound, up-to-date knowledge. The result of our work in response to what Thieme had learned is this atlas. The *Thieme Atlas of Anatomy*, unlike most other atlases, is a comprehensive educational tool that combines illustrations with explanatory text and summary tables, introducing clinical applications throughout, and presenting anatomic concepts in a step-by-step sequence that includes system-by-system and topographical views.

For the first edition we had hoped that our *Atlas of Anatomy* would help the medical student to understand the anatomical basis of clinical medicine. This indeed was accepted by the students all over the world and soon a second edition had to come on the market in Germany, which was extensively extended and revised, followed by further third, fourth, fifth, and sixth editions. More and more information has been added over these editions, including spreads on important foundational information on the common imaging planes for plain film, MRI, and CT scans, the structure of and function of muscle fasciae, the structure and chemical composition of hyaline cartilage, and the regeneration of peripheral nerves, bone marrow, and paraganglia, as well as new graphical summaries in neuroanatomy and new sections describing internal female genitalia and surgical interventions with respect to female genitalia. This atlas takes special care to cover variants in human anatomy such as blood vessels whose courses deviate from the "norm", or anomalous positions of organs. Hence the sixth German edition looks ever more distinctly different from the first one. Of course, we have also checked, corrected, and updated all of the information in this atlas.

We are grateful to the American branch of Thieme that they have made this fourth English edition possible. We hope that this updated version will serve the medical students and practitioners of medicine alike in helping them to understand human morphology which is indispensable for diagnosis and therapy.

Michael Schünke, Erik Schulte, Udo Schumacher,
Markus Voll, and Karl Wesker

Acknowledgments

First we wish to thank our families. This atlas is dedicated to them.

Since the publication of the first volume of the *Thieme Atlas of Anatomy* in 2006, we have received numerous suggestions for refinements and additions. We would like to take this opportunity to express our sincere thanks to all those who through the years have helped us to improve the *Thieme Atlas of Anatomy* in one way or another. Specifically, this includes Kirsten Hattermann, Ph.D.; Runhild Lucius, D.D.S.; Prof. Renate Lüllmann-Rauch, M.D.; Prof. Jobst Sievers, M.D.; Ali Therany, D.D.S.; Prof. Thilo Wedel, M.D. (all at the Anatomic Institute of Christian Albrecht University of Kiel); as well as Prof. Christoph Düber, M.D. (University of Mainz Medical School); Christian Friedrichs, D.D.S. (Practice for Tooth Preservation and Endodontics, Kiel); Prof. Reinhart Gossrau, M.D. (Charité Berlin, Institute of Anatomy); Prof. Daniel Haag-Wackernagel, Ph.D. (Basel); Johannes-Martin Hahn, M.D., (Tübingen); Prof. Stefan Müller-Hülsbeck, M.D. (DIAKO Krankenhaus gGmbH Flensburg); Róbert Késmárszky, M.D.; Prof. Susanne Klutmann (UKE Hamburg); Michael Kriwat (Kiel); Prof. Paul Peter Lunkenheimer, M.D. (Westphalian Wilhelm University Münster); Dr. Janos Mester, professor (UKE Hamburg); Dr. Jörg Detlev Moritz, associate professor (Klinik für Radiologie und Neuroradiologie Kiel); Dr. Thomas Müller, associate professor (University of Mainz Medical School); Dan mon O'Dey, M.D., associate professor (Luisenhospital Aachen); Dr. Kai-Hinrich Olms, Fußchirurgie Bad Schwartau, Daniel Paech, M.D., M.S. (physics) (Deutsches Krebsforschungszentrum Heidelberg); Dr. Thilo Schwalenberg, senior physician (Urologische Klinik des Universitätsklinikums Leipzig); Hans-Peter Sobotta, M.D. (Stiftung Herzogin Elisabeth Hospital Braunschweig); Dr. Katharina Spanel-Borowski, professor emeritus (University of Leipzig); Dr. Jürgen Specht (Orthopaedicum Frankfurt); Prof. Christoph Viebahn, Ph.D. (Georg-August-Universität Göttingen); Imke Weyers, M.D. (Univ. Lübeck).

For extensive corrections, especially in the first edition, we thank Gabriele Schünke, M.S. biology; Jakob Fay, M.D.; as well as medical students Claudia Dücker, Simin Rassouli, Heike Teichmann, Susanne Tippmann, and dental student Sylvia Zilles; also, Julia Jörns-Kuhnke, M.D., especially for her assistance with the figure labels.

We extend special thanks to Stephanie Gay and Bert Sender, who prepared the layouts. Their ability to arrange the text and illustrations on facing pages for maximum clarity has contributed greatly to the quality of the atlas.

We particularly acknowledge the efforts of those who handled this project on the publishing side:

Jürgen Lüthje, M.D., Ph.D., executive editor at Thieme Medical Publishers, has "made the impossible possible." He not only reconciled the wishes of the authors and artists with the demands of reality but also managed to keep a team of five people working together for years on a project whose goal was known to us from the beginning but whose full dimensions we only came to appreciate over time. For this he has earned our deepest heartfelt gratitude. Since Dr. Jürgen Lüthje retired in 2018, Dr. Jochen Neuberger has taken over and has shown great initiative in continuing and further developing the *Thieme Atlas of Anatomy*.

Sabine Bartl, developmental editor, became a touchstone for the authors in the best sense of the word. She has, as a humanities scholar, not a doctor, reviewed the text and, in connection with the illustrations, was able to determine whether a beginning student, and thus one who is not (yet) a physician, could clearly appreciate the logic of the presentation. Based on her suggestions topics were also reformulated and redesigned. It is not only the authors who are indebted to her, but also the reader benefits from her didactic talent.

We owe a great debt of thanks to Martin Spencker, who was at the time of publication of the first edition managing director of Educational Publications at Thieme, especially to his ability to make quick and unconventional decisions when dealing with problems and uncertainties. His openness to all the concerns of the authors and artists and the transparency and fairness in all discussions gave the project momentum and established conditions for a cooperative partnership. We also owe him a huge debt of gratitude.

We are grateful to Antje Bühl, who was there from the beginning as project assistant, working "behind the scenes" on numerous tasks such as repeated proofreading and helping to arrange the figure labels.

We are also indebted to Yvonne Strassburg, Michael Zepf, and Laura Diemand who saw to it that the *Thieme Atlas of Anatomy* was printed and bound on schedule, and that the project benefited from the best practical expertise throughout the entire process of publication. We also thank Susanne Tochtermann-Wenzel and Anja Jahn for their assistance with technical issues involving every aspect of the illustrations; Julia Fersch who ensured that the *Thieme Atlas of Anatomy* is also accessible via eRef; Almut Leopold and Dr. Wilhelm Kuhn for the exceptional index; Marie-Luise Kürschner and Nina Jentschke for the appealing cover design; as well as Dr. Thomas Krimmer, Liesa Arendt, Birgit Carlsen, Stephanie Eilmann, Marion Hamm, and Anne Döbler, representing all those now and previously involved in the marketing, sale, and promotion of the *Thieme Atlas of Anatomy*.

The authors, August 2022

As consulting editor, I was asked to review, for accuracy and appropriateness, the English translation of the *Thieme Atlas of Anatomy: General Anatomy and Musculoskeletal System*, fourth edition. My work involved the review—and, where necessary, adaptation and correction to the standards of North American anatomy programs—of approximately 600 pages, including approximately 2,600 illustrations. The clear organization and structure of the prior edition (3rd edition) made this task almost effortless. Throughout this process, I tried diligently to remain faithful to the intentions and insights of the authors and illustrators, whom I wish to thank for this brilliant revision. I would also like to thank the team at Thieme Medical Publishers who patiently worked with me to complete this edit in a timely manner. First, I wish to thank Andrew Moyer, Executive Content Portfolio Manager, for his patience, knowledge and cordial support. Next, I would like to thank Torsten Scheihagen, Sewnior Content Service Manager, for his insightful feedback and patience. I would also like to thank translator John Grossman for his work with the new and altered text. Team Thieme has been an absolute pleasure to work with.

Finally, sincerest thanks go to Brian R. MacPherson, Ph.D., Vice-Chair for Educational Programs at the University of Kentucky, for inviting me to contribute to Thieme's mission of educating medical professionals.

Nathan F. Johnson

As consulting editor I was asked to review the English translation of *Thieme Atlas of Anatomy: General Anatomy and Musculoskeletal System*, fourth edition. My work involved reviewing and editing the translation and nomenclature conversion of the German text to currently used English terms. In addition, some minor changes in presentation were made to reflect commonly accepted approaches in North American educational programs. This task was facilitated by the clear organization of the original text and the exceptional work of the translators. Throughout this process I have tried to remain faithful to the intentions and insights of the authors and illustrators, whom I thank for this outstanding revision.

I would also like to thank the entire team at Thieme Medical Publishers who worked on this edition. In particular I would like to give special thanks to Andrew Moyer, Executive Content Portfolio Manager, for inviting me to work on this atlas and for his continued support, Torsten Scheihagen, Senior Content Service Manager, for his support and help throughout the process of completing this atlas, and translator John Grossman for his work with the new and altered text.

Wayne A. Cass

The people behind the Thieme Atlas of Anatomy

A work such as the *Thieme Atlas of Anatomy* can only arise when the people involved in the project work hand in hand. The integrated educational and artistic work you now hold in your hands is the product of an intensive discourse between anatomy professors Michael Schünke, Erik Schulte, and Udo Schumacher and anatomic illustrators Markus Voll and Karl Wesker.

Creating learning units that comprehensively treat a topic on a two-page spread is a challenge in itself. The authors must carefully select the content, assemble it, and add explanatory legends. Yet how this content is presented in the atlas, how appealing and memorable it is, depends largely on the illustrations. And the *Thieme Atlas of Anatomy* now includes a good 5,000 of them. In creating them, Markus Voll and Karl Wesker drew on many years of experience in anatomic illustration, visited anatomic collections, studied specimens, and immersed themselves in old and new works of anatomy. This was the foundation on which the *Thieme Atlas of Anatomy* arose.

Photo: private

Michael Schuenke, MD, PhD, professor

Institute of Anatomy of the University of Kiel; studied biology and medicine in Tübingen and Kiel; extensive teaching of medical students and physical therapists; author and translator of other textbooks.

Photo: Kristina Schäfer

Erik Schulte, MD, professor

Institute of Functional and Clinical Anatomy of the Johannes Gutenberg University of Mainz; studied medicine in Freiburg; extensive teaching of medical students; award for excellence in teaching in Mainz.

Photo: private

Udo Schumacher, MD, professor

MSB Medical School Berlin; studied medicine in Kiel, and a one-year sojourn at the Wistar Institute of Anatomy and Biology, Philadelphia; intensive experience in teaching medical students, physical therapists, and specialty candidates (FRCS). Spent several years in Southampton and gained experience in integrated interdisciplinary instruction.

It guides the reader through anatomy step by step, revealing what a crucial role anatomy will later play in medical practice. This was a particularly important consideration for the authors. Whether performing bowel surgery for a tumor, puncturing the tympanic membrane in a middle ear infection, or examining a pregnant patient, no physician lacking knowledge of anatomy is a good physician.

Even the *Thieme Atlas of Anatomy* cannot spare you the effort of learning, yet the authors and illustrators can assure you that it will make it a lot more pleasant.

Photo: private

Markus Voll

Freelance illustrator and graphic artist in Munich, trained as an artist at the Blocherer School of Design in Munich, studied medicine at the University of Munich. He has decades of experience as a scientific illustrator for numerous book projects.

Photo: private

Karl Wesker

Freelance painter and graphic artist in Berlin. Apprenticeship as a plate etcher and lithographer, studied visual communication at the University of Applied Sciences in Münster and at the Berlin University of the Arts and art science at the Technical University of Berlin. He has worked for decades as a freelance painter and graphic scientific artist, including book projects in anatomy.

Contents

A General Anatomy

B Trunk Wall

C Upper Limb

D Lower Limb

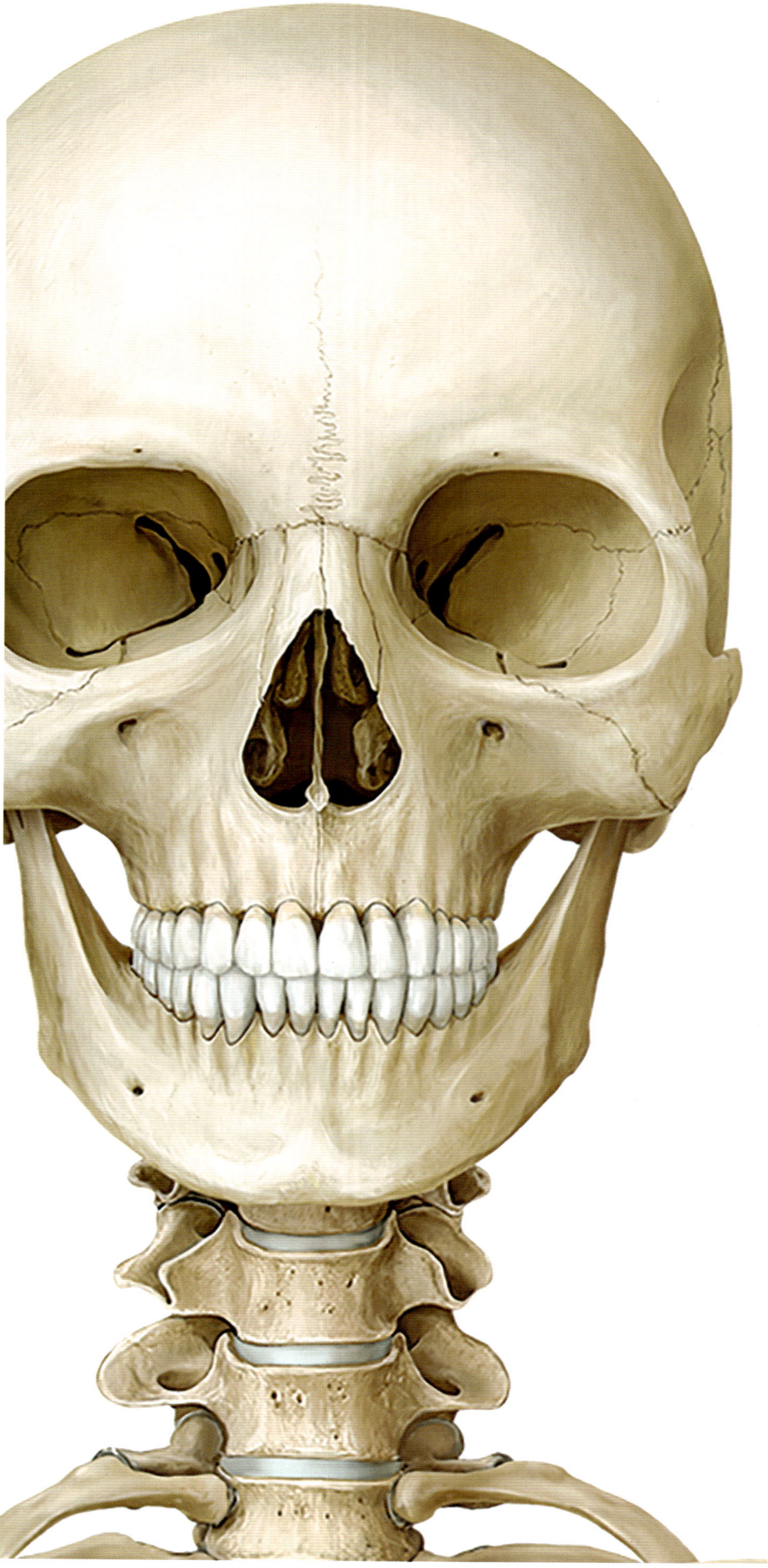
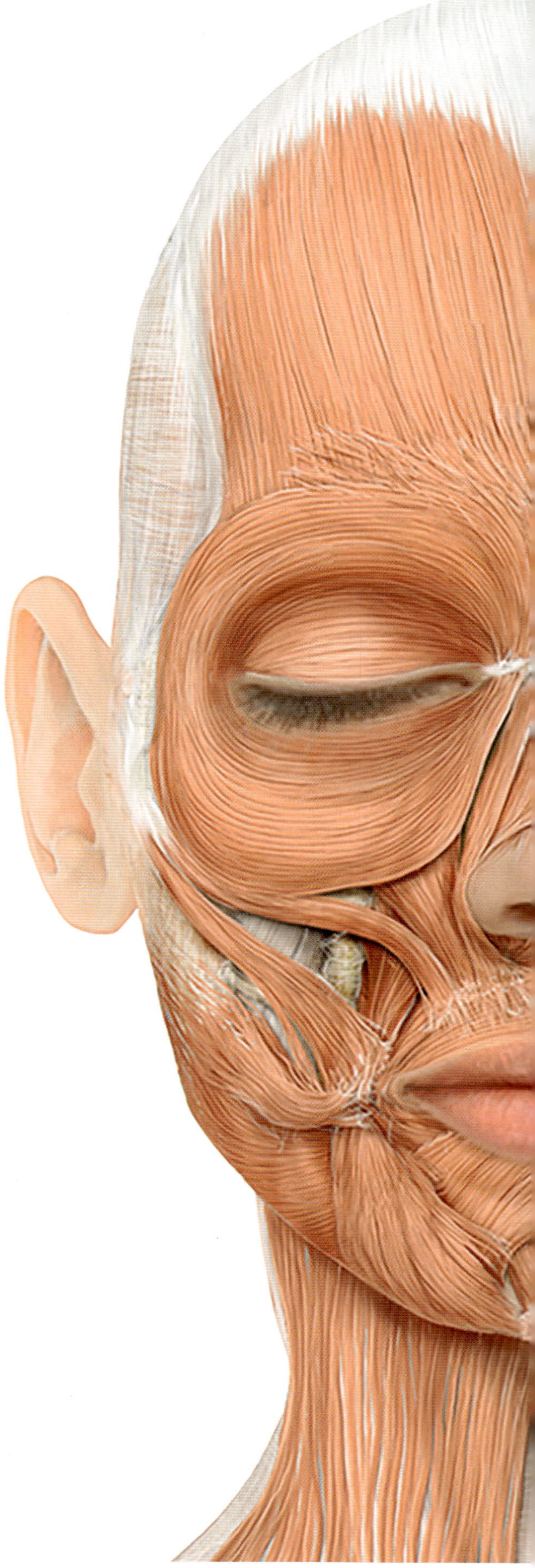

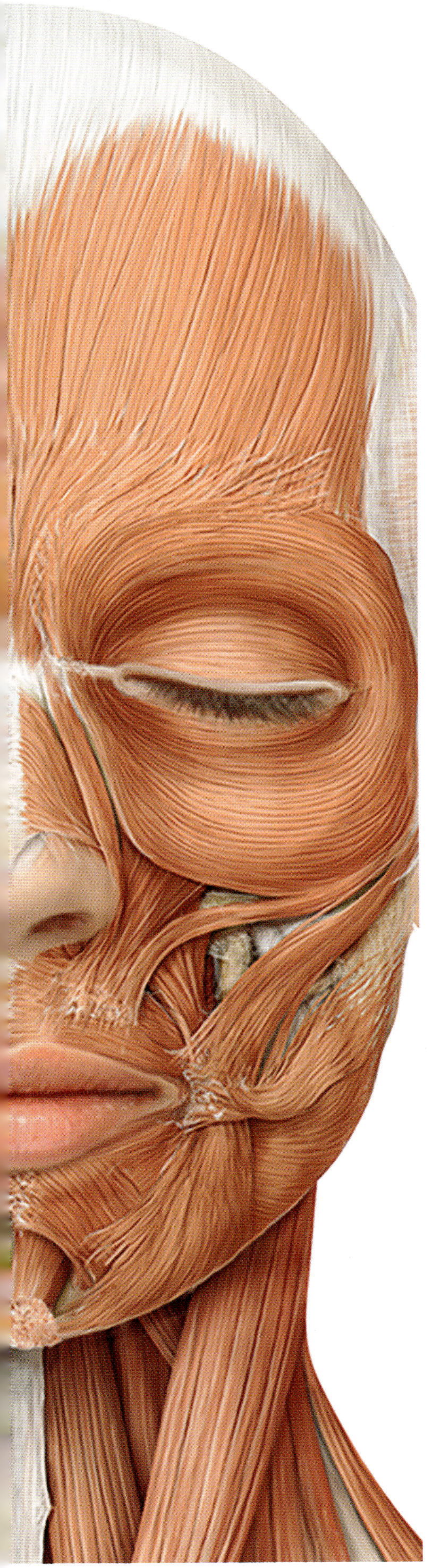

A General Anatomy

1.1 Human Phylogeny

A Brief overview of human phylogenetic development

To better understand the evolution of the human body, it is helpful to trace its phylogenetic development. Humans and their closest relatives belong to the **phylum Chordata**, which includes approximately 50,000 species. It consists of two subphyla:

- Invertebrata: the tunicates (Tunicata) and chordates without a true skull (Acraniata or Cephalochordata)
- Vertebrata: the vertebrates (animals that have a vertebral column)

Although some members of the chordate phylum differ markedly from one another in appearance, they are distinguished from all other animals by characteristic morphological structures that are present at some time during the life of the animal, if only during embryonic development (see **G**). Invertebrate chordates, such as the cephalochordates and their best-known species, the lancelet (*Branchiostoma lanceolatum*) are considered the *model of a primitive vertebrate* by virtue of their organization. They provide clues to the basic structure of the vertebrate body and thus are important in understanding the general organization of vertebrate organisms (see **D**).

All the **members of present-day vertebrate classes** (jawless fish, cartilaginous fish, bony fish, amphibians, reptiles, birds, and mammals) have a number of characteristic features in common (see **H**), including a row of vertebrae arranged in a *vertebral column*, which gives the subphylum its name (Vertebrata). The evolution of an *amniotic egg* (i. e., the development of the embryo within a fixed shell inside a fluid-filled amniotic cavity) was a critical evolutionary breakthrough that helped the vertebrates to survive on land. This reproductive adaptation enabled the terrestrial vertebrates (reptiles, birds, and mammals) to live out their life cycles entirely on land and sever the final ties with their marine origin. When we compare the embryos of different vertebrate classes, we observe a number of morphological and functional similarities, including the formation of branchial arches (see **B**).

Mammals comprise **three major groups**: monotremes (egg-laying mammals), marsupials (mammals with pouches), and placentals (mammals with a placenta). The placental mammals, which include humans, have a number of characteristic features (see **I**), including a tendency to invest much greater energy in the care and rearing of their young. Placental mammals complete their embryonic development inside the uterus and are connected to the mother by a placenta. Humans belong to the mammalian order of **primates**, whose earliest members were presumably small tree-dwelling mammals. Together with lemurs, monkeys, and the higher apes, human beings have features that originate from the early adaptation to an arboreal way of life. For example, primates have movable shoulder joints that enable them to climb in a hanging position while swinging from branch to branch. They have dexterous hands for grasping branches and manipulating food, and they have binocular, broadly overlapping visual fields for excellent depth perception.

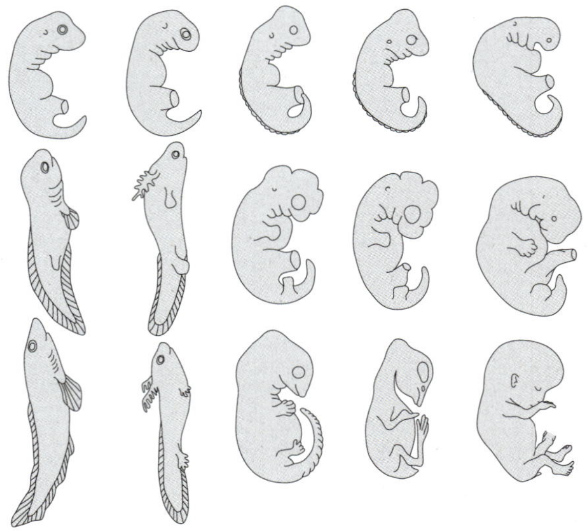

B Different stages in the early embryonic development of vertebrates

The early developmental stages (top row) of fish, amphibians, reptiles, birds, and mammals (as represented by humans) present a series of striking similarities that suggest a common evolutionary origin. One particularly noteworthy common feature is the set of branchial or pharyngeal arches in the embryonic regions that will develop into the head and neck. Although it was once thought that the developing embryo of a specific vertebrate would sequentially display features from organisms representing every previous step in its evolution ("Ontogeny recapitulates phylogeny," the "biogenetic law" of Ernst Haeckel [1834–1919]), subsequent work has shown that the vertebrates share common embryonic components that have been adapted to produce sometimes similar (fins and limbs) and sometimes radically different (gills vs. neck cartilages) adult structures.

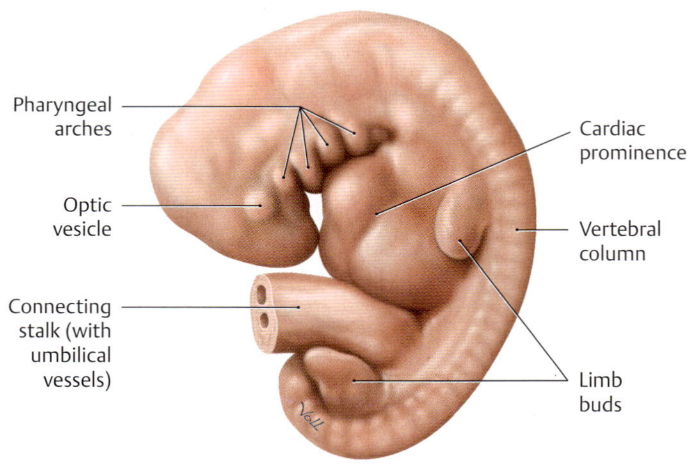

C Formation of the branchial or pharyngeal arches in a 5-week-old human embryo

Left lateral view. The branchial or pharyngeal arches of the vertebrate embryo have a *metameric* arrangement (similar to the somites, the primitive segments of the embryonic mesoderm); this means that they are organized into a series of segments that have the same basic structure. Among their other functions, they provide the raw material for the species-specific development of the visceral skeleton (maxilla, mandible, middle ear, hyoid bone, and larynx), the associated facial muscles, and the pharyngeal gut (see p. 11).

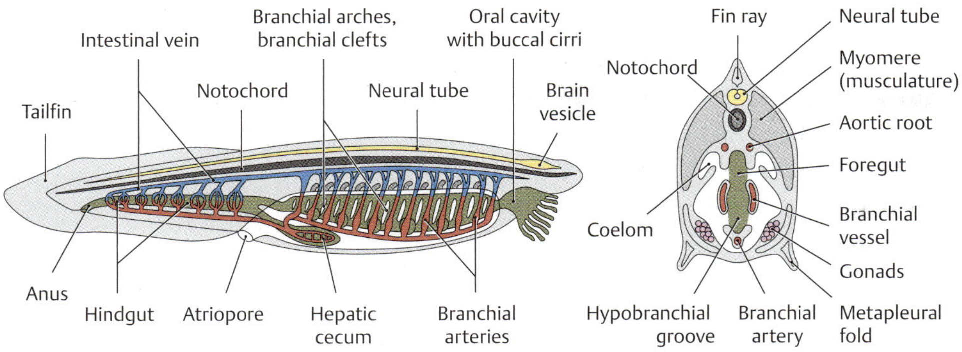

D Basic chordate anatomy, illustrated for the lancelet *(Branchiostoma lanceolatum)*

The vertebrates (including humans) are a subphylum of the chordates (Chordata), of which the lancelet is a typical representative. Its anatomy displays relatively simple terms of structures common to all vertebrates. The characteristic features of chordates include the development of an axial skeleton called the *notochord*. The human body still has remnants of the notochord, such as the nucleus pulposus of the intervertebral disks. The notochord is present in humans only during embryonic life, however, and is not a fully developed structure. Its remnants may give rise to developmental tumors called *chordomas*. Chordates have a *tubular nervous system* lying dorsal to the notochord. The body, particularly the muscles, is composed of multiple segments called *myomeres*. In humans, this myomeric pattern of organization is most clearly apparent in the trunk. Another distinguishing feature of chordates is the presence of a closed circulatory system.

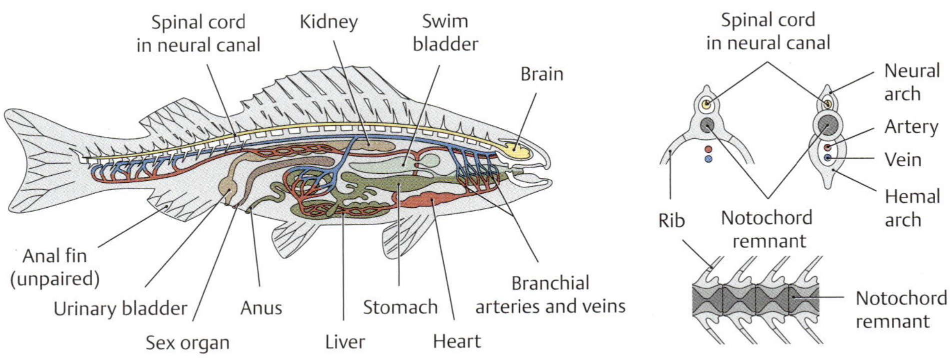

E Basic vertebrate anatomy, exemplified by the bony fish

The vertebrates are the *subphylum of chordates* from which humans evolved. With the evolution of fish, the notochord was transformed into a vertebral column (spinal column). The segmentally arranged bony vertebrae of the spinal column encircle remnants of the notochord and have largely taken its place. Dorsal and ventral arches arise from the vertebral bodies. The dorsal arches (vertebral or neural arches) in their entirety make up the neural canal, while the ventral arches (hemal arches) form a caudal "hemal canal" that transmits the major blood vessels. The ventral arches in the trunk region are the origins of the ribs.

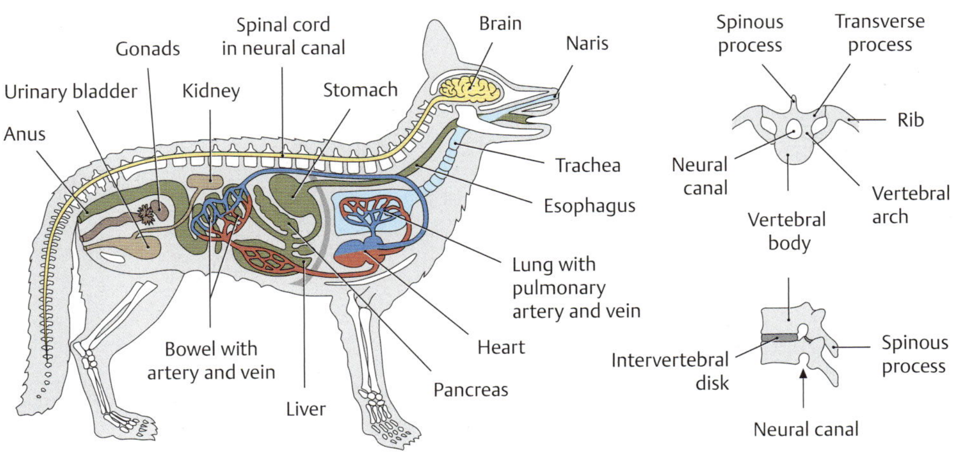

F Basic vertebrate anatomy, the dog

G Characteristic features of chordates

- Development of an axial skeleton (notochord)
- Dorsal neural tube
- Segmental arrangement of the body, particularly the muscles
- Foregut pierced by slits (branchial gut)
- Closed circulatory system
- Postanal tail

H Characteristic features of vertebrates

- Nerve cells, sensory organs, and oral apparatus concentrated in the head (cephalization)
- Multipart brain with a pituitary gland
- Replacement of the notochord by the vertebral column
- Generally, two pairs of limbs
- Development of branchial arches
- Presence of neural crest cells
- Closed circulatory system with a ventral, chambered heart
- Labyrinthine organ with semicircular canals
- Stratified epidermis
- Liver and pancreas always present
- Complex endocrine organs such as the thyroid and pituitary
- Complex immune system
- Sexes almost always separate

I Characteristic features of mammals

- Highly glandular skin covered with true hair (terminal hair)
- Females always have mammary glands for nursing offspring, which are usually born live (viviparous)
- Well-developed cerebrum
- Well-developed cutaneous muscles
- Diaphragm is the major respiratory muscle and separates the thoracic and abdominal cavities
- Heterogeneous and specialized teeth
- Four-chambered heart with a (left-sided) aortic arch
- Constant body temperature (homeothermy)

1.2 Human Ontogeny: Overview, Fertilization, and Earliest Developmental Stages

Besides gross and microscopic anatomy, the developmental history of the individual organism (ontogeny) is of key importance in understanding the human body. Ontogeny is concerned with the formation of tissues (*histogenesis*), organs (*organogenesis*), and the shape of the body (*morphogenesis*).

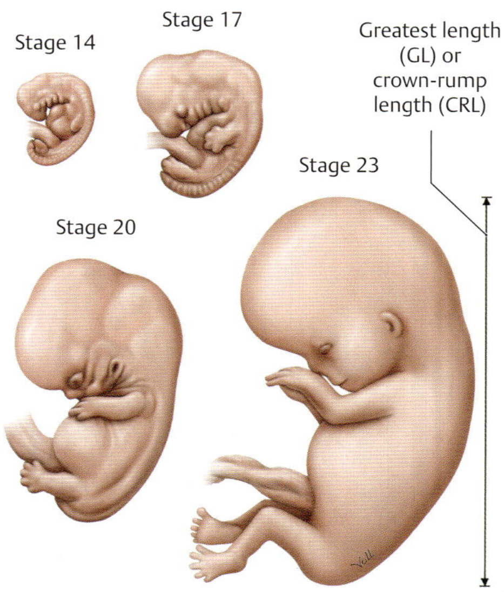

A 5- to 8-week-old human embryos
Streeter (1942) and O'Rahilly (1987) classified early human development and the embryonic period into 23 stages based on specimens from the Carnegie Collection. The Carnegie stages are defined by morphological characteristics that can be closely correlated with specific age (postovulatory days or weeks) and size (measured as the greatest length, excluding lower limb [GL], or crown-rump length [CRL], see **C**).

Stage 14: 5th week, GL 5-7 mm, future cerebral hemispheres become identifiable.
Stage 17: 6th week, GL 11-14 mm, digital rays become visible.
Stage 20: 7th week, GL 18-22 mm, upper arms bent at the elbow, hands in a pronated position.
Stage 23: 8th week, GL 27-31 mm, eyelids fuse, external genitalia begin differentiation.

B Longitudinal growth and weight gain during the fetal period

Age (weeks)	Crown-rump length, CRL (cm)	Weight (g)
9–12	5–8	10–45
13–16	9–14	60–200
17–20	15–19	250–450
21–24	20–23	500–820
25–28	24–27	900–1300
29–32	28–30	1400–2100
33–36	31–34	2200–2900
37–38	35–36	3000–3400

C Timetable of antenatal human development
(The Carnegie stages are shown in parentheses.)

Weeks 1–3: Early development	
Week 1:	Tubal migration, segmentation, and blastocyst formation (stages 1–3)
Week 2:	Implantation and bilaminar embryonic disc, yolk sac (stages 4–5)
Week 3:	Trilaminar embryonic disc, start of neurulation (stages 6–9)
Weeks 4–8: Embryonic period	
Week 4:	Folding of the embryo, neurulation concluded, axial organs, basic body shape (stages 10–13)
Weeks 5–8:	Organogenesis (formation of all essential external and internal organs, elongated limb buds) (stages 14–23)
Weeks 9–38: Fetal period	
Weeks 9–38:	Organ growth and functional maturation (sex-specific differentiation of the external genitalia)
Length of gestation	
• p.o. = postovulatory age	266 days = 38 weeks
• p.m. = postmenstrual age	280 days = 40 weeks

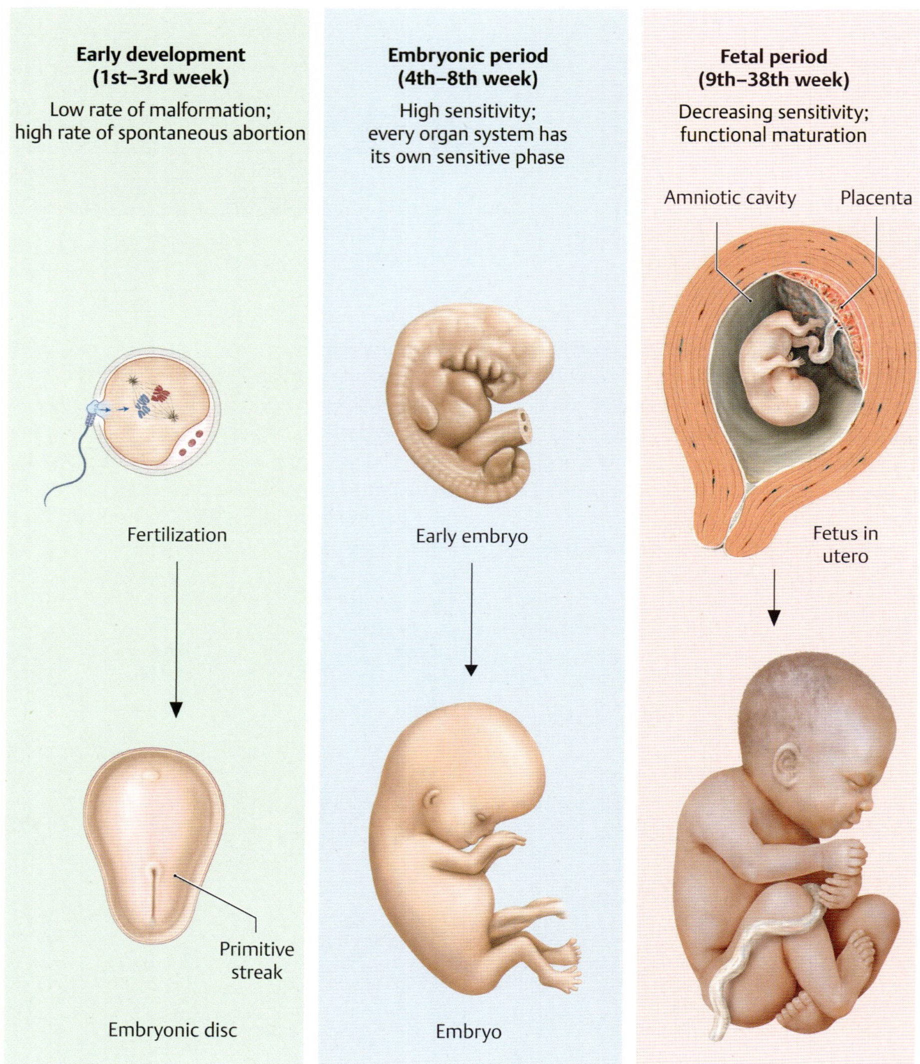

D Stages sensitive to teratogenic influences

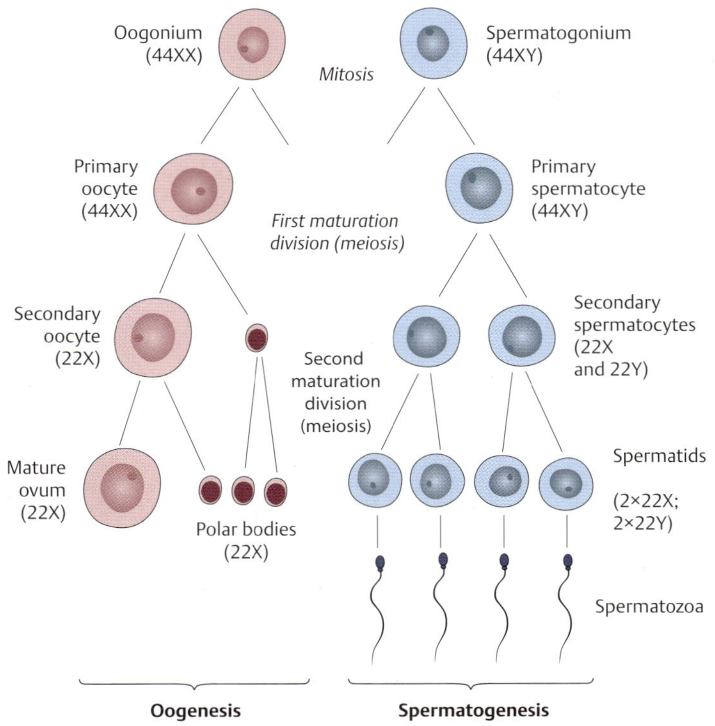

E Formation of the ovum and sperm

During the formation of the gametes (sex cells), two successive cell divisions occur (the first and second meiotic maturation divisions). This results in cells having a chromosome set that is reduced by one half (haploid). When fertilization occurs, a diploid (full) chromosome set is restored. During meiosis, extensive chromosomal rearrangement occurs, thus recombining the internal genetic information into new and different subsets.

Oogenesis: The initial oogonia first undergo a mitotic division to form primary oocytes, which still have a diploid chromosome number (44XX). Later the primary oocytes undergo a first and second maturation division by meiosis, resulting in four haploid cells (22X): one mature ovum and three polar bodies.

Spermatogenesis: Diploid spermatogonia undergo mitosis to form primary spermatocytes (44XY). These cells then divide meiotically to form four haploid spermatids, two of which have an X chromosome (22X) and two a Y chromosome (22Y). The spermatids develop into motile spermatozoa (spermatohistogenesis).

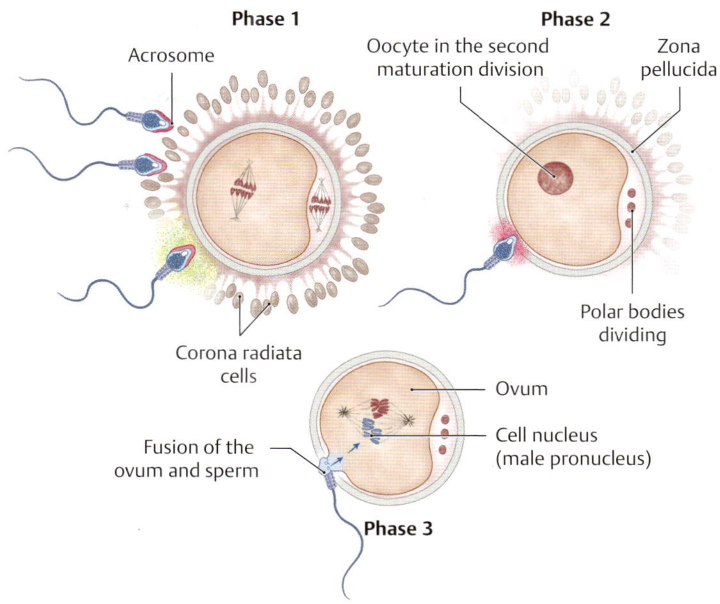

F Schematic representation of the fertilization process

In *phase 1,* the spermatozoon penetrates the corona radiata cells. In *phase 2,* the acrosome dissolves, releasing enzymes that digest the zona pellucida. In *phase 3,* the cell membranes of the ovum and sperm fuse, and the spermatozoon enters the egg.

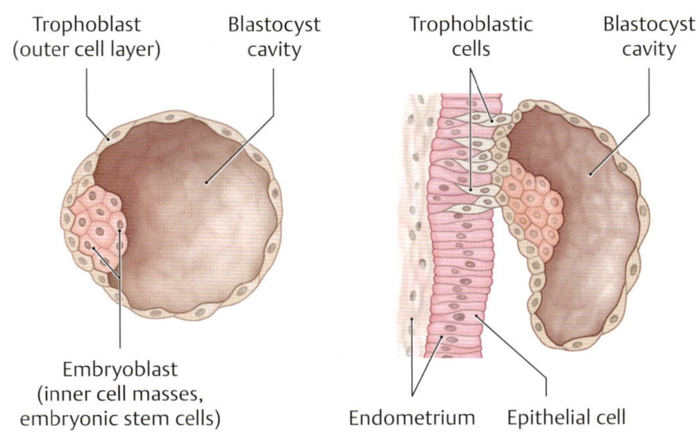

G Implantation of the blastocyst in the uterine mucosa on postovulatory day 5–6

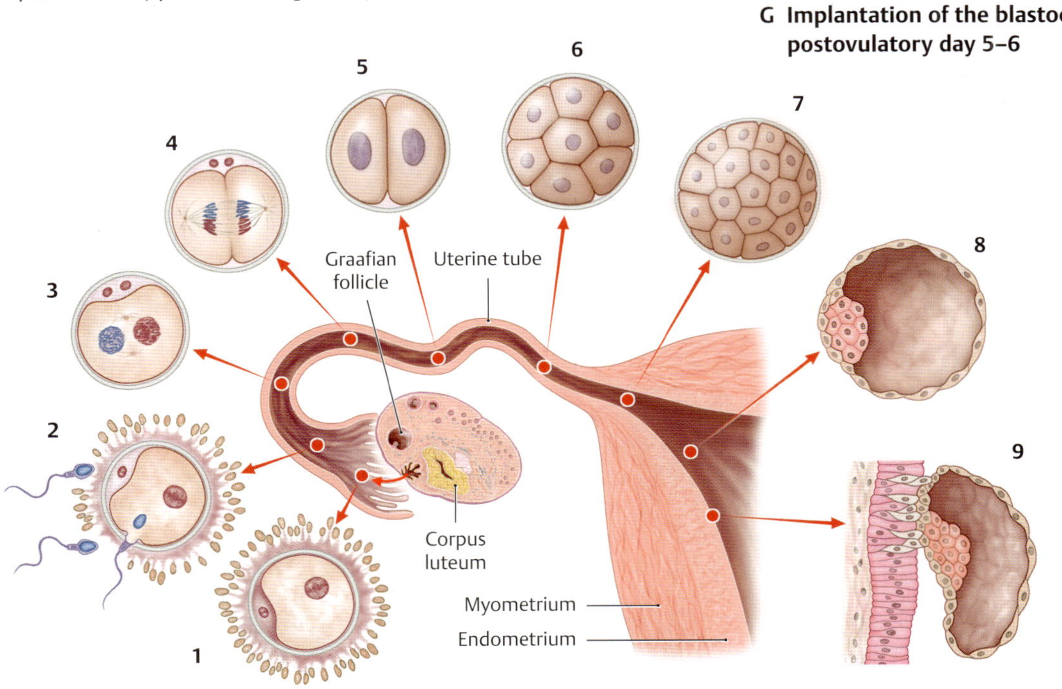

H Developmental processes during the first week of development

1. Ovum immediately after ovulation
2. Fertilized within approximately 12 hours
3. Male and female pronucleus with subsequent zygote formation
4. First segmentation
5. Two-cell stage (30 hours)
6. Morula stage (3 days)
7. Entry into the uterine cavity (4 days)
8. Blastocyst (4–5 days)
9. Early implantation (5–6 days)

1.3 Human Ontogeny: Gastrulation, Neurulation, and Somite Formation

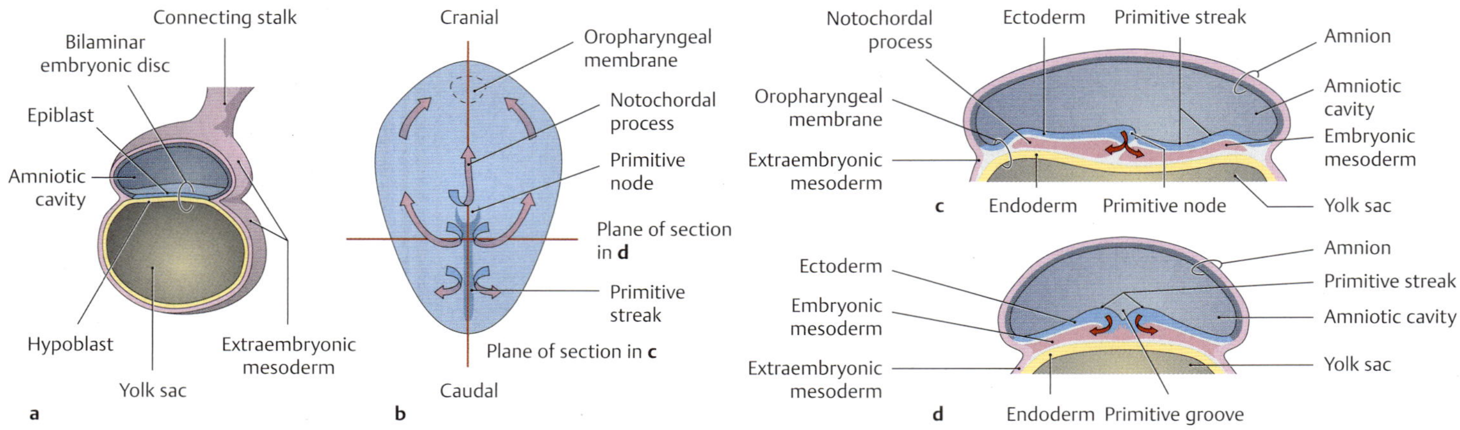

A Formation of the trilaminar human embryonic disc (gastrulation) at the start of the third postovulatory week

As a result of gastrulation, the cell layers become differentiated into an *ectoderm*, *endoderm*, and *mesoderm*, from which all structures of the human body are derived (e.g., the ectoderm gives rise to the central nervous system and the sensory organs). Gastrulation also establishes the primary axes of the body (ventral–dorsal, cranial–caudal, and left–right).

a Sagittal section through a conceptus at 2 postovulatory weeks. The embryonic disc is still bilaminar and is stretched between the amniotic cavity and yolk sac. The extraembryonic mesoderm, whose formation commences at the posterior pole of the embryonic disc, already covers the entire conceptus, which is attached to the chorionic cavity by a connecting stalk.

b Dorsal view of a human embryonic disc at the start of gastrulation. The amnion has been removed. At the start of gastrulation, the epiblast develops a primitive streak, where the embryonic mesoderm is generated and migrates between epiblast and hypoblast. Shortly afterward, at the level of the primitive node (the cranial tip of the primitive streak), epiblast cells migrate cranially to form the notochordal process and radially to form the definitive endoderm. In the process, the definitive endoderm sequentially replaces the hypoblast, while the notochordal process only temporarily fuses with the hypoblast layer. The notochordal process expands cranially from the primitive node to the oropharyngeal membrane.

c Sagittal section: embryonic disc along the notochordal process.

d Cross section: embryonic disc at the level of the primitive groove (arrows in **c** and **d:** direction of gastrulation movements by the mesoderm).

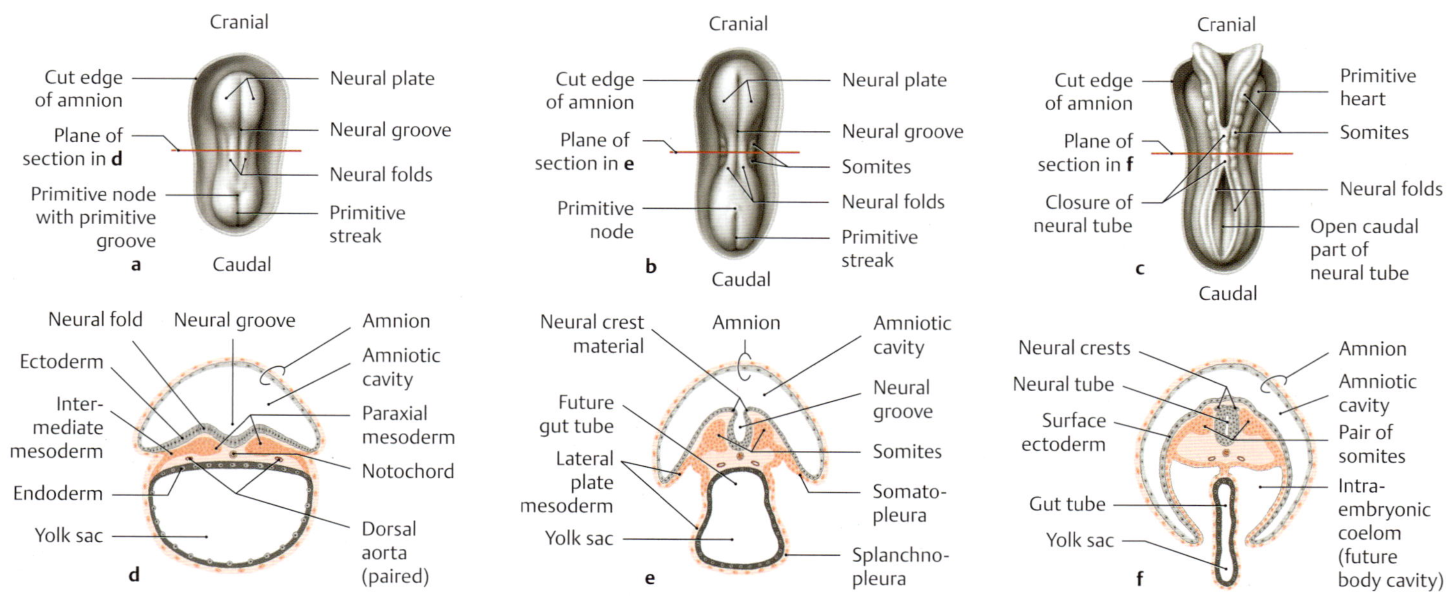

B Neurulation during early human development

a–c Dorsal view after removal of the amnion.

d–f Schematic cross sections of the corresponding stages at the planes of section marked in **a–c**. During neurulation, the neuroectoderm differentiates from the surface ectoderm due to inductive influences from the notochord.

a, d Embryonic disc at 19 days. The neural tube is developing in the area of the neural plate.

b, e Embryonic disc at 20 days. The first somites have formed, and the neural groove is beginning to close to form the neural tube, with initial folding of the embryo.

c, f Embryo at 22 days. Eight pairs of somites are seen flanking the partially closed neural tube, which has sunk below the ectoderm. At the sites where the neural folds fuse to close the neural tube, cells form a bilateral neural crest that detaches from the surface and migrates into the mesoderm.

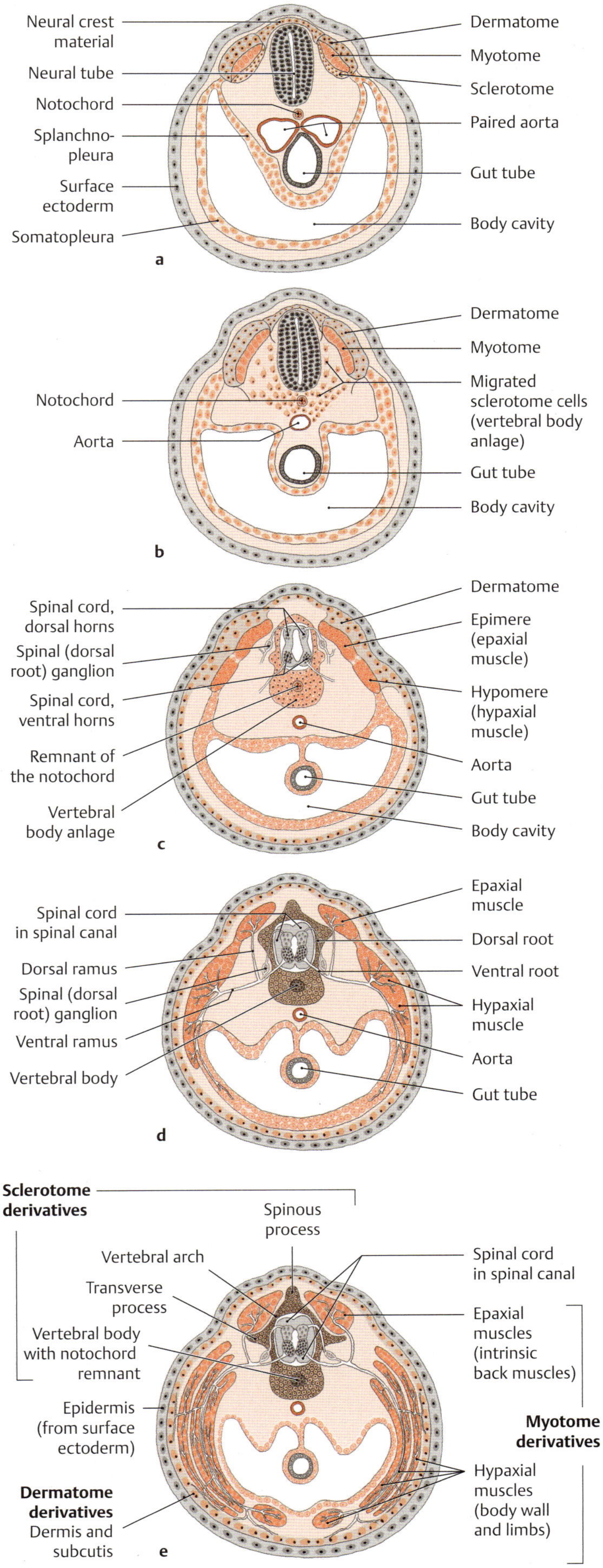

C Somite derivatives and spinal nerve formation during the embryonic period (weeks 4–8), shown in schematic cross sections

D Differentiation of the germ layers

Germ layer			Derivatives
Ectoderm	Neural tube		Brain, retina, spinal cord
	Neural crest	Neural crest of the head	Sensory and parasympathetic ganglia, intramural nervous system of the bowel, parafollicular cells, smooth muscle, pigment cells, carotid body, bone, cartilage, connective tissue, dentin and cementum of the teeth, dermis and subcutaneous tissue of the head
		Neural crest of the trunk	Sensory and autonomic ganglia, peripheral glia, adrenal medulla, pigment cells, intramural plexuses
	Surface ecto-derm	Ectodermal placodes	Anterior pituitary, cranial sensory ganglia, olfactory epithelium, inner ear, lens
			Enamel organ of the teeth, epithelium of the oral cavity, salivary glands, nasal cavities, paranasal sinuses, lacrimal passages, external auditory canal, epidermis, hair, nails, cutaneous glands
Mesoderm	Axial	Notochord, prechordal mesoderm	Extraocular muscles
	Paraxial		Spinal column, ribs, skeletal muscle, connective tissue, dermis and subcutis of the back and part of the head, smooth muscle, blood and lymph vessels
	Intermediate		Kidneys, gonads, renal and genital excretory ducts
	Lateral plate meso-derm	Visceral (splanchno-pleura)	Heart, blood vessels, smooth muscle, bowel wall, blood, adrenal cortex, visceral serosa
		Parietal (somato-pleura)	Sternum, limbs (cartilage, bones, and ligaments), dermis and subcutaneous tissue of the anterolateral body wall, smooth muscle, connective tissue, parietal serosa
Endoderm			Epithelium of the bowel, respiratory tract, digestive glands, pharyngeal glands, eustachian tube, tympanic cavity, urinary bladder, thymus, parathyroid glands, thyroid gland

(For clarity, the surrounding amnion is not shown.) The first pairs of somites appear at approximately 20 postovulatory days. All 34 or 35 of the somites ("primitive segments") have formed by day 30.

a When differentiation begins, each of these somites subdivides into a dermatome, myotome, and sclerotome (i.e., a cutaneous, muscular, and vertebral segment).

b At the end of 4 weeks, the sclerotome cells migrate toward the notochord and form the anlage of the spinal column.

c The neural tube—the precursor of the spinal cord and brain—differentiates to form a rudimentary spinal cord with dorsal and ventral horns. Cells within the ventral horn differentiate into motor neurons that sprout axons that form the *ventral root*. The neural crest has multiple derivatives, including sensory neurons that form dorsal root (spinal) ganglia, which send central processes into the spinal cord via the *dorsal root*. The myotomes become segregated into a dorsal part (epimere = epaxial muscles) and a ventral part (hypomere = hypaxial muscles).

d Each pair of dorsal and ventral roots unites to form a spinal nerve, which then divides into two main branches (a dorsal ramus and a ventral ramus). The epaxial muscles are supplied by the dorsal ramus, the hypaxial muscles by the ventral ramus.

e Cross section at the level of the future abdominal muscles. The epaxial muscles become the intrinsic back muscles, while the hypaxial muscles develop into structures that include the lateral abdominal muscles (external and internal oblique, transversus abdominis) and the anterior abdominal muscles (rectus abdominis).

1.4 Human Ontogeny: Development of the Fetal Membranes and Placenta

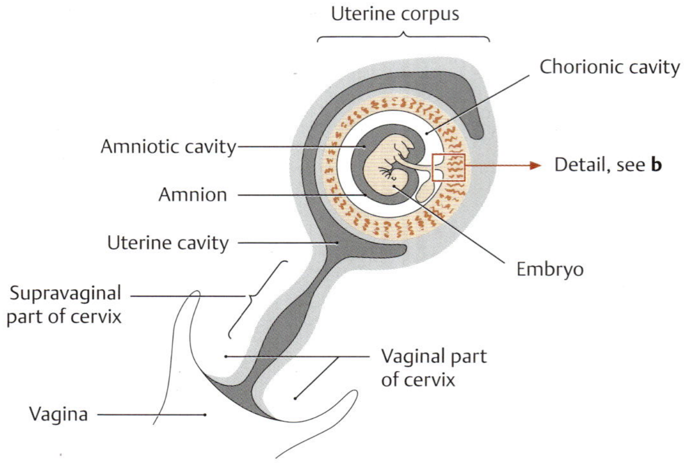

a

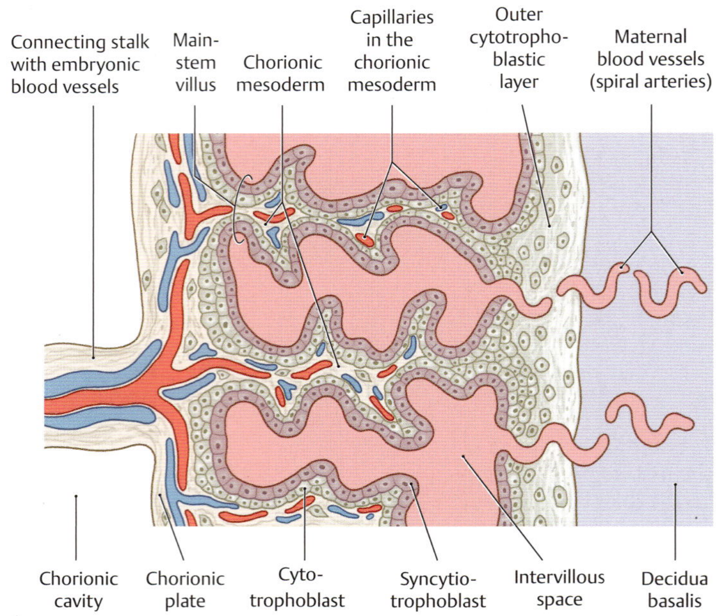

b

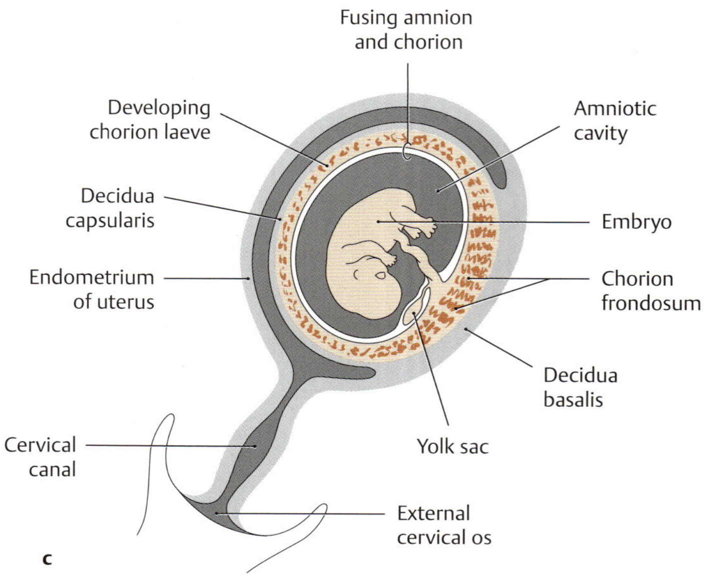

c

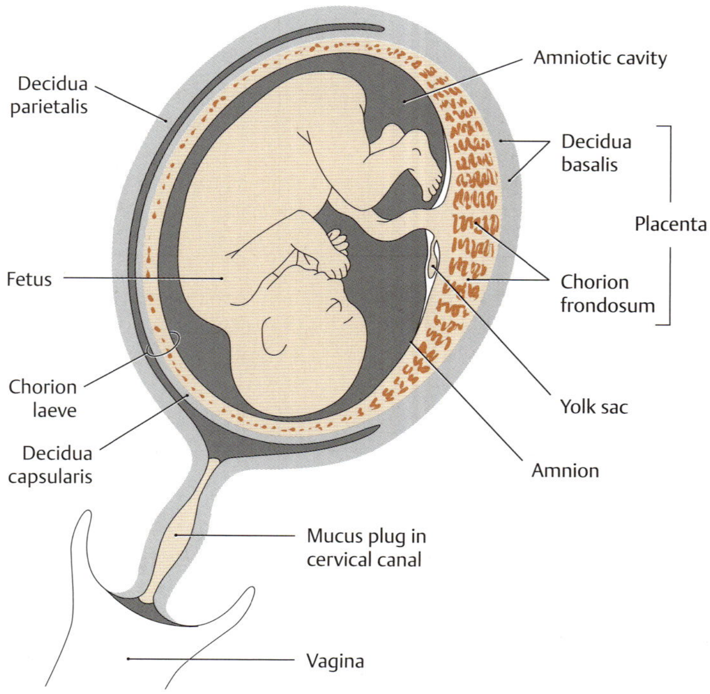

d

A Development of the fetal membranes and placenta
a, **c**, and **d** Schematic sections through a pregnant uterus at different points in gestation.
b Detail from **a**.

a Embryo at 5 weeks: After the blastocyst has implanted in the uterine mucosa, the embryo initially derives its nutrition through the developing trophoblast and chorionic mesoderm. Chorionic villi are formed that surround the entire chorionic sac and embryo. They develop from primary to secondary villi and finally to tertiary villi (see close-up in **b**).

b Detail from a: The mainstem villi of the chorionic plate are attached on the maternal side to the basal plate of the decidua basalis by compact columns of trophoblastic cells. Like the small villous trees that sprout and branch from them, these mainstem villi have a syncytial covering (syncytial trophoblast), which in turn rests on a continuous layer of trophoblastic cells. Inside the villi, capillaries develop in the chorionic mesoderm and communicate with the vessels in the connecting stalk. Maternal blood flows through spiral arteries into the intervillous spaces.

c Embryo at 8 weeks: While the chorionic villi continue to grow and arborize at the embryonic pole, forming the chorion frondosum, the villi outside of this zone begin to regress, forming the nonvillous chorion laeve directly below the decidua capsularis. The amniotic cavity has enlarged at the expense of the chorionic cavity, and the amnion fuses with the chorion.

d Fetus at 20 weeks: The placenta is fully formed and consists of two parts: a fetal part formed by the chorion frondosum and a maternal part, the decidua basalis.

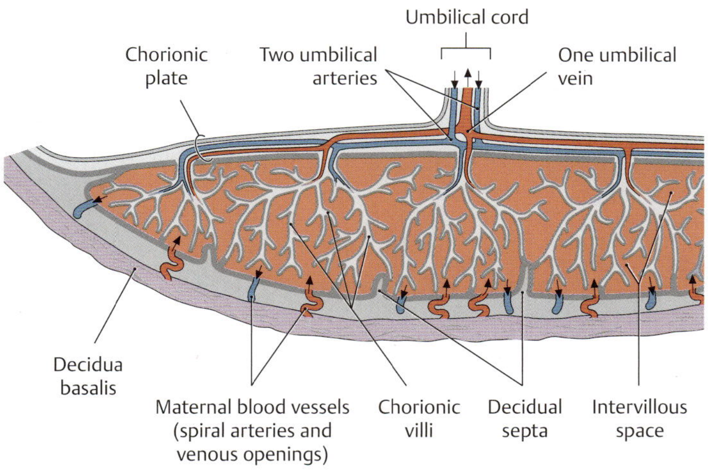

B Schematic cross section of a mature human placenta

The mature placenta is shaped like a frying pan, the maternal decidua basalis (basal plate) forming the base of the pan and the fetal chorionic plate forming the "lid." Some 40 arborizing villous trees containing fetal vessels project from the chorionic plate into the portions of the placenta that are filled with maternal blood (intervillous spaces). The maternal blood flows through approximately 80 to 100 spiral arteries into the intervillous spaces, which are divided into cotyledons by incomplete decidual septa. After the blood has bathed the villi, it is collected by irregularly distributed venous openings in the basal plate and returned to the maternal circulation.

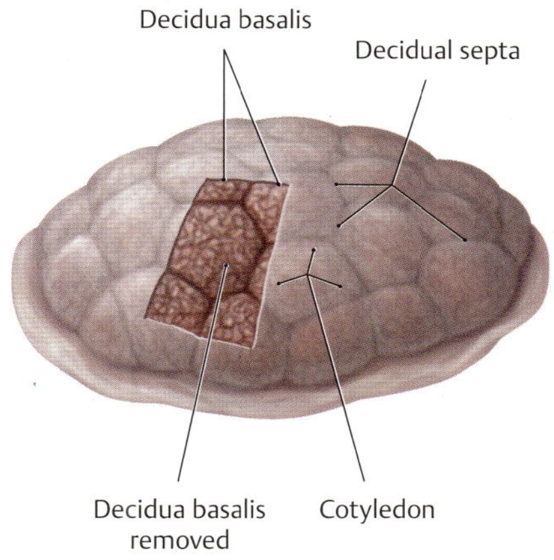

C The postpartum placenta

View of the maternal side of the delivered placenta (with a piece of the decidua basalis removed). The bulging cotyledons on the maternal surface are separated from one another by decidual septa.

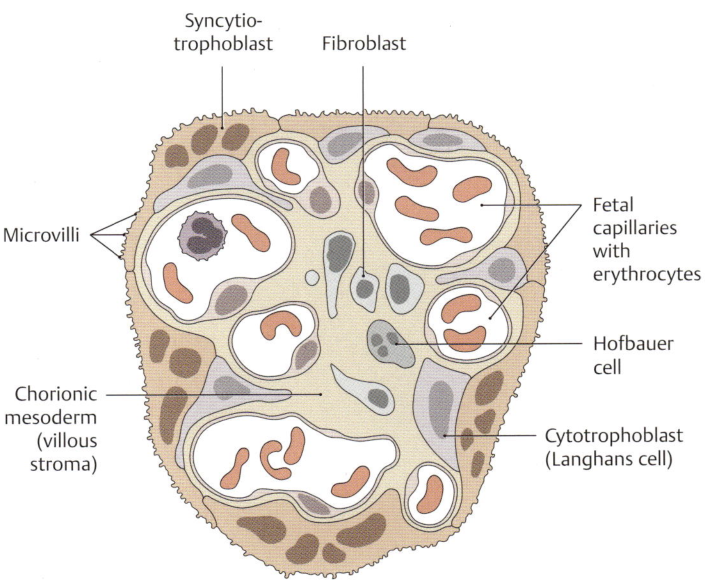

D Cross section through a terminal villus from a mature human placenta

E Characteristics of a mature human placenta

Size:	18–23 cm in diameter 2–3 cm in thickness
Weight:	450–500 g
Total placental volume:	approximately 500 mL
Volume of the intervillous spaces:	approximately 150 mL
Villous surface area:	approximately 11–13 m^2
Blood circulation on the maternal side:	500–600 mL/min

Structure of the placental barrier

- Endothelium of the fetal capillaries and basal lamina
- Fibrous villous stroma
- Syncytiotrophoblast and basal lamina
- Continuous trophoblast cell layer (becomes discontinuous after 20 weeks of gestation)

Diffusion distance approx. 5 µm (initially approx. 50 µm)

Primary functions of the mature placenta

1. *Transport of substances and exchange of metabolic products*

Mother-to-fetus	Fetus-to-mother
O_2, water, electrolytes, carbohydrates, amino acids and lipids, hormones, antibodies, vitamins and trace elements, but also drugs, toxins, and certain viruses	CO_2, water, electrolytes, urea, uric acid, bilirubin, creatinine, hormones

2. *Hormone production (syncytiotrophoblast)*
   - Human chorionic gonadotropin (HCG) → maintenance of the corpus luteum
   - Estrogens → growth of the uterus and breasts
   - Progesterone → inhibits uterine muscle contractions

Clinical note: The HCG formed in the syncytiotrophoblast prevents premature breakdown of the corpus luteum and sustains the pregnancy. HCG can be detected in the maternal urine at an early stage, providing the basis for early pregnancy testing.

1.5 Development of the Pharyngeal (Branchial) Arches in Humans

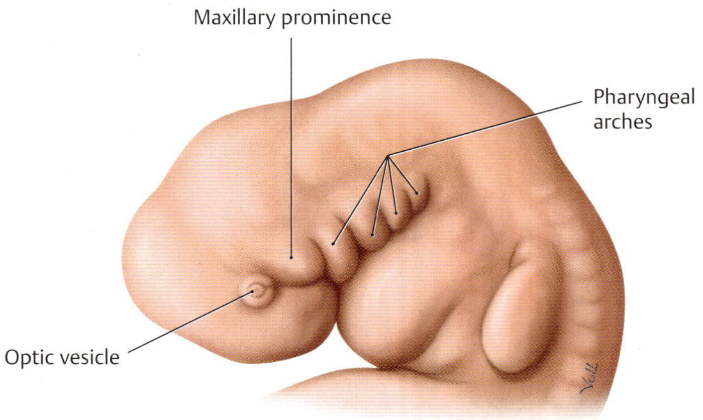

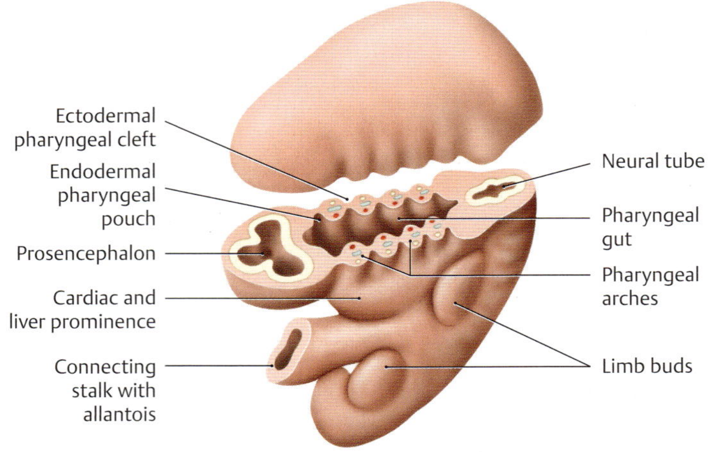

A Head and neck region of a 5-week-old human embryo, demonstrating the pharyngeal arches and clefts

Left lateral view. The pharyngeal arches are instrumental in the development of the neck and face. In *fish* and *amphibians*, the branchial arches develop into a respiratory organ (gills) for exchanging oxygen and carbon dioxide between the blood and water. *Land-dwelling vertebrates* (including humans) have *pharyngeal arches* rather than true branchial arches. Development of the pharyngeal arches begins in the 4th week of embryonic life as cells migrate from the neural crest to the future head and neck region. Within 1 week, a series of four oblique ridges (first through fourth pharyngeal arches) form that are located at the level of the cranial segment of the foregut and are separated externally by four deep grooves (pharyngeal clefts). The pharyngeal arches and grooves are prominent features of the embyro at this stage. Although the human embryo has no equivalent to the fifth and sixth branchial arches of other vertebrates, some of their components are incorporated into the human fourth pharyngeal arch.

B Cross section through a human embryo at the level of the pharyngeal gut

Left superior oblique view. Due to the craniocaudal curvature of the embryo, the cross section passes through the pharyngeal arches and pharyngeal gut, as well as the prosencephalon and spinal cord.
The pharyngeal gut is bounded on both sides by the pharyngeal arches (see also **A**), which contain a mesodermal core. They are covered externally by ectoderm and internally by endoderm. Ectodermal pharyngeal *clefts* and endodermal pharyngeal pouches lie directly opposite one another. Because the embryo is curved craniocaudally, the pharyngeal gut and pharyngeal arches overlie the prominence of the rudimentary heart and liver.

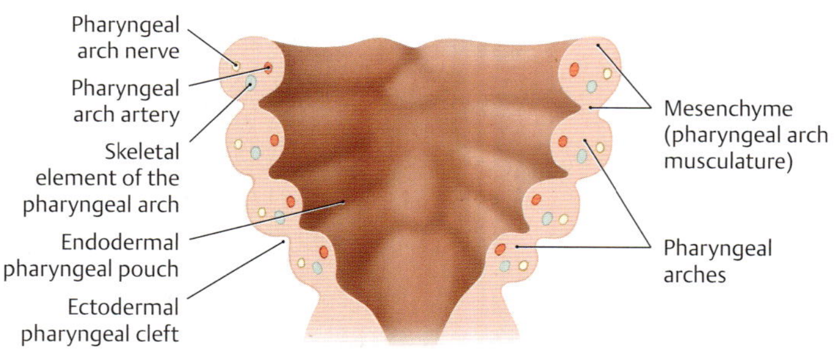

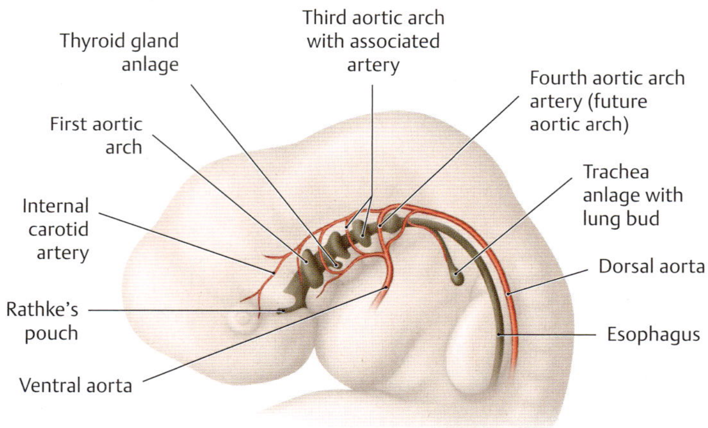

C Structure of the pharyngeal arches

View of the floor of the pharyngeal gut and the transversely sectioned pharyngeal arches. The typical components of a pharyngeal arch are easily identified: the aortic arch, musculature and associated nerves, and a cartilaginous skeletal element of each pharyngeal arch. The derivatives of these structures are of key importance in the formation of the face, neck, larynx, and pharynx. Because the developmental transformation of pharyngeal arch structures is complex, it is readily disrupted, causing malformations that may involve a cluster of related derivatives. Defects in pharyngeal arch development result in branchial and lateral cervical cysts and fistulas and in a group of "first arch syndromes" involving mandibulofacial deformities.

D Location of the aortic arch and pharyngeal pouches

The aortic arches (branchial arch arteries) *arise* from the paired embryonic *ventral aorta* and run between the pharyngeal pouches. They *open* dorsally into the *dorsal aorta*, which is also paired. The definitive aortic arch develops from the fourth aortic arch on the left side (the development of the aortic arch is described on p. 12). The pharyngeal pouches are paired, diverticula-like outpouchings of the endodermal pharyngeal gut. A total of four distinct pharyngeal pouches develop on each side; the fifth is often absent or rudimentary.
Note: The pouch protruding from the roof of the oral cavity is called *Rathke's pouch* (precursor of the anterior pituitary). Note also the *lung bud* extending ventrally from the pharyngeal gut and the anlage of the *thyroid gland*.

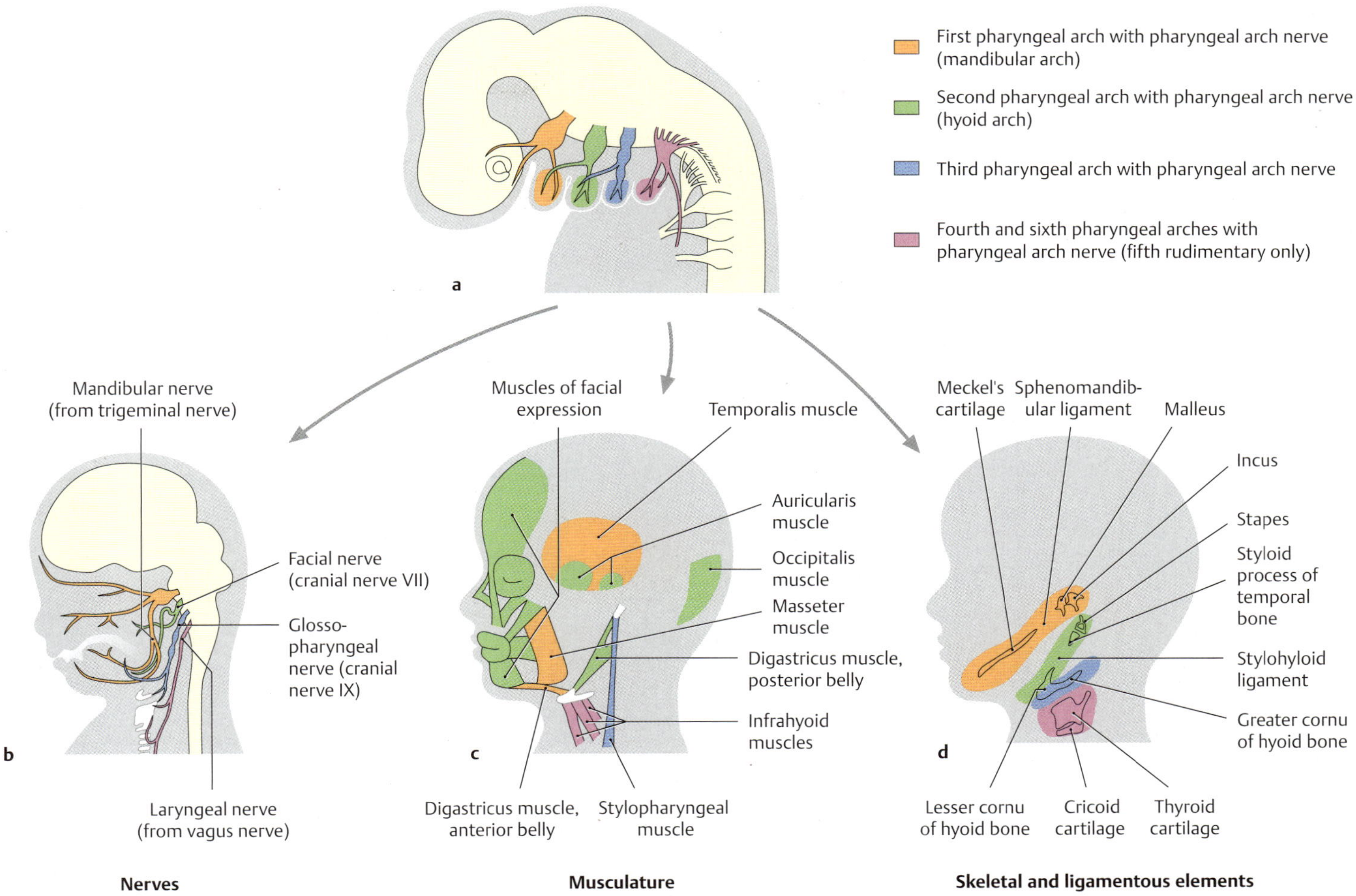

E The system of pharyngeal or branchial arches

a Anlage of the embryonic pharyngeal arches with the associated pharyngeal arch nerves.
b Definitive arrangement of the future cranial nerves V, VII, IX, and X.
c Muscular derivatives of the pharyngeal arches
d Skeletal derivatives of the pharyngeal arches.

F Derivatives of the pharyngeal (branchial) arches in humans

Pharyngeal arch	Nerve	Muscles	Skeletal and ligamentous elements
First (mandibular arch)	Cranial nerve V (mandibular nerve from the trigeminal)	Masticatory muscles – Temporalis – Masseter – Lateral pterygoid – Medial pterygoid Mylohyoid Digastric (anterior belly) Tensor tympani Tensor veli palatini	Malleus and incus Portions of the mandible Meckel's cartilage Sphenomandibular ligament Anterior ligament of malleus
Second (hyoid arch)	Cranial nerve VII (facial nerve)	Muscles of facial expression Stylohyoid muscle Digastric muscle (posterior belly) Stapedius	Stapes Styloid process of the temporal bone Lesser cornu of hyoid bone Upper part of hyoid body
Third	Cranial nerve IX (glossopharyngeal nerve)	Stylopharyngeus muscle	Greater cornu of hyoid bone Lower part of hyoid body
Fourth and sixth	Cranial nerve X (superior and recurrent laryngeal nerve)	Pharyngeal and laryngeal muscles	Laryngeal skeleton (thyroid cartilage, cricoid cartilage, arytenoid cartilage, corniculate and cuneiform cartilages)

1.6 Early Embryonic Circulation and the Development of Major Blood Vessels

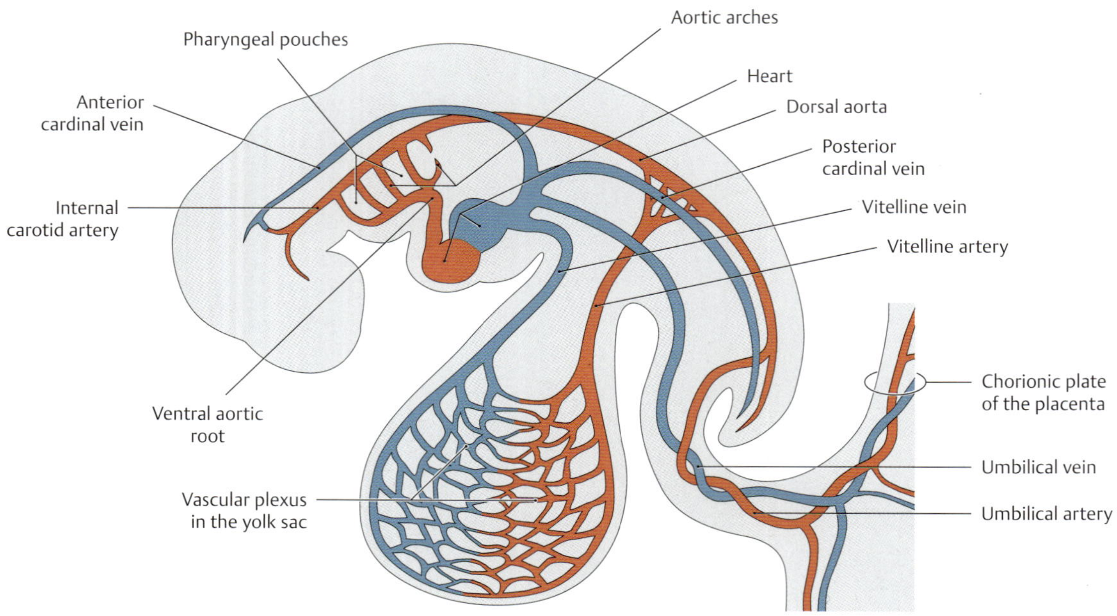

A Circulatory system of a 3- to 4-week-old human embryo
Lateral view. The cardiovascular system of a 3- to 4-week-old human embryo consists of a well-functioning two-chambered heart and three distinct circulatory systems:

1. An **intraembryonic systemic circulation** (ventral and dorsal aorta, branchial arch and aortic arches, anterior and posterior cardinal veins)
2. An **extraembryonic vitelline circulation** (omphalomesenteric arteries and veins)
3. A **placental circulation** (umbilical arteries and veins)

The vascular pathways still show a largely symmetrical arrangement at this stage.

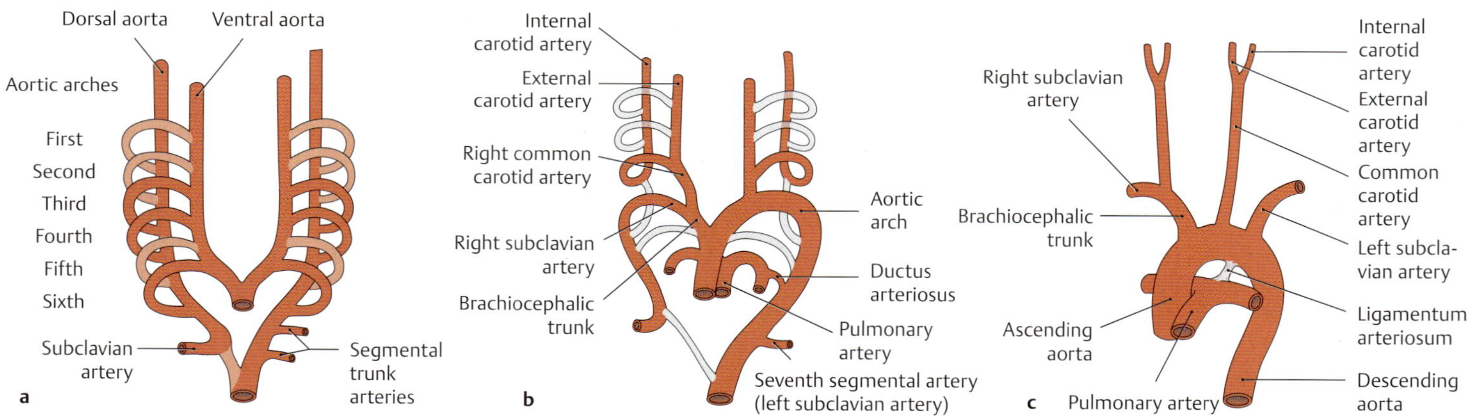

B Development of the arteries derived from the aortic arch (after Lippert and Pabst)

a Initial stage (4-week-old embryo, ventral view). An artery develops in each of the pharyngeal arches, proceeding in the craniocaudal direction. These arteries arise from the paired ventral aortic roots, course through the mesenchyme of the pharyngeal arches, and open into an initially paired dorsal aorta. These vessels give rise to segmental trunk arteries. The six aortic arches are not all present at any one time, however. For example, while the fourth arch is forming, the first two arches are already beginning to regress. The development proceeds in such a way that the original symmetry is lost in favor of a preponderance on the left side.

b Structures that regress or persist: The first, second, and fifth aortic arches on both sides regress with continued development. The third aortic arch gives rise to a common carotid artery on each side and the proximal portion of the internal carotid artery. The left fourth aortic arch later becomes the definitive *aortic arch*, while the artery on the right side becomes the proximal right subclavian artery. The left subclavian artery is derived from the seventh segmental artery. The trunk of the pulmonary arteries and the ductus arteriosus are derived from the sixth aortic arch.

c Variants in the adult: Besides the typical case pictured here (77%), there are numerous variants of the brachiocephalic trunk that occur with different frequencies. In the second most common pattern (13%), the left common carotid artery also arises from the brachiocephalic trunk. A right-sided aortic arch and a duplicated aortic arch each occur with a frequency of about 0.1%.

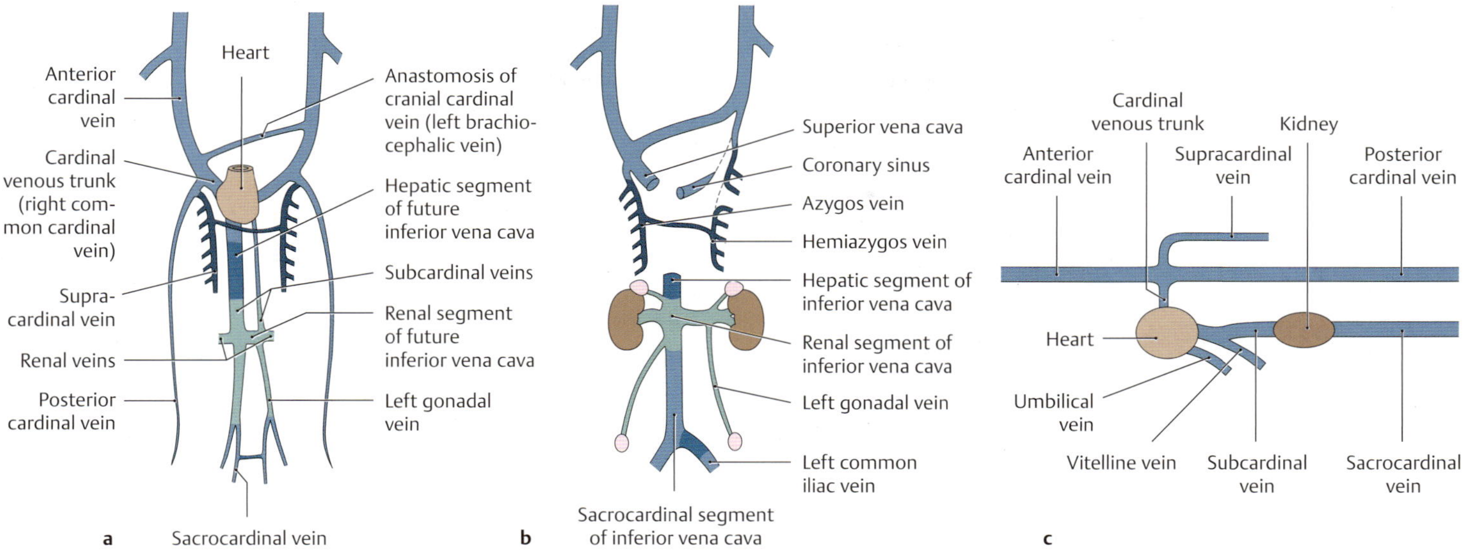

C Development of the cardinal venous system from weeks 5–7 to birth

a At 5–7 weeks (ventral view), **b** at term (ventral view), **c** lateral view at 5–7 weeks.

Up until the 4th week of development, three paired venous trunks return the blood to the heart: the vitelline, umbilical, and cardinal veins. The cardinal venous system at this stage consists of the anterior, posterior, and common cardinal veins. The following additional cardinal venous systems are formed between weeks 5 and 7:

- **Supracardinal veins:** These vessels replace the posterior cardinal veins and receive blood from the intercostal veins (future azygos system: azygos and hemiazygos veins).
- **Subcardinal veins:** These vessels develop to drain the kidneys—the right subcardinal vein becoming the middle part of the inferior vena cava, and the transverse anastomosis becoming the left renal vein. The distal segment of the left subcardinal vein persists as a gonadal vein (testicular vein or left ovarian vein).
- **Sacrocardinal veins:** These vessels develop during the formation of the lower limbs, their transverse anastomosis becoming the left common iliac vein.

Characteristic transverse anastomoses are formed between the individual cardinal venous systems. These connections transfer blood from the right to the left side, channeling it to the inflow tract of the heart. The transverse anastomosis between the anterior cardinal veins, for example, forms the future left brachiocephalic vein. The future superior vena cava develops from the right anterior and common cardinal veins, while the left common cardinal vein contributes to the venous drainage of the heart (coronary sinus).

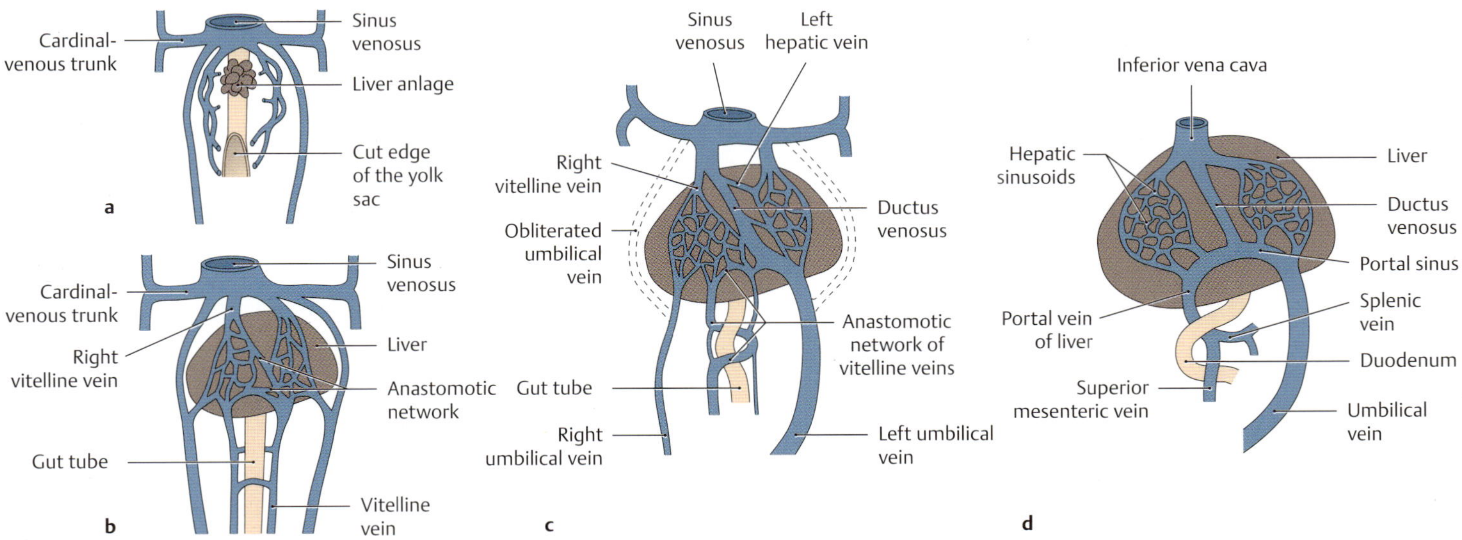

D Development of the vitelline and umbilical veins

a 4th week, **b** 5th week, **c** 2nd month, **d** 3rd month (ventral view).

Before the vitelline veins (omphalomesenteric veins) open into the venous sinus, they form a venous plexus around the duodenum, perfuse the embryonic liver anlage, and form the first hepatic sinusoids. At this stage the two umbilical veins still course on both sides of the liver anlage. With further development, however, they establish a connection with the hepatic sinusoids. While the right umbilical vein regresses completely during the 2nd month, the left umbilical vein assumes the function of transporting all blood back from the placenta to the fetus. The blood flows through a shunt (ductus venosus) into the *proximal* trunk of the right vitelline vein (the future posthepatic part of the inferior vena cava) and back to the sinus venosus. The *distal* portion of the right vitelline vein develops into the future portal vein, by which blood is conveyed from the unpaired abdominal organs to the liver (superior and inferior mesenteric veins, splenic vein).

1.7 Primordial Skeleton, Prenatal Skeletal Development, and Ossification Centers

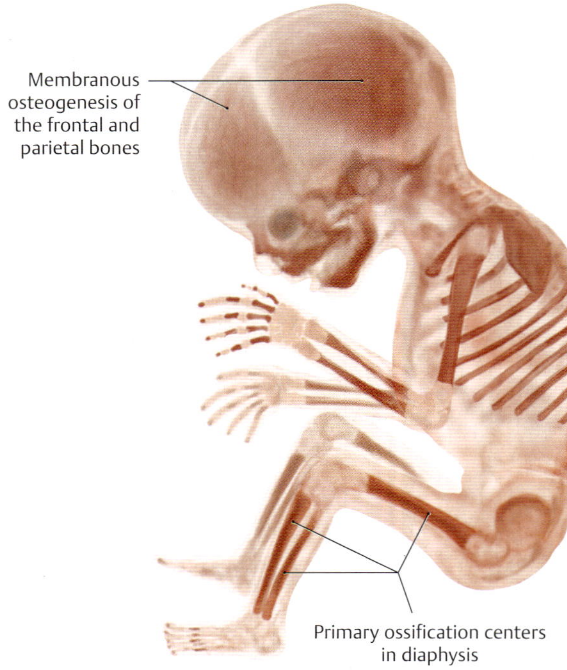

A Primordial skeleton, prenatal skeletal development, and ossification centers (Transparent specimen of an 11-week-old fetus's ossification centers stained with alizarin red).

The supporting tissue of the human skeleton (mainly cartilage and bone tissue) arises from embryonic connective tissue (= mesenchyme), derived from the middle germ layer (= mesoderm, see p. 7, **D**). Mesenchymal cells initially differentiate into chondroblasts (endochondral osteogenesis), which form a miniature model of the skeleton in its final form (primordial skeleton) from hyaline cartilage. It is only at a later developmental stage that the hyaline cartilage is replaced by bone tissue. Endochondral osteogenesis leads to the formation of the largest part of the human skeleton (torso, limbs, and skull base). The ossification of the primordial skeleton begins at the end of the embryonic period (8th week). It extends from the secondary ossification centers in the diaphyseal region of the long bones (perichondral ossification = bone formation directly from the mesenchyme, thus desmal osteogenesis). Soon after, with the start of endochondral ossification (= bone formation around the cartilage, thus endochondral osteogenesis), the formation of primary ossification centers in the diaphysis begins. By the 12th week, there are ossification centers in all long bones. In the epiphysis, endochondral ossification (secondary ossification centers) only begins some time after birth (exception: distal femur and proximal tibial epiphysis, which already exist at birth). At birth, many short bones (e.g., most of the tarsal bones and all of the carpal bones) consist entirely of cartilage and form an ossification center only in the ensuing months and years. Only the bones of the cranial roof, parts of the facial bones, and the collar bone (clavicle) ossify directly (membranous osteogenesis, which means mesenchyme cells develop into osteoblasts [see p. 17]).

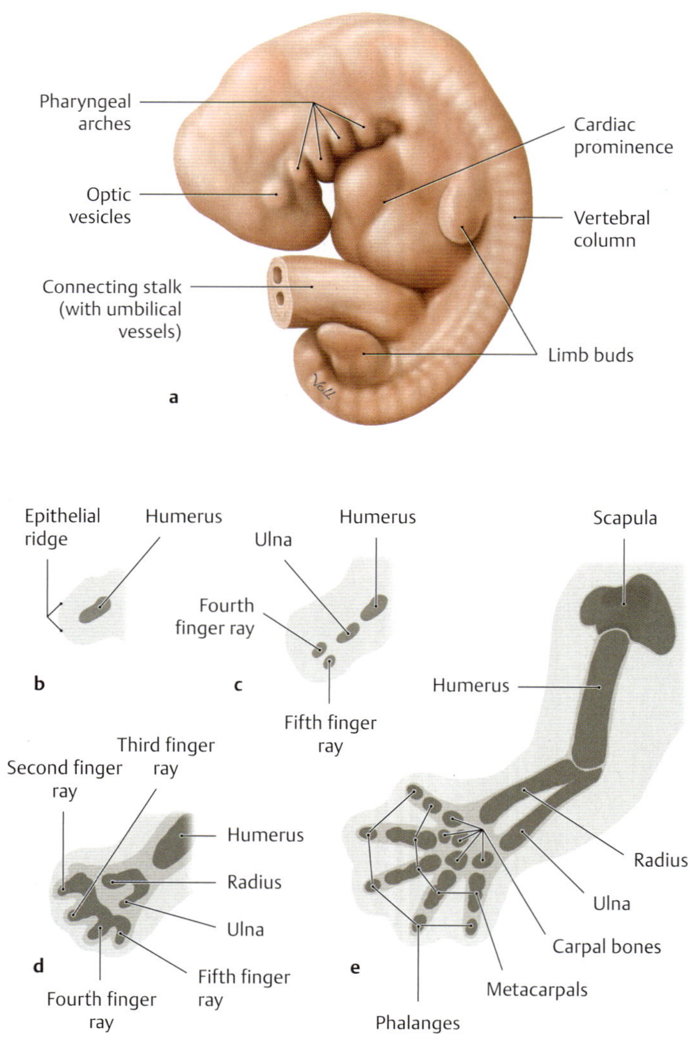

B Development of the limbs illustrated for the arm

At the end of the 4th week of embryonic development, the limb anlage (primordia) appear as paddle-shaped outpouchings around the lateral trunk regions (**a**). These limb buds consist of mesenchymal cells (so-called limb bud blastema, see p. 153) and an ectoderm covering with an apical epithelial thickening, the epithelial ridge (**b**). The growth and arrangement of the limb buds occurs within a three-dimensional coordinate system, which has both proximal-distal and cranial-caudal axes of differentiation. The humerus of the upper arm develops first, followed by the ulna and adjacent skeletal elements of the carpal roots (**c**). The ulna and the fourth and fifth finger rays are referred to as the postaxial (caudal) section; the radius and first, second, and third finger rays, as the preaxial (cranial) section (**d**). The finger and foot rays, for example, form when the cells of the epithelial ridge divide into five segments through programmed cell death (interdigital apoptosis [**e**]). In case of a disruption of apoptosis, adjacent fingers or toes fuse together (syndactyly). The absence of entire limbs is called *amelia*; the absence of parts of the arm or leg is called *meromelia*.

Note: Achondroplasia is a genetic disorder that causes a disruption in the endochondral ossification and is one of the most frequent causes of disproportionate short stature (short limbs, short trunk, too large of a skull relative to the rest of the body). Due to a disruption in cell division and maturation of the chondrocytes in the growth plates, all bones generated by endochondral osteogenesis remain too short. Membranous osteogenesis, however, proceeds normally.

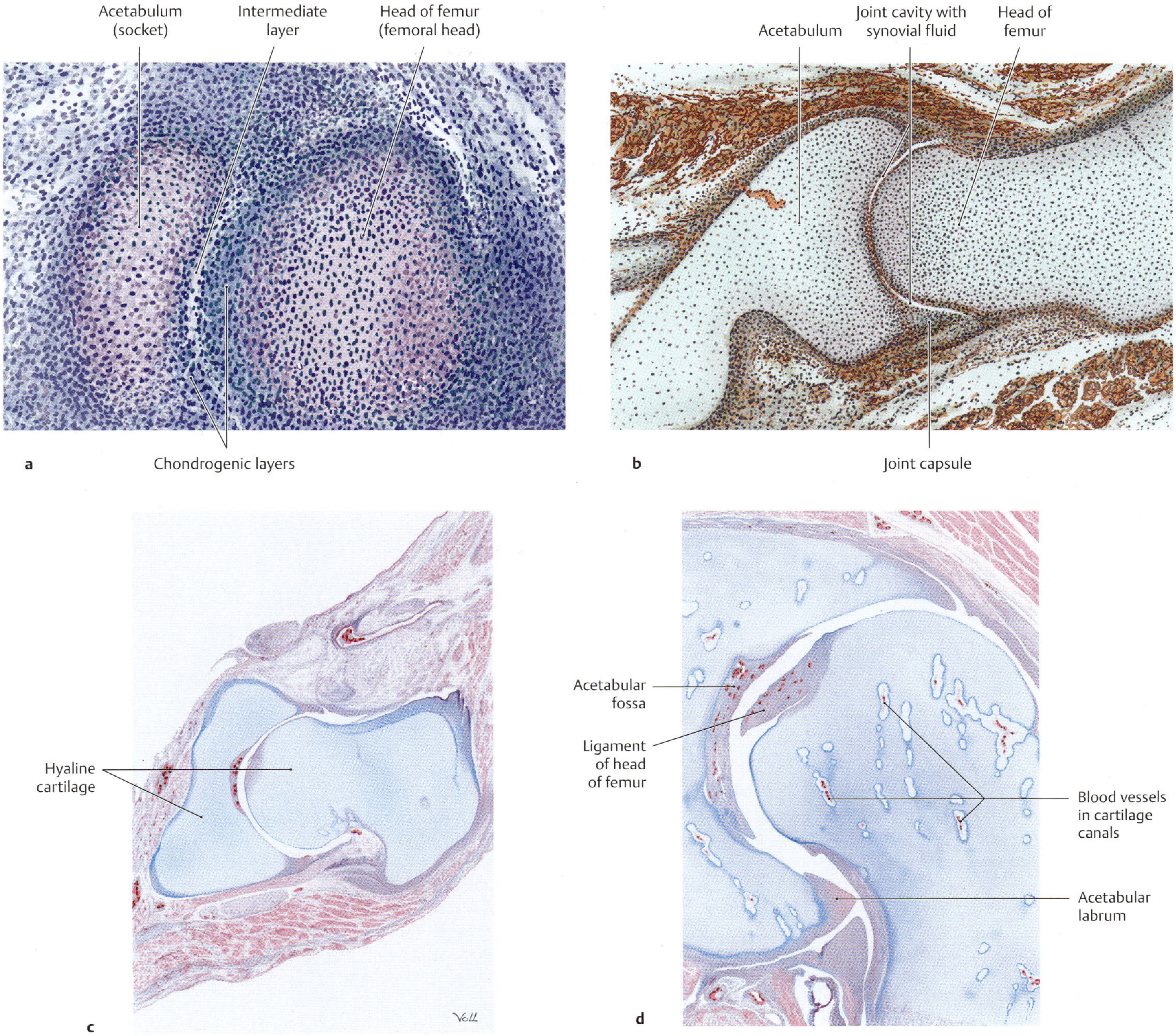

C Development of joints illustrated for the hip joint

a In the 6th week of embryonic development, cell condensation occurs where the joints will eventually develop. A three-layered joint interzone forms. It consists of two chondrogenic layers covering the cartilage anlage (primordia) and an intermediate layer containing fewer cells.

b Around the 8th week of embryonic development, programmed apoptopic cell death leads to the formation of the joint cavity around the intermediate layer. The joint capsule arises from the peripheral region of the joint interzone and begins to produce synovial fluid.

c After the joint cavity has been formed, hyaline cartilage develops in chondrogenic layers. At the end of the 12th week of embryonic development, the development of joints is complete. The final structure of the joint is determined by functional loads (e.g., muscle force), although the joint has a genetically predetermined shape.

d The joint continues to grow (interstitial and appositional growth). With the start of the 13th week of embryonic development, nutrition through diffusion from both the perichondrium and synovia of the joint cavity is no longer sufficient, which leads to the formation of intrachondral vessels along the so-called cartilage canals. Only one area close to the joint cavity remains blood vessel-free. The vascularization of the cartilaginous epiphysis, however, stands in no relation to the development of the secondary ossification center. At the femoral head of the hip joint, for instance, there is a delay of about 12 months between the first vascularization (3rd month of embryonic development) and the appearance of the ossification center in the proximal femoral epiphysis (6th postnatal month).

Note: There are essentially two ways in which joints can develop:

- Through segmentation (most common form), meaning cleavage of a continuous skeletal condensation (applies to almost all joints: hip joint, shoulder joint, elbow joint, etc.).
- Through appositional growth, meaning two initially separate skeletal elements grow toward one another (e.g., temporomandibular, sternoclavicular, and iliosacral joints). First, a synovial bursa forms at the junction between those elements, which then transforms into the joint cavity. In addition, intra-articular disks are typical for this type of joints (exception: iliosacral joint).

1.8 Bone Development and Remodeling

Bone development and bone remodeling are closely interrelated. During growth, for instance, bone undergoes a constant remodeling process in which immature woven bone tissue is replaced by "mature" lamellar bone. Continual remodeling also occurs in the mature skeleton, particularly in cancellous bone (called also trabecular or spongy bone, see **F**). In this way, on average, approximately 10% of the entire adult skeleton is remodeled each year, meaning that the skeleton is completely renewed over about a 10-year period. This process is basically a functional adaptation of the bones to the dominant stress patterns to which they are exposed (and which vary over time). It also serves to prevent material fatigue, repair microinjuries to bone, and provide a rapidly available source of calcium.

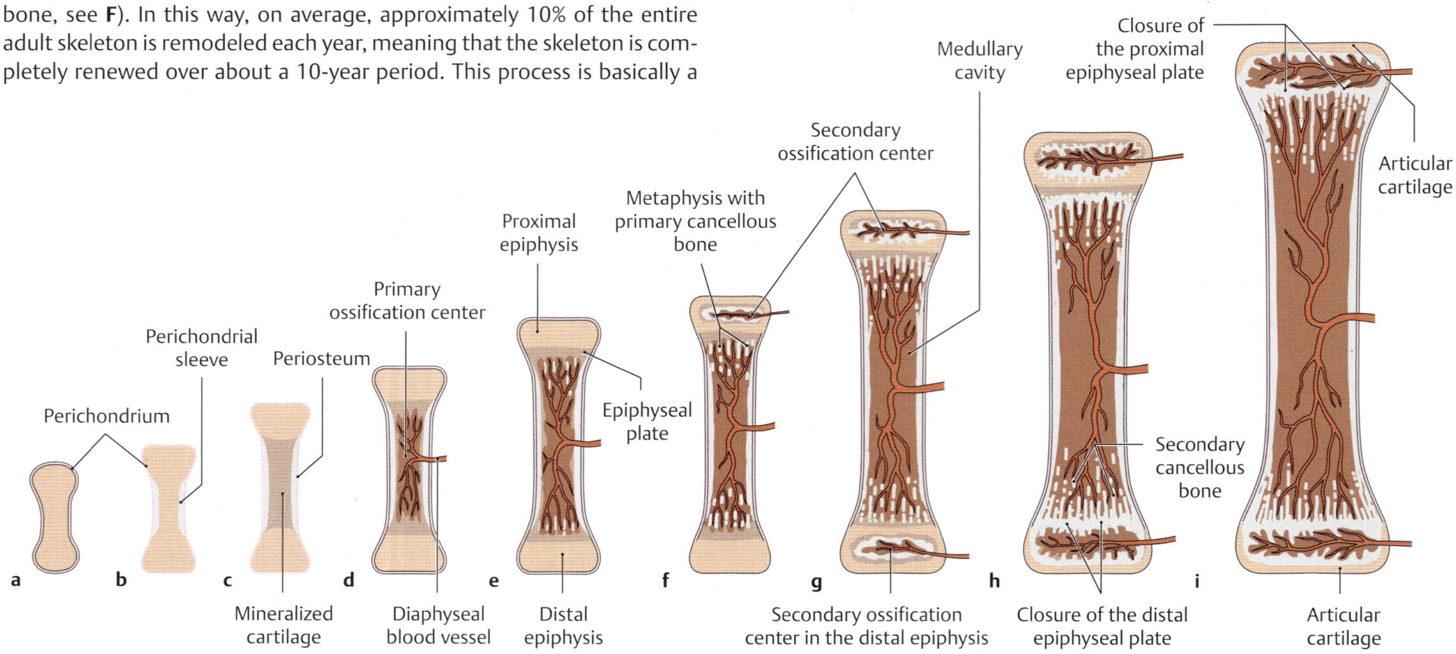

A Development of a long bone

The long bones (humerus, tibia, etc.) are mainly a product of indirect bone formation; i. e., they form by replacing a preexisting cartilaginous model of the bone (*endochondral osteogenesis*).But portions of the long bones (the perichondrial bone collar, which allows the bone to grow in thickness) are a product of *direct* bone formation (i. e., they form from the direct transformation of condensed mesenchyme [*membranous osteogenesis*, see **E**]).

a Cartilaginous model of a bone in the embryonic skeleton. **b** Formation of a perichondrial bone collar (directly from mesenchyme). **c** Differentiation to hypertrophic chondrocytes and mineralization of the cartilaginous extracellular matrix. **d** Ingrowth of a diaphyseal vessel and formation of a primary ossification center. **e** Development of the proximal and distal growth centers (epiphyseal plates). **f** Appearance of the proximal epiphyseal ossification center (secondary ossification center). **g** Formation of the distal epiphyseal ossification center. **h** Closure of the distal epiphyseal plate. **i** Closure of the proximal epiphyseal plate (occurs at the end of skeletal growth, between about 18 and 23 years of age for most tubular bones).

Note: Osteogenesis = the formation of an individual bone; ossification = the formation of bone tissue.

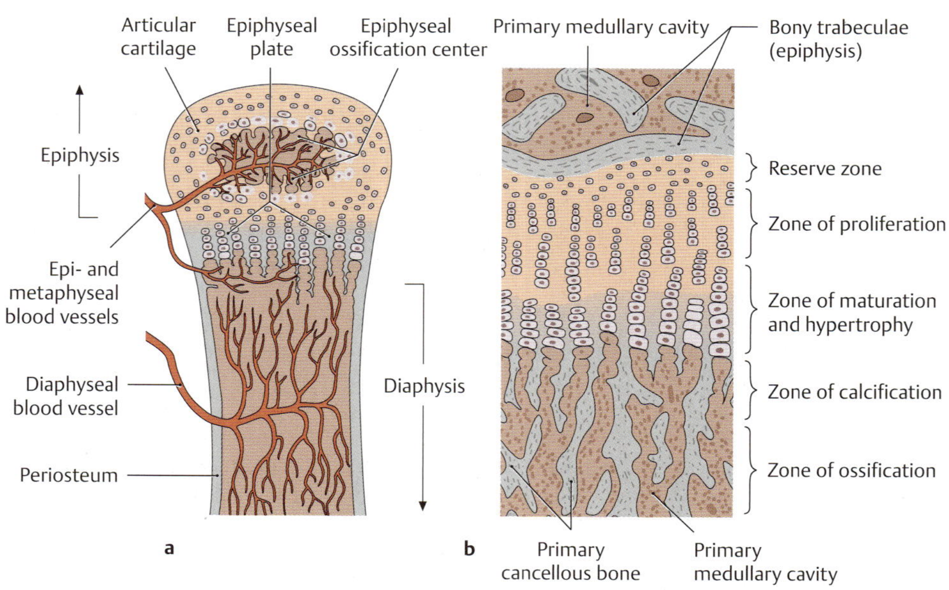

B Structure of the epiphyseal plate

a Blood supply, **b** detail from **a:** zones of the epiphyseal plate.

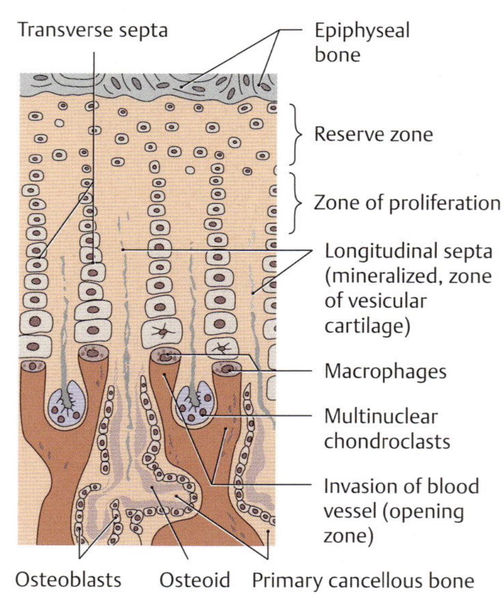

C Schematic representation of cellular processes within the epiphyseal plate

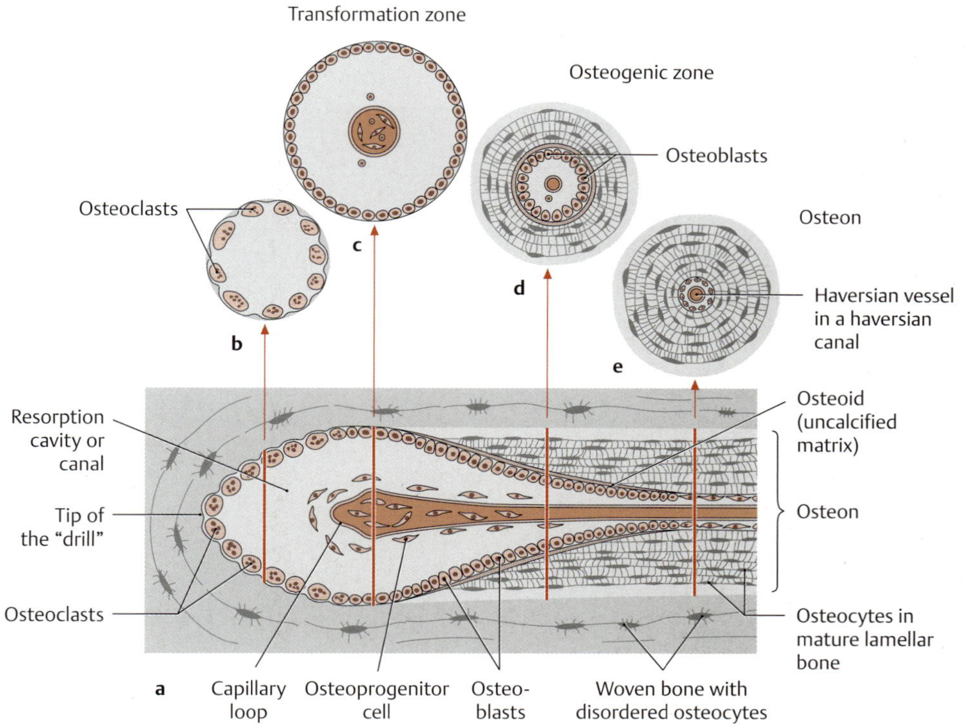

D Development of an osteon

The process of functional remodeling (see upper left page) begins with the invasion of blood vessels and accompanying osteoclasts ("bone-eating cells") into woven bone. They burrow through the woven bone like a drill, cutting a vascularized channel (resorption canal or cavity) that is equal in diameter to the future osteon.

a Longitudinal section through a resorption canal.
b Cross section at the level of the resorption canal.
c Transformation zone: osteoprogenitor cells (a kind of precursor for bone-forming cells) are transformed into osteoblasts.
d Osteogenic zone (osteoblasts produce bony lamellae).
e Newly formed osteon.

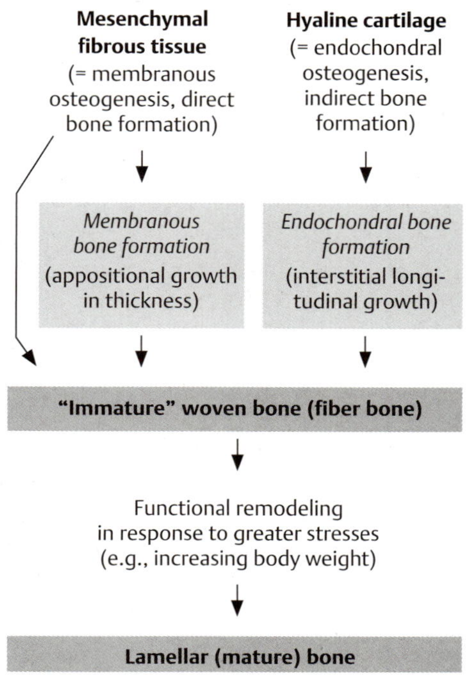

E Types of bone development (osteogenesis)

Note: Most bones are formed by indirect osteogenesis (the few exceptions include the clavicle and certain bones of the calvaria). But portions of these bones still develop from the direct transformation of mesenchyme, i.e., by direct osteogenesis.

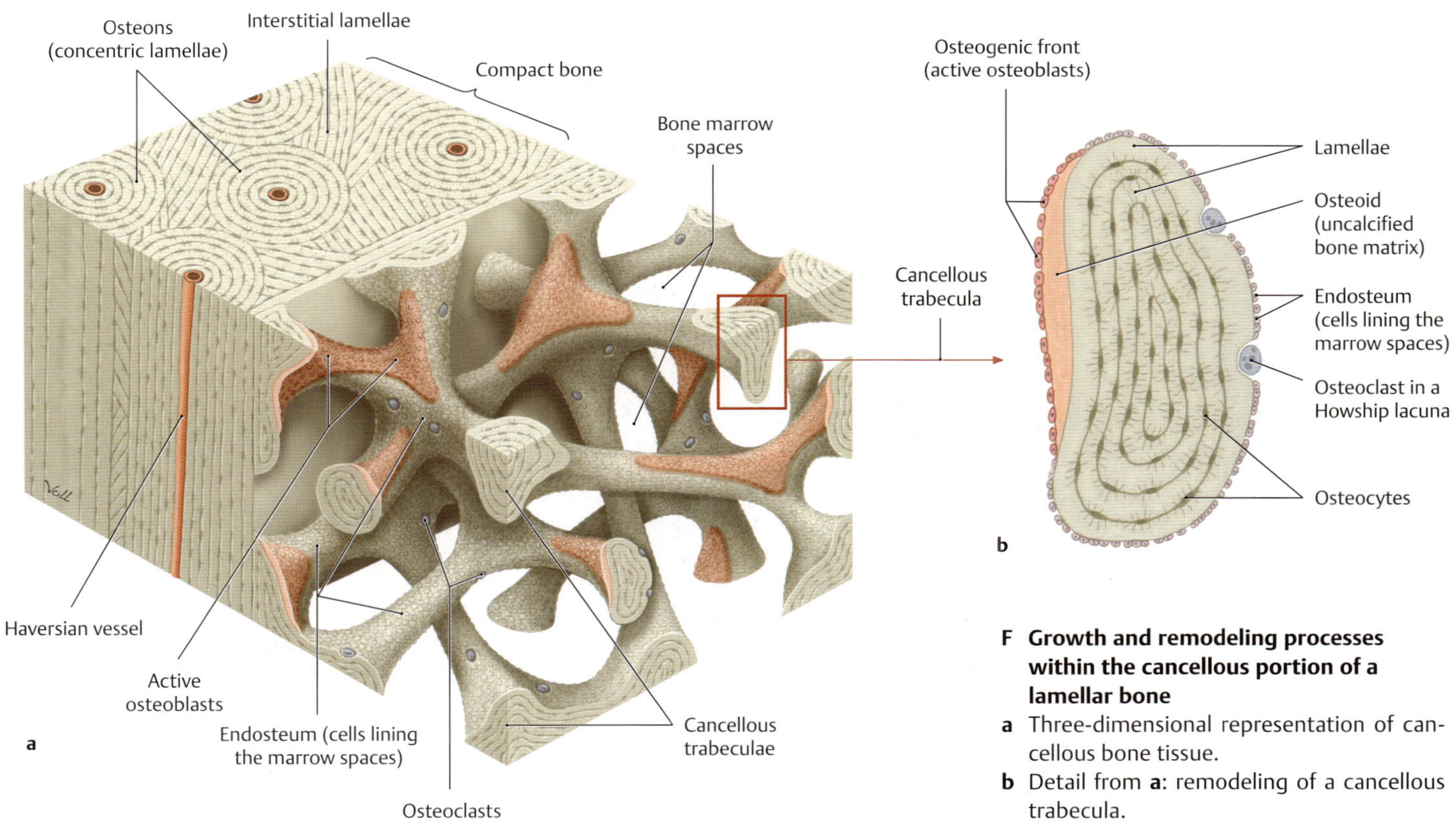

F Growth and remodeling processes within the cancellous portion of a lamellar bone

a Three-dimensional representation of cancellous bone tissue.
b Detail from **a**: remodeling of a cancellous trabecula.

1.9 Ossification of the Limbs

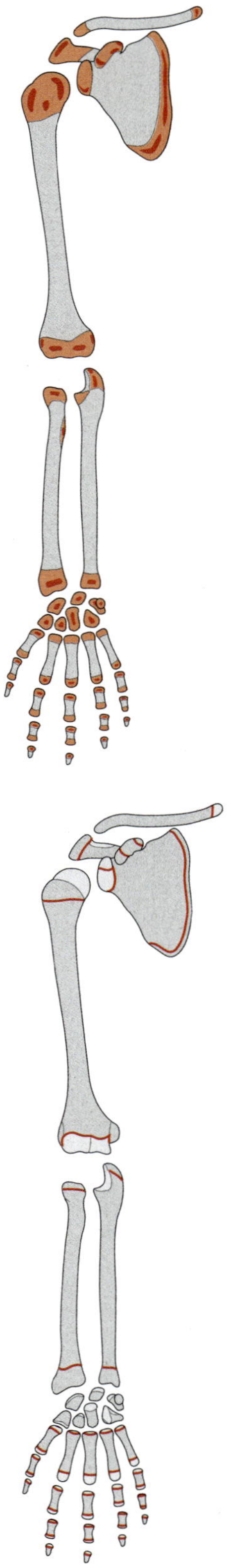

a

b

B Ossification of the skeleton of the upper limb

a Location of the epi- and apophyseal ossification centers.

b Location of the epi- and apophyseal plates.

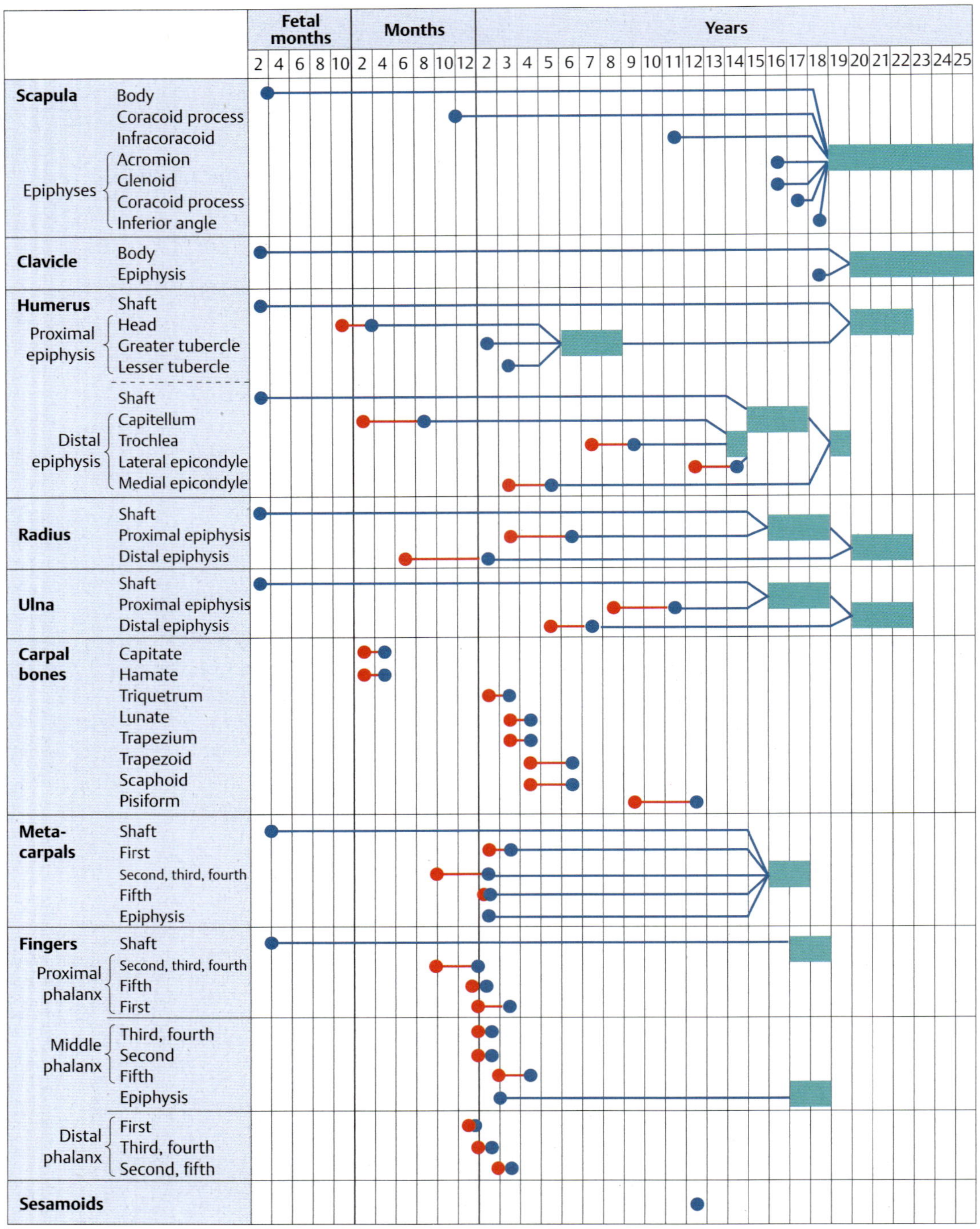

Times of appearance of ossification centers:
- ● Female
- ● Male
- —— Period of ossification
- ▬ Period of synostosis

A Fetal months and first year of life are set off from the rest of the table.

A and C Timetable of regional bone growth in the upper limb (A) and lower limb (C)

The current stage of skeletal development, and thus the individual skeletal age, can be estimated from the times of appearance of the ossification centers. *Primary ossification centers*, which generally appear in the shaft region of bones *during the fetal period* (diaphyseal ossification), are distinguished from *secondary ossification centers*, which form *after birth* within the cartilaginous epiphysis and apophysis (epi- and apophyseal ossification). Longitudinal growth ceases with the closure of the epiphyseal plate (synostosis). The greater tubercle of the humerus, for example, begins to ossify at 2 years of age. A period of synostosis follows from 6 to 8 years of age, and after that the greater tubercle shows only external, appositional growth. With the cessation of longitudinal growth, the ossification centers disappear and are no longer visible on X-ray films. The relationship between maturation and the appearance of secondary ossification centers is most clearly demonstrated in the bones of the carpus (see also **B**). The eight carpal bones ossify

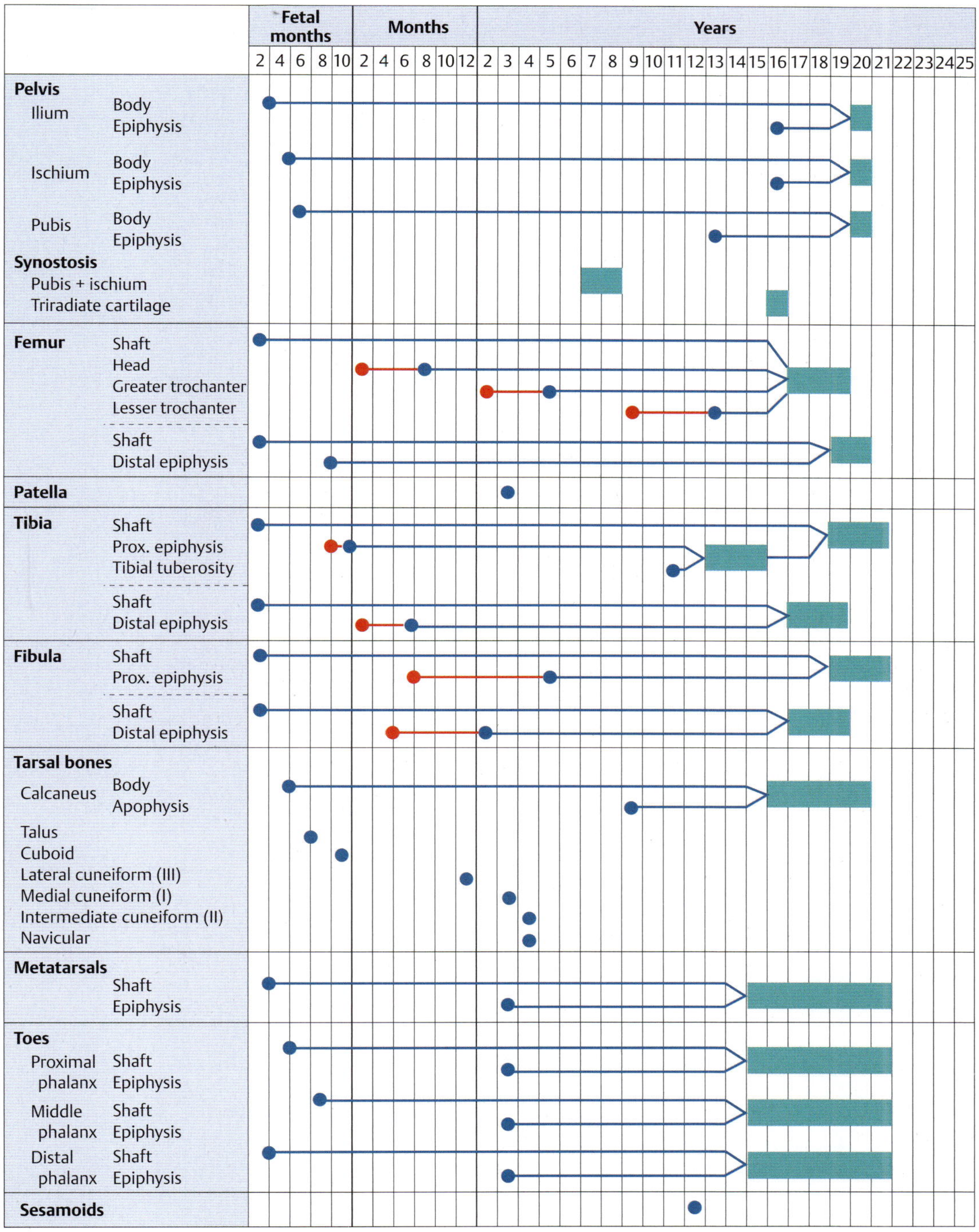

Times of appearance of ossification centers:
- ● Female
- ● Male
- — Period of ossification
- ▬ Period of synostosis

C Fetal months and 1st year of life are set off from the rest of the table.

gradually over a period of approximately 9 years. The first ossification center is that of the capitate bone, which appears during the 1st year of life; the last is that of the pisiform bone, which ossifies at 9 years of age. It is standard practice to use the left or nondominant hand for radiographic examinations. The skeletal age reflects the biological maturity of the organism more than chronological age. The estimation of skeletal age, and thus of growth potential, is of key importance, for example, in the prognosis and treatment of orthopedic diseases and deformities in children. Also, given the relationship that exists between skeletal maturity and body height, the definitive adult height can usually be predicted with reasonable accuracy after 6 years of age based on the skeletal age and longitudinal measurements.

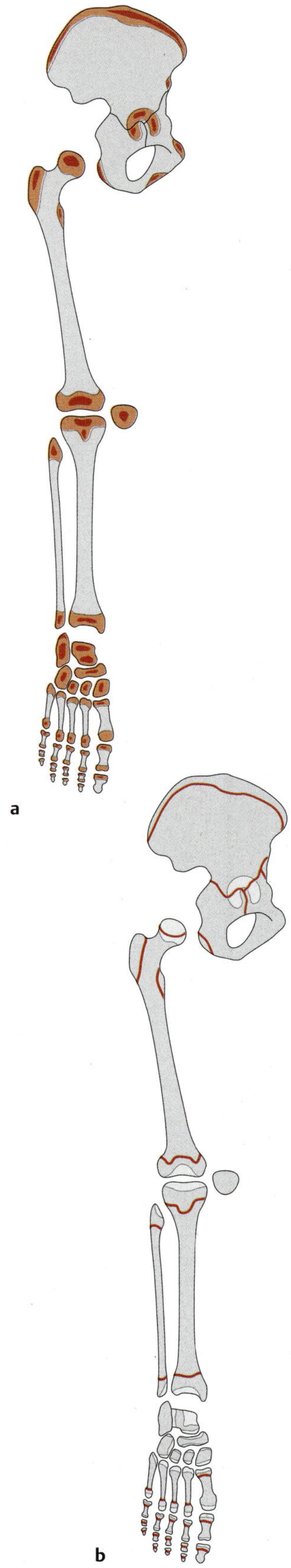

D Ossification of the skeleton of the lower limb

a Location of the epi- and apophyseal ossification centers.

b Location of the epi- and apophyseal plates.

1.10 Basic Skeletal Structure and Development and Position of the Limbs

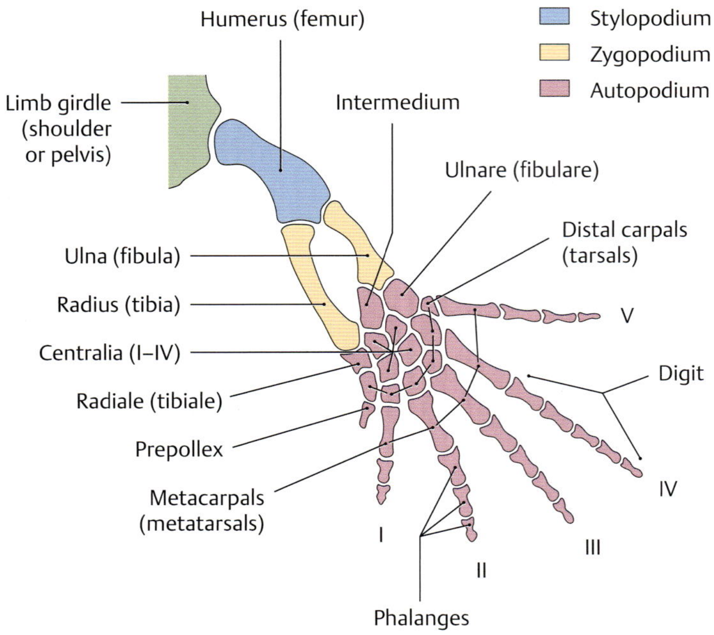

A Basic skeletal structure of a five-ray (pentadactyl) tetrapod limb
Both the forelimb and hindlimb of a free-ranging terrestrial vertebrate have the same basic, three-part structure consisting of a proximal, middle, and distal segment (called the *stylopodium*, *zygopodium*, and *autopodium*). The elbow or knee joint is placed between the stylopodium, which consists of a single bone (humerus or femur), and the zygopodium, which consists of two bones (the radius and ulna or the tibia and fibula). The five-ray autopodium (hand or foot) is also made up of proximal, middle, and distal units (called the *basipodium*, *metapodium*, and *acropodium*; see **C**). Additionally, there are some vertebrate classes that depart from this basic structure by showing a reduction or fusion of various bony units.

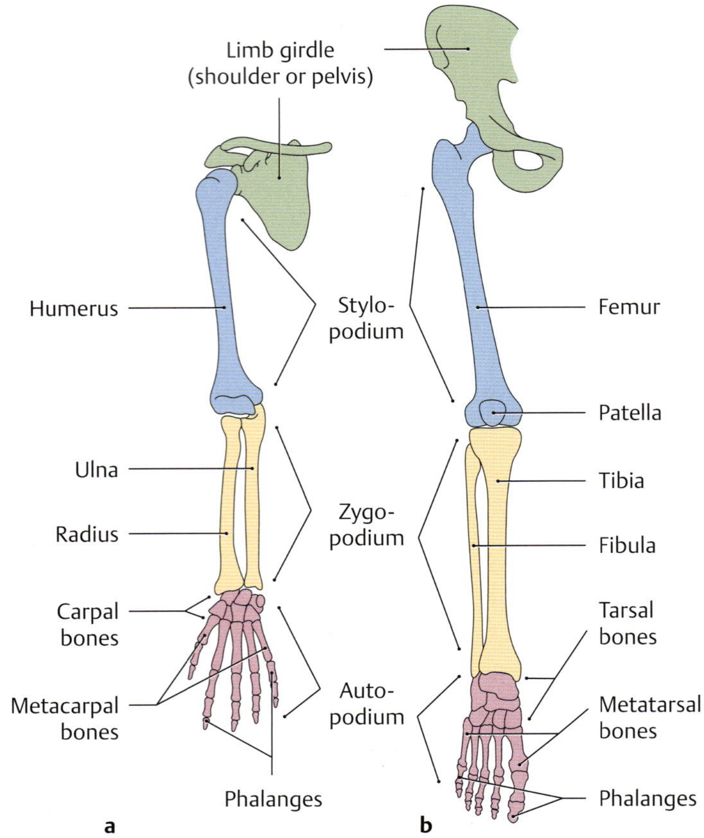

B Basic skeletal structure of the human limbs
Anterior view. **a** Right upper limb, **b** right lower limb.
The skeletal elements of the human upper and lower limbs have been colored to show how they are homologous with the tetrapod limb segments shown in **A** (stylo-, zygo-, and autopodium). Congenital malformations such as polydactyly or syndactyly (the presence of supernumerary fingers or toes or their fusion) are not uncommon.

C Bony constituents of the pentadactyl tetrapod limb

Segments	Paired forelimbs	Paired hindlimbs
Limb girdle	Shoulder girdle – Scapula and clavicle	Pelvic girdle – Hip bone
Free limbs		
Stylopodium	Arm (brachium) – Humerus	Thigh (femur) – Femur
Zygopodium	Forearm (antebrachium) – Radius – Ulna	Leg (crus) – Tibia – Fibula
Autopodium	Hand	Foot
– Basipodium	Carpus – Proximal row: radiale, intermedium, ulnare – Central group: centralia I–IV – Distal row: carpals I–V	Tarsus – Proximal row: tibiale, intermedium, fibulare – Central group: centralia I–IV – Distal row: tarsals I–V
– Metapodium	Metacarpus – Metacarpals I–V	Metatarsus – Metatarsals I–V
– Acropodium	Fingers – Digits I–V (with different numbers of phalanges)	Toes – Digits I–V (with different numbers of phalanges)

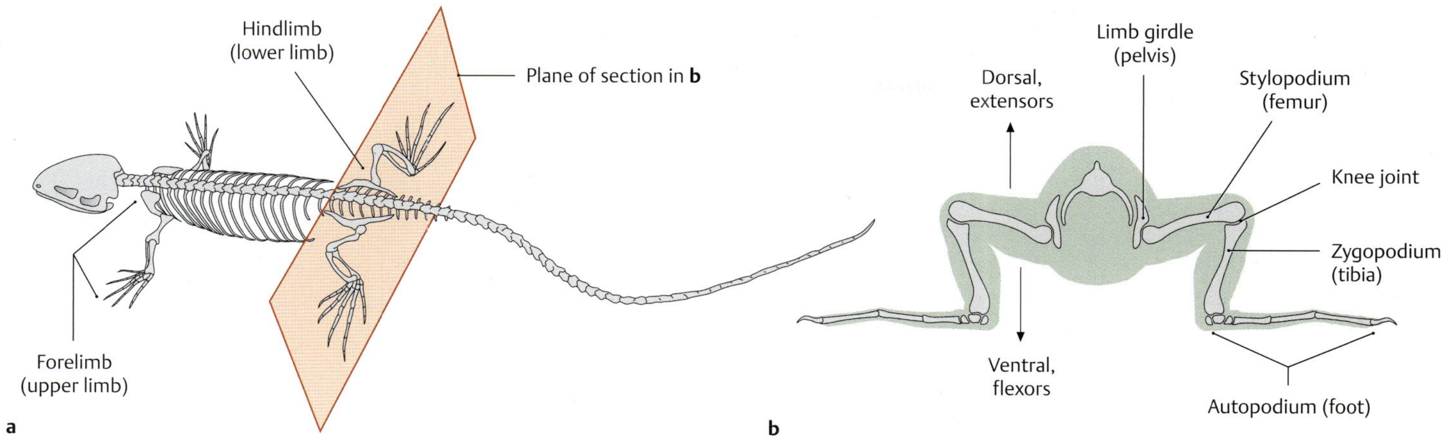

D Limb positions in a primitive terrestrial tetrapod (the lizard *Lacerta viridis*)

a Dorsal view, **b** cross section at the level of the hindlimbs.

In amphibian and reptilian tetrapods (e.g., salamanders, turtles, and lizards), the trunk is slung between the limbs and frequently touches the ground. The limbs are set almost at right angles to the body, so that the arm and thigh are nearly horizontal, and the elbow and knee point outward. The radius and ulna and the tibia and fibula are flexed at right angles at the elbow and knee. The volar surface of the hand and the plantar surface of the foot are in contact with the ground. The axes of all the joints are directed parallel to the spinal column (see **E**).

Note that the *extensor* muscles are placed *dorsally*, while the *flexor* muscles are *ventral*. Thus, the location of the extensors and flexors relative to the bone does not change with evolution—the bone merely assumes a different alignment (see also **F**).

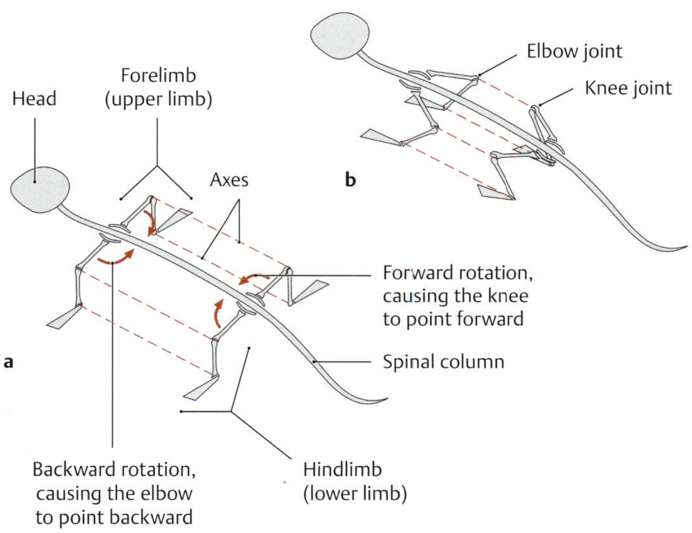

E Rotation of the limbs in mammalian evolution

a Before rotation, **b** after rotation.

An important feature of mammalian evolution involved rotation of the tetrapod limb. The limb was reoriented, placed parallel to the body, and moved closer to or beneath the body. This improved locomotion and supported the body more efficiently. The hindlimb rotated *forward* (with the knee pointing cephalad), while the forelimb rotated *backward* against the body (the elbow pointing caudad). As a result, both sets of limbs assumed a sagittal orientation under or alongside the trunk (see **F**).

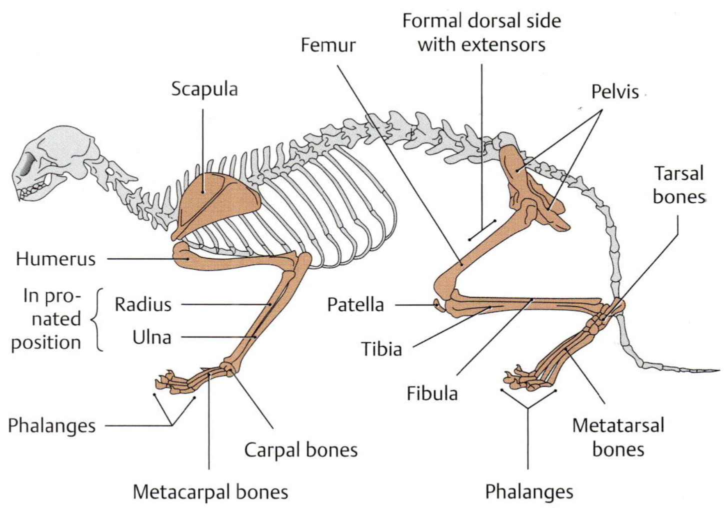

F Skeleton of a cat *(Felis catus)*

Left lateral view. In order for the volar surfaces of the forelimbs to rest on the ground despite the backward angulation of the elbows, the forearm bones must cross to a pronated position. In the hindlimbs, there is no need for pronation of the leg bones because the thigh is rotated forward.

This arrangement of the skeletal elements in the various limb segments is essentially preserved in humans. Because the lower limb has been rotated forward, the *former dorsal side* of this limb faces forward in a human standing upright. As a result, the extensors of the thigh and leg (the genetically "dorsal" muscles) are on the anterior side of the limb, placed in front of the corresponding limb bones. This is one reason why the terms "anterior" and "posterior" are preferred over "dorsal" and "ventral" in the human lower limb. By contrast, the extensors and flexors of the arm and forearm have maintained their original dorsal and ventral positions, respectively.

2.1 The Human Body: Proportions, Surface Areas, and Body Weights

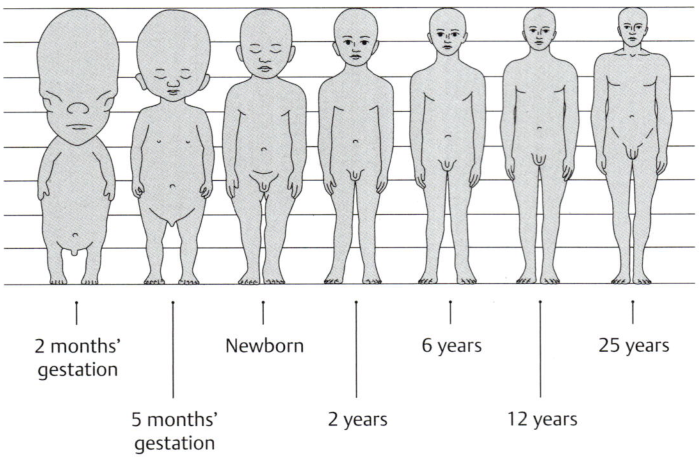

A Change in body proportions during growth
While the head height in embryos at 2 months' gestation is equal to approximately half the total body length, it measures approximately one fourth of the body length in newborns, one sixth in a 6-year-old child, and one eighth in an adult.

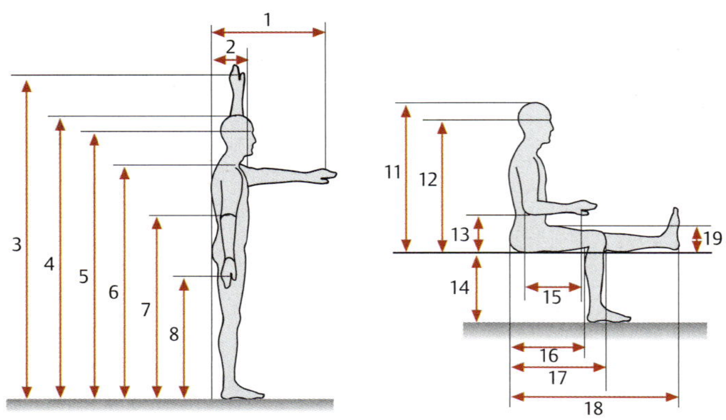

Measurements (in cm) (points 9, 10, 20, and 21 not shown)	Percentiles					
	Male			Female		
	5th	50th	95th	5th	50th	95th
1 Forward reach	68.5	74.0	81.5	62.5	69.0	75.0
2 AP body thickness	26.0	28.5	38.0	24.5	29.0	34.5
3 Overhead reach (with both arms)	197.5	207.5	220.5	184.0	194.5	202.5
4 Body height	165.0	175.0	185.5	153.5	162.5	172.0
5 Ocular height	153.0	163.0	173.5	143.0	151.5	160.5
6 Shoulder height	134.5	145.0	155.0	126.0	134.5	142.5
7 Elbow-to-floor distance	102.5	110.0	117.5	96.0	102.0	108.0
8 Hand-to-floor distance	73.0	76.5	82.5	67.0	71.5	76.0
9 Shoulder width	44.0	48.0	52.5	39.5	43.5	48.5
10 Hip width, standing	34.0	36.0	38.5	34.0	36.5	40.0
11 Sitting body height (trunk height)	85.5	91.0	96.5	81.0	86.0	91.0
12 Ocular height while sitting	74.0	79.5	85.5	70.5	75.5	80.5
13 Elbow to sitting surface	21.0	24.0	28.5	18.5	23.0	27.5
14 Height of leg and foot (height of sitting surface)	41.0	45.0	49.0	37.5	41.5	45.0
15 Elbow to gripping axis	32.5	35.0	39.0	29.5	31.5	35.0
16 Sitting depth	45.0	49.5	54.0	43.5	48.5	53.0
17 Buttock–knee length	56.5	61.0	65.5	54.5	59.0	64.0
18 Buttock–leg length	96.5	104.5	114.0	92.5	99.0	105.5
19 Thigh height	13.0	15.0	18.0	12.5	14.5	17.5
20 Width above the elbow	41.5	48.0	55.5	39.5	48.5	55.5
21 Hip width, sitting	35.0	37.5	42.0	36.0	39.0	46.0

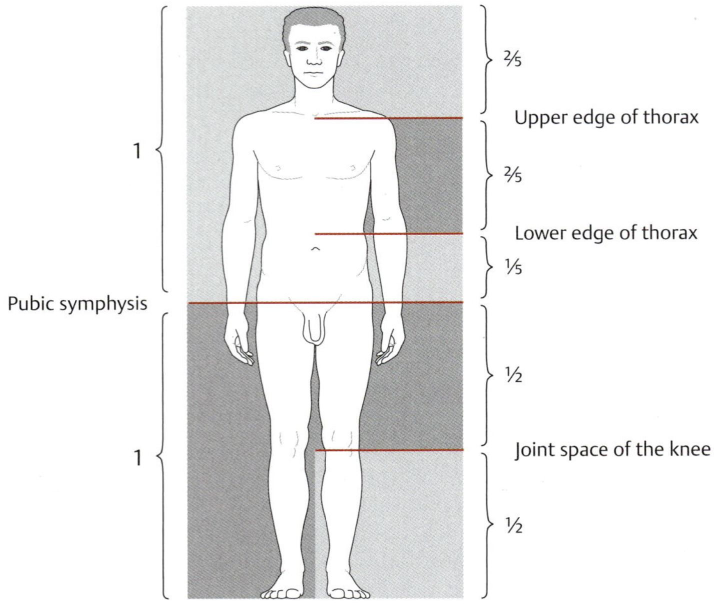

B Normal body proportions
In an adult, the midpoint of the total body height lies approximately at the level of the pubic symphysis; i.e., there is a 1:1 ratio of upper to lower body height at that level. The pelvis accounts for one fifth of the upper body height, the thorax for two fifths, and the head and neck for two fifths. The lower body height is distributed equally between the thigh and leg (plus heel) at the joint space of the knee.

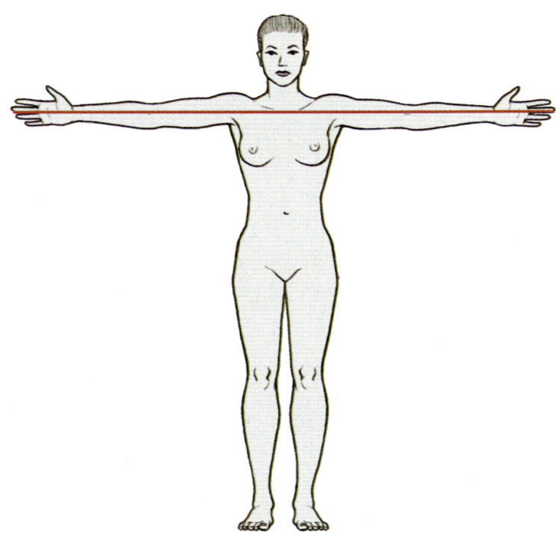

C Span of the outstretched arms
The arm span from fingertip to fingertip is slightly *greater* than the body height (approximately 103% in women and 106% in men).

D Selected body measurements in the standing and sitting human being (unclothed, 18–65 years of age) after DIN 33402-2 (from Ergonomie—Körpermaße des Menschen—Teil 2: Werte. Berlin: Beuth; 2005)
The percentile values indicate what percentage in a population group are below the value stated for a particular body measurement (in this case, all those living in Germany, thus including migrants in the years 1999–2002). For example, the 95th percentile for body height in males 18 to 65 years of age with a body height of 185.5 cm means that 95% of this population group are shorter than 185.5 cm, and 5% are taller.

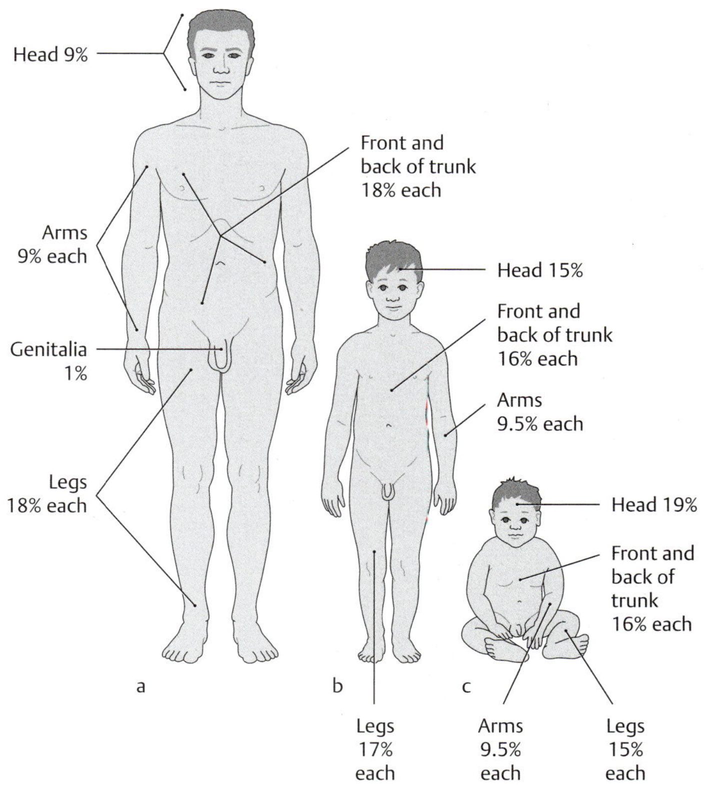

E Distribution of body surface area in adults, children, and infants

According to the "rule of nines" described by Wallace (1950), the body surface area of adults over about 15 years of age (**a**) can be divided into units that are a *multiple* of 9%: the head and each arm account for 9% each, the front and back of the trunk and each leg account for 18% (2 x 9) each, and the external genitalia comprise 1%. In children (**b**) and infants (**c**), the rule of nines must be adjusted for age.

Note: The rule of nines can be used in burn victims to provide a quick approximation of the area of skin that has been burned.

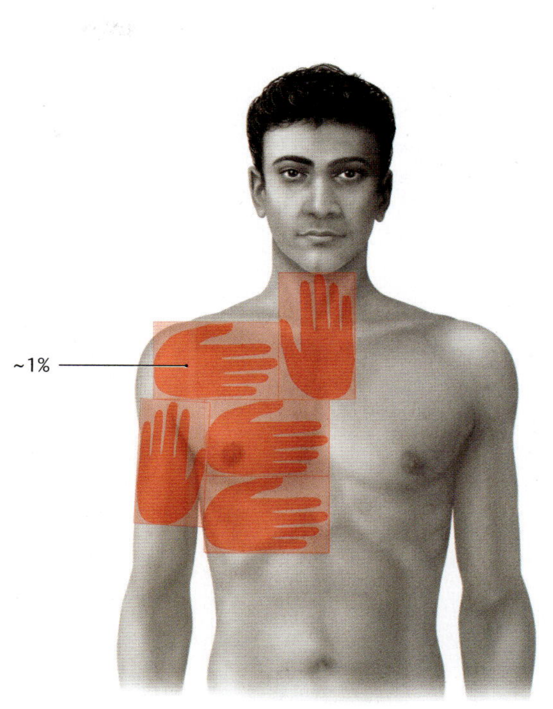

F Hand area rule

The percentage of the body surface affected by burns can be accurately estimated with the hand area rule, which states that the area of the patient's hand is approximately 1% of the patient's own total body surface area. The hand rule also applies to children, whose hands and total surface area are both proportionately smaller than in adults.

G Dependence of relative body surface area (skin surface area) on age, and consequences

For progressively larger solid bodies, the surface area increases as the square of the radius, but the volume increases as the cube of the body's radius. Because of this basic geometrical relationship, smaller animals generally have a larger relative surface area than larger animals. A higher ratio of surface area to volume causes smaller animals to radiate relatively more body heat. As a result, small animals, such as mice and children, tend to have a higher metabolic rate than larger animals, such as elephants and human adults.

Age	Body weight (kg)	Body surface area (cm^2)	Body surface area over body weight (cm^2/kg)
Newborn	3.4	2100	617.6
6 months	7.5	3500	466.7
1 year	9.3	4100	440.9
4 years	15.5	6500	419.4
10 years	30.5	10500	344.3
Adult	70.0	18100	258.6

Height in meters

Weight in kilograms	1.48		1.52		1.56		1.60		1.64		1.68		1.72		1.76		1.80		1.84		1.88		1.92		1.96		2.00
120	55	53	52	51	49	48	47	46	45	44	43	42	41	40	39	38	37	36	35	35	34	33	33	32	31	31	30
118	54	52	51	50	48	47	46	45	44	43	42	41	40	39	38	37	36	36	35	34	33	33	32	31	31	30	30
116	53	52	50	49	48	46	45	44	43	42	41	40	39	38	37	37	36	35	34	34	33	32	31	31	30	30	29
114	52	51	49	48	47	46	45	43	42	41	40	39	39	38	37	36	35	34	34	33	32	32	31	30	30	29	29
112	51	50	48	47	46	45	44	43	42	41	40	39	38	37	36	35	35	34	33	32	32	31	30	30	29	29	28
110	50	49	48	46	45	44	43	42	41	40	39	38	37	36	36	35	34	33	32	32	31	30	30	29	29	28	28
108	49	48	47	46	44	43	42	41	40	39	38	37	37	36	35	34	33	33	32	31	31	30	29	29	28	28	27
106	48	47	46	45	44	42	41	40	39	38	38	37	36	35	34	33	33	32	31	31	30	29	29	28	28	27	27
104	47	46	45	44	43	42	41	40	39	38	37	36	35	34	34	33	32	31	31	30	29	29	28	28	27	27	26
102	47	45	44	43	42	41	40	39	38	37	36	35	34	34	33	32	31	31	30	29	29	28	28	27	27	26	26
100	46	44	43	42	41	40	39	38	37	36	35	35	34	33	32	32	31	30	30	29	28	28	27	27	26	26	25
98	45	44	42	41	40	39	38	37	36	36	35	34	33	32	32	31	30	30	29	28	28	27	27	26	26	25	25
96	44	43	42	40	39	38	38	37	36	35	34	33	32	32	31	30	30	29	28	28	27	27	26	26	25	24	24
94	43	42	41	40	39	38	37	36	35	34	33	33	32	31	30	30	29	28	28	27	27	26	25	25	24	24	24
92	42	41	40	39	38	37	36	35	34	33	33	32	31	30	30	29	28	28	27	27	26	25	25	24	24	23	23
90	41	40	39	38	37	36	35	34	33	33	32	31	30	30	29	28	28	27	27	26	25	25	24	24	23	23	23
88	40	39	38	37	36	35	34	34	33	32	31	30	30	29	28	28	27	27	26	25	25	24	24	23	23	22	22
86	39	38	37	36	35	34	34	33	32	31	30	30	29	28	28	27	27	26	25	25	24	24	23	23	22	22	22
84	38	37	36	35	35	34	33	32	31	30	30	29	28	28	27	27	26	25	25	24	24	23	23	22	22	21	21
82	37	36	35	35	34	33	32	31	30	30	29	28	28	27	26	26	25	25	24	24	23	23	22	22	21	21	21
80	37	36	35	34	33	32	31	30	30	29	28	28	27	26	26	25	25	24	24	23	23	22	22	21	21	20	20
78	36	35	34	33	32	31	30	30	29	28	28	27	26	26	25	25	24	24	23	23	22	22	21	21	20	20	20
76	35	34	33	32	31	30	30	29	28	28	27	26	26	25	25	24	23	23	22	22	22	21	21	20	20	19	19
74	34	33	32	31	30	30	29	28	28	27	26	26	25	24	24	23	23	22	22	21	21	20	20	20	19	19	19
72	33	32	31	30	30	29	28	27	27	26	26	25	24	24	23	23	22	22	21	21	20	20	20	19	19	18	18
70	32	31	30	30	29	28	27	27	26	25	25	24	24	23	23	22	22	21	21	20	20	19	19	19	18	18	18
68	31	30	29	29	28	27	27	26	25	25	24	24	23	22	22	21	21	21	20	20	19	19	18	18	18	17	17
66	30	29	29	28	27	26	26	25	25	24	23	23	22	22	21	21	20	20	19	19	19	18	18	18	17	17	17
64	29	28	28	27	26	26	25	24	24	23	23	22	22	21	21	20	20	19	19	18	18	18	17	17	17	16	16
62	28	28	27	26	25	25	24	24	23	22	22	21	21	20	20	20	19	19	18	18	18	17	17	16	16	16	16
60	27	27	26	25	25	24	23	23	22	22	21	21	20	20	19	19	19	18	18	17	17	17	16	16	16	15	15
58	26	26	25	24	24	23	23	22	22	21	21	20	20	19	19	18	18	18	17	17	16	16	16	15	15	15	15
56	26	25	24	24	23	22	22	21	21	20	20	19	19	18	18	18	17	17	17	16	16	16	15	15	15	14	14
54	25	24	23	23	22	22	21	21	20	20	19	19	18	18	17	17	17	16	16	16	15	15	15	14	14	14	14
52	24	23	23	22	21	21	20	20	19	19	18	18	18	17	17	16	16	16	15	15	15	14	14	14	14	13	13
50	23	22	22	21	21	20	20	19	19	18	18	17	17	17	16	16	15	15	15	14	14	14	14	13	13	13	13
48	22	21	21	20	20	19	19	18	18	17	17	17	16	16	15	15	15	14	14	14	14	13	13	13	12	12	12

Extremely obese · Obese · Overweight · Normal · Underweight

H Body mass index

In anthropometry, the body mass index (BMI) has become the international standard for evaluating body weight because it correlates relatively well with total body fat. BMI is defined as the body weight in kilograms divided by the square of the height in meters:

$$BMI = \frac{kg}{m^2}$$

2.2 The Structural Design of the Human Body

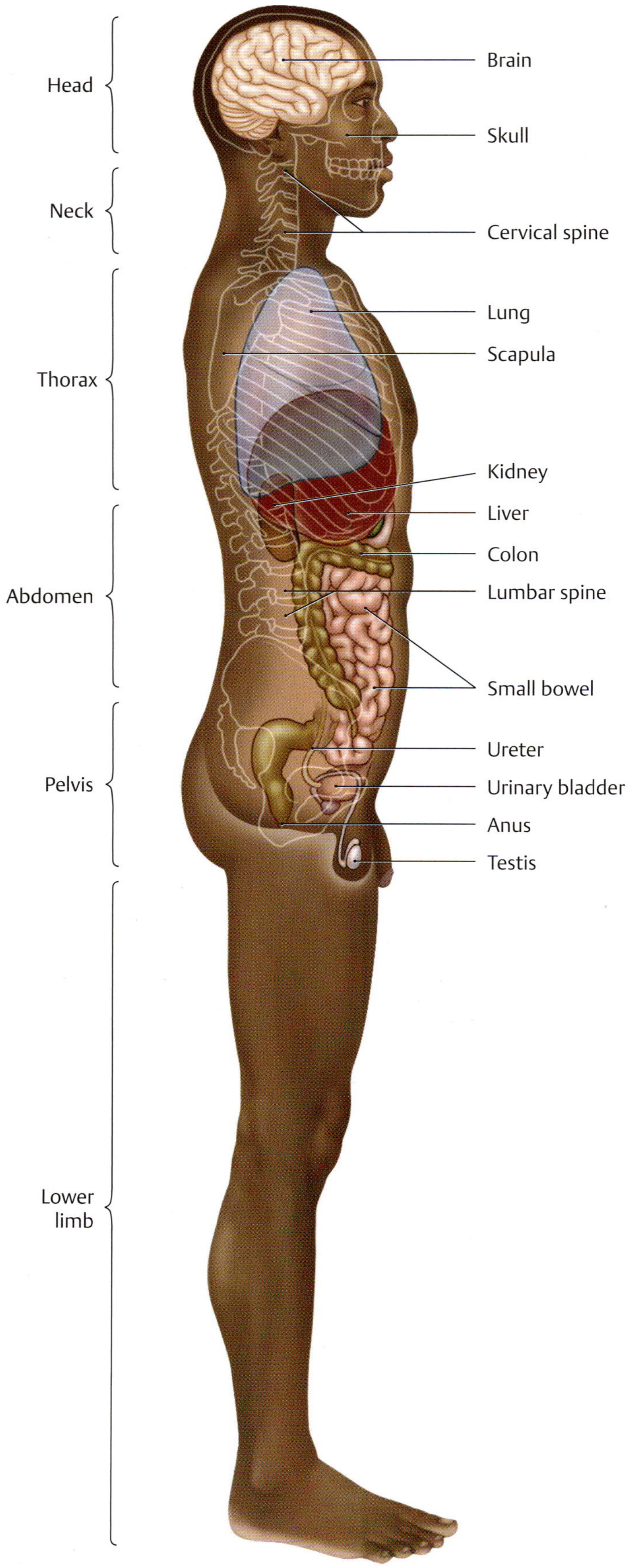

A Location of the internal organs
Lateral view.

B Regional subdivisions of the body

Head

Neck

Trunk
- Thorax (chest)
- Abdomen
- Pelvis

Upper limb
- Shoulder girdle
- Free upper limb

Lower limb
- Pelvic girdle
- Free lower limb

C Functional subdivisions by organ system

Locomotor system (musculoskeletal system)
- Skeleton and skeletal connections (passive part)
- Striated skeletal musculature (active part)

Viscera
- Cardiovascular system
- Hemolymphatic system
- Endocrine system
- Respiratory system
- Digestive system
- Urinary system
- Male and female reproductive system

Nervous system
- Central and peripheral nervous system
- Sensory organs

The skin and its appendages

D Topography of body cavities, serous cavities, and connective-tissue spaces

Organs and organ systems are embedded either in serous cavities or in connective-tissue spaces of varying size. A serous cavity is a fully enclosed potential space that is lined by a shiny membrane (serosa) and contains a small amount of fluid. The serosa consists of two layers that are *usually* apposed (both layers are not *necessarily* in direct contact, as in the abdominal cavity): a visceral layer that directly invests the organ, and a parietal layer that lines the wall of the serous cavity.

Enclosed serous cavities
- Thoracic cavity:
  - Pleural cavity
  - Pericardial cavity
- Abdominal cavity
  - Peritoneal cavity of the abdomen
- Pelvic cavity
  - Peritoneal cavity of the pelvis

Connective-tissue spaces
- Space between the middle and deep layers of cervical fascia
- Mediastinum
- Extraperitoneal space with
  - the retroperitoneal space (retroperitoneum) and
  - the subperitoneal space
- Bursa and synovial cavities

Note: The peritoneal cavities of the abdomen and pelvis are connected.

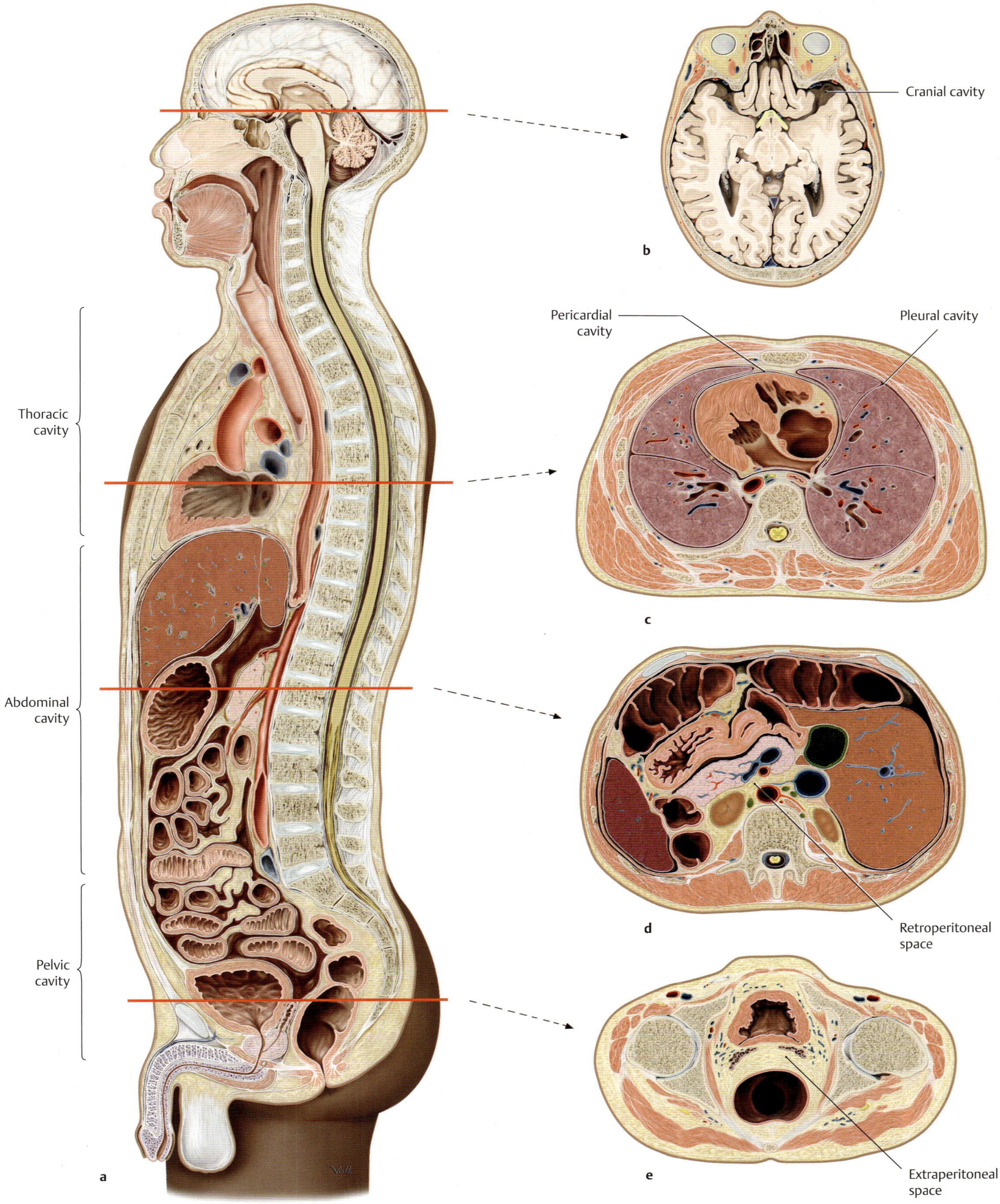

E Selected planes of section through the body
Superior view

a Midsagittal section.
b Cross section at the level of the head.
c Cross section through the thorax.
d Cross section through the abdomen.
e Cross section through the lesser pelvis (see also Terms of Location and Direction, Cardinal Planes and Axes, p. 26).

3.1 Terms of Location and Direction, Cardinal Planes and Axes

A General terms of location and direction

Upper body (head, neck, and trunk)	
Cranial	Pertaining to or located toward the head
Cephalad	Directed toward the head
Caudal	Pertaining to or located toward the tail
Caudad	Directed toward the tail
Anterior	Pertaining to or located toward the front Synonym: Ventral (used for all animals)
Posterior	Pertaining to or located toward the back Synonym: Dorsal (used for all animals)
Superior	Upper or above
Inferior	Lower or below
Medius	Located in the middle
Flexor	Pertaining to a flexor muscle or surface
Extensor	Pertaining to an extensor muscle or surface
Axial	Pertaining to the axis of a structure
Transverse	Situated at right angles to the long axis of a structure
Longitudinal	Parallel to the long axis of a structure
Horizontal	Parallel to the plane of the horizon
Vertical	Perpendicular to the plane of the horizon
Medial	Toward the median plane
Lateral	Away from the medial plane (toward the side)
Median	Situated in the median plane or midline
Central	Situated at the center or interior of the body
Peripheral	Situated away from the center
Superficial	Situated near the surface
Deep	Situated deep beneath the surface
External	Outer or lateral
Internal	Inner or medial
Apical	Pertaining to the tip or apex
Basal	Pertaining to the bottom or base
Occipital	Pertaining to the back of the head
Temporal	Pertaining to the lateral region of the head (the temple)
Sagittal	Situated parallel to the sagittal suture
Coronal	Situated parallel to the coronal suture (pertaining to the crown of the head)
Rostral	Situated toward the nose or brow
Frontal	Pertaining to the forehead
Basilar	Pertaining to the skull base

Limbs	
Proximal	Close to or toward the trunk
Distal	Away from the trunk (toward the end of the limb)
Radial	Pertaining to the radius or the lateral side of the forearm
Ulnar	Pertaining to the ulna or the medial side of the forearm
Tibial	Pertaining to the tibia or the medial side of the leg
Fibular	Pertaining to the fibula or the lateral side of the leg
Palmar (volar)	Pertaining to the palm of the hand
Plantar	Pertaining to the sole of the foot
Dorsal	Pertaining to the back of the hand or top of the foot

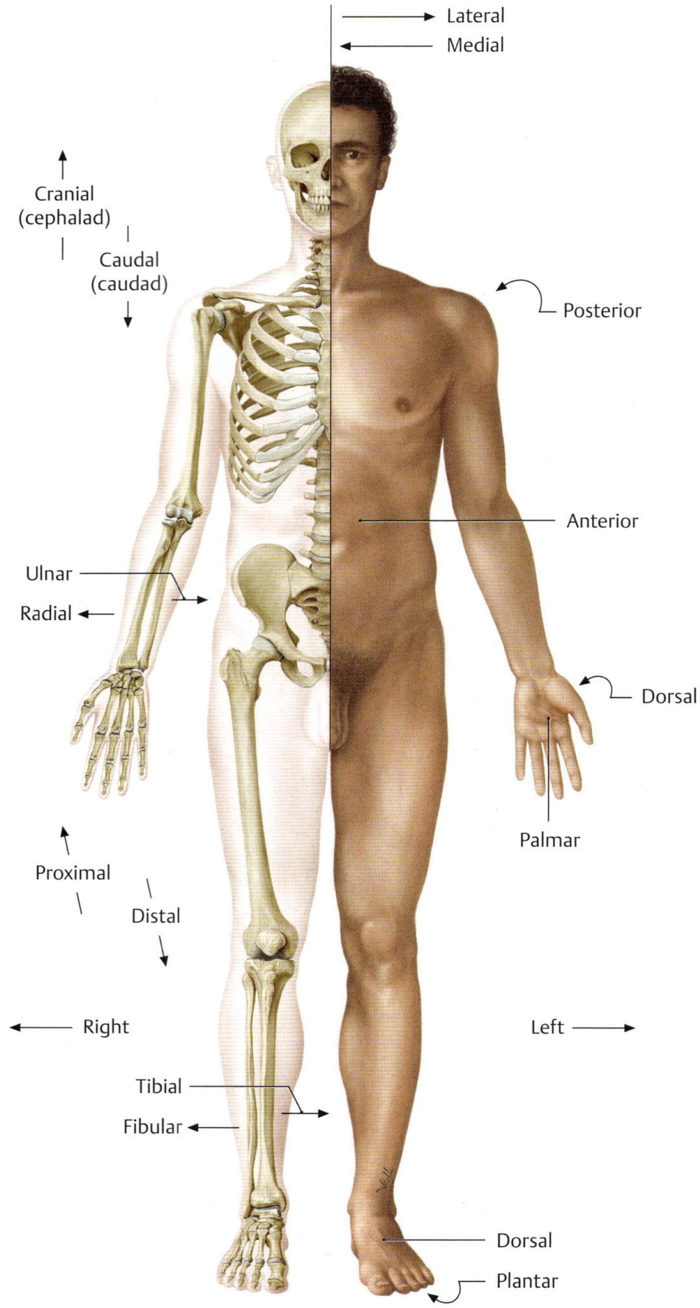

B The anatomic body position
The gaze is directed forward, the hands are supinated. The *right* half of the body is shown in light shading to demonstrate the skeleton. *Note* that the designations “left” and “right” always refer to the patient.

C Abbreviations

a.	artery (aa. = arteries)
v.	vein (vv. = veins)
m.	muscle (mm. = muscles)
lig.	ligament (ligs. = ligaments)
l. n.	lymph node(s)

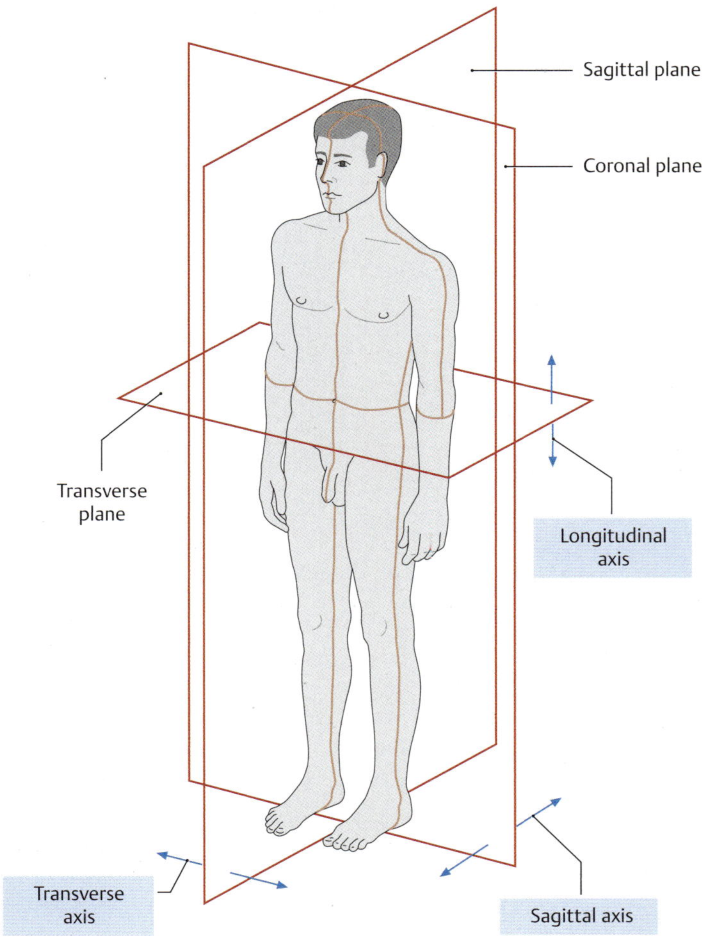

D Cardinal planes and axes in the human body (neutral position, left anterolateral view)

Although any number of planes and axes can be drawn through the human body, it is standard practice to designate *three cardinal planes and axes*. They are perpendicular to one another and are based on the three spatial coordinates.

The cardinal body planes:

- **Sagittal plane:** Any *vertical* plane that is parallel to the sagittal suture of the skull, passing through the body from front to back. The *midsagittal plane* (= median plane) divides the body into equal left and right halves.
- **Frontal (coronal plane):** Any plane that is *parallel to the forehead* or to the coronal suture of the skull. In the standing position, it passes vertically through the body from side to side.
- **Transverse plane (axial plane):** Any *horizontal*, cross-sectional plane that divides the body into upper and lower portions. It is perpendicular to the longitudinal body axis.

The cardinal body axes:

- **Vertical or longitudinal axis:** In the standing position, this axis runs through the body *craniocaudally* and is perpendicular to the ground. It lies at the intersection of the coronal and sagittal planes.
- **Sagittal axis:** This axis runs *anteroposteriorly* from the front to back surface of the body (or from back to front) and lies at the intersection of the sagittal and transverse planes.
- **Transverse or horizontal axis:** This axis runs from side to side and lies at the intersection of the coronal and transverse planes.

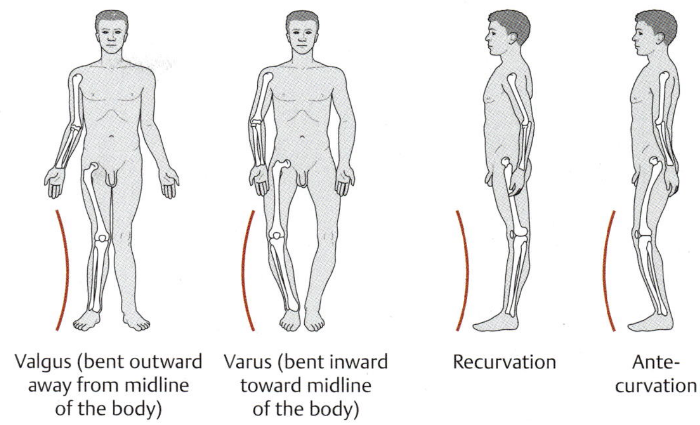

E Axis deviation of upper and lower limbs

Joint deformities at the limbs can lead to axis deviation in the two articulating bones in both frontal and sagittal planes. In accordance with international guidelines, axis deviation in the frontal plane is called *varus* or *valgus deformity* and *recurvation* or *antecurvation* in the sagittal plane. A varus deformity in the knee joint exists if the axis deviation is concave to the vertical body axis or if the distal bone (the tibia) points to the median line (e.g., *genu varum* = bow-leggedness). In a valgus deformity of the knee joint, the axis deviation is convex to the vertical body axis, or the tibia points away from the median line (e.g., *genu valgum* = knock knee).

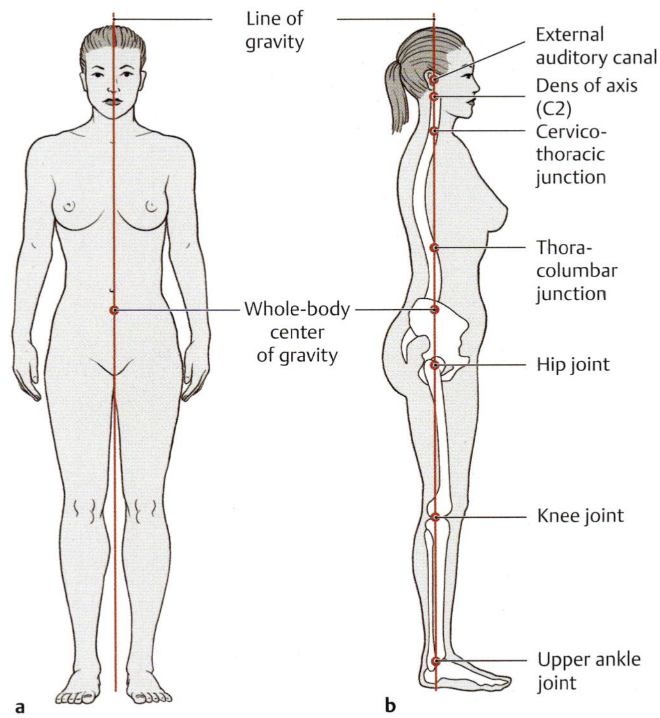

F The whole-body center of gravity and the line of gravity

a Anterior view. The line of gravity is directed vertically along the midsagittal plane, passing through the whole-body center of gravity below the sacral promontory at the level of the second sacral vertebra.

b Lateral view. The line of gravity passes through the external auditory canal, the dens of the axis (second cervical vertebra), the anatomic and functional junctions within the spinal column, the whole-body center of gravity, and through the hip, knee, and ankle joints.

3.2 Location and Designation of Imaging Planes

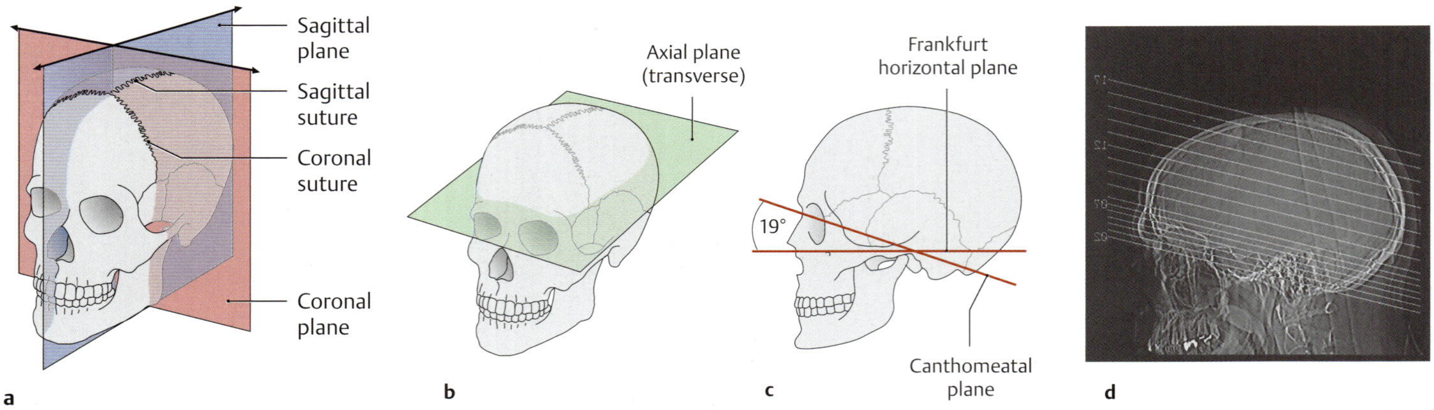

A Radiographic planes, illustrated for the skull

a Just as there are three standard anatomic sectional planes (frontal, sagittal, and transverse planes, see p. 27), certain standard sectional planes have been defined for the most important imaging techniques (computed tomography [CT] and magnetic resonance imaging [MRI]).

Radiographic plane		Anatomic plane
Coronal plane	=	Frontal plane (along or parallel to *coronal suture*)
Sagittal plane	=	Sagittal plane (along or parallel to *sagittal suture*)
Axial plane	=	Transverse plane (at the head, parallel to the *Frankfurt horizontal plane*, or at the trunk and limbs, vertical to the corresponding longitudinal axes)

b The Frankfurt horizontal plane (transverse plane) extends from the upper margin of the external auditory canal to the lower margin of the orbital cavity.

c and **d** As a general rule, CT scans of the skull or the brain are slanted at a 19-degree angle from anterior to posterior (along the so-called canthomeatal line). This line extends from the upper margin of the external auditory canal to the canthus (corner of the eye) and protects the contents of the orbit from heavy X-ray exposure.

Note: Both CT and MRI allow visualization in any desired imaging plane. Both modalities acquire a volume of data that can be used to generate not only two-dimensional images but also spatial representations using 3D reconstruction. In vascular diagnostics, for example, this technique can visualize complicated anatomy or vascular pathology.

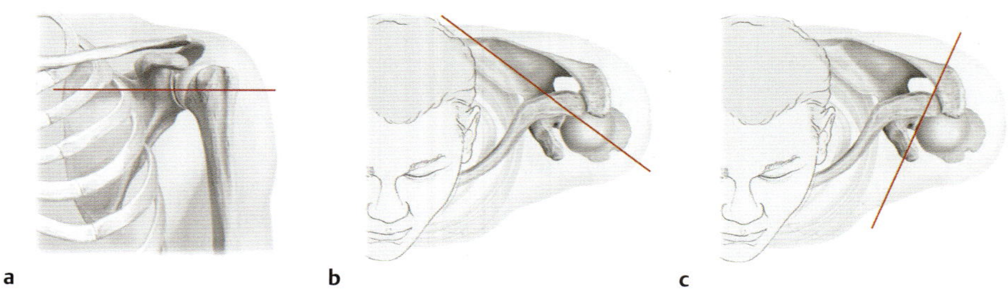

B Standard planes in MRI shoulder examination

The patient lies on his or her back with the arm of the shoulder joint that is to be examined positioned parallel to the body in arm outward rotation or neutral (null) position.

a Position of the axial (transverse) plane of the left shoulder, anterior view.

b Position of the angled coronal plane (parallel to the supraspinatus and vertical to the glenoid cavity), superior view.

c Position of the angled sagittal plane (parallel to the glenoid cavity), superior view.

Thanks to the imaging of the body in multiple sectional planes (or multiplane imaging) and the particularly good differentiation of the periarticular soft tissues, MRI is very well suited for joint exams. Depending on the sectional plane selected and the relaxation time (T1- and T2-weighted scans), the tissues can be differentiated according to their anatomic position and their fat and water percentages. *Note:* Axial (transverse) planes (see **D**) are always viewed from below (distal).

C Optimal MRI planes to show important anatomic structures of shoulder joint

MRI plane	Important visible anatomic structures
Axial (transverse) plane	• Supraspinatus • Glenoid labrum • Joint capsule • Glenohumeral ligaments • Tendon of biceps femoris, long head
Angled coronal plane	• Supraspinatus • Infraspinatus • Subacromial bursa • Acromioclavicular joint • Upper and lower glenoid labrum
Angled sagittal plane	• Rotator cuff • Coracoacromial ligament • Acromion • Glenohumeral ligament • Subacromial space

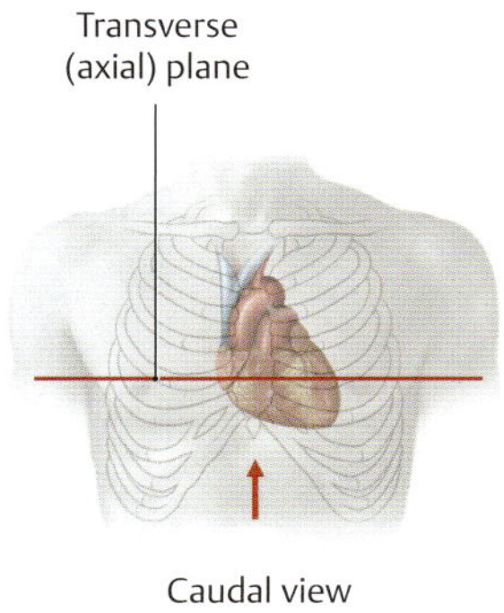

a

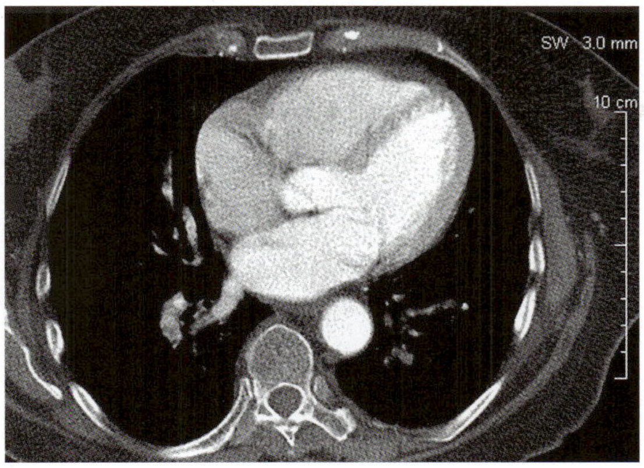

b

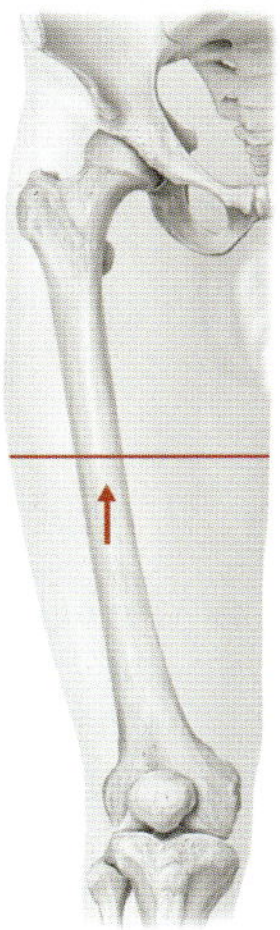

c

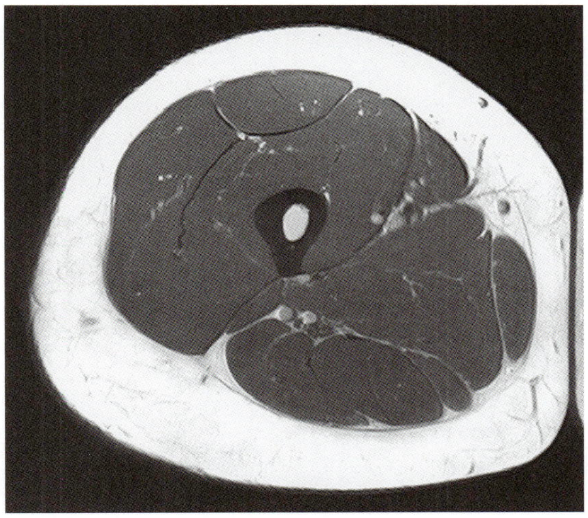

d

D Examining axial images

a and **b** trunk; **c–e** lower limb.

a Anterior view of thorax and location of the transverse (axial) plane at the level of T8.

b Axial CT of thorax (inferior view) in the so-called soft-tissue window, which means the soft tissue (the heart) can be evaluated in high resolution; the lungs, however, because they are filled with air, are overexposed (black), and the bones are underexposed (white) (from: Duale Reihe Radiologie. Reiser M, Kuhn F, Debus J, eds. 2nd ed. Stuttgart: Thieme; 2006).

c Right upper thigh, anterior view, and position of the transverse (axial) plane.

d Axial, T1-weighted MRI of right thigh, distal view (from Möller T, Reif E, eds. Taschenatlas Schnittbildanatomie, Vol. III: Extremitäten, Gelenke, Wirbelsäule – CT und MRT. 2nd ed. Stuttgart: Thieme; 2019).

e Anatomic cross section with description of truncated structures.

Axial (transverse) CT and MRI images are viewed from below and distally for limbs: the patient lies on his or her back. Thus, axial images always show the posteriorly positioned spinal cord pointing downward, the anteriorly positioned skeleton of the thorax pointing upward. Correspondingly, the front of the upper thigh points upward and the back, downward; structures on the right side are displayed on the left side, and structures on the left side are displayed on the right side.

Note: Coronal (frontal) cross sections are viewed as if the patient stands in front of the examiner.

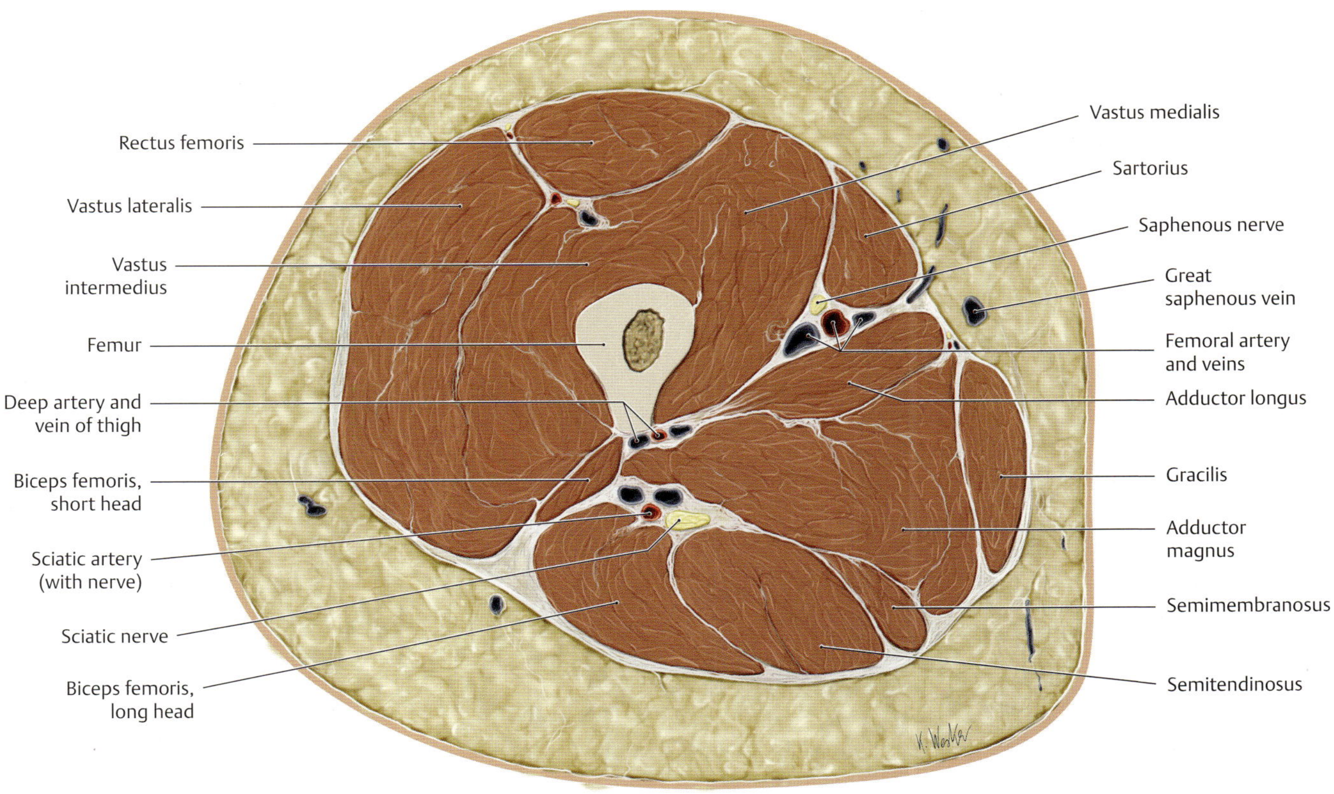

e

3.3 Body Surface Anatomy

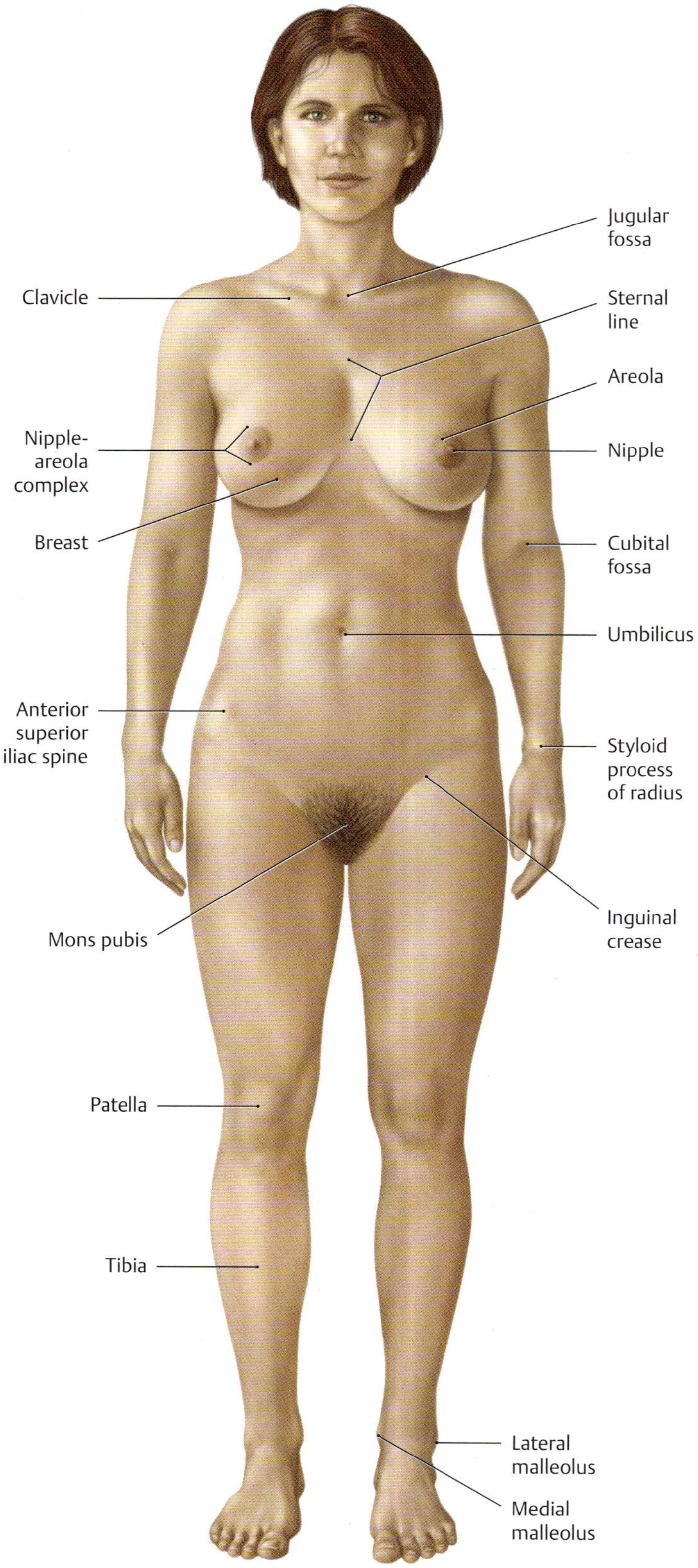

A Body surface anatomy of the female
Anterior view. Body surface anatomy deals with the surface anatomy of the living subject. It plays an important role in classic methods of examination (inspection, palpation, percussion, auscultation, function testing), and so it has particular significance in clinical examination courses. To avoid repetition, identical structures such as the olecranon have not been labeled on both the female and male bodies.

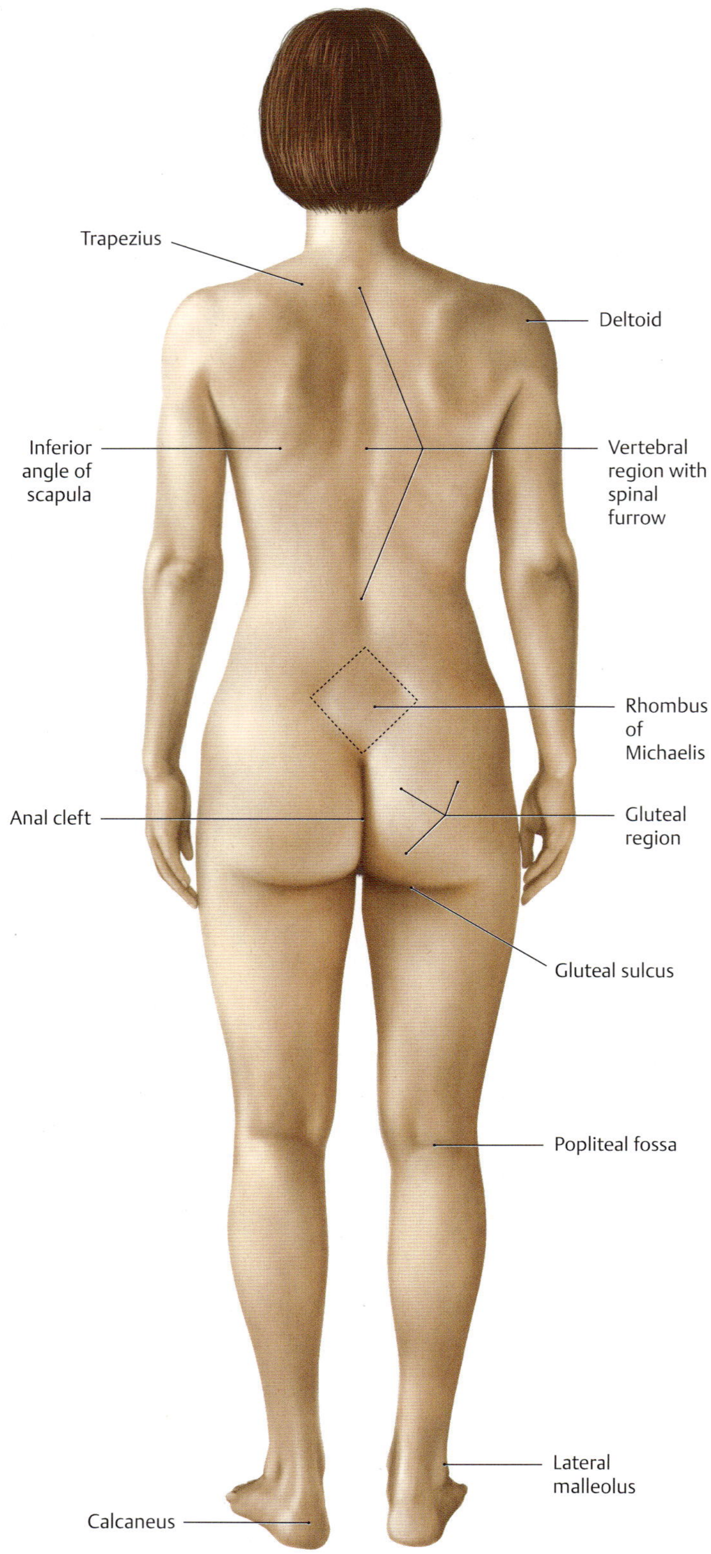

B Body surface anatomy of the female
Posterior view.

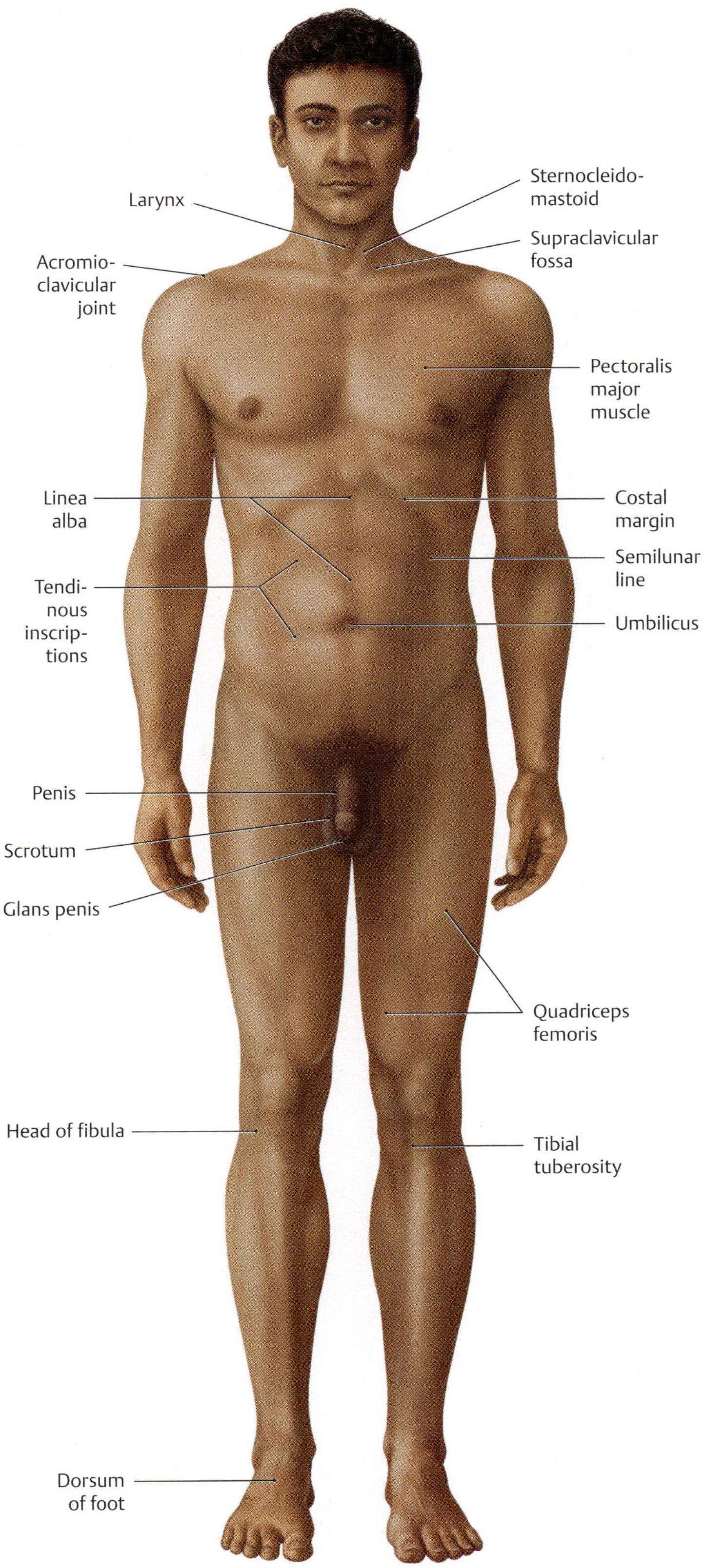

C Body surface anatomy of the male
Anterior view.

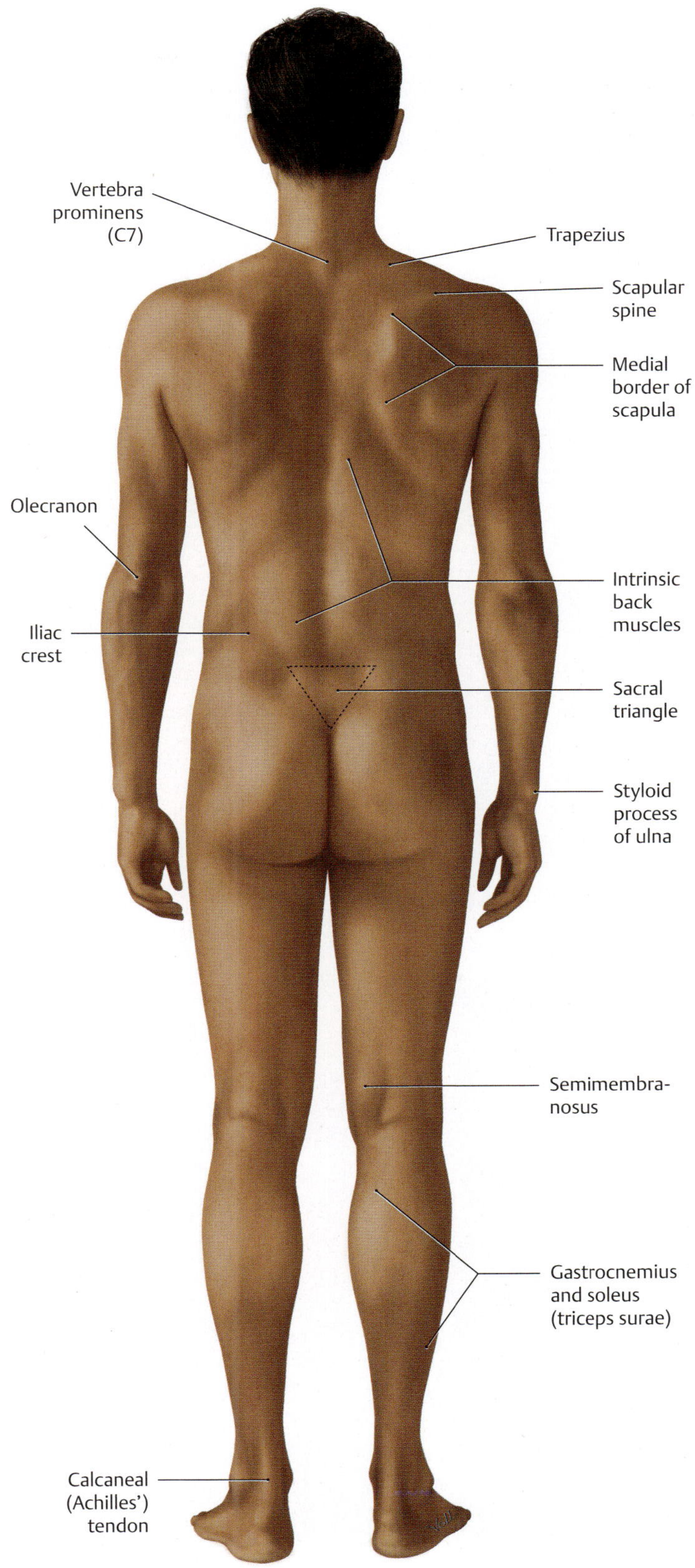

D Body surface anatomy of the male
Posterior view.

3.4 Body Surface Contours and Palpable Bony Prominences

Palpable bony prominences are important landmarks for anatomic orientation in the skeleton, as it is not always possible to palpate articulating skeletal structures (e.g., the hip joint). In these cases, the examiner must rely on palpable bony prominences as an indirect guide to the location of the inaccessible structure.

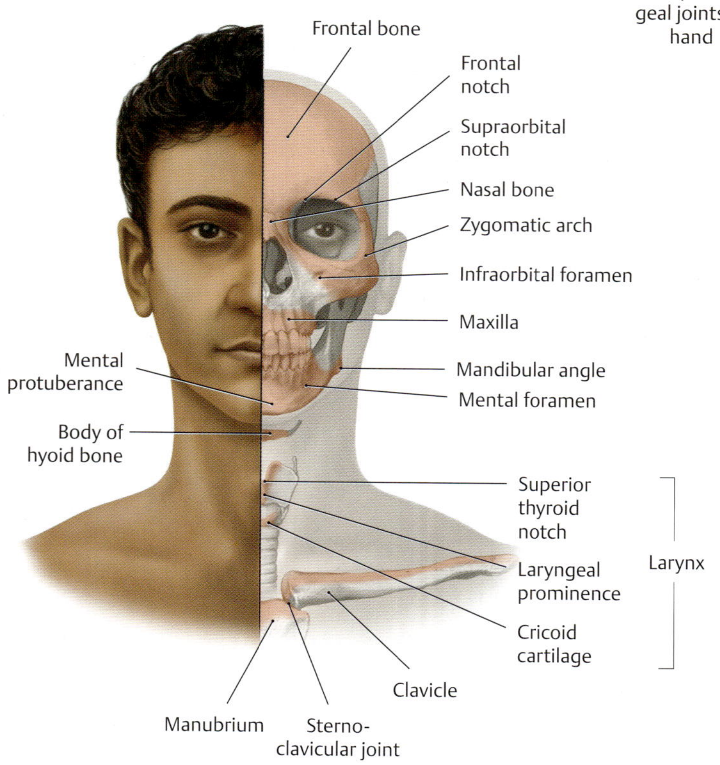

A Surface contours and palpable bony prominences of the face and neck
Anterior view.

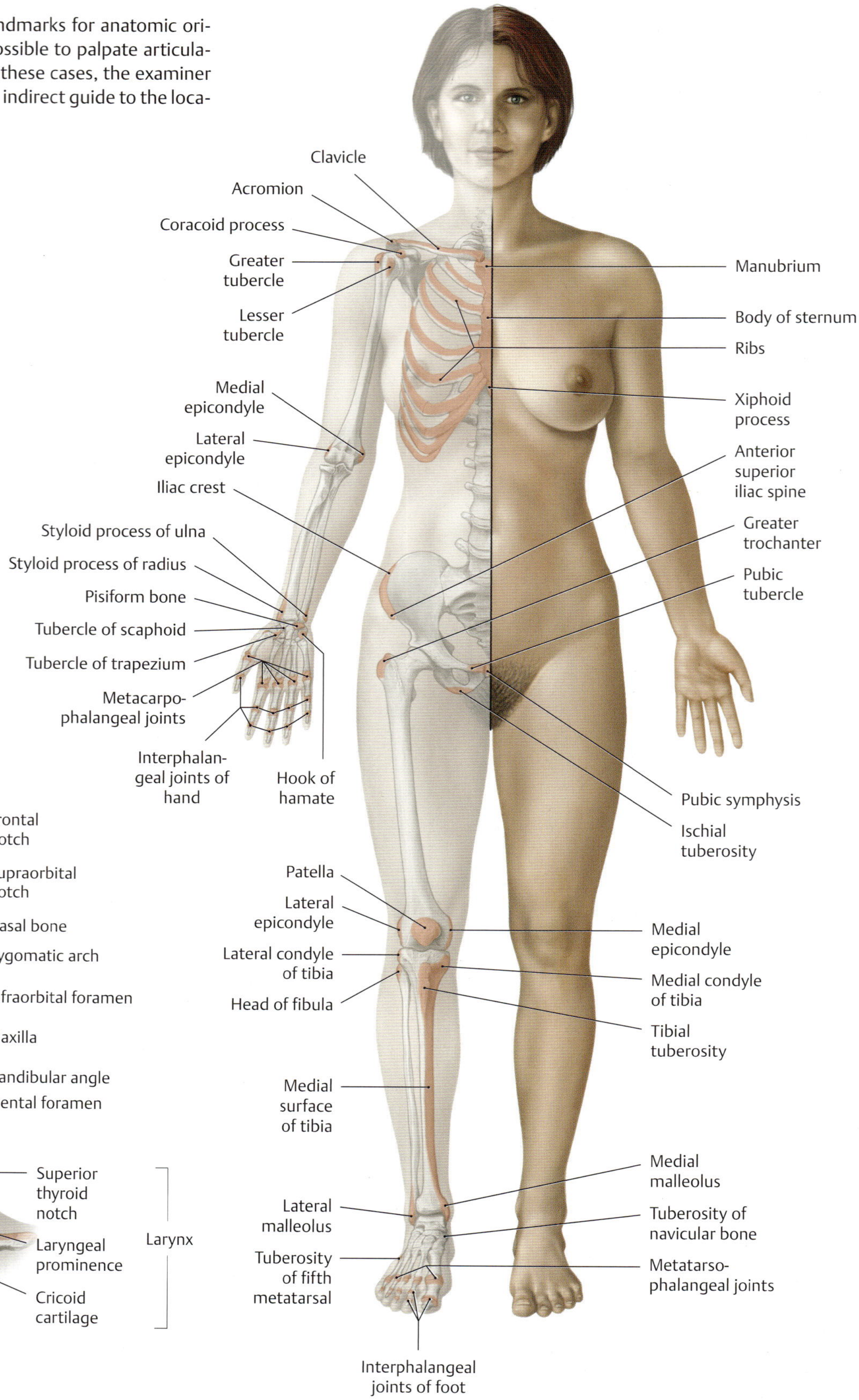

B Surface contours and palpable bony prominences of the trunk and upper and lower limbs in the female
Anterior view.

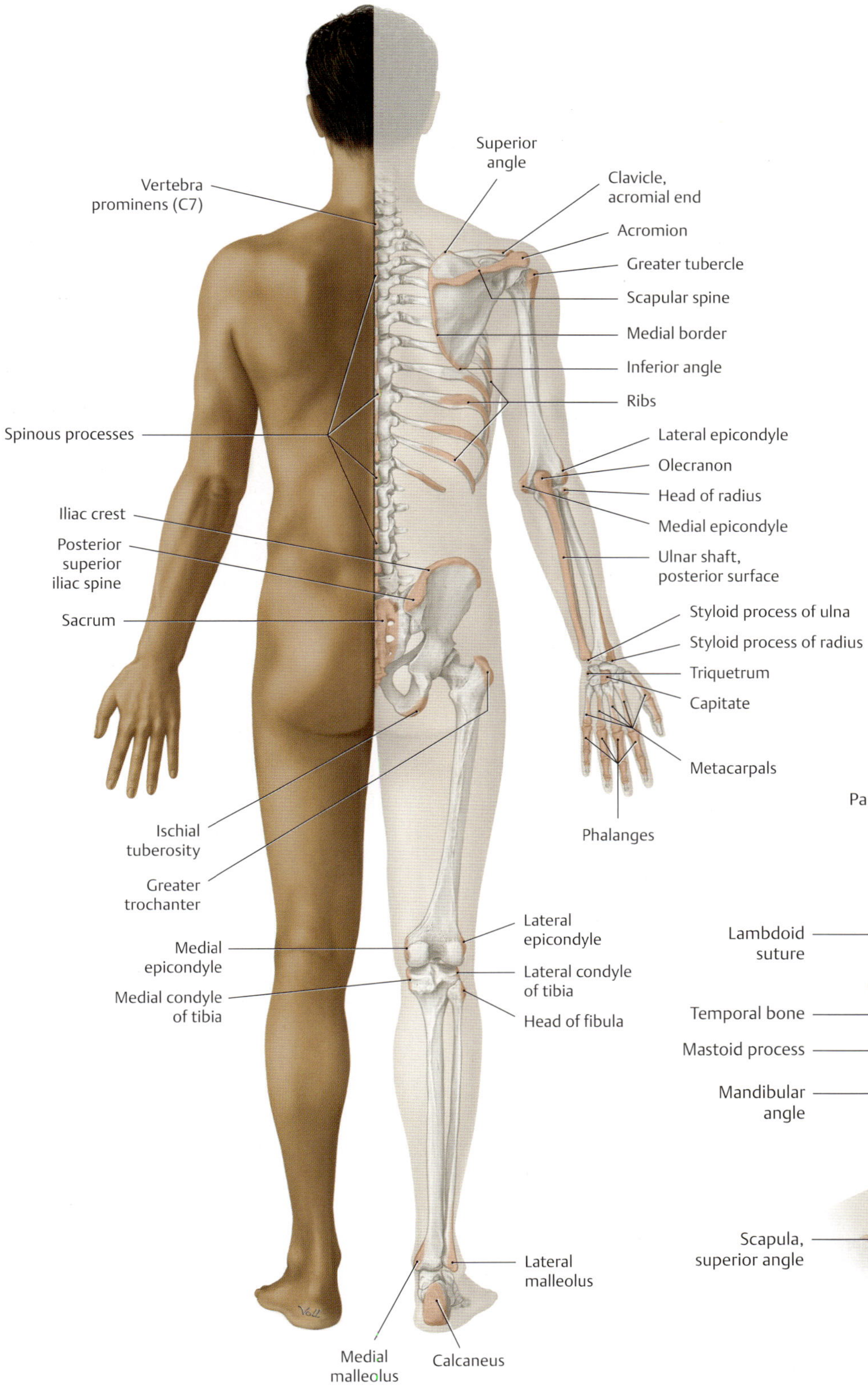

C Surface contours and palpable bony prominences of the trunk and upper and lower limbs in the male

Posterior view.

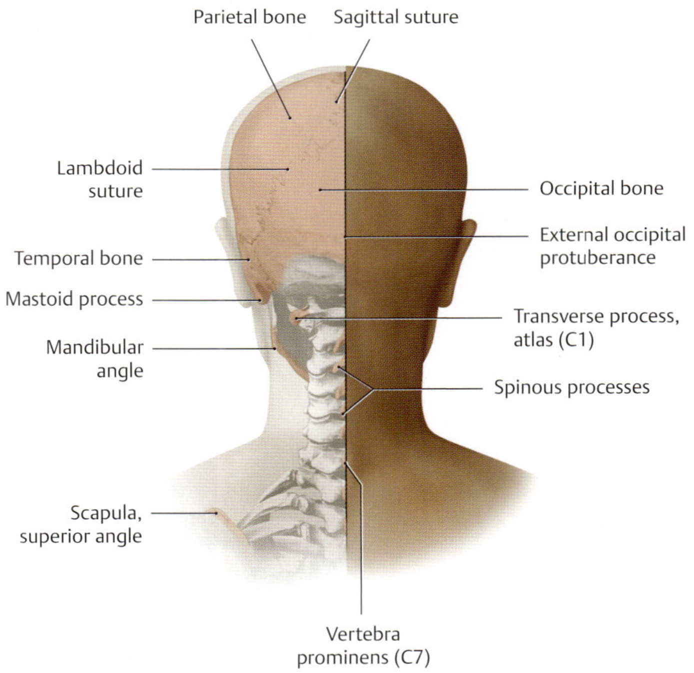

D Surface contours and palpable bony prominences of the head and neck

Posterior view.

3.5 Landmarks and Reference Lines on the Human Body

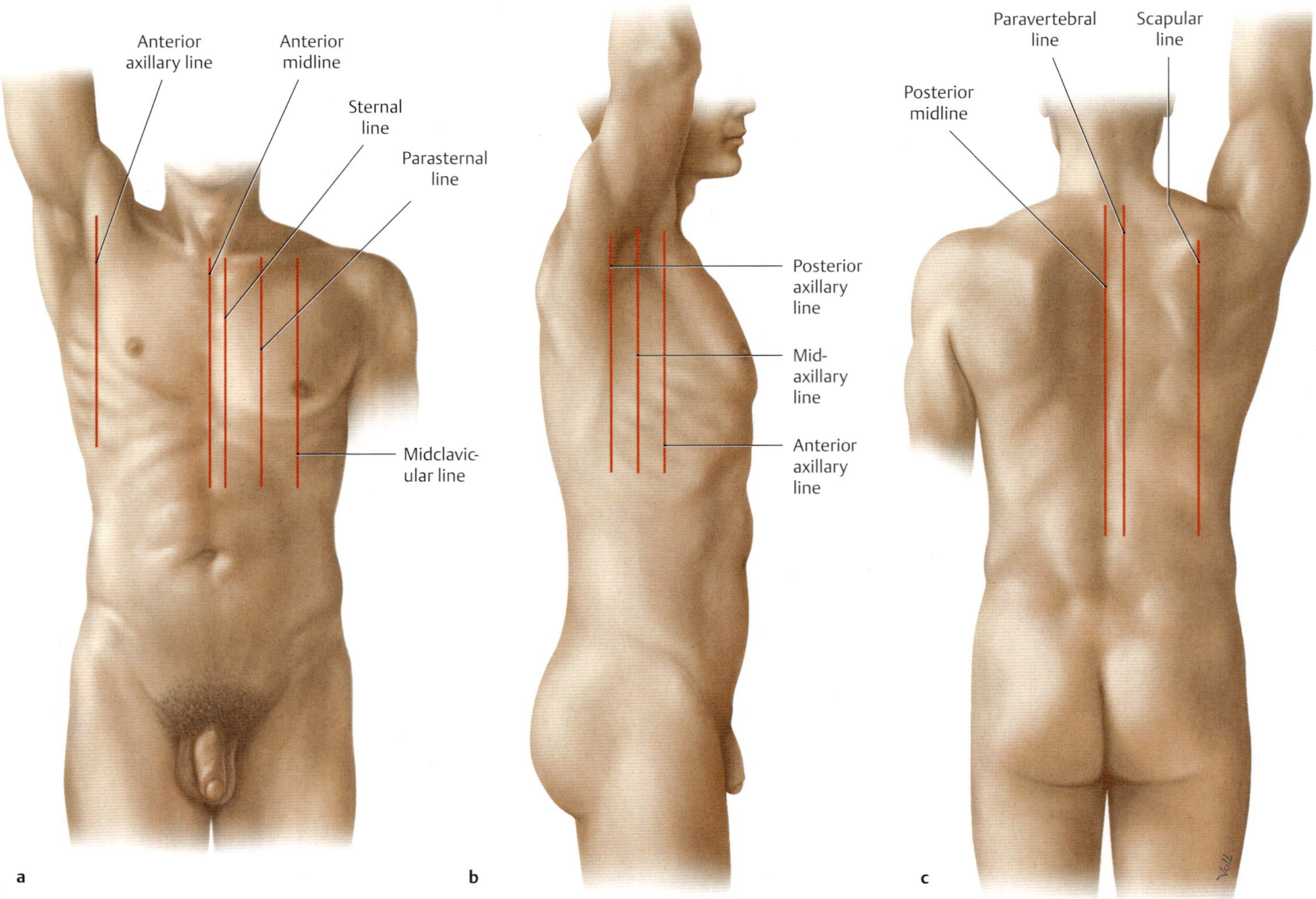

A Vertical reference lines on the trunk
a Anterior view, **b** right lateral view, **c** posterior view.

Anterior midline	Anterior trunk midline passing through the center of the sternum
Sternal line	Line along the sternal margin
Parasternal line	Line midway between the sternal line and midclavicular line
Midclavicular line	Line through the midpoint of the clavicle (often identical to the nipple line)
Anterior axillary line	Line at the level of the anterior axillary fold (pectoralis major muscle)
Midaxillary line	Line midway between the anterior and posterior axillary lines
Posterior axillary line	Line at the level of the posterior axillary fold (latissimus dorsi muscle)
Posterior midline	Posterior trunk midline at the level of the spinous processes
Paravertebral line	Line at the level of the transverse processes
Scapular line	Line through the inferior angle of the scapula

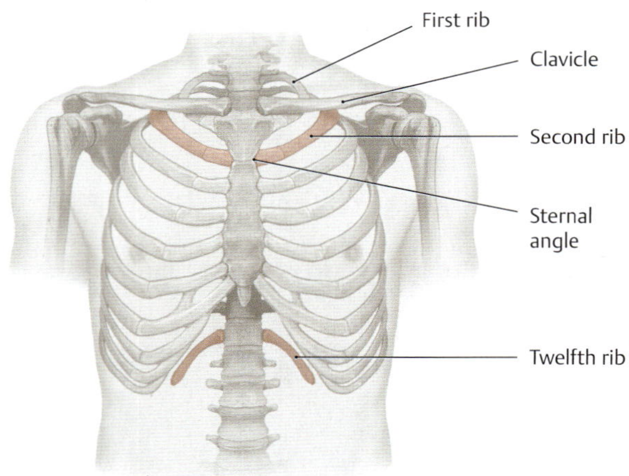

B "Rib counting" for anatomic orientation in the thorax
The first rib is covered by the clavicle. The first palpable rib is the second rib; therefore, the count begins at that level. The second rib attaches to the sternum at the level of the sternal angle. At the lower end of the rib cage, it is best to start at the twelfth rib, which is palpable only in its posterior portion.

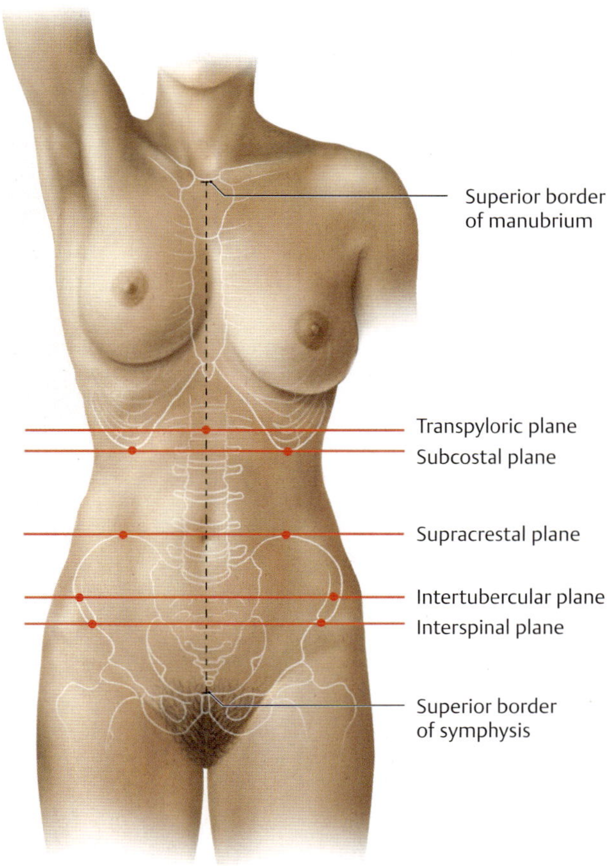

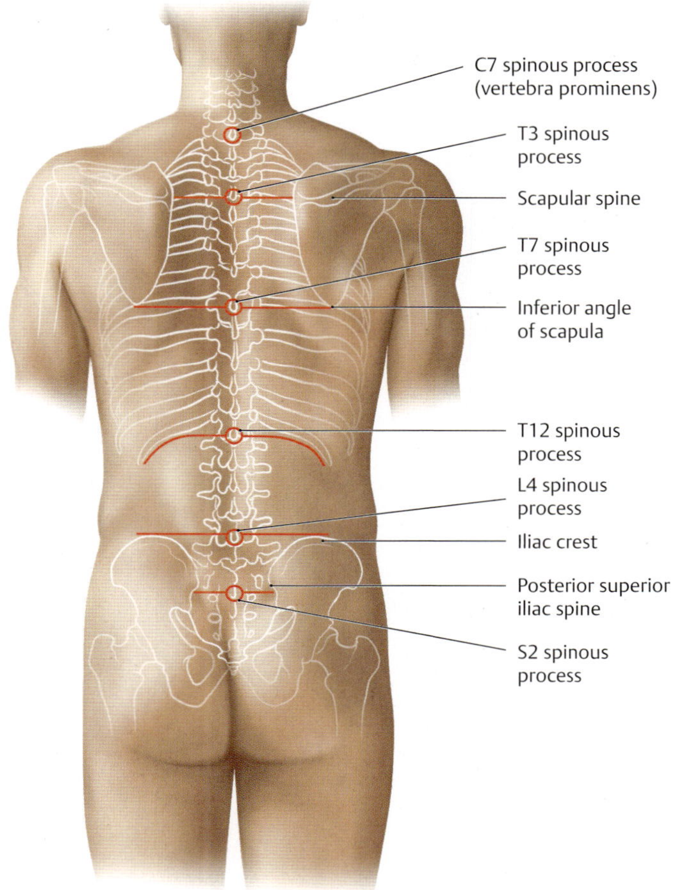

C Standard transverse planes through the abdominal cavity (see also p. 207)
Anterior view.

Transpyloric plane	Transverse plane midway between the superior borders of the symphysis and manubrium
Subcostal plane	Plane at the lowest level of the costal margin (inferior margin of the tenth costal cartilage)
Supracrestal plane	Plane passing through the summits of the iliac crests
Intertubercular plane	Plane at the level of the iliac tubercles (the iliac tubercle lies approximately 5 cm posterolateral to the anterior superior iliac spine)
Interspinal plane	Plane at the level of the anterior superior iliac spine

D Spinous processes that provide useful posterior landmarks
Posterior view.

C7 spinous process	Vertebra prominens (the projecting spinous process of C7 is clearly visible and palpable)
T3 spinous process	At the level of the line connecting the two scapular spines
T7 spinous process	At the level of the line connecting the inferior angles of both scapulae
T12 spinous process	Just below the twelfth rib
L4 spinous process	At the level of the line connecting the summits of the iliac crests
S2 spinous process	At the level of the line connecting the posterior superior iliac spine (recognized by small skin depressions directly over the iliac spines)

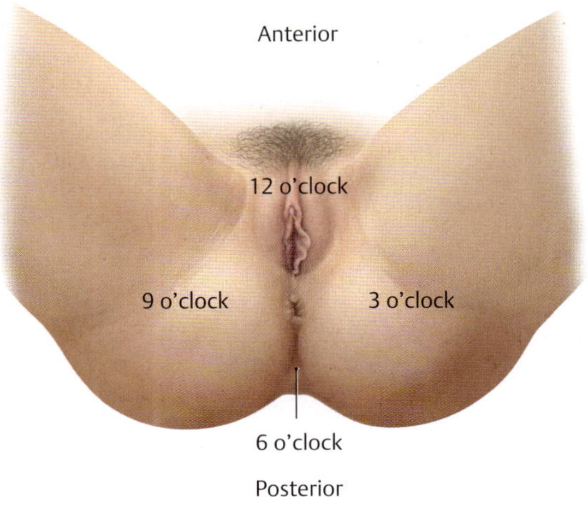

E Lithotomy position (supine with the legs, hips, and knees flexed and the thighs abducted)
The position of choice for proctological examinations. Clock-face notation is used for anatomic orientation (e.g., to describe the location of a lesion):

- Top = toward the pubis = 12 o'clock
- Bottom = toward the sacrum = 6 o'clock
- Right = 3 o'clock
- Left = 9 o'clock

3.6 Body Regions (Regional Anatomy)

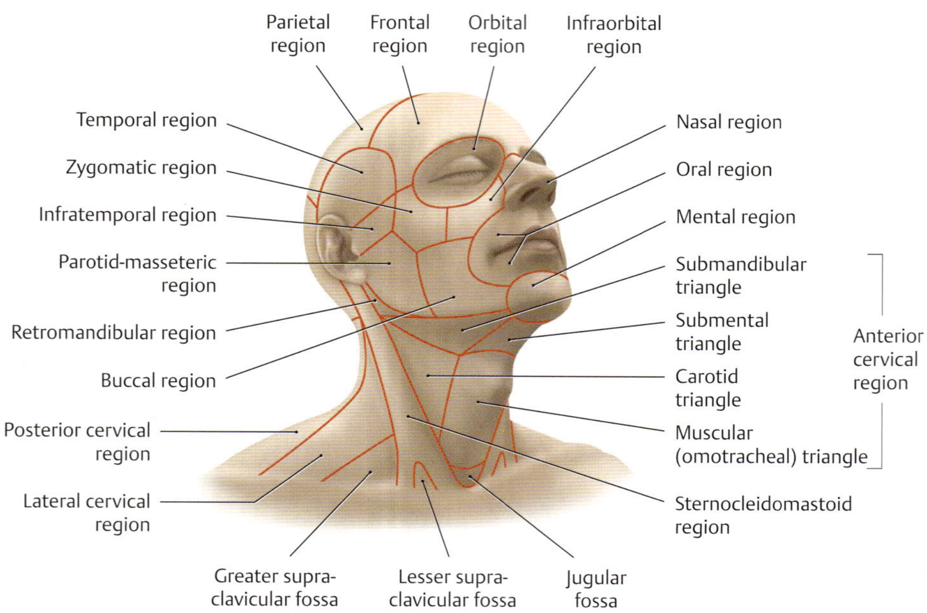

A Regions of the head and neck
Right anterolateral view.

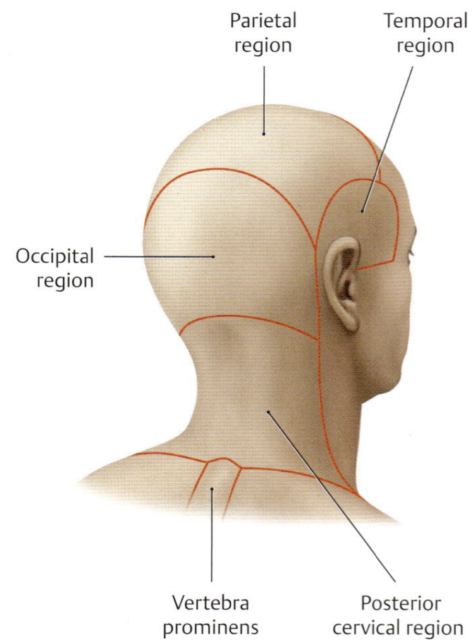

B Regions of the head and neck
Right posterolateral view.

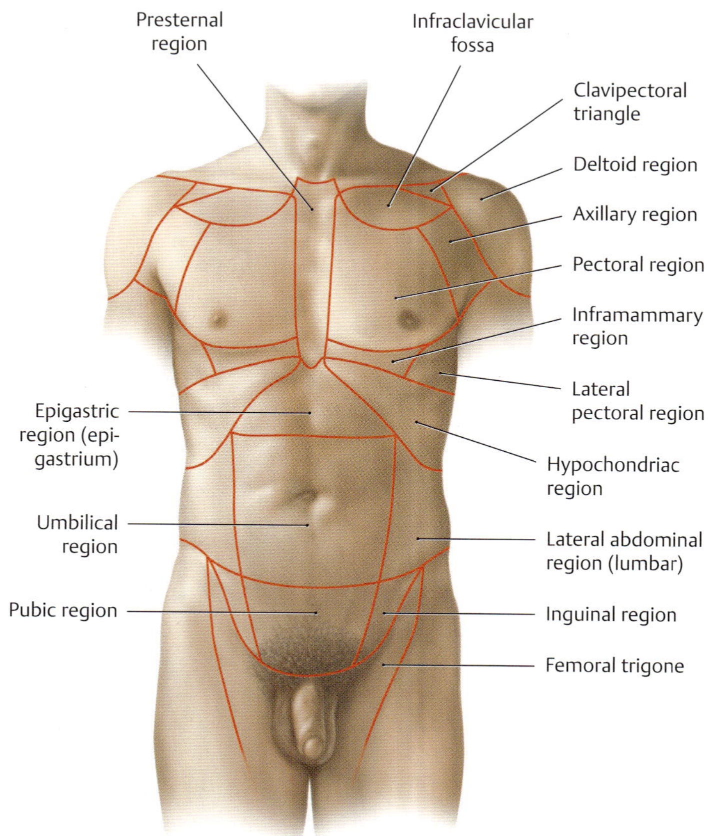

C Regions of the thorax and abdomen
Anterior view.

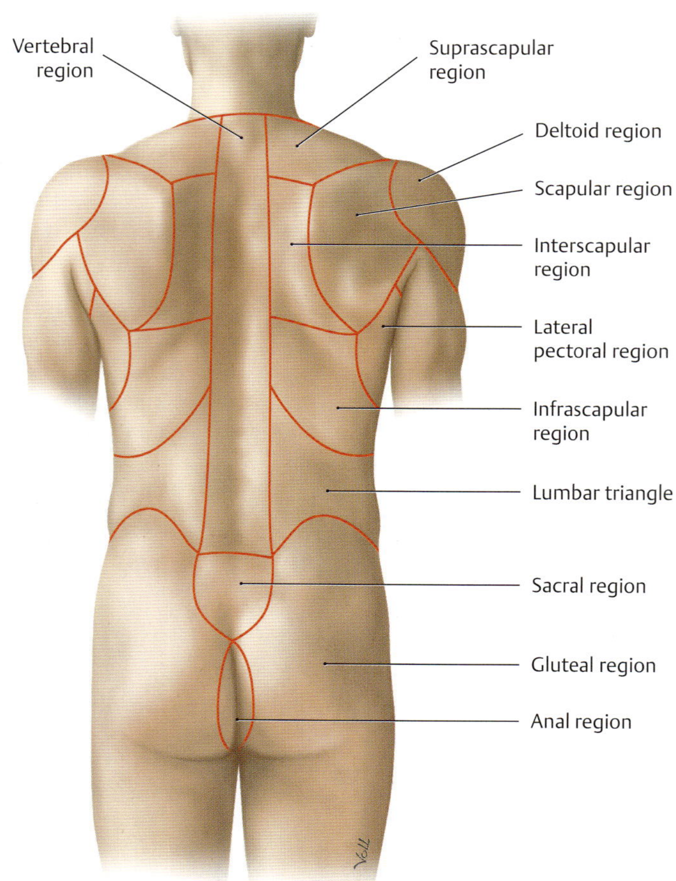

D Regions of the back and buttocks
Posterior view.

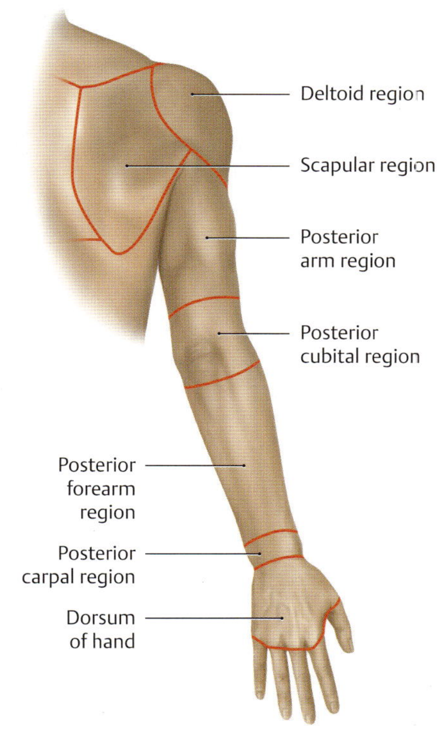

E Regions of the upper limb
Posterior view.

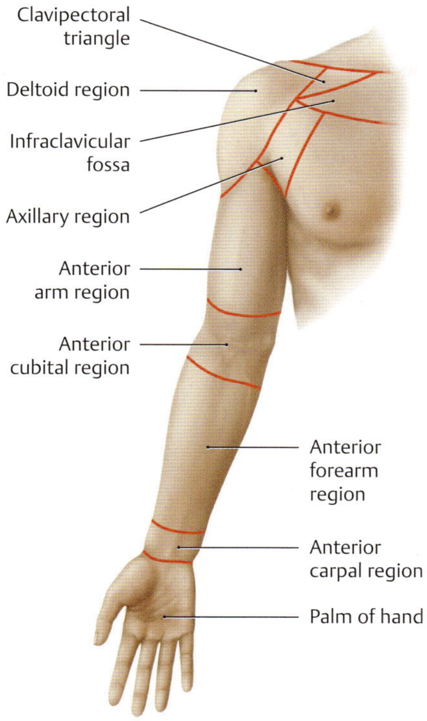

F Regions of the upper limb
Anterior view.

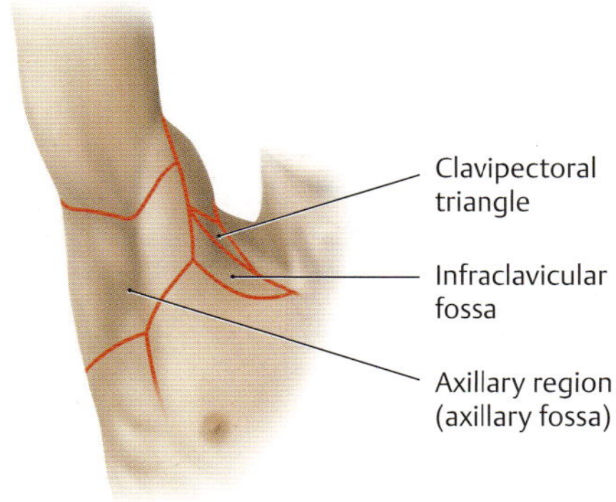

G Regions about the axilla
Anterior view.

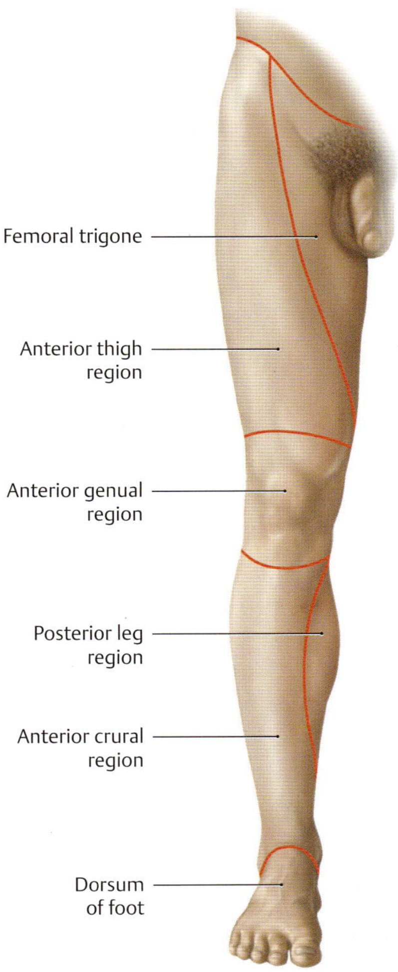

H Regions of the lower limb
Anterior view.

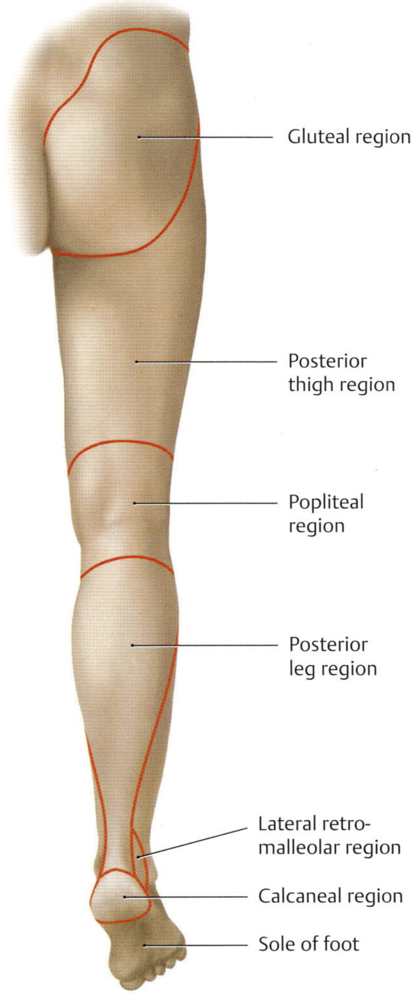

I Regions of the lower limb
Posterior view.

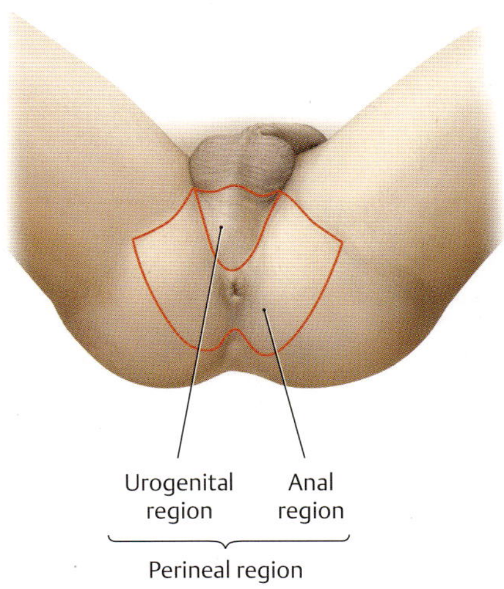

J Perineal region (lithotomy position)

3.7 Skin

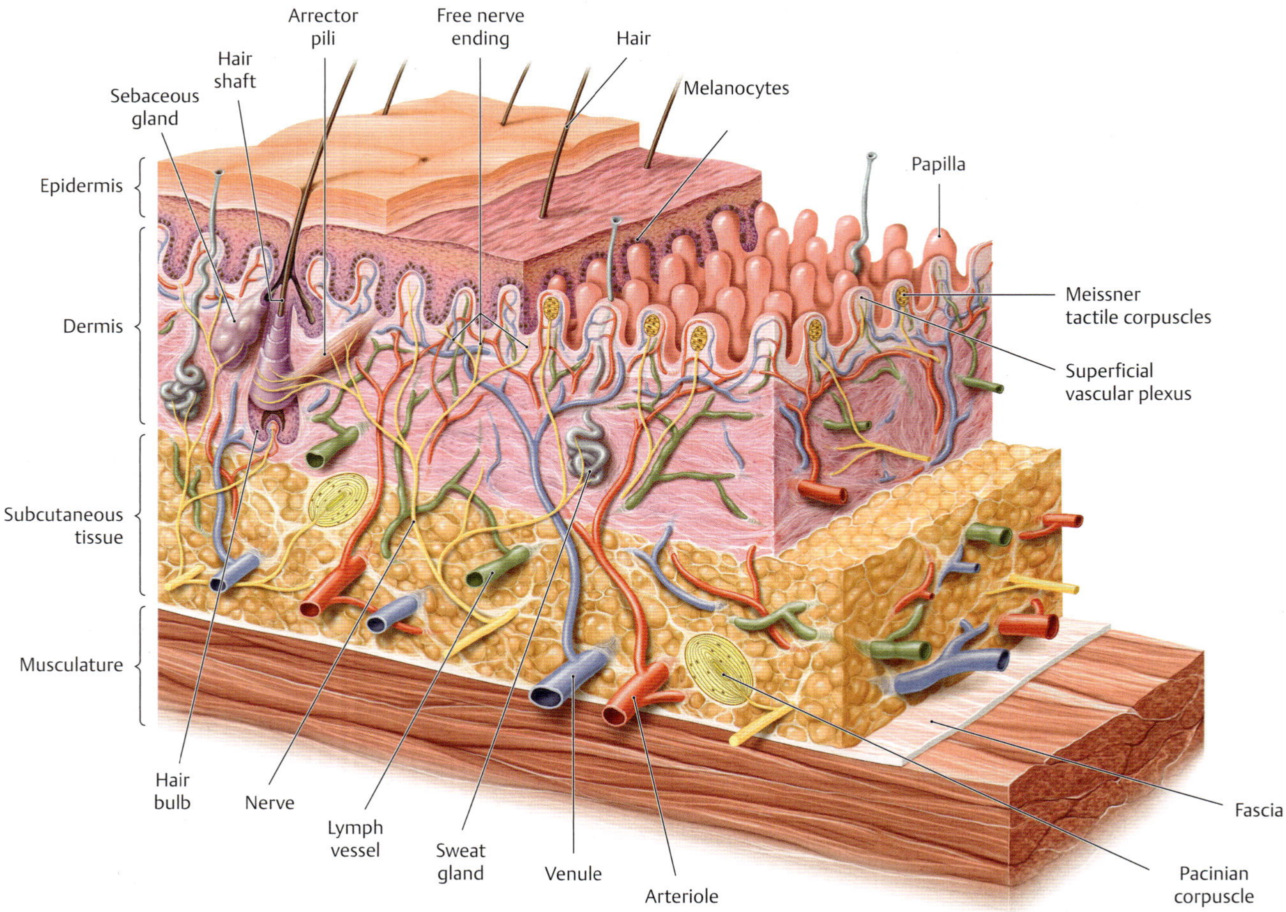

A Structure of the skin

The skin is the outer barrier that protects human beings against the environment. It covers the entire body and in terms of area (approximately 1.8m^2) it is the largest organ. Depending on its surface texture (diamond-shaped fields or dermal ridges), one refers to meshed skin or ridged skin. The dermal ridges form the fingerprints (see p. 388). Ridged skin is found only on the palmar (volar) aspect of the hands and the plantar aspect of the feet. Both types of skin are composed of layers from superficial to deep or, respectively, from outside to inside (see Table **C**):

- Epidermis = epithelial layer of the skin.
- Dermis (also cutis or corium = the connective-tissue layer of the skin; this layer of the skin of domestic animals is used to produce leather).
- Subcutis or subcutaneous tissue.

The epidermis interdigitates with the dermis in a pattern resembling an egg carton. This increases the area of contact between the two structures. Dissection of the skin reveals indentations on the underside of the epidermis (papillary loops of the dermal papillae). Like all other epithelia, the epidermis is completely avascular. The blood vessels and sensory receptors lie in the dermis (see **B** and **D**). The subcutaneous tissue consists largely of fatty tissue. It ends at the general body fascia.

B Vascular supply to the skin

Intact vascular supply is essential not only for ensuring the supply of nutrients to the skin but also for thermoregulation. The blood vessels are located in the dermis. The superficial plexus lies at the border between the papillary dermis and reticular dermis; the deep dermal plexus lies at the border between the dermis and the subcutaneous tissue. Vertical vessels connect the two plexuses. The upper plexus gives off capillary branches that supply the papillae and the overlying epidermis. Using a special cutting device (dermatome), one can divide the dermis between these two plexuses and use the skin flap created in this manner to cover larger wounds such as burns. This is known as a split-thickness skin graft.

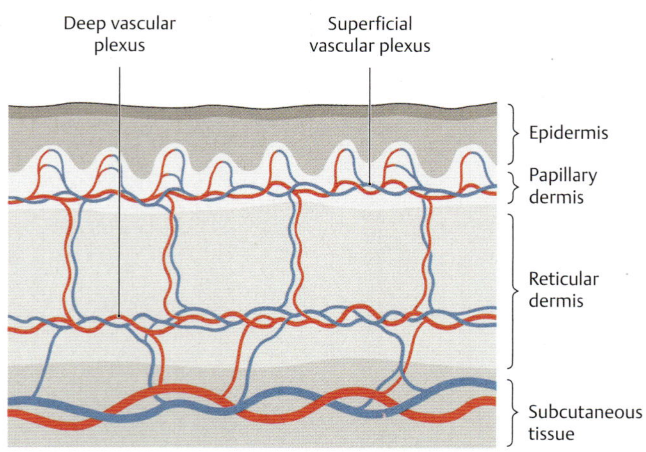

C Skin layers and their components

Skin layer	Applicable subdivisions	Description	Main functions
Epidermis (epithelial layer of the skin) *Note:* In histology one invariably examines the layers of the epidermis from *deep* to *superficial*. Here the individual components, analogous to the other layers of the skin, are listed in order from *superficial* to *deep* or, respectively, from the outside in.	• Stratum corneum	Cornified layer of dead keratinized cells	Mechanical protection against agents such as germs, radiation, and chemical substances (especially acids)
	• Stratum lucidum	"Bright layer" occurring only in the *ridged skin* (= sole of the foot and palm of the hand)	Mechanical protection
	• Stratum granulosum	Granular layer; cells contain keratohyalin granules that create a physical barrier	Protection against dessication
	• Stratum spinosum	Spinous layer; usually the thickest layer of living cells in the skin	• Mechanical protection by many desmosomes and their interconnections by means of cytokeratins • Immune response by Langerhans cells
	• Stratum basale	Lowest epithelial layer of the epidermis; its basement membrane borders on the dermis	• Protection against UV radiation by secretion of melanin on the melanocytes • Merkel discs (= slowly adapting pressure receptors consisting of Merkel cells and its associated nerve endings)
Dermis (connective-tissue layer of the skin)	• Stratum papillare (also known as papillary layer) • Stratum reticulare (also known as reticular layer)	Papillary layer; papillae extend into the epidermis so that dermis and epidermis interdigitate Reticular layer with taut collagenous connective tissue; the reticular structure creates the skin cleavage lines	• Sweat glands: sweat secretion, thermoregulation, secretion of antimicrobial peptides • Scent glands: axilla, nipple, groin region, perianal region, pubic region • Sebaceous glands: fat secretion, protection against evaporation • Blood vessels: supply of nutrients, thermoregulation • Lymph vessels: drainage of interstitial fluid • Hair: protection against sunlight and heat loss, perception of stimuli • Pressure receptors: Meissner tactile corpuscles in glabrous skin and in the lips; less often: Ruffini corpuscles (lie in the deeper dermis) • Free nerve endings (everywhere): perception and transmission of stimuli
Subcutaneous tissue (fat cells and loose connective tissue)		Layer providing mobility between the skin and adjacent general body fascia	• Fat cells: fat storage providing an energy reservoir • Mechanical cushion • Large blood vessels: supply of nutrients, thermoregulation • Vater–Pacini corpuscles: rapidly adapting mechanoreceptors for vibration sensations; smaller form: Krause corpuscles (also known as Krause end bulbs)

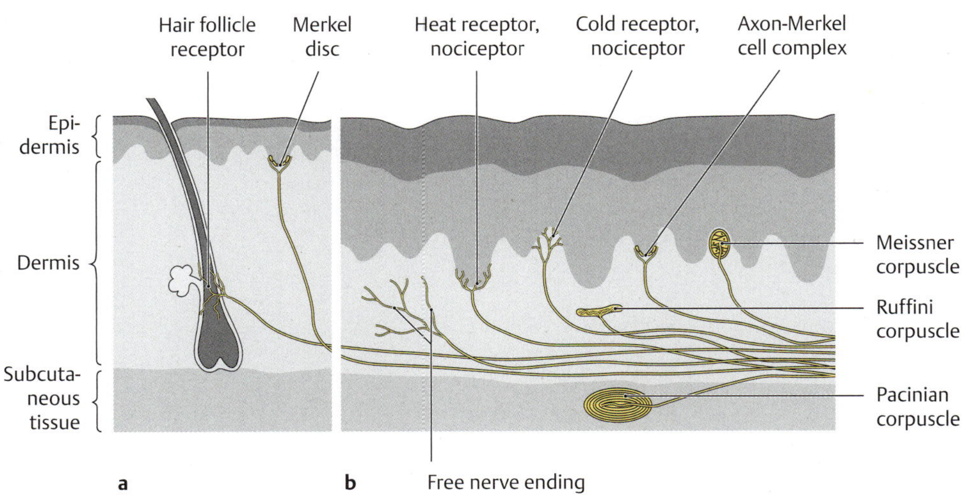

D Skin as a sensory organ

a Hair-bearing skin; **b** glabrous skin.

The skin is an important sensory organ. The dermal papillae of the glabrous ridged skin (sole of the foot, palm of the hand) and the lips contain:

- *Meissner tactile corpuscles* (rapidly adapting pressure receptors that register changes in pressure),
- *Merkel discs* (slowly adapting pressure receptors),
- and *Ruffini corpuscles* (slowly adapting pressure and stretching receptors).

In the hair-bearing skin, hair follicle receptors replace the Meissner tactile corpuscles. The Vater–Pacini corpuscles (rapidly adapting mechanoreceptors for vibration sensations) lie in the subcutaneous tissue.

4.1 The Bony Skeleton and the Structure of Tubular Bones

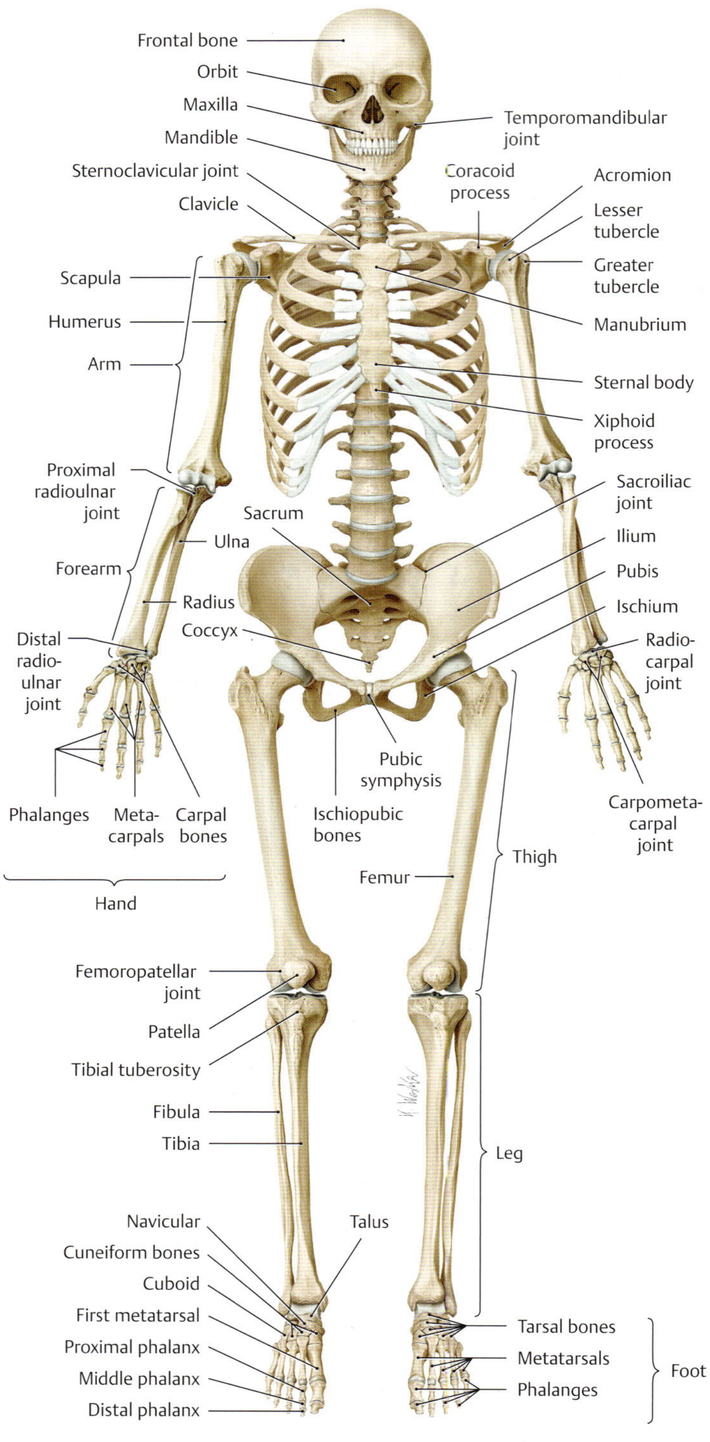

A Female skeleton from the anterior view
The left forearm is pronated, and both feet are in plantar flexion.

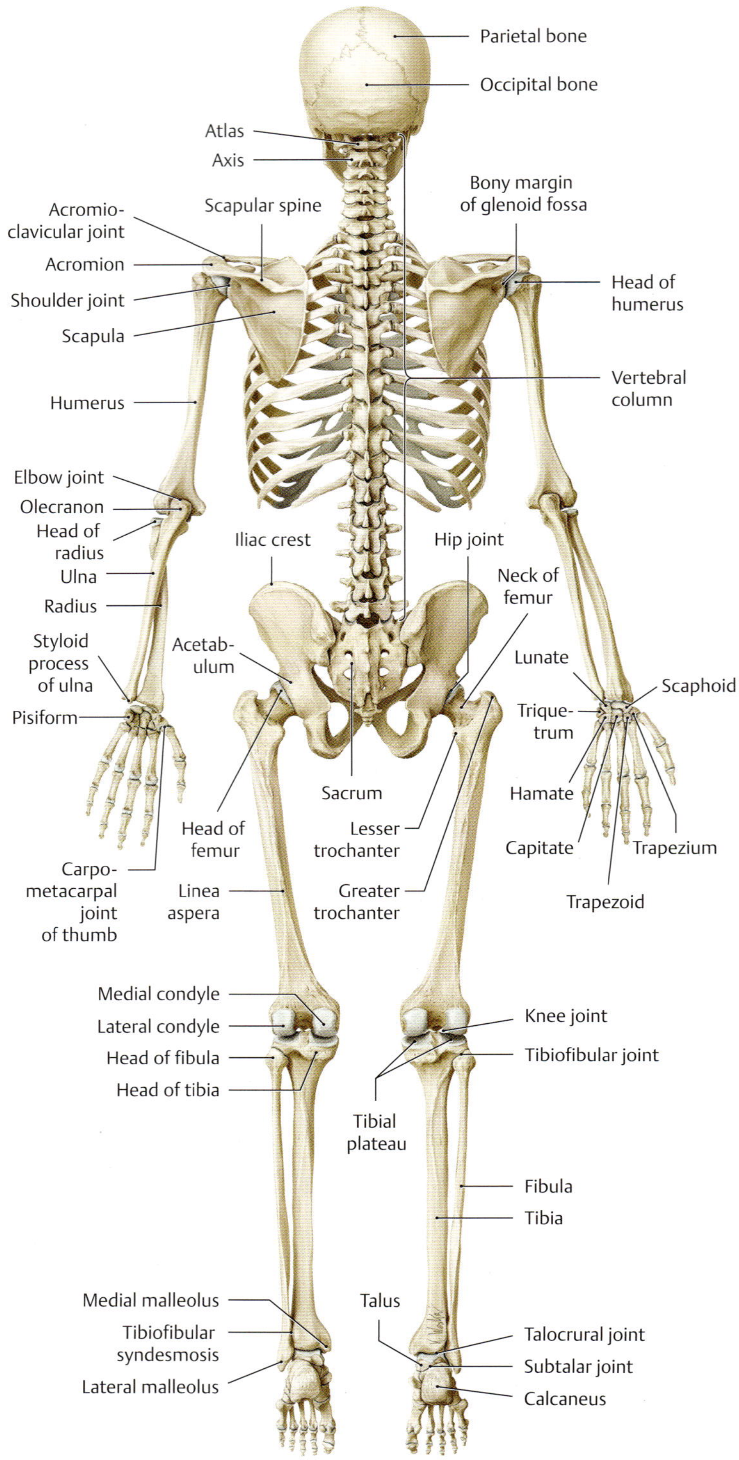

B Male skeleton from the posterior view
The left forearm is pronated, and both feet are in plantar flexion.

C Types of bone

- **Long bones**, e.g., tubular bones of the limbs
- **Short bones**, e.g., carpal and tarsal bones
- **Flat bones**, e.g., scapula, ilium, and bones of the calvaria
- **Irregular bones**, e.g., vertebrae anomalous, supernumerary bones not consistently present, as in the skull base
- **Pneumatic bones** (containing air-filled spaces), e.g., bones of the facial skeleton and paranasal sinuses
- **Sesamoid bones** (bones incorporated in tendons), e.g., the patella
- **Accessory bones** (anomalous, supernumerary bones), as in the calvaria and foot (generally result from the failure of fusion of certain adjacent ossification centers)

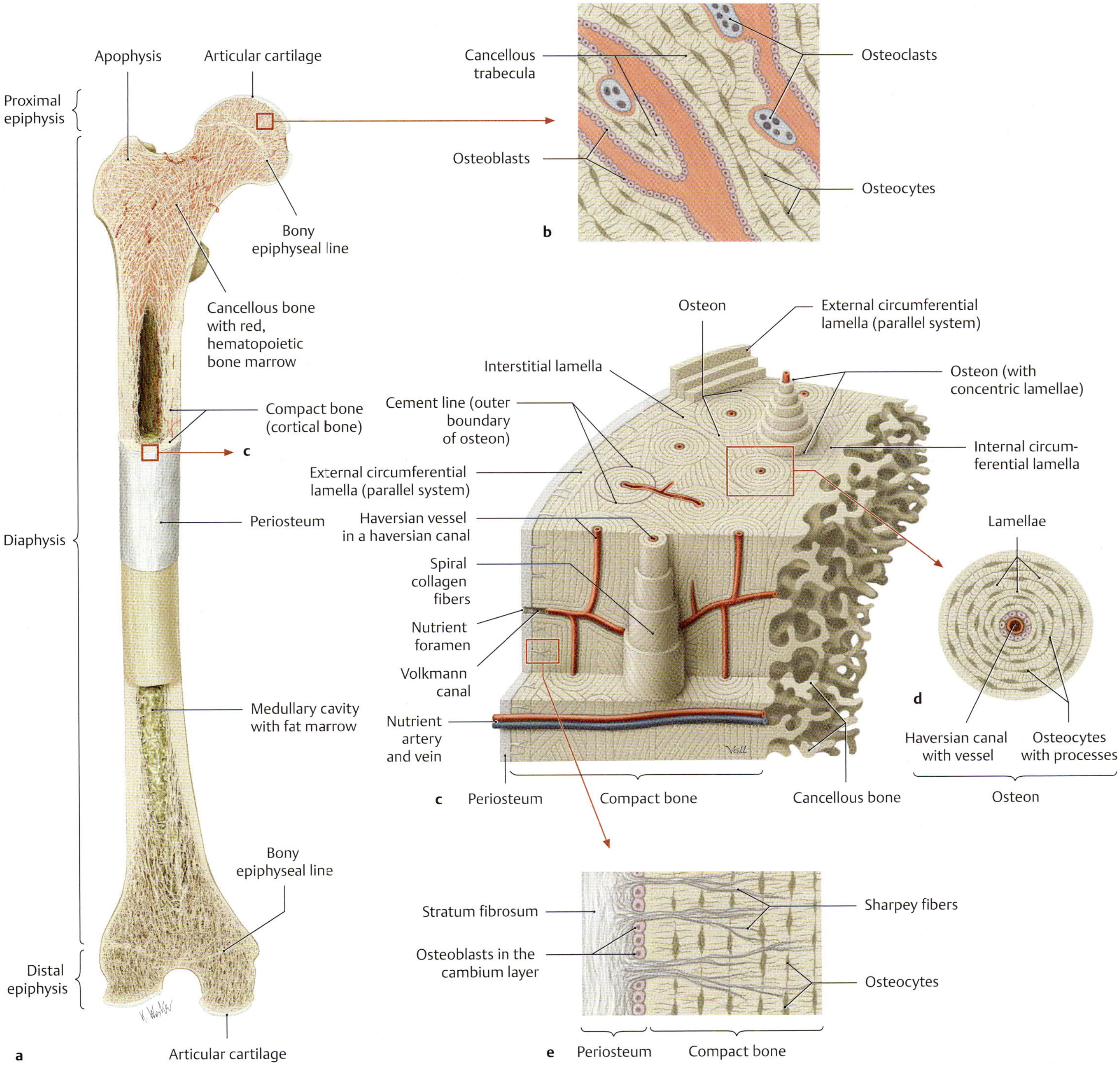

D Structure of a typical tubular bone, illustrated for the femur

a Coronal saw cuts have been made through the proximal and distal parts of an adult femur (without sectioning the midshaft region).

b Detail from **a**: The sectioned areas display the lamellar architecture ("lamellar bone") of the cancellous trabeculae. The lamellae are arranged in contiguous plates, similar to plywood. Since the cancellous trabeculae do not have an actual vascular supply and are nourished by diffusion from the adjacent medullary cavity, the trabeculae attain a thickness of only about 200–300 µm.

c Detail from **a**: Three-dimensional representation of compact bone, whose structural units consist of vascularized osteons approximately 1 cm long and 250–350 µm in diameter. The haversian canals, which tend to run longitudinally in the bone, are connected to one another by short transverse and oblique Volkmann canals and also to the vessels of the periosteum and medullary cavity.

d Detail from **c**, demonstrating the microstructure of an osteon. The haversian canal at the center is surrounded by approximately 5 to 20 concentric lamellar systems composed of osteocytes and extracellular matrix. The osteocytes are interconnected by numerous fine cytoplasmic processes.

e Detail from **c**, showing the structure of the periosteum.

4.2 Continuous and Discontinuous Joints: Overview and False Joints (Synarthroses)

A Different types of joints

False joints	True joints
(Synarthroses; continuously connected by connective tissue, cartilage, or bone; low to moderate degree of mobility; see right side) • **Syndesmoses** (fibrous joints, e.g., newborn fontanelles) • **Synchondroses** (cartilaginous joints, e.g., spinal disks; if the intervening tissue is mostly fibrocartilaginous, the joint is called a symphysis, e.g., pubic symphysis) • **Synostoses** (sites of bony fusion, e.g., sacral bone)	(Separated by a joint space; mobility is variable, depending on the attached ligaments = discontinuous connection) • **Diarthroses** (joints with varying degrees of freedom, see p. 44); classified according to – Shape and arrangement of articular surfaces (spheroidal, ellipsoidal, and saddle joints) – Number of joint axes – Number of degrees of freedom • **Amphiarthroses** (stiff joints, e.g., iliosacral joint)

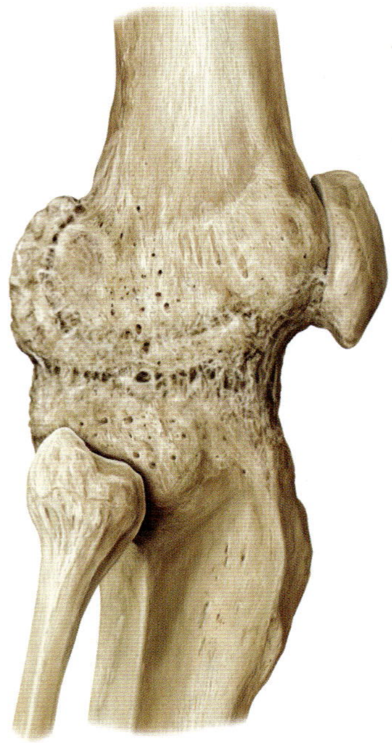

C Ankylosis of the knee joint
Illustration based on a specimen from the anatomic collection of the University in Kiel. Following the complete loss of cartilage, bone gradually fills the joint cavity, resulting in complete stiffness of the joint.

B Short glossary

- **Pseudarthrosis:** So-called false joint (instability due to missing wedge of bone) after abnormal fracture healing (see p. 55).
- **Ankylosis:** Abnormal bony fixation of a true joint (see **C**).
- **Arthrodesis (called spondylodesis in the spinal cord):** Surgically induced joint ossification by performing an osteosynthesis procedure (see p. 55 bottom **D**).
  - Main indicators: infectious arthritis, post-traumatic joint destruction, degenerative joint diseases, and paralytic instability.
  - Principle: loss of mobility in favor of absence of pain and stability.
  - Important for good joint motion postsurgery: stabilization in mode of function, which means in the position that is crucial for the most important joint functions, such as grasping for the arm and standing and walking for the leg.
- **Arthrotomy:** Surgical opening of a joint.
- **Arthrography:** Examination of the joint cavity using X-rays with the aid of a contrast agent (has increasingly lost relevance after rise in the use of MRI).
- **Arthroscopy:** Endoscopic examination of the joint, often associated with ensuing endoscopic therapy, e.g., arthroscopic reconstruction of injured ligament and capsule structures, removal of loose bodies, and treatment of joint cartilage defects (e.g., osteochondrosis dissecans).
- **Synovectomy:** Removal of the synovial membrane, e.g., in chronic polyarthritis.
- **Joint puncture (intra-articular injection):** Puncture of a joint space to remove fluid (e.g., in an articular effusion) or application of substances (medication).
- **Joint endoprosthesis:** Joint replacement surgery through total or partial endoprosthesis in advanced joint disease (arthritis) (see p. 48).

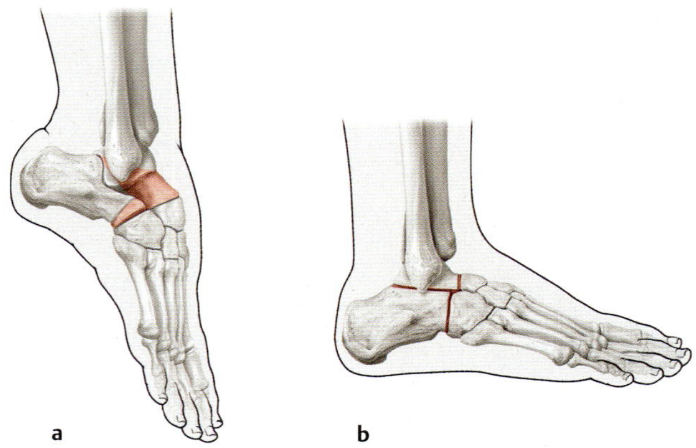

D T-arthrodesis after Lambrinudi for treatment of a club foot deformity
Right foot before (**a**) and after (**b**) arthrodesis treatment, lateral view. The surgical goal is to repair the club foot position caused by stiffness (arthrodesis) to achieve a plantigrade foot (lower leg and foot form a 90-degree angle) while at the same time maintaining mobility in the upper ankle joint. Since three joints (subtalar, talonavicular, and calcaneocuboid joints, see p. 470) are fused or stiffened, the procedure is called T-arthrodesis for "triple arthrodesis." It involves resecting the talar head and interlocking it with the navicular bone by performing osteosynthesis.
Note: In case of instability and ankle joint misalignment, arthrodesis is the safest treatment method.

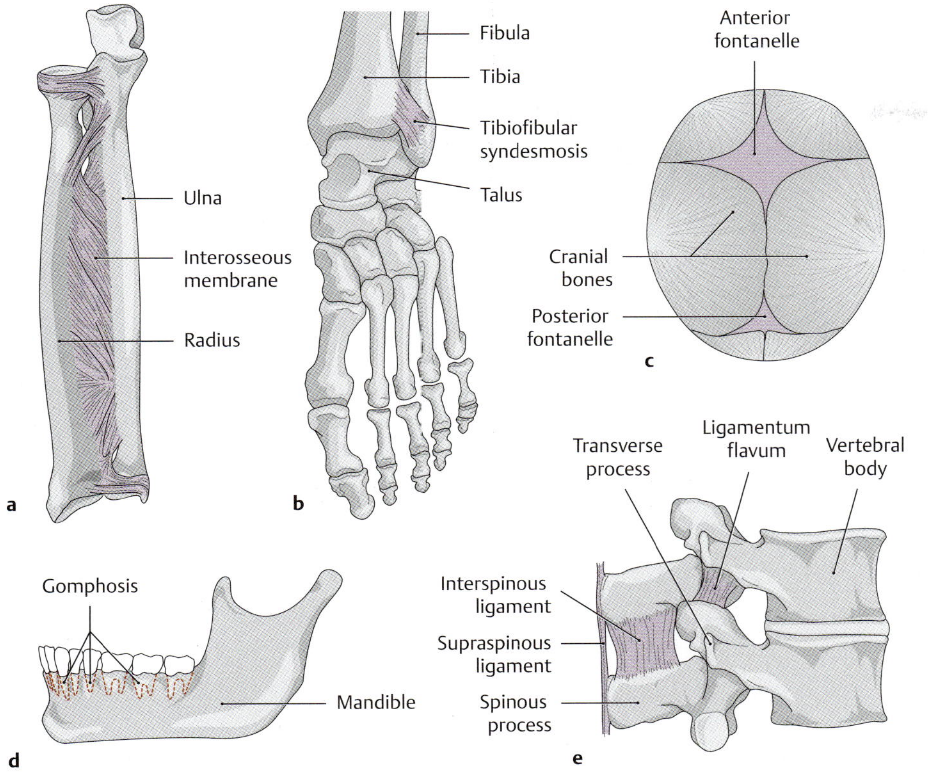

E Syndesmoses (fibrous joints)
a Interosseous membrane.
b Tibiofibular syndesmosis.
c Fontanelles.
d Gomphosis (immovable articulation).
e Ligamentum flavum, interspinous ligament, and supraspinous ligament.

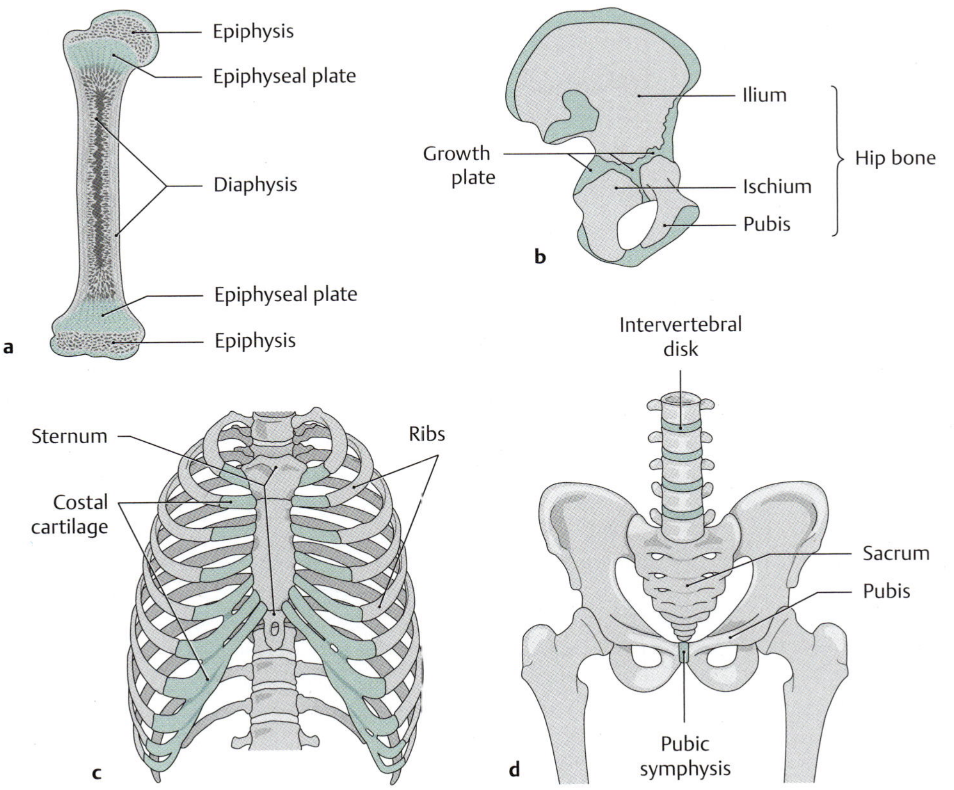

F Synchondroses (cartilaginous joints)
a Epiphyseal plates prior to closure.
b Hip bone before closure of the growth plates.
c Costal cartilage.
d Pubic symphysis and intervertebral disks (intervertebral symphysis).

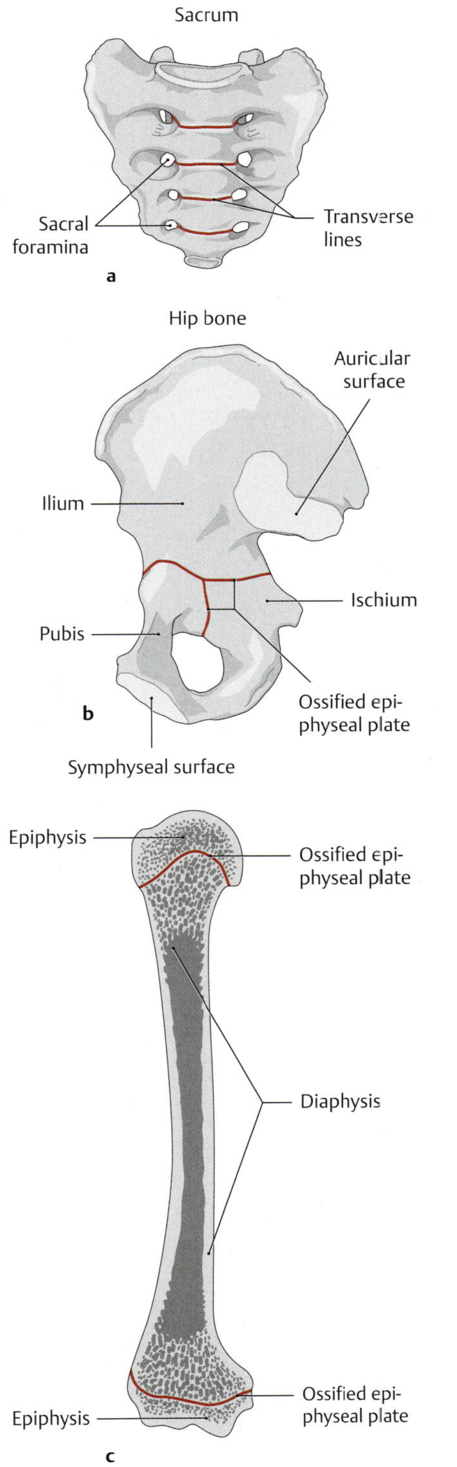

G Synostoses (sites of bony fusion)
a Sacrum (fused sacral vertebrae).
b Hip bone (fusion of the ilium, ischium, and pubis).
c Closed and ossified epiphyseal plates.

4.3 True Joints: Components and Intra- and Extra-articular Structures

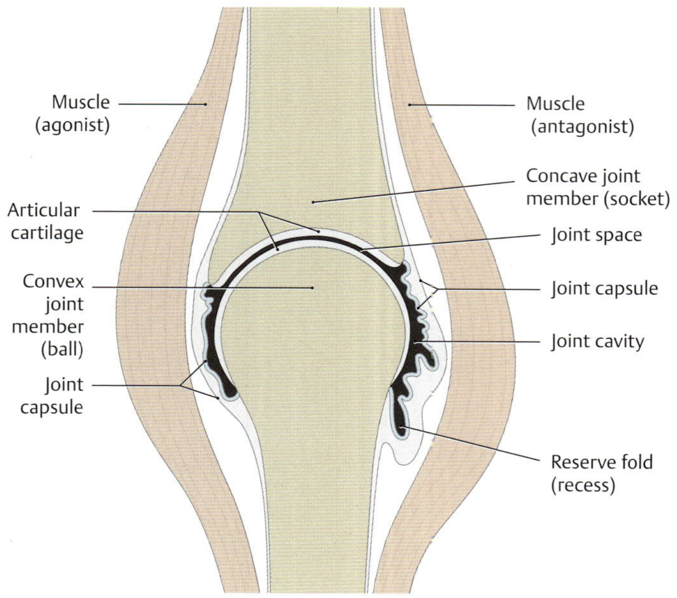

A Structure of a true joint
In true joints, the articulating joint surfaces are separated by cavities measuring no more than a few millimeters in width (*di*arthroses), unlike false joints, in which the joint surfaces are connected (*syn*arthroses). True joints are also called synovial joints, as the joint capsule produces synovial fluid. In addition to the cavity, true joints have the following characteristics:

- Variously shaped articular surfaces covered by hyaline cartilage.
- Joint cavity with articular recesses, the width of which vary.
- All-around closed joint capsule increased by variously shaped alar folds, synovial folds, and synovial villi.
- Highly viscous synovial fluid produced by joint capsule.
- Intra-articular components to improve joint congruency and increase force-absorbing surface, e.g., articular menisci, disks, and labra.
- Various forms of ligaments, with intra- and extracapsular ligaments acting as primary joint stabilizers.
- Joint crossed by muscles that move it in opposite directions (agonist/antagonist).
- Synovial bursa, often located near the joint and thus articulating with the joint cavity.

* Not all of the above are present in all true joints.

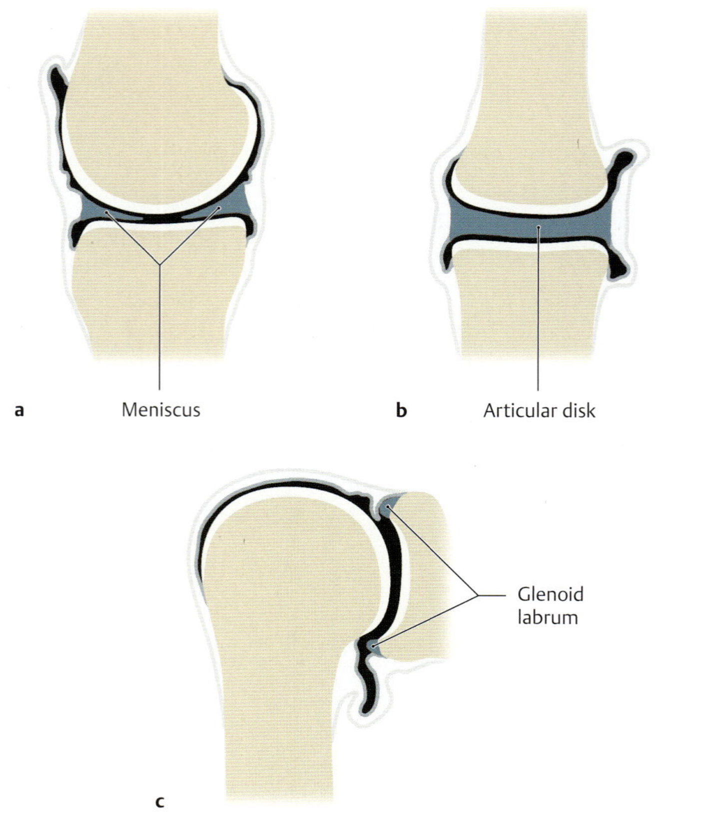

B Intra-articular structures and their importance to functional integrity of joints
Some joints contain intra-articular structures that ensure stability and function. For example, they increase the force-transmitting surface by balancing out incongruencies between the different parts of the joint, thus considerably reducing the compressive force on the cartilage. By definition, intra-articular structures are located in the articular cavity and float within synovial fluid, which means they are in direct contact with the synovial fluid through which they are primarily nourished (menisci, disks, articular labra).

a Menisci: Crescent-shaped structures, wedge-shaped in cross section, which are formed regularly in the knee only. They consist of collagen fibers, connective tissue, and fibrous cartilage. While their peripheral components are supplied by blood vessels of the joint capsule with which they are conjoined, the interiorly located fibrocartilaginous components are nourished by the synovia.

b Articular disks partially consist of connective tissue and fibrous, cartilaginous, disk-like structures, which divide the joint into two separate chambers. Articular disks are regularly found in the jaw, sternoclavicular joint, and proximal wrist joint (see p. 301).

c Articular labra are wedge-shaped structures at the outer margins of the osseous hip and shoulder joint sockets (glenoid and acetabular labra, see pp. 278 and 444). They consist mainly of fibrocartilage, and their outer layer is composed of connective tissue, the fibers of which are joined to the capsules. The articular labra enlarge the articulating surfaces of the shoulder and hip joint.

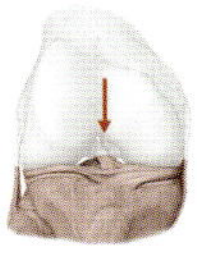

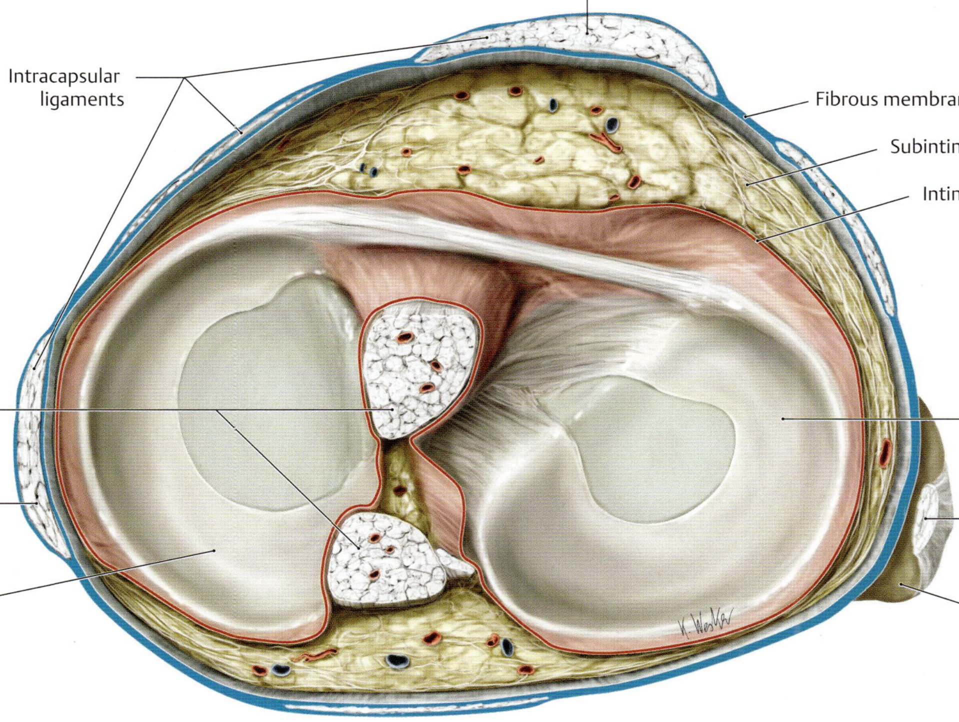

C The joint ligaments: extra-articular structures, which are usually located intracapsular (inside the joint capsule)

Intracapsular ligaments can run along both in the subintima (cruciate ligaments) and in the fibrous membrane (e.g., iliofemoral, ischiofemoral, and pubofemoral ligaments of the hip joint). They strengthen the capsule and thus serve mainly mechanical functions. For example, they are responsible for stabilizing and guiding the joint (guiding ligaments) or, to varying degrees, limiting or restricting motion (restraining ligaments). Occasionally, the joint ligaments extend without any direct contact to the actual joint capsule (extracapsular ligaments), such as the fibular collateral ligament in the knee joint. Often, ligament structures, which at first seem to run in the articular cavity (e.g., the cruciate ligaments in the knee joint, see p. 458, and the ligament of the head of the femur in the hip joint), are incorrectly called intra-articular ligaments. However, unlike the meniscus and the articular labrum, they are always covered by a thin synovial intima and thus run subintimally. Strictly speaking, they are located inside the capsule, which means they are intracapsular, and thus outside the actual joint capsule, which means they are extra-articular (see p. 46).

Note: Ligaments must have the "right" length in order to be able to guide and stabilize the joint. If they are overstretched or tear (as a result of chronic overuse or injuries, e.g., as sprains or dislocation), the joint becomes unstable. If ligaments shorten because the joint is not moved or held permanently in an unfavorable position (e.g., knee joint in bent position, proximal phalanges in stretched position), the joint becomes more or less immobile (joint contracture).

D Composition and function of synovial fluid

Synovial fluid is a clear, slightly yellowish, viscous fluid (product of secretion of the synovial cysts; pH value 7.4–7.7). Even in large joints, such as the knee joint, its volume is only 3–5 mL (less than 1 mL in small joints).

Compositon	Function
• Hyaluronate (or hyaluronic acid), lubricin (mucin-like glycoprotein), and phospholipids • Blood plasma components (mainly blood proteins and glucose) • Defense cells 60–150/µL (mainly macrophages and lymphocytes)	① Nourishing the hyaline cartilage, which is devoid of blood vessels, through diffusion and convection ② Lubricating the cartilaginous articular surfaces, thus enabling frictionless motion ③ Functioning as a buffer by evenly distributing pressure across the articular surfaces

4.4 True Joints: Structure of Joint Capsule and Hyaline Cartilage

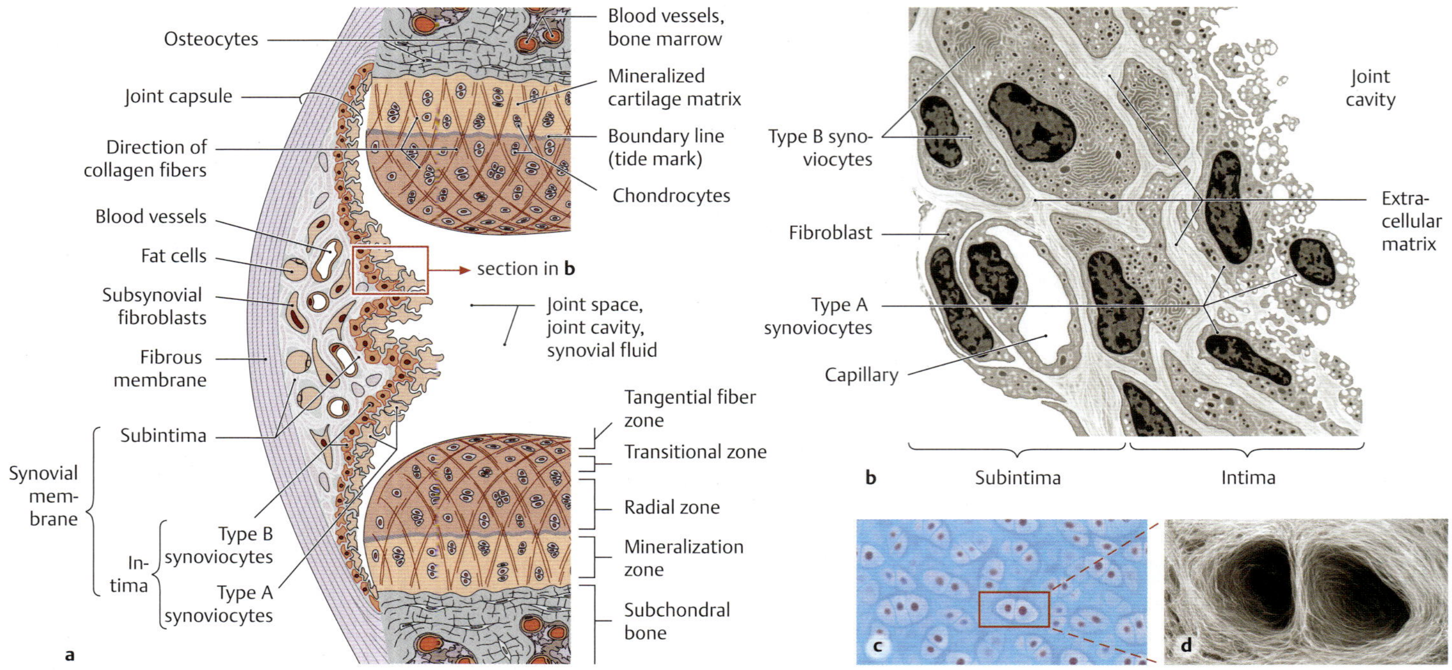

A Structure of joint capsule and hyaline cartilage

The joint cavity of true joints is fully enclosed by a joint capsule, the two layers of which differ morphologically from one another and serve different functions:

Joint capsule

- **Fibrous membrane** – WALL (firm, collagenous connective tissue; fibers extend to the periosteum). Thickness varies both within one joint and in different joints, in many joints reinforced by ligaments (intracapsular ligaments), see p. 45
- **Synovial membrane** – WALLPAPER (attached to the margin of the joint capsule)
  - **Synovial intima** – lining cells (1 to 3 layers of epithelial-like [not epithelial cells because no basal membrane] covering cells); **production and resorption of synovial fluid.**
    - **Type A synoviocytes** (boundary layer to the joint capsule): macrophage-like cells with vacuoles, Golgi apparatus, mitochondria, lysosomes; resorption of used synovial fluid and phagocytosis of bacteria and cell debris
    - **Type B synoviocytes** (underneath the type A synoviocytes): fibroblast-like cells with rough endoplasmic reticulum, secretory granules; mainly production of synovial fluid
  - **Synovial subintima** (connective tissue with blood and lymph vessels, pain receptors [free nerve endings], and mechanoreceptors [Pacini and Ruffini bodies = proprioceptors], which react to changes in lengths or tension in the joint capsule = helps guide joint movement sequences), structure differs based on topography
    - **Areolar synovial membrane** (loose, richly vascularized connective tissue)
    - **Adipose synovial membrane** (mainly fat cells in connective tissue)
    - **Fibrous synovial membrane** (vessel-poor and collagen-rich connective tissue)

Note: The **synovial membrane** can regenerate into old age, even after complete removal (= synovectomy; necessary, e.g., in cases of chronic arthritis due to rheumatism).

The synovial membrane reacts to stimuli by increased secretion, which appears as articular effusion and swelling in the entire joint region. Depending on the type of stimuli (mechanical, allergic, infectious), the effusion is either clear and thin or turbid and pus-filled. A true bloody effusion (hemarthrosis) develops in the wake of injuries (e.g., cruciate ligament rupture). The ensuing pain is the result of overstretching the joint capsule and the release of inflammatory mediators (e.g., prostaglandins, histamines, bradykinin, and cytokines). With the exception of the jaw and sternoclavicular joint, which are covered by fibrous cartilage, all articular surfaces are covered by **hyaline cartilage**. It is malleable and reacts to pressure (elastic compression). Depending on the amount of pressure, its thickness varies—ranging from 1–2 mm in the phalangeal joints to 5–7 mm in the femoropatellar joints. Hyaline cartilage consists of extracellular matrix (ECM) and chondrocytes (approximately 5% volume fraction).

Note: While the synovial membrane can regenerate through the adjoining connective tissue, the hyaline cartilage (except in the primordial skeleton) does not have any surrounding connective tissue (perichondrium) and thus can hardly regenerate. In addition, it is devoid of vessels. The cartilage cells (chondrocytes) must exclusively be nourished by synovial fluid (diffusion and convection, see **D**, on the right).

Hyaline cartilage can be divided into zones. They are determined by the arcade-like course the collagen fibrils follow (**a**). From zone IV (deepest zone, where the collagen fibrils are connected to the bone underneath the cartilage), the collagen fibrils extend almost vertically (radially) to the articular surface (zone III), then turn to follow a course parallel to the surface (zone II), and again after a short tangential course extend downward (zone I).

The collagen fibrils are visible only under the transmission electron microscope (see **d**). In a light microscope image, they are "masked" as they break the light just as their surroundings do. After enzymatically liberating chondrocytes and proteoglycans using hyaluronidase, looking through the transmission electron microscope creates the impression of empty "cartilage holes" that are surrounded by a thick mesh of collagenous fibrils (for composition of collagen fibrils, see **d**). Proteoglycans, a component of the extracellular matrix, however, are clearly visible under the light microscope because they carry many negative charges, which can be visualized by using dyes such as hematoxylin (see **c**).

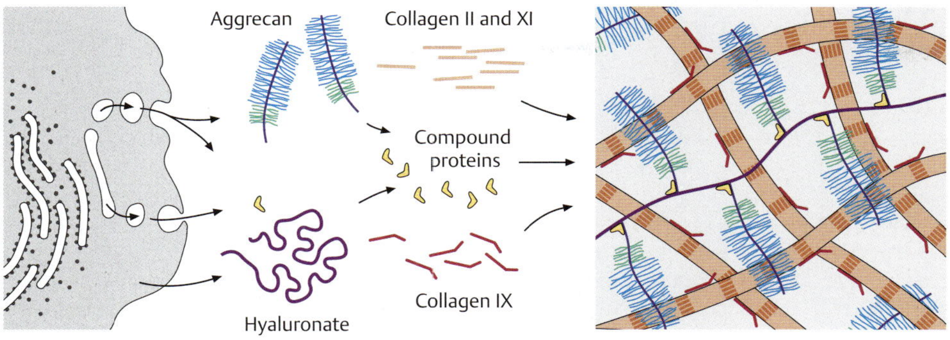

B Structure and function of extracellular matrix (ECM)
The extracellular matrix (all macromolecules that are secreted by cartilage cells and immobilized when interacting with other molecules in the extracellular space) primarily consists of

- Collagen fibers or fibrils (mainly collagen type II, IX, and XI).
- Proteoglycans (mainly aggrecan) with glycosaminoglycans (e.g., hyaluronate, chondroitin, and keratan sulfate), as well as core and compound proteins.
- Glycoproteins (e.g., adhesion proteins such as chondronectin).
- Interstitial fluid (water) and electrolytes (mainly cations, Ca^+, K^+, Na^+).

The collagen fibrils (diameter of 15–130 nm) consist of collagen molecules, each of which are formed by three polypeptide chains (α-chains) and have the shape of a right-handed screw (triple helix). Individual triple helices (tropocollagens) are cross-linked by covalent connections, which give them their extraordinary tensile strength.

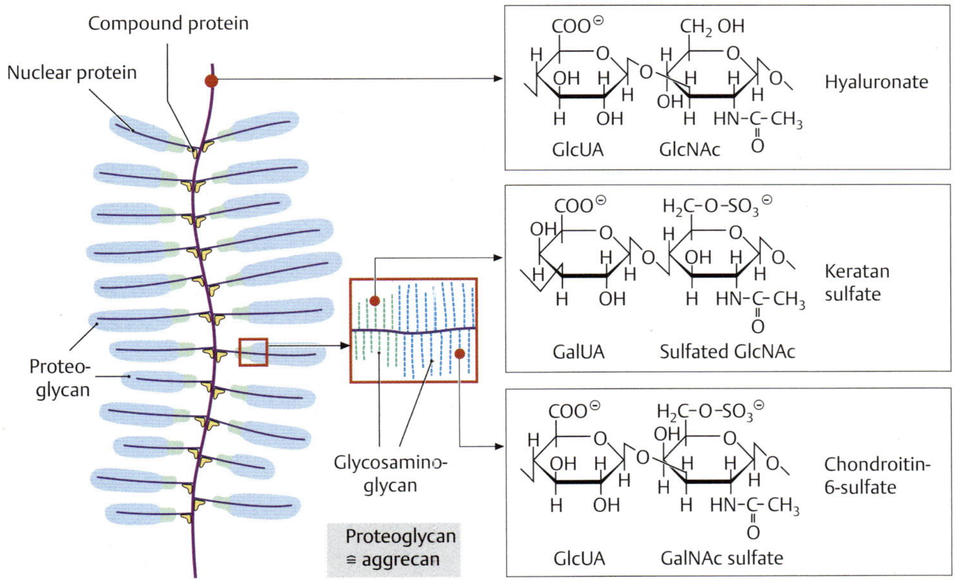

C Aggrecan, the most important proteglycan of the hyaline cartilage (after Koolman and Roehm)
Proteoglycans such as aggrecan are very large molecular complexes that consist of more than 100 glycosaminoglycan chains. They look like a bottle brush, with the individual glycosaminoglycan molecules as "bristles," affixed to a central hyaluronate molecule through so-called compound proteins. Disaccharide units, which mostly consist of a uric acid (e.g., glucuronic acid, GlcUA) and a type of amino sugar (e.g., N-acetylgalactosamine, GalNAc) and usually carry two negative charges (one each on the carboxyl and on the sulfate groups), form the basic structure of the glycosaminoglycan chains. For reasons of electroneutrality, the proteoglycans bind cations, which thanks to osmosis can reversibly bind a lot of water. Due to their strong water-binding capabilities, proteoglycan molecules take up huge volumes of water and thus create large, space-filling structures. GalUA, galacturonic acid; GlcNAc, N-acetylglucosamine.

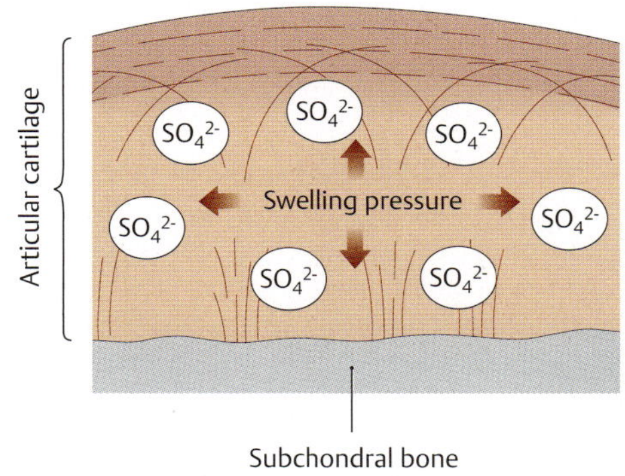

D Biphasic viscoelastic model of the joint cartilage (after Mow and colleagues)
Firmness and elasticity during compression of the cartilage are crucial for the proper functioning of the joints. They are both ensured by the swelling pressure (the principle of the elastic spring). It arises from the interaction of proteoglycans and collagen fibrils in the extracellular matrix (ECM), where they collaborate to carry out particular functions. The extraordinary tensile strength of the collagen fibrils allows for mechanical stability and structural integrity of the joint cartilage. The polyanionic proteoglycans ensure its physiochemical properties (e.g., reversible water-binding capacity). In free, watery fluid, proteoglycans, which are negatively charged, would normally greatly expand by repelling each other. They are prevented from doing so only by the tensile strength of the collagen fibrils, so they can take up only one fifth of the space they actually require. There is only a limited possibility for further compressing the proteoglycans. If cartilaginous tissue is subjected to mechanical pressure (normal pressure), interstitial fluid and cations escape into the joint cavity, and the ECM increasingly thickens. As a result, it is harder for fluid to drain from the cavity, the hydrostatic pressure in the tissue rises (*viscid components* of the viscoelastic model), until the hydrostatic pressure from inside and mechanical pressure from outside are in balance, and no further compression occurs. With the thickening of the ECM, the negatively charged glycosaminoglycans approach each other, leading to an increased repulsion force as a result of which negative pressure builds up. When the pressure is relieved, the tissue swells up to its original size due to proteoglycans repelling each other, and interstitial fluid and free cations migrate anew into the tissue (*elastic component* of the viscoelastic model) until the collagen network again puts a stop to it. The flow of interstitial fluid is also called *convection*. It is a crucial precondition for the nourishment of the joint cartilage, which does not contain blood vessels.

4.5 Degenerative Joint Diseases as Exemplified by Hip Joint Arthrosis

A Etiology and pathogenesis of osteoarthritis

For many years now, degenerative joint diseases—mainly osteoarthritis—have been the top cause of disability worldwide. Due to the high cost of treatment and the large number of days employees are absent at work, osteoarthritis represents a major economic problem. Elderly people are particularly affected. The seeming increase in cases of this disease over the past few decades is due to people reaching ever higher ages, which allows them to actually experience their osteoarthritis (higher age and degenerative joint disease don't necessarily go hand in hand).

The main cause of osteoarthritis is an *imbalance between the load and the capacity of the articular cartilage. Secondary* osteoarthritis (in contrast to *primary* osteoarthritis, the cause of which is not pathologic) develops almost exclusively due to suboptimal biomechanics, e.g., in cases of incongruencies of the articulating surfaces. Over- or incorrectly loaded joints play a major role, e.g., caused by post-traumatic deformities, joint dysplasia, axis deviation, and instabilities. In particular, an increasing number of sports injuries lead to osteoarthritis. Excess weight, lack of exercise, and bad nutrition further contribute to the development of osteoarthritis. Early detection and correction of axis deviations or osteotomies to correct joint incongruencies are of particular significance for the prevention of osteoarthritis. If osteoarthritis has already developed, and conservative therapies have been exhausted, in addition to performing arthrodesis (joint stiffening, see p. 42), there is a possibility to replace the joint (endoprosthesis, see **C**).

One of the main causes of the disease is the gradual attrition, meaning the wear and tear and destruction, of the articular cartilage (see **D**). If the hyaline cartilage is too heavily loaded over the long term (see etiology), it disrupts the regular flow of interstitial fluid into the articular cartilage (see p. 42). This affects not only the elasticity but also the nourishment of the bradytrophic cartilage (no blood vessels), which, due to its limited metabolism and a missing perichondrium, is barely able to regenerate (see p. 46). As a result of increased mechanical overload of articular cartilage areas, both cartilage matrix and chondrocytes are damaged (see **D**). This damage is irreversible; complete restoration or repair is not possible. There are, however, endogenic compensatory mechanisms, which are supposed to make up for the loss of cartilage in the early stages (I and II, see **D**) and the increased pressure in the subchondral bone in the late stages (III and IV). The body's attempts to repair itself explain the shift between pain-free periods and periods marked by heavy pain. Heavily loaded joints such as hip or knee joints or the intervertebral joints of the spinal column are per se affected by osteoarthritis (see **B**).

* Synonyms: osteoarthrosis—in contrast to arthritis, in the process of which the cartilage disappears only secondarily as a result of inflammation of the articular cartilage.

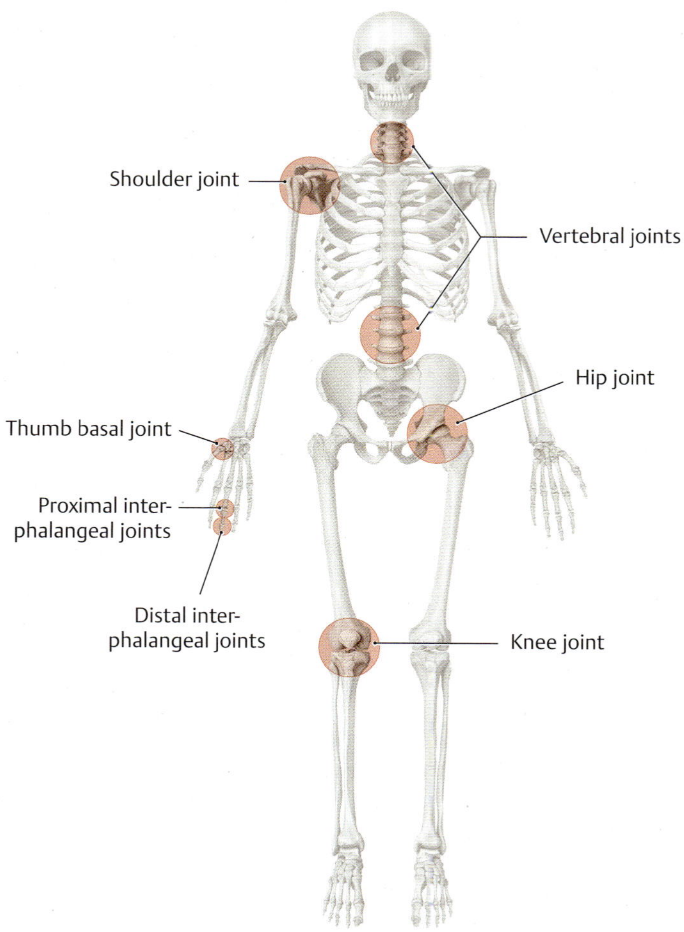

B Preferred locations of degenerative joint diseases

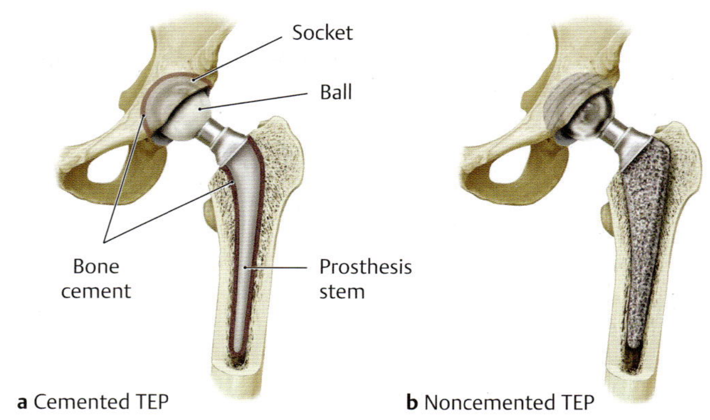

C Total endoprosthesis (TEP) of the hip joint

Endoprosthetic joint replacement has become a possibility for almost all joints. The best results, particularly in terms of long-term success, have been achieved in the hip joint (replacement of head and socket). The implants—usually made of metal, synthetic material, or ceramic—are either attached to the joint with a special cement or implanted "cement free." The cemented TEP is immediately stressable postsurgery; the noncemented TEP requires a little more time (several weeks) to firmly connect with the surrounding bone tissue. However, a noncemented prosthesis is easier to replace should it be necessary if it becomes loose, since there is no cement to remove. Which prosthesis to implant depends on several factors, such as age, primary bone stability, and possible preexisting bone diseases (osteoporosis). Complications of artificial joint replacement are foremost loosening of the joint and infection. Both cemented and noncemented prostheses currently have a 90% probability of lasting significantly longer than 10 years.

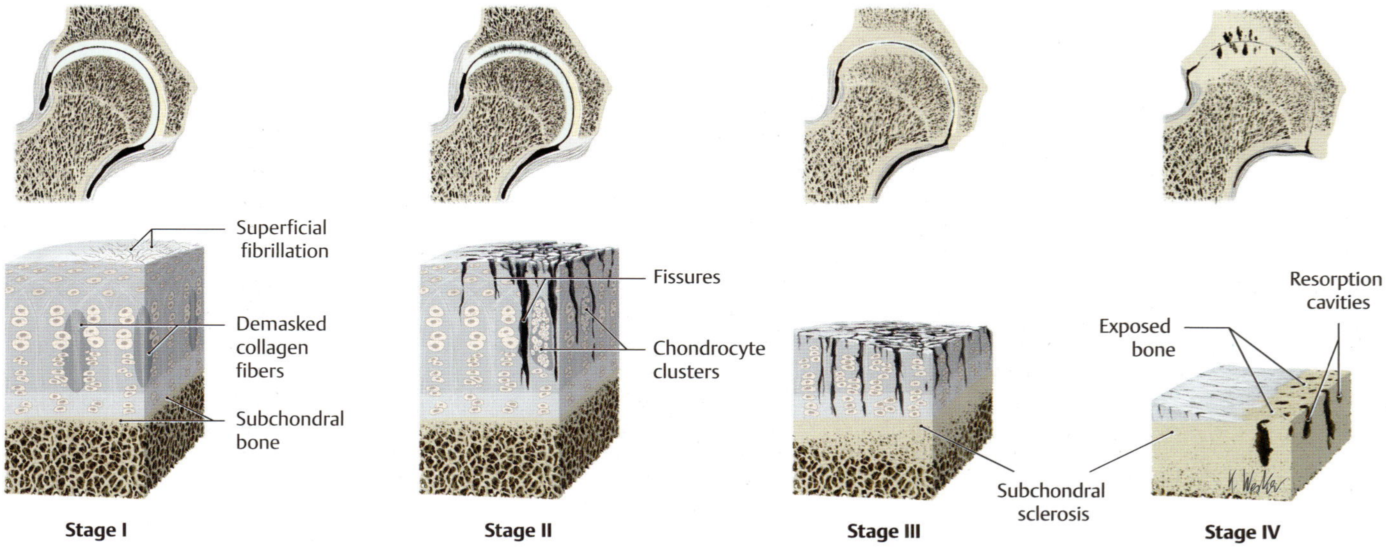

D Stages, compensatory mechanisms, and clinical symptoms of osteoarthritis

Stage I	Stage II	Stage III	Stage IV
• Articular surface is roughed up and begins to fray (surface fibrillations) • In the deeper lying cartilaginous layers, collagen fibers are exposed (demasked) by decreased proteoglycan synthesis	• Development of fissures that extend deep into the cartilage • Loss of cartilage matrix • In the fissure area, individual cartilage cells divide again (chondrocyte clusters develop)	• Increasing density of the subchondral bone as a result of increasing cartilage attrition • First osteophytes appear at the bone bridge • Inflammatory response at the articular capsule (reactive synovitis, usually with effusion) as a result of increased accumulation of deposits caused by cartilage wear	• The bone is entirely exposed; the bone surfaces rub on each other • Collapse of the subchondral bone roof (bone necrosis) • Development of large resorption cavities in the articulating joint surfaces (see **E**) • Increasingly strong growth of osteophytes

Compensatory mechanisms in the early stage	Compensatory mechanisms in the advanced stage	Goal of the compensatory mechanisms
• Limited proliferation (cell division) of the previously post-mitotic cartilage cells • Formation of so-called chondrocyte clusters, or brood capsules, with newly developed extracellular matrix	• Activation of mesenchymal progenitor cells in opened subchondral cavities within the spinal cord (subchondral cysts) • Followed by formation of a new articular surface through meta-plastic fibrocartilage	Compensate for cartilage loss
Reactive thickening of the subchondral bone with increasing bone density (subchondral sclerosis)	Enlargement of the articular force-transmitting surface through formation of marginal osteophytes (so-called marginal exostosis)	Equalize pressure in the subchondral bone

Clinical symptoms: After an asymptomatic early stage of the disease (so-called latent arthrosis) that can last for years, patients initially report discrete pain, which gets worse with increased joint load and is accompanied by the first signs of referred muscle pain. In a more advanced stage of the disease, in addition to stress-related pain, patients experience pain when in neutral position (so-called activated osteoarthritis). Its cause lies in the inflammatory response of the articular capsule to the increased accumulation of deposits caused by cartilage wear. The reactive synovitis is almost always accompanied by effusions. As the disease progresses, symptoms intensify, with simultaneous increased joint instability. The final stage is characterized by constant pain, heavily limited load-carrying capacity, strong muscle contraction, drastic malalignment, and stiffening of the joint.

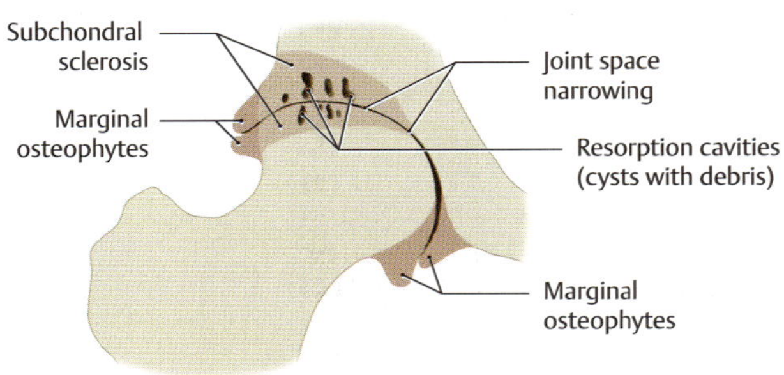

E Synopsis of the results of X-ray examination coxarthrosis of hip joint arthrosis

Schematic representation showing the radiologic changes seen in advanced of hip joint arthrosis. The destruction and degeneration of the hyaline articular cartilage leads to typical signs in the conventional X-ray, including narrowing of the cartilage gap in the joint parts that are most heavily loaded as a result of increasing loss of cartilage, the subchondral thickening (subchondral sclerosis) and lightening (subchondral cyst formation) of the bone, and the presence of osteophytes (from reactive bone changes) both in loaded and unloaded areas. This also results in deformation of the femoral head.

4.6 Basic Principles of Joint Mechanics: Movements

A Joints as functional units

In order to understand joints as functional units, it is best to draw on physical laws, for although joints primarily enable *motion*, at the same time they guarantee *stability* of the musculoskeletal system and assist in transmitting *force* between the articulating bones.

- Movements are changes of location in space (translational and rotational motion), which have both temporal (uniform or nonuniform motion in time) and spatial aspects (variable capacity for motion along the three axes) (see **B–D**).
- In order to balance the body and stand erect against gravity, almost every joint, according to their location relative to the body's center of gravity, has to be stabilized mainly through ligaments, muscles, and tendons (see p. 52).
- The forces acting on a joint (muscle and ligament force, body weight, or gravity) create joint pressure and cause movement to occur (via the torsional moment) if they are applied outside the center of rotation, e.g., via a lever arm (see p. 53).

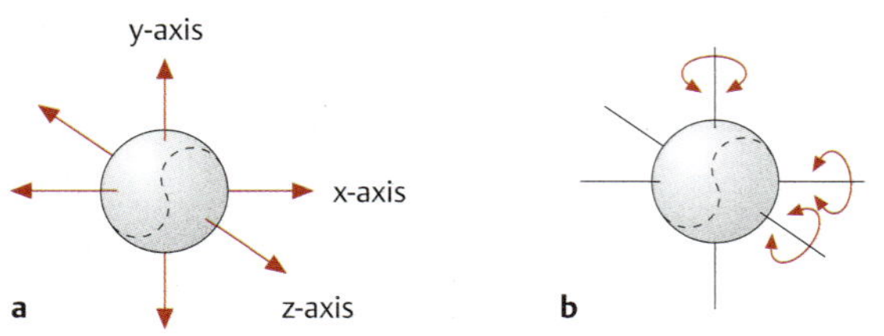

B Degrees of freedom, illustrated for the possible movements of a tennis ball in space

a Three degrees of freedom in translation (one each along the x-, y-, and z-axis).
b Three degrees of freedom in rotation (one each around the x-, y-, and z-axis).

With its similarity to a tennis ball, the spheroidal joint (see **Fc** and **Fd**) has the greatest freedom of movement.

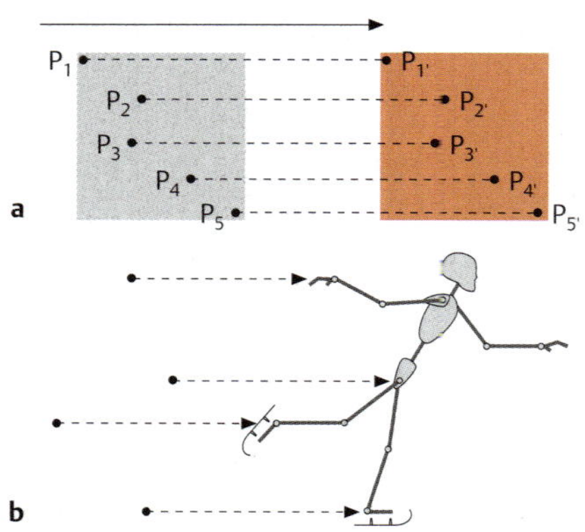

C Translation

Translation means that a body is sliding on a straight or curved path *without rotating*. As a result, all points on the moving body travel an *equal distance* in the same direction.
a All the points move in parallel lines.
b The gliding ice skater illustrates movement in translation.

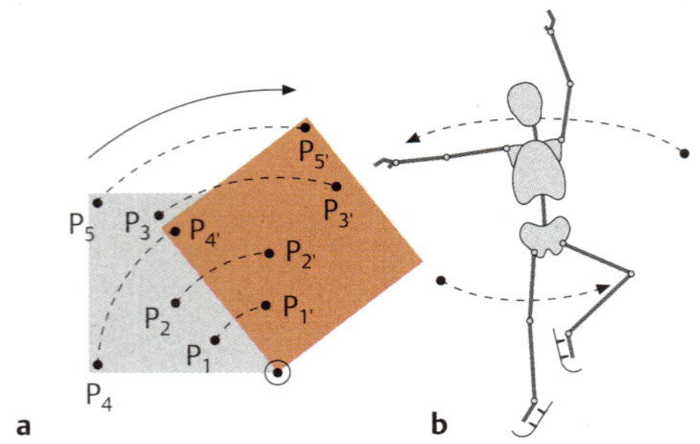

D Rotation

When a body is rotating, different points on the body move in concentric circles and travel *different distances*.
a All the points move in circular arcs. **b** The spinning ice skater illustrates movement in rotation.

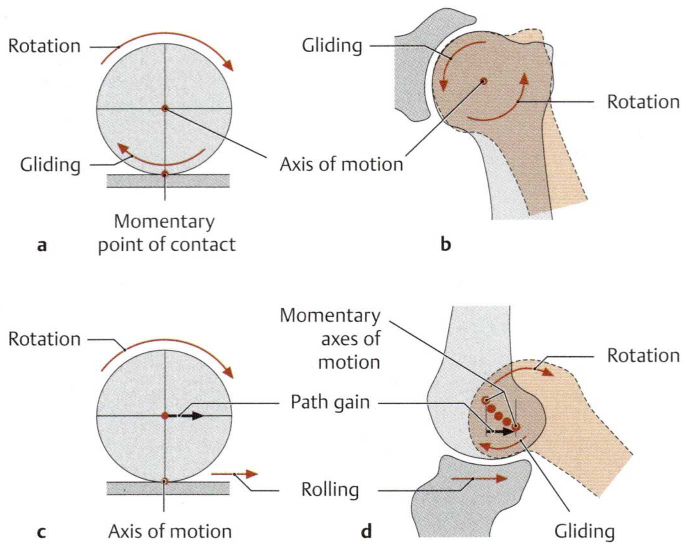

E Sliding and rolling in rotational motions

When rotating, articular surfaces can glide onto each other or roll off each other. In most cases, however, it is a combination of both (roll-slide).
Note: The more congruent the articular surfaces (roughly the same radius of curvature, e.g., shoulder joint; **a**, **b**), the better they slide onto each other; the less congruent the articular surfaces, the more they roll off each other (e.g. knee joint; **c**, **d**). Rolling and sliding differ in relation to the path gain of the axis of rotation.

a, b Rotation without path gain of the axes of motion (= sliding), meaning one point of the articular surface successively touches different points on the other articular surface (**a**), as happens with abduction (= rotational motion) in the shoulder joint (**b**).

c, d Rotation with path gain of the axes of motion (= rolling), meaning the surface of the rotating body rolls off the articular surface of the other body, and so each point of the articular surface touches one point of the other surface, with the distance covered being the same for both joints. An example is flexion in the knee joint (**d**): rotation here is a combination of rolling and sliding, because the femur and the condyles of the tibia have markedly different radii of curvature. Thus, when rolling, the axis of motion moves back on a particular path (evolute), indicating only where the movement of flexion occurs at that instant. Such an axis is also called an instantaneous axis.

F Correlation between form and movement capabilities of a joint
The movements a joint can perform are most notably determined by the form, meaning the geometry, of the articulating surfaces. In general, there are two basic movements:

- Translational movement (**a**, **b**) and
- Rotational movement (**c**–**h**).

Whereas translational movements in the joints are usually limited to one or two axes of motion (e.g., femoropatellar joint in the knee [**a**], vertebral joints [**b**] in the spinal column), rotational movements that occur, for instance, in spheroidal joints (**c**, **d**) have up to three axes of motion. In the femoropatellar joint, for example, when bending the knee, the patella slides up or down in the femoral groove, meaning the joint has one degree of freedom in translation with, accordingly, two main movements. In vertebral joints with their mostly flat surfaces, translational movements usually occur along one straight line, albeit in different directions (often two degrees of freedom in translation with four main movements). A spheroidal joint, however, has three degrees of freedom in rotation relative to the number of axes of motion and thus six main movements (abduction/adduction, flexion/extension, internal/external rotation). In joints with rotational movements, depending on the form of the articulating surfaces, the number of axes of motion and thus the number of degrees of freedom can be limited, e.g., ellipsoidal or saddle joint (two axes of motion, thus four degrees of freedom, **e**, **f**), hinge or pivot joint (one axis of motion, thus two degrees of freedom, **g**, **h**).

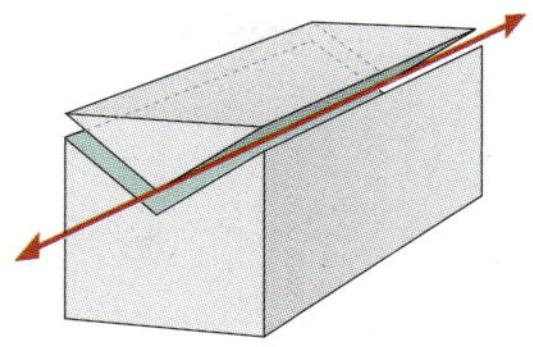

a Femoropatellar joint

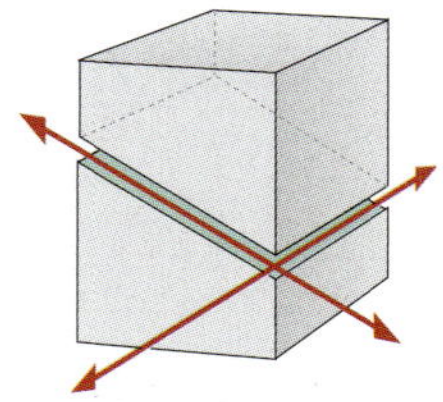

b Vertebral joint

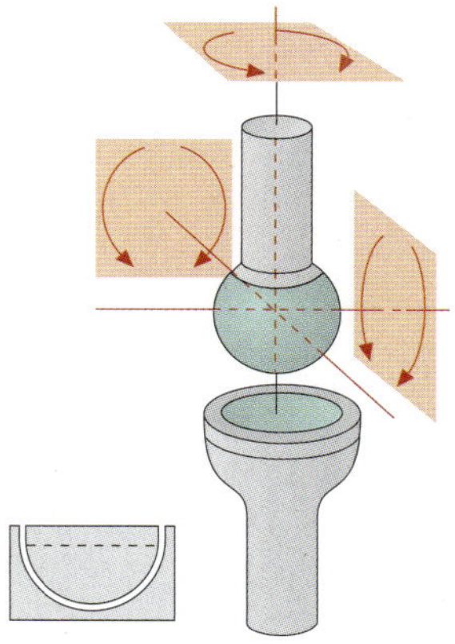

c Spheroidal joint
A joint with three mutually perpendicular axes of motion, resulting in six primary movements (i.e., hip joint with deep socket).

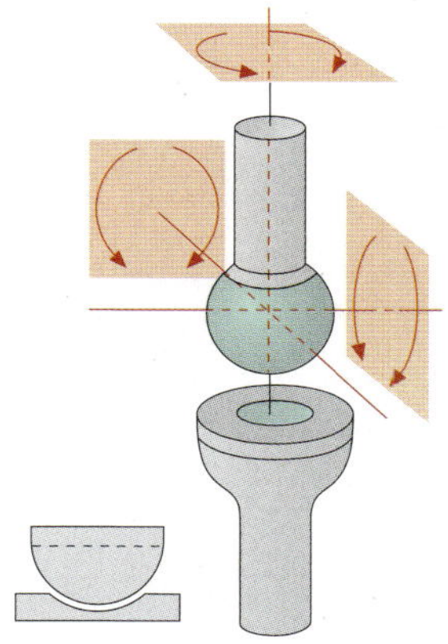

d Spheroidal joint
A joint with three mutually perpendicular axes of motion, resulting in six primary movements (i.e., shoulder joint with shallow socket).

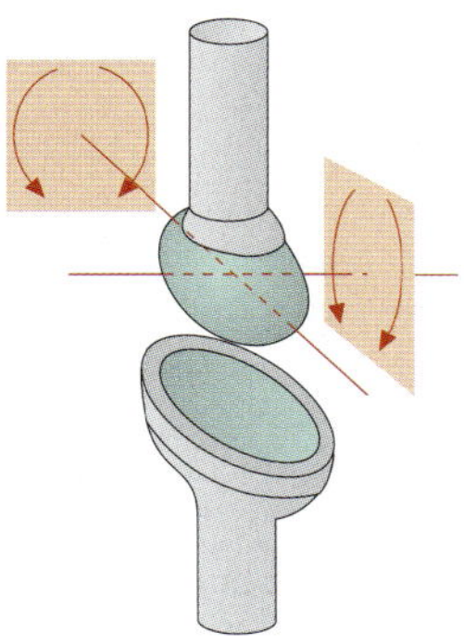

e Ellipsoidal joint
A biaxial joint with four primary movements (i.e., the radiocarpal joint).

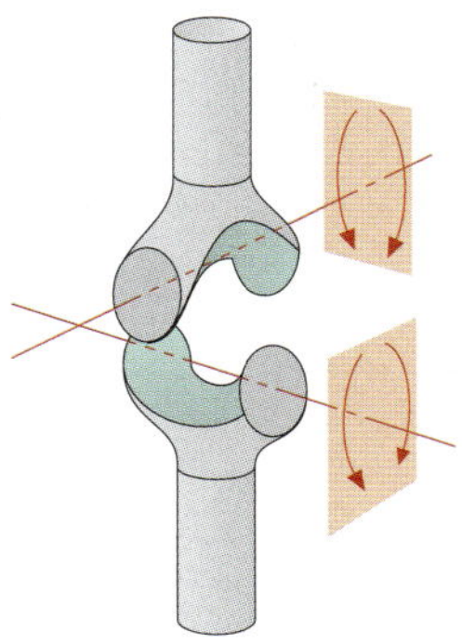

f Saddle joint
A biaxial joint with four primary movements (i.e., the carpometacarpal joint of the thumb).

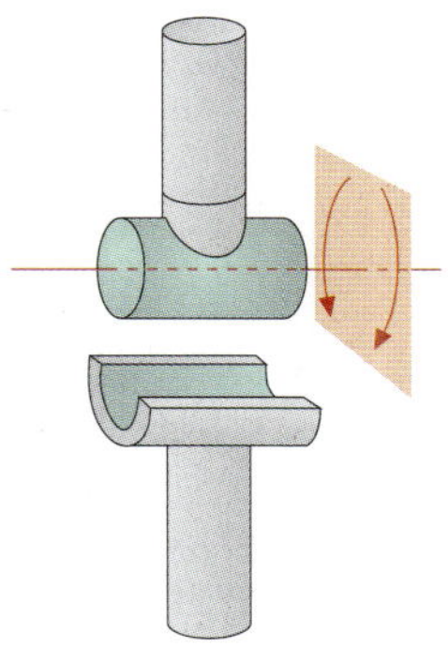

g Hinge joint
A joint with one axis of motion and two primary movements (i.e., parts of the elbow joint).

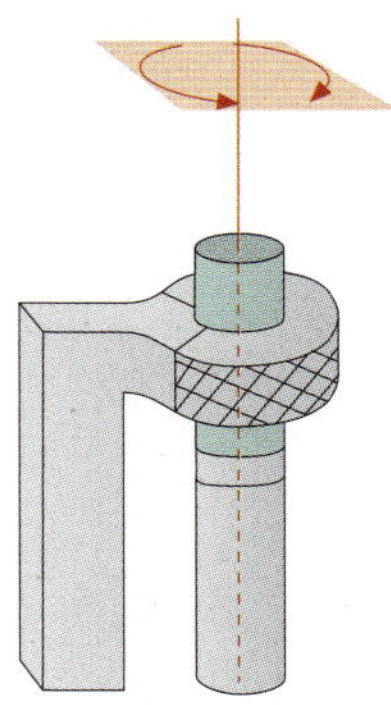

h Pivot joint
A uniaxial joint with two primary movements (i.e., the proximal radioulnar joint).

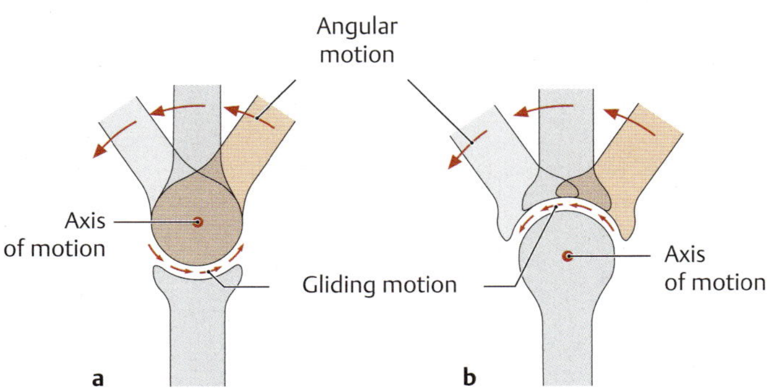

G Convex-concave rule
The convex-concave rule describes the sliding motion in dependence of the angular rotation motion.

a If the *convex joint partner* moves in an angular direction around the axis of motion, its rotation motion inside the joint is in the opposite direction of the angular motion.
b If the *concave joint partner* moves in an angular direction around the axis of motion, its rotational motion inside the joint is in the same direction as the angular motion.

4.7 Basic Principles of Joint Mechanics: Stability and Transmission

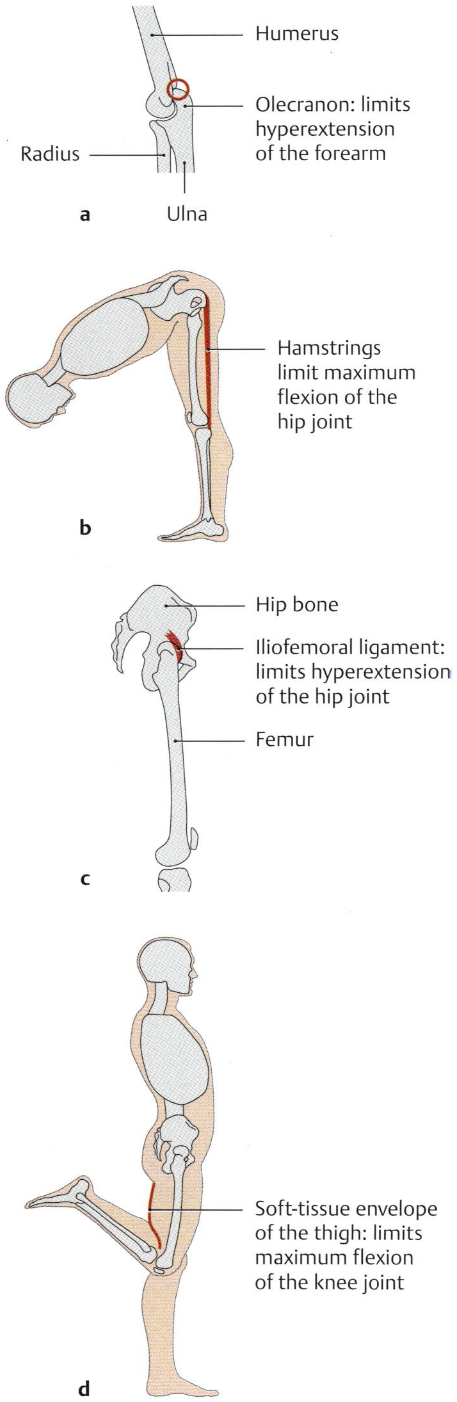

A Constraints on joint motion
The range of joint motion depends not only on the shape of the bony joint members but also on the muscles, ligaments, and soft-tissue envelope that surround the joint. Accordingly, all of these factors determine the total range of joint motion:

a Bony constraint.
b Muscular constraint.
c Ligamentous constraint.
d Soft-tissue constraint.

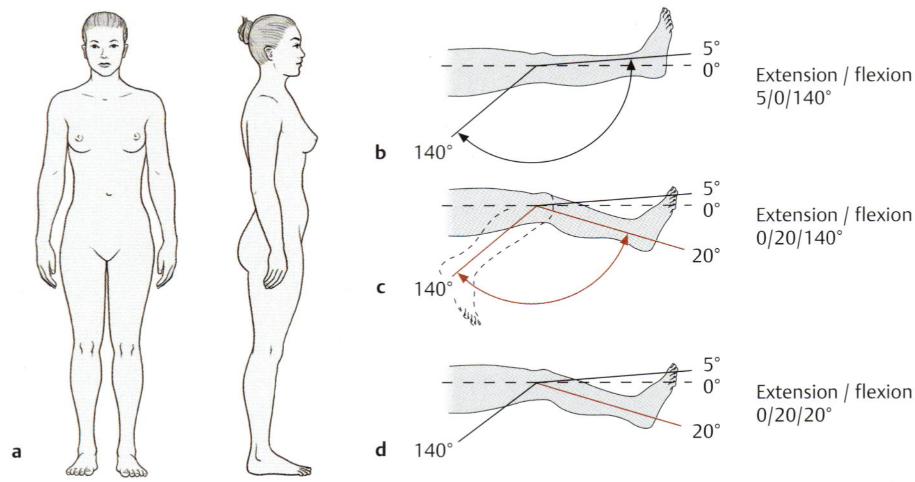

B The neutral zero method
The neutral zero method is a standardized method for measuring the range of joint motion.

a Zero-degree starting position, viewed from the anterior and lateral aspects.
b Range of motion of a normal knee joint.
c Limitation of motion caused by a flexion contracture.
d Ankylosis of the knee joint in 20° of flexion.

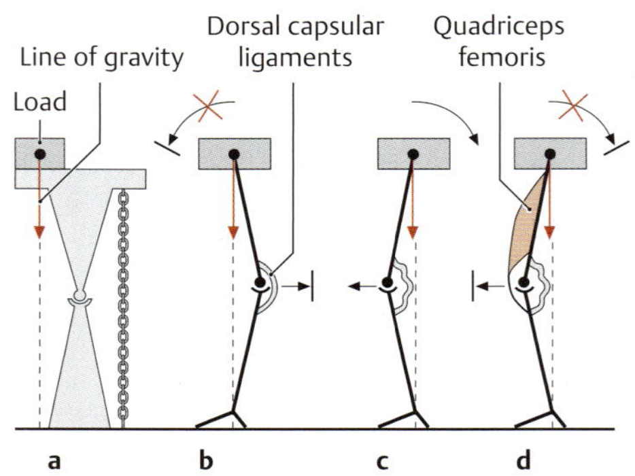

C Active and passive joint stabilization as demonstrated in the knee joint
The knee joint is actively stabilized by the quadriceps femoris and passively stabilized by the posterior ligaments (oblique popliteal and arcuate ligaments).

a Mechanical model: A load (body weight) causes a torsional moment in the joint as the line of gravity is in front of the middle of the joint. To prevent the upper joint partner from tilting forward, a chain acts as a counterforce.
b Passive stabilization through the popliteal and arcuate ligaments: The knee joint needs to shift behind the line of gravity posterior to the capsular ligaments so that the ligaments can stabilize the knee joint.
c If the line of gravity extends behind the knee joint, the ligaments are useless; the body tilts backward, and the knee joint gives way.
d Active stabilization of the knee joint by the quadriceps femoris extending on its front side.

D The most important groups of muscles and ligaments responsible for the maintenance of erect posture

Muscles	Ligaments
• The lower leg muscles, mainly triceps surae and tibialis anterior, stabilize the upper ankle joint in the sagittal plane • The upper leg muscles (quadriceps femoris) stabilize the knee joint • The gluteal muscles stabilize the hip joint in the sagittal plane • The smaller gluteal muscles (glutei medius and minimus) stabilize the hip joint in the frontal plane • The intrinsic back muscles stabilize the spinal column	• The anterior ligaments stabilize the knee joint • The iliofemoral ligament stabilizes the hip joint • The collateral ligaments stabilize the knee joint and the upper ankle joint in the frontal plane

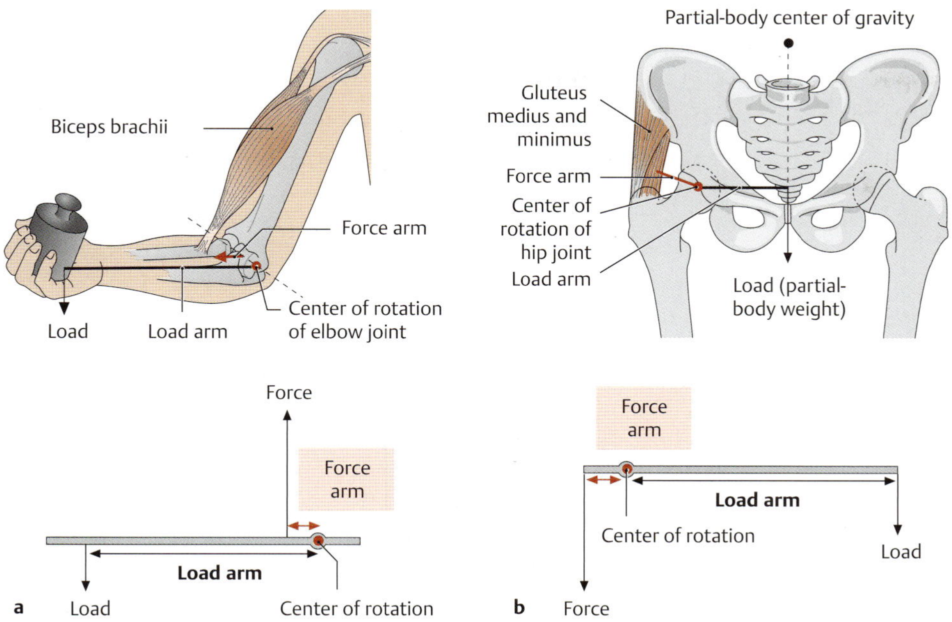

E One- and two-arm levers

a One-arm lever (elbow joint), **b** two-arm lever (hip joint).

Joint mechanics is based on the principles of the lever. The amount of force that a muscle can transmit to a joint depends on the length of the associated lever arm. This depends on the perpendicular distance from the muscle and its tendon to the center of rotation (= *force arm*) and is opposed by the force of the *load arm*. In the case of the elbow joint in **a**, the load arm is the distance from the joint axis (center of rotation) to the load. The *magnitude of each of the three active forces* is determined by multiplying the force by the force arm and the load by the load arm. This product is called the *torque* (= moment of rotational force) because the active forces produce a rotational movement of the associated lever. If the product of the load times the load arm equals the product of the force times the force arm, both torques are identical, and the joint is at rest. The lever in **a** is classified as a *one-arm lever* because the muscular force and load act *on the same side*—in this case, to the left of the center of joint rotation. The lever in **b** is a *two-arm lever* because the muscular force acts to the left of the center of joint rotation, while the force of the body weight acts to the right of the joint center.

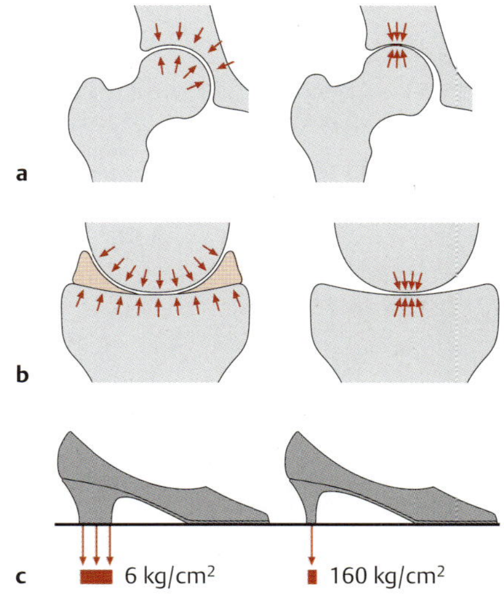

F Dependence of joint load on the size of the force-transmitting surface on the joint load

The larger the force-transmitting surface, the more evenly the force is distributed and the lower the joint pressure, or joint load, per unit area.

- **a** Normal and limited covering of the head of the femur.
- **b** Knee joint with and without meniscus.
- **c** Example of a square stacked heel and a stiletto heel: if body weight is the same, but the force-transmitting surface is less, the load on the bottom of the foot is many times higher.

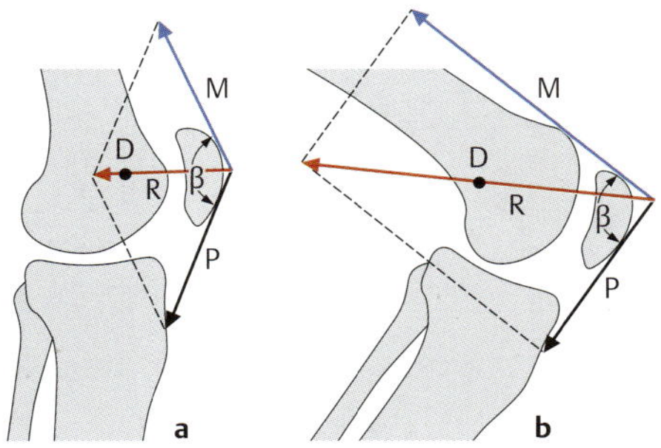

G Dependence of magnitude of the resultant force vector on the net joint force

- **a** The resultant force vector (red) of the femoropatellar joint runs through the center of rotation of the joint (D) and is the summation of individual muscle force vectors (M) (quadriceps femoris) and the ligament force vector (P) (patellar ligament). It creates the joint pressure and is a measure of the forces acting on the joint.
- **b** Due to the increased intensity of contraction of the quadriceps femoris when bending the knee, the resultant force in the bend position is significantly higher than in the stretching position, which means that with increased bending, the joint load in the femoropatellar joint increases.

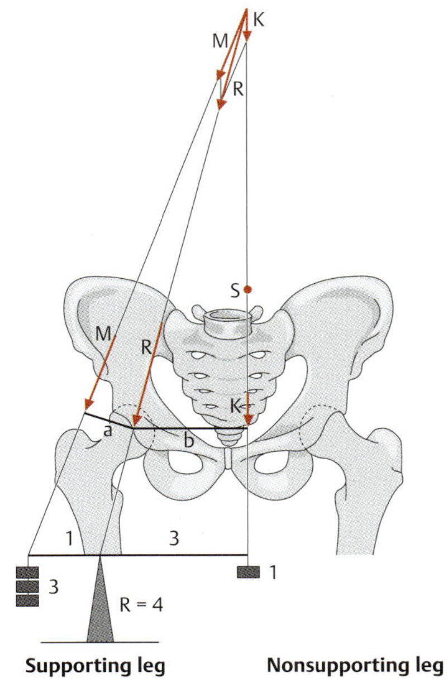

H Load on the right hip joint in the support phase

Anterior view. In a one-legged stance or while walking in the support phase, the partial-body center of gravity (S) shifts, which results in a turning moment of the load, which tends to tilt the body above the joint toward the supporting leg.

In order to remain stable, the turning moment has to be balanced by a counterforce (e.g., a muscle or ligament force). In the hip joint, it is provided by the muscle force (M) of the hip abductors (glutei medius and minimus). However, unlike the force of the partial-body weight (K) at the hip joint, it acts at a lever arm that is three times shorter, which means the ratio between the lever arm of the muscle force (**a**) to the lever arm of the partial-body weight (**b**) is 1:3. The *muscle force* needed to stabilize the hip therefore is approximately three times that of the partial-body weight *when standing on one leg*. From this it follows that based on the different length of the lever arms, the *compressive force* that the hip joint has to withstand when walking, (resultant force R) has to be four times higher than the partial-body weight (after Pauwels). That means that the hip joint has to be able to permanently withstand extreme loads and is thus very often affected by osteoarthritis.

4.8 Fractures: Classification, Healing, and Treatment

A Fracture classification

Fractures can be classified according to different characteristics; those mentioned below are practical and have proven successful in daily use:

- According to **formation and injury mechanism**:
  - Traumatic fracture caused by *direct* force (transverse or comminuted fracture, see **Ba** and **Bg**, e.g., as a result of being hit while walking) or *indirect* force (lever arm and rotational motion directed away from the force effect, e.g., torsion fracture to the lower leg as a result of a ski crash, see **Bf**).
  - Nontraumatic fracture ("spontaneous fracture") as a result of preexisting localized or generalized bone disease, such as tumors, osteomyelitis, or osteoporosis (= pathologic fracture, can occur without any external forceful impact) or as a result of microtrauma, which is caused by repeated moderate application of force at the same spot over and over again eventually leading to a fracture (= fatigue fracture, e.g., metatarsal fracture such as a so-called march fracture).
- According to **fracture morphology** (e.g., degree of break in continuity [fissure or fracture]), course of fracture lines, number of fragments, see **B** and **C**).
- According to severity of **soft-tissue damage**:
  - Closed fracture or
  - Open fracture: generally, it can be assumed to give direct access to germs; however, what is important in the end is the severity of soft-tissue damage, meaning to what extent muscles, vessels, or nerves have also been damaged

Furthermore, with regard to **long bones**, when considering the location of the fracture (epiphyseal, metaphyseal, and diaphyseal), a distinction is made between proximal fractures, fractures of the long bones, and distal fractures with or without joint involvement. In addition to pain, clinical symptoms of a bone fracture include obvious deformation, abnormal mobility, and palpable and distinguishable crepitus (creaking, crackling as a result of crepitation).

Note: Fractures suffered during childhood show distinct characteristics and have to be disinguished from those suffered in adulthood. The most important anatomic difference involves the growth plates and the periosteum, which is thicker and more robust in children. If, for example, a fracture causes damage in the proliferative zone in the growth plate, cellular demise and callous bridges can lead to a premature closure of the epiphyseal plate. In so-called greenstick fractures, the cortex fractures on one side while maintaining the thick periosteal tube (intraperiosteal fracture).

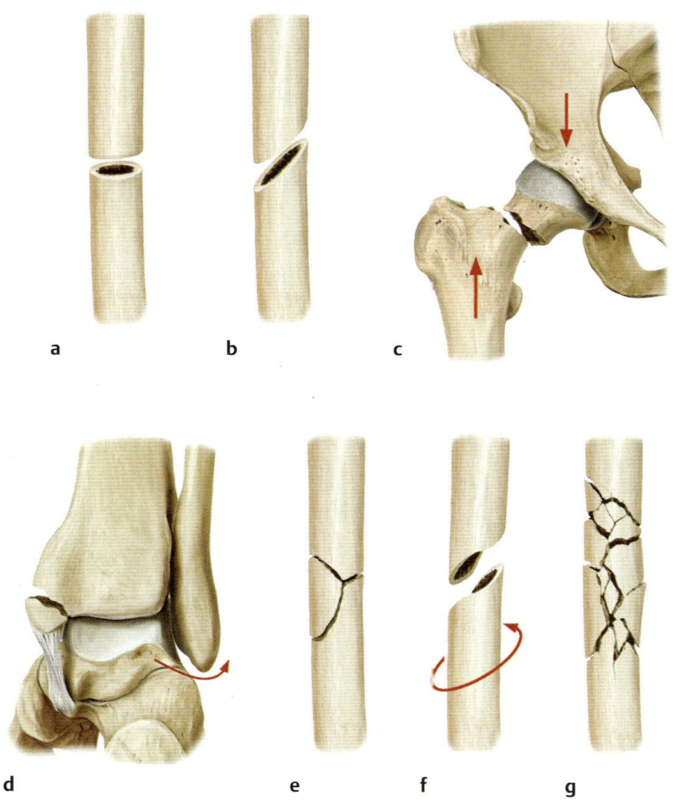

B The most important types of fractures

a Transverse fracture; **b** oblique fracture; **c** shearing fracture (e.g., fracture of the femoral neck); **d** avulsion fracture (e.g., avulsion fragment in the intra-articular ligament of the medial malleolus); **e** bending fracture with bending wedge as third fragment; **f** torsion, or spiral, fracture; **g** comminuted fracture with more than six fragments.

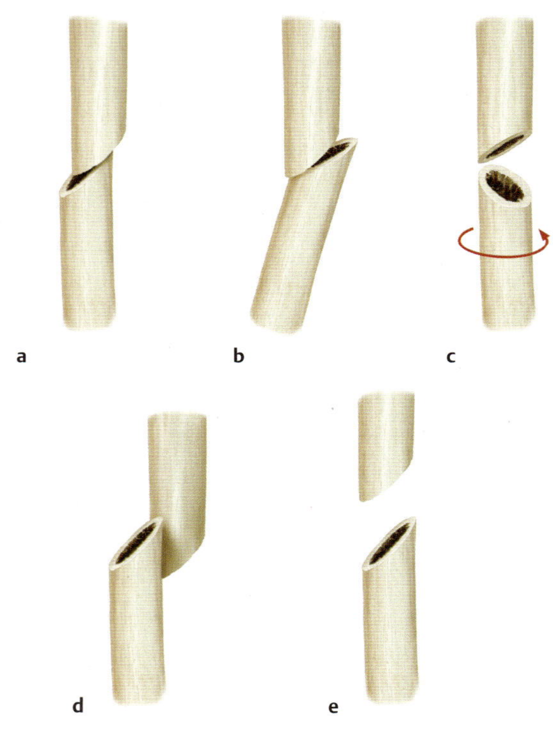

C Typical fracture dislocations

Fractures can be primarily (by acting force) or secondarily (by a disrupted muscle balance, e.g., muscular traction at the fragments) more or less distinctly displaced or dislocated.

a Sideways dislocation; **b** axial dislocation; **c** angular malalignment; **d**, **e** shifting direction axis and overlapping.

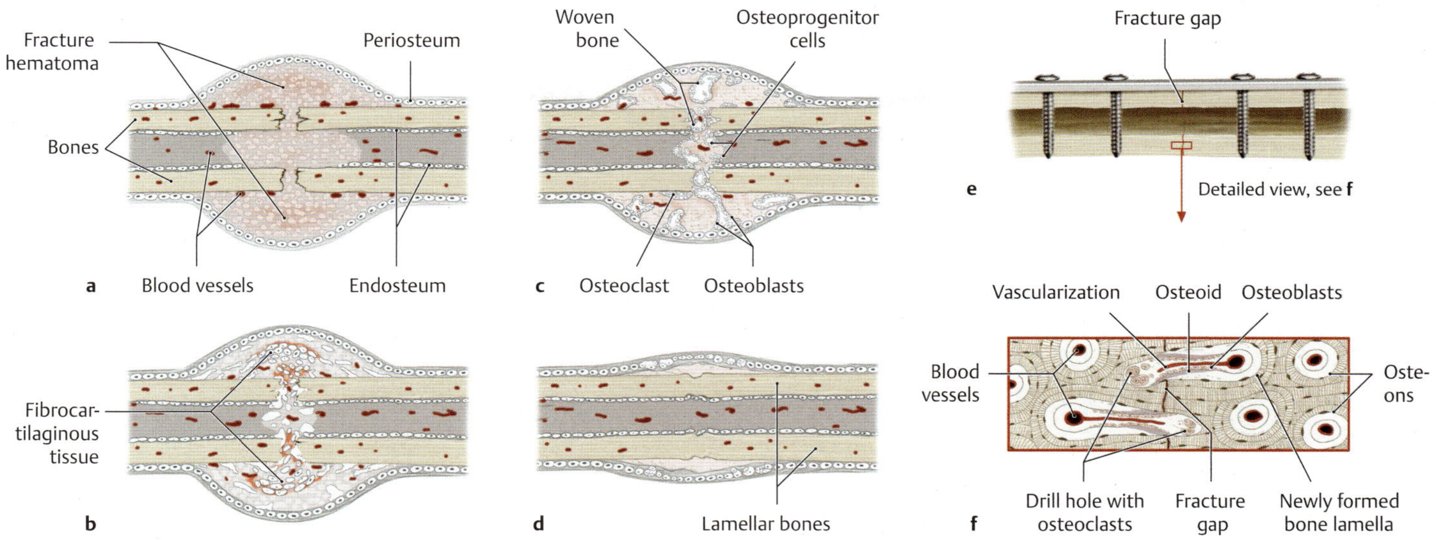

D Fracture healing

Preconditions for optimal fracture healing are repositioning and retaining the bone fragments of the fracture. In principle, this can be achieved both conservatively (e.g., with a cast) and surgically (see **E**). In addition, fracture healing requires cells with osteogenic potential, sufficient vascularization, and absolute mechanical rest in the fracture area. During the healing process, active movement therapy is crucial in order to prevent muscle atrophy, due to osteoporosis inactivity, stiffening of the joints, and circulatory disturbances. Endochondral and membranous ossification are part of the healing process. Thus, these processes, which occur in the course of the normal growth and the development of the skeleton, are repeated.

- **Indirect (natural or secondary) fracture healing (a–d)**, which has the following physiologic stages:
  - Development of a fracture hematoma (inflammatory phase, **a**) and organization of the fracture hematoma with the help of loose, vessel-rich connective tissue (granulation stage).
  - Development of a fibrocartilaginous cartilage while involving cells of the periosteum and the endosteum (so-called fibrocartilaginous callus or fixation callus, **b**).
  - Mineralization of the cartilage (callus hardening within 6 to 8 weeks after fracture occurrence; so-called osseous callus or fracture callus; at first thick and bulky).
  - Loss of cartilage substance and new bone formation (woven bone) with the help of osteoprogenitor cells (**c**).
  - Bone restructuring in the lamellar bone, which takes months (**d**).
- **Direct (or primary) fracture healing** (**e**, **f**). In this case, no chondral fracture callus develops, but ideally (though only rarely) direct angiogenic ossification occurs, meaning the sprouting osteons grow directly from one fragment into another (**e**, **f**). However, this *requires* surgical osteosynthesis and that the fracture gap is no more than 0.5-mm wide (contact healing). Conservative stabilization using synthetic dressing will *not* result in direct fracture healing.

Note: Pseudarthrosis (failure of bone formation after 6 months) is among the most common complications of fracture healing. It arises due to both excessive mechanical load on the regenerating bone and absence of callus formation as a result of poor circulation.

E Basic principles of fracture fixation

Surgical fixation creates a temporary mechanical bone union using screws, plates, wires, or nails. The procedure neither replaces nor expedites bone healing. However, there are a number of **advantages compared to a conservative cast treatment**:

- Exact repositioning of bone fragments in joint fractures.
- Immediate mobilization (prevents thrombosis, emboli, decubitus, and fracture disease, e.g., edema and dystrophy).
- Timely stabilization exercises (mobilization of joints without load) and partial load-bearing capacity to allow for healing.

Disadvantages include the risks associated with anesthesia and surgery, as well as the possibility of infection. There are five different osteosynthetic procedures (which can be combined as needed):

- Screw osteosynthesis (e.g., compression with cortical-traction screw, **a**).
- Plate osteosynthesis (e.g., compression plate, **b**).
- Intramedullary nail, rod, or fixation (e.g., in diaphyseal fracture, **c**).
- Tension-band wiring (e.g., using a wire loop, **d**).
- External fixation (e.g., three-dimensional frame construction, **e**).

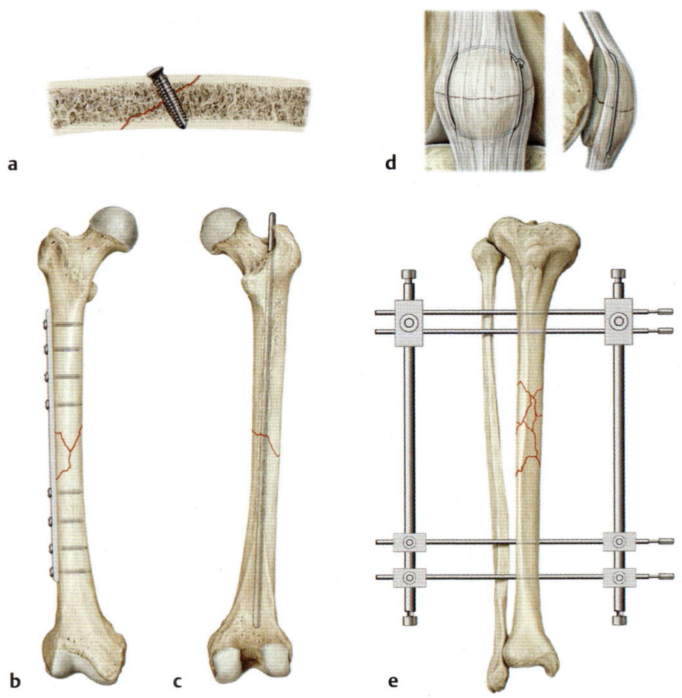

5.1 Overview: Female

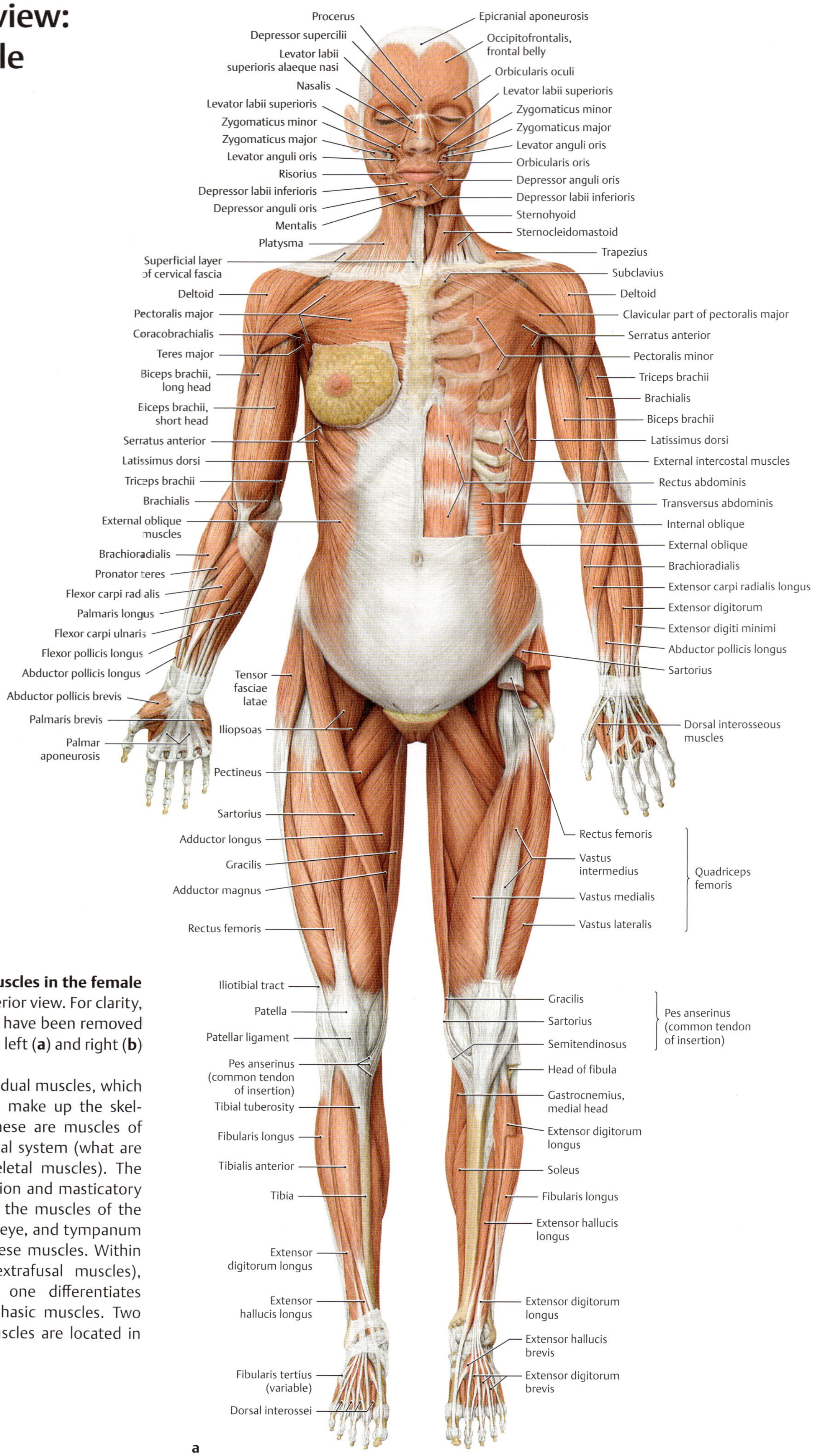

A Skeletal (striated) muscles in the female
a Anterior view, **b** posterior view. For clarity, some superficial muscles have been removed or windowed on both the left (**a**) and right (**b**) side of the body.

Approximately 220 individual muscles, which differ in shape and size, make up the skeletal muscles. Overall, these are muscles of the active musculoskeletal system (what are often referred to as skeletal muscles). The muscles of facial expression and masticatory muscles of the face and the muscles of the tongue, pharynx, larynx, eye, and tympanum are similar in form to these muscles. Within the skeletal muscles (extrafusal muscles), for functional reasons, one differentiates between postural and phasic muscles. Two thirds of the skeletal muscles are located in

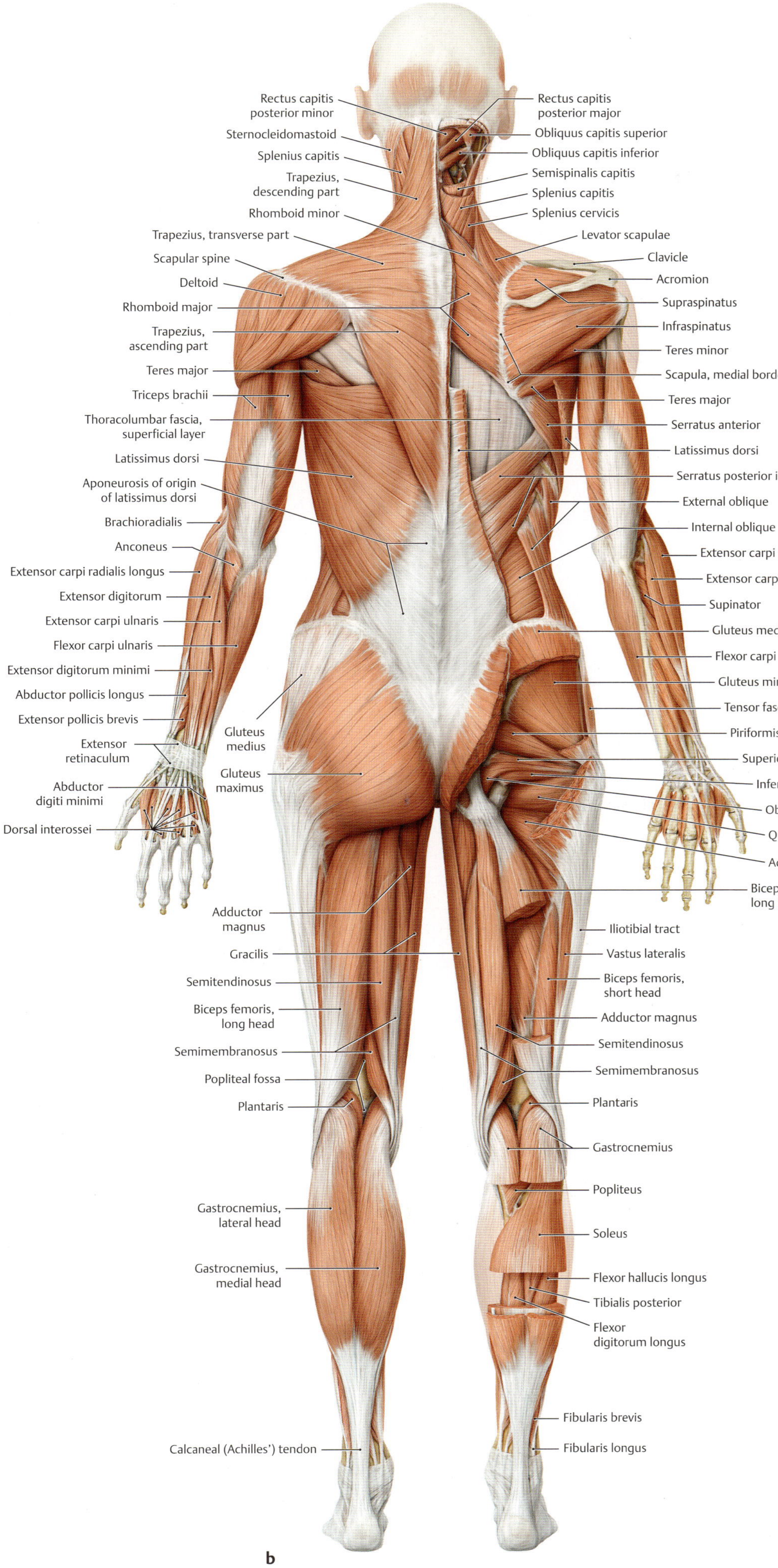

b

the lower limb region and mainly help to hold the body in an erect position in the field of gravity (= antigravity = postural muscles) and in movement (= locomotion = phasic muscles). Depending on gender, age, and physical condition, the skeletal muscles account on average for 40% of the total weight of a person and are thus the heaviest organ. They consist of up to 75% water, 20% proteins (mainly contractile proteins of muscles: myosin, actin, troponin, and tropomyosin), 2% low molecular weight organic components (e.g., triglycerides), and 3% inorganic components (e.g., calcium ions). As people age, their percentage of muscle mass decreases, and their percentage of fat increases (due to physical inactivity and the influence of hormonal factors, see p. 60).

5.2 Overview: Male

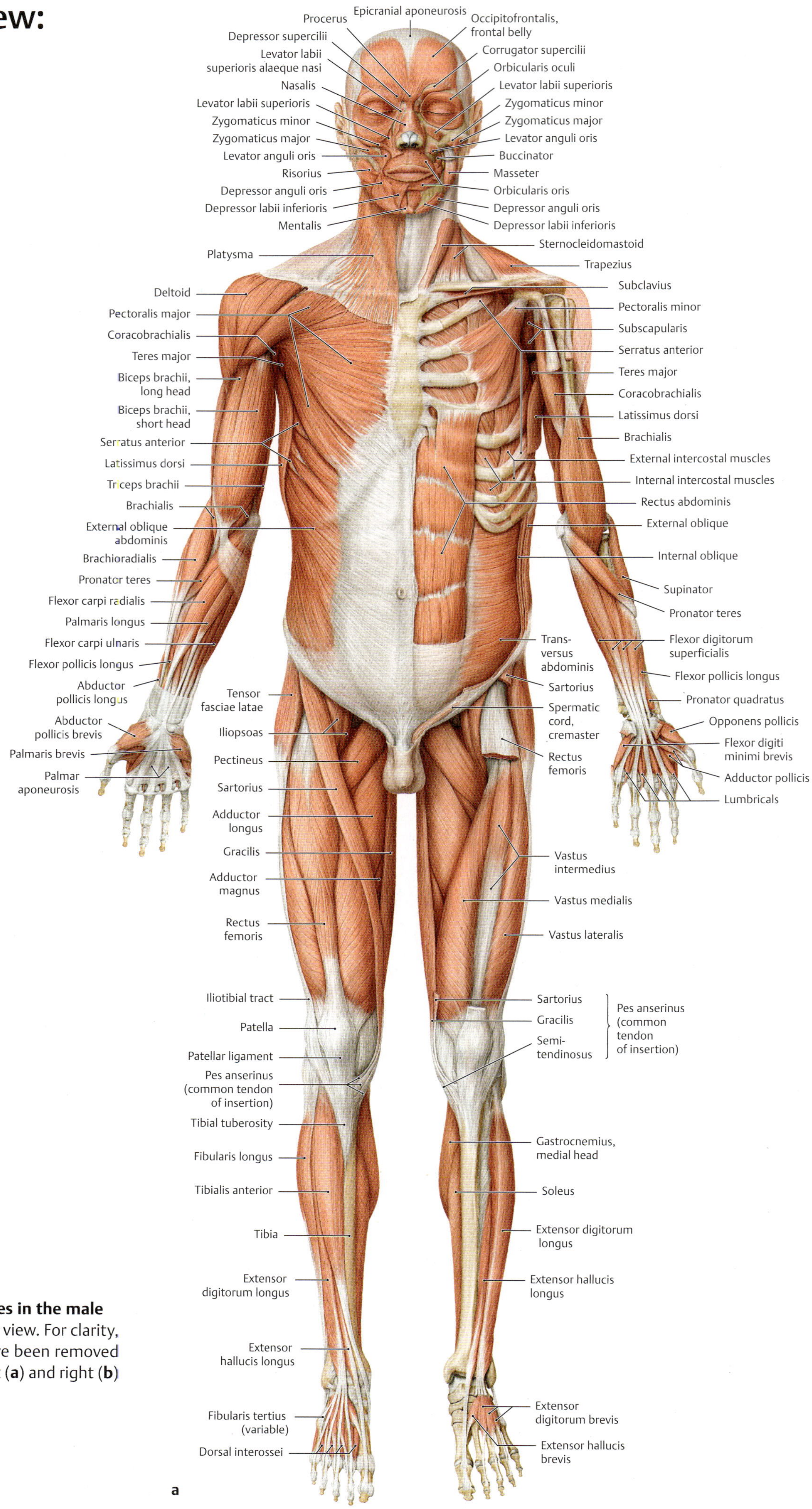

A Skeletal (striated) muscles in the male
a Anterior view, **b** posterior view. For clarity, some superficial muscles have been removed or windowed on both the left (**a**) and right (**b**) side of the body.

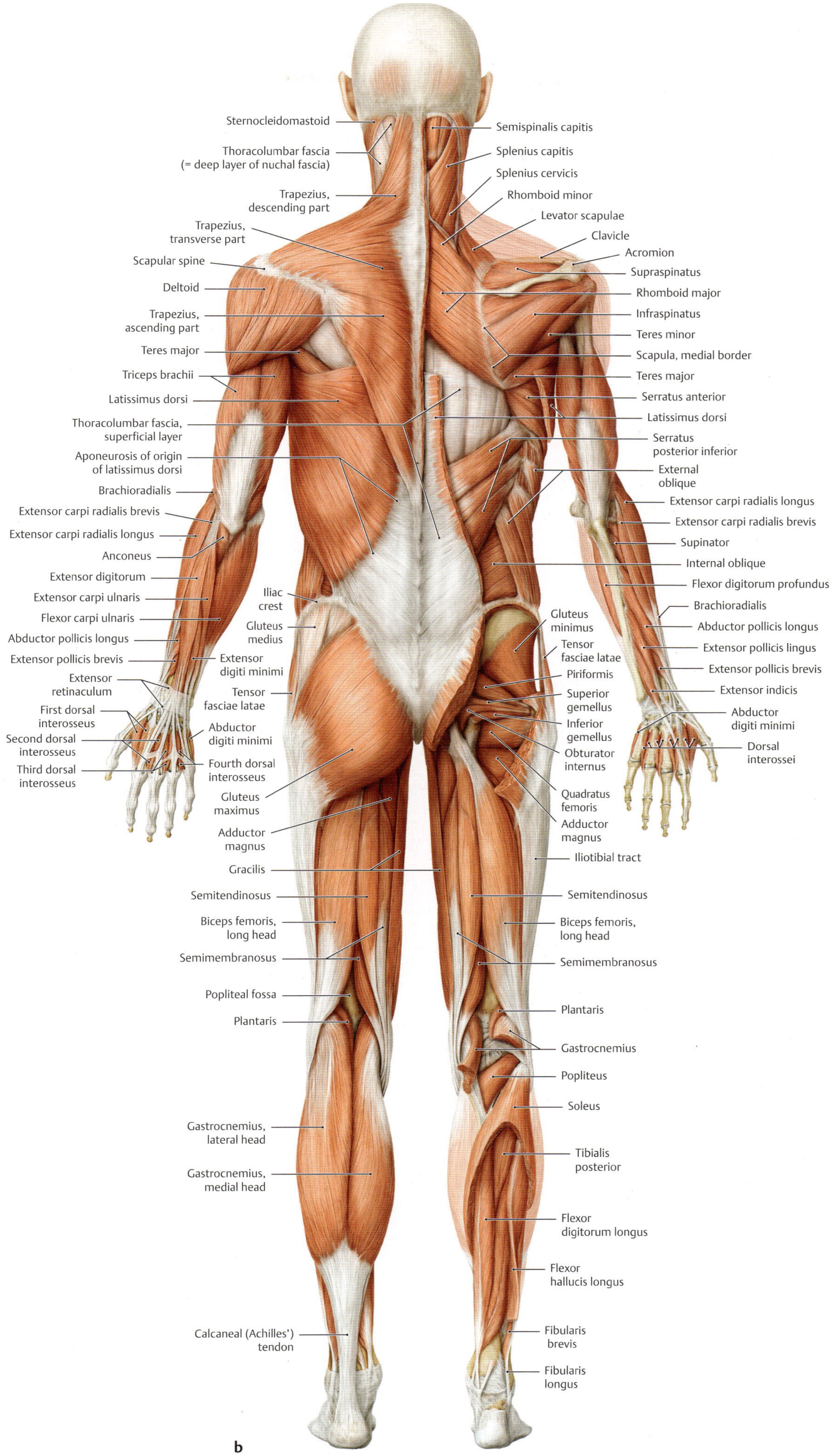
Sternocleidomastoid
Thoracolumbar fascia (= deep layer of nuchal fascia)
Trapezius, descending part
Trapezius, transverse part
Scapular spine
Deltoid
Trapezius, ascending part
Teres major
Triceps brachii
Latissimus dorsi
Thoracolumbar fascia, superficial layer
Aponeurosis of origin of latissimus dorsi
Brachioradialis
Extensor carpi radialis brevis
Extensor carpi radialis longus
Anconeus
Extensor digitorum
Extensor carpi ulnaris
Flexor carpi ulnaris
Abductor pollicis longus
Extensor pollicis brevis
Extensor retinaculum
First dorsal interosseus
Second dorsal interosseus
Third dorsal interosseus
Iliac crest
Gluteus medius
Extensor digiti minimi
Tensor fasciae latae
Abductor digiti minimi
Fourth dorsal interosseus
Gluteus maximus
Adductor magnus
Gracilis
Semitendinosus
Biceps femoris, long head
Semimembranosus
Popliteal fossa
Plantaris
Gastrocnemius, lateral head
Gastrocnemius, medial head
Calcaneal (Achilles') tendon
Semispinalis capitis
Splenius capitis
Splenius cervicis
Rhomboid minor
Levator scapulae
Clavicle
Acromion
Supraspinatus
Rhomboid major
Infraspinatus
Teres minor
Scapula, medial border
Teres major
Serratus anterior
Latissimus dorsi
Serratus posterior inferior
External oblique
Extensor carpi radialis longus
Extensor carpi radialis brevis
Supinator
Internal oblique
Flexor digitorum profundus
Brachioradialis
Abductor pollicis longus
Extensor pollicis lingus
Extensor pollicis brevis
Extensor indicis
Abductor digiti minimi
Dorsal interossei
Gluteus minimus
Tensor fasciae latae
Piriformis
Superior gemellus
Inferior gemellus
Obturator internus
Quadratus femoris
Adductor magnus
Iliotibial tract
Semitendinosus
Biceps femoris, long head
Semimembranosus
Plantaris
Gastrocnemius
Popliteus
Soleus
Tibialis posterior
Flexor digitorum longus
Flexor hallucis longus
Fibularis brevis
Fibularis longus
b

5.3 Muscle Fiber Types and Pennate and Nonpennate Skeletal Muscles

A Muscle fiber types for postural muscles and muscles of movement: a comparison

All striated skeleton muscles basically consist of two different muscle fiber types (type I and type II fibers), which differ in their metabolic, physiological, histochemical, and biochemical characteristics (see **B**). The type II muscle fibers can be further divided into type IIA and type IIB fibers based on weight of isoforms in their myosin chains. Since the type I and type II fibers respond to a single action potential from the innervated axon with a single twitch in an *all-or-none response*, they are also known as *twitch fibers*. In contrast, *tonic fibers*, which are only found in muscle spindles and external eye muscles, *gradually depolarize*, leading to gradual fiber shortening.

The twitch duration of type I and II fibers differs. Type I fibers twitch more slowly (slow-twitch, or ST, fibers) than type II fibers (fast-twitch, or FT, fibers). The *basic distribution pattern* of type I and type II fibers in individual skeletal muscles is apparently genetically determined. In each case, the muscle fiber type of a motor unit is determined by the innervating neuron, so that all fibers of a motor unit belong to the same fiber type. Muscles with small motor units (< 100) have mostly fast-twitch (type II) fibers; muscles with large motor units (several thousand muscle fibers) have mostly slow-twitch (type I) fibers (see p. 63). The fiber type and its twitch duration are linked to a specific form of capacity and thus a particular type of twitch duration. Due to their high oxidative metabolism, type I fibers have endurance capacity and are thus mainly found in the postural muscles. Due to their high glycolytic metabolism, type II fibers can perform short, forceful contractions and are thus mainly found in phasic or elastic muscles. However, as skeletal muscles are highly adaptable, the genetically programmed distribution of type I and type II fibers can be influenced through neuromuscular activity (exercise). Accordingly, muscles that predominantly consist of type I fibers are found in endurance athletes, such as long-distance runners, cyclists, rowers, and cross-country skiers. Muscles that predominantly consist of type II fibers are found in explosive athletes, such as sprinters, high or long jumpers, and weight lifters (Pette and Saron, 2001). This *phenotypic plasticity* of the muscle is based on qualitative and quantitative changes in gene expression and mainly involves contractile and regulatory proteins of the myofibril apparatus, as well as the enzymes in metabolism (Pette, 1999).

	Postural muscles (red muscles)	**Muscles of movement** (white muscles)
Characteristics:	• Phylogenetically older • Predominantly slow-twitch fibers (type I fibers, approximately 100 ms) • Function best in endurance • Fatigue slowly • Large motor units • Rich in myoglobin • Abundant mitochondria • Energy derived from oxidative (aerobic) metabolism • Little glycogen (PAS-negative) • Relatively highly vascularized • Prone to shortening (increased resting tonus) and require regular stretching	• Phylogenetically more recent • Predominantly fast-twitch fibers (type II fibers, approximately 30 ms) • Brief periods of intense activity • Fatigue more rapidly • Small motor units • Scant myoglobin • Few mitochondria • Energy derived mainly from anaerobic glycolysis • Abundant glycogen (PAS-positive) • Much smaller capillary supply • Prone to atrophy and require regular strengthening
Examples:	Intercostal muscles, masticatory muscles, trapezius (descending part), hamstrings, iliopsoas, adductors, rectus femoris, soleus, intrinsic back muscles (mainly the cervical and lumbar part)	Biceps brachii, vastus lateralis and medialis, tibialis anterior, serratus anterior, gluteus maximus, gastrocnemius *Abbreviation:* PAS, periodic acid–Schiff.

Note: The loss of muscle mass (**sarcopenia**) in old age and the associated loss of muscular force and endurance are the most important causes for impaired physical mobility and, as a result, impaired ability to live independently. Sarcopenia is prevalent among the old and is a heavy financial burden on the health care system. Between the ages of 25 and 75, a person loses almost 40% of muscle mass (mainly type II muscle fibers). Especially dramatic is the loss of muscle mass after the age of 50 (15% per decade). Taking into account that muscle weakness is the most common factor in the risk of falling, exercises that target a specific muscle group (especially for elasticity) can efficiently counteract the loss of strength and in combination with balance exercises significantly reduce the risk of falling.

B Histochemical differences between type I and type II skeletal muscle fibers

Microscopic cross section of a skeletal muscle (tibialis anterior, rat, 200X magnification, 8-µm-thick frozen section). Succinated dehydrogenase (SDH), a mitochondrial enzyme of muscle metabolism, catalyzes the transformation of succinate into fumarate (reduction of fumarate leads to a brownish black reaction product, histochemical proof for the presence of SDH). The activity of mitochondrial SDH makes the varying percentage of mitochondria in the individual muscle fibers visible: heavily colored type I fibers with numerous mitochondria between the myofibrils and underneath the sarcolemma, and pale-colored type II fibers with few mitochondria.

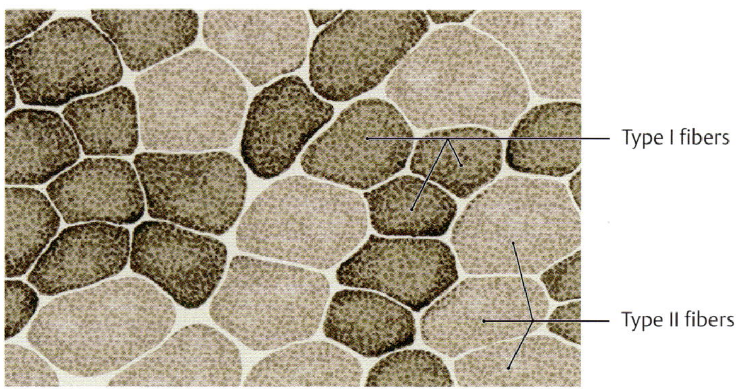

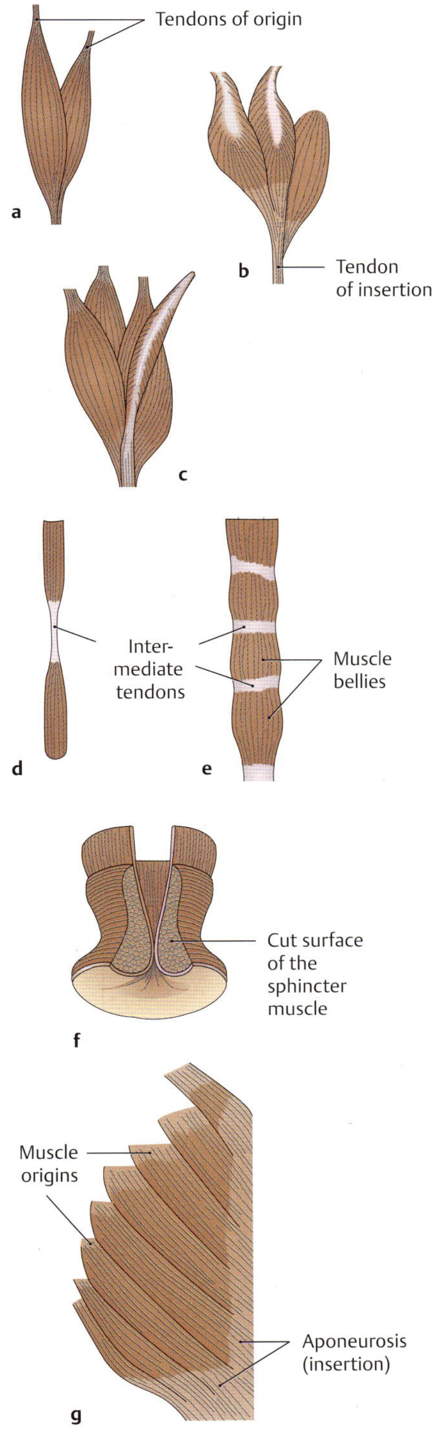

C Morphological forms of muscles

a Two heads = bicipital (e.g., the biceps brachii).
b Three heads = tricipital (e.g., the triceps surae).
c Four heads = quadricipital (e.g., the quadriceps femoris).
d Two bellies = digastric (e.g., the digastric muscle).
e Multiple bellies = multigastric (e.g., the rectus abdominis).
f Radial (e.g., the external anal sphincter).
g Flat (e.g., the external oblique).

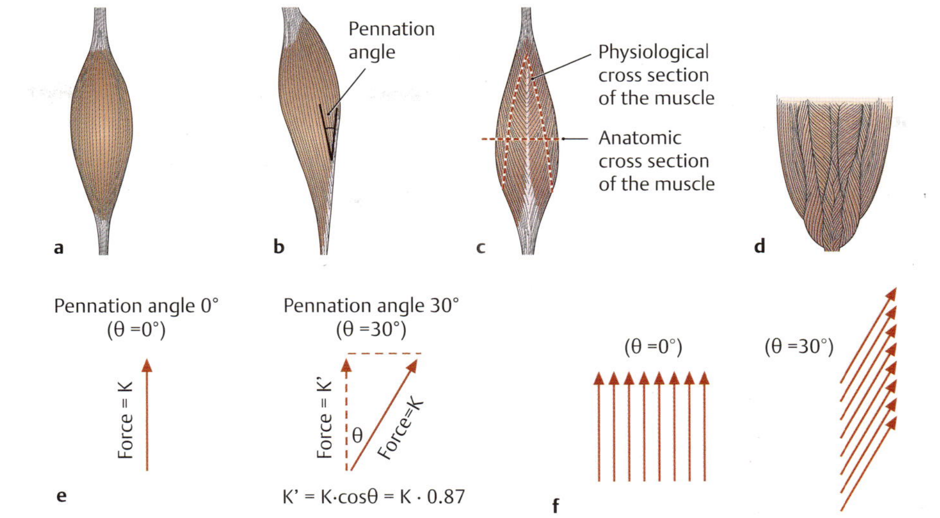

D Pennate and nonpennate skeletal muscles

Muscle fibers may have a parallel arrangement or may show varying degrees of convergence, or pennation, at their attachment.

a Nonpennate muscle (with parallel fibers); **b** unipennate muscle; **c** bipennate muscle; **d** multipennate muscle; **e** effect of pennation angle on force (at a 30° pennation angle, the tendon force K' amounts to 87% of the force generated in the muscle fibers); **f** effect of fiber arrangement on the number of fibers in a given cross section of muscle.

Whereas almost all fibers of skeletal muscles are of the same thickness (average diameter approximately 60 µm, see p. 62), their *length* varies considerably (ranging from a few millimeters to up to 20 cm), as does their *ratio of fiber length to muscle length* (between 0.2 and 0.6, meaning the fiber length amounts to between 20 and 60% of the length of the muscle) and their *pennation angle*. All three factors play a significant role in lifting *force* and lifting *height* and thus in muscle "power output" (power = force × velocity or lifting force × lifting height):

- The longer the fibers, the greater the maximum shortening and thus the maximum lifting height of the muscle.
- The longer the fibers relative to the length of the muscle, the smaller the physiological cross-sectional area of the muscle and thus the lifting *force* of the muscle.
- The greater the pennation angle, the greater the physiological cross-sectional area and thus the lifting *force*.

E Nonpennate and pennate muscles: a comparison

Nonpennate (parallel-fibered) muscle	Pennate muscle
• Fibers run roughly along the longitudinal direction of the tendon (= muscle line of action) and are thus able to transmit almost all of their force to the tendon, see **f** • Maximum fiber shortening (= lifting height) and extent of actual fiber shortening during muscle activity are almost identical • Anatomic cross section (perpendicular to the longitudinal axis of the muscle, at its thickest part) and physiological cross section (perpendicular to the longitudinal axis of the fibers, see **c**) are roughly the same size	• The fibers form a (pennation) angle (up to 30°) with the longitudinal axis of the tendon and thus can transmit only a portion of their force to the tendon • Due to pennation angle, maximum fiber shortening (= lifting height) is greater than actual fiber shortening during muscle activity (= path gain) • Physiological cross section is larger than anatomic cross section, meaning (due to the angle) more muscle fibers can insert into the tendon in a given cross section as compared to a parallel-fibered muscle, thus increasing the lifting height, which depends on the size of the physiological cross section

Note: In nonpennate (parallel-fibered) muscles, force *production* is greater (direct force transmission from muscle to tendon, as there is no pennation angle); in pennate muscles, force *development* is greater (there are more fibers in a given cross section, since fibers are arranged obliquely to a tendon, resulting in greater lifting height, see above). The lesser force production of pennate muscles is compensated for by the greater force development. A huge advantage of pennate muscles compared to nonpennate muscles is their smaller size for similar size power output. If there were only nonpennate muscles, the body, in a lot of places, would not have space for a sufficient number of strong muscles.

5.4 Structure and Function

A Structure of a skeletal muscle

a Cross section of a skeletal muscle; **b** detail from **a** (cross section); **c** detail from a (longitudinal section); **d** structure of a muscle fiber (= muscle cell); **e** structure of a myofibril.

In striated skeletal muscle, muscle fibers and connective tissue are tightly interwoven and collaborate to carry out particular functions. The connective tissue is arranged in sheaths, which perform the following functions:

- Endomysium: most interior connective tissue sheaths (important for the tensile strength of the muscles); surrounds and condenses 200 to 250 muscle fibers (= muscle cells) into primary bundles. Each primary bundle, or muscle cell, contains the end branches of the motor axons to the motor end plates, as well as numerous (300–400 per mm^2) heavily gyrose capillaries, see **c** (important for the supply of the muscles).
- Perimysium (important for the transmission of muscle tensile force to tendons): condenses muscle fibers into finer primary bundles (with an average medial cross-sectional area of 1 mm^2; approximately 200–250 muscle fibers = muscle cells; formerly known as perimysium internum) and primary bundles into into secondary bundles (formerly known as perimysium externum; see **a**), which are several millimeters thick and thus are clearly visible to the naked eye ("meat fibers"). In some muscles, the primary bundles do not appear to be clearly organized into secondary bundles and instead seem random. Additionally, certain very small muscles that perform fine motor functions such as the stapedius and outer ocular muscles exhibit only fine primary bundles.
- Epimysium: loose connective tissue layer directly beneath the muscle fascia (see **b**); establishes the connection to the muscle.

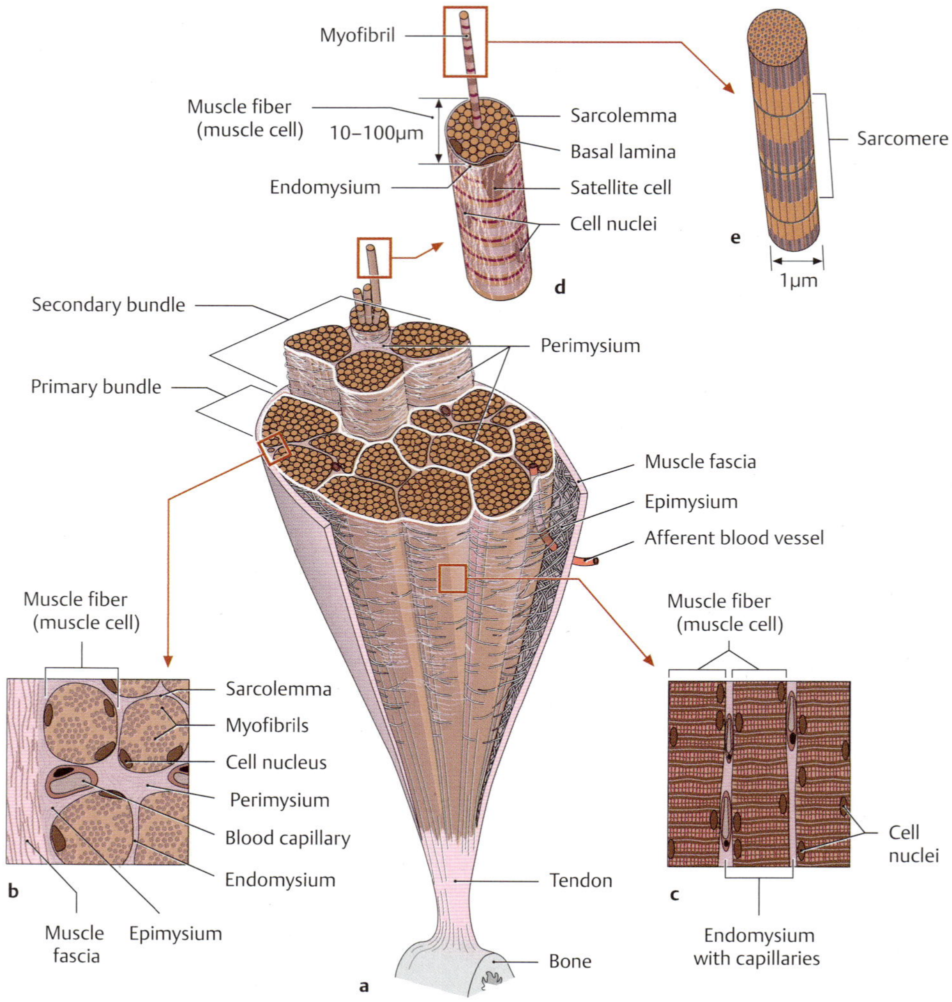

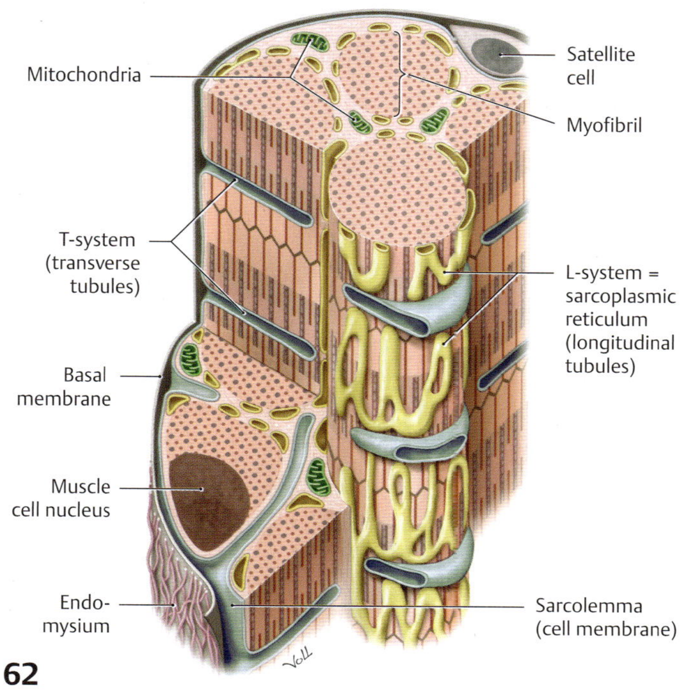

B Structure of a skeletal muscle fiber

Skeletal muscle fibers are especially large cells, with average diameters of roughly 60 µm (10–100 µm) and lengths up to 20 cm. The dominant structures in the cytoplasm are myofibrils, mitochondria, and the L-and T-systems. The *L-*, or *longitudinal, system* is a cavity system (sarcoplasmic reticulum, longitudinal tubules), which is arranged lengthwise to the myofibrils and offers a pool for calcium ions. The *T-*, or *transverse, system* is created when a cell membrane periodically invaginates deeply into the interior of the muscle cell and as a result develops transverse tubules. It enlarges the membrane surface of a muscle fiber by 5 to 10 times, and the extracellular space can extend over the entire muscle fiber cross section. It guarantees the fast spread of an action potential deep into the muscle fiber.

Characteristic for the skeletal muscle cell are the numerous cell nuclei (approximately 50 nuclei/mm length), which are located directly underneath the cell membrane (sarcolemma). This very large number of cell nuclei arises during embryonic development, when the progenitor cells (myoblasts), arranged in a chainlike pattern, fuse with each other. Scattered between the sarcolemma and basal membrane are satellite cells (approximately 800/mm^3 muscle tissue), which as dormant myoblasts (stem cells) represent some sort of cell pool.

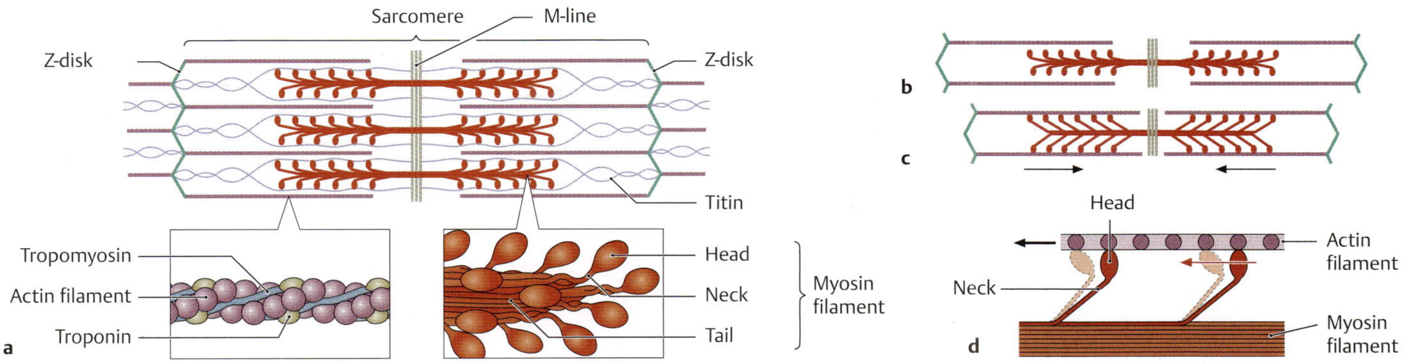

C Structure of a sarcomere

a Schematic representation of a sarcomere (= structural unit of a striated muscle cell); **b** myosin heads at rest; **c** myosin heads during contraction; **d** interaction between myosin heads and actin.

The arrangement of myofibrils in skeletal muscle fibers (see **B**) shows a cross-striation under the light microscope. It arises as the result of an alternating arrangement of thin actin (diameter: 7 nm) and thick myosin filaments (diameter: 15 nm). While actin filaments and their accompanying proteins (tropomyosin and troponin) of two adjacent sarcomeres are connected with each other at the Z-disks, the myosin filaments in the M-line among others are cross-linked by the myosin protein and in addition are kept in place by the elastic protein titin.

With the bipolar structure of the myosin filaments, a distinction is made between three parts: the head, neck, and tail. During muscle contraction, the myosin heads migrate along the actin filaments (so-called filament sliding) toward the Z-disks. Thereby, when at rest, each sarcomere shortens maximally to 70% of their optimal length of 2.2 µm; however, the individual filaments maintain their original length. This sliding mechanism is based on a quick succession of reaction cycles, during which cross-bridges between myosin heads and actin filaments are developed and detached again. It is associated with a tilting and rowing movement of the myosin head toward the middle of the sarcomere, with the myosin heads snapping off in an articulated manner (red arrow), and the actin filaments shoving on (black arrow). A single tilting movement shifts the actin filaments by about 10 to 20 mm, meaning the sarcomere shortens by about 1% of its original length. Greater changes in length are enabled by numerous repetitions of the interaction between actin and myosin. The basis for these cyclical processes is the fission of adenosine triphosphate (ATP) through the myosin head ATPase and the increase in cytosolic Ca^{+} concentration (electromechanical coupling; for more details, see books about physiology).

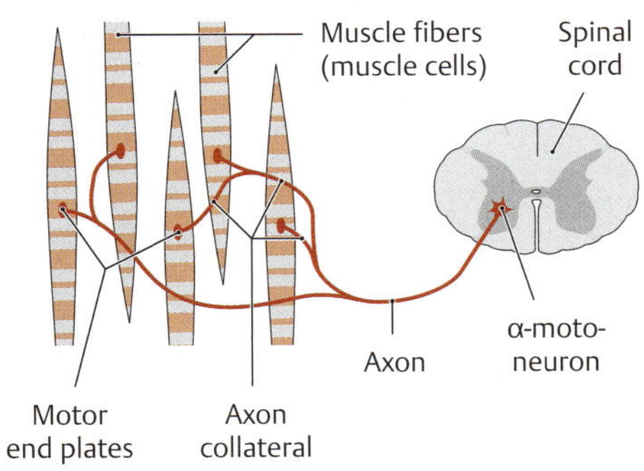

D Motor Unit

A motor unit is the sum of all muscle fibers that are innervated by a motor nerve fiber (the axon of a motor nerve cell of the spinal cord = α-motoneuron). There are *small motor units* (< 100 fibers) with mostly *fast*-twitching muscle fibers and *large motor units* (several thousand muscle fibers) with mostly *slow*-twitching muscle fibers. The fewer the muscle fibers combined in a motor unit, the more precise and finely tuned the movements of the muscle (e.g., finger muscles and external eye muscles). However, in muscles whose main function is to maintain posture or to produce movement (e.g., gluteal and back muscles), several thousand muscle fibers are innervated by one motor nerve cell.

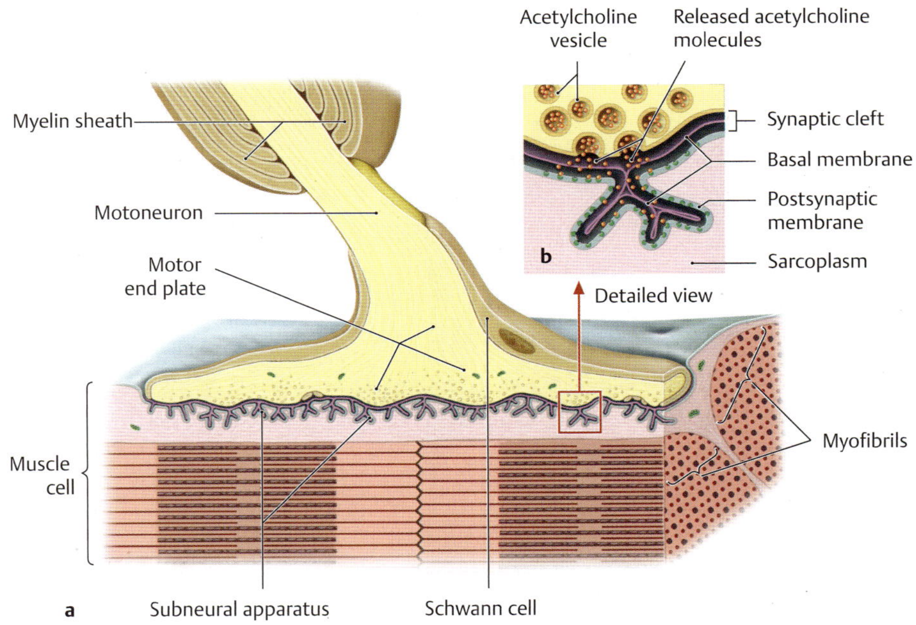

E Motor end plate

a Schematic representation of a motor end plate; **b** components of the synaptic contact (detailed view of **a**).

The motor axon divides at its end into several branches (one branch per muscle fiber), loses its myelin sheath, and develops on each of the branches a motor end plate (neuromuscular synapse), where the synaptic transmission of action potentials of the motoneuron occurs. Acetylcholine serves as a neurotransmitter and is stored in synaptic vesicles in the axoplasm of the end plate. As a result of folds in the sarcolemma (see **B**), the postsynaptic membrane with its acetylcholine receptors is greatly enlarged (subneural folding apparatus). In the entire 100-nm-wide system, there is one basal lamina, to which the acetylcholinesterase (an enzyme responsible for breaking down acetylcholine) is attached. The functional sequence at the motor end plate is generally identical to that at other synapses.

5.5 The Tendons and Mechanisms That Assist Muscle Function

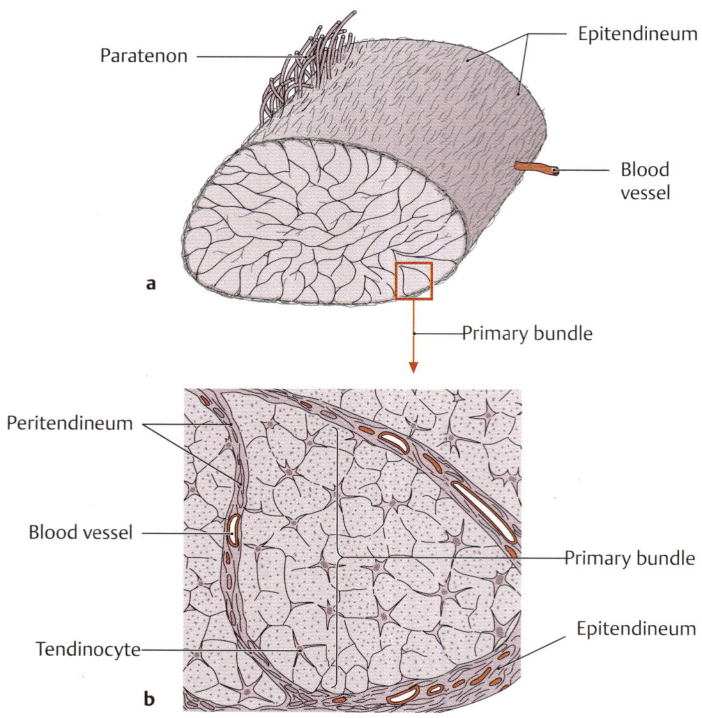

A Structure of a tendon (after Kristic)

a The tendon is connected to its surroundings by the loose, richly vascularized paratenon.

b Detail from **a**: The individual primary bundles are surrounded by peritendineum and are grouped into the actual tendon by the epitendineum. The function of a tendon is to transmit force from the muscle to the bone.

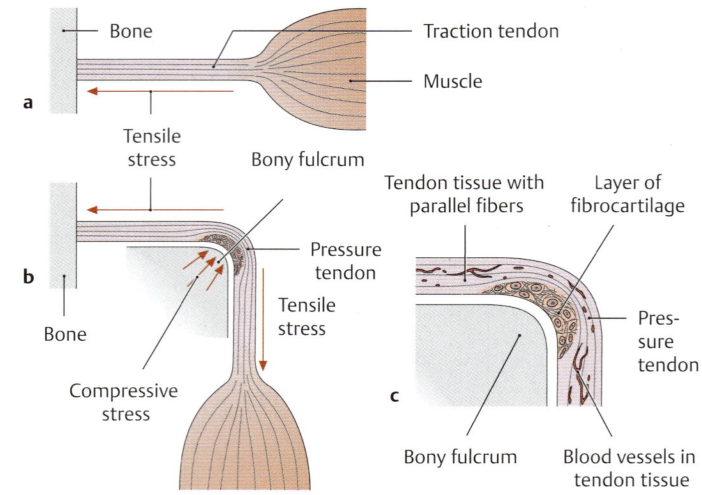

B Pressure tendons and traction tendons

a Traction tendons are subject to tensile stresses and consist of strong connective tissue with parallel fibers.

b Pressure tendons are strained by pressure and change their direction by running around the bone (unlike traction tendons). They consist of fibrocartilage on the side in contact with the bone, which acts as a fulcrum.

c Detail from **b**: The fibrocartilage layer in the compressed area, unlike the strong connective tissue in a traction tendon, is not vascularized.

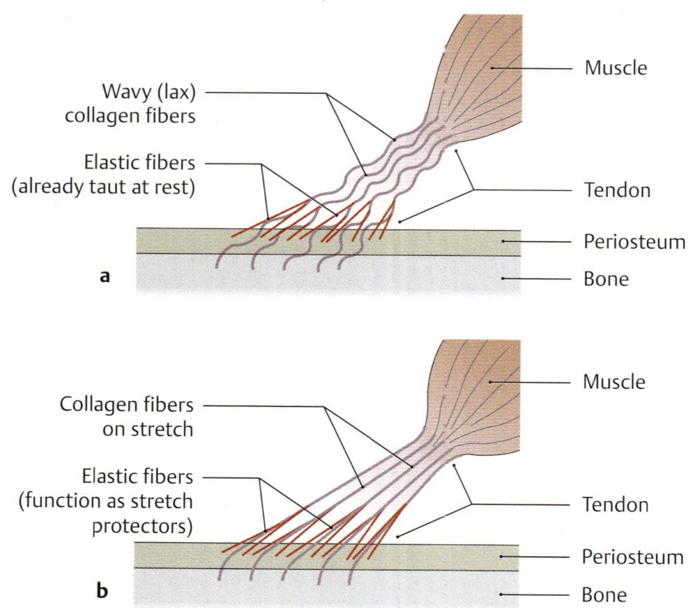

C Structure and function of a periosteal diaphyseal tendon insertion

a Tendon in the lax condition.

b Tendon on stretch.

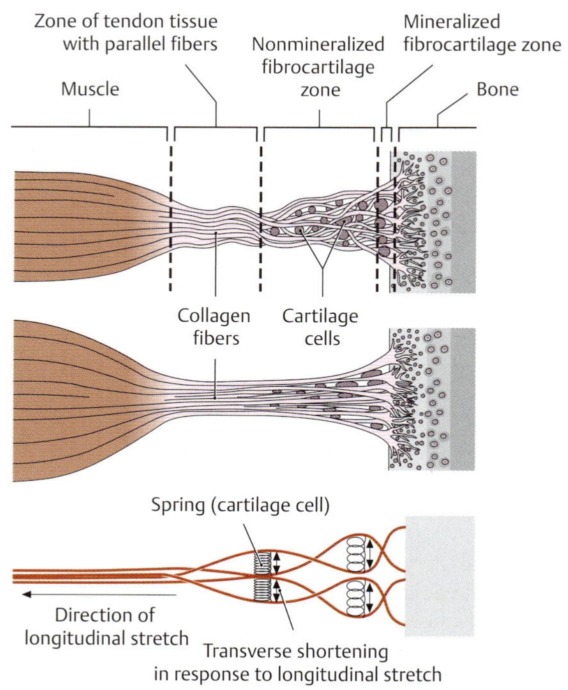

D Structure and function of a chondral apophyseal tendon insertion

a Tendon in the lax condition (muscle relaxed).

b Tendon on stretch (muscle contracted).

c The principle of stretch protection: cartilage cells in the nonmineralized fibrocartilage zone act like taut springs to resist transverse shortening.

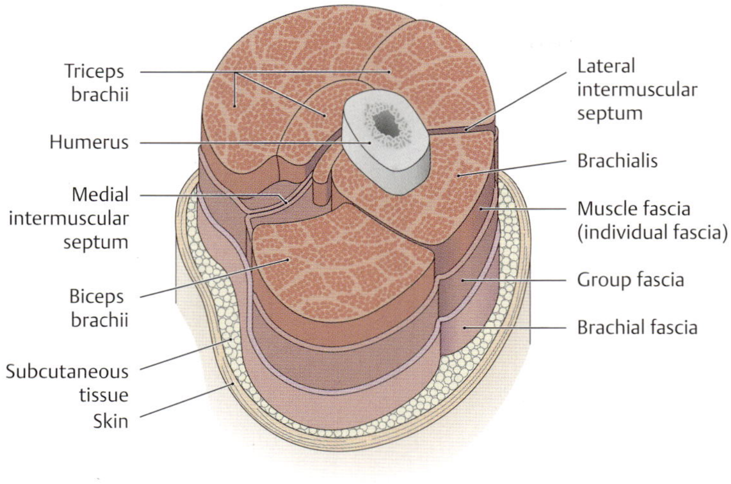

E Muscle fasciae

Cross section through the middle third of the right arm, proximal view.

Except for muscles of facial expression, all individual muscles are sheathed in *individual fasciae*, functional muscle groups in *group fasciae*, and the muscles of every limb in a *common fascia* (such as the brachial fascia). An *intermuscular septum* attached to the bone develops at the intersection of two group fasciae. This septum combines with the muscle fascia and bone to form what are known as osteofibrous channels containing muscles and neurovascular structures (these are also referred to as compartments; see also p. 66).

Note: A compartment can contain only a single muscle or several muscles.

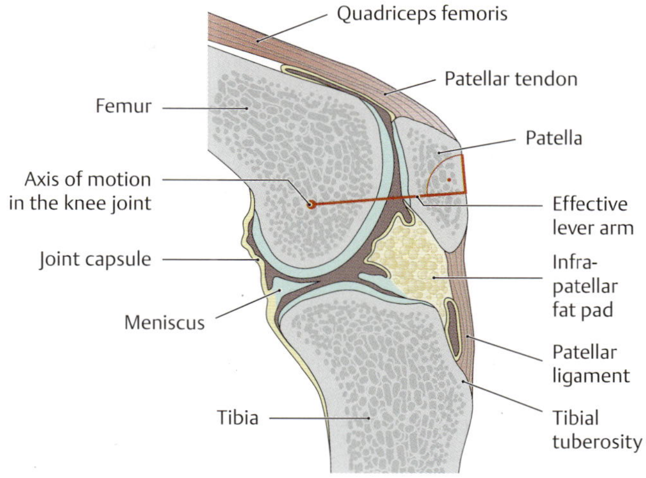

F Sesamoid bones

Sagittal section through the knee joint. Sesamoid bones are bones that are embedded in tendons and protect the tendons from excessive friction. Not everyone has an equal number of sesamoid bones. Sesamoid bones develop where bone deflects the course of the tendon (fulcrum). The compression initially leads to the development of cartilage and later bone. Sesamoid bones lengthen the effective lever arm of a muscle, increasing its mechanical efficiency. The diagram illustrates this principle for the patella, which is the largest sesamoid bone in the body. The patella significantly lengthens the effective lever arm, represented by a perpendicular line from the joint axis to the tendon of insertion of the quadriceps femoris.

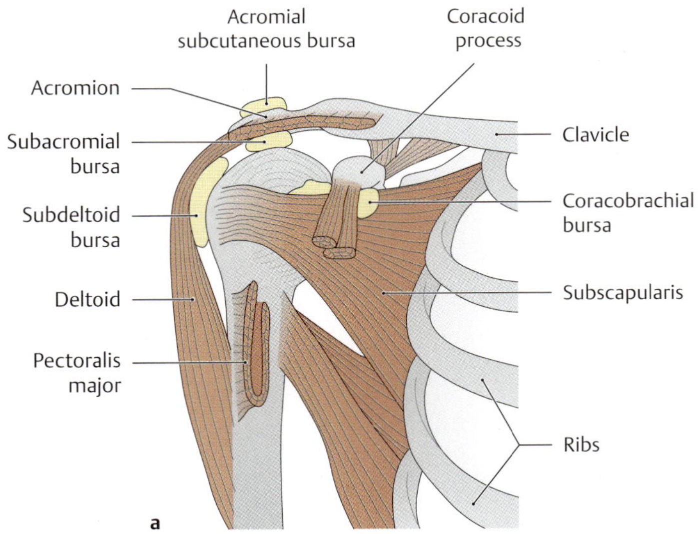

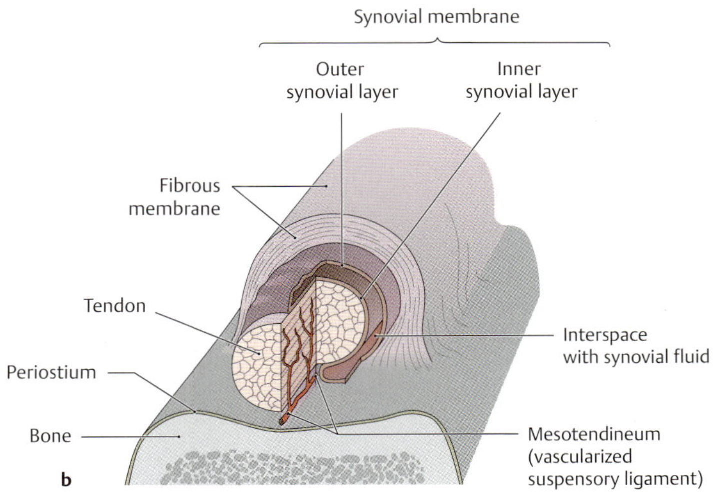

G Synovial bursae and tendon sheaths – mechanisms that protect tendons and muscles

It is thought that synovial bursae (**a**) protect the tendons at those locations where they course in direct contact with the bone. A tendon sheath is a slightly specialized form of synovial bursa (**b**).

a Synovial bursae in the shoulder region. Right shoulder viewed from the anterior aspect with some of the muscles removed. Bursae are pouchlike structures of varying size, usually flattened, that contain synovial fluid. Their wall structure is similar to that of a joint capsule. The bursae may become inflamed (bursitis), causing severe pain.

b Tendon sheath (synovial sheath): The structure of the sheath, consisting of an outer fibrous membrane and an inner synovial membrane, resembles that of a joint capsule. The inner layer of the synovial membrane is firmly attached to the tendon, while its outer layer is attached to the fibrous membrane of the tendon sheath. The mesotendineum (sometimes referred to as vinculum) transmits blood vessels to the tendon.

6.1 Muscle Fasciae: Structure and Function

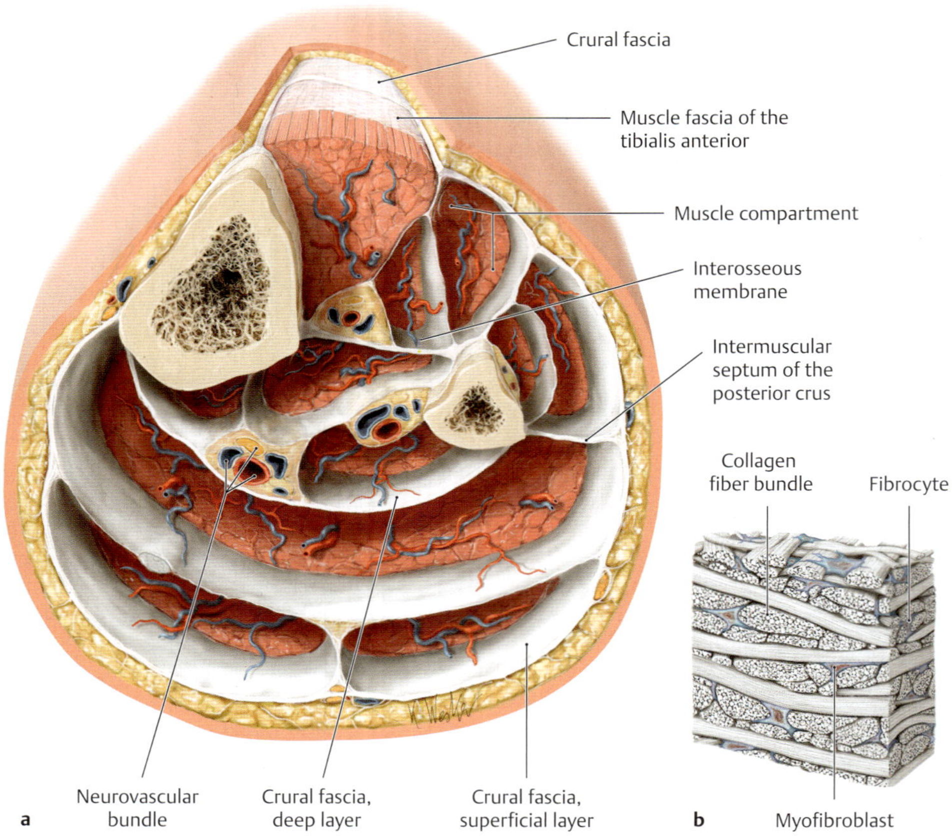

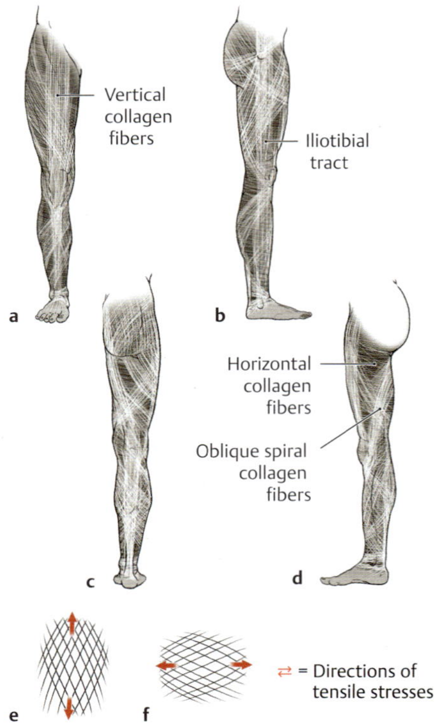

B Lattice structure illustrated by the fascias of the leg

a–d Anterior, lateral, posterior, and medial views; **e** and **f** lattice structure of fascial tissue.

A Definition of the term "fascia" and structure of muscle fasciae

a Fascia of the lower leg (crural fascia); **b** taut, weblike collagenous connective tissue.

Definition: Fasciae are "divisible accumulations of connective tissue" (see Nomina anatomica 1978) that form envelopes and layers and occur as sheetlike or tubular structures. Aside from muscle fasciae, one also differentiates fasciae of the trunk wall, body cavity fasciae, and subcutaneous fasciae (see p. 68). In muscles, "muscle fascia" in the strict sense refers only to the outermost layer of connective tissue overlying the epimysium (see p. 62).

Note: Current usage of the term "fascia" varies widely, and the term can describe nearly all types of dense and loose connective tissue that form a three-dimensional network investing the body and extending into its innermost structures (see 1st International Fascia Research Congress, Harvard Medical School, Boston, 2007).

Structure: **a** shows how the superficial layer of the crural fascia separates muscle and subcutaneous tissue whereas the deep layer separates the individual muscles and functional muscle groups of the lower leg (for the distinction between individual and group fasciae, see p. 65). An *intermuscular septum* attached to the bone develops at the intersection of two common fasciae. This septum combines with the muscle fascia and bone to form osteofibrous channels (compartments) containing muscles and neurovascular structures. Where fasciae and intermuscular septa are reinforced by an aponeurosis, they provide *surfaces for muscle origins*; this may be in the form of retinacula such as in the ankle or wrist regions or as lateral reinforcement of the fascia lata in the thigh (iliotibial tract, see pp. 496 and 524).

Muscle fasciae are taut, weblike collagenous connective tissue structures. They form a web (see **b**) with a lattice structure (see **B**), that is to say layers consisting of parallel type I collagen fibers (90% of all collagen) alternate with adjacent layers of collagen fibers exhibiting different angles of inclination. Collagen fibers are about 5% elastic and because of their undulating course can be elongated about 3%. They resist tissue deformation and are invariably aligned in the direction of tensile forces. They absorb tensile stresses (providing tensile strength). An extended period of relaxation (reduced tensile loading) can lead to shortening of the fibers, which can then become overstretched with increased tension. However, collagen fibers are very resistant to tearing (50–100 N/mm^2) and functionally adapt to increased loading (see **B**).

The fascial system of the lower leg as dissected by Gerlach and Lierse (1990), primarily including the fascia lata of the thigh and crural fascia of the lower leg, shows under polarized light multiple successive layers with collagen fibers coursing vertically, horizontally, and obliquely in spirals (see **Ab**). The varying angles of inclination of the collagen fibers, that is the lattice structure, not only increases the fascial tube's strength and resilience but also its functionality:

- optimal tensile loading in various directions,
- fascia adapts to the changing shape of the muscle as it contracts.

Note: The vertical reinforcement along the lateral aspect of the fascia lata in particular (the iliotibial tract) reduces the bending stress that body weight places on the femur in the coronal plane (Pauwel tension banding principle, see p. 429). This reduces the required amount of bone tissue which in turn reduces body weight. Together with the various connective tissues of muscle (endomysium, perimysium, and epimysium; see p. 62), fasciae allow free mobility between the fibers of a muscle and between adjacent muscles, see **A**).

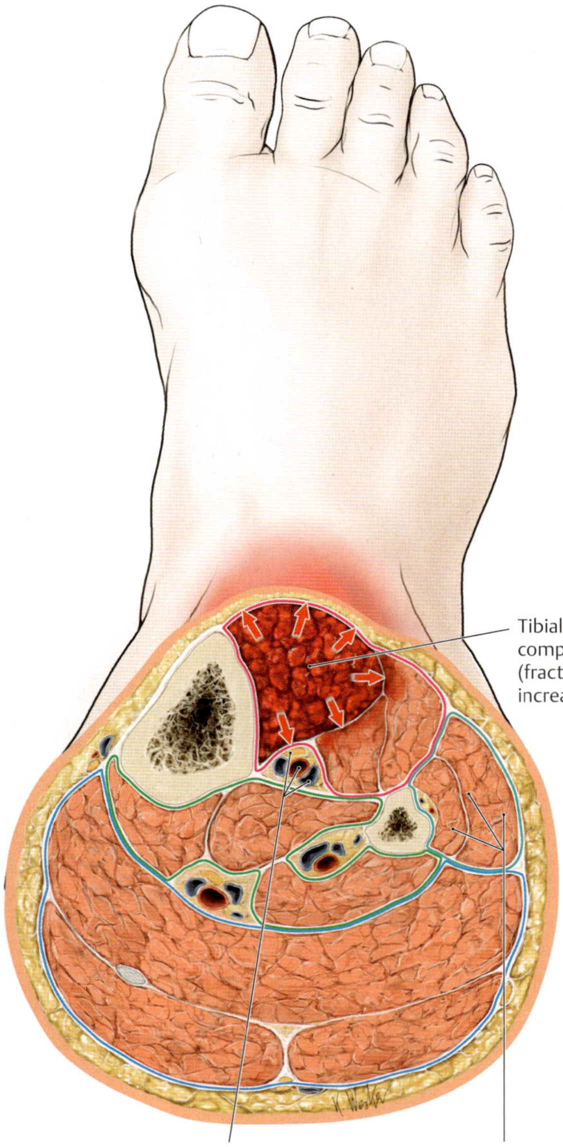

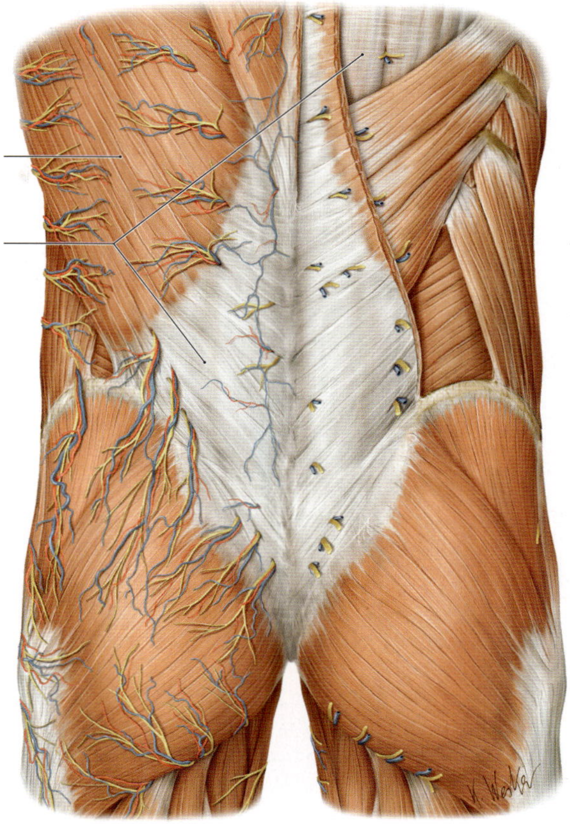

C Compartment syndrome in the lower leg

Eighty percent of all compartment syndromes involve the lower leg or forearm. The common tibialis anterior syndrome (anterior compartment, see p. 585) involves increased tissue pressure such as can occur secondary to a fracture hematoma or muscle edema. This impairs neuromuscular function in the extensor compartment. In terms of pathophysiology, the loss of capillary blood supply leads to ischemia and consequently to metabolic disorders of muscles and nerves (irreversible damage such as ischemic muscle necrosis occurs after about 4 hours).

Surgical principle: Immediate decompression by surgical division of the fascia (skin incision and fasciotomy).

Clinical symptoms: Following an accident such as an impact injury in football, patients often present with rapidly increasing pain in the affected compartment with subsequent compartmental swelling (local pressure over 40 mmHg suggests a manifest compartment syndrome). Additionally, there is pain on passive muscle extension accompanied by motor weakness and sensory deficits in the innervated region (ischemia of the nerves coursing through the compartment).

Note: Distal pulses are generally palpable as increases in pressure of 30–60 mmHg are not generally sufficient to compress major vessels.

D Innervation of fasciae

Most findings on the innervation of fascia pertain to the large fascia of the back, the thoracolumbar fascia (Mense, 2021) which invests the erector spinae, the intrinsic back musculature, while also providing a strong aponeurosis for the origin of several muscles, particularly the latissimus dorsi. Since Staubesand first demonstrated myofibroblasts in muscle fasciae in 1997, we have known that the thoracolumbar fascia contains not only numerous nerve fibers but also free nerve endings and encapsulated receptor organs: nociceptors and proprioceptors as well as numerous afferent and efferent nerve fibers of the autonomic nervous system. However, one must assume that not all muscle fascias are uniformly innervated with various types of receptors. The free nerve endings all primarily contain the two neuropeptides substance P and CGRP (calcitonin gene-related peptide), which indicates that the fascia probably has a nociceptive function. Compared with the density of the CGRP fibers in the erector spinae, the innervation of the fascia is three times more dense and it is thus significantly more sensitive to pain than the musculature of the back. What was surprising was that the thoracolumbar fascia was found to have particularly dense innervation with sympathetic nerve fibers (40% of total innervation). This would mean that the blood supply to the fascia depends heavily on the activity of the sympathetic nervous system, meaning that increased activity of the sympathetic nervous system could lead to reduced perfusion of the fascia.

Note: It is possible that the intrafascial myofibroblasts serve to regulate fascial prestressing for the autonomic nervous system; that is to say, that fascial tone and the autonomic nervous system are tightly coupled and mutually influence each other. This in turn means that involuntary autonomic influences such as stress, fear, mental disorders as well as physiologic parameters such as cold and heat result not only in changes in fascial tone but can also lead to long-term fascial indurations and adhesions. Techniques such as fascia mobilization can therefore lead to improved posture and greater mobility.

6.2 Classification of the Fasciae of the Trunk and Body Cavities

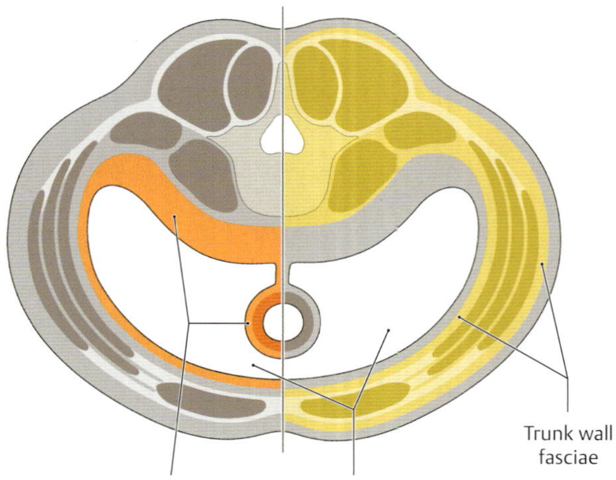

a

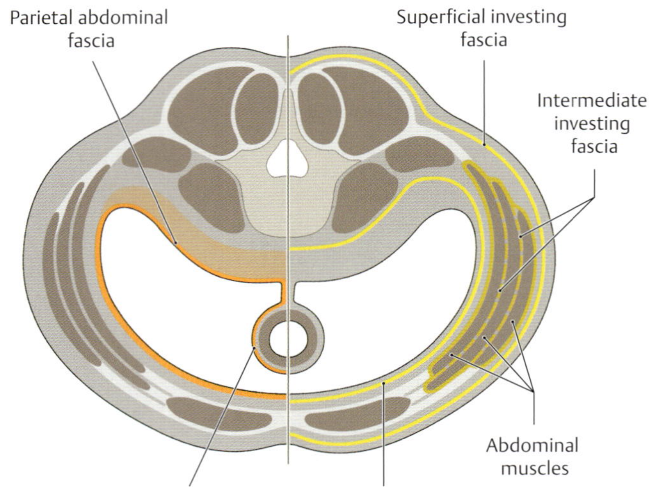

b

A Categorization and configuration of the trunk fasciae illustrated by the abdomen

Superior view of horizontal sections through the abdomen. The fasciae of the abdomen (abdominal fasciae), thorax (thoracic fasciae), and pelvis (pelvic fasciae) belong to the fasciae of the trunk. Like all fasciae they invest structures and spaces and separate functionally distinct layers. For the general structure of fasciae, see p. 66.

a Categorization of the trunk fasciae: In terms of their topography and function, the fasciae of the trunk are divided into two major groups:

- **Trunk wall fasciae:** They invest and separate the muscular trunk wall in the abdomen or the muscular and bony trunk wall in the thorax (thorax and pelvis, see **B** and **C**) and "clothe" them (Latin *vestis* = garment): investing fascia.
- **Body cavity fasciae:** They line the respective body cavity, subdivide it into chambers, and enclose organs and groups of organs in connective tissue. They are named according to their location with respect to the body cavity wall or organ.

b Arrangement of the trunk fasciae

Trunk wall fasciae: The fasciae of the trunk wall can be represented in three layers from superficial to deep:

- Outer (= superficial) layer, superficial investing fascia: separates the trunk wall musculature from the skin (cutis) and subcutaneous tissue (subcutis) and allows the skin and muscle wall to move independently of one another in opposite directions.
- Middle (= intermediate) layer, intermediate investing fascia: covers the trunk wall musculature and separates the individual muscle layers, allowing them to move independently of one another. The middle layer thus invests a muscle group and is therefore often referred to as a muscle group fascia.
- Inner (= deep) layer, deep investing fascia: separates the muscle wall from the actual body cavity.

Note: In the abdomen, which here serves to illustrate the categorization of fasciae in general, the three layers mentioned are the layers of a single muscle wall. Contrast this with the two functionally distinct muscle walls in the thorax (see **B**).

Body cavity fasciae: They exhibit the same structure in the abdomen, thorax, and pelvis but are significantly less prominent in the thorax. Here, too, we differentiate **three parts** but this time **as viewed from the cavity:**

- *Parietal* (= mural) fascia: parietal fascia (of the abdomen): forms the outermost margin of the cavity and lies immediately adjacent to the fascial layer of the inner trunk wall, to which it often adheres. The parietal fascia lies within the peritoneal cavity as a serous membrane overlying the parietal peritoneum.
- *Visceral* (= organ) fascia: visceral fascia (of the abdomen): surrounds the organs to varying degrees (particularly pronounced in the pelvis around the urinary bladder and rectum) and is fused directly with the outer organ tissue (tunica adventitia, tela subserosa) and/or the organ fascia (see note). On the cavity side it is covered by the visceral peritoneum.
- *Extraperitoneal* (= intermediate) fascia: extraperitoneal fascia (of the abdomen or pelvis): connective tissue that connects the parietal and visceral fascia to each other and usually contains neurovascular structures, more pronounced in the pelvis than in the abdomen.

Note: The individual muscles of the trunk wall are each covered by their own separate muscle fascia (proper muscle fascia, see p. 66); the individual organs are often also covered by their separate fascia (proper organ fascia). These proper fasciae are named according to the muscle or organ that they invest. In the classification above, the proper muscle fasciae belong topographically to the intermediate trunk wall fasciae, the proper organ fasciae to the visceral body cavity fascia. Structurally, deep trunk wall fascia is not clearly distinguishable from parietal body cavity fascia as is also the case with visceral body cavity fascia and organ fascia. They are nonetheless differentiated for purposes of classification. In the interest of clarity, the illustration does not show proper muscle and organ fasciae.

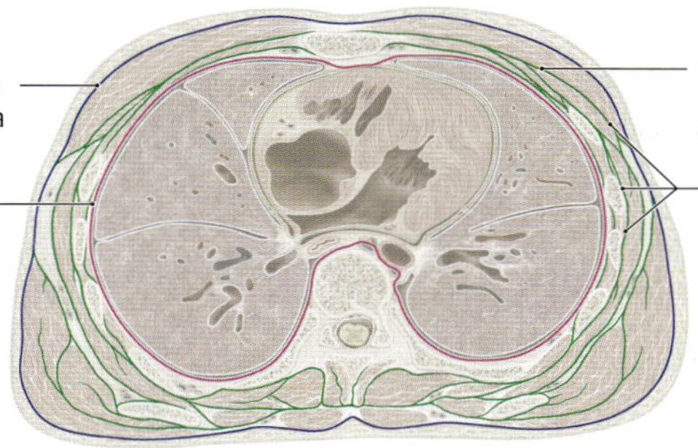

B Fasciae in the thorax

In the thorax, **two** topographically separate **muscle walls** are distinguishable (in the abdomen only **one** 3-layer muscle wall is present).

- Intercostal muscles: in systematic and evolutionary terms, they correspond to the musculature of the abdominal wall, are attached to the ribs, and have three layers;
- Muscles of arm and shoulder girdle motion: located "farther out" and have their origins in the thorax.

Therefore, we distinguish the following **thoracic wall fascias:**

- superficially, deep to the subcutaneous tissue, a superficial investing fascia (blue, corresponds to the superficial investing fascia in the abdomen);
- deep, a deep investing fascia (pink) which defines the innermost layer of the trunk wall fasciae in the thorax (corresponds to the deep investing fascia);
- between them, a complex intermediate investing fascia (green), which is separated into layers of different classifications (and different names as well) because, unlike in the abdomen, they separate two different muscle groups from each other and their surroundings.

The body cavity fasciae of the thorax correspond to those in the abdomen but are significantly less prominent. In the thorax as well, there is extensive fascia-like connective tissue outside the serosa (extraserosal fascia). This corresponds to the extraperitoneal fascia in the abdomen and pelvis. *Note:* Because of its central location in the mediastinum of the thorax, the fascia-like connective tissue outside the serosa is not usually referred to as fascia (although that is what it represents) but very generally as mediastinal connective tissue. For the terms endothoracic fascia, etc., see **D**.

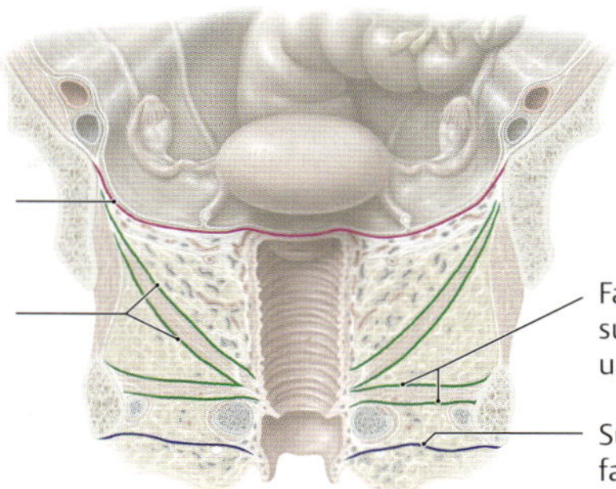

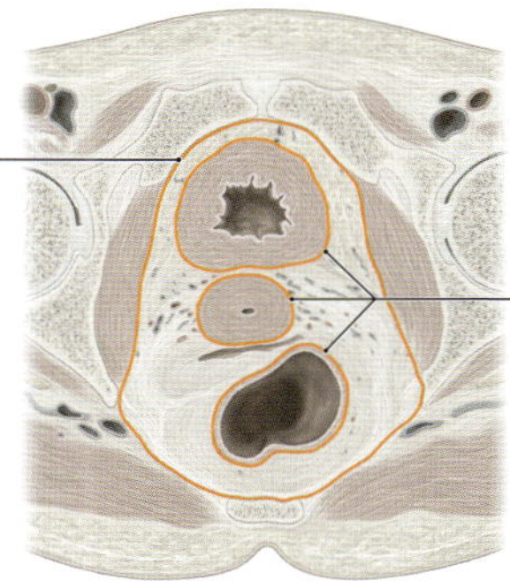

C Fasciae of the pelvis

a Fasciae of the pelvic wall; **b** fasciae of the pelvic cavity.

Fasciae of the pelvic wall: Large parts of the pelvis are bounded by the bony pelvic girdle. Therefore the gluteal muscles on the posterior aspect of the pelvis are almost the only structures invested in fasciae; they have no connection to the pelvic cavity. The pelvic floor with the pelvic and urogenital diaphragms is an important muscle wall defining the pelvic cavity. Both diaphragms consist of several muscles and are surrounded by intermediate investing fascia. A superficial investing fascia also exists below the subcutaneous tissue, especially in the perineal region: the perineal fascia. Current nomenclature does not describe a separate deep investing fascia yet it exists in principle as the pelvic continuation of the transversalis fascia of the abdomen.

Body cavity fasciae in the pelvis: The extensive parietal, visceral, and pelvic extraperitoneal fasciae correspond to the analogous fasciae of the abdomen.

D Specific nomenclature of several fasciae

All fasciae of the trunk can be described with the systematic nomenclature. However, for historical reasons several fasciae have other names (see table) that have long been established and are therefore important.

Systematic name of fascia	Specific designation in thorax	Specific designation in abdomen	Specific designation in pelvis
Superficial investing fascia	*In the region of the pectoralis major:* pectoralis fascia	• Superficial abdominal fascia • *Specifically in the inguinal canal:* external spermatic fascia	Superficial perineal fascia
Intermediate investing fascia	• *For the intercostal muscles:* thoracic fascia • *For the pectoralis minor and subclavius:* clavipectoral fascia	None	• *Specifically in the pelvic diaphragm:* fascia of the superior/inferior pelvic diaphragm • *Specifically in the urogenital diaphragm:* fascia of the superior/inferior urogenital diaphragm
Deep investing fascia	None (because the fasciae in the thorax are poorly distinguishable this is occasionally known as endothoracic fascia)	• Transversalis fascia • *Specifically in the inguinal canal:* internal spermatic fascia	None
Parietal fascia	• Endothoracic fascia • *Specifically in the diaphragm:* phrenicopleural fascia	Endoabdominal fascia	None
Visceral fascia	None or organ specific	None or organ specific	None or organ specific
Extraperitoneal fascia	None	None	None

7.1 Overview of the Human Cardiovascular System

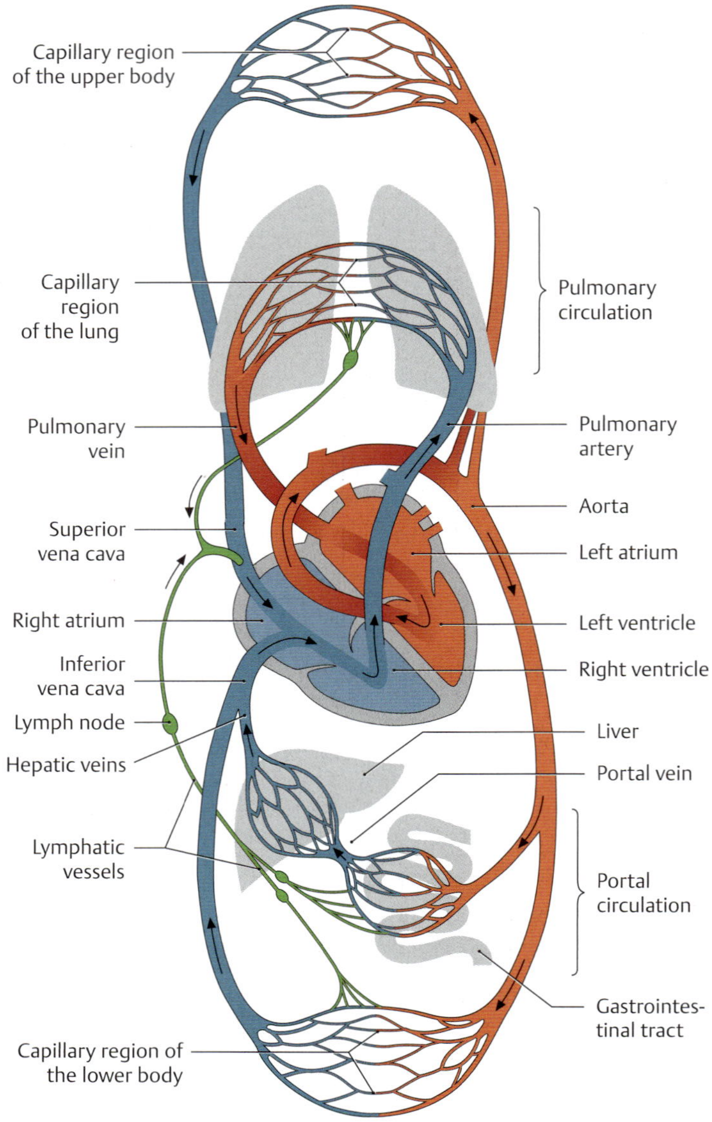

A Schematic representation of the circulatory system

The **circulatory system** is a closed system of arteries, veins, and capillaries in which blood flow is maintained by the pumping action of the heart. Arteries carry the blood from the heart and veins return it to the heart. The exchange of substances and gases (oxygen, carbon dioxide) takes place in the capillaries.

Note: A vessel is designated an "artery" or "vein" according to the **direction of blood flow**. This says nothing about the oxygen content of the blood it conducts. In contrast to all other figures in the book, oxygenated blood is shown here in red and deoxygenated blood in blue to illustrate the transport of oxygen. Functionally, the circulatory system is divided into two main circuits:

- The **pulmonary circulation** with the pulmonary arteries and veins with all their branches.
- The **systemic circulation** with the aorta and the superior and inferior vena cava with all their branches.

Note: These circuits include only the vessels mentioned; they do not include the chambers of the heart.

- The **pulmonary circulation**: Deoxygenated blood flows through the pulmonary arteries away from the heart to the lungs where it is enriched with oxygen; oxygenated blood is returned via the pulmonary veins to the left atrium of the heart and from there to the left ventricle.
- The **systemic circulation**: From the left ventricle the aorta conducts the oxygenated blood away from the heart to the organs, which require oxygen. The deoxygenated blood flows via the superior and inferior vena cava back to the right atrium which conducts it to the right ventricle.

A special part of the systemic circuit is the **portal circulation**, which includes two successive capillary beds. Before venous blood returns to the inferior vena cava from the capillary beds of the unpaired abdominal organs (stomach, bowel, pancreas, and spleen), it is carried by the portal vein to the capillary bed of the liver. This ensures that nutrient-rich blood from the digestive organs initially flows into the liver. Only after undergoing metabolic processes in the liver is the blood returned to the inferior vena cava via the hepatic veins

The **lymphatic system** runs parallel to the venous half of the circulatory system. It begins in the extracellular capillary regions, takes in the extracellular fluid found there and transports it back to the venous blood. This return transport occurs through the lymph vessels, which are interspersed with lymph nodes that act as biological filters.

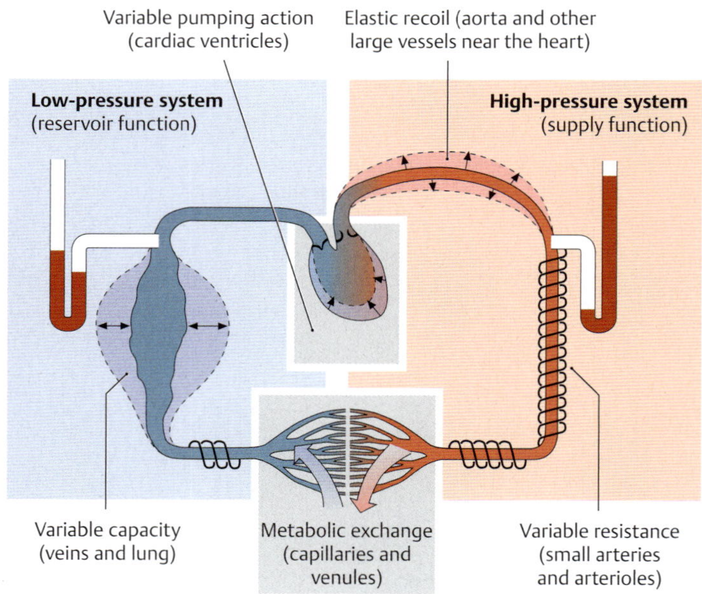

B Basic functional diagram of the circulatory system

(no distinction is made between the systemic and pulmonary systems in the diagram; after Klinke, Silbernagl)

Blood is transported through the circulatory system along a *pressure gradient* created by the different pressure levels in the arterial and venous systems. While the average blood pressure in the *arterial high-pressure system* is approximately 100 mmHg (13.3 N), the pressure in the *venous low-pressure system* generally does not exceed 20 mmHg (2.6 N). The two systems meet in the capillary region of the terminal vascular bed, where metabolic exchange takes place. When the heart expels blood during systole, the **arteries surrounding the heart** (*elastic*-type arteries) can temporarily expand to accommodate the ejected blood volume. During the diastole that follows, the vessel lumen undergoes an *elastic recoil* that transforms the intermittently ejected blood volumes into continuous flow. **Arteries distant from the heart** (*muscular*-type arteries) can actively expand (*vasodilation*) and contract (*vasoconstriction*), providing a very effective means of controlling vascular resistance and regulating local blood flow. The veins are also called *capacitance vessels* because of the high volume of blood contained within the veins. They can accommodate 80% of the total blood volume and thus serve an important reservoir function.

Note: In the pulmonary circulation, even the arteries are part of a low-pressure system. Nonetheless there is an arteriovenous pressure gradient even here.

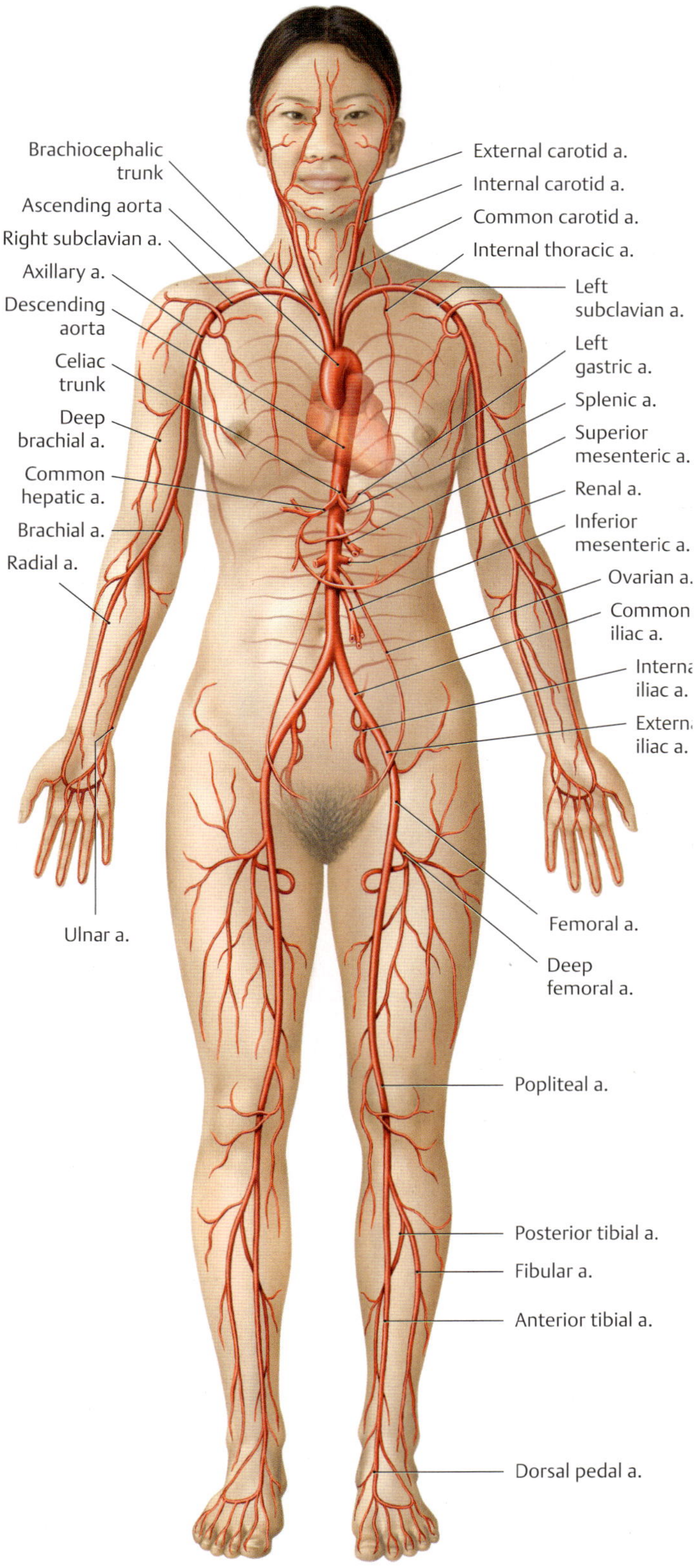

C Overview of the principal arteries in the systemic circulation

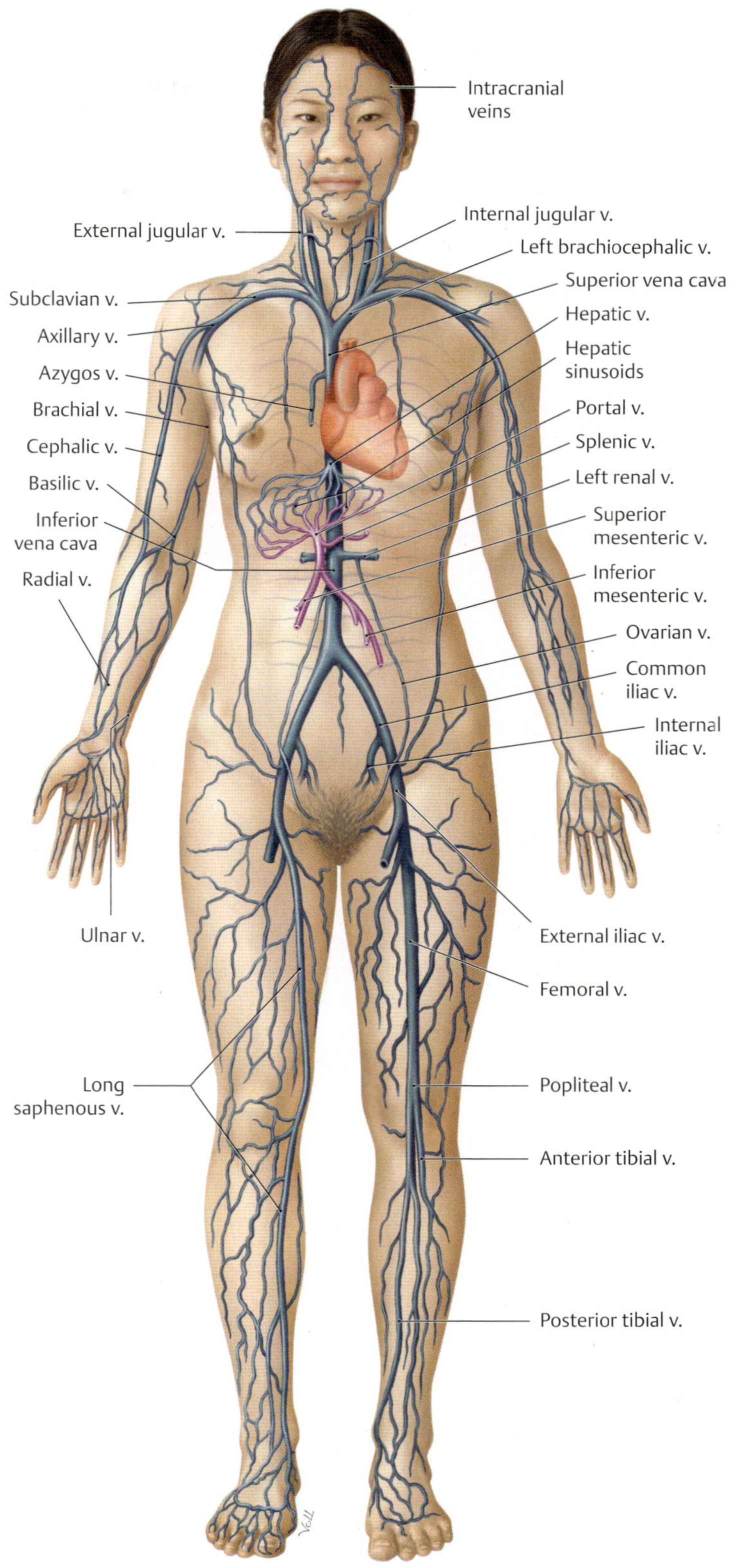

D Overview of the principal veins in the systemic circulation

The venous system is comprised of superficial veins, deep veins, and perforator veins, which interconnect the superficial and deep venous systems.

Note the portal circulation (portal vein), which carries nutrient-rich blood (shown here in purple) from the digestive organs directly to the liver (compare to the left side of **A**).

7.2 The Structure of Arteries and Veins

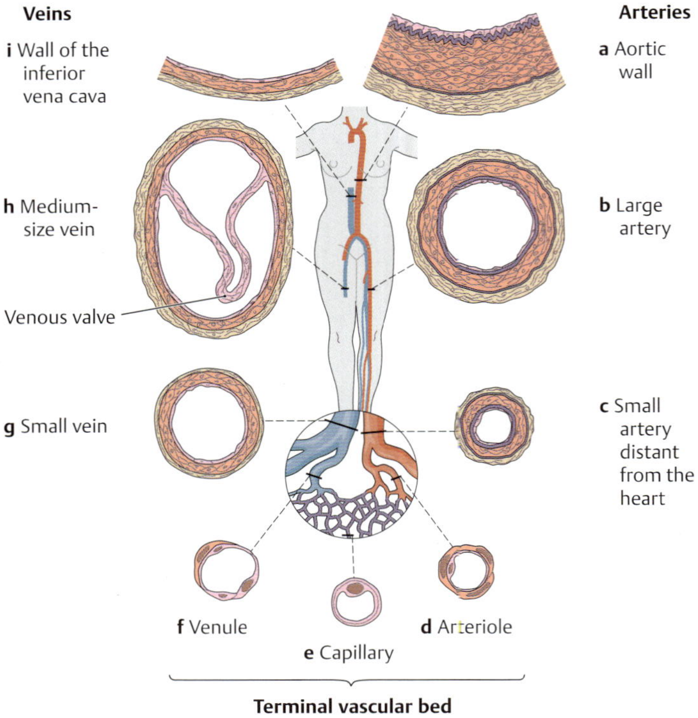

B Organization of the blood vessel system

Arteries (high-pressure system = supply function)
• Elastic-type arteries • Muscular-type arteries
Terminal vascular bed (microcirculation = exchange function)
• Arterioles • Capillaries • Venules
Veins (low-pressure system = reservoir function)
• Small and medium-size veins (with valves) • Large venous trunks

	Arteries		Terminal vascular bed		Veins	
	Aorta	Small artery	Arteriole	Venule	Vein	Vena cava
Wall thickness (w)	2.5mm	1mm	20µm	5µm	0.5mm	1.5mm
Luminal radius (ri)	12.5mm	2mm	20µm	20µm	2.5mm	15mm

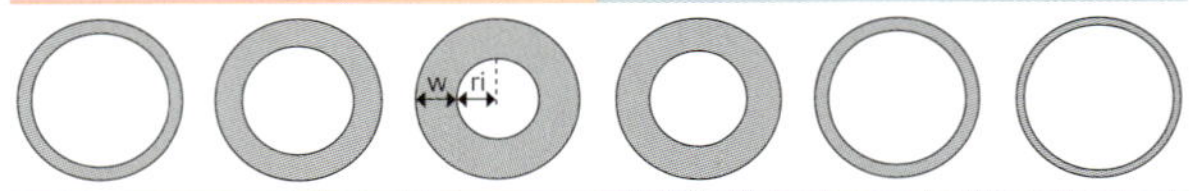

A Structure of the blood vessels in different regions of the systemic circulation (after Frick/Leonhardt/Starck)

Consistent with changing demands, the vessels in different regions of the systemic circulation (high- and low-pressure systems, microcirculation) show significant local structural differences despite a basic similarity in the arrangement of their wall layers. Whereas a relatively high internal pressure prevails throughout the *arterial system*, and the arterial vessels have correspondingly thick walls, the *veins* have a considerably lower intravascular pressure, resulting in thinner walls and larger luminal diameters than in the arteries. In the terminal vascular bed, on the other hand, the vessel wall layers are reduced to permit the exchange of gases, fluids, and other substances.

a–c Arteries, **d–f** terminal vascular bed, **g–i** veins. **a** Close-up view of the aortic wall (elastic-type artery). **b, c** Large and small arteries distant from the heart (muscular-type arteries). **d** Arteriole. **e** Capillary. **f** Venule. **g, h** Small and medium-size veins (some with venous valves). **i** Close-up view of the wall of the inferior vena cava.

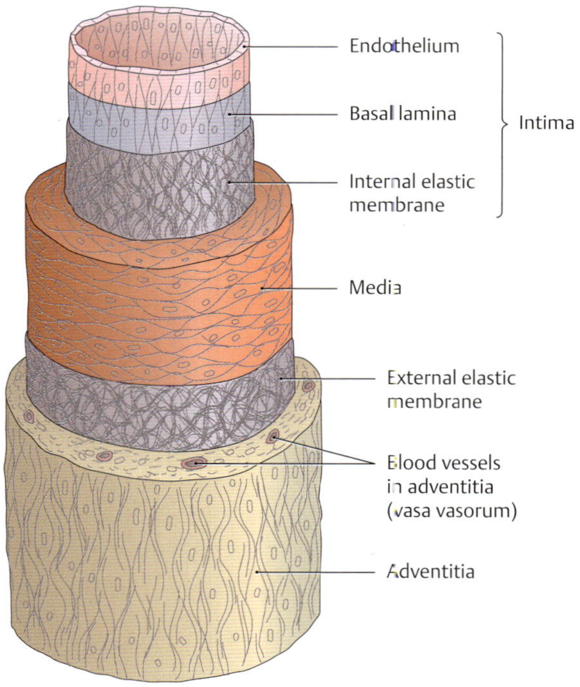

C Wall structure of a blood vessel, illustrated for a muscular-type artery

The wall of a blood vessel basically consists of *three layers*: the *intima*, *media*, and *adventitia*. The three-layered structure is clearly apparent in the walls of arteries and is less conspicuous in veins (see **D**).

- The **intima** consists of a layer of spindle-shaped endothelial cells that are aligned along the vessel axis and rest upon a basement membrane and a thin layer of subendothelial connective tissue. In muscular-type arteries, the intima is consistently separated from the media by an internal elastic membrane.
- The **media** consists of an approximately circular arrangement of smooth muscle cells, elastic and collagenous fibers, and proteoglycans. Muscular-type arteries may have an external elastic membrane that separates the media from the adventitia.
- The **adventitia**, like the intima, is composed of longitudinally aligned elements, mostly connective tissue. The adventitia of veins may additionally contain smooth muscle. The adventitia transmits autonomic nerves to the muscle of the vessel wall, and, especially in larger vessels, it also transmits the vasa vasorum, which supply blood to the outer third of the vessel wall.

Specific functions can be assigned to all three layers. The intima is concerned with the exchange of gases, fluids, and other substances through the vessel wall; the media regulates blood flow, and the adventitia integrates the blood vessel into its surroundings.

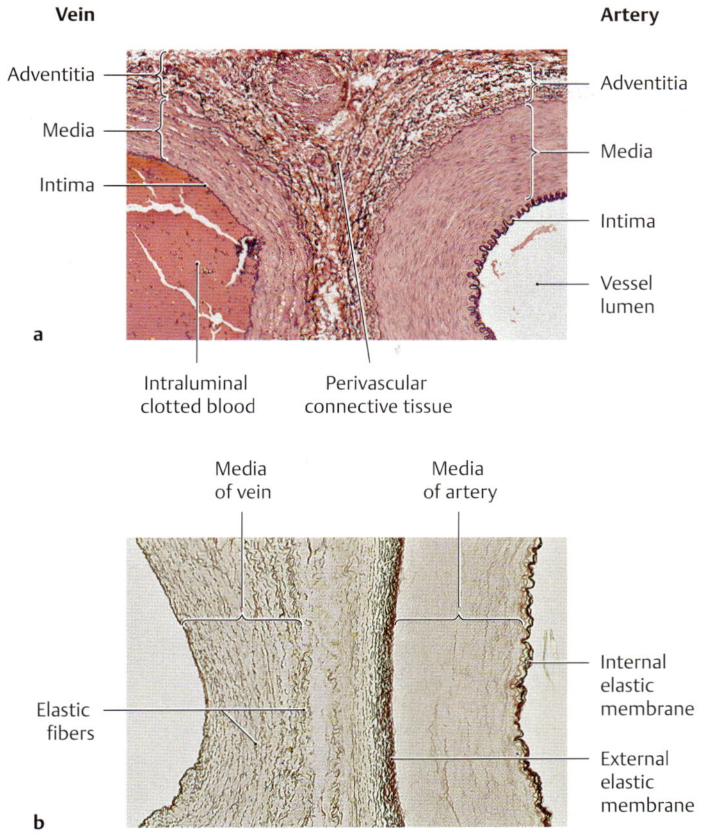

D Differences in the wall structure of arteries and veins
Wall sections from a muscular-type artery and an accompanying vein. Comparison of tissue cross sections treated with different stains. **a** H & E-resorcin-fuchsin stain of the posterior tibial artery and vein. **b** Resorcin-fuchsin stain of the femoral artery and vein.
Note the characteristic structural differences in the media. While the arterial media consists of densely packed layers of smooth muscle cells, the venous media contains a far greater amount of connective tissue elements (collagenous and elastic fibers), giving it a much looser structure. The veins also lack a conspicuous layered structure and an internal elastic membrane (from Lüllmann-Rauch: Histologie, 2nd ed. Stuttgart: Thieme; 2006).

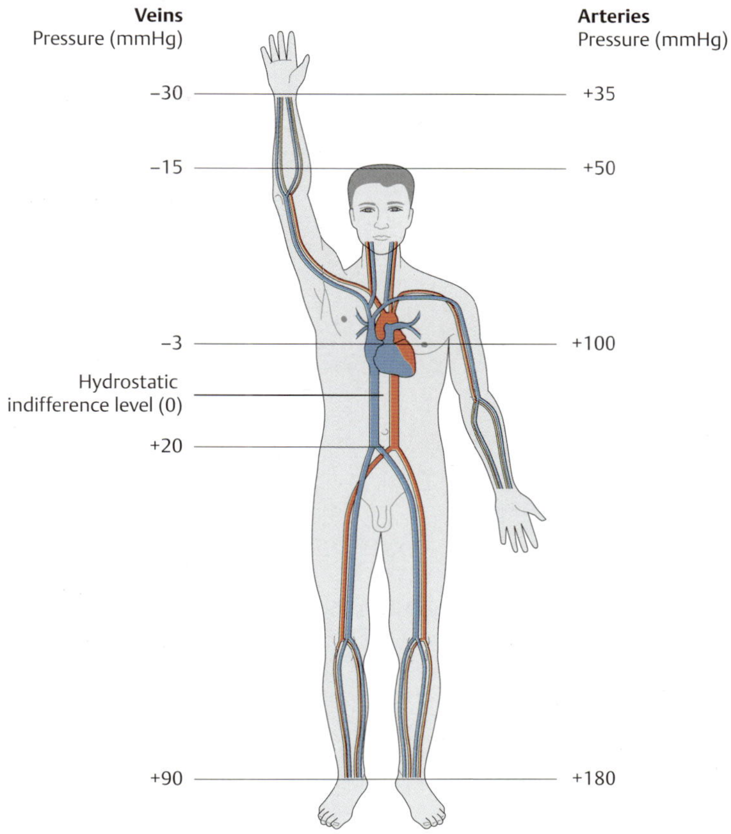

E Arterial and venous pressure changes in the standing position
Changing from a recumbent to a standing position radically alters the pressure relationships in the circulatory system. The hydrostatic effects of this change cause the pressure to rise sharply in the lower parts of the body, while the pressures in the upper body decrease (the pressures remain unchanged at the "hydrostatic indifference level" just below the diaphragm). Along with the hydrostatic pressure changes, approximately 500 mL of blood volume is shifted into the lower limb veins. This rise of venous pressure greatly increases the transmural pressure in the lower limb veins, while the pressure in the head and neck veins may fall so low that the veins collapse. This explains why analogous veins in the lower and upper body regions have wall layers of different thickness; i.e., the veins on the dorsum of the foot are much more muscular than on the dorsum of the hand. The wall of the inferior vena cava, however, is paper-thin due to the low venous pressures at that level.

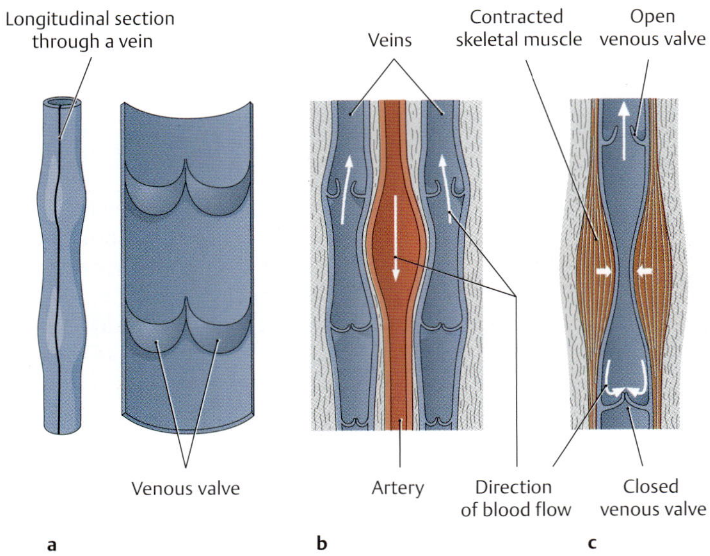

F Venous return to the heart
The following factors promote the return of venous blood to the heart: **a** opening and closing of the venous valves; **b** arteriovenous coupling (the pulse wave in the artery is transmitted to the accompanying vein); and **c** the muscle pump.
Venous return is also aided by the "suction effect" of the heart, i. e., the negative pressure produced when the valve plane moves toward the cardiac apex during systole. A lack of muscular movement due to prolonged standing or sitting, for example, can cause the damming back of venous blood, leading to a raised intravascular pressure and incompetence of the venous valves. This can result in edema, varicose veins, and circulatory impairment.

7.3 The Terminal Vascular Bed

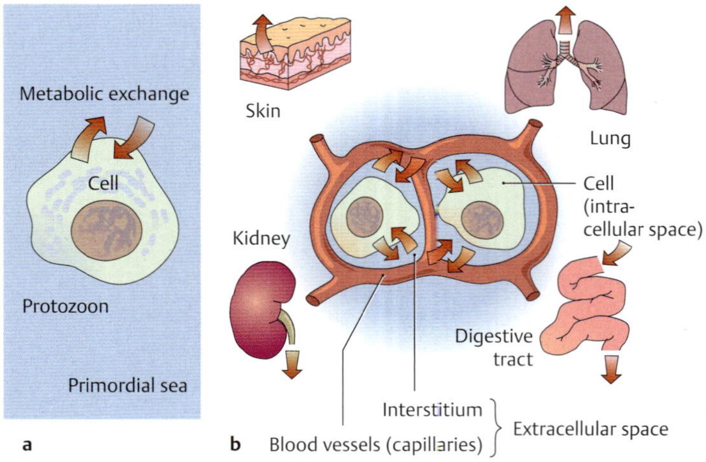

A The milieu in which a cell lives (after Silbernagl, Despopoulos)

a Protozoan: The first single-cell organisms lived in an environment, the primordial sea, that provided a milieu of constant composition. The internal and external milieus were the same, and so neither of them changed during metabolism.

b Human: The cells of a multicellular organism are bathed by extracellular fluid, whose volume is substantially smaller than the intracellular volume. The extra- and intracellular fluids also have a different composition. In this situation the internal milieu would change very quickly if the intercellular space (interstitium) were not linked via the bloodstream to organs such as the lungs, kidneys, and digestive tract, which absorb nutrients and excrete metabolic products. Nutrients absorbed from the bowel are distributed to the cells of the various organs (interstitium of the capillary beds) via the bloodstream. The blood also transports the metabolic products of the cells to the organs that are responsible for their excretion (e.g., the lungs and kidneys).

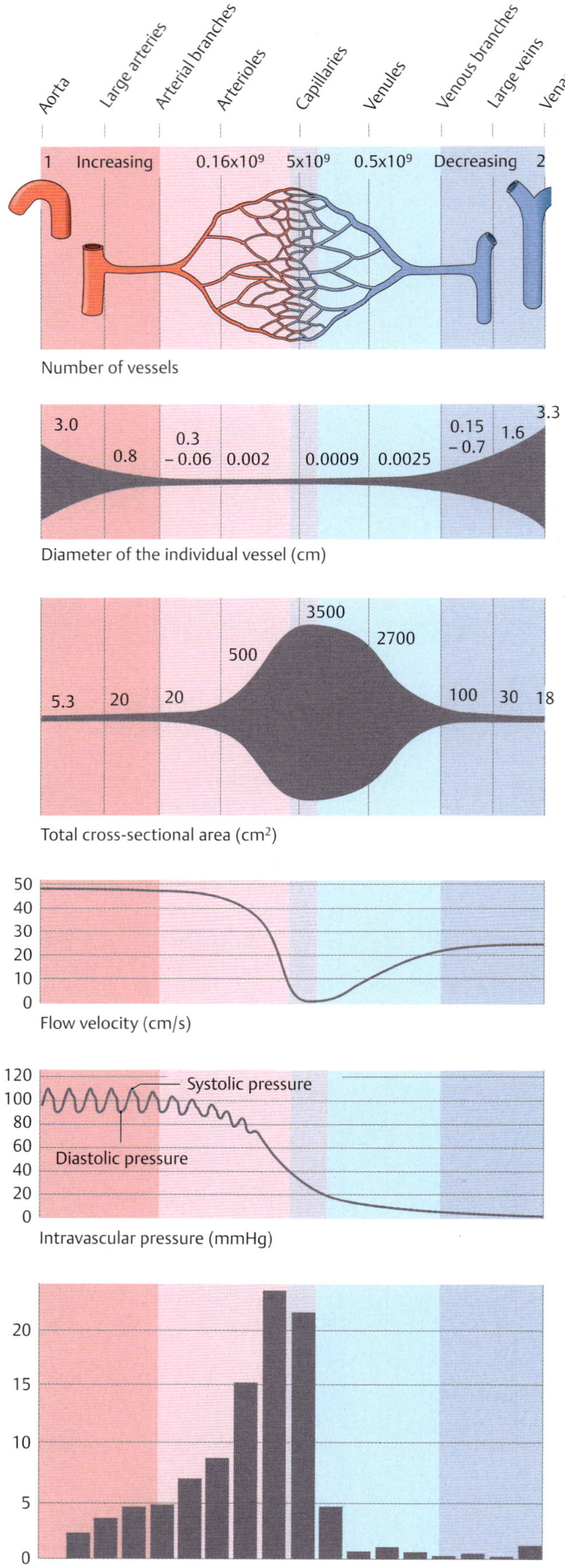

B Characteristics of different vascular regions (after Silbernagl, Despopoulos)

The terminal vascular bed is the site of the microcirculation and therefore is the site where gases, fluids, and other substances are exchanged. It consists of

- an *afferent arterial limb* (precapillary arterioles),
- the *capillary bed* itself, and
- an *efferent venous limb* (postcapillary venules).

The smallest vessels, the capillaries, consist only of an endothelial layer and a basal lamina to which pericytes may be externally attached (contrast with the more complex structure of large vessels, p. 72). Owing to the extensive branching of the vessels in the capillary bed, the total vascular cross section is greatly increased (approximately 800 times), while the flow velocity is correspondingly reduced (from 50 cm/s in the aorta to 0.05 cm/s in the capillaries). With an average capillary length of 0.5 mm, a time of approximately 1 second is available for metabolic exchange. The increased vascular resistance in the arterioles and capillaries caused by contact of the blood with the large endothelial surface area (increased friction) lowers the blood pressure and eliminates pressure spikes. Thus, the capillaries provide ideal conditions for exchange processes to occur between the blood and the interstitial fluid that bathes the body's cells.

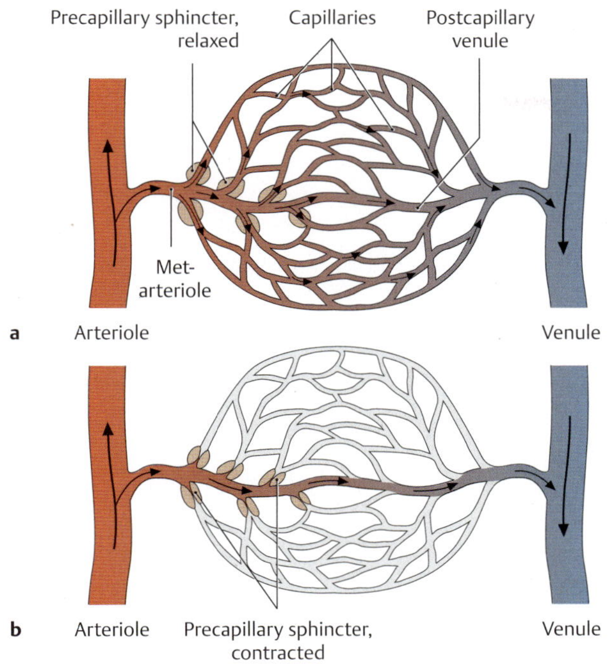

C Blood flow in the capillary bed

a Sphincter relaxed, **b** sphincter contracted.

Precapillary sphincters, with their circular array of muscle cells, are located at the junction of the metarterioles and capillaries and regulate blood flow within the capillary network. When the sphincters contract, the branching capillaries are closed, and the capillary bed is unperfused except for the metarterioles (e.g., only about 25 to 35% of all capillaries are perfused under resting conditions). The arterioles and venules may also be interconnected by shunts called *arteriovenous anastomoses*.

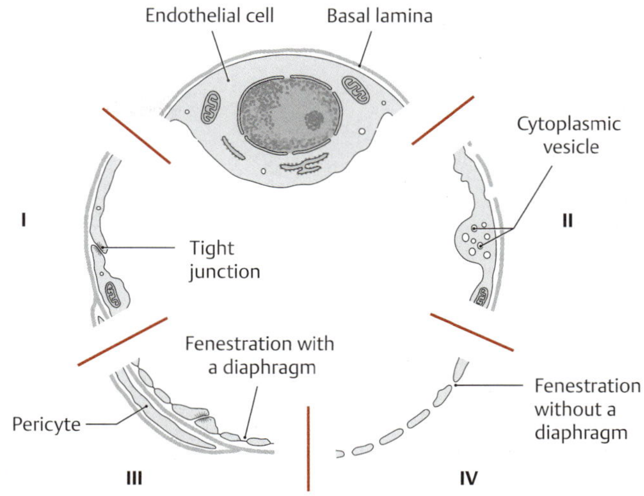

D Different forms of capillary endothelial cells (scheme of ultrastructural features)

Capillaries range from 5 to 15 µm in diameter and consist of endothelial cells, basal lamina, and external pericytes. Pericytes have various properties and functions, including a role in vascular development and angiogenesis. The individual endothelial cells are connected to one another by adhesion contacts, tight junctions, and gap junctions, largely preventing any metabolic exchange between individual endothelial cells. The endothelia of different capillaries have varying degrees of permeability, and several types of endothelium are distinguished on that basis:

- **I** Closed endothelium without fenestrations and with a continuous basal lamina (e.g., nervous system).
- **II** Closed endothelium with pinocytotic activity (e.g., cardiac and skeletal muscle).
- **III** Endothelial cells fenestrated by a diaphragm (e.g., gastrointestinal tract).
- **IV** Endothelial cells with intercellular gaps (large fenestrations) and without a continuous basal lamina (e.g., liver).

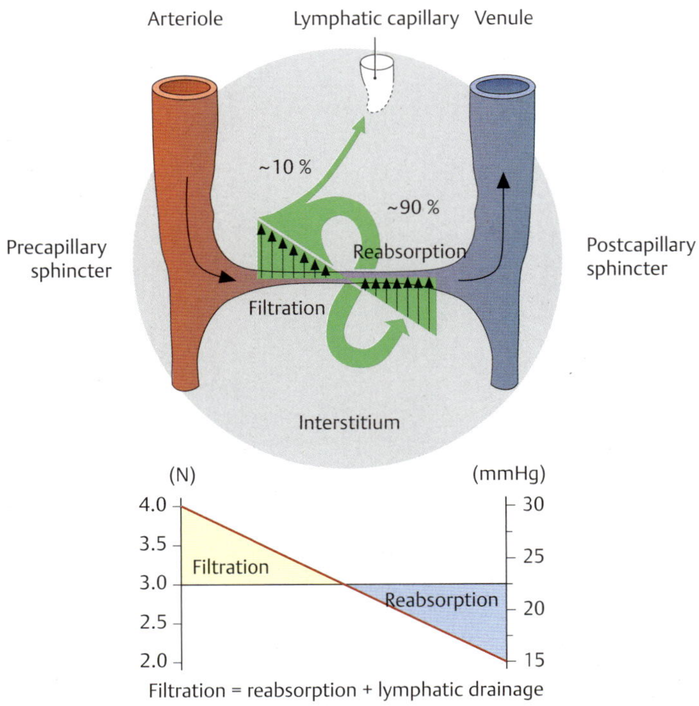

E Mechanism of fluid exchange in a capillary (after Silbernagl, Despopoulos)

Fluid exchange between capillaries and the surrounding tissue (interstitium) is regulated by a changing pressure gradient between the blood pressure in the capillaries (hydrostatic pressure) and the intravascular colloid osmotic pressure. The driving force behind the fluid exchange is the hydrostatic blood pressure. The pressure at the arterial end of the capillary is 35 mmHg (= 4.6 N), which is 10 mmHg higher than the colloid osmotic pressure of approximately 25 mmHg (= 3.3 N). This positive pressure differential makes it possible for fluid as well as dissolved particles to filter out of the capillaries and into the surrounding tissue. These relationships are reversed at the venous end of the capillary—there, the hydrostatic blood pressure falls to approximately 15 mmHg (2.0 N), while the colloid osmotic pressure remains essentially unchanged at about 25 mmHg. As a result of this, the hydrostatic pressure on the *venous side of the capillary* is 10 mmHg lower than the colloid osmotic pressure (15–25 = –10 mmHg), causing fluid with its solute particles to flow back into the vessel (*reabsorption*).

Of the 20 L of fluid that leave the capillaries each day, only about 18 L (90%) are reabsorbed. Approximately 2 L (10%) of the filtered volume is removed by lymphatic vessels in the form of lymph. If this fluid exchange does not occur in the manner described, edema may develop (i. e., a persistent accumulation of fluids in the interstitium). The reason for this may be an *elevated hydrostatic pressure* (due to blood pooling on the venous side of the capillaries) or a *decreased colloid osmotic pressure* (due to a decrease in plasma proteins). In both cases there is an imbalance of fluid exchange, allowing fluid to accumulate in the tissue, known as edema.

8.1 The Human Lymphatic System

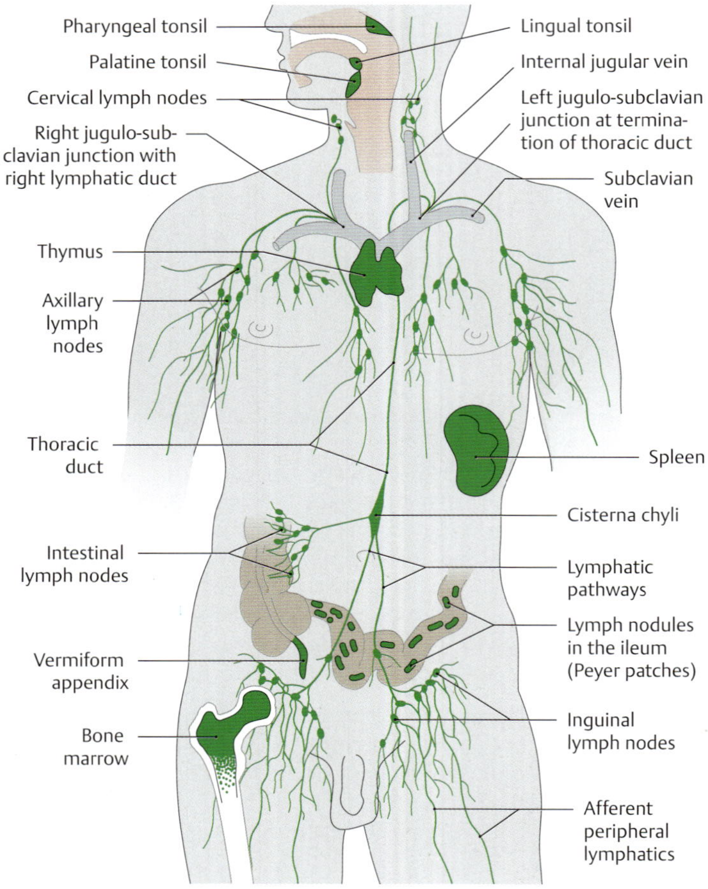

A The human lymphatic system

This system includes the lymphatic vessels and the lymphatic organs (immune organs, see **B**). The **lymphatic vascular system** runs parallel to the venous system and performs several functions:

- Its primary function is to clear the interstitial spaces of tissue fluid and substances that cannot be reabsorbed in the venous capillary bed. The composition of the lymph varies in different regions and is similar to that of the surrounding interstitial fluid.
- It carries away food lipids (chylomicrons) that are absorbed in the bowel.
- It returns lymphocytes from the lymphatic organs to the blood.

The lymphatic vascular system consists of

- *lymphatic capillaries*, which begin peripherally as blind-ended vessels,
- the *lymphatic vessels* and interposed lymph nodes, and
- the *major lymphatic trunks* (thoracic duct and right lymphatic duct).

The lymphatic capillaries collect fluid from the interstitium and transport it via the lymphatic vessels and lymph nodes to the major lymphatic trunks. The fluid reenters the venous system from these trunks at the junctions of the left and right subclavian and internal jugular veins. The lymph drained from three body quadrants enters the left jugulo-subclavian venous junction, while only lymph from the right upper quadrant enters the right jugulo-subclavian venous junction.
The **lymphatic organs** are part of the specific immune system and, as such, are situated at likely portals of entry for infectious microorganisms. The spleen is the only immune organ that is directly integrated into the bloodstream.

B Primary and secondary lymphatic organs

The functions of the lymphatic organs include mounting a specific immune response. A distinction is drawn between primary and secondary lymphatic organs. The primary lymphatic organs are concerned with the production, maturation, and selection of immune cells. The secondary lymphatic organs are subsequently populated by the immunocompetent lymphocytes and are sites for various processes, such as antigen presentation, lymphocyte proliferation, and antibody formation.

- **Primary lymphatic organs:**
  - Thymus (selection of T-lymphocytes)
  - Bone marrow (selection of B-lymphocytes)
- **Secondary lymphatic organs:**
  - Spleen
  - Lymph nodes
  - Mucosa-associated lymphatic tissue (MALT) and the pharyngeal lymphatic (Waldeyer's) ring—the pharyngeal, palatine, and lingual tonsils.
  - Bronchus-associated lymphatic tissue (BALT)
  - Gut-associated lymphatic tissue (GALT), such as Peyer patches and the vermiform appendix

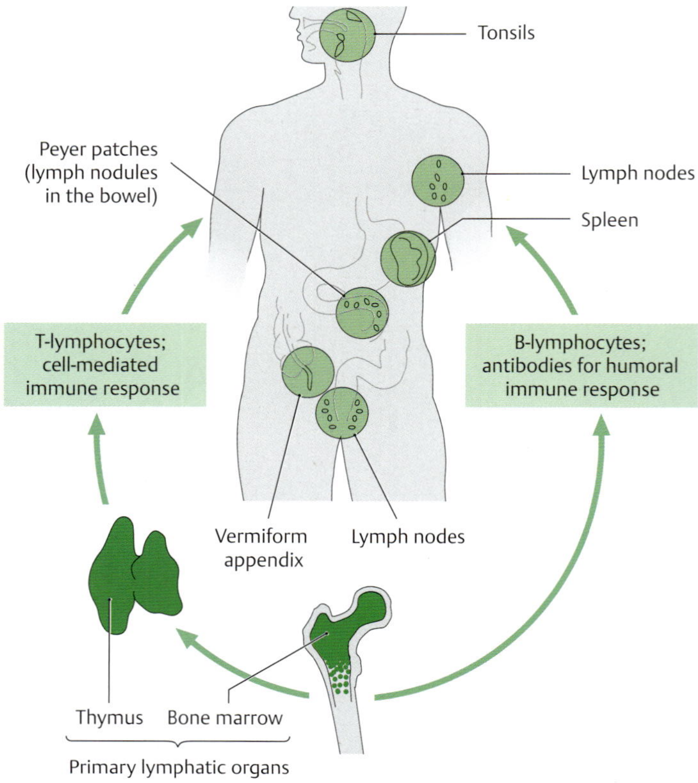

C Organization of the lymphatic vascular system

Three compartments can be distinguished in the lymphatic vascular system based on **topographical and functional** criteria:

1. A superficial system → drains the skin and subcutaneous tissue.
2. A deep system → drains lymph from the muscles, joints, tendon sheaths, and nerves.
3. An organ-specific system → drains the organs and shows organ-specific differences.

A system of perforator vessels interconnects the superficial and deep systems, conveying lymphatic fluid from the surface to the deeper tissues.
The lymphatic vascular system can be subdivided into four *different regions* based on the **histologic structure of the vessel walls**:

1. Lymphatic capillaries
2. Precollectors
3. Collectors
4. Lymphatic trunks

Lymphatic capillaries and precollectors are also known as *initial lymphatics*.

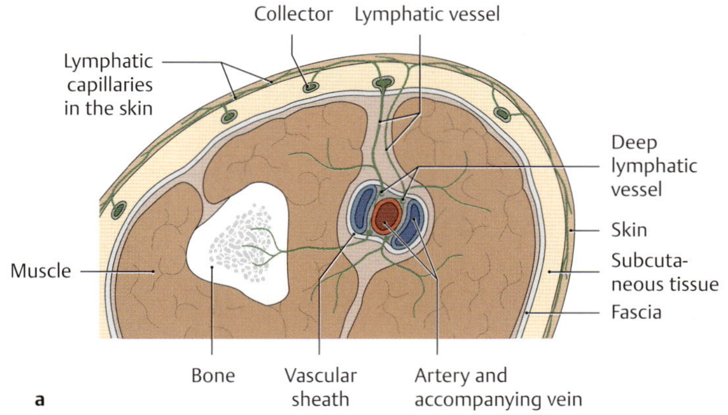

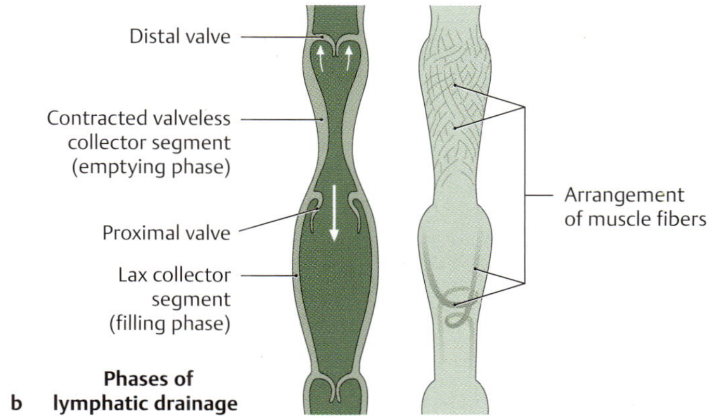

D Organization and structure of the different lymphatic regions

a Lymphatics in the skin and muscles.

b Detail from **a**, showing the structure and function of a collector segment.

Both the superficial and deep lymphatics originate with the extremely thin-walled **lymphatic capillaries**, which are approximately 50 µm in diameter. Their endothelium is bounded by an incomplete basal lamina, and they are attached by collagenous "anchoring filaments" to elastic fibers and collagen fibers in their surroundings. The network of lymphatic capillaries opens into larger **precollectors** approximately 100 µm in diameter. Unlike the lymphatic capillaries, these vessels contain valve cusps, and their walls are reinforced by a layer of connective tissue. They open into **collectors**, which also contain valves and have a transverse diameter of 150 to 600 µm. Like the larger lymphatic vessels and the lymphatic trunks, the collectors have a venous-type wall structure divided indistinctly into an intima (endothelium and basement membrane), a smooth-muscle media, and a fibrous adventitia. Lymph transport is effected by a series of rhythmic contractile waves (10–12/min) that are generated in the smooth-muscle, valveless collector segments. The direction of lymph flow is controlled by closing the distal valves and opening the proximal valves of the precollectors and collectors.

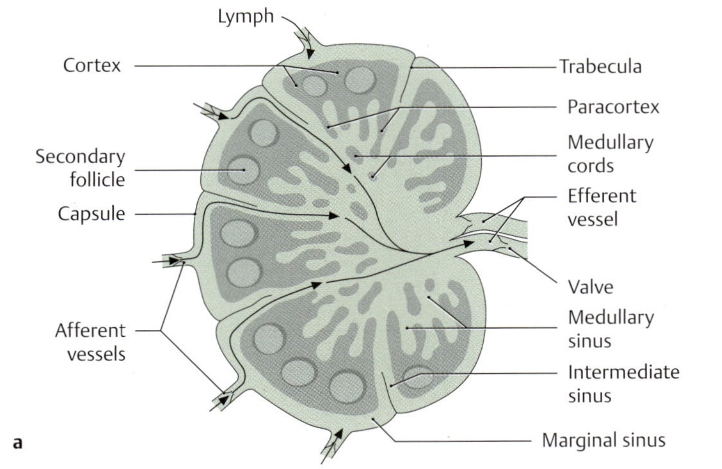

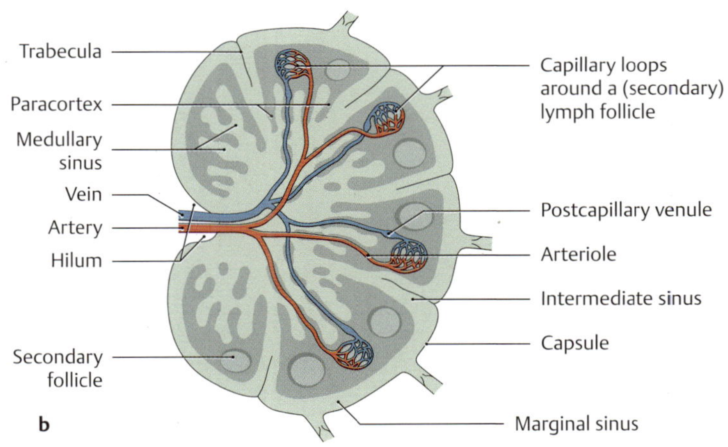

E Structure of a lymph node

a Lymph circulation, **b** blood supply to the lymph node.

Lymph nodes are small filtering stations located in the course of lymphatic vessels and are components of the specific immune response (they contain T- and B-lymphocytes). *Regional lymph nodes* are distinguished from the *collecting lymph nodes* that receive lymph from multiple regional nodes. The lymph enters the lymph node through multiple afferent vessels. As the fluid passes along the various lymph sinuses to the efferent vessels, it comes into contact with the lymph node tissue over a broad surface area. From outside to inside, a lymph node consists of the cortex, paracortex, and medulla. The numerous secondary follicles in the *cortex* form the *B-lymphocyte region*, and the lymphocyte-rich areas between and below the secondary follicles are the *T-lymphocyte regions* (paracortex). Lymphocytes leave the bloodstream in the high-endothelial postcapillary venules of the T-lymphocyte region; then, after differentiating, they leave the lymph node with the draining lymph via efferent lymph vessels, which often become the afferent vessel of another lymph node of a lymph node group.

8.2 Exocrine and Endocrine Glands

A Development and classification of glands

Glands are epithelial aggregations of highly specialized single cells (goblet cells, multicellular intraepithelial glands) or of larger cell groups that have migrated to deeper levels. Their function is to synthesize and release secretions. Glands fall into two main categories:

- **Exocrine glands** (e.g., salivary glands, sweat glands): These glands release their secretion *externally* to the skin or mucosa, either directly or through excretory ducts.
- **Endocrine glands:** Their secretions (in this case hormonal messengers) are released *internally*, i.e., into the bloodstream, lymphatics, or intercellular spaces. Endocrine glands do not have excretory ducts (see **F** for mechanisms of hormone release). Once released into the bloodstream, the hormones are distributed throughout the body and are transported to their target cells, where they bind to specific receptors and exert their effect.

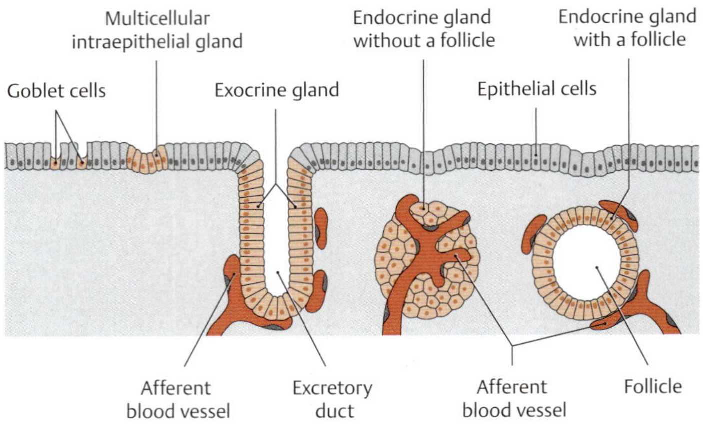

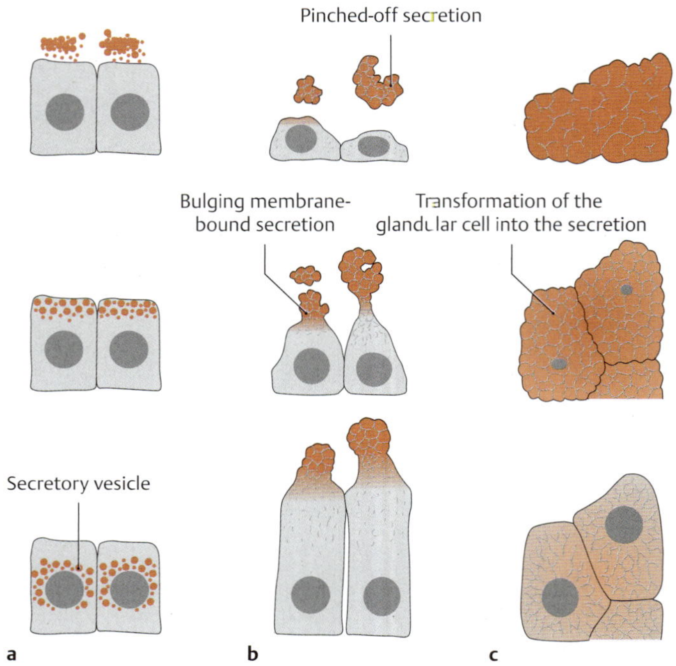

B Mechanisms by which exocrine glands release their secretions (light-microscopic scale)

a **Exocytosis:** In this mechanism, the secretion is released *without an enclosing membrane* (merocrine or eccrine secretion). The membrane-bound vesicles containing the secretion fuse with the apical cell membrane and discharge their contents to the outside with no loss of membranous material (the secretory mechanism of most glands, see also **C**).

b **Apocytosis:** The membrane-bound vesicles form a bulge in the apical cell membrane and are finally *pinched off* by it (apocrine secretion). The pinched-off secretory products are enclosed within a membrane. This mechanism is necessary in the secretion of fats. The membrane encloses the fats and keeps them emulsified (e.g., scent glands, mammary glands).

c **Holocytosis:** In this mechanism the *entire glandular cell disintegrates and becomes the secretory product* (holocrine secretion). As a result, the glandular cells must be constantly replaced by a basal regenerative cell layer (e.g., sebaceous glands in the skin).

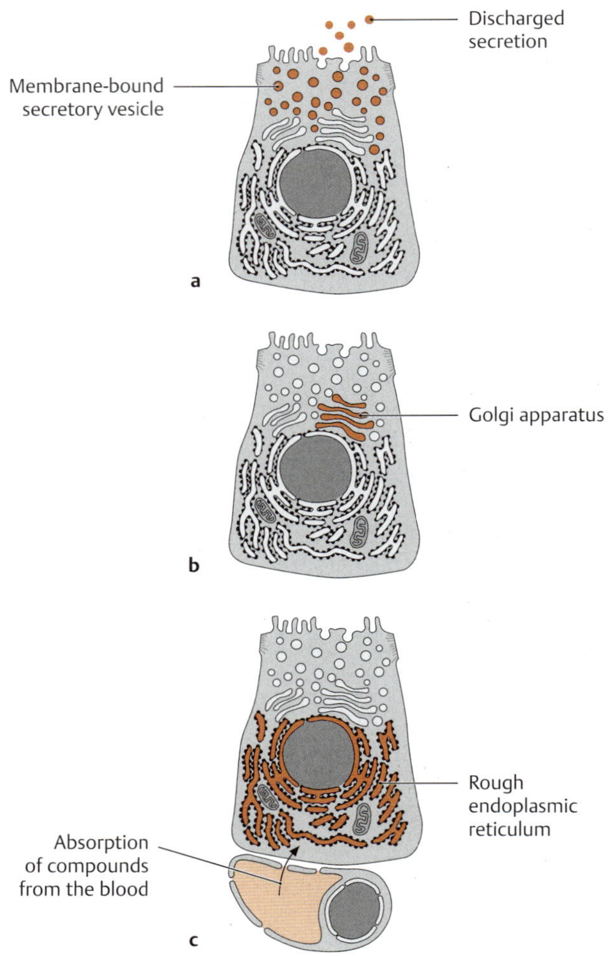

C Production and release of secretions by exocytosis (electron-microscopic scale)

After the glandular cell has absorbed essential compounds from the blood and synthesized necessary materials such as secretory proteins in the rough endoplasmic reticulum (**a**), the secretions are transported by the Golgi apparatus (**b**) to the apical part of the cell, where they are discharged by exocytosis (**c**).

D Principal sites where hormones and hormonelike substances are formed

Hormones are vitally important chemical messengers that enable cells to communicate with one another. Usually, very small amounts of these messengers act on metabolic processes in their target cells. Different hormones can be classified on the basis of their

- site of formation,
- site of action,
- mechanism of action, or
- chemical structure.

Examples are steroid hormones (e.g., testosterone, aldosterone), amino acid derivatives (e.g., epinephrine, norepinephrine, dopamine, serotonin), peptide hormones (e.g., insulin, glucagon), and fatty acid derivatives (e.g., prostaglandins).

Principal sites of formation	Hormones and hormonelike substances
Classic endocrine hormonal glands	
Pituitary gland (anterior and posterior lobes)	ACTH (adrenocorticotropic hormone, corticotropin) TSH (thyroid-stimulating hormone, thyrotropin) FSH (follicle-stimulating hormone, follitropin) LH (luteinizing hormone, lutropin) STH (somatotropic hormone, somatotropin) MSH (melanocyte-stimulating hormone, melanotropin) PRL (prolactin) ADH (antidiuretic hormone or vasopressin) Oxytocin (formed in the hypothalamus and secreted by the posterior pituitary)
Pineal gland	Melatonin
Thyroid gland	Thyroxine (T4) and triiodothyronine (T3)
C cells of the thyroid gland	Calcitonin
Parathyroid glands	Parathyroid hormone
Adrenal glands	Mineralocorticoids and glucocorticoids Androgens Epinephrine and norepinephrine
Pancreatic islet cells (Langerhans cells)	Insulin, glucagon, somatostatin, and pancreatic polypeptide
Ovary	Estrogens and progestins
Testis	Androgens (mainly testosterone)
Placenta	Chorionic gonadotropin, progesterone
Hormone-producing tissues and single cells	
Central and autonomic nervous system	Neuronal transmitters
Parts of the diencephalon (e.g., the hypothalamus)	Releasing and inhibitory hormones (liberins and statins)
System of gastrointestinal cells in the GI tract	Gastrin, cholecystokinin, secretin
Cardiac atria	Atrial natriuretic peptide
Kidney	Erythropoietin, renin
Liver	Angiotensinogen, somatomedins
Immune organs	Thymus hormones, cytokins, lymphokines
Tissue hormones	Eicosanoids, prostaglandins, histamine, bradykinin

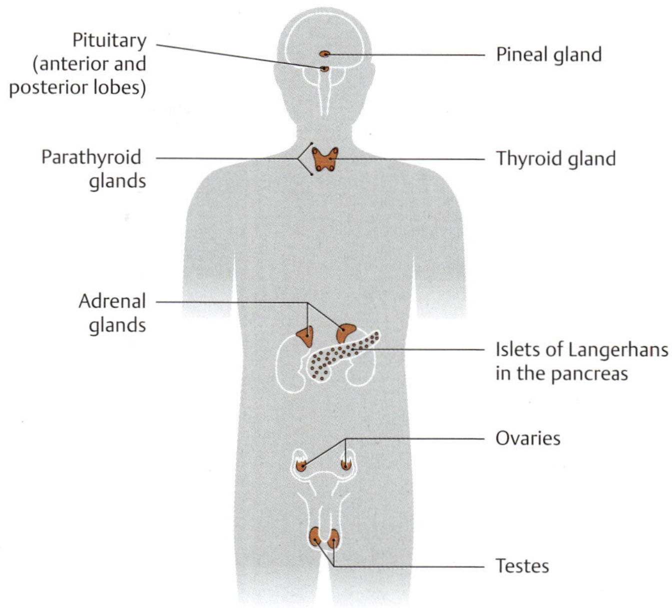

E Overview of the human endocrine glands

The diffuse or disseminated endocrine cell system (individual endocrine cells dispersed among the cells of the surface epithelium) in the gastrointestinal tract is not shown.

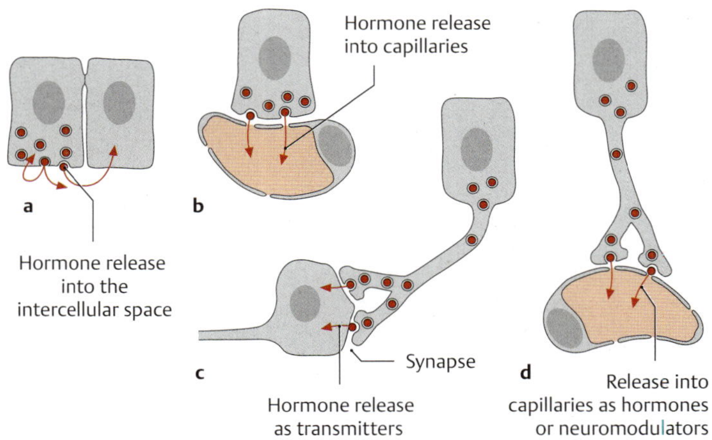

F Types of hormone-mediated information transmission

The endocrine system is closely linked to the autonomic nervous system and immune system in terms of its biological tasks. It functions as a kind of wireless communication system that coordinates the functions of target tissues and target organs, which may be located at distant sites.

a Paracrine and autocrine secretion: The hormones are not released into the bloodstream but into the intercellular space. Hence they act only in close proximity to their site of synthesis.

b Endocrine secretion: The hormones are synthesized and released into the bloodstream (fenestrated capillaries).

c Neurocrine secretion: Hormones of the neurocrine system (neurotransmitters) act in the form of synaptic transmitter substances and are concerned with local information transmission.

d Neurosecretion: Hormones or neuromodulators (neurohormones) are produced in specialized nerve cells and released to blood vessels in neurohemal regions (e.g., the pituitary). This enables them to act on distant organs.

9.1 Development of the Central Nervous System (CNS)

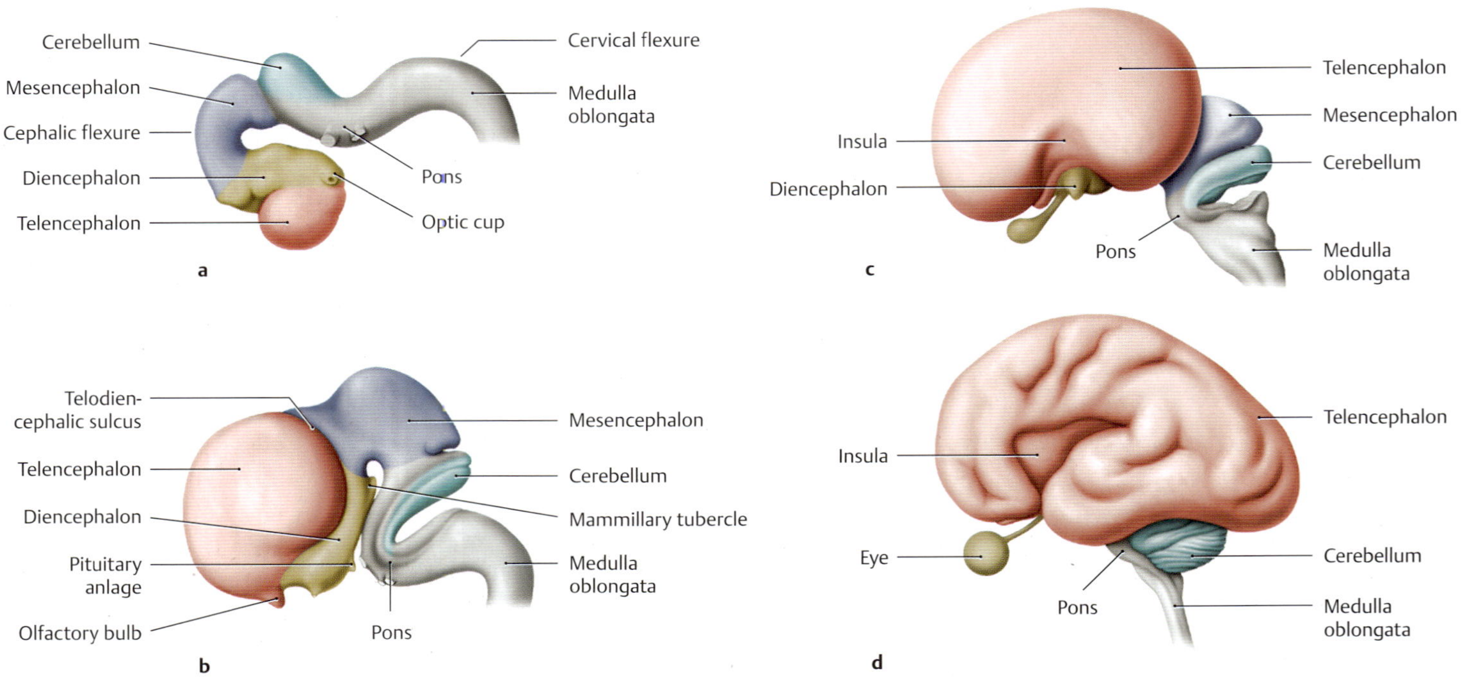

A Development of the brain

a Embryo with a greatest length (GL, see p. 4) of 10 mm, *at the beginning of the 2nd month of development*. Even at this stage we can see the differentiation of the neural tube into segments that will generate various brain regions (see **C**):

- Medulla oblongata (gray)
- Pons (gray)
- Cerebellum (light blue)
- Midbrain (mesencephalon, dark blue)
- Interbrain (diencephalon, yellow)
- Forebrain (telencephalon, red)

Note: The telencephalon grows over all the other brain structures as development proceeds.

b Embryo with a GL of 27 mm, *near the end of the 2nd month of development* (end of the embryonic period). The olfactory bulb is developing from the telencephalon, part of the pituitary anlage (*neurohypophysis*) from the diencephalon.

c Fetus with a GL of 53 mm, in approximately the *3rd month of development*. By this time the telencephalon has begun to overgrow the other brain areas. The insula is still on the surface but will subsequently be covered by the cerebral hemispheres (compare with **d**).

d Fetus with a crown–rump length (CRL, see p. 4) of 27 cm (270 mm), in approximately the *7th month of development*. The brain has begun to develop conspicuous gyri and sulci.

B Brain vesicles and their derivatives

The cranial end of the neural tube expands to form three primary brain vesicles: the forebrain (prosencephalon), the midbrain (mesencephalon), and the hindbrain (rhombencephalon). The forebrain (telencephalon) and interbrain (diencephalon) develop from the prosencephalon. The mesencephalon gives rise to the superior and inferior colliculi and related structures. The rhombencephalon differentiates into the pons, cerebellum, and medulla oblongata. The pons and cerebellum are also known collectively as the metencephalon. Some important structures of the adult brain are listed at far right to illustrate the derivatives of the brain vesicles. They can be traced back in the diagram to their developmental precursors.

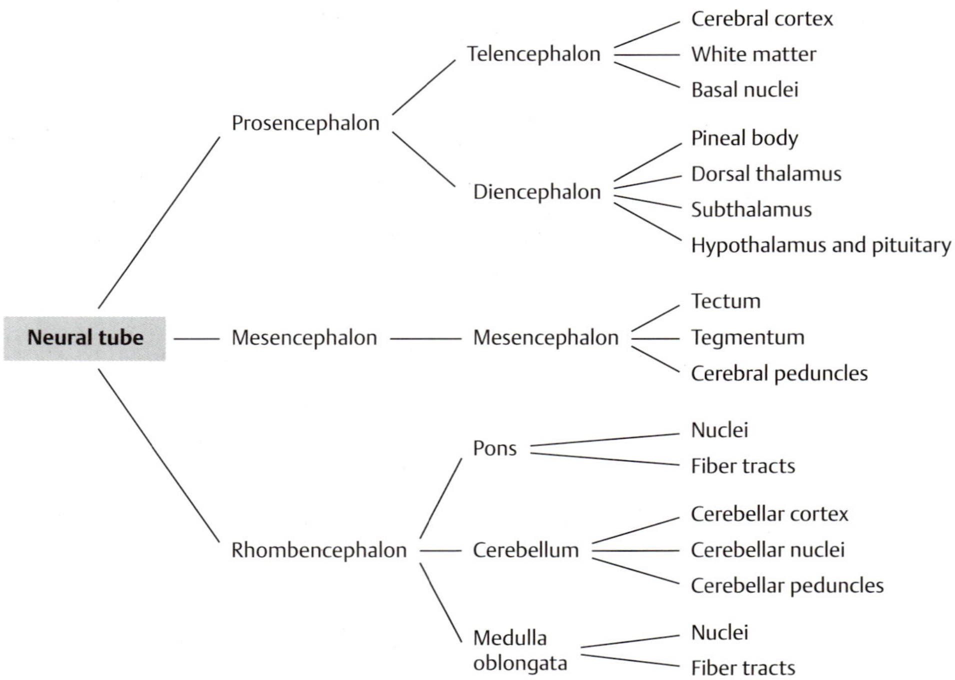

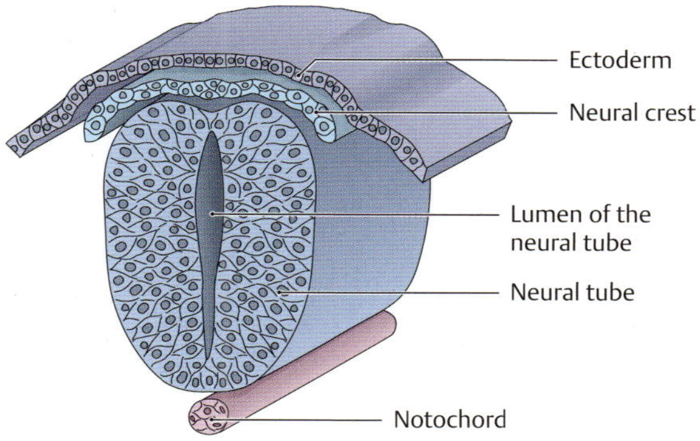

C Development of the nervous system: cross section through the neural tube, neural crest, and dorsal ectoderm

During development, the neural groove folds away from the overlying dorsal ectoderm and closes to form the *neural tube*. Cells migrate from the lateral portions of the neural groove to form the *neural crest* on each side. The *central* nervous system (brain and spinal cord) develops from the neural *tube*, while the *peripheral* nervous system develops from derivatives of the neural crest (see p. 82).

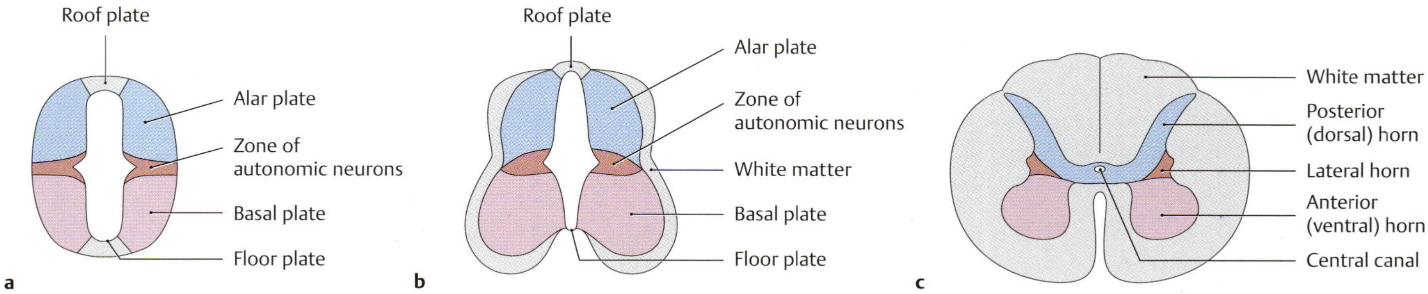

D Differentiation of the neural tube in the spinal cord region during development

Cross section, cranial view.

a Early neural tube, **b** intermediate stage, **c** adult spinal cord.

The neurons that form in the *basal plate are efferent* (motor) neurons, while those that form in the *alar plate are afferent* (sensory) neurons. The area between them—the future thoracic, lumbar, and sacral cord—is another zone that gives rise to preganglionic autonomic neurons. The roof plate and floor plate do not form neurons. A knowledge of how these neuron populations are distributed is helpful in understanding the structure of the hindbrain (rhombencephalon, see **E**).

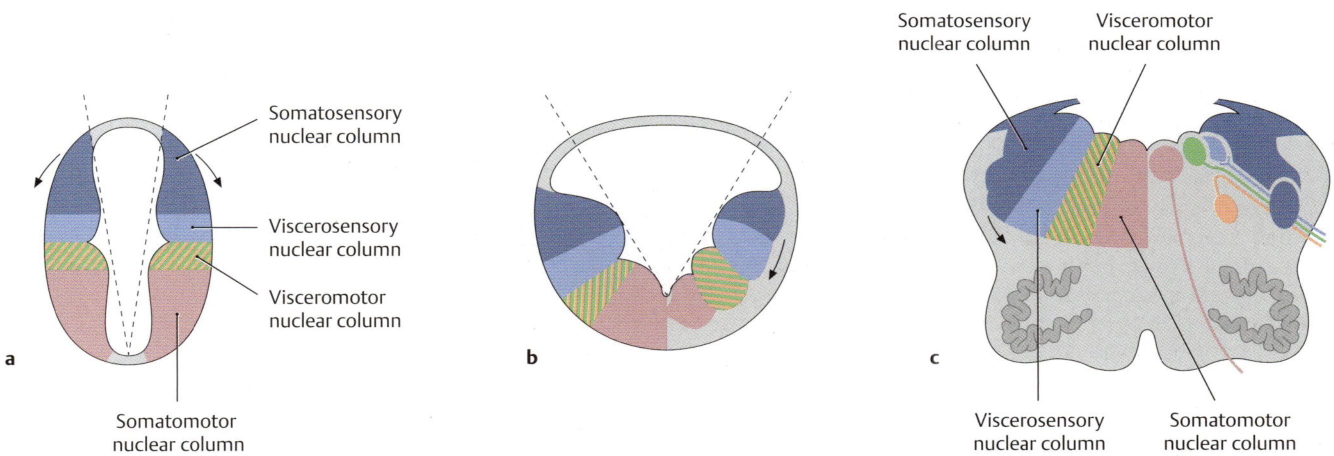

E Embryonic migratory movements of neuron populations and their effect on the location of the cranial nerve nuclei

Cross section, cranial view. (Visual aid: If we compare the spinal cord to a book, it would be closed in **a** and open in **b** and **c**.)

a In the **initial stage**, the motor neurons are ventral (anterior), and the sensory neurons are dorsal (posterior). The arrows indicate the directions of migration.

b In the **early embryonic stage**, the neurons of the alar plate migrate laterally and ventrally (anteriorly).

c In the adult brain (medulla oblongata and pons, derivatives of the rhombencephalon), we can distinguish **four nuclear columns** (after His and Herrick) that contain functionally distinct cranial nerve nuclei (from medial to lateral):

1. Somatomotor column (lilac)
2. Visceromotor column (orange and green stripes)
3. Viscerosensory column (light blue)
4. Somatosensory column (dark blue)

9.2 Neural Crest Derivatives and the Development of the Peripheral Nervous System (PNS)

A Development of the neural crest cells

At 3 weeks' development, the notochord induces surface ectoderm in the medial embryonic disc to thicken and form the neural plate (neuroectoderm). The neural plate differentiates to form the primordia of the nervous system. Neural folds are raised on each side of the neural plate, and a median groove develops between them—the neural groove. This groove subsequently deepens and closes to form the neural tube, which sinks below the ectoderm. Portions of the folds that do not contribute to neural tube formation differentiate to form the neural crest. While neural crest cells in the future head region start to migrate even before the neural tube is closed, crest migration in the trunk is delayed until tube closure. Cells destined to form the neural crest detach from the ectoderm at the fusing margins of the neural tube and undergo an *epithelio-mesenchymal* transition, diving into underlying mesoderm, where they begin a long and tortuous migration. As the neural tube differentiates into the central nervous system, migrating neural crest cells settle in different locations and develop into sensory and autonomic ganglia, endocrine glands, melanocytes, cartilage, and other structures (see **B** and **C**) (after Wolpert).

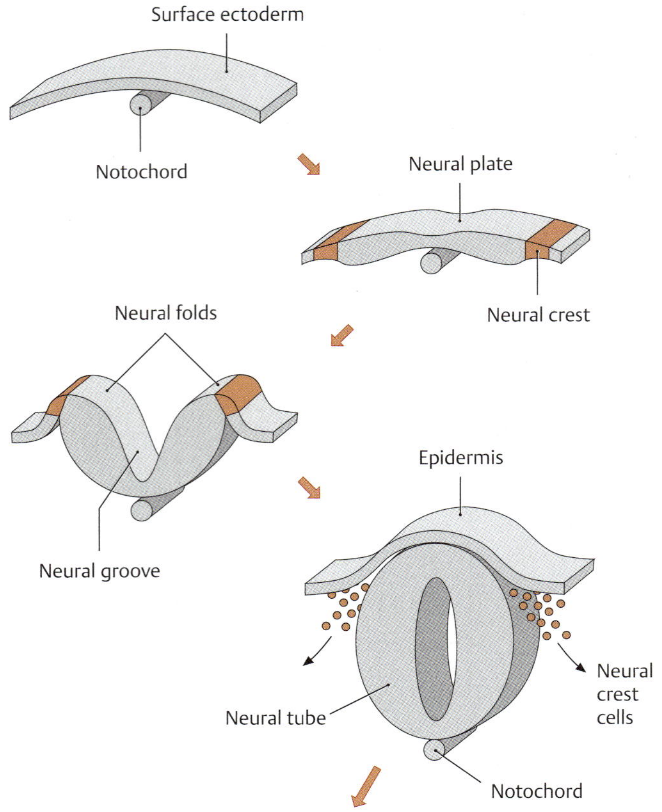

B Main migratory pathways and derivatives of the neural crest

Neural crest cells originated in different regions have different migration paths and fates. Those from cranial levels contribute to cartilage and bone in the head and neck, and to cranial parasympathetic ganglia (see **C**). Those from thoracolumbar levels do not become skeletal cells, but generate peripheral neurons, endocrine cells, melanocytes, and Schwann cells. This diagram shows the migration of the neural crest cells *in the trunk of the early embryo* (4 weeks' development, see p. 7). They follow **three main migratory pathways:**

① Dorsolateral pathway (melanoblasts, which differentiate into melanocytes)
② Ventrolateral pathway (ganglioblasts, which differentiate into sensory nerve cells in the dorsal (posterior) root [spinal] ganglia)
③ Ventral pathway (cells differentiate into neurons and associated cells of the paravertebral sympathetic ganglia, into chromaffin cells of the adrenal medulla, and into autonomic plexuses in the gastrointestinal tract)

Thus, the neural crest can develop into a variety of seemingly unrelated nonneuronal cells as well as peripheral ganglion cells. These unusual characteristics of the neural crest—its pluripotential capacity and wide-ranging migration—have consequences when its differentiation or migration is defective. Disruption of neural crest development may deprive organs of their autonomic innervation (Hirschsprung disease). Tumors derived from neural crest cells tend to be highly malignant and difficult to treat. (See **D**.)

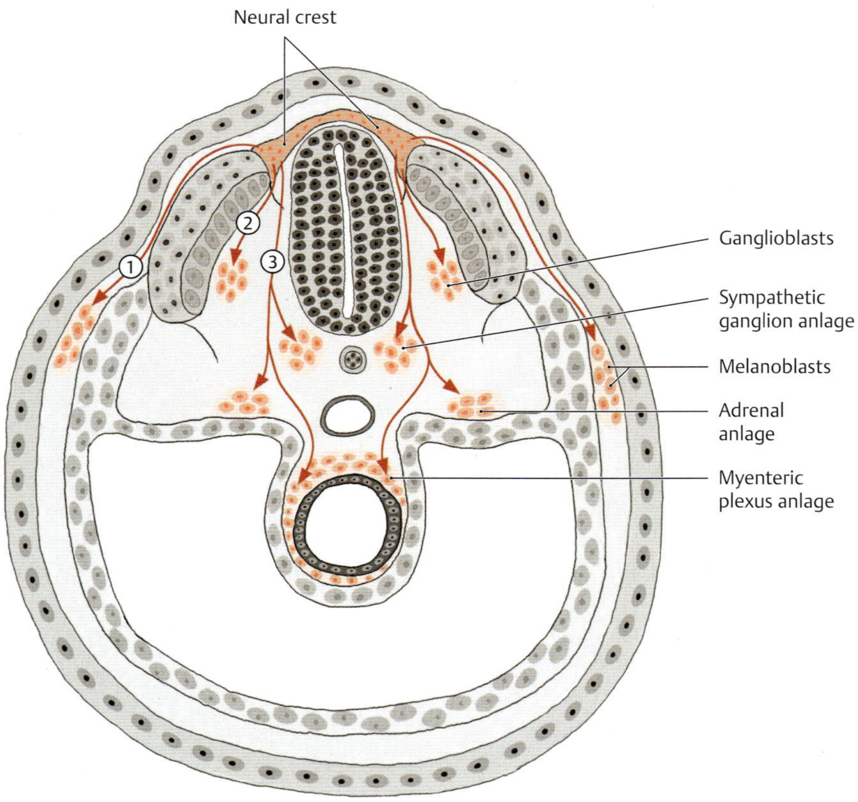

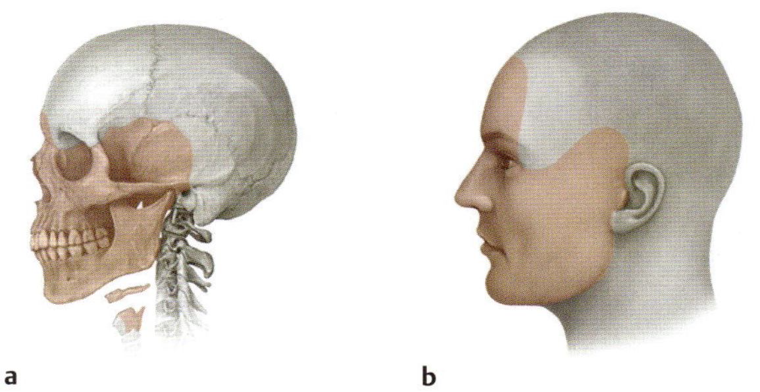

C Neural crest derivatives in the head and neck region
Besides the equivalents of the structures named in **B** (e.g., melanocytes), other structures in the head and neck region that originate from the cranial neural crest are skeletal and cartilaginous muscles and muscles of facial expression.

a Cranial neural crest derivatives in the adult skeleton: facial bones, hyoid bone, portions of the thyroid cartilage.
b Most of the facial skin is derived from the neural crest.

D Diseases of neural crest derivatives (selected examples)

Neural crest	Disease
Parasympathetic visceral ganglia	Neuroblastoma (malignant childhood tumor)
Enteric nervous system	Hirschsprung disease (aganglionic colon)
Glial cells (Schwann cells, satellite cells)	Neurofibromatosis (Recklinghausen disease)
Melanocytes	Malignant melanoma, albinism
Adrenal medulla	Pheochromocytoma (adrenal gland tumor)
Endocrine cells of the lung and heart	Carcinoids (malignant tumors with endocrine activity)
Parafollicular cells (C cells) of the thyroid gland	Medullary thyroid carcinoma

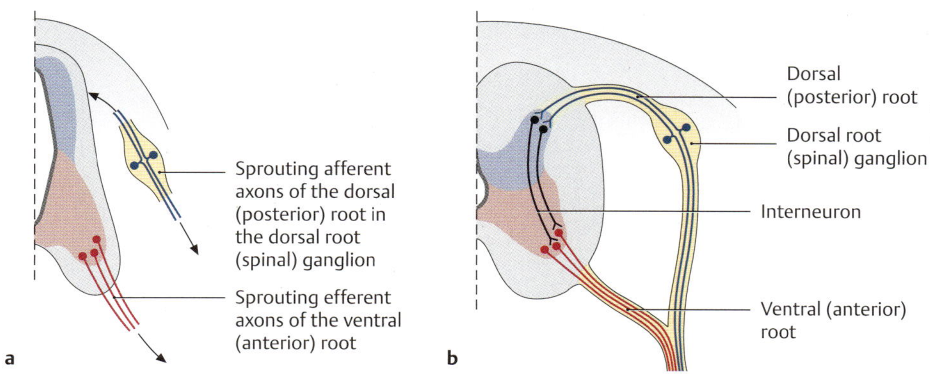

E Development of a peripheral nerve
Afferent (blue) and efferent (red) axons sprout *separately* from the neuron cell bodies during early development (**a**). Primary afferent (sensory) neurons develop in the dorsal (posterior) root (spinal) ganglia, and primary motor neurons develop from the basal plate of the spinal cord (**b**). *Interneurons* (black), which connect sensory ganglia and motor neurons, develop later.

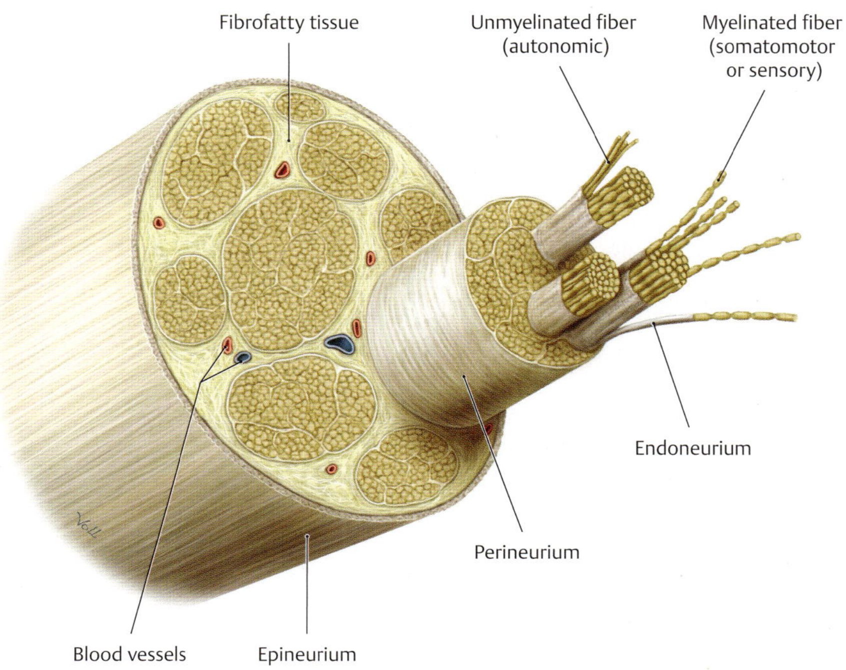

F Structure of a peripheral nerve
A peripheral nerve consists entirely of axons (also called neurites) and sheath tissue (Schwann cells, fibroblasts, blood vessels). The axons transmit information either from the periphery to the CNS (*afferents*) or in the opposite direction from the CNS to the periphery (*efferents*). Axons may be myelinated or unmyelinated. The latter have a much slower conduction velocity and are usually fibers of the autonomic nervous system (see p. 103). Among the investing layers of the nerve, the *perineurium* ensheaths the nerve fascicles and provides an important tissue barrier (see p. 101).

9.3 Topography and Structure of the Nervous System

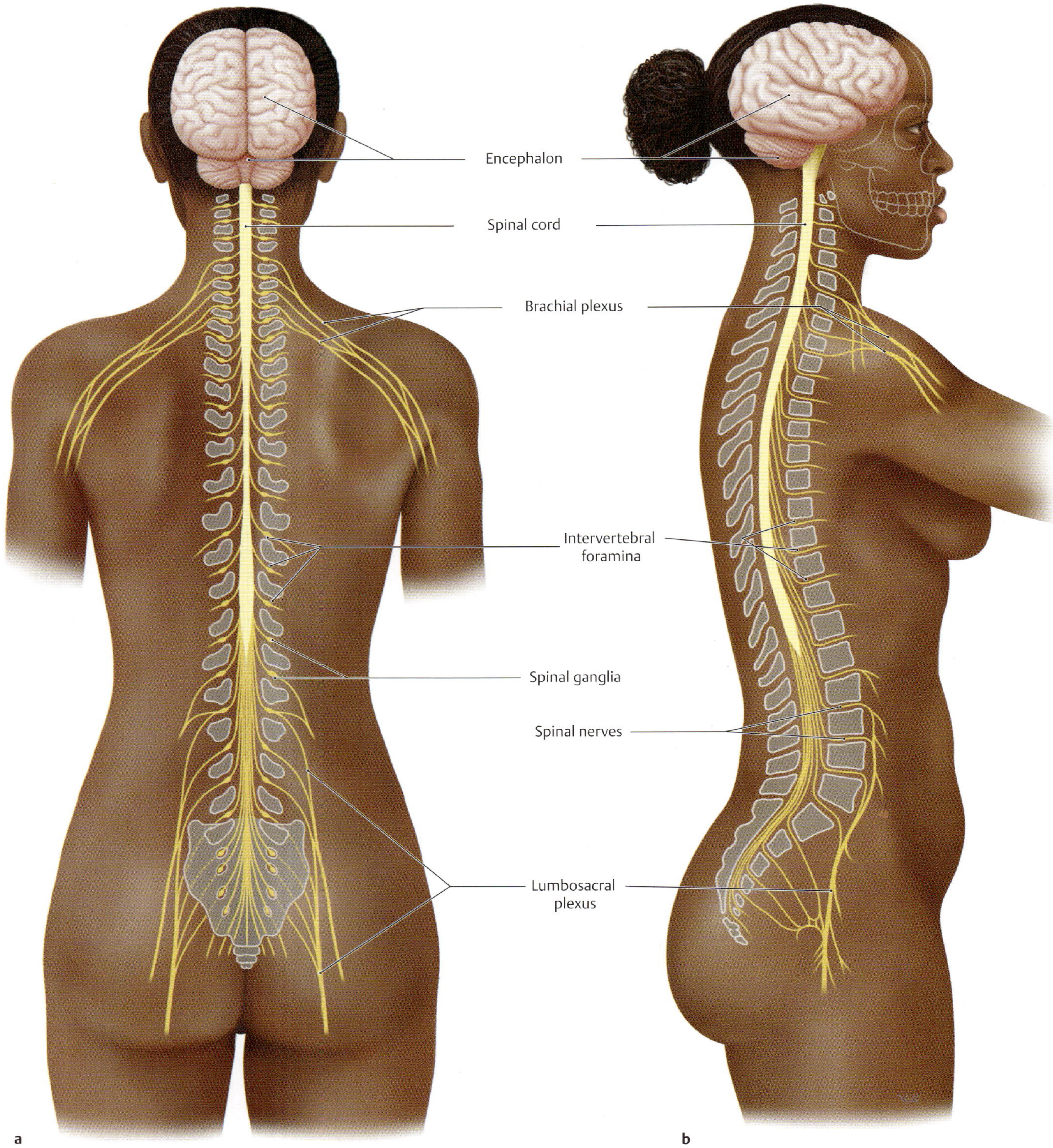

A Topography of the nervous system
a Posterior view, **b** right lateral view.
The *central nervous system* (CNS), consisting of the brain (encephalon) and spinal cord, is shown in pink. *The peripheral nervous system* (PNS), consisting of nerves and ganglia, is shown in yellow. The nerves arising from the spinal cord leave their bony canal through the *intervertebral foramina* and are distributed to their target organs. The *spinal nerves* are formed in the foramina by the union of their dorsal (posterior) roots and ventral (anterior) roots (see p. 89). The small *spinal ganglion* in the intervertebral foramen appears as a slight swelling of the dorsal root (visible only in the posterior view; its function is described on p. 89).
In the limbs, the ventral rami of the spinal nerves come together to form plexuses. These plexuses then give rise to the peripheral nerves that supply the limbs.

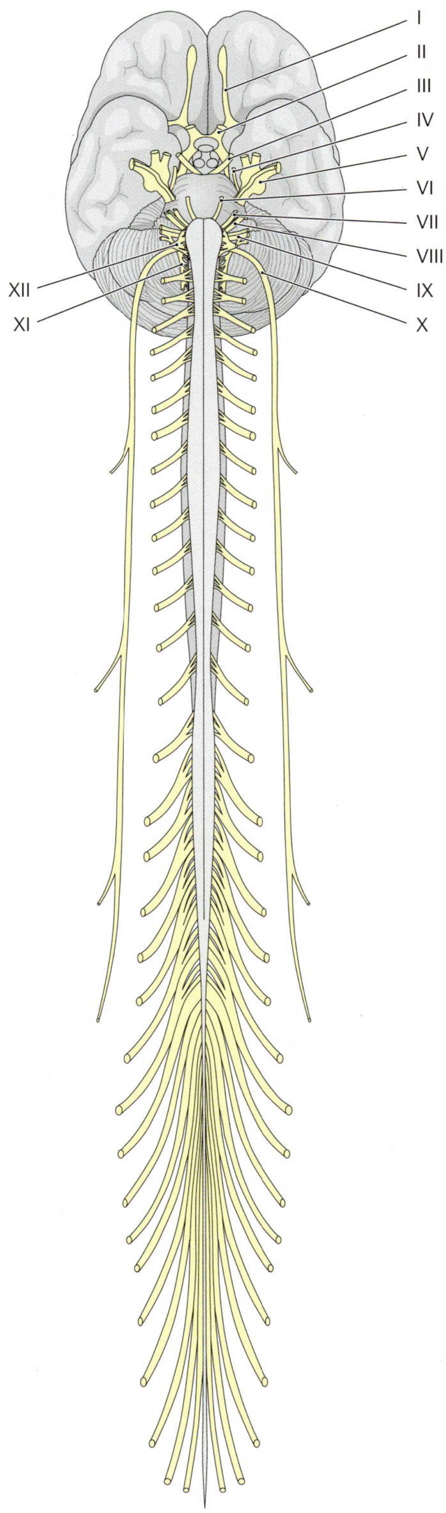

B Spinal nerves and cranial nerves
Anterior view. *Thirty-one pairs of spinal nerves* arise from the spinal cord in the peripheral nervous system, compared to *12 pairs of cranial nerves* that arise from the brain. The cranial nerve pairs are traditionally designated by roman numerals.
Note: The first two cranial nerves, the olfactory (I) and optic (II) nerves, are not peripheral nerves in the strict sense, but outpouchings from the brain, which means they are pathways of the central nervous system. They are enclosed within the meninges and contain cells that are exclusively found in the central nervous system, oligondendrocytes and microglial cells.

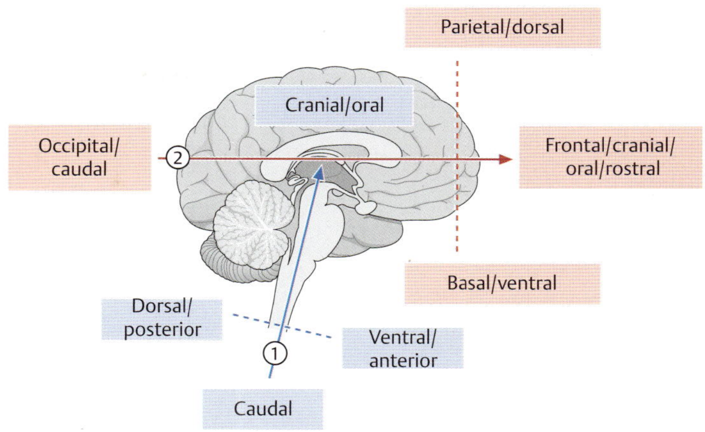

C Terms of location and direction in the CNS
Midsagittal section, right lateral view.
Note two important axes:

① The almost vertical brainstem axis (corresponds approximately to the body axis).
② The horizontal axis through the diencephalon and telencephalon.

Keep these reference axes in mind when using directional terms in the CNS.

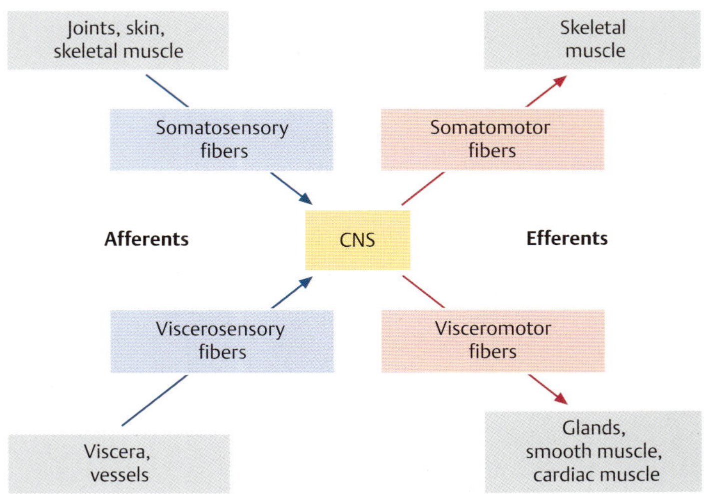

D Schematic representation of information flow in the nervous system
The information encoded in nerve fibers is transmitted either *to the CNS* (brain and spinal cord) or *from the CNS* to the periphery (PNS, including the peripheral parts of the autonomic nervous system, see p. 102). Fibers that carry information to the CNS are called afferent fibers or *afferents* for short; fibers that carry signals away from the CNS are called efferent fibers or *efferents*.

9.4 Cells of the Nervous System

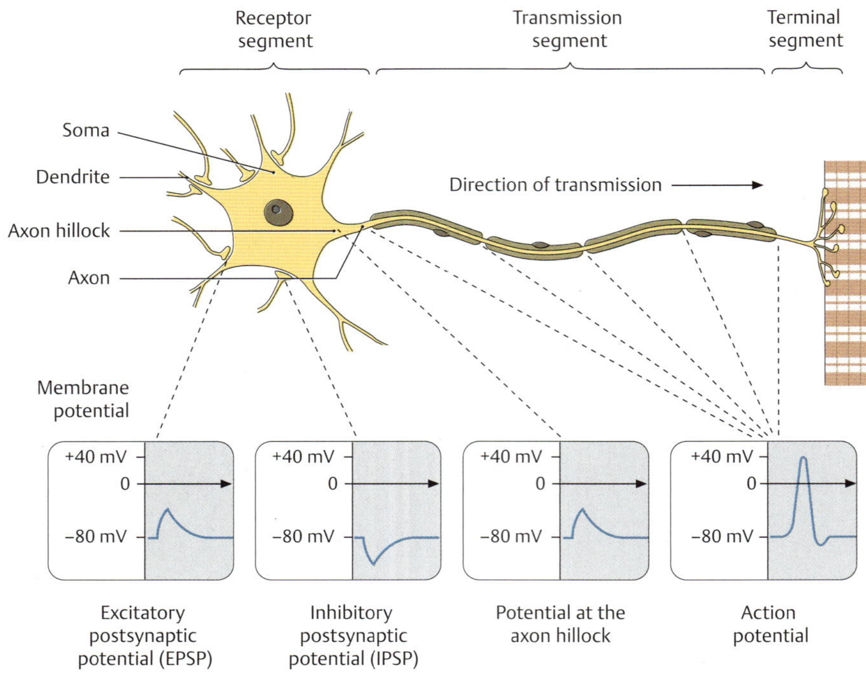

A The nerve cell (neuron)

The neuron is the smallest functional unit of the nervous system. Neurons communicate with other nerve cells through synapses. The *synapses that end at nerve cells* usually do so at dendrites (as seen here). The transmitter substance that is released at the synapses to act on the dendrite membrane may have an *excitatory* or *inhibitory* action, meaning that the transmitter either increases or decreases the local action potential at the nerve cell membrane. All of the excitatory and inhibitory potentials of a nerve cell are integrated in the axon hillock. If the excitatory potentials predominate, the stimulus exceeds the excitation threshold of the neuron, causing the axon to fire (transmit an impulse) according to the all-or-nothing rule.

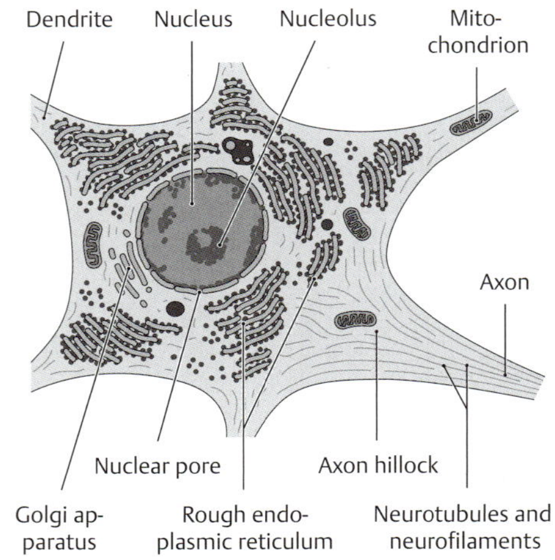

B Electron microscopy of the neuron

Neurons are rich in *rough endoplasmic reticulum* (protein synthesis, active metabolism). This endoplasmic reticulum (known also as *Nissl substance*) is easily demonstrated by light microscopy using cationic dyes, which bind to the phosphodiester backbone of the ribosomal RNAs. The distribution pattern of the Nissl substance is used in neuropathology to evaluate the functional integrity of neurons. Neurotubules and neurofilaments are referred to collectively as *neurofibrils* in light microscopy, as they are too fine to be identified as separate structures under a light microscope. Neurofibrils can be demonstrated in light microscopy by impregnating the nerve tissue with silver salts. This is of interest in neuropathology because the clumping of neurofibrils is an important histologic feature of Alzheimer disease.

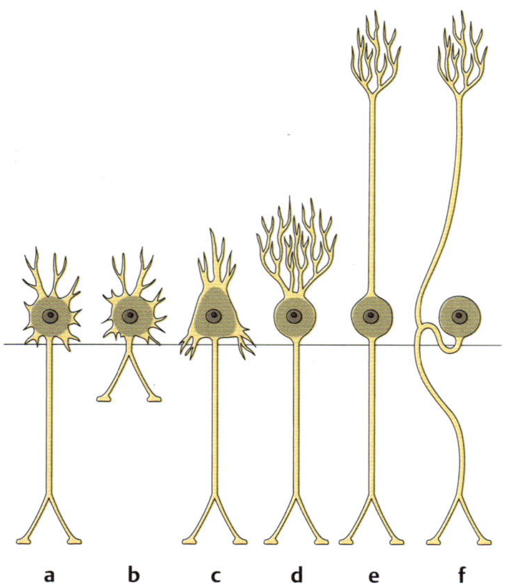

C Basic forms of the neuron and its functionally adapted variants

The horizontal line marks the region of the axon hillock, which represents the initial segment of the axon. (The structure of a peripheral nerve, consisting only of axons and sheath tissue, is shown on p. 83.)

- **a** Multipolar neuron (multiple dendrites) with a *long* axon (= long transmission path). Examples are projection neurons such as α-motoneurons in the spinal cord.
- **b** Multipolar neuron with a *short* axon (= short transmission path). Examples are interneurons like those in the gray matter of the brain and spinal cord.
- **c** Pyramidal cell: Dendrites are present only at the apex and base of the *tridentate* cell body, and the axon is long. Examples are efferent neurons of the cerebral motor cortex.
- **d** Purkinje cell: An elaborately branched dendritic tree arises from a circumscribed site on the cell body. The Purkinje cell receives many synaptic contacts from afferents to the cerebellum and is also the efferent cell of the cerebellar cortex.
- **e** Bipolar neuron: The dendrite branches in the periphery. Examples are bipolar cells of the retina.
- **f** Pseudounipolar neuron: The dendrite and axon are not separated by a cell body. An example is the primary afferent (= first sensory) neuron in the spinal ganglion (see p. 97).

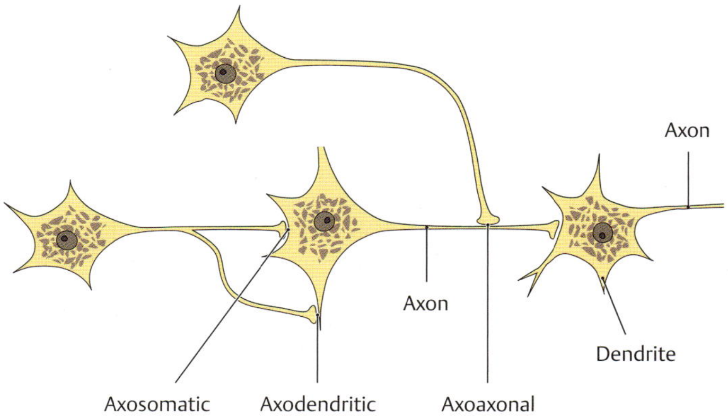

D Synaptic patterns in a small group of neurons
Axons can terminate at various sites on the target neuron and form synapses there. The synaptic patterns are described as axodendritic, axosomatic, or axoaxonal. Axodendritic synapses are the most common (see also **A**).

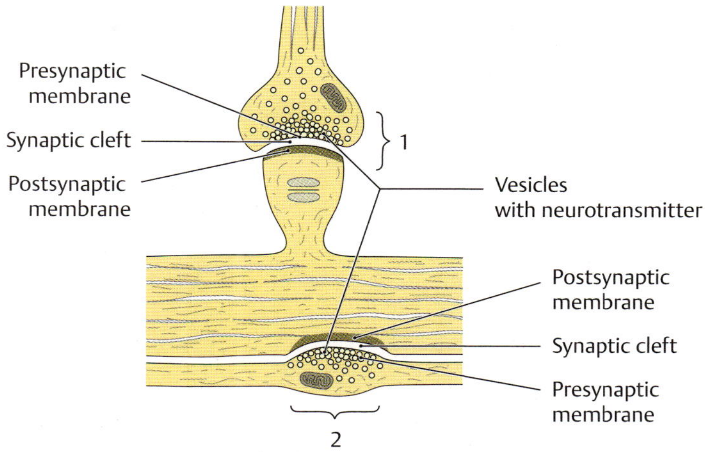

E Electron microscopy of synapses in the CNS
Synapses are the functional connection between two neurons. They consist of a presynaptic membrane, a synaptic cleft, and a postsynaptic membrane. In a *spine synapse* (1), the presynaptic knob (bouton) is in contact with a specialized protuberance (spine) of the target neuron. The side-by-side synapse of an axon with the flat surface of a target neuron is called a parallel contact or *bouton en passage* (2). The vesicles in the presynaptic expansion contain the neurotransmitters that are released into the synaptic cleft by exocytosis when the axon fires. From there the neurotransmitters diffuse to the postsynaptic membrane, where their receptors are located. A variety of drugs and toxins act upon synaptic transmission (antidepressants, muscle relaxants, toxic gases, botulinum toxin).

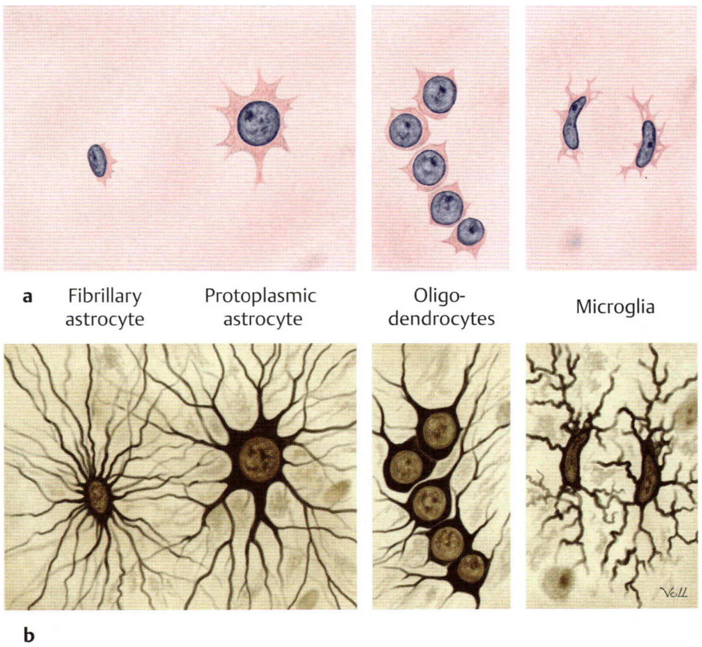

F Cells of the neuroglia in the CNS
Neuroglial cells surround the neurons, providing them with structural and functional support (see **G**). Various staining methods are available in light microscopy for selectively demonstrating different portions of the neuroglial cells:

a Cell nuclei demonstrated with a basic stain.
b Cell body demonstrated by silver impregnation.

G Summary: cells of the CNS and PNS and their functional importance

Type of cell	Function
Neurons (CNS and PNS)	1. Impulse formation 2. Impulse conduction 3. Information processing
Glial cells	
Astrocytes (CNS only)	1. Maintain a constant internal milieu in the CNS 2. Contribute to the structure of the blood–brain barrier (see p. 101) 3. Phagocytize dead synapses 4. Form scar tissue in the CNS (e.g., in multiple sclerosis or following a stroke)
Microglial cells (CNS only)	Phagocytosis ("macrophages of the brain")
Oligodendrocytes (CNS only)	Myelin sheath formation in the CNS
Schwann cells (PNS only)	Myelin sheath formation in the PNS
Satellite cells (PNS only)	Modified Schwann cells; surround the cell body of neurons in PNS ganglia

9.5 Structure of a Spinal Cord Segment

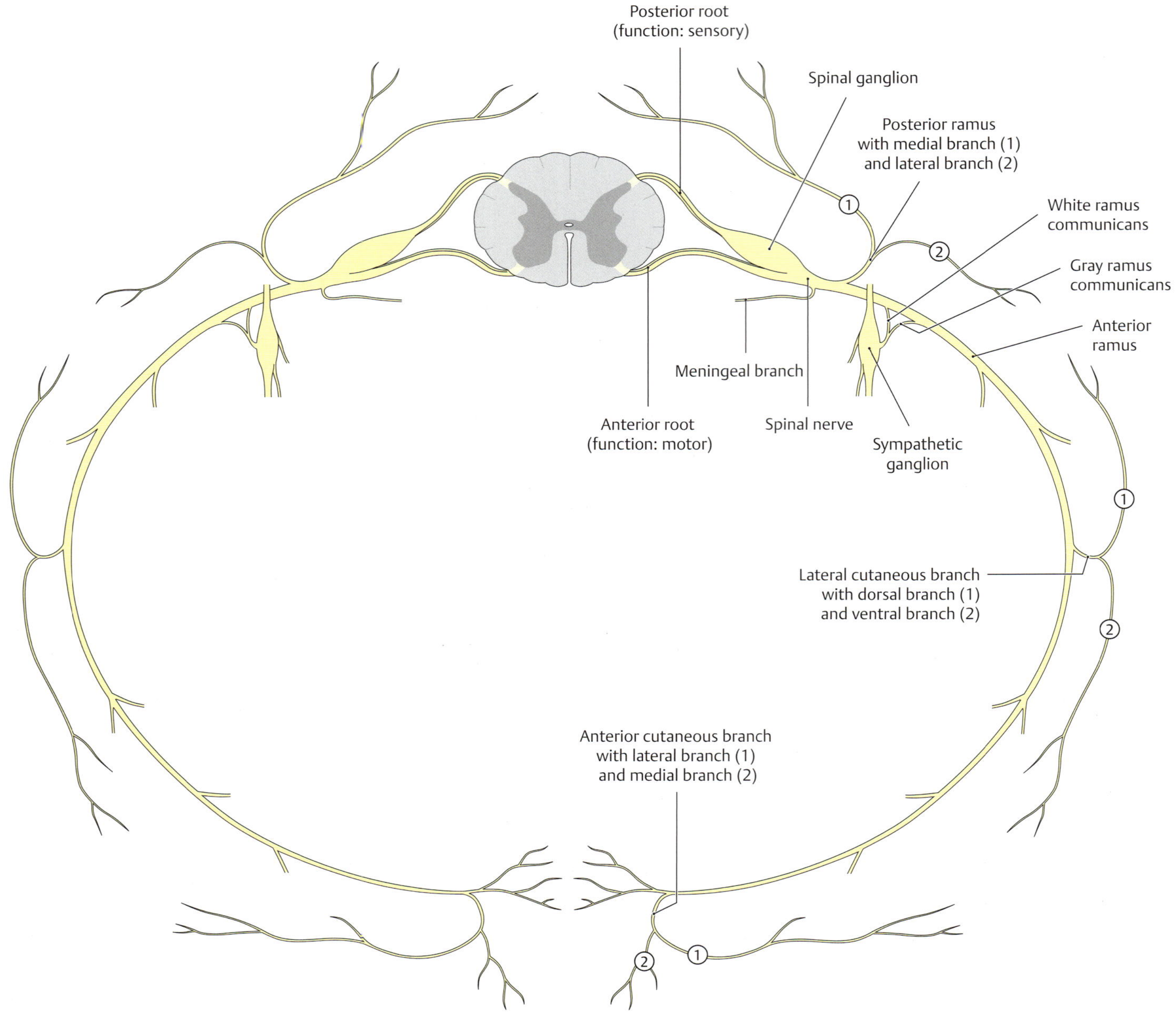

A Structure of a spinal cord segment with its spinal nerve

Superior view. The spinal cord is made up of 31 consecutive segments arranged one above the other (see **B**). An *anterior root* and a *posterior root* emerge from the sides of each segment. The anterior root consists of *efferent* (motor) fibers, while the *posterior* root consists of *afferent* (sensory) fibers. Both roots from one segment unite in the intervertebral foramen to form the *spinal nerve*. The afferent (sensory) fibers and efferent (motor) fibers intermingle at this junction, so that the branches into which the spinal nerve divides (see below) contain motor and sensory elements (except for the meningeal branch, which is purely sensory). This division into branches (rami) occurs shortly after the anterior and posterior roots unite to form the spinal nerve. Consequently, the spinal nerve itself is only about 1 cm long.

Note: This segment of the spinal nerve, the spinal nerve in the strict sense, is referred to as the trunk of the spinal nerve.

The principal branches of the spinal nerve have the following functions:

- The anterior ramus innervates the anterior and lateral body wall and the limbs.
- The posterior ramus innervates the skin of the back and the intrinsic back muscles.
- The meningeal branch reenters the spinal canal, providing sensory innervation to the spinal membranes and other structures.
- The white ramus communicans carries white (= myelinated) fibers to the ganglion of the sympathetic trunk.
- The gray ramus communicans carries gray (= unmyelinated) fibers *from* the sympathetic ganglion back to the spinal nerve (the functional significance of this is described on p. 103). The anterior and posterior rami of the spinal nerves subdivide into further branches.

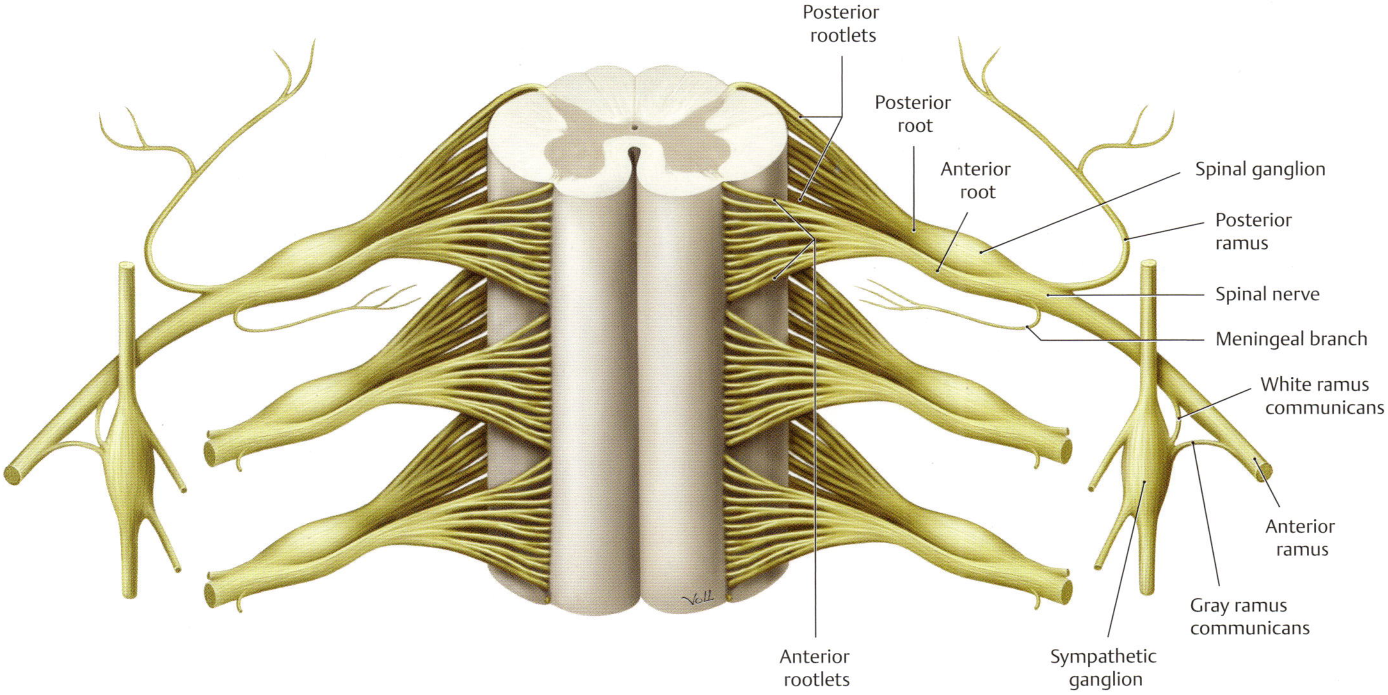

B Spinal cord segments
Anterior view. The spinal cord is composed of many segments, each of which gives rise to one pair of spinal nerves (for clarity, only the upper segment of the spinal cord is shown). Several rootlets unite to form the anterior and the posterior roots of a spinal nerve, which then divides into five branches (see **A**). The segmental structure of the spinal cord is visible externally only at the rootlets; the spinal cord itself does not show any external segmentation.

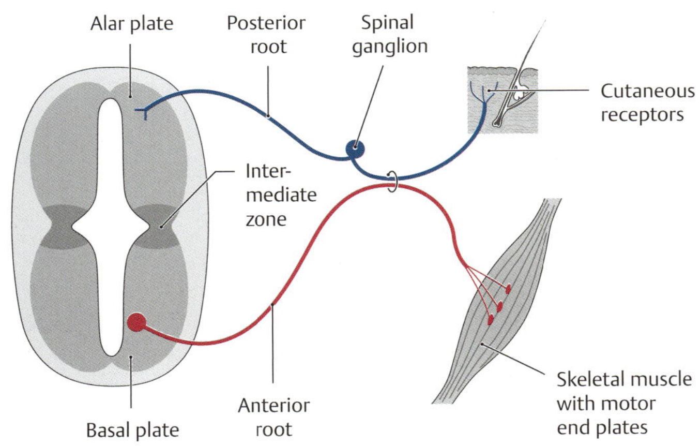

C Embryological origins of the topographical and functional anatomy of a spinal cord segment
The afferent fibers (e.g., from cutaneous receptors) pass through the *posterior root* into the posterior horn of the spinal cord, which is derived embryologically from the alar plate. The efferent fibers arise from neurons located in the anterior horn of the spinal cord, which is a derivative of the basal plate. They leave the spinal cord by the *anterior root* and are distributed to their target organ, such as a skeletal muscle. The intermediate zone develops into the lateral horn containing autonomic neurons that control organs.

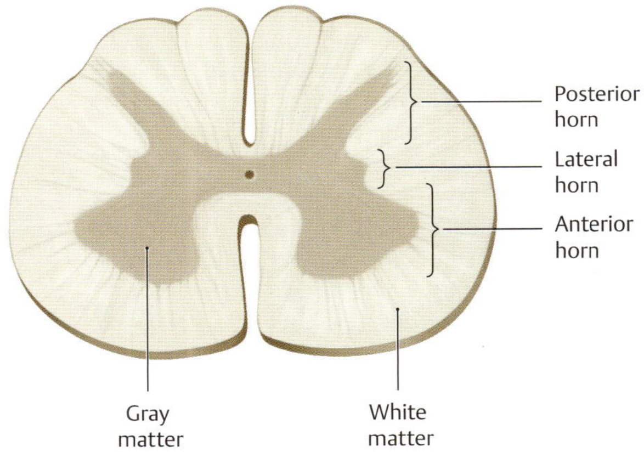

D Cross section of a spinal cord segment
Superior view. A cross section of the spinal cord shows the centrally located gray matter arranged in a butterfly-shaped pattern and the white matter surrounding it. Cell bodies of neurons are located in the gray matter; axons move up and down in the white matter. The gray matter is divided into anterior, posterior, and lateral horns (the latter not in all segments). The cell bodies of efferent neurons are located in the anterior horns, those of afferent neurons in the posterior horns, and the autonomic neurons in the lateral horn (controlling the organs). Thus, the spinal cord is functionally arranged horizontally from anterior (motor) to posterior (sensory). For the relationship between spinal cord segments and dermatomes (vertical arrangement of the spinal cord), see p. 90.

9.6 Sensory Innervation: An Overview

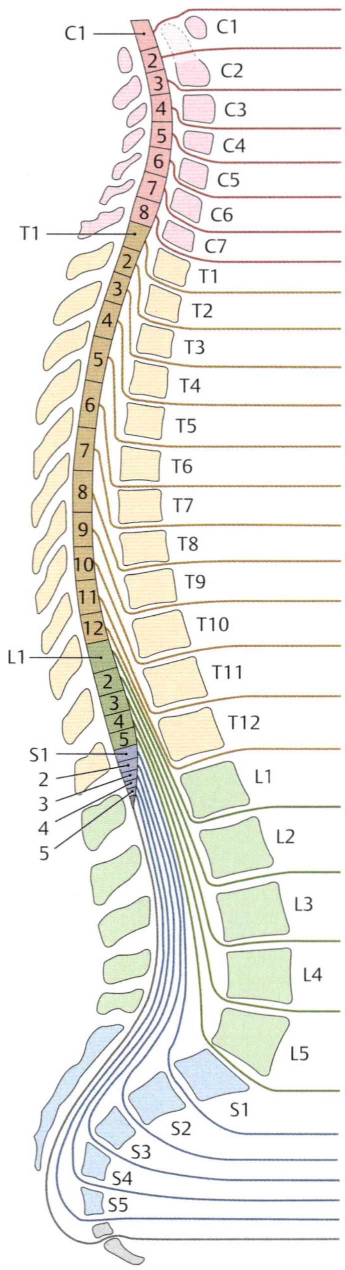

A Location and designation of spinal cord segments in relation to the vertebral canal

Right lateral view. The spinal cord consists of 31 segments that are arranged from top to bottom

- 8 cervical segments,
- 12 thoracic segments,
- 5 lumbar segments,
- 5 sacral segments, and
- 1 coccygeal segment (no clinical significance)

The growth of the spinal cord lags behind that of the spinal column, so in an adult it extends only to the first lumbar vertebra, as shown here (compare p. 136). The spinal cord segments are equivalent to the dermatomes on the body surface (see **C**). In other words, a dermatome is an area on the body surface, the sensory receptors of which (e.g., for pressure, temperature, pain, vibration) are drawn into the periphery, thus enabling a 1:1 relationship between body surface and spinal cord segment. In clinical terms, it means that in the case of sensory disorder (dermatome malfunction), it is possible to determine at which level of the spinal cord the malfunction is located. A myotome (see p. 7) is the equivalent of a dermatome, as regards motor function.

Note: All the pairs of cervical nerves exit the vertebral canal superior to the respective cervical vertebra, with the exception of the eighth pair of cervical nerves (C8). This is because there are seven cervical vertebrae (C1–C7) but eight pairs of cervical nerves (C1–C8). As a result, the eighth pair of cervical nerves must exit superior to the first thoracic vertebra (T1). This in turn means that the first thoracic nerve pair (T1) must exit inferior to the respective thoracic vertebra. The remaining pairs of spinal nerves also exit inferior to the respective vertebral body.

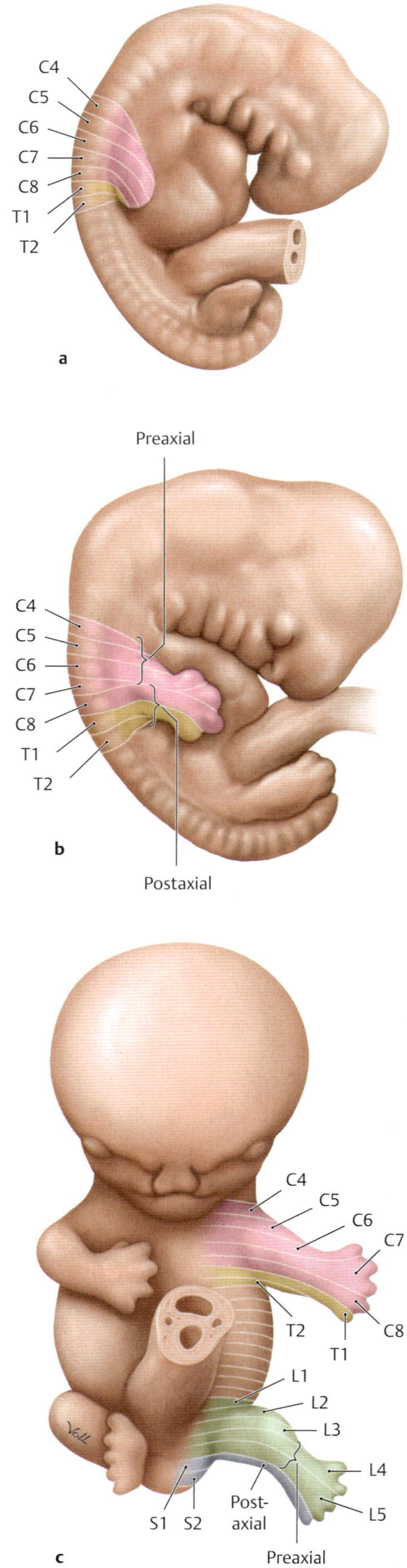

B Dermatomes and limb buds

From the somites, which give rise to limbs, cells migrate, and limb buds form. These migrating cells carry their original segmental innervation distally as the limbs develop. Thus, dermatomes reflect the segmental arrangement, which is based on the somites (see p. 7).

a In a 5-week-old embryo, the dermatomes are still segmentally arranged.

b In a 6-week-old embryo, migrations have already taken place: the preaxial segments are positioned cranially, the postaxial segments caudally. The segments lying in between have shifted in a distal direction toward the hand.

c In a 7-week-old embryo, the proportions are similar to those postbirth.

This dermatome diagram, similar to the diagram in **C**, is simplified, because the cells, which form the dermatomes, migrate (for details, see p. 14).

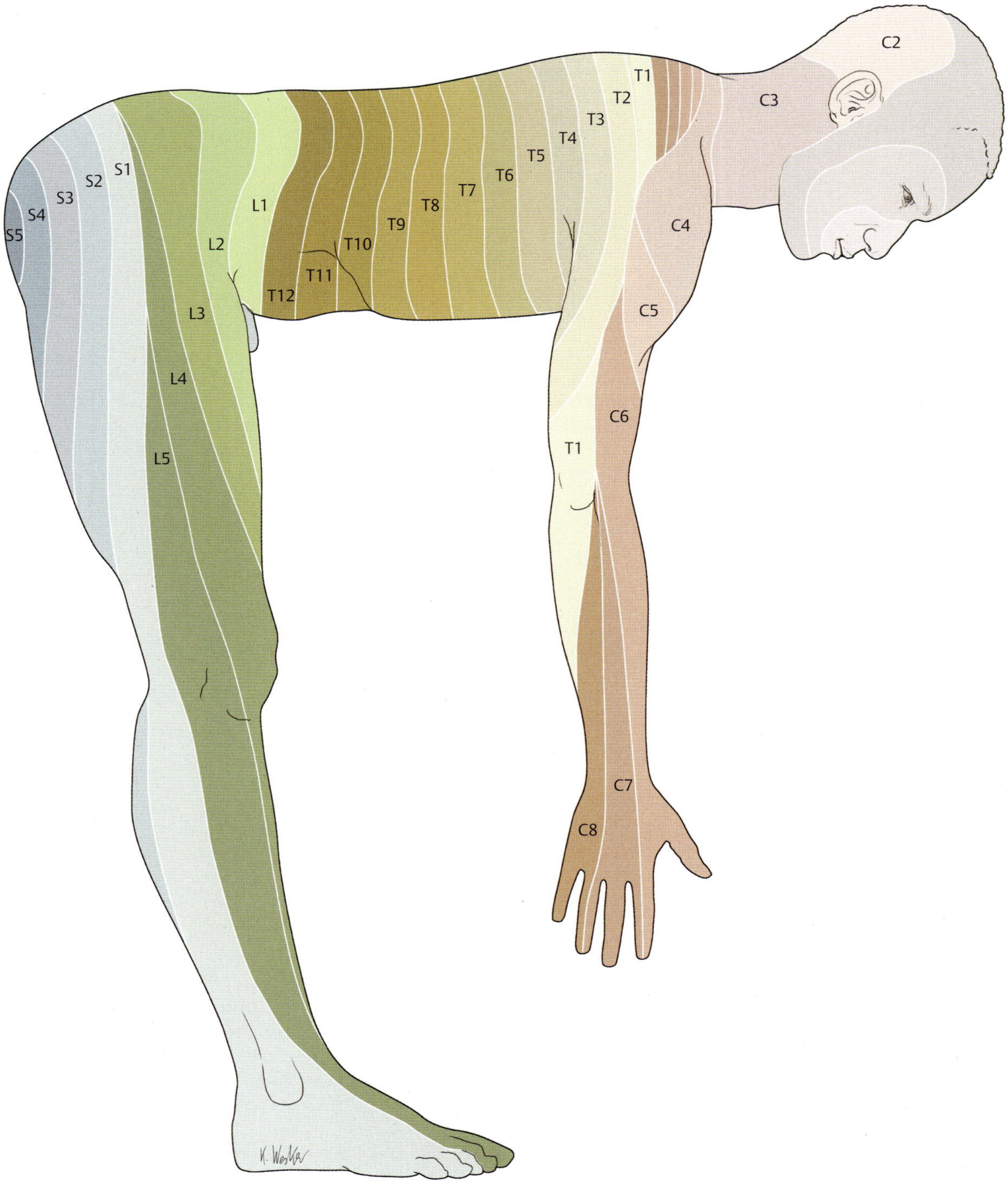

C Simplified scheme for learning the dermatomes

The distribution of dermatomes on the human limbs results from the sprouting of the limb buds that occurs during embryonic development. When the limbs are positioned at right angles to the body as in a quadruped, it is easier to appreciate the pattern of innervation. Compare the location of the dermatomes to those of the spinal cord sections in **A**. The more complex pattern of dermatomes resulting from cell migration is shown on p. 94.

Note: The C1 segment contains only motor fibers; thus, there is not a C1 dermatome.

9.7 Sensory Innervation: Principles of Dermatome and Plexus Formation

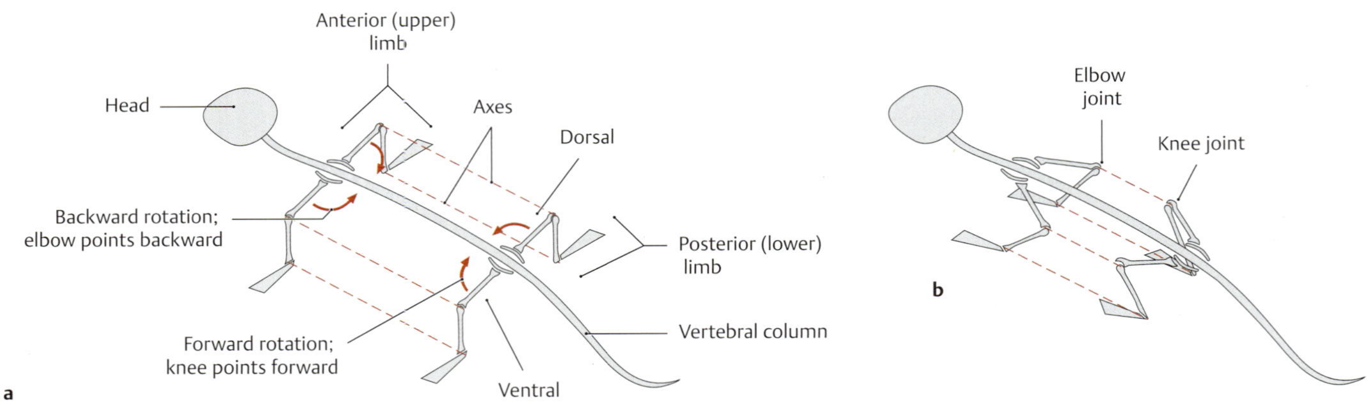

A Phylogenetic development of dermatomes
In lower tetrapods, the trunk is located between the limbs (**a**). In mammals, the limbs are also rotated (**b**). The posterior (= lower) limbs are rotated forward, and the original posterior position of the limb muscles shifts anteriorly when picturing the mammal walking erect (for details, see p. 20). This is not the case for the anterior (= upper) limbs, which are rotated backward. That is why the cells that form dermatomes in the lower limb region migrate more than similar cells in the upper limb region. That leads to a screwlike distribution of dermatomes in the lower limbs, unlike that in the upper limbs (see p. 94).

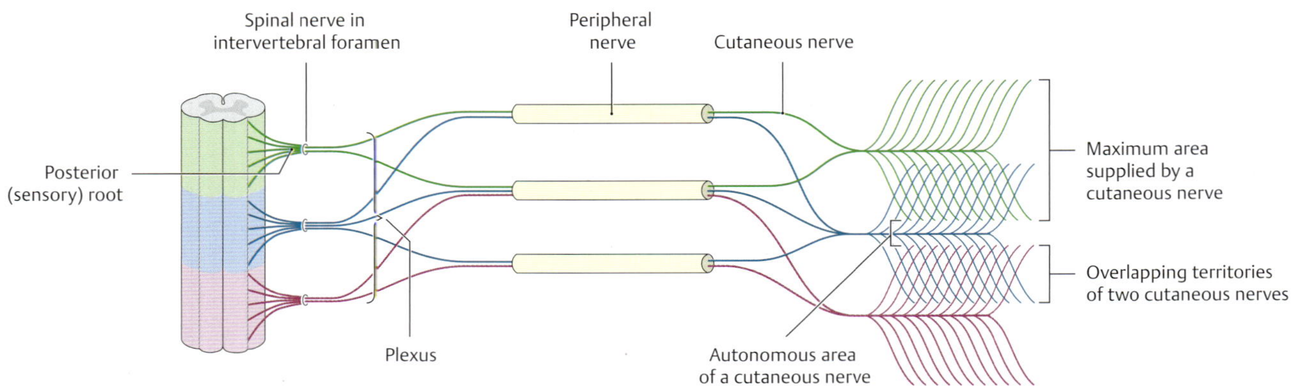

B Course of sensory fibers from the posterior root to the dermatome
Sensory fibers pass from the posterior root to the intervertebral foramen, where they unite with motor fibers to form the spinal nerve. The sensory fibers are then distributed to the anterior and posterior rami of the spinal nerve. The simple segmental arrangement of the sensory territories that occurs on the trunk is not found in the limbs (see **D**). This results from the migratory movements of various muscular and cutaneous primordia, or precursors, during the development of the limbs. Because these primordia carry their segmental innervation with them, the sensory fibers from different segments become intermingled in the limbs (plexus formation, see **C**). After fibers have been distributed in the plexus, the fibers in peripheral nerves travel to their destinations, their terminal portions often consisting entirely of sensory cutaneous nerves. The area of skin that is innervated by one spinal cord segment is called a *dermatome*. The dermatomes of adjacent spinal cord segments are often located so close together that their territories broadly overlap. This explains why the clinically detectable area of sensory loss caused by a segmental nerve lesion may be considerably smaller than the dermatome itself. The area that receives all of its sensory supply from one cutaneous nerve is called the autonomous area of that nerve. When a segmental nerve lesion occurs, the cutaneous nerves from the two adjacent spinal cord segments are still available to supply at least the peripheral part of the affected skin area.
This helps in understanding the difference between radicular (= segmental) and peripheral sensory innervation. When a nerve root becomes damaged (e.g., due to a herniated intervertebral disk), the resulting sensory loss will follow a radicular innervation pattern (see p. 94). But when a peripheral nerve is damaged (e.g., due to a limb injury), the sensory loss will conform to a peripheral innervation pattern (see p. 95).

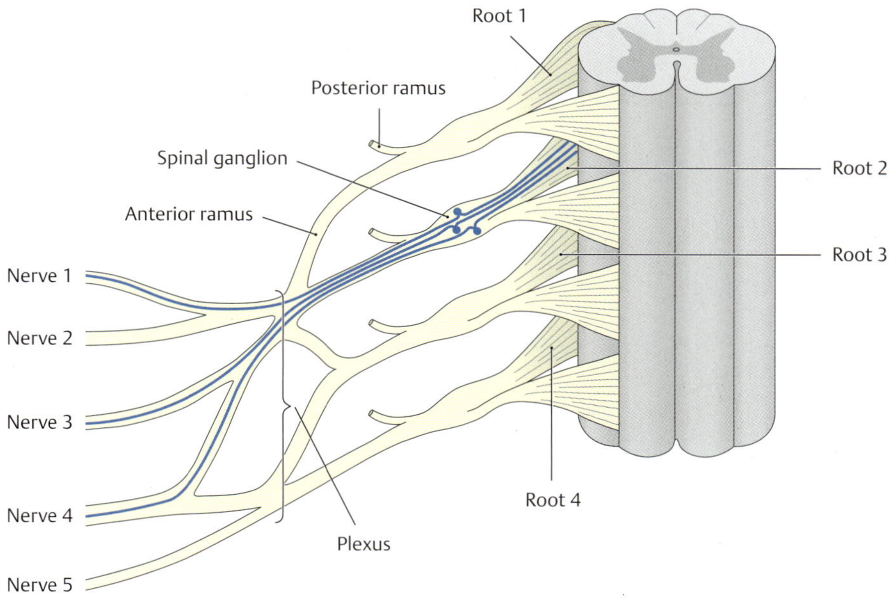

C Principles of plexus formation
The axons, which form the afferents from a dermatome, extend from that dermatome via several peripheral nerves to a single root in the spinal cord. As part of embryonic plexus formation, afferent axons of that one root spread via several peripheral nerves to the dermatome, in response to signals from the primordia of the dermatome.

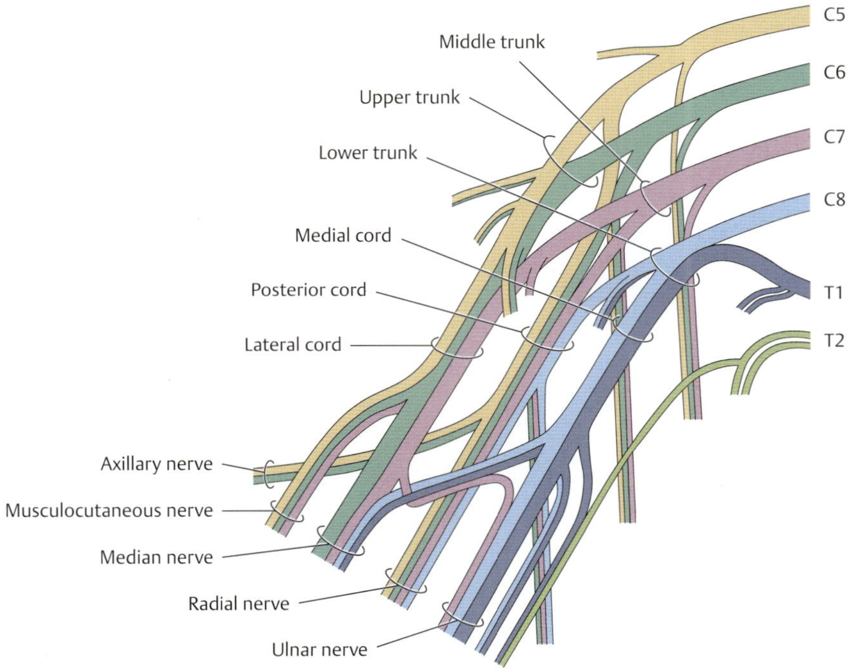

D Plexus formation as illustrated by the brachial plexus
During embryonic development, migration and, associated with it, mixing of dermatome and myotome primordia occur with the primordia dragging their innervation along behind them. Dermatomes and myotomes have an identical developmental pattern. In the following, we will discuss only dermatomes (for more on myotomes, see p. 96). During their development, the primordia of dermatomes transmit signals to the budding sensory axons (see p. 83). This leads to every dermatome receiving its axons from its spinal cord segment. (To exhibit features more clearly, every spinal cord segment is individually colored). In order for the axons to leave their spinal cord segment and reach "their" dermatome, they have to be paired with the various peripheral nerves. The place where the shifting and mixing of axons (= fibers) occurs is called a plexus. With the brachial plexus, a distinction is made between different parts of the plexus. Prior to plexus formation, axons from one spinal cord segment combine to form roots. The axons from the roots of C5 and C6 merge to form the upper trunk, the axons from the root of C7 form the middle trunk, and those of C8 and T1 form the lower trunk. The anterior branches of the upper and middle trunks form the lateral and medial cords, and the posterior branches of all three trunks form the posterior cord. Finally, from the cords arise the large nerves of the arm and shoulder, where the axons travel on their way to the dermatomes.

9.8 Sensory Innervation: Dermatomes and Cutaneous Nerve Territories

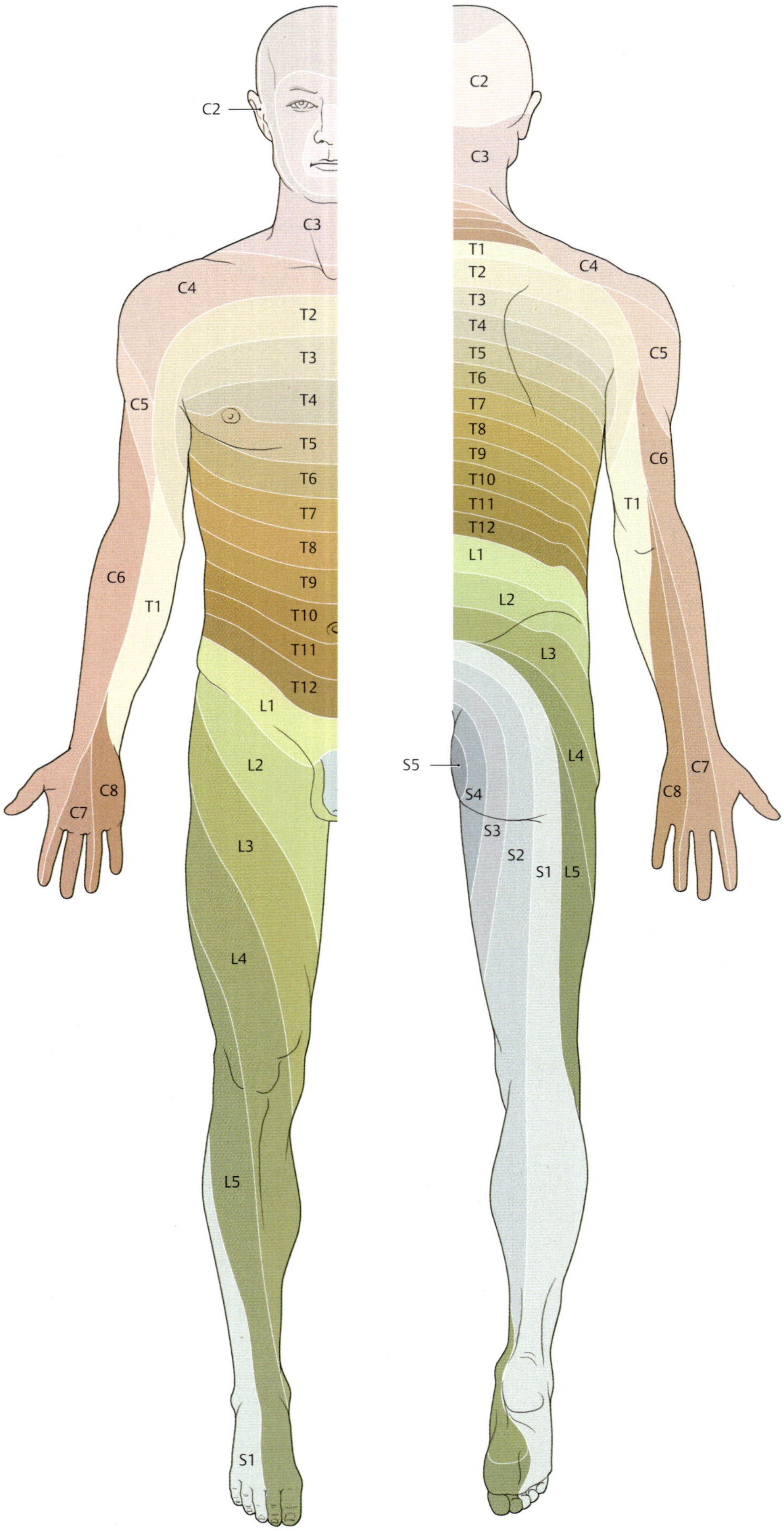

A Pattern of radicular (segmental) sensory innervation (dermatomes)

The skin area supplied by a posterior spinal nerve root is called a dermatome. Because the C 1 segment consists entirely of motor fibers, it lacks a corresponding sensory field. A knowledge of radicular innervation is very important clinically. For example, when a herniated intervertebral disk is impinging on a sensory root, it will cause sensory losses in the affected dermatome. The area of sensory loss can then be used to locate the level of the lesion: Which intervertebral disk is affected? In a patient with shingles (herpes zoster inflammation of a spinal ganglion), the dermatome supplied by that ganglion will be affected.

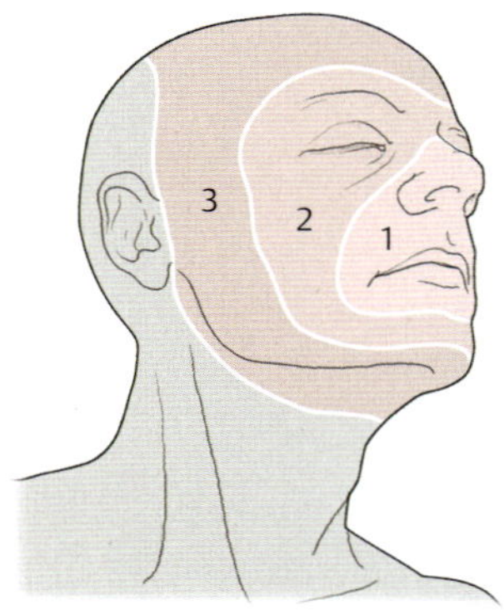

B Pattern of nuclear sensory innervation in the head region

The head receives its sensory innervation from the trigeminal nerve (cranial nerve V). A lesion of the sensory nucleus of the trigeminal nerve within the brain (= central lesion) causes an onion-skin pattern of sensory alteration along the concentric *Sölder lines* that encircle the mouth and nostrils. The pattern of these lines corresponds to the distribution of the neurons in the sensory nucleus of the trigeminal nerve (somatotopic organization, i. e., certain groups of neurons in the CNS are linked to certain territories in the periphery). Territory 1 is supplied by the cranial nuclear column, territory 2 by the middle column, and territory 3 by the caudal column. This pattern of sensory loss is like that associated with radicular neuropathy affecting a peripheral nerve.

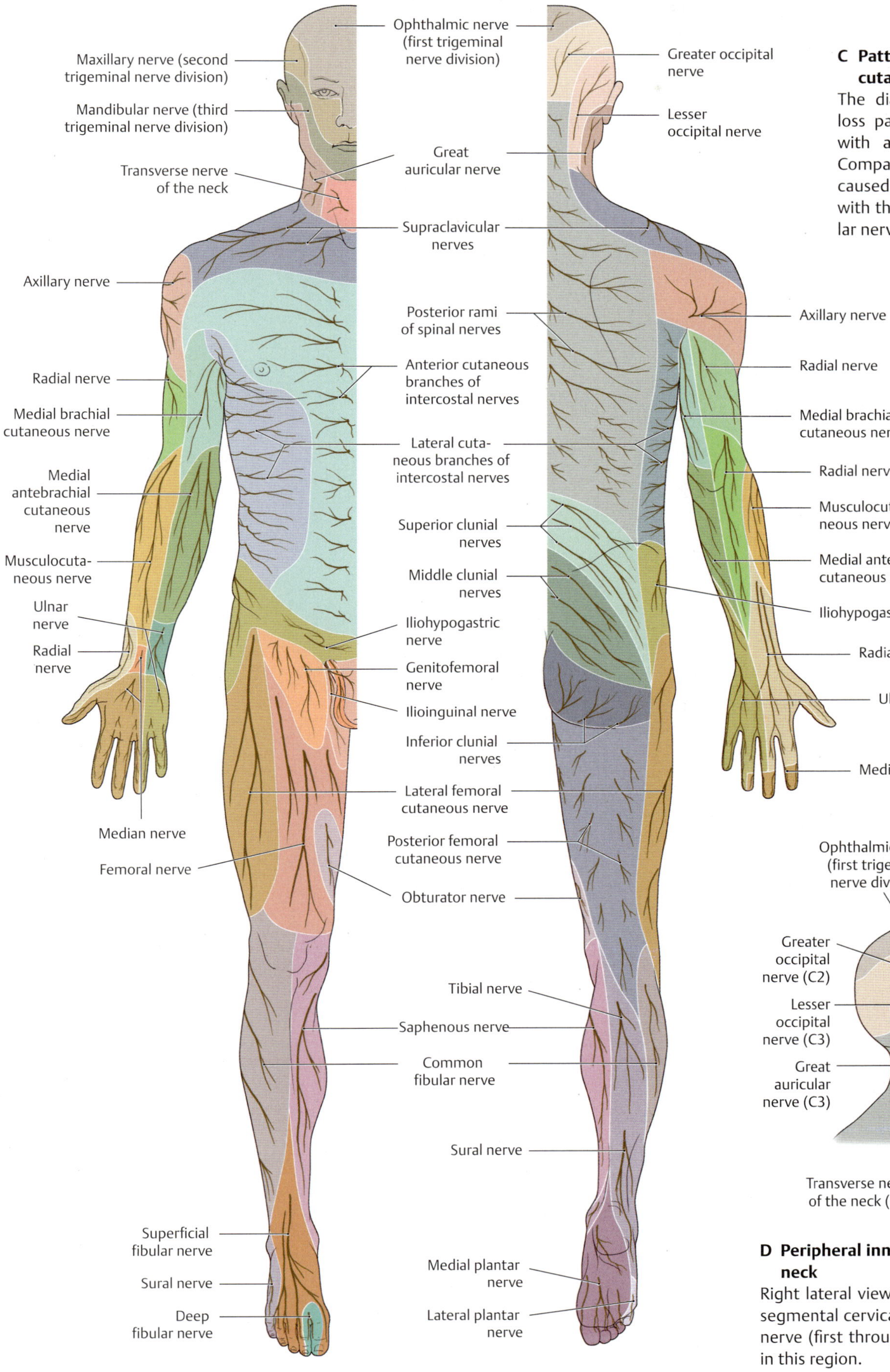

C Pattern of peripheral sensory cutaneous innervation

The diagram shows the sensory loss patterns that are associated with a peripheral nerve lesion. Compare the area of sensory loss caused by a peripheral nerve lesion with the area affected by a radicular nerve lesion in **A**.

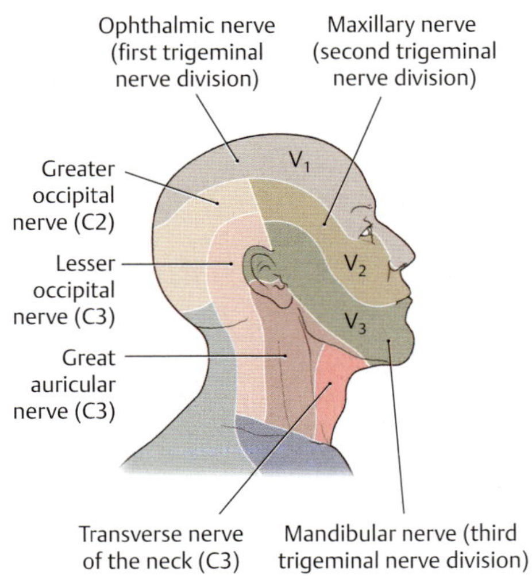

D Peripheral innervation of the head and neck

Right lateral view. The areas supplied by the segmental cervical nerves and the trigeminal nerve (first through third divisions) intersect in this region.

9.9 Motor Innervation: Organization of the Spinal Cord and Reflexes

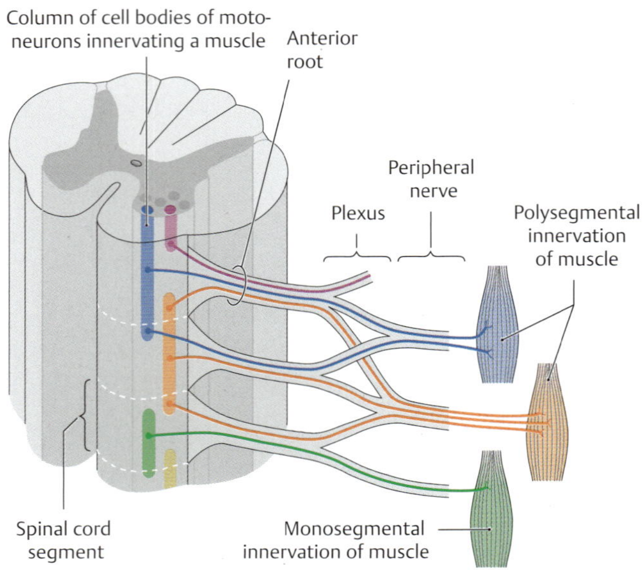

A Organization of anterior horns in the spinal cord

Whereas the dermatomes are areas of sensory innervation (= afferents), myotomes are areas of motor innervation of the skeletal muscles (= efferents). Just as in the posterior horn, in the anterior horn, too, the spinal cord segments are stacked on top of each other. With regard to innervation, a distinction is generally made between two types of skeletal muscles:

- Those with monosegmental innervation
- Those with polysegmental innervation

In monosegmentally innervated muscles (green muscle), the cell bodies of the motoneurons innervating these muscles are located at the level of a single spinal cord segment. In polysegmentally innervated muscles (blue and orange muscle), the cell bodies of the motoneurons reside inside a column that extends over several spinal cord segments. Muscles that are exclusively or predominantly innervated by one spinal cord segment are called segment-indicating muscles. They can be clinically evaluated by testing a corresponding reflex.

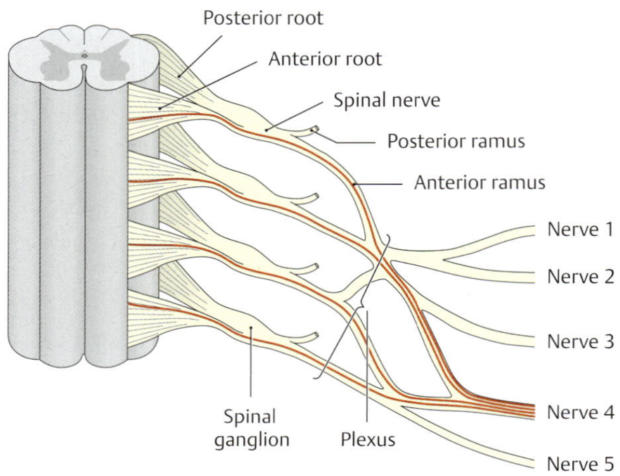

B Principles of plexus formation

In the plexus region, efferent axons from various spinal cord roots that innervate a polysegmentally innervated muscle combine into one peripheral nerve, which then extends to the muscle.

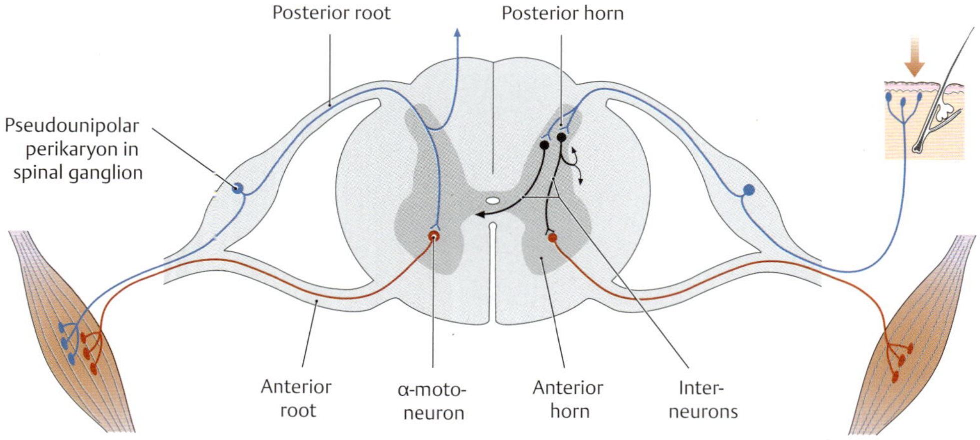

C Reflexes

The gray matter in the spinal cord, among other things, helps to coordinate muscle action on the spinal cord level (without involvement of the cortex), that is, reflexes. Broadly speaking, a distinction is made between a monosynaptic proprioceptive reflex (shown on the left) and a polysynaptic exteroceptive reflex (shown on the right).

Monosynaptic proprioceptive reflex: Receptors in a muscle transmit information about the condition of the muscle (length, muscle tension) via neurons in the spinal ganglia, the synapses of which end on the motor neuron of the muscle in the spinal cord, thereby influencing its activity.

Polysynaptic exteroceptive reflex: Receptors in the skin, rather than a muscle, transmit information to an interneuron; the interneuron synapses with the motoneuron of the muscle in the spinal cord. Since there is more than one neuron involved in the transmission of sensory information, the reflex is referred to as polysynaptic.

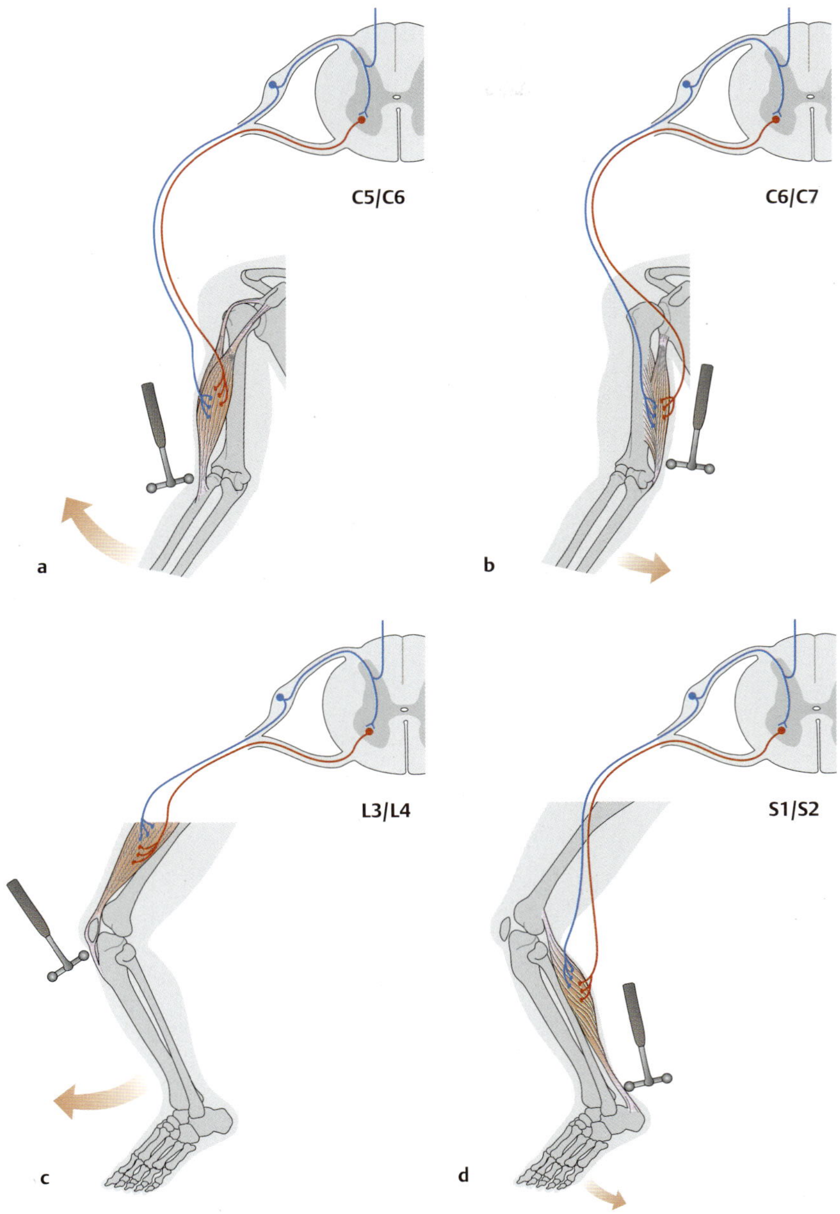

D Clinically important reflexes

a Biceps reflex; **b** triceps reflex; **c** patellar tendon reflex (quadriceps reflex); **d** calcaneal (Achilles') tendon reflex.

For each reflex, the muscles, reflex trigger points, involved nerves (afferents in blue, efferents in red), and corresponding spinal cord segments are shown. The most important proprioceptive reflexes should be tested during every clinical examination. The reflex is triggered by a brief blow with the reflex hammer, in most cases hitting the tendon of the muscle. As a result, the muscle stretches. If the muscle then contracts in reaction to the expansion, the reflex arc is intact. Although the reflex involves one muscle and the nerve supplying it, several spinal cord segments are involved in the innervation (multisegmental muscles, see **A**). As part of reflex testing, it is important to compare each reflex to its contralateral counterpart in order to detect one-sided augmentation, impairment, or pathologic deviations.

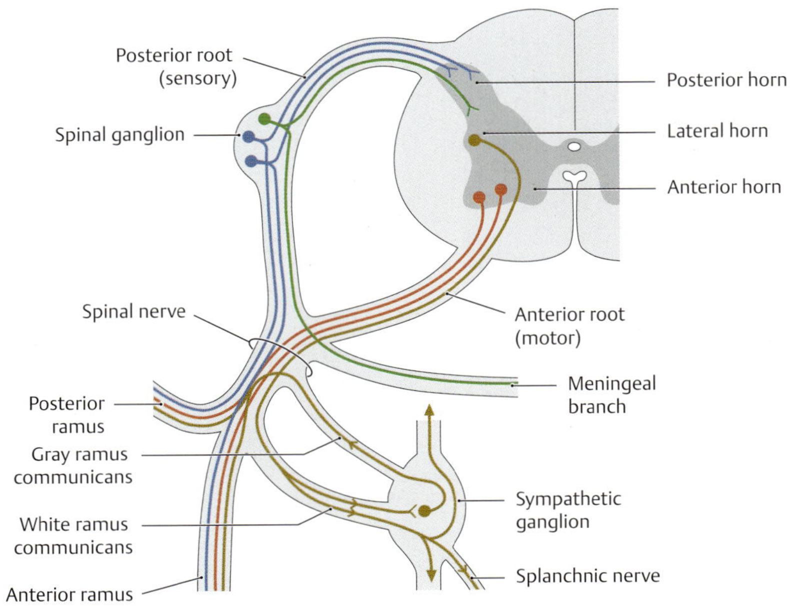

E Topographical and functional organization of a spinal cord segment

The **afferent fibers** from the skin, muscles, and joints (somatosensory, blue) and from the viscera (viscerosensory, green) pass through the posterior root into the spinal cord and terminate in the posterior horn. Both fibers arise from pseudounipolar cells in the posterior root (spinal) ganglion.

The **efferent fibers** for the skeletal musculature (somatomotor, red) and for the viscera (visceromotor, brown) pass through the anterior root to their respective target organs, i.e., the skeletal muscles, the vascular smooth muscles, and the smooth muscles of the internal organs. The fibers differ in their origin. Fibers for the skeletal muscles originate in the anterior horn of the spinal cord, and fibers for the viscera originate in the lateral horn. The axons of somatic motoneurons in the anterior horn synapse directly on skeletal muscle fibers. Visceral motoneurons in the lateral horn, however, provide indirect innervation to most of their target organs. Their axons form synapses with autonomic neurons in larger discrete sympathetic ganglia or in scattered clusters embedded in the visceral organs.

9.10 Motor Innervation: Upper (First) and Lower (Second) Motor Neurons

A Simplified scheme of motor innervation

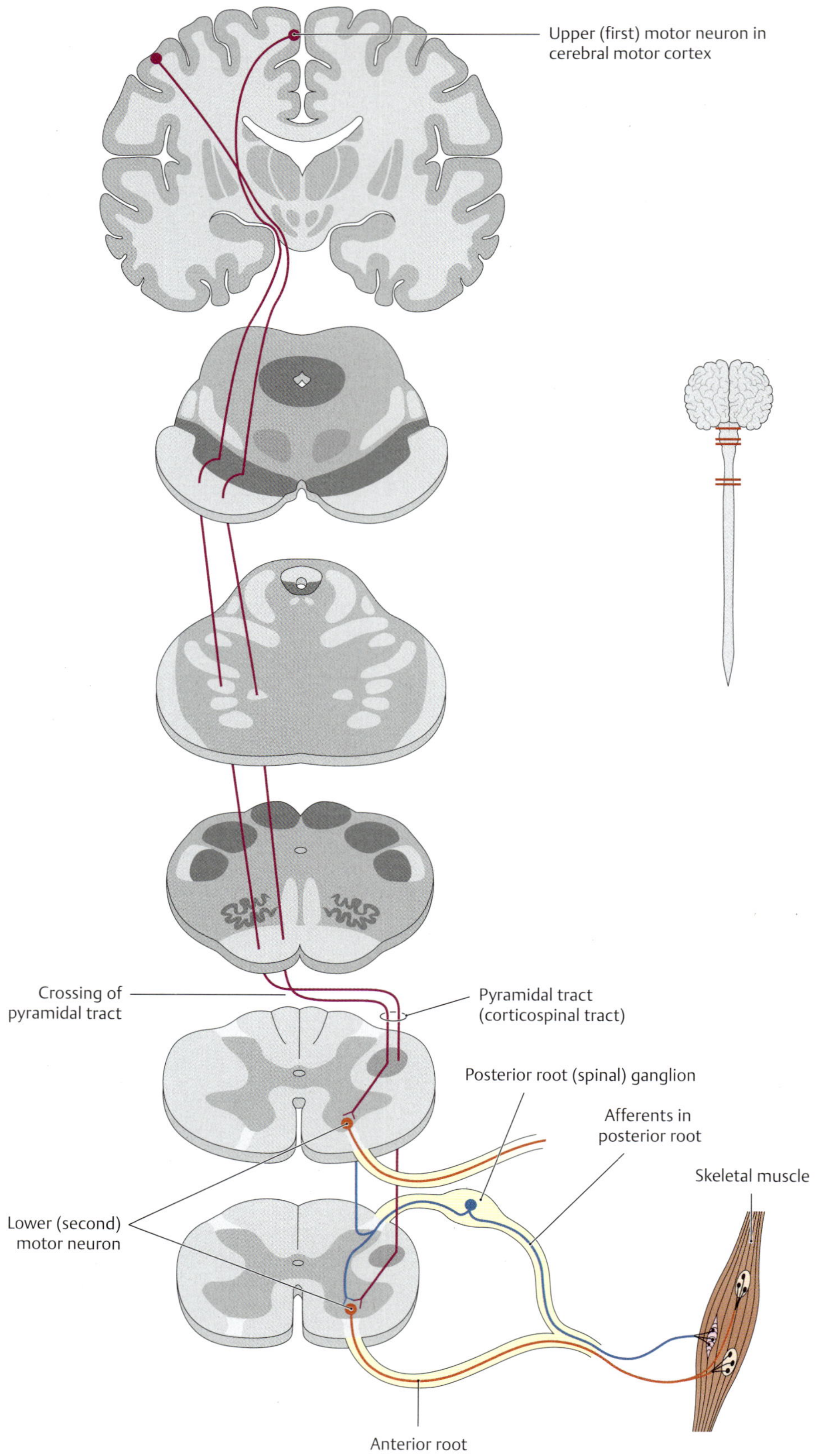

The simplest common path of motor innervation originates in the primary cerebral motor cortex. Large neurons (dark red) contribute axons to bundles collected into massive *corticospinal* tracts. Many axons pass uninterrupted to the spinal cord, where they synapse upon motor neurons in the anterior horn (orange). In the brainstem the bundles of corticospinal axons are designated *pyramids or pyramidal tracts*. In the medulla oblongata, most corticospinal axons cross the midline in the *pyramidal decussation* and continue on the opposite side. Axons from motor neurons in the anterior horn of the spinal cord exit in *anterior roots* and reach their targets, skeletal muscles, via peripheral nerves, where they synapse directly at *neuromuscular junctions*.

The typical spinal motor neuron receives inputs from multiple sources. The simplest local circuit involves afferent (sensory) synapses from fibers from posterior root (spinal) ganglion cells (blue). Although most afferent information is relayed to spinal motor neurons through multiple intermediate *interneurons*, some sensory input—particularly from tendon stretch receptors—is transmitted directly via a single synapse, as depicted. This chain of connections is called a *reflex arc*. One such arc produces the knee-jerk or patellar tendon reflex.

This pattern of motor connections is responsible for two fundamental principles in the diagnosis of neurologic disorders. Because of their positions in the pathway, the cerebral cortical cells are referred to as *upper motor neurons*, those in the spinal cord *as lower motor neurons*. Damage to lower motor neurons or their axons leads to denervation of muscles, with *flaccid paralysis* and eventual muscle atrophy. In contrast, interruption of corticospinal axons or destruction of the upper motor neurons themselves, resulting, for instance, from a cerebral infarction (stroke), leads to a loss of voluntary control over the lower motor neurons. With such an upper motor neuron lesion, the lower motor neuron is controlled solely by local spinal circuits. Local reflexes remain or are enhanced, but muscles show sustained contractions and increased tone, with *spastic paralysis*.

The second fundamental correlation between the anatomy of the motor pathway and neurologic diagnosis involves the crossing of corticospinal axons in the pyramidal decussation. Upper motor lesions that occur above (cranial to) this decussation will produce spastic paralysis on the *contralateral* side of the body; lesions in corticospinal axons below (caudal to) this point will cause spastic paralysis on the same side as (*ipsilateral* to) the lesion.

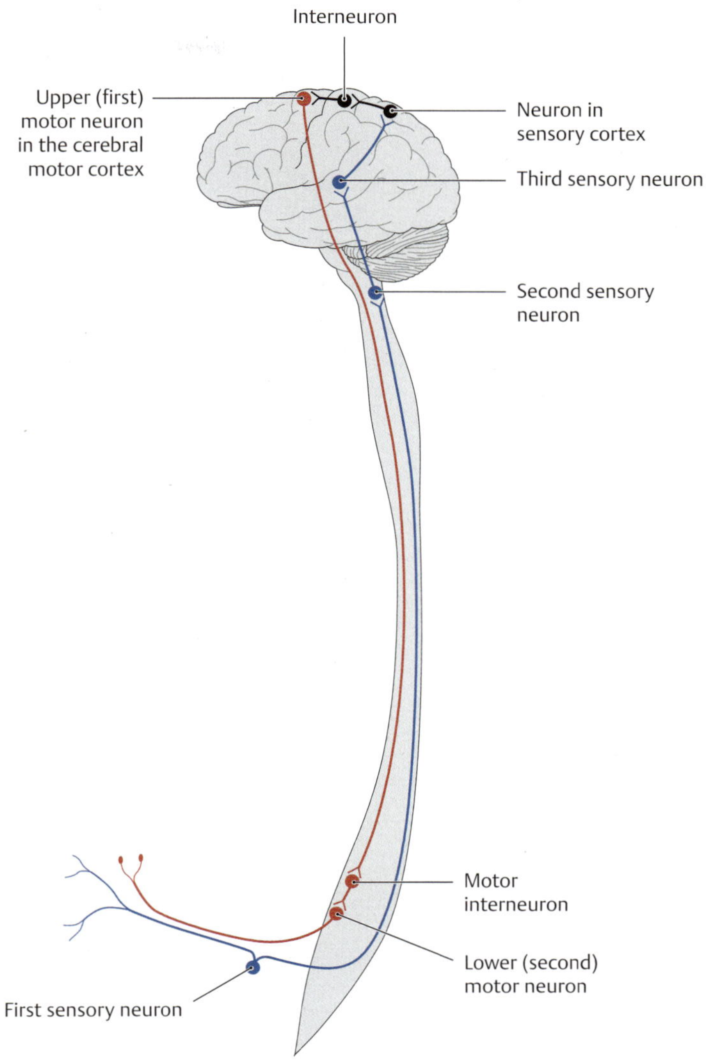

B Neural circuit for sensory and motor innervation

Left lateral view. Information processing in the CNS is actually more complex than shown in **A** because motor innervation is influenced by afferents of a kind not illustrated here (= sensorimotor circuit). Although a great many neurons are active at each step in the circuit, the diagram shows only one at each level. Sensory information, encoded as electrical impulses, enters the spinal cord through a primary afferent neuron (first sensory neuron) whose cell body is in the spinal ganglion (afferent neurons shown in blue). The information is relayed synaptically to the second and third afferent (sensory) neurons, which transfer it to the sensory cortex (black). Associative neurons (interneurons, black) transmit the information to the upper motor neuron in the cerebral motor cortex, where the information is modified by many associated inputs. The cortical motor neuron (upper motor neuron) redirects the information to the spinal cord, where it reaches the spinal (lower) motor neuron, either directly or via interneurons and additional synaptic relays. Finally, the lower motor neuron in the spinal cord transmits the impulses to the voluntary skeletal muscle. Corticospinal projections are most direct and numerous for lower motor neurons that innervate muscles, like those in the human hand, which are often under fine control and coordinated cognitive direction. Other motor pathways, not depicted here, are more important for controlling posture and balance. Because the axons in some of these other pathways are not in the corticospinal and pyramidal tracts, they are sometimes referred to as *extrapyramidal*.

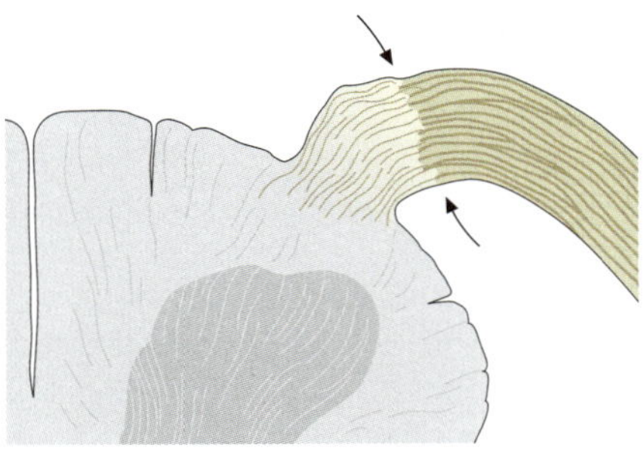

C Obersteiner–Redlich zone in a posterior root

The Obersteiner–Redlich zone marks the morphological junction between the CNS and PNS (arrows). Oligodendrocytes in the CNS form a myelin sheath around the axons only as far as this zone, which is located just past the emergence of the posterior root from the spinal cord. The myelin sheaths in the PNS are formed by Schwann cells (see p. 100 for details). The myelin is so thin at these sites that the fibers appear almost unmyelinated. This creates a site of predilection for immunologic diseases such as the immune reactions that occur in the late stages of syphilis.

9.11 Differences between the Central and Peripheral Nervous Systems

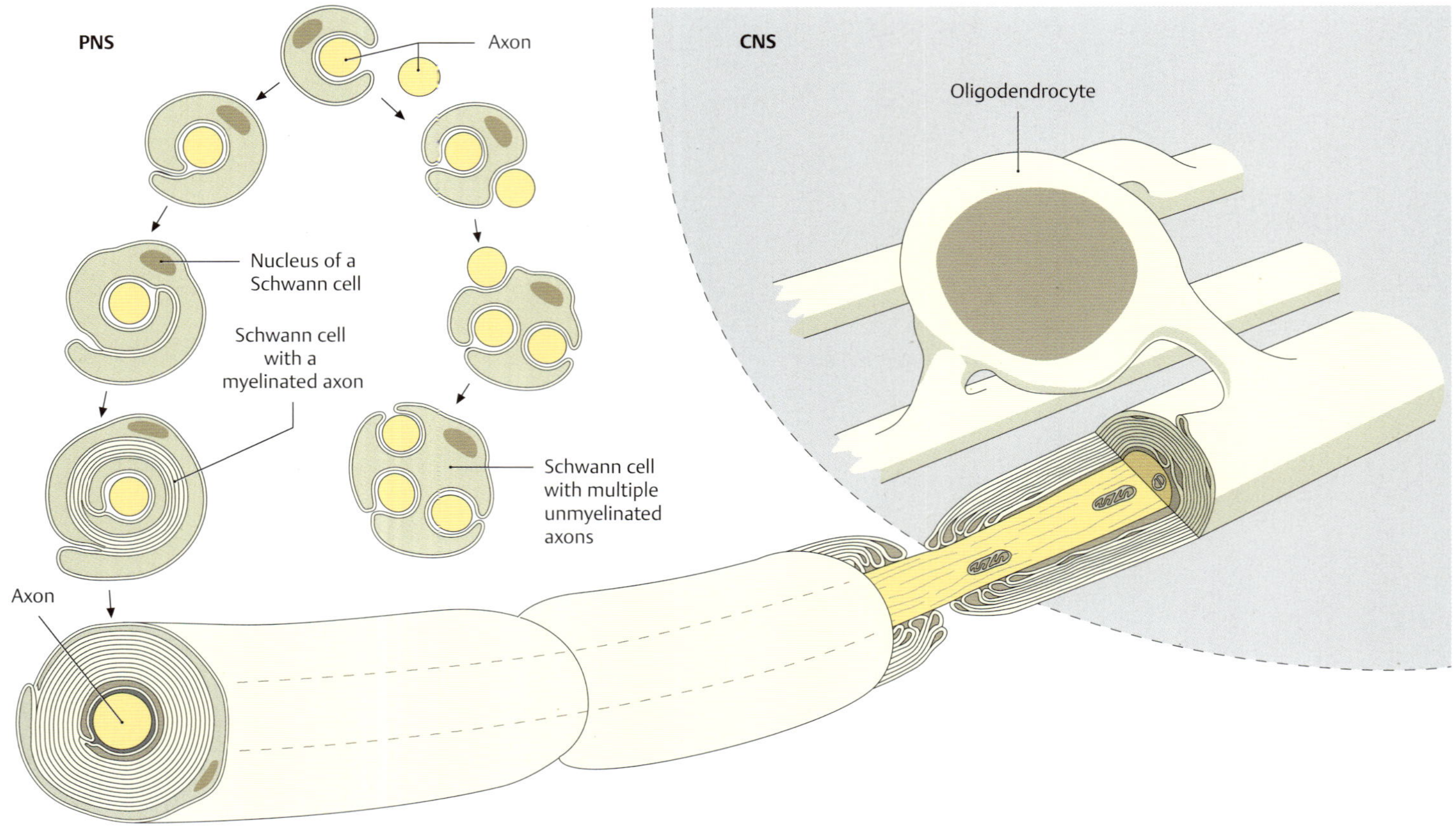

A Myelination differences in the PNS and CNS
The purpose of myelination is to provide the axons with electrical insulation, which significantly increases the nerve conduction velocity. The very lipid-rich membranes of myelinating cells are wrapped around the axons to produce this insulation. Schwann cells (left) myelinate the axons in the *PNS*, while oligodendrocytes (right) form the myelin in the *CNS*. *Note:* In the *CNS*, one oligodendrocyte always wraps around *multiple* axons. In the *PNS*, one Schwann cell *may* ensheath *multiple* axons if the peripheral nerve is *unmyelinated*. If the peripheral nerve is *myelinated*, *one* Schwann cell *always* wraps around *one* axon. Owing to this improved insulation, the nerve conduction velocity is higher in myelinated nerves than in unmyelinated nerves. Myelinated fibers occur in areas where fast reaction speeds are needed (muscular contractions), while unmyelinated fibers occur in areas that do not require rapid information transfer, as in the transmission of visceral pain. Because of the different cell types, myelin has a different composition in the CNS and PNS. This difference in myelination is also important clinically. An example is multiple sclerosis, in which the oligodendrocytes are damaged, but the Schwann cells are not, so that the central myelin sheaths are disrupted, while the peripheral myelin sheaths remain intact.

B Myelination

Principles
• Myelinated axons are generally thicker than unmyelinated axons. • Myelinated axons transmit impulses faster than unmyelinated axons. • Unmyelinated axons reside only in the PNS.

CNS
• Oligodendrocytes myelinate axons, with the oligodendrocyte enveloping from several up to 50 axons.

PNS
• With myelinated axons, Schwann cells wrap many layers around a single axon. • Somatomotor and somatosensory axons are usually myelinated (exception: small pain fibers). • Axons of the autonomic nervous system are often unmyelinated (exception: e.g., white rami communicantes of preganglionic fibers). • With unmyelinated axons, Schwann cells wrap one layer around several axons. • The transition from Schwann cell to oligodendrocyte occurs in the Obersteiner-Redlich zone (see **C**, p. 99).

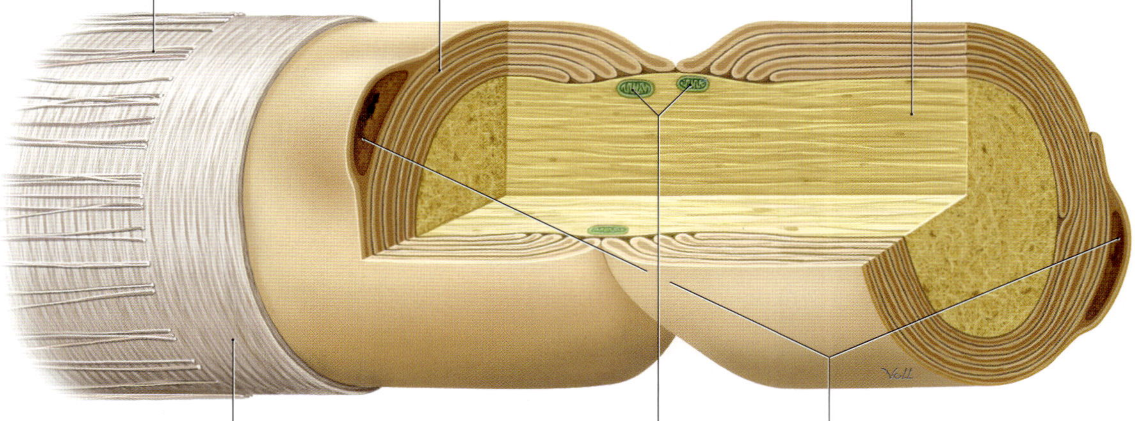

C Structure of a node of Ranvier in the PNS

In the PNS, the *node of Ranvier* is the site where two Schwann cells come together. That site is marked by a small gap in the myelin sheath, which forms the morphological basis for *saltatory nerve conduction*, allowing impulses to be transmitted at a higher velocity. Mitochondria concentrate in the nodes of Ranvier to ensure the energy supply of Na/K ATPase.

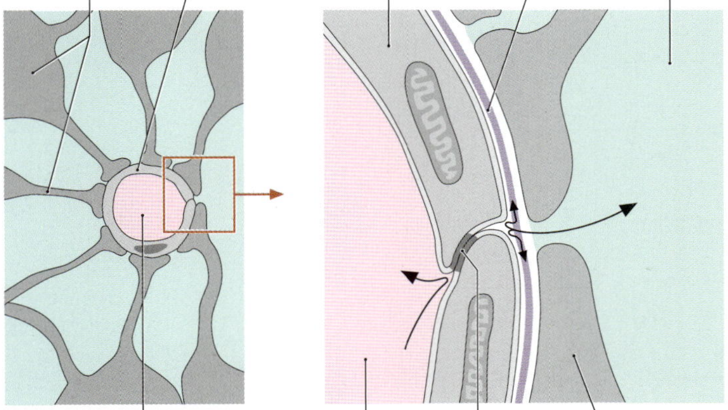

D Structure of the blood–brain barrier in the CNS

Besides the type of myelination, there is also a difference in tissue barriers between the CNS and PNS. The CNS is isolated from surrounding tissues by the blood–brain barrier. The components of the blood–brain barrier include (1) most importantly, a continuous capillary endothelial cell layer, sealed by tight junctions; (2) a continuous basal lamina surrounding the endothelial cells; and (3) enveloping astrocytic processes surrounding the brain capillary. This barrier serves to exclude macromolecules, as well as many small molecules that are not actively transported by the endothelial cells, thus protecting the delicate environment of the CNS. The barrier is vulnerable, however, to lipid-soluble molecules that can traverse the endothelial cell membranes. The perineurial sheath creates a similar barrier in the PNS (see **E**).

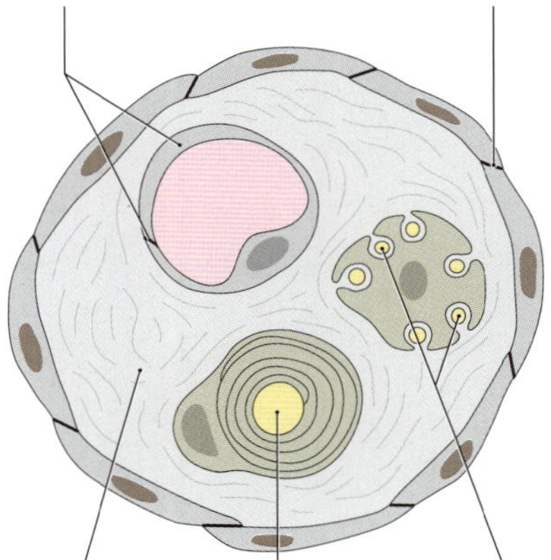

E Structure of the perineurial sheath in the PNS

The perineurial sheath, like the blood–brain barrier, is formed by tight junctions between the epithelium-like fibroblasts (perineurial cells; the perineurium is described on p. 83). It isolates the milieu of the axon from that of the surrounding endoneural space (endoneurium), thereby preventing harmful substances from invading the axon. This tissue barrier must be surmounted by drugs that are designed to act on the axon, such as local anesthetic agents.

9.12 The Autonomic Nervous System

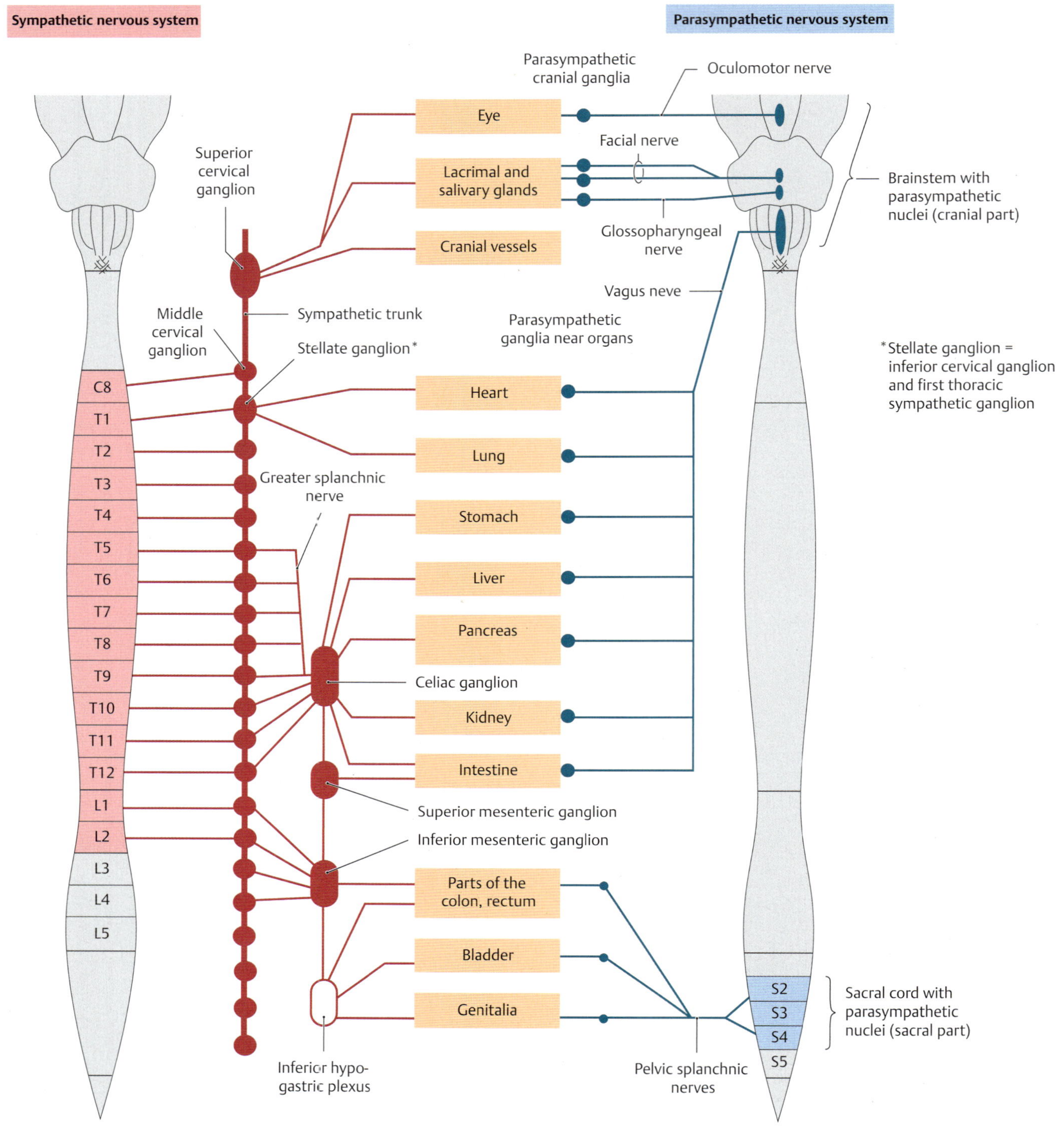

A Structure of the autonomic nervous system

The system of motor innervation of skeletal muscle is complemented by the *autonomic nervous system*, with two divisions: sympathetic (red) and parasympathetic (blue). Both divisions have a two-neuron path between the CNS and their targets: a *preganglionic* CNS neuron and a ganglion cell in the PNS that is close to the target. The preganglionic neurons of the sympathetic system are in the lateral horns of the cervical, thoracic, and lumbar spinal cord. Their axons exit the CNS via the anterior roots and synapse in sympathetic ganglia in bilateral *paravertebral* chains (*sympathetic trunks*), or as single midline *prevertebral ganglia* (see **E**). Axons from these ganglion cells course in unmyelinated bundles on blood vessels or in peripheral nerves to their targets. The preganglionic neurons of the parasympathetic system are located in the brainstem and sacral spinal cord. Their axons exit the CNS via cranial and pelvic splanchnic nerves to synapse with parasympathetic ganglion cells. In the head, these cells are in discrete ganglia associated with the cranial nerves. In other locations the parasympathetic ganglion cells are in tiny clusters embedded in their target tissues. The sympathetic and parasympathetic systems regulate blood flow, secretions, and organ function; the two divisions often act in antagonistic ways on the same target (see **B**). Although this basic dichotomy of visceral motor activity was identified early, by Langley (1905) and others, it has been shown more recently that autonomic control of various organs, particularly in the gastrointestinal and urogenital tracts, is highly sophisticated, dependent upon feedback from local visceral afferents that relay pain and stretch information, etc., through complex local circuits.

B Synopsis of the sympathetic and parasympathetic nervous systems

1. The sympathetic nervous system can be considered the excitatory part of the autonomic nervous system that prepares the body for a "fight or flight" response.
2. The parasympathetic nervous system is the part of the autonomic nervous system that coordinates the "rest and digest" responses of the body.
3. Although there are separate control centers for the two divisions in the brainstem and spinal cord, they have close anatomic and functional ties in the periphery.
4. The principal transmitter at the target organ is acetylcholine in the parasympathetic nervous system and norepinephrine in the sympathetic nervous organ.
5. Stimulation of the sympathetic and parasympathetic nervous systems produces the following different effects on specific organs:

Organ	Sympathetic nervous system	Parasympathetic nervous system
Eye	Pupillary dilation	Pupillary constriction and increased curvature of the lens
Salivary glands	Decreased salivation (scant, viscous)	Increased salivation (copious, watery)
Heart	Elevation of the heart rate	Slowing of the heart rate
Lungs	Decreased bronchial secretions and bronchial dilation	Increased bronchial secretions and bronchial constriction
Gastrointestinal tract	Decreased secretions and motor activity	Increased secretions and motor activity
Pancreas	Decreased secretion from the endocrine part of the gland	Increased secretion
Sex organs	Ejaculation	Erection
Skin	Vasoconstriction, sweat secretion, piloerection	No effect

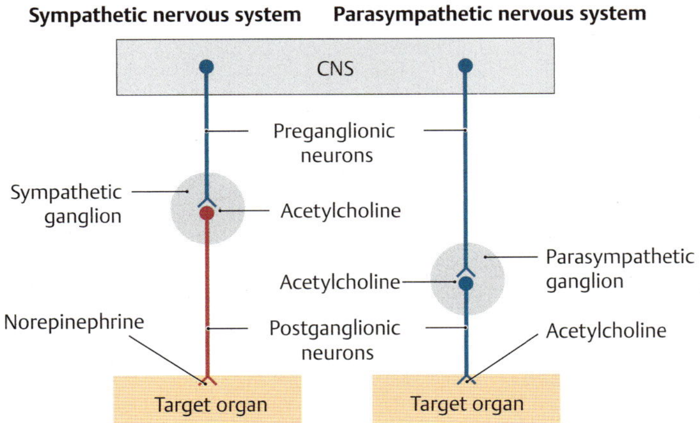

C Circuit diagram of the autonomic nervous system

The synapse of the central, preganglionic neuron uses *acetylcholine* as a transmitter in both the sympathetic *and* parasympathetic nervous systems (cholinergic neuron, shown in blue). In the sympathetic nervous system, the transmitter changes to *norepinephrine* at the synapse of the postganglionic neuron with the target organ (adrenergic neuron, shown in red), while the parasympathetic system continues to use acetylcholine at that level.

Note: Various types of receptors for acetylcholine (= neurotransmitter sensors) are located in the membrane of the target cells. As a result, acetylcholine can produce a range of effects depending on the receptor type.

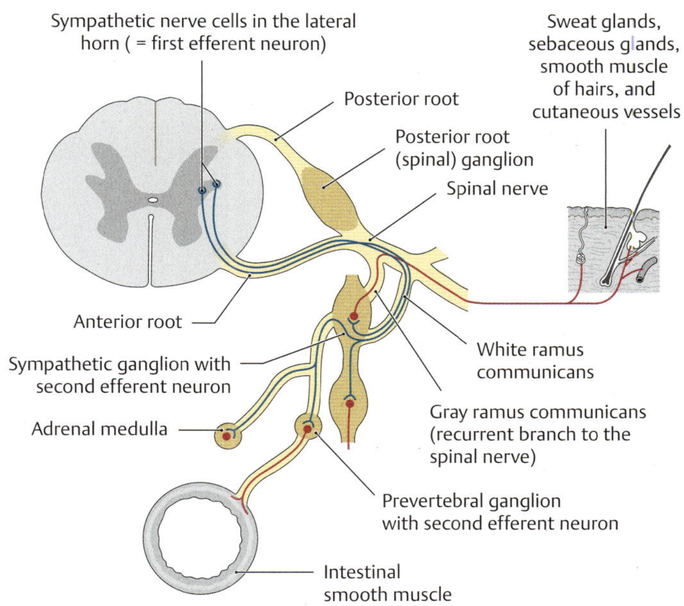

D Distribution of sympathetic fibers in the periphery

Skin (blood vessels, cutaneous glands, and smooth muscle cells attached to hair follicles) and blood vessels of skeletal muscles are only innervated sympathetically. For these, synapses between the first and second neurons occur in the ganglia of the sympathetic trunk. For the sympathetic fibers to the organs, the synapses occur in the prevertebral ganglia or in the organ itself (see **E**).

Note: Skin and skeletal muscles are not parasympathetically innervated.

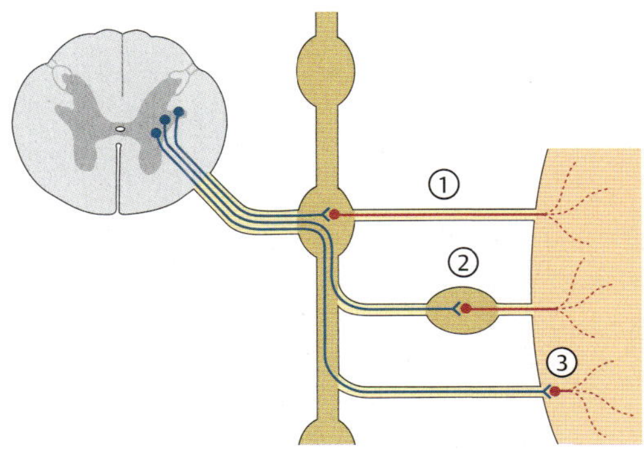

E Location of synapses in the sympathetic nervous system

Acetylcholine is the neurotransmitter released by both the first and second neurons in the parasympathetic system, but in the sympathetic system the second neuron releases norepinephrine. Synapses of the first (preganglionic cholinergic) neuron (blue) with the second neuron (red) may be located:

① In the paravertebral ganglia on either side of the vertebral column, which are linked to one another as the sympathetic trunk. The synapses for the blood vessels of the skin and skeletal muscles mainly reside in these ganglia (see **D**).
② iIn prevertebral ganglia located in front of the spinal column. The synapses for visceral fibers of the abdomen and pelvis reside here.
③ In the target organ, or effector, itself, i.e., the adrenal medulla.

9.13 Lesions of Peripheral Nerves

A Damage to a peripheral nerve

A peripheral nerve is a cluster of afferent (pain, temperature, pressure, vibration, sense of position) and efferent (somatomotor and visceromotor) axons. Pressure (left) or cuts (right) can damage the occasionally long course of the nerve.

Damage through pressure: It only takes slight pressure for the peripheral nerves to respond sensitively, with the afferent system being the first to be affected. Already a mild stimulus leads to paresthesia (e.g., tingling when the ulnar nerve is stimulated due to a slight tap on the medial elbow); longer and heavier pressure leads to loss of sensation in the area innervated by the receptive field. Finally, the efferent system is damaged, causing motor loss. There are numerous anatomic bottlenecks (e.g., tunnel or slit-shaped [osteo] fibrous canals), which can occasionally lead to compression syndromes. They are first perceived as pain and later as loss of function in the affected muscles. Some of the important syndromes are listed under **B**.

Damage through cutting: As a result of a cut (a mixed nerve has been completely severed), the afferent and efferent systems fail at the same time. The muscles innervated by this nerve can no longer be moved, resulting in flaccid paralysis (for more about the difference between flaccid and spastic paralysis, see p. 98). In addition, the loss of afferents leads to a loss of sensation, and the loss of sympathetic fibers leads to vegetative symptoms (increased sweat secretion and increased cutaneous perfusion) in the skin area innervated by the nerve.

B Selected nerve compression (bottleneck) syndromes

This provides an overview of the syndromes or symptoms that can be caused by the compression of individual nerves (for more details about the syndromes, see textbooks of neurology).

Affected nerve	Anatomic region	Syndrome/symptoms
Shoulder girdle and upper limb		
Brachial plexus	Upper thoracic aperture	Thoracic outlet syndrome (e.g., scalene syndrome, cervical rib syndrome, costoclavicular syndrome)
Suprascapular nerve	Scapular notch	Scapular notch syndrome
Axillary nerve	Lateral axillary hiatus	Lateral axillary hiatus syndrome
Ulnar nerve	Ulnar sulcus Flexor carpi ulnaris tendon of origin, ulnar palmar branches	Ulnar nerve entrapment (ulnar sulcus syndrome) Cubital tunnel syndrome Ulnar tunnel syndrome (Guyon -Logen syndrome)
Radial nerve	Groove for radial nerve	Park bench paralysis
– Deep branch	Supinator (arcade of Frohse)	Supinator syndrome (distal radial compression syndrome)
– Superficial branch	Distal radial forearm	Wartenberg disease
Median nerve	Passage through pronator teres Carpal tunnel	Pronator teres syndrome Carpal tunnel syndrome
Pelvic girdle and lower limb		
Sciatic nerve	Gluteal region	Piriformis syndrome
Femoral nerve	Inguinal region	Pain/weakness of the front upper leg
Obturator nerve	Anterior pubic bone area	Howship–Romberg syndrome
Saphenous nerve	Adductor canal	Dysesthesia at the interior lower leg
Lateral femoral cutaneous nerve	Lateral inguinal/upper leg region	Meralgia paresthetica (inguinal ligament syndrome)
Common fibular nerve	Head and neck of the fibula	Fibular tunnel syndrome
Deep fibular nerve	Anterior upper ankle joint	Anterior tarsal tunnel syndrome
Tibial nerve/plantar nerves	Inner malleolus region	Posterior tarsal tunnel syndrome
Plantar digital nerves	Toe ball area 3/4	Interdigital Morton neuralgia

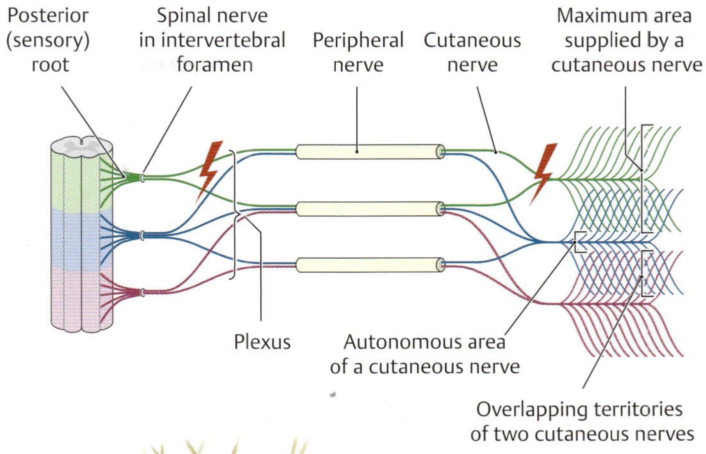

C Damage to a peripheral nerve exemplified by loss of sensitivity

Functionally speaking, in the innervated area of an afferent nerve, the nerve is only a part of a conduit cable for stimuli, which ends in the sensory area of the cerebral cortex, where conscious perception of the respective sensory stimulus occurs (see p. 99). The brain always locates damage to an afferent nerve in the area innervated by that cutaneous nerve, regardless of whether the damage occurs close to the spinal cord or not. In case of sensitivity loss, the damage is not necessarily located where sensation is lost.

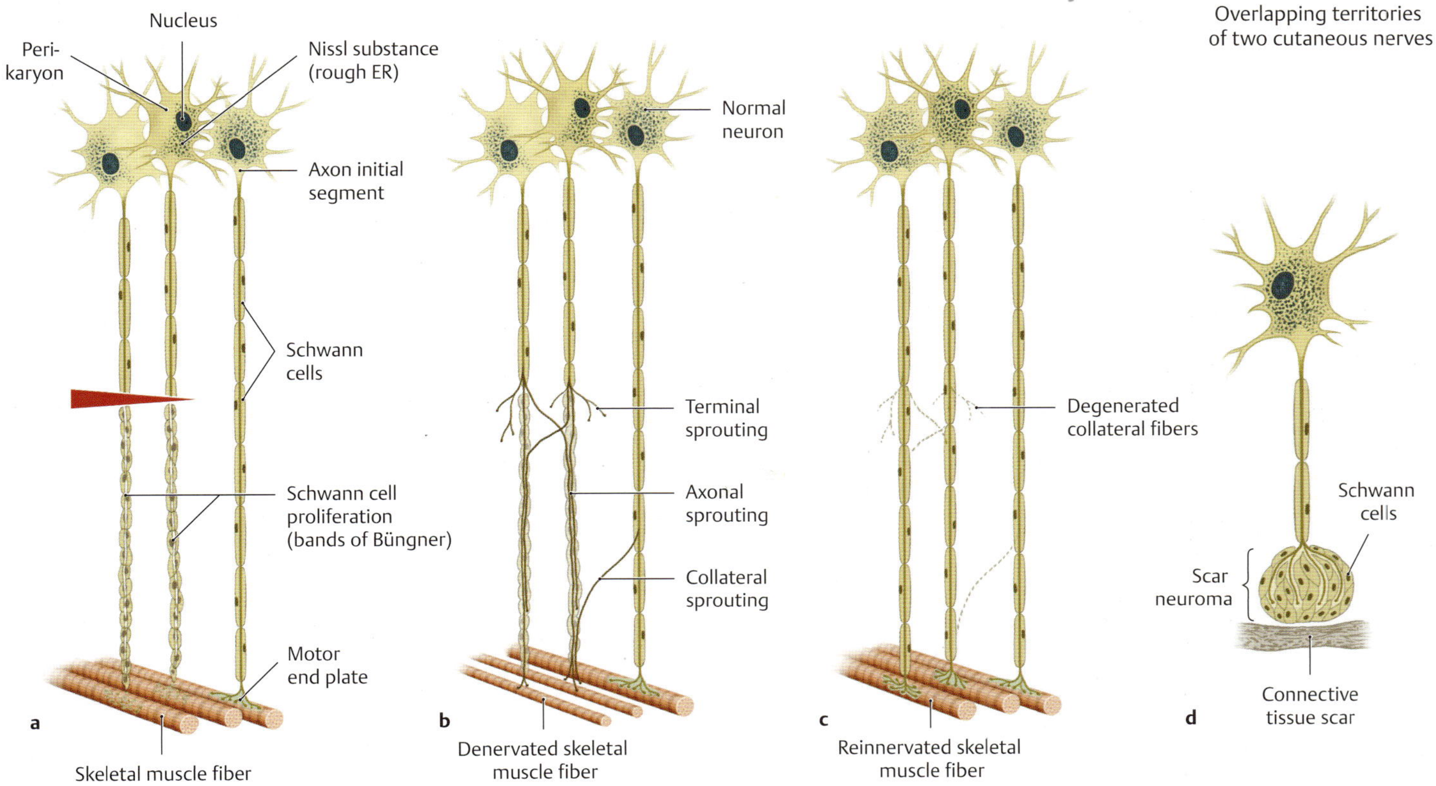

D Regeneration of a peripheral nerve after having been cut

In the peripheral nervous system (PNS), a nerve can usually regenerate after its axon has been severed.

- **a** A few days after the axon has been severed, it first degenerates distally to the lesion (so-called Waller degeneration). Its myelin sheath disintegrates, and the axon dissipates, as it is no longer nourished through the perikaryon (the so-called trophic function of the perikaryon is no longer required). In the affected perikarya, the cell nucleus shifts to the margin, and the Nissl substance partially dissolves (tigrolysis). At the same time, the Schwann cells divide, forming a cluster of cells from which the so-called Büngner ligaments arise.
- **b** Weeks later, the severed axons begin to sprout again proximally (axonal sprouting) and extend by 1 mm/day into the areas they originally innervated. From adjacent axons, which have not been affected by the lesions, axon collaterals can also extend into the affected area. The process of axonal sprouting can be tested with the help of the Hoffmann–Tinel sign. When percussing over the course of the nerve, patients experience a tingling around the area where the axons sprout.
- **c** Months later, the regeneration is complete. The axons have reached the skeletal muscles again and innervate them. Axons that have not made it to their relevant skeletal muscles and axons that have sprouted from the axon collaterals perish. As the Schwann cells divide distally along the lesion, the same segment of axons is now being myelinated by more Schwann cells than previously. This leads to an increased number of nodes of Ranvier, thus slowing down the saltatory conduction. This process can be measured by determining the nerve conduction velocity. The position of the nuclei and the texture of the Nissl substance in the perikarya have normalized.

After the peripheral nerve has been severed, its ends are sutured back together end-to-end. The goal of the nerve suture is to maintain the continuity of the connective tissue tube with its Schwann cell tubes as the anatomic landmark for the sprouting axons. In case of an injury to a large segment of the nerve, a nerve transplant may be performed. This transplant bridges the gap; it, too, only provides a conduction for the spouting axons. Such a conduction is crucial for sprouting axons because axons cannot penetrate into the scar tissue. If, because of such a scar, axons cannot find their way back to their muscle, it can lead to amputation or scar neuroma (see **d**), which is a conglomerate of Schwann cells and axonal sprouts.

B Trunk Wall

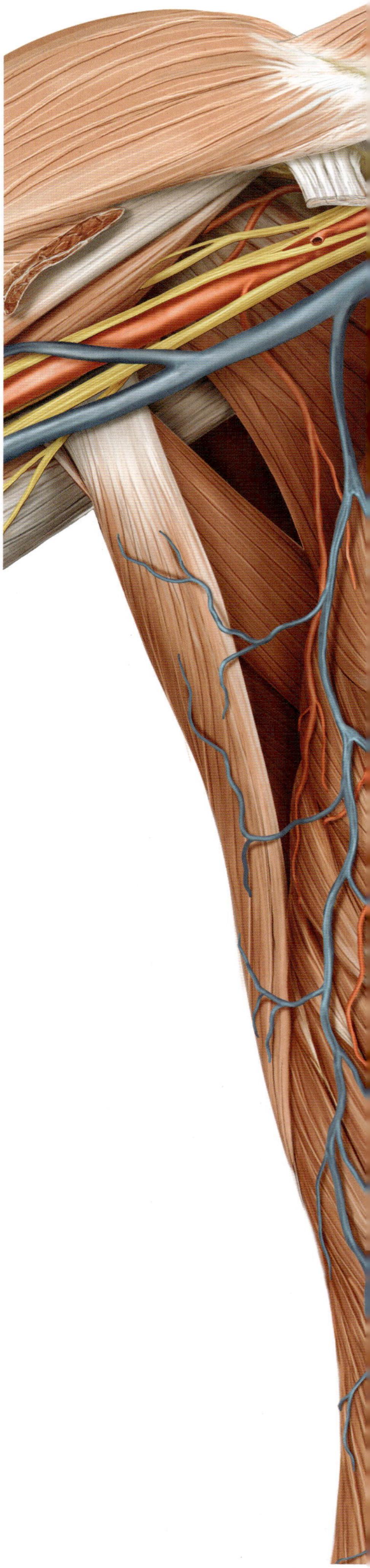

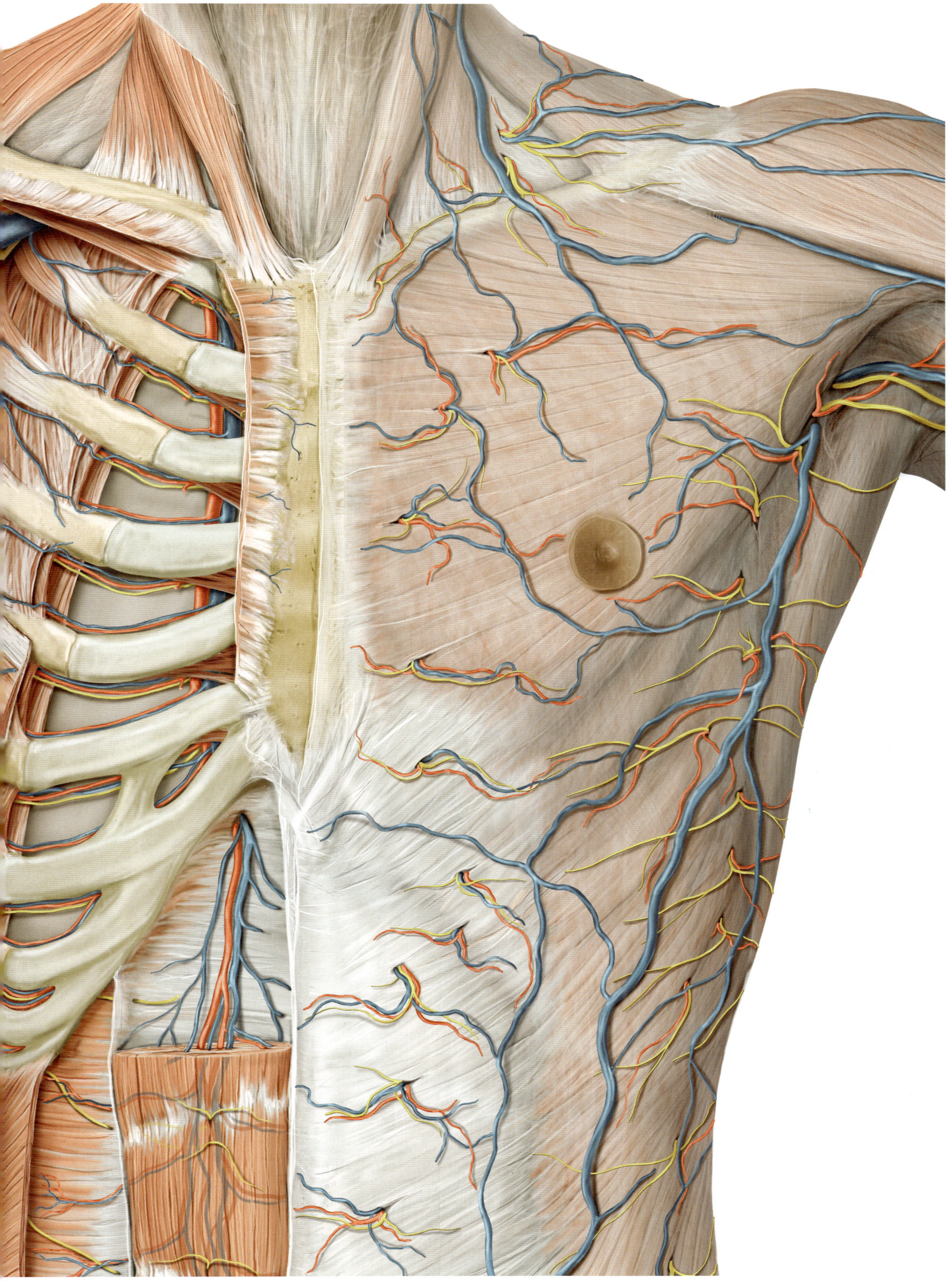

1.1 The Skeleton of the Trunk

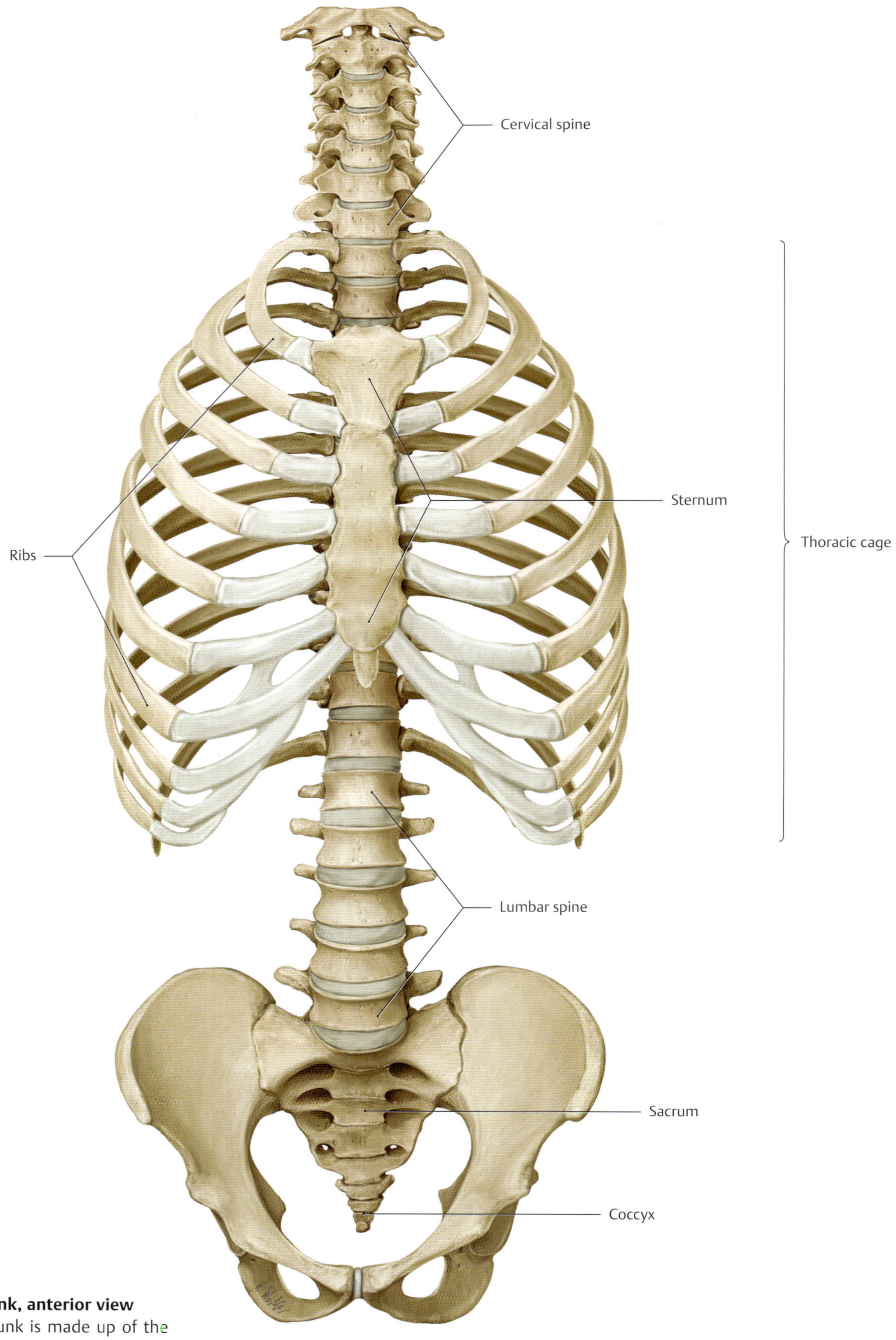

A Skeleton of the trunk, anterior view
The skeleton of the trunk is made up of the vertebral column, ribs, and sternum. Details on the ribs and sternum are given on p. 140.

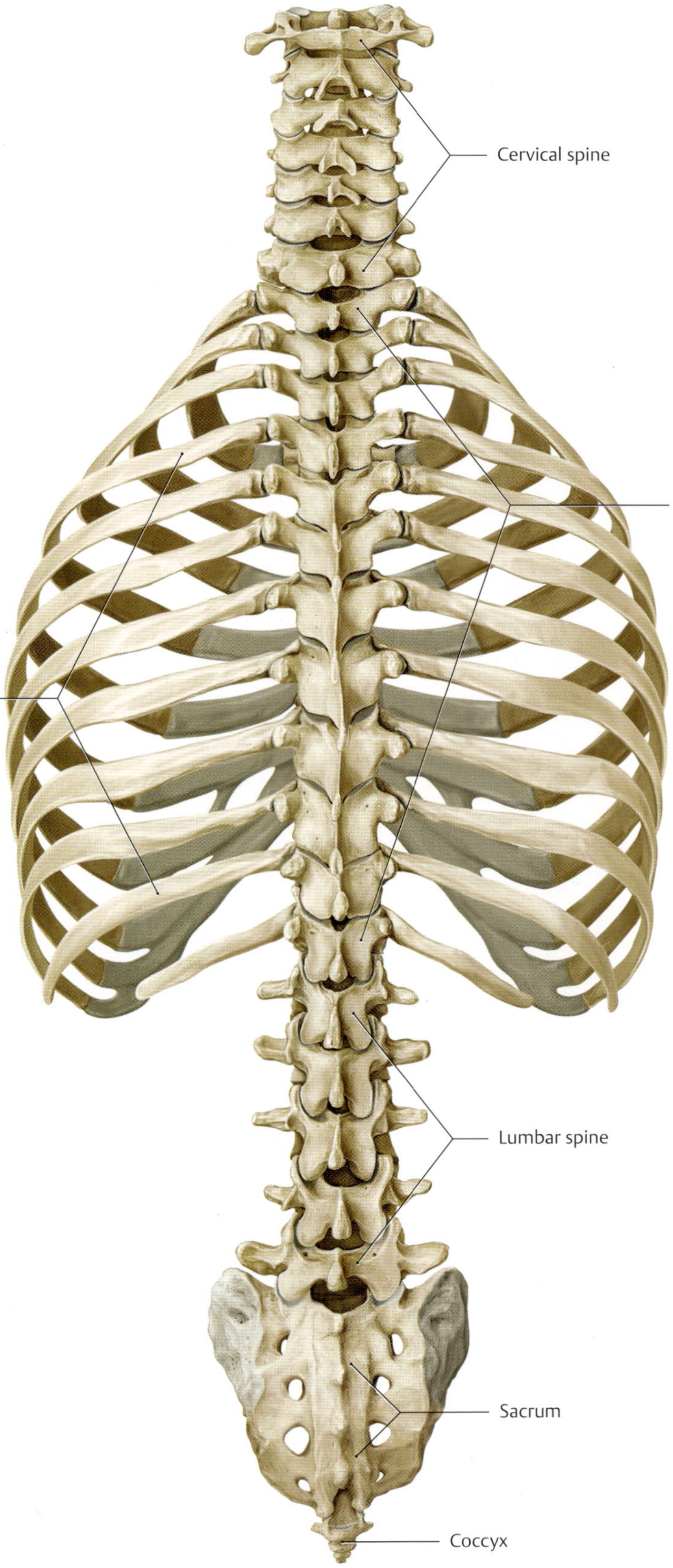

B Skeleton of the trunk, posterior view

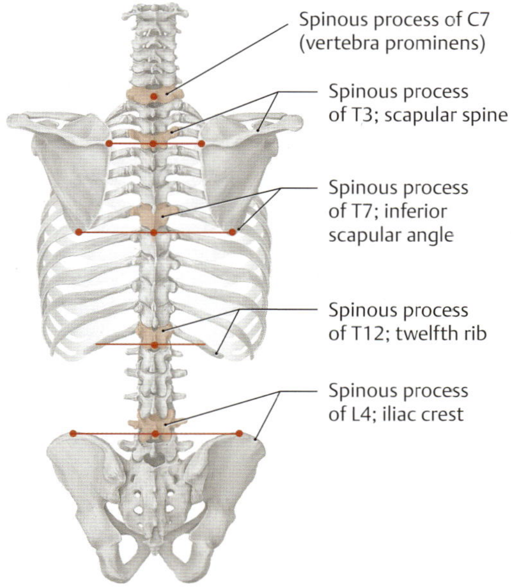

C The spinous processes as anatomical landmarks Posterior view. The spinous processes of the vertebrae appear as variable prominences beneath the skin and provide important landmarks during the physical examination. With few exceptions, they are easily palpated.

- The spinous process of the seventh cervical vertebra, located at the junction of the cervical and thoracic spine. Usually it is the most prominent of the spinous processes, causing the seventh cervical vertebra to be known also as the *vertebra prominens*.
- The spinous process of the third thoracic vertebra, located on a horizontal line connecting the scapular spines
- The spinous process of the seventh thoracic vertebra, located at the level of the inferior angles of the scapulae
- The spinous process of the twelfth thoracic vertebra, located slightly below the attachment of the last rib
- The spinous process of the fourth lumbar vertebra, located on a horizontal line connecting the highest points of the iliac crests

Note: The spinous processes of the thoracic vertebrae are angled downward (see p. 118), so the spinous process of the fifth thoracic vertebra, for example, is at the level of the sixth thoracic vertebral body.

1.2 The Bony Spinal Column

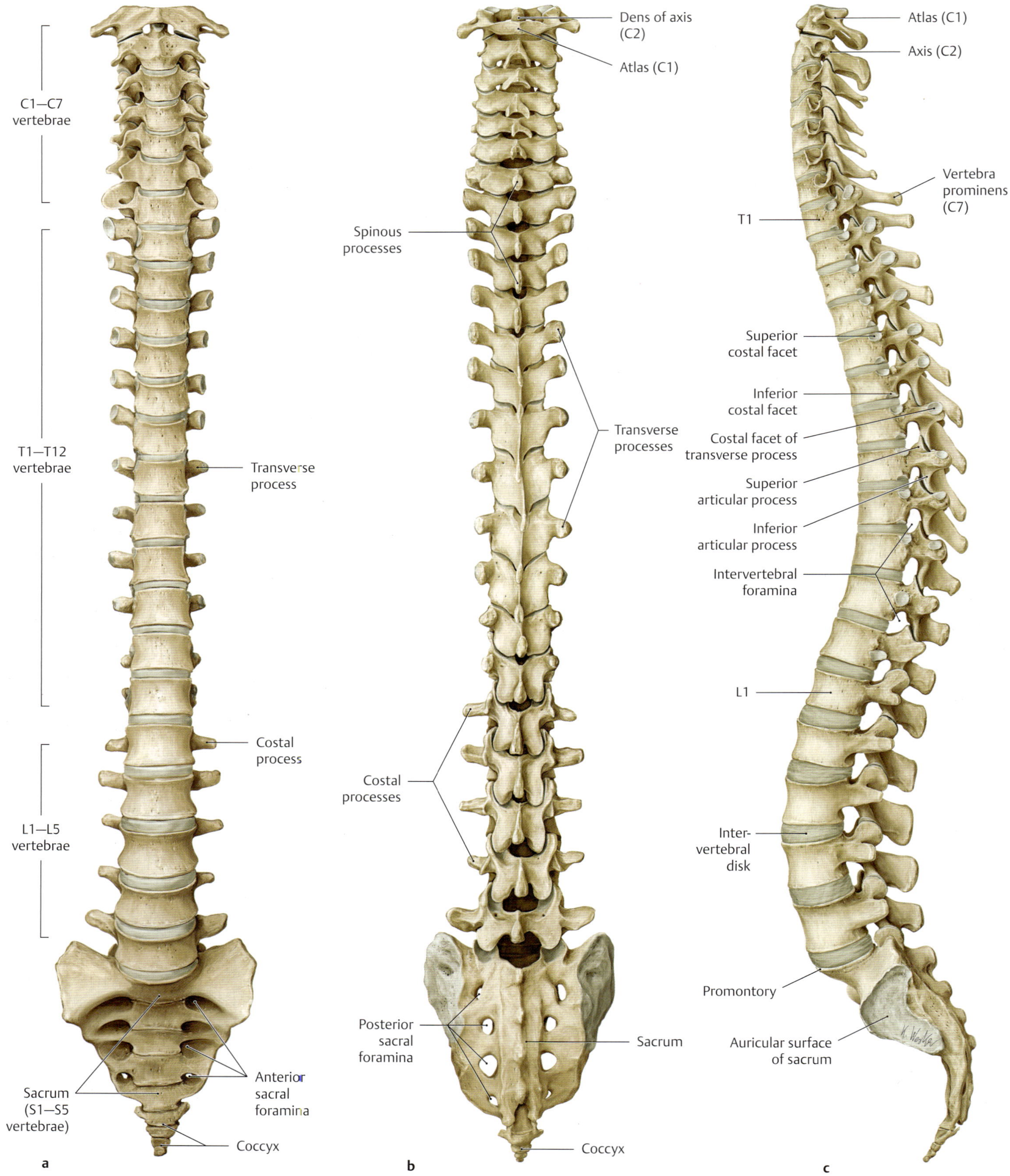

A The bony spinal column
a Anterior view, **b** posterior view, **c** left lateral view.

Note: Phylogenetically, the transverse processes of the lumbar vertebrae are rudimentary ribs. They therefore are known as costal processes (see also p. 114).

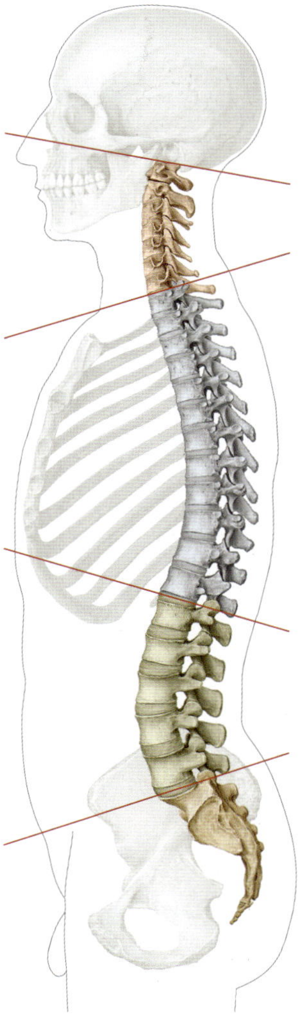

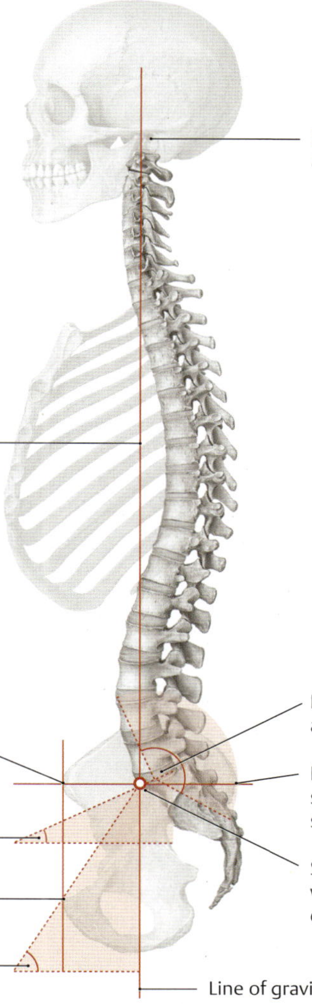

B Regions and curvatures of the spinal column

Left lateral view. The spinal column of an adult is divided into four regions and presents four characteristic curvatures in the sagittal plane. These curves are the result of human adaptation to upright bipedal locomotion, acting as springs to cushion axial loads. The following regions and curvatures are distinguished in the craniocaudal direction:

- Cervical spine – cervical lordosis
- Thoracic spine – thoracic kyphosis
- Lumbar spine – lumbar lordosis
- Sacral spine – sacral kyphosis

The cervical, thoracic, and lumbar regions of the spinal column are also known collectively as the *presacral spine*. The transitional areas between the different regions are of clinical importance because they are potential sites for spinal disorders (e.g., herniated disks). Occasionally, the vertebrae in these transitional areas have an atypical morphology that identifies them as *transitional vertebrae*. This is particularly common at the lumbosacral junction, where *sacralization* or *lumbarization* may be seen, depending on the appearance of the atypical vertebra. With lumbarization, the first sacral vertebra is not fused to the sacrum and constitutes an extra lumbar vertebra. With sacralization, there are only four lumbar vertebrae, the fifth being "sacralized" by fusion to the sacrum. These *assimilation disorders* are often unilateral (hemilumbarization, hemisacralization).

C Integration of the spinal column into the pelvic girdle

Skeleton of the trunk with the skull and pelvic girdle, left lateral view. Normally, the spinal column is curved and integrated into the pelvic girdle in such a way that characteristic angles are formed between certain imaginary lines and axes.These angles and lines are useful in the radiographic evaluation of positional abnormalities and deformities of the spine and trunk.

Lumbosacral angle: angle formed by the axes of the L 5 and S 1 vertebrae, averaging 143°. It results from the fact that the sacrum is a fixed component of the pelvic ring (see p. 146) and thus contributes little to straightening the vertebral column. The result is a characteristic sharp angle at the junction of the presacral part of the spinal column with the sacrum.

Sacral angle: angle between the horizontal plane and the superior surface of the sacrum, averaging approximately 30°.

Pelvic inclination angle: angle formed by the pelvic inlet plane (connecting the sacral promontory to the upper border of the symphysis) with the horizontal. It measures approximately 60° in upright stance. The pelvic inclination angle increases or decreases as the pelvis is tilted forward or backward (see p. 165). With an ideal pelvic position in upright stance, the anterior and posterior superior iliac spines are at the same horizontal level, and the anterior superior iliac spine is directly above the pubic symphysis. By knowing this, the examiner can easily evaluate the position of the pelvis by using palpable bony landmarks.

Line of gravity: the line of gravity passes through landmarks that include the external auditory canal, the dens of the axis (C2), the functional-anatomical transition points in the spinal column (between lordosis and kyphosis), and the whole-body center of gravity just anterior to the sacral promontory.

1.3 Development of the Spinal Column

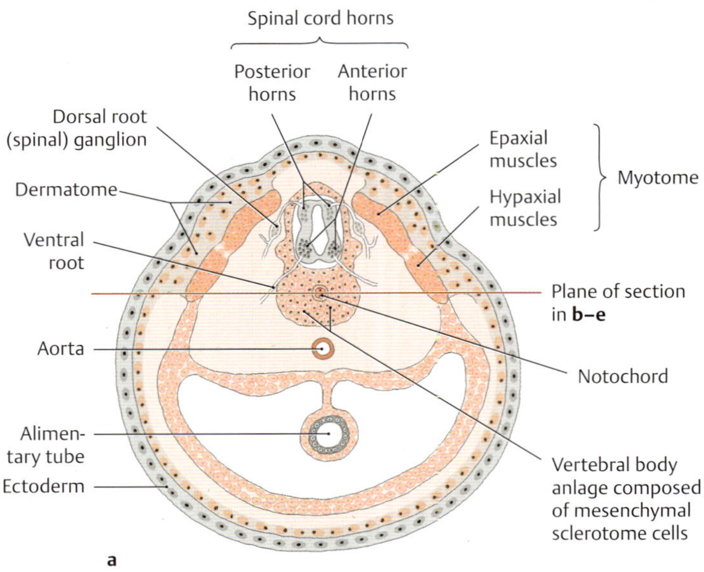

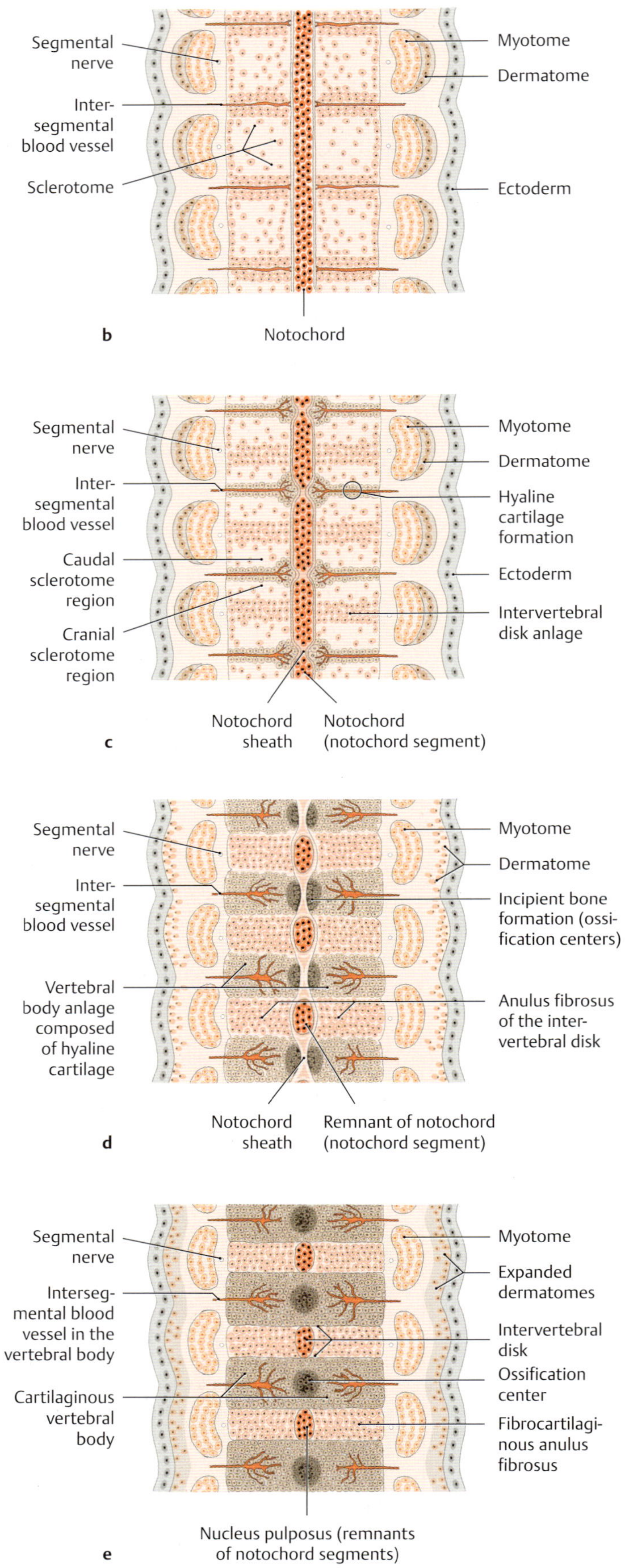

A Development of the spinal column (weeks 4–10)

a Schematic cross section, **b–e** schematic coronal sections (the plane of section in **b–e** is indicated in **a**).

a, b The former somites have differentiated into the myotome, dermatome, and sclerotome. The sclerotome cells separate from the other cells at 4 weeks, migrate toward the notochord, and form a cluster of mesenchymal cells around the notochord (anlage of the future spinal column).

c Adjacent cranial and caudal sclerotome segments above and below the intersegmental vessels join together and begin to chondrify in the sixth week, displacing the notochord material superiorly and inferiorly (notochord segments).

d The intervertebral disks with their nucleus pulposus and anulus fibrosus develop between the rudimentary vertebral bodies. Ossification begins at the center of the vertebral bodies in the eighth week of development.

e By fusion of the caudal and cranial sclerotome segments, the segmentally arranged myotomes interconnect the processes of two adjacent vertebral anlages, bridging the gap across the intervertebral disks. This is how the *motion segments* are formed (see p. 132). The segmental spinal nerve courses at the level of the future intervertebral foramen, and the intersegmental vessels become the nutrient vessels of the vertebral bodies (week 10).

Clinical aspects: If the neural tube or the dorsal portions of the vertebral arches fail to close normally during embryonic development, the result is *spina bifida*—a cleft spine. In this anomaly, the spinal column is open posteriorly, and the spinous processes are absent (the various forms and manifestations are described in textbooks of embryology). Usually there is a bilateral defect in the vertebral arches (generally affecting the L 4 and L 5 region), known as *spondylolysis*. This defect may be congenital or acquired (e.g., due to trauma). Acquired cases are common in sports that pose a risk of vertebral arch fractures (javelin throwing, gymnastics, high jumping). If the associated intervertebral disk is also damaged, the vertebral body will begin to slip forward (*spondylolisthesis*). In cases of *congenital spondylolysis* (which are associated with varying degrees of spondylolisthesis), the slippage progresses slowly during growth, and the condition tends to stabilize after 20 years of age.

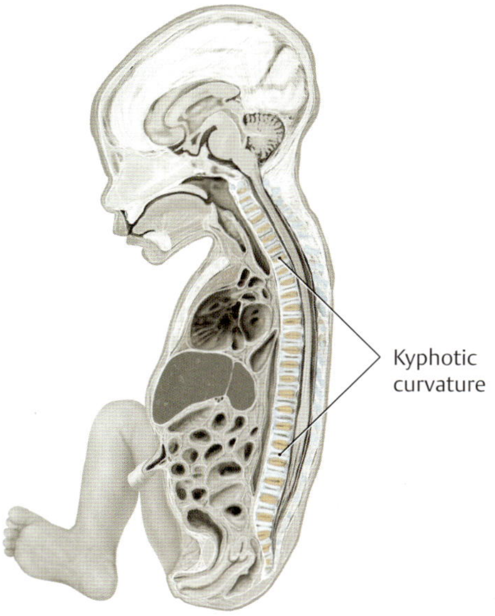

B Neonatal kyphosis

Midsagittal section through a newborn, left lateral view. Owing to the curved intrauterine position of the fetus, the newborn has a "kyphotic" spinal curvature with no lordotic straightening of the cervical and lumbar spine.

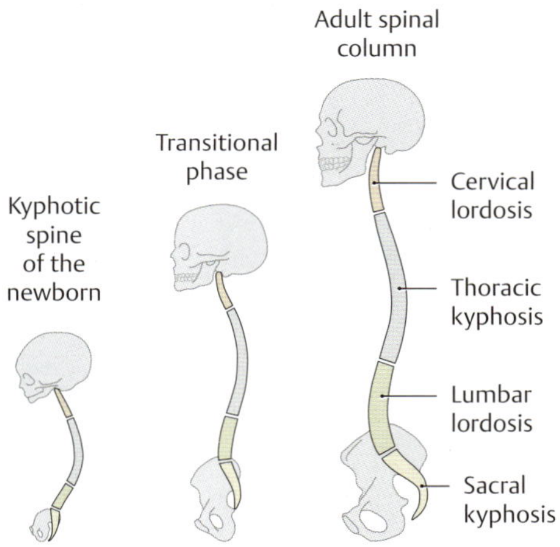

C Straightening of the spine during normal development

The characteristic curvatures of the adult spine are only partially present in the newborn (compare with **B**) and appear only during the course of postnatal development. First cervical lordosis develops to balance the head in response to the growing strength of the posterior neck muscles. Lumbar lordosis develops later as the child learns to sit, stand, and walk. The degree of lordosis increases until the legs can be fully extended at the hips, and it finally becomes stable during puberty. A similar transformation of the spinal column is observed in the phylogenetic transition from quadrupedal to bipedal locomotion.

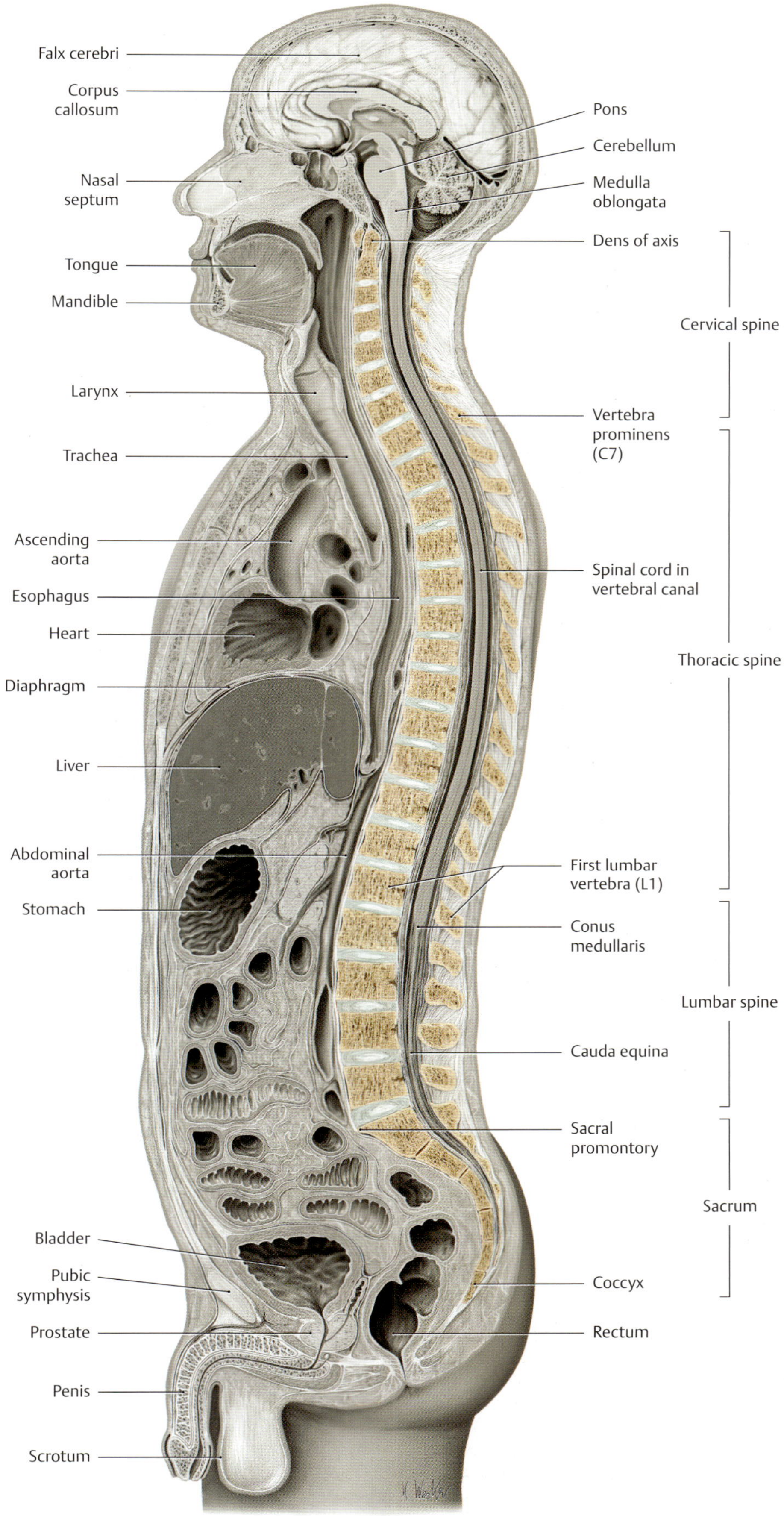

D Physiological curvatures of the adult spinal column

Midsagittal section through an adult male, left lateral view.

1.4 The Structure of a Vertebra

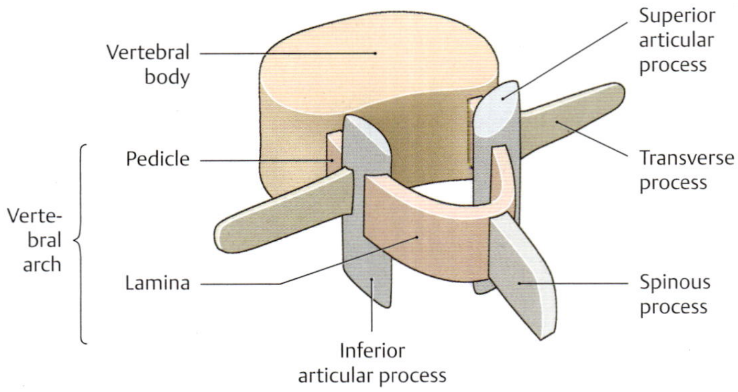

A Structural elements of a vertebra
Left posterosuperior view. All vertebrae except the atlas and axis (see p. 115) consist of the same basic structural elements:

- A vertebral body
- A vertebral arch
- A spinous process
- Two transverse processes (called costal processes in the lumbar vertebrae)
- Four articular processes

The processes give attachment to muscles and ligaments, and the bodies of the thoracic vertebrae have costovertebral joints. The vertebral bodies and arches enclose the vertebral foramen, and all of the vertebral foramina together constitute the vertebral (spinal) canal.

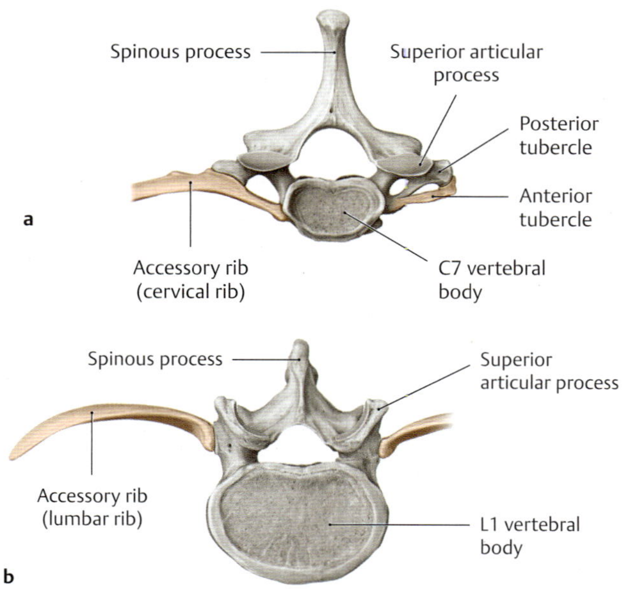

B Accessory ribs
Superior view. **a** Cervical rib, **b** lumbar rib.
The presence of anomalous cervical ribs can narrow the scalene interval, causing compression of the brachial plexus and subclavian artery (scalenus syndrome or cervical rib syndrome, see also p. 376). An accessory lumbar rib, on the other hand, has no adverse clinical effects.

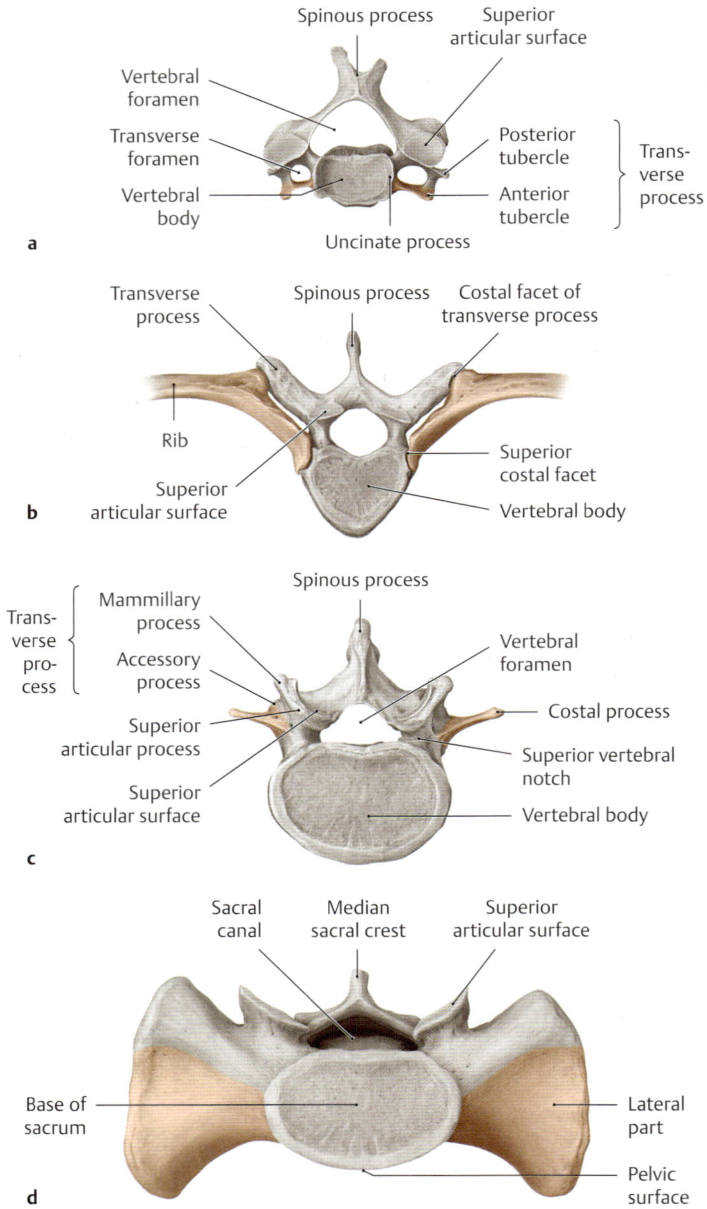

C Costal elements in different regions of the spinal column
Superior view. The shape and configuration of the vertebrae are closely related to the development of the ribs and their rudiments (indicated here by color shading).

- **a** Cervical vertebrae: Here the rudimentary rib forms a process called the anterior tubercle. It unites with the posterior tubercle to form the transverse foramen.
- **b** Thoracic vertebrae: Because these vertebrae give attachment to the ribs, their bodies and transverse processes bear corresponding cartilage-covered articular surfaces (costal facets on the transverse processes, also superior and inferior costal facets).
- **c** Lumbar vertebrae: The costal elements in the lumbar spine take the form of "transverse processes," which are much larger than in the cervical spine. Because of their size, they are also known as costal processes.
- **d** Sacrum: Here the rudimentary rib forms the anterior portion of the lateral part of the sacral vertebra. It is fused to the transverse processes.

D Typical vertebrae from different regions of the spinal column

Superior and left lateral view.

a, b First cervical vertebra (atlas).
c, d Second cervical vertebra (axis).
e, f Fourth cervical vertebra.
g, h Sixth thoracic vertebra.
i, j Fourth lumbar vertebra.
k, l Sacrum.

The vertebrae in different regions of the spinal column differ not only in their size but also in their special features. While the vertebral bodies gradually become larger from superior to inferior to accommodate the increasing stresses imposed by the gravity and body weight, the vertebral *foramina* gradually become smaller to match the decreasing diameter of the spinal cord. The arrangement of the vertebral arches and adjacent processes also varies at different levels in the spine (for details see pp. 117, 119, and 121).

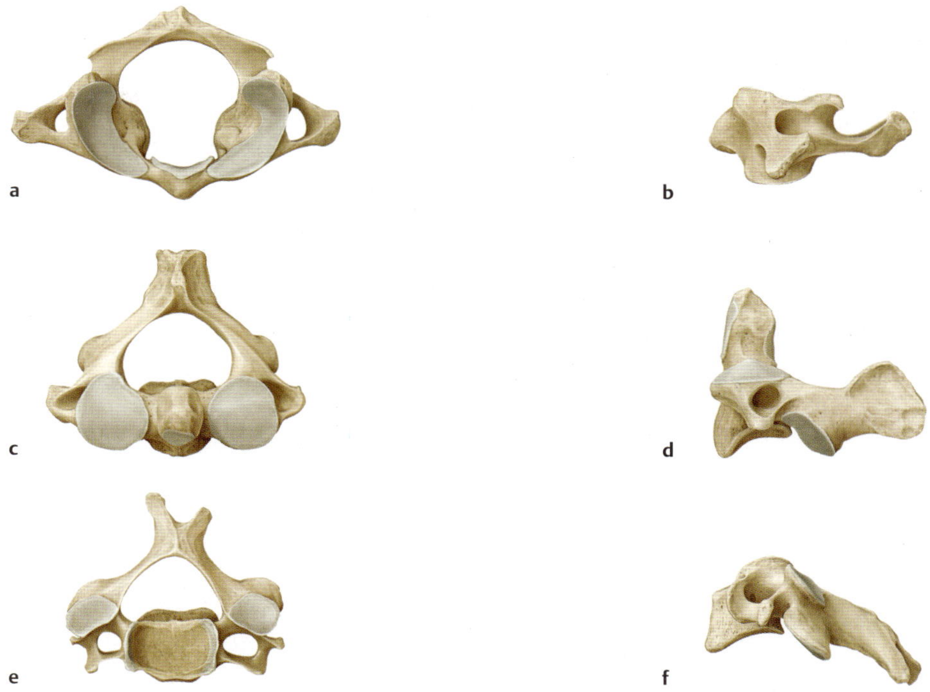

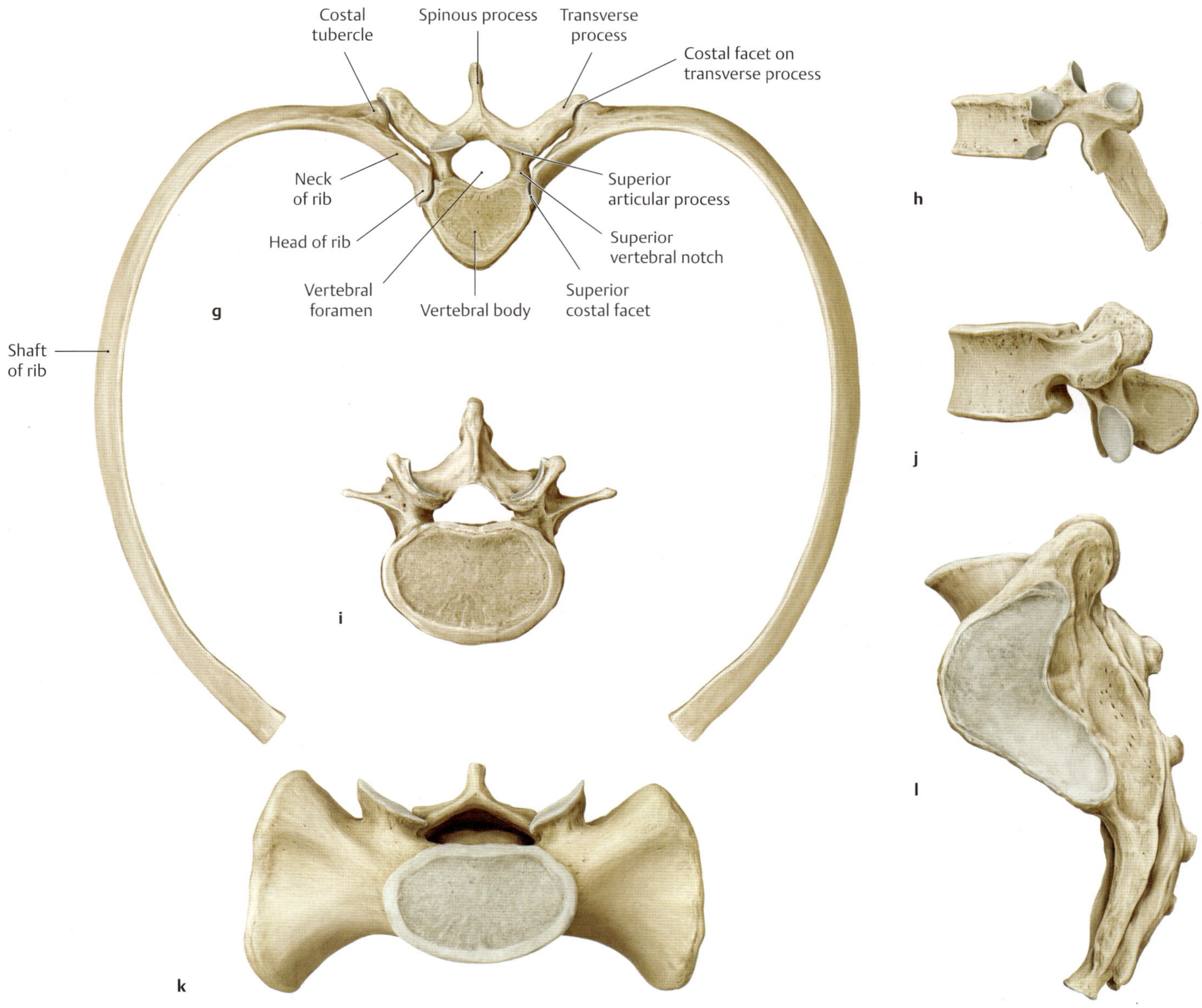

1.5 The Cervical Spine

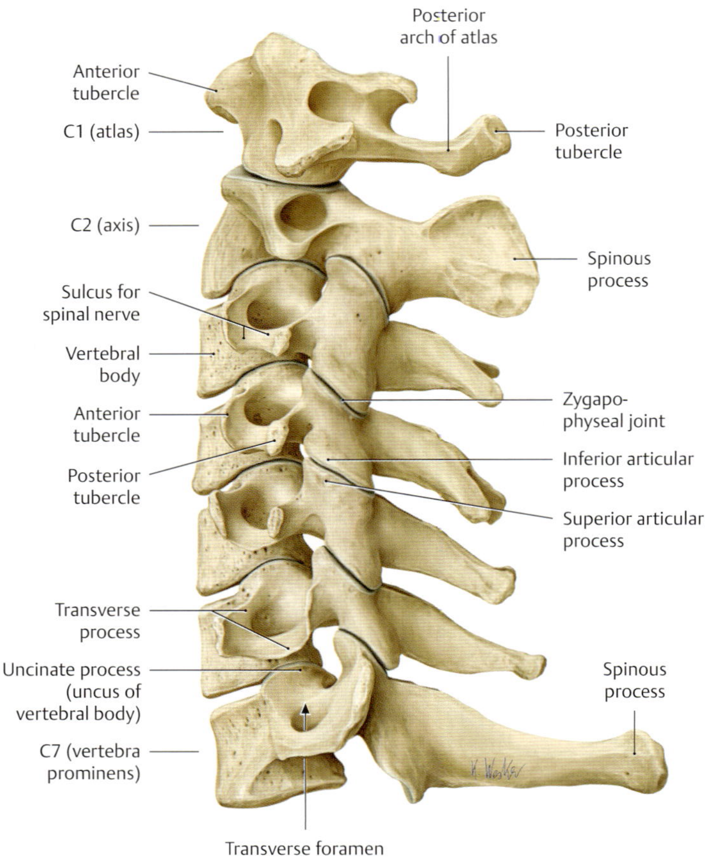

A Cervical spine, left lateral view

Of the cervical vertebrae, which number seven in all, the **first and second cervical vertebrae** (C1 and C 2, atlas and axis) differ most conspicuously from the common vertebral morphology. They are specialized for bearing the weight of the head and allowing the head to move in all directions, similar to a ball-and-socket joint. Each of the **remaining five cervical vertebrae** (C3–C7) has a relatively small body, which presents a more or less square shape when viewed from above, and a large, triangular vertebral foramen (see **Cc**). The superior and inferior surfaces of the vertebral bodies are saddle-shaped, the superior surfaces bearing lateral uncinate processes that do not appear until about the tenth year of life (see p. 134). The transverse process consists of an anterior and a posterior bar, which terminate laterally in two small tubercles (anterior and posterior tubercles). These bars enclose the transverse foramen, through which the vertebral artery ascends from the C6 to the C1 level. The superior surface of the transverse process of the first three cervical vertebrae bears a broad, deep notch (spinal nerve sulcus), in which lies the emerging spinal nerve at that level. The superior and inferior articular processes are broad and flat. Their articular surfaces are flat and are inclined approximately 45° from the horizontal plane. The spinous processes of the third through sixth cervical vertebrae are short and bifid. The spinous process of the seventh cervical vertebra is longer and thicker than the others and is the first of the spinous processes that is distinctly palpable through the skin (vertebra prominens).

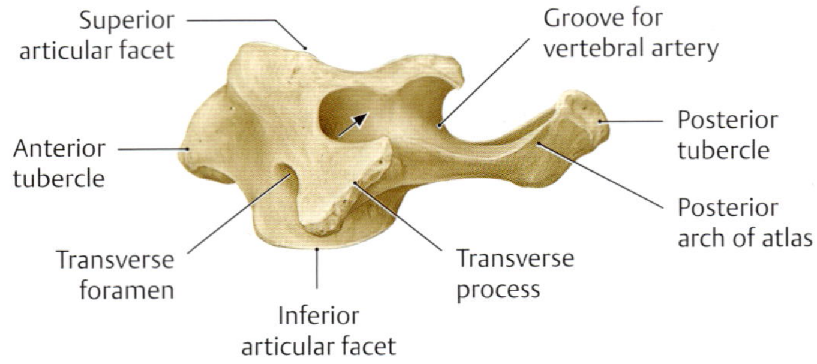

a First cervical vertebra (C1, atlas)

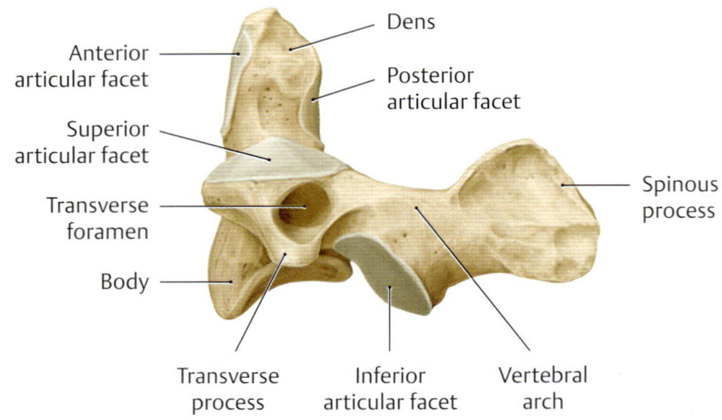

b Second cervical vertebra (C2, axis)

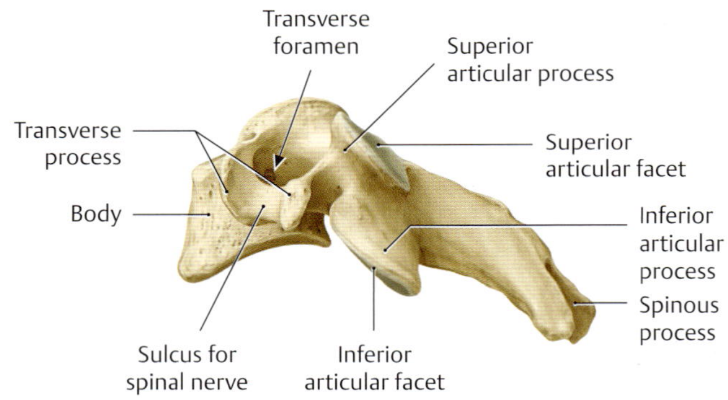

c Fourth cervical vertebra (C4)

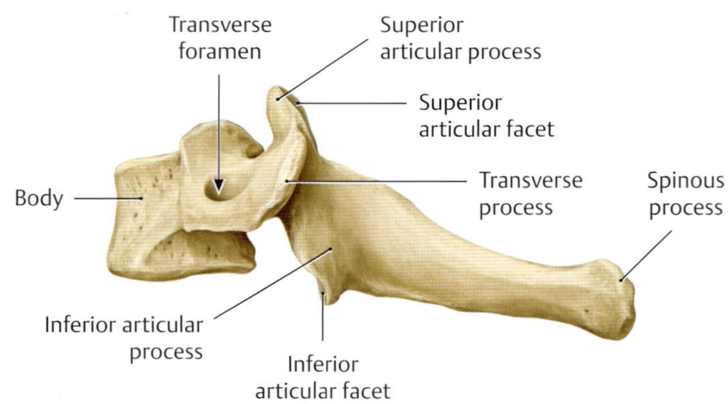

d Seventh cervical vertebra (C7, vertebra prominens)

B Cervical vertebrae, left lateral view

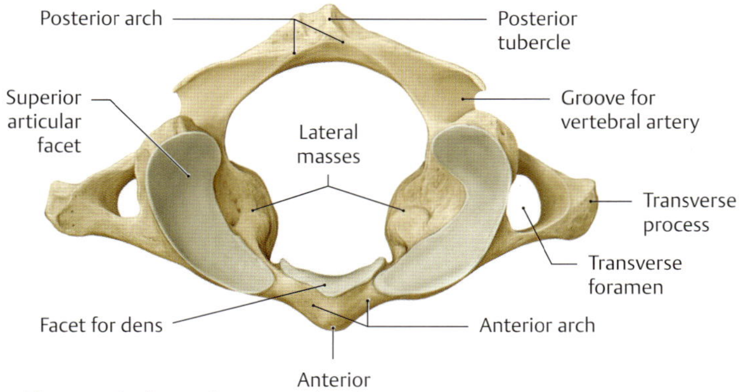

a First cervical vertebra (atlas)

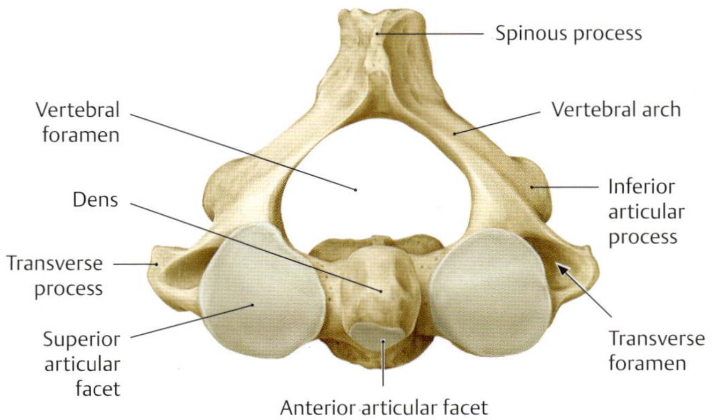

b Second cervical vertebra (axis)

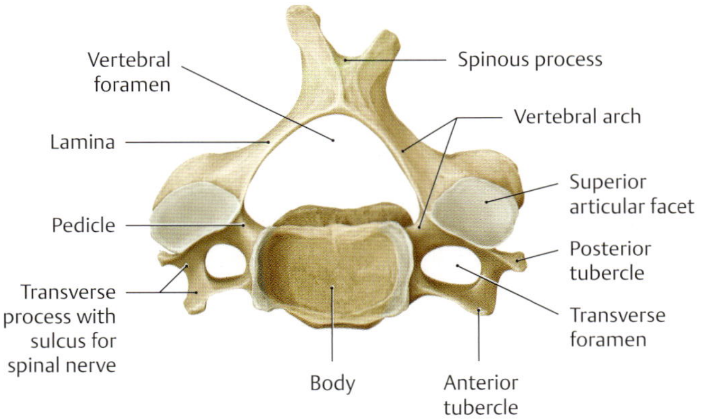

c Fourth cervical vertebra

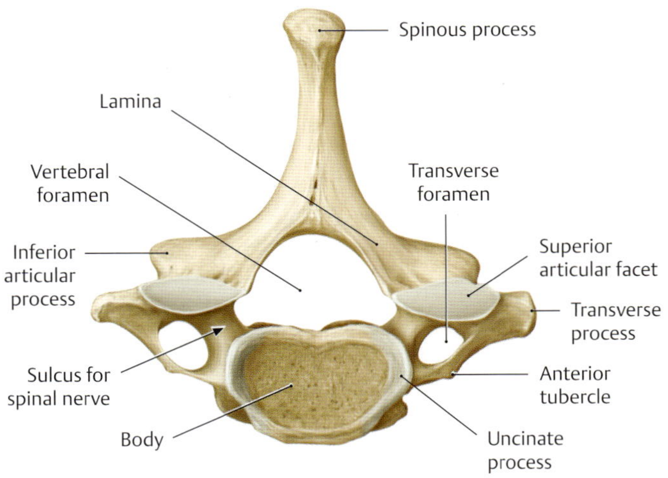

d Seventh cervical vertebra (vertebra prominens)

C Cervical vertebrae, superior view

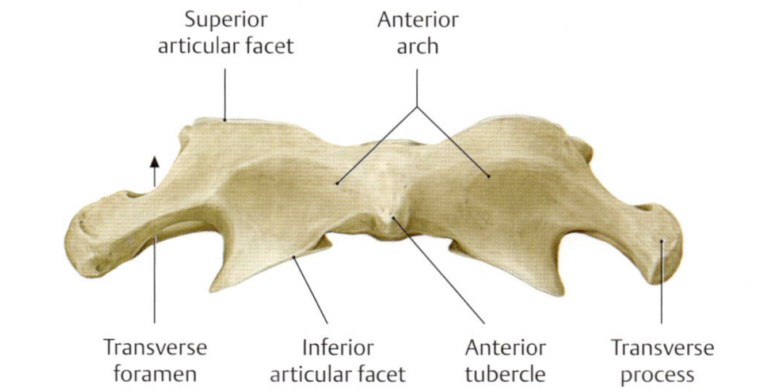

a First cervical vertebra (atlas)

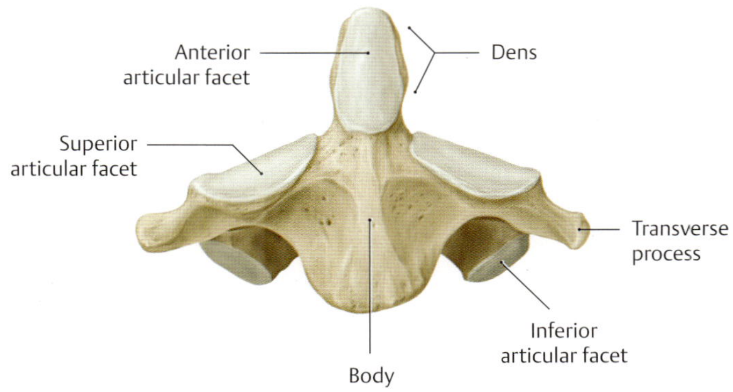

b Second cervical vertebra (axis)

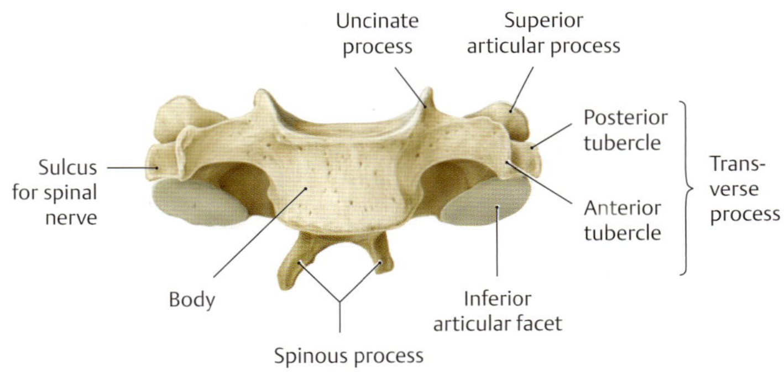

c Fourth cervical vertebra

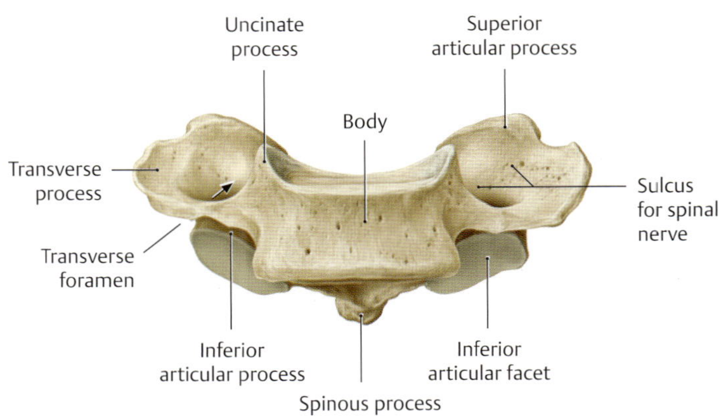

d Seventh cervical vertebra (vertebra prominens)

D Cervical vertebrae, anterior view

1.6 The Thoracic Spine

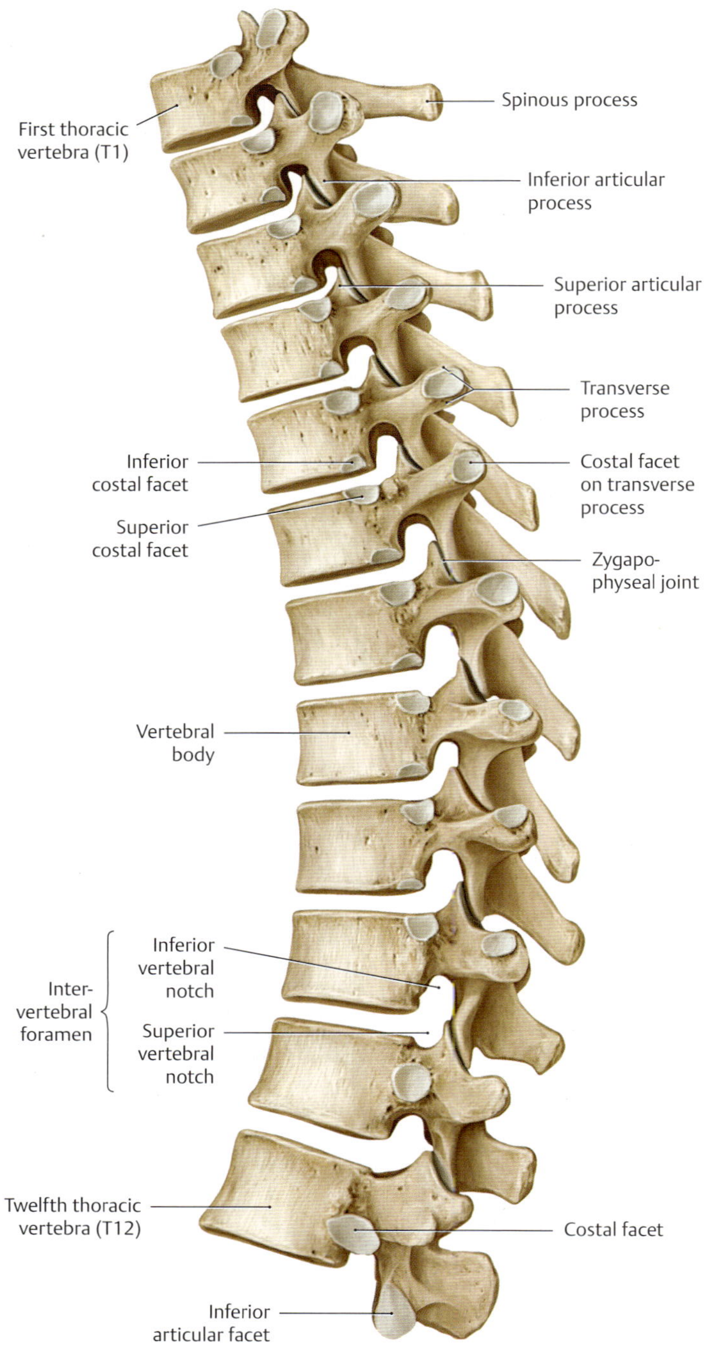

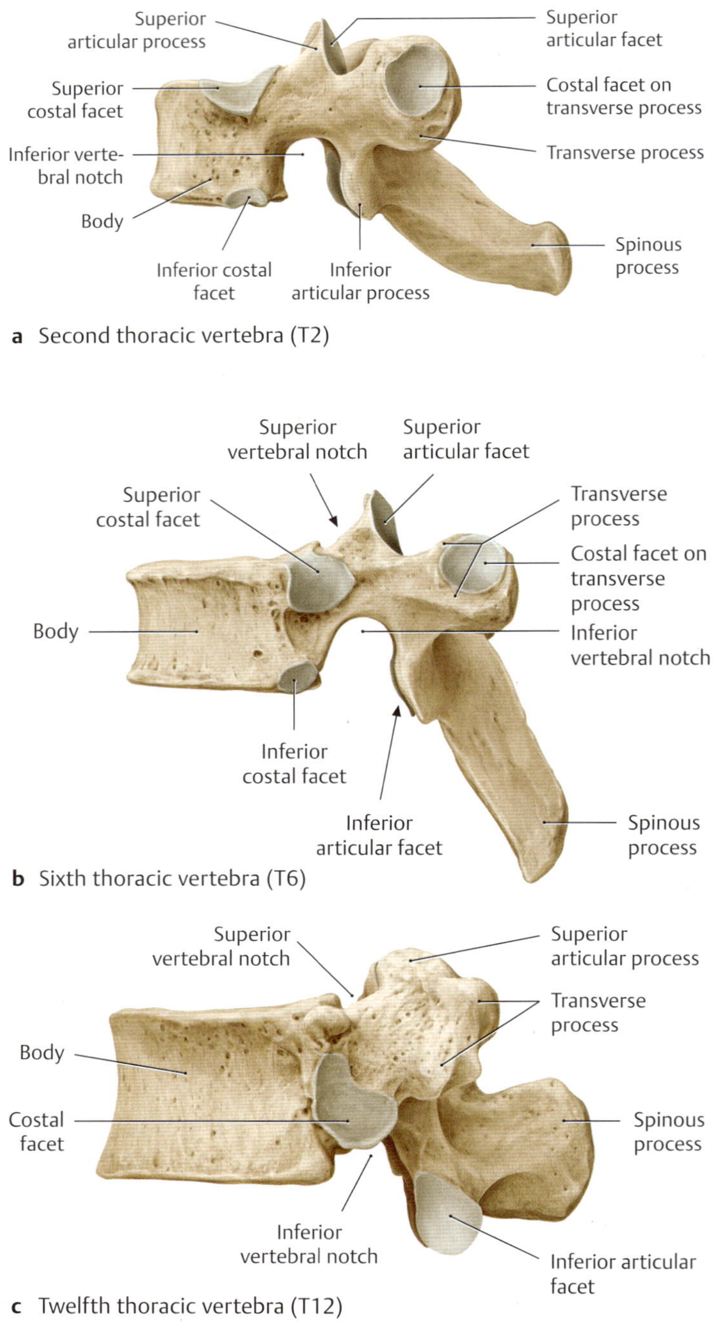

a Second thoracic vertebra (T2)

b Sixth thoracic vertebra (T6)

c Twelfth thoracic vertebra (T12)

A Thoracic spine, left lateral view

The thoracic vertebral bodies gradually become taller and broader from superior to inferior, the lower vertebral bodies assuming a transverse oval shape like that of the lumbar vertebrae. The vertebral foramen is roughly circular and is smaller than in the cervical and lumbar vertebrae. The end plates are rounded and triangular. The spinous processes are long and angled sharply inferiorly, creating an overlapping arrangement that interlinks the thoracic vertebrae. The facets of the *inferior* articular processes are directed anteriorly, while the facets of the *superior* articular processes face posterior so that they can articulate with the inferior facets to form the zygapophyseal or facet joints (p. 126). Another feature of the thoracic vertebrae is that their transverse processes are angled backward to allow for articulations with the ribs.

B Thoracic vertebrae, left lateral view

The costal facets are cartilage-covered surfaces that articulate with the corresponding ribs (see p. 145). The bodies of the second through ninth thoracic vertebrae (T1–T9) bear two articular facets on each side—a superior costal facet and an inferior costal facet—such that two adjacent vertebrae combine to articulate with the head of a rib. Thus, a given numbered rib articulates with its own numbered vertebra and the vertebra above. Exceptions to this scheme are ribs 1 and 10 to 12, which articulate only with the vertebral body of the same number. The body of the tenth thoracic vertebra has only superior articular facets, and the tenth rib usually articulates only with the tenth thoracic vertebrae. As mentioned, the transverse processes of the thoracic vertebrae (except for T11 and T12) also bear articular facets for the ribs.

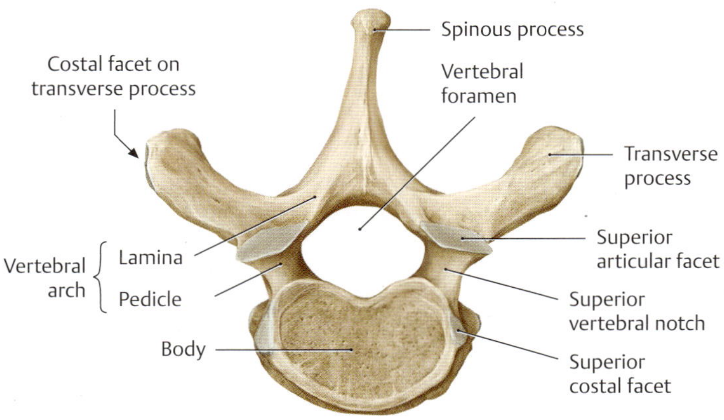

a Second thoracic vertebra (T2)

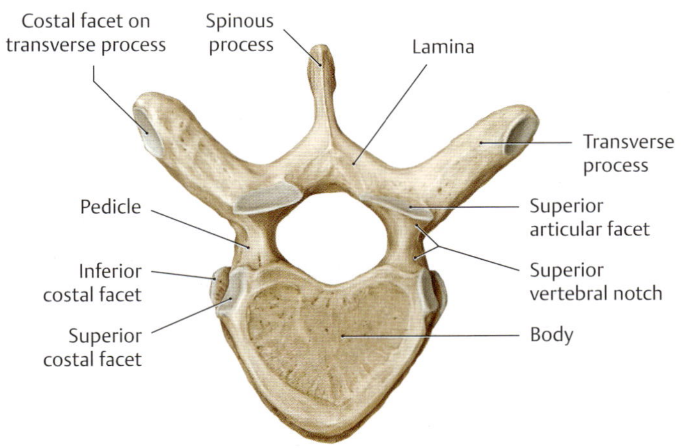

b Sixth thoracic vertebra (T6)

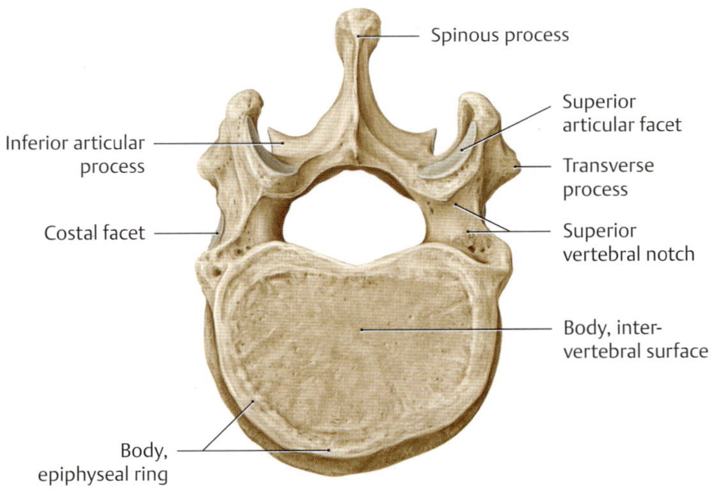

c Twelfth thoracic vertebra (T12)

C Thoracic vertebrae, superior view
The laminae and pedicles make up the vertebral arch.

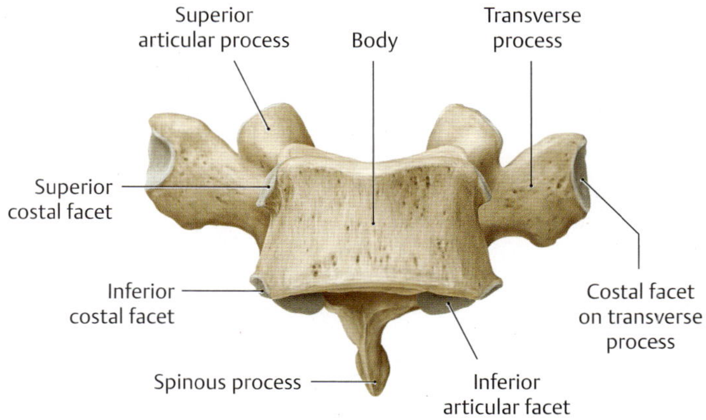

a Second thoracic vertebra (T2)

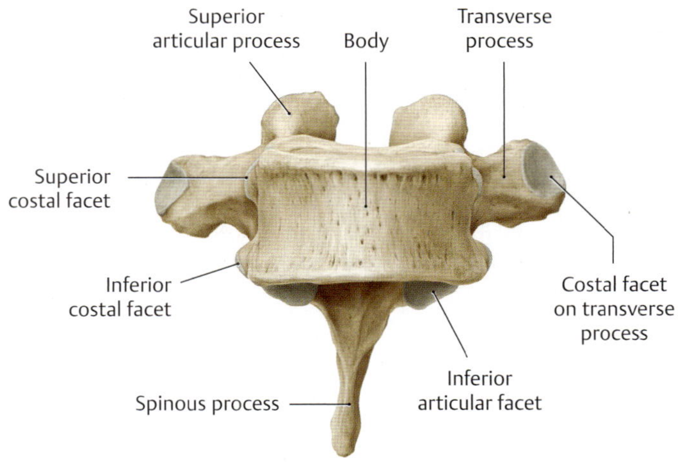

b Sixth thoracic vertebra (T6)

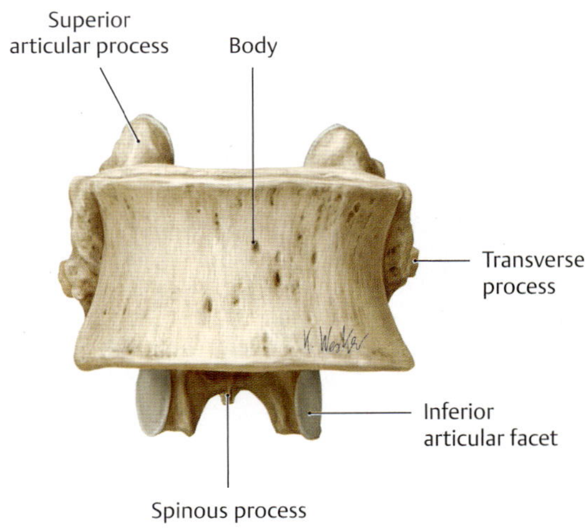

c Twelfth thoracic vertebra (T12)

D Thoracic vertebrae, anterior view

1.7 The Lumbar Spine

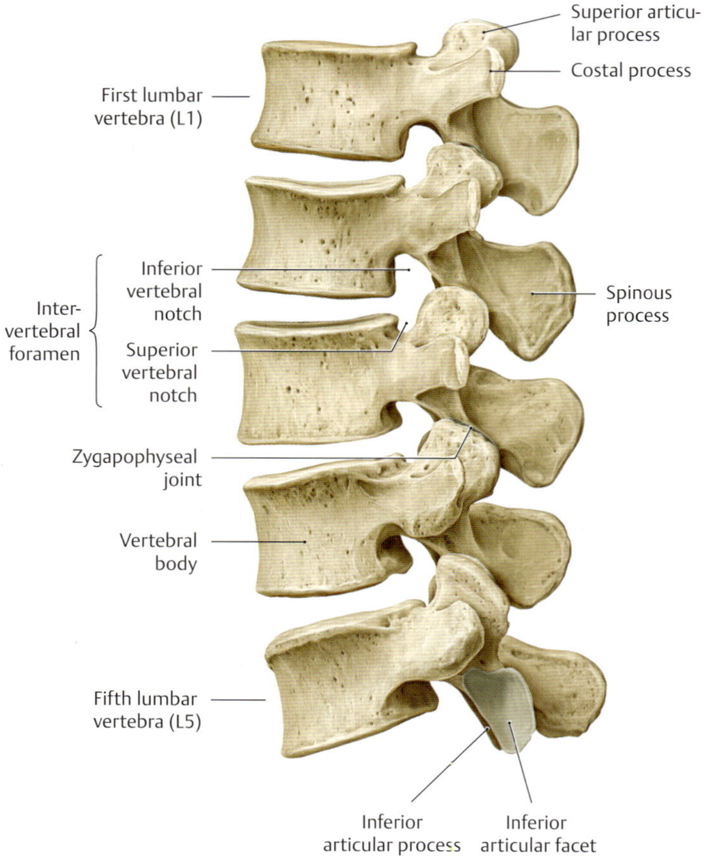

A Lumbar spine, left lateral view

The bodies of the lumbar vertebrae are large and have a transverse oval shape when viewed from above (see **C**). The massive vertebral arches enclose an almost triangular vertebral foramen, and they unite posteriorly to form a thick spinous process that is flattened on each side. The "transverse processes" of the lumbar vertebrae correspond phylogenetically to rudimentary ribs (see p. 114). Thus, they are more accurately termed costal processes and are not homologous with the transverse processes of the other vertebrae. The thick costal process is fused with the actual transverse process, which is a small, pointed eminence located at the root of each costal process (accessory process, see **Cb**). The relatively massive superior and inferior articular processes bear slightly inclined articular facets that have a vertical, almost sagittal orientation. The articular facets of the superior articular processes are slightly concave and face medially, while those of the inferior articular processes are slightly convex and face laterally. The mammillary processes on the lateral surfaces of the superior articular processes serve as origins and insertions for the intrinsic back muscles (see **Bb** and **Ca**).

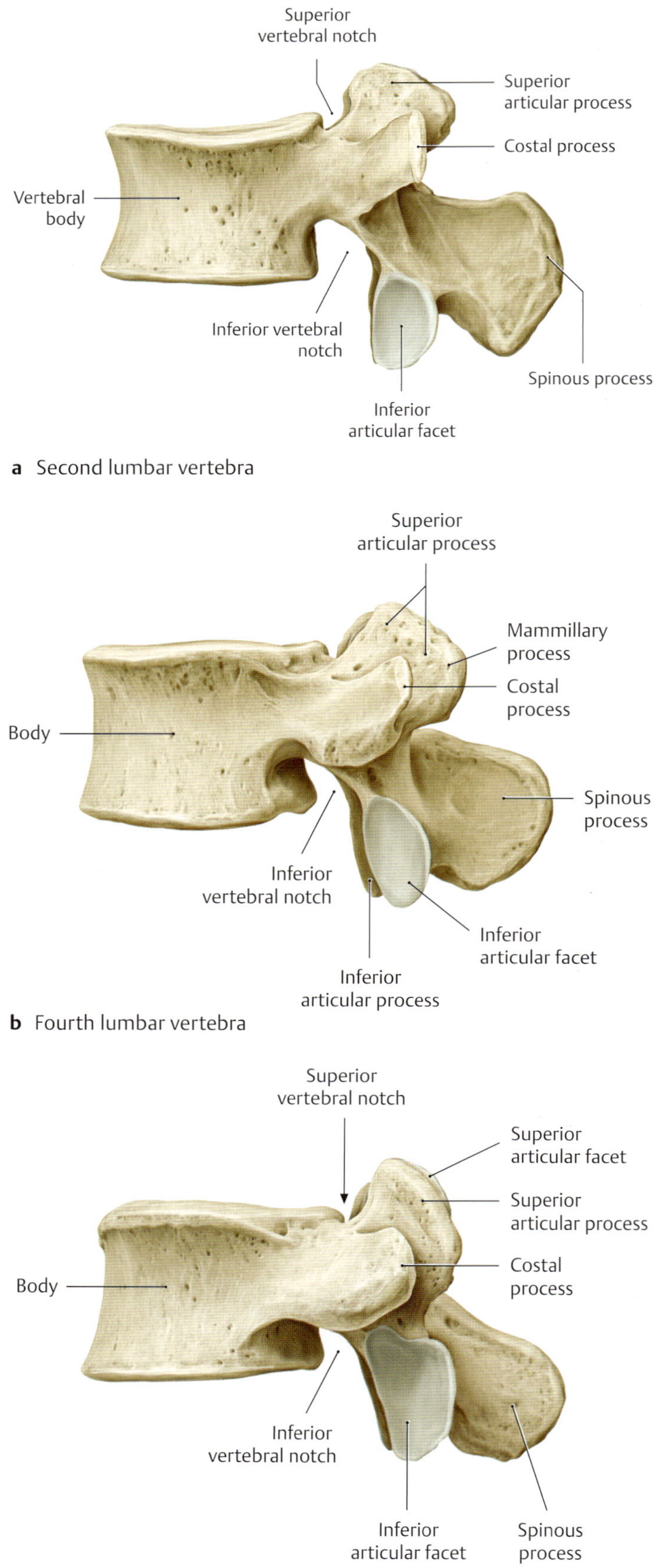

a Second lumbar vertebra

b Fourth lumbar vertebra

c Fifth lumbar vertebra

B Lumbar vertebrae, left lateral view

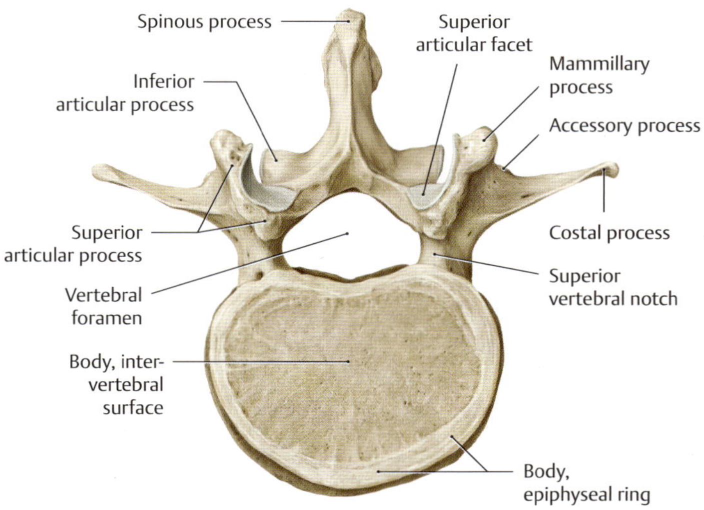

a Second lumbar vertebra

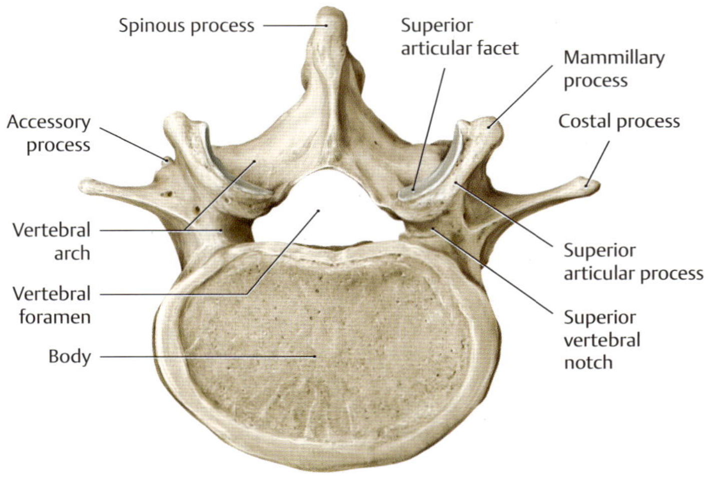

b Fourth lumbar vertebra

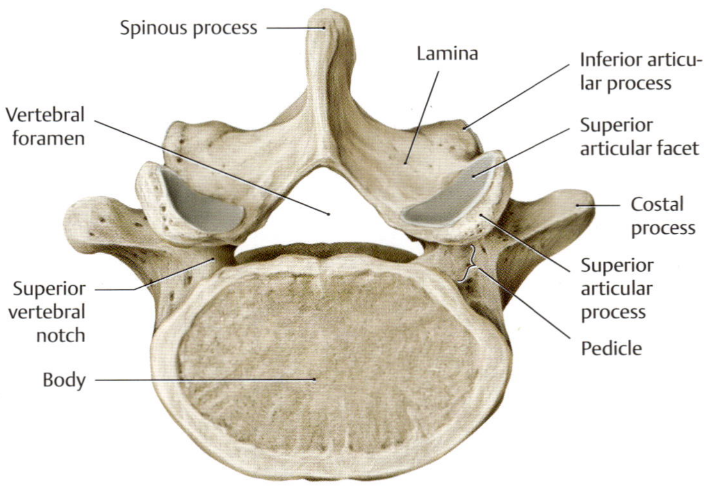

c Fifth lumbar vertebra

C Lumbar vertebrae, superior view

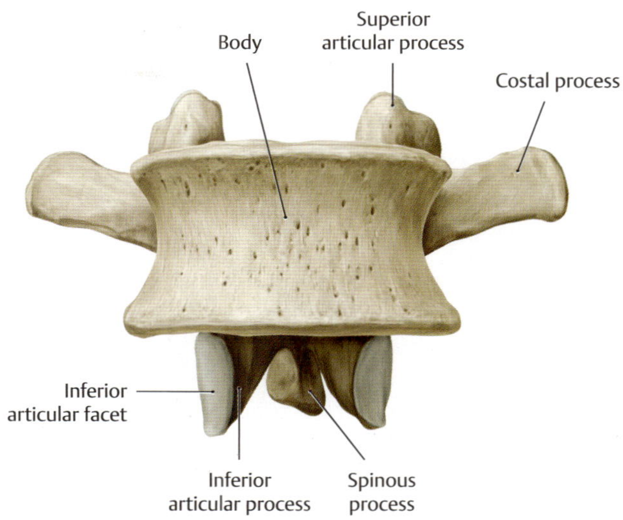

a Second lumbar vertebra

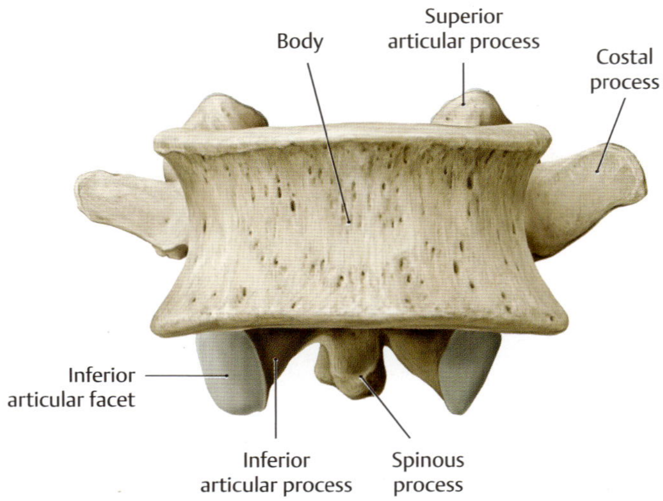

b Fourth lumbar vertebra

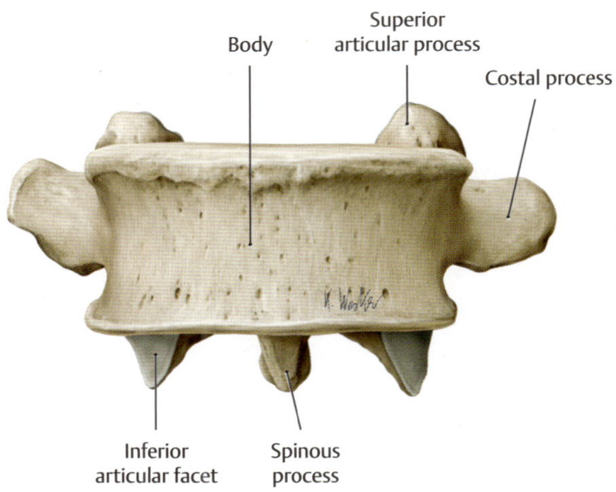

c Fifth lumbar vertebra

D Lumbar vertebrae, anterior view

1.8 The Sacrum and Coccyx

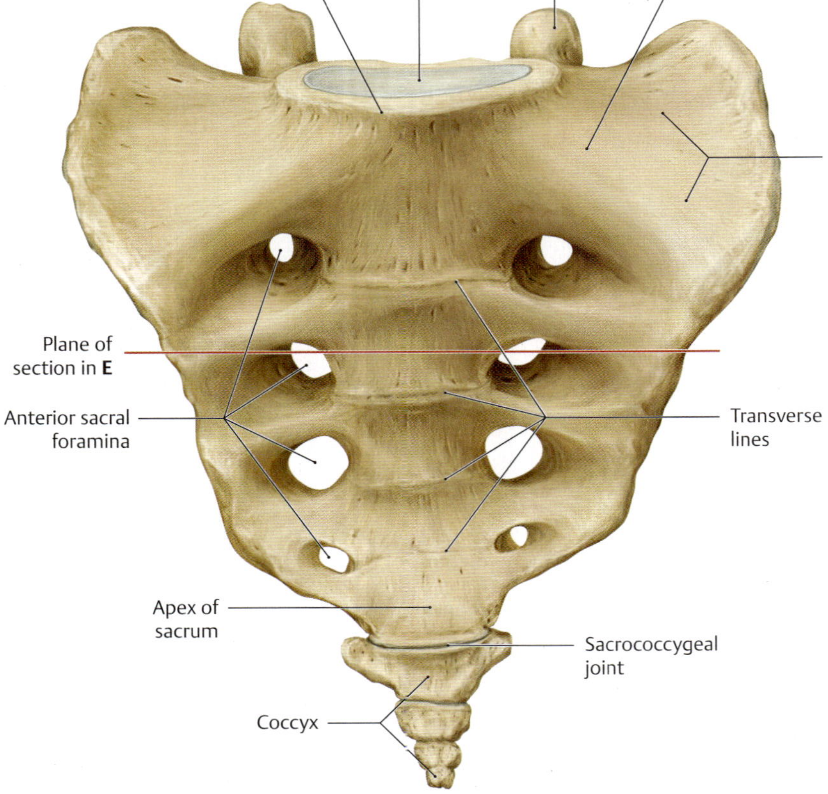

A Sacrum and coccyx, anterior (pelvic) view

The sacrum at birth is composed of five separate sacral vertebrae. Postnatally, they fuse to form a single bone that is flattened anteroposteriorly and has a triangular shape when viewed from the front. The *base of the sacrum*, located at the superior end of the bone, articulates with the body of the fifth lumbar vertebra by a wedge-shaped intervertebral disk. The *apex of the sacrum* is at the inferior end of the bone and articulates with the coccyx. The anterior or *pelvic surface* of the sacrum is concave in both the sagittal and transverse planes (see **C**). Between the anterior sacral foramina are four transverse ridges (transverse lines), which mark the sites of fusion between the five sacral vertebrae. The coccyx consists of three or four rudimentary vertebrae. Only the first coccygeal vertebra exhibits some of the typical structural elements of a fully formed vertebra. A cartilaginous disk usually connects the base of the coccyx to the apex of the sacrum (sacrococcygeal joint). This joint allows passive forward and backward motion of the coccyx, increasing the anteroposterior diameter of the pelvic outlet during childbirth (see p. 147).

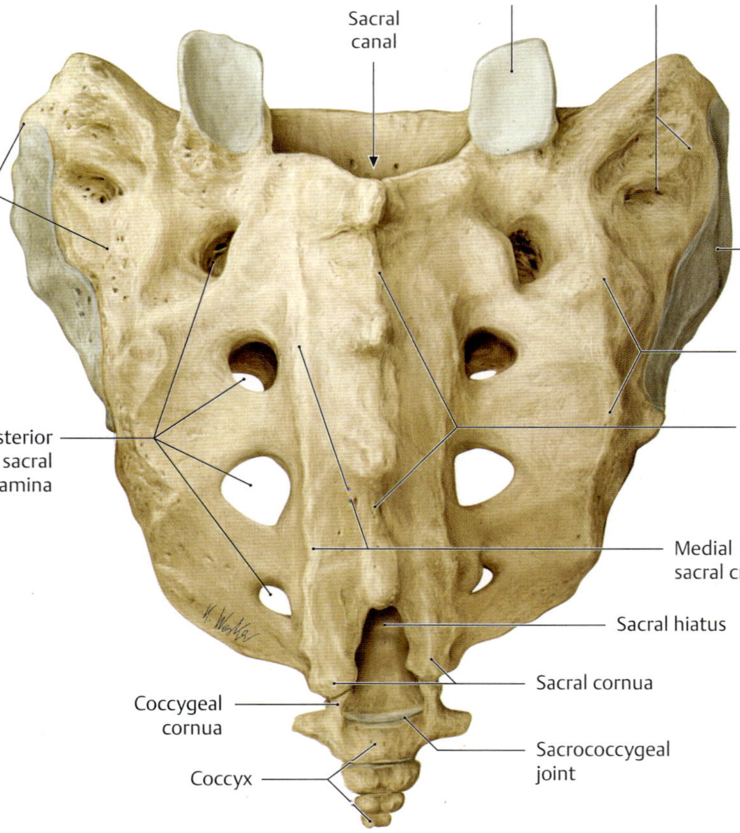

B Sacrum and coccyx, posterior view

The fused spinous processes form a jagged bony ridge on the convex dorsal surface of the sacrum, the *median sacral crest*. This ridge is flanked on each side by the paired medial sacral crests, formed by fusion of the articular processes. The medial crests are continuous below with the rudimentary inferior articular processes of the fifth sacral vertebra (the sacral cornua); they are continuous above with the two superior articular processes that face posteriorly. Between the sacral cornua is an aperture, the sacral hiatus, which is formed by the incomplete vertebral arch of the fifth sacral vertebrae and provides access to the sacral canal (e.g., for anesthesia). Lateral to the posterior sacral foramina are another pair of longitudinal ridges, the *lateral sacral crests*, formed by the fused transverse processes. The bony union of the transverse processes with the rudimentary ribs forms the thick *lateral* parts of the sacral wing that flank the body of the sacrum. Each lateral part bears an ear-shaped ("auricular") surface that articulates with the ilium (see **C**).

C Sacrum, left lateral view

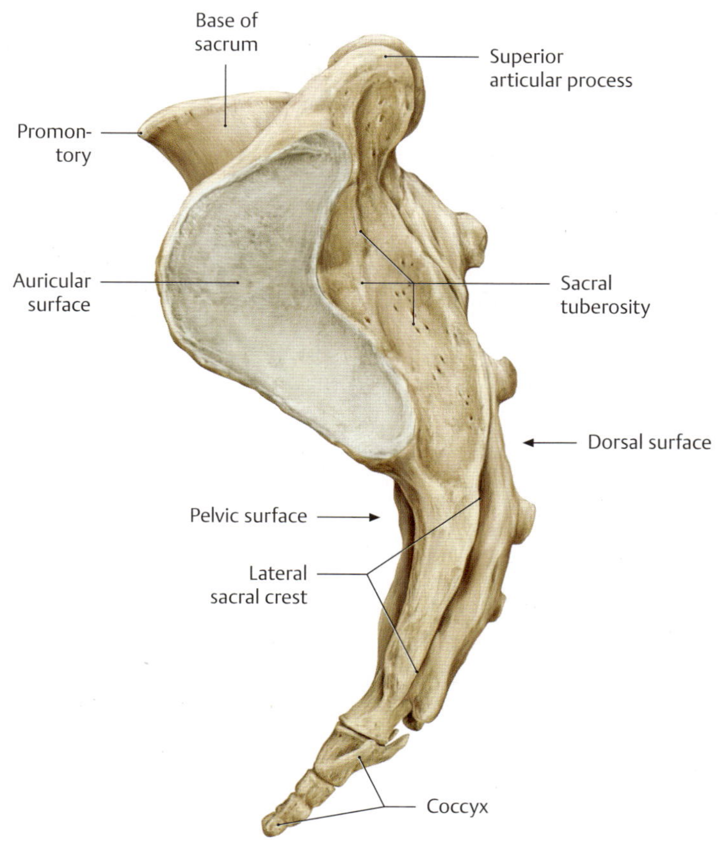

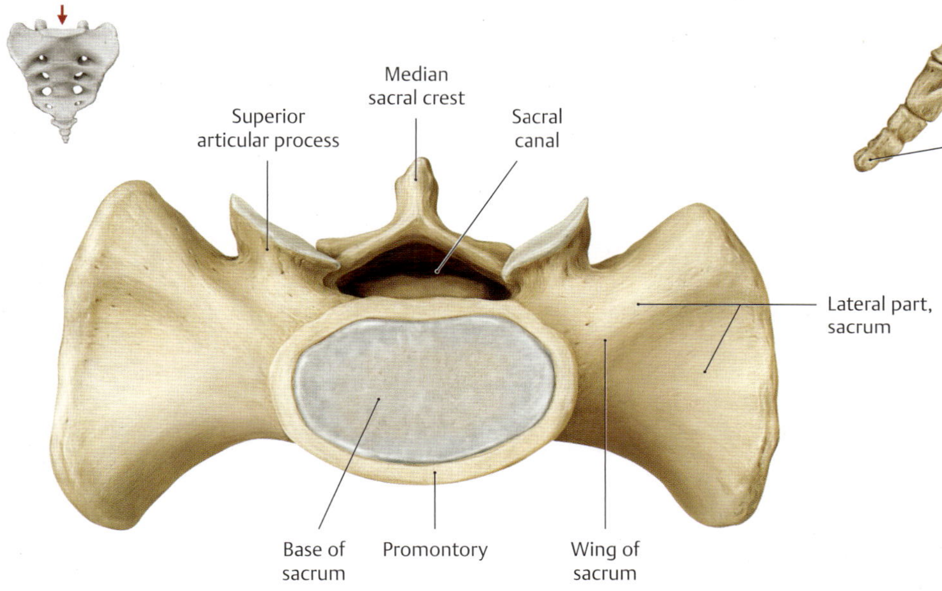

D Sacrum, superior view

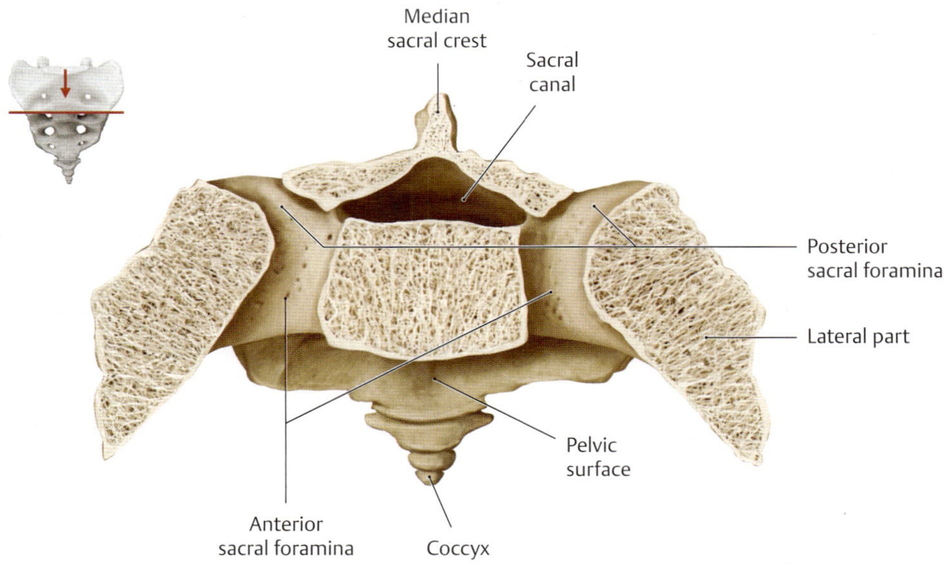

E Transverse section through the sacrum
Superior view (the level of the section is shown in **A**). Fusion of the upper four sacral vertebrae creates four T-shaped bony canals at the level of the intervertebral foramina on each side, providing sites of emergence for the first through fourth sacral nerves. The corresponding ventral and dorsal rami of the spinal nerves exit the bony canals through the anterior and posterior sacral foramina (see p. 555).

1.9 The Intervertebral Disk: Structure and Function

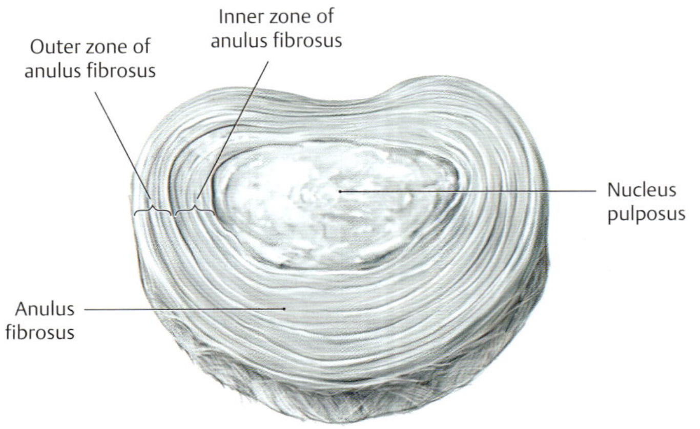

A Structure of an intervertebral disk
Isolated lumbar intervertebral disk, anterosuperior view.
The intervertebral disk consists of an external fibrous ring, the *anulus fibrosus*, and a gelatinous core, the *nucleus pulposus*. The anulus fibrosus consists of an outer zone and an inner zone. The outer zone is a fibrous sheath that possesses high tensile strength and is made up of concentric laminae of type I collagen fibers. Its fiber systems crisscross due to their varying obliquity and interconnect the marginal ridges of two adjacent vertebrae (see **B**), to which they are attached. At the junction with the inner zone of the anulus fibrosus, the tough fibrous tissue of the outer zone blends with a fibrocartilaginous tissue whose type II collagen fibers are attached to the hyaline cartilage end plates of the vertebral bodies (see **Da** and **Ea**).

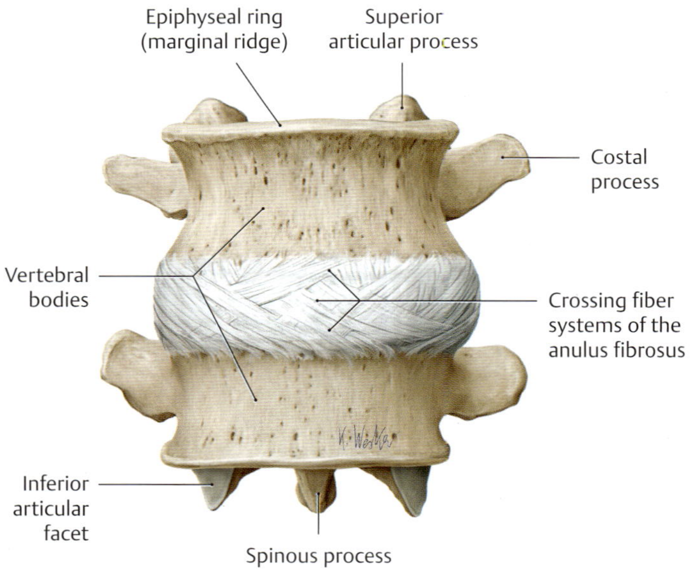

B Outer zone of the anulus fibrosus
Intervertebral disk between the third and fourth lumbar vertebrae, anterior view.
The connective-tissue fiber bundles in the outer zone of the anulus fibrosus cross one another at various angles, interconnecting the bony marginal ridges of two adjacent vertebral bodies.

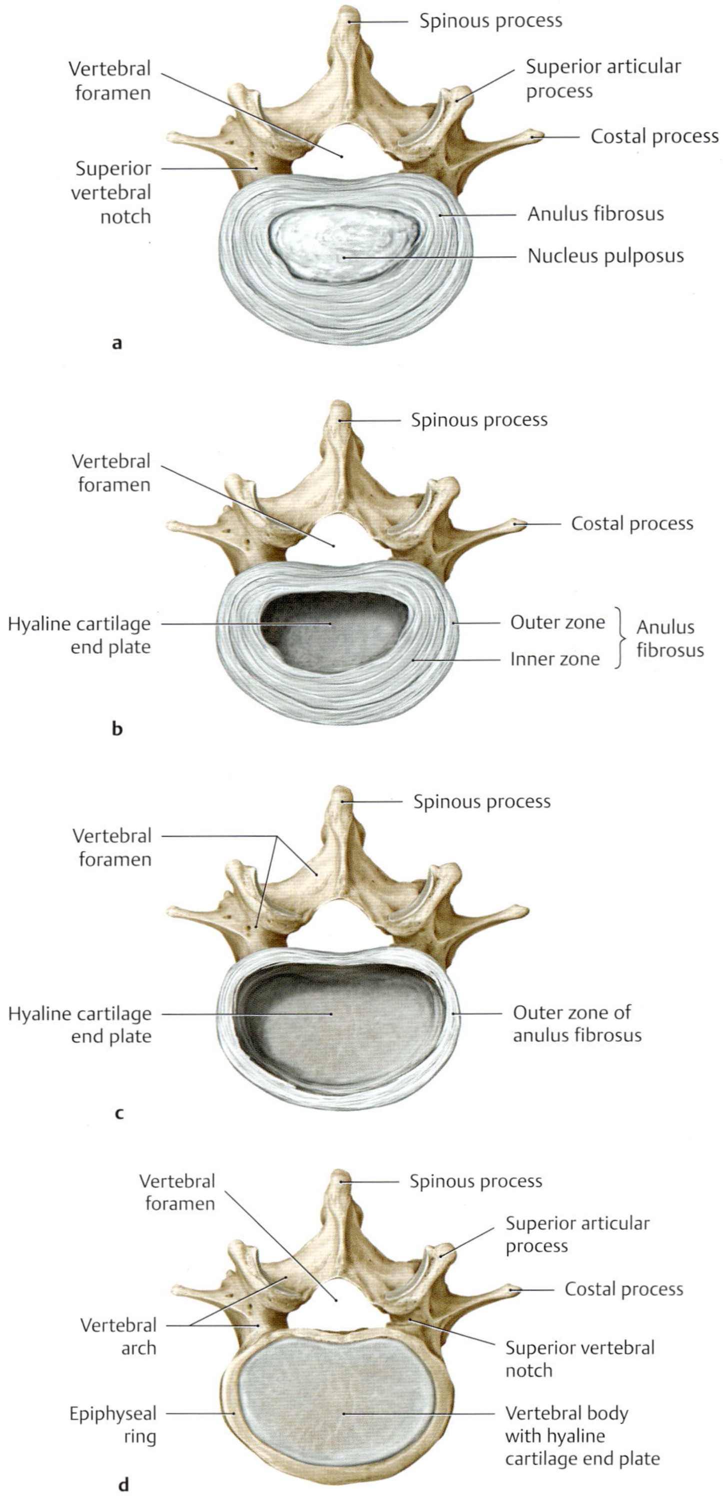

C Main structural components of an intervertebral disk
Fourth lumbar vertebra with its associated upper disk, superior view.

a Intervertebral disk with the anulus fibrosus and nucleus pulposus.
b Anulus fibrosus (with the nucleus pulposus removed).
c Outer zone of the anulus fibrosus (with the inner zone removed).
d Hyaline cartilage end plate within the bony marginal ridge (entire intervertebral disk removed).

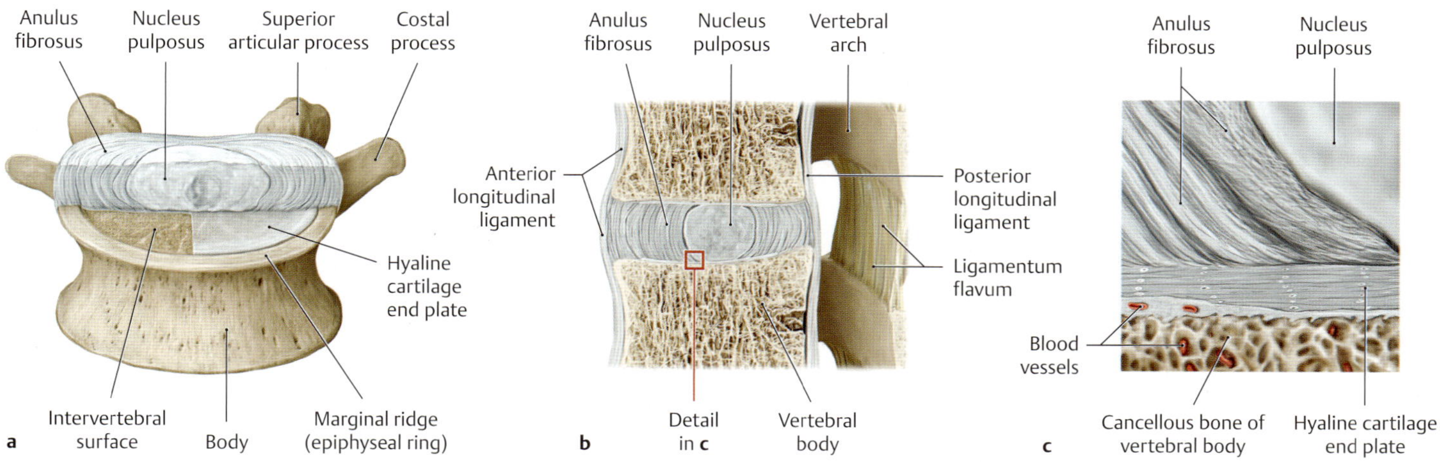

D Position of the intervertebral disk in the motion segment

a Hyaline cartilage end plate, anterosuperior view (the anterior half of the disk and right half of the end plate have been removed).

b Sagittal section through a motion segment (see p. 132), left lateral view.

c Detail from **b**.

Except for the outer zone of the anulus fibrosus, the entire disk is in contact superiorly and inferiorly with the hyaline cartilage layer of the adjacent vertebral endplates. The subchondral, bony portion of the end plate consists of compact bone (intervertebral surface) permeated by myriad pores (see **c**) through which the vessels in the bone marrow spaces of the vertebral bodies can supply nutrients to the disk tissue.

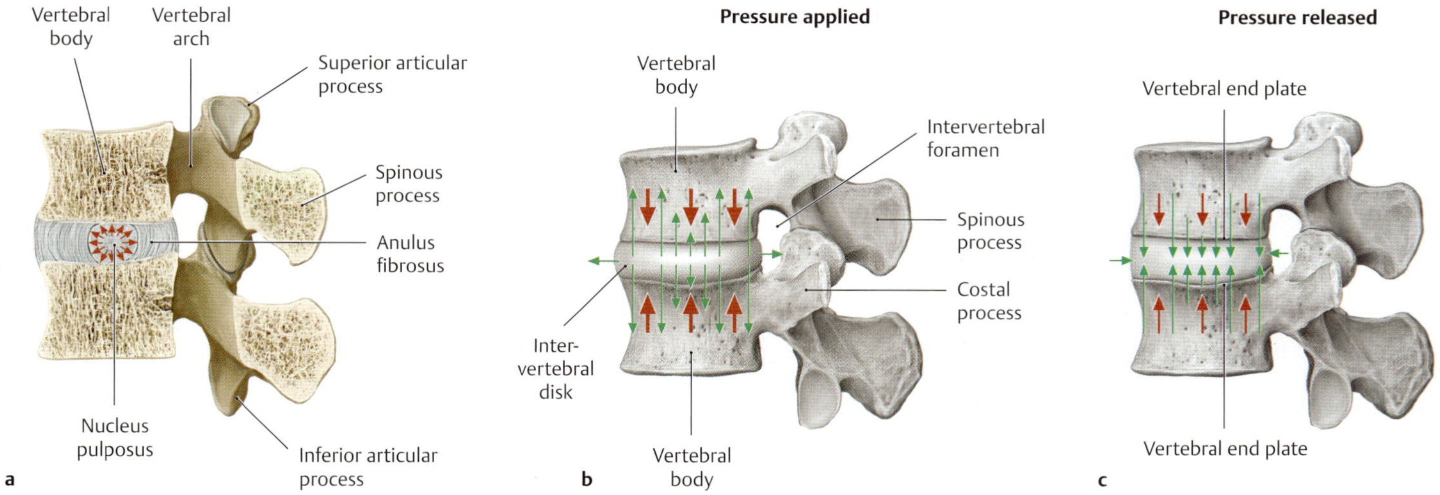

E Load-dependent fluid shifts in the intervertebral disk

a The nucleus pulposus functions as a "water cushion" to absorb transient axial loads on the intervertebral disk. Mechanically, the disk represents a hydrostatic system that is resilient under pressure. It is composed of a tension-resistant sheath (the anulus fibrosus) and a watery, incompressible core, the nucleus pulposus. This core consists of 80 to 85% water, which it can reversibly bind in its paucicellular, gelatinous, mucoviscous tissue (owing to its high content of glycosaminoglycans). The nucleus pulposus is under a very high hydrostatic pressure, particularly when acted upon by gravity and other forces. This pressure can be absorbed by the adjacent cartilaginous end plates and also by the anulus fibrosus (which transforms compressive forces into tensile forces). In this way the nucleus pulposus functions as a "water cushion" or hydraulic press between two adjacent vertebral bodies. Combined with the anulus fibrosus, it acts as an effective shock absorber that can distribute pressures uniformly over the adjacent vertebral end plates.

b Fluid outflow from the intervertebral disk (green arrows) in response to a sustained pressure load (thick red arrows). While transient loads are cushioned by the shock-absorber function of the nucleus pulposus and anulus fibrosus (see **a**), sustained loads cause a gradual but permanent outflow of fluid from the disk. The turgor and height of the disk are reduced, while the end plates and eventually the bony vertebral elements move closer together (see p. 139 for further details on disk degeneration).

c Fluid uptake by the intervertebral disk (green arrows) when pressure is released (thin red arrows). The process described in **b** is reversed when the pressure on the disk is released, and the height of the disk increases. This increase is caused by fluid uptake from the subchondral blood vessels in the bone marrow spaces, which are instrumental in disk nutrition (see **Dc**). As a result of pressure-dependent fluid shifts (convection) in the intervertebral disks, the overall body height decreases temporarily by approximately 1% (1.5–2.0 cm) relative to the initial body height during the course of the day.

1.10 The Ligaments of the Spinal Column: Overview and Thoracolumbar Region

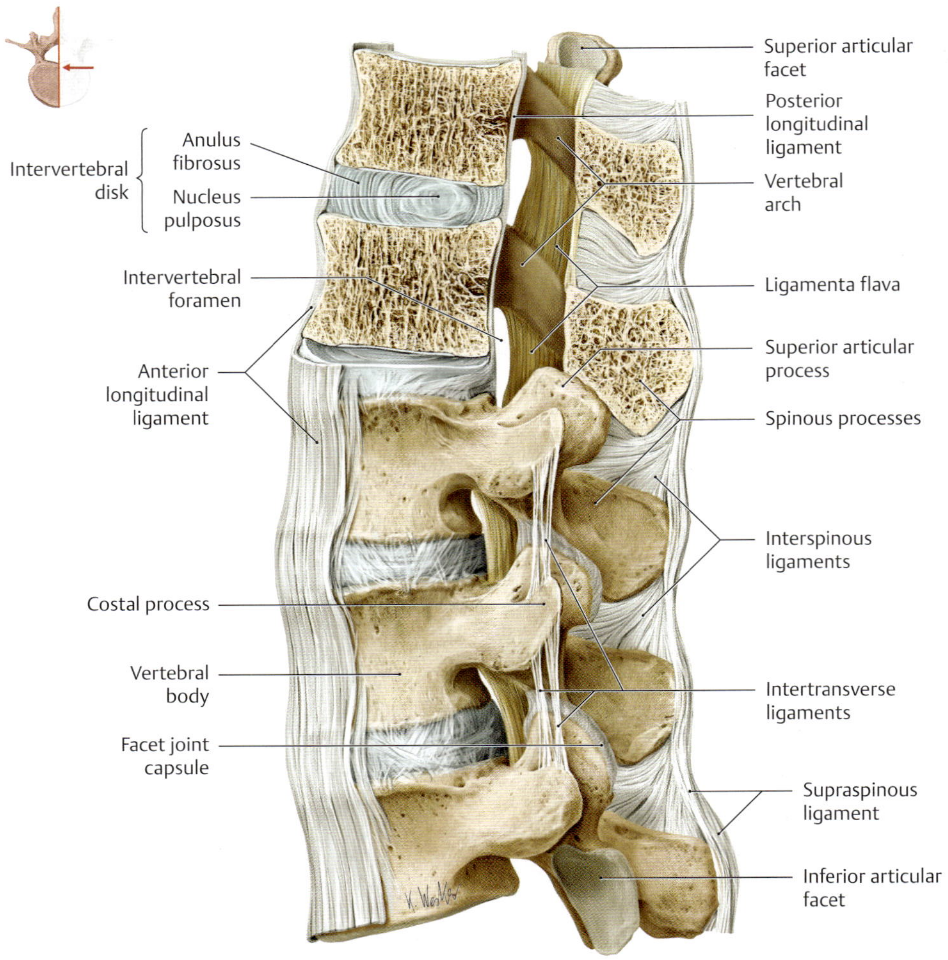

A The ligaments of the spinal column at the level of the thoracolumbar junction (T11–L3)
Left lateral view. The two uppermost thoracic vertebrae have been sectioned in the midsagittal plane.

B The ligaments of the spinal column
The ligaments of the spinal column bind the vertebrae securely to one another and enable the spine to withstand high mechanical loads and shearing stresses. The ligaments are subdivided into vertebral body ligaments and vertebral *arch* ligaments.

Vertebral body ligaments
• Anterior longitudinal ligament • Posterior longitudinal ligament
Vertebral arch ligaments
• Ligamenta flava • Interspinous ligaments • Supraspinous ligament • Nuchal ligament* • Intertransverse ligaments
* The sagittally oriented nuchal ligament runs between the external occipital protuberance and the seventh cervical vertebra; it corresponds to a supraspinal ligament that is broadened superiorly (see p. 129).

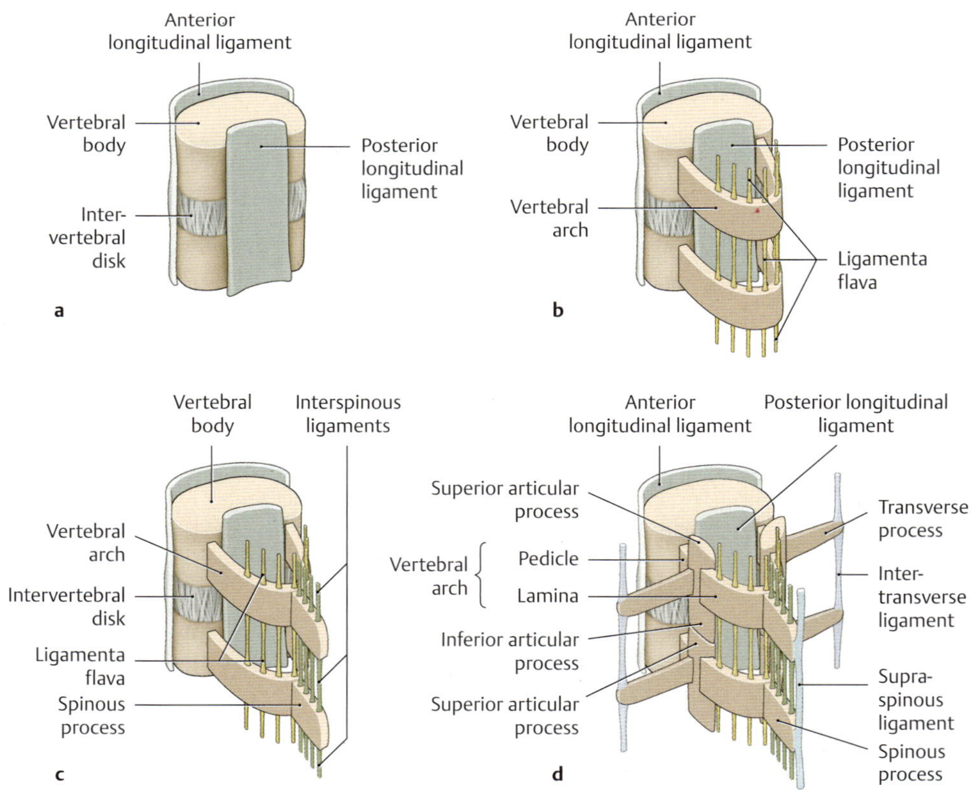

C Schematic representation of the vertebral body and vertebral arch ligaments
Viewed obliquely from the left posterior view.

a Vertebral body ligaments.
b–d Vertebral arch ligaments.

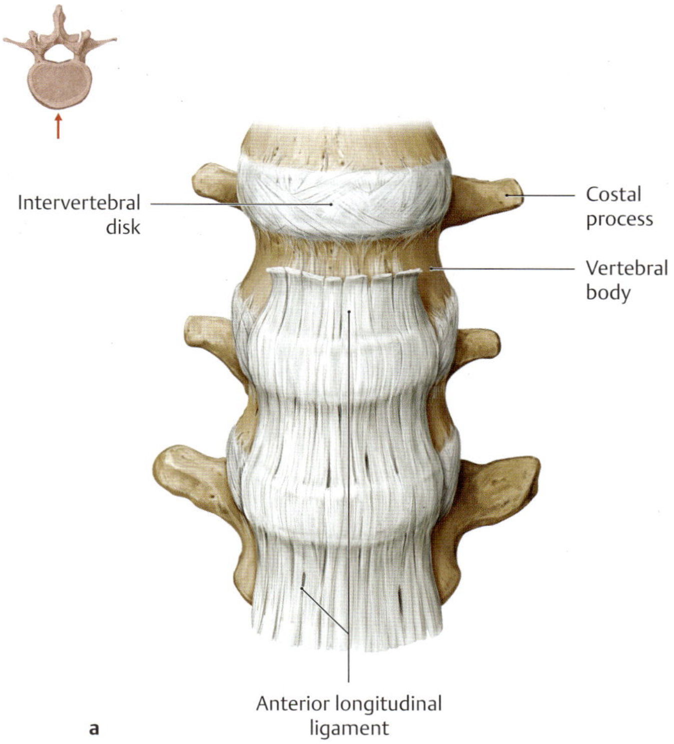

a

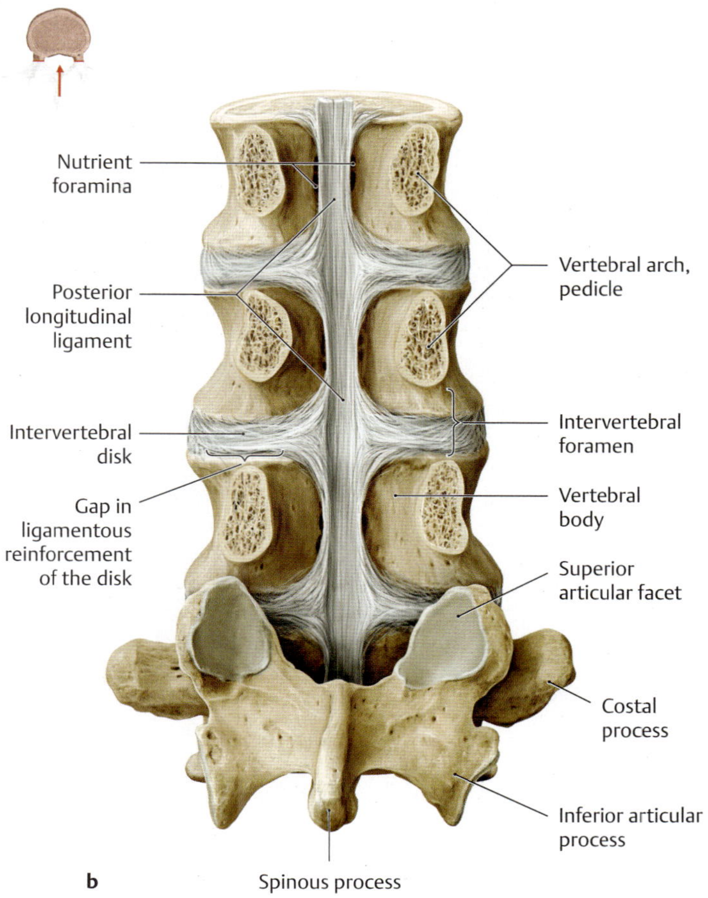

b

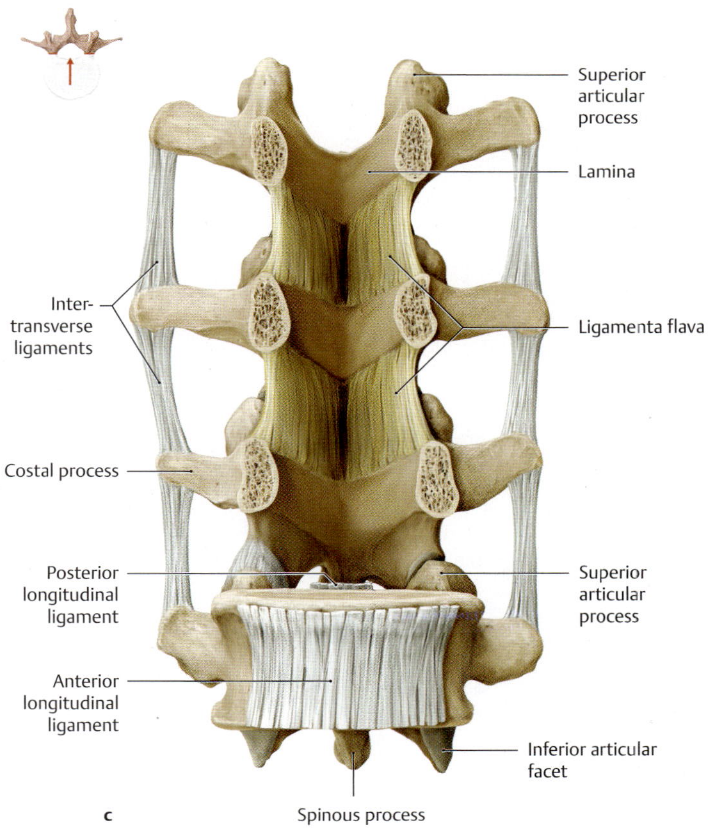

c

D The ligaments surrounding the lumbar spine

a Anterior longitudinal ligament, anterior view.

b Posterior longitudinal ligament, posterior view after removal of the vertebral arches at the pedicular level.

c Ligamenta flava and intertransverse ligaments, anterior view (after removal of the L2–L4 vertebral bodies). (The other vertebral arch ligaments are not visible in this view.)

The **anterior longitudinal ligament** runs broadly on the anterior side of the vertebral bodies, extending from the skull base to the sacrum. Its deep fibers bind adjacent vertebral bodies together, while its superficial fibers span multiple segments. Its collagenous fibers are attached firmly to the vertebral bodies but have only a loose attachment to the intervertebral disks. The thinner **posterior longitudinal ligament** descends from the clivus along the posterior surface of the vertebral bodies, passing into the sacral canal. The portion over the vertebral bodies is narrow and is attached to their superior and inferior margins. The ligament broadens at the levels of the intervertebral disks, to which it is firmly attached by tapered lateral extensions. Despite the attachment of the ligament to the anulus fibrosus of the intervertebral disks (not seen clearly here, being hidden by the posterior longitudinal ligament), a large portion of the intervertebral disk has no ligamentous reinforcement, especially laterally (predisposing to lateral disk herniation, see p. 139). Both longitudinal ligaments contribute to maintaining the normal curvature of the spinal column. The **ligamenta flava** consist mainly of elastic fibers, which give these ligaments their characteristic yellow color. They are thick, powerful ligaments that connect the laminae of adjacent vertebral arches and reinforce the wall of the vertebral canal posterior to the intervertebral foramina (see **A**). When the spinal column is erect, the ligamenta flava are under tension and help the back muscles to stabilize the spine in the sagittal plane. They also act as reins to limit forward flexion of the spinal column, thereby helping to maintain the position of the flexed spine. The tips of the transverse processes are connected on each side by the **intertransverse ligaments**, which serve to limit rocking movements of one vertebra upon another.

1.11 Overview of the Ligaments of the Cervical Spine

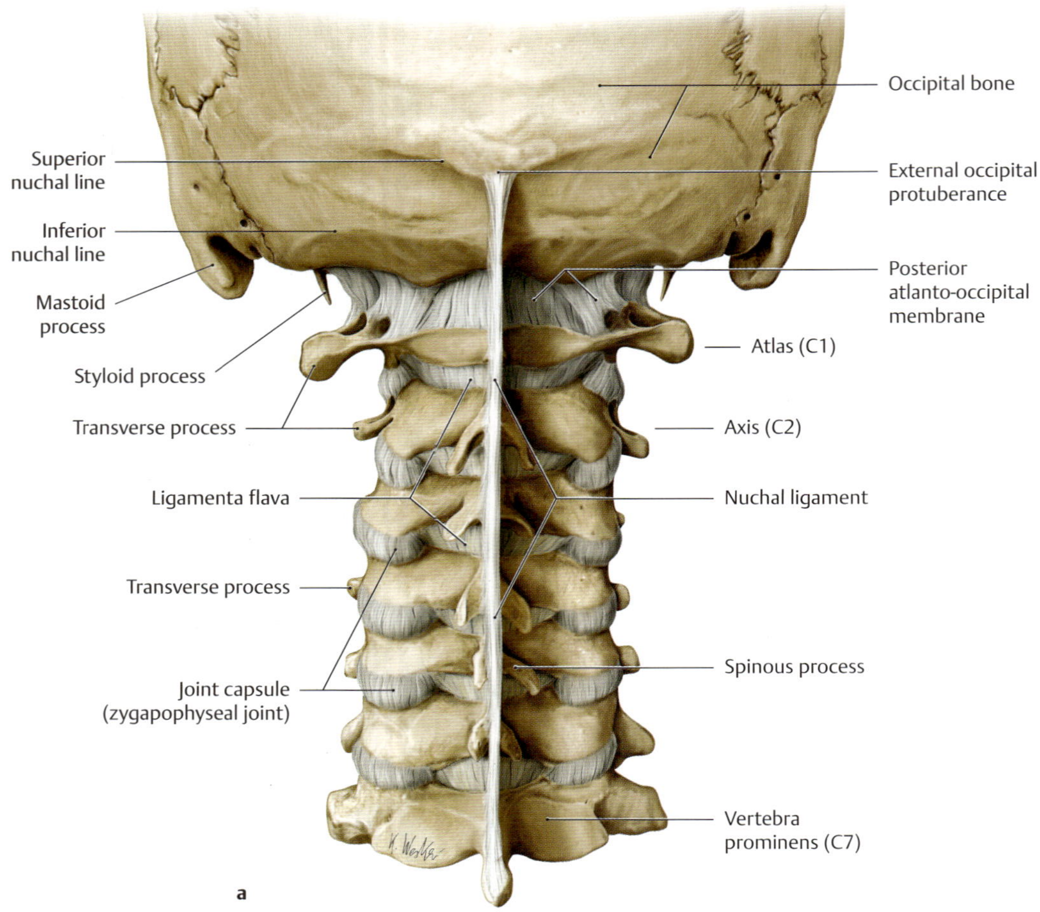

A The ligaments of the cervical spine

a Posterior view.

b Anterior view after removal of the anterior skull base (see p. 130 for the ligaments of the upper cervical spine, especially the craniovertebral joints).

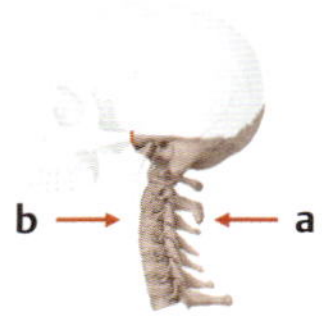

B The craniovertebral joints

The craniovertebral joints are the articulations between the atlas (C1) and occipital bone (atlanto-occipital joints) and between the atlas and axis (C2, atlantoaxial joints). While these joints, which number six in all, are anatomically distinct, they are mechanically interlinked and comprise a functional unit (see p. 133).

Atlanto-occipital joints
Paired joints where the oval, slightly concave superior articular facets of the atlas articulate with the convex occipital condyles.
Atlantoaxial joints
• *Lateral atlantoaxial joint* = paired articulation between the inferior articular facets of the atlas and the superior articular facets of the axis • *Median atlantoaxial joint* = unpaired articulation (comprising an anterior and posterior compartment) between the dens of the axis, the fovea of the atlas, and the cartilage-covered anterior surface of the transverse ligament of the atlas (see p. 131)

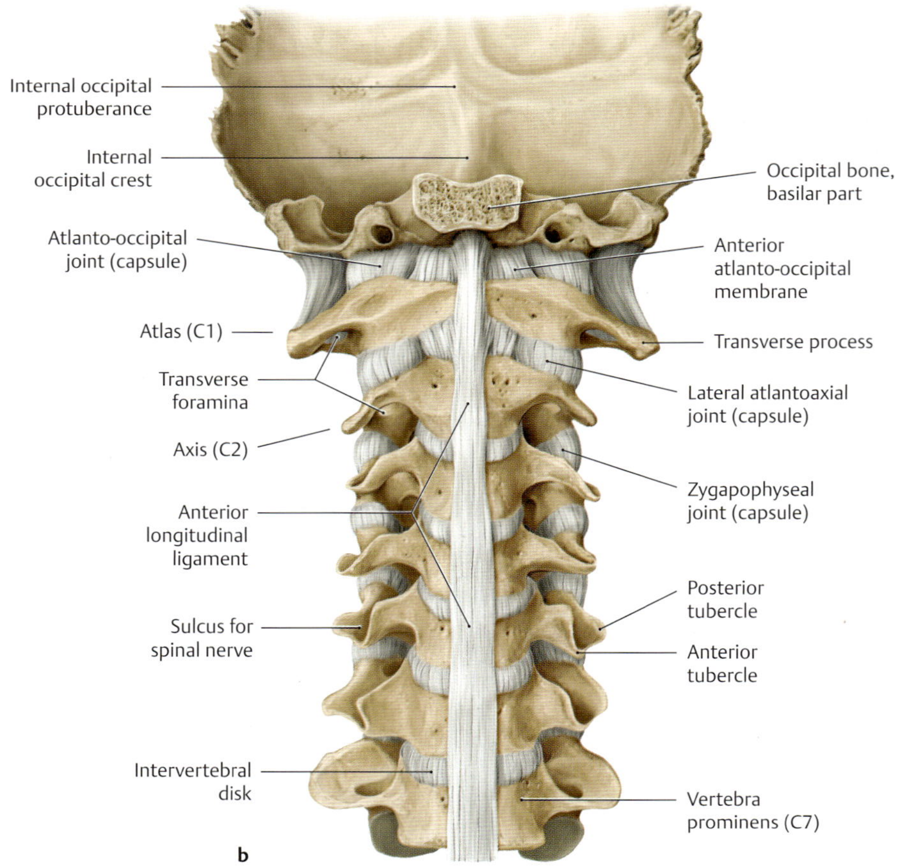

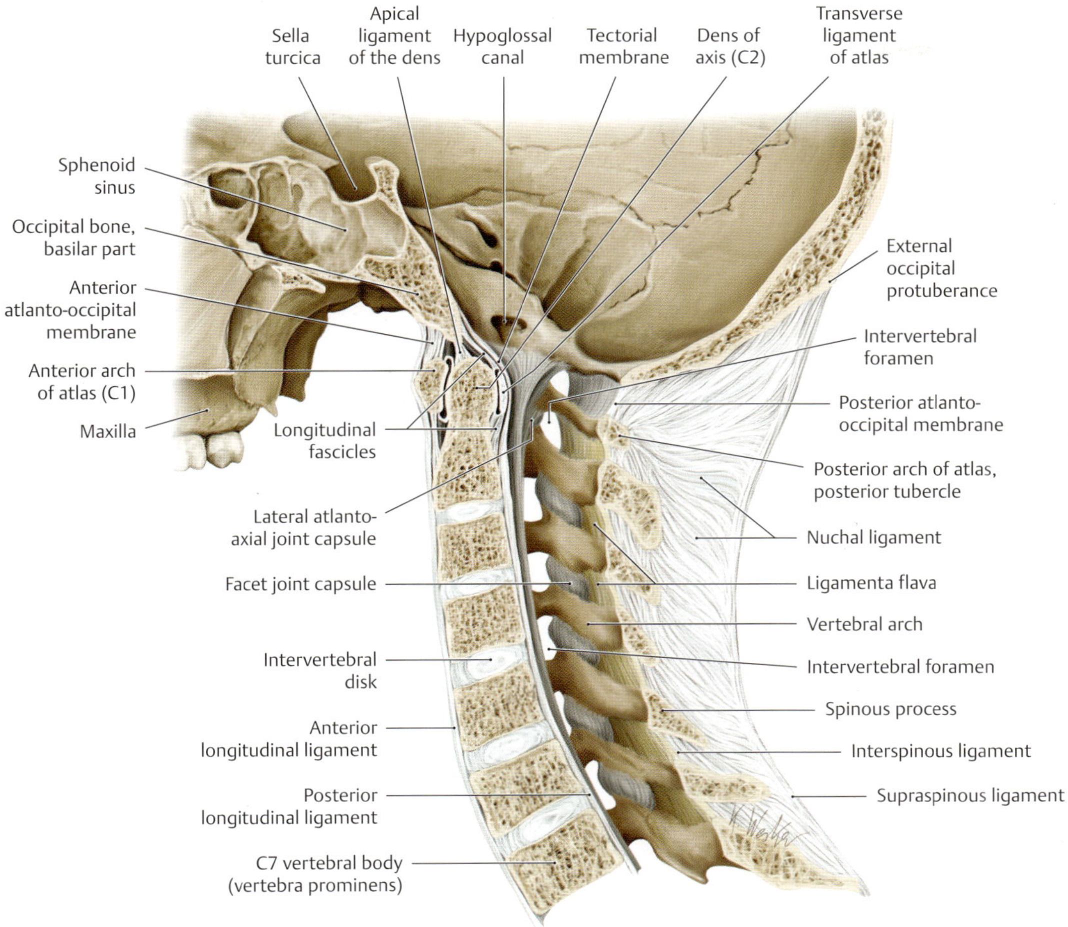

C The ligaments of the cervical spine: nuchal ligament

Midsagittal section, left lateral view. The nuchal ligament is the broadened, sagittally oriented part of the supraspinous ligament that extends from the vertebra prominens (C1) to the external occipital protuberance (see **A**; see also p. 130 for the ligaments of the atlanto-occipital and atlantoaxial joints).

Note: The intervertebral foramen lies posterior to the joint capsule between the atlas and axis.

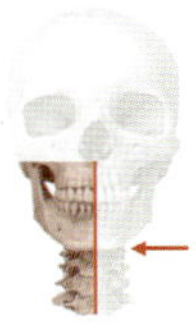

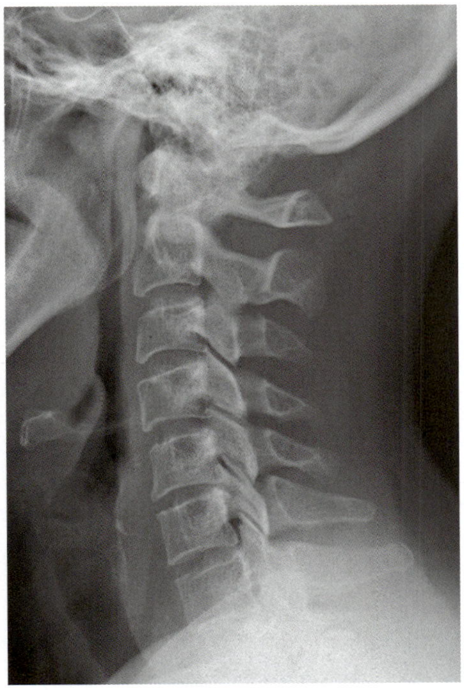

D Plain lateral radiograph of the cervical spine

Left lateral view.

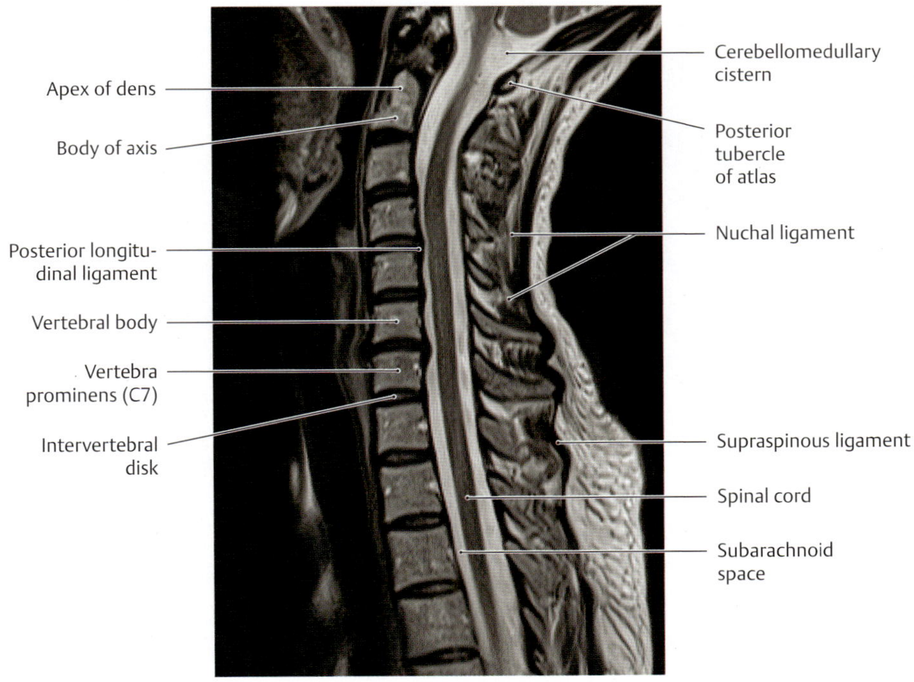

E Magnetic resonance image of the cervical spine

Midsagittal section, left lateral view T2-weighted TSE sequence. (Original image: Prof. S. Müller-Hülsbeck, M.D., Inst. für Diagnostische und Interventionelle Radiologie/Neuroradiologie, DIAKO Krankenhaus GmbH Flensburg).

1.12 The Ligaments of the Upper Cervical Spine (Atlanto-occipital and Atlantoaxial Joints)

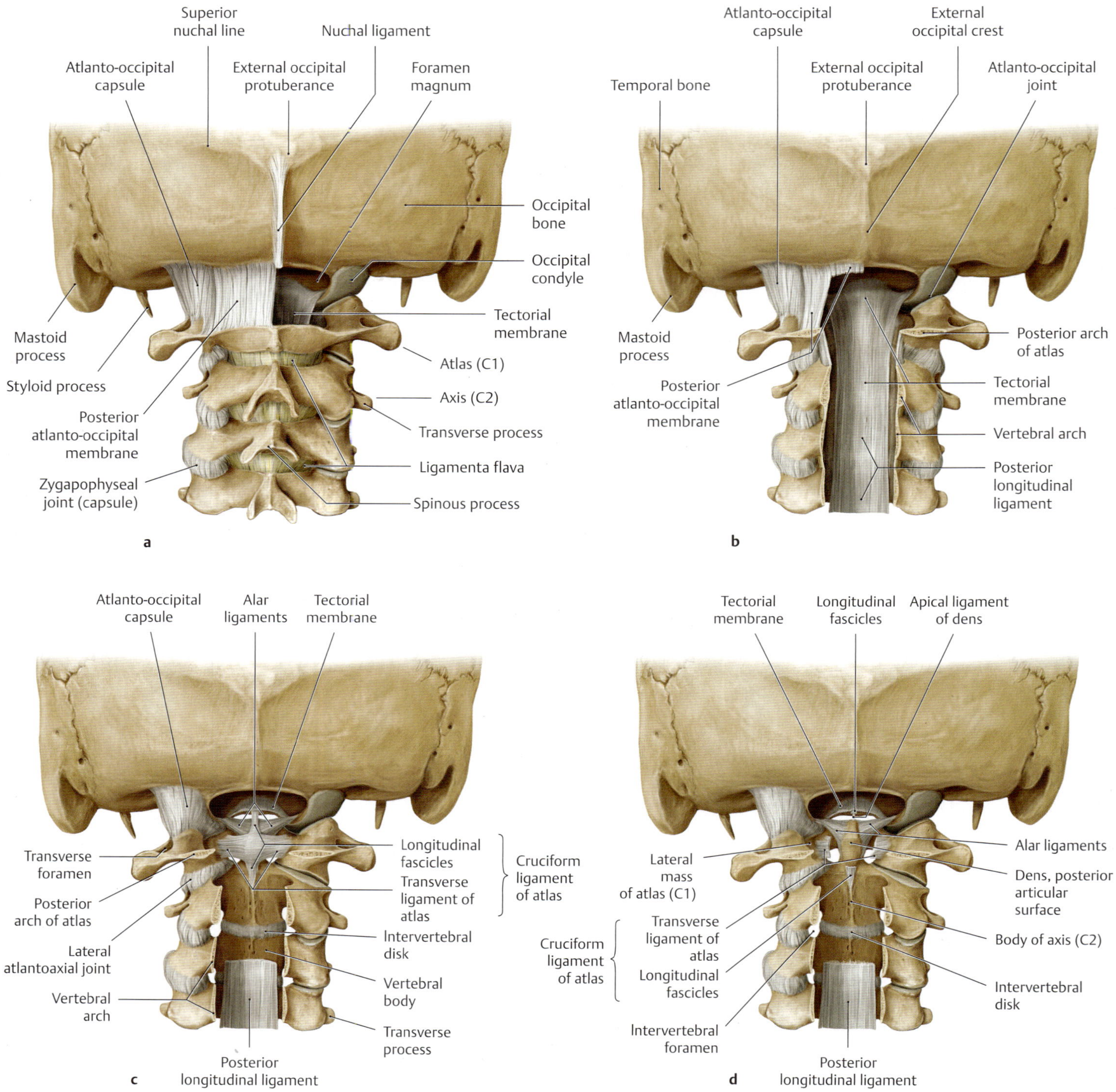

A The ligaments of the craniovertebral joints
Skull and upper cervical spine, posterior view.

a The posterior atlanto-occipital membrane—the "ligamentum flavum" between the atlas and occipital bone (see p. 128)—stretches from the posterior arch of the atlas to the posterior rim of the foramen magnum. This membrane has been removed on the right side.

b With the vertebral canal opened and the spinal cord removed, the tectorial membrane, a broadened expansion of the posterior longitudinal ligament, is seen to form the anterior boundary of the vertebral canal at the level of the craniovertebral joints.

c With the tectorial membrane removed, the cruciform ligament of the atlas can be seen. The transverse ligament of the atlas forms the thick horizontal bar of the cross, and the longitudinal fascicles form the thinner vertical bar.

d The transverse ligament of the atlas and longitudinal fascicles have been partially removed to demonstrate the paired alar ligaments, which extend from the lateral surfaces of the dens to the corresponding inner surfaces of the occipital condyles, and the unpaired apical ligament of the dens, which passes from the tip of the dens to the anterior rim of the foramen magnum.

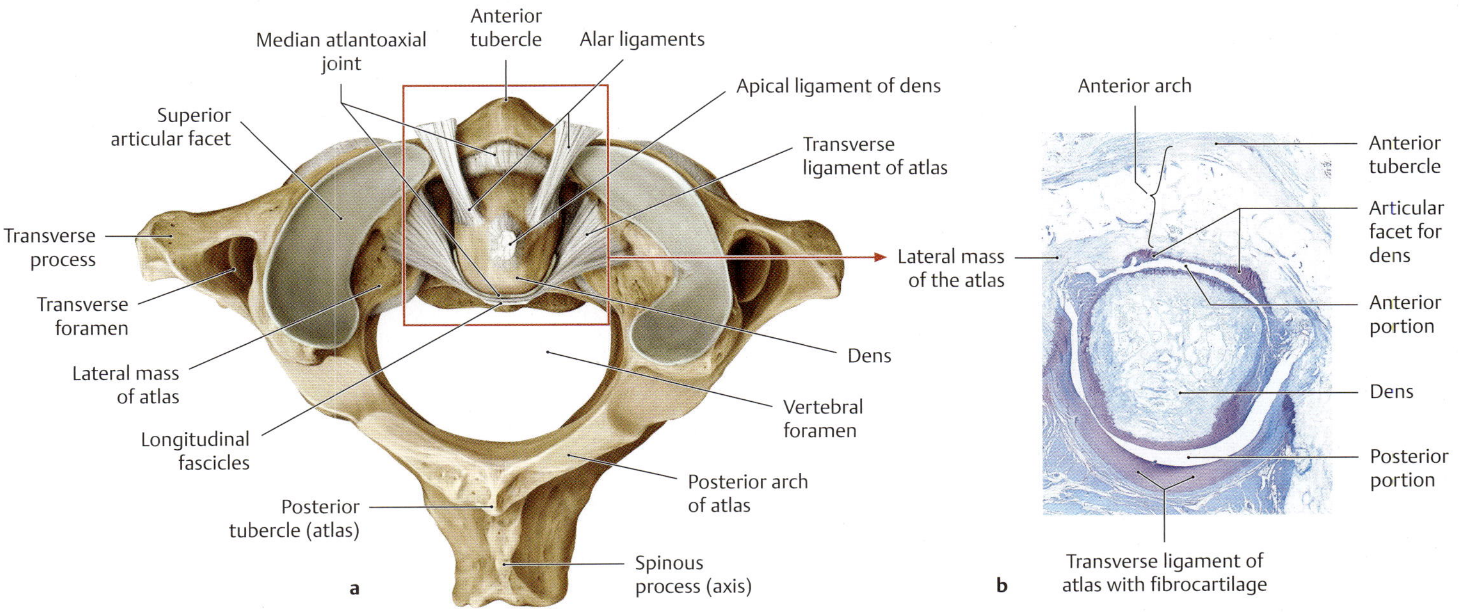

B Ligaments and histology of the median atlantoaxial joint

a Atlas and axis, superior view. The fovea, while part of the median atlantoaxial joint, is hidden by the joint capsule.

b Toluidine blue histological section (thickness 10 µm) (detail of **a**). See markings there. Horizontal section through the anterior and posterior portions of the median atlantoaxial joint, superior view (human male, 85 years old).

Note: 1. The fibrous structure of the transverse ligament of the atlas due to compressive stresses and 2. the advanced degeneration of the anterior median atlantoaxial joint (significant loss of articular cartilage). This may be attributable to lifelong excessive stress in this joint (which accounts for nearly 70% of the rotation in the cervical spine).

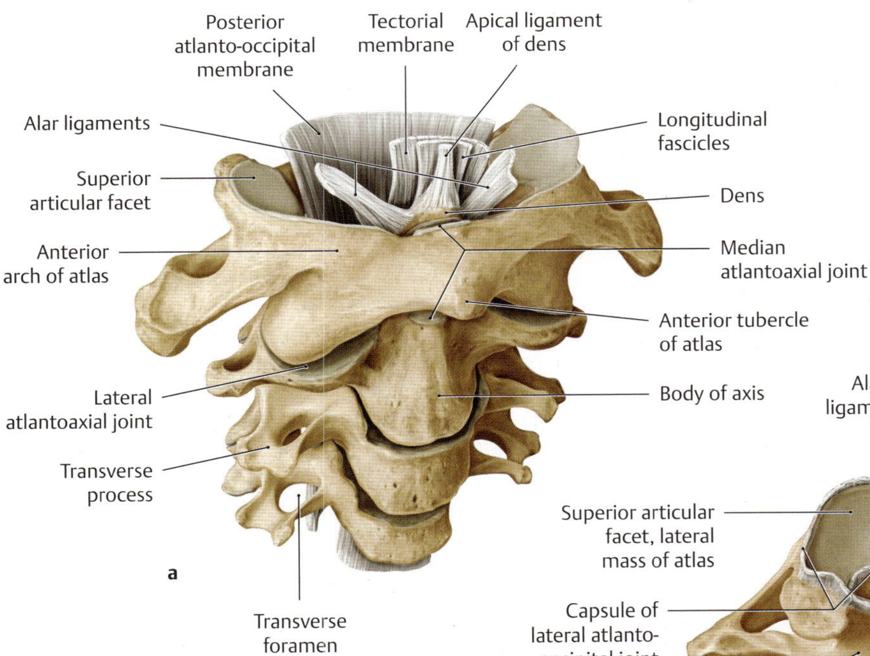

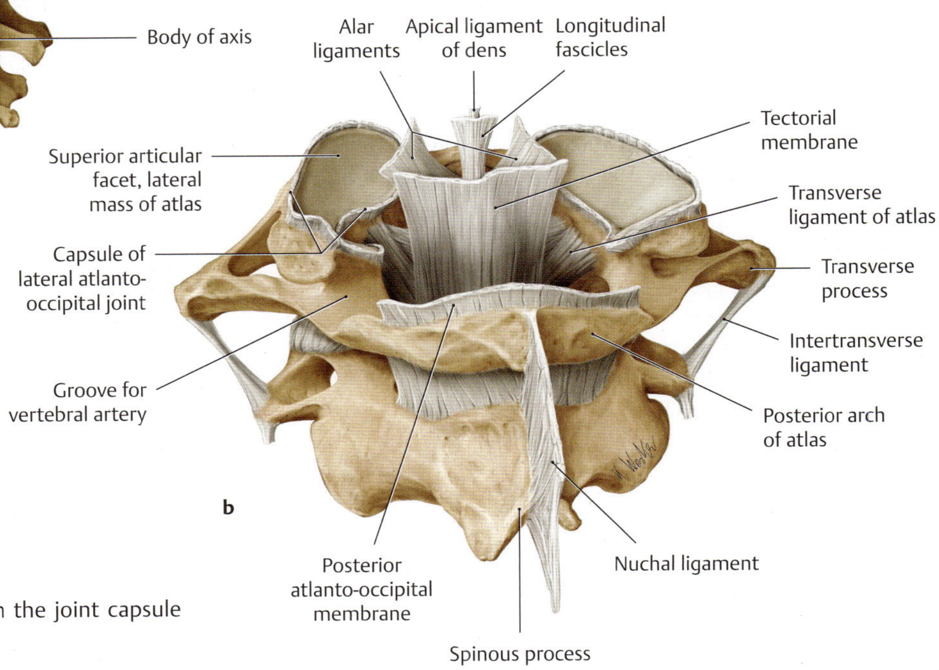

C The ligaments of the craniovertebral joints (joint capsule removed)

a Proximal cervical spine, anterosuperior view with the joint capsule removed.

b Atlas and axis, posterosuperior view.

1.13 The Intervertebral Facet Joints, Motion Segments, and Range of Motion in Different Spinal Regions

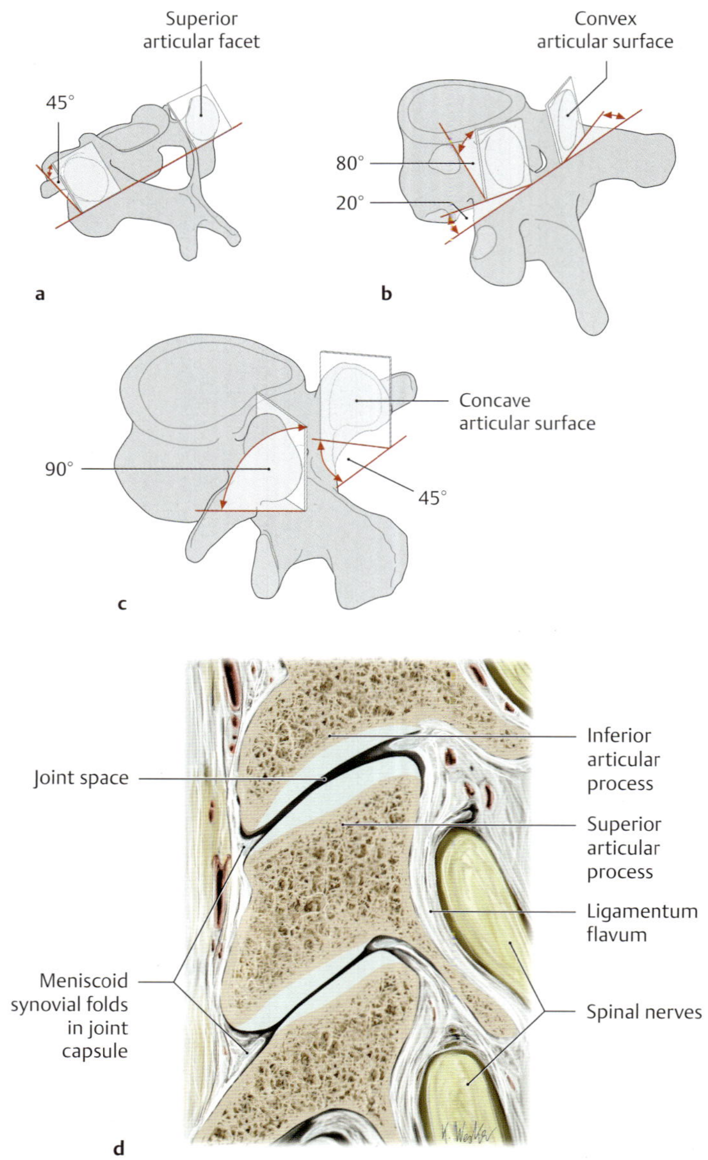

A The intervertebral facet joints (zygapophyseal joints)

The diagrams show the position of the articular surfaces of the intervertebral facet joints in different regions of the spinal column, shown from the left posterosuperior view: **a** cervical spine, **b** thoracic spine, **c** lumbar spine. Panel **d** is a sagittal section through the facet joints at the level of the third, fourth, and fifth cervical vertebrae, shown from the lateral view (drawing based on a specimen from the Anatomical Collection at Kiel University).

The paired facet joints are true synovial joints formed by the articular processes of the vertebral arches (see p. 114). Their articular surfaces, called **facets**, show varying degrees of inclination from the horizontal (and vertical) in different spinal regions and are therefore specialized for certain directions and ranges of motion (the movements that can occur in different spinal regions are shown in **D**). The **joint capsule** of the facet joint is inserted into the margins of the articular facets and is often firmly attached to the ligamentum flavum (see **d**). The joint capsule tends to be broad and lax in the cervical spine but is narrower in the thoracic and lumbar regions. Almost all the intervertebral joints contain *meniscoid synovial folds*, which form a crescent-shaped projection into the joint space. The folds consist of some loose but mostly firm connective tissue that has a rich vascular supply. The function of these folds is to fill the spaces around the articular surfaces (**d**).

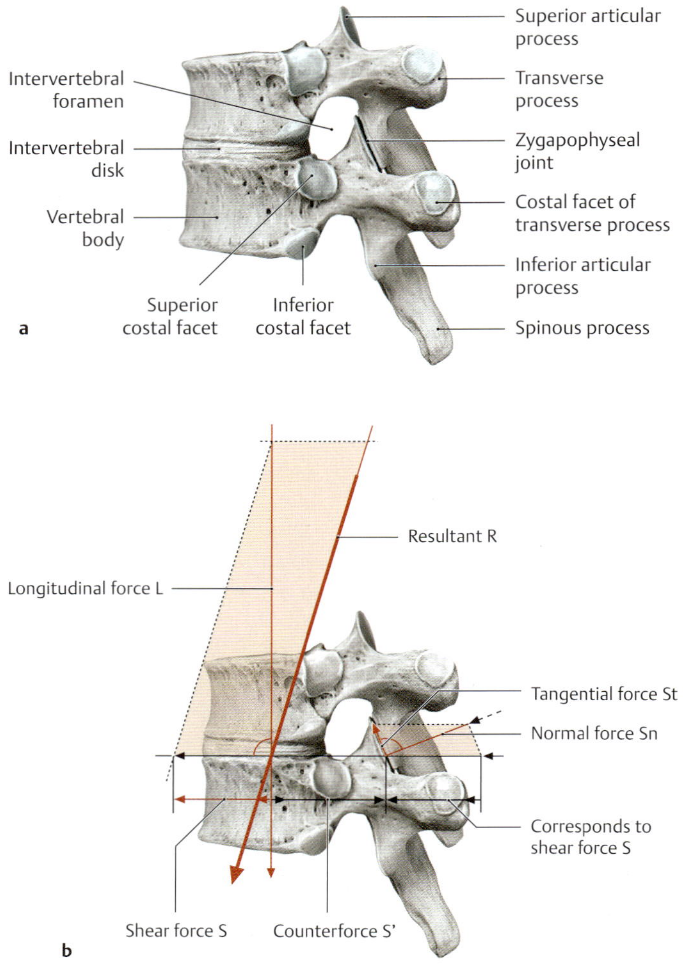

B Structure and loading of a motion segment, illustrated for two thoracic vertebrae

Lateral view. "Motion segment" is the term applied to the articular and muscular connection between two adjacent vertebrae (**a**). It consists of the intervertebral disk, the paired intervertebral facet joints (zygapophyseal joints), and the associated ligaments and muscles (not depicted here). For clinical purposes, the motion segment is also considered to include the contents of the intervertebral foramina (nerves and blood vessels, see pp. 198 and 204) and the contents of the vertebral canal. In all, the spinal column contains 25 of these motion segments, which constitute distinct functional and morphological units. However, because of the interdependence of these units, abnormalities in a circumscribed portion of the spinal column tend to affect the other spinal segments as well. Each of these motion segments is subjected to certain loads that can be represented as applied forces (**b**): a forward-directed *shear force* and a downward-directed *longitudinal force*, which combine to give the resultant force R. The longitudinal force acts on the vertebral body and intervertebral disk, while the shear force is absorbed mainly by the ligaments and facet joints (counterforce S'). The shear force can be subdivided into a *normal force* (Sn) and a tangential force (St). Because the shearing forces are not perpendicular to the articular surfaces of the facet joints, these surfaces are loaded by the axially directed normal force (Sn), which is weaker than the original shear force. Displacement of the vertebrae by the upward-directed *tangential force* (St) is prevented by the ligaments and intrinsic back muscles (after Kummer).

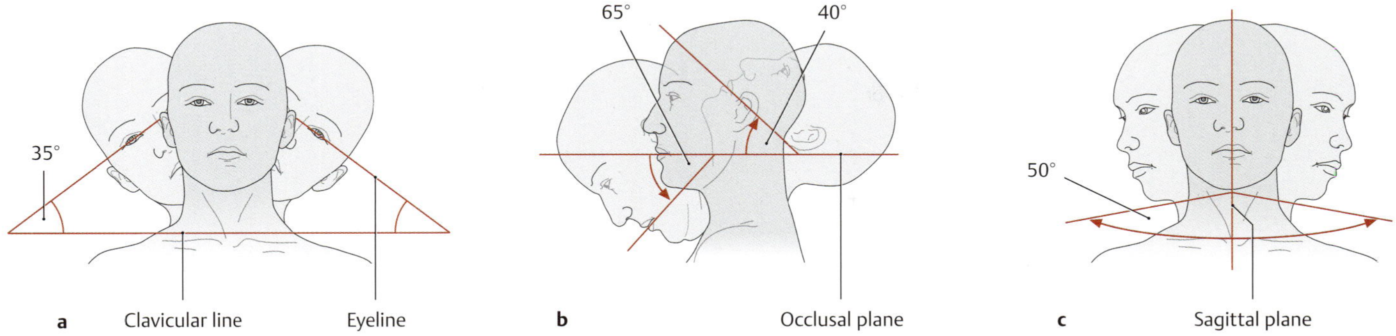

C Total range of motion of the cervical spine
a Lateral flexion, **b** flexion/extension, **c** rotation.

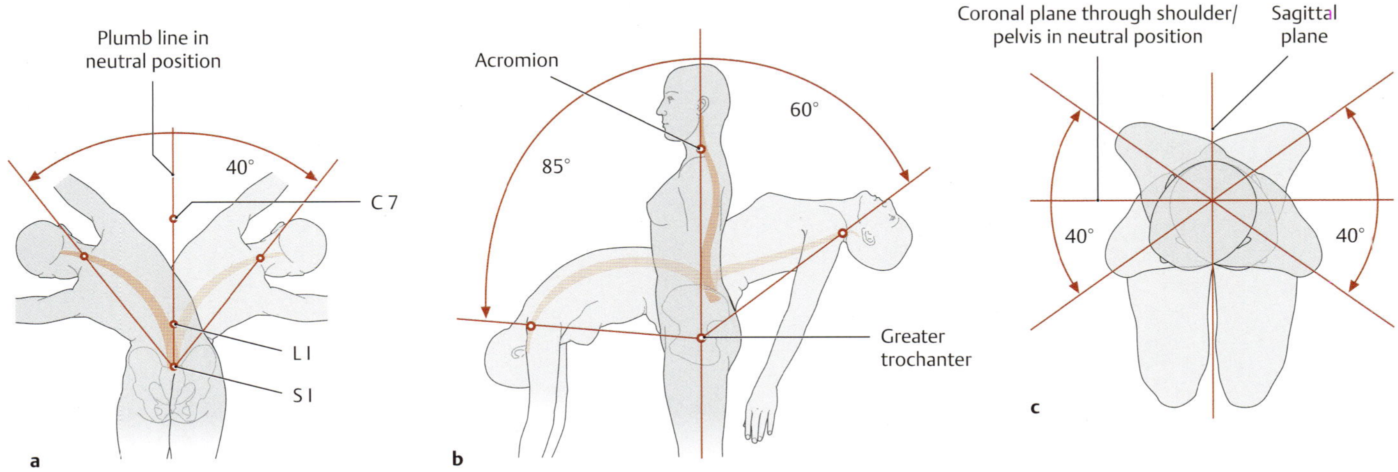

D Total range of motion of the thoracic and lumbar spine
a Lateral flexion, **b** flexion/extension, **c** rotation.
The clinical examination, particularly function testing, is of key importance in examinations of the spine. Because the total range of spinal motion is comprised of the movements of a total of 25 motion segments, generally the examiner can detect only movement disorders that affect individual regions. For example, clinical testing can show clear evidence of ankylosis affecting particular spinal segments. The examiner makes use of standard reference lines (e.g., the clavicular line or occlusal plane) in determining whether the range of spinal motion is normal or decreased.

E Average ranges of motion in different spinal regions (degrees)

	Cervical spine			Thoracic spine	Lumbar spine	Cervical + thoracic + lumbar
	A-o joint	A-a joint	Entire cervical spine			
Flexion	20	–	65	35	50	150
Extension	10	–	40	25	35	100
Lateral flexion*	5	–	35	20	20	75
Rotation*	–	35	50	35	5	90

A-o joint = Atlanto-occipital joint
A-a joint = Atlantoaxial joint

*To each side

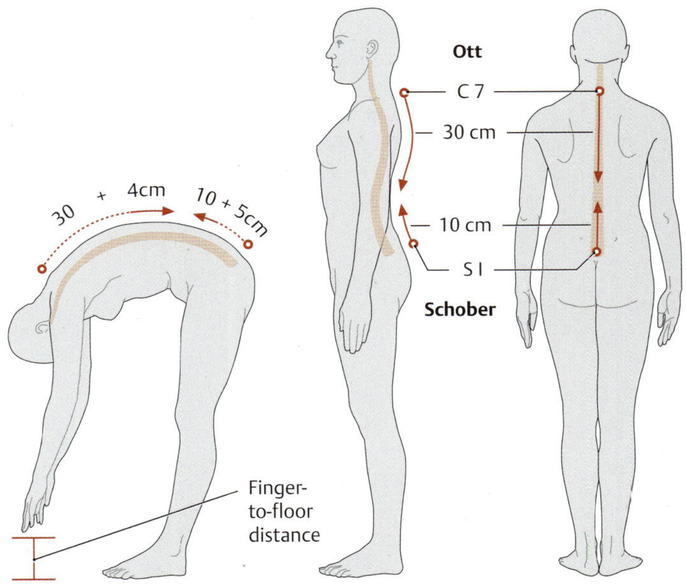

F Measurement of the range of thoracic and lumbar spinal flexion by the method of Schober and Ott
In the method of Schober and Ott, the patient stands erect while the examiner marks the S1 spinous process and a second point 10 cm higher. When the patient bends as far forward as possible, the distance between the two skin markings will increase to approximately 15 (10 + 5) cm (range of motion of the lumbar spine). The thoracic range of motion is determined by measuring 30 cm down from the spinous process of the C7 vertebra (vertebra prominens) and marking that point on the skin. When the patient bends forward, the distance between the markings may increase by up to 4 cm. An alternative method is to measure the smallest finger-to-floor distance (FFD) with the knees extended.

1.14 The Uncovertebral Joints of the Cervical Spine

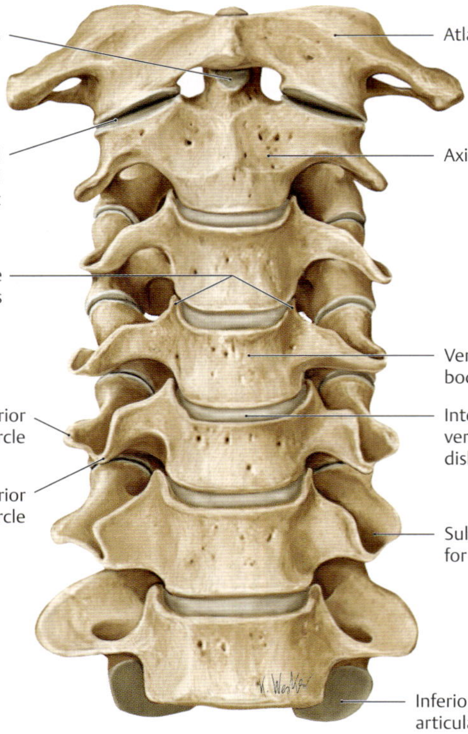

a

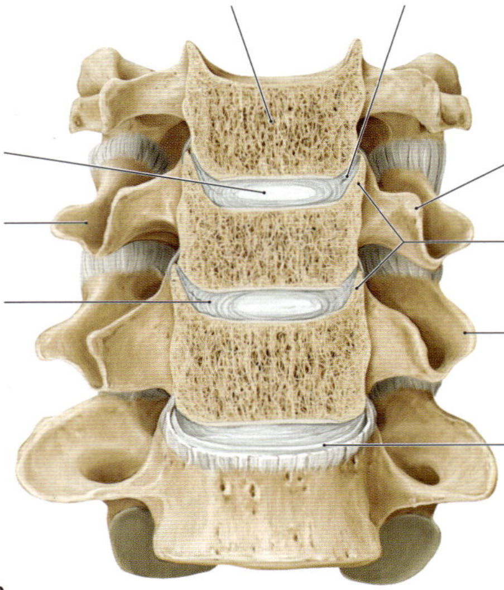

b

A The uncovertebral joints in a young adult
Cervical spine of an 18-year-old man, anterior view.

a The upper end plates of the C3 through C7 vertebral bodies have lateral projections (uncinate processes) that develop during childhood. Starting at about 10 years of age, the uncinate processes gradually come into contact with the oblique, crescent-shaped margin on the undersurface of the next higher vertebral body. This results in the formation of lateral clefts (uncovertebral clefts or joints, see **b**) in the outer portions of the intervertebral disks.

b C4 through C7 vertebrae. The bodies of the C4–C6 vertebrae have been sectioned in the coronal plane to demonstrate more clearly the uncovertebral joints or clefts. These clefts are bounded laterally by a connective tissue structure, a kind of joint capsule, which causes them to resemble true joint spaces. These clefts or fissures in the intervertebral disk were first described by the anatomist Hubert von Luschka in 1858, who called them *lateral hemiarthroses*. He interpreted them as primary mechanisms designed to enhance the flexibility of the cervical spine and confer a functional advantage (drawings based on specimens from the Anatomical Collection at Kiel University).

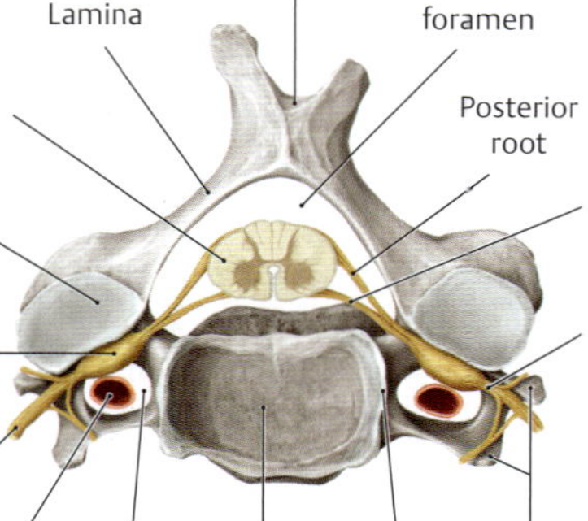

a

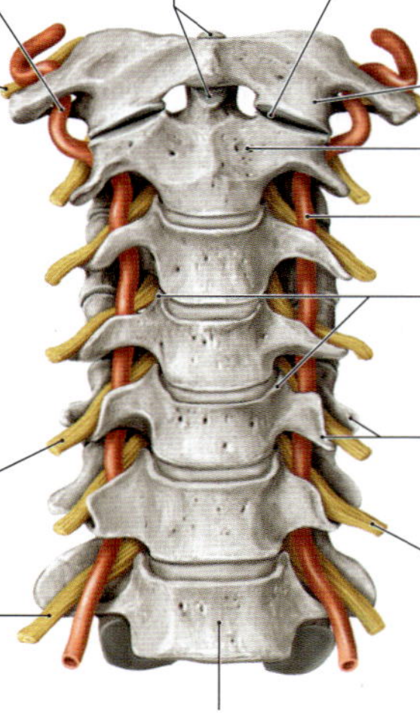

b

B Topographical relationship of the spinal nerve and vertebral artery to the uncinate process

a Fourth cervical vertebra with spinal cord, spinal roots, spinal nerves, and vertebral arteries, superior view.

b Cervical spine with both vertebral arteries and the emerging spinal nerves, anterior view.

Note the course of the vertebral artery through the transverse foramina and the course of the spinal nerve at the level of the intervertebral foramina. Given their close proximity, both the artery and nerve may be compressed by osteophytes (bony outgrowths) caused by uncovertebral arthrosis (see **D**).

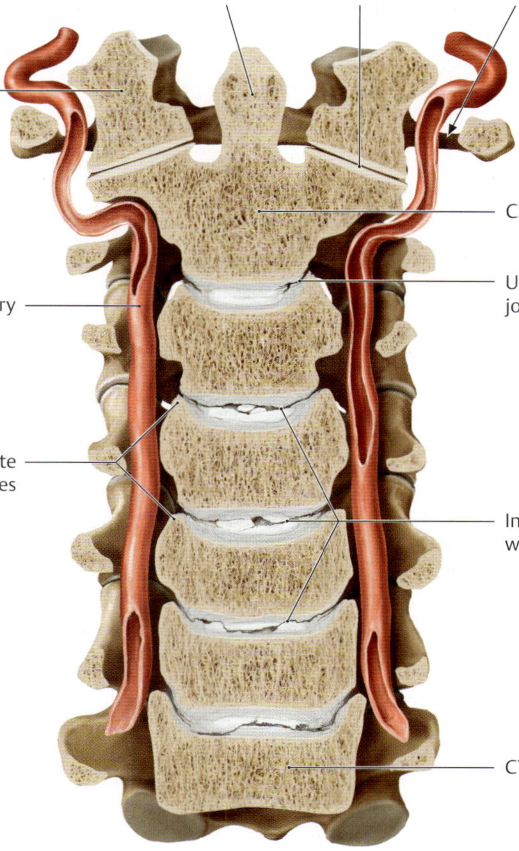

C Degenerative changes in the cervical spine (uncovertebral arthrosis)
Coronal section through the cervical spine of a 35-year-old man, anterior view. *Note* the course of the vertebral arteries on both sides of the vertebral bodies.
The development of the uncovertebral joints at approximately 10 years of age initiates a process of cleft formation in the intervertebral disks. This process spreads toward the center of the disk with aging, eventually resulting in the formation of complete transverse clefts that subdivide the intervertebral disks into two slabs of roughly equal thickness. The result is a progressive degenerative process marked by flattening of the disks and consequent instability of the motion segments (drawing based on specimens from the Anatomical Collection at Kiel University).

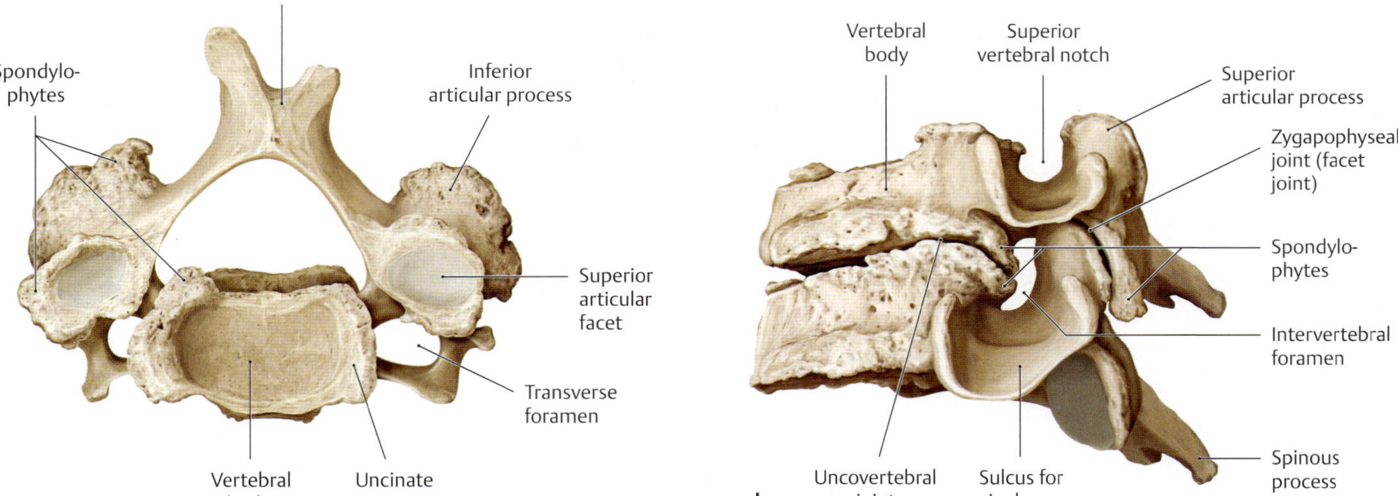

D Advanced uncovertebral arthrosis of the cervical spine
a Fourth cervical vertebra, superior view.
b Fourth and fifth cervical vertebrae, lateral view (drawings based on specimens from the Anatomical Collection at Kiel University).

The uncovertebral joints undergo degenerative changes comparable to those seen in other joints, including the formation of osteophytes (called spondylophytes when they occur on vertebral bodies). These sites of new bone formation serve to distribute the imposed forces over a larger area, thereby reducing the pressure on the joint. With progressive destabilization of the corresponding motion segment, the facet joints undergo osteoarthritic changes leading to osteophyte formation. Osteophytes of the uncovertebral joints have major clinical importance because of their relation to the intervertebral foramen and vertebral artery (uncovertebral arthrosis). They cause a progressive narrowing of the intervertebral foramen, with increasing compression of the spinal nerve and often of the vertebral artery as well (see **C**). Meanwhile, the spinal canal itself may become significantly narrowed (spinal stenosis) by the same process.

1.15 Cross-sectional Anatomy of the Lumbar Spine

A Midsagittal section through the lower part of the spinal column
Left lateral view.
Note: The caudal end of the spinal cord, the conus medullaris, terminates in most people at the level of the first lumbar vertebra but in a few at the level of the second lumbar vertebra. The spinal cord and spinal canal are approximately the same length until the 12th week of prenatal development, so that each spinal cord segment and its pair of spinal nerves lie adjacent to the vertebra of the same number. With further growth, however, the vertebral column lengthens more rapidly than the spinal cord, resulting in an increasing cephalad displacement of the conus medullaris. At birth the conus medullaris has already reached the level of the third lumbar vertebra, and it continues its gradual upward shift until about the 10th year of life. Because of these disparate growth rates, the spinal roots run obliquely downward from their segment of origin in the cord to reach their corresponding intervertebral foramen. The spinal roots that descend from the lower end of the cord are collectively termed the *cauda equina* ("horse's tail"). Because the membranes that enclose the spinal cord (the meninges) extend into the sacral canal (lumbar cistern), a needle can be safely introduced into the subarachnoid space below the conus medullaris to sample cerebrospinal fluid without injuring the cord (*lumbar puncture*). This site is also used for *lumbar spinal anesthesia* to block both the afferent nerve roots (for analgesia) and the efferent nerve roots (for muscular paralysis) that supply the pelvic region and lower limbs.

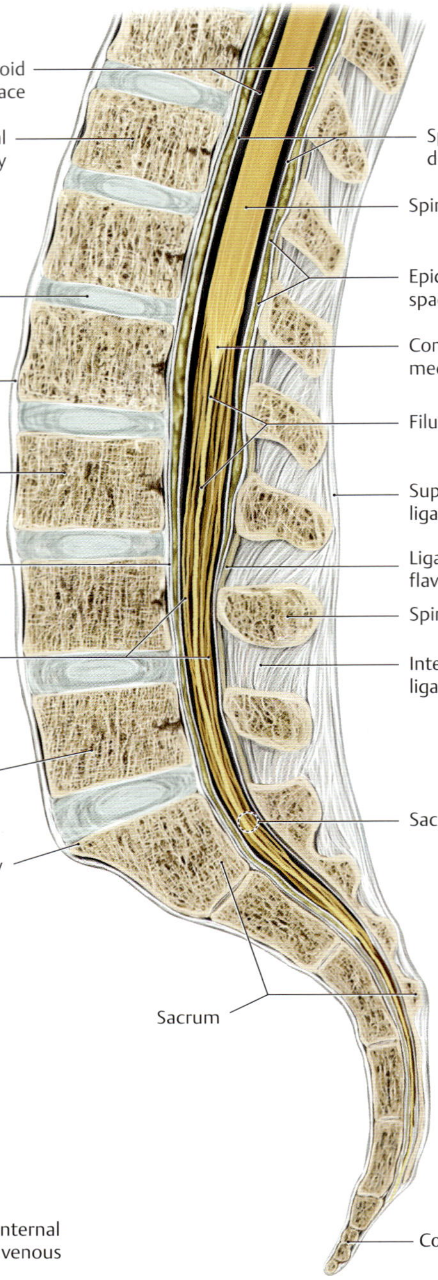

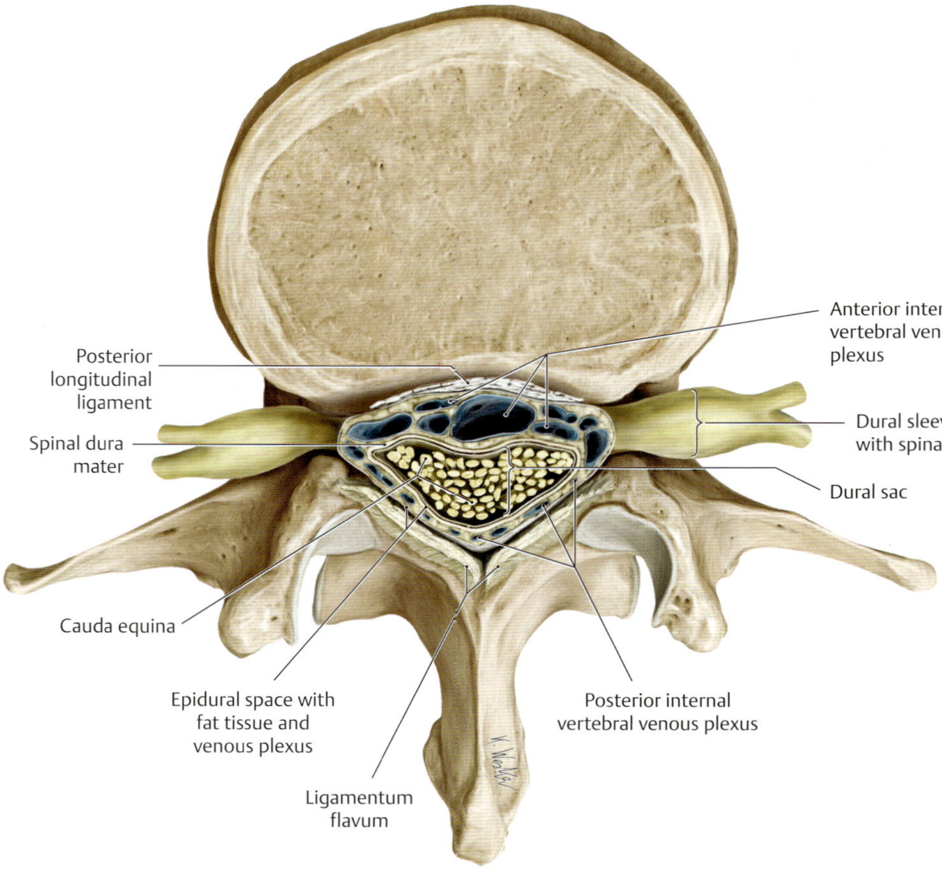

B Cauda equina at the level of the second lumbar vertebra
Transverse section, superior view. Because the spinal cord ends at the level of the first lumbar vertebra in most people, at the level of the second lumbar vertebra, the dural sac, which ends at the level of the second sacral vertebrae, consists only of the cauda equina and the filum terminale (see **A**). Here the epidural space is enlarged and filled with extended venous plexus and fatty tissue.

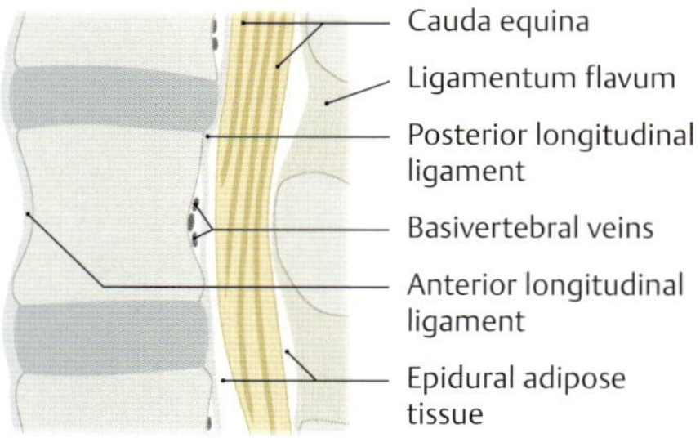

a

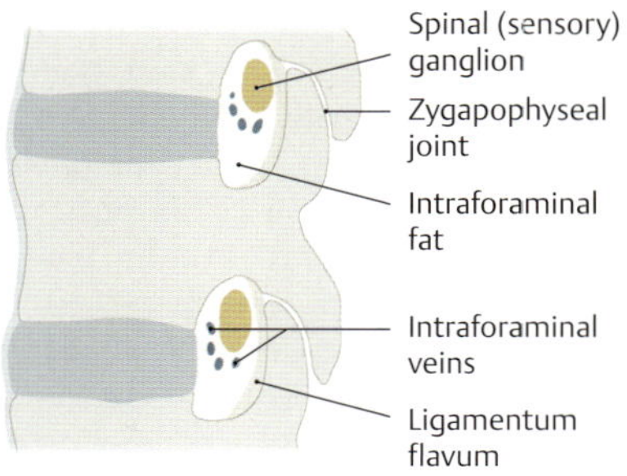

b

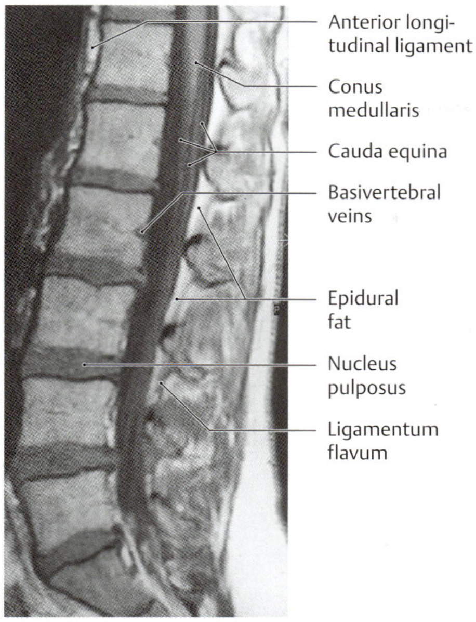

c

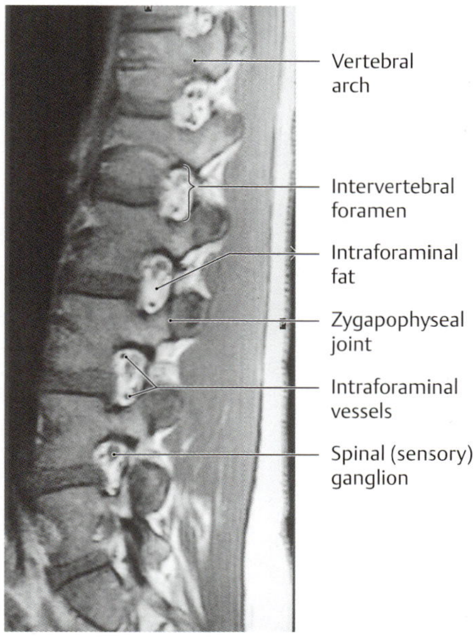

d

C MRI of the normal lumbar spine: sagittal and transverse sections (photos from Vahlensieck M, Reiser M. MRT des Bewgungsapparates. 3rd ed. Stuttgart: Thieme; 2006)

a, b structures visible in the median and parasagittal section.

c, d T1-weighted SE sequence of the lumbar spine in median and parasagittal section.

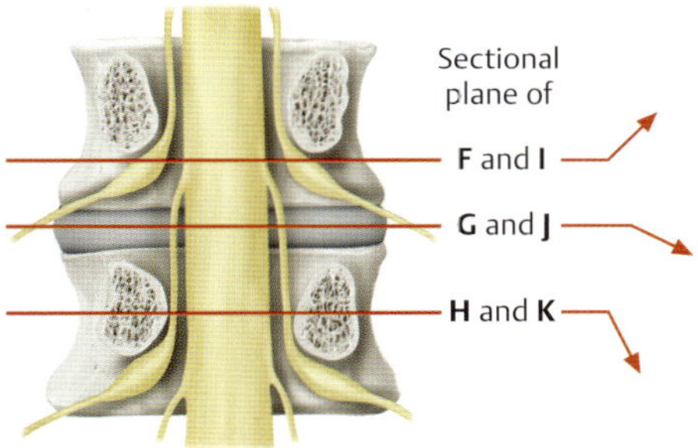

e

e motion segment L4–L5 after removal of vertebral arch, dorsal view; position of the transverse planes:

f supradisk level.

g disk level L4–L5.

h infradisk level.

In **f** and **g**, the intervertebral foramen is shown; in **h**,the pedicle is shown.

i–k T1-weighted SE sequences of the motion segment L4–L5 in the corresponding axial (transverse) planes.

Note the position of the spinal roots in relation to the dural sac. The spinal roots extend in so-called dural bags, which reach into the intervertebral foramen, so the roots and the spinal ganglion are completely bathed in fluid (see Fig. **B**).

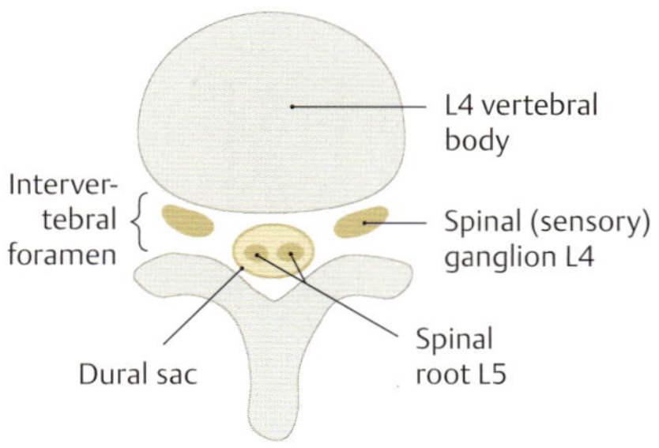

f

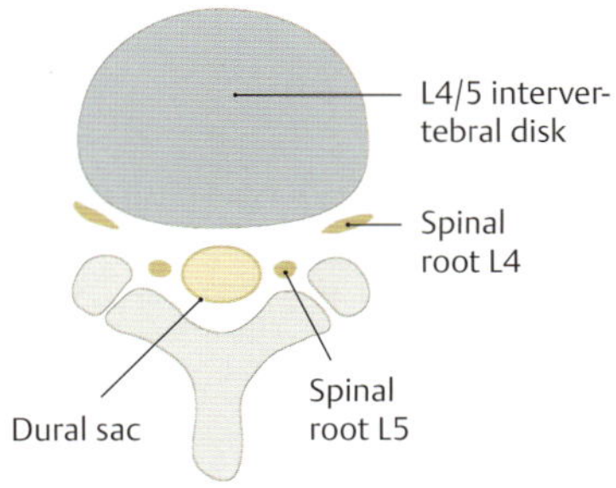

g

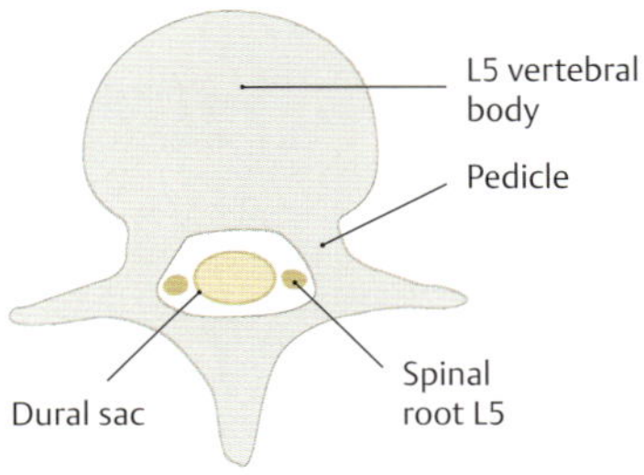

h

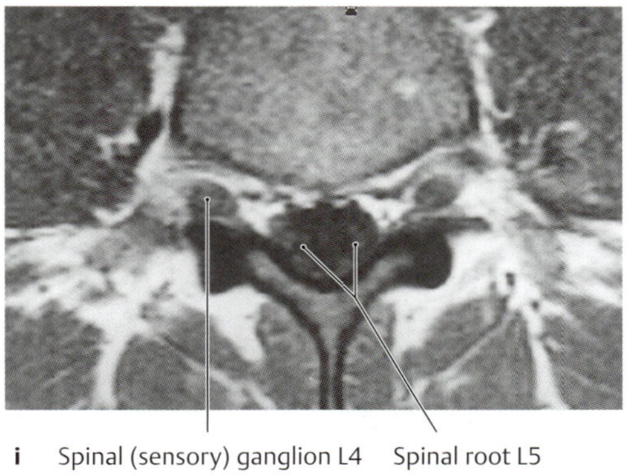

i

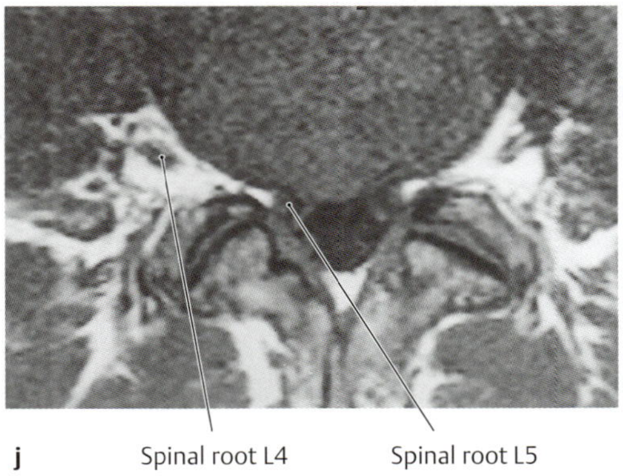

j

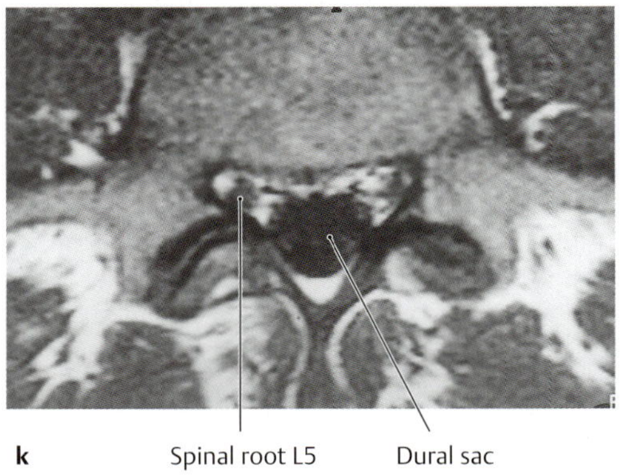

k

1.16 Degenerative Changes in the Lumbar Spine

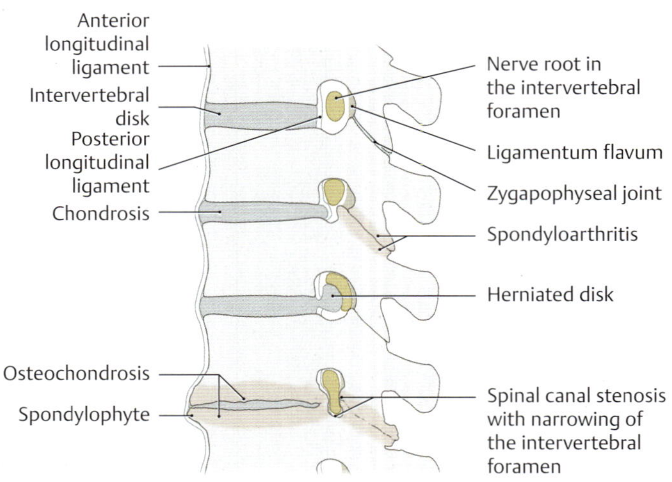

A Degenerative spine changes at a glance

Degenerative spine changes are among the most common disorders. They are age-dependent and increase dramatically after the age of 30. The intervertebral disk, the adjacent osseous end plates, the vertebrae, and the ligaments of the participating motion segment are most often affected.

Degeneration of the intervertebral disks and disk herniation: As a result of fluid depletion in the nucleus pulposus (decreased water-binding capacity—a typical symptom of old age), the intervertebral disk space increasingly narrows, and the affected motion segment becomes increasingly unstable. The resulting locally increased mechanical loading of the disk causes fraying and cracking in the anulus fibrosus (chondrosis) and eventually leads to a herniated disk (protrusion or prolapse, see **D**).

Lumbar spinal stenosis: The body tries to withstand disk degeneration through reactive bone changes, similar to what happens with osteoarthritis in the limb joints (see p. 48). At the vertebral bodies, marginal spikes (osteophytes = spondylophytes) form, which are supposed to enlarge the force-transmitting surface, thereby relieving the relevant vertebral joint and stabilizing the motion segment, causing end plate sclerosis (osteochondrosis). Similar processes occur at the small vertebral joints (spondylarthrosis). The increasing narrowing of the vertebral canal and the intervertebral foramina leads to degenerative lumbar spinal stenosis (see **B**). Later, the marginal spikes grow toward each other until they finally touch and as a result bridge and ossify the motion segment ("formation of bone bridges," see **Cc**). This increasingly impairs spine movement, often decreasing discomfort. A herniated disk is no longer possible, which is why herniated disks do not occur in old age.

Vertebral body fractures and deformities: In old age, the vertebral body is often the weakest link. Advanced osteoporosis or osteolytic bone metastasis accompanied by loss of stability in the bone can lead to vertebral body fractures and deformities. The symptoms include initially localized pain, which can spread if, at the same time, nerve roots are compressed.

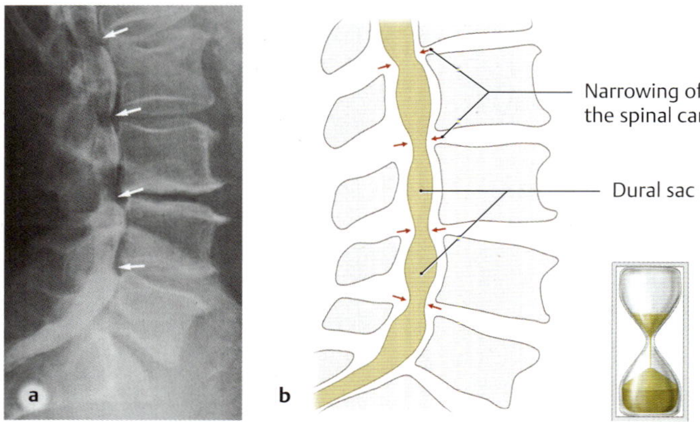

B Degenerative lumbar spinal stenosis

Symptoms of this disorder include the concentric narrowing of the vertebral canal at the level of the relevant motion segment. It develops based on a congenital narrowing and/or acquired degenerative changes, such as dorsal spondylophytes and arthrotically transformed vertebral joints (spondylarthrosis). Although rare, ligamentous hypertrophy, too, particularly of the posterior longitudinal ligament and the ligamenta flava, can contribute to narrowing of the vertebral canal. Typical symptoms include load-dependent neurogenic pain in the lumbar region and legs. It is caused by walking for long periods of time but also by standing, and subsides only by supporting the body (e.g., by propping oneself up) and by kyphosis of the lumbar spine (sitting with the upper body bent forward). Pain, sensory disturbance, and paralysis lead to drastically impaired walking distance (spinal claudication). The diagnosis is made using MRI and/or with the help of lumbar myelography with the spine positioned in lateral projection (**a**). (From: Niethard F, Pfeil J, Biberthaler P, eds. Duale Reihe Orthopädie und Unfallchirurgie. 8th ed. Stuttgart: Thieme; 2017.) Note the hourglass constriction of the contrasted dural sac (**b**).

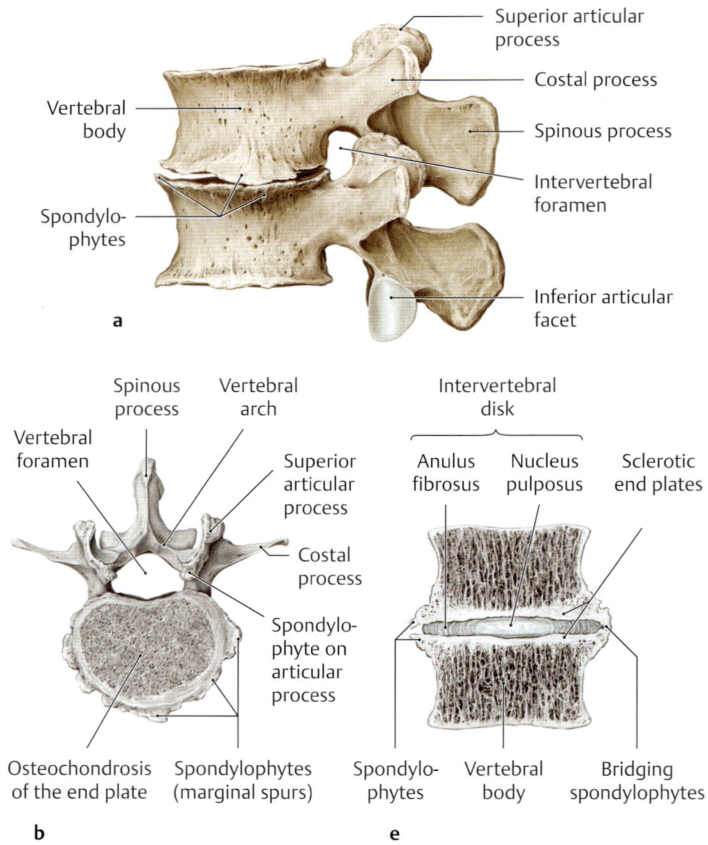

C Lumbar disk herniation

a Posterolateral herniation, superior view
b Posterior herniation, superior view
c Posterolateral herniation, posterior view (the vertebral arches have been removed to demonstrate the lumbar dural sac and corresponding nerve roots)

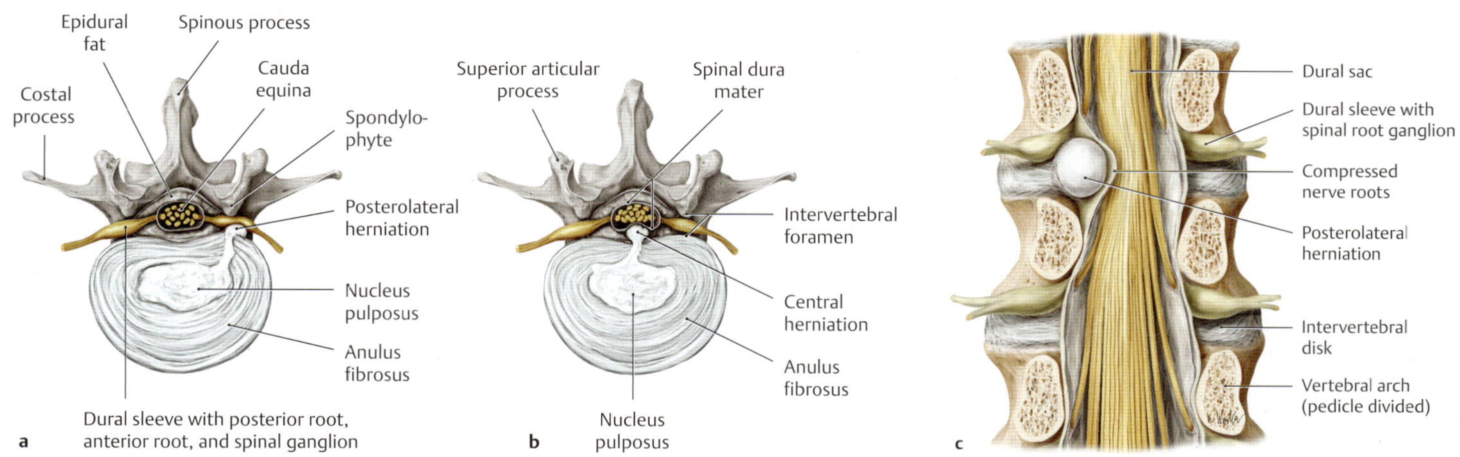

D Lumbar disk herniation

- **a** Posterolateral herniation, superior view.
- **b** Posterior herniation, superior view.
- **c** Posterolateral herniation, posterior view (the vertebral arches have been removed at the level of the pedicle to demonstrate the lumbar dural sac and corresponding nerve roots).

In old age, not only does the intervertebral disk get thinner (due to loss of water in the nucleus pulposus, see **A**), it also shifts more easily, as a result of a decrease in resistance in the intervertebral disk sleeve, the anulus fibrosus, which begins to tear and fray. The tissue of the jelly-like nucleus pulposus shifts initially toward the weak spots of the anulus fibrosus (intervertebral disk protrusion). If it tears as a result of continuous strain, it protrudes (disk prolapse) and compresses the content of the intervertebral foramen. Sequestration occurs when a part of the intervertebral disk that has shifted forward no longer shows any connection to the rest of the intervertebral disk. A posterolateral herniation generally compresses the subjacent lying nerve root (**c**), causing pain and paralysis that affect inferiorly adjacent dermatomes and corresponding muscles (see **E**).

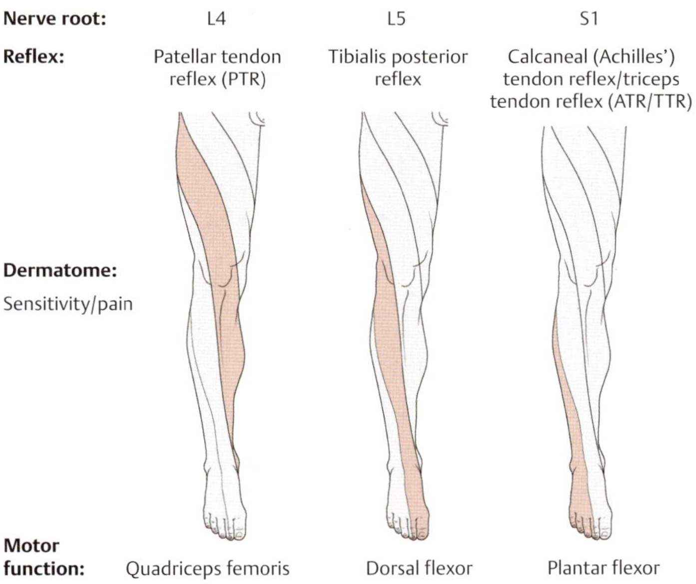

E Schematic representation of neurologic disorders in lumbar root compression syndromes as a result of herniated disks (L4, L5, and S1)

In over 90% of all cases, herniated lumbar disks affect the L5 and S1 roots. Patients complain about pain and sensitivity loss in the affected dermatomes, as well as weakness or paralysis in the muscles innervated by these nerve roots. Patients experience the symptoms of root compression syndrome in the following order with increasing severity: pain, sensitivity loss, and paralysis. Often the cause is a weakening or abnormal proprioceptive reflex associated with the affected root.

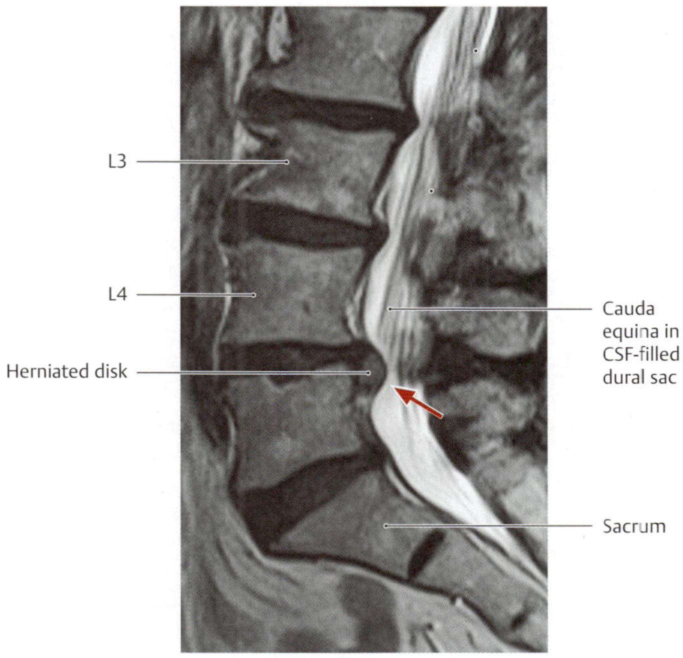

F Posterior disk herniation

Seventy-four-year-old woman with compression of L5 on the right side. **Findings:** Caudally sequestered extruded nucleus pulposus tissue at the L4–L5 level. The midsagittal T1-weighted (fat-saturated) magnetic resonance image shows peripheral contrast enhancement in this tissue (red arrow) (from: Stäbler A, Ertl-Wagner B. editors. Radiologie-Trainer: Bewegungsapparat. 4th ed. Stuttgart: Thieme; 2022).

1.17 The Thoracic Skeleton

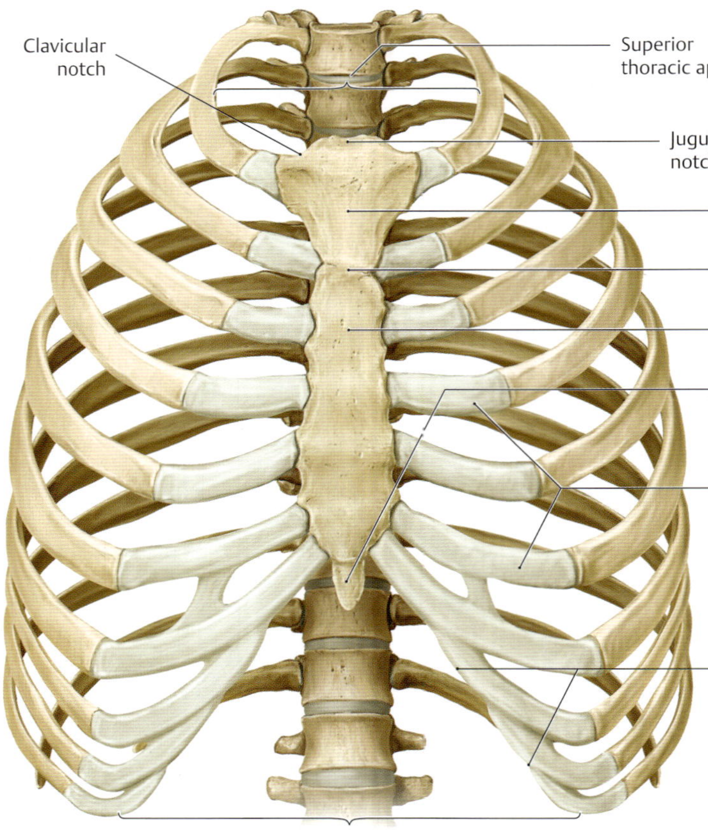

a

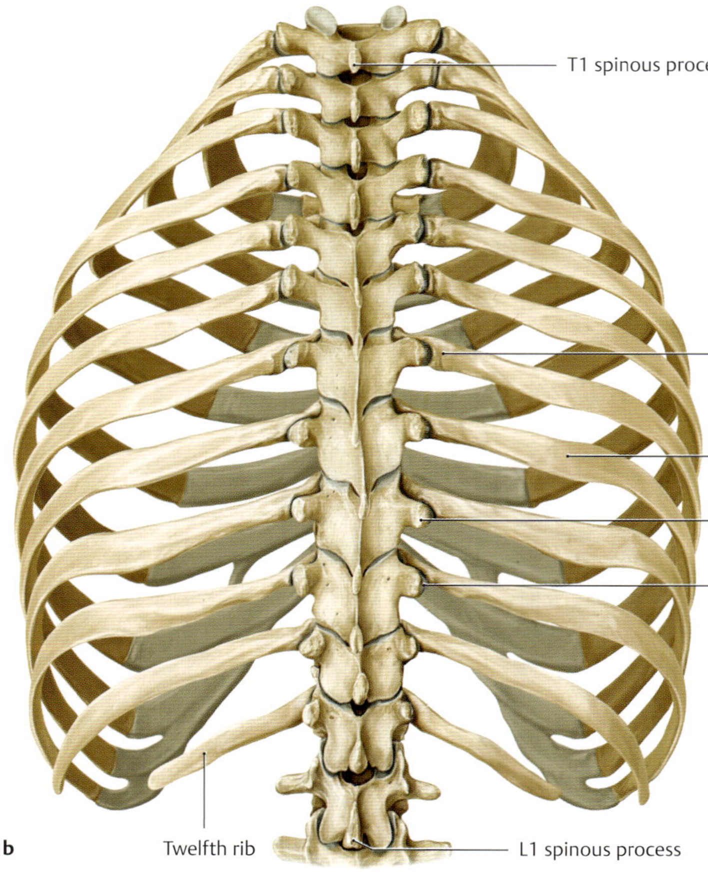

b

A Thoracic skeleton

a Anterior view, **b** posterior view.

The thoracic skeleton (thoracic cage, thorax) consists of the vertebral column, the 12 pairs of ribs, and the sternum. These structures are movably interconnected by ligaments, true joints, and synchondroses. Tension is imparted to the thorax by the intercostal muscles. The thoracic cage encloses the chest cavity and has a superior aperture (the thoracic inlet) and an inferior aperture (the thoracic outlet). Its shape is subject to marked variations relating to age and gender. The ribs of an infant have very little obliquity and are approximately horizontal. With increasing age, the ribs incline downward, while the thoracic cage becomes flattened anteroposteriorly. This is associated with a relative decrease in the size of the thoracic outlet. Generally, the female thorax is narrower and shorter than in the male. From a functional standpoint, the thoracic skeleton and its muscular wall structures form a rugged, stable enclosure that permits the respiratory excursions that are necessary for normal breathing. This is apparent in patients with severe chest injuries such as multiple rib fractures due to blunt trauma, where instability of the chest wall leads to paradoxical respiration: the affected side of the rib cage moves inward during inspiration and outward during expiration. This results in a *pendelluft* effect (from the German *Pendel*, pendulum, and *Luft*, air): air streams back and forth between the lungs, leading to increased dead space ventilation, decreased alveolar gas exchange, and respiratory failure. Patients thus affected generally require intubation.

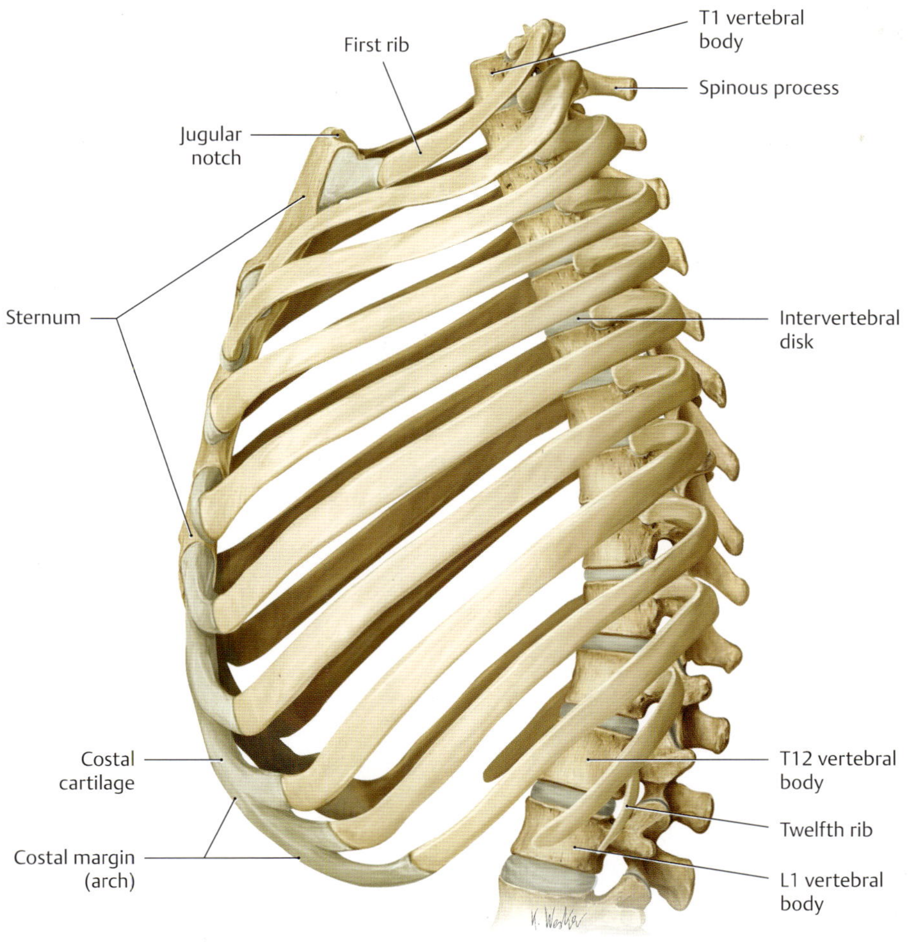

B Thoracic skeleton, lateral view

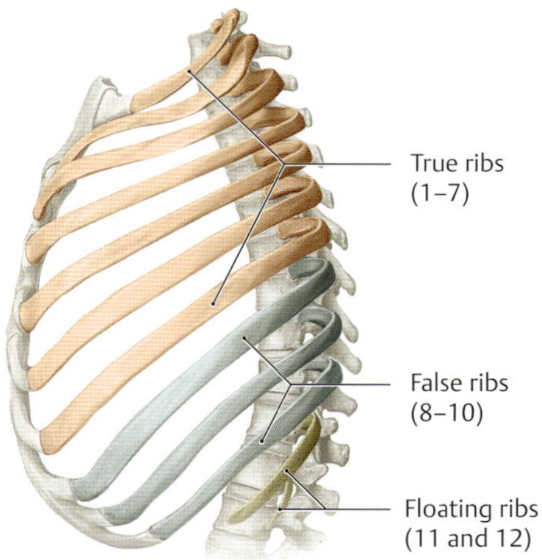

C True, false, and floating ribs

Lateral view. Each of the 12 pairs of ribs is bilaterally symmetrical, but the shapes of the ribs vary at different levels. The first seven pairs of ribs, called the *true ribs*, are normally connected anteriorly to the sternum. Of the remaining five, called the *false ribs*, the eighth, ninth, and tenth ribs are joined by their costal cartilage to the cartilage of the rib directly above, contributing to the structure of the costal margin (arch) (see **Aa**). The last two pairs of "false" ribs, called the *floating ribs*, usually terminate freely between the muscles of the lateral abdominal wall.

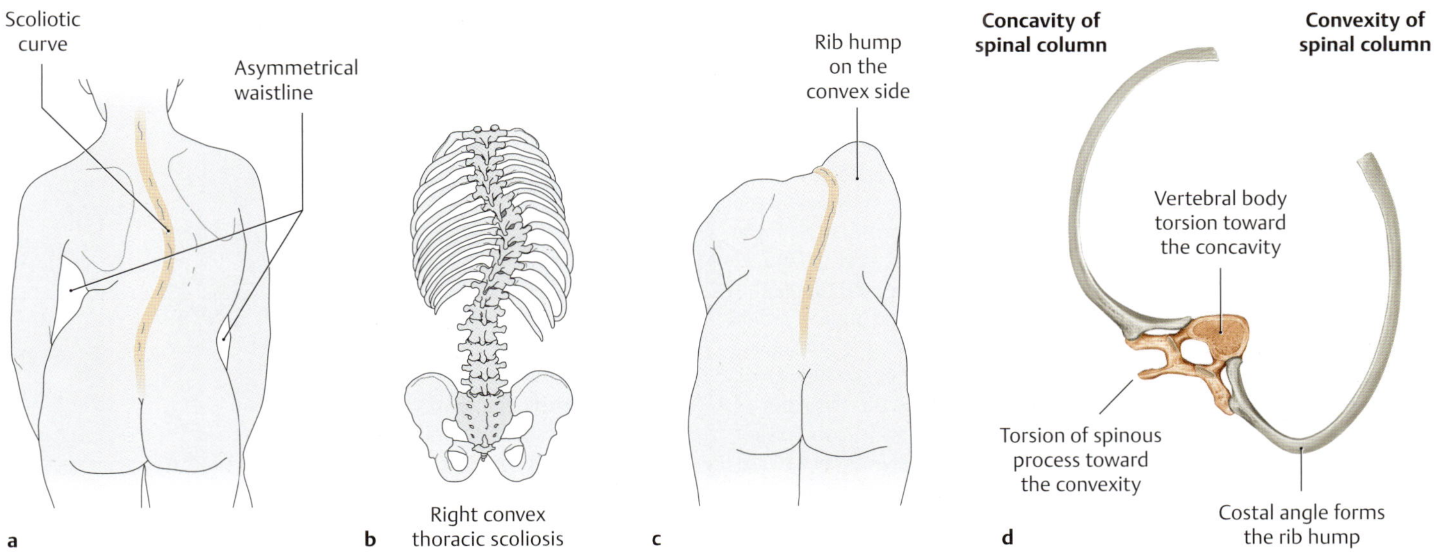

D Lateral curvature of the spinal column (scoliosis)

a, b Posterior view. Scoliosis commonly presents as a right convex curve of the spinal column at the level of the T8–T9 vertebrae (**b**). It is manifested by a typical postural deformity in upright stance (**a**).

c, d When a patient with right convex scoliosis bends forward, a typical rib hump appears on the convex side of the curve (**c**). This happens because the adjacent ribs also occupy an abnormal position due to torsion of the vertebral bodies (**d**, superior view).

1.18 The Sternum and Ribs

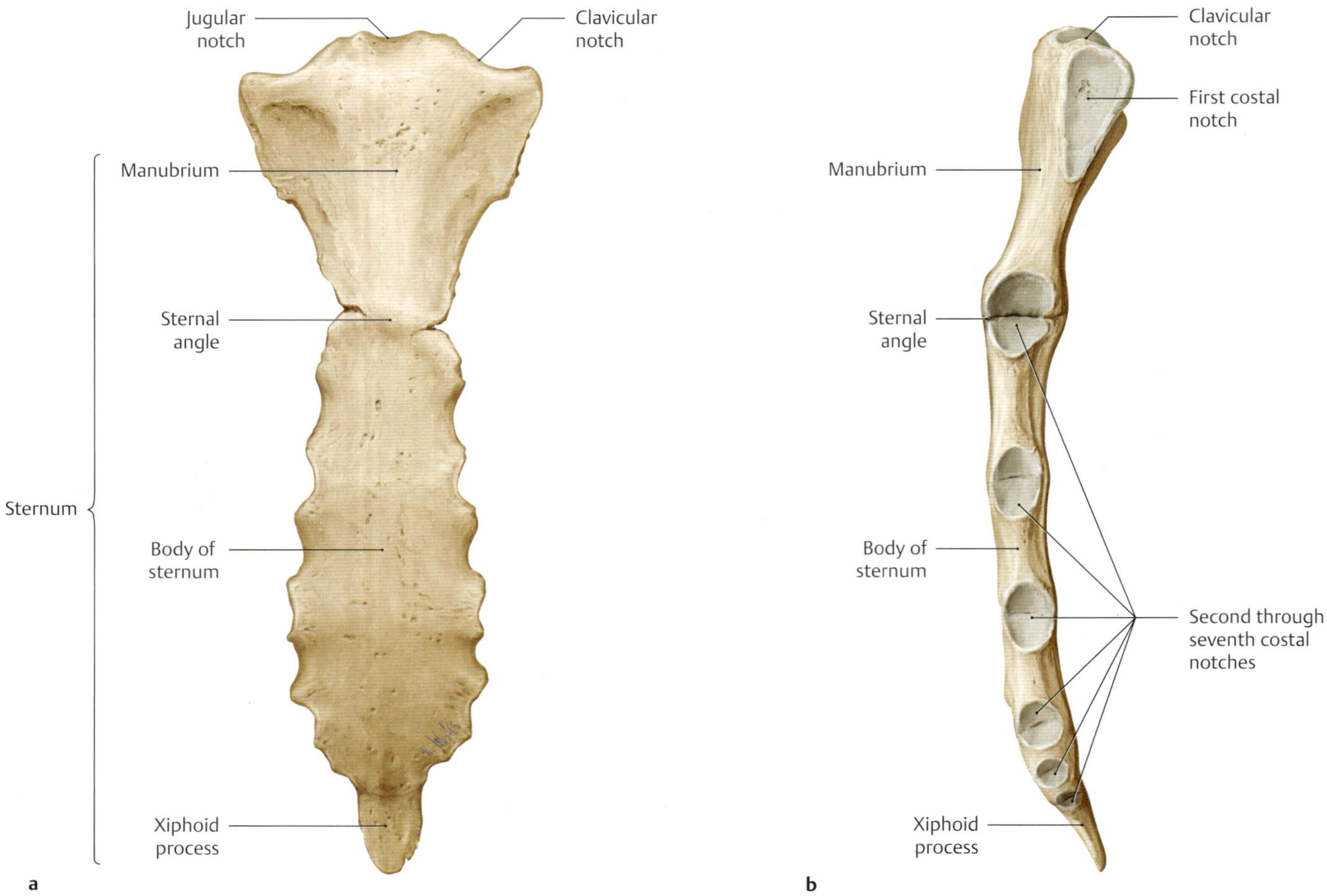

A The sternum

a Anterior view. **b** Lateral view.

The sternum is a flattened bone, slightly convex anteriorly, with multiple indentations along its lateral borders (costal notches). The adult sternum consists of three bony parts:

- The manubrium
- The body of the sternum
- The xiphoid process

The manubrium, body, and xiphoid process in adolescents and young adults are connected to one another by cartilaginous plates (manubriosternal synchondrosis and xiphosternal synchondrosis), which gradually ossify with increasing age. The fully ossified, mature adult form is depicted here. The depression in the superior border of the manubrium (jugular notch) is clearly palpable through the skin and marks the inferior margin of the jugular fossa. On each side of the jugular notch is a depression for articulation with the clavicle on that side (clavicular notch), and just below that is a shallow concavity (first costal notch) for the synchondrosis with the first rib. At the junction of the sternal manubrium and body is an articular facet for the second rib (second costal notch). At that site the manubrium is usually angled slightly backward in relation to the body of the sternum (sternal angle). The lateral borders of the sternal body bear additional costal notches that articulate with the third through seventh costal cartilages, the notches for the sixth and seventh costal cartilages being placed very close together. The sometimes bifid and perforated xiphoid process itself is devoid of costal attachments and is highly variable in its form. Frequently, the xiphoid process is still cartilaginous in adults

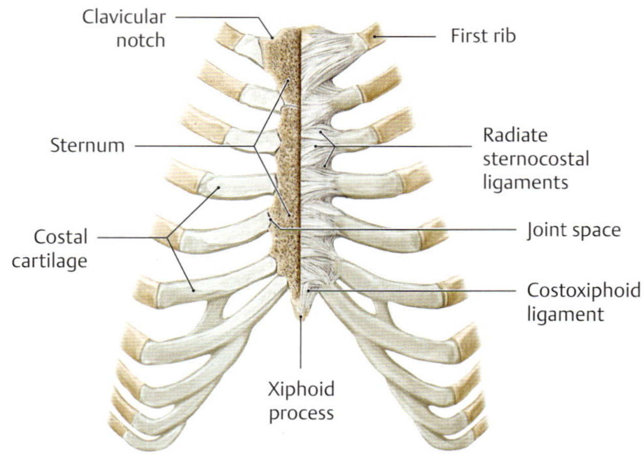

B The sternocostal joints

Anterior view (the right half of the sternum has been sectioned frontally to show the sternocostal joints). The connections between the costal cartilages for the first through seventh ribs and the costal notches of the sternum consist partly of synchondroses and partly of true joints. Generally, a joint space is found only at the second through fifth ribs, while the first, sixth, and seventh ribs are attached to the sternum by synchondroses. In the true joints as well as the synchondroses, ligaments (radiate sternocostal ligaments) radiate from the perichondrium of the costal cartilage to the anterior surface of the sternum, where they blend with the periosteum to form a dense fibrous membrane (the sternal membrane). The second rib is typically the most superior palpable rib.

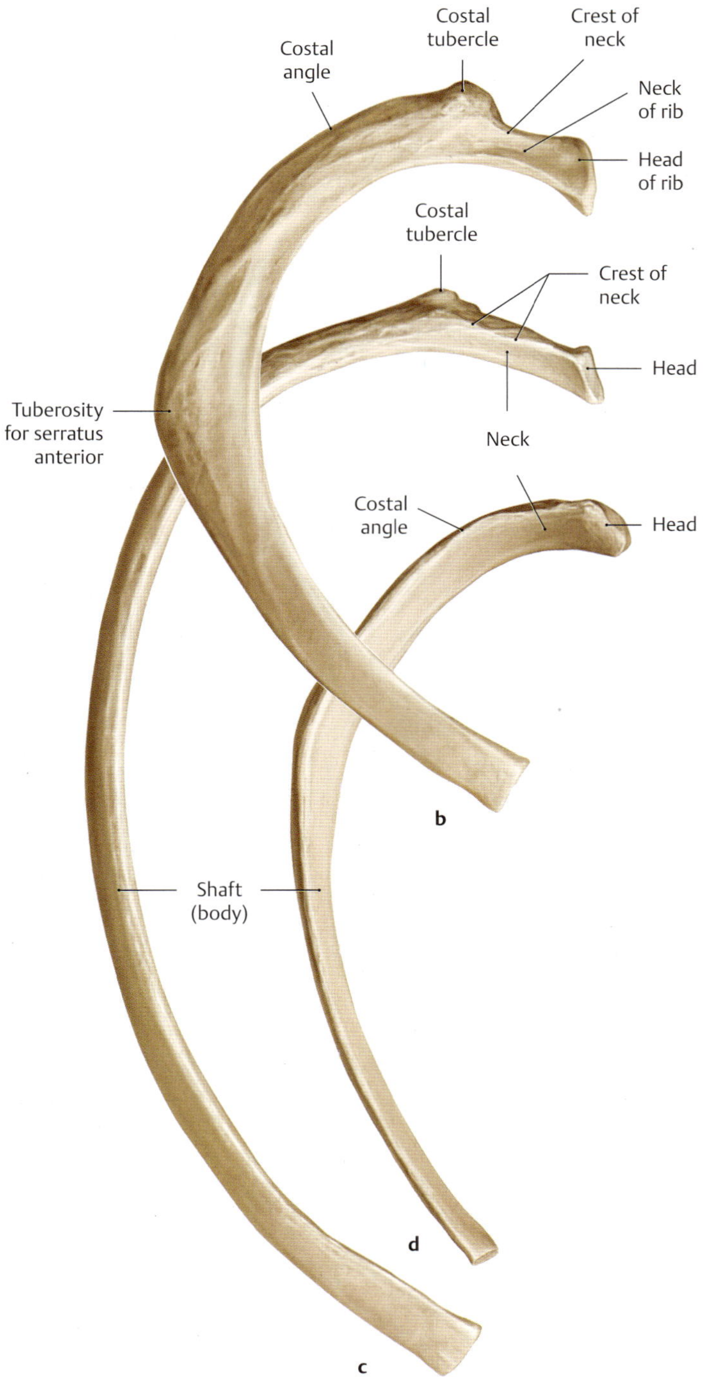

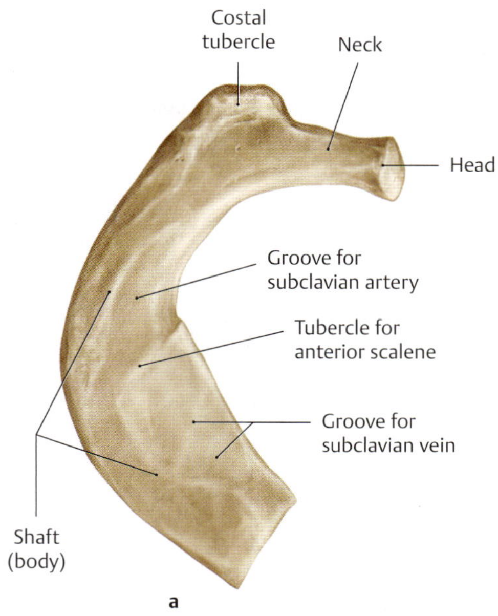

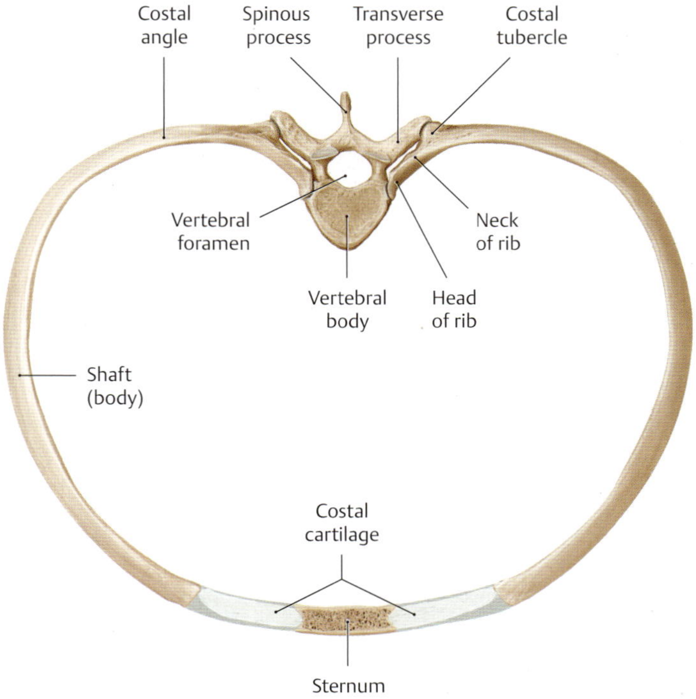

C Variable size and shape of the ribs

a First rib, **b** second rib, **c** fifth rib, **d** eleventh rib (all are right ribs, viewed from above).

The neck of the rib extends from the head of the rib to the costal tubercle. Except on the first rib, the neck bears a sharp superior ridge (crest of the neck of the rib). Lateral to the costal tubercle is the shaft (body) of the rib, which curves forward to form the costal angle. The shafts of the second through twelfth ribs in particular show irregular curvatures (on the flat and on the edges) and are also twisted about their long axis. Because of this torsion, the external surfaces of the ribs face slightly downward at their vertebral end and slightly upward at their anterior end. Normally, the first and twelfth ribs are the shortest, while the seventh rib is the longest. The costal cartilage increases in length from the first to the seventh rib and shortens again past the eighth rib. Every rib except the first, eleventh, and twelfth has a groove along its inferior border (costal groove) that affords some protection for the intercostal vessels and nerves (see pp. 181 and 213).

D Segments of the rib and structure of a thoracic segment

Sixth pair of ribs, superior view. Each rib consists of a bony part (costal bone) and a cartilaginous part (costal cartilage). The bony part consists of the following segments, starting from its vertebral end:

- Head
- Neck
- Costal tubercle
- Shaft (body), which includes the costal angle

1.19 The Costovertebral Joints and Thoracic Movements

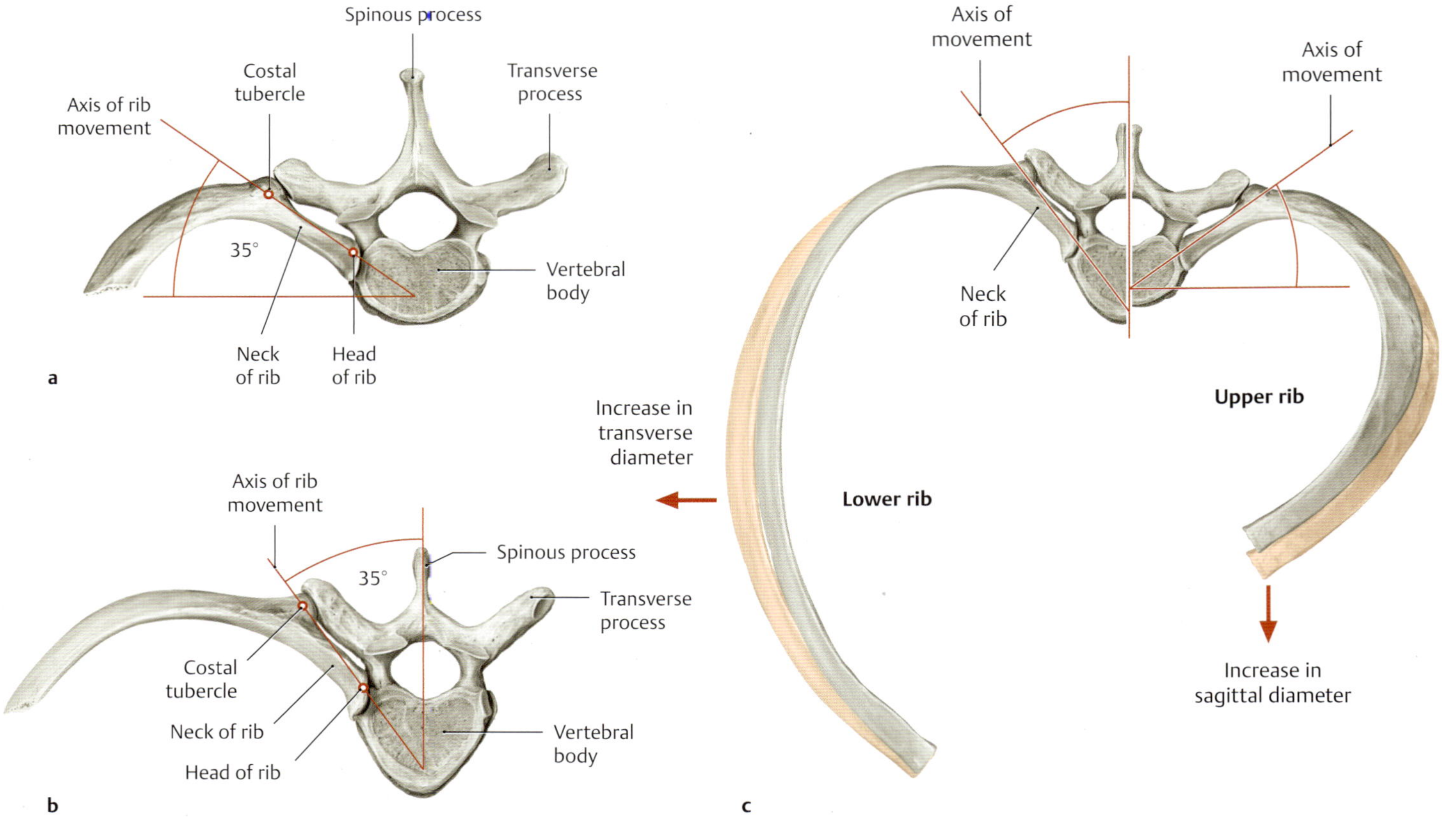

A Axes for movements of the costovertebral joints and ribs
Superior view. **a** Axis of upper rib movements. **b** Axis of lower rib movements. **c** Direction of rib movements (see **C** for the costovertebral joints).
The axes of rib movements are directed parallel to the necks of the ribs. The axes for the upper ribs are closer to the coronal plane (**a**), while those for the lower ribs are closer to the sagittal plane (**b**). For this reason, a rib excursion in the upper part of the rib cage mainly increases the sagittal thoracic diameter, while a lower rib excursion increases the transverse diameter (see **B**).

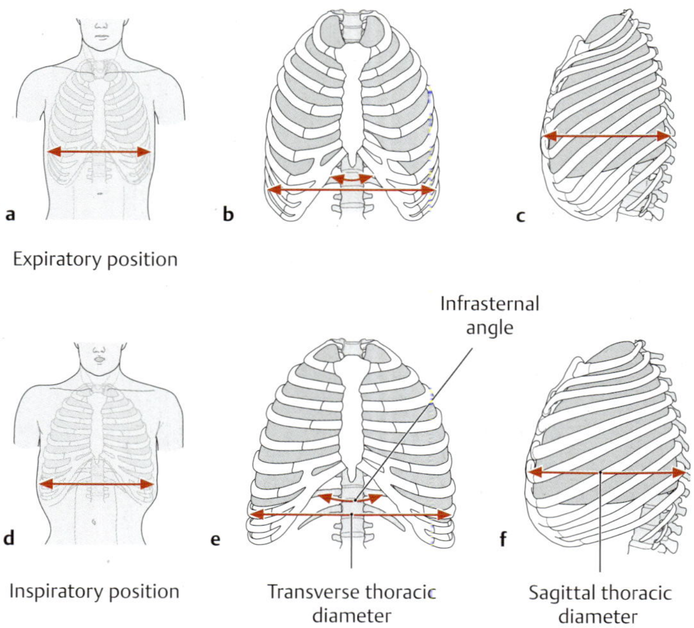

B Movements of the rib cage during costal or chest breathing (sternocostal breathing)
Respiration (ventilation) is effected through changes in the thoracic volume. The increase in thoracic volume that is necessary for inspiration can be accomplished in two ways:

1. By lowering the diaphragm (costodiaphragmatic breathing or abdominal breathing, see p. 168).
2. By elevating the ribs (sternocostal breathing, also known as costal or chest breathing).

While breathing at rest is almost entirely abdominal, respiration during physical effort is augmented by chest breathing with the intercostal muscles and the auxiliary muscles of respiration. The drawings illustrate the thoracic volume changes that occur during chest or costal breathing, in which the thoracic volume decreases and increases in both the coronal and sagittal planes. Panels **a–c** show the decrease in the transverse and sagittal chest diameters at end-expiration, and **d–f** show the increase at end-inspiration.

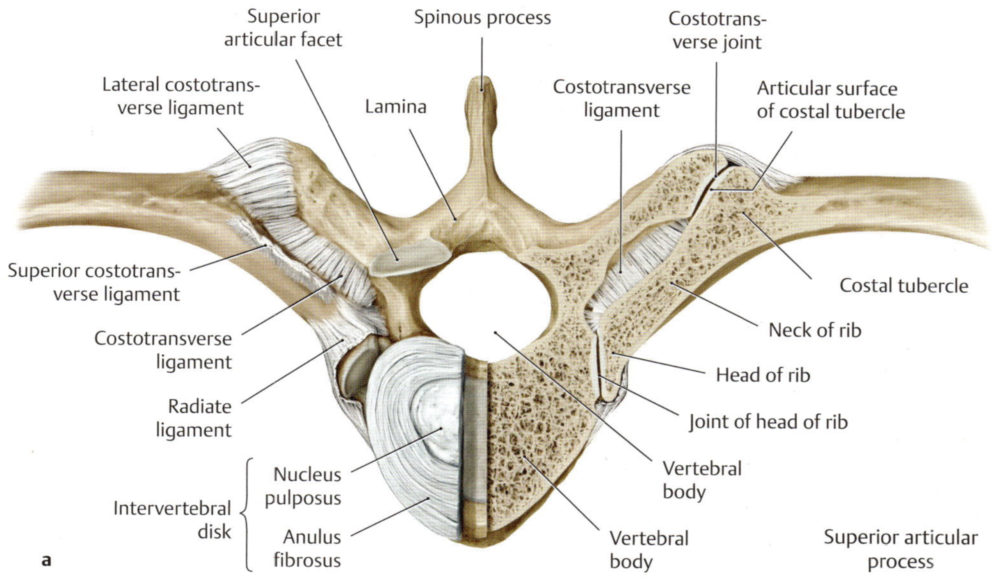

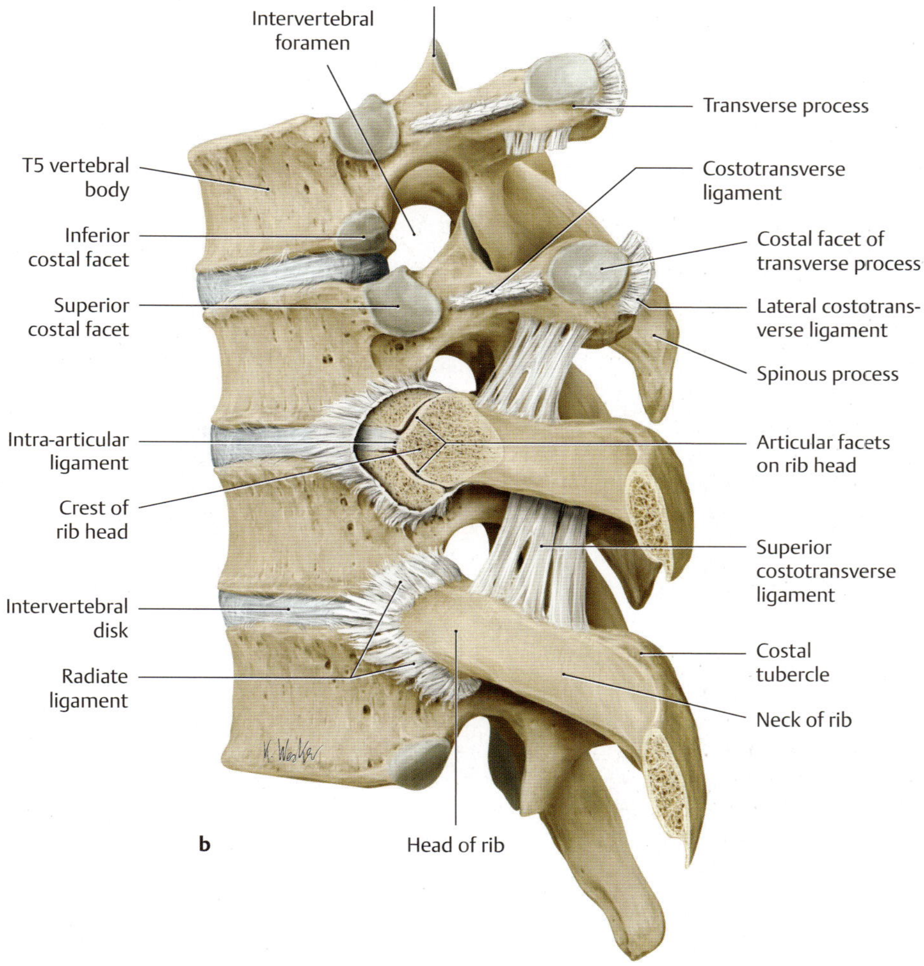

C Ligaments of the costovertebral joints

The costovertebral joints are the joints by which the ribs articulate with the vertebrae. They consist of two types: the joints of the heads of the ribs and the costotransverse joints. Though morphologically distinct, these different joint types are functionally interrelated.

a Articulation of the eighth rib with the eighth thoracic vertebra, superior view (the joint of the head of the rib and costotransverse joint on the left side have been transversely sectioned).

b The fifth through eighth thoracic vertebrae and associated ribs (seventh and eighth ribs), left lateral view (the joint of the head of the seventh rib has been transversely sectioned).

Joint of the head of the rib: This joint consists of two articular surfaces:

1. An articular facet on the head of the rib.
2. A costal facet on the vertebral body.

The articular facets on the head of the second through tenth ribs (defined by the crest of the rib head) articulate with the fossa formed by the superior and inferior costal facets on two adjacent vertebral bodies and the intervertebral disk between them. The intra-articular ligament, which extends from the crest of the rib head to the intervertebral disk, separates the joint cavity of the second through tenth rib heads into two compartments. By contrast, the heads of the first, eleventh, and twelfth ribs each articulate with only one thoracic vertebral body (see **A**, p. 118). In all the joints of the heads of the ribs, the joint capsule is reinforced by the radiate ligament.

Costotransverse joint: In the costotransverse joints of the first through tenth ribs, the articular surface of the costal tubercle articulates with the costal facet on the transverse process of the corresponding thoracic vertebra. The eleventh and twelfth ribs do not have a costotransverse joint because their transverse processes do not have articular facets (see **A**, p. 118). Three ligaments stabilize the costotransverse joint and also strengthen the joint capsule:

1. The lateral costotransverse ligament (from the tip of the transverse process to the costal tubercle).
2. The costotransverse ligament (between the neck of the rib and the transverse process).
3. The superior costotransverse ligament (between the neck of the rib and the transverse process of the vertebra above it).

1.20 The Bony Pelvis

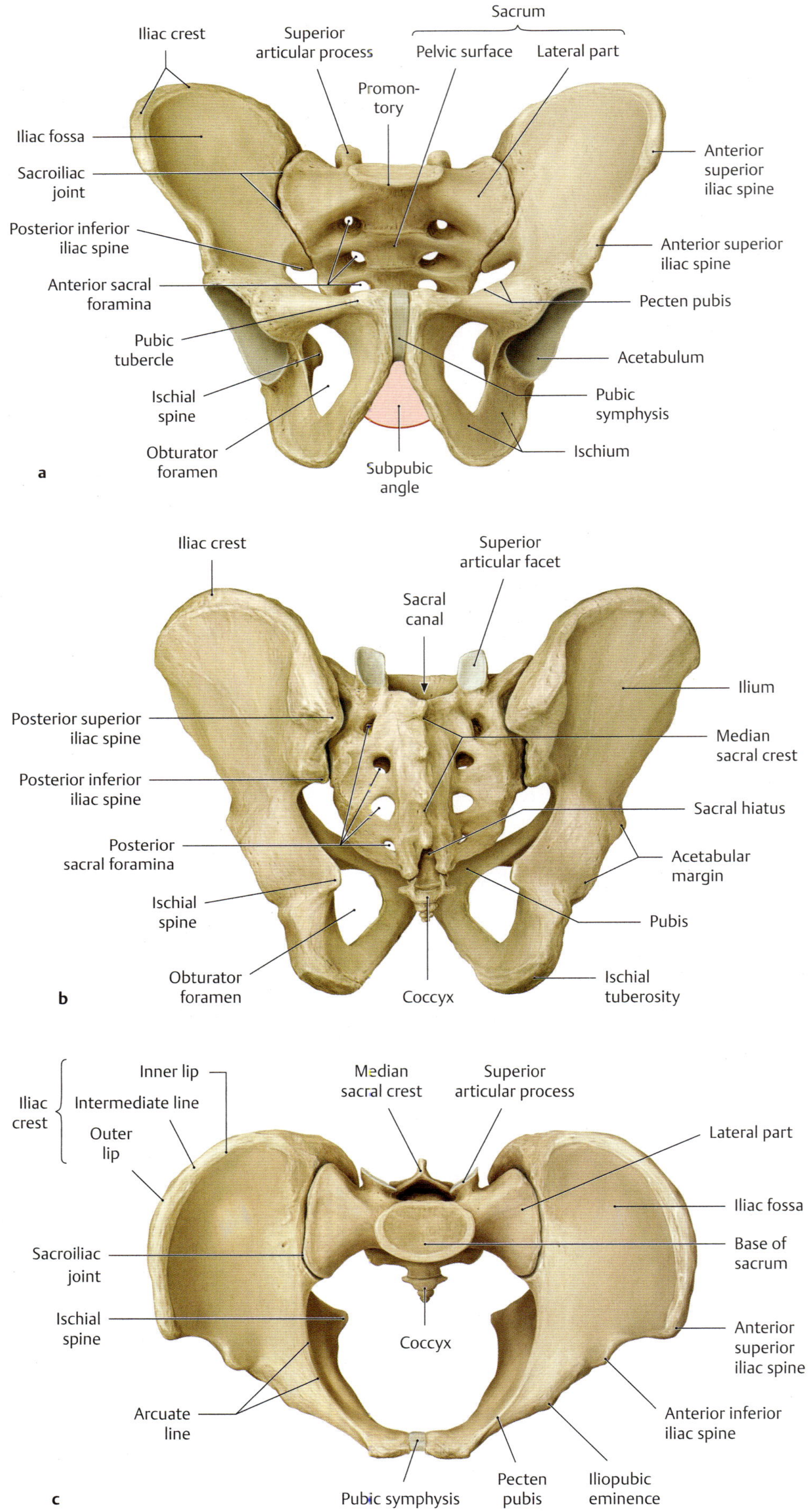

A The male pelvis

a Anterior view.
b Posterior view.
c Superior view.

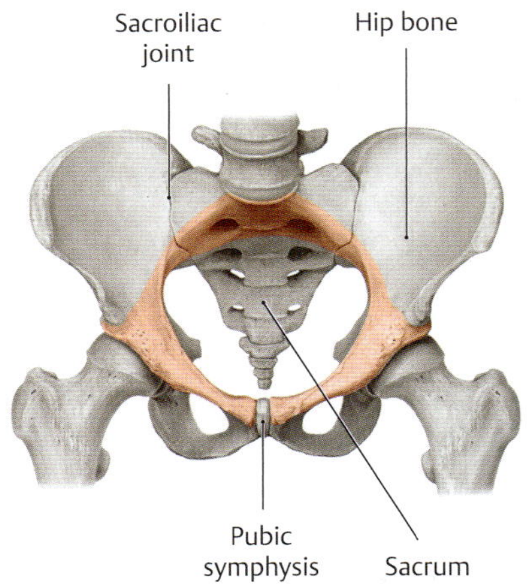

B The pelvic girdle and pelvic ring

Anterosuperior view. The pelvic girdle consists of the two hip bones (coxal bones). The sacroiliac joints and the cartilaginous pubic symphysis unite the bony parts of the pelvic girdle with the sacrum to form a stable ring called the pelvic ring (indicated by color shading). It allows very little mobility, because stability throughout the pelvic ring is an important prerequisite for transmitting the trunk load to the lower limbs.

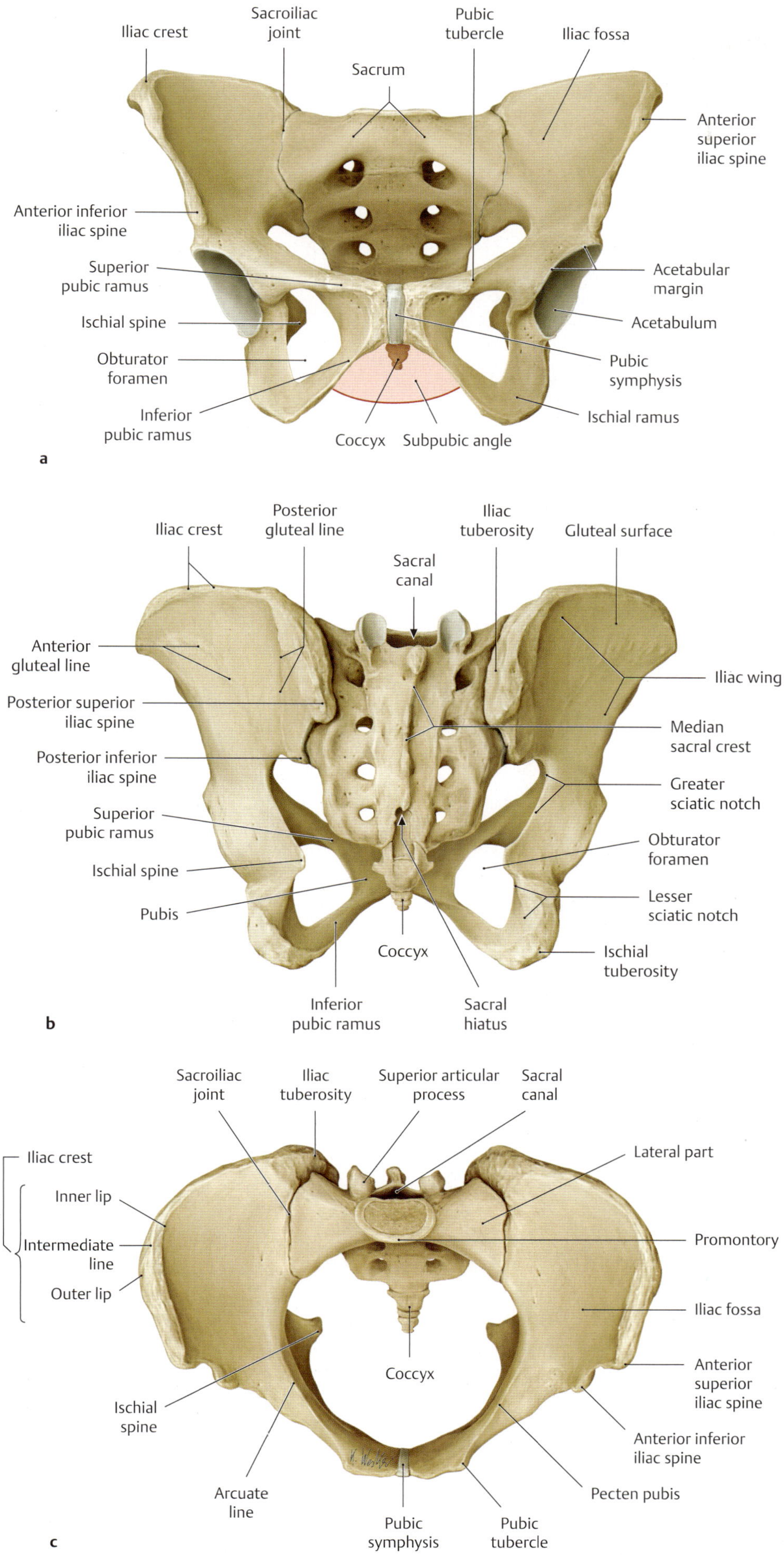

C The female pelvis
a Anterior view.
b Posterior view.
c Superior view.

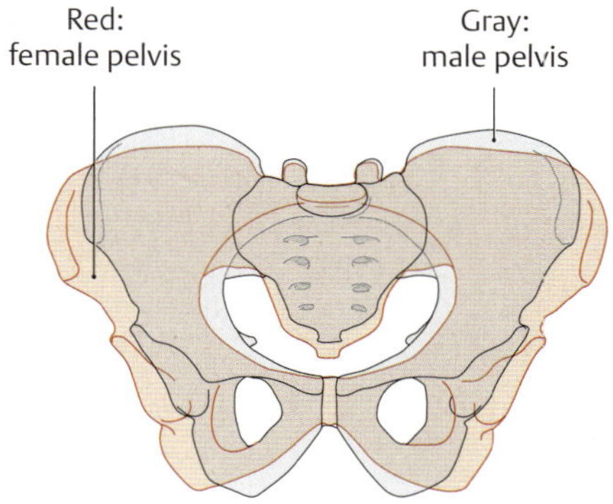

D Gender-specific features of the pelvis
Anterosuperior view. A male and female pelvis have been superimposed to illustrate the gender-specific differences. Comparison reveals that the female pelvis is larger and broader than the male pelvis, while the latter is taller, narrower, and more massive than the female pelvis. The inlet of the female pelvis is larger and has an almost oval shape, while the sacral promontory shows a greater projection in the male pelvis (**Cc**). Sex differences are also noted in the angle of the pubic arch formed by the inferior pubic rami, which is acute in men (70°) but significantly larger in women (almost 90–100°). Accordingly, this angle is called the subpubic angle (see **D**, p. 149). The sacrum also exhibits differences between the sexes. In women the sacrum is angled at the level of the third and fourth vertebrae (see p. 122), while in men it presents a uniform curvature.

1.21 The Pelvic Ligaments and Pelvic Measurements

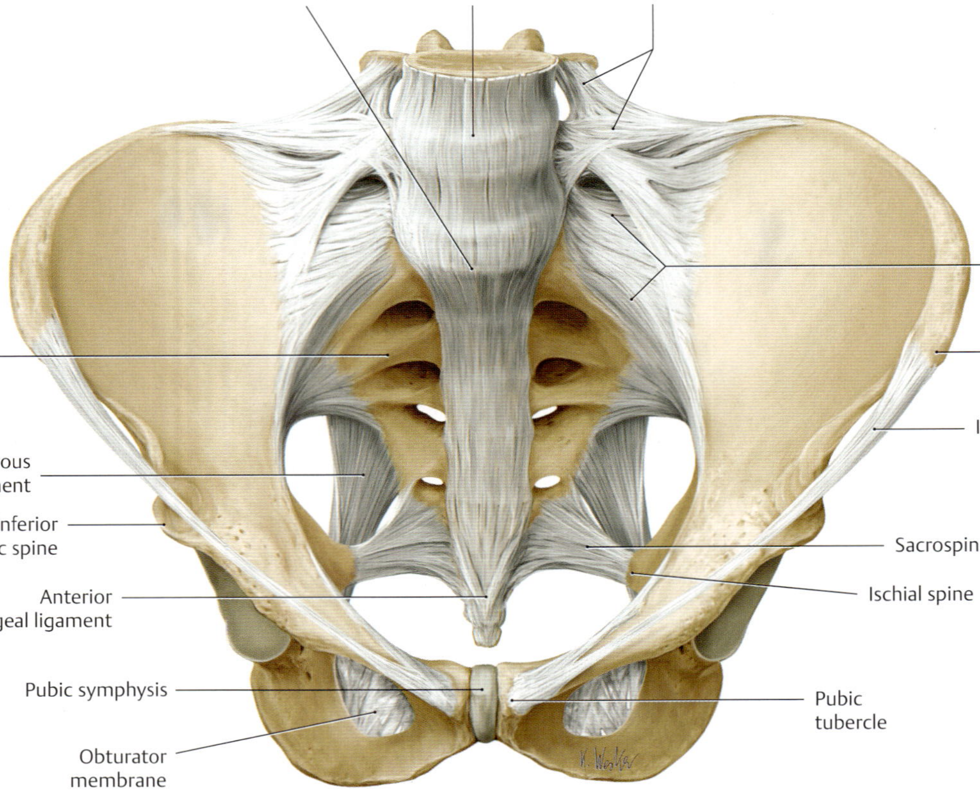

a

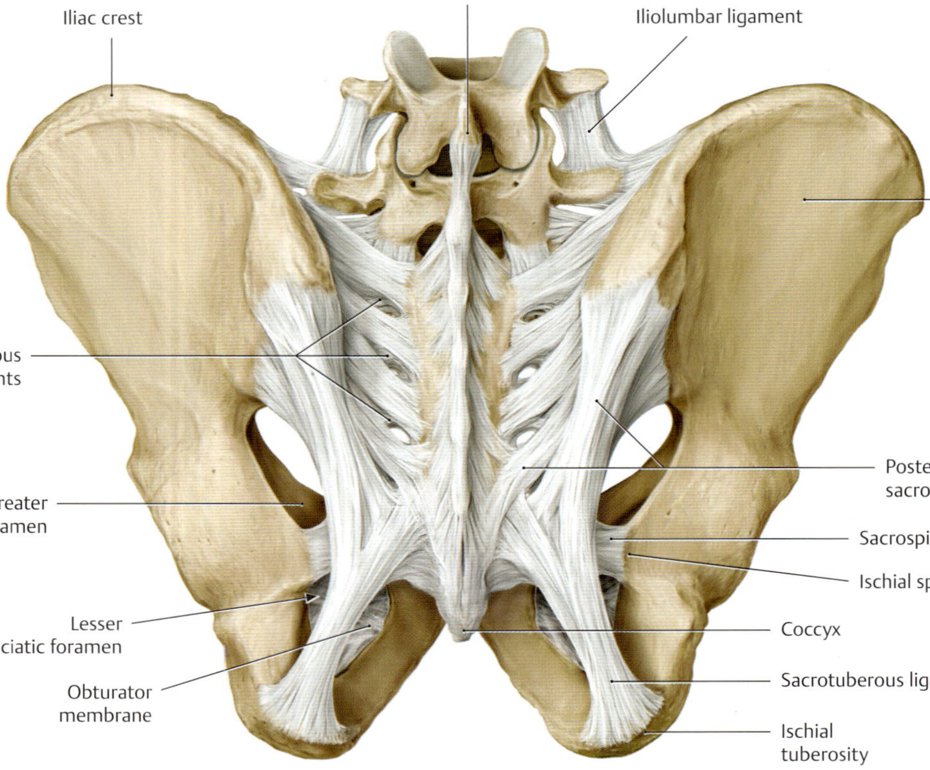

b

A Ligaments of the male pelvis
a Anterosuperior view.
b Posterior view.

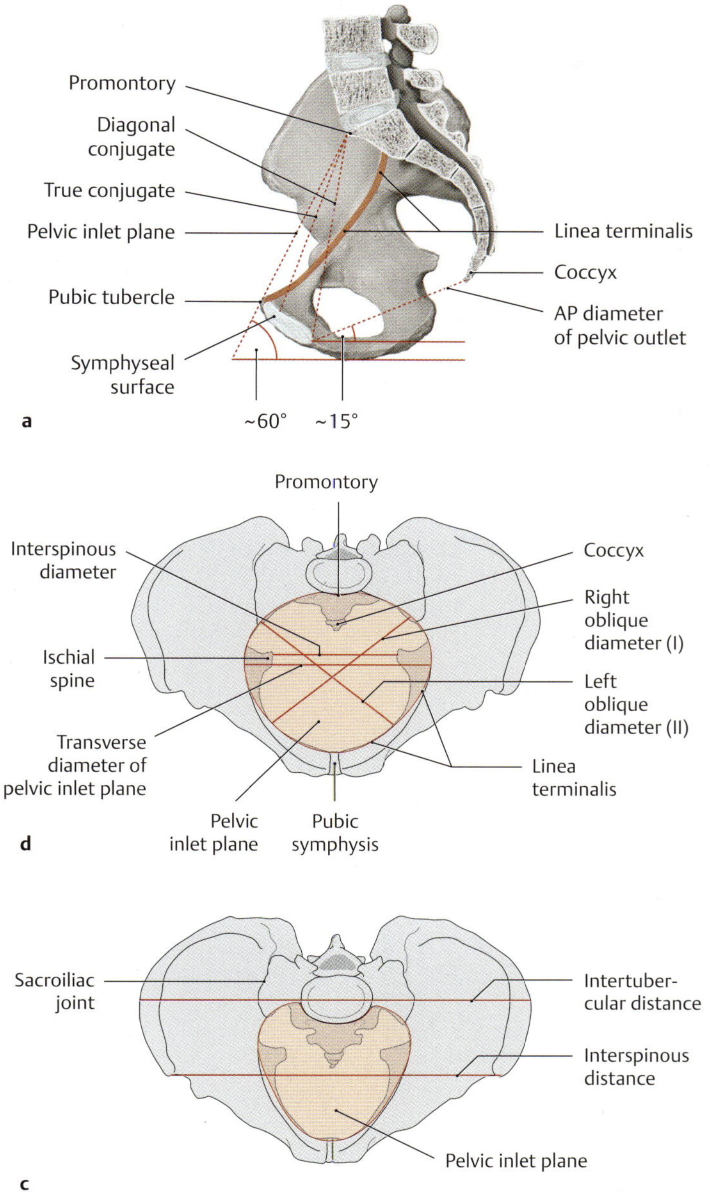

B B Internal and external pelvic measurements

a Right half of a female pelvis, medial view.
b Female pelvis, superior view.
c Male pelvis, superior view.

The linea terminalis is marked in red in **a**. The pelvic inlet plane is color-shaded in **b** and **c**.

C Internal and external pelvic measurements, linea terminalis, and pelvic inlet plane

The internal and external pelvic measurements provide direct or indirect information on the size and shape of the bony boundaries of the lesser pelvis. Because the lesser pelvis functions as the birth canal, the internal and external pelvic dimensions have special practical significance in obstetrics, determining whether the pelvic cavity is broad enough to allow for a vaginal delivery. A particularly important measurement is the true conjugate of the pelvic inlet (the obstetric conjugate), which is the smallest anteroposterior dimension of the lesser pelvis. With *pelvimetry*, a method of measuring pelvic dimensions, potential obstructions to labor can be identified prior to the delivery. Generally, the measurements are performed by transvaginal sonography. Some pelvic dimensions, such as the diagonal conjugate, can be accurately determined by bimanual examination.

Internal pelvic measurements in women (see **Ba** and **Bb**)

- True conjugate = 11 cm (distance from the sacral promontory to the posterior border of the symphysis)
- Diagonal conjugate = 12.5–13 cm (distance from the sacral promontory to the lower border of the symphysis)
- Anteroposterior (AP) diameter of the pelvic outlet = 9 (+2) cm (distance from the lower border of the symphysis to the tip of the coccyx)
- Transverse diameter of the pelvic inlet plane = 13 cm (greatest distance between the lineae terminales)
- Interspinous diameter = 11 cm (distance between the ischial spines)
- Right (I) and left (II) oblique diameter = 12 cm (distance from the sacroiliac joint at the level of the linea terminalis to the iliopubic eminence [formerly the "iliopectineal eminence"] on the opposite side)

External pelvic measurements in men (see **Bc**)

- Intertubercular distance = 25–26 cm (distance between the anterior superior iliac spines)
- Intercristal distance = 28–29 cm (greatest distance between the left and right iliac crests in the coronal plane)
- External conjugate = 20–21 cm (distance from the upper border of the symphysis to the spinous process of the L5 vertebra)

Linea terminalis (see **B**)

Boundary line between the greater and lesser pelvis, consists of the pubic symphysis and crest + pecten pubis + the arcuate line and sacral promontory.

Pelvic inlet plane (see **Bb** and **Bc**)

Plane through the pelvic inlet at the level of the linea terminalis, below which is the lesser pelvis.

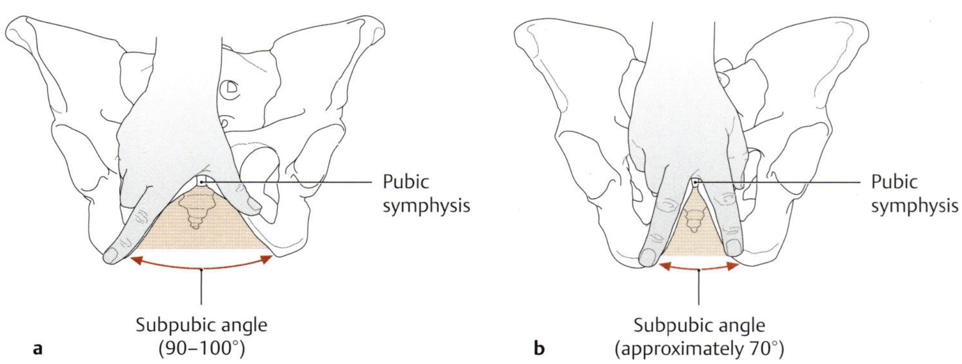

D Subpubic angle
Anterior view.

a Female pelvis
b Male pelvis.

1.22 The Sacroiliac Joint

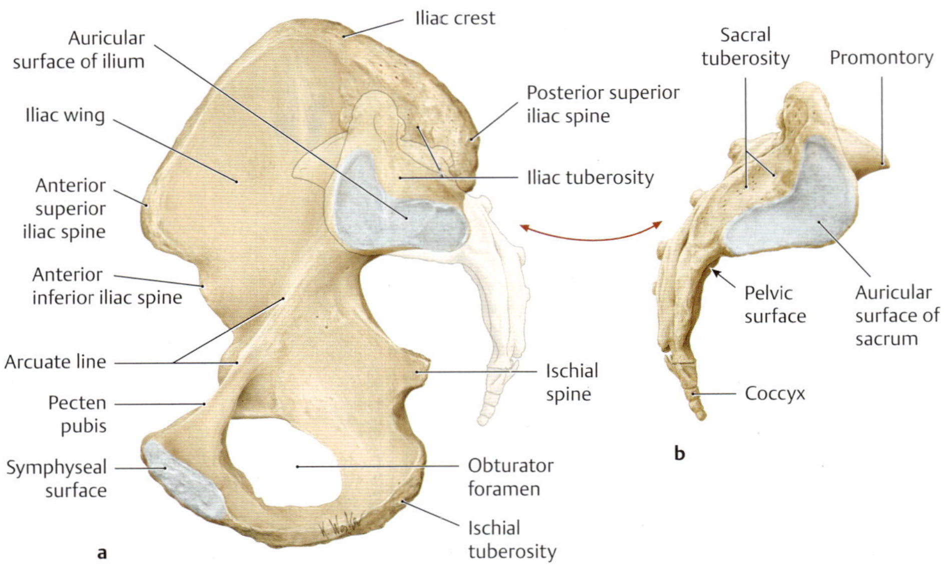

A Articular surfaces of the sacroiliac joint

a Auricular surface of the ilium, right hip bone, medial view (the sacrum is transparent in this view)

b Auricular surface of the sacrum, right lateral view.

The two ear-shaped articular surfaces of the ilium and sacrum (auricular surfaces) are brought together at the sacroiliac joint. The auricular surfaces of the ilium and sacrum are reciprocally irregular. The shape and size of both articular surfaces show considerable individual variation—more so than in other joints. Their cartilaginous covering is generally irregular, the articular cartilage on the sacral side being approximately twice as thick as on the iliac side.

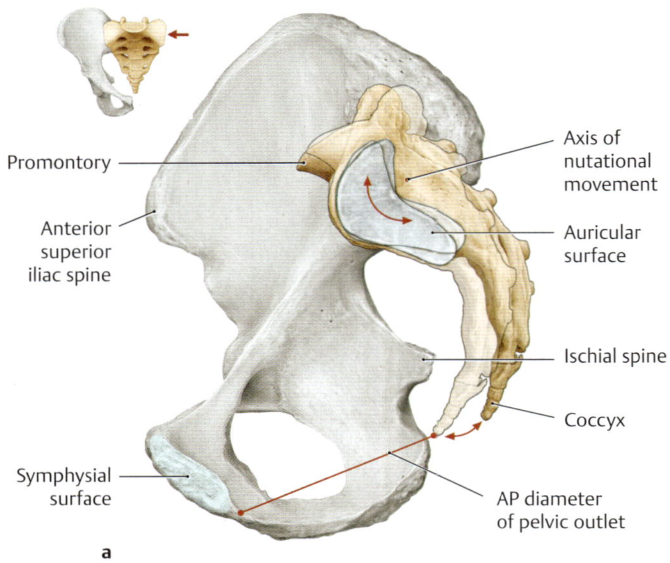

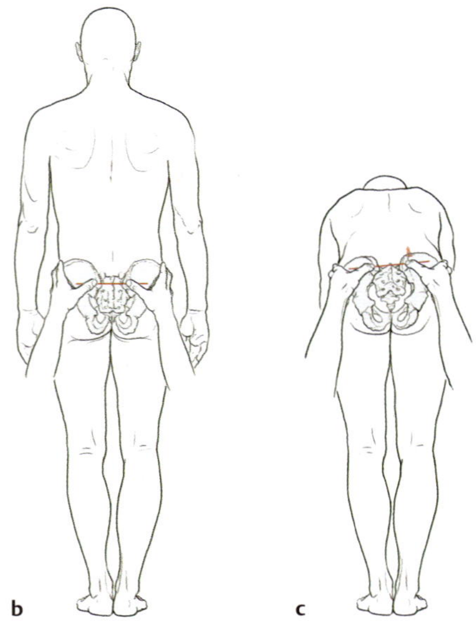

B Motion and blockage in the sacroiliac joint (pathologic anterior motion).

Movement in the sacroiliac joints is normally severely limited (see **C**). However, it can vary considerably among individuals and between the sexes. Movement in the sacroiliac joints can alter the width of the pelvic ring and thus has practical importance in obstetrics. Minimal rotational and translational movements are possible In a rotational movement, a **nutation** or **"tilting" movement** (see **a**), the sacrum rotates about an axis located at the attachments of the interosseous sacroiliac ligaments. With anterior rotation of the sacrum, the promontory moves forward and downward, while the coccyx moves upward and backward, thereby increasing the AP diameter of the pelvic outlet. With posterior rotation of the sacrum, the AP diameter of the pelvic outlet is decreased and the AP diameter of the pelvic inlet increases.

A sports injury such as a hard landing when jumping can jam the sacrum, rendering movement within the sacroiliac joints impossible (**sacroiliac joint blockage**). This is apparent in the **pathologic anterior motion** (**b** and **c**) seen in the standing flexion test. The patient stands with his or her back to the examiner and legs about hip width apart. The examiner sits on a stool behind the patient and palpates both posterior superior iliac spines (**b**) simultaneously. The examiner asks the patient to fully extend the knee and slowly bend forward (see **c**). During this motion the examiner observes whether both posterior superior iliac spines simultaneously move the same distance superiorly and anteriorly. In normally mobile sacroiliac joints, the nutation of the sacrum with respect to the iliac wings is smooth and uniform; both the examiner's thumbs (and thus both the patient's posterior superior iliac spines) will remain at the same level at the end of the trunk motion as at its beginning. If the sacroiliac joint is blocked on one side, the iliac spine on the affected side will be drawn superiorly faster than the contralateral side (a positive finding on the affected side). Blockage of the sacroiliac joints leads to severe stretching of the joint capsule, causing severe pain with practically every movement of the body. However, pain in the sacroiliac joint can also be caused by inflammatory or degenerative disorders such as spondylitis deformans or osteoarthritis or by ligament weakness (see **C**). Generalized ligament weakness or ligament loosening due to pregnancy or hormonal influence can lead to hypermobility and pain in the sacroiliac joint(s).

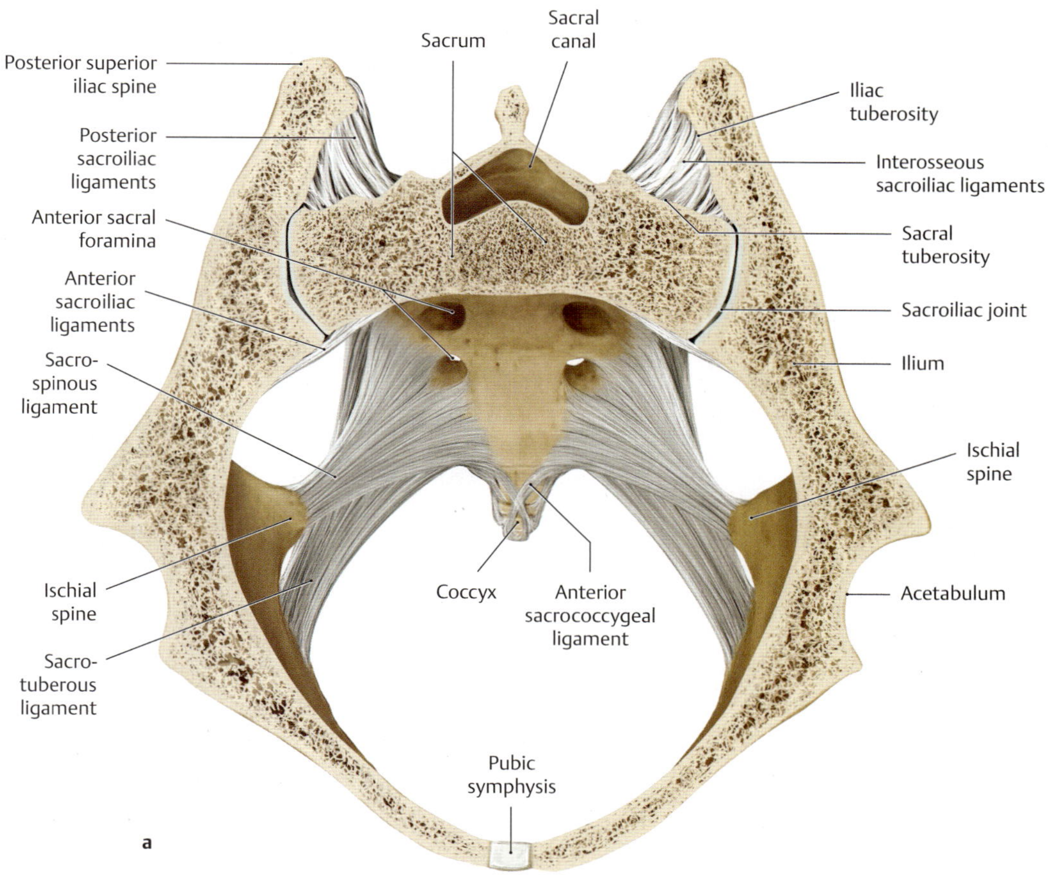

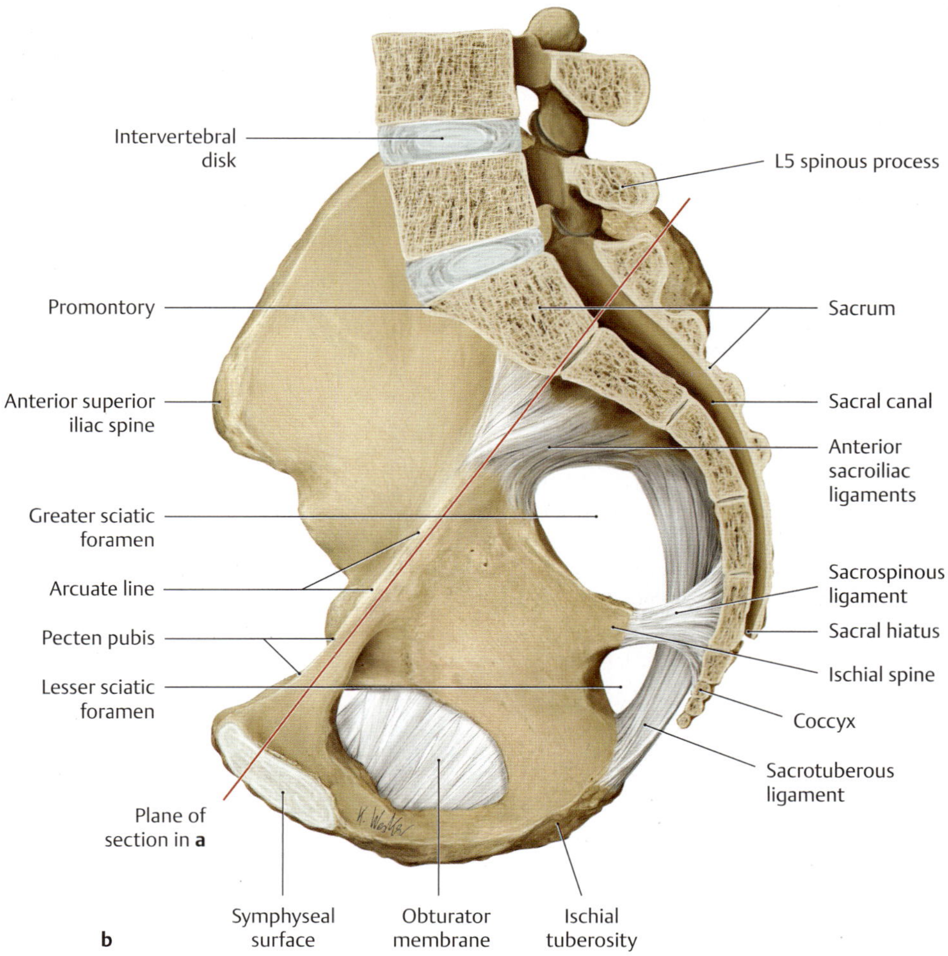

C Ligaments of the sacroiliac joint

a Oblique section through the pelvis at the level of the pelvic inlet plane, superior view (plane of section indicated in **b**).

b Right half of pelvis, medial view.

Although the sacroiliac joint is a true joint, its movements are greatly restricted by a tight joint capsule and powerful ligaments (amphiarthrosis). The anterior sacroiliac ligaments stabilize the joint on the anterior side of the pelvis, while the interosseous and posterior sacroiliac ligaments and iliolumbar ligaments are the main stabilizing elements on the dorsal side (see p. 148). The powerful interosseous sacroiliac ligaments are deep bands that run medially just behind the sacroiliac joint from the iliac tuberosity to the tuberosity of the sacrum. They are completely covered by the posterior sacroiliac ligaments. The ligament complex helps to anchor the sacrum in the pelvic ring during upright stance and keep it from sliding into the pelvic cavity. Additionally, the sacrotuberous ligament and sacrospinous ligament (see **b**) stabilize both sacroiliac joints and prevent posterior tilting of the pelvis about the transverse axis.

2.1 The Muscles of the Trunk Wall: Their Origin and Function

Overview of the Trunk Wall Muscles
The muscles of the trunk wall in the strict sense consist of the intrinsic back muscles and the muscles of the chest and abdominal wall. In the broad sense, they include the muscles of the pelvic floor (which form the inferior boundary of the abdominopelvic cavity) and the diaphragm (which separates the thoracic and abdominal cavities). Besides the muscles of the trunk wall, the back and thorax also contain muscles of the shoulder girdle and upper limb that migrated to the trunk during the course of development (*nonintrinsic back and thoracic muscles*). Examples are the thoracohumeral muscles anteriorly, the *spinohumeral* muscles laterally and posteriorly, and the *spinocostal* muscles. Other muscles that have migrated to the trunk, such as the trapezius, are derived from the mesenchyme of the branchial arches (branchial musculature). They are innervated by cranial nerves (the trapezius by the accessory nerve) and were incorporated secondarily into the trunk wall (see p. 312 ff).

A Trunk wall muscles in the strict sense

Intrinsic back muscles
Lateral tract • Sacrospinal system – Iliocostalis – Longissimus • Spinotransverse system – Splenius • Intertransverse system – Intertransversarii – Levatores costarum **Medial tract** • Spinal system – Interspinales – Spinalis • Transversospinal system – Rotatores breves and longi – Multifidus – Semispinalis **Short nuchal and craniovertebral joint muscles (suboccipital muscles)*** • Rectus capitis posterior major • Rectus capitis posterior minor • Obliquus capitis superior • Obliquus capitis inferior
Prevertebral neck muscles (belong topographically to the group of deep neck muscles but act mainly on the cervical spine)
• Longus capitis • Longus colli • Rectus capitis lateralis • Rectus capitis anterior
Muscles of the thoracic cage and wall
• Intercostal muscles • Transversus thoracis • Subcostal muscles • Scalene muscles (belong topographically to the group of deep neck muscles but are functionally related to thoracic breathing)
Muscles of the abdominal wall
Anterolateral abdominal muscles • External oblique • Internal oblique • Transversus abdominis **Anterior (straight or strap) abdominal muscles** • Rectus abdominis • Pyramidalis **Posterior (deep) abdominal muscles** • Quadratus lumborum • Psoas major (belongs functionally to the hip muscles, see p. 494)

B Trunk wall muscles in the broad sense

Muscles of the pelvic floor and perineum
Pelvic diaphragm • Levator ani muscle – Puborectalis – Pubococcygeus – Iliococcygeus – Coccygeus **Deep perineal muscles** • Deep transverse prineal (in male) • External urethral sphincter • Compressor urethra (in female) • Urethrovaginal sphincter (in female) **Superficial perineal muscles** • Superficial transverse perineal • Bulbospongiosus • Ischiocavernosus • External anal sphincter
Diaphragm
• Costal part • Lumbar part (crura) • Sternal part

C Muscles that migrated secondarily to the trunk wall
(described in the Upper Limb unit, p. 250)

Spinocostal muscles (see also p. 166)
• Serratus posterior superior • Serratus posterior inferior
Spinohumeral muscles between the trunk and shoulder girdle
• Rhomboid major and minor • Levator scapulae • Serratus anterior • Subclavius • Pectoralis minor • Trapezius
Spinohumeral muscles between the trunk and arm
• Latissimus dorsi
Thoracohumeral muscles
• Pectoralis major

* The suboccipital muscles in the strict sense are the short or deep nuchal muscles that are counted among the intrinsic back muscles (because they are innervated by a posterior ramus). The rectus capitis anterior and lateralis muscles are *not* classified as intrinsic back muscles because they are innervated by anterior rami, even though they are also suboccipital in their location.

Origin of the Trunk Wall Muscles

The striated muscles of the trunk wall (including the muscles of the diaphragm and pelvic floor), like the limb muscles, develop embryologically from the myotomes of the somites (see p. 6) and are therefore called the *somatic muscles*. In all, approximately 42 to 44 pairs of segmental somites are formed in the paraxial mesoderm between the 20th and 30th day of development. Five occipital, 7 cervical, 12 thoracic, 5 lumbar, 5 sacral, and 8 to 10 coccygeal somites are formed in a craniocaudal sequence (see **D**). Some of these somites regress with further development, particularly the first occipital somites and most of the coccygeal somites, so that the number of original somites is greater than the number of subsequent vertebral segments. The boundary between the head and neck runs through the fifth pair of occipital somites. At the end of the sixth week of development, the somite myotomes migrate in the dorsoventral direction and become separated into a dorsal part (epimere or epaxial muscles) and a ventral part (hypomere or hypaxial muscles) (**E**). While the epaxial muscles develop into intrinsic (local) back muscles and retain their original location, the hypomere develops into the anterolateral muscles of the chest and abdominal wall and the limb musculature (**F**). As the myotomes become segregated, the spinal nerves undergo a corresponding division into a *dorsal (posterior) ramus* for the epaxial muscles and a *ventral (anterior) ramus* for the hypaxial muscles (see **Ea**). The original segmental (metameric) arrangement of the trunk muscles mostly disappears with further development. It persists only in the deep layers of the intrinsic back muscles (e.g., the rotator, interspinal, and intertransverse muscles) and the thoracic muscles (e.g., the internal and external intercostal muscles), while the superficial portions of the myotomes fuse together to form long, continuous muscles ("polymerization") in which only the neurovascular supply still exhibits the original segmental arrangement (see **F**).

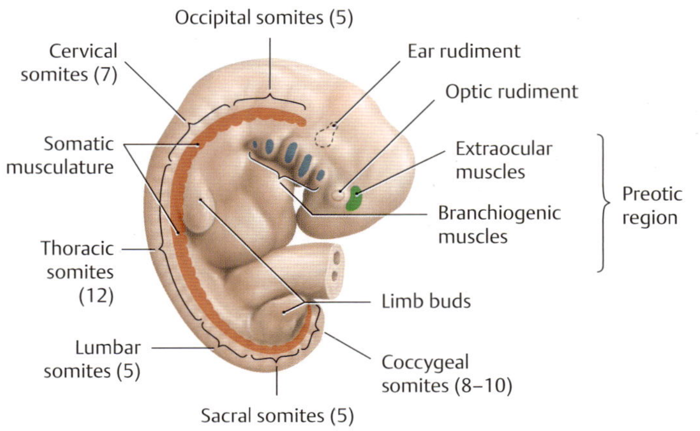

D Somites in a 5-week-old human embryo

Right lateral view. The somites formed from the paraxial mesoderm are classified as *preotic* (shown in blue and green) or *postotic* (shown in red), meaning that they are located cranial or caudal to the ear rudiment. The somatic muscles develop from postotic somites. Segmentation into different somites is not observed in the preotic region, which contains the rudiments for the branchiogenic pharyngeal arch muscles and the extraocular muscles. The structures in this region are innervated by cranial nerves.

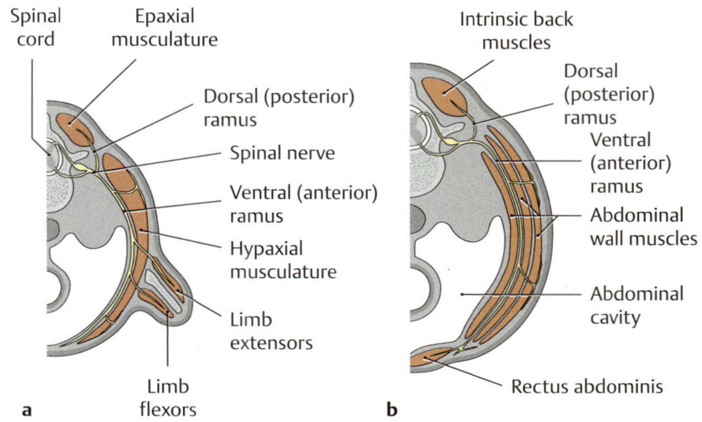

E Transverse sections through a 6-week-old human embryo

a Transverse section at the level of a limb bud. **b** Transverse section through the abdominal wall.

Muscle precursor cells with replicative capacity located at the level of the limb buds migrate from the myotomes into the limb buds, and the myotome that remains in these regions develops into intrinsic back muscles. With further growth of the limb bud, the muscle tissue differentiates into a dorsal rudiment (blastema) for the extensor muscles and a ventral rudiment for the flexor muscles of the upper and lower limbs (see p. 20). As in the hypomere, the limb muscles are innervated by ventral (anterior) spinal nerve rami (brachial plexus and lumbosacral plexus, see pp. 374 and 548).

Note the different innervation of the epaxial muscles (dorsal [posterior] ramus) and hypaxial muscles (ventral [anterior] ramus).

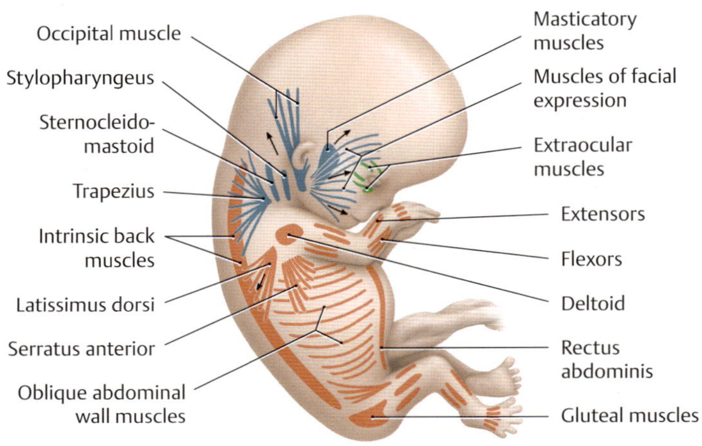

F Diagram of the principal muscle groups in an 8-week-old human embryo

Right lateral view. Red = somatic muscles, blue = branchiogenic (branchial arch) muscles, green = extraocular muscles.

2.2 The Intrinsic Back Muscles: Lateral Tract

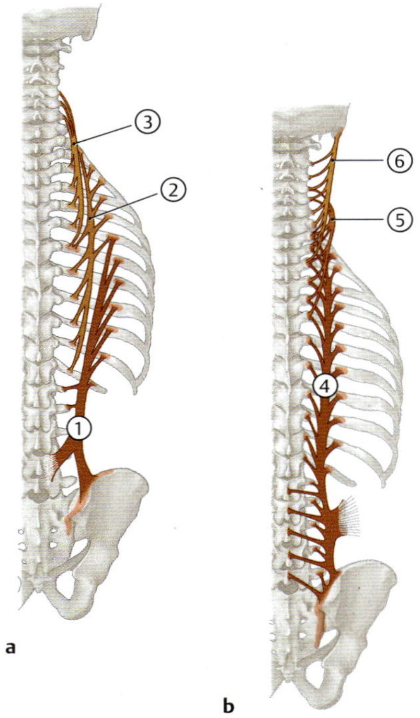

A Lateral tract: schematic of the sacrospinal system
a Iliocostalis muscles.
b Longissimus muscles.

Iliocostalis muscles (*see p. 155, opposite, lower right)

Origin:	① Iliocostalis lumborum: sacrum, iliac crest, superficial layer of the thoracolumbar fascia ② Iliocostalis thoracis: seventh to twelfth ribs ③ Iliocostalis cervicis: third to seventh ribs
Insertion:	• MIliocostalis lumborum: sixth to twelfth ribs, deep layer of thoracolumbar fascia, transverse processes of upper lumbar vertebrae • Iliocostalis thoracis: first to sixth ribs • Iliocostalis cervicis: transverse processes of C4–C6 vertebrae
Action:	Entire muscle: bilateral contraction extends the spine, unilateral contraction bends the spine laterally to the same side.
Innervation:	Lateral branches of posterior rami of spinal nerves C8–L1

Longissimus muscles

Origin:	④ Longissimus thoracis: sacrum, iliac crest (common tendon of origin with iliocostalis), spinous processes of lumbar vertebrae, transverse processes of lower thoracic vertebrae ⑤ Longissimus cervicis: transverse processes of T1–T6 vertebrae ⑥ Longissimus capitis: transverse processes of T1–T3 vertebrae and transverse and articular processes of C2–C7 vertebrae
Insertion:	• Longissimus thoracis: second to twelfth ribs, costal processes of lumbar vertebrae, transverse processes of thoracic vertebrae • Longissimus cervicis: transverse processes of C4–C5 vertebrae • Longissimus capitis: mastoid process of temporal bone
Action:	• Entire muscle: bilateral contraction extends the spine, unilateral contraction bends the spine laterally to the same side. • Longissimus capitis: bilateral
Innervation:	Lateral branches of posterior rami of spinal nerves C1–L5

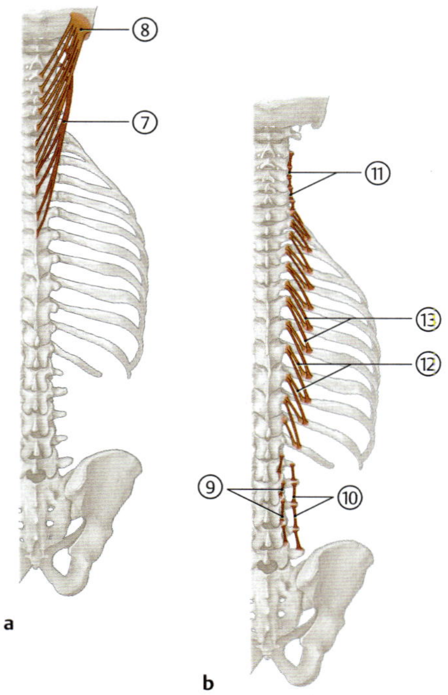

B Lateral tract: schematic of the spinotransverse and intertransverse systems
a Splenius muscles.
b Intertransversarii and levatores costarum muscles.

Splenius muscles

Origin:	⑦ Splenius cervicis: spinous processes of the T3–T6 vertebrae ⑧ Splenius capitis: nuchal ligament, spinous processes of the C7–T3 vertebrae
Insertion:	• Splenius cervicis: transverse processes of C1 and C 2 • Splenius capitis: lateral superior
Action:	Entire muscle: bilateral contraction extends the cervical spine and head, unilateral contraction flexes and rotates the head to the same side.
Innervation:	Lateral branches of dorsal rami of spinal nerves C1–C6

Intertransversarii

Origin and insertion:	⑨ Intertransversarii mediales lumborum: course between adjacent mammillary processes of all lumbar vertebrae ⑩ Intertransversarii laterales lumborum: course between adjacent costal processes of all lumbar vertebrae ⑪ Intertransversarii posteriores cervicis: course between adjacent posterior tubercles of the C2–C7 vertebrae
Action:	• Intertransversarii anteriores cervicis: course between adjacent anterior tubercles of the C2–C7 vertebrae • Bilateral contraction stabilizes and extends the cervical and lumbar spine. • Unilateral contraction bends the cervical and lumbar spine laterally to the same side.
Innervation:	Posterior rami of the spinal nerves for intertransversarii mediales lumborum, intertransversarii laterales lumborum, and intertransversarii posteriores cervicis; anterior rami of the spinal nerves for intertransversarii anteriores cervicis

Levatores costarum

Origin:	⑫ Levatores costarum breves: transverse processes of the C7 and T1–T11 vertebrae ⑬ Levatores costarum longi: transverse processes of the C7 and T1–T11 vertebrae
Insertion:	• Levatores costarum breves: costal angle of the next lower rib • Levatores costarum longi: costal angle of second lower rib
Action:	• Bilateral contraction extends the thoracic spine. • Unilateral contraction: ipsilateral lateral flexion and contralateral rotation.
Innervation:	Posterior rami of the spinal nerves

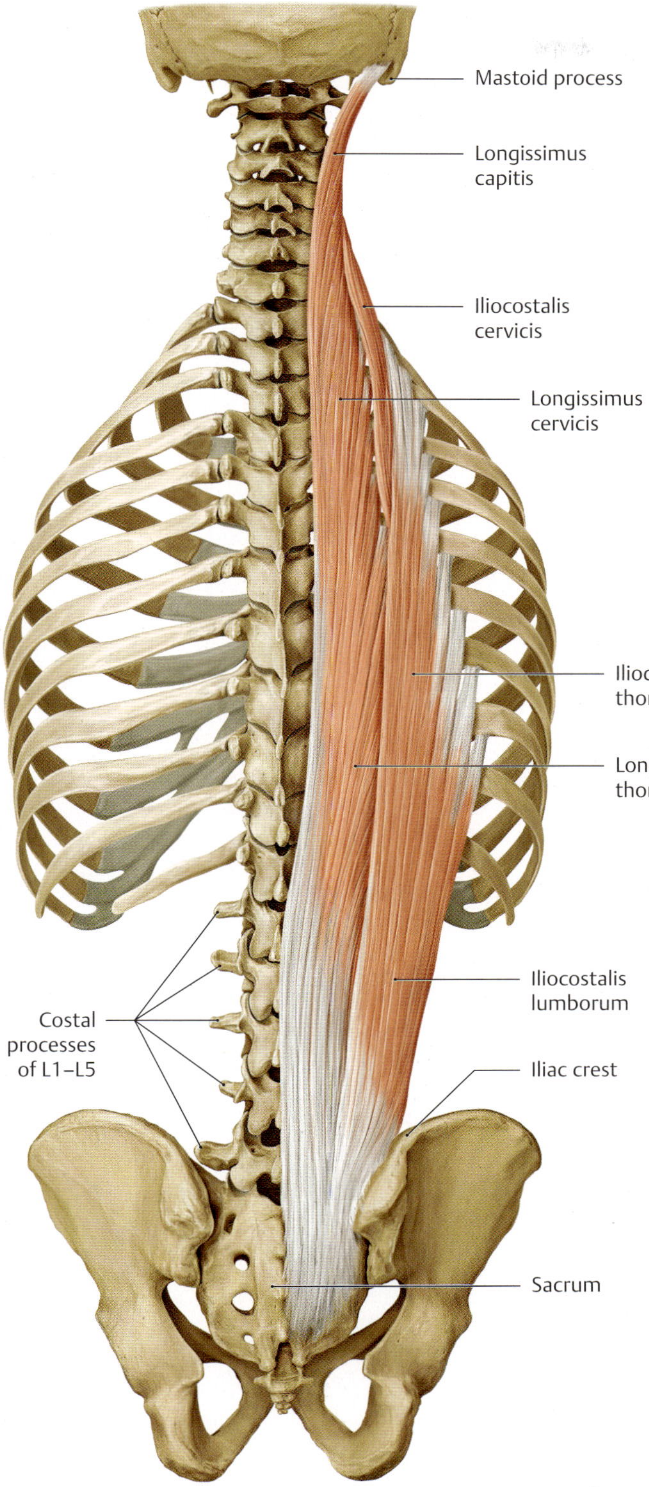

C Lateral tract of the intrinsic back muscles: the sacrospinal system (iliocostalis and longissimus muscles)

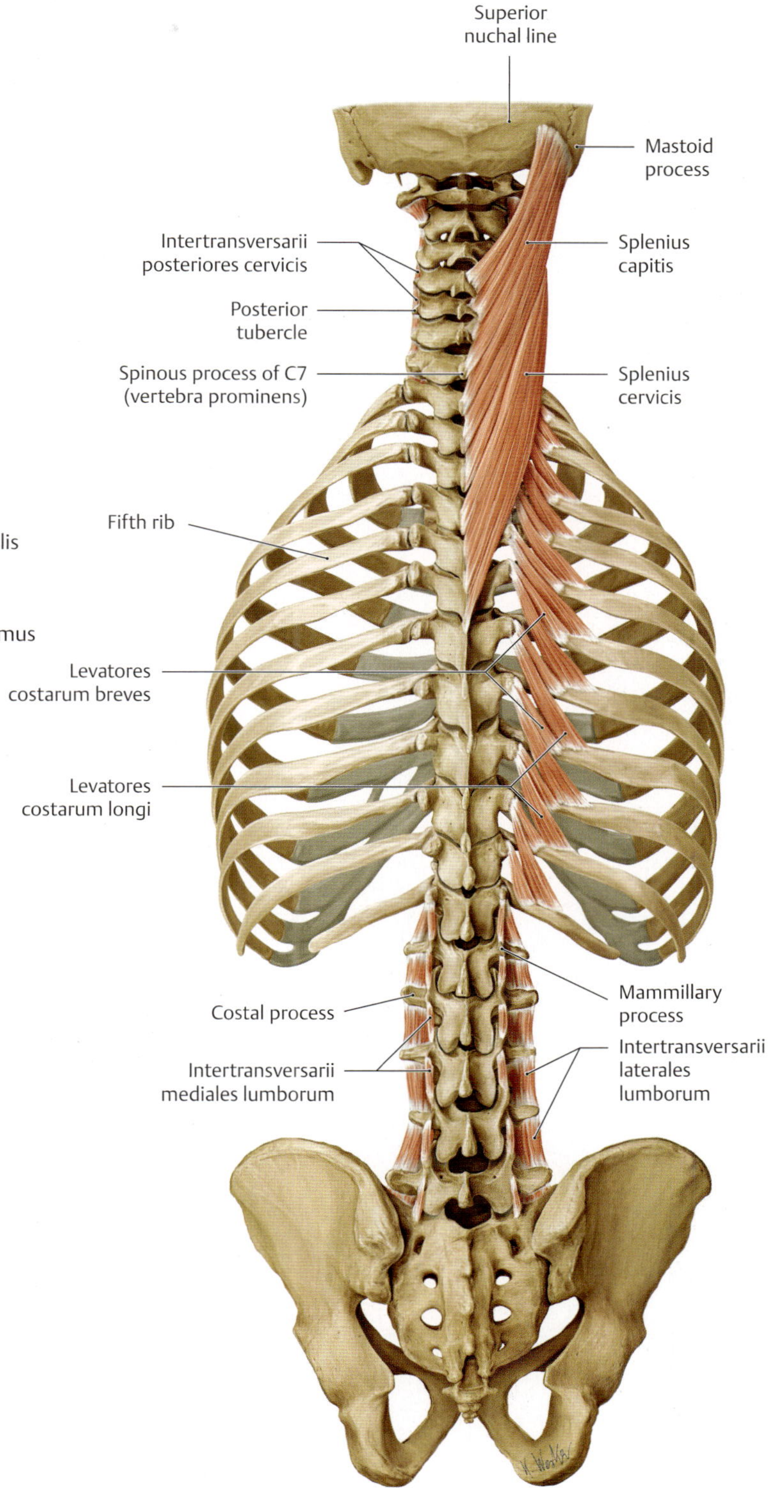

D Lateral tract of the intrinsic back muscles: the spinotransverse system (splenius muscles) and intertransverse system (intertransversarii and levatores costarum muscles)

* The schematic muscle diagrams with accompanying tables on p. 154 are intended to give a systematic overview of the muscles and their actions. The drawings on the right-hand page are intended to display the muscles as they would appear in a dissection. Not all the structures listed in the tables on p. 154 are labeled in the drawings above, as they are not all visible in the views shown.

2.3 The Intrinsic Back Muscles: Medial Tract

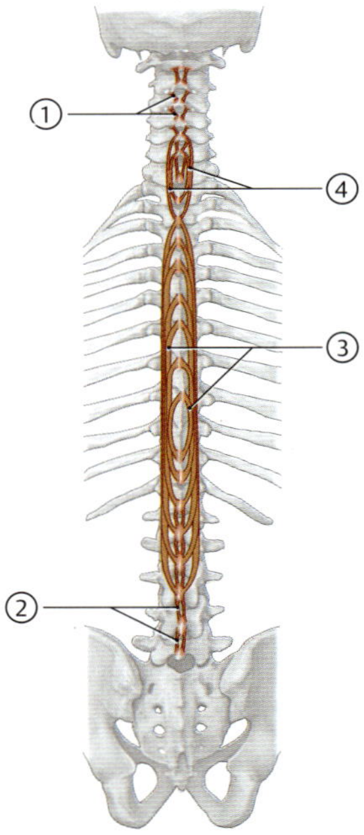

A Medial tract of the intrinsic back muscles: schematic of the spinal system
The interspinales and spinalis muscles.

Interspinales muscles

Origin and insertion:
① Interspinales cervicis: course between the spinous processes of adjacent cervical vertebrae
② Interspinales lumborum: course between the spinous processes of adjacent lumbar vertebrae

Action: Extends the cervical and lumbar spine

Innervation: Posterior rami of the spinal nerves

Spinalis muscles

Origin:
③ Spinalis thoracis: lateral surface of the spinous processes of the T10–T12 and L1–L3 vertebrae
④ Spinalis cervicis: spinous processes of the C5–C7 and T1–T2 vertebrae

Insertion:
- Spinalis thoracis: lateral surface of the spinous processes of the T2–T8 vertebrae
- Spinalis cervicis: spinous processes of the C2–C5 vertebrae

Action:
- Bilateral contraction extends the cervical and thoracic spine.
- Unilateral contraction bends the cervical and thoracic spine to the same side.

Innervation: Posterior rami of the spinal nerves

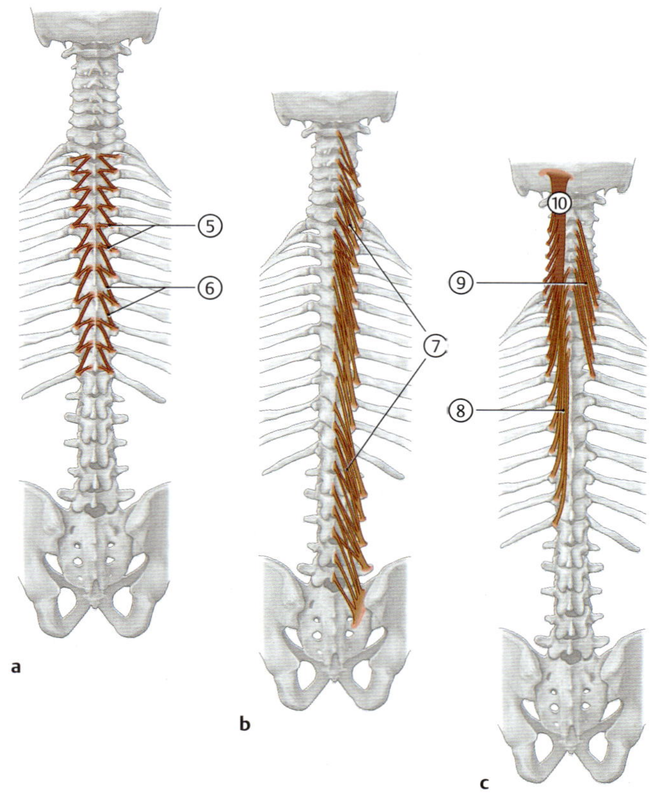

B Medial tract of the intrinsic back muscles: schematic of the transversospinal system
a Rotatores breves and longi.
b Multifidus.
c Semispinalis.

Rotatores breves and longi

Origin and insertion:
⑤ Rotatores breves: from transverse process of thoracic vertebrae to spinous process of the next higher vertebra
⑥ Rotatores longi: from transverse process of thoracic vertebrae to spinious process of the vertebra two levels higher

Action:
- Bilateral contraction extends the thoracic spine.
- Unilateral contraction rotates it to the opposite side.

Innervation: Posterior rami of the spinal nerves

⑦ Multifidus (most fully developed in lumbar spine)

Origin and insertion: Sacrum, ilium, mammillary processes of L1–L5, transverse and articular processes of T1–T4, C4–C7
Superomedially to spinous process of vertebra two to four levels higher

Action:
- Bilateral contraction extends the spine.
- Unilateral contraction flexes the spine to the same side and rotates it to the opposite side.

Innervation: Posterior rami of the spinal nerves

Semispinalis muscles

Origin:
⑧ Semispinalis thoracis: transverse processes of the T6–T12 vertebrae
⑨ Semispinalis cervicis: transverse processes of the T1–T6 vertebrae
⑩ Semispinalis capitis: transverse processes of the C3–T6 vertebrae

Insertion:
- Semispinalis thoracis: spinous processes of the C6–T4 vertebrae
- Semispinalis cervicis: spinous processes of the C2–C5 vertebrae
- Semispinalis capitis: occipital bone between the superior and inferior nuchal lines

Action:
- Bilateral contraction extends the thoracic spine, cervical spine, and head (stabilizes the craniovertebral joints).
- Unilateral contraction bends the head, cervical spine, and thoracic spine to the same side and rotates them to the opposite side.

Innervation: Posterior rami of the spinal nerves

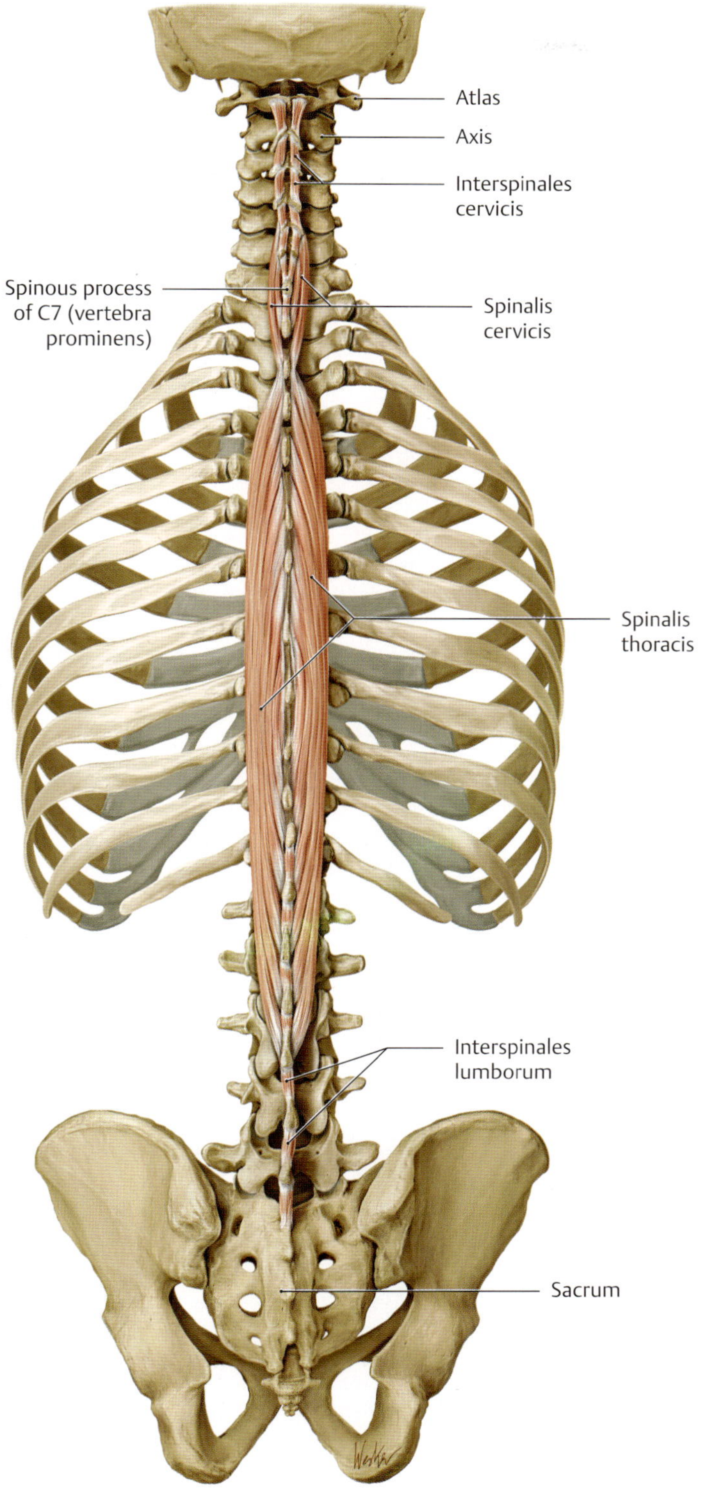

C Medial tract of the intrinsic back muscles: the spinal system (interspinales and spinalis muscles)

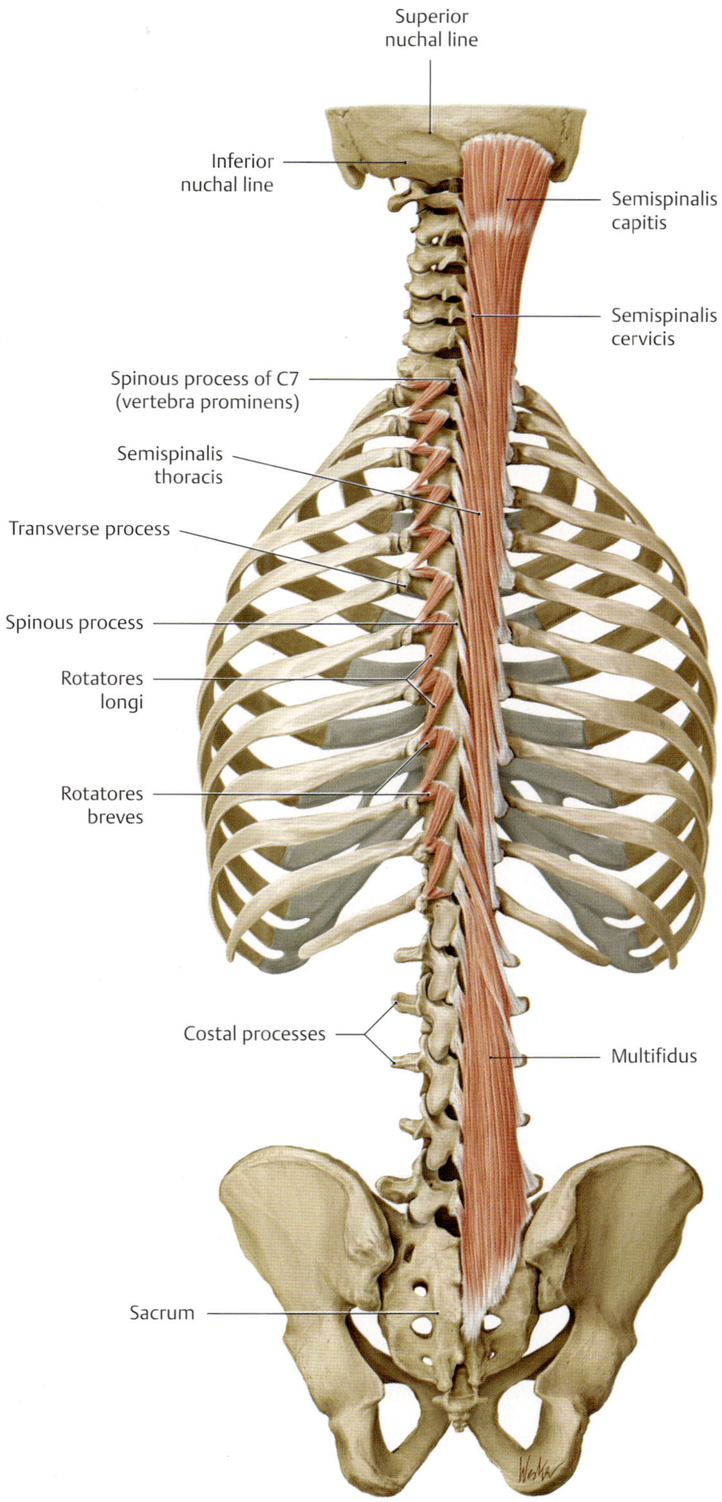

D Medial tract of the intrinsic back muscles: the transversospinal system (rotatores breves and longi, multifidus, and semispinalis muscles)

2.4 The Intrinsic Back Muscles (Short Nuchal and Craniovertebral Joint Muscles) and the Prevertebral Muscles

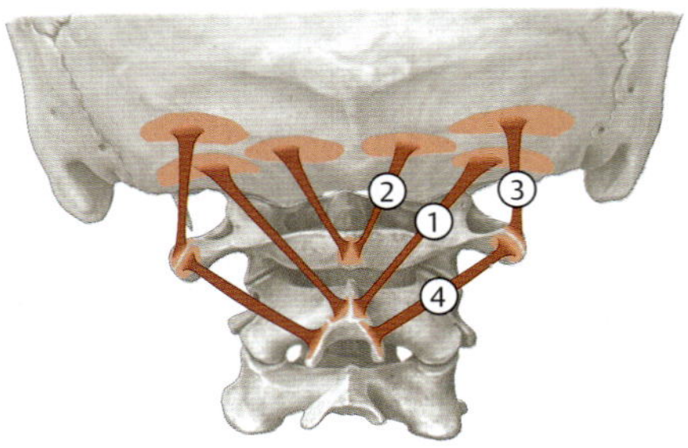

A Schematic of the short nuchal and craniovertebral joint muscles (suboccipital muscles): recti capitis posterior major and minor and obliquii capitis superior and inferior
Posterior view.

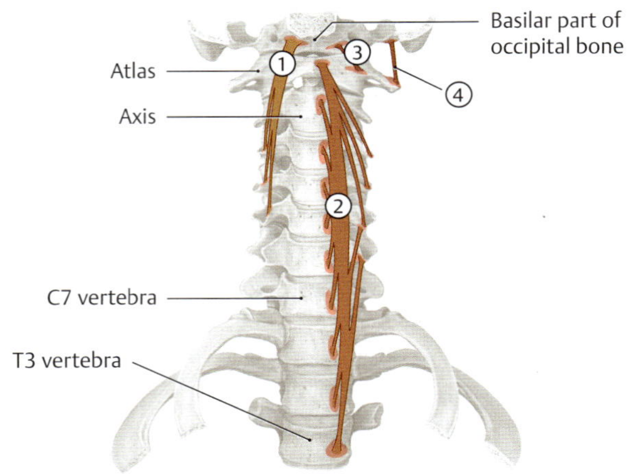

B Schematic of the prevertebral neck muscles (colli and cervicis muscles): longi capitis and colli and recti capitis anterior and lateralis
Anterior view.

① **Rectus capitis posterior major**

Origin: Spinous process of the axis
Insertion: Middle third of the inferior nuchal line
Action:
- Bilateral: extends the head
- Unilateral: rotates the head to the same side

Innervation: Posterior ramus of C1 (suboccipital nerve)

② **Rectus capitis posterior minor**

Origin: Posterior tubercle of the atlas
Insertion: Inner third of the inferior nuchal line
Action:
- Bilateral: extends the head
- Unilateral: rotates the head to the same side

Innervation: Posterior ramus of C1 (suboccip tal nerve)

③ **Obliquus capitis superior**

Origin: Transverse process of the atlas
Insertion: Above the insertion of the rectus capitis posterior major or middle third of nuchal line
Action:
- Bilateral: extends the head
- Unilateral: tilts the head to the same side and rotates it to the opposite side

Innervation: Posterior ramus of C1 (suboccipital nerve)

④ **Obliquus capitis inferior**

Origin: Spinous process of the axis
Insertion: Transverse process of the atlas
Action: Querfortsatz des Atlas
- Bilateral: extends the head
- Unilateral: rotates the head to the same side

Innervation: Posterior ramus of C1 (suboccipital nerve)

Note: The prevertebral muscles are not included among the intrinsic back muscles as they are innervated by the a terior rami of the spinal nerves.

① **M. longus capitis**

Origin: Anterior tubercle of the transverse processes of the C3–C6 vertebrae
Insertion: Basilar part of the occipital bone
Action:
- Unilateral: tilts and slightly rotates the head to the same side
- Bilateral: flexes the head

Innervation: Direct branches from the cervical plexus (C1–C3)

② **Longus colli (cervicis)**

Origin:
- Vertical (medial) part: anterior sides of the C5–C7 and T1–T3 vertebral bodies
- Superior oblique part: anterior tubercle of the transverse processes of the C3–C5 vertebrae
- Inferior oblique part: anterior sides of the T1–T3 vertebral bodies

Insertion:
- Vertical part: anterior sides of the C2–C4 vertebrae
- Superior oblique part: anterior tubercle of the atlas
- Inferior oblique part: anterior tubercle of the transverse processes of the C5 and C6 vertebrae

Action:
- Unilateral: tilts and rotates the cervical spine to the same side
- Bilateral: flexes the cervical spine

Innervation: Anterior rami of C2–C6

③ **Rectus capitis anterior**

Origin: Lateral mass of the atlas
Insertion: Basilar part of the occipital bone
Action:
- Unilateral: lateral flexion at the atlanto-occipital joint
- Bilateral: flexion at the atlanto-occipital joint

Innervation: Anterior ramus of the C1 nerve

④ **Rectus capitis lateralis**

Origin: Transverse process of the atlas
Insertion: Lateral part of the occipital bone (lateral to the occipital condyles)
Action:
- Unilateral: lateral flexion at the atlanto-occipital joint
- Bilateral: flexion at the atlanto-occipital joint

Innervation: Anterior ramus of the C1 nerve

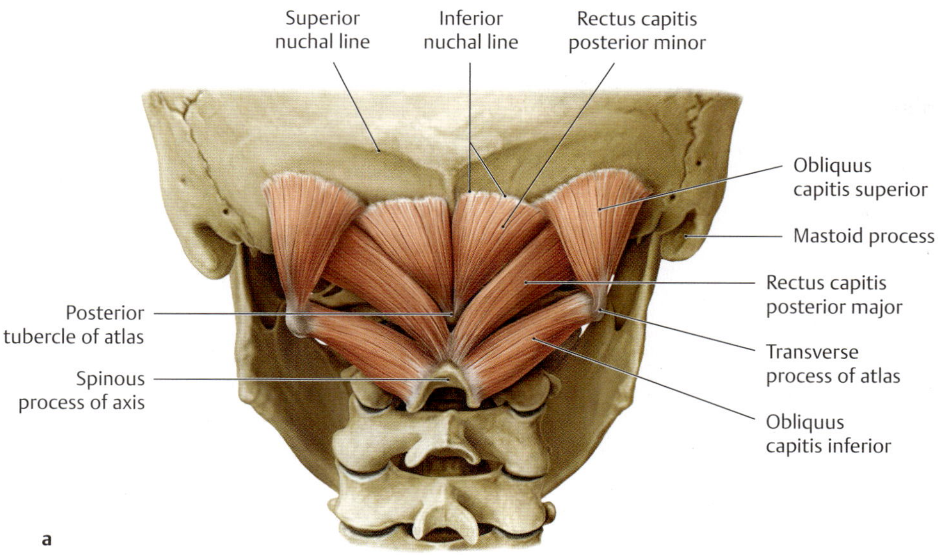

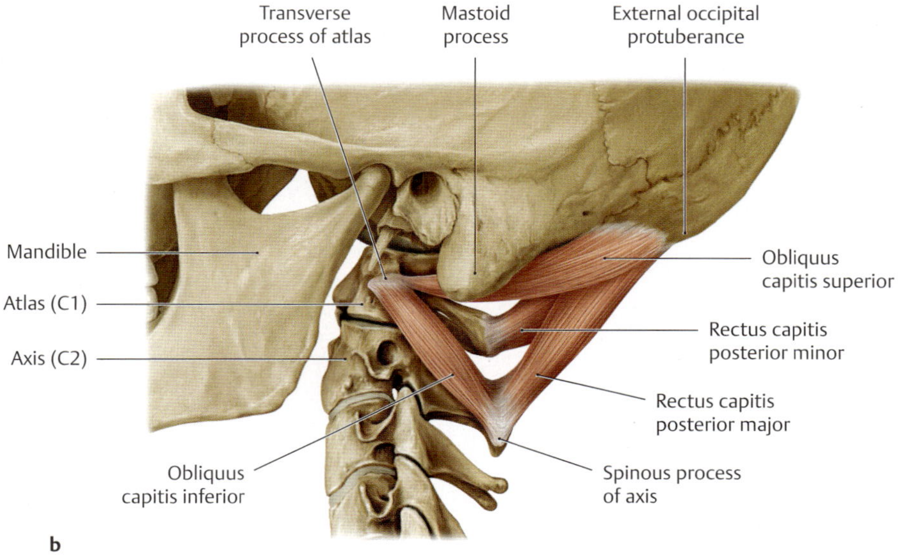

C The short nuchal and craniovertebral joint muscles: rectus capitis posterior and obliquus capitis

a Posterior view, **b** lateral view.

In a strict sense, the short nuchal muscles consist only of the muscles innervated by the posterior ramus of the first spinal nerve (suboccipital nerve). They include representatives of the lateral tract (obliquus capitis inferior) and the medial tract (obliquus capitis superior and recti capitis posterior major and minor). The anterior group of short nuchal muscles (recti capitis lateralis and anterior) is innervated by the anterior rami, placing them among the prevertebral neck muscles (see **D**).

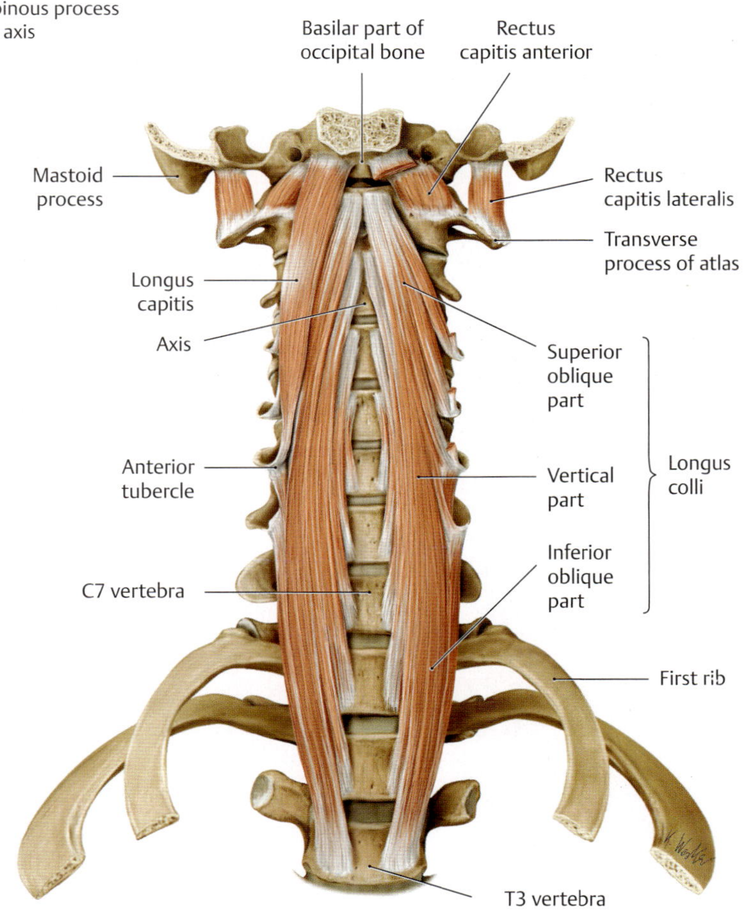

D The prevertebral neck muscles: longi capitis and colli and recti capitis anterior and lateralis

Anterior view after removal of the cervical viscera. The longus capitis muscle has been partially removed on the left side.

2.5 The Muscles of the Abdominal Wall: Anterolateral Muscles

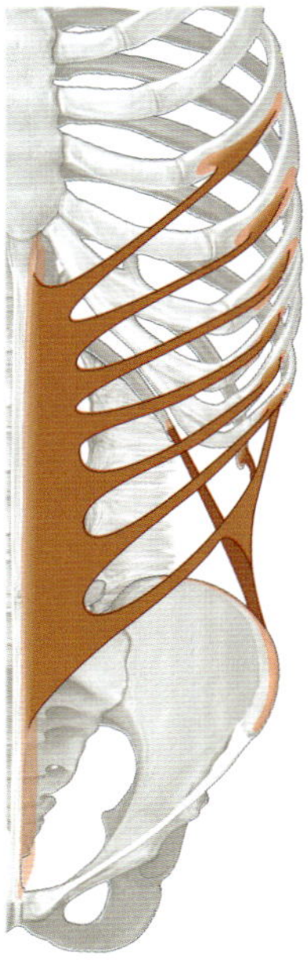

A Schematic of the external oblique

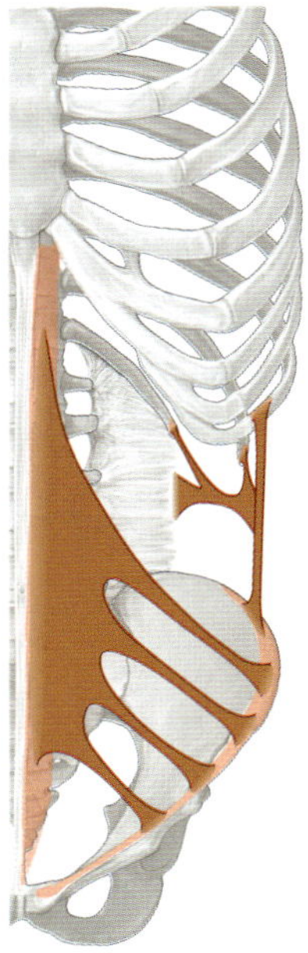

B Schematic of the internal oblique

External oblique

Origin:	Outer surface of the fifth to twelfth ribs
Insertion:	• Outer lip of the iliac crest • Anterior layer of the rectus sheath, linea alba
Action:	• Unilateral: bends the trunk to the same side, rotates the trunk to the opposite side • Bilateral: flexes the trunk, stabilizes the pelvis, active in expiration, compresses the abdomen
Innervation:	Intercostal nerves (T5–T11), subcostal nerve (T12), iliohypogastric nerve

Internal oblique

Origin:	Deep layer of the thoracolumbar fascia, intermediate line of the iliac crest, anterior superior iliac spine, iliopsoas fascia
Insertion:	• Lower borders of the tenth to twelfth ribs • Anterior and posterior layers of the rectus sheath, linea alba • Junction with the cremaster muscle
Action:	• Unilateral: bends the trunk to the same side, rotates the trunk to the same side • Bilateral: flexes the trunk, stabilizes the pelvis, active in expiration, compresses the abdomen
Innervation:	• Intercostal nerves (T8–T11), subcostal nerve (T12), iliohypogastric nerve, ilioinguinal nerve • Cremaster muscle (genital branch of genitofemoral nerve)

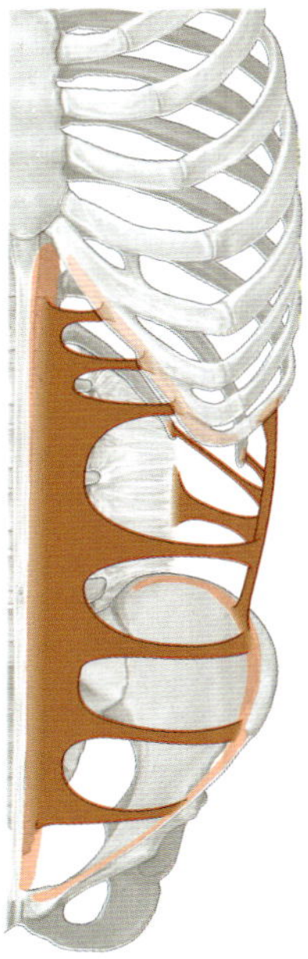

C Schematic of the transversus abdominis

Transversus abdominis

Origin:	• Inner surfaces of the seventh to twelfth costal cartilages • Deep layer of the thoracolumbar fascia • Inner lip of the iliac crest, anterior superior iliac spine • Lateral part of the inguinal ligament • Posterior layer of the rectus sheath, linea alba
Insertion:	• Pubic crest
Action:	• Unilateral: rotates the trunk to the same side • Bilateral: active in expiration, compresses the abdomen
Innervation:	Intercostal nerves (T5–T11), subcostal nerve (T12), iliohypogastric nerve, ilioinguinal and genitofemoral nerves

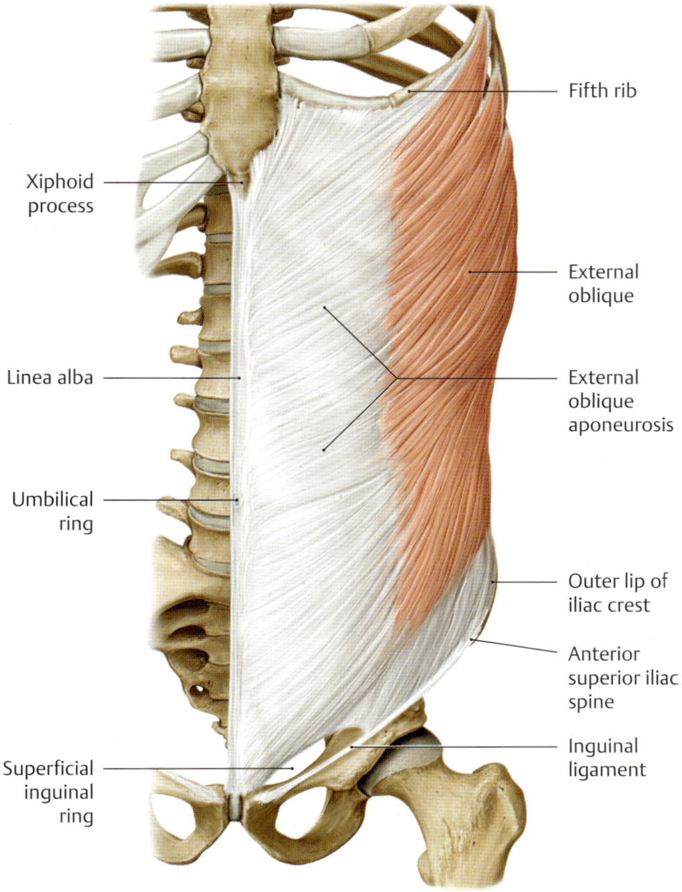

D External oblique
Left side, anterior view.

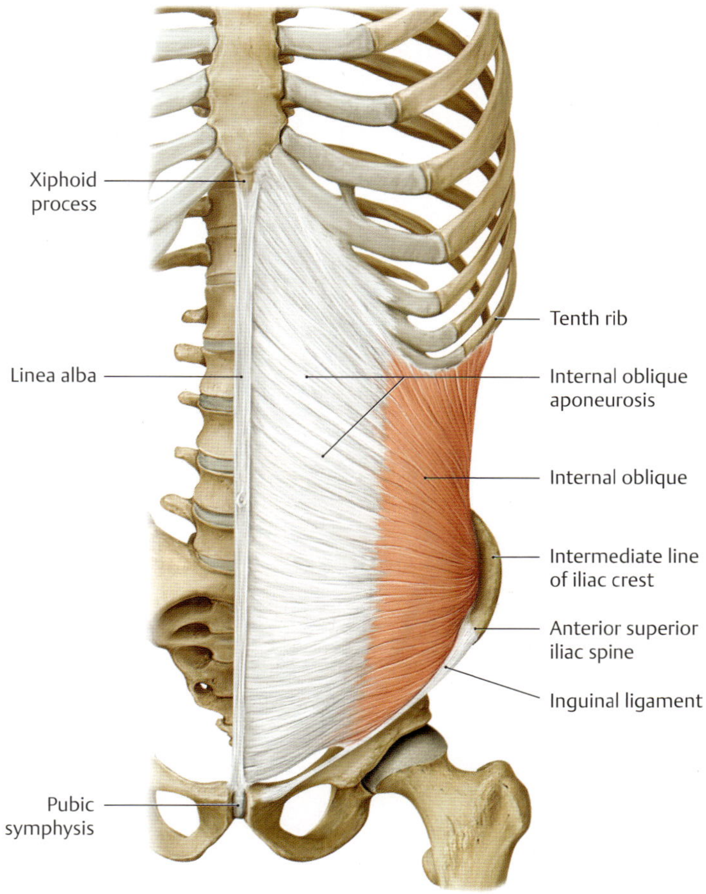

E Internal oblique
Left side, anterior view.

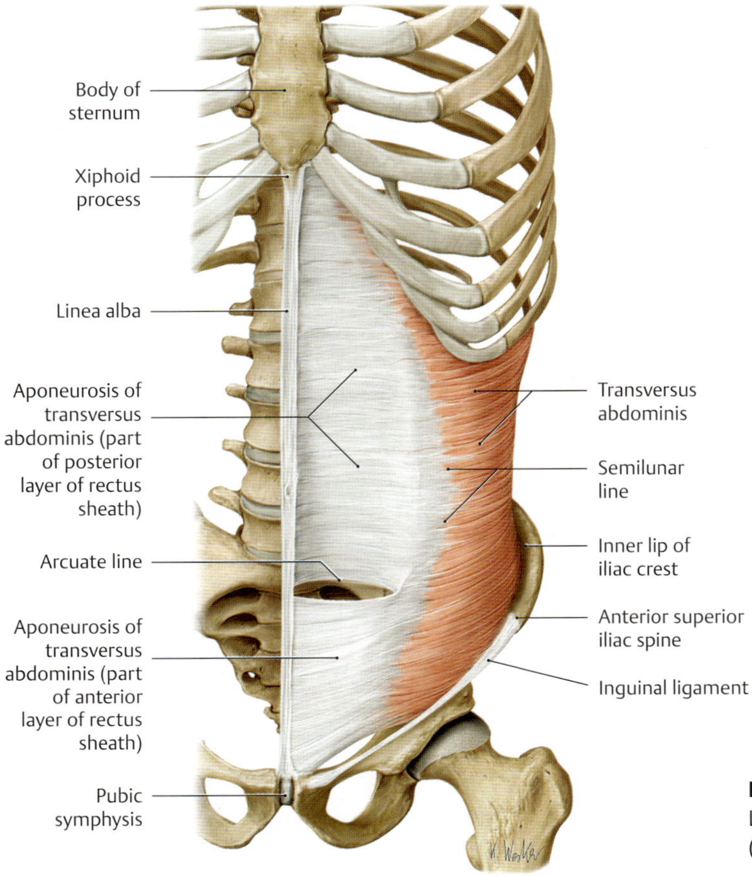

F Transversus abdominis
Left side, anterior view.
(For structure of the rectus sheath, see p. 187)

2.6 The Muscles of the Abdominal Wall: Anterior and Posterior Muscles

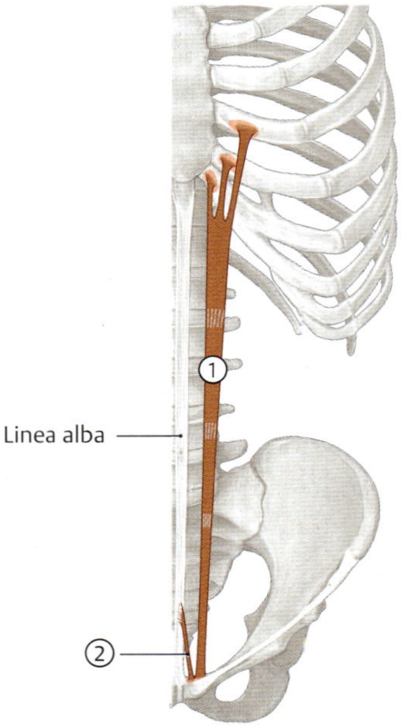

A Schematic of the anterior (straight, or strap) muscles of the abdominal wall: rectus abdominis and pyramidalis

① Rectus abdominis

Origin: Cartilages of the fifth to seventh ribs, xiphoid process of the sternum
Insertion: Pubis (between the pubic tubercle and symphysis)
Action: Flexes the lumbar spine, stabilizes the pelvis, active in expiration, compresses abdomen
Innervation: Intercostal nerves (T5–T11) and subcostal nerve (T12)

② Pyramidalis*

Origin: Pubis (anterior to insertion of the rectus abdominis)
Insertion: Linea alba (runs within the rectus sheath)
Action: Tenses the linea alba
Innervation: Subcostal nerve (T12)

*Variable, absent in 10–25% of all people

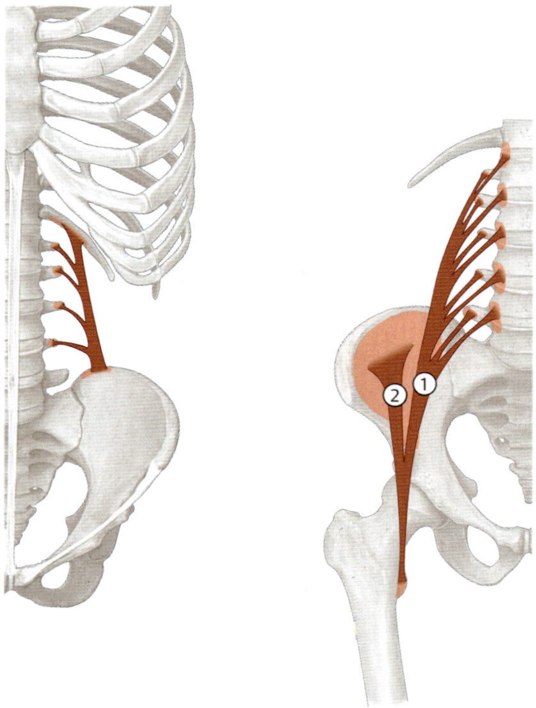

B Schematic of the posterior (deep) muscles of the abdominal wall: quadratus lumborum and psoas major

Quadratus lumborum

Origin: Iliac crest
Insertion: Twelfth rib, costal processes of the L1–L4 vertebrae
Action:
- Unilateral: bends the trunk to the same side
- Bilateral: bearing down and expiration

Innervation: Subcostal nerve (T12)

Iliopsoas (① psoas major and ② iliacus)*

Origin:
- Psoas major (superficial layer): lateral surfaces of the T12 vertebral body, the L1–L4 vertebral bodies, and the associated intervertebral disks
- Psoas major (deep layer): costal processes of the L1–L5 vertebrae
- Iliacus: iliac fossa

Insertion: Insert jointly as the iliopsoas muscle on the lesser trochanter of the femur
Action:
- Hip joint: flexion and external rotation
- Lumbar spine: unilateral contraction (with the femur fixed) bends the trunk laterally, bilateral contraction raises the trunk from the supine position

Innervation: Direct branches from spinal nerves (L1–L4) and femoral nerve (L2–L4)

* Of these two muscles, only the psoas major belongs topographically to the posterior abdominal muscles. It is classified functionally as a hip muscle (see p. 488).

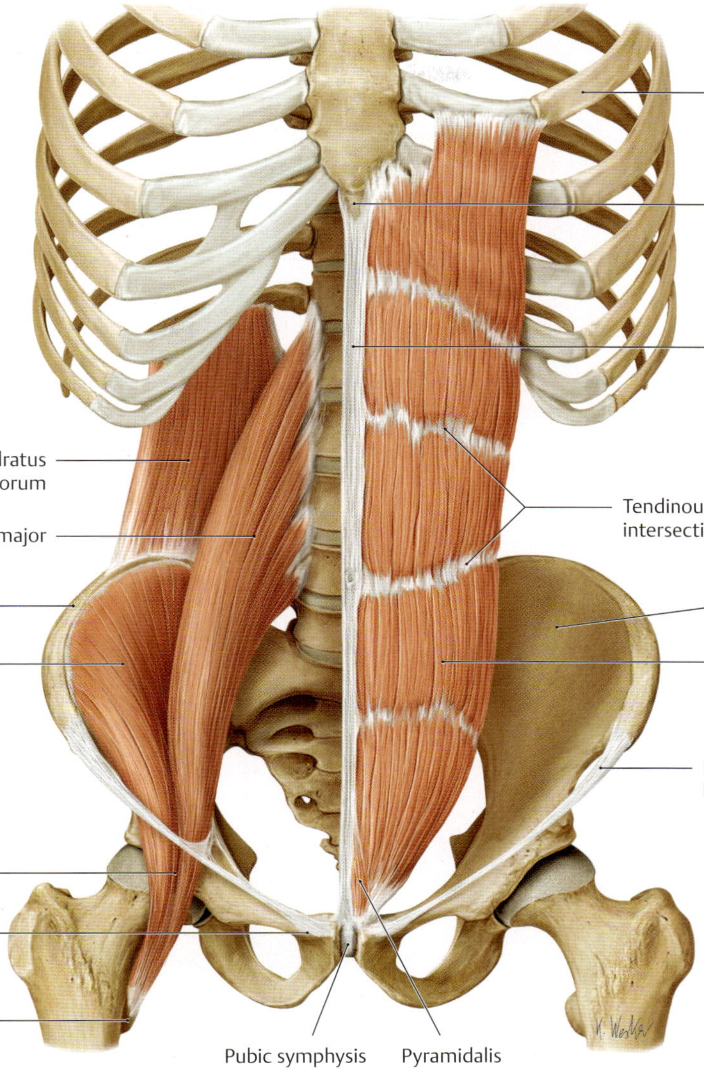

C Anterior muscles (rectus abdominis and pyramidalis) and posterior muscles (quadratus lumborum and iliopsoas) of the abdominal wall

Anterior view. The anterior abdominal wall muscles are shown on the left side and the posterior muscles on the right side.

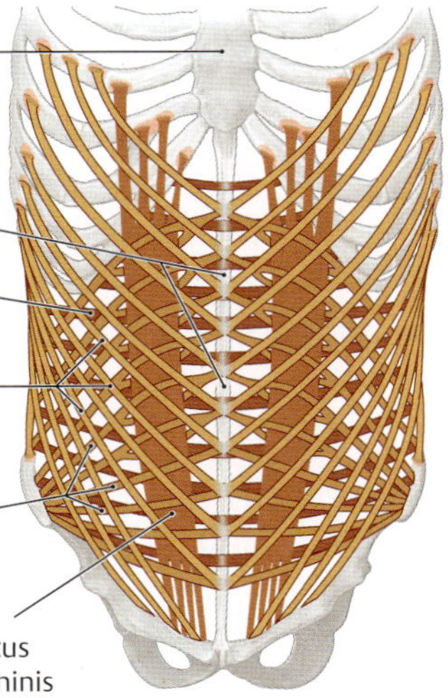

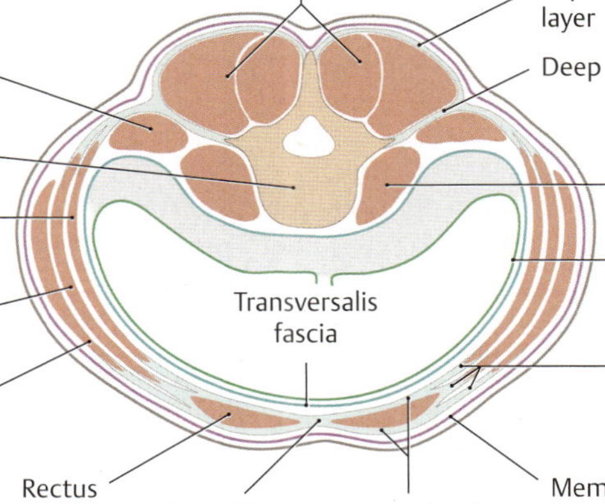

D Arrangement of the abdominal wall muscles and rectus sheath

The anterior (rectus abdominis) and lateral (obliques) abdominal wall muscles and their aponeuroses comprise a functional unit. The muscles' course and the fusion of their broad aponeuroses in the region of the linea alba create functional muscle bands that serve to maintain tension in the abdominal wall. Hence, the aponeuroses of the lateral muscles of the abdominal wall enclose the anterior rectus muscles like a quiver, and together with the fasciae of the abdominal wall, form the rectus sheath (see p. 187). The course of the muscles and their tendons produce oblique, transverse, and vertical systems of reinforcement. This reinforcement is important to keep the viscersa "in place" as this region is subject to particularly severe stresses.

2.7 Functions of the Abdominal Wall Muscles

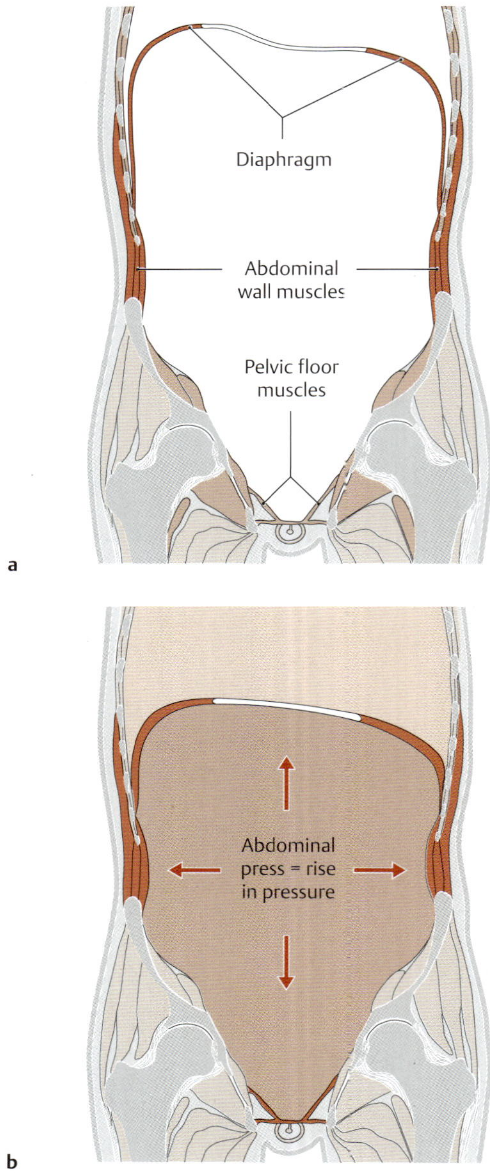

Functions of the Abdominal Wall Muscles

The different abdominal wall muscles perform numerous functions, which very often are carried out in concert with other muscle groups (e.g., the back and gluteal muscles and the diaphragm). The principal actions of the abdominal wall muscles are as follows:

- Maintenance of abdominal tone: tensing the abdominal wall and compressing the abdominal viscera (abdominal press).
- Stabilizing the vertebrae and reducing stresses on the spinal column.
- Moving the trunk and pelvis.
- Assisting in respiration

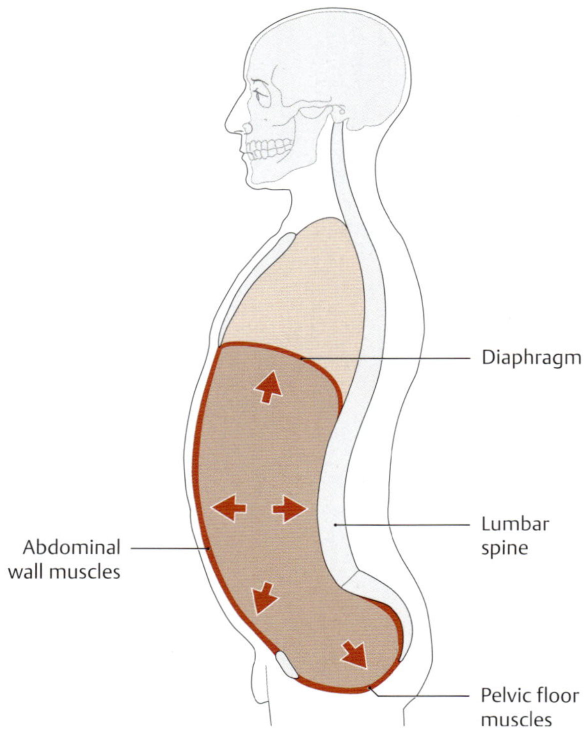

A Abdominal press = raising the intra-abdominal pressure by tensing the abdominal wall and pelvic floor muscles and the diaphragm

Schematic coronal section through the abdominal cavity, anterior view.

- **a** The walls of the abdominal and pelvic cavity are formed by bony structures (spinal column, thoracic cage, pelvis) and also by muscles (diaphragm, muscles of the abdominal wall and pelvic floor).
- **b** When the muscles about the abdomen contract (abdominal press), they reduce the volume of the abdominal cavity, thereby raising the intra-abdominal pressure and actively compressing the abdominal viscera. This action is important, for example, in expelling stool from the rectum (defecation), expelling urine from the bladder (micturition), and emptying the gastric contents (vomiting). The abdominal press is also an essential part of maternal pushing during the expulsive phase of labor.

B Abdominal press = stabilizing the spinal column by raising the intra-abdominal pressure

Schematic midsagittal section through the trunk, left lateral view. Simultaneous contraction of the diaphragm and the muscles of the abdominal wall and pelvic floor raises the pressure in the abdominal cavity (abdominal press). The hydrostatic effect of this maneuver stabilizes the trunk, reduces stresses on the spinal column (especially at the lumbar level), and stiffens the trunk wall like the wall of an inflated ball. This action is performed automatically during the lifting of heavy loads. The "inflatable space" of the trunk can be employed in this way to lighten the pressure load on the intervertebral disks by up to 50% in the upper lumbar spine and by approximately 30% in the lower lumbar spine. Meanwhile, the forces exerted by the intrinsic back muscles are reduced by more than 50%. This explains the importance of well-conditioned abdominal muscles in preventing and treating diseases of the spinal column.

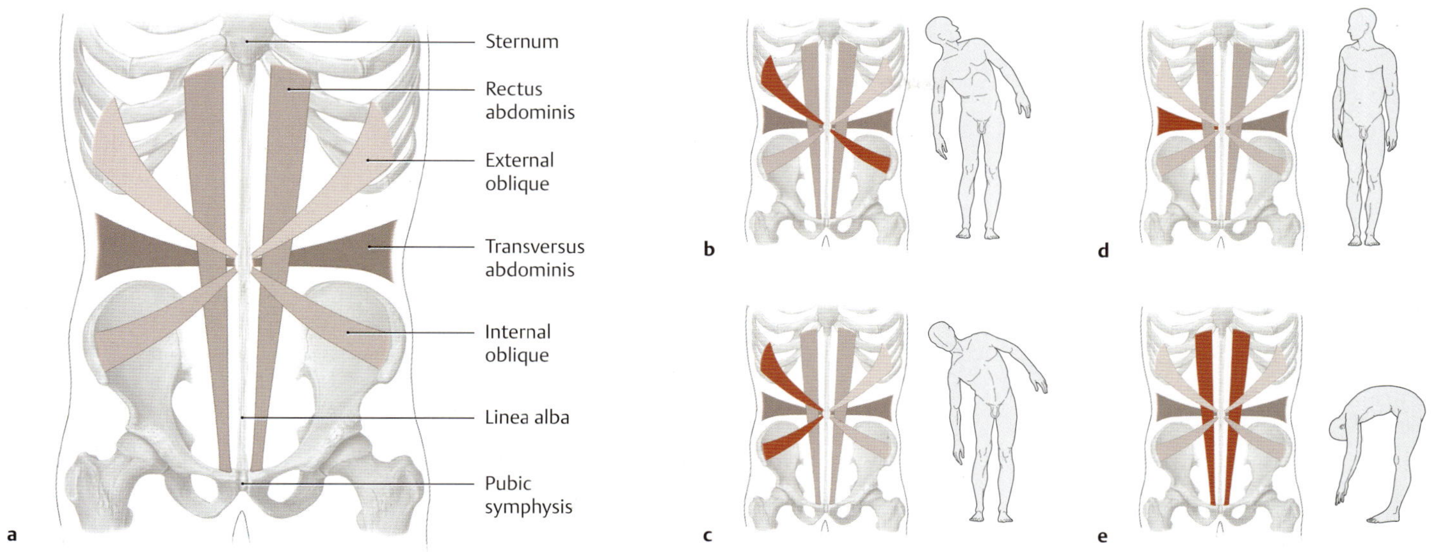

C Trunk movements aided by the straight and oblique muscles of the abdominal wall

a Course and arrangement of the straight and oblique abdominal wall muscles.

b Trunk flexion to the right with simultaneous rotation of the trunk to the left side by contraction of the external oblique muscle on the right side and the internal oblique muscle on the left side.

c Trunk flexion to the right by contraction of the right lateral abdominal muscles (aided by the right intrinsic back muscles).

d Rotating the trunk to the right side is effected by the right lateral abdominal muscles and the left intrinsic back muscles.

e Flexion of the trunk is effected by bilateral contraction of the rectus abdominis, lateral abdominal and psoas muscles.

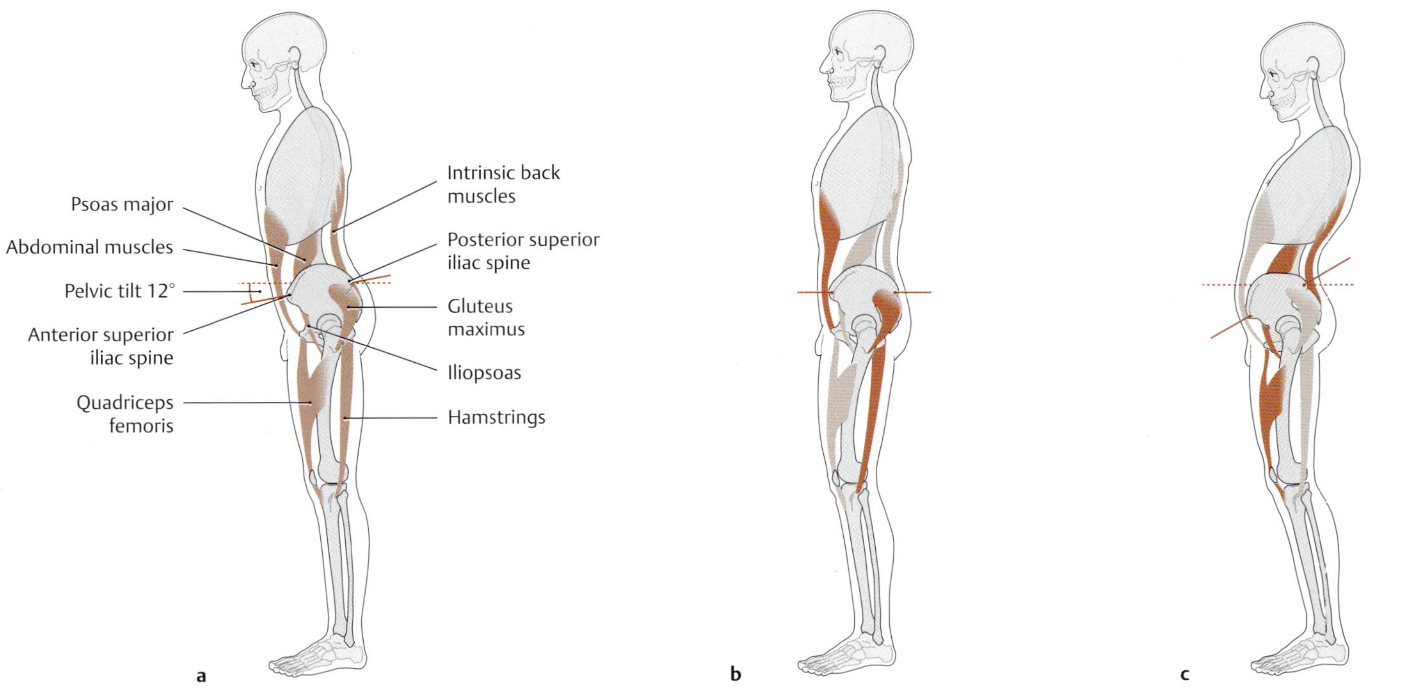

D Effect of the abdominal wall muscles on pelvic movements: active and passive posture

a Normal active posture, **b** active rigid posture, **c** passive slumped posture.

An imbalance between the intrinsic back muscles and abdominal muscles is particularly evident in the curvature of the lower spine and in the degree of pelvic tilt. In a normal active posture, the pelvis is tilted forward by approximately 12° (**a**). When a rigid posture is assumed ("stomach in, chest out"), the pelvis is held in a more upright position so that the anterior superior iliac spine and the posterior superior iliac spine are at the same level (**b**). The most active muscles are the abdominal wall muscles, the gluteal muscles, and the hamstrings. When the abdominal muscles are lax and are not well conditioned, the result is a passive slumped posture (**c**) with an excessive degree of anterior pelvic tilt. Also, the lordotic curvature of the lumbar spine is accentuated due to progressive shortening of the intrinsic back muscles. This posture is reinforced by the tendency of the iliopsoas muscles (psoas major and iliacus) to become shortened.

2.8 The Muscles of the Thoracic Cage: Scalenes, Intercostals, Subcostals, and Transversus Thoracis

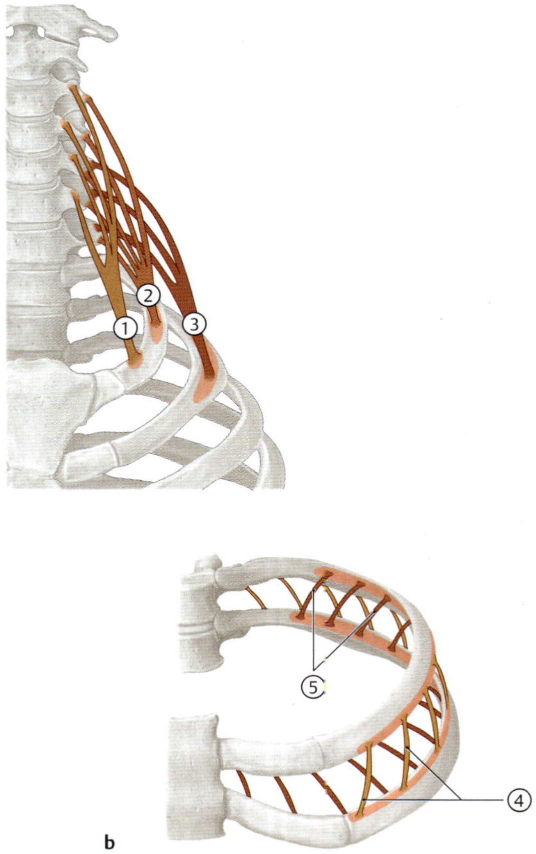

A Overview of the thoracic wall muscles: (a) scalenes and **(b)** intercostals
Anterior view.

Scalene muscles

Origin:	① Anterior scalene: anterior tubercle of the transverse processes of the C3–C6 vertebrae ② Middle scalene: posterior tubercle of the transverse processes of the C3–C7 vertebrae ③ Posterior scalene: posterior tubercle of the transverse processes of the C5–C7 vertebrae
Insertion:	• Anterior scalene: anterior scalene tubercle of first rib • Middle scalene: first rib (posterior to the groove for the subclavian artery) • Posterior scalene: outer surface of the second rib
Action:	• With the ribs mobile: raises the upper ribs (in inspiration) • With the ribs fixed: bends the cervical spine toward the same side (with unilateral contraction) and flexes the neck (with bilateral contraction) • Anterior scalene: anterior rami of C4–C6 • Middle scalene: anterior rami of C3–C8 • Posterior scalene: anterior rami of C6–C8
Innervation:	Directly from the plexus cervicalis and the plexus brachialis (C3–6)

Intercostal muscles

Origin and insertion:	④ External intercostal muscles (costal tubercle to chondro-osseous junction): arise at the lower margin of a rib and insert on the upper margin of the next lower rib (course obliquely forward and downward) ⑤ Internal intercostal muscles (costal angle to sternum): arise at the lower margin of a rib and insert on the upper margin of the next lower rib (course obliquely backward and downward) • Innermost intercostal muscles: division of the internal intercostals (same course and action)
Action:	• External intercostals: elevate (raise) the ribs (in inspiration), support the intercostal spaces, stabilize the chest wall • Internal intercostals: depress (lower) the ribs (in expiration), support the intercostal spaces, stabilize the chest wall
Innervation:	First through eleventh intercostal nerves

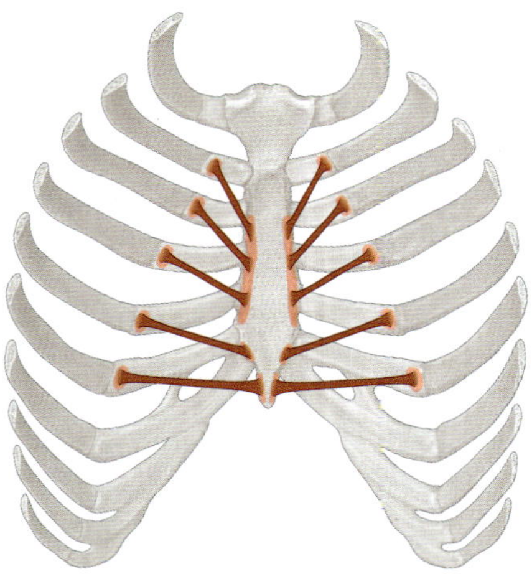

B Overview of the thoracic wall muscles: transversus thoracis
Posterior view.

Transversus thoracis

Origin:	Inner surface of the sternum and xiphoid process
Insertion:	Inner surface of the costal cartilage of the second through sixth ribs
Action:	Lowers the ribs (in expiration)
Innervation:	Second through seventh intercostal nerves

C Anterior, middle, and posterior scalenes and the internal and external intercostals

Anterior view with anterior thoracic cage partially removed. Topographically, the scalene muscles are included among the deep neck muscles, but functionally they play an important role in thoracic breathing. Subcostal muscles have the same orientation as internal intercostals but skip over one or two ribs to form continuous sheets, especially at the angle of the sixth through eleventh ribs.

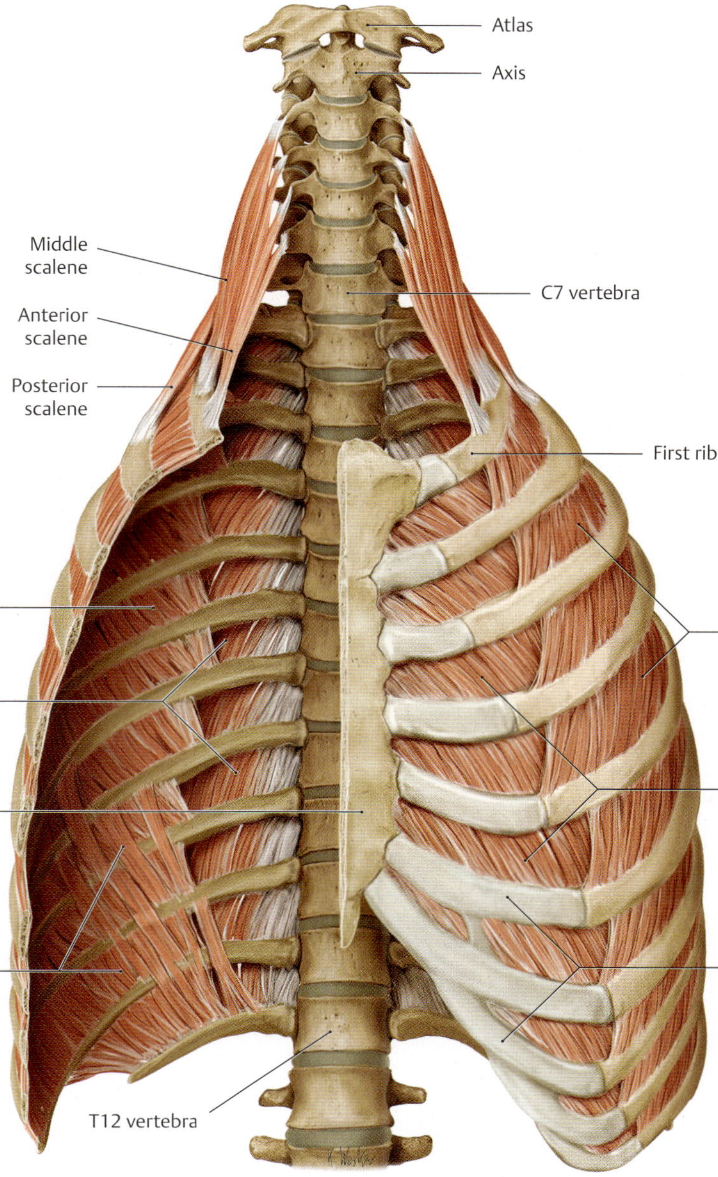

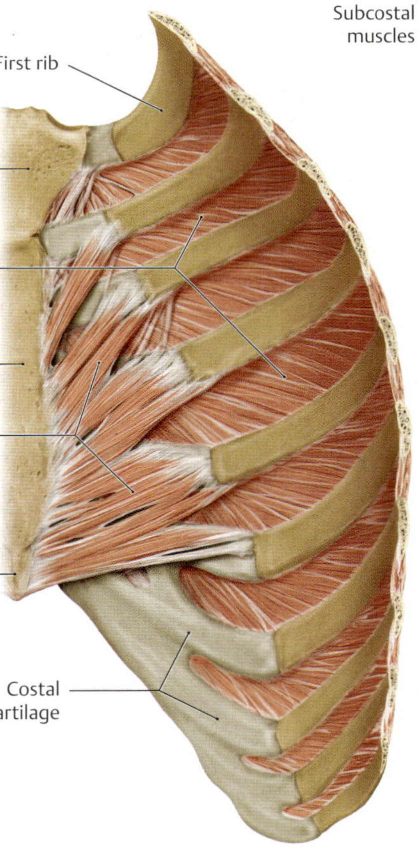

D Transversus thoracis

Posterior surface of the anterior thoracic cage removed in **C** (right half of the transversus thoracis muscle).

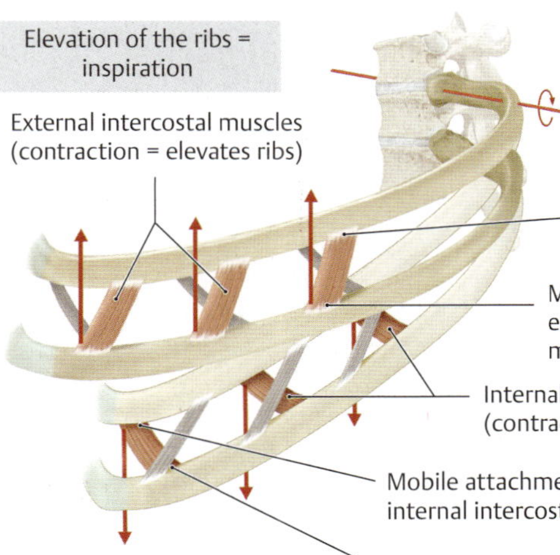

E Action of the external and internal intercostals

Note the position of the axis of motion along the neck of the rib.

2.9 The Muscles of the Thoracic Cage: The Diaphragm

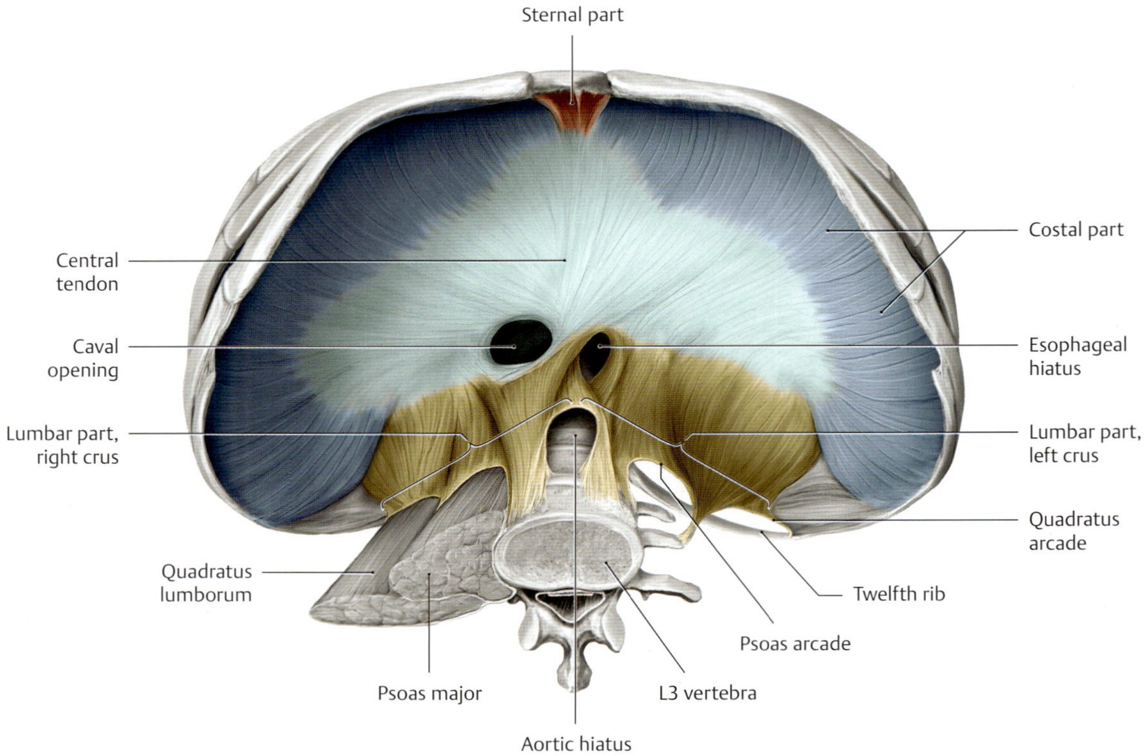

A Overview of the diaphragm

Origin:	• Costal part: lower margin of costal arch (inner surface of seventh through twelfth ribs) • Lumbar part (right and left crura): – Medial parts: L1–L3 vertebral bodies, second and third intervertebral disks, anterior longitudinal ligament – Lateral parts: first tendinous arch of the abdominal aorta (median arcuate ligament) at L1 associated with its anterior surface; second tendinous arch of the psoas arcade (medial arcuate ligament) from L2 vertebral body to associated costal process; third tendinous arch of the quadratus lumborum arcade (lateral arcuate ligament) from the costal process of L2 to the tip of rib 12 • Sternal part: posterior surface of the xiphoid process
Insertion:	Central tendon
Action:	Principal muscle of respiration (diaphragmatic and thoracic breathing), aids in compressing the abdominal viscera (abdominal press)
Innervation:	Phrenic nerve from the cervical plexus (C3–C5)

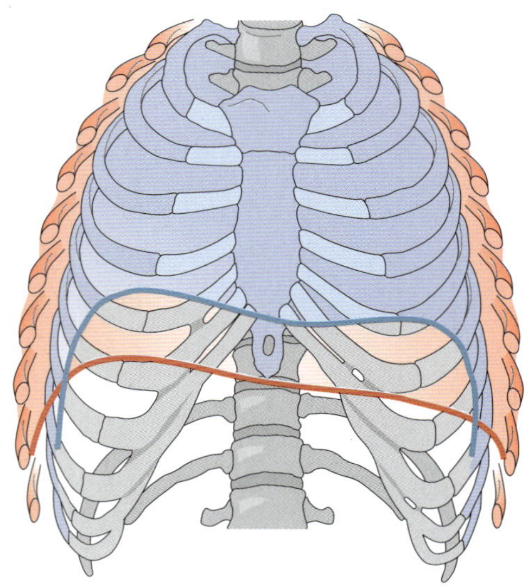

B Position of the diaphragm and ribs at full inspiration and expiration

Thoracic cage, anterior view.

Note the different positions of the diaphragm at full inspiration (red) and full expiration (blue). During a physical examination, the posterior lung boundaries can be identified by percussion (tapping the body surface). The respiratory movement of the diaphragm from end-expiration to end-inspiration should be determined; it is approximately 4 to 6 cm (see p. 182f).

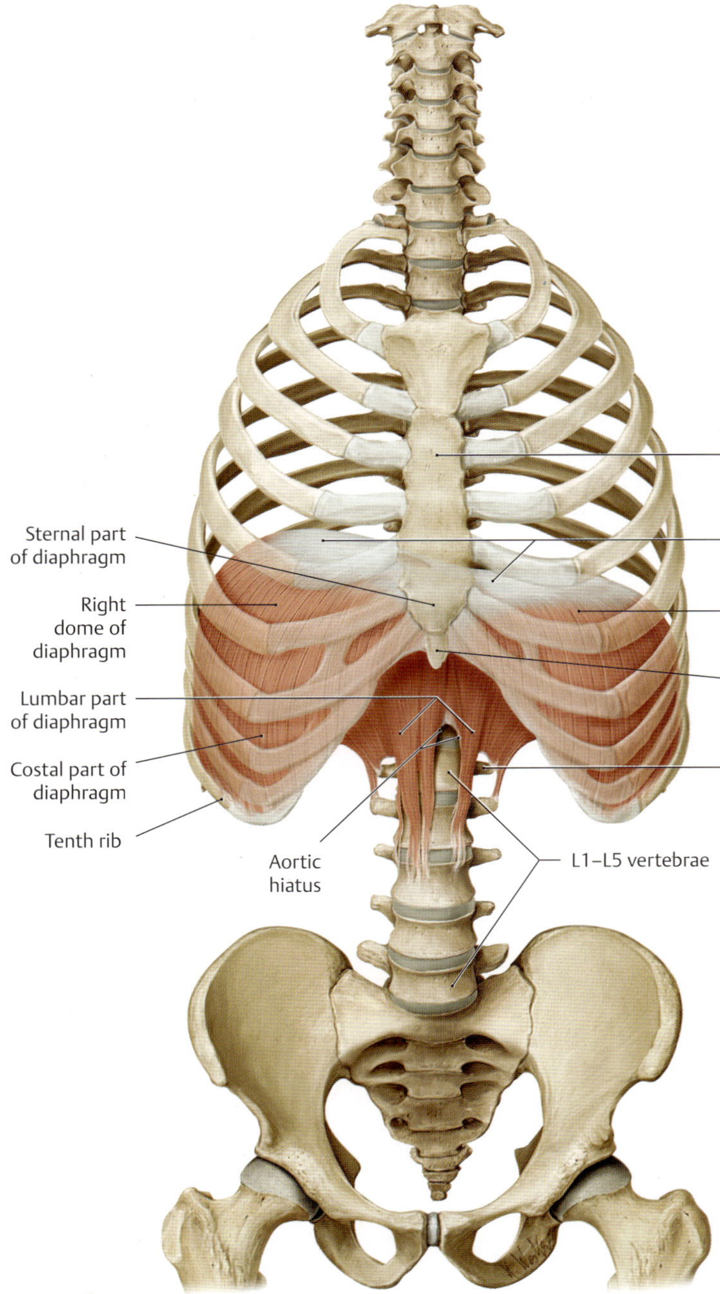

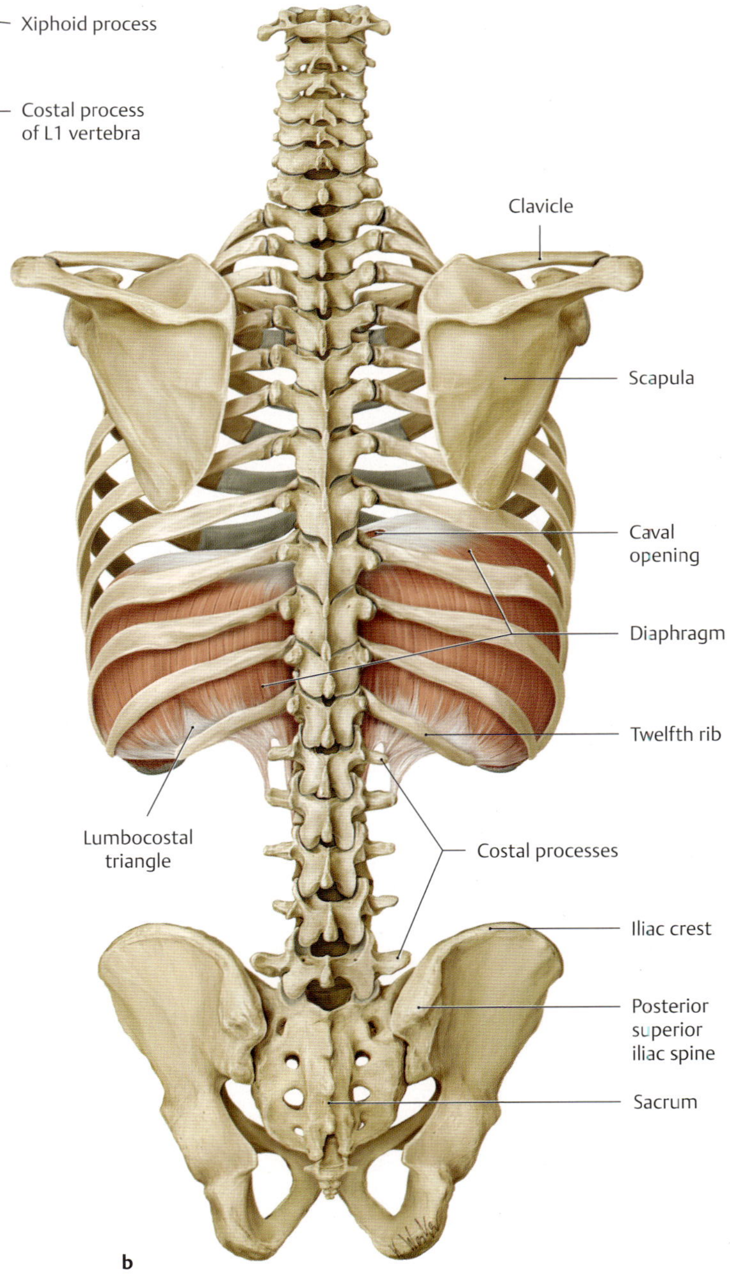

C The diaphragm

a Anterior view, **b** posterior view.

2.10 The Muscles of the Pelvic Floor and Perineum: Pelvic Diaphragm and Deep and Superficial Perineal Muscles

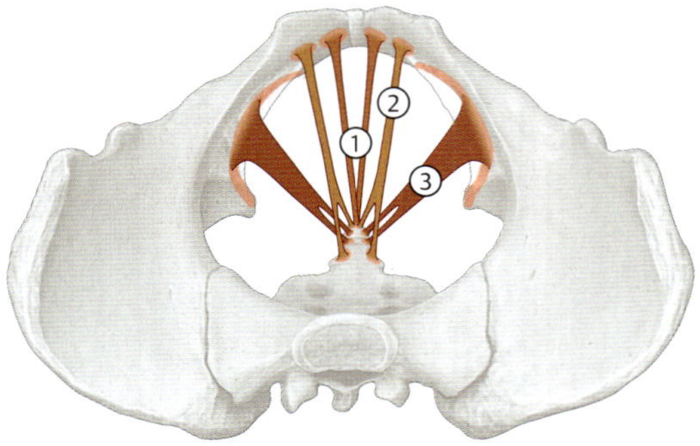

A Schematic of the pelvic diaphragm: the levator ani (puborectalis, pubococcygeus, and iliococcygeus)
Superior view.

Levator ani

① Puborectalis

Origin: Superior pubic ramus on both sides of the pubic symphysis
Insertion: Loop-shaped around the anorectal junction, interwoven with the deep part of the external anal sphincter
Innervation: Pudendal nerve and direct branches of sacral plexus (S2–S4)

② Pubococcygeus

Origin: Pubis (lateral at origin of the puborectalis)
Insertion: Anococcygeal ligament, coccyx
Innervation: Pudendal nerve (S2–S4)

③ Iliococcygeus

Origin: Tendinous arch of the internal obturator fascia (of the levator ani)
Insertion: Levator (iliococcygeal) raphe, anococcygeal ligament, coccyx
Function of the pelvic diaphragm: Holds the pelvic organs in place
Innervation: Pudendal nerve (S2–S4)

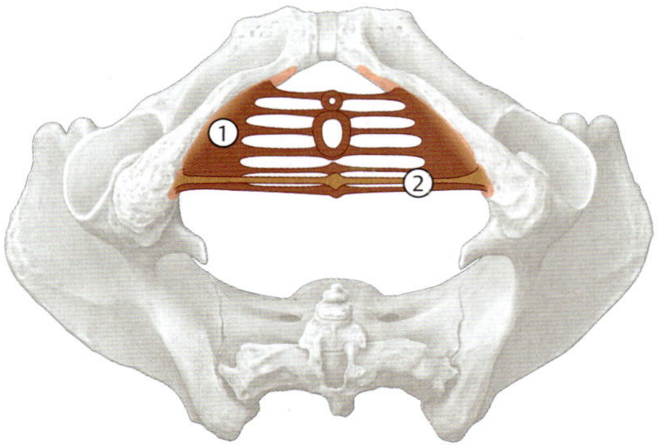

B Schematic of the deep and superficial transverse perineal
Inferior view.

① Deep transverse perineal

Origin: Inferior pubic ramus, ischial ramus
Insertion: Wall of vagina or prostate, urethra
Innervation: Pudendal nerve (S2–S4)

② Superficial transverse perineal

Origin: Ischial ramus
Insertion: Perineal body
Function of the transverse perineal muscles: Holds the pelvic organs in place, closes the urethra
Innervation: Pudendal nerve (S2–S4)

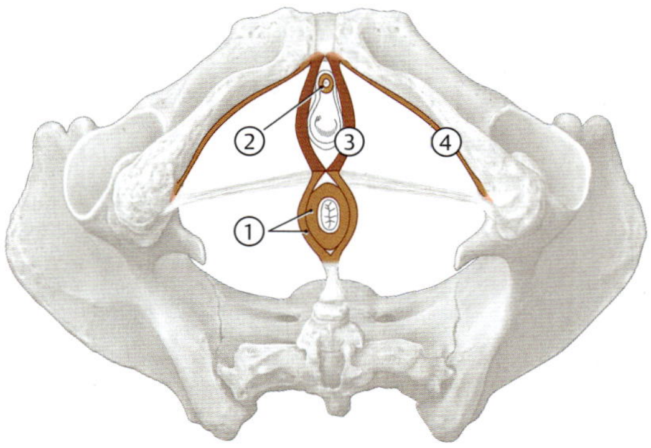

C Schematic of the external anal sphincter, external urethral sphincter, bulbospongiosus, and ischiocavernosus
Inferior view.

① External anal sphincter
The external anal sphincter is a circular muscle that closes the orifice around the anal canal, extends from the perineal body to the anococcygeal ligament, and is divided into subcutaneous, superficial, and deep parts.

Action: Closes the anus
Innervation: Pudendal nerve (S2–S4)

② External urethral sphincter
This consists of two parts: an inner, ring-shaped portion of smooth muscle, and an outer striated portion in the shape of an omega.

Action: Closes the urethra
Innervation: Pudendal nerve (S2–S4)

③ Bulbospongiosus
Runs anteriorly from the perineal body to the clitoris in females or to the penile raphe in males.

Action: Narrows the vaginal introitus in females, surrounds the corpus spongiosum in males
Innervation: Pudendal nerve (S2–S4)

④ Ischiocavernosus

Origin: Ischial ramus
Insertion: Crus of penis or clitoris
Action: Compresses the crura of the clitoris or penis; helps to promote and maintain erection
Innervation: Pudendal nerve (S2–S4)

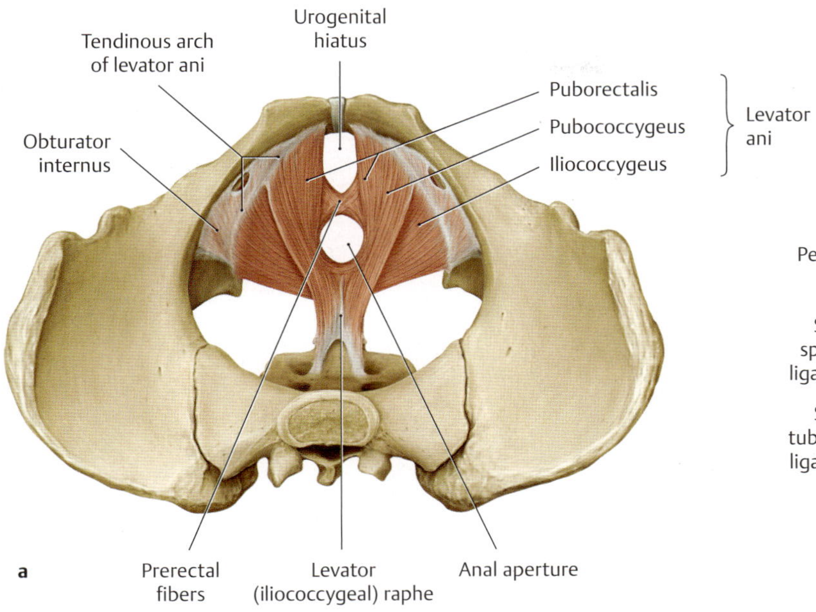

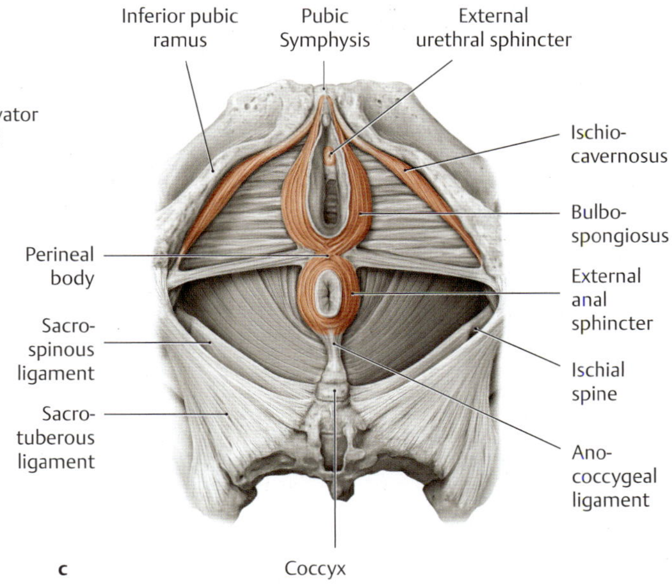

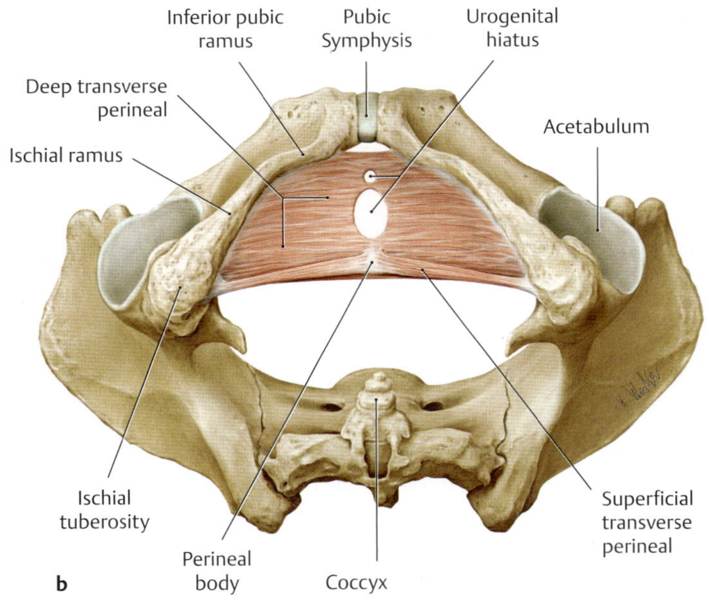

D The muscles of the female pelvic floor
a Pelvic diaphragm, superior view.
b Transverse perineal muscles, inferior view.
c Sphincter and erectile muscles, inferior view.

Note: The existence of the deep transverse perineal muscle in the female has been called into question. This is because the deep transverse perineal becomes heavily permeated by connective tissue with aging and especially after vaginal deliveries. In older women, the deep perineal space (see p. 191) basically contains connective tissue that completely occupies the urogenital hiatus at the openings of the urethra and vagina (see p. 196). We have noted this replacement by smooth muscle in each illustration of the female pelvis.

Important Functions of the Pelvic Floor Muscles
The pelvic floor performs a dual function:

- It supports the abdominal and pelvic organs by closing the abdominal and pelvic cavities inferiorly, bearing the bulk of the visceral load.
- It controls the openings of the rectum and urogenital passages (sphincter functions), which mechanically weaken the pelvic floor by piercing it.

To accomplish these inherently conflicting functions (sealing off the pelvic cavity while maintaining several apertures), the pelvic floor is lined by overlapping sheets of funnel-shaped muscle and connective tissue. This complex structure, however, also makes the pelvic floor highly susceptible to damage, especially in women.
Repetitive, extreme fluctuations in intra-abdominal pressure and other stresses, particularly at the end of pregnancy, can weaken the connective tissue apparatus and damage the pelvic floor muscles. Stretching and other injuries of the pelvic floor or the nerves supplying it, as during labor and delivery (in multiparous women), can eventually lead to pelvic floor insufficiency and its various clinical sequelae:

- Descent of the pelvic floor and thus the pelvic organs (e.g., *uterine descent*).
- In extreme cases, prolapse of the uterus with eversion of the vagina (*uterine prolapse*).

Visceral descent is generally associated with urinary or fecal incontinence in response to coughing or other acts (*stress incontinence*). Mild degrees of descent often respond well to regular pelvic floor exercises, but more serious cases may require surgical treatment with *pelvic floor repair* (surgical exposure and approximation of the two levator crura = puborectalis muscle) (e.g., approximating the levator muscles by tightening the puborectales) or surgically attaching the pelvic organs (e.g., colpopexy or rectopexy) to the pelvic wall or sacrum.

2.11 Secondarily Incorporated Trunk Muscles: Spinocostal, Spinohumeral, and Thoracohumeral

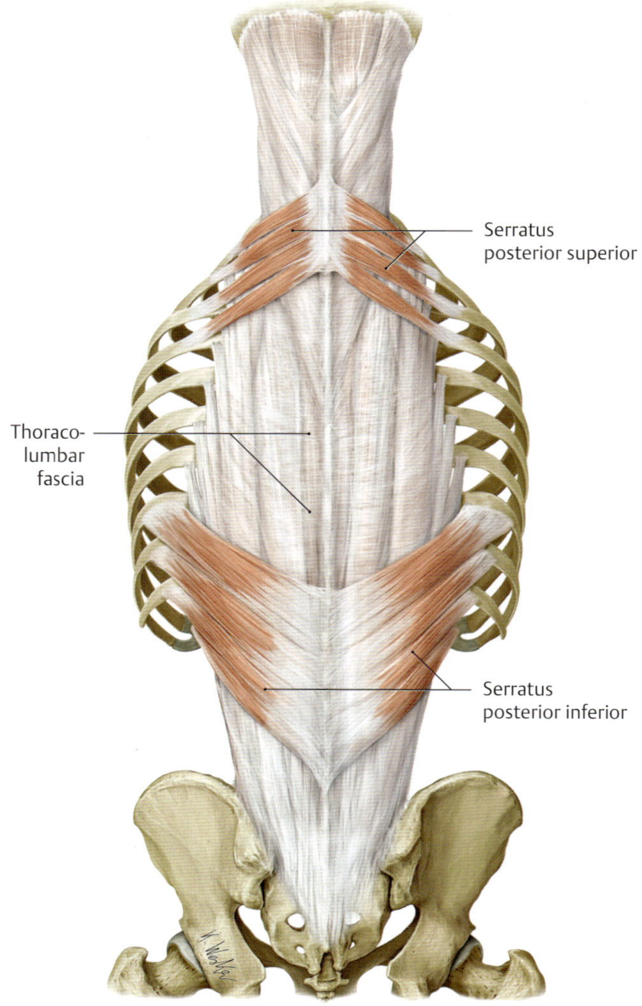

A Overview of spinocostal muscles

Due to their position above the thoracolumbar fascia and their innervation by the anterior rami of the spinal nerves, the two posterior serratus muscles are counted among the secondarily incorporated trunk muscles. These muscles, which are mostly very thin and often markedly segmented, act as auxiliary inspiratory muscles of respiration (see below).

Serratus posterior superior	
Origin:	Nuchal and supraspinous ligaments, spinal processes of C7–T3 vertebrae
Insertion:	Second through fifth ribs lateral to costal angle
Funktion:	Lifts the ribs, thus supporting inspiration
Innervation:	Intercostal nerves (T2–T5)
Serratus posterior inferior	
Origin:	Spinal processes of the T11–L2 vertebrae and the thoracolumbar fascia
Insertion:	Toward the lower margin of the ninth through twelfth ribs
Action:	Also aids inspiration; by drawing the lower ribs backward and downward, it assists in preventing further narrowing of the thoracic aperture, thus providing a fixed stable base of the diaphragm
Innervation:	Intercostal nerves (T9–T11), subcostal nerve (T12)

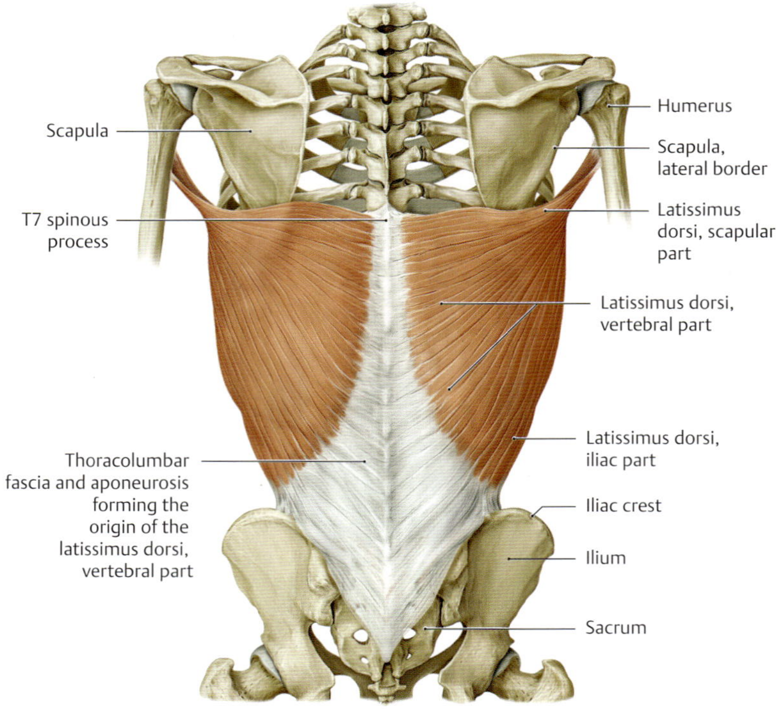

B Latissimus dorsi

Posterior view. *Note:* The aponeurosis forming the origin of the latissimus dorsi is fused with the thoracolumbar fascia. The two structures cannot be separated.

C List of secondarily incorporated trunk muscles

(Discussed with the exception of the serratus muscles on p. 312.) All of the intrinsic back muscles are covered by large muscles, which in part extend all the way to the pelvis. In the course of phylogenesis, they have extended their origin to the trunk (incorporated trunk muscles, compare p. 153). These muscles initially migrated from the back as hypaxial muscles (innervation: anterior rami) toward the upper limb (thus shoulder girdle muscles) in order to later relocate again to the back. This probably happened as part of the development of tetrapods, when the pectoral girdle muscles grew and had to carry more weight. A distinction is made between the following:

Spinocostal or trunk rib muscles (between spinal column and ribs)	
• Serratus posterior superior • Serratus posterior inferior	
Spinohumeral or muscles of the trunk and shoulder girdle or trunk and arm (between spinal column and shoulder girdle or upper arm)	
• Muscles of the trunk and shoulder girdle – Rhomboid major and rhomboid minor – Levator scapulae – Serratus anterior – Subclavius – Pectoralis minor – Trapezius	• Muscles of the trunk and arm – Latissimus dorsi • Muscles of the thorax and arm – Pectoralis major
Note: All incorporated muscles are innervated by anterior rami of the spinal nerves.	

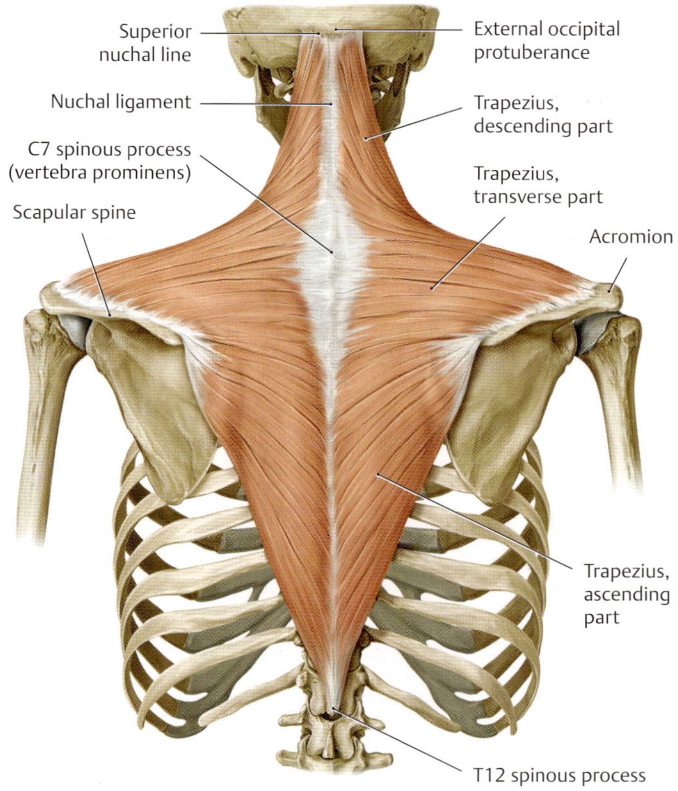

D Trapezius
Posterior view.

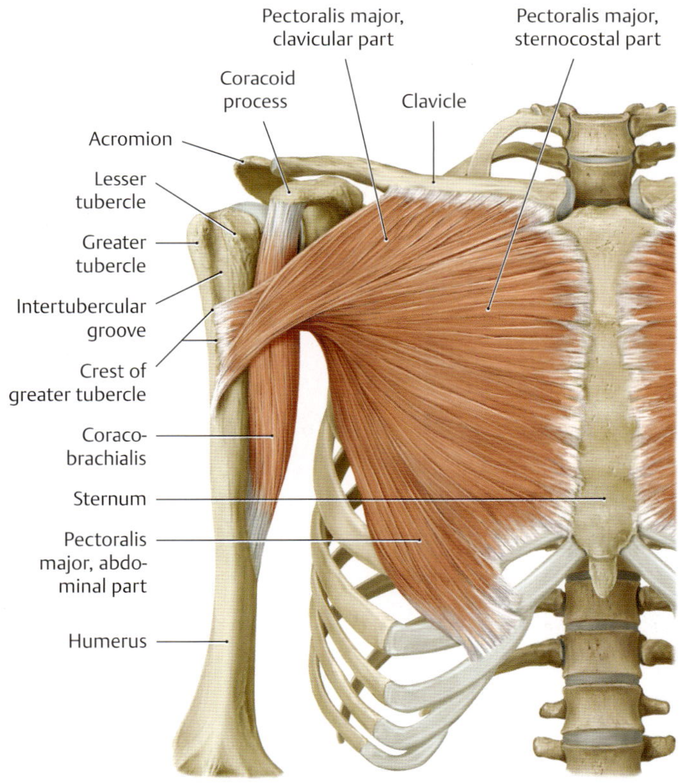

E Pectoralis major and coracobrachialis
Right side, anterior view.

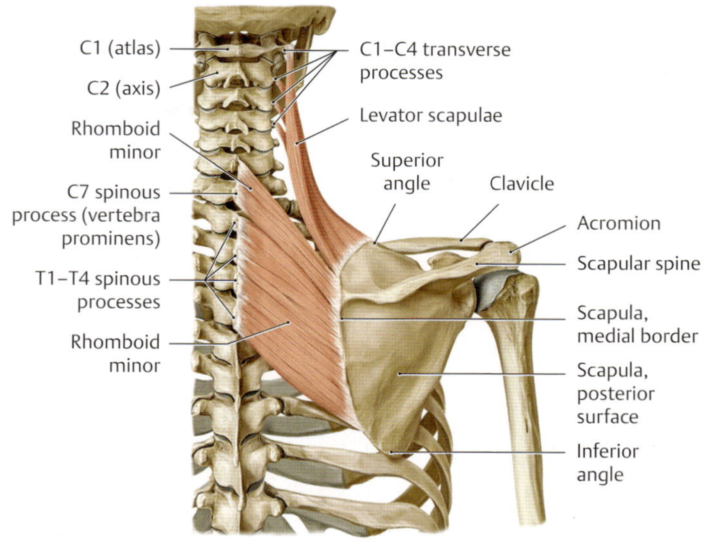

F Levator scapulae, rhomboid major, and rhomboid minor
Right side, posterior view.

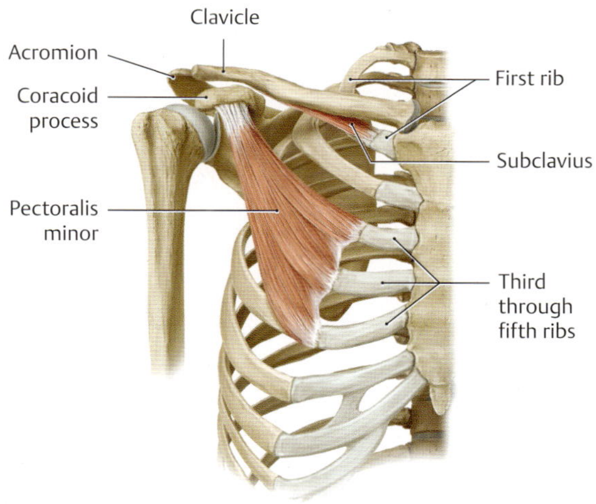

G Pectoralis minor and subclavius
Right side, anterior view.

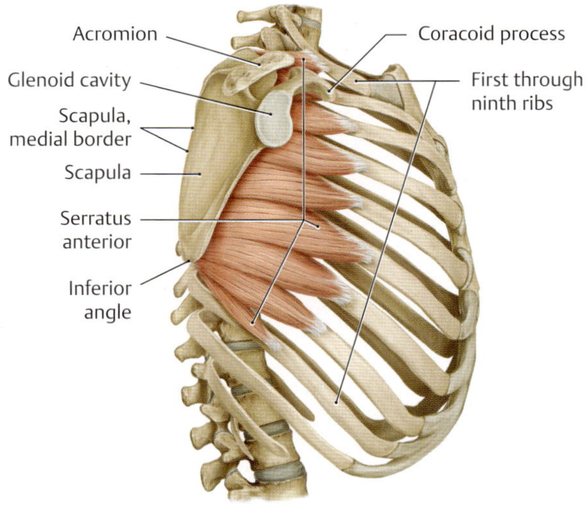

H Serratus anterior
Right side, lateral view.

3.1 The Back Muscles and Thoracolumbar Fascia

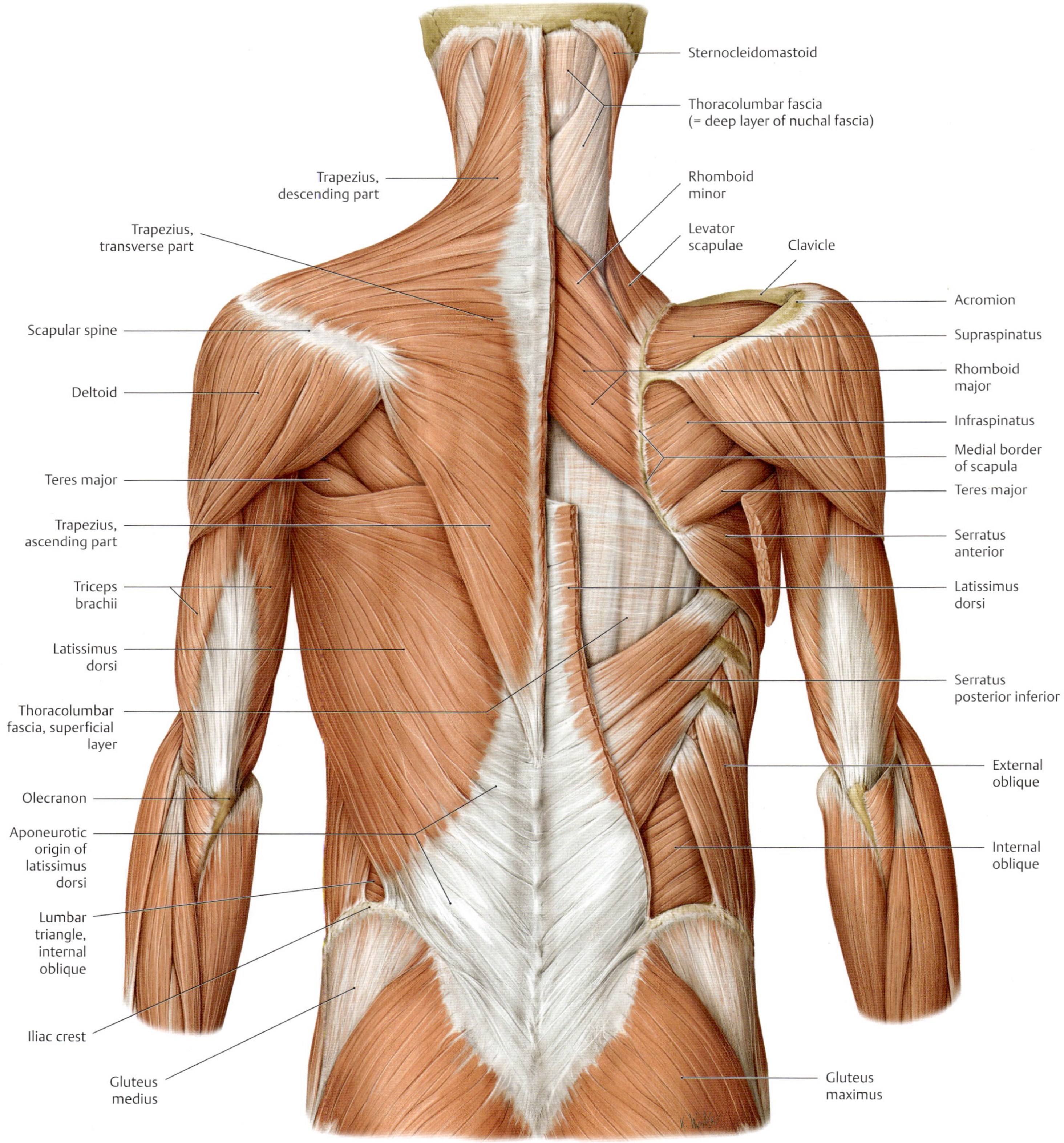

A The thoracolumbar fascia as a partition between the intrinsic and nonintrinsic back muscles

The trapezius muscle has been completely removed and the latissimus dorsi has been partially removed on the right side to reveal the thoracolumbar fascia. The superficial layer of the thoracolumbar fascia separates the intrinsic back muscles from the nonintrinsic muscles that have migrated to the back.

Note: The superficial layer of the thoracolumbar fascia is located close to the spinal column just beneath the skin, where it is heavily reinforced and serves as the origin for the extensive tendon of the latissimus dorsi. Lateral to the origin, the superficial layer is pushed beneath both the tendon and the muscle, thereby losing direct contact to the subcutaneous tissue. Along its entire width, medial to lateral, the superficial layer covers the intrinsic back muscles.

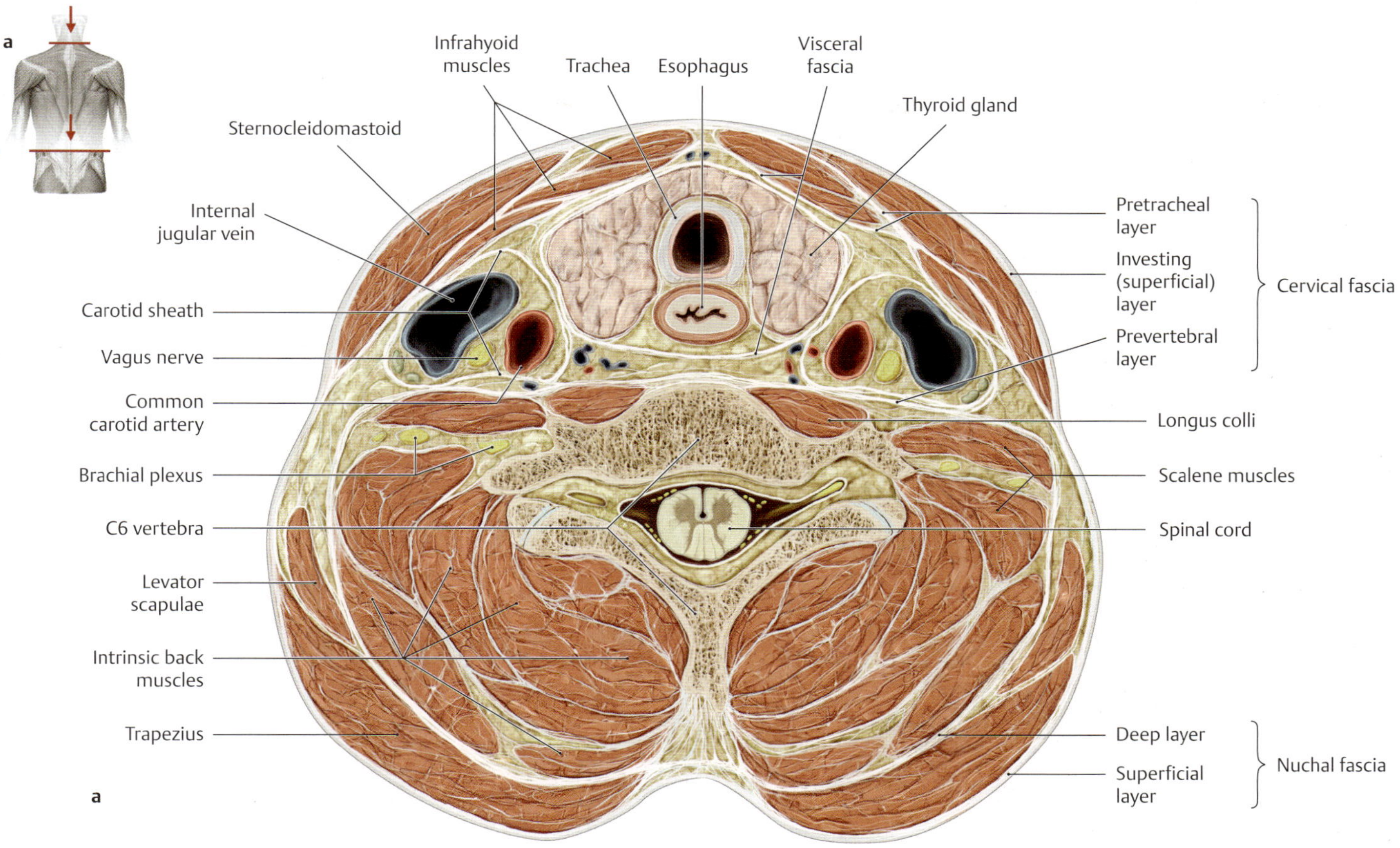

a

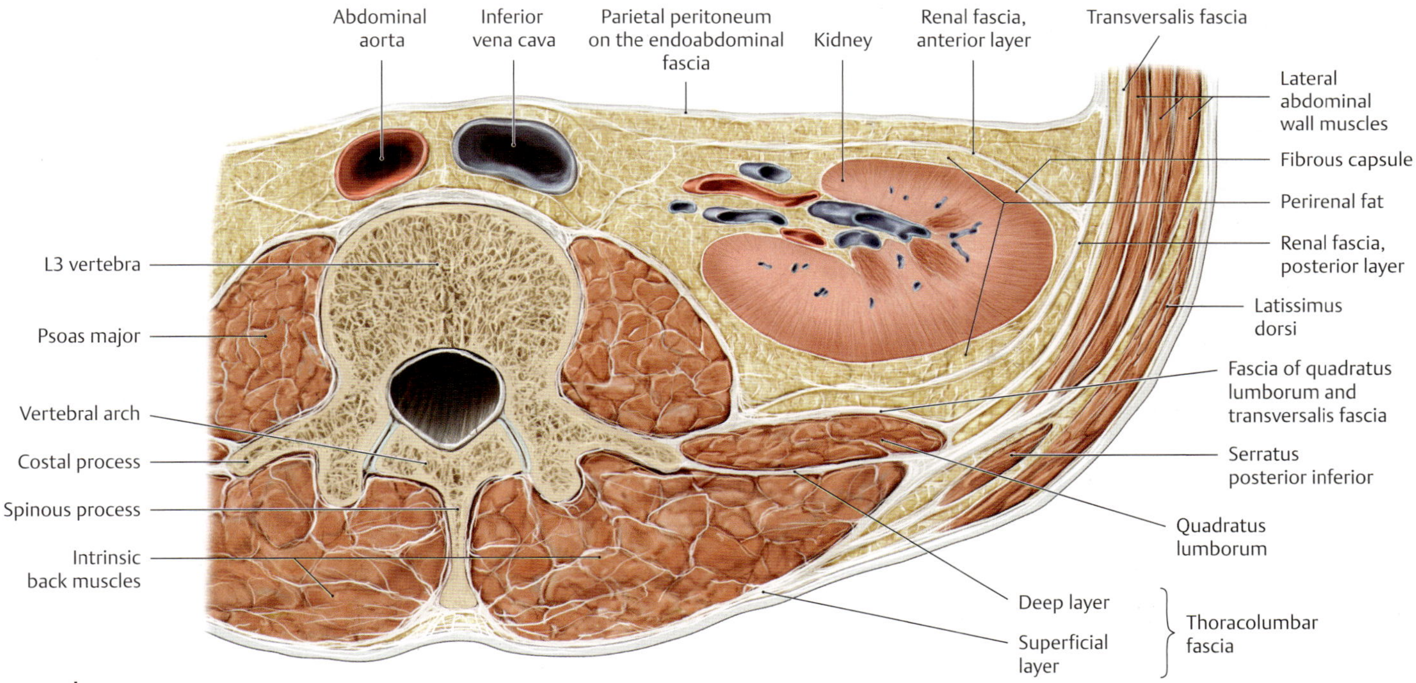

b

B Thoracolumbar fascia

a Transverse section through the neck at the level of the C6 vertebra, superior view.

b Transverse section through the posterior trunk wall at the level of the L3 vertebra (cauda equina removed), superior view.

The thoracolumbar fascia forms the lateral portion of an osseofibrous canal that encloses all of the *intrinsic back muscles*. In addition to the thoracolumbar fascia, this canal is formed by the vertebral arches and the spinous and costal processes of the associated vertebrae. The thoracolumbar fascia consists of a superficial and a deep layer, especially in the lumbar region; both layers unite at the lateral margin of the intrinsic back muscles. At the back of the neck, the superficial layer of the thoracolumbar fascia blends with the nuchal fascia (deep layer), becoming continuous with the prevertebral layer of the cervical fascia.

Note: According to the new anatomic nomenclature, the thoracolumbar fascia consists of three layers: What was previously the deep layer is now the middle layer (lamina media); the new deep layer is the fascia of the quadratus lumborum. The superficial layer remains unchanged.

3.2 The Intrinsic Back Muscles: Lateral and Medial Tracts

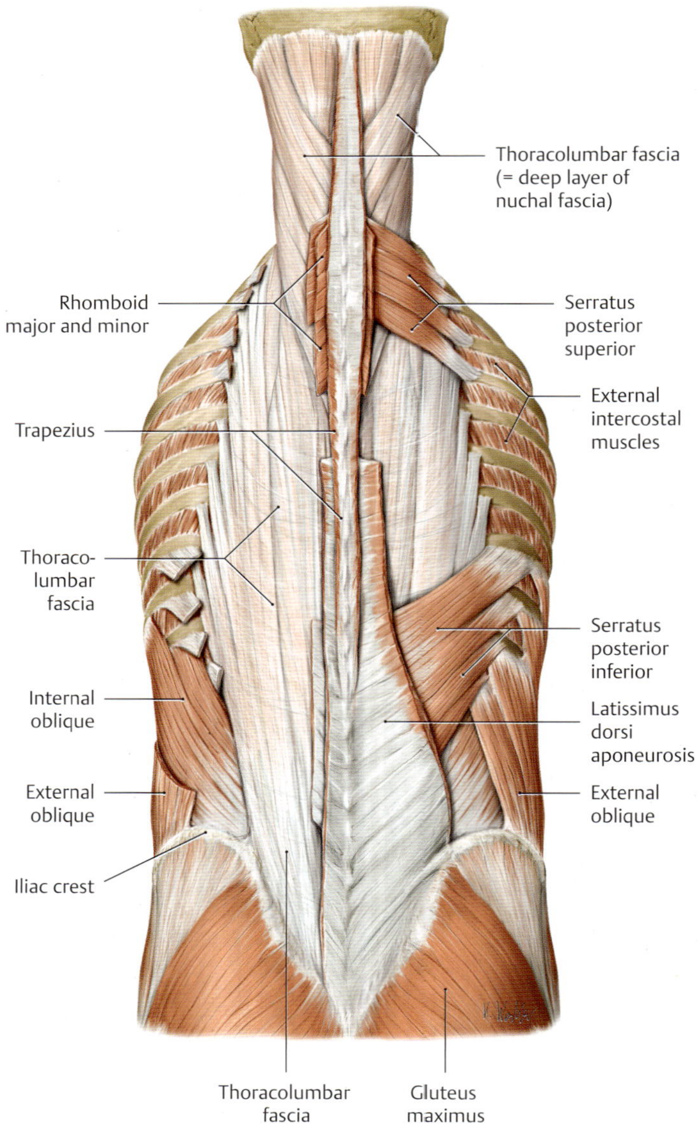

A Course of the thoracolumbar fascia
Posterior view. To reveal the thoracolumbar fascia, both shoulder girdles and the extrinsic back muscles have been removed (except for the serratus posterior superior and inferior and the aponeurotic origin of the latissimus dorsi on the right side).

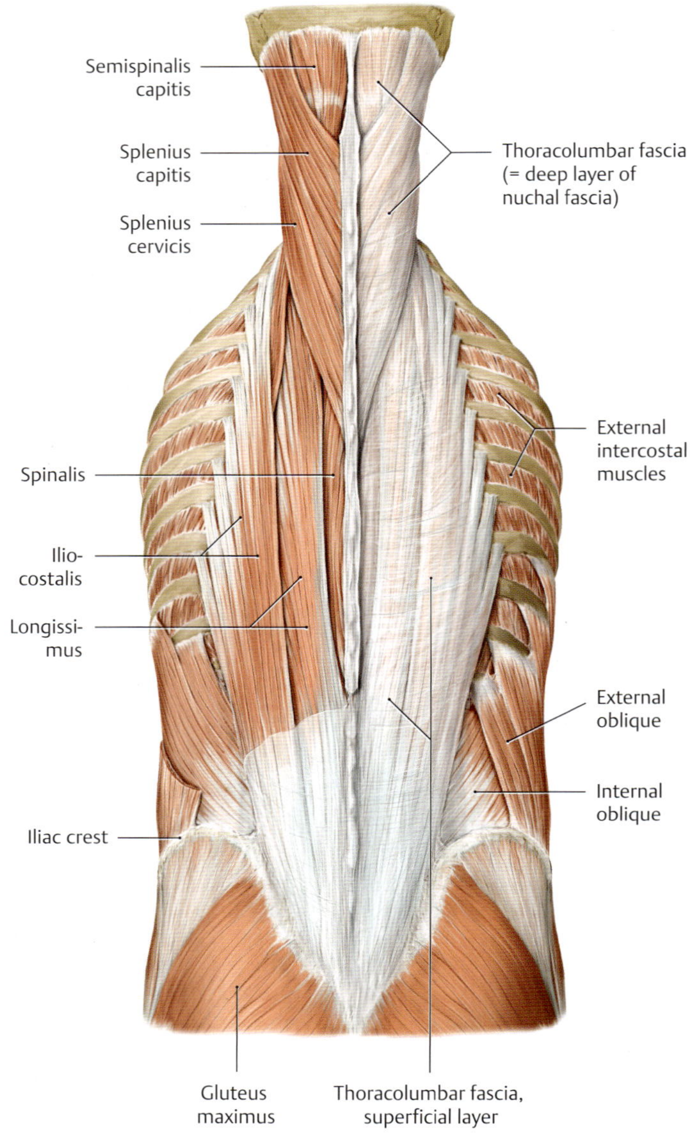

B Lateral tract of the intrinsic back muscles
Posterior view. Portions of the superficial layer of the thoracolumbar fascia have been removed on the left side of the back to expose the lateral tract muscles (iliocostalis, longissimus, and splenius muscles). The levatores costarum and intertransversarii muscles, also part of the lateral tract, are covered here by the iliocostalis and longissimus muscles (see **C** and **D**).
Note that the thoracolumbar fascia on the back of the neck is continuous with the deep layer of the nuchal fascia.

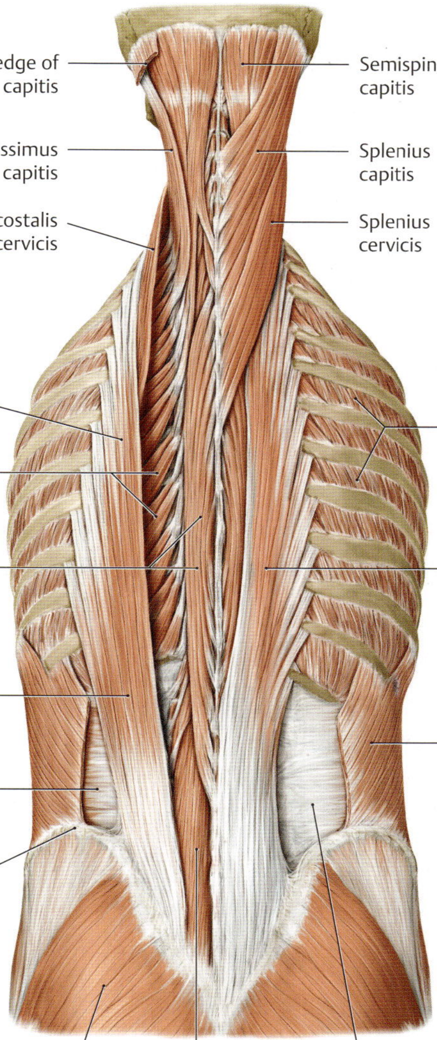

C Medial tract of the intrinsic back muscles (portions of the lateral tract left in place)

Posterior view. The longissimus (except for longissimus capitis) and splenius muscles have been removed on the left side of the back, and the iliocostalis muscles have been removed on the right side (see **D** for the rotatores muscles).

Note the deep layer of the thoracolumbar fascia from which both the internal oblique and the transversus abdominis muscles originate (see **D**).

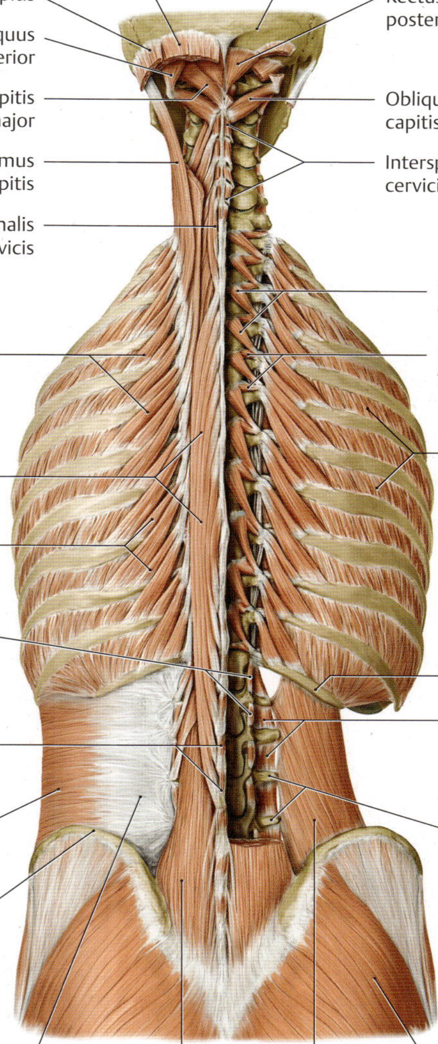

D Medial tract of the intrinsic back muscles (with the entire lateral tract removed)

Posterior view. The entire lateral tract (except for the intertransversarii and levatores costarum) has been removed, along with portions of the medial tract, to demonstrate the various individual muscles of the medial tract.

Note the origin of the transversus abdominis from the deep layer of the thoracolumbar fascia in the lumbar region (left side).

On the right side, the deep fascial layer and multifidus muscle have been removed to display the intertransversarii (lateral tract) and the quadratus lumborum (posterior [deep] abdominal muscle).

3.3 The Intrinsic Back Muscles: Short Nuchal Muscles

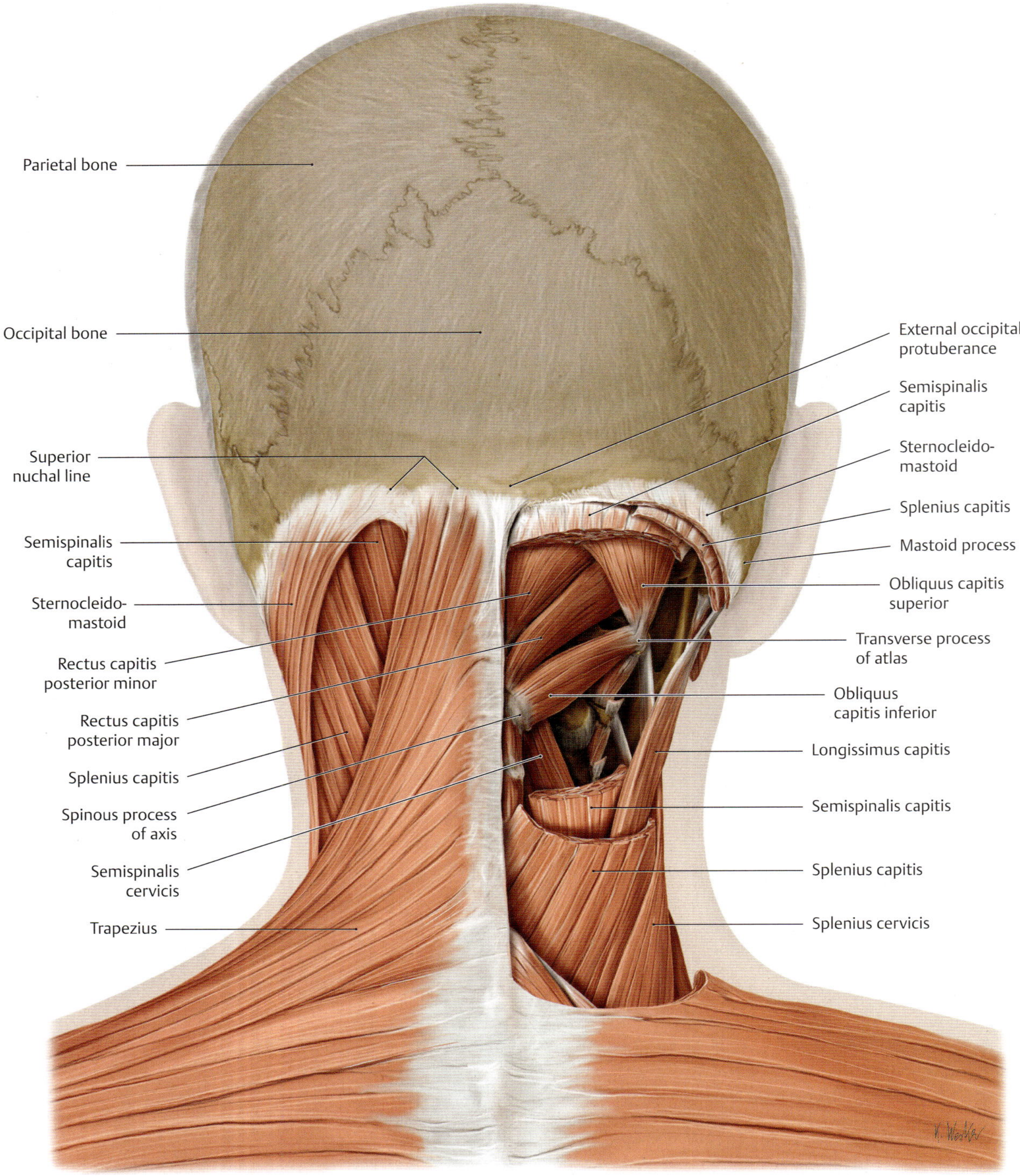

A Location of the short nuchal muscles (suboccipital muscles)
Nuchal region, posterior view. The suboccipital muscles in the strict sense are the short (deep) nuchal muscles that belong to the intrinsic back muscles (recti capitis posterior major and minor and obliquii capitis superior and inferior). They meet the criterion of being innervated by a posterior ramus—in this case the C1 posterior ramus, the suboccipital nerve. The recti capitis *anterior* and *lateralis* are *not* classified as intrinsic back muscles, despite their suboccipital location, because they are innervated by anterior rami. The short nuchal muscles lie within the thoracolumbar fascia deep at the back of the neck and course between the occiput and the first two cervical vertebrae. They act mainly on the craniovertebral joints (see p. 128) and support differentiated head movements (e.g., for fine adjustments of head position). The following muscles have been partially removed to reveal their location in the right nuchal region: trapezius, sternocleidomastoid, splenius capitis, and semispinalis capitis. An important landmark in the deep nuchal region is the spinous process of the axis.

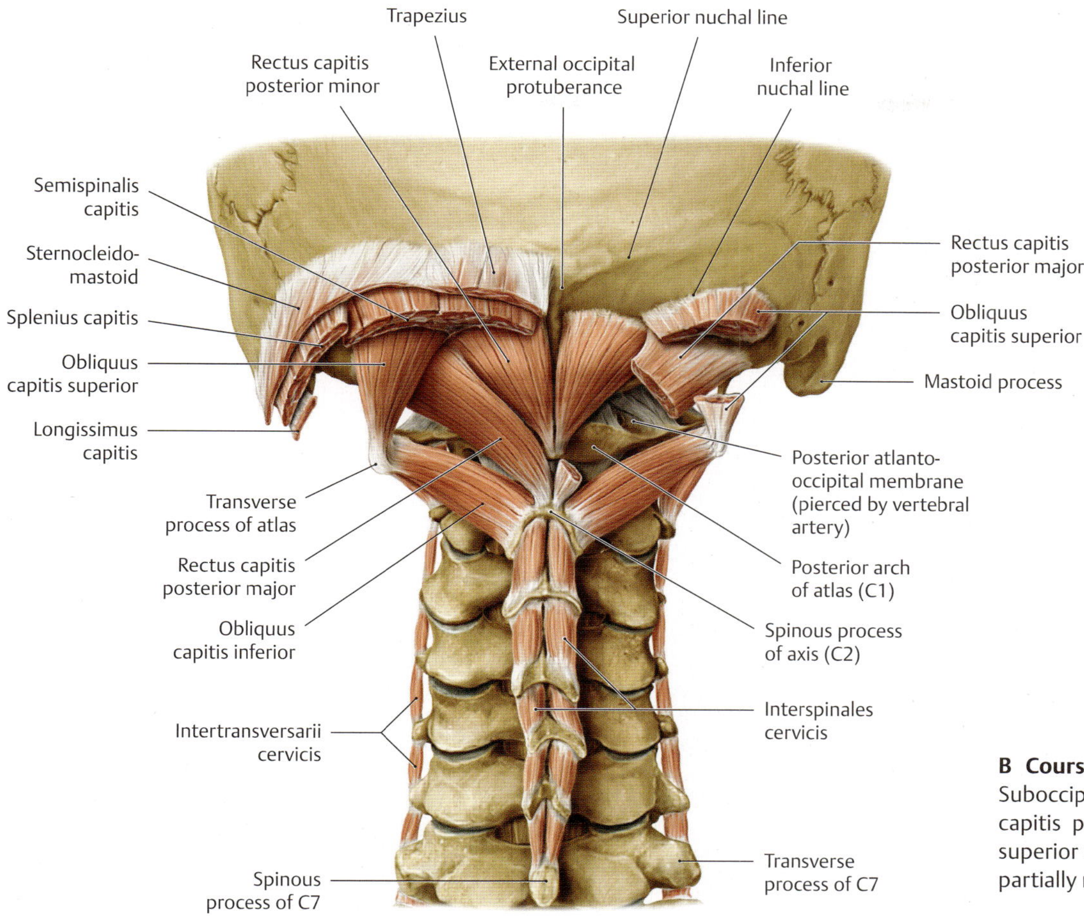

B Course of the short nuchal muscles
Suboccipital region, posterior view. The rectus capitis posterior major and obliquus capitis superior muscles on the right side have been partially removed.

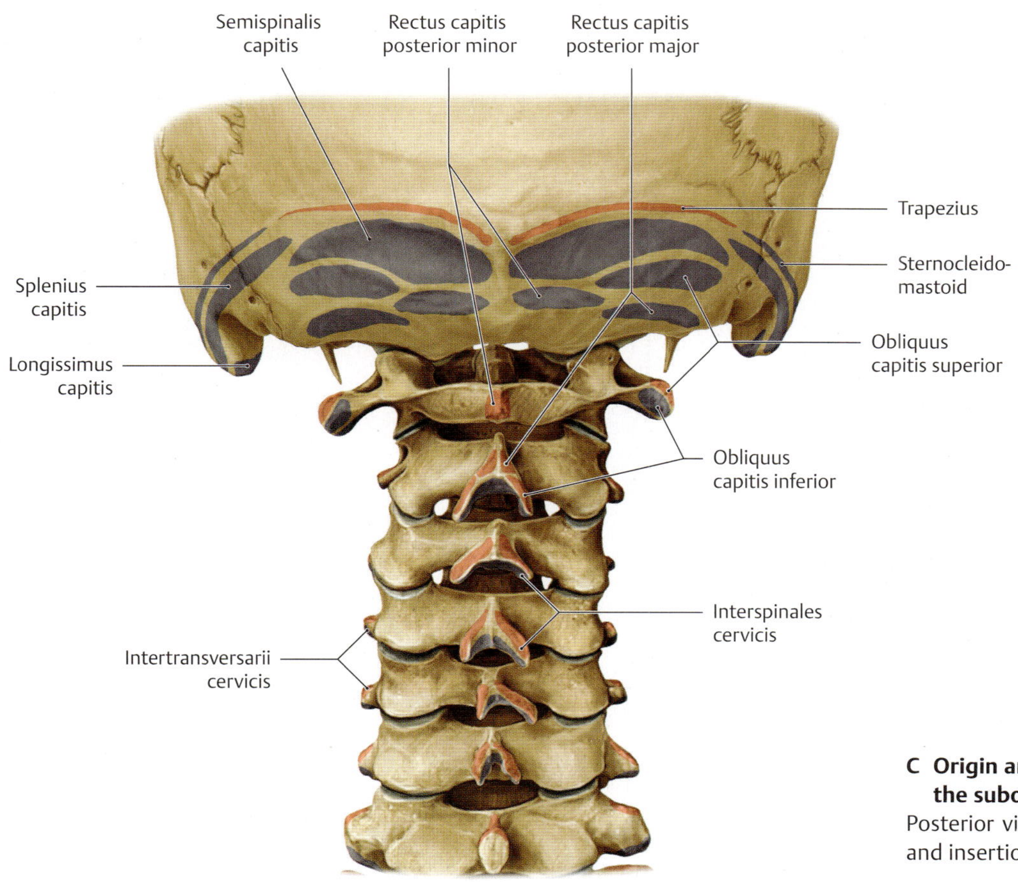

C Origin and insertion of the muscles of the suboccipital region
Posterior view. The origins are shown in red and insertions in blue.

3.4 The Thoracic Wall Muscles and Endothoracic Fascia

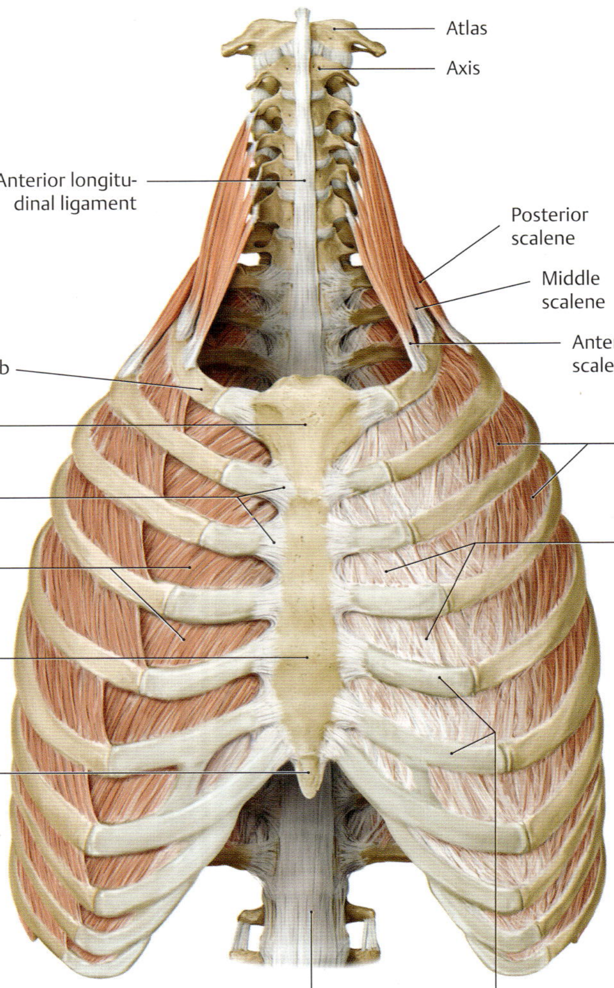

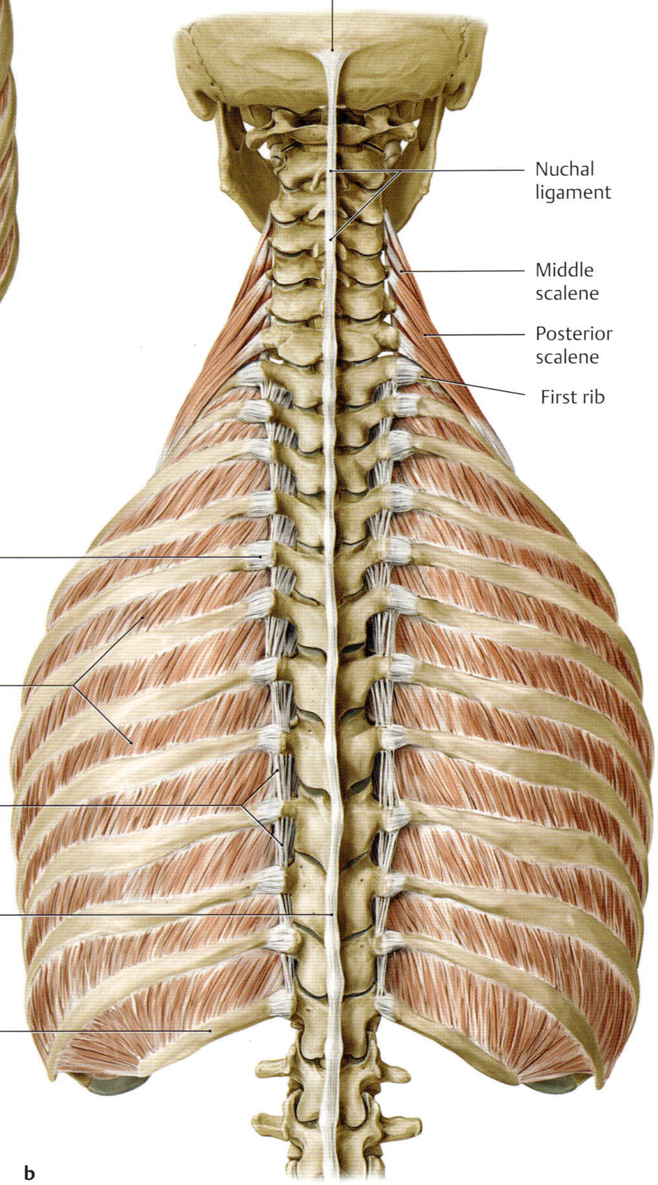

A Muscles of the thoracic wall

a Anterior view, **b** posterior view.

Topographically, the intercostal, transversus thoracis (see **Ba**), and subcostal (see **Bb**) muscles are included in the category of thoracic wall muscles.

Note: The scalene muscles, which belong to the deep neck muscles topographically, are classified functionally as thoracic muscles owing to their involvement in thoracic breathing. Hence they are included in the drawings.

The diaphragm, also considered a thoracic muscle, is shown in **Bc** and on p. 182. The nonintrinsic back muscles are described on p. 152.

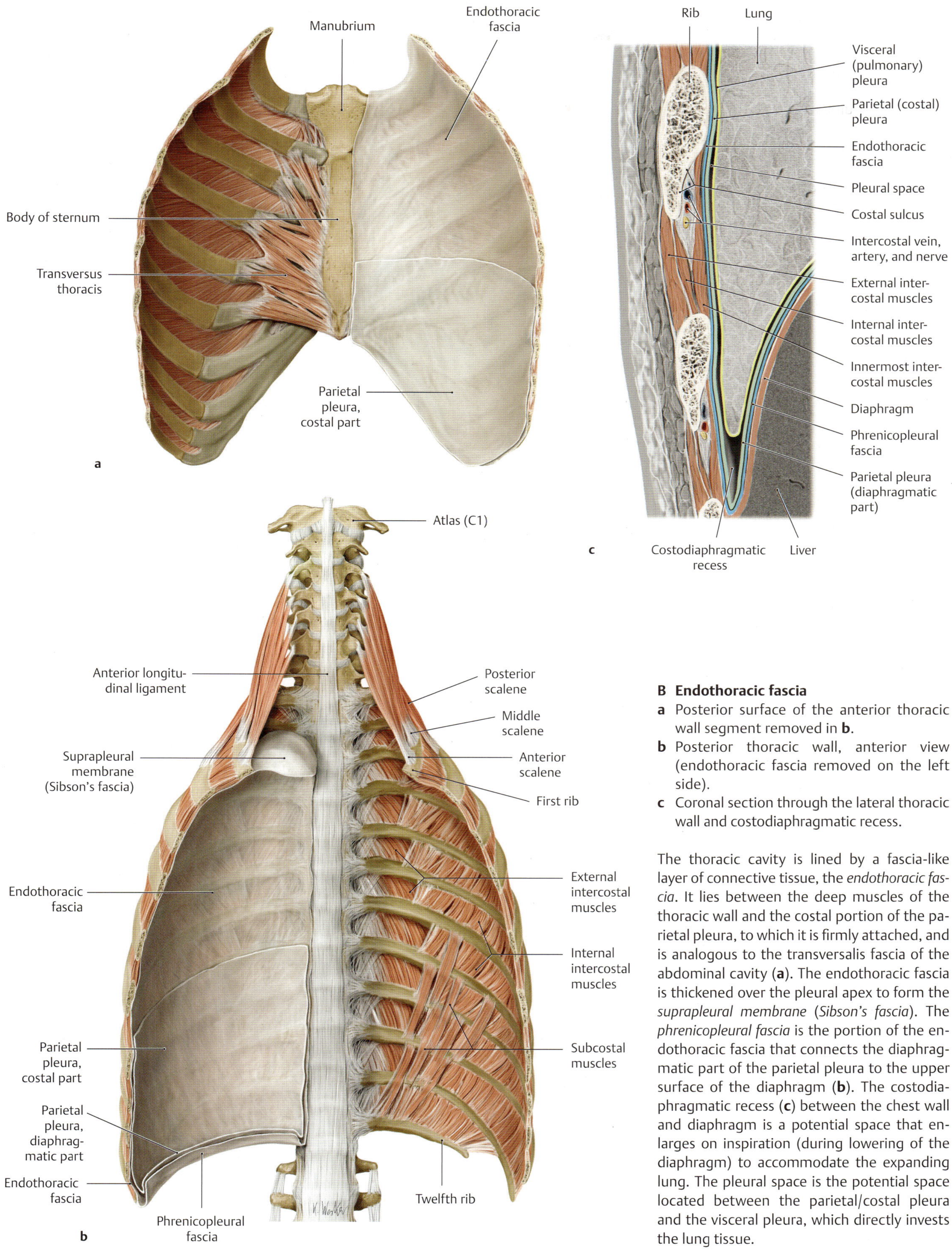

B Endothoracic fascia

a Posterior surface of the anterior thoracic wall segment removed in **b**.

b Posterior thoracic wall, anterior view (endothoracic fascia removed on the left side).

c Coronal section through the lateral thoracic wall and costodiaphragmatic recess.

The thoracic cavity is lined by a fascia-like layer of connective tissue, the *endothoracic fascia*. It lies between the deep muscles of the thoracic wall and the costal portion of the parietal pleura, to which it is firmly attached, and is analogous to the transversalis fascia of the abdominal cavity (**a**). The endothoracic fascia is thickened over the pleural apex to form the *suprapleural membrane* (*Sibson's fascia*). The *phrenicopleural fascia* is the portion of the endothoracic fascia that connects the diaphragmatic part of the parietal pleura to the upper surface of the diaphragm (**b**). The costodiaphragmatic recess (**c**) between the chest wall and diaphragm is a potential space that enlarges on inspiration (during lowering of the diaphragm) to accommodate the expanding lung. The pleural space is the potential space located between the parietal/costal pleura and the visceral pleura, which directly invests the lung tissue.

3.5 Thoracoabdominal Junction: The Diaphragm

A The diaphragm, superior view
The diaphragm consists of three parts: costal, lumbar, and sternal. As the muscle that separates the thoracic and abdominal cavities, the diaphragm has characteristic apertures for the passage of the esophagus, inferior vena cava, and aorta (see **Cb** and **Db**).

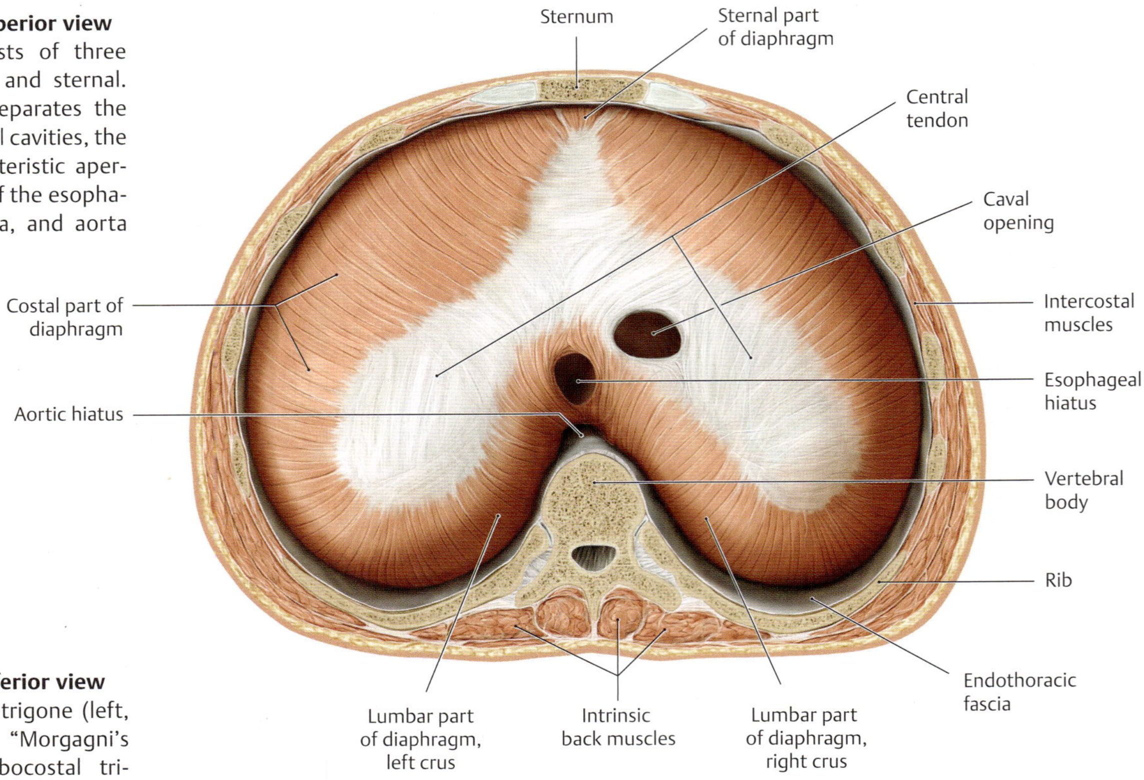

B The diaphragm, inferior view
Both the sternocostal trigone (left, "Larrey's cleft"; right, "Morgagni's cleft") and the lumbocostal trigone (both sides also referred to as "Bochdalek's triangle") represent places of lower mechanical resistance as connective tissue-filled muscle gaps in the diaphragm (*locus minoris resistantiae*, the place of least resistance). In these places, hernias can form from the abdomen into the thoracic cavity. Hernias are often referred to by proper names based on their anatomical location (e.g. "Bochdalek hernia").

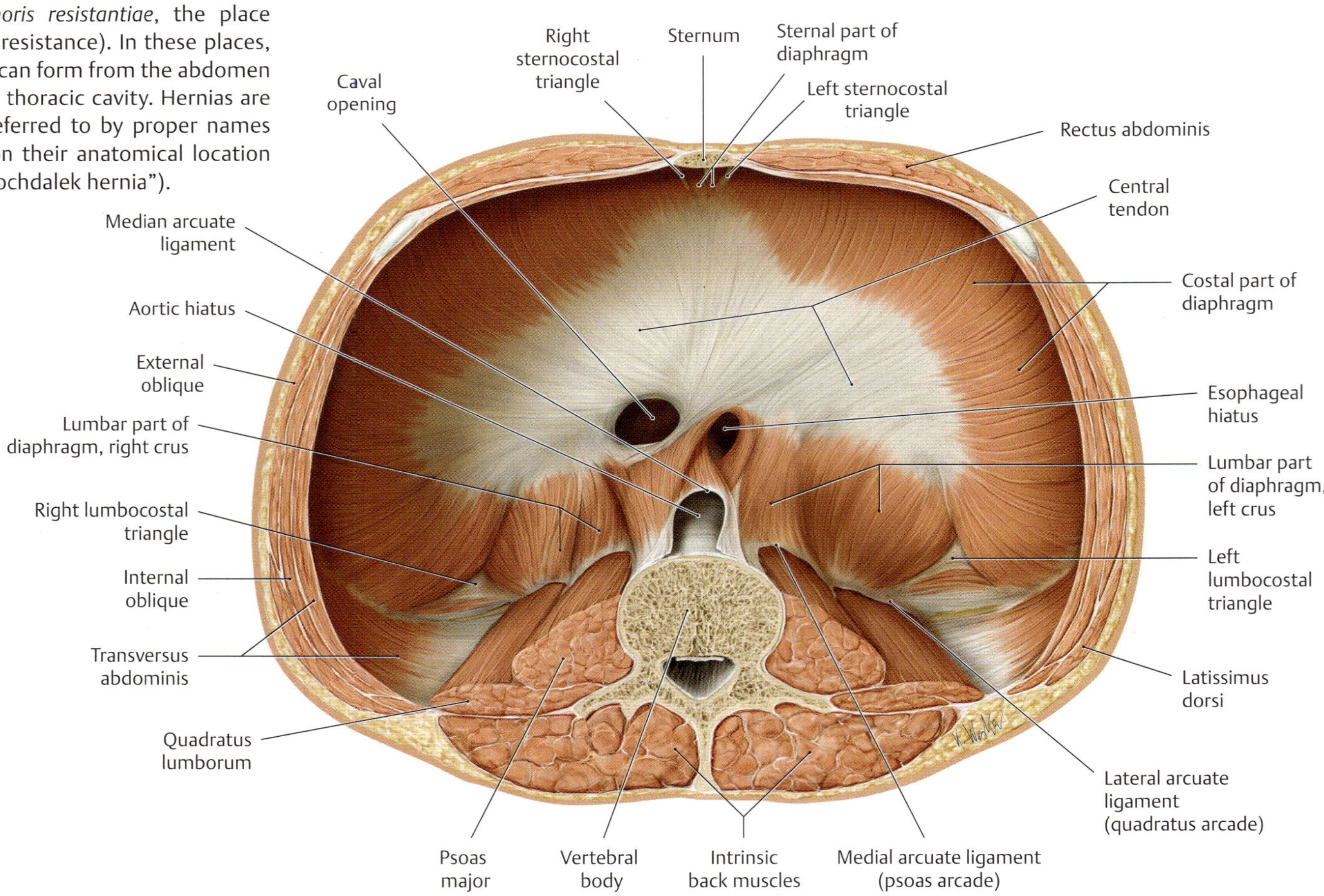

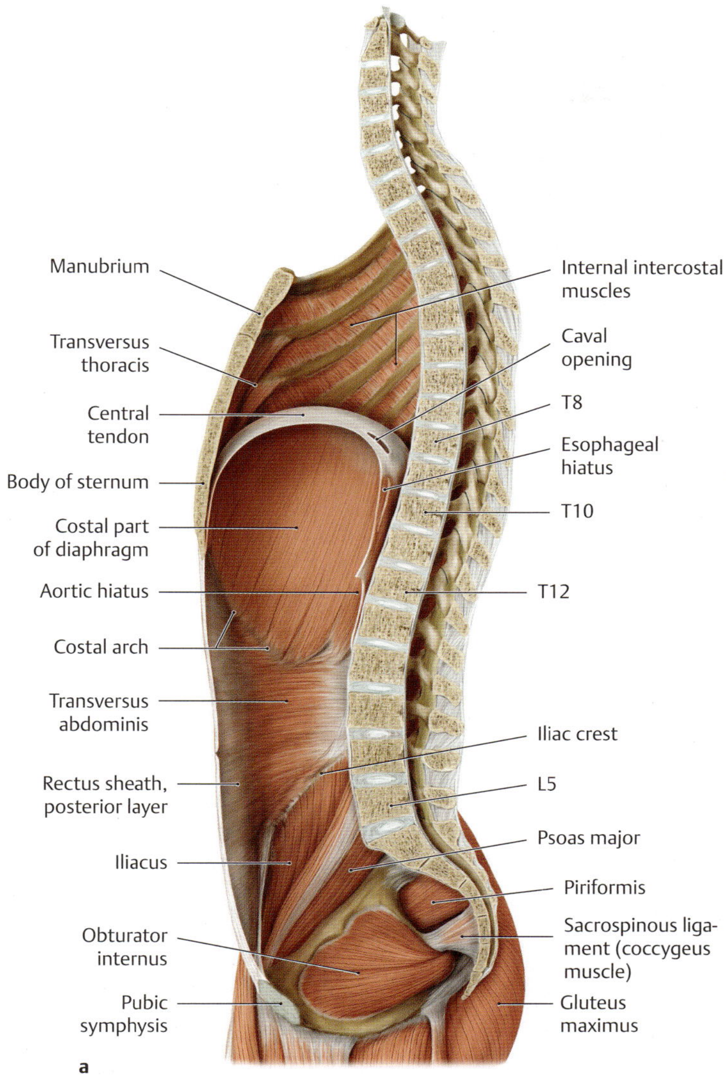

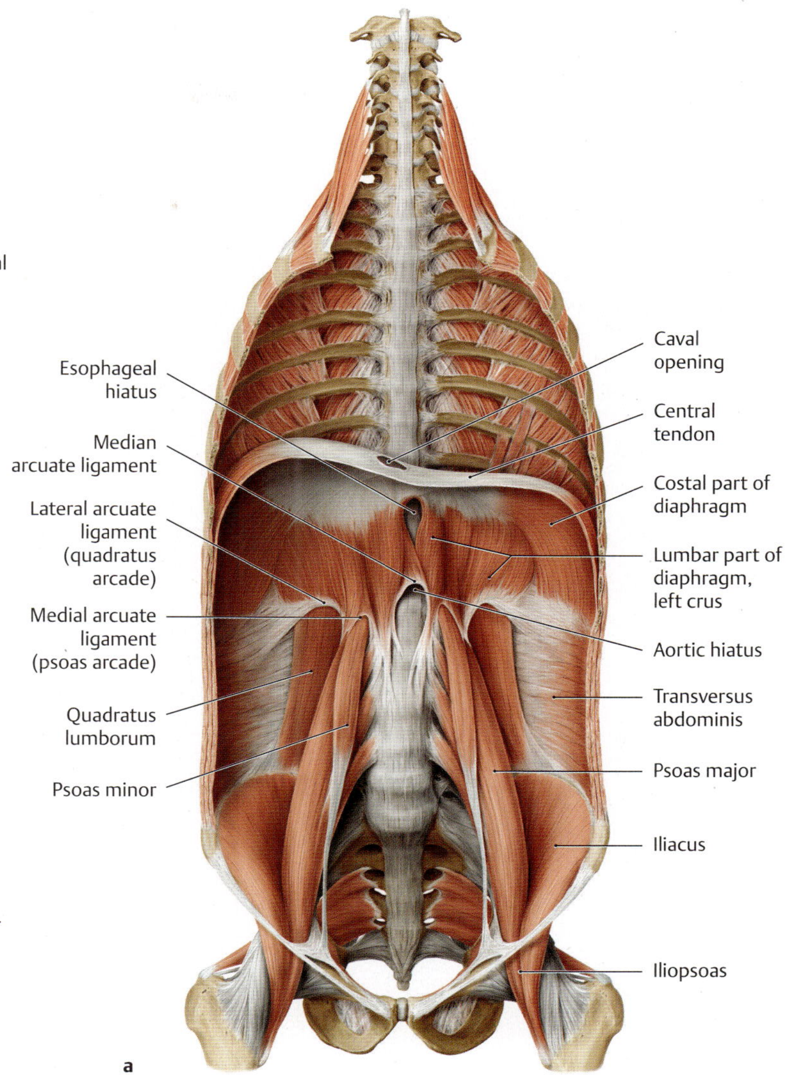

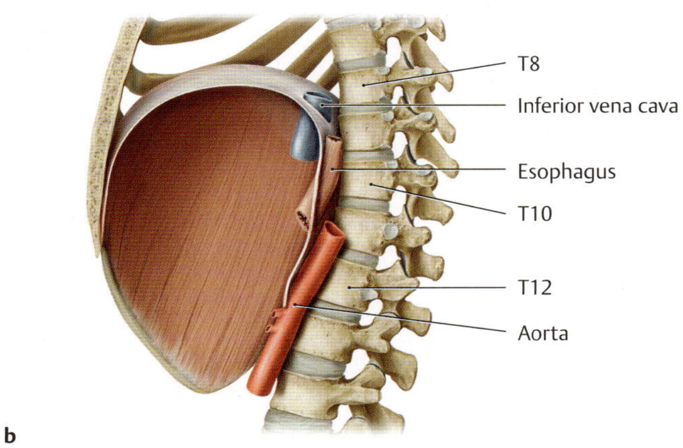

C Position and shape of the diaphragm, viewed from the left side
Midsagittal section demonstrating the right half of the body. The diaphragm is in an intermediate position at end-expiration.

a The apertures in the diaphragm are depicted at vertical positions corresponding to the following landmarks in the lower thoracic spine: caval opening = T 8 vertebral body, esophageal hiatus = T 10 vertebral body, aortic hiatus = T 12 vertebral body.
b The diaphragmatic apertures and the structures that they transmit (see p. 213).

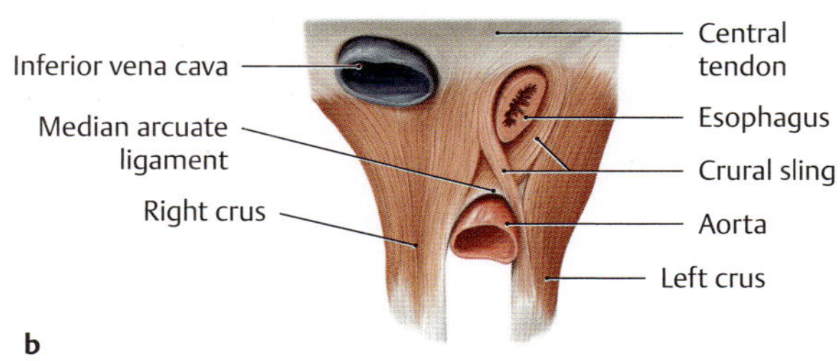

D Position and shape of the diaphragm, anterior view
Coronal section with the diaphragm in an intermediate position.

a The apertures are located in the region of the central tendon (inferior vena cava) and in the lumbar part of the diaphragm (esophageal and aortic hiatus).
b Enlarged view of the diaphragmatic apertures, with vessels transected. The caval opening is located to the right of the midline, the esophageal and aortic hiatus to the left.

In a diaphragmatic hernia (diaphragmatic rupture), abdominal viscera prolapse into the chest cavity through a congenital or acquired area of weakness in the diaphragm. By far the most common herniation site is the esophageal hiatus, accounting for 90% of cases. Typically, the distal end of the esophagus and the gastric cardia (gastric inlet) "slide" upward through the esophageal hiatus into the chest (axial hiatal hernia or sliding hernia; approximately 85% of all hiatal hernias). Typical symptoms are acid reflux, heartburn, and a feeling of retrosternal pressure after meals. More severe cases may present with nausea, vomiting, and functional cardiac complaints.

3.6 The Anterolateral and Anterior Abdominal Wall Muscles*

* The posterior (deep) abdominal wall muscles, most notably the psoas major, are actually hip muscles in a functional sense because they act predominantly on the hip joint. For this reason they are described in the Lower Limb unit (see p. 492).

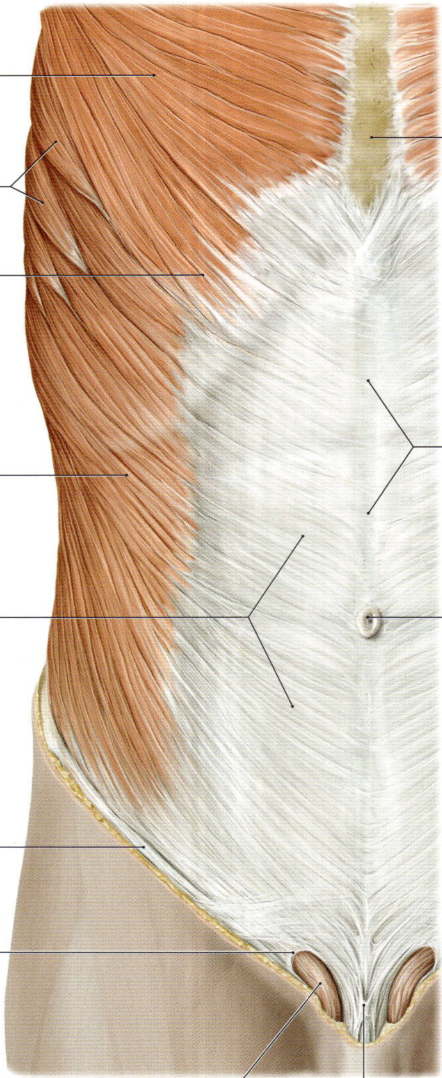

a

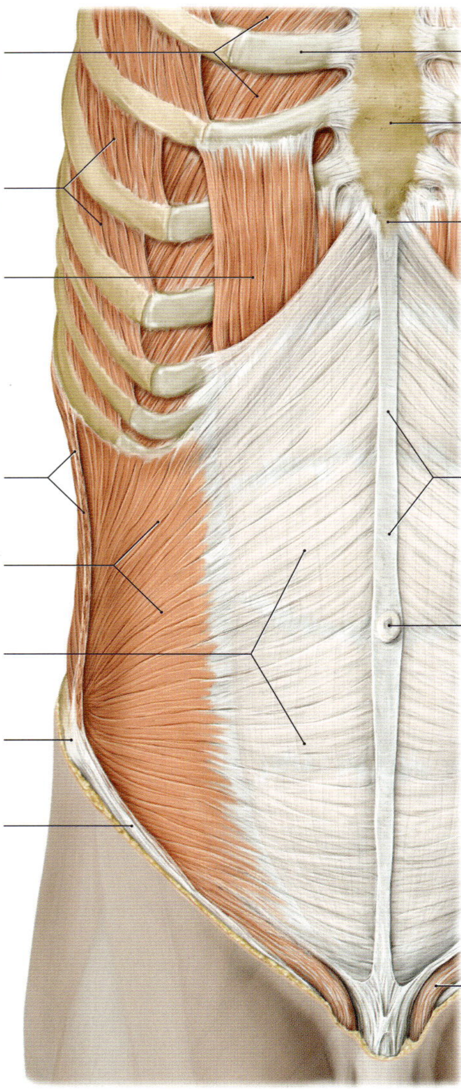

b

A Anterolateral abdominal wall muscles in the male

Right side, anterior view. The oblique muscles of the abdominal wall consist of the external and internal oblique and the transversus abdominis.

a The external oblique aponeurosis borders the superficial inguinal ring, its inferior margin forming the inguinal ligament.

b The external oblique, pectoralis major, and serratus anterior muscles have been removed. The inferior border of the internal oblique forms the roof of the inguinal canal (see p. 219) and is continued onto the spermatic cord in the male as the cremaster muscle and cremasteric fascia.

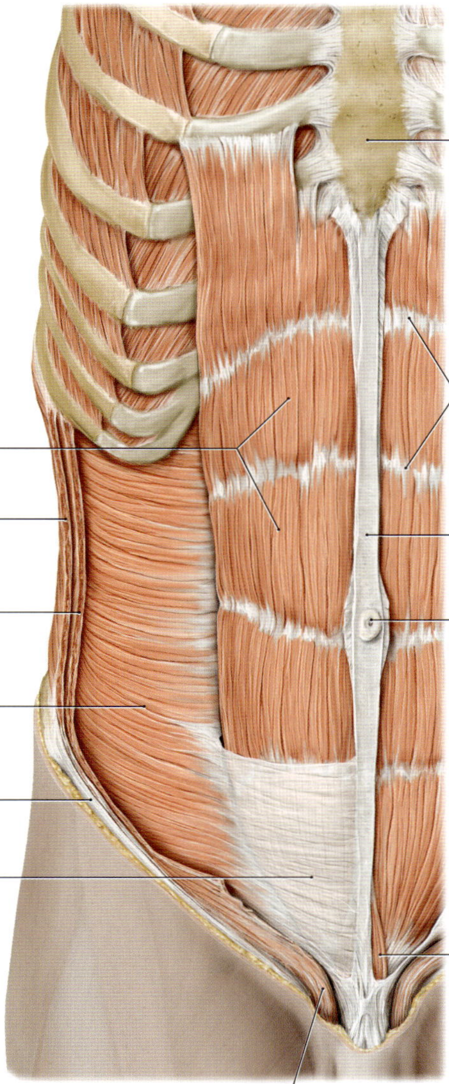

B Anterior (straight) abdominal wall muscles in the male

Right side, anterior view. The straight, or strap, muscles of the abdominal wall include the rectus abdominis and pyramidalis.

a The internal oblique has been removed.
b In addition, the upper portion of the rectus abdominis has been removed.

Note: Below the arcuate line, the transversus abdominis aponeurosis and the inner layer of internal oblique aponeurosis lie anterior to the rectus abdominis (see also p. 187).

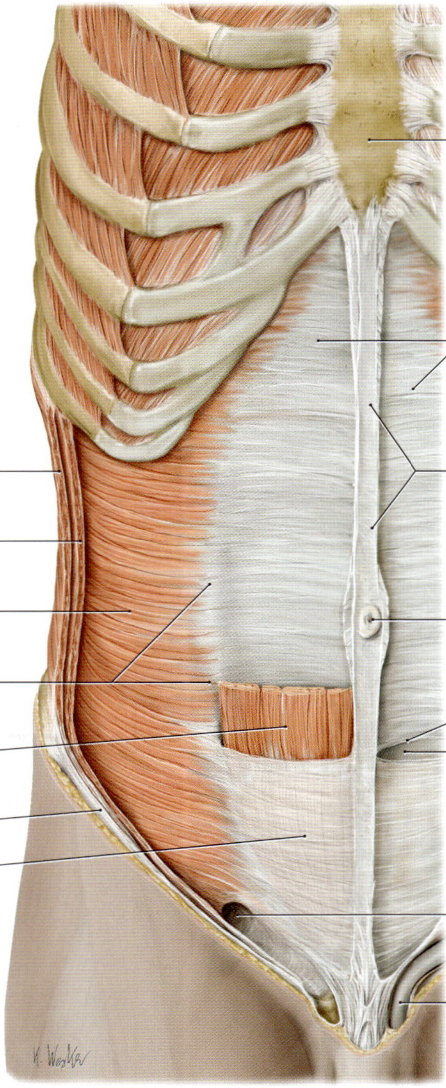

3.7 Structure of the Abdominal Wall and Rectus Sheath

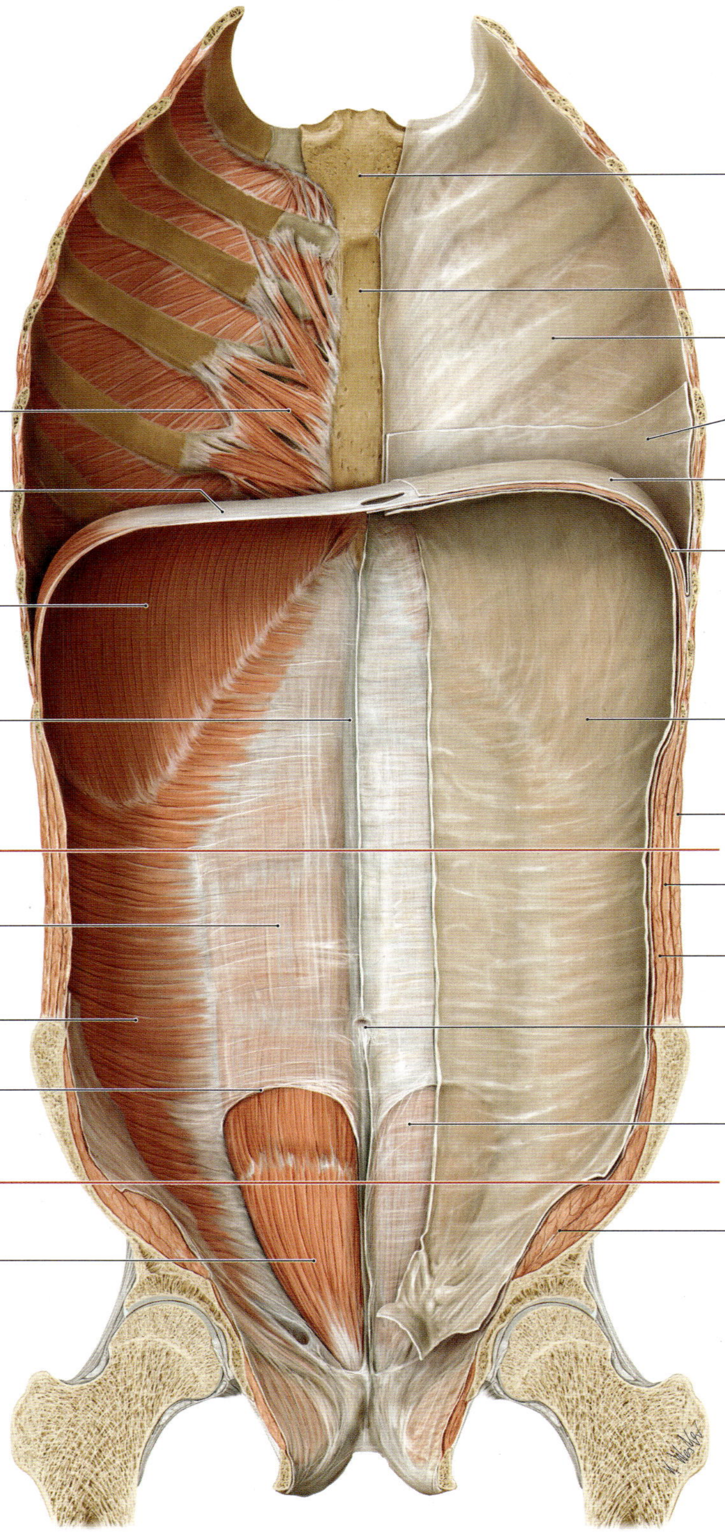

A Overview of the abdominal wall and rectus sheath
Posterior view with the viscera removed. To show how the diaphragm separates the thoracic cavity from the abdominal cavity, the transversalis fascia and parietal peritoneum have been removed from the left abdominal wall, while the endothoracic fascia and parietal pleura have been removed from the left chest wall. The rectus sheath (enclosing the rectus abdominis muscle) plays a special role in the abdominal wall, because its structure changes below the arcuate line to accommodate the increasing pressure of the viscera against the body wall (see **C**). The rectus sheath is formed by the aponeuroses of the lateral abdominal muscles (of which only the transversus is visible here; the others are hidden) and is divided into an anterior and a posterior layer.

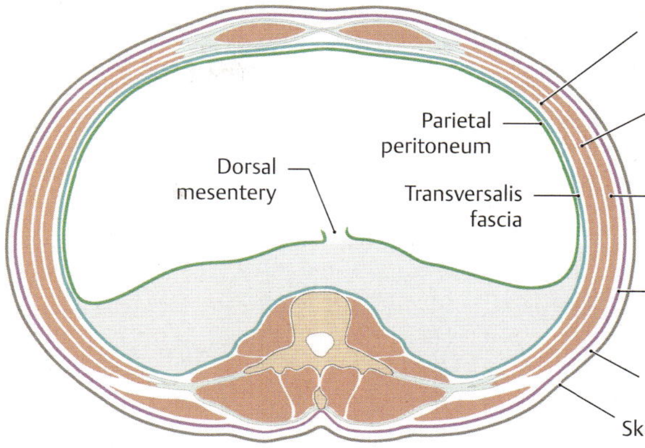

B Structure of the abdominal wall

Cross section through the abdominal wall above the umbilicus, superior view. The following layers are distinguished in the lateral abdominal wall, from deep to superficial:

- Parietal peritoneum
- Transversalis fascia
- Transversus abdominis muscle
- Internal oblique muscle
- External oblique muscle
- Superficial abdominal fascia
- Subcutaneous tissue and skin

* *Note:* The previous anatomic nomenclature divided the superficial fascia of the anterior abdominal wall into a superficial layer and deep layer:

- The superficial layer is the subcutaneous fascia of Camper, named after Peter Camper, 1722–1789, professor of anatomy, surgery, and medicine in Amsterdam.
- The deep layer (the actual superficial body fascia) is the superficial abdominal fascia of Scarpa, named after Antonio Scarpa, 1747–1832, professor of anatomy and surgery in Modena.

The current anatomic nomenclature identifies these fascia structures as the fatty layer and the membranous layer of the abdomen. However, these terms are not yet commonly used, especially in a clinical setting.

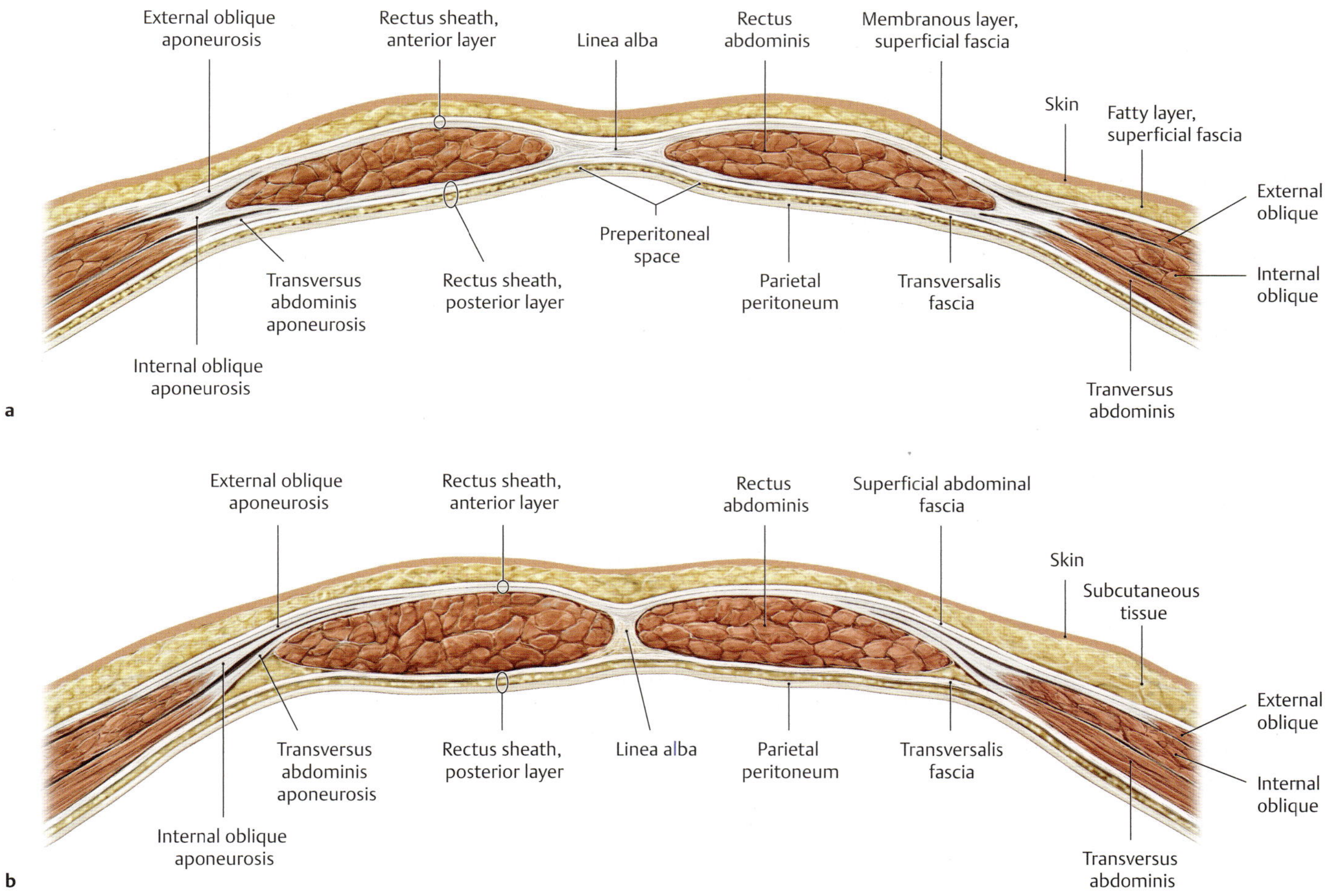

C Structure of the rectus sheath

Cross sections through the rectus sheath superior (**a**) and inferior (**b**) to the arcuate line, superior view. The aponeuroses of the lateral abdominal wall muscles ensheath the anterior strap muscles to form the rectus sheath. This creates a muscular compartment consisting of an anterior layer and a posterior layer. While the aponeuroses of the three lateral abdominal muscles contribute equally to the anterior and posterior layers of the sheath above the umbilicus, the two layers blend together approximately 3 to 5 cm below the umbilicus (at the level of the arcuate line) to form a single (and consequently more stable) sheet that passes in front of the rectus abdominis muscle. *Below the arcuate line*, the posterior layer of the rectus sheath is absent. The preperitoneal space, a fine layer of loose connective tissue and fat (see also p. 224), is located between transversalis fascia and parietal peritoneum.

3.8 The Pelvic Floor Muscles: Overview of the Perineal Region and Superficial Fasciae

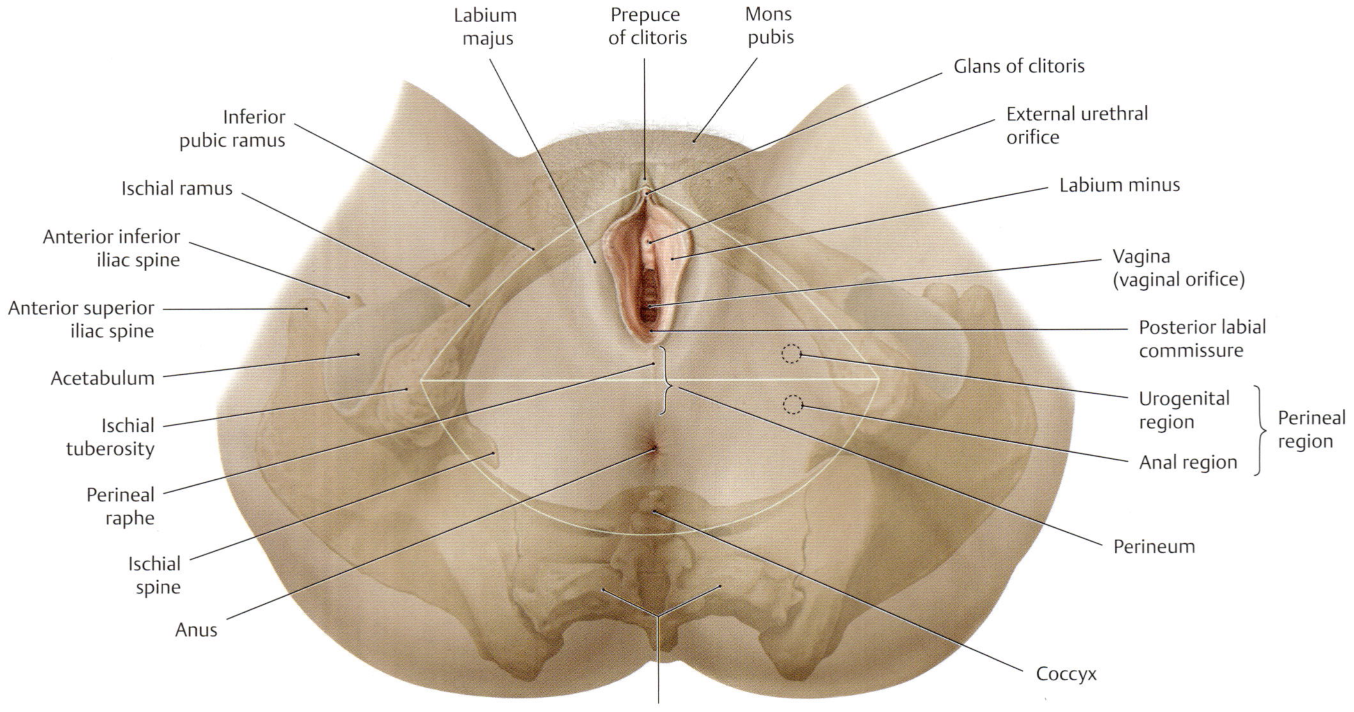

a

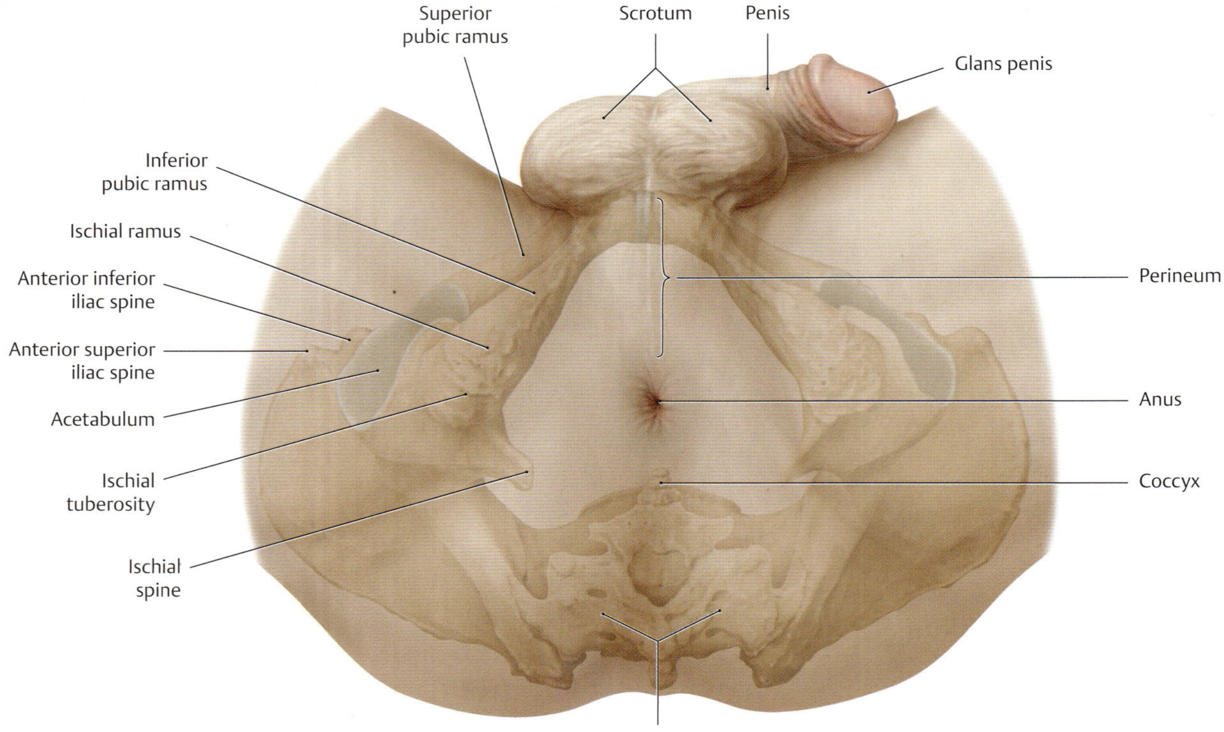

b

A Perineal region in the female (a) and male (b)
Lithotomy position, inferior view. The perineal region in both sexes consists of the urogenital region anteriorly and the adjacent anal region posteriorly. The two regions are separated by a line that runs between the ischial tuberosities. The perineum denotes the soft-tissue area between the thighs and the buttocks. In the female, obstetricians refer to the perineum as the region from the anterior margin of the anus to the posterior commissure of the vagina. It is considerably longer in the male, extending from the anal margin to the root of the scrotum. The perineum is a region containing fibrous and fatty tissue, and the perineal body (see **Ba**) is a fibromuscular mass, also referred to as the central tendon of the perineum. A more detailed structural description of the perineum is given on p. 192.

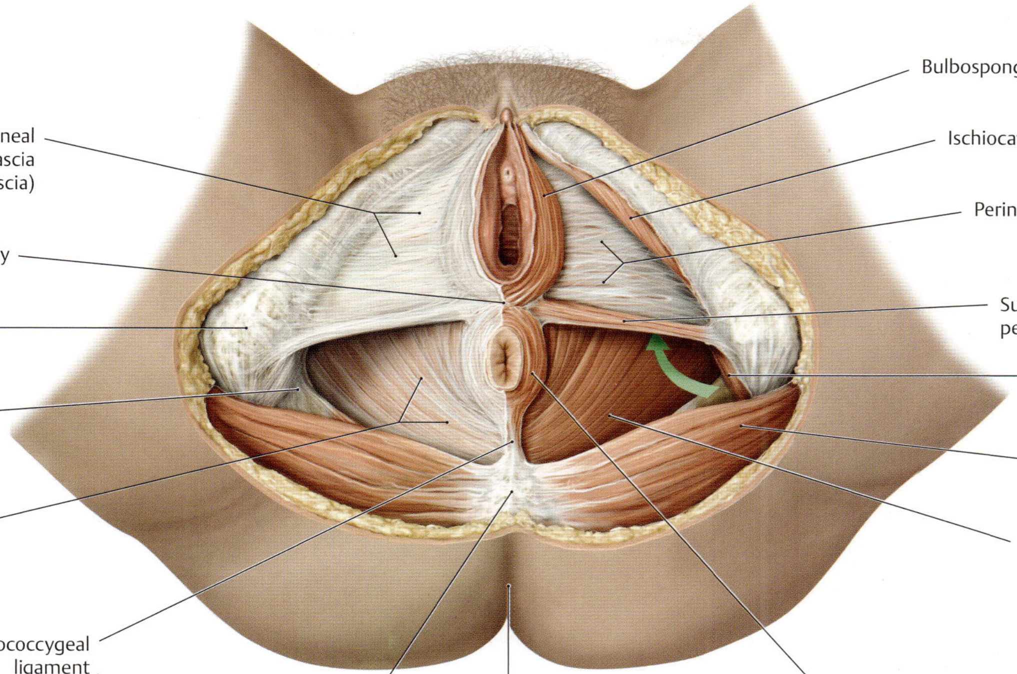

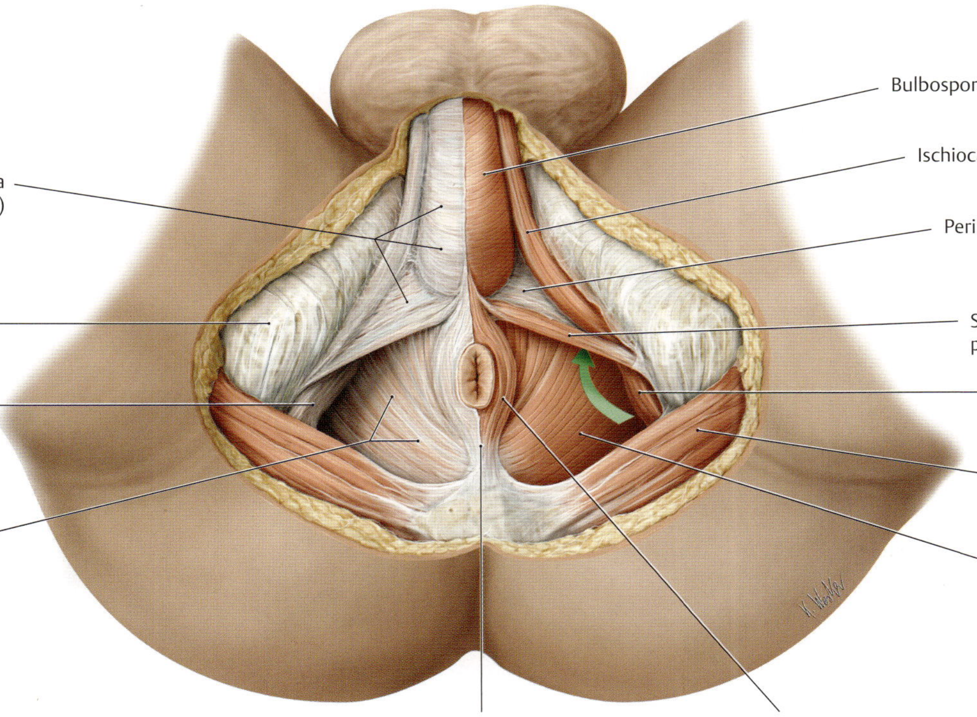

B Superficial fasciae of the female (a) and male pelvic floor (b)
Lithotomy position, inferior view. The superficial perineal fascia (urogenital region) and the inferior pelvic diaphragmatic fascia (anal region) are intact on the right side but have been removed on the left side. In this way the superficial perineal space has been opened in the urogenital region on the left side of each dissection, and the levator ani muscle has been exposed in the anal region. The superficial perineal space is bounded posteriorly by the superficial transverse perineal muscle and superiorly by the perineal membrane. The bulbospongiosus muscle is located in the medial part of this superficial perineal space; the ischiocavernosusis muscle is located laterally. The green arrow in each figure points to the anterior recess of the left ischioanal fossa (see also p. 574).

3.9 Structure of the Pelvic Floor and Pelvic Spaces: Female versus Male

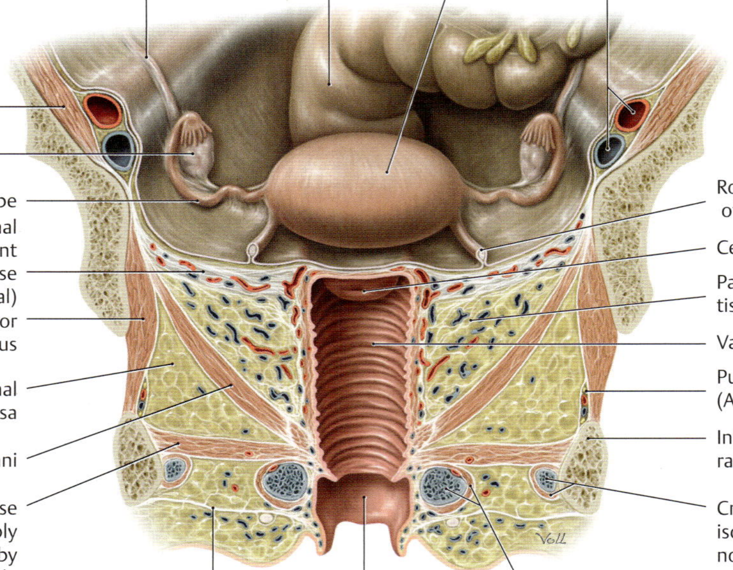

A Coronal section through the female pelvis
Anterior view.

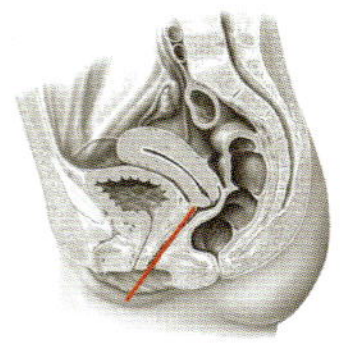

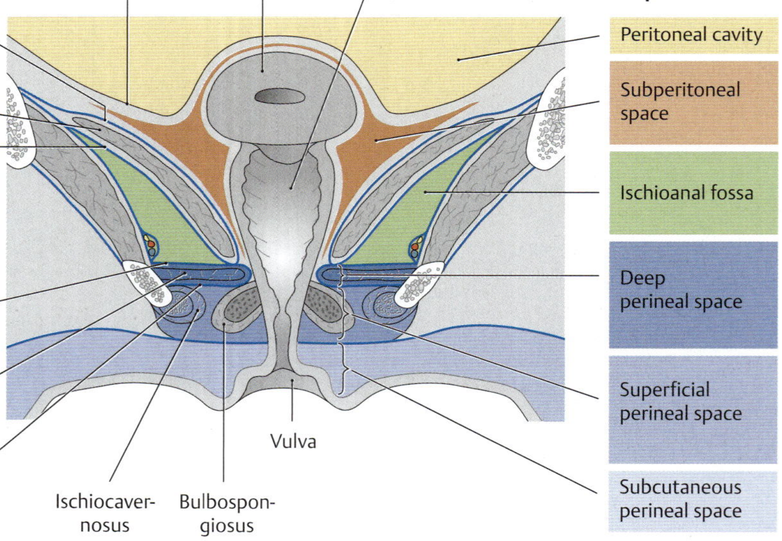

B Pelvic spaces, fasciae, and arrangement of the pelvic floor muscles in a female pelvis
Coronal section at the level of the vagina (see small diagram above for the exact location of the plane). The levels of the pelvis are shown in different colors.

C Subdivisions of the pelvis and structure of the pelvic floor (in both sexes)

Subdivisions of the pelvis

The pelvis is the portion of the abdominal cavity located in the lesser pelvis. It is surrounded by the skeleton of the lesser pelvis, which meets the greater pelvis at the *linea terminalis* (see p. 149). The pelvic cavity lies above the pelvic floor (supralevator) and is subdivided into upper and lower levels by the peritoneum:

- **Upper level:** peritoneal cavity of the lesser pelvis
- **Lower level:** subperitoneal space

The perineum lies below the pelvic floor (infralevator) and is separated into

- Deep perineal space with ischioanal fossa
- Superficial perineal space
- Subcutaneous perineal space

Subjacent to the pelvis are three more spaces that are considered to be separate from the pelvis itself: the *deep perineal space*, the *superficial perineal space*, and the *subcutaneous perineal space* (see **B** and **F**).

Structure of the pelvic floor

The three muscular and connective-tissue sheets that contribute to the structure of the pelvic floor are arranged in three levels:

- **Upper level:** pelvic diaphragm
- **Middle level:** deep urogenital muscles
- **Lower level:** sphincters and erectile muscles of the urogenital and intestinal tract

The funnel-shaped pelvic diaphragm is formed by the levator ani and coccygeus muscles and the superior and inferior fasciae of the pelvic diaphragm. The deep urogenital muscles (see p. 192) lie in the **deep perineal pouch**. They include the deep transverse perineal muscle in the male, a possible sheet of smooth muscle in the female, and the inferior portion of the external urethral sphincter. The perineal membrane separates the deep from the **superficial perineal pouch**, the muscles of which are the superficial transverse perineal, bulbospongiosus, ischiocavernosus, and the external anal sphincter more posteriorly.

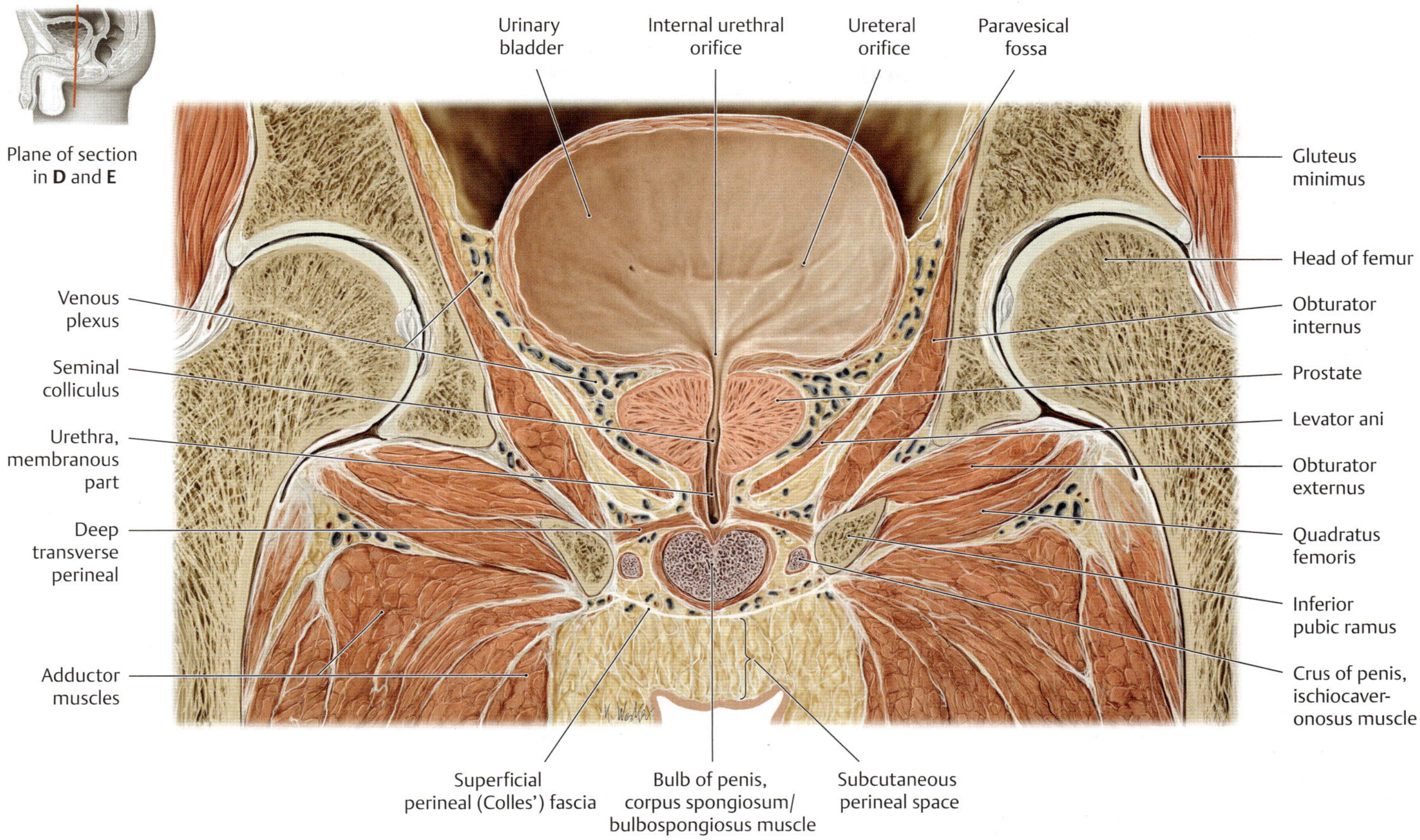

D Coronal section through the male pelvis
Ansicht von ventral.

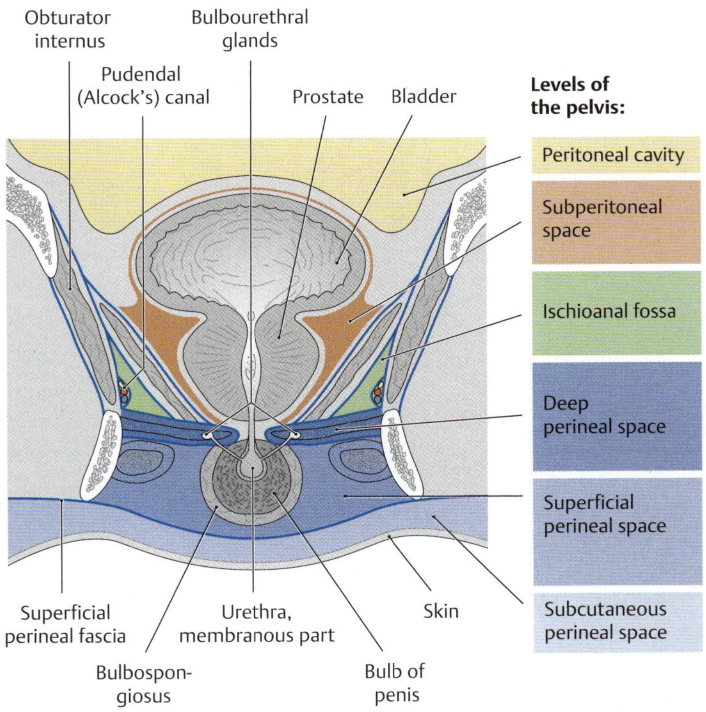

E Pelvic spaces, fasciae, and arrangement of the pelvic floor muscles in a male pelvis
Coronal section at the level of the prostate (see small diagram above for the exact location of the plane). Different colors indicate the various levels of the pelvis and the perineal spaces.

F Boundaries and contents of the deep perineal space and superficial perineal space (Colles' space) in the male [and female]

Deep perineal space

- Boundaries:
  - Perineal membrane
  - Inferior fascia of pelvic diaphragm
  - Inferior portion of inner fascia of obturator internus
- Contents:
  - Deep transverse perineal muscle
  - External urethral sphincter
  - Compressor urethra muscles (in female)
  - Anterior extension of ischioanal fat pad
  - Intermediate (membranous) portion of urethra (in male)
  - Bulbourethral glands (in male)
  - Branches of related neurovascular structures

Superficial perineal space

- Boundaries:
  - Superficial perineal (Colles') fascia
  - Perineal membrane
  - Ischiopubic rami
- Contents:
  - Superficial transverse perineal muscle
  - Bulbospongiosus muscles
  - Ischiocavernosus muscles
  - Root of penis and proximal portion of spongy urethra (in males)
  - Clitoris, bulb of vestibule, and greater vestibular glands (in females)
  - Branches of related neurovascular structures

The subcutaneous perineal space is located between the superficial perineal fascia and skin and contains mostly fatty tissue.

3.10 The Muscles of the Female Pelvic Floor and Wall

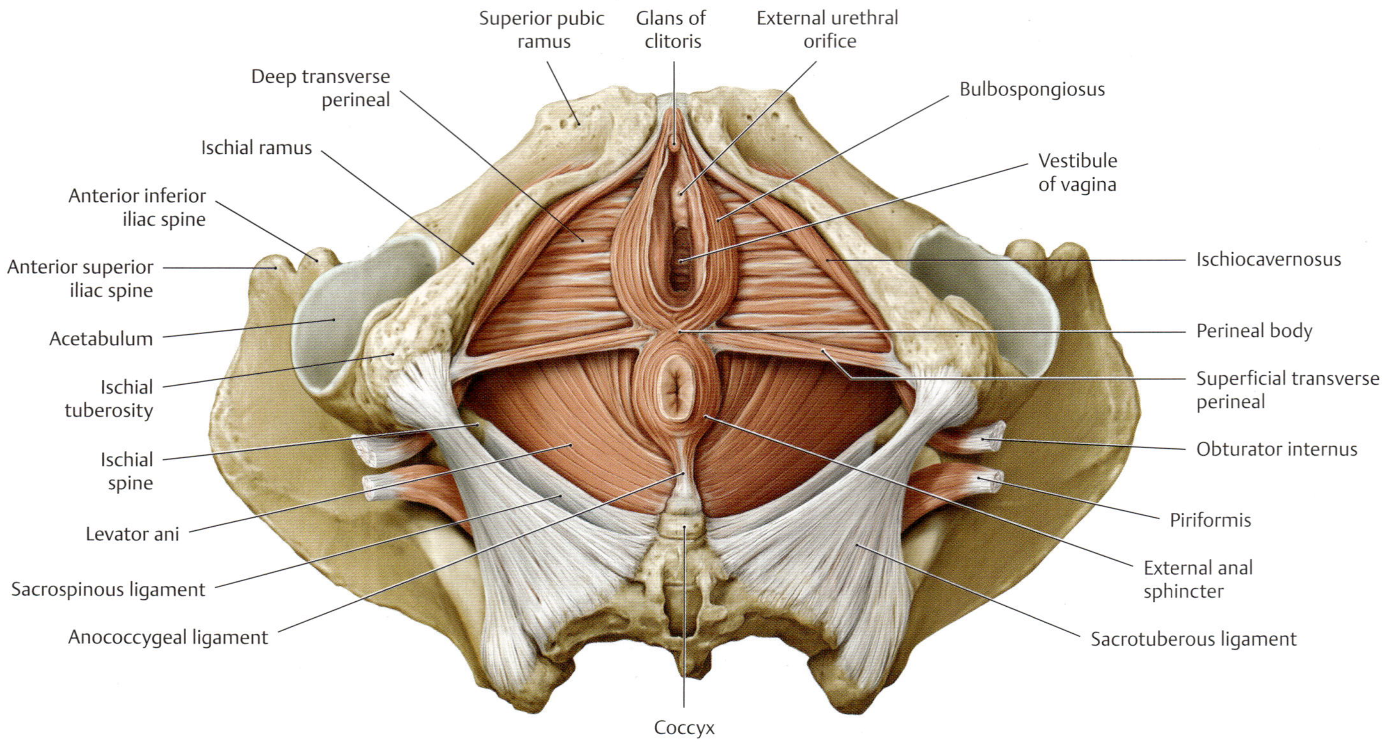

A Muscles of the pelvic floor with the fasciae removed
Female pelvis, inferior view. The muscular layers are progressively removed from **B** through **D** to demonstrate the underlying muscles from a consistent perspective. The levator ani muscle is described more fully on pp. 194–195.

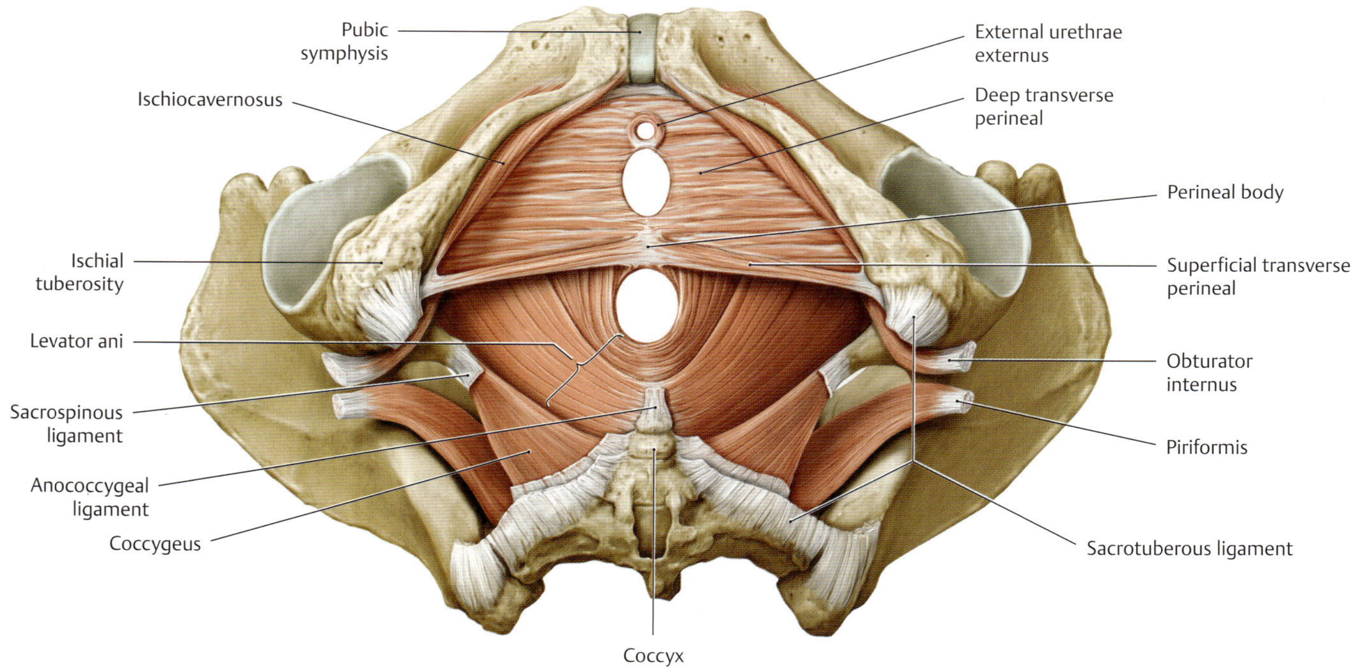

B Muscles of the pelvic floor with the sphincters removed
Female pelvis, inferior view. The sphincter muscles have been removed from the urogenital and intestinal tracts (= bulbospongiosus and external anal sphincter), leaving the external urethral sphincter intact.

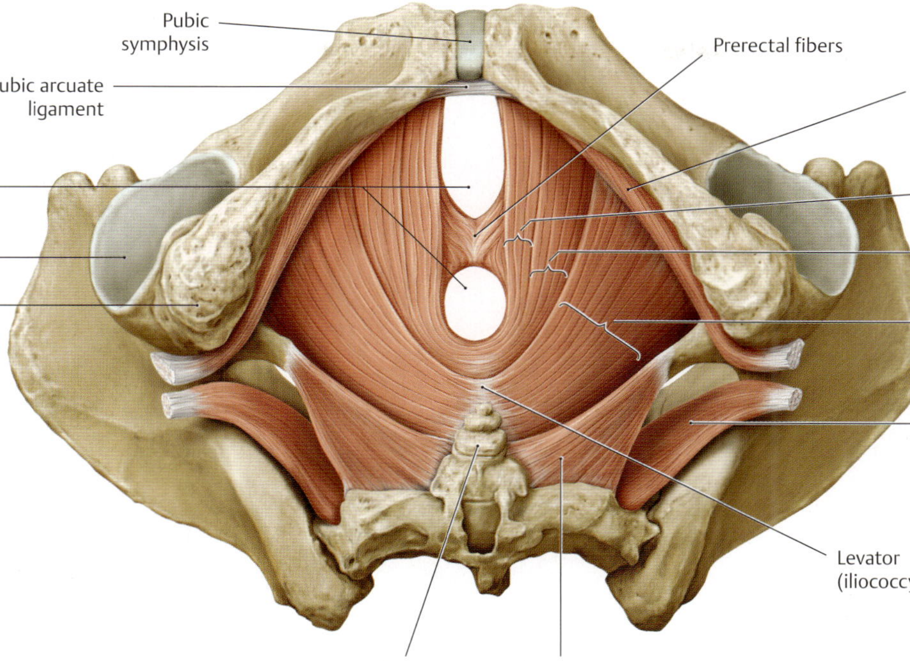

C Muscles of the pelvic floor with the urogenital muscles removed
Female pelvis, inferior view. The urogenital muscles (= superficial and deep transverse perineal and the ischiocavernosi) have been removed. *Note* the opening of the urogenital hiatus, which is bounded by the two crura of the puborectalis muscle (the levator crura).

Note also the prerectal fibers that have split off from the puborectalis muscle. The prerectal fibers are interwoven with connective-tissue fibers and smooth muscle to form the fibromuscular framework of the perineum (see also p. 188).

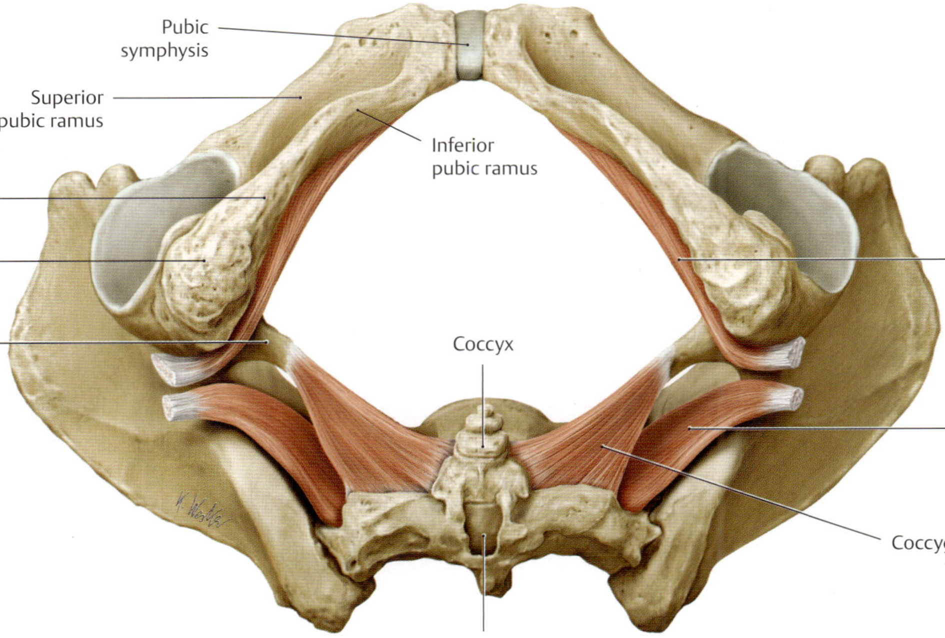

D Muscles of the pelvic wall (parietal muscles of the lesser pelvis)
Female pelvis, inferior view. All of the pelvic floor muscles have been removed, leaving the parietal muscles intact (obturator internus, coccygeus, and piriformis). Along with the skeleton of the lesser pelvis, these muscles contribute structurally to the pelvic wall and assist in closing the posterior pelvic outlet. The obturator internus and its fascia provide a tendon of origin for the iliococcygeus, which is part of the levator ani complex (tendinous arch of the levator ani, see p. 195).

3.11 Pelvic Floor Muscles: The Levator Ani

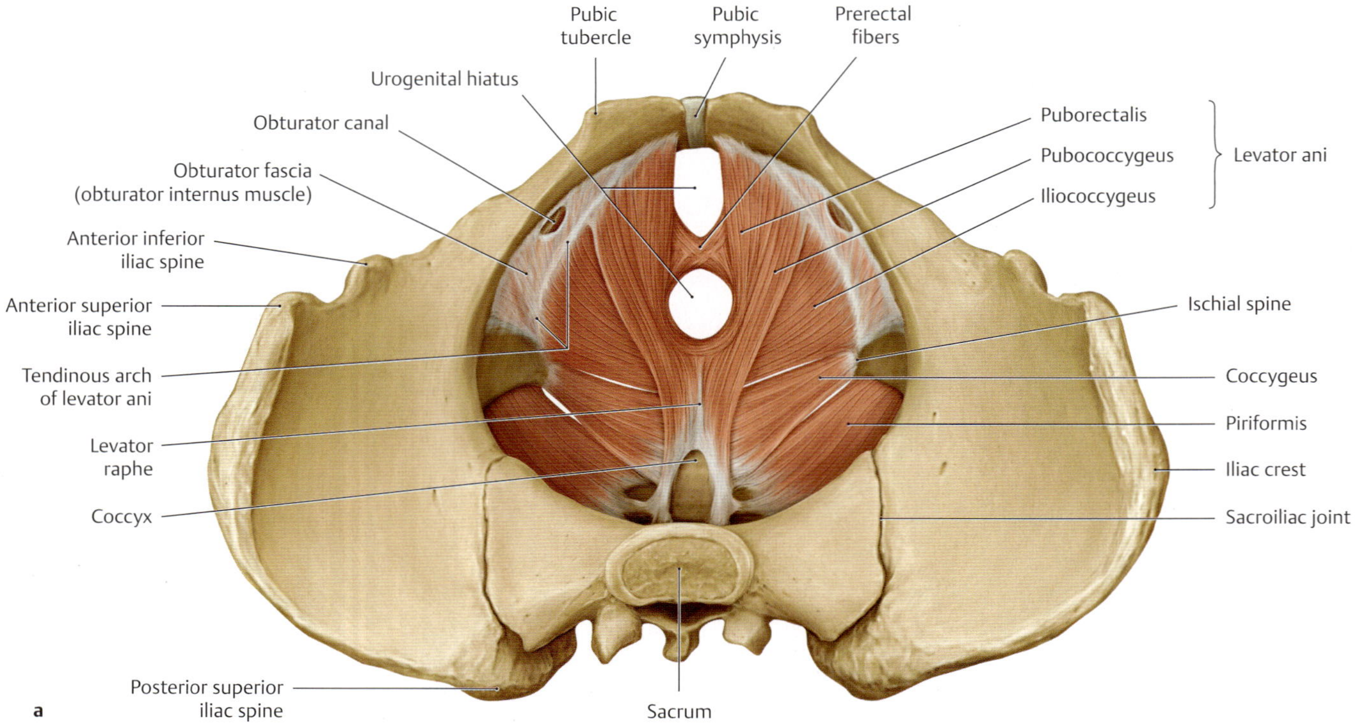

a

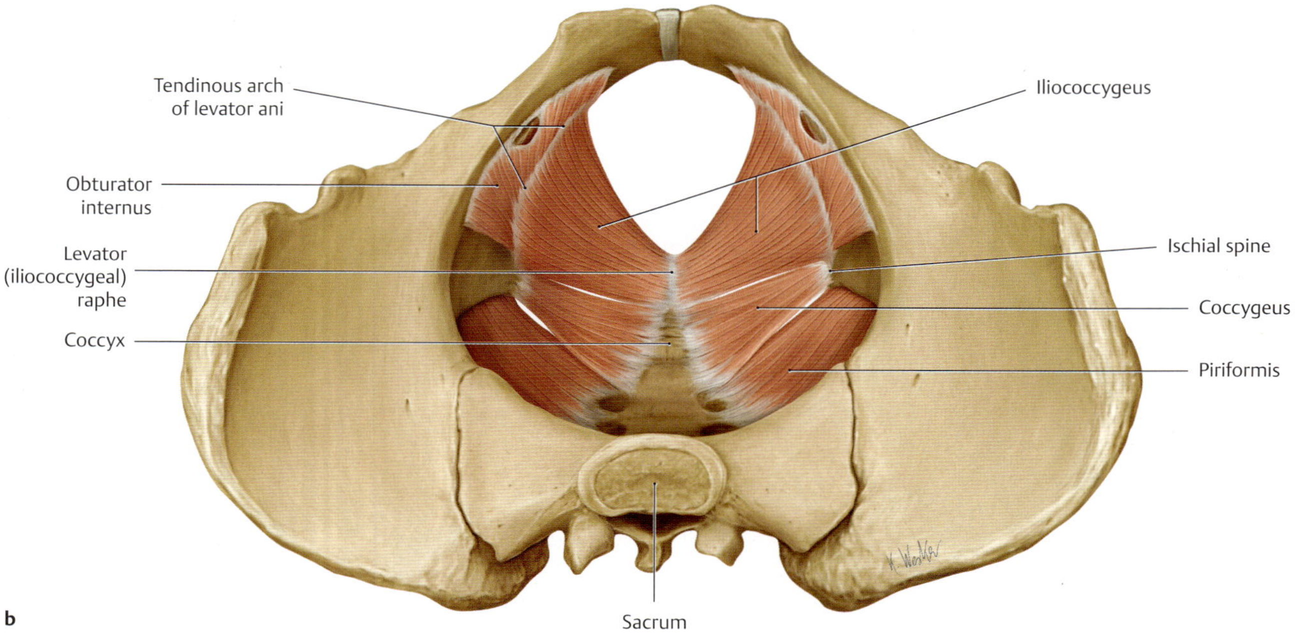

b

A Parts of the levator ani and the parietal muscles of the pelvic wall

Female pelvis, superior view.

a The levator ani muscle consists of three parts: the puborectalis, pubococcygeus, and iliococcygeus. It arises from the anterior and lateral pelvic wall on a line running from the center of the pubic symphysis to the ischial spine (= tendinous arch of the levator ani). As a contributor to continence, the puborectalis muscle assists the external anal sphincter (not visible here) in keeping the anus closed. It arises from the superior pubic ramus on both sides of the pubic symphysis and runs in the shape of a loop around the rectum, where the fibers are connected to the deep part of the external anal sphincter. It has the shape of an arched gateway, with its two crura (the levator crura) forming the boundaries of the urogenital hiatus. In contraction of the puborectal sling, the rectal wall is pulled forward toward the pubic bone, which increases the so-called anorectal angle (angle between rectum and anal canal) and tightens the anal canal by squeezing it.

b The puborectalis and pubococcygeus muscles have been removed. The coccygeus (muscle fibers on the sacrospinous ligament) and the piriformis complete the pelvic outlet posteriorly on both sides of the sacrum.

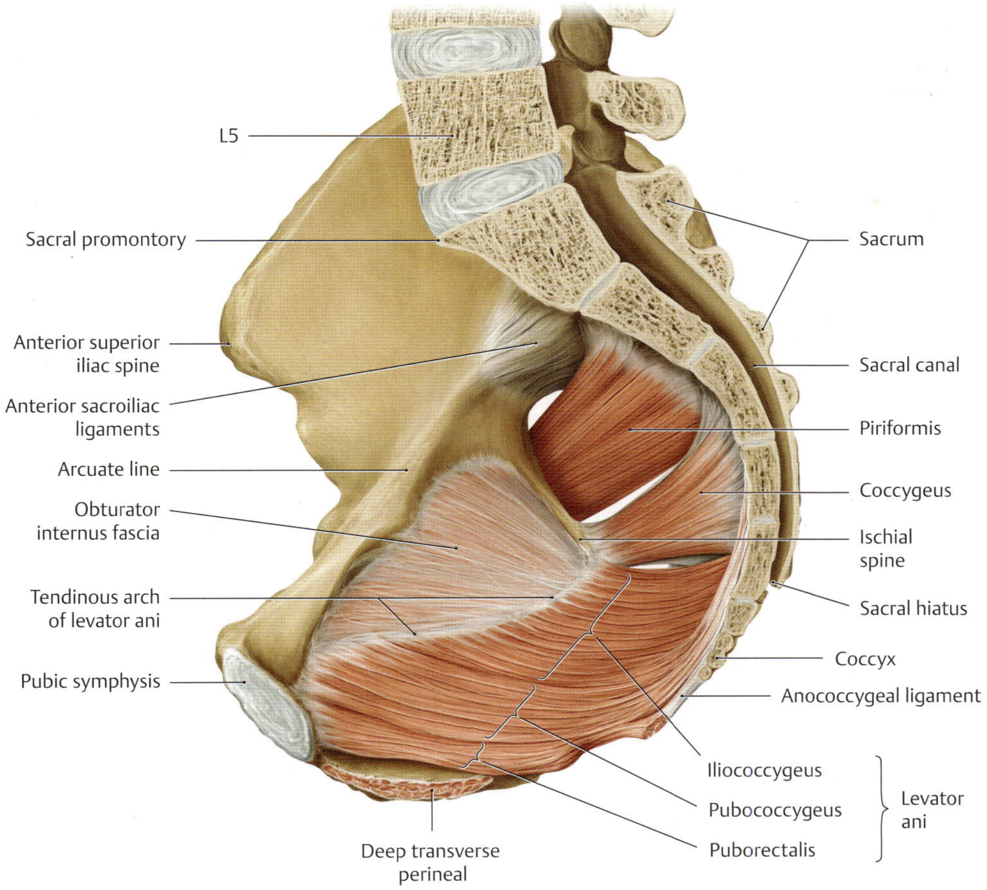

B Tendinous arch of the levator ani
Right hemipelvis, medial view. The tendinous arch of the levator ani is a thickening of the obturator internus fascia that serves as the origin of the iliococcygeus muscle.

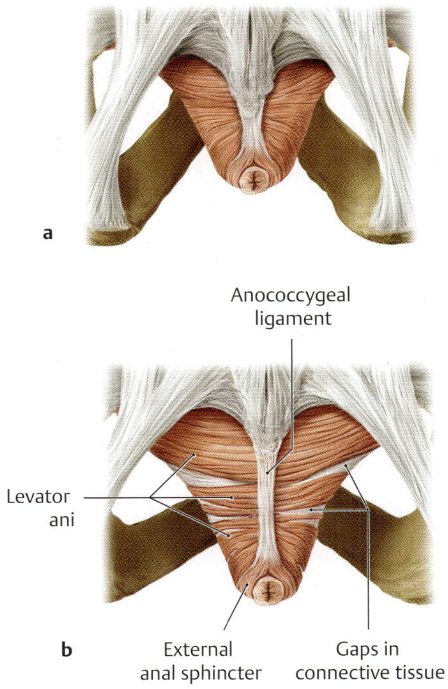

D Gender-related differences in structure of the levator ani (**a** male, **b** female)
Posterior view.
Note the gaps in the connective tissue between the muscular parts of the levator ani in female (**b**).

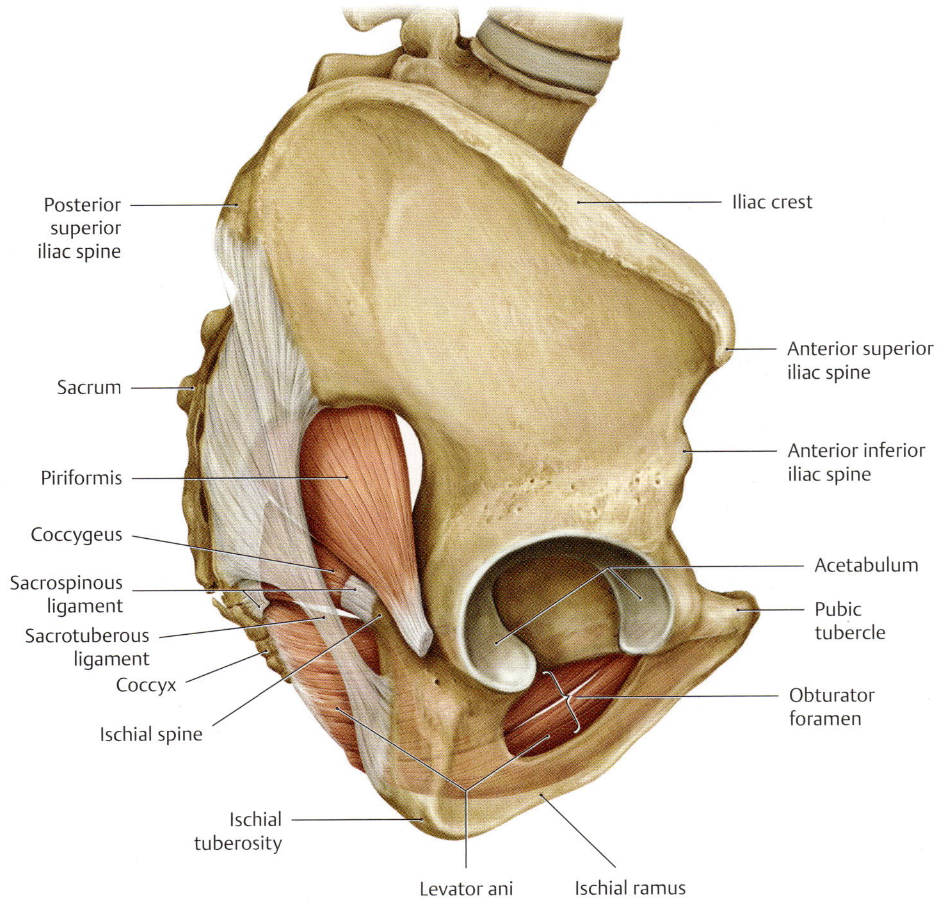

C Funnel shape of the levator ani
Pelvis viewed from the right side. A portion of the ischium is shown transparent. A contraction of the funnel-shaped levator ani leads to lifting of the anus (levatore = to rise) and to forward motion caused by extension of the puborectalis. At its neutral position, the levator funnel stands at a steep angle but flattens when contracted. Because the muscle has a cone or funnel shape, during defecation, only the peripheral parts of the levator ani contract; the four areas close to the sphincter, including the puborectalis, go limp so that the anus descends, and the stool is pushed through the opened anal canal.

3.12 Pelvic Floor Muscles: Their Relation to Organs and Vessels in Males and Females

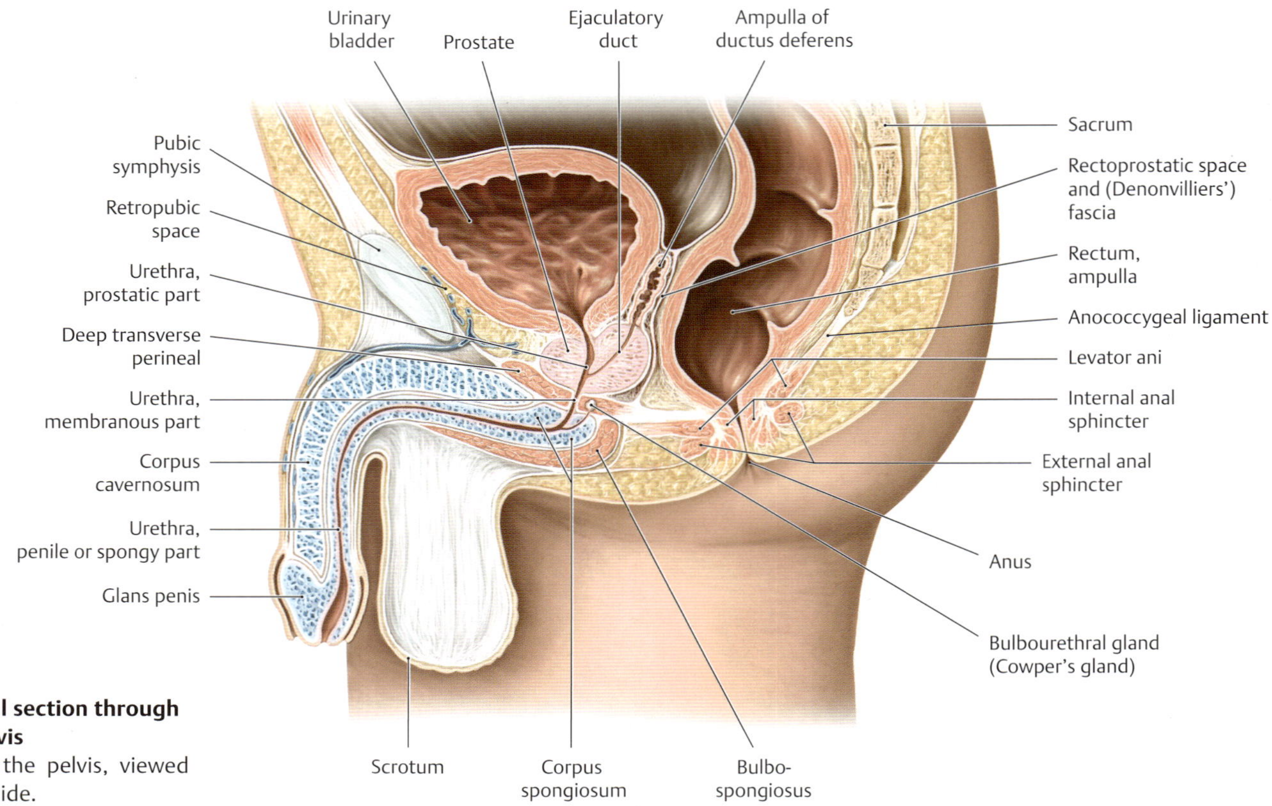

A Midsagittal section through a male pelvis
Right half of the pelvis, viewed from the left side.

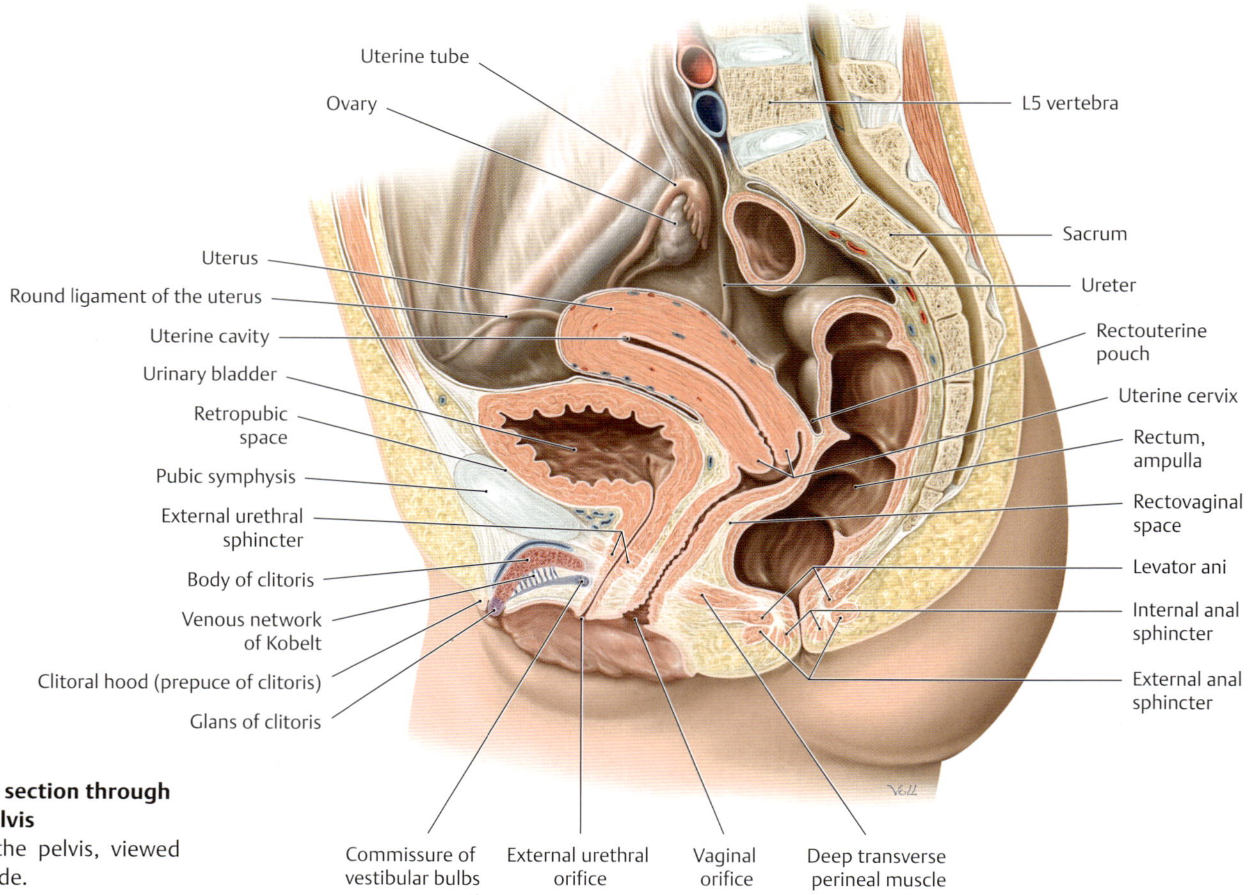

B Midsagittal section through a female pelvis
Right half of the pelvis, viewed from the left side.

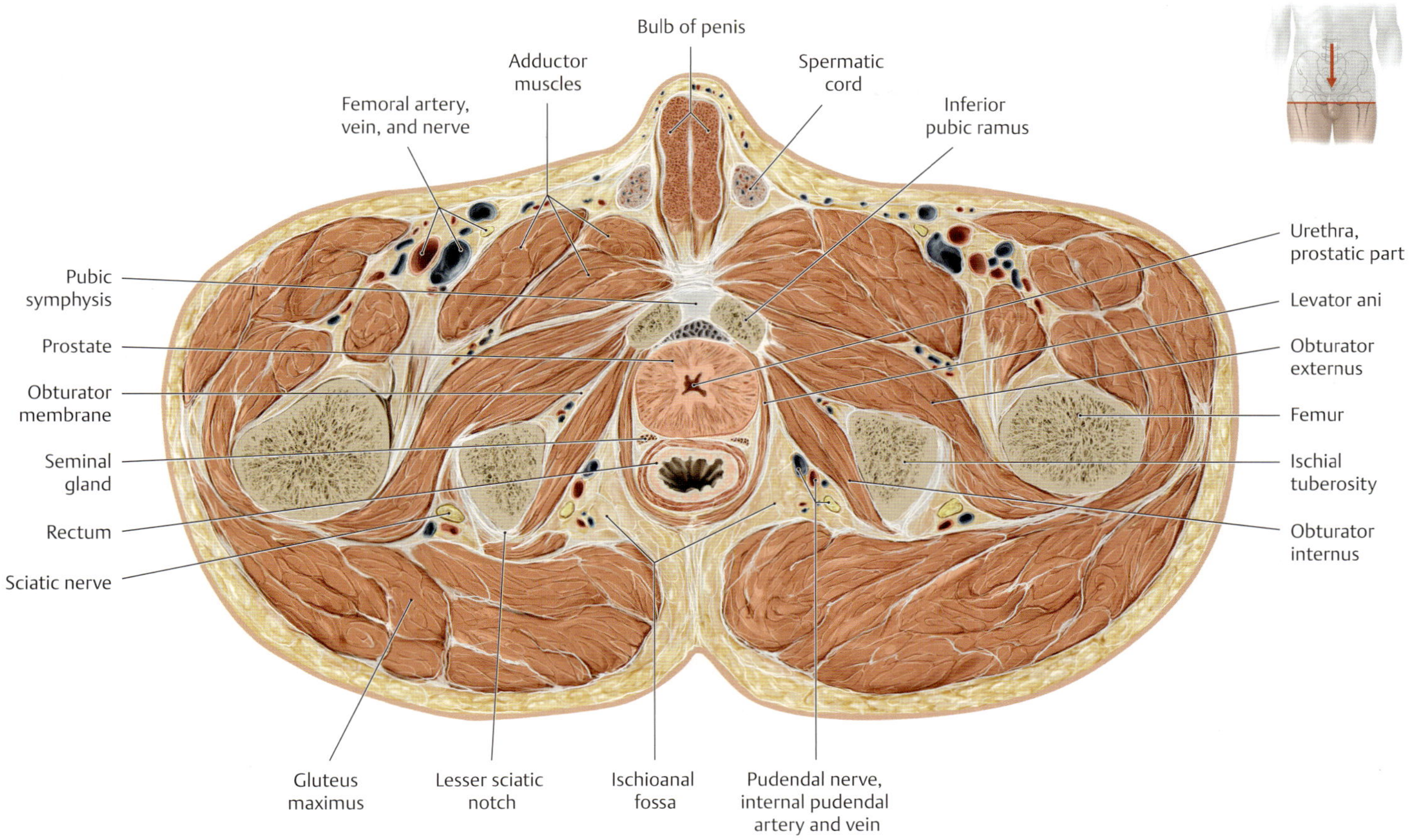

C Cross section through a male pelvis
Superior view.

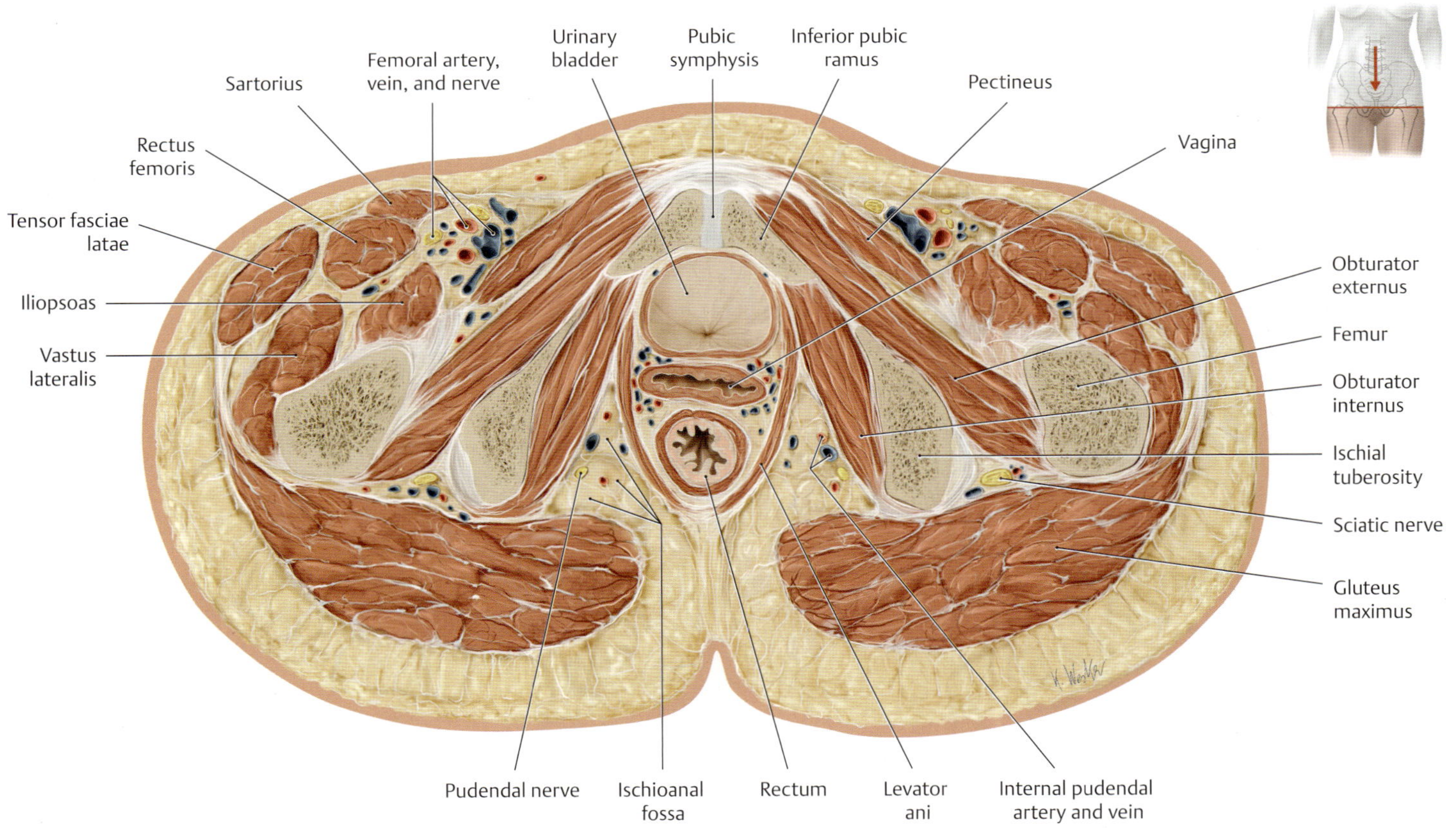

D Cross section through a female pelvis
Superior view.

4.1 The Arteries

A Overview of the arteries of the trunk wall

The arrangement of the neurovascular structures in the trunk reflects the segmental anatomy of the trunk wall, particularly in the thoracic region. Accordingly, each of the intercostal spaces is traversed by an intercostal artery, vein, and nerve. All *arteries* supplying the trunk wall ultimately come from the aorta.

The chest wall is supplied principally by the posterior intercostal arteries, which arise from the aorta, and by anterior intercostal branches arising from the internal thoracic artery:

- The first and second posterior intercostal arteries, which are given off by the superior intercostal artery (= branch of the costocervical trunk, see **Da**)
- The third through eleventh posterior intercostal arteries (each giving off a dorsal, a collateral, and a lateral cutaneous branch, see **Db**)
- The musculophrenic artery (one of the two terminal branches of the internal thoracic artery), which runs behind the costal arch, see **B**
- The subcostal artery (twelfth intercostal artery), see **B**
- The anterior intercostal arteries, which arise from the internal thoracic artery, see **B**

Many other "regional" arteries supply the anterior, lateral, and posterior trunk wall.

Anterior trunk wall

- The perforating branches (from the internal thoracic artery, e.g., the medial mammary branches that supply the breast), see **Db**
- The superior epigastric artery (continuation of the internal thoracic artery, see **B** and **C**)
- The inferior epigastric artery (from the external iliac artery, see **B** and **C**)
- The superficial epigastric artery, see **B**
- The superficial circumflex iliac artery, see **B**
- The deep circumflex iliac artery, see **B**

Posterior trunk wall

- Dorsal branches (from the posterior intercostal arteries), each with a medial, a lateral cutaneous, and a spinal branch, see **Dc**
- The first through fourth lumbar arteries (each with a dorsal and spinal branch), see **B**
- The median sacral artery, see **B**

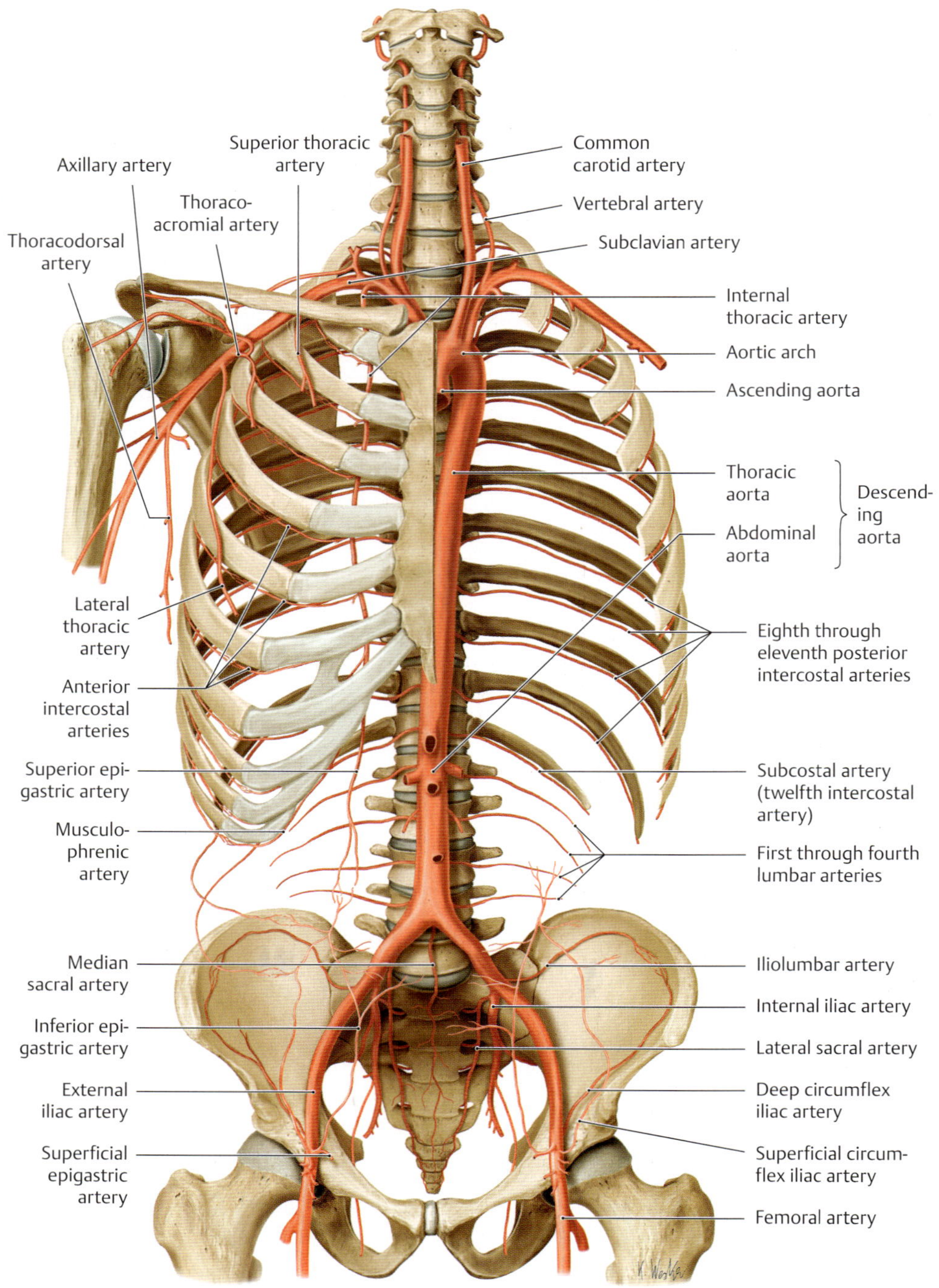

Lateral trunk wall

- The superior thoracic artery, see **B**
- The thoracoacromial artery, see **B**
- The lateral thoracic artery, see **B**
- The lateral cutaneous branches (from the intercostal arteries), which distribute branches mainly to the breast (lateral mammary branches, see **Db**)
- The iliolumbar artery (from the internal iliac artery), which gives off an iliac, a lumbar, and a spinal branch, see **B**

B Arteries of the trunk wall

Anterior view. The anterior portions of the ribs have been removed on the left side.

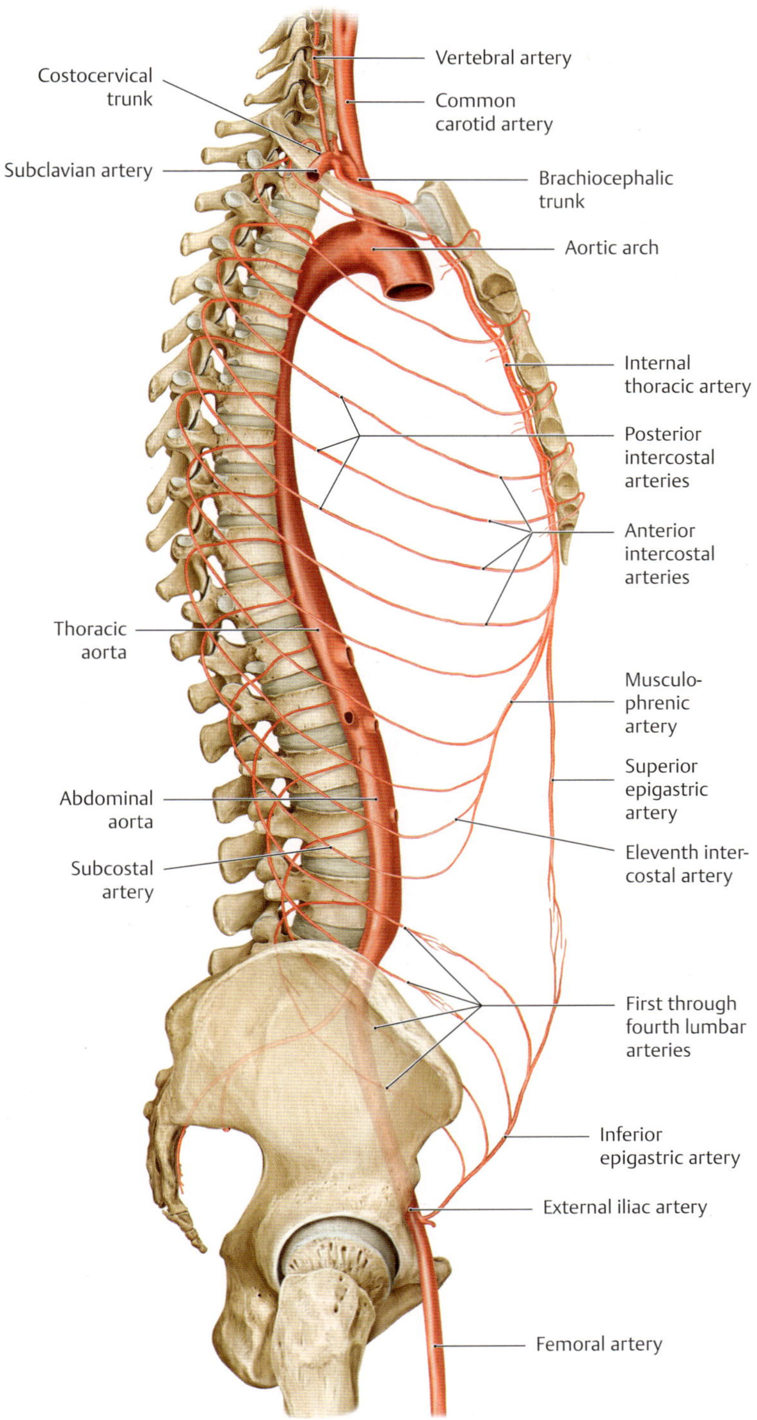

C Arteries of the trunk wall
Right lateral view.

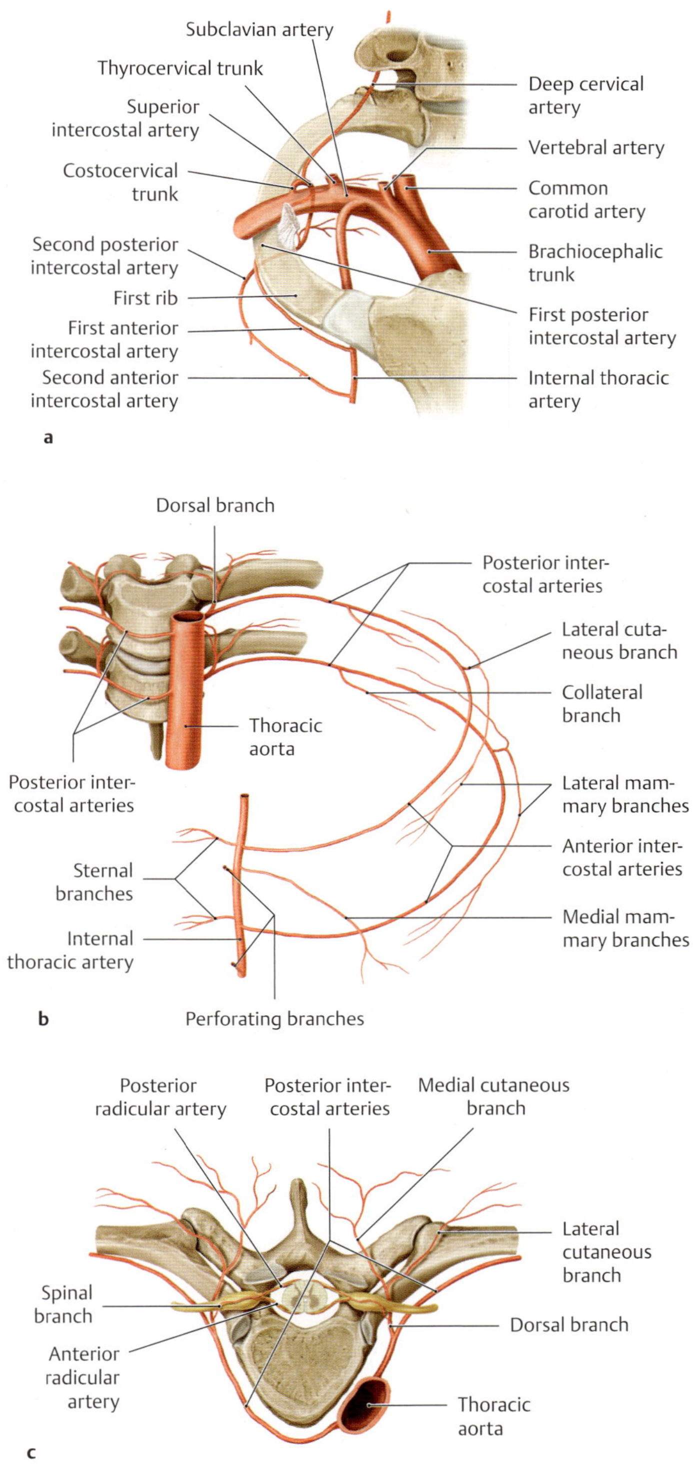

D Course and branches of the intercostal arteries

a Anterior view of the superior intercostal artery, which gives off the first two intercostal arteries.
Note: The first and second posterior intercostal arteries are not branches of the thoracic aorta but arise from the superior intercostal artery (branch of the costocervical trunk), which are branches of the subclavian artery.

b Anterior view of the posterior intercostal arteries that are segmental branches of the thoracic aorta.
Note: The *anterior* intercostal arteries arise from the subclavian artery (via the internal thoracic artery), while the *posterior* intercostal arteries arise directly from the thoracic aorta.

c Branches of the posterior intercostal arteries, viewed from the superior view.

4.2 The Veins

A Overview of the veins of the trunk wall

The veins of the trunk wall drain into both the vena caval and azygos systems (see **B**). Within the vena caval system, we can distinguish between the tributary regions of the inferior vena cava, which ends at the junction of the thorax and abdomen, and the superior vena cava. Connections between the superior and inferior vena cavae are called *cavocaval anastomoses* (collateral channels).

Tributaries of the superior vena cava

- Superior intercostal vein (brachiocephalic vein) (see **B**)
- Anterior intercostal veins (internal thoracic vein, subclavian vein) (see **D**)
- Superior epigastric vein (internal thoracic vein, subclavian vein)
- Lateral thoracic vein (axillary vein) (see **C**)
- Thoracoepigastric vein (axillary vein) (see **C**)

Tributaries of the inferior vena cava (see **B**)

Tributaries	Drains into
• Posterior intercostal veins (only inferior thoracic wall) • Subcostal vein • First through fourth lumbar veins	Ascending lumbar vein
• Iliolumbar vein • Median sacral vein • Lateral sacral vein	Common iliac vein
• Deep circumflex iliac vein • Inferior epigastric vein	External iliac vein
• Obturator vein (see p. 220) • Internal pudendal vein (see p. 238)	Internal iliac vein
• External pudendal veins • Superficial circumflex iliac vein • Superficial epigastric vein	Femoral vein

Tributaries of the azygos vein (see **B**)

- Superior intercostal veins (the majority)
- Posterior intercostal veins
- Hemiazygos vein
- Accessory hemiazygos vein
- Veins of the vertebral column, see **Ea**

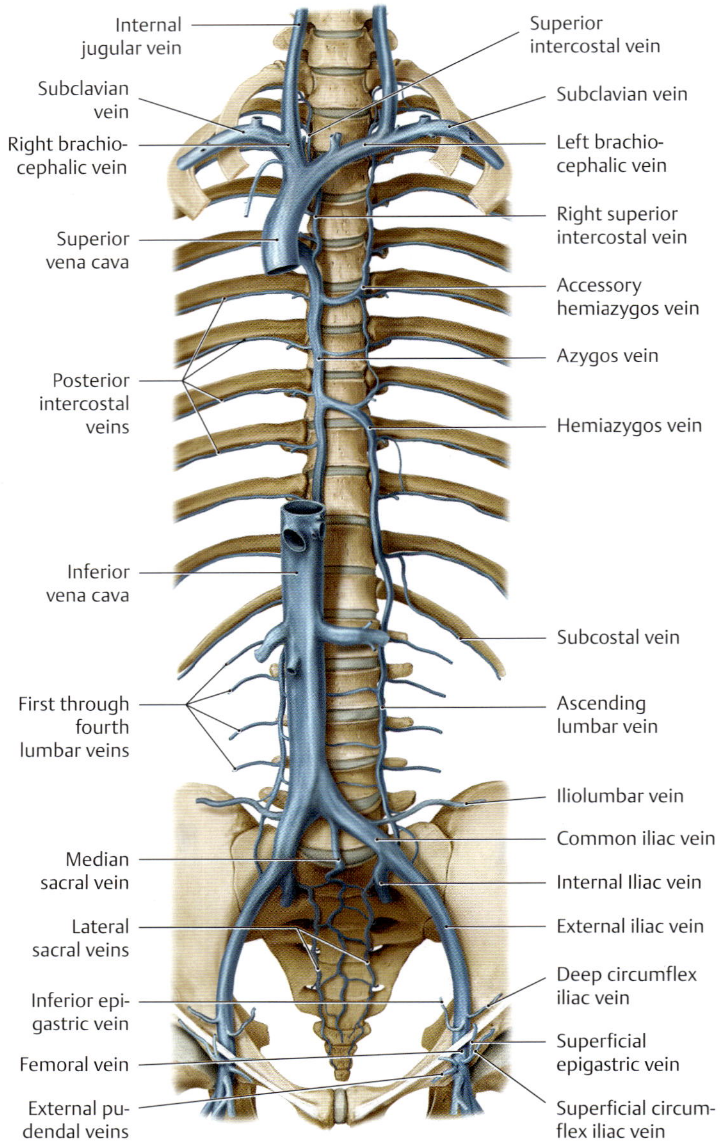

B Major veins in the trunk
Anterior view.

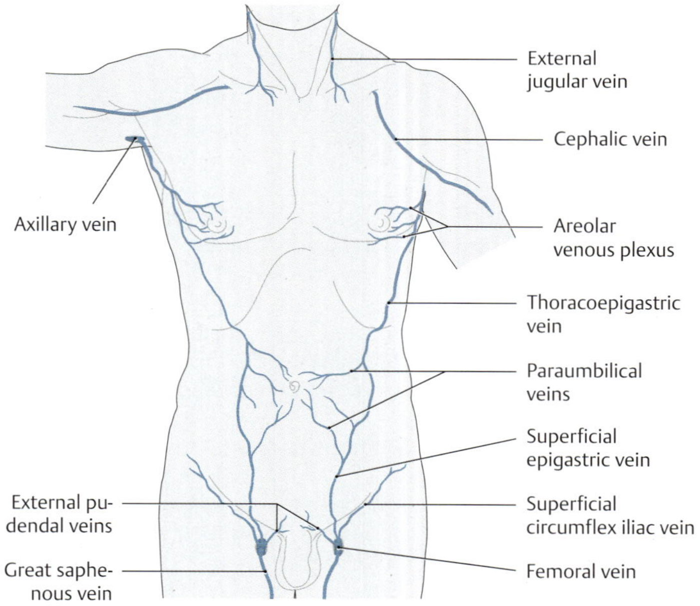

C Superficial veins of the anterior trunk wall

Anterior view. Normally, these veins are not palpable, but they are of key importance in the development of portocaval anastomoses, in which the former umbilical vein connects the portal vein of the liver to the superior and inferior venae cavae. In cases where portal hypertension develops as a result of liver disease (hepatic cirrhosis due to alcohol abuse), the portal venous blood must partially bypass the liver, flowing through the paraumbilical veins (see p. 218) to the superficial truncal veins in the umbilical region (periumbilical veins) and finally to the heart. Since the superficial veins must carry considerably more blood in this situation, they undergo varicose dilation, becoming visible and palpable on the abdomen. This is also called a "Medusa's head" (*caput medusae*) because the serpentine, dilated veins around the umbilicus resemble the snake-haired head of the Gorgon.

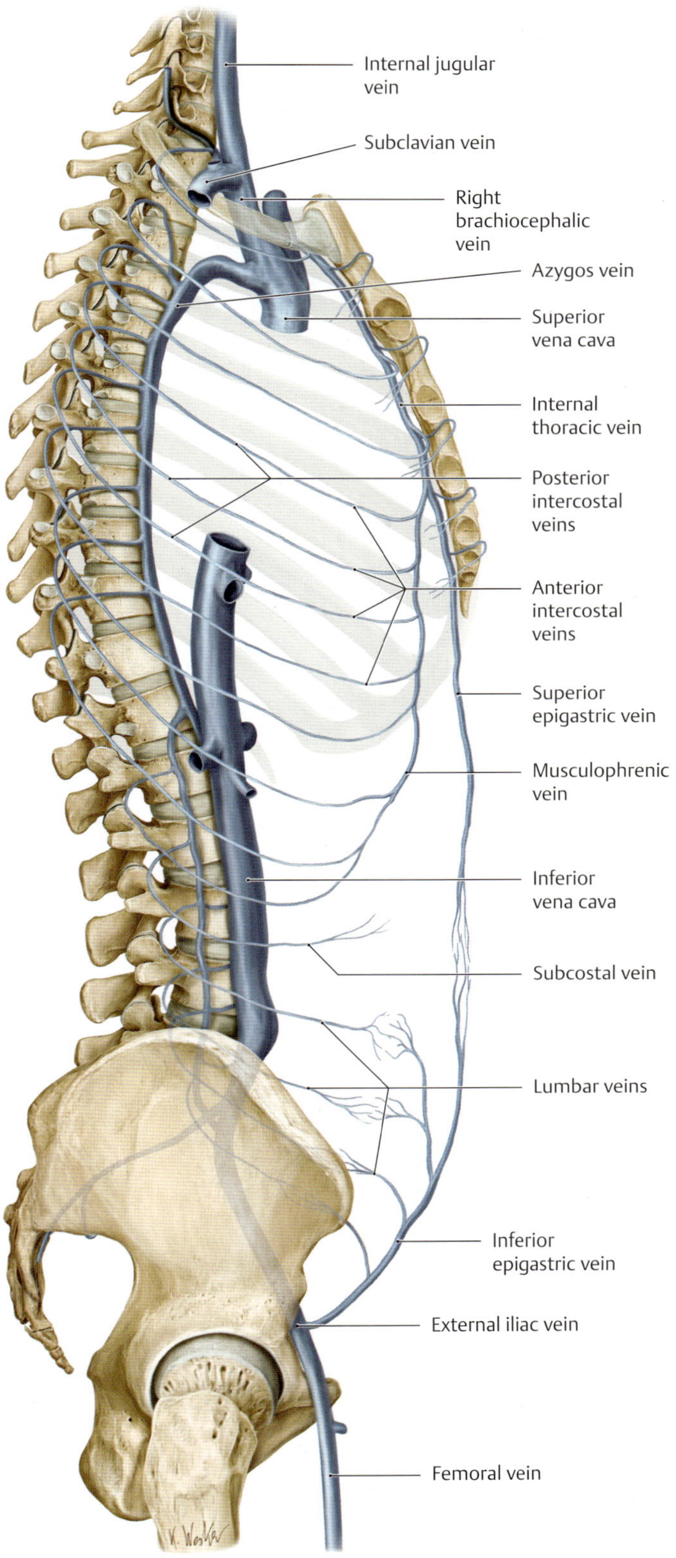

D Veins of the trunk wall
Right lateral view.

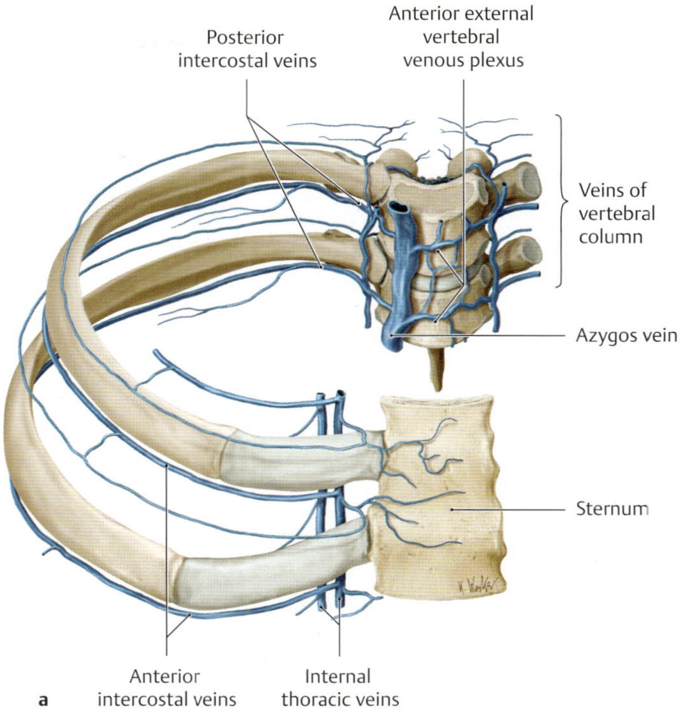

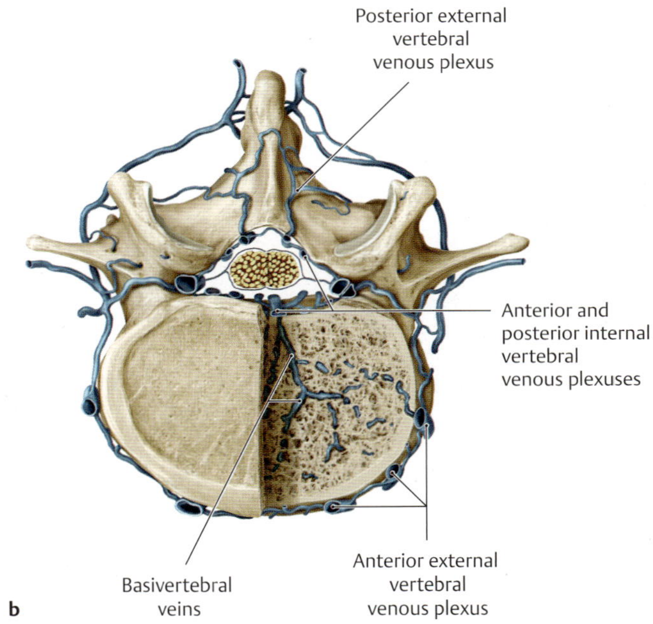

E Intercostal veins and venous plexuses of the vertebral canal
a Vertebral column and rib segment, anterosuperior view.
b Lumbar vertebra, superior view.

4.3 The Lymphatic Vessels and Lymph Nodes

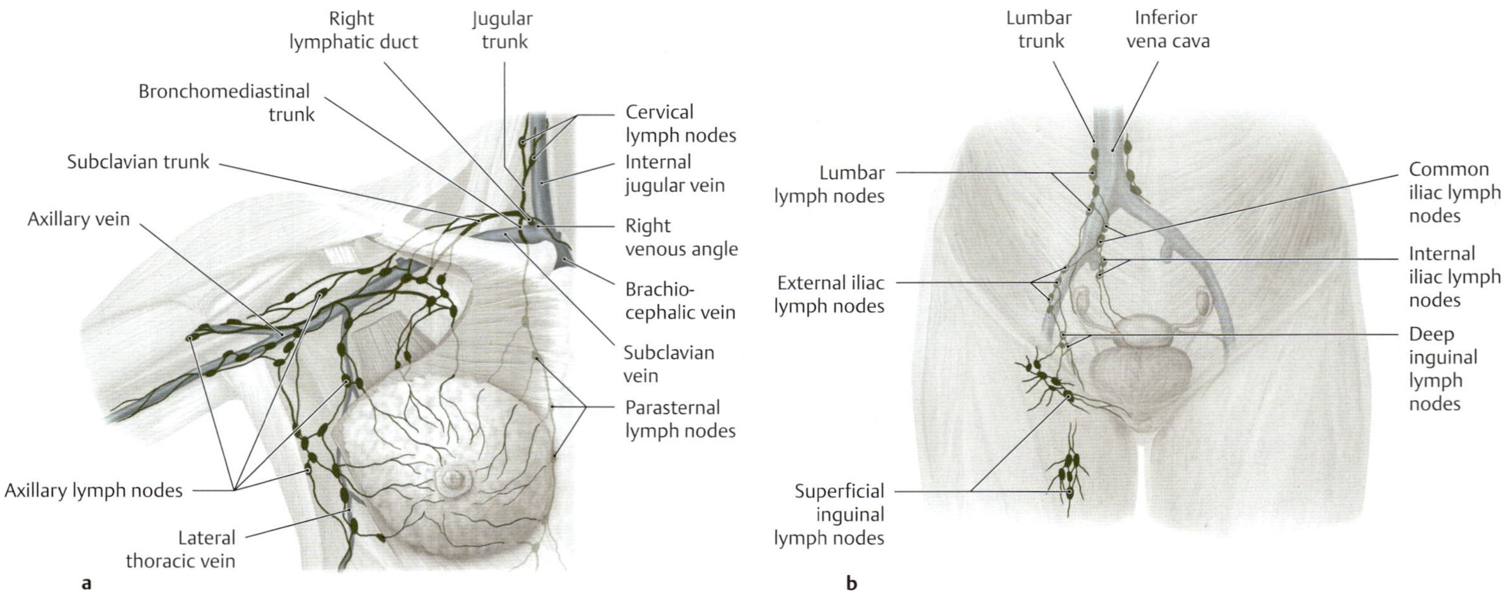

A Regional lymph nodes and their associated lymphatics
Anterior view.

- **a** Axillary, parasternal, and cervical lymph nodes (right thoracic and axillary region with the arm abducted). The lymph node levels are described on p. 217.
- **b** Lymph nodes of the inguinal region and lesser pelvis.

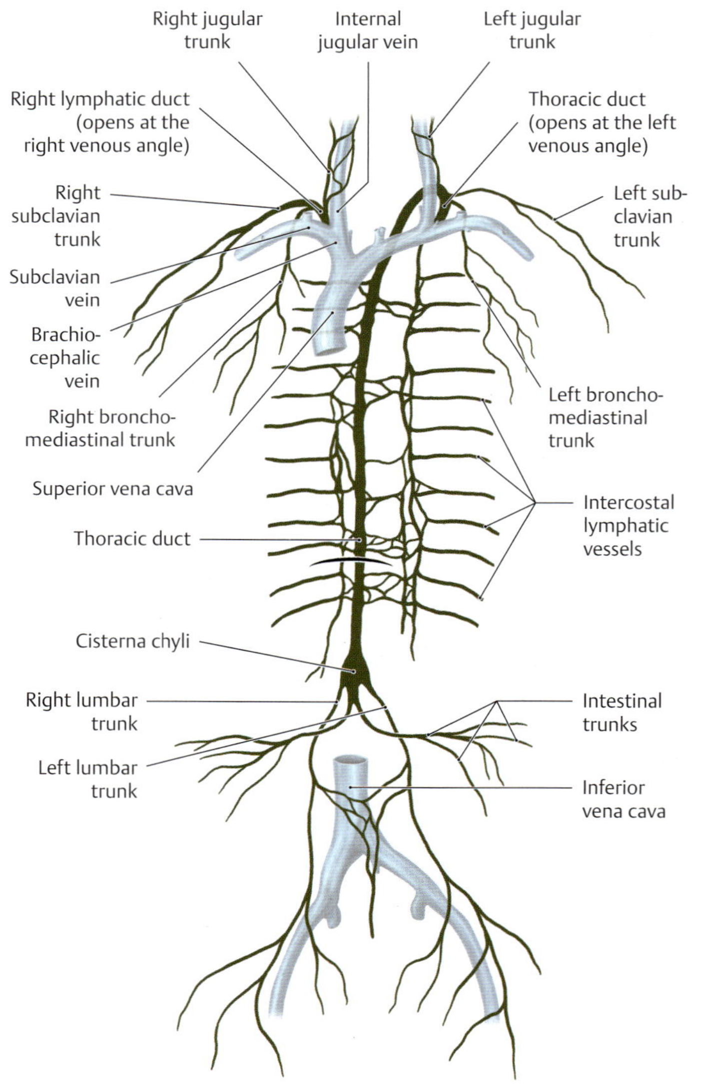

B Left and right venous angles
Anterior view. The approximately 1-cm-long *right lymphatic duct* collects lymph from the right upper quadrant of the body (see **Ca**) and empties into the **right venous angle** at the junction of the right internal jugular vein with the right subclavian vein. Its major tributaries are

- the right jugular trunk (right half of the head and neck),
- the right subclavian trunk (right upper limb, right side of the chest and back wall), and
- the right bronchomediastinal trunk (organs of the right thoracic cavity).

The thoracic duct is approximately 40 cm long and transports lymph from the entire lower half of the body and left upper quadrant. It empties into the **left venous angle** between the left internal jugular vein and left subclavian vein. Its main tributaries are

- the left jugular trunk (left half of the head and neck),
- the left subclavian trunk (left upper limb, left side of the chest and back wall),
- the left bronchomediastinal trunk (organs of the left thoracic cavity),
- the intestinal trunks (abdominal organs), and
- the right and left lumbar trunks (right and left lower limb; pelvic viscera; right and left pelvic, abdominal, and back wall).

The intercostal lymphatic vessels transport lymph from the upper left and right intercostal spaces to the corresponding ipsilateral lymphatic duct. The lower intercostal spaces from both sides drain to the cisterna chyli.

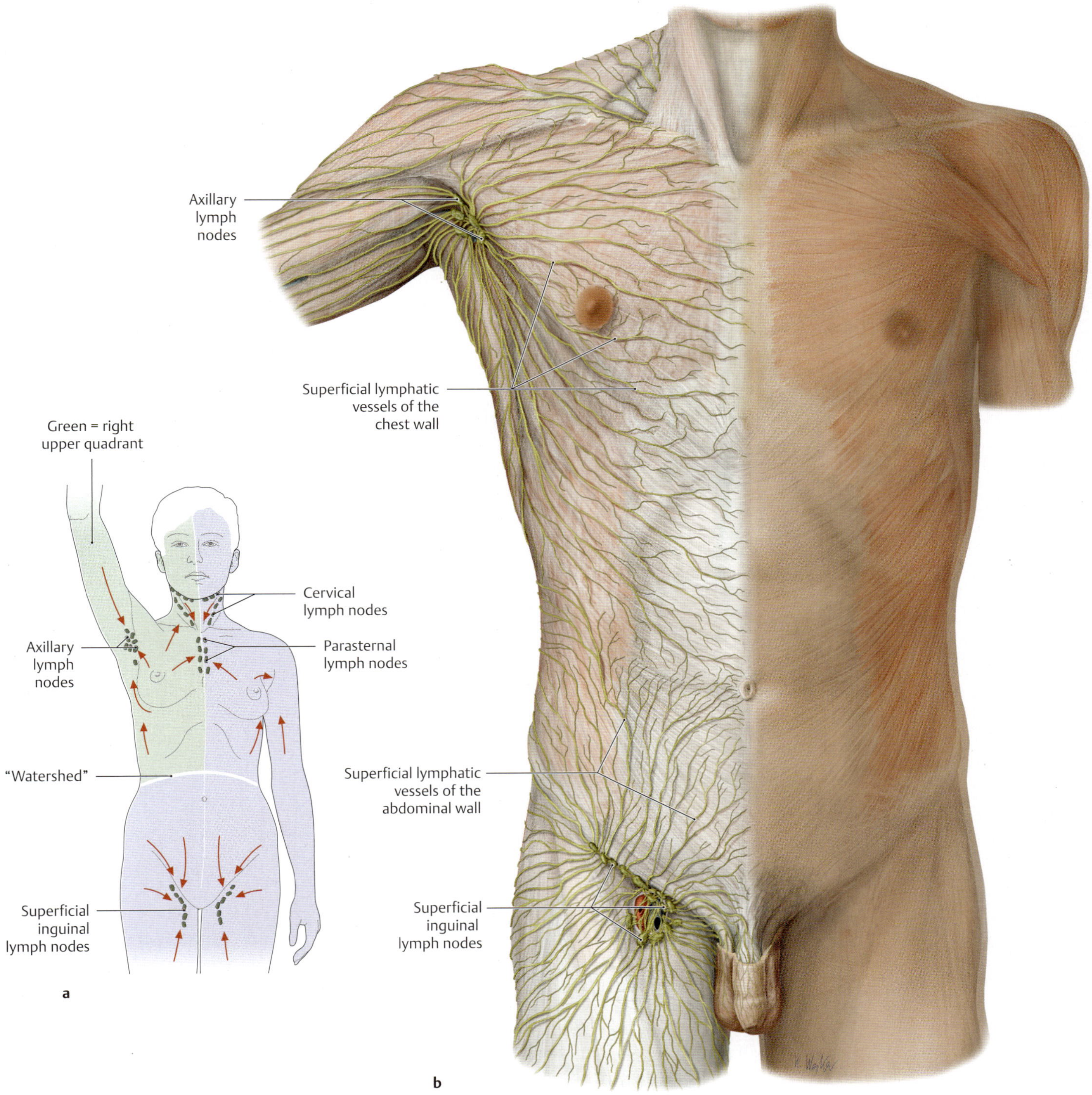

C Areas drained by the superficial lymphatic vessels of the anterior trunk wall

Anterior view.

a Lymphatic pathways and regional lymph nodes of the anterior trunk wall (arrows indicate the direction of lymph flow).

b Superficial network of lymphatic vessels in the right anterior trunk wall.

Lymph from the skin of the trunk wall is collected mainly by the axillary and superficial inguinal lymph nodes, following the general pattern of venous drainage in the anterior trunk wall. The "watershed" zone between the two drainage regions is defined by a curved line above the umbilicus and below the costal arch. Lymph from the regional axillary and inguinal lymph nodes is finally collected by two lymphatic trunks, each of which drains into the jugulosubclavian venous junction on the corresponding side of the body (right or left venous angle, see **B**). Lymphatic fluid from the right upper quadrant (green) is returned to the venous system by the *right lymphatic duct*, while lymph from the other three body quadrants (violet) is returned to the veins by the *thoracic duct*.

4.4 The Nerves

A Anterior and posterior branches (rami) of the spinal nerves

The trunk wall receives most of its sensory nerve supply from the anterior and posterior rami of the T 1–T 12 spinal cord segments (intercostal nerves and posterior rami of the spinal nerves) (see also p. 208).

Spinal cord segment	Anterior branches (anterior rami)	Posterior branches (posterior rami)
C1	Cervical plexus (C1–C4)	Suboccipital nerve
C2		Greater occipital nerve
C3		Third occipital nerve (see p. 208)
C4		Posterior rami* (C4–Co2)
C5	Brachial plexus (C5–T1)	
C6		
C7		
C8		
T1	Intercostal nerves and subcostal nerve (T1–T12)	
T2		
T3		
T4		
T5		
T6		
T7		
T8		
T9		
T10		
T11		
T12		
L1	Lumbar plexus (L1–L4)	
L2		
L3		
L4		
L5	Sacral plexus (L5–S4)	
S1		
S2		
S3		
S4		
S5	Coccygeal plexus (see p. 560) (S5–Co2)	
Co1		
Co2		

* The posterior rami of the L1–L3 spinal nerves are also known as the superior clunial nerves, those from S1–S3 as the middle clunial nerves (see **C**).

Note: The *inferior* clunial nerves are anterior rami from the sacral plexus; see also p. 554.

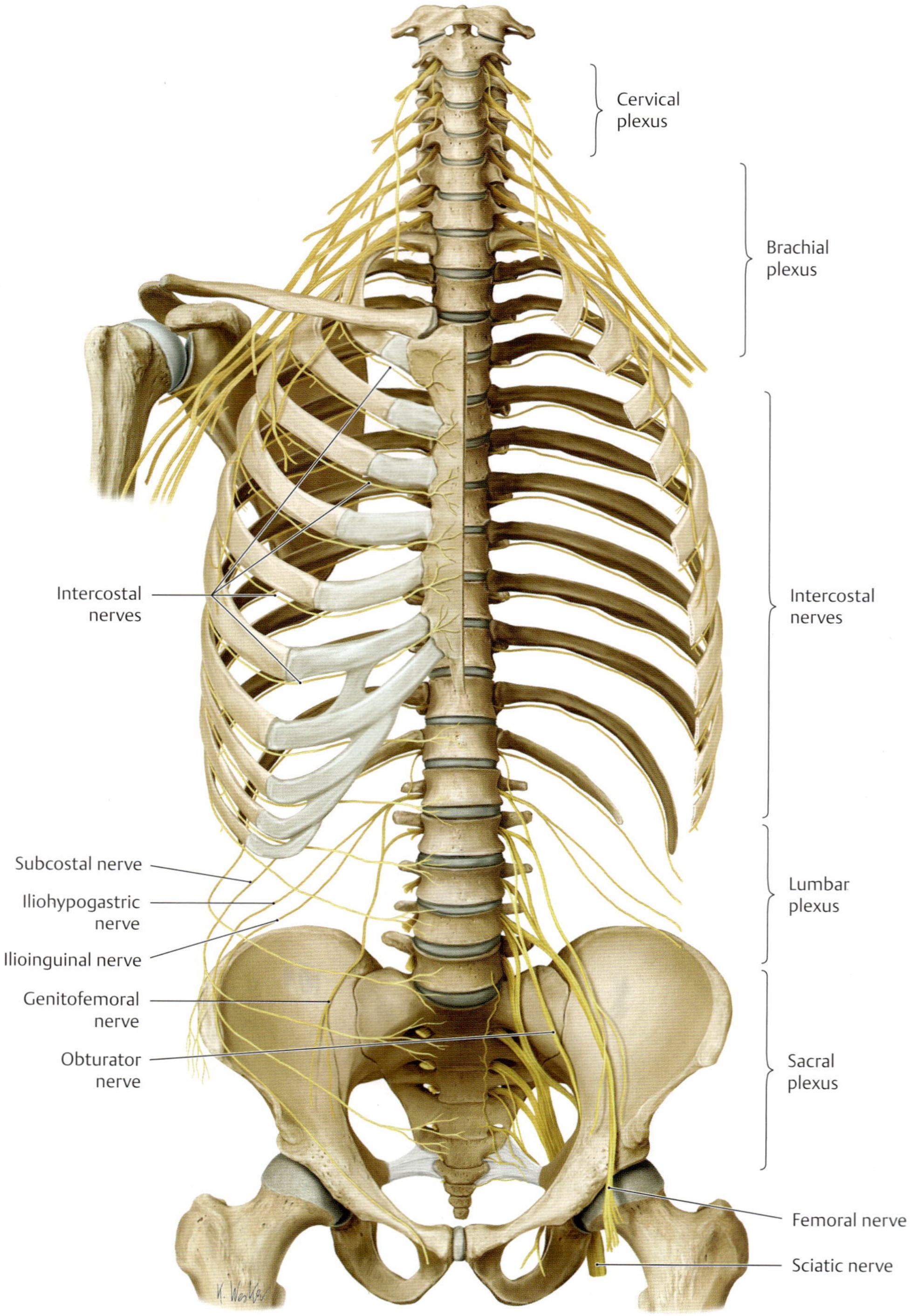

B Nerves of the trunk wall

Anterior view. The anterior part of the left half of the thoracic cage has been removed. The trunk wall receives most of its motor and sensory innervation from the 12 thoracic spinal nerves. Of all the spinal nerves, these 12 pairs most clearly reflect the original segmental (metameric) organization of the body. The anterior rami run foward in the intercostal spaces to become the intercostal nerves, while the posterior rami are distributed to the intrinsic back muscles and to the skin of the back. Portions of the trunk wall are also supplied by nerves from the cervical plexus (supraclavicular nerves), brachial plexus (e.g., the long thoracic nerve), and lumbar plexus (e.g., the ilioinguinal nerve).

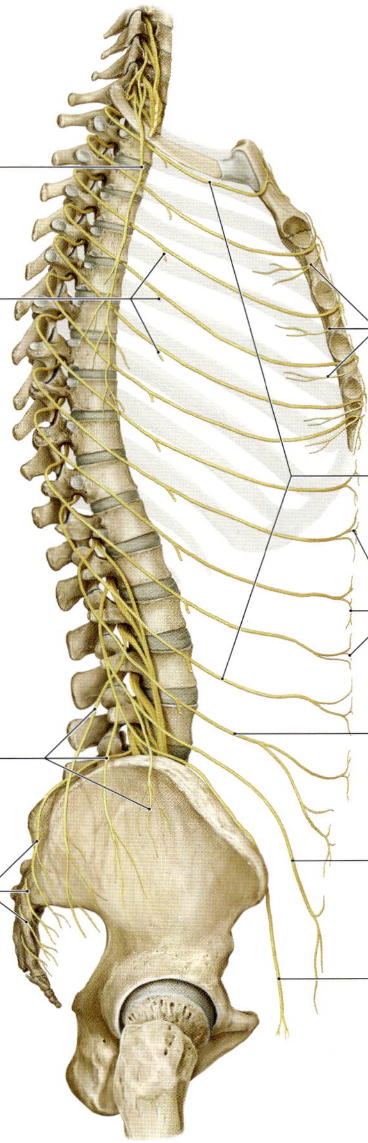

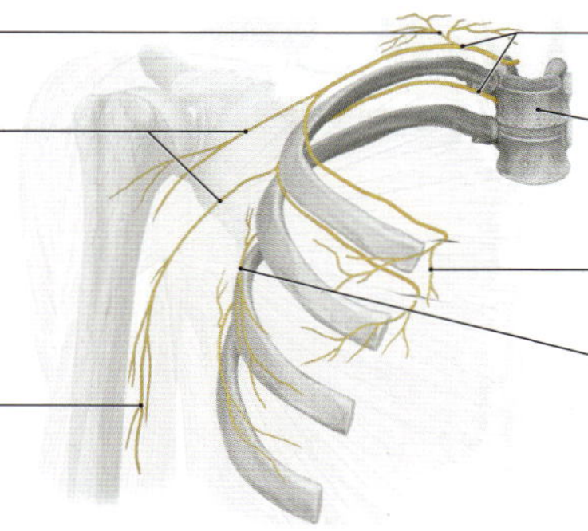

D Course of the intercostal nerves
Right side, anterior view.

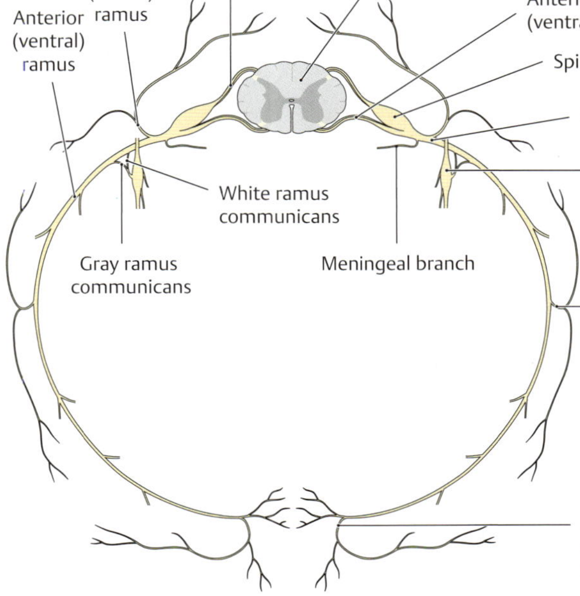

C Course of the nerves on the lateral trunk wall
Right lateral view.
Note the segmental arrangement of the intercostal nerves (compare with the segmental arrangement of the arteries and veins, pp. 199 and 201).

E Branches of a spinal nerve
Formed by the union of the posterior (sensory) and anterior (motor) roots, the approximately 1-cm-long spinal nerve courses through the intervertebral foramen and divides into five branches after exiting the vertebral canal (see **F**).

F Spinal nerve branches and the territories they supply

Spinal nerve branch	Motor or visceromotor territory	Sensory territory
① Anterior (ventral) ramus	All somatic muscles except for the intrinsic back muscles	Skin of the lateral and anterior trunk wall and of the upper and lower limbs
② Posterior (dorsal) ramus	Intrinsic back muscles	Posterior skin of the head and neck, skin of the back and buttock
③ Meningeal branch		Spinal meninges, ligaments of the spinal column, capsules of the facet joints
④ White ramus communicans	Carries preganglionic fibers from the spinal nerve to the sympathetic trunk ("white" because the preganglionic fibers are myelinated)	
⑤ Gray ramus communicans*	Carries postganglionic fibers from the sympathetic trunk back to the spinal nerve ("gray" because the fibers are unmyelinated)	

* Strictly speaking, the gray ramus communicans is not a spinal nerve branch but a branch passing from the sympathetic trunk to the spinal nerve.

5.1 Anterior Trunk Wall: Surface Anatomy and Superficial Nerves and Vessels

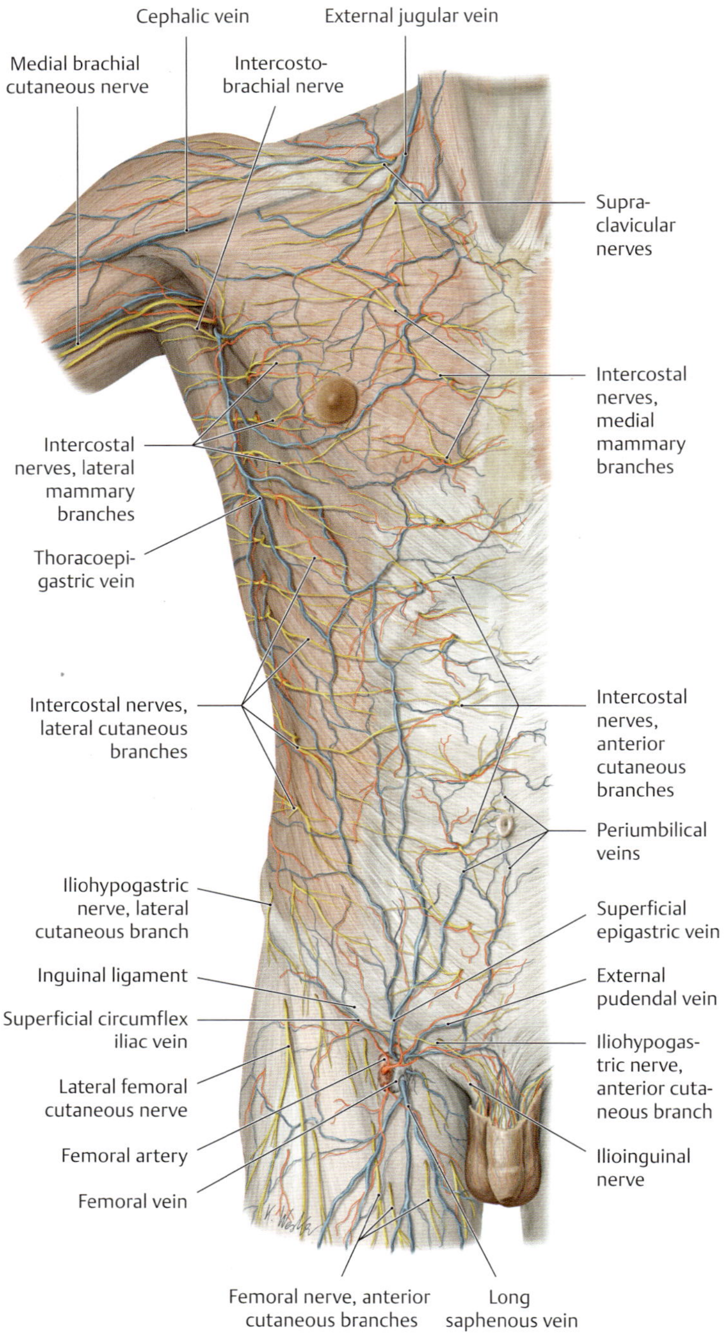

A Superficial cutaneous vessels and nerves of the anterior trunk wall

Anterior view.

Superficial vessels: Most of the *arterial supply* to the anterior trunk wall comes from two sources: the internal thoracic artery and the superficial epigastric artery. The superficial *veins* drain chiefly into the axillary vein (via the thoracoepigastric vein) and into the femoral vein (via the superficial epigastric and superficial circumflex iliac veins). The peri- and paraumbilical veins provide the main communication between the superficial veins of the trunk wall and the portal veins (portocaval anastomoses).

Superficial nerves: The *sensory supply* to the anterior trunk wall has a largely segmental arrangement (provided, for example, by lateral and anterior cutaneous branches from the intercostal nerves). The cervical plexus (supraclavicular nerves) is additionally involved in the thoracic region, as is the lumbar plexus (e.g., iliohypogastric and ilioinguinal nerves) in the lower abdominal region.

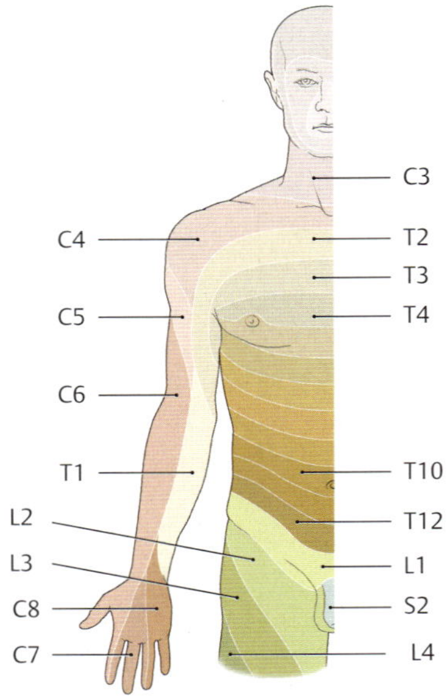

B Segmental (radicular) cutaneous innervation of the anterior trunk wall (dermatomes)

Right half of the trunk and adjacent upper limb, anterior view. Every sensory nerve root (dorsal root) innervates a specific skin area with its fibers. These "dermatomes" (see p. 94), then, correspond to associated spinal cord segments. The dermatomes are arranged in bandlike patterns that encircle the chest wall and upper abdomen. Below the umbilicus, the dermatomes become angled slightly downward toward the median plane. A "segmental gap" exists between the C 4 and T 2 dermatomes because the phylogenetic outgrowth of the human upper limb has removed the sensory fibers of C 5–C 8 and T 1 from the trunk wall.

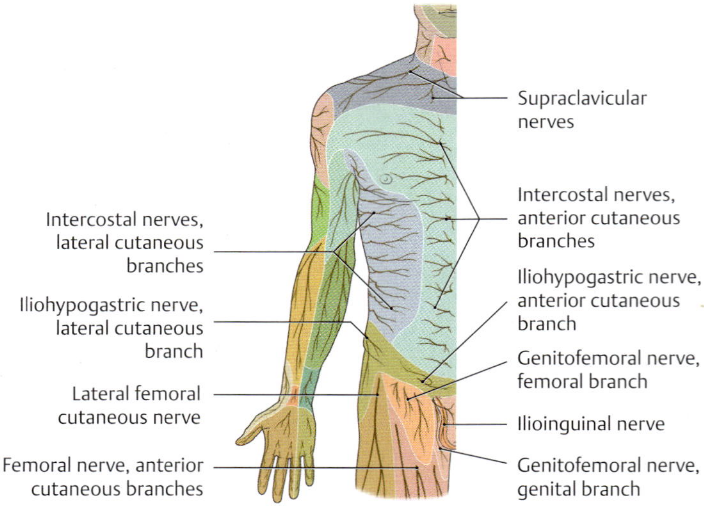

C Peripheral sensory cutaneous innervation of the anterior trunk wall

Right half of the trunk and adjacent upper limb, anterior view. The color-coded map of the peripheral cutaneous nerve territories follows the branching pattern of the cutaneous nerves in the subcutaneous connective tissue. Besides the cutaneous branches of the intercostal nerves (anterior and lateral cutaneous branches), it is chiefly the supraclavicular nerves and the iliohypogastric and ilioinguinal nerves that supply the skin of the anterior trunk wall.

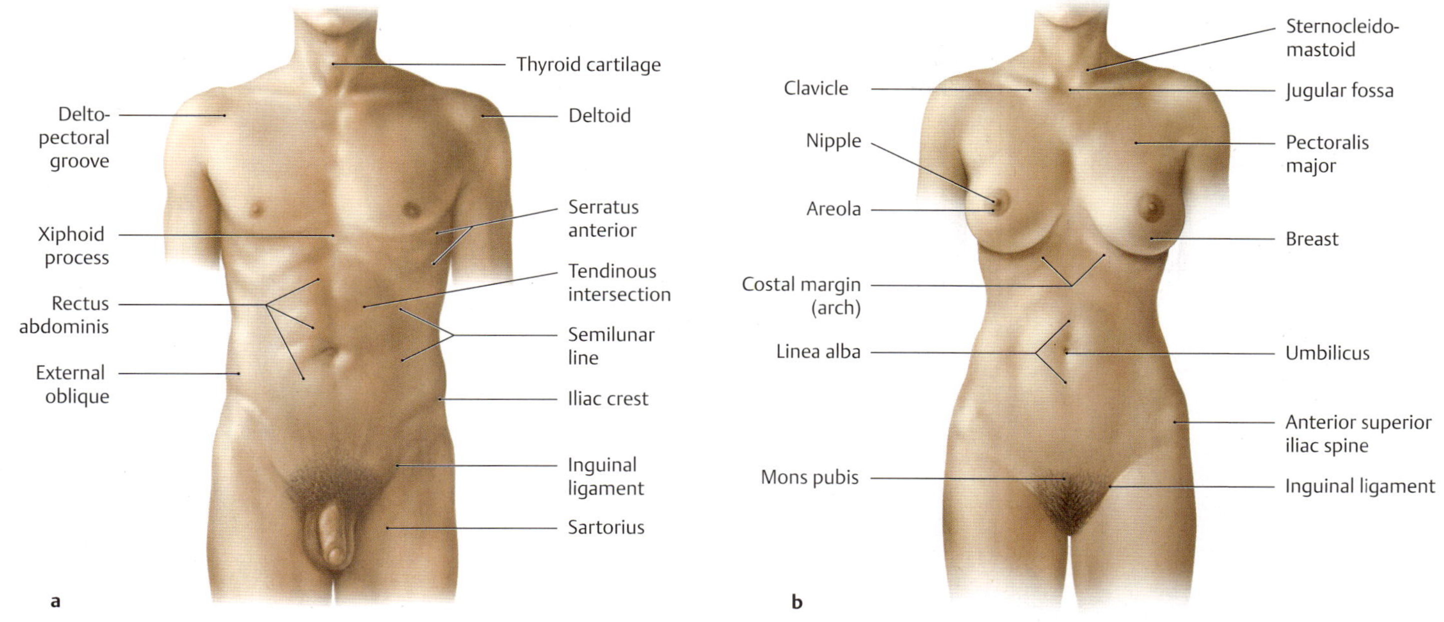

D Surface anatomy of the anterior trunk wall

a Male, **b** female.

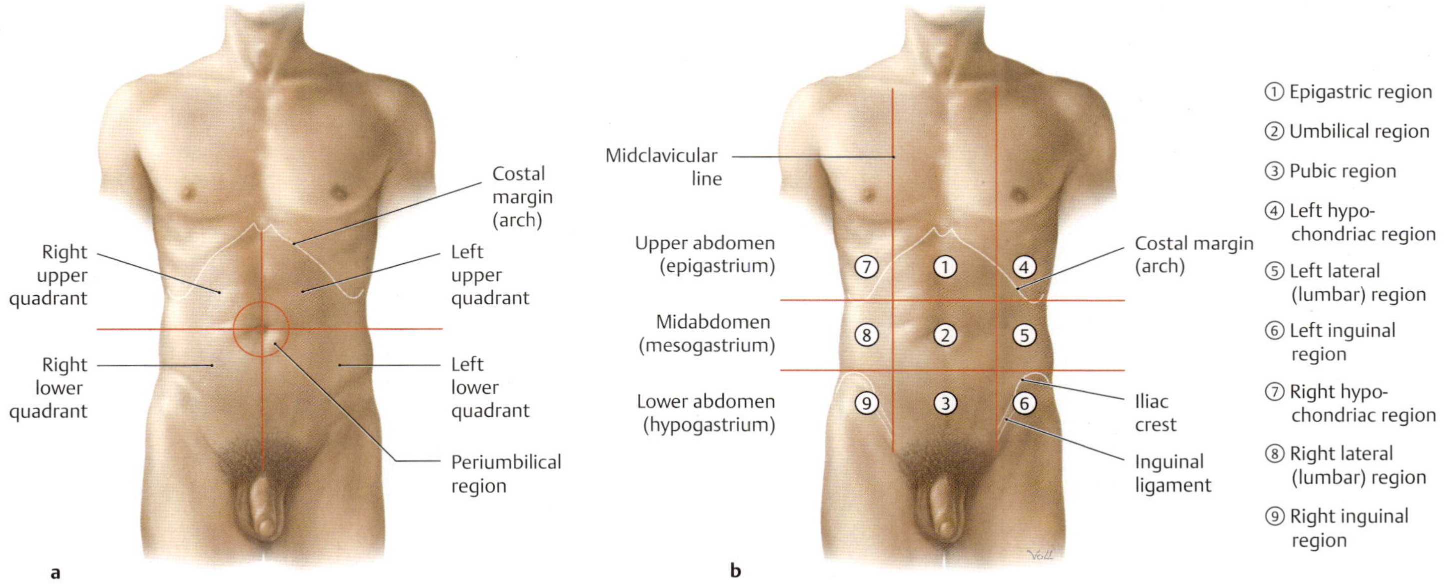

E Criteria for dividing the abdomen into regions

a The abdomen is divided into four quadrants by two perpendicular lines that intersect at the umbilicus.

b Coordinate system composed of two vertical and two horizontal lines. They divide the abdomen into nine regions, each located in either the upper, middle, or lower abdomen. The two vertical lines represent the left and right midclavicular lines. The two horizontal lines pass through the lowest point of the tenth ribs or the summit of the two iliac crests (see p. 35).

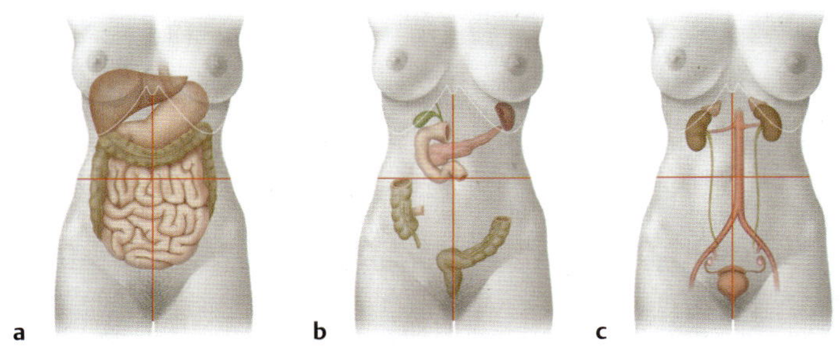

F Projection of the abdominal organs onto the four quadrants of the anterior abdominal wall

a Organs of the anterior layer, **b** organs of the middle layer, **c** organs of the posterior layer.

The organs of the *anterior* layer abut the anterior abdominal wall. The organs of the *middle* layer are located in the posterior part of the abdominal cavity (some are partially retroperitoneal), and those of the *posterior* layer are located outside or behind the actual abdominal cavity (i.e., they are retroperitoneal).

5.2 Posterior Trunk Wall: Surface Anatomy and Superficial Nerves and Vessels

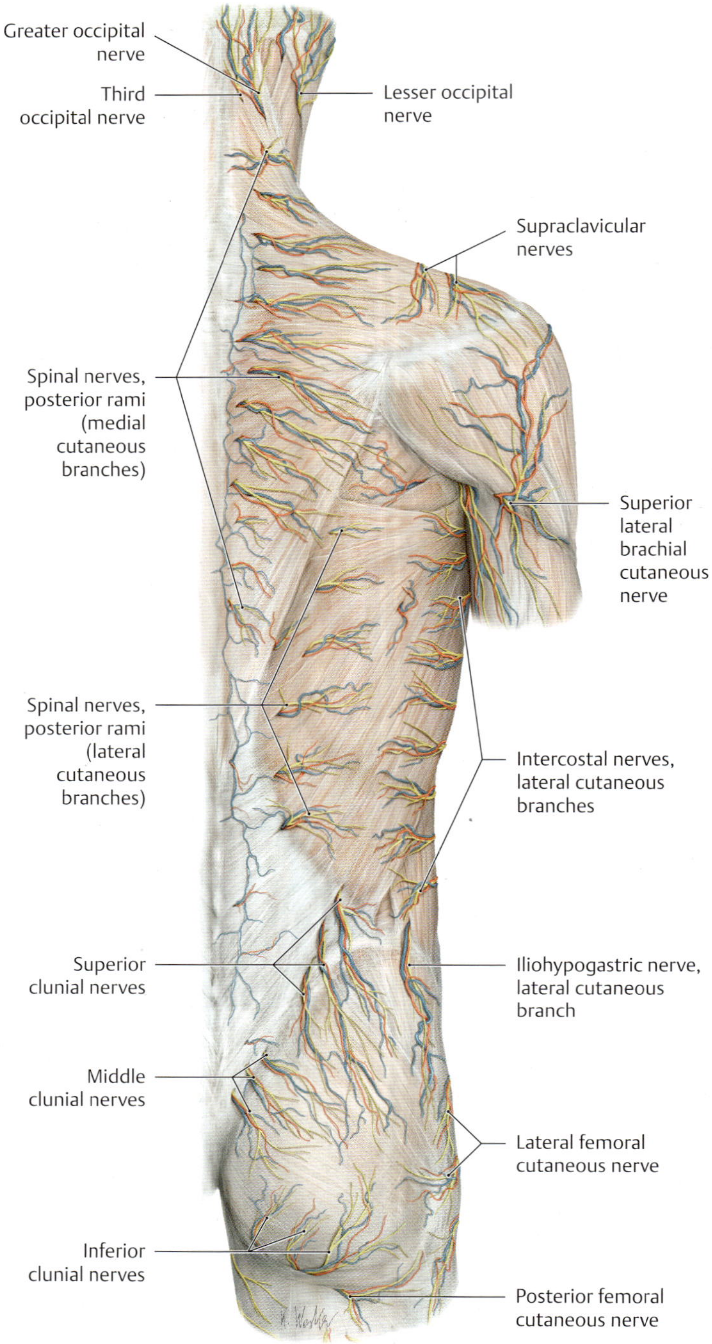

A Superficial cutaneous vessels and nerves of the posterior trunk wall

Posterior view. Except for the lower buttocks and lateral portions of the trunk wall, the posterior trunk wall derives its sensory innervation from posterior rami of the spinal nerves and from lateral cutaneous branches of the intercostal nerves. This is a predominantly segmental innervation pattern, analogous to that described in the anterior trunk wall. Both the medial and lateral cutaneous nerve branches pass with the cutaneous vessels through the intrinsic back muscles to the skin of the back. The skin of the buttocks is supplied by lateral branches from the three cranial lumbar and sacral nerves (superior and middle clunial nerves).

Note: The lower part of the buttock is supplied by the inferior clunial nerves, which are branches of the sacral plexus; thus, they are derived from the *anterior* rami of the spinal nerves.

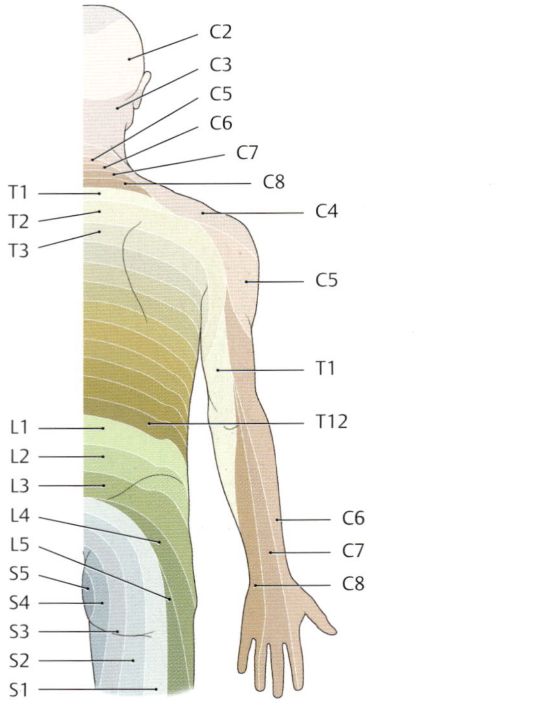

B Segmental (radicular) cutaneous innervation of the posterior trunk wall (dermatomes)

Right half of the trunk and adjacent upper limb, posterior view.

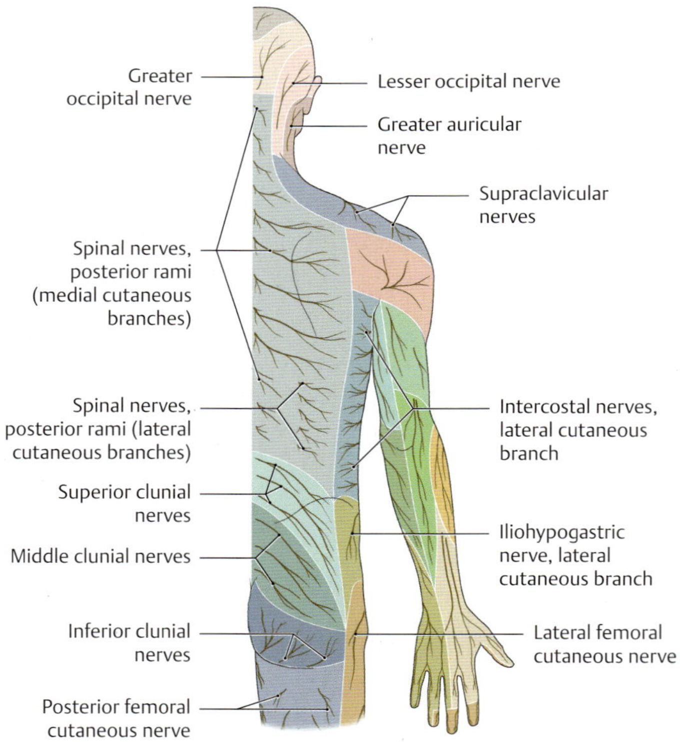

C Peripheral sensory cutaneous innervation of the posterior trunk wall

Right half of the trunk and adjacent upper limb, posterior view.

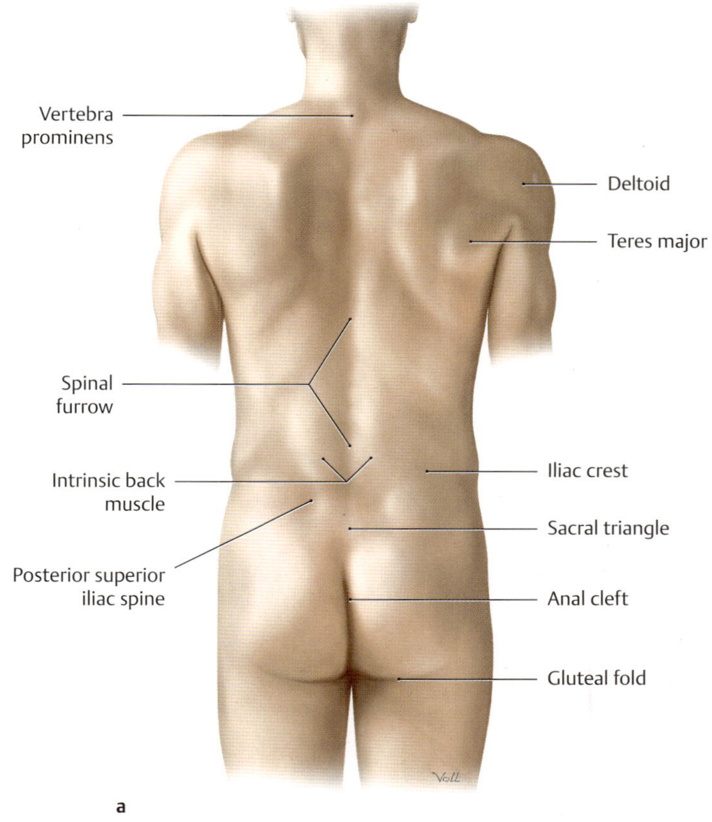

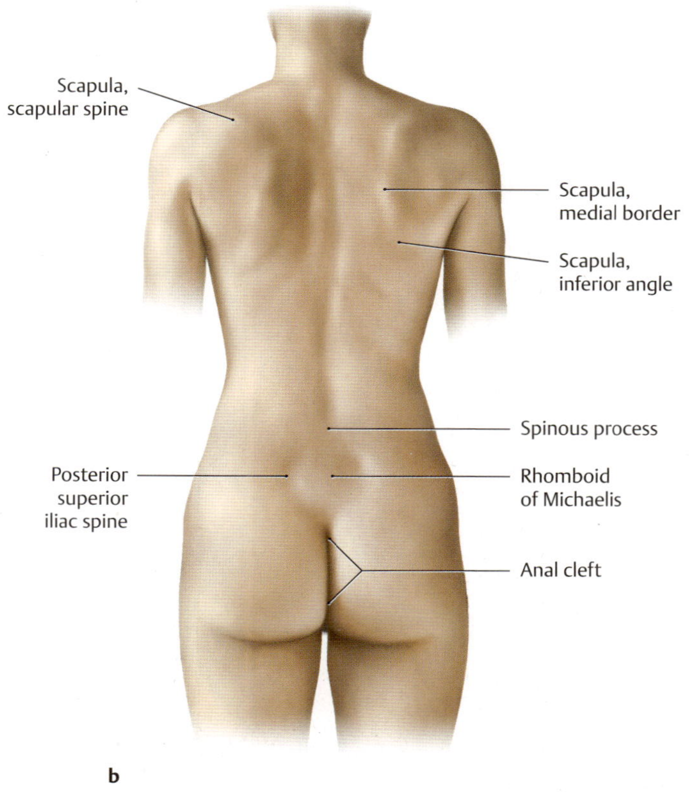

D Surface anatomy of the posterior trunk wall

a Male, **b** female.

In both sexes a *spinal furrow* runs vertically in the posterior midline of the trunk below the C 7 spinous process. It is formed by the fixation of the subcutaneous tissue to the corresponding spinous processes. At the sacral level in males, the furrow widens to form the *sacral triangle* (bounded by the right and left posterior superior iliac spines and the upper part of the anal cleft). The corresponding diamond-shaped area in females is called the *rhomboid of Michaelis* (see **F**).

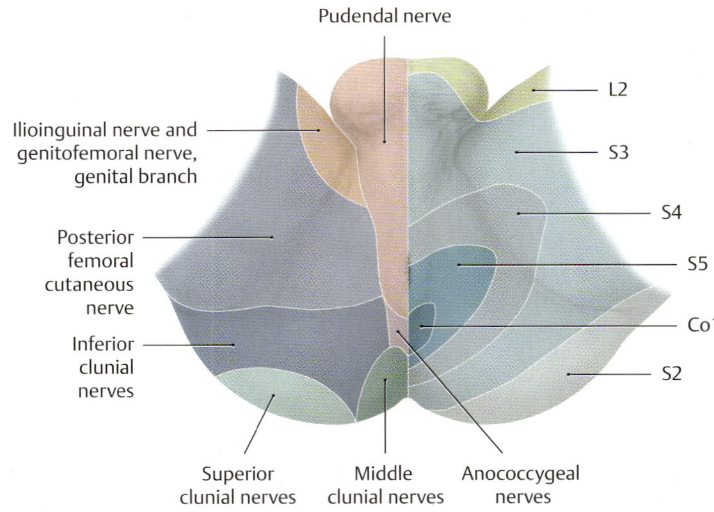

E Segmental and peripheral cutaneous innervation of the male perineal region

Lithotomy position. The segments or dermatomes have been mapped on the left side of the body, and the areas supplied by the peripheral cutaneous nerves are shown on the right side.

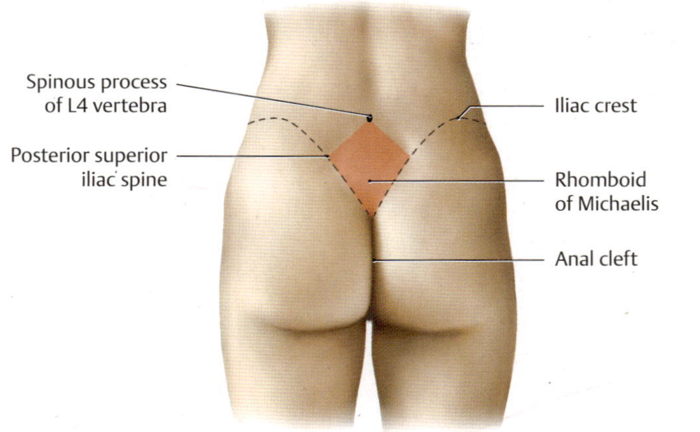

F Anatomic boundaries of the Michaelis rhomboid

Female gluteal region, posterior view. In women the sacral triangle is expanded to form a diamond-shaped figure with the following boundaries: the left and right posterior superior iliac spines, the spinous process of the L 4 vertebra, and the upper part of the anal cleft. With a normal female pelvis, the vertical and horizontal dimensions of the rhomboid are approximately equal. The shape of the Michaelis rhomboid (named for the German gynecologist G. A. Michaelis, 1798–1848) reflects the width of the female pelvis, providing an indirect indicator of the size of the birth canal.

5.3 Posterior Trunk Wall, Posterior View

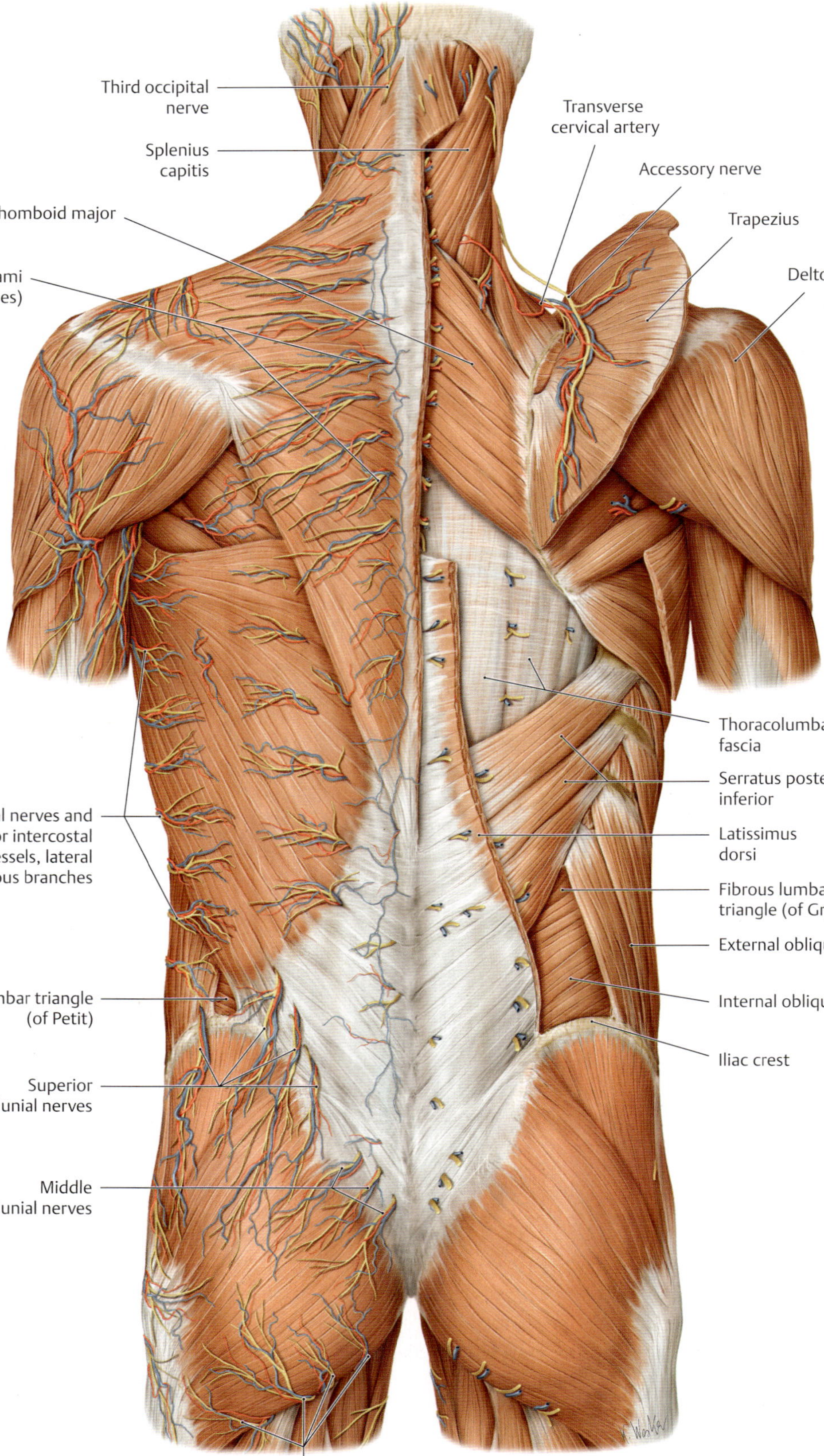

A Neurovascular structures of the posterior trunk wall and nuchal region

Posterior view. The segmentally arranged neurovascular structures of the posterior trunk wall (posterior rami of the spinal nerves and posterior branches of the posterior intercostal and lumbar vessels) are demonstrated on the left side of the trunk (all muscle fasciae have been removed except for the superficial layer of the thoracolumbar fascia). On the right side, the trapezius muscle has been detached from its origins and reflected laterally to show the course of the transverse cervical artery in the deep scapular region (compare with **B**).

Note: On the posterior trunk wall, only the lateral nuchal region (lesser occipital nerve, see **C**) and the lower gluteal region (inferior clunial nerves) receive their sensory supply from anterior spinal nerve rami.

The latissimus dorsi muscle has been partially removed on the right side to demonstrate the upper costolumbar triangle (of Grynfeltt). The fibrous lumbar triangle (boundaries: twelfth rib, intrinsic back muscles, and internal oblique) is similar to the lower iliolumbar triangle (of Petit, bounded by the iliac crest, latissimus dorsi, and external oblique) in that it creates a site of predilection for rare, usually acquired lumbar hernias (Grynfeltt or Petit hernia, see also p. 229).

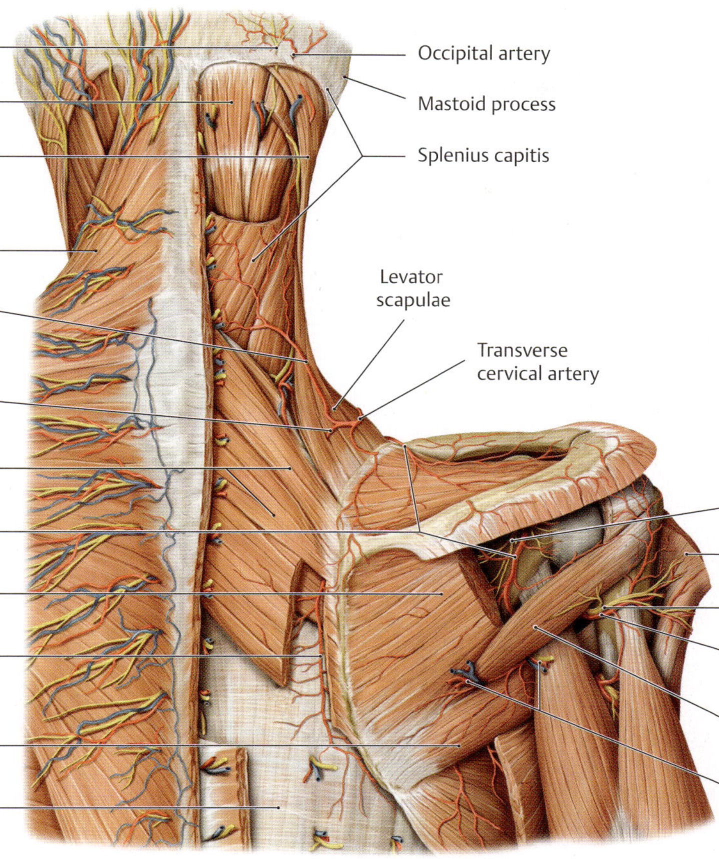

B Arteries of the deep scapular region

Right scapular region, posterior view. The trapezius, splenius capitis, deltoid, infraspinatus, and rhomboid major and minor muscles have been completely or partially removed on the right side. The deep scapular region is supplied by the transverse cervical artery, deep cervical artery (see **C**), suprascapular artery, circumflex scapular artery, and posterior circumflex humeral artery. All of these vessels arise directly or indirectly—via the thyrocervical trunk—from the subclavian artery (neither is visible here). The suprascapular, circumflex scapular, dorsal scapular, and thoracodorsal arteries form the "*scapular arcade*" (see p. 403). Medial to the mastoid process, the occipital artery appears below the tendon of insertion of the sternocleidomastoid muscle and runs upward with the sensory greater occipital nerve to the skin of the occiput. The greater occipital nerve pierces both the trapezius and semispinalis capitis muscles in the area of their firm tendinous attachments. It may become compressed at these sites, leading to *occipital neuralgia*.

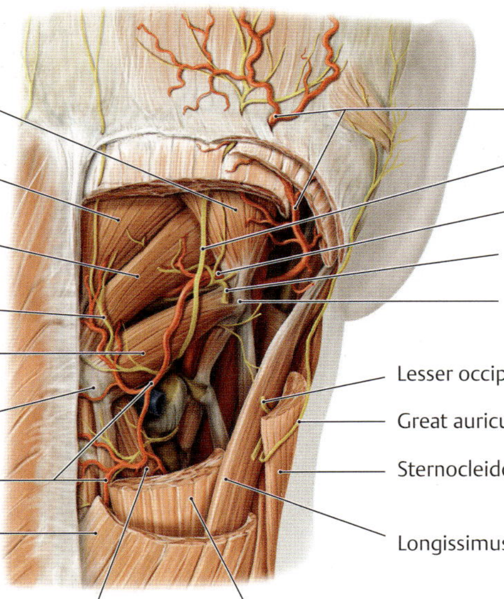

C Suboccipital triangle (vertebral artery triangle)

Posterior view. The trapezius, sternocleidomastoid, splenius capitis, and semispinalis capitis muscles have been removed to display the suboccipital region on the right side. The suboccipital triangle is bounded by the suboccipital muscles (rectus capitis posterior major and the obliquus capitis superior and inferior). In the deep portion of the triangle, the vertebral artery runs through its groove in the atlas. The suboccipital nerve (C 1), which is purely motor, emerges above the posterior arch of the atlas to supply the short muscles of the head. The greater occipital nerve (C 2) and, at a lower level, the third occipital nerve (C 3) wind posteriorly as they pass the lower margin of the obliquus capitis inferior. The deep cervical artery, a branch of the costocervical trunk, runs between the semispinalis capitis and cervicis muscles.

5.4 Posterior Trunk Wall, Anterior View

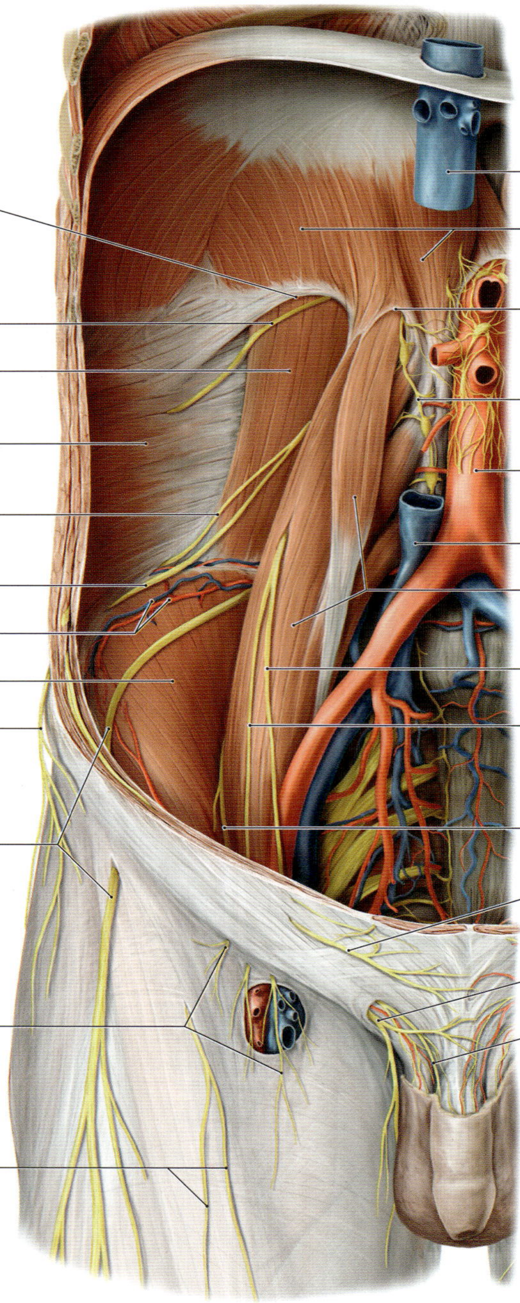

a

A Neurovascular structures on the anterior side of the posterior trunk wall
Anterior view.

a Lumbar fossa on the right side after removal of the anterior and lateral trunk wall, the intra- and retroperitoneal organs, the peritoneum, and all the fasciae of the trunk wall. The inferior vena cava has been partially removed.

b Lumbar fossa with the lumbar plexus of the right side after removal of the superficial layer of the psoas major.

The lumbar plexus (see p. 548) is formed by the anterior rami of the T 12–L 4 nerves lateral to the lumbar spine and is partially covered by the psoas major muscle. The nerves run laterally and obliquely downward to the abdominal wall and thigh, except for the obturator nerve (see **b**), which runs through the lateral wall of the lesser pelvis and the obturator foramen (not visible here) to the medial part of the thigh.
Note: The sites of emergence of the lumbar and iliolumbar vessels are located below the psoas major muscle. They run almost horizontally across the quadratus lumborum and iliacus muscles. Medial to the psoas major (and covered by the inferior vena cava) is the lumbar sympathetic trunk.

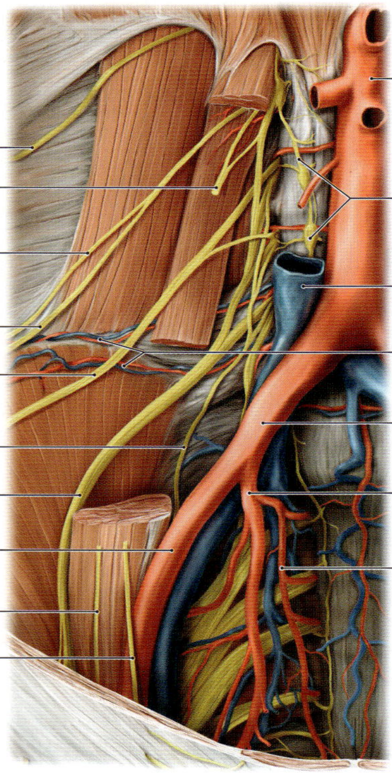

b

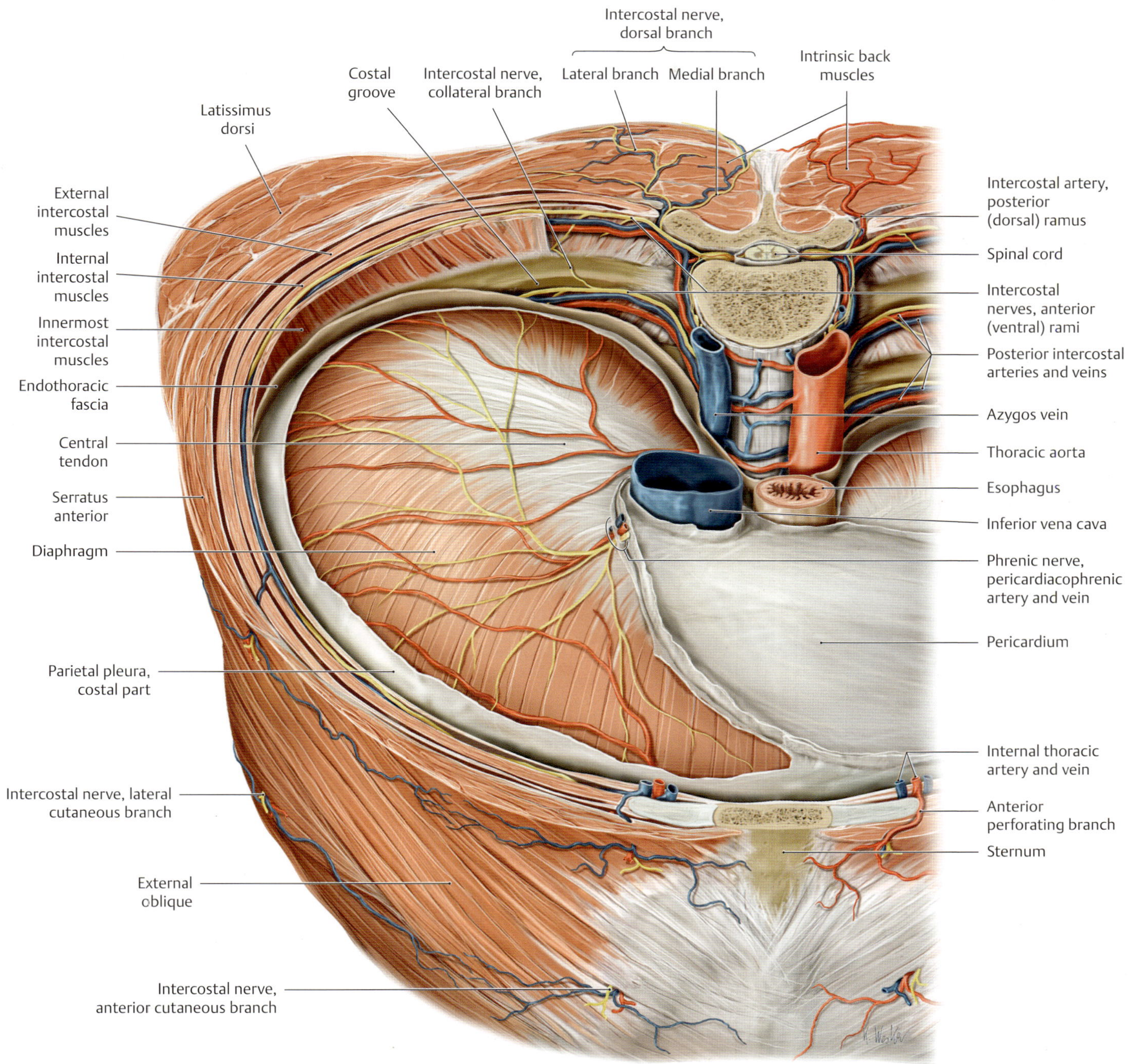

B Neurovascular structures of the posterior trunk wall at the thoracic level

Transverse section through the thorax after removal of the thoracic organs, parietal pleura, and part of the endothoracic fascia, anterosuperior view. The chest wall receives its arterial blood supply from the posterior intercostal arteries and is drained by the intercostal veins, which empty into the azygos system. The intercostal vessels run with the intercostal nerves along the inferior border of the associated rib, lodged in the costal sulcus.

5.5 Anterior Trunk Wall: Overview and Location of Clinically Important Nerves and Vessels

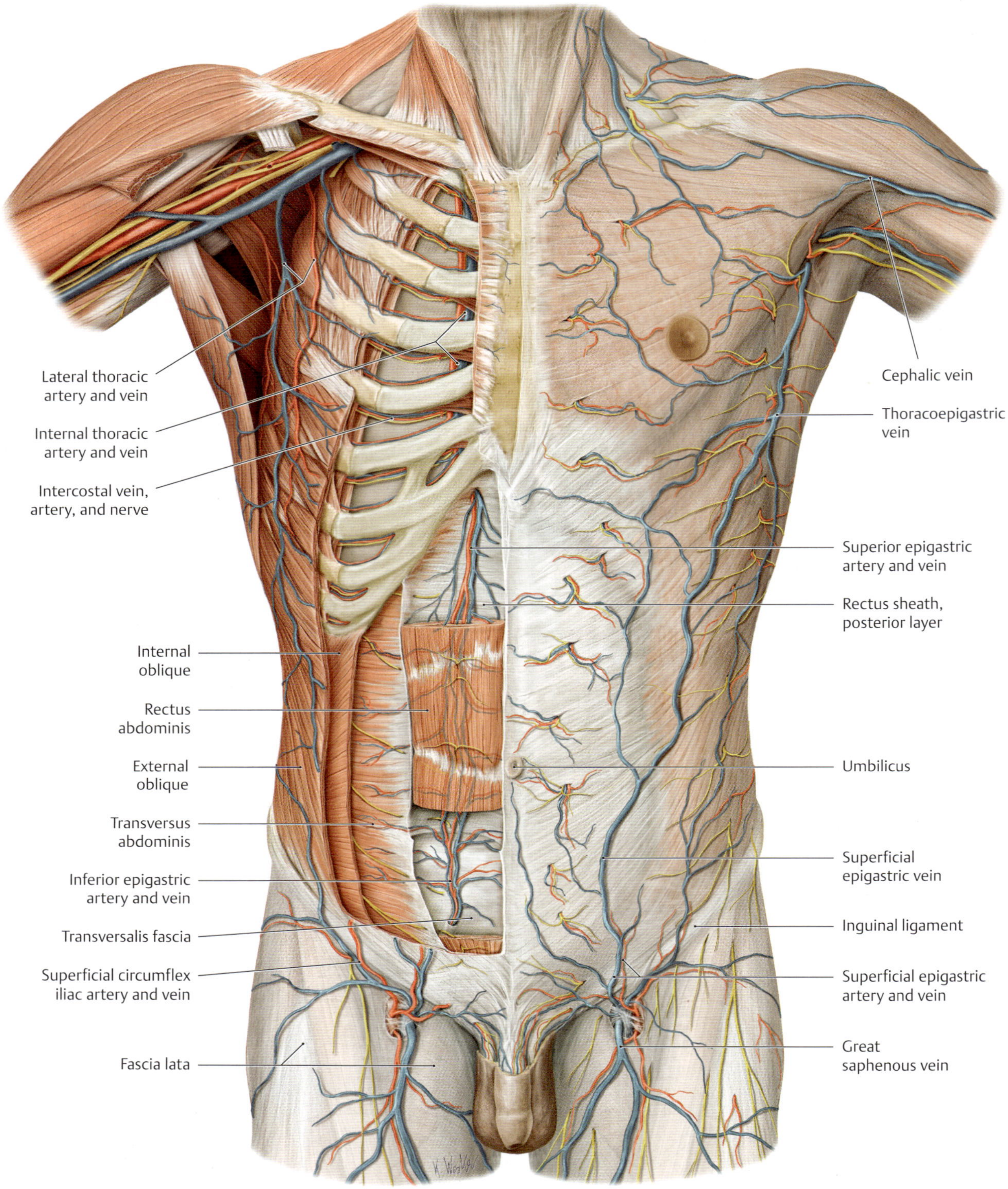

A Neurovascular structures on the anterior side of the anterior trunk wall

Anterior view. The superficial (subcutaneous) neurovascular structures are demonstrated on the left side of the trunk and the deep neurovascular structures on the right side. For this purpose the pectoralis major and minor muscles have been completely removed on the right side, and the external and internal oblique muscles have been partially removed. Portions of the right rectus abdominis muscle have been removed or rendered transparent to demonstrate the inferior epigastric vessels. Finally, the intercostal spaces have been exposed to display the course of the intercostal vessels and nerves.

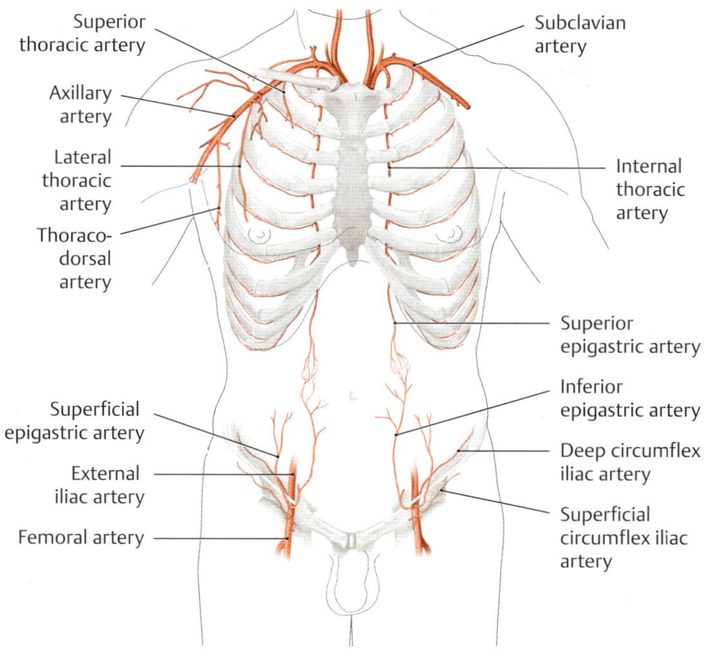

B The arterial supply of the anterior trunk wall

Anterior view. The anterior trunk wall receives its blood supply from two main sources: the internal thoracic artery, which arises from the subclavian artery, and the inferior epigastric artery, which arises from the external iliac artery. It is also supplied by smaller vessels arising from the axillary artery (superior thoracic artery, thoracodorsal artery, and lateral thoracic artery) and from the femoral artery (superficial epigastric artery and superficial circumflex iliac artery).

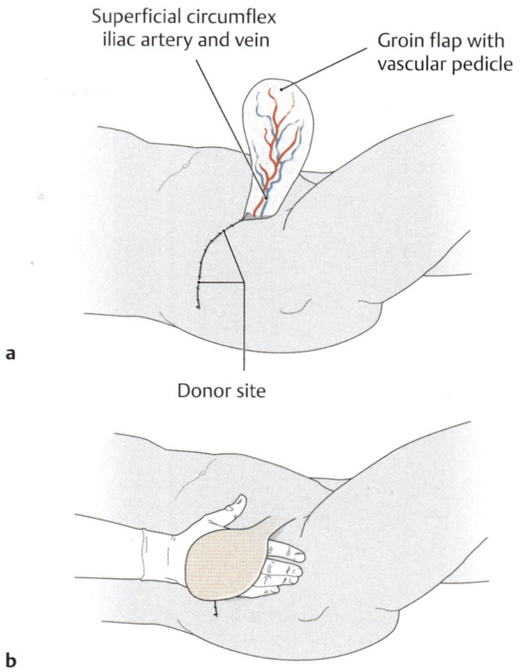

C Importance of the superficial circumflex iliac artery in harvesting skin flaps for plastic surgery

a Dissection of a skin flap based on the superficial circumflex iliac artery.

b The groin flap transferred to the dorsum of the right hand.

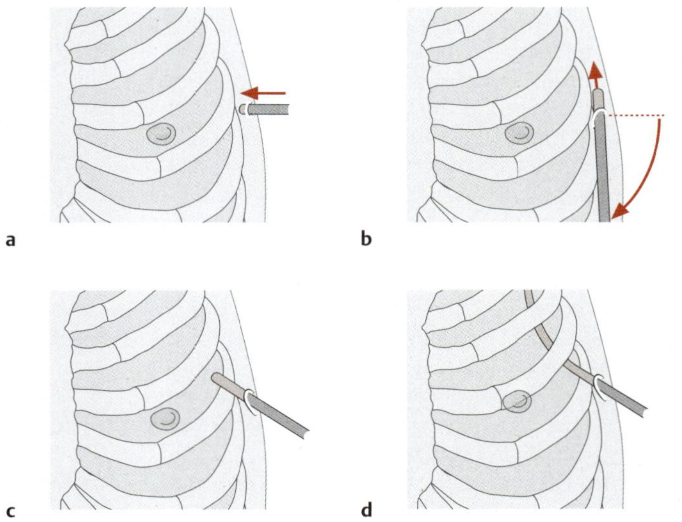

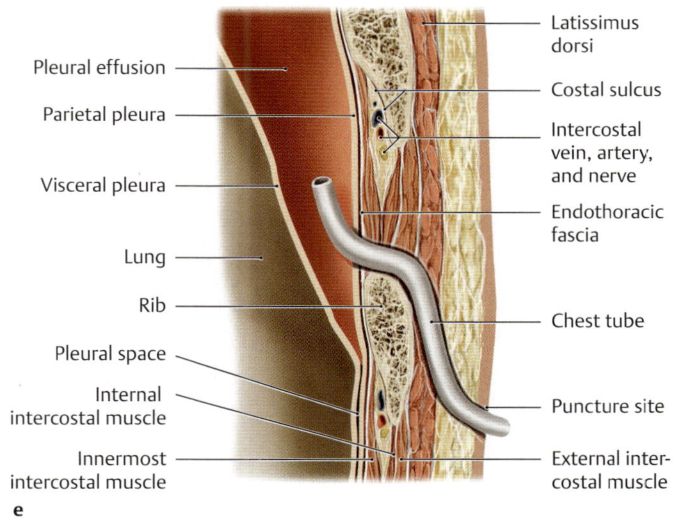

D Preserving the intercostal vein, artery, and nerve during the insertion of a chest tube

A chest tube (Bülau drain) may be inserted or a pleural puncture performed to drain an abnormal fluid collection from the pleural space, such as a pleural effusion due to bronchial carcinoma. The best site for placing the chest tube can be determined by percussion or ultrasound examination. Generally, one optimum puncture site in the sitting patient is at the level of the seventh or eighth intercostal space on the posterior axillary line (see **e** and p. 34). The drain should always be introduced at the *upper margin of a rib* to avoid injuring the intercostal vein, artery, and nerve. For additional puncture sites, see textbooks on surgery.

a–d Steps in the placement of a chest tube (anterior view):

a A skin incision is made under local anesthesia, and the drainage tube is introduced perpendicular to the chest wall.

b On reaching the ribs, the tube is angled 90° and advanced cephalad in the subcutaneous plane, parallel to the chest wall.

c On reaching the next higher intercostal space, the tube is passed through the intercostal muscles above the rib.

d The tube is then advanced into the pleural cavity.

e **Longitudinal section through the chest wall at the level of the posterior axillary line, after placement of the chest tube in the presence of pleural effusion.**

5.6 Anterior Trunk Wall: Nerves, Blood Vessels, and Lymphatics in the Female Breast

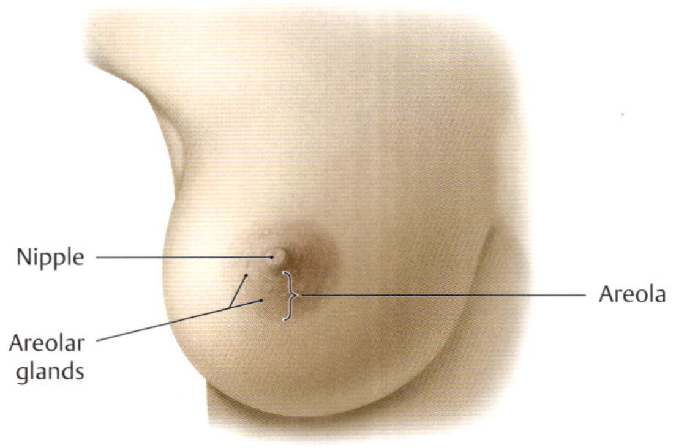

A Shape and appearance of the female breast
Right breast, anterior view. The female breast is shaped like a cone that is more rounded in its lower half than in the upper quadrants. It consists of the glandular tissue (mammary gland) and a fibrous stroma that contains fatty tissue. The excretory ducts of the glandular tissue open on the cone-shaped nipple, which lies at the center of the more heavily pigmented areola. Numerous small protuberances mark the openings of apocrine sweat glands and free sebaceous glands (areolar glands).

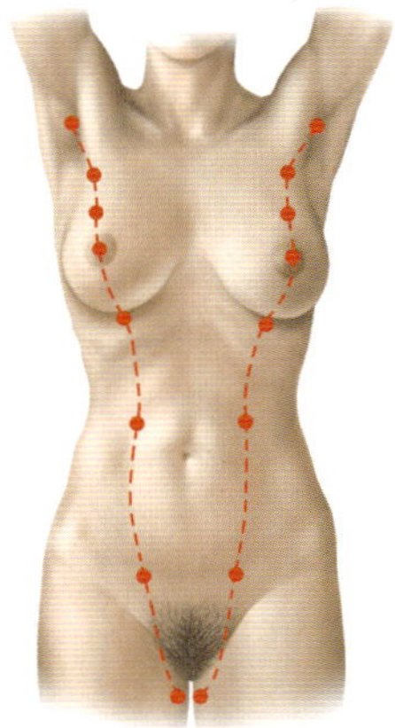

B The mammary ridges
The rudiments of the mammary glands form in both sexes along the mammary ridges, appearing as an epidermal ridge extending from the axilla to the inguinal region on each side. Although rarely the mammary ridges may persist in humans to form accessory nipples (polythelia), normally all the rudiments disappear except for the thoracic pair. By the end of fetal development, lactiferous ducts have sprouted into the subcutaneous tissue from the two remaining epithelial buds. After menarche, breast development in females is marked by growth of the fibrous stroma and proliferation of the glandular tree in response to stimulation by sex hormones.

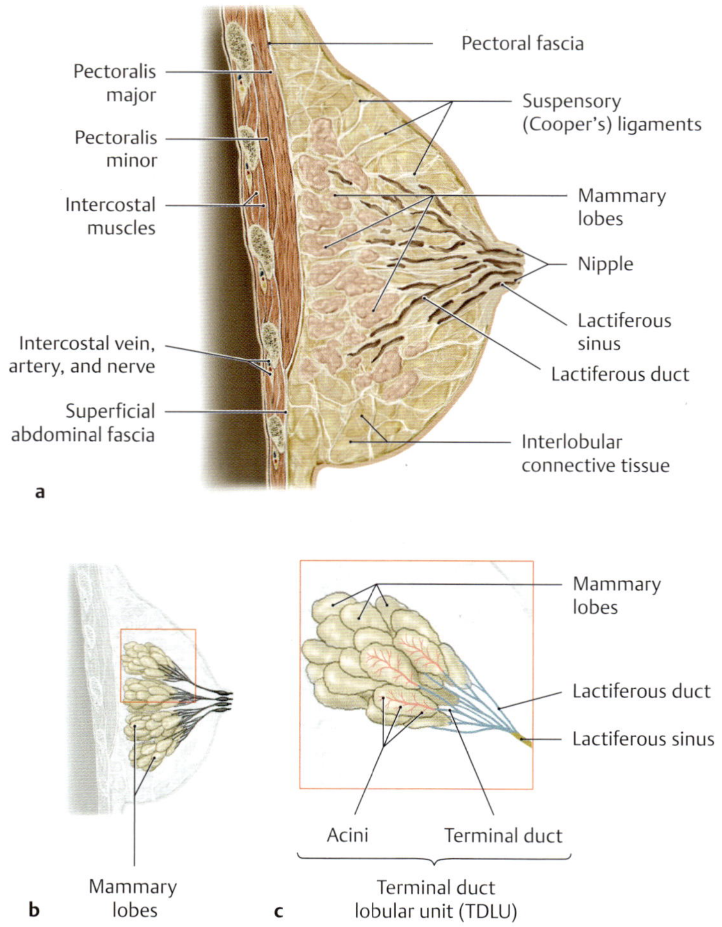

C Gross and microscopic anatomy of the breast

a The base of the adult female breast extends from the second to the sixth rib along the midclavicular line and directly overlies the pectoralis major, serratus anterior, and external oblique muscles. It is loosely attached to the pectoralis fascia and adjacent fascial planes (axillary and superficial abdominal fascia) by connective tissue. The breast is additionally supported, especially in its upper portion, by permeating bundles of connective tissue (the suspensory ligaments of the breast, or *Cooper's ligaments*). The glandular tissue is composed of 10 to 20 individual lobes, each of which has its lactiferous major duct that opens on the nipple by way of a dilated segment, the lactiferous sinus (structure of the lobe shown in **b**). The glands and lactiferous ducts are surrounded by firm, fibrofatty tissue that has a rich blood supply.

b Sagittal section of the duct system and portions of a lobe. A mammary lobe resembles a tree composed of branching lactiferous ducts, which terminate in smaller lobules (approximately 0.5 mm in diameter). In the *nonlactating breast* (as shown here), these lobules contain rudimentary acini that are arranged in clusters of small epithelial buds without a visible lumen.

c The terminal duct lobular unit (TDLU). One lobule and its terminal duct make up the basic secretory unit of the female breast. Each lobule is composed of acini that empty into a terminal ductule. The associated intralobular connective tissue (mantle tissue) contains stem cells that give rise to the tremendous cell growth (proliferation of the duct system and differentiation of the acini) that occurs during the transformation to the *lactating breast*. The TDLU is of key importance in pathohistology because it is the site where most malignant breast tumors originate (after Lüllmann).

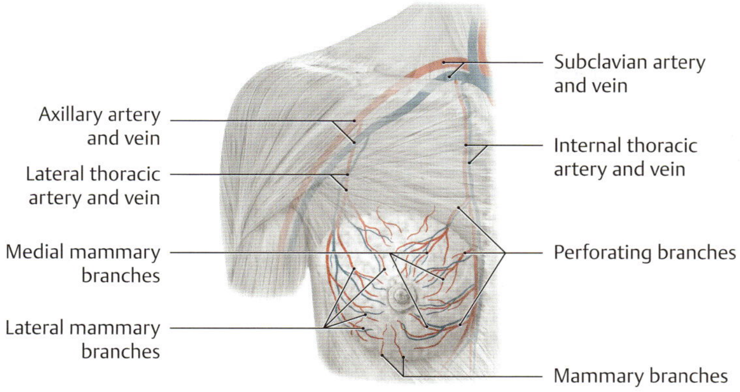

D Blood supply to the breast
The breast derives its blood supply from perforating branches of the internal thoracic artery (= medial mammary branches from the second through fourth intercostal spaces), branches of the lateral thoracic artery (lateral mammary branches), and direct branches from the second through fifth intercostal arteries (mammary branches). The breast is drained by the internal and lateral thoracic veins.

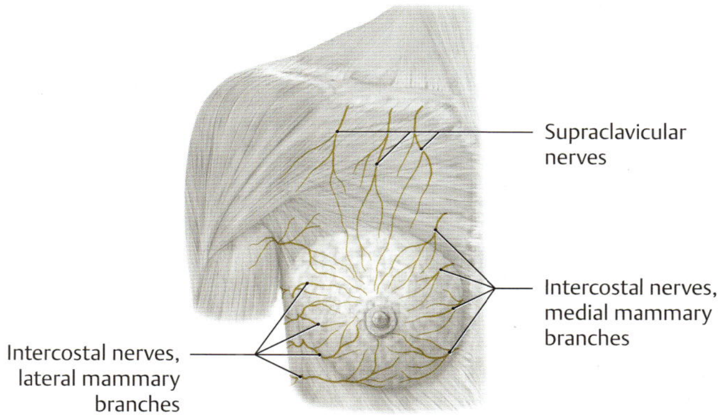

E Nerve supply to the breast
The sensory innervation of the breast has a segmental arrangement and is supplied by branches of the second through sixth intercostal nerves (lateral and medial mammary branches). Branches of the cervical plexus (supraclavicular nerves) also supply the upper portion of the breast.

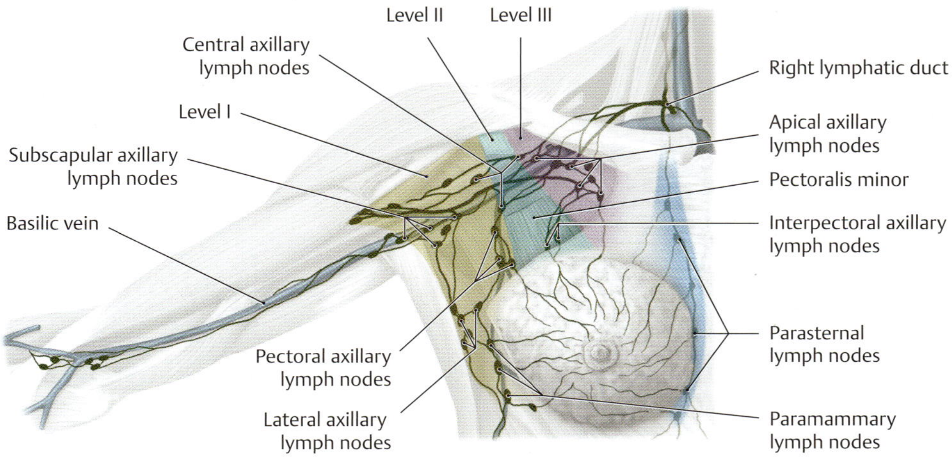

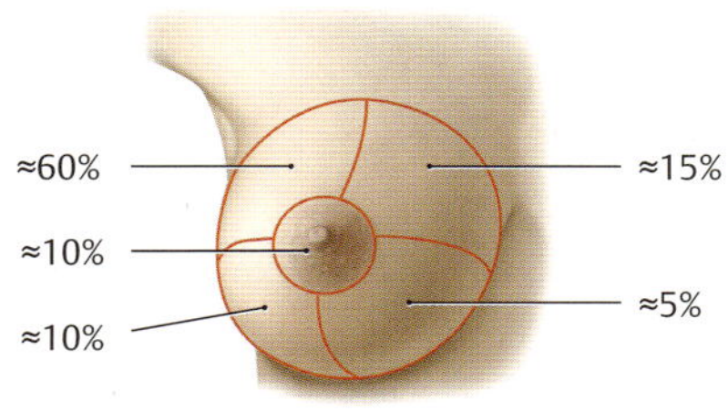

G Distribution of malignant tumors by quadrant in the female breast
The numbers indicate the average percentage location of malignant breast tumors.

F Lymphatic drainage of the breast
The lymphatic vessels of the breast can be divided into a superficial, subcutaneous, and deep system. The deep system begins with lymphatic capillaries at the acinar level (see **Cb** and **c**) and is particularly important as a route for tumor metastasis. The main regional filtering stations are the axillary and parasternal lymph nodes, the approximately *30 to 60 axillary lymph nodes* receiving most of the lymphatic drainage. They are the first nodes to be affected by metastasis (see **G**) and therefore have major oncological significance. The axillary lymph nodes are subdivided into levels (see p. 372):

- **Level I: lower axillary group** (lateral to the pectoralis minor):
  - Pectoral axillary lymph nodes
  - Subscapular axillary lymph nodes
  - Lateral axillary lymph nodes
  - Paramammary lymph nodes
- **Level II: middle axillary group** (at the level of the pectoralis minor):
  - Interpectoral axillary lymph nodes
  - Central axillary lymph nodes
- **Level III: upper infraclavicular group** (medial to the pectoralis minor):
  - Apical axillary lymph nodes

The *parasternal lymph nodes*, which are distributed along the thoracic vessels, chiefly drain the medial portion of the breast. From there, tumor cells may spread across the midline to the opposite side. The survival rate in breast cancer patients correlates most strongly with the number of involved lymph nodes at the various *axillary* nodal levels. The parasternal lymph nodes are rarely important in this regard. According to Henne-Bruns, Dürig, and Kremer, the 5-year survival rate is approximately 65% with metastatic involvement of level I, 31% with involvement of level II, but approaches 0% in patients with level III involvement. This explains the key prognostic importance of a *sentinel lymphadenectomy* (removal of the sentinel lymph node). This technique is based on the assumption that every point in the integument drains via specific lymphatic pathways to a particular lymph node, rarely draining to more than one. Accordingly, the lymph node that is the first to receive lymph from the primary tumor will be the first node to contain tumor cells that have spread from the primary tumor by lymphogenous metastasis. The specific lymphatic drainage path, and thus the sentinel node, can be identified by scintigraphic mapping with radiolabeled colloids (^{99m}TC sulfur microcolloid), which has superseded the older technique of patent blue dye injection. The first lymph node to be visualized is the sentinel node. That node is selectively removed and histologically examined for the presence of tumor cells. If the sentinel node does not contain tumor cells, generally the rest of the axillary nodes will also be negative. This method is 98% accurate in predicting the level of axillary nodal involvement prior to surgery.

5.7 Anterior Trunk Wall: The Inguinal Canal

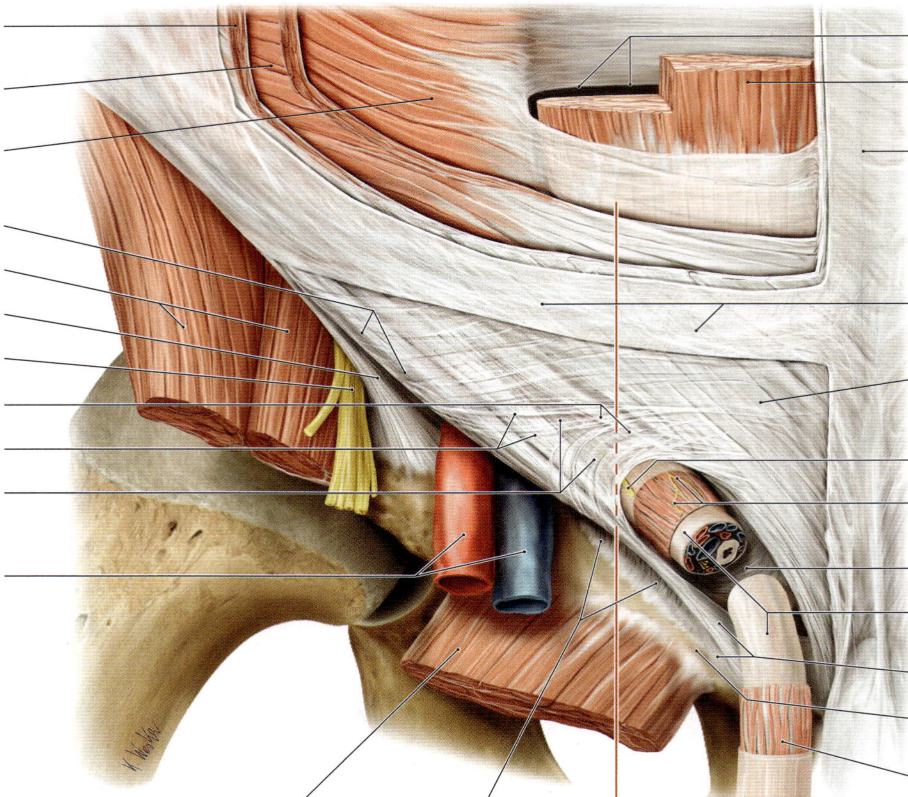

A Location of the inguinal canal in the male
Right inguinal region, anterior view. Approximately 4 to 6 cm long, the inguinal canal passes obliquely forward, downward, and medially above the inguinal ligament to pierce the anterior abdominal wall. It begins internally at the deep inguinal ring (**D** and **E**) in the lateral inguinal fossa (see p. 220) and opens externally at the superficial inguinal ring, lateral to the pubic tubercle. With the superficial abdominal fascia removed, this "external opening" of the canal can be identified as a slitlike orifice in the aponeurosis of the external oblique muscle (external oblique aponeurosis). It is bounded by the *medial crus* superomedially and by the *lateral crus* inferolaterally. Both crura are interconnected by the *intercrural fibers*. The superficial inguinal ring is completed internally by arched fibers from the inguinal ligament (reflected ligament), forming a deep groove. The inguinal canal in the male provides a pathway for the descent of the testis during fetal life (see p. 232). Its contents in the male (after testicular descent) include the spermatic cord, and its contents in the female include the round ligament of the uterus.

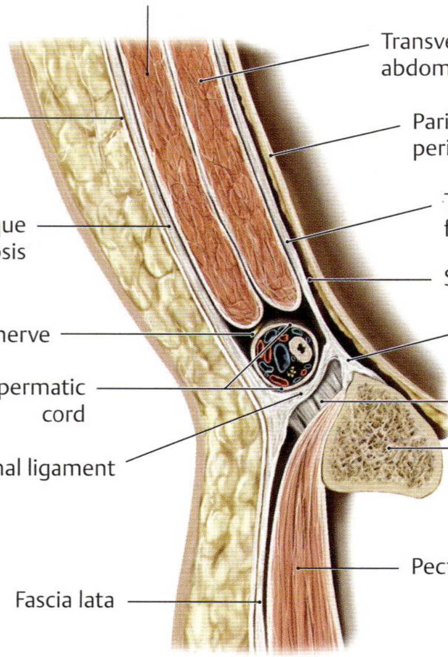

B Sagittal section through the inguinal canal in the male
Medial view. Note the structures that form the walls of the inguinal canal above and below the spermatic cord and on the anterior and posterior sides (compare with **C**). The openings and wall structures of the inguinal canal bear an important relationship to the pathophysiology of hernias.

C Openings and wall structures of the inguinal canal

The inguinal canal resembles a flattened tube with an internal and external opening (see below), a floor, a roof, and anterior and posterior walls. A lumen is present only after its contents have been removed (the spermatic cord in males, the uterine round ligament and its artery in females, the ilioinguinal nerve and lymphatic vessels in both sexes). The inguinal canal remains patent for life, especially in males, and thus forms a path for potential herniation through the abdominal wall (see p. 221).

Openings of the inguinal canal (see **A**)	
Superficial inguinal ring	Opening in the external oblique aponeurosis bounded by the medial crus, lateral crus, intercrural fibers, and reflected inguinal ligament
Deep inguinal ring	Opening between the interfoveolar ligament, inguinal ligament, and lateral umbilical fold; formed by an outpouching of the transversalis fascia (becomes the internal spermatic fascia) (see p. 220)
Wall structures of the inguinal canal (see **B**)	
Floor	Inguinal ligament (densely interwoven fibers of the lower external oblique aponeurosis and adjacent fascia lata of the thigh)
Roof	Transversus abdominis and internal oblique muscles
Anterior wall	External oblique aponeurosis
Posterior wall	Transversalis fascia and peritoneum (partially thickened by the interfoveolar ligament)

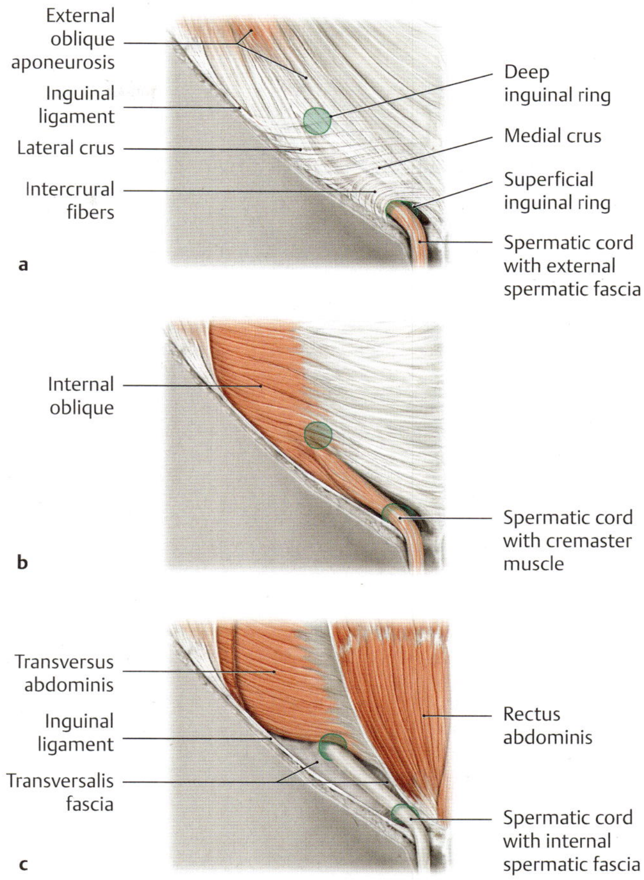

D Contribution of the oblique abdominal muscles to the structure of the male inguinal canal

Right inguinal region, anterior view.

a–c Progressive removal of the abdominal wall muscles.

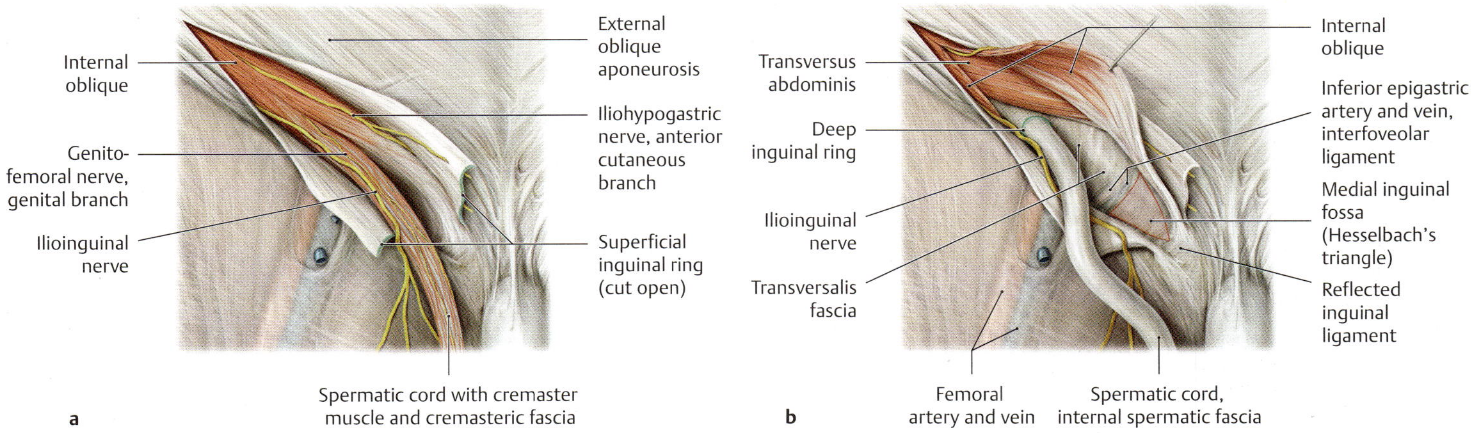

E The inguinal canal, progressively opened to expose the spermatic cord

a Division of the external oblique aponeurosis reveals the internal oblique muscle, some of whose fibers are continued onto the spermatic cord as the cremaster muscle. The genital branch of the genitofemoral nerve runs with it below the cremasteric fascia (see p. 550). The ilioinguinal nerve runs through the inguinal canal on the spermatic cord. Its sensory fibers pass through the superficial inguinal ring to the skin over the pubic symphysis and are distributed to the lateral portion of the scrotum or labia majora and medial thigh.

b With the internal oblique muscle divided and the cremaster muscle split, the full course of the spermatic cord through the inguinal canal can be displayed. The spermatic cord appears at the deep inguinal ring, where the transversalis fascia is invaginated into the inguinal canal (and encloses the spermatic cord on its way to the testis as the internal spermatic fascia). It runs below the transversus abdominis along the posterior wall of the inguinal canal (transversalis fascia and peritoneum). The wall at the midportion of the canal is formed by the interfoveolar ligament and is reinforced medially by the reflected inguinal ligament. Medial to the interfoveolar ligament, deep to which run the epigastric vessels, and superior to the inguinal ligament is the medial inguinal fossa, a weak spot in the abdominal wall that is a common site for direct inguinal hernias (see also p. 221).

5.8 Anterior Abdominal Wall: Anatomy and Weak Spots

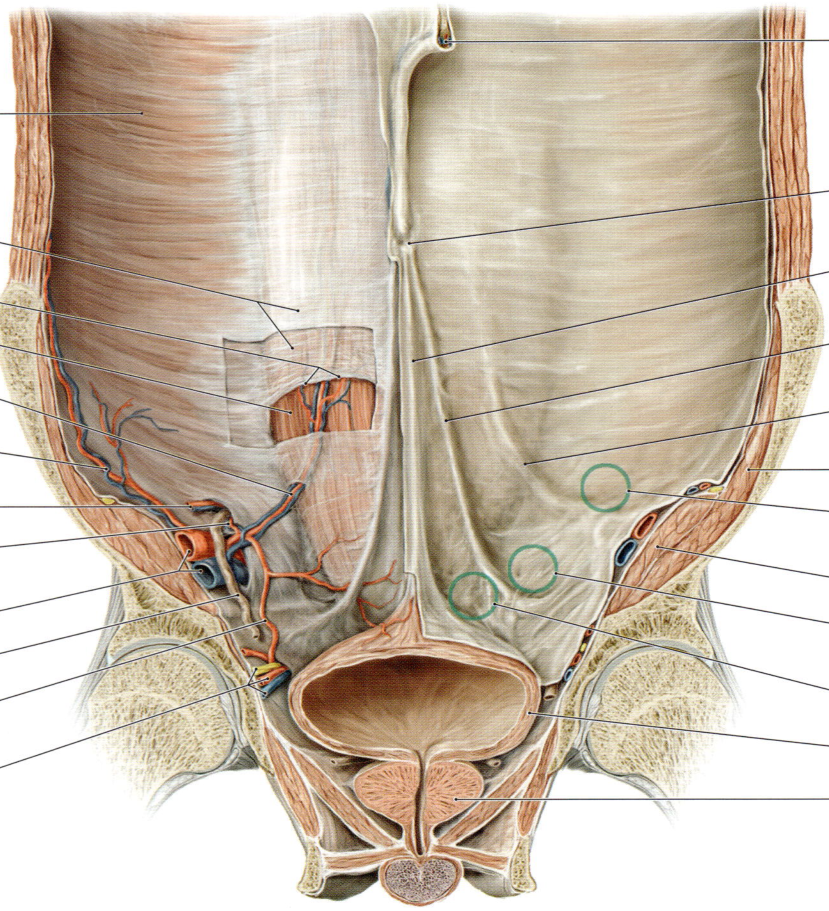

A Internal surface anatomy of the anterior abdominal wall in the male

Coronal section through the abdominal and pelvic cavity at the level of the hip joints, posterior view. All of the abdominal and pelvic organs have been removed except for the urinary bladder and prostate. Portions of the peritoneum and transversalis fascia have also been removed on the left side. The internal surface anatomy of the lower abdominal wall is marked by five peritoneal folds, which extend toward the umbilicus:

- An unpaired *median umbilical fold* on the midline (contains the obliterated urachus).
- Paired left and right *medial umbilical folds* (contain the left and right obliterated umbilical artery).
- Paired left and right *lateral umbilical folds* (contain the left and right inferior epigastric vessels)

Located between the peritoneal folds on each side are three more or less distinct fossae, which are sites of potential herniation through the anterior abdominal wall:

- The *supravesical fossa*, located between the median and medial umbilical folds above the apex of the bladder
- The *medial inguinal fossa* (Hesselbach's triangle), located between the medial and lateral umbilical folds
- The *lateral inguinal fossa*, located lateral to the lateral umbilical fold (site of the deep inguinal ring)

B Internal and external openings for abdominal hernias

Above the inguinal ligament, the median, medial, and lateral umbilical folds (see **A**) form three sites of weakness on each side of the abdominal wall where indirect and direct inguinal hernias and suprapubic hernias typically occur. Another weak spot is located *below the inguinal ligament* and medial to the femoral vein in the femoral ring. At that site the femoral ring is covered only by loose, compliant connective tissue, the femoral septum, which is permeated by numerous lymphatic vessels. The sharp-edged lacunar ligament forms the medial border of the femoral ring and can contribute to the incarceration of a femoral hernia (see p. 223).

Internal opening*	Hernia	External opening*
Above the inguinal ligament:		
Supravesical fossa	Supravesical hernia	Superficial inguinal ring
Medial inguinal fossa (Hesselbach's triangle)	Direct inguinal hernia	Superficial inguinal ring
Lateral inguinal fossa (deep inguinal ring)	Indirect inguinal hernia	Superficial inguinal ring
Below the inguinal ligament:		
Femoral ring	Femoral hernia	Saphenous opening

* The internal opening is where the hernia starts and the external opening is where it is observed clinically.

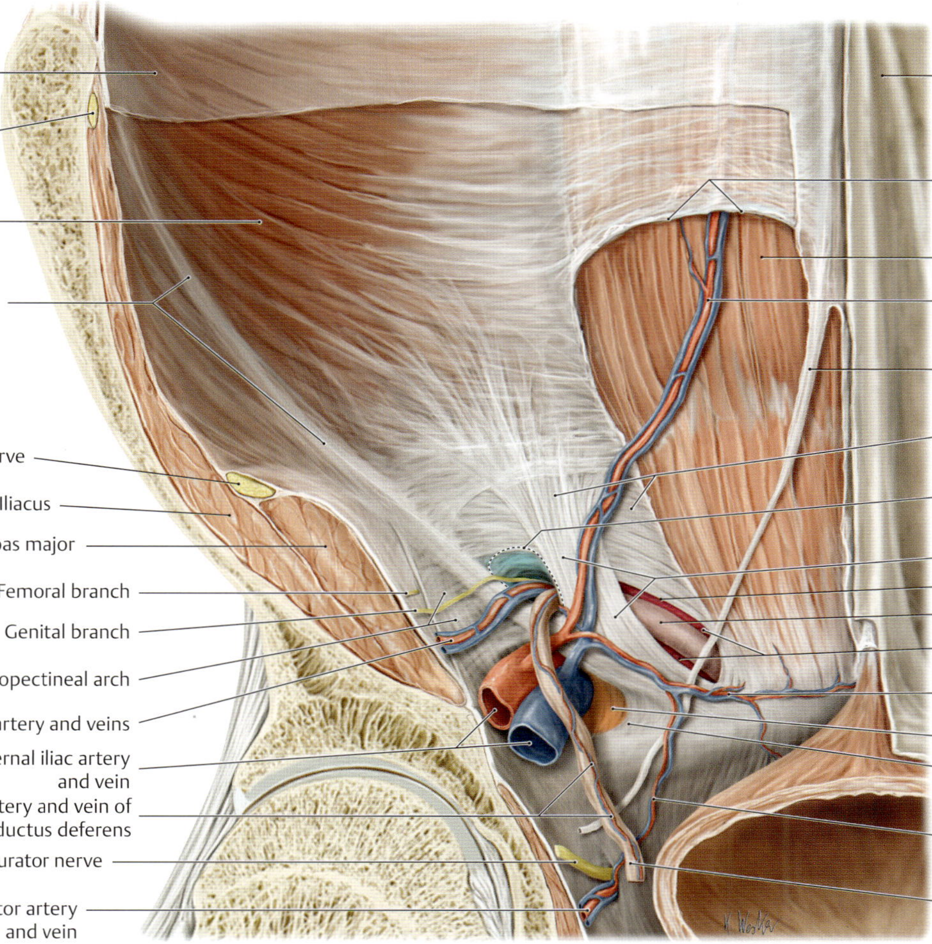

C Internal hernial openings in the male inguinal and femoral region

Detail from **A**, posterior view. The peritoneum and transversalis fascia have been partially removed to reveal the hernial openings more clearly. The interfoveolar ligament, which reinforces the transversalis fascia, was not removed. It forms the lateral boundary of the Hesselbach Triangle. The internal openings (see **A** and **B**) for indirect and direct inguinal hernias, femoral hernias, and suprapubic (= supravesical) hernias are indicated by color shading.

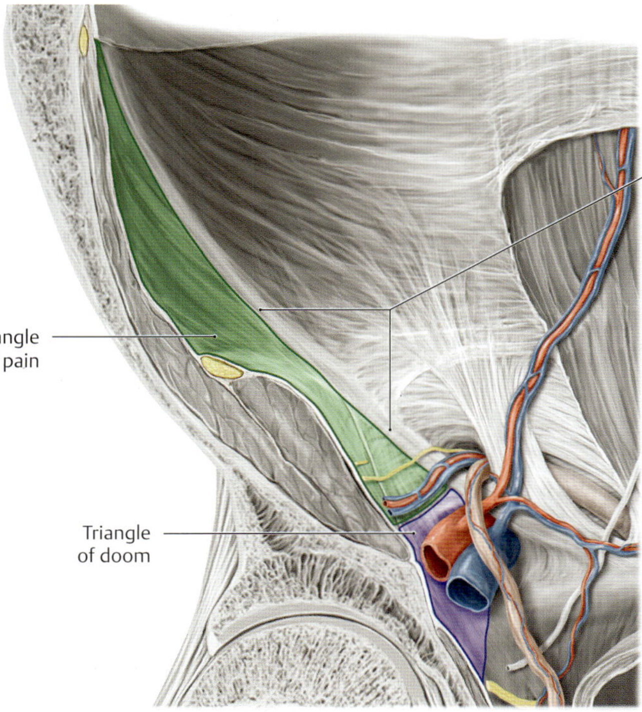

D Triangle of doom and triangle of pain

In two key regions in the preperitoneal space known respectively as the "triangle of pain" and the "triangle of doom," it is important to avoid placing clips to secure mesh implants, particularly in laparoscopic hernia repairs (see p. 227). The iliopubic tract (the union of the transverse fascia and the inguinal ligament) defines the superior border of both triangles. The triangle of pain lies lateral to the testicular vessels, the triangle of doom medial to them. The triangle of pain contains both branches of the genitofemoral nerve, the femoral nerve, and the lateral femoral cutaneous nerve. Injury to these nerves can result in pain and dysesthesia. The testicular vessels and the major vessels of the leg (the external iliac vessels) course through the triangle of doom. Injury to these vessels can lead to poorly controllable bleeding.

5.9 Inguinal and Femoral Hernias

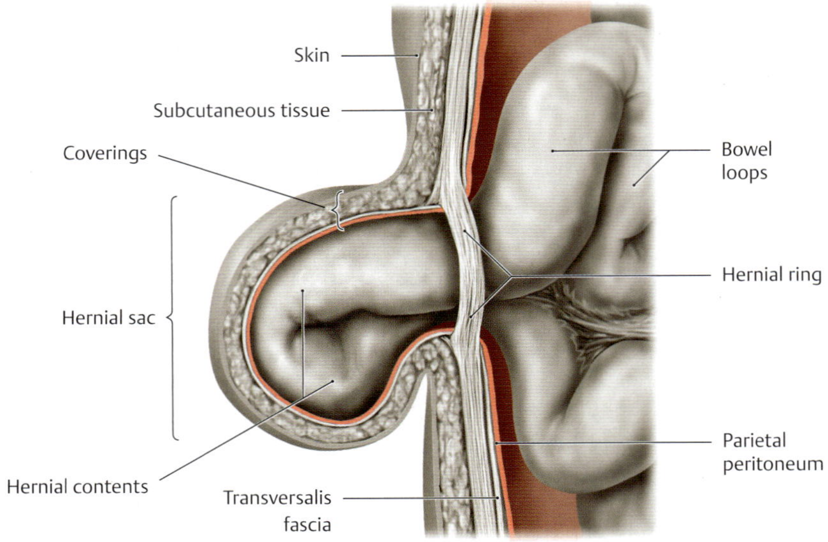

A Definition, occurrence, and structure of a hernia
Between the thorax and the bony pelvis lies an extensive gap in the skeleton which is bridged by a multi-layered abdominal wall of broad muscles, fasciae, aponeuroses, and peritoneum. The muscular layer is absent at certain specific locations in the abdominal wall. At these locations, the abdominal wall is composed of connective tissue structures such as the transverse fascia, which also forms the posterior aspect of the inguinal canal. These areas are weak points in the abdominal wall. Especially where the connective tissue is weakened, as occurs in elderly persons, they occasionally fail to resist intra-abdominal pressure and become sites of hernias. Triggering events include lifting heavy objects or even coughing. The term hernia (Latin hernia = rupture) refers to the protrusion of the parietal peritoneum through a preexisting gap (inguinal or femoral hernia) or an acquired gap (hernia through a scar). A hernia that exits the abdominal space and is visible on the surface of the body is referred to as an external hernia; a protrusion into a peritoneal pocket is referred to as an internal hernia. As to the time of occurrence, one differentiates congenital hernias (such as an umbilical hernia or indirect inguinal hernia in a patent processus vaginalis) and acquired hernias (such as a direct inguinal hernia or femoral hernia). The following components of a hernia are important from a surgical standpoint:

- **Hernial opening:** the orifice or defect through which the viscus herniates (inguinal, femoral, or scar hernia).
- **Hernial sac:** the pouch, generally lined by parietal peritoneum, that contains the herniated viscus. Its size is highly variable, depending on the extent of the hernia.
- **Hernial contents:** may be almost any intra-abdominal viscus but usually consist of greater omentum or loops of small bowel.
- **Coverings:** the tissue layers surrounding the hernial sac. The composition of the coverings depends on the location and mechanism of the hernia.

B Hernias of the groin region (inguinal and femoral hernias)*

Hernia	Openings and course
• **Direct (medial) inguinal hernia** Always acquired	• *Internal opening:* medial inguinal fossa (Hesselbach's triangle), i.e., above the inguinal ligament and medial to the inferior epigastric artery and vein • *Course:* sac is perpendicular to the abdominal wall • *External opening:* superficial inguinal ring
• **Indirect (lateral) inguinal hernia** Congenital (patent processus vaginalis) or acquired	• *Internal opening:* deep inguinal ring, i.e., above the inguinal ligament and lateral to the inferior epigastric artery and vein • *Course:* sac passes through the inguinal canal • *External opening:* superficial inguinal ring
• **Femoral hernia** Always acquired	• *Internal opening:* femoral ring and septum, i.e., below the inguinal ligament and medial to the femoral artery and vein • *Course:* sac passes through the femoral canal below the fascia lata • *External opening:* saphenous opening

*80% of all hernias are inguinal hernias (90% of which occur in males), and approximately 10% are femoral hernias (more common in females) (see also p. 224). Inguinal hernias are among the most common structural defects in humans, accounting for some 20% of all surgical operations.

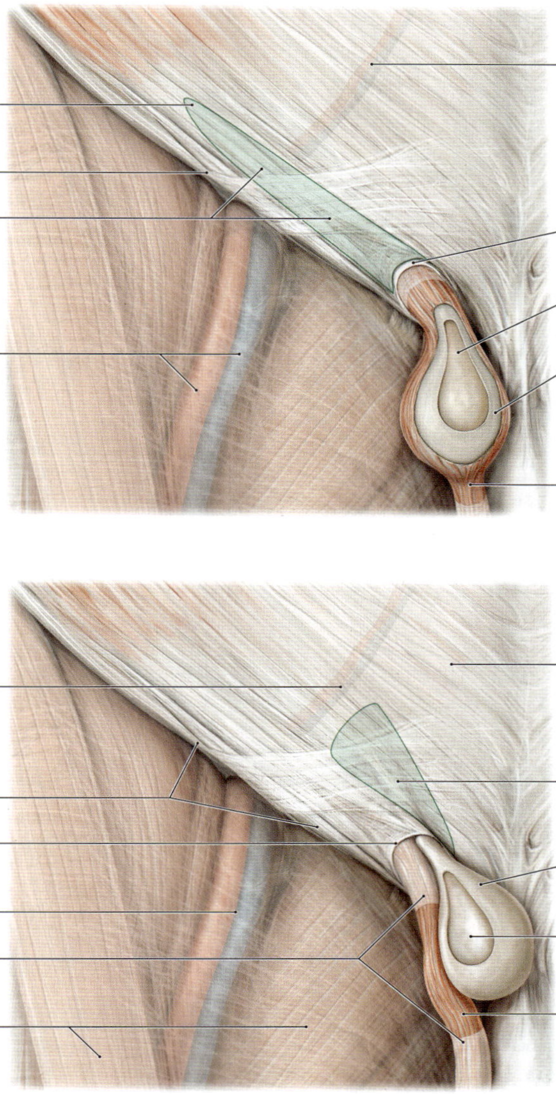

C Indirect and direct inguinal hernia

Right male inguinal region with the skin and superficial body fascia removed, anterior view. The fascia lata of the thigh is shown transparent, and the spermatic cord is windowed.

a Indirect (lateral) inguinal hernia (congenital or acquired): The hernial sac passes parallel to the abdominal wall through the inguinal canal and can descend into the scrotum. The internal hernial ring is the expanded deep inguinal ring lying lateral to the epigastric vessels; the external hernial ring is the superficial inguinal ring.

b Direct (medial) inguinal hernia (always acquired): The hernial sac courses directly through and perpendicular to the abdominal wall without the "detour" through the inguinal canal (therefore "direct hernia"). The internal hernial ring in this case is located in the triangle of Hesselbach, which lies medial to the epigastric vessels. The external hernial ring is the superficial inguinal ring, as in the indirect inguinal hernia. The exposed hernial sac usually emerges medial to the spermatic cord.

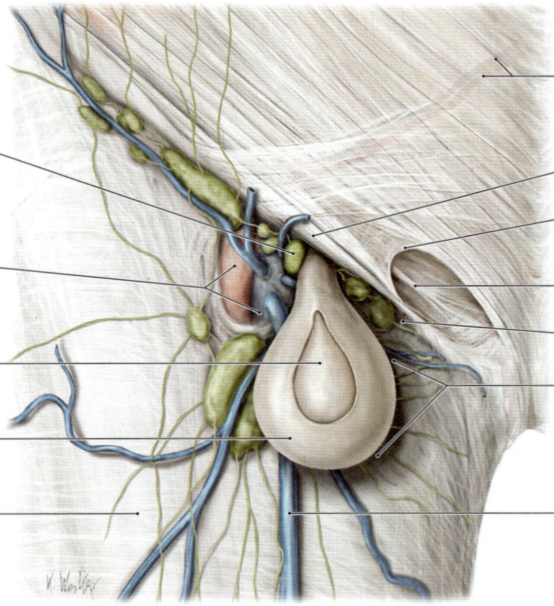

D Acquired femoral hernia

Right female inguinal region with the skin and superficial fascia removed, anterior view. Femoral hernias are always acquired and are more common in women (broader pelvis and larger femoral ring). They always pass inferior to the inguinal ligament and medial to the femoral vein, extending through the femoral ring to enter the femoral canal (not visible here). The funnel-shaped canal begins at the femoral ring (the internal opening of the hernia, see p. 220), ends approximately 2 cm inferiorly at the saphenous opening, and lies anterior to the pectineal fascia. The sharp-edged lacunar ligament forms the medial boundary of the femoral ring (risk of incarceration). Typically, the femoral canal is occupied by loose fatty connective tissue and deep inguinal lymph nodes. Femoral hernias may emerge at the saphenous opening (external opening), which is covered by the thin cribriform fascia, and thus become subcutaneous.

5.10 Topographic Anatomy of Inguinal Hernias

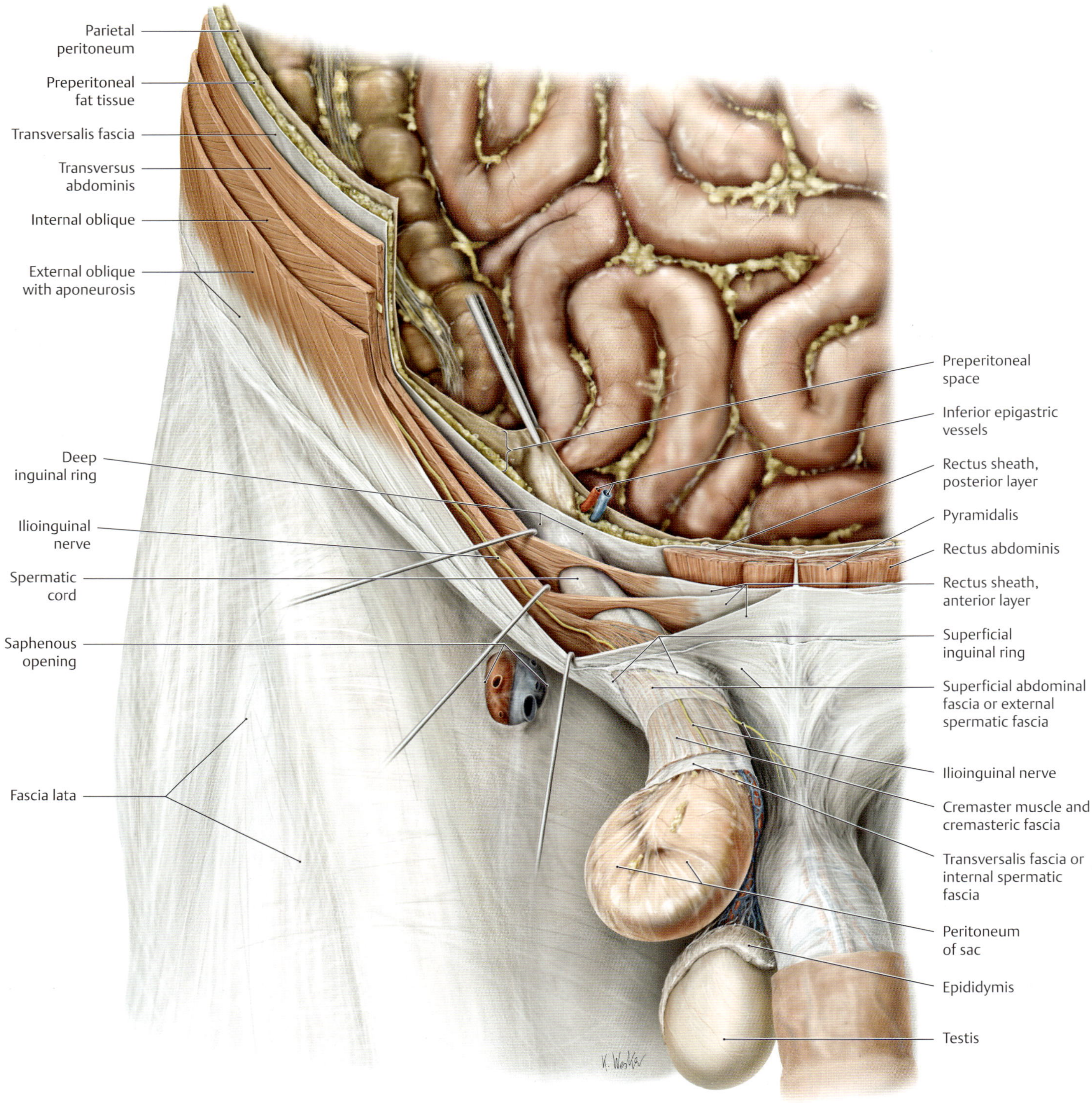

A Acquired, indirect inguinal hernia in a male (right side)
Anterior view, skin and abdominal walls largely removed. In the area of the inguinal canal that runs from upper lateral to lower medial, the individual abdominal wall muscles have been spread apart to reveal the relationship of structures. The hernial sac of the indirect inguinal hernia (a hernia that runs through the inguinal canal) containing an intestinal loop is visible. Because the hernial sac is within the spermatic cord, the hernia is covered by parietal peritoneum, transversalis fascia, and the cremaster muscle and cremasteric fascia.

Note: The position of the internal and external hernial rings and what is known as the preperitoneal space between the peritoneum and the transverse fascia, which is filled with loose fatty connective tissue. This space is surgically opened in an inguinal or femoral hernia repair in order to perform a mesh reconstruction to reinforce the posterior wall of the inguinal canal (see p. 227).

B Schematic representation of the inguinal canal, the abdominal wall layers, and their continuation into the covering of the testes

Anterior view; several section planes have been combined in this schematic. The two arrows point to the lateral inguinal fossa (= deep inguinal ring) and the medial inguinal fossa (Hesselbach's triangle). The inferior epigastric vessels extend between them in the lateral umbilical fold. The lateral inguinal fossa is the internal hernial orifice for indirect inguinal hernias (see **Da**); the medial inguinal fossa, for direct inguinal hernias (see **Db**). Both types of inguinal hernias use the superficial inguinal ring as the external hernial orifice.

Note: The tunica vaginalis with its visceral layer and parietal layer is the remnant of the obliterated processus vaginalis (**b**) (compare p. 232).

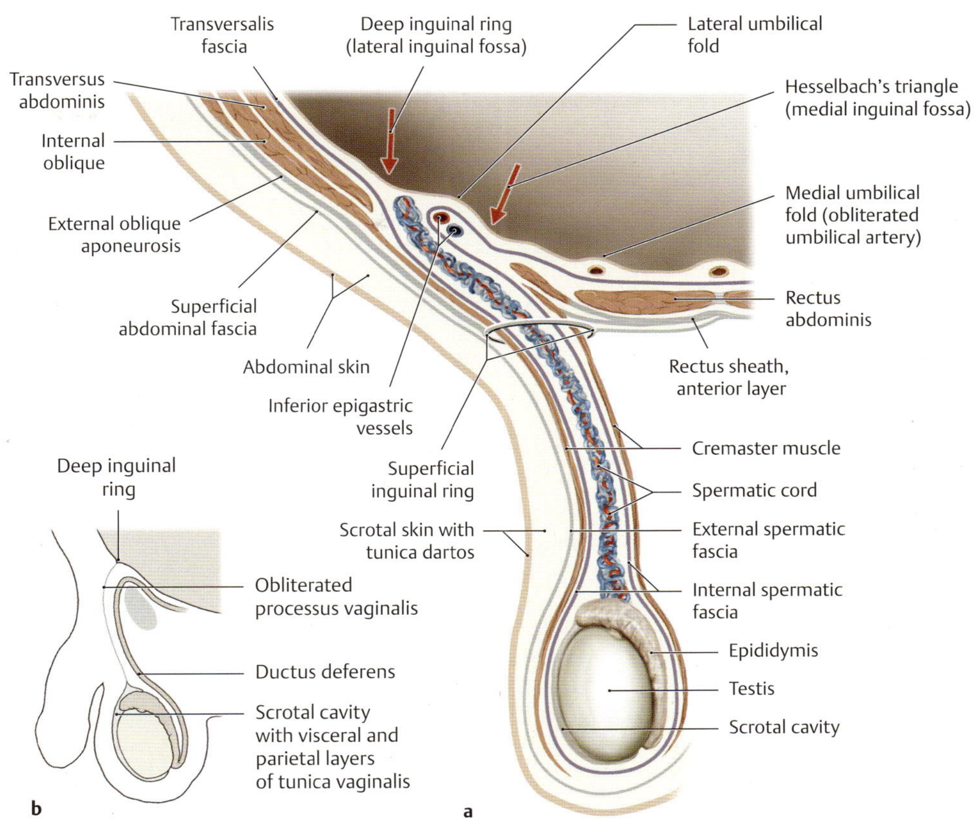

C Comparison of layers of the abdominal wall and the corresponding coverings of the spermatic cord and testes

The coverings of the spermatic cord and testes are derivatives of muscles and fasciae of the abdominal wall and envelop the spermatic cord and testes in a sac (scrotum), whose superficial layer is formed from the abdominal skin.

Abdominal wall layers	Coverings of the spermatic cord and testes
• Abdominal skin and membranous layer of superficial fascia	→ Scrotal skin with tunica dartos
• External oblique fascia	→ External spermatic fascia
• Internal oblique muscle and fascia	→ Cremaster muscle with cremasteric fascia
• Transversalis fascia	→ Internal spermatic fascia
• Peritoneum	→ Tunica vaginalis, parietal layer and visceral layer

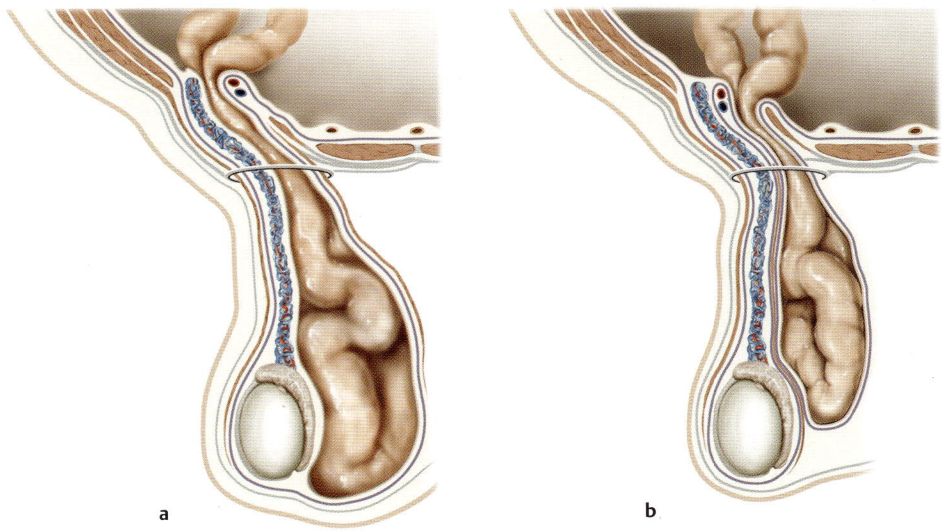

D Comparison of indirect and direct inguinal hernias

a Indirect inguinal hernia (congenital or acquired): The inner hernial orifice (lateral inguinal fossa = deep inguinal ring) is lateral to the inferior epigastric vessels. The hernial contents (e.g., intestinal loops), which are enclosed by the parietal peritoneum, either push through the inguinal canal to the scrotum (acquired hernia) or move through the still-open processus vaginalis (see **Bb**) to the scrotum (congenital hernia). The hernial sac is made of the same elements in both cases: parietal peritoneum, transversalis fascia, and cremaster muscle.

b Direct inguinal hernia (acquired): The inner hernial orifice (medial inguinal fossa = Hesselbach's triangle) is medial to the inferior epigastric vessels. The hernial contents push directly through the abdominal wall and via the superficial inguinal ring to the scrotum. In comparison to indirect inguinal hernias, the hernial sac consists only of parietal peritoneum and transversalis fascia.

5.11 Diagnosis and Treatment of Hernias

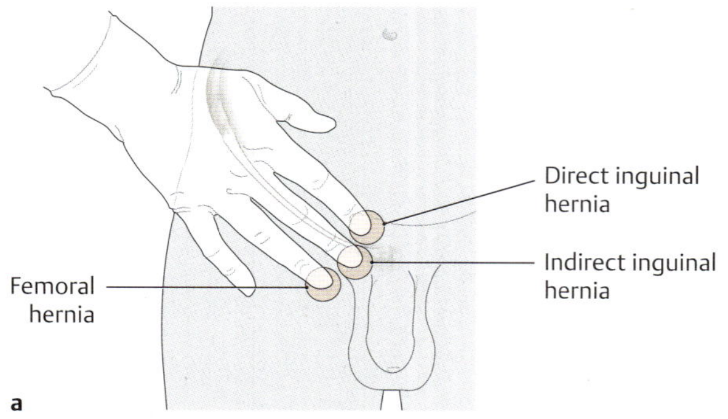

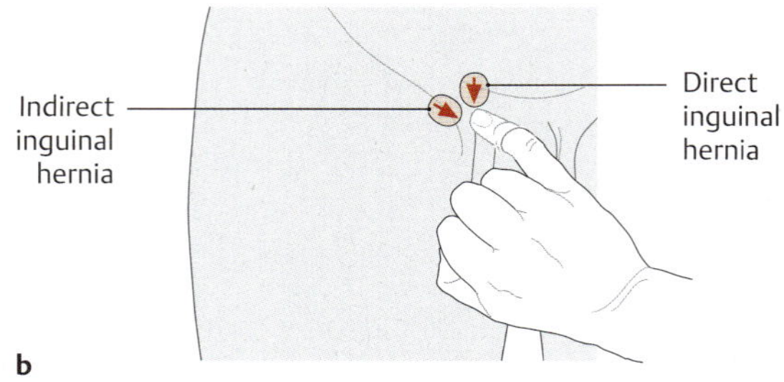

A Technique for the examination of inguinal and femoral hernias
Hernias of the groin region, like most hernias, are typically precipitated by a rise of intra-abdominal pressure (e.g., due to coughing, sneezing, or straining) and present as a palpable bulge or swelling in the inguinal region. Usually this swelling regresses spontaneously when the patient lies down; thus, the clinical examination is performed with the patient in a standing position. Spontaneous pain is generally absent with an uncomplicated hernia, and a foreign body sensation is more common. Persistent pain accompanied by a feeling of pressure at the hernia site, nausea, and vomiting are signs of incarceration (see **C**). In patients with a bulge in the inguinal region or a scrotal mass (see p. 235), the differential diagnosis should include hydrocele, varicocele, ectopic testis, lymphoma, and other tumors of the testis or epididymis. Since inguinal and femoral hernias are of the external type and are easily accessible to inspection and palpation, the diagnosis is generally made clinically.

a **Palpation from the iliac spine (the three-finger rule):** The "three-finger rule" makes it easier to appreciate the topographical anatomy of inguinal and femoral hernias and differentiate among direct and indirect inguinal hernias and femoral hernias. When the examiner places the thenar eminence on the anterior superior iliac spine, the index finger points to a direct hernia, the middle finger to an indirect hernia, and the ring finger to a femoral hernia. Thus, when a hernia is felt below the index finger, for example, this means that the patient has a direct inguinal hernia.
Caution: The hernial sac protrudes through the superficial inguinal ring in both direct and indirect hernias, so there is no difference between them based on inspection or palpation.

b **Palpation from the scrotum:** This technique is particularly useful for palpating smaller inguinal hernias in the standing patient. By invaginating the scrotum and groin skin, the examiner palpates along the spermatic cord to the superficial inguinal ring and touches the posterior wall of the inguinal canal with the pad of the finger. When the patient coughs, an experienced examiner can distinguish a direct hernia, which strikes the pad of the finger, from an indirect hernia, which strikes the distal tip of the finger.

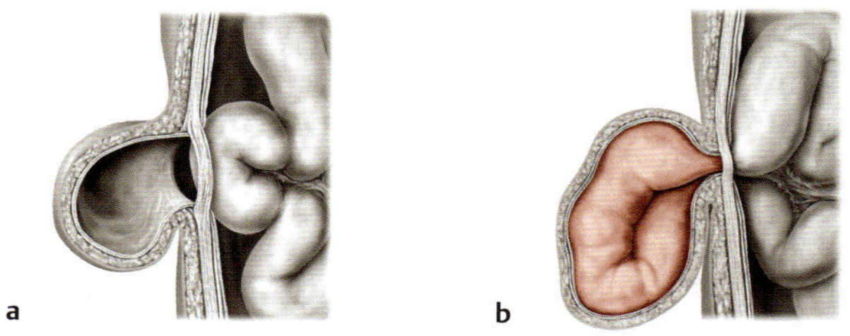

B Complete reduction of a hernia and an incarcerated hernia (strangulation)

a When the herniated viscus can move freely within the hernial sac and at the hernial opening, generally the hernia can be reduced spontaneously (e.g., by lying down) or by manual manipulation. As a result, there is no risk of acute incarceration.

b Incarceration is the most serious complication of a hernia. Strangulation at the neck of the hernia restricts blood flow to the herniated bowel, with effects ranging from ischemia to necrosis. The patient may develop symptoms of a functional or mechanical bowel obstruction with a life-threatening interruption of intestinal transit due to narrowing or obstruction of the bowel lumen. Given the risk of bowel perforation and peritonitis in these cases, immediate surgery is indicated (see **C**).

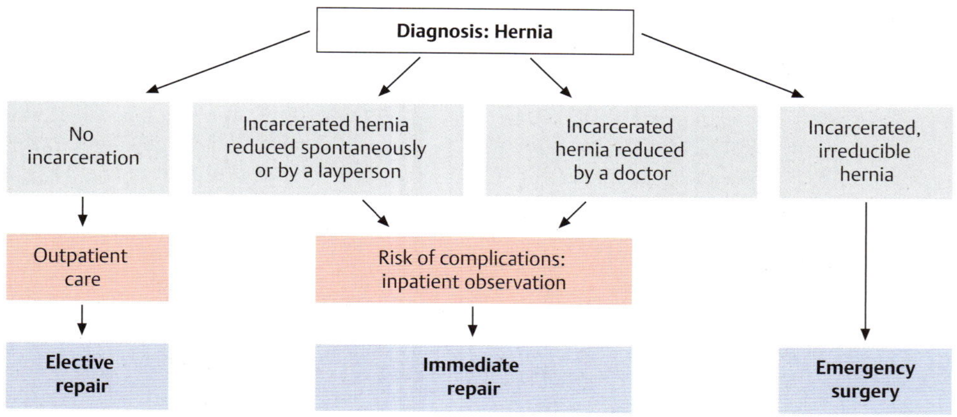

C Hernia symptoms and timing of the repair (after Henne-Bruns, Dürig, and Kremer)
As a rule, hernias do not respond definitively to conservative treatment (e.g., trusses or binders), and a permanent reduction is achieved only by surgical closure of the hernial opening (see **E**). The timing of the repair is based on the clinical presentation, i.e., the presence of a reducible, irreducible, or incarcerated hernia.

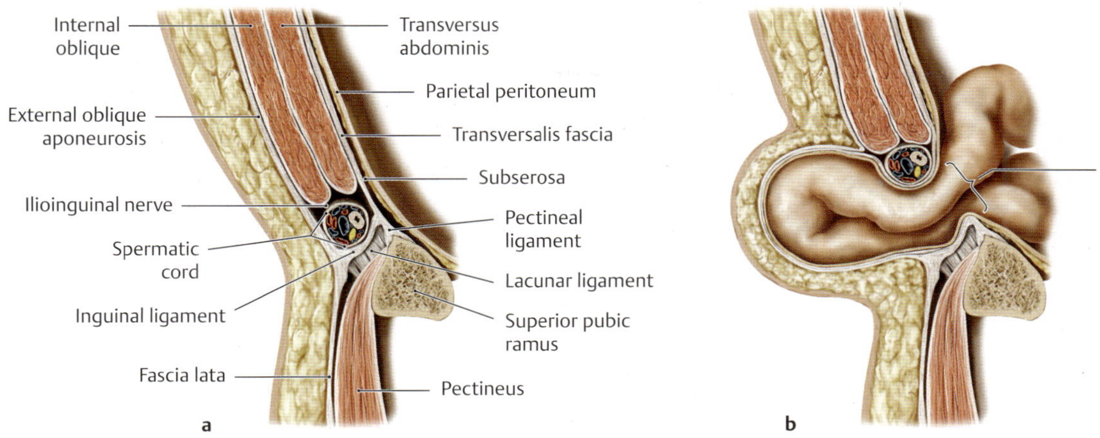

D Condition after direct, acquired inguinal hernia
Sagittal sections through the male inguinal region: **a** normal; **b** direct, acquired inguinal hernia.

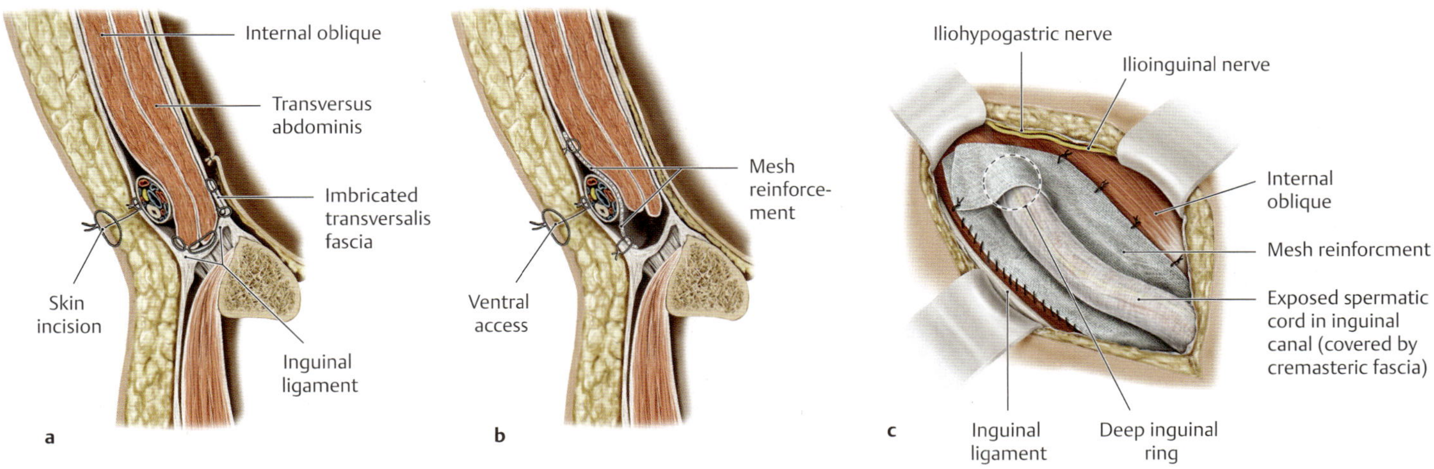

E Inguinal hernia repair
Sagittal section through the male inguinal region.
Various surgical methods are available for hernia repair. They differ mainly in the technique used to reinforce the posterior wall of the inguinal canal. In all methods, the steps are as follows:

- Expose the hernial sac.
- Reduce the hernial contents.
- Close the hernial orifice and thus restore the stablility of the abdominal walls.

In younger patients, in which there is a strong relationship between fascia and muscle, reconstruction usually is performed without the use of foreign material to reinforce the abdominal wall (e.g., the Shouldice technique). In older patients or in the case of a recurring hernia, the abdominal wall is usually reinforced with a mesh patch (e.g., the Lichtenstein technique) during surgery.

a Shouldice technique: The posterior wall of the inguinal canal is reinforced by overlapping the transversalis fascia and suturing the internal oblique and transversus abdominis muscles to the inguinal ligament in two layers.
b, c Lichtenstein technique: With this tension-free repair, the abdominal wall is reinforced by placing a synthetic mesh patch posterior to the internal oblique and transversus abdominis.
d–f Mesh reinforcement through totally extraperitoneal (TEP) or transabdominal preperitoneal (TAPP) repair. The abdominal wall is reinforced by placing a mesh patch through a peritoneal incision anterior to the internal oblique and transversus abdominis in the preperitoneal space between the transversalis fascia and parietal peritoneum (**d**). Whereas in TEP the peritoneal cavity is not entered (extraperitoneal, **e**), in TAPP the parietal peritoneum has to be entered from inside the peritoneal cavity (transabdominal, **f**).

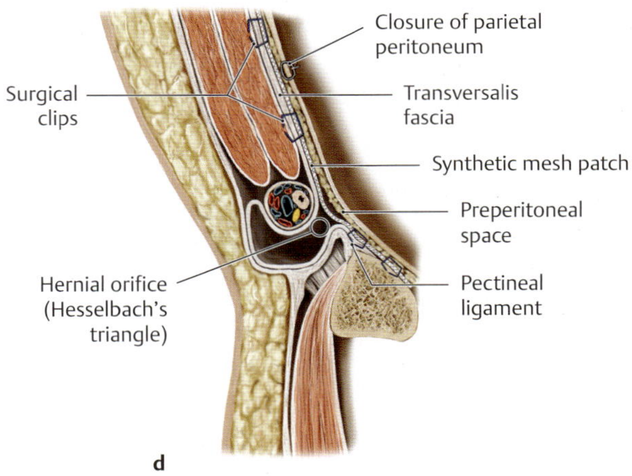

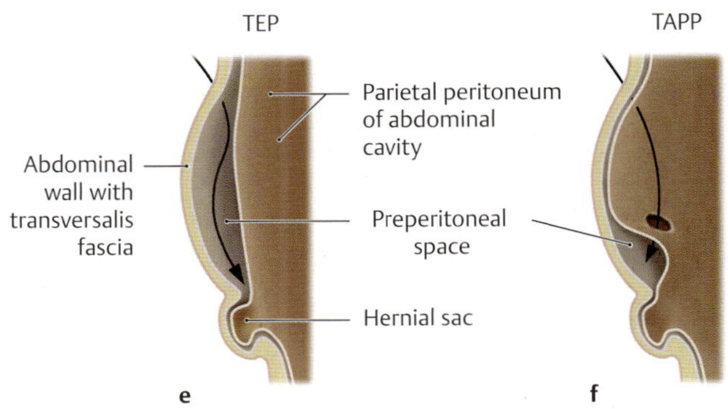

5.12 Rare External Hernias

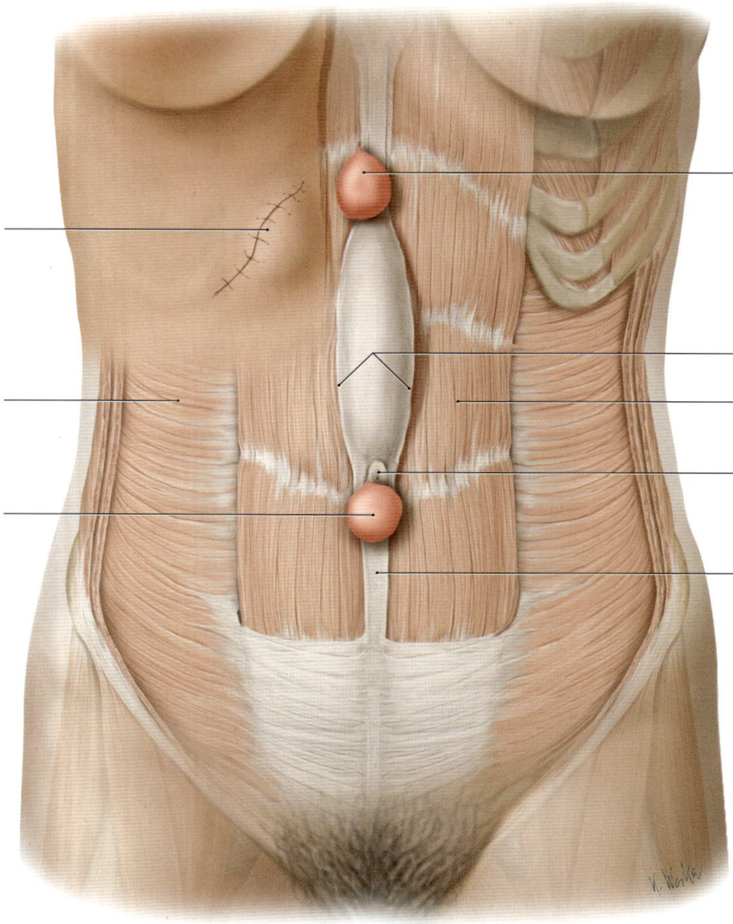

A Location of hernias in the anterior abdominal wall

B Hernias of the anterior abdominal wall*

Hernia	Location, occurrence, and typical features
• **Umbilical hernia**	• Umbilical region, passing through the umbilical ring: – Congenital umbilical hernia: incomplete regression of the normal fetal umbilical hernia due to scarring of the umbilical papilla (hernial sac: amnion and peritoneum) – Acquired umbilical hernia: common after multiple pregnancies, also in association with obesity, hepatic cirrhosis, or ascites (secondary widening of the umbilical ring)
• **Omphalocele**	• Congenital persistence (1 : 6,000) of an abdominal wall defect with an incomplete reduction of abdominal viscera during fetal life; unlike an umbilical hernia, the omphalocele is not covered by skin or subcutaneous tissue but only by peritoneum, mucous connective tissue (Wharton jelly), and amniotic epithelium (so the contents are easily recognized)
• **Epigastric hernia**	• The hernial openings are gaps in the linea alba excluding the umbilicus (on a continuum with rectus diastasis, see below)
• **Rectus diastasis****	• The rectus muscles separate at the linea alba when the abdominal muscles are tightened, creating a site for potential herniation (the hernia reduces on relaxation, and complaints are rare)
• **Incisional hernia**	• Occurs at a previous incision site (usually in the upper abdominal midline)

* Umbilical and epigastric hernias comprise approximately 10% of all hernias.
** The typical hernias of the anterior abdominal wall do not include diastasis recti.

C Other rare hernias occurring elsewhere on the trunk*

Hernia	Location
• **Spigelian hernia**	• Anterior abdominal wall between the semilunar line and lateral rectus sheath, usually at the level of the arcuate line
• **Lumbar hernia**	• Between the twelfth rib and iliac crest: – Superior lumbar hernia (superior costolumbar triangle, Grynfeltt's triangle): between the twelfth rib and iliocostalis – Inferior lumbar hernia (inferior iliolumbar triangle, Petit's triangle): between the iliac crest, latissimus dorsi, and externus oblique muscles
• **Obturator hernia**	• Through the obturator foramen and then between the pectineus, adductor longus, and obturator externus muscles
• **Sciatic hernia**	• Through the greater sciatic foramen: – Suprapiriform hernia (above the piriformis muscle) – Infrapiriform hernia (below the piriformis muscle) – Spinotuberous hernia (in front of the sacrotuberous ligament)
• **Perineal hernia**	• Through the pelvic floor: – Anterior perineal hernia (in front of the deep transverse perineal) – Posterior perineal hernia (behind the deep transverse perineal) – Ischiorectal hernia (through the levator ani into the ischioanal fossa)

* Less than 1% of all hernias, generally acquired (after Schumpelick).

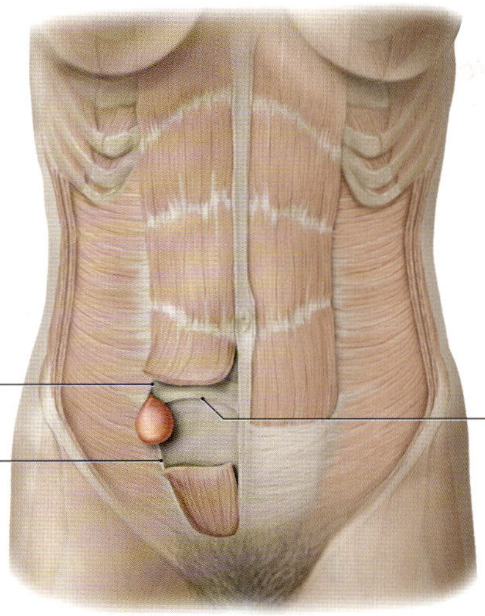

D Spigelian hernia

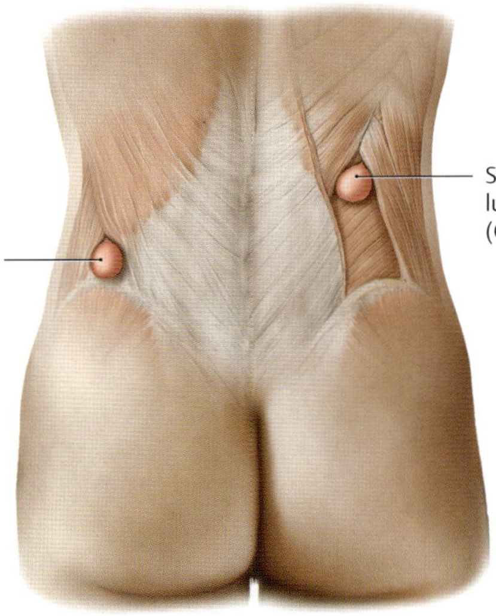

E Lumbar hernia

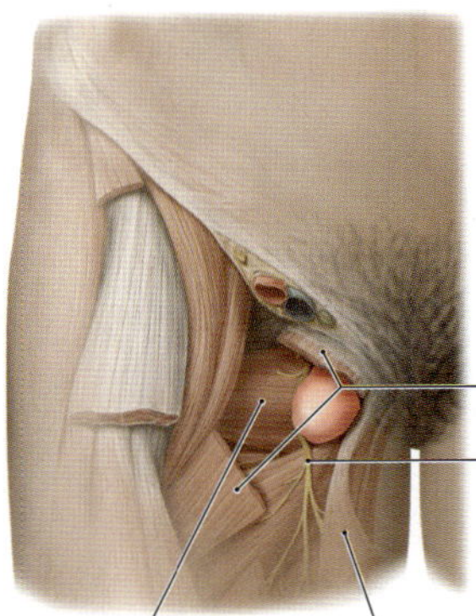

F Obturator hernia

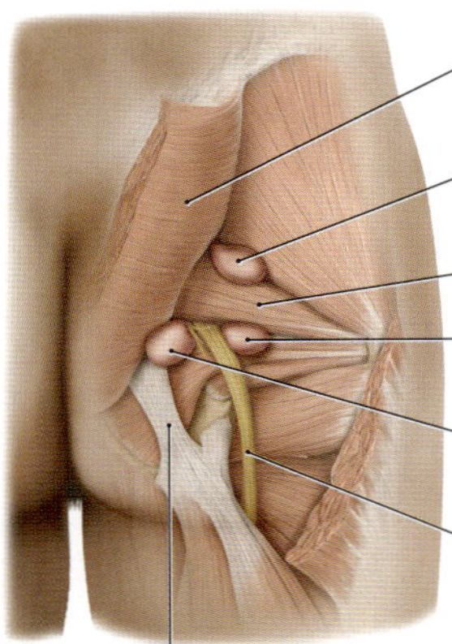

G Sciatic hernia

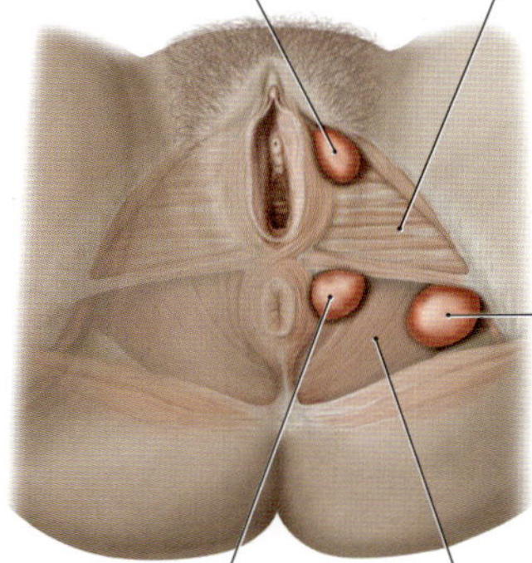

H Perineal hernia

5.13 Development of the External Genitalia

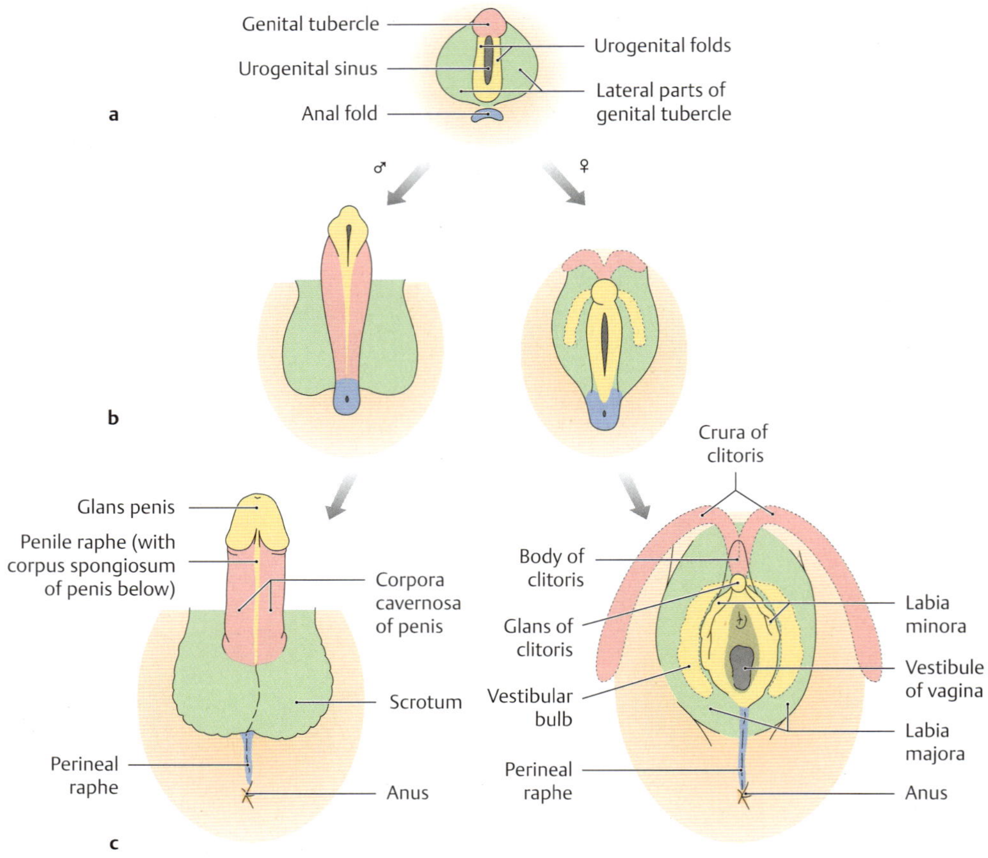

A Development of the external genitalia

a Rudimentary, undifferentiated external genitalia in a 6-week-old embryo.
b Differentiation of the external genitalia along male or female lines in a 10-week-old fetus.
c Differentiated external genitalia in the newborn.

The external genital organs develop from an undifferentiated mesodermal primordium in the cloaca and, like the gonads, pass through an initial **indifferent stage**. The anorectal area and urogenital sinus (cloaca) are not yet separated from each other and are closed externally by a common cloacal membrane. The following elevations develop around the cloacal membrane due to intensive mesodermal cell divisions:

- Anterior: the genital tubercles
- Lateral: the urogenital folds
- Posterior: the anal folds
- Lateral to the urogenital folds: the genital swellings (labioscrotal swellings)

Later, between the sixth and seventh weeks of development, the urorectal septum divides the cloaca into an anterior part (urogenital sinus) and a posterior part (anus and rectum). The cloacal membrane disappears, and the urogenital ostium forms anteriorly. The early perineum forms at the level of the urorectal septum (by fusion of the paired anal folds to the perineal raphe). **Differentiation of the genital organs** begins approximately in the eighth to ninth week of fetal development. Sexual differentiation is clearly evident by the 13th week and is fully developed by the 16th week.

- In the *male fetus*, the genital tubercles enlarge *under the influence of testosterone* to form the phallus and future penis. The urogenital sinus closes completely by fusion of the genital folds and forms the spongy part of the urethra. The genital (scrotal) swellings unite to form the scrotum
- In the *female fetus* (*absence of testosterone*), the genital tubercles give rise to the body and crura of the clitoris. The urogenital sinus persists as the vestibule of the vagina, and the two urogenital folds form the labia minora. The genital swellings enlarge to form the labia majora, vestibular bulbs, and what is known as the RSP or infracorporeal **r**esidual **s**pongy **p**art.

Male sex organs develop only in the presence of the factors listed below:

- A functionally competent SRY (sex-determining region of the Y) gene on the Y chromosome (otherwise, ovaries and a female phenotype will develop). The SRY gene ensures that anti-müllerian hormone and Leydig cells are produced (see below).
- Among its other functions, the *anti-müllerian hormone* induces regression of the müllerian ducts. It is formed in the somatic cells of the testicular cords (future Sertoli cells) starting in the eighth week of fetal life.
- *Leydig cells* begin to form in the fetal testes by the ninth week and produce large amounts of androgens (testosterone) until birth. They stimulate differentiation of the *wolffian duct* into seminiferous tubules and the development of the male external genitalia.

Alteration or interruption of this process of differentiation at any stage can lead to incomplete midline fusion, which leaves persistent clefts (hypospadias, epispadias, see **C**), or to more extensive external genital anomalies (see **E**).

B Derivatives of the undifferentiated embryonic genital anlage (primordia) during development of the external genital organs* **(Baskin et al. 2018)**

Undifferentiated anlage	♂ Male	♀ Female
Genital tubercles	Corpora cavernosa of penis	Body of clitoris with ascending and descending parts, angle of clitoris, crura of clitoris
Urogenital folds	Corpus spongiosum of penis, glans penis	Labia minora, vestibular bulb, glans of clitoris, infracorporeal residual spongy part (RSP)
Genital swellings	Scrotum	Labia majora
Urogenital sinus	Spongy part of urethra	Vestibule of vagina
Anal folds	Perineal raphe	Perineal raphe

* Details on the development of the gonads and genital tracts can be found in textbooks of embryology.

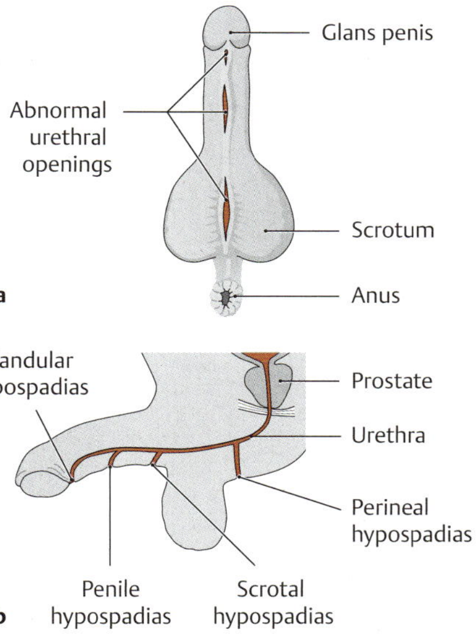

C Hypospadias: a urethral anomaly in boys

a Cleft anomalies affecting the underside of the penis and scrotum.

b Possible sites of emergence of the urethra in hypospadias (penis viewed from the lateral view).

If the genital folds do not fuse completely during sexual differentiation (see **A**), the result is a cleft anomaly of the urethra, which may open on the underside of the penis (*hypospadias*) or on its dorsal surface (*epispadias*). Hypospadias is much more common, with an incidence of 1 : 3,000 compared to 1 : 100,000 for epispadias. It is most common to find an abnormal urethral orifice in the glans region of the penis (glandular hypospadias). Additionally, the shaft of the penis is usually shortened and angled downward by the presence of ventral fibrous bands. Surgical correction is generally performed between the sixth month and second year of life.

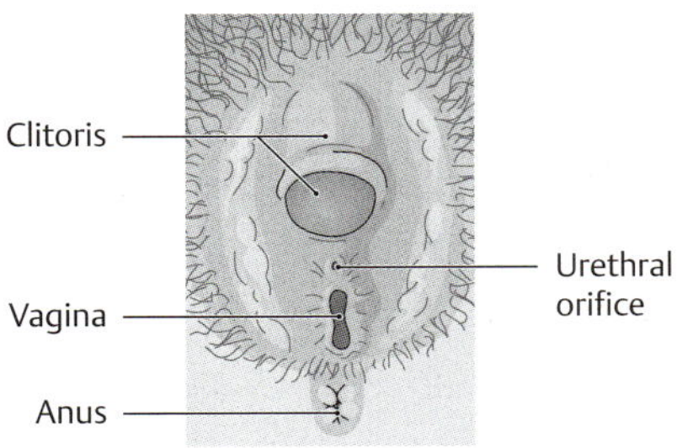

D External genitalia of a woman with adrenogenital syndrome

Anterior view. The external genitalia show definite signs of masculinization. The clitoris is markedly enlarged. The labia majora and minora are partially fused, and the urogenital sinus forms an undersized vestibule (see **E**, female pseudohermaphroditism).

E Various forms of intersexuality*

Condition	Features
• **True hermaphroditism****	• Very rare form of hermaphroditism (approximately 70% of cases have a female karyotype: 46,XX). The gonads contain both testicular and ovarian tissue (*ototestis*), but with a preponderance of ovarian tissue. Hence the external genitalia tend to have a female appearance with a markedly enlarged clitoris. A uterus is frequently present. Most hermaphrodites are raised as girls.
• **Pseudohermaphroditism**	• The pseudohermaphrodite has a definite chromosomal sex (female: 46,XX or male: 46,XY) but a phenotype of the opposite gender. The condition is termed male pseudohermaphroditism when a testis is present and female pseudohermaphroditism when an ovary is present.
– *Male pseudohermaphroditism* → Chromosomal sex: male (46,XY) → Phenotype: female	• **Etiology and pathogenesis** The female phenotype results from a lack of fetal androgen exposure: 1. Disturbance of testosterone synthesis 2. Disturbance of testosterone conversion 3. Androgen receptor defect 4. Testicular dysgenesis • **Example: testicular feminization** (1: 20,000 live births): – 46,XY chromosome complement – Female phenotype (estrogen synthesis present) but lacks pubic and axillary hair ("hairless woman"). The upper vagina and uterus are also absent. – *Cause:* androgen receptor defect or a disturbance of androgen metabolism (5α-reductase-2 defect) – *Result:* absence of spermatogenesis – *Treatment:* removal of the testes, which are usually in the inguinal region (risk of malignant transformation) and estrogen replacement for life
– *Female pseudohermaphroditism* → Chromosomal sex: female (46,XX) → Phenotype: male	• **Etiology and pathogenesis** The male phenotype results from fetal androgen exposure: 1. Congenital enzyme defect 2. Diaplacental androgen exposure • **Example: congenital adrenogenital syndrome** (1: 5 000 live births): – 46,XX chromosome complement – Female internal genital organs with masculinized external genitalia (enlarged clitoris, partial fusion of the labia majora, small urogenital sinus, see **D**) – Cause: adrenocortical hyperplasia with impaired steroid synthesis based on a genetic enzyme defect (most commonly a 21-hydroxylase deficiency). The low hormone level causes increased adrenocorticotropic hormone secretion, leading to the overproduction of androgens. – Treatment: hydrocortisone therapy for life, which may be combined with a mineralocorticoid.

* Intersexuality refers to a condition marked by contradictions in the development of general external sex characteristics, the gonads, and the chromosomal sex.

** Named after Hermaphroditos, the androgynous son of Hermes and Aphrodite from Greek mythology.

5.14 Male External Genitalia: Testicular Descent and the Spermatic Cord

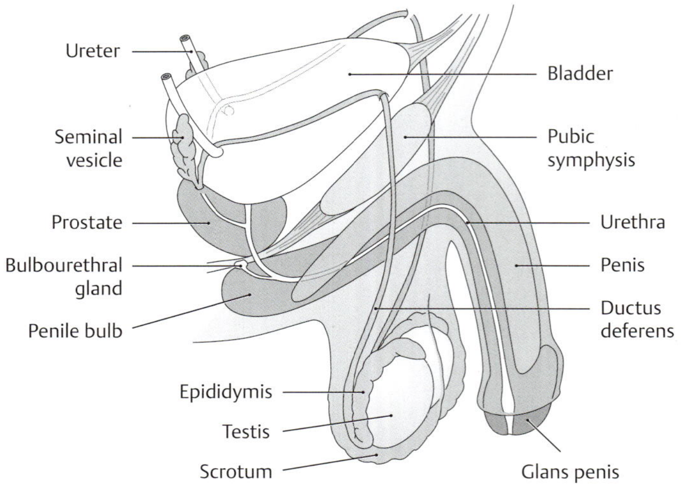

A Overview of the male genital organs

The internal and external male genitalia are distinguished by their *origins*. The internal reproductive organs originate from the two urogenital ridges located above the pelvic floor (except for the prostate and bulbourethral glands, which develop from urethral epithelium and thus are derivatives of the urogenital sinus). By contrast, the external genital organs develop around the urogenital sinus from a genital anlage (primordium) located below the pelvic floor (see p. 230).

Male internal genital organs	Male external genital organs
• Testes • Epididymis • Ductus deferens (vas deferens) • Accessory sex glands – Prostate – Seminal glands – Bulbourethral glands (Cowper's glands)	• Penis • Scrotum • Coverings of the testis
Topographically, however, the testis, epididymis, and a portion of the ductus deferens are classified among the external genital organs because they migrate from the abdominal cavity into the scrotum during fetal development (testicular descent).	

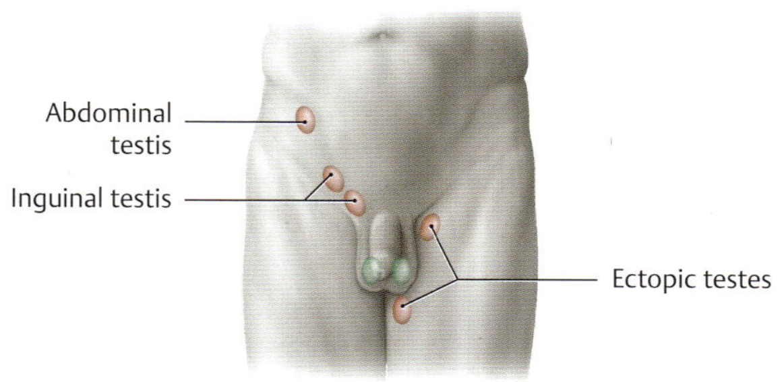

C Anomalous position of the testes

Abnormalities in the descent of the tests occur in approximately 3% of all newborns. A testis may be retained in the abdominal cavity or in the inguinal canal (cryptorchidism or retained testis). A deficiency of androgen production is the presumed cause. An ectopic testis is one that strays from the normal tract and occupies an abnormal position. The principal results are infertility due to the higher ambient temperature and an increased risk of malignant transformation.

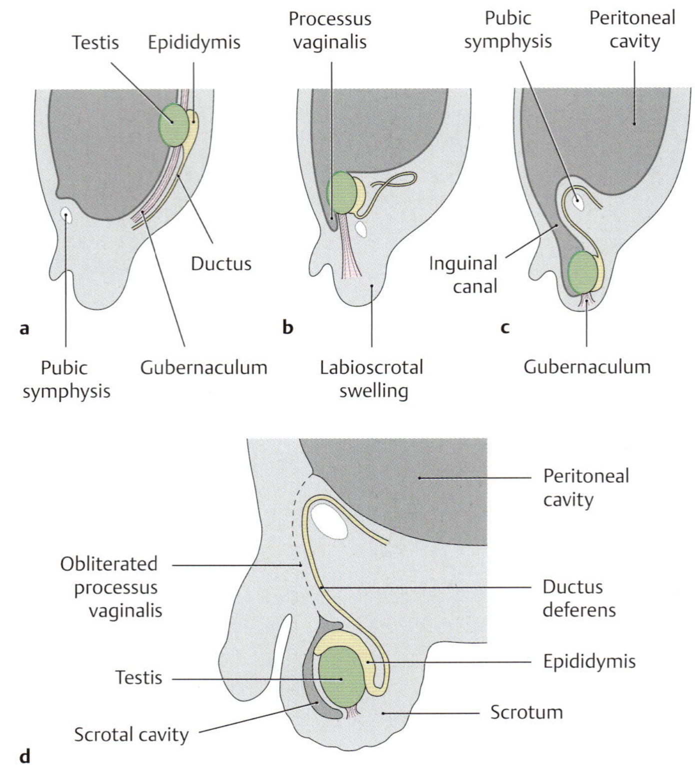

B Descent of a testis

Lateral view.

a Second month, **b** third month, **c** at birth, **d** after obliteration of the processus vaginalis of the peritoneum.

Near the end of the second month of development, the gonads and the rest of the mesonephros lie in a common peritoneal fold (urogenital fold), from which the "gonadal ligaments" are derived after regression of the mesonephros. The lower gonadal ligament, called the gubernaculum, is important for the descent of the testis. It passes below the genital ducts, pierces the abdominal wall in the area of the inguinal canal, and ends in the labioscrotal swelling, an outpouching of the anterior abdominal wall. Traction from this gonadal ligament (a consequence of body growth, which is more rapid than the growth of the genital organs) causes the testis and epididymis to slide downward along the posterior trunk wall external to the peritoneum (**transabdominal descent**). By the start of the third month, the testis has already reached the entrance of the future inguinal canal. The processus vaginalis, a funnel-shaped outpouching of peritoneum, forms anterior to the gubernaculum and is continued into the scrotal swelling with the other layers of the abdominal wall. It gives rise to the coverings of the spermatic cord and testis after the testis has completed its descent. A second phase, which is completed shortly before birth (**transinguinal descent**), culminates in passage of the testis through the inguinal canal into the scrotum. After testicular descent is completed (by birth), the processus vaginalis is obliterated except for a small space that partially surrounds the testis as the scrotal cavity (tunica vaginalis of the testis with a visceral layer and a parietal layer, see p. 234). Failure of this process of obliteration results in a persistent communication between the abdominal cavity and testicular cavity (congenital indirect inguinal hernia, see p. 222) (after Starck).

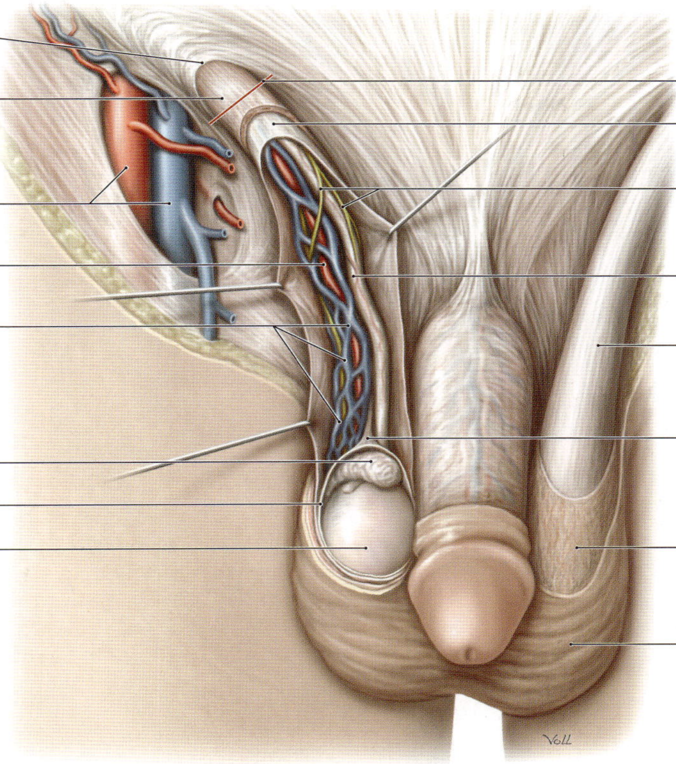

D The penis, scrotum, and spermatic cord
Anterior view. The skin has been partially removed from over the scrotum and spermatic cord. The tunica dartos and external spermatic fascia are exposed on the left side, and the spermatic cord on the right side has been opened in layers. The skin of the scrotum differs in many respects from the skin of the abdominal wall. It is more pigmented, markedly thinner, more mobile, and devoid of subcutaneous fat. It also contains a network of myofibroblasts (tunica dartos), whose contraction causes a wrinkling of the skin. This reduces the surface area of the scrotum and, when accompanied by vasoconstriction of the cutaneous vessels, reduces heat loss from the testes. This mechanism regulates the temperature to optimize spermatogenesis.

E Contents of the spermatic cord
Cross section through the spermatic cord (see **D** for section level). The neurovascular structures that supply the testes converge at the level of the deep inguinal ring, forming a bundle the thickness of the small finger that is held together by loose connective tissue and by the coverings of the spermatic cord and testes. The coverings consist of the following structures:

- Artery and vein of the ductus deferens
- Ilioinguinal nerve*
- Ductus deferens
- Genital branch of the genitofemoral nerve
- Testicular artery
- Cremaster muscle with cremasteric fascia
- Cremasteric artery and vein
- Testicular veins (= pampiniform plexus)
- Autonomic nerves (testicular plexus)
- Lymph vessels
- The obliterated processus vaginalis

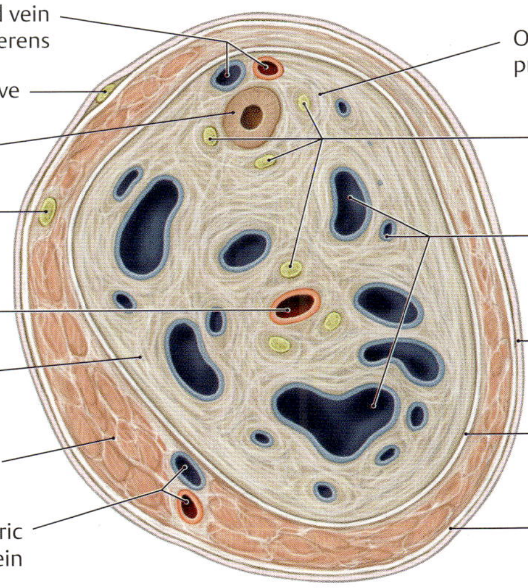

Special features: The large veins of the pampiniform plexus are exceptionally thick-walled, with a three-layered structure, and hence are easily mistaken for arteries. In life the ductus deferens, with its compact muscular wall, is palpable through the skin as a firm cord the thickness of a knitting needle. The ease of surgical access to this site is utilized in *vasectomy*, or ligation of the ductus deferens, to interrupt the transport of sperm (sterilization).

* When the spermatic cord is observed outside the inguinal canal as shown in this figure, the external spermatic fascia (the continuation of the superficial body fascia) belongs to the spermatic cord so that the ilioinguinal nerve courses a short distance within this fascia and later takes an epifascial course. As there is no external spermatic fascia in the inguinal canal, the ilioinguinal nerve courses outside the spermatic cord in that region. When the structure is demonstrated in the anatomy lab, the superficial body fascia (external spermatic fascia) has usually been removed. This creates the false impression that the entire length of the ilioinguinal nerve courses outside the spermatic cord.

5.15 Male External Genitalia: The Testis and Epididymis

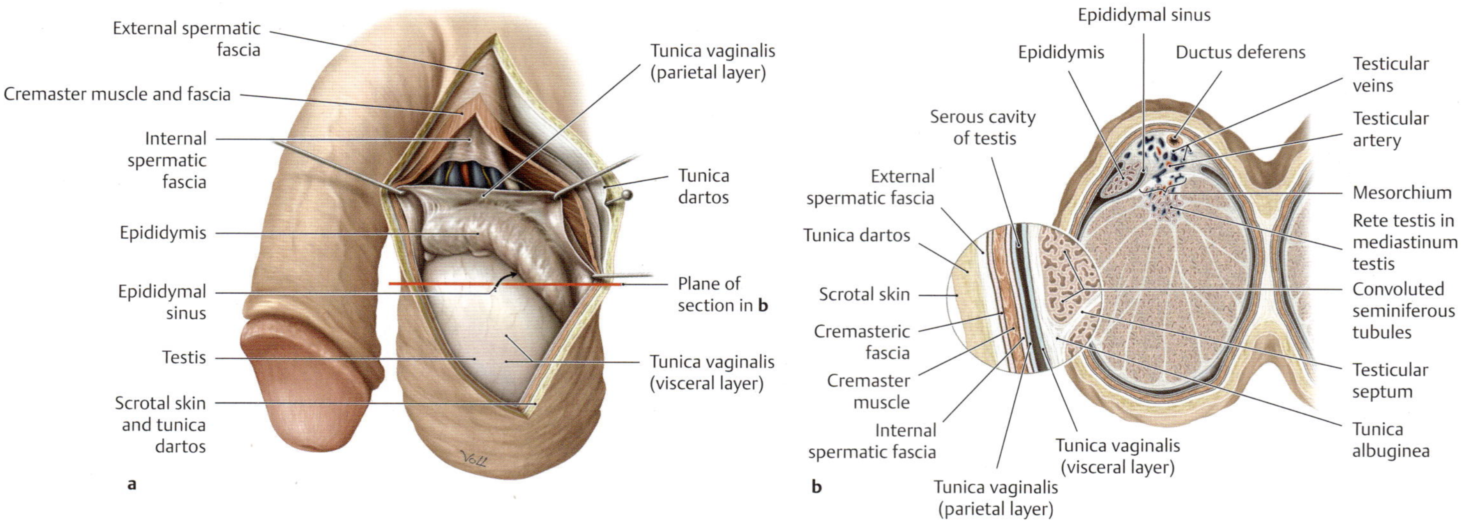

A Tunica vaginalis and serous cavity of the testis

a Opened tunica vaginalis of the left testis, lateral view.

b Cross section through the testis, epididymis, and scrotum, superior view.

The tunica vaginalis (unobliterated end of the processus vaginalis, see p. 232) forms a serous coat surrounding the testis and epididymis. Its visceral layer is fused to the tunica albuginea of the testis. At the mediastinum testis (a suspensory ligament where nerves and vessels enter and leave the testis), the tunica vaginalis is reflected to form the parietal layer, which is covered externally by the internal spermatic fascia. Between the two layers is a slitlike mesothelium-lined space (serous cavity of the testis) that contains a scant amount of fluid and is partially continuous between the testis and epididymis (epididymal sinus). An abnormal fluid collection in this cavity is called a *testicular hydrocele* (see **Fb**).

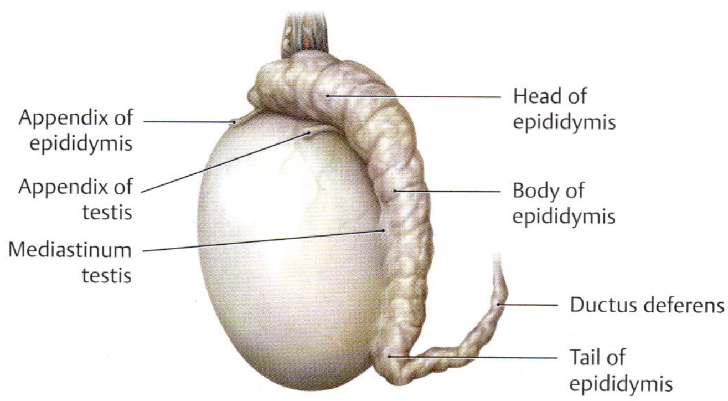

B Surface anatomy of the testis and epididymis

Left testis and epididymis, lateral view. The combined weight of the testis and epididymis at sexual maturity is approximately 20 to 30 g. The testis has an ovoid shape (approximately 5 cm long and 3 cm wide) and an average volume of approximately 18 mL (12–20 mL). The testicular tissue is enclosed in a tough fibrous capsule (tunica albuginea) and has a rubbery consistency. The *epididymis* consists of a head, which is attached to the upper pole of the testis, and a body and tail that curve down along the mediastinum on the posterior side of the testis. The tail of the epididymis becomes continuous with the ductus deferens at the lower pole of the testis.

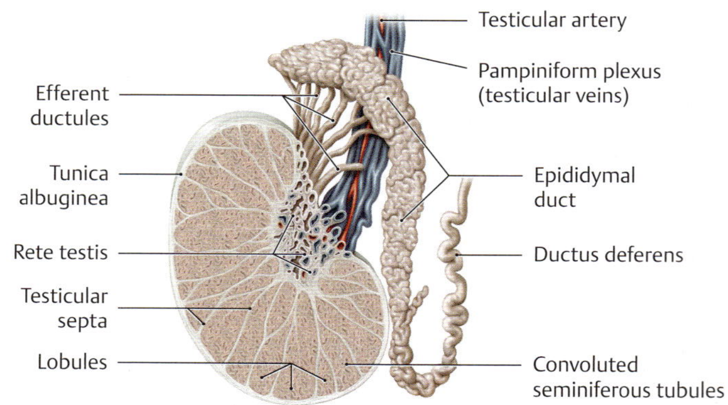

C Structure of the testis and epididymis

Section through the testis (epididymis intact), lateral view. Fibrous testicular septa extend radially from the tunica albuginea of the testis toward the mediastinum testis, subdividing the testicular tissue into approximately 370 wedge-shaped lobules. Each lobule contains one or more convoluted seminiferous tubules, in whose epithelium the spermatocytes are formed (spermatogenesis, see p. 4) and which open into the rete testis. From there, approximately 10 to 15 efferent ductules pass to the head of the epididymis, where the epididymal duct begins. This single duct is continuous distally with the ductus deferens, which passes through the inguinal canal in the spermatic cord to enter the abdominal cavity and opens into the prostatic part of the urethra via a short intervening segment, the ejaculatory duct (see p. 237).

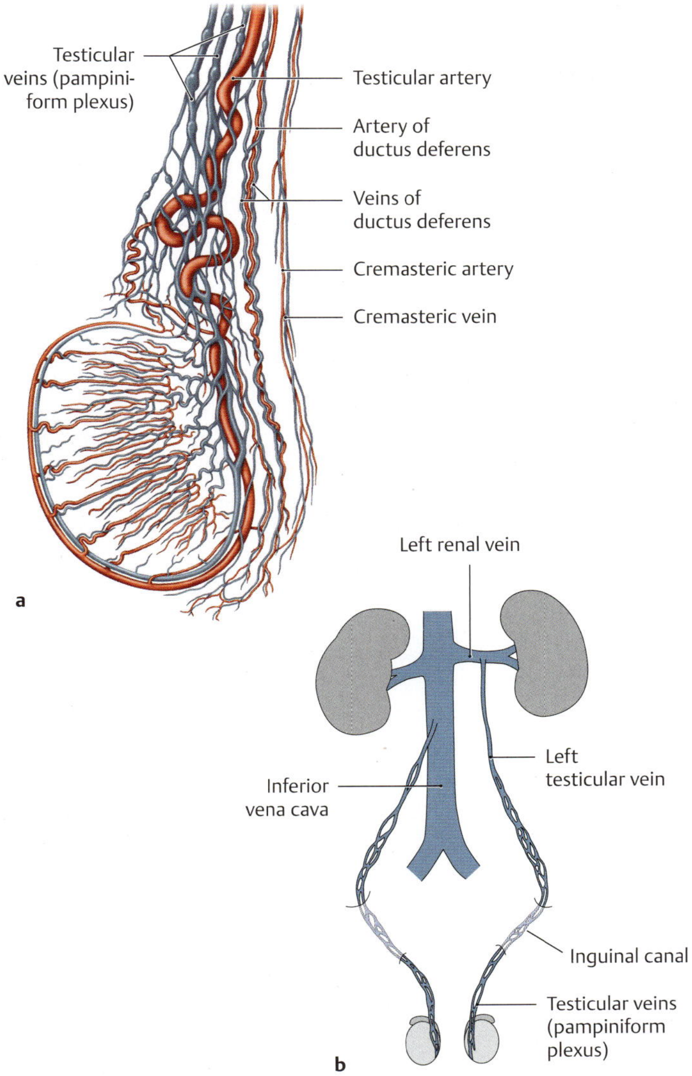

D Blood supply of the testis

a Arterial supply: The testis, epididymis, and their coverings are supplied by three different arteries, which anastomose with one another:

- Testicular artery: arises directly from the aorta.
- Artery of the ductus deferens: arises from the internal iliac artery.
- Cremasteric artery: arises from the inferior epigastric artery.

The vessels supplying the scrotum arise from the internal pudendal artery (see p. 574).

b Different venous drainage patterns of the right and left testes: Venous blood from the testis and epididymis flows into the testicular veins in the area of the mediastinum testis. These veins form an elongated venous network, especially distally, called the pampiniform plexus. It surrounds the branches of the testicular artery and ascends with it through the inguinal canal into the retroperitoneum. There the *right* testicular vein empties into the inferior vena cava, while the *left* testicular vein opens into the left renal vein. This asymmetry of venous drainage has major clinical relevance: the left testicular vein enters the left renal vein at a right angle. This creates a physiologically significant constriction that can obstruct venous outflow from the left testicular vein and thus from the pampiniform plexus (varicocele, see **Fd**). In this case, the pampiniform plexus can no longer perform its "thermostat" function (cooling venous blood returning from the testicular artery), resulting in a local heat buildup that may compromise the fertility of the left testis.

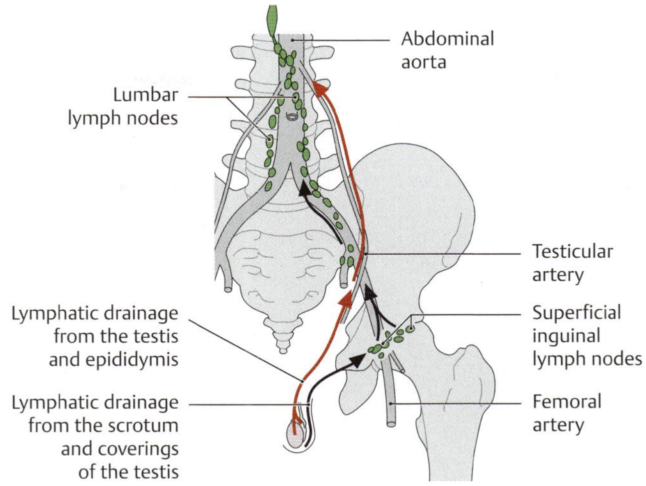

E Lymphatic drainage and regional lymph nodes of the testis, the epididymis, the coverings of the testis, and the scrotum

The lymphatic vessels of the testis and epididymis drain to the lumbar lymph nodes, accompanied by the testicular vessels. The lymphatic vessels from the scrotum and the coverings of the testis drain to the superficial inguinal nodes (see p. 546).

Note: Advanced testicular tumors tend to metastasize to retroperitoneal lymph nodes because they serve as the primary lymphatic conduit from the testis and epididymis.

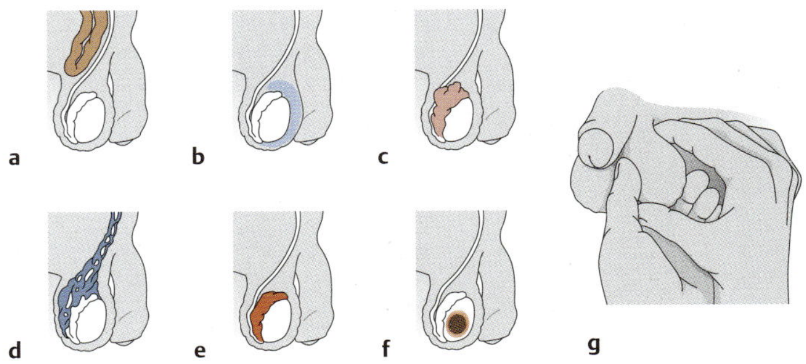

F Abnormal findings on clinical examination of the external genitalia

a–f Diseases that may present with scrotal swelling: **a** inguinal hernia, **b** testicular hydrocele (serous fluid collection in the serous cavity of the testis), **c** spermatocele (retention cyst in the epididymis), **d** varicocele (painful, varicose dilation of the pampiniform plexus), **e** epididymitis (painful bacterial inflammation of the epididymis), **f** testicular tumor (painless, usually unilateral induration of the testis).

g Bimanual examination of the testis and epididymis: Clinical examination of the external genitalia should include palpation of the testis and epididymis (bimanual examination). Based on the disease features noted above, the following questions should be addressed during the clinical examination:

- Is the mass confined to the scrotum?
- Is there transient enlargement of the mass when the patient coughs?
- Is the mass translucent when examined by *transillumination* (illumination with a flashlight)?
- Is the mass painless or tender to pressure?

Note: A painless induration of the testis, especially in young men, should always raise suspicion of a testicular tumor.

5.16 Male External Genitalia: The Fasciae and Erectile Tissues of the Penis

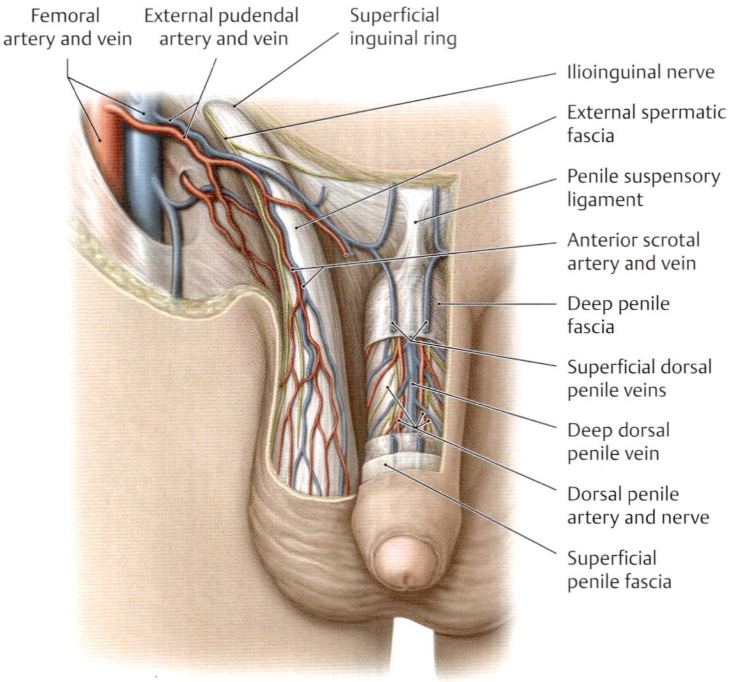

A Arrangement of the penile fasciae
a Anterior view of the penis (skin and fasciae partially removed).
b Right lateral view of the penis (skin and fasciae partially removed).
c Cross section through the shaft of the penis.

The penis is covered by thin, mobile skin that is devoid of fatty tissue. The skin over the glans penis is duplicated to form the prepuce (foreskin), which is attached to the undersurface of the glans by the median fold of the frenulum (see **b**). The erectile tissues of the penis are surrounded by a common, strong envelope of collagenous fibers, the *tunica albuginea*. The two layers of the penile fascia (superficial and deep) also surround the corpus spongiosum and corpora cavernosa. The erectile tissues, their fibrous sheaths, and the way in which the vessels are incorporated into these fibrous sheaths are of key interest in understanding the function of the penis (see p. 239).

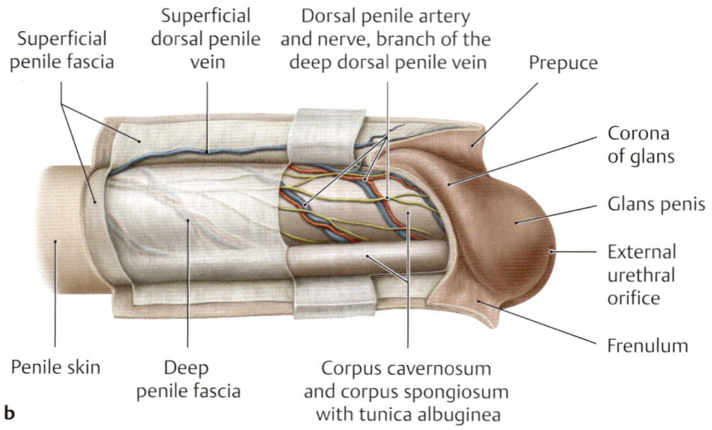

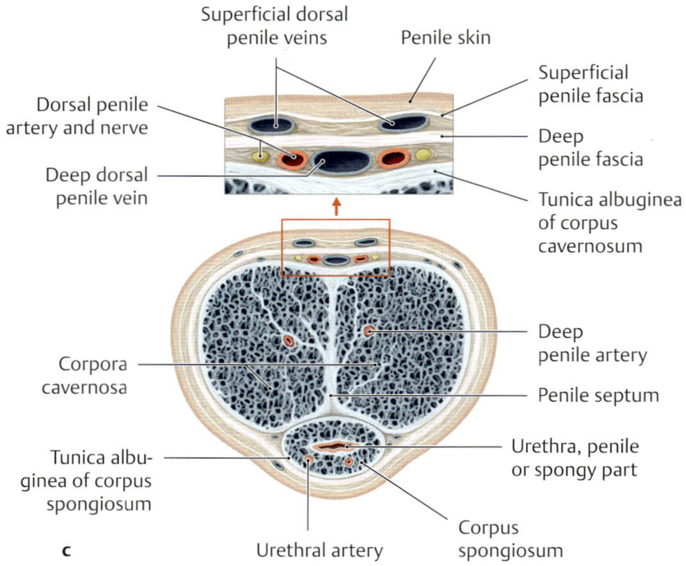

B Constriction of the prepuce (phimosis)
a Constriction of the prepuce in a 3-year-old boy.
b Appearance following circumcision.

The epithelium of the inner layer of the prepuce is adherent to the surface epithelium of the glans penis in newborns and infants. Because of this, the distal junction of the outer and inner layers of the prepuce is normally constricted, a condition characterized as **physiological phimosis**. During the first 2 years of life, the epithelial attachments become separated due to enlargement of the glans and the secretion of smegma (cellular debris sloughed from the stratified keratinized epithelium). If the prepuce still cannot slide over the glans by 3 years of age due to a functional stenosis (e.g., persistent epithelial attachments due to an absence of smegma secretion), the phimosis should be surgically corrected by circumcision. This procedure may be conservative or radical, resecting all the foreskin (as shown here), depending on the severity of the phimosis. Immediate surgical intervention (before 3 years of age) is necessary for **paraphimosis**—an emergency situation in which the glans is strangulated by the narrowed foreskin (painful, livid swelling of the glans due to decreased blood flow, with risk of necrosis) (after Sökeland, Schulze, and Rübben).

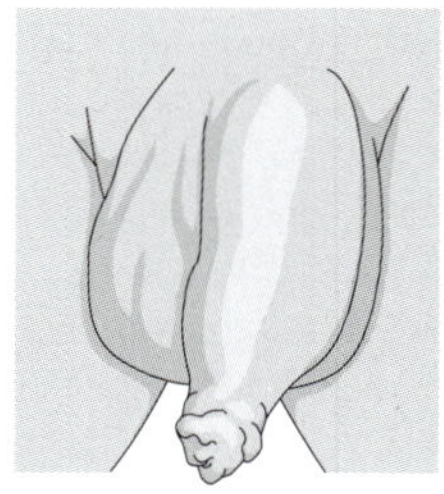

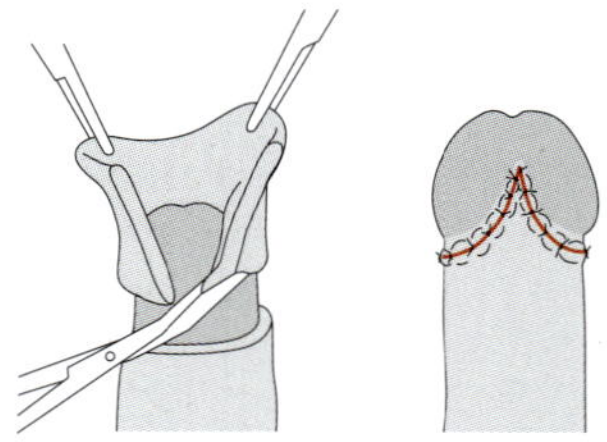

C The erectile tissues and erectile muscles of the penis

a Inferior view. The corpus spongiosum is partially mobilized, and the skin and fasciae have been removed. The ischiocavernosus and bulbospongiosus muscles have been removed on the left side, along with the inferior fascia of the urogenital diaphragm.

b Cross section through the root of the penis. The root of the penis is firmly attached to the perineal membrane and pelvic skeleton. It is distinguished from the freely mobile shaft (body) of the penis, with its dorsal and urethral surfaces, and from the glans penis, which bears the external urethral orifice. The penis contains two types of erectile tissue:

- The paired corpora cavernosa
- The unpaired corpus spongiosum

At the root of the penis, each of the corpora cavernosa tapers to form a crus penis. Between the two crura lies the thickened end of the corpus spongiosum, the bulb of the penis. The glans penis forms the distal end of the corpus spongiosum. Its posterior margin is broadened to form the corona, which is turned over the ends of the corpora cavernosa. The erectile tissues receive their blood supply from branches of the internal pudendal artery, which branches in the deep perineal space (see pp. 191 and 238).

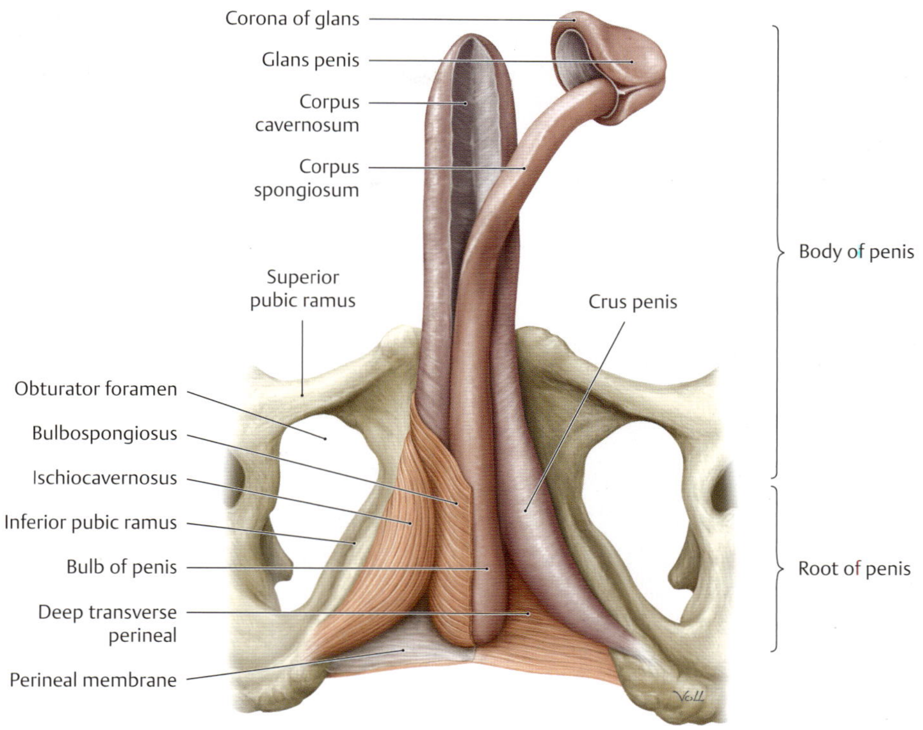

a

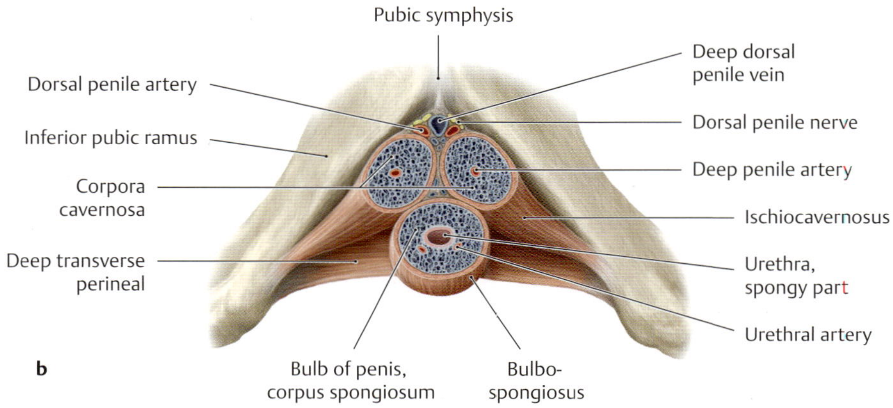

b

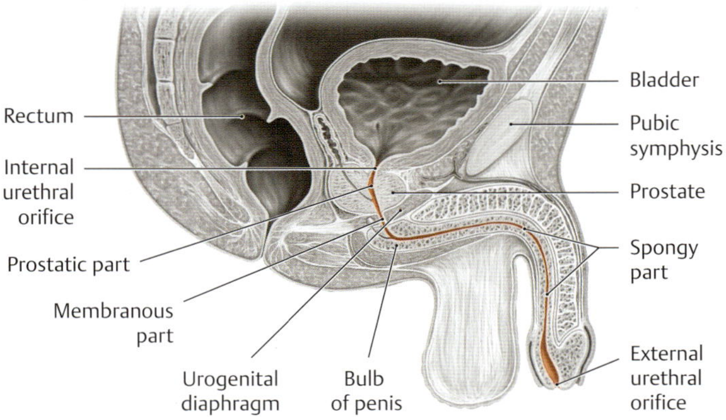

D Course of the male urethra

Midsagittal section through a male pelvis. The male urethra consists of a prostatic part, a membranous part, and a penile or spongy part named for different regions of the pelvis and external genitalia (see p. 232). The spongy part begins where it enters the bulb of the penis and terminates at the external urethral orifice.

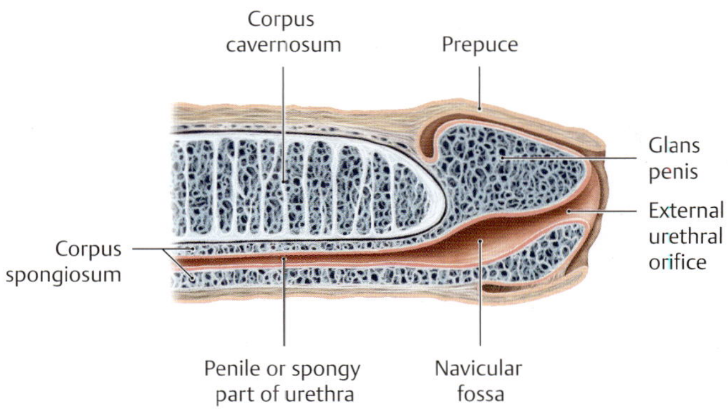

E Midsagittal section through the distal penis

The spongy part of the urethra undergoes an approximately 2-cm-long fusiform dilation within the glans penis. In this area, the navicular fossa, the stratified columnar epithelium of the urethra, gives way to stratified, nonkeratinized squamous epithelium. The upper cell layers of this epithelium are rich in glycogen, which—as in the vaginal milieu in females—provides a culture medium for the lactic acid bacteria that thrive there (acidic pH protects against pathogenic organisms).

5.17 Male External Genitalia: Nerves and Vessels of the Penis

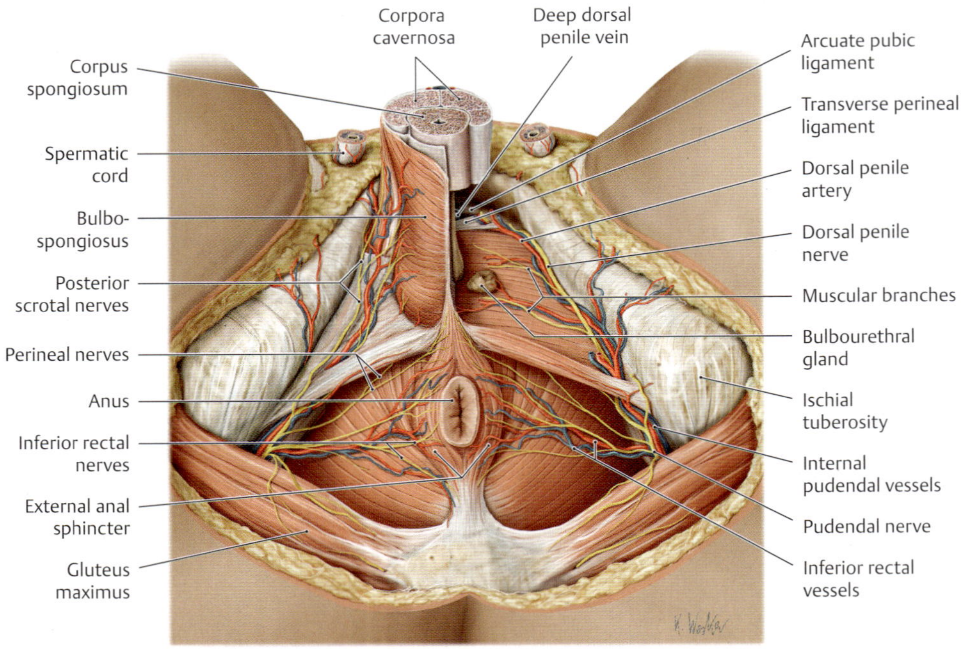

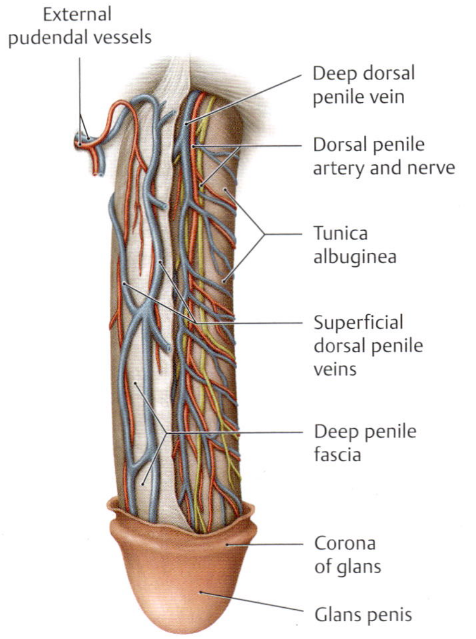

A Neurovascular structures of the male perineal region
Lithotomy position with the scrotum removed, inferior view. The superficial perineal space on the right side has been opened by removing the superficial perineal fascia. The erectile muscles and root of the penis have been removed on the left side, and the deep perineal space is partially exposed. The penis has been transected across the shaft, and the spermatic cords have been divided.

B Dorsal vessels and nerves of the penis
The prepuce, skin, and superficial fascia have been completely removed from the penile shaft. The deep penile fascia has also been removed from the left dorsum.

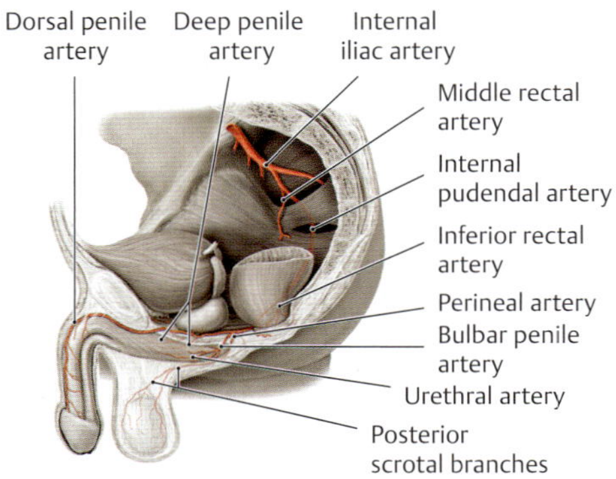

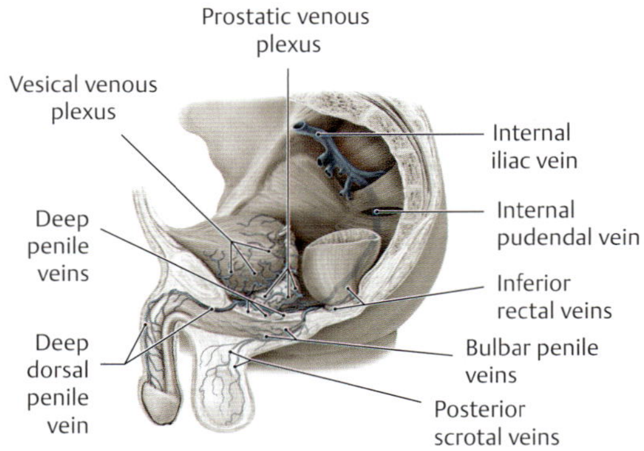

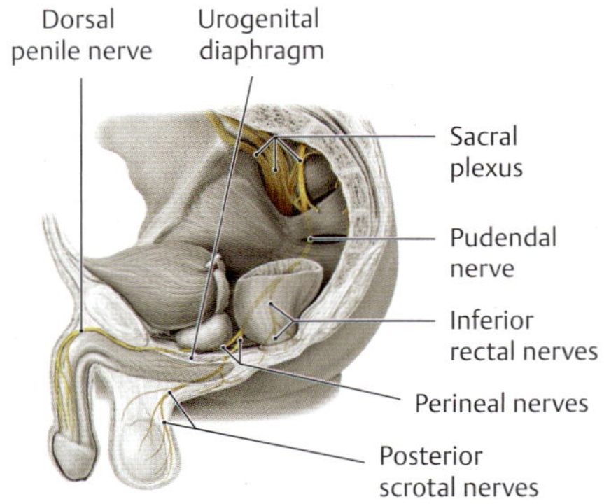

C Arterial supply to the penis and scrotum
Left lateral view. The penis and scrotum derive their arterial supply from the internal pudendal artery. This vessel enters the ischioanal fossa and, after giving off the inferior rectal artery to the anus, courses to the posterior border of the urogenital diaphragm. Then, after giving off the perineal artery, it passes through the deep perineal space into the superficial perineal space (see p. 191), where it divides into its terminal branches: the dorsal penile artery, deep penile artery, bulbar penile artery, and urethral artery.

D Venous drainage of the penis and scrotum
Left lateral view. The veins of the penis (especially the deep dorsal penile vein and its tributaries, the deep penile and bulbar penile veins) open initially into the internal pudendal vein and then into the prostatic venous plexus. Exceptions are the superficial dorsal penile veins (not seen here), which drain via the external pudendal veins into the long saphenous vein. On its way to the prostatic venous plexus, the deep dorsal penile vein passes through a narrow space just below the symphysis between the arcuate pubic ligament and the transverse perineal ligament (see ligaments in **A**).

E Nerve supply to the penis and scrotum
Left lateral view. The pudendal nerve enters the ischioanal fossa and, after giving off the inferior rectal nerves, courses to the external anal sphincter and to the skin of the anus at the posterior border of the urogenital diaphragm. There it divides into its terminal branches, the perineal nerves. The superficial branches pass through the superficial perineal space to the skin of the perineum and posterior scrotum (posterior scrotal branches). The deep branches course in the deep perineal space. They innervate the erectile muscles (via muscular branches), the skin of the penis, and the erectile bodies (via the dorsal penile nerve). The course of the autonomic fibers is shown in **F**.

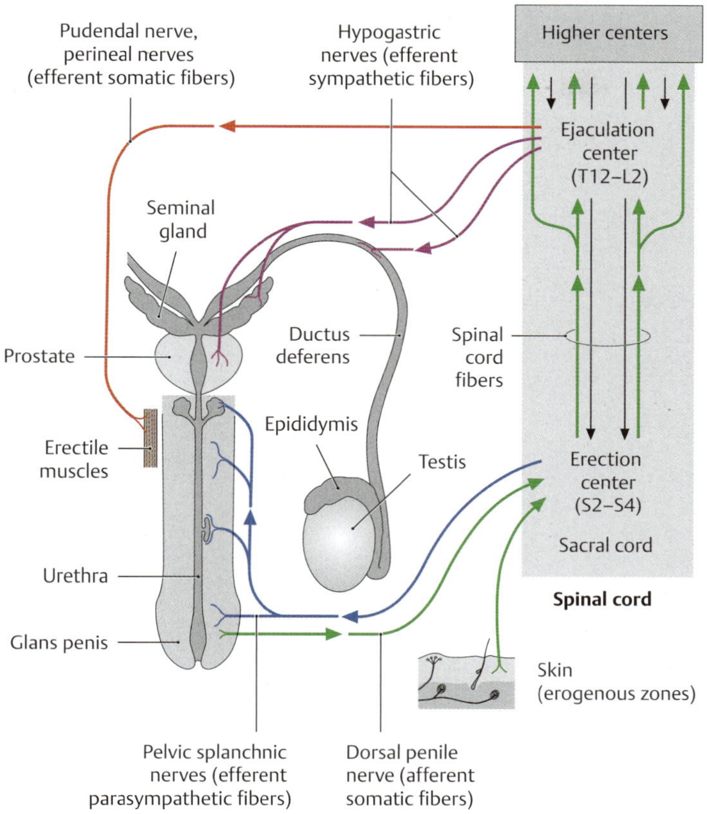

F Overview of the male sexual reflexes

The sexual reflexes in males are evoked by a variety of stimuli (e.g., tactile, visual, olfactory, acoustic, and psychogenic). Somatic and autonomic nerve pathways transmit the stimulus to the erection and ejaculation centers in the thoracolumbar and sacral spinal cord, from which it is relayed to higher centers (e.g., the hypothalamus and limbic system). For example, *tactile cutaneous stimuli* to the genitalia are transmitted to the sacral cord by *afferent somatic* fibers (dorsal penile nerve from the pudendal nerve, shown in green) and are relayed in the erection center (S2–S4) to *efferent parasympathetic* fibers (pelvic splanchnic nerves, shown in blue). These impulses, which stimulate vasodilation of the arteries supplying the erectile tissues (see **G**), are critically influenced by descending pathways from higher centers. Conversely, the excitatory impulses evoked by increasing mechanical stimulation of the glans penis ascend from the sacral cord to the ejaculation center located at the T12–L2 levels. There they are relayed to *efferent sympathetic fibers* (hypogastric nerves, shown in purple) and stimulate smooth muscle contractions in the epididymis, ductus deferens, prostate, and seminal glands. Simultaneous stimulation of the erectile muscles by efferent *somatic* nerve fibers (perineal nerves from the pudendal nerve, shown in red) produces rhythmic contractions that expel the ejaculate from the urethra (emission). Failure to achieve an erection despite an active libido (psychological interest in sexual activity) is defined as erectile dysfunction. The recent development of successful medical treatment of *erectile dysfunction* with sildenafil (Viagra [Pfizer]) is based on its modulation of the second messenger cyclic guanosine monophosphate (cGMP). When the primary messenger nitrous oxide (NO) is released by neural stimulation, it activates the enzyme guanylate cyclase in penile erectile tissues. This enzyme generates cGMP as a second messenger, which in turn induces vasodilation and produces an erection. Sildenafil selectively inhibits cGMP breakdown by a specific phosphodiesterase (PDE5) that is prominent in erectile tissue. cGMP accumulates, vessels remain dilated, and penile erection is sustained. Sildenafil treatment thus effectively amplifies the initial neural stimuli, with the potential of prolonging normal erections and overcoming other, inhibitory physiological problems (after Klinke and Silbernagl).

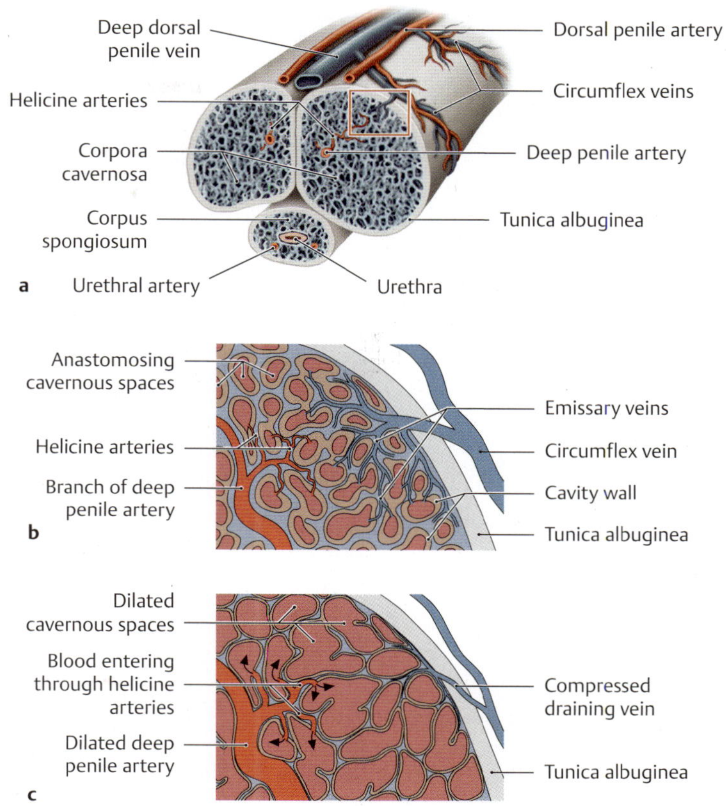

G Mechanism of penile erection

a Penis in cross section, showing the blood vessels involved in erection (enlarged views in **b** and **c**).
b Corpus cavernosum in the flaccid state.
c Corpus cavernosum in the erect state.

Penile erection is based essentially on *maximum engorgement* and pressure elevation in the cavities (cavernous spaces) of the corpora cavernosa combined with a *constriction of venous outflow*. This mechanism raises the intracavernous blood pressure to approximately 10 times the normal systolic blood pressure (approximately 1200 mmHg in young men). Microscopically, the erectile tissue of the penis consists of an arborized trabecular meshwork of connective-tissue and smooth-muscle cells that is connected to the tunica albuginea. Among the trabeculae are interanastomosing cavities that are lined with endothelium. Branches of the deep penile artery, called the helicine arteries, open into these cavities. In the flaccid state, the helicine arteries are more or less occluded by "intimal pads." When an erection occurs, the afferent arteries dilate, and the helicine arteries open under the influence of the autonomic nervous system. The result is that with each pulse wave, blood is forced into the cavernous spaces, increasing the volume of the erectile tissue and raising the intracavitary pressure. The tunica albuginea, which has a limited capacity for distension, becomes taut and compresses the veins that pass through it. This mechanism, aided by the occlusion of emissary veins, causes a constriction of venous outflow, enabling the penis to remain stiff and hard. Meanwhile, the dense venous plexuses in the corpus spongiosum and glans prevent excessive compression of the urethra. The flaccid phase begins with vasoconstriction of the afferent arteries. An undesired, prolonged, and painful erection is called *priapism* (from Priapus, the Greco-Roman god of procreation) and may occur, for example, in certain blood diseases or metabolic disorders. The initial treatment of this condition is medical; one option is the use of vasoconstrictors (etilefrine or norepinephrine). Surgical treatment involves making "punch anastomoses" to promote the outflow of blood.

5.18 Female External Genitalia: Overview and Episiotomy

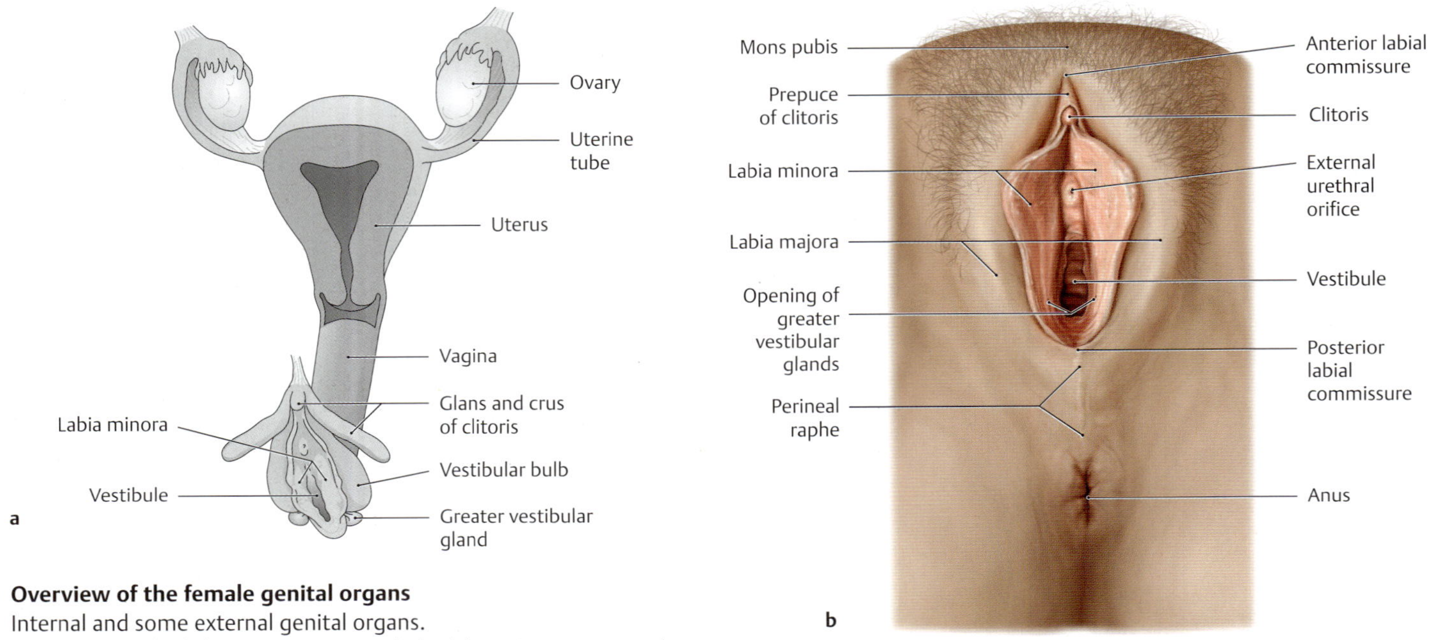

A Overview of the female genital organs
a Internal and some external genital organs.
b External genitalia, lithotomy position with the labia minora separated.

As in males, a distinction is drawn between the development and the topography of the female internal and external reproductive organs. The homology in the development of the male and female genital organs is reflected chiefly in the comparable histological features of the corresponding parts (see textbooks of histology). The female external genitalia (pudendum) is also known in clinical parlance as the *vulva*. It is separated from the internal genital organs by the *hymen* (not shown here). The outer boundaries of the vulva are formed by the mons pubis, a fatty-fleshy prominence over the pubic symphysis, and the labia majora, two pigmented ridges of skin that contain smooth muscle cells as well as sebaceous glands, sweat glands, and scent glands. The labia majora are interconnected anteriorly and posteriorly by a bridge of tissue called the anterior or posterior labial commissure. The area between the posterior commissure and anus is the perineal raphe. Specific structures are listed in the table at right.

Female internal genital organs	**Female external genital organs** (vulva)
• Ovary • Uterine tube • Uterus • Vagina	• Mons pubis • Labia majora • Labia minora • Vestibule of vagina • Vestibular bulb • Clitoris • Vestibular glands – Greater vestibular glands (Bartholin's glands) – Lesser vestibular glands

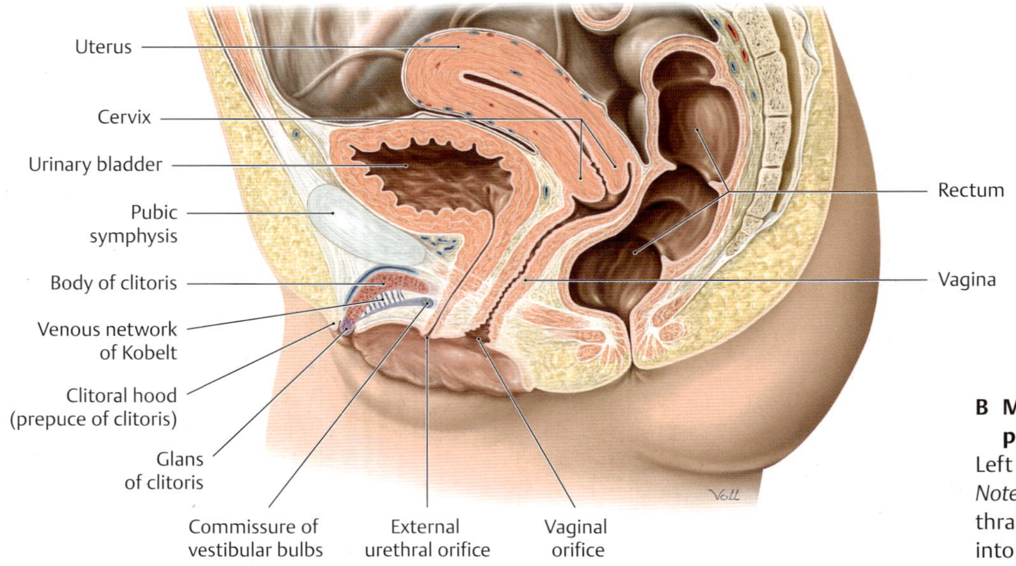

B Midsagittal section through a female pelvis
Left lateral view.
Note the close proximity of the external urethral orifice to the vaginal orifice, which opens into the vestibule.

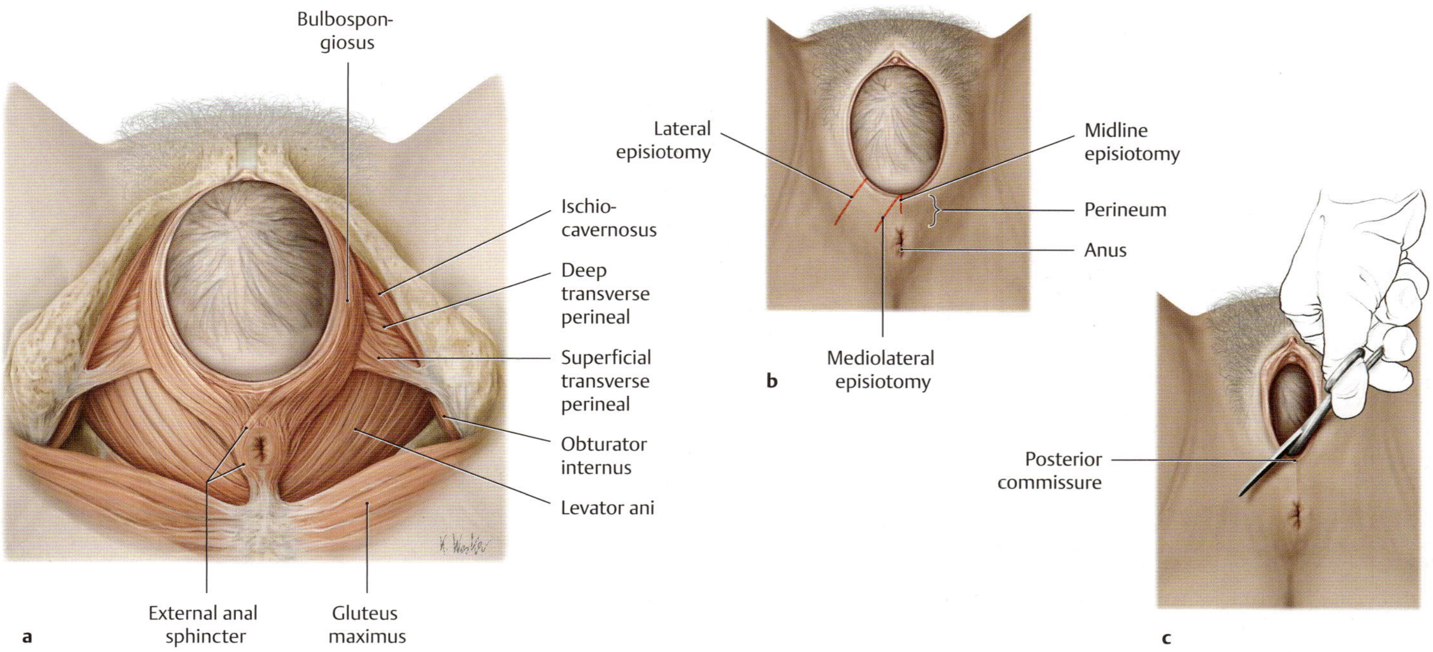

C Episiotomy: indications and technique

a Pelvic floor with crowning of the fetal head.
b Types of episiotomy: midline, mediolateral, lateral.
c Mediolateral episiotomy performed at the height of a contraction.

Episiotomy is a common obstetric procedure utilized to enlarge the birth canal during the expulsive stage of labor (see p. 560). When the fetal head crowns through the pelvic floor, the levator ani muscle in particular is passively stretched, forced downward, and rotated approximately 90°. The "levator plate" thus helps to form the wall of the distal birth canal along with the urogenital diaphragm and bulbospongiosus muscle. As such, it comes under considerable tension at the perineal body during the pushing stage of labor. To protect the perineal muscles from tearing, the obstetrician counteracts this tension by supporting the perineum with two fingers (*perineal protection*). An *episiotomy* is often performed to prevent uncontrolled laceration of the perineum (maternal indication). There is an imminent danger of perineal laceration during the delivery when the perineal skin is stretched to the point that it turns white, indicating diminished blood flow. The primary purpose of an episiotomy, however, is to expedite the delivery of a baby that is at risk for hypoxia during the expulsive stage. An *early episiotomy* is one that is made before the head crowns (the head is visible with contractions and pushing but recedes between contractions). A timely episiotomy is made after the head has crowned, when there is maximum tension on the perineal skin. Three types of episiotomy are available (see **D** for advantages and disadvantages):

- **Midline episiotomy:** straight down from the vagina toward the anus
- **Mediolateral episiotomy:** oblique incision from the posterior commissure
- **Lateral episiotomy:** lateral incision from the lower third of the vulva

After the placenta is delivered, the episiotomy is usually closed in at least three layers (vaginal suture, deep perineal suture, and cutaneous suture). Local anesthesia is generally required, especially when an early episiotomy has been done. If the episiotomy was made at the height of a contraction after crowning of the head, anesthesia is unnecessary. Local infiltration anesthesia and the pudendal block (PDB) are described on p. 560.

D Advantages and disadvantages of the different types of episiotomy (after Goerke)

Episiotomy	Divided muscles	Advantages	Disadvantages
• Midline	• None	• Easy to repair • Heals well	• May lengthen to a grade III perineal laceration
• Mediolateral	• Bulbospongiosus • Superficial transverse perineal	• Gains more room • Low risk of laceration	• Heavier bleeding • More difficult to repair • More difficult healing
• Lateral*	• Bulbospongiosus • Superficial transverse perineal • Levator ani (puborectalis)	• Gains the most room	• Heaviest bleeding • Potential complications (e.g., anal incontinence) • Greatest postpartum complaints

* Very rarely used.

5.19 Female External Genitalia: Neurovascular Structures, Erectile Tissues, Erectile Muscles, and Vestibule

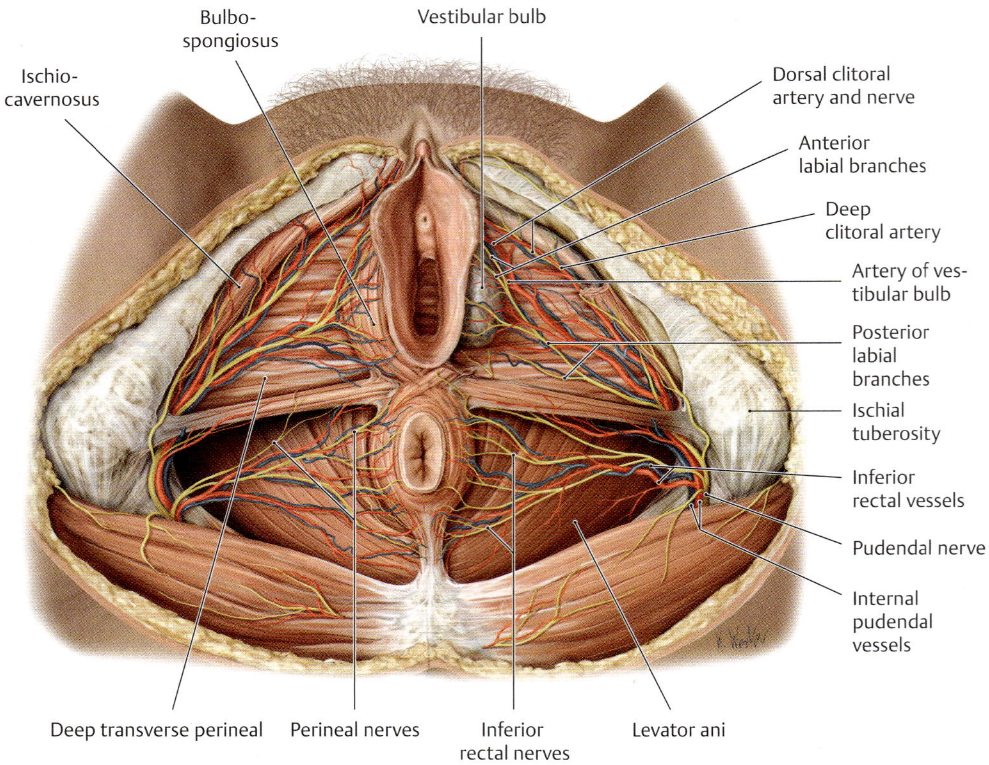

A Nerves and vessels of the female perineal region
Lithotomy position. The labia majora, skin, superficial perineal fascia, and fatty tissue in the ischioanal fossa have been removed to demonstrate the neurovascular structures. The bulbospongiosus and ischiocavernosus muscles and the perineal membrane have also been dissected away on the left side.

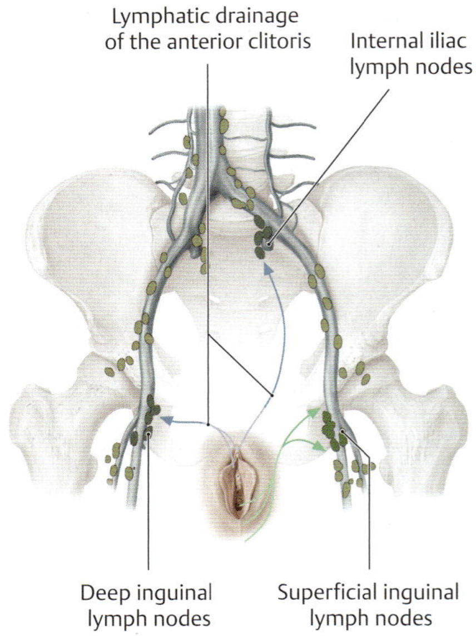

B Lymphatic drainage of the female external genitalia
Female pelvis, anterior view. Lymph from the female external genitalia drains to the superficial inguinal lymph nodes. The only exceptions are the anterior portions of the clitoris (body and glans), which drain to the deep inguinal and internal iliac lymph nodes.

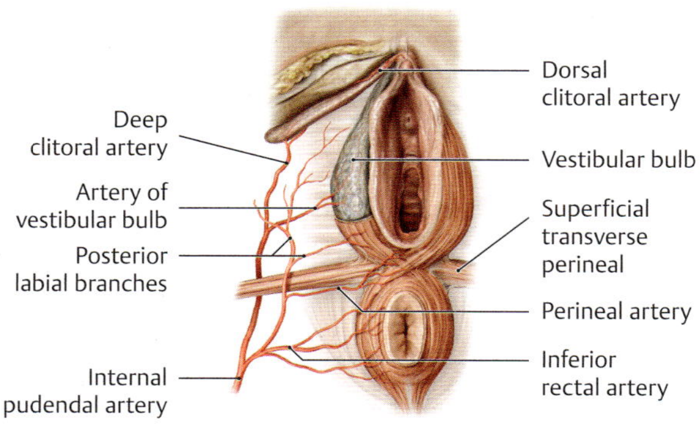

C Arterial supply to the female external genitalia
Perineal region, inferior view. The female external genitalia, like the penis and scrotum, are supplied by the internal pudendal artery, which enters the ischioanal fossa (not seen here). After giving off the inferior rectal artery to the anus, the internal pudendal artery passes anteriorly into the superficial perineal space. Another branch, the perineal artery, supplies the perineal region, the erectile muscles, and the *posterior part* of the labia majora (posterior labial branches). In the superficial perineal space (not visible here, see p. 191), the internal pudendal artery divides into its terminal branches, the artery of the vestibular bulb and the deep and dorsal clitoral arteries (which supply the corpus cavernosum of the clitoris). The *anterior part* of the labia majora is supplied by the external pudendal arteries (anterior labial branches), which arise from the femoral artery (not seen here).

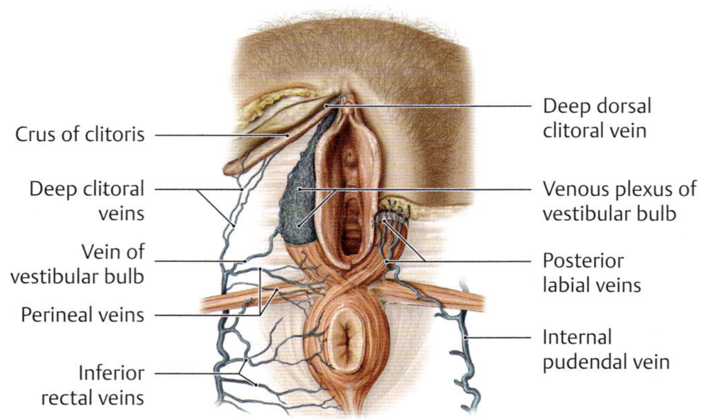

D Venous drainage of the female external genitalia
Venous drainage is handled by the following vessels:

- The deep clitoral veins, posterior labial veins, and vein of the vestibular bulb, which drain into the *internal pudendal vein*.
- The superficial dorsal clitoral veins and anterior labial veins, which drain into the *external pudendal veins* (not seen here).
- The deep dorsal clitoral vein, which drains into the *vesical venous plexus*.

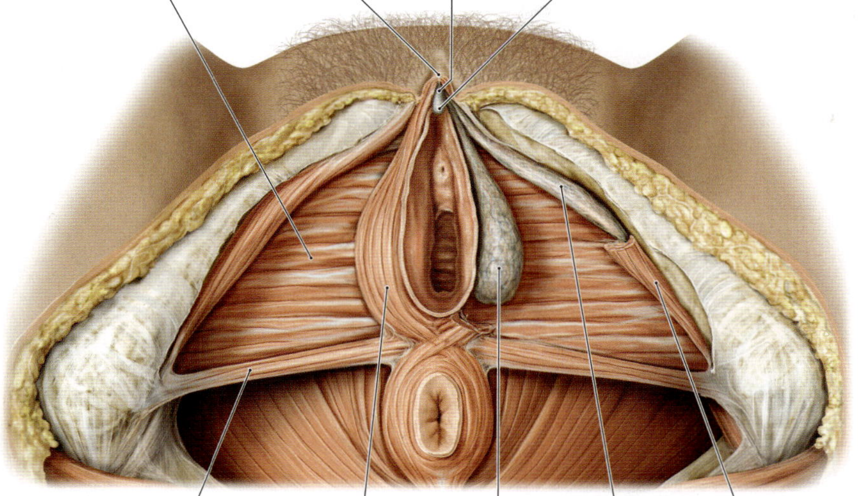

E Erectile tissues and erectile muscles in the female

Perineal region, lithotomy position. The labia majora and minora, skin, and superficial perineal fascia have been removed, as well as the erectile muscles on the left side.

Erectile tissues: The erectile tissues of the *clitoris* are distributed around both of its crura and its short shaft (body). They correspond to the erectile tissues in the male and are named accordingly: the right and left *corpora cavernosa of the clitoris*, homologous to the corpora cavernosa of the penis. The swelling at the end of the clitoral shaft is called the glans of clitoris, homologous to the glans penis. Its sensory innervation is like that of the penile glans, and it is mostly covered by a clitoral prepuce. The erectile tissue of the *labia minora* is located in the hairless, fat-free skin folds of the labia minora and is termed the *vestibular bulb*, which is the homologue of the corpus spongiosum in the male.

Erectile muscles: The two crura of the clitoris, one on each side, by which the clitoris arises from the inferior pubic ramus are covered by the *ischiocavernosus* muscle, and the dense erectile venous plexus at the base of the labia is covered by the *bulbospongiosus*.

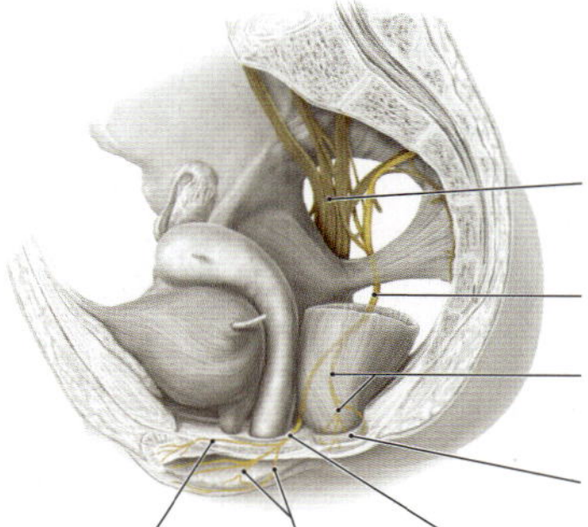

F Nerve supply to the female external genitalia

Lesser pelvis, left lateral view. The pudendal nerve enters the ischioanal fossa. After giving off the inferior rectal nerves to the external anal sphincter and anal skin, it courses to the posterior border of the urogenital diaphragm, where it divides into its terminal branches (perineal nerves). The *superficial branches* pass through the superficial perineal space (not seen here) to the skin of the perineum and the posterior portions of the labia majora (posterior labial branches). The *deep branches* course in the deep perineal space, distributing muscular branches to the erectile muscles and the dorsal clitoral nerve to the clitoris. The anterior portions of the labia majora are supplied by anterior labial branches from the ilioinguinal nerve (not seen here).

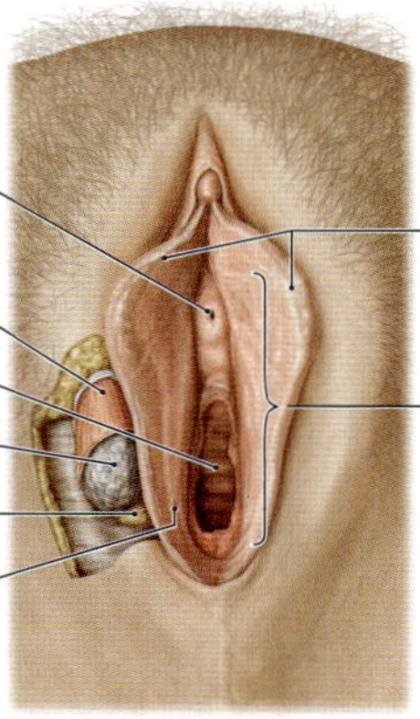

G The vestibule and vestibular glands

Lithotomy position with the labia separated. The vestibule, bounded by the labia minora, contains the external openings of the urethra and vagina (external urethral orifice and vaginal orifice) and the vestibular glands. The *lesser vestibular glands* (not seen here) have numerous openings near the external urethral orifice, while the *paired greater vestibular glands* each open by a 1-cm-long duct at the posterior border of the vestibular bulb on the inner surface of the labia minora. The lesser vestibular glands are homologous to the urethral glands in the male, and the greater vestibular glands to the bulbourethral glands. The vestibular glands produce a mucous secretion that moistens the vestibule and reduces friction during coitus, preventing epithelial injury. Next to the external urethral orifice are two short, blind, rudimentary excretory ducts called paraurethral ducts (not shown here). They correspond developmentally to the male prostate and are therefore also referred to as the female prostate (*prostata feminina*) (examined and described in great detail by the Dutch gynecologist Regnier de Graaf in the 17th century). During orgasm, the secretion of the female prostate is released as female ejaculate through the urethral opening into the vestibule of the vagina. There is still speculation about the functional significance of the female ejaculate (antimicrobial properties, environmental optimization for the sperm?). The paraurethral ducts, like the vestibular glands, are susceptible to bacterial colonization. Bacterial colonization of the greater vestibular glands (Bartholin's glands) can lead to bartholinitis, a painful inflammation with swelling and redness. An inflammation that occludes the excretory ducts of the glands can produce a painful retention cyst, which should be opened or removed.

5.20 Location, Structure, and Innervation of the Bulboclitoral Organ

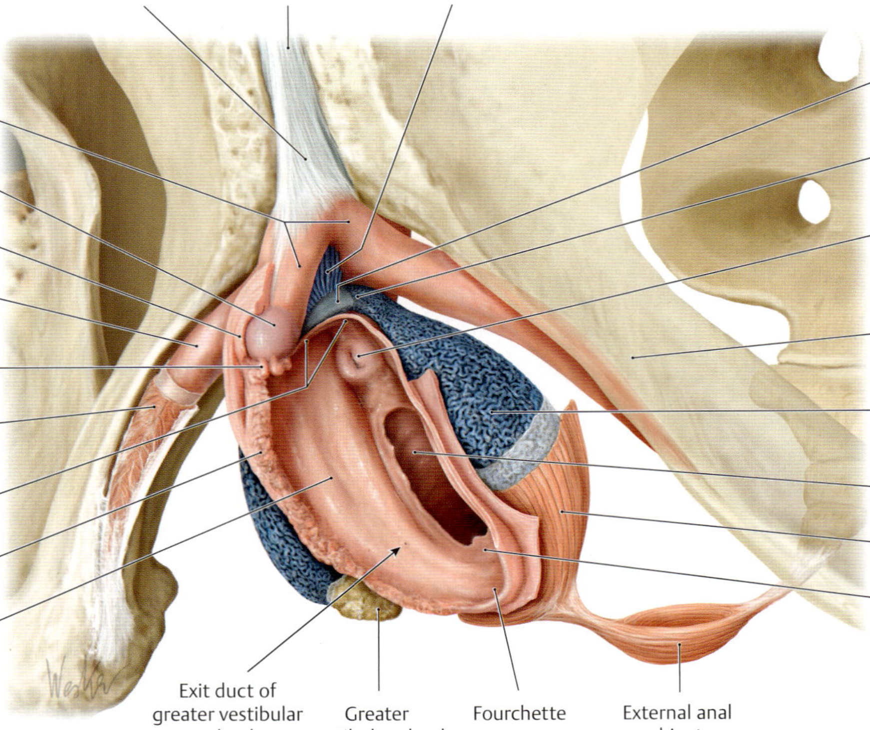

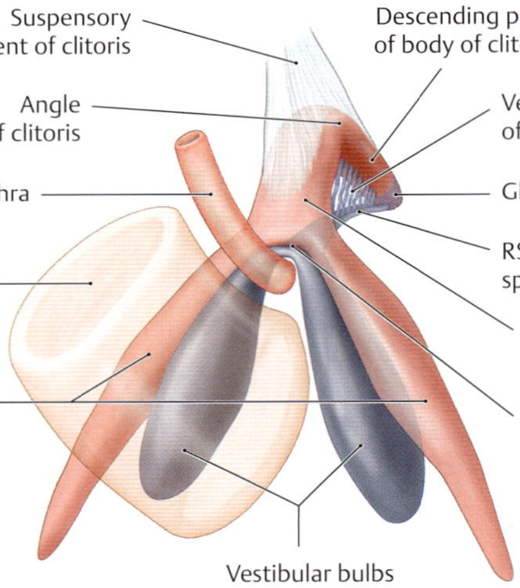

A Structure and location of the bulboclitoral organ

a Bulboclitoral organ in situ, left anterior view of partially windowed specimen; **b** Right posterior schematic view of the isolated bulboclitoral organ showing location relative to the urethra and vagina.

Marino and Lepidi (2014) describe the closely interconnected cavernous and spongy structures of the external female genitalia (corpora cavernosa and corpora spongiosa) as the *bulboclitoral organ*. As erectile tissues these structures play a crucial role in sexual arousal and are important sources of pleasure in the female. The **corpora cavernosa** include

- Ascending and descending parts of the body of the clitoris,
- Angle of the clitoris, and
- Left and right crura of the clitoris.

The **corpora spongiosa** consist of

- Glans of the clitoris,
- Vestibular bulbs (paired), and
- The infracorporeal residual spongy part (RSP).

The two corpora lie inferior to the two inferior rami of the pubic symphysis in what is known as the pubic arch. The bulboclitoral organ is attached to the small pelvis by both the suspensory ligament of the clitoris, which extends over a broad area from the entire body of the clitoris to the symphysis (see **b**), and via the retrocrural fasciae of the clitoral crura to the inferior public rami.

Note: The bulboclitoral organ is about 9 cm long from the glans of the clitoris to the ends of the clitoral crura.

The fourchette (frenulum of the labia minora) joins the posterior extensions of the two labia minora and forms the boundary of the fossa of the vestibule of the vagina. The frenulum of the clitoris is a paired structure extending from the glans of the clitoris on either side into the respective labium minus. The anterior portion of the glans of the clitoris, the only externally visible structure, is discernible under the clitoral hood (prepuce of the clitoris). Because they develop from a common embryonic primordium (see p. 230), the glans of the clitoris is conjoined with the two vestibular bulbs via the residual spongy part. The urethra and vagina (see **b**) course immediately inferior to the commissure of the vestibular bulbs. The eminence visible and palpable from the vestibule between the urethral meatus and the glans of the clitoris corresponds to the connective tissue union of the two urogenital folds and is referred to as the raphe (see **a**). Between the RSP and the descending body of the clitoris lies a complex of 8-10 communicating veins known as the venous network of Kobelt after Georg Ludwig Kobelt (1844). When the woman is sexually aroused, the muscles of the crura of the clitoris (ischiocavernosus) and the vestibular bulbs (bulbospongiosus) rhythmically press blood into the body and glans of the clitoris via the venous network of Kobelt.

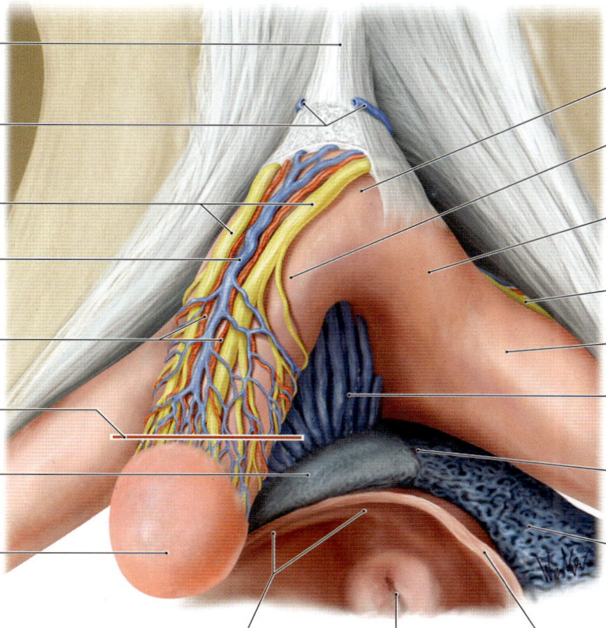

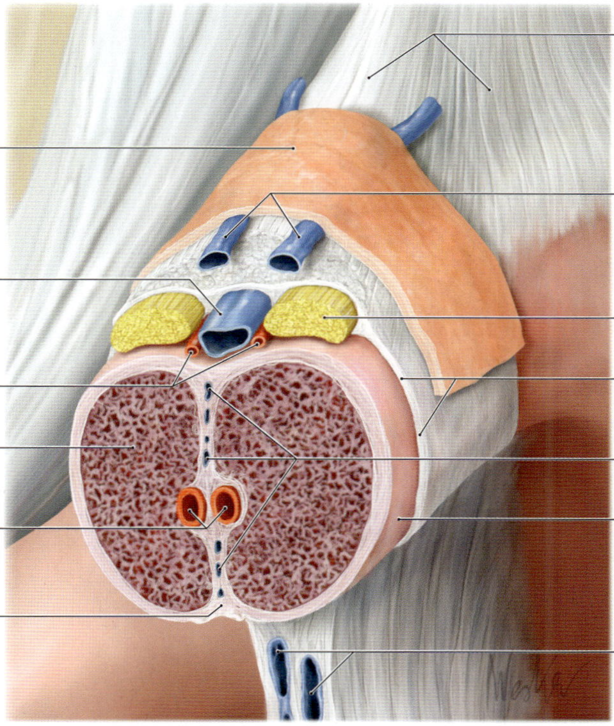

B Innervation of the bulboclitoral organ

a Left anterior view of the course of the dorsal nerve of the clitoris on the descending body of the clitoris; **b** Cross section through the descending body of the clitoris (level of section see **a**). The dorsal nerve of the clitoris extends from lateral to medial along the respective crus to the dorsal aspect of the body of the clitoris. It initially courses within the suspensory ligament of the clitoris, later between the tunica albuginea and clitoral fascia on to the glans of the clitoris (see **b**). At the margin of the glans of the clitoris it gives off the actual terminal branch of the dorsal nerve of the clitoris, which is still about 2 mm thick (i. e., 95% of the afferent nerve fibers begin in the glans; before this only a few lateral branches split off). With over 8,000 sensory nerve endings the glans of the clitoris is one of the most sensitive structures of the female body. In addition to free nerve endings, corpuscular mechanoreceptors (essentially genital and Vater-Pacini corpuscles, not shown here) play an important role in causing sexual arousal and triggering orgasm (Haag-Wackernagel, 2021a, 2021b, and 2022).

5.21 Forms of Female Circumcision and Vulvar Reconstruction

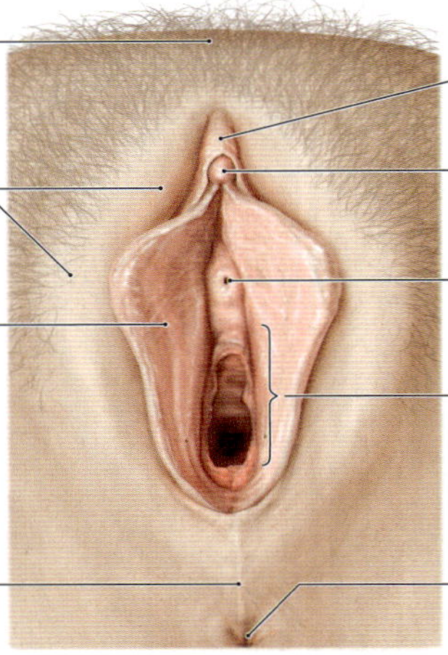

A Intact uncircumcised female genitals

Regarding forms of circumcision see **B**; with respect to legal aspects and acute and chronic physical and psychological complications, see also the recommendations of your country's national medical association on treating patients who have undergone female genital mutilation (FGM).

Note: The term "vaginal introitus" is used nearly exclusively in daily clinical practice. It denotes both the vestibule of the vagina and the vaginal orifice.

B Forms of circumcision* according to WHO classification

Type I: The partial or total removal of the clitoral glans (the external and visible part of the clitoris, which is a sensitive part of the female genitals), and/or the prepuce/clitoral hood (the fold of skin surrounding the clitoral glans).

Type II: The partial or total removal of the clitoral glans and the labia minora (the inner folds of the vulva), with or without removal of the labia majora (the outer folds of skin of the vulva).

Type III: Also known as infibulation, this is the narrowing of the vaginal opening through the creation of a covering seal. The seal is formed by cutting and repositioning the labia minora, or labia majora, sometimes through stitching, with or without removal of the clitoral prepuce/clitoral hood and glans.

Type IV: All other harmful procedures to the female genitalia for non-medical purposes, e.g., pricking, piercing, incising, scraping and cauterizing the genital area.

* According to the WHO over 200 million women worldwide in about 30 countries (most frequently in Egypt, Ethiopia, Nigeria, and Sudan) are affected by female genital mutilation/cutting (FGM/C). (*Source:* https://www.who.int/news-room/fact-sheets/detail/female-genital-mutilation, current as of 31 Jan 2023.)

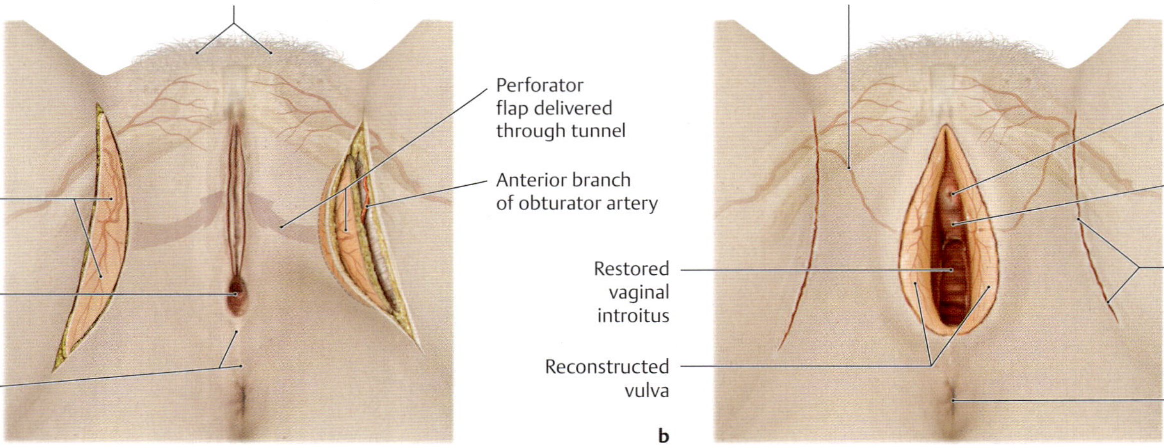

C Functional anatomic reconstruction of the external female genitals after Type III circumcision

Special reconstructive and microsurgical operative procedures allow largely anatomic reconstruction of the external genitals (O'Dey 2018, 2019, 2020). Treatments range from the comparatively simple reopening of the vaginal introitus (defibulation) to complex plastic surgical procedures (such as reconstruction of the labia, see **a** and **b**, surgical reconstruction of the glans and clitoral prepuce, see **c–g**, right). Whereas defibulation serves to allow sexual intercourse, excretion of urine, and menstruation, the primary goal of reconstruction of the clitoris is to reduce chronic pain and improve sexual sensation. The following section explains surgical reconstruction of the external genitals as is necessary after Type III circumcision and is possible as of today.

a and b Reconstruction of the labia majora and vaginal introitus (Anterior obturator artery perforator or aOAP flap); the patient is placed in the lithotomy position. A perforator flap (a cutaneous flap without muscle tissue) supplied by the *anterior branch of the obturator artery* (*anterior obturator artery perforator flap* or aOAP flap) is raised. The donor region is the genitofemoral sulcus, which lends itself to the procedure due to its topographical proximity.

- The perforator flap is advanced medially to the recipient site subcutaneously (tunneled transposition, see **a**) and the labia majora on either side of the reopened vestibule are reconstructed from the perforator flap (see **b**).
- The wound is closed over the donor site which restores the genitofemoral sulcus, resulting in an invisible scar.

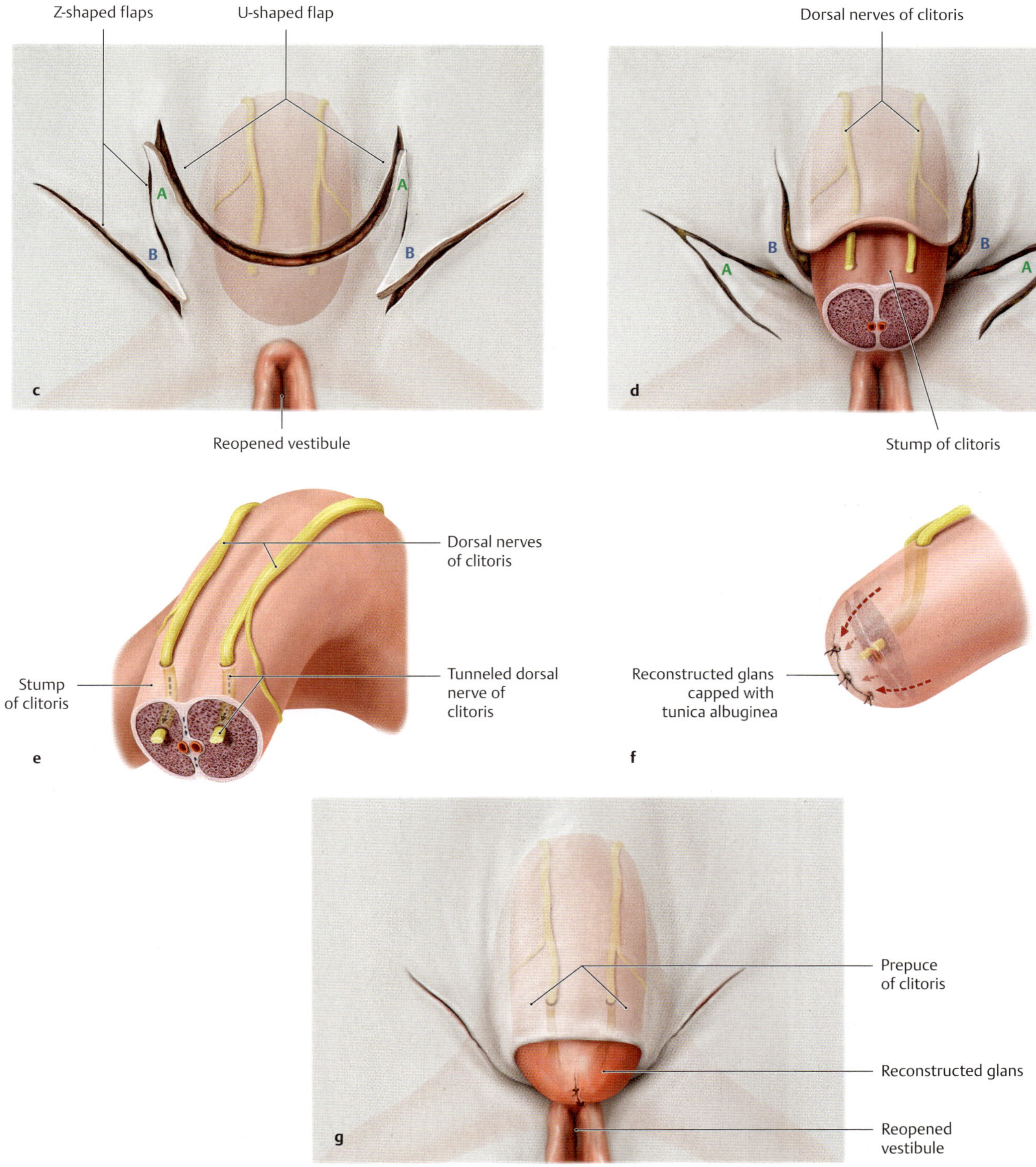

c–g Reconstruction of the glans of the clitoris (neurotizing and molding of the clitoral stump or NMCS procedure) and clitoral hood (omega domed flap or OD flap)

This part of the reconstruction begins with what is known as an omega incision (see **c**), i.e.:

- An incision is made in the form of a central U-shaped flap (see **A** in the figure; in the official nomenclature this is known as an anteriorly pedicled, well perfused transposition flap) to allow reconstruction of the clitoral hood. Additionally, bilateral incisions are made to raise two lateral Z-shaped flaps (see **B** in the figure) for the subsequent reconstruction of the clitoral hood.
- The U-shaped flap is reflected superiorly and the clitoral stump is located in the deep plane.
- The suspensory ligament is released to mobilize the clitoral stump and to allow pulling the stump forward for placement and fixation of a reconstructed glans in place of the glans removed by circumcision, see **d**.
- The dorsal nerves of the clitoris are identified and both nerves are reintegrated into the newly fashioned tip of the clitoris (**e**); the anterior cut surface is then closed with a tunica albuginea flap ("capping"; **f**);
- Finally, a new clitoral hood is reconstructed by means of a bilateral Z-plasty (**g**): The Z-plasty allows tension-free transposition of skin and for this reason is often used when correcting scars. The two triangular partial flaps (A and B) are shifted medially and sutured together beneath the newly formed glans. Shifting the two Z-plasties medially draws in the edges of the base of the curved central flap without creating tension; that is, it narrows it. The excess skin creates a roof or hood over the donor site, the reconstructed clitoral hood.

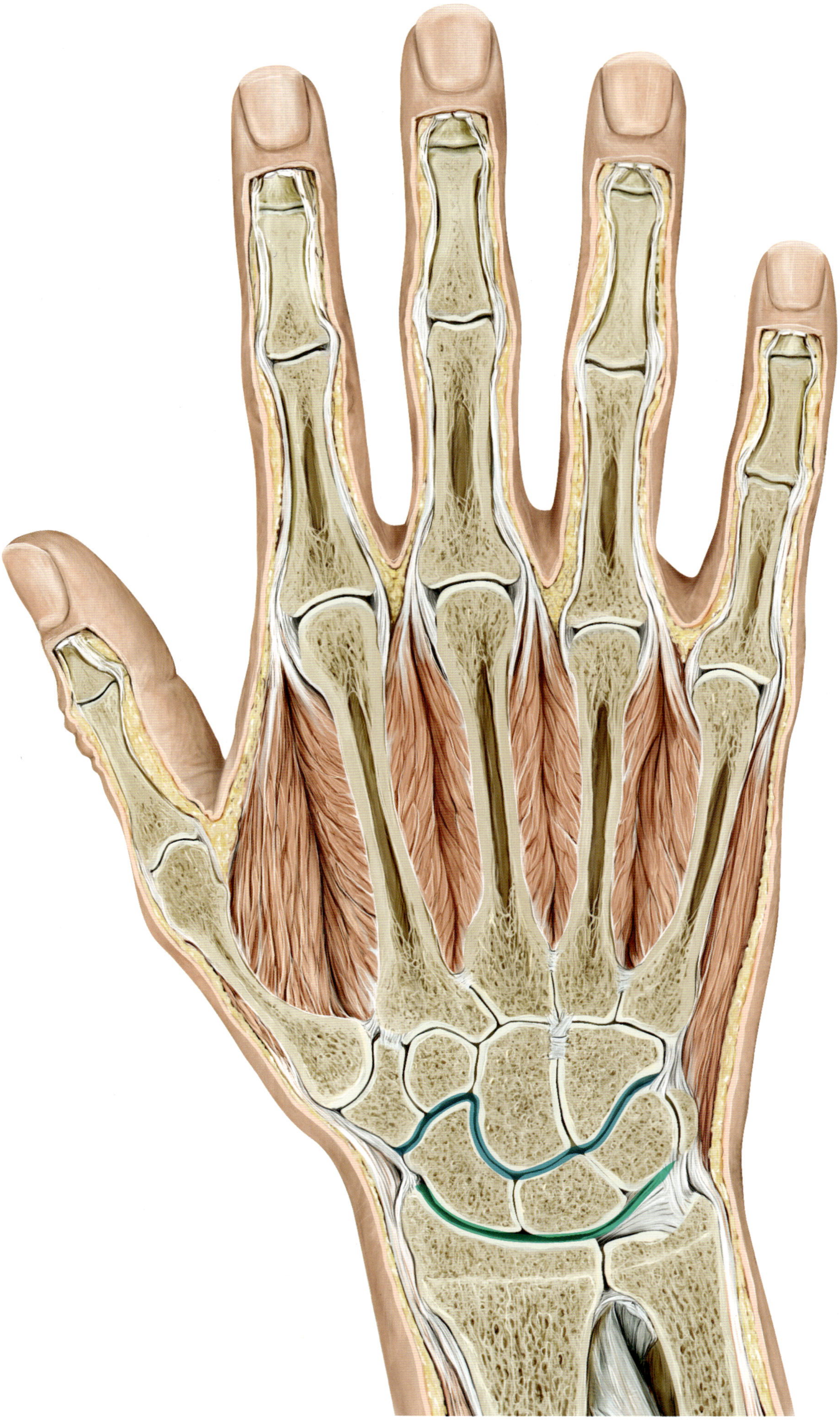

C Upper Limb

1.1 The Upper Limb as a Whole

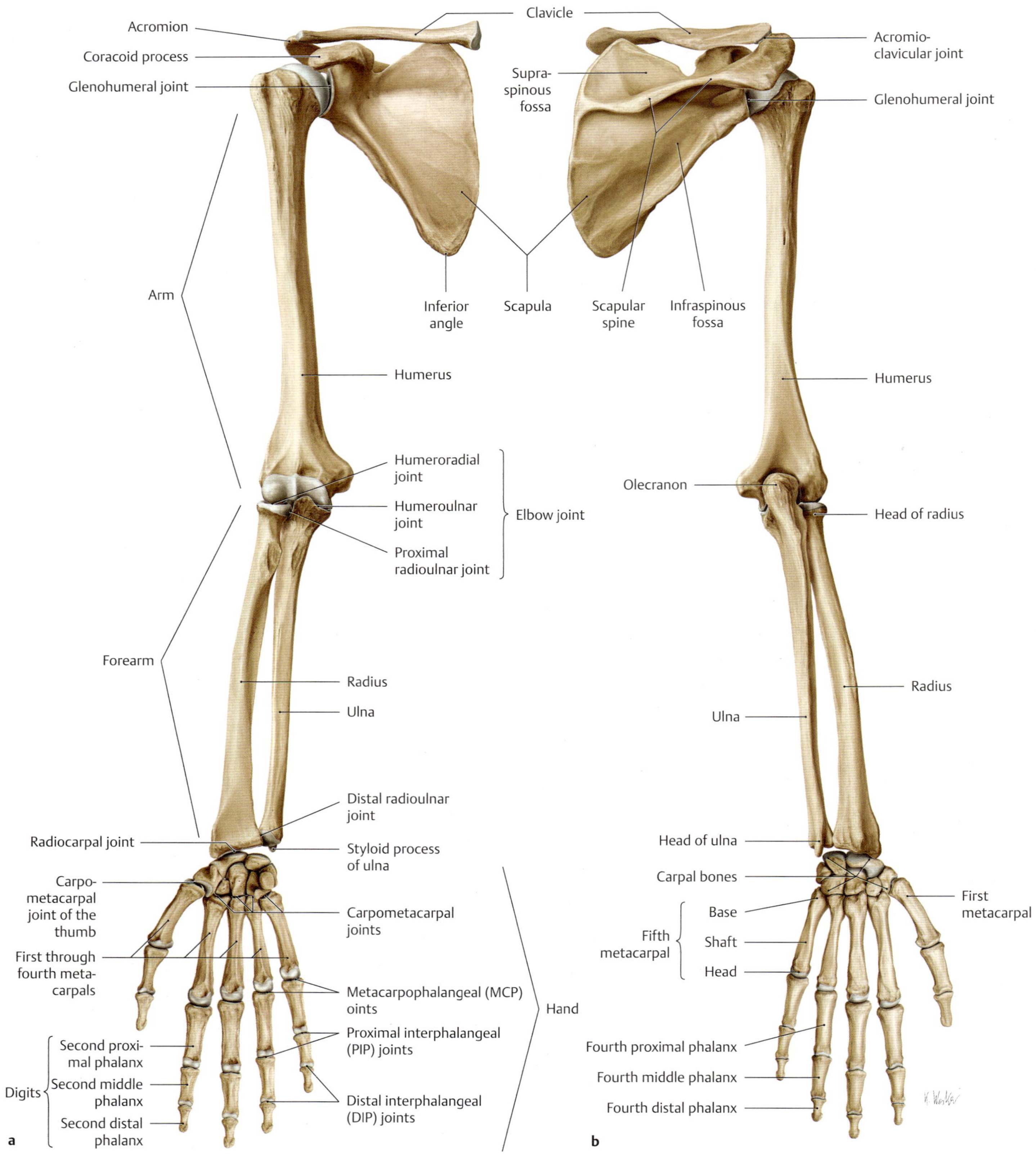

A Skeleton of the right upper limb

a Anterior view, **b** posterior view.

The skeleton of the upper limb consists of the shoulder girdle, arm, forearm, wrist, and hand The shoulder girdle (clavicle and scapula) is joined to the upper limb at the shoulder joint, and it joins the upper limb to the thorax at the sternoclavicular joint (see p. 271). The limb itself consists of the

- arm,
- forearm, and
- hand.

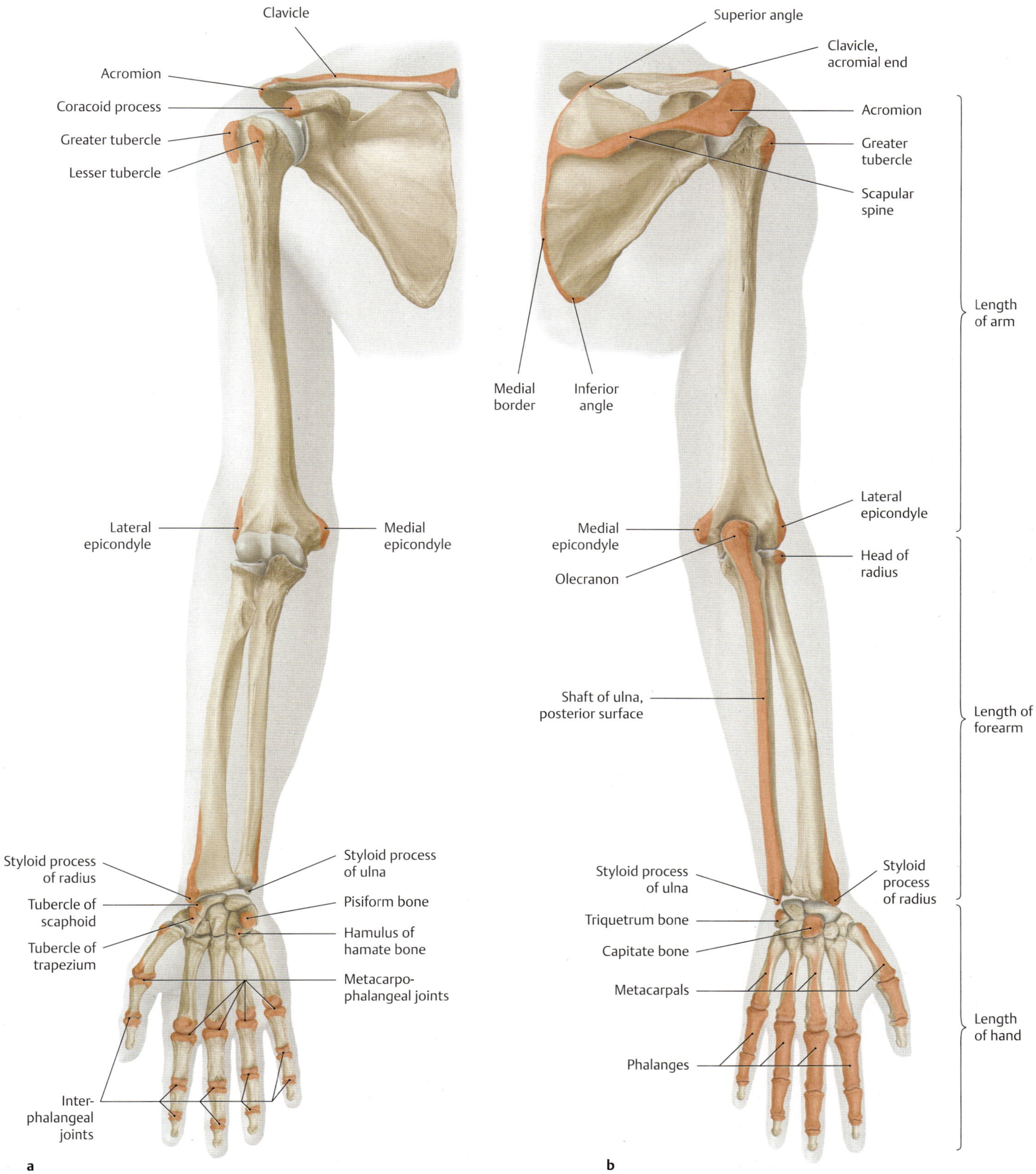

B Palpable bony prominences on the right upper limb

a Anterior view, **b** posterior view.

Except for the lunate and trapezoid bones, all of the bones in the upper limb are palpable to some degree through the skin and soft tissues. By consensus, standard reference points have been defined for use in measuring the lengths of the segments of the dependent limb (with the palm turned forward):

- Length of the arm = distance from the acromion to the lateral epicondyle.
- Length of the forearm = distance from the lateral epicondyle to the styloid process of the radius.
- Length of the hand = distance from the styloid process of the radius to the tip of the third finger.

The segment lengths of the limb may be measured, for example, to aid in the precise evaluation of isolated pediatric growth disturbances that are confined to a particular bone.

1.2 Integration of the Shoulder Girdle into the Skeleton of the Trunk

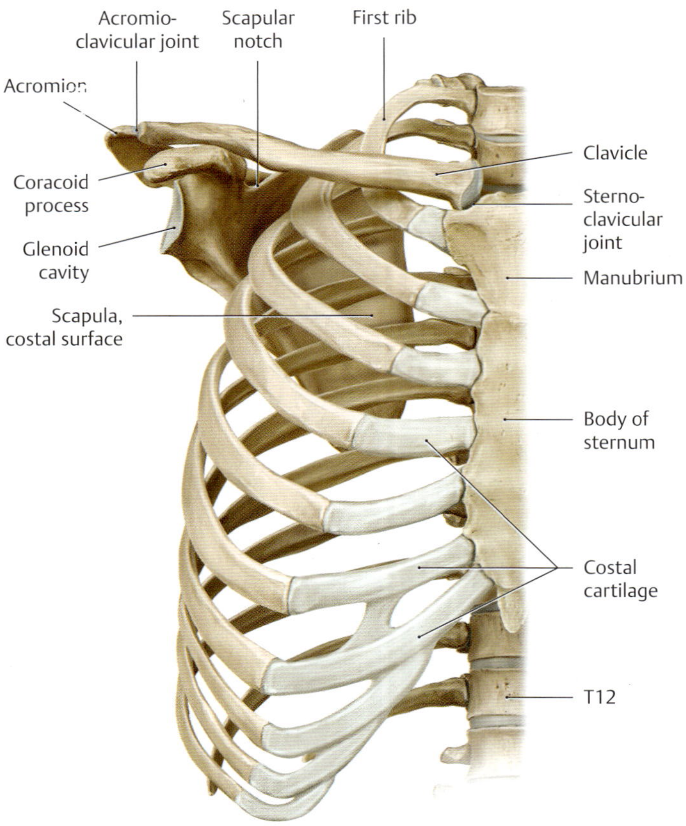

a

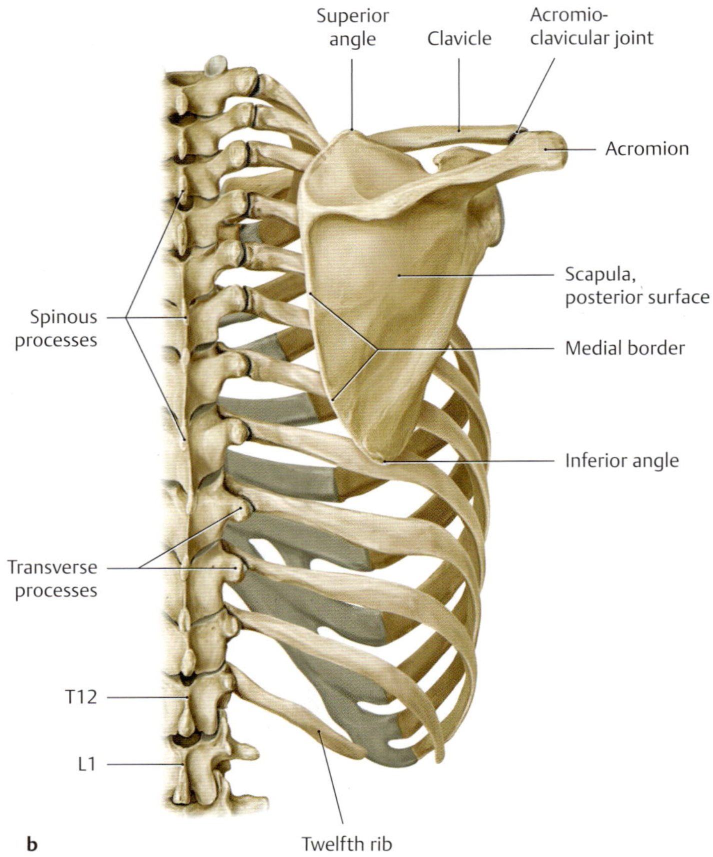

b

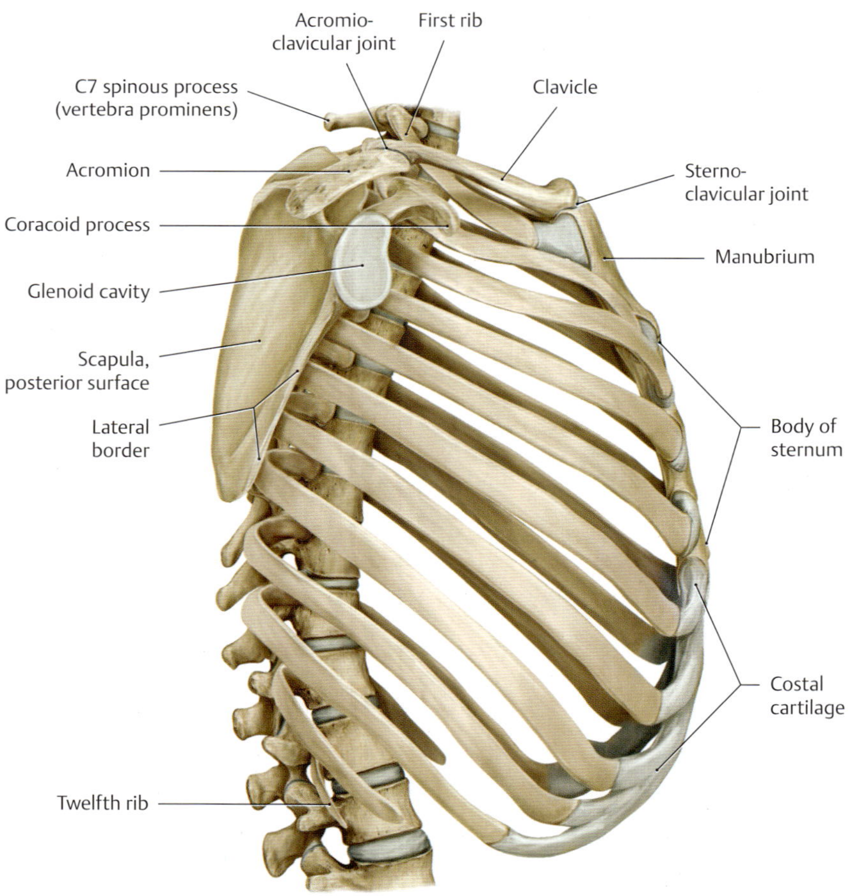

c

A Bones of the right shoulder girdle in relation to the trunk

a Anterior view, **b** posterior view,
c lateral view.

The two bones of the shoulder girdle (the clavicle and scapula) are connected at the acromioclavicular joint (see p. 271). In its normal anatomic position, the scapula extends from the second to the seventh rib. The inferior angle of the scapula is level with the spinous process of the seventh thoracic vertebra, and the scapular spine is level with the spinous process of the third thoracic vertebra. When the scapula occupies a normal position, its long axis is angled slightly laterally, and its medial border forms a 3 to 5° angle with the midsagittal plane.

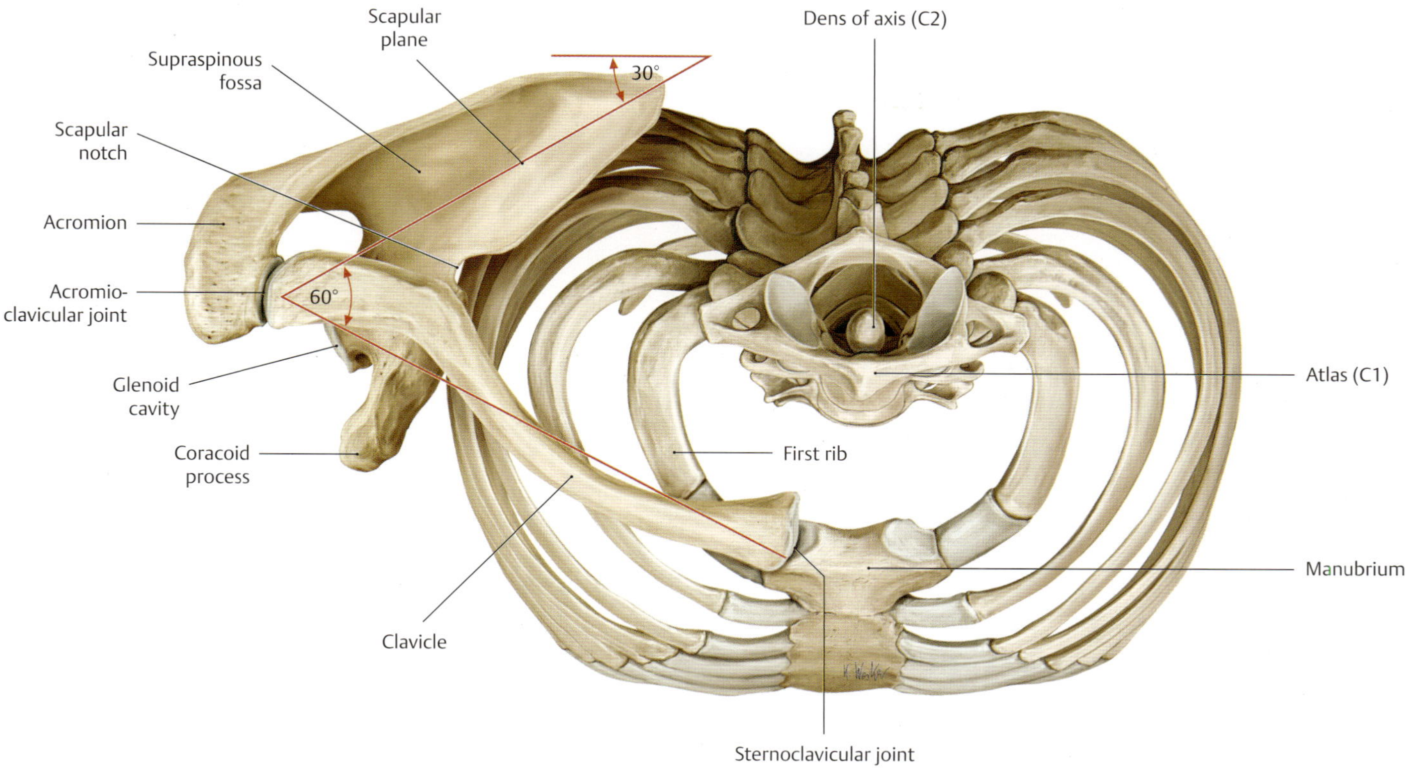

B Right shoulder girdle

Superior view. With the transition to a bipedal mode of locomotion, the human scapula moved from the more lateral placement in quadruped mammals to a more posterior position, and also a more frontal orientation, on the back of the thorax. Viewed from above, the scapula forms a 30° angle with the coronal plane. The scapula and clavicle subtend an angle of approximately 60°. Because of this arrangement, the two shoulder joints are angled slightly forward, shifting the range of arm movements forward into the field of vision and action. This reorientation in humans creates the opportunity for visual control of manual manipulations (hand–eye coordination).

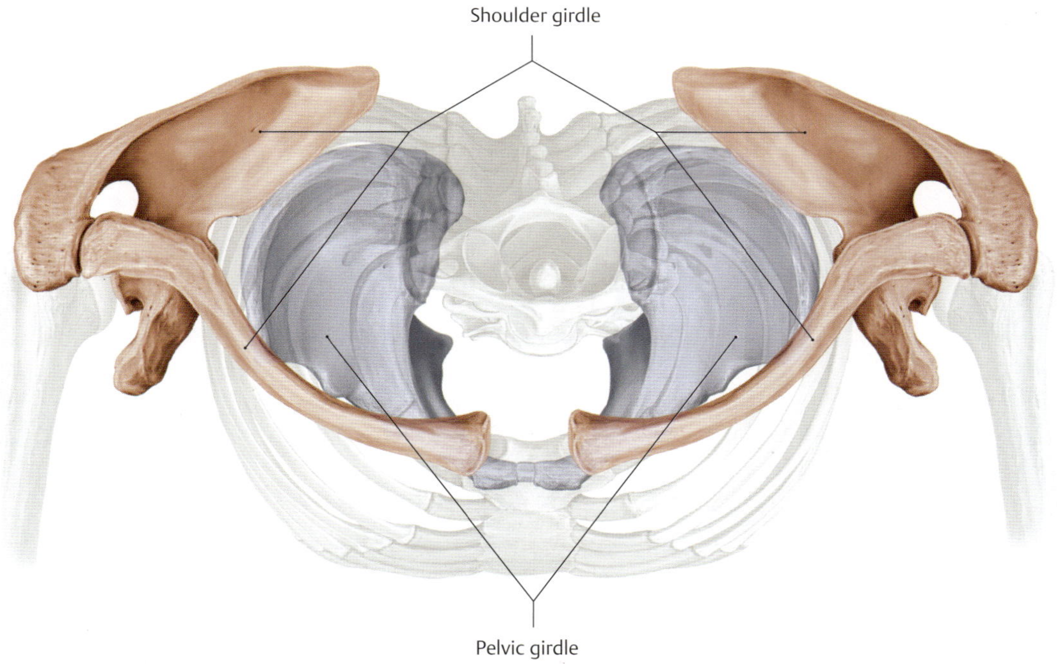

C Comparison of the shoulder girdle and pelvic girdle in their relation to the skeleton of the trunk

Superior view. Unlike the very mobile shoulder girdle, the pelvic girdle, consisting of the paired hip (coxal) bones, is firmly integrated into the axial skeleton. As the trunk assumes an upright position, the pelvis moves over the weight-bearing surface of the feet, making it necessary for the pelvis to support the total weight of the trunk. This basically limits the lower limbs to functions of locomotion and support while freeing the upper limbs from these tasks and making them a versatile organ of movement and expression that is particularly useful for touching and grasping.

1.3 The Bones of the Shoulder Girdle

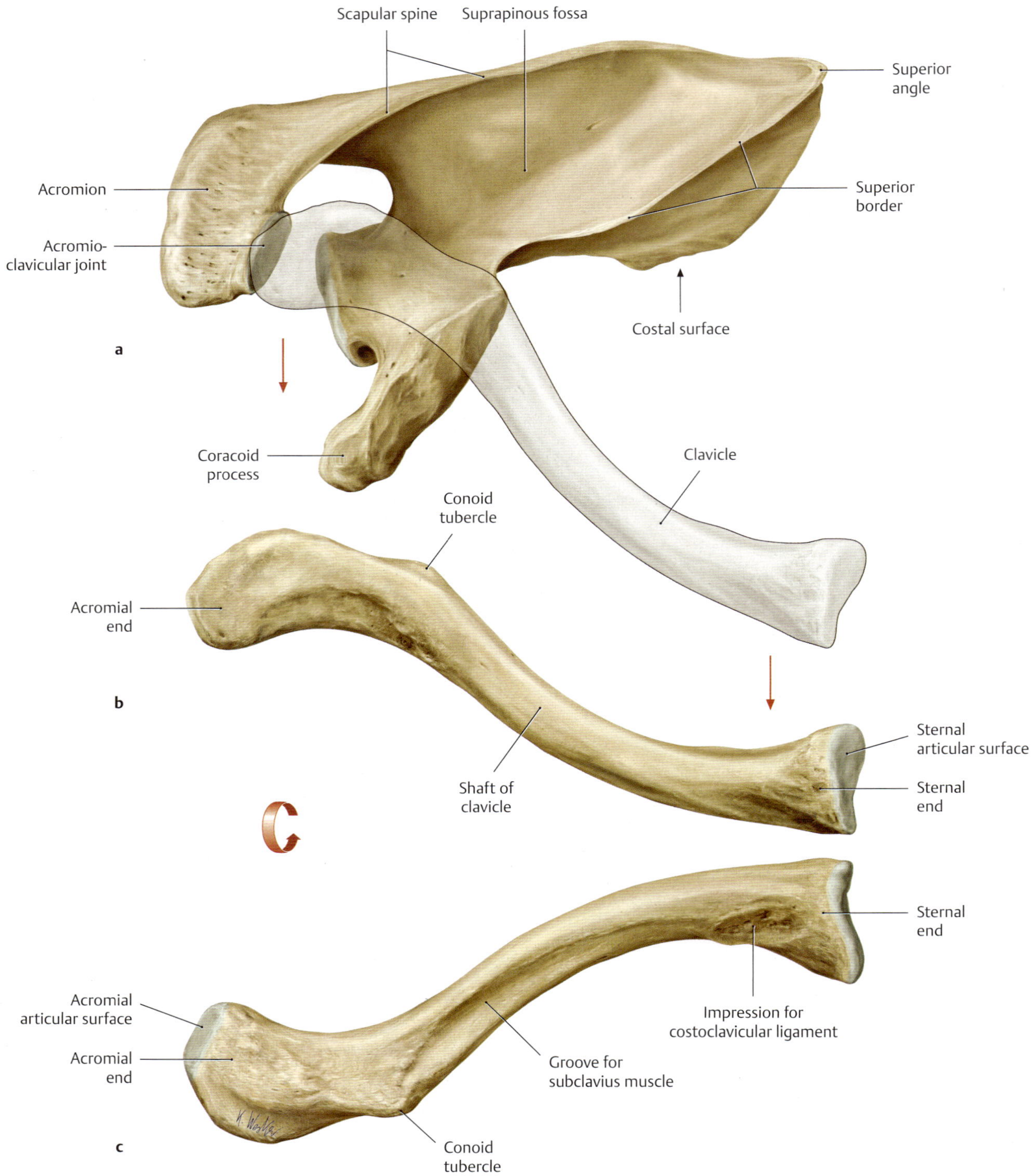

A Position and shape of the right clavicle
a Clavicle in its normal relation to the scapula, superior view.
b Isolated clavicle, superior view.
c Isolated clavicle, inferior view.

The clavicle is an S-shaped bone, approximately 12 to 15 cm long in adults, that is visible and palpable beneath the skin along its entire length. The medial or *sternal end* of the clavicle bears a saddle-shaped articular surface, while the lateral or *acromial end* has a flatter, more vertical articular surface. The clavicle is the only bone in the limbs that is not preformed in cartilage during embryonic development; instead, it ossifies directly from *connective tissue* (membranous ossification). A congenital failure or abnormality in the development of this connective tissue results in an anomaly called *cleidocranial dysostosis*. There may be associated ossification defects in the cranial vault, which are also formed by membranous ossification (*craniofacial dysostosis*). Besides fractures due to obstetric trauma (1–2% of all newborns), fractures of the middle third of the clavicle are one of the most common fractures that are sustained by children and adults (in children, some 50% of all clavicular fractures occur before 6 years of age).

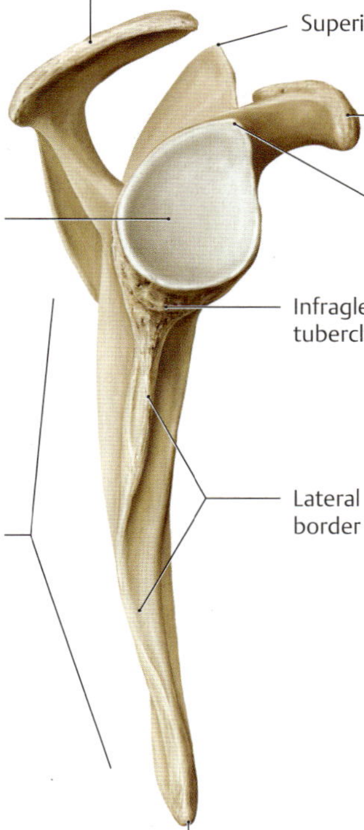

B The right scapula. Lateral view

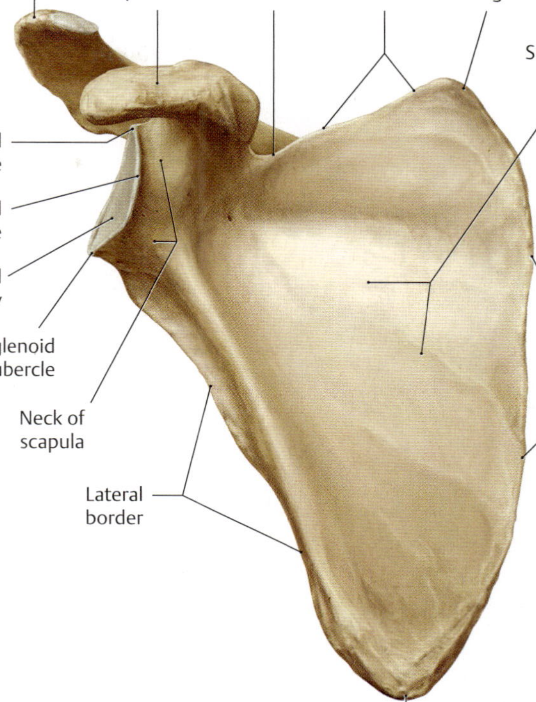

C The right scapula. Anterior view

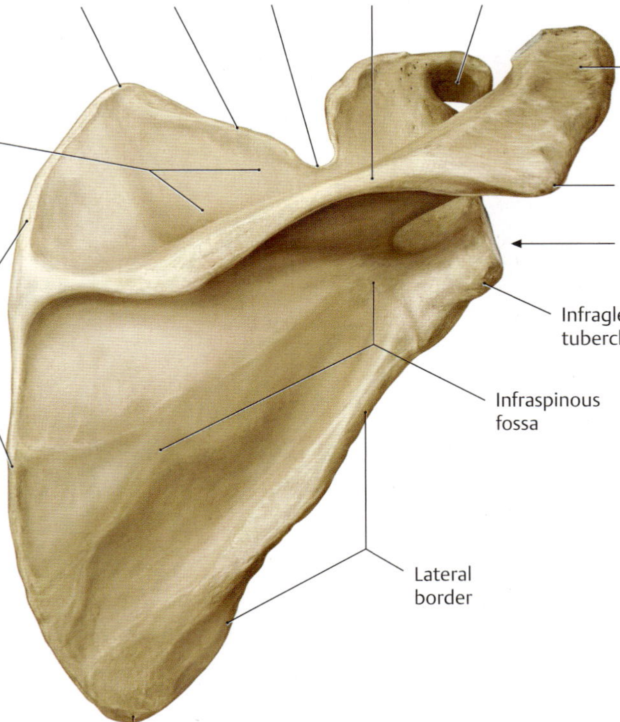

D The right scapula. Posterior view

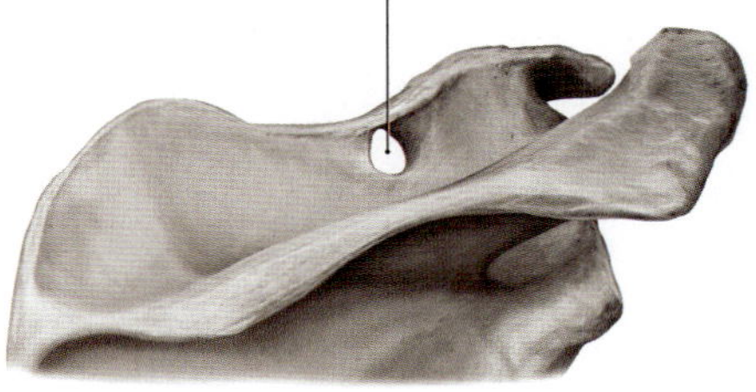

E The scapular foramen

The superior transverse ligament of the scapula (see p. 279) can become ossified, transforming the scapular notch into an anomalous bony canal referred to as a *scapular foramen*. This can lead to compression of the suprascapular nerve as it passes through this canal (see p. 402). Active rotational movements of the shoulder aggravate the nerve, leading to significant symptoms (*scapular notch syndrome*). A common result is weakness and atrophy of the muscles—the supraspinatus and infraspinatus—that the suprascapular nerve innervates (see p. 317).

1.4 The Bones of the Upper Limb: The Humerus

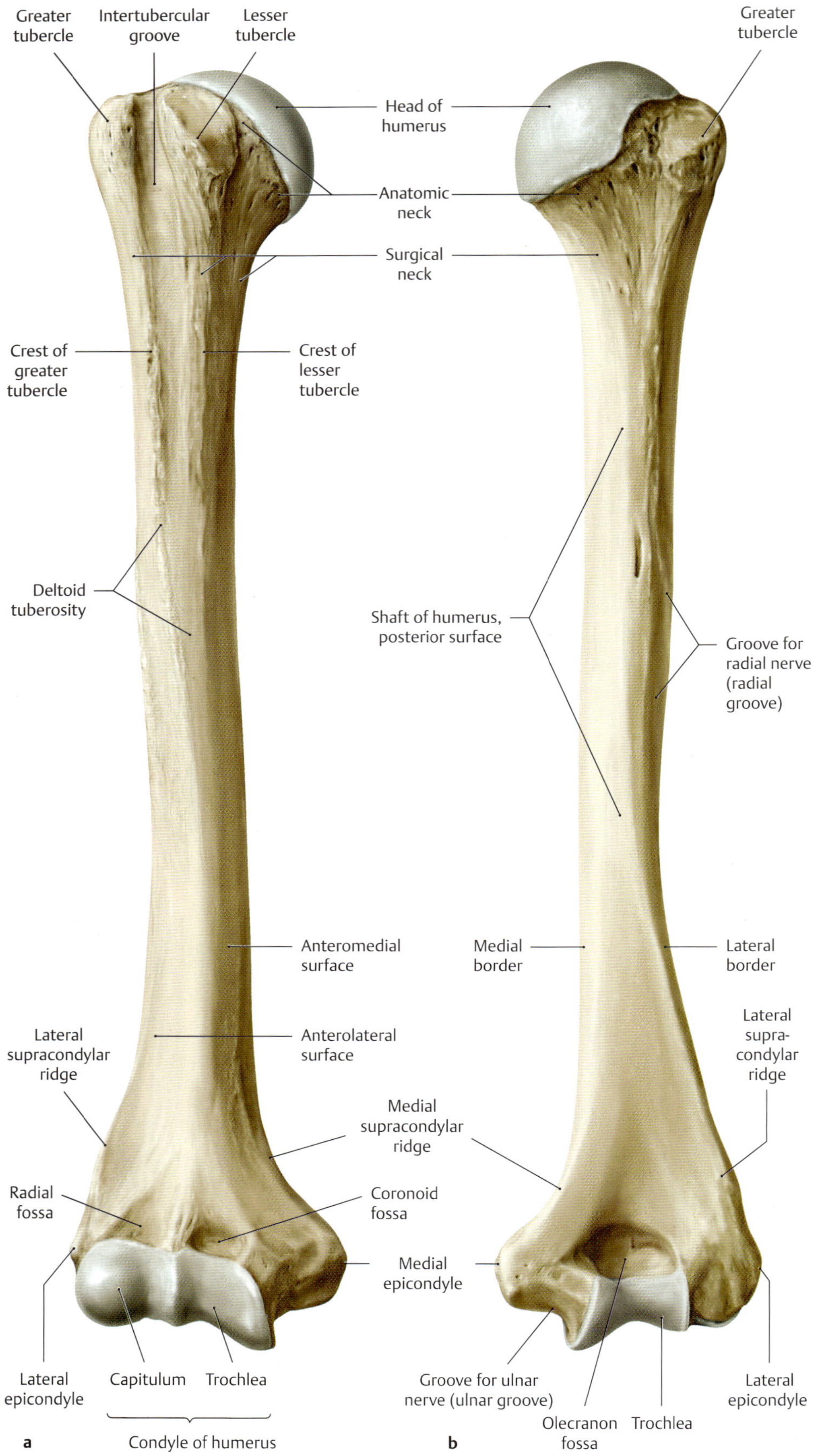

A The right humerus
a Anterior view, **b** posterior view.

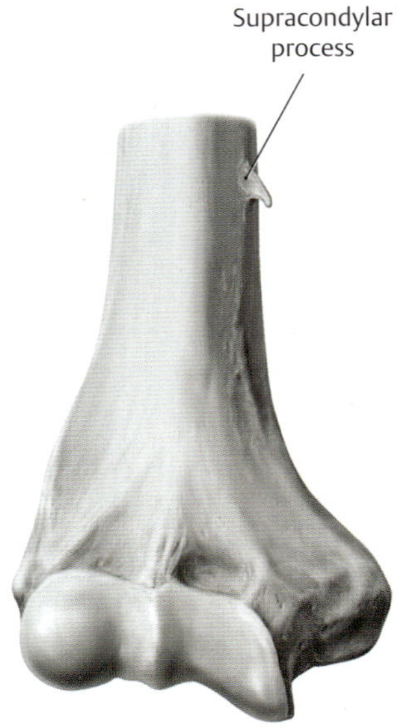

B Supracondylar process
An anomaly sometimes found on the distal humerus above the medial epicondyle is referred to as a supracondylar process. This bony outgrowth is a relatively rare atavistic feature in humans that corresponds to a normal structure in other vertebrates, in which it forms part of a supracondylar canal (see p. 407).

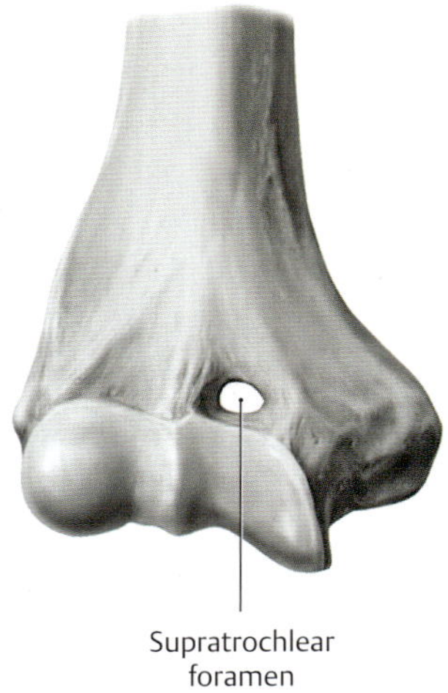

C The supratrochlear foramen
The presence of a supratrochlear foramen is another rare variant in which the two opposing olecranon and coronoid fossae communicate through an opening.

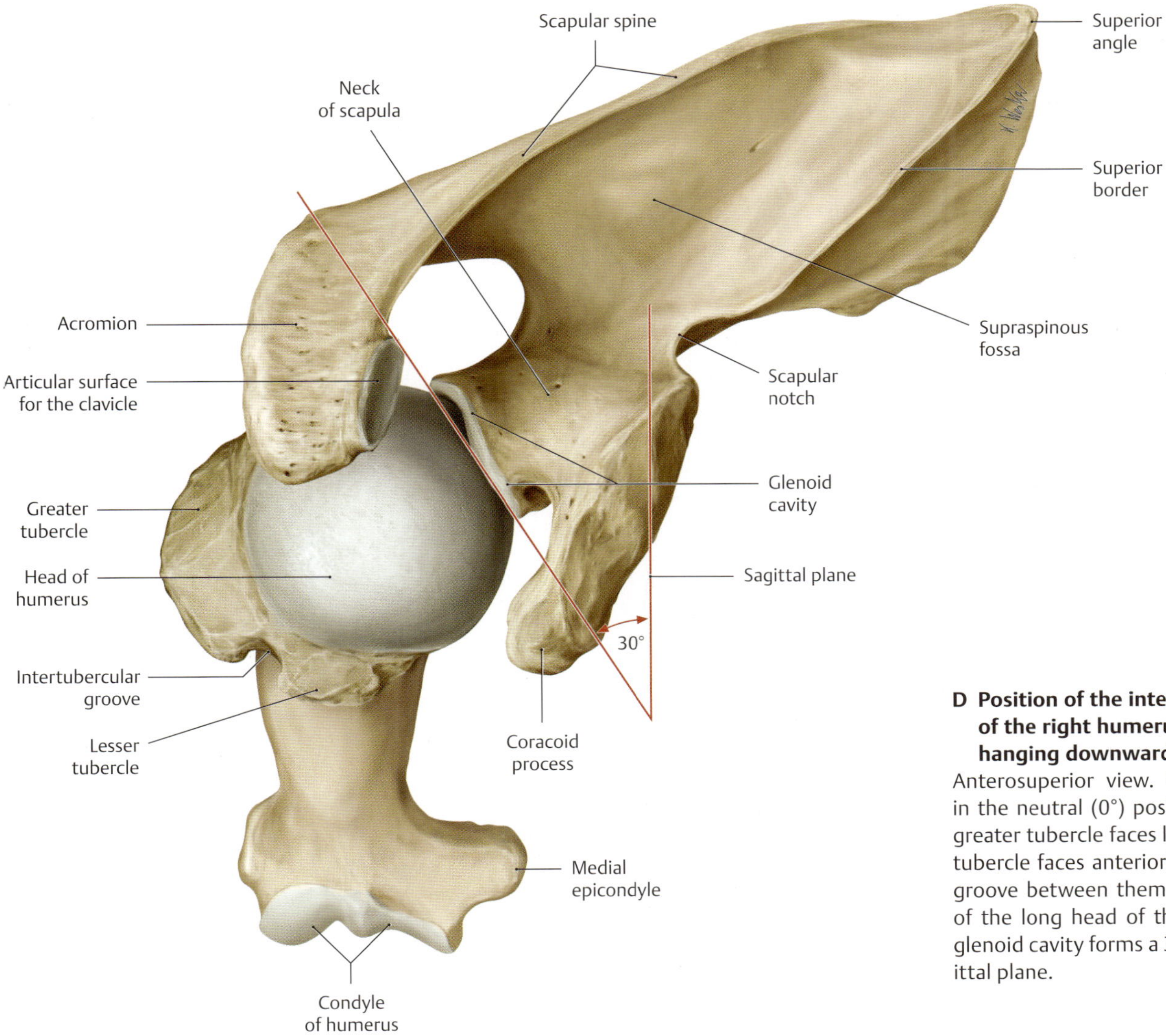

D Position of the intertubercular groove of the right humerus with the limb hanging downward

Anterosuperior view. With the upper limb in the neutral (0°) position (see p. 287), the greater tubercle faces laterally, and the lesser tubercle faces anteriorly. The intertubercular groove between them transmits the tendon of the long head of the biceps muscle. The glenoid cavity forms a 30° angle with the sagittal plane.

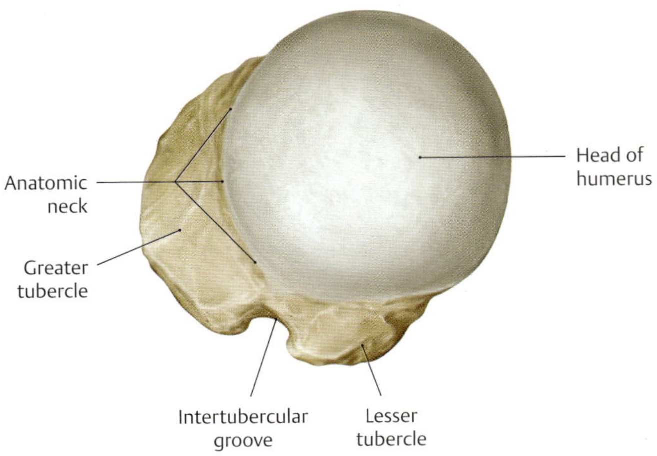

E The proximal right humerus. Superior view

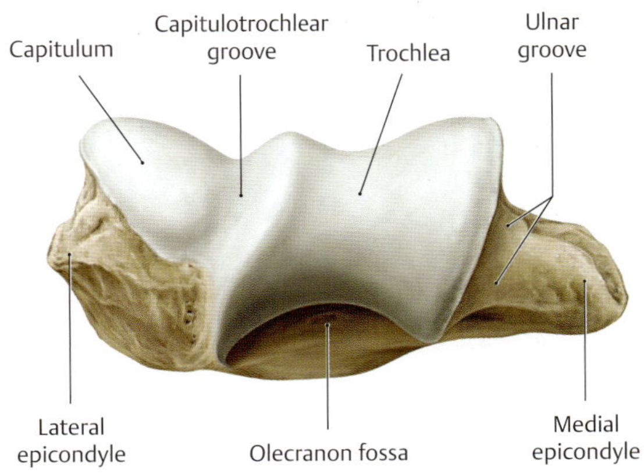

F The distal right humerus. Inferior view

1.5 The Bones of the Upper Limb: Torsion of the Humerus

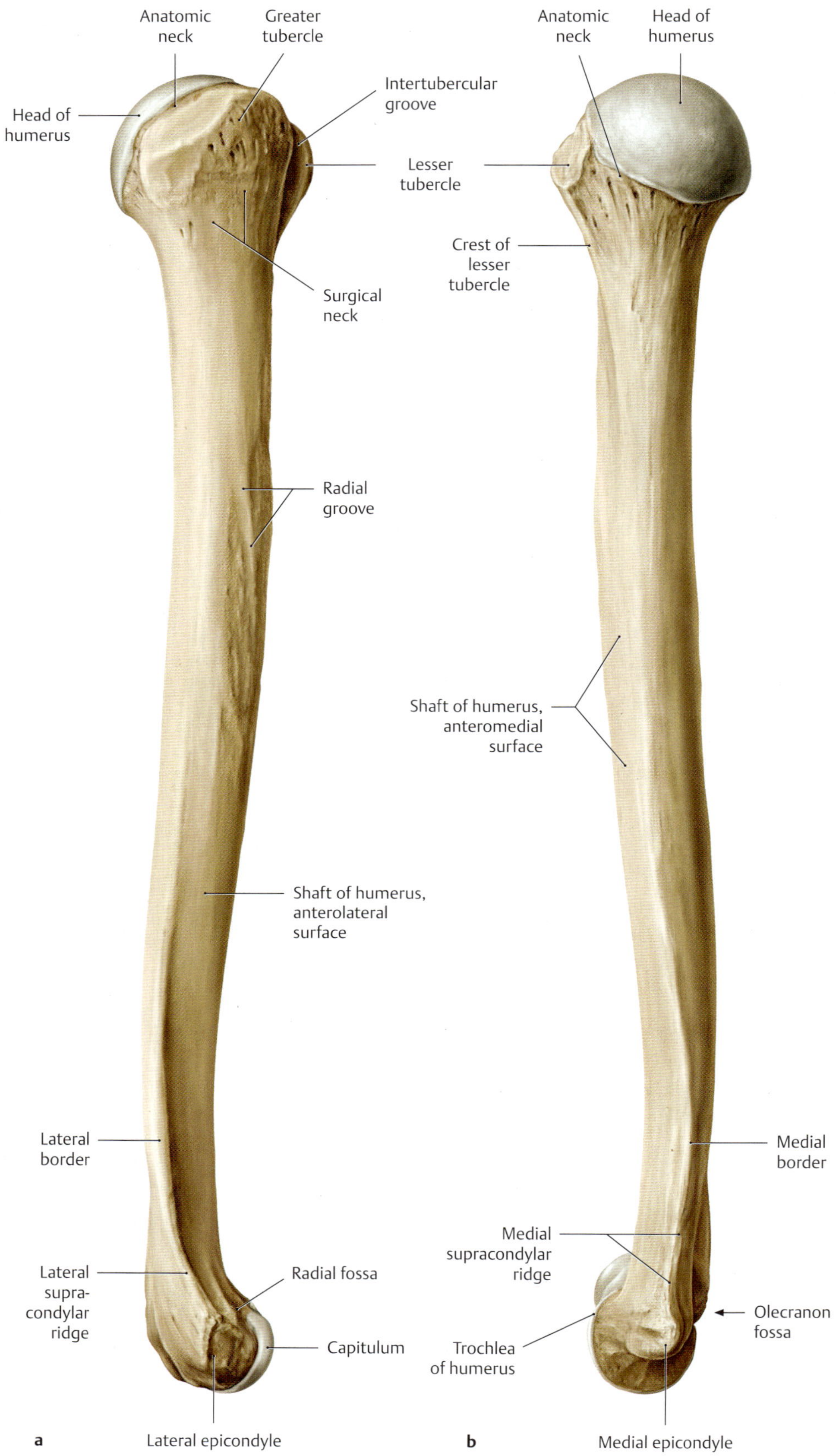

A The right humerus
a Lateral view, **b** medial view.

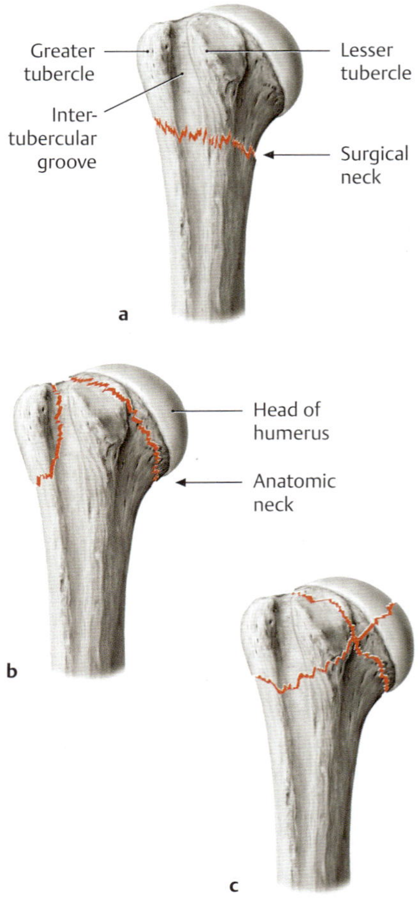

B Fractures of the proximal humerus
Anterior view. Fractures of the proximal humerus comprise approximately 4 to 5% of all fractures. They occur predominantly in older patients who sustain a fall onto the outstretched arm or directly onto the shoulder. Three main types are distinguished:

- Extra-articular fractures (**a**)
- Intra-articular fractures (**b**)
- Comminuted fractures (**c**)

Not infrequently, extra-articular fractures at the level of the *surgical neck* (site of predilection for extra-articular fractures of the proximal humerus) and intra-articular fractures at the level of the *anatomic neck* are accompanied by injuries of the blood vessels that supply the humeral head (anterior and posterior circumflex humeral arteries, see p. 369), with an associated risk of post-traumatic avascular necrosis. Besides proximal humeral fractures, other important injuries are fractures of the humeral shaft and fractures of the distal humerus (e.g., supracondylar fractures). Fractures of the humeral shaft are frequently associated with damage to the radial nerve in its groove (see p. 383 for neurologic deficits following a radial nerve lesion).

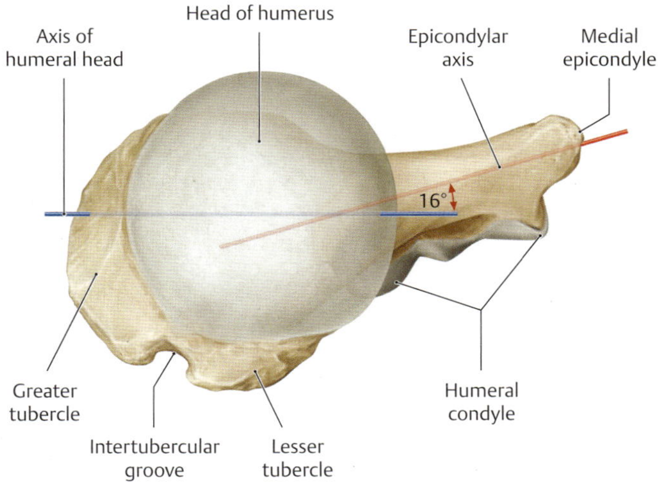

C Torsion of the humerus
Right humerus, superior view. The shaft of the adult humerus normally exhibits some degree of torsion, i.e., the proximal end of the humerus is rotated relative to its distal end. The degree of this torsion can be assessed by superimposing the axis of the humeral head (from the center of the greater tubercle to the center of the humeral head) over the epicondylar axis of the elbow joint. This *torsion angle* equals approximately 16° in an adult, compared to about 60° in a newborn. The decrease in the torsion angle with body growth correlates with the change in the position of the scapulae. Thus, while the glenoid cavity in the newborn still faces anteriorly, it is directed much more laterally in the adult (see p. 253). As the position of the scapula changes, there is a compensatory decrease in the torsion angle to ensure that hand movements will remain within the visual field of the adult.

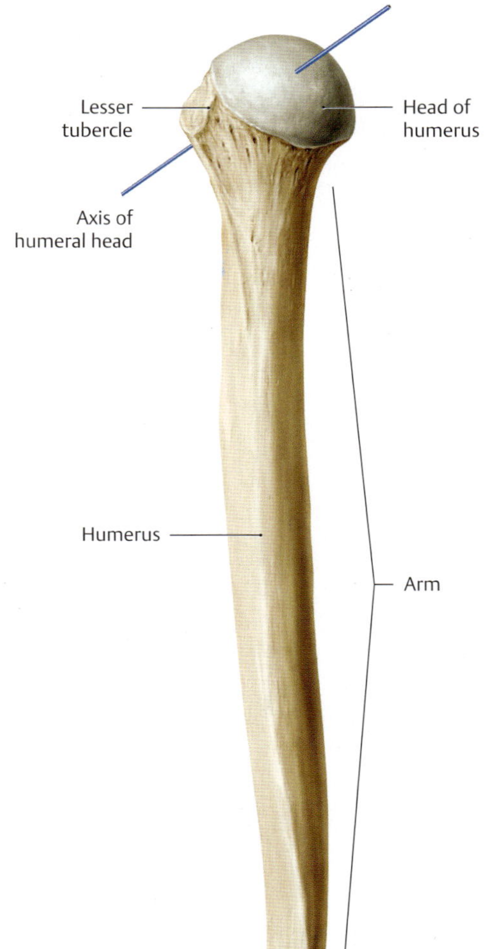
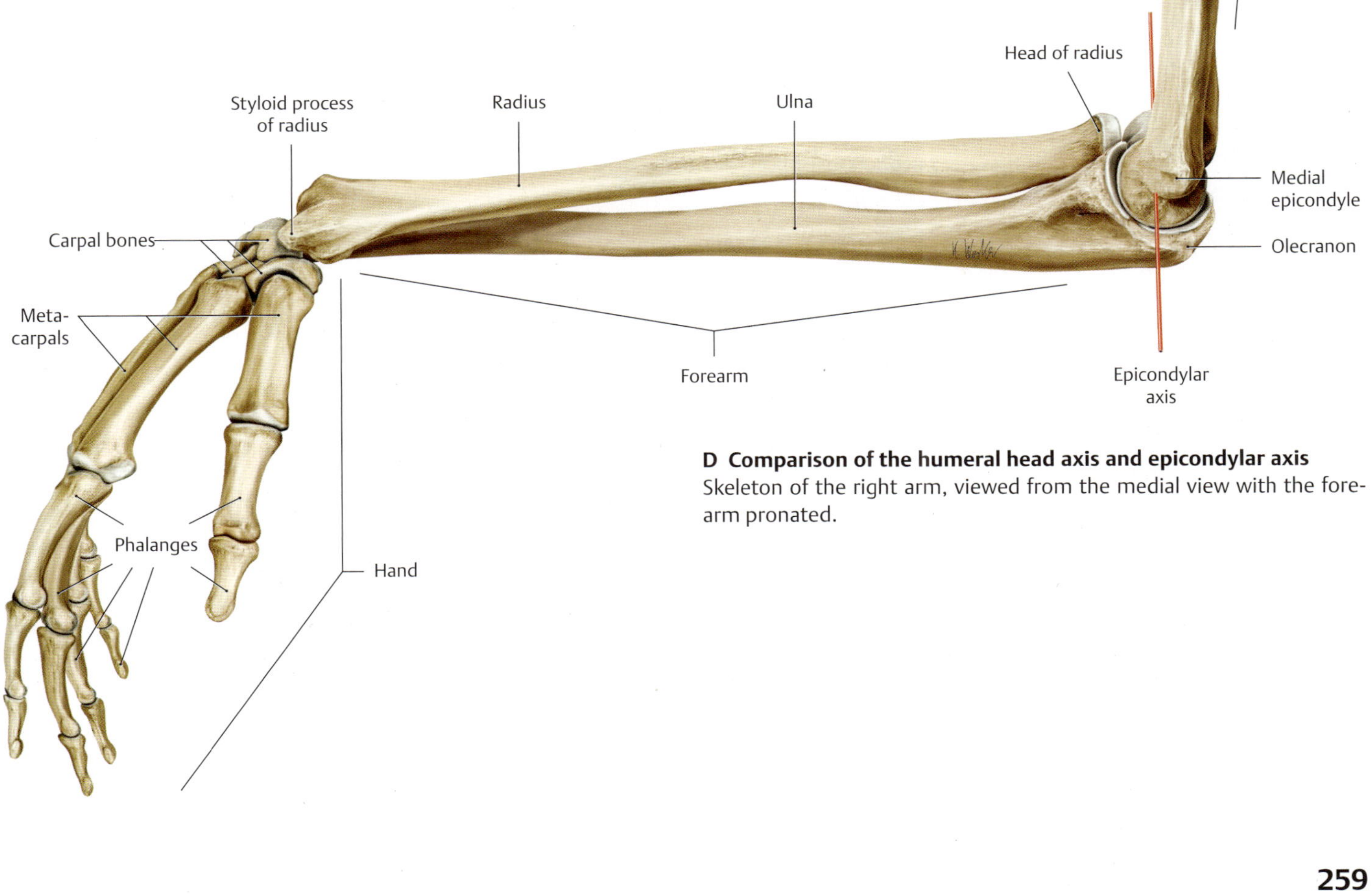

D Comparison of the humeral head axis and epicondylar axis
Skeleton of the right arm, viewed from the medial view with the forearm pronated.

1.6 The Bones of the Upper Limb: The Radius and Ulna

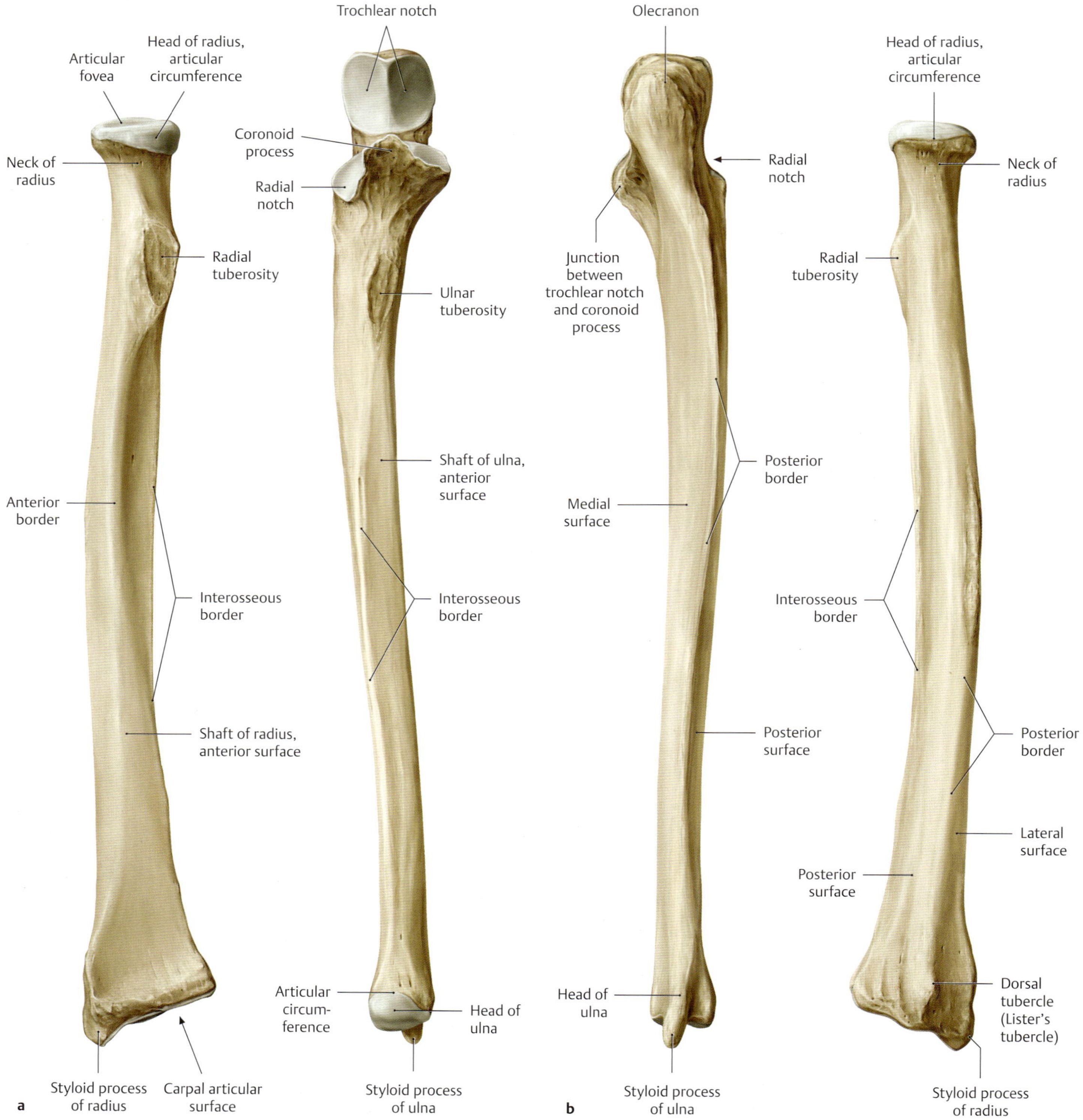

A The radius and ulna of the right forearm
a Anterior view, **b** posterior view.
The radius and ulna are not shown in their normal relationship; they have been separated to demonstrate the articular surfaces of the proximal and distal radioulnar joints.

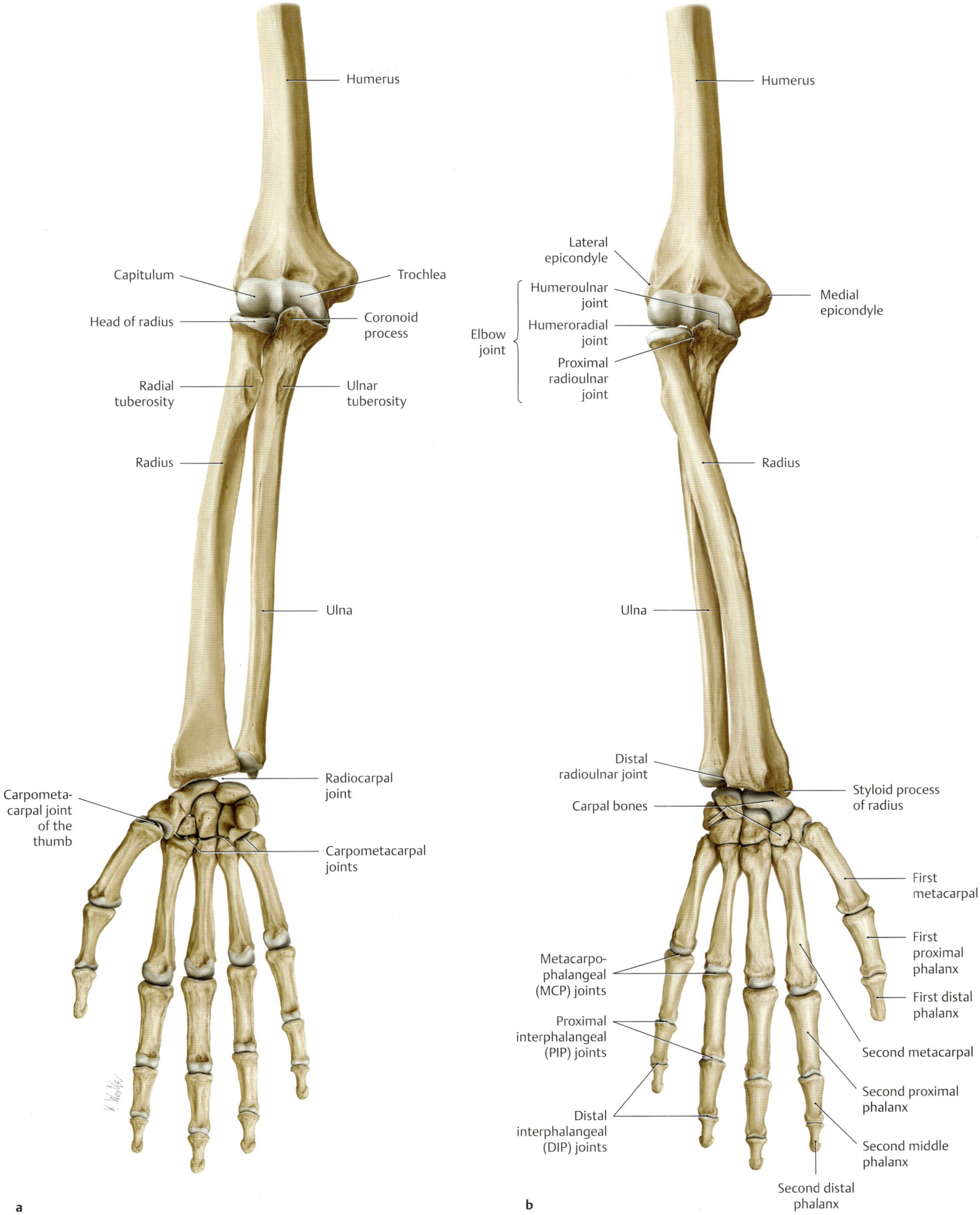

B The radius and ulna of the right arm in (a) supination and (b) pronation

The radius and ulna are parallel to each other in supination, whereas in pronation the radius crosses over the ulna. The movement of turning the palm upward or downward (supination/pronation) takes place at the proximal and distal radioulnar joints (see p. 294).

1.7 The Bones of the Upper Limb: The Articular Surfaces of the Radius and Ulna

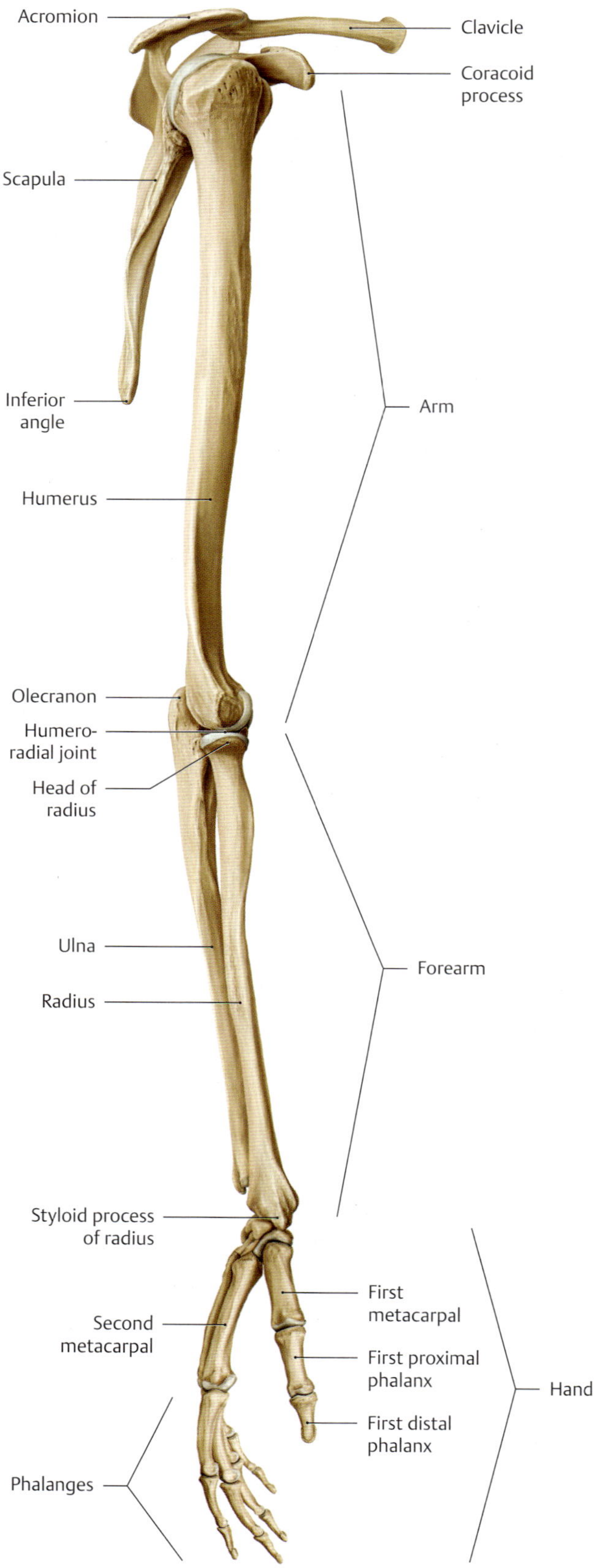

A The right upper limb
Lateral view. The forearm is supinated (the radius and ulna are parallel).

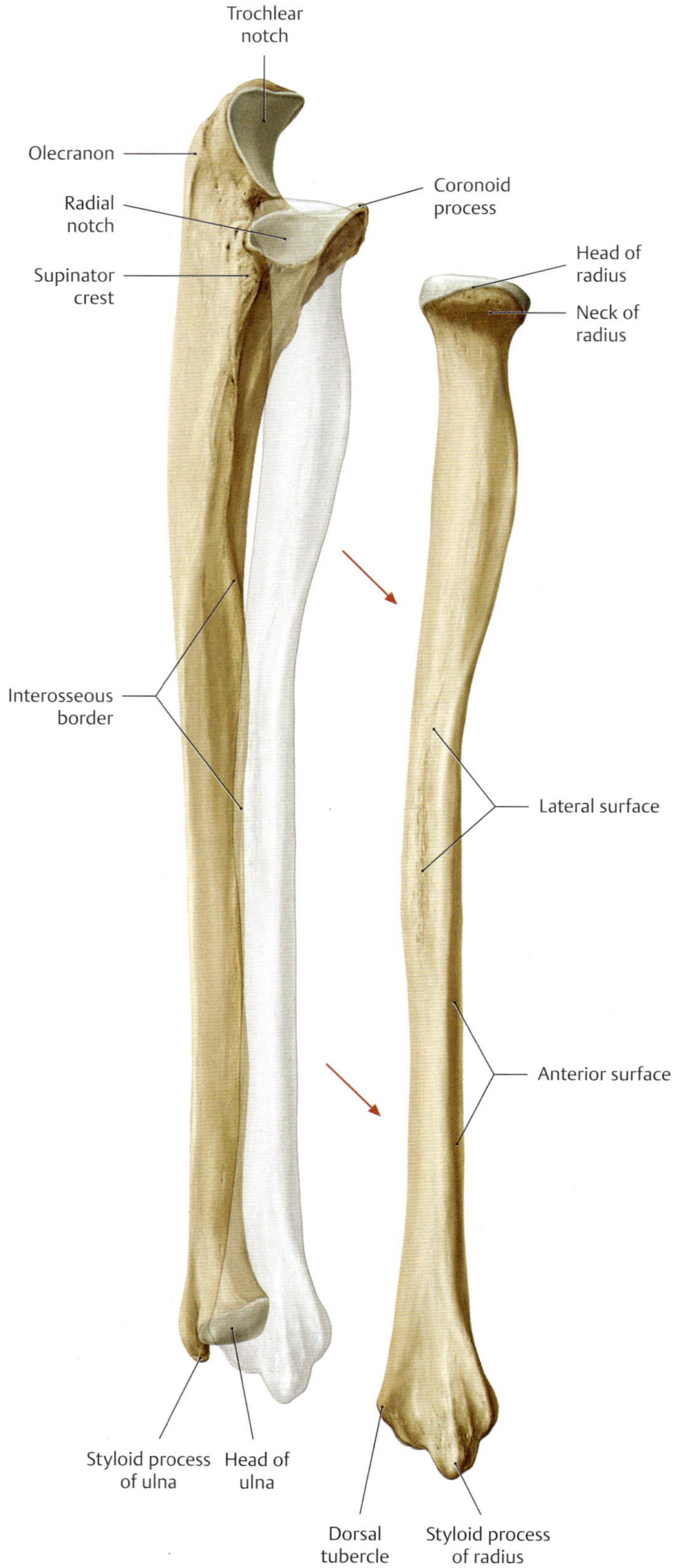

B Right forearm
Lateral view. The radius and ulna are shown in a disarticulated position to demonstrate the articular surfaces of the ulna for the proximal and distal radioulnar joints (see **C**).

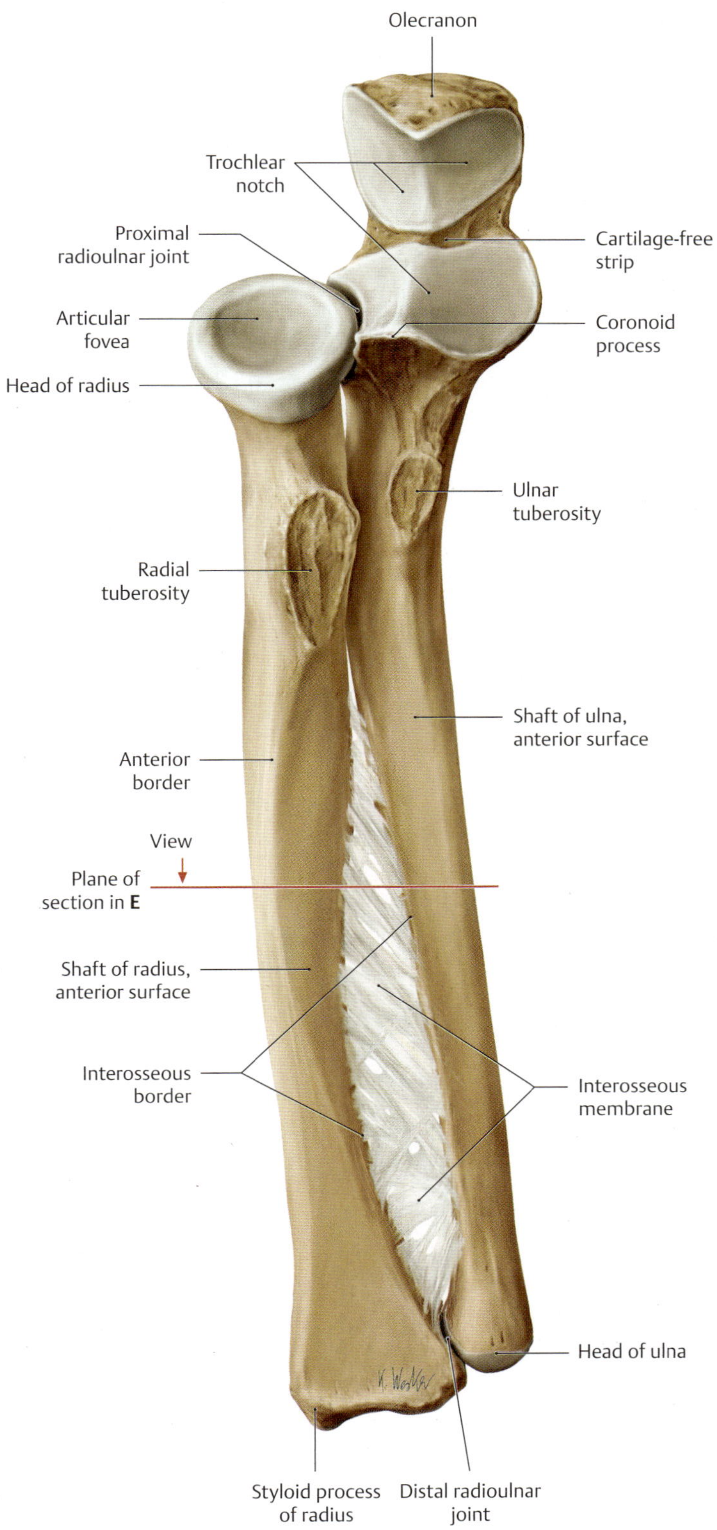

C The radius and ulna of the right forearm
Anterosuperior view. The proximal and distal radioulnar joints are functionally interlinked by the interosseous membrane between the radius and ulna. As a result, motion in one joint is invariably combined with motion in the other (see p. 296).

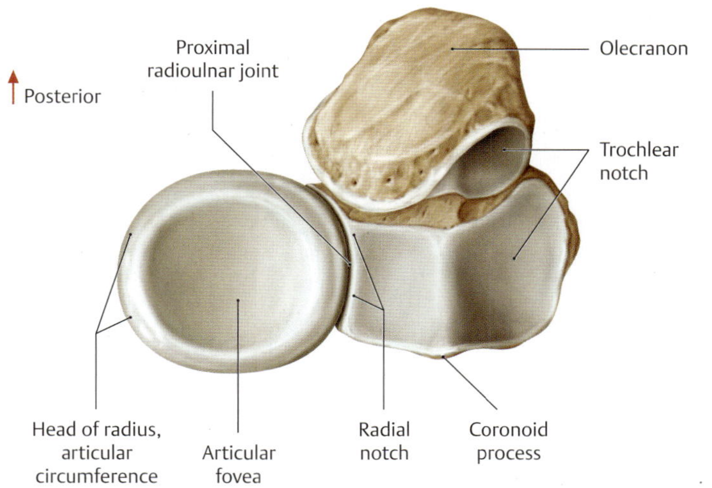

D The proximal articular surfaces of the radius and ulna of the right forearm
Proximal view.

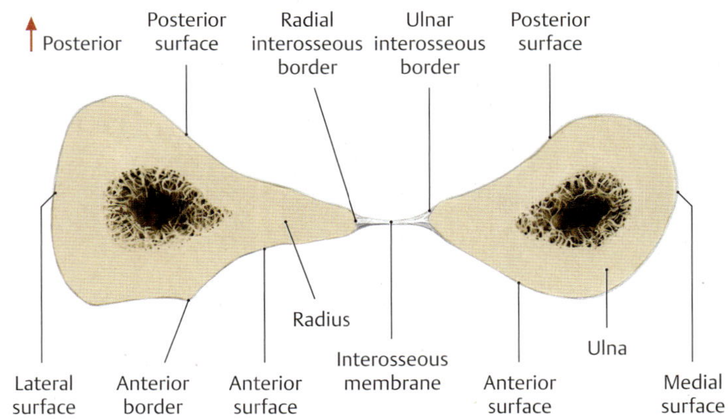

E Cross section through the right radius and ulna
Proximal view.

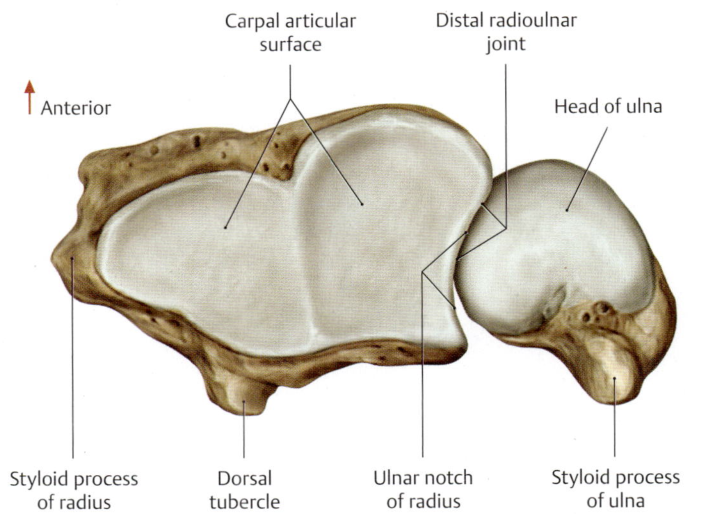

F The distal articular surfaces of the radius and ulna of the right forearm
Distal view.

1.8 The Bones of the Upper Limb: The Hand

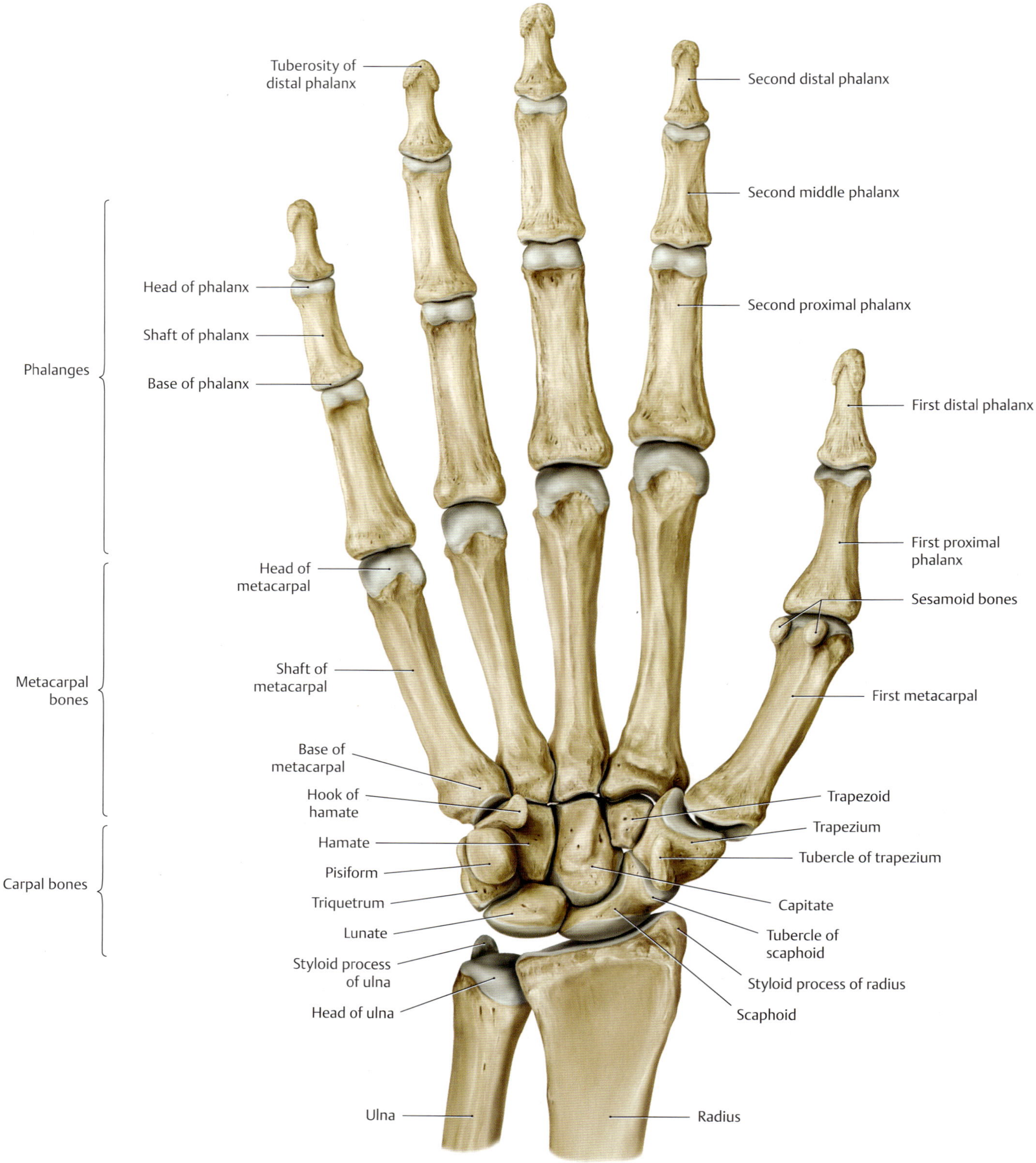

A The bones of the right hand. Palmar view
The skeleton of the hand consists of

- the carpal bones,
- the metacarpal bones, and
- the phalanges.

The palm refers to the anterior (flexor) surface of the hand, the dorsum to the posterior (extensor) surface. The terms of anatomic orientation in the hand are palmar or volar (toward the anterior surface), dorsal (toward the posterior surface), ulnar (toward the ulna or small finger), and radial (toward the radius or thumb).

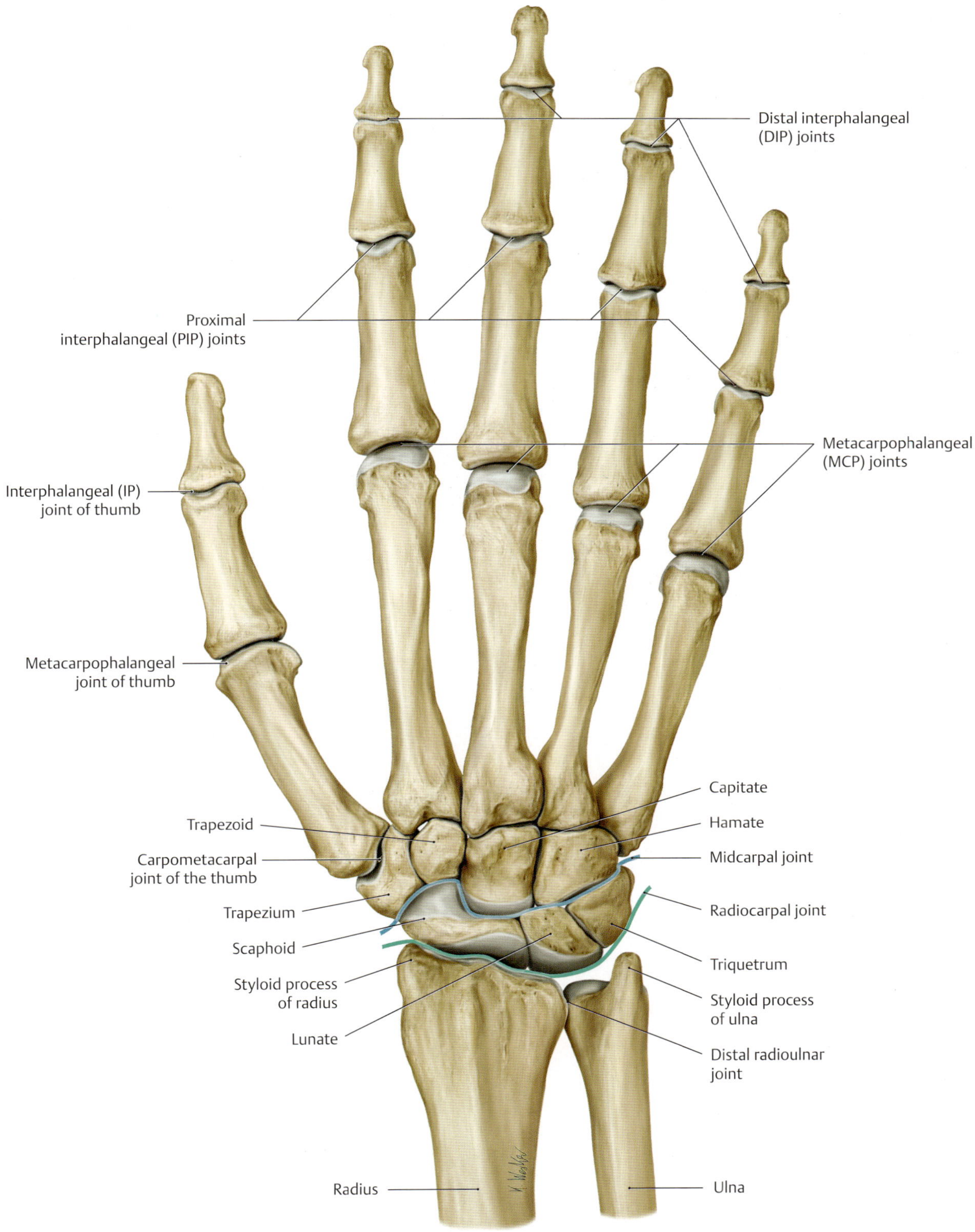

B The bones of the right hand. Dorsal view
The radiocarpal and midcarpal joints are indicated by blue lines.

1.9 The Bones of the Upper Limb: The Carpal Bones

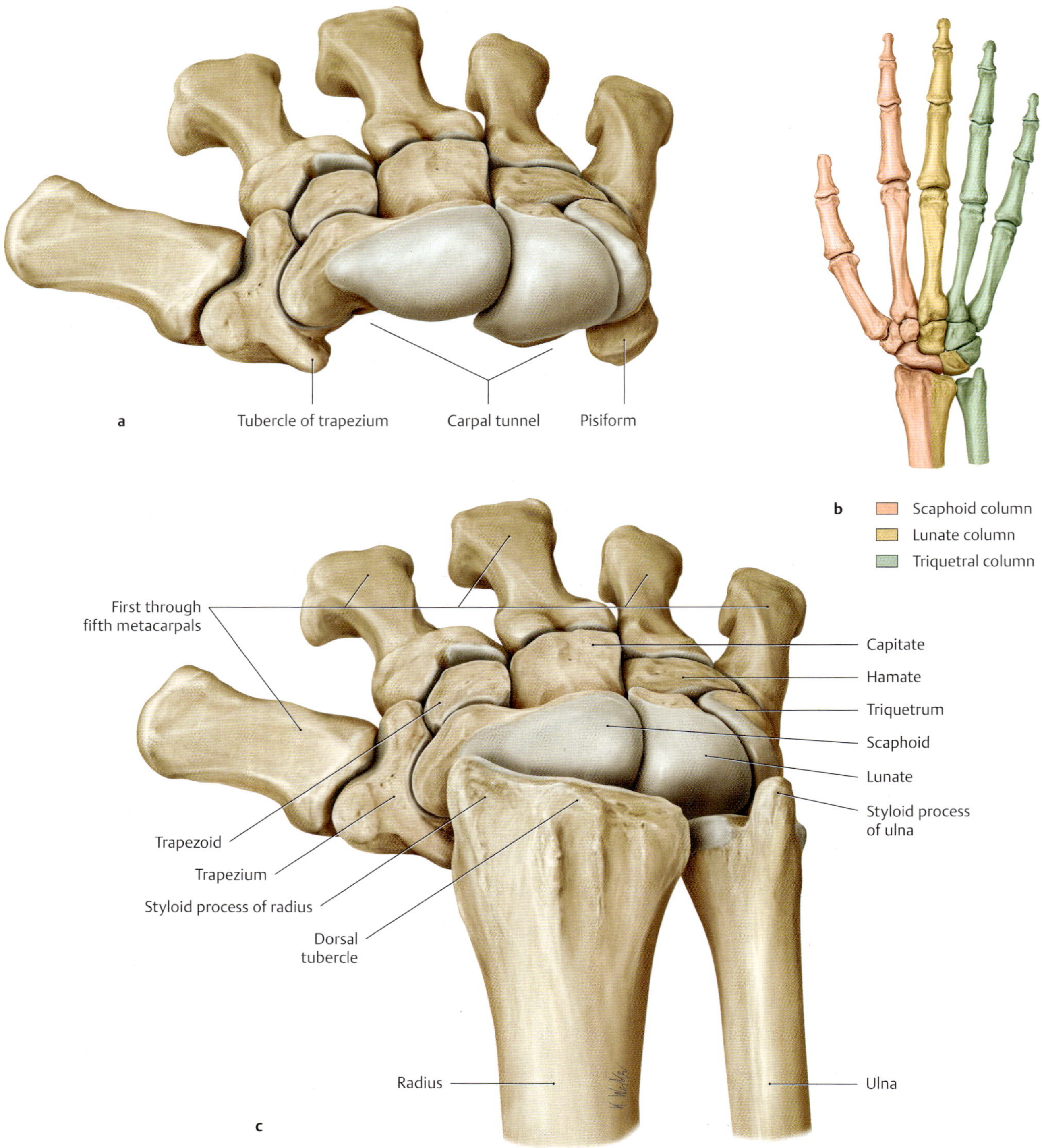

A Carpal bones of the right hand
a After removal of radius and ulna, proximal view; **b** column model of the hand, dorsal view; **c** with the wrist in flexion, proximal view.

The carpal bones are arranged in two rows of four bones each—one proximal row and one distal row (see also **B**). From a biomechanical and clinical standpoint, the carpal bones do not form two transverse rows but are arranged in three longitudinal columns: a radial scaphoid column (consisting of the scaphoid, trapezium, and trapezoid bones), a central lunate column (consisting of the lunate and capitate), and an ulnar triquetral column (consisting of the triquetrum and hamate). In this functional classification, the pisiform is regarded as a sesamoid bone embedded in the tendon of the flexor carpi ulnaris muscle (see p. 418). The bones in each row are interconnected by tight joints, their surfaces exhibiting a palmar concavity and a dorsal convexity. This creates the carpal tunnel on the palmar surface (see p. 302), which is bounded by a bony eminence on the radial and ulnar sides.

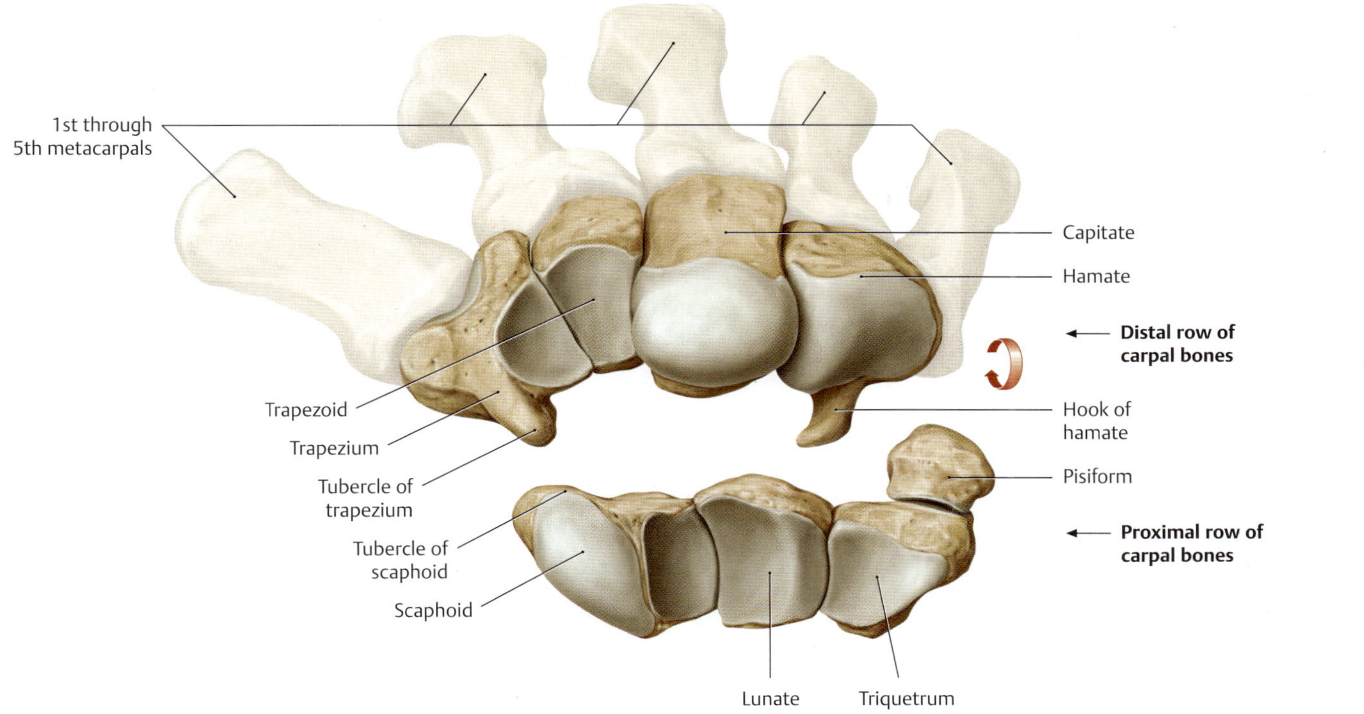

B Articular surfaces of the midcarpal joint of the right hand
The *distal* row of carpal bones is shown from the *proximal view*. The proximal row is shown from the *distal* view.

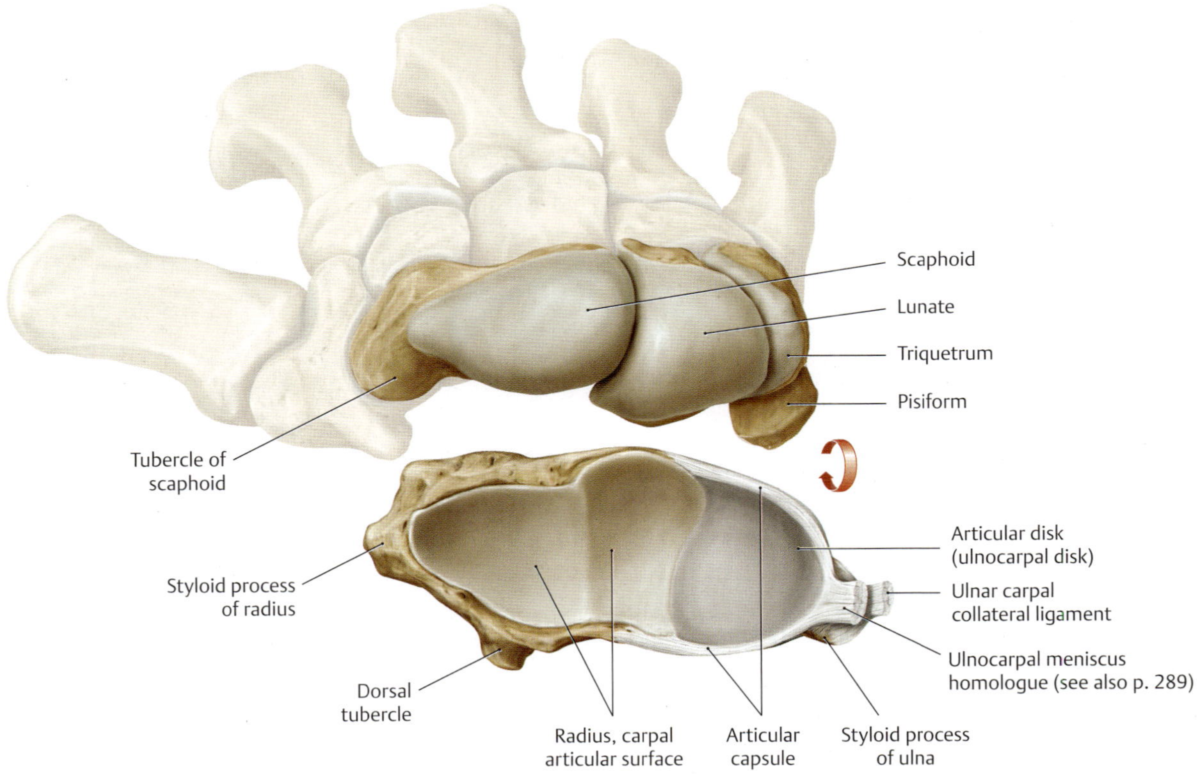

C Articular surfaces of the radiocarpal joint of the right hand
The proximal row of carpal bones is shown from the proximal view. The articular surfaces of the radius and ulna and the articular disk (ulnocarpal disk) are shown from the distal view.
Clinically, the radiocarpal joint is subdivided into a radial compartment and an ulnar compartment. This takes into account the presence of the interposed ulnocarpal disk, which creates a second, ulnar half of the radiocarpal joint in addition to the radial half. Accordingly, the radius articulates with the proximal row of carpal bones in the radial compartment, while the head of the ulna and ulnocarpal disk articulate with the proximal row of carpal bones in the ulnar compartment.

1.10 Architecture of the Radiocarpal Junction and the Metacarpus; Distal Radius and Scaphoid Fractures

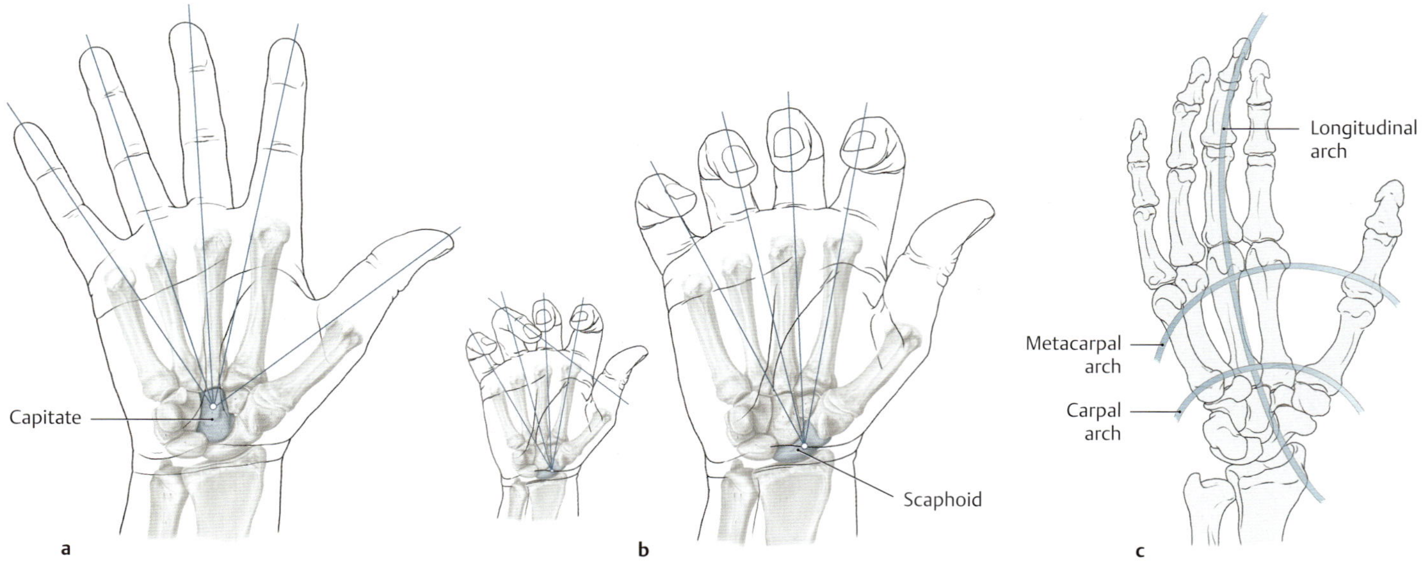

A Architecture of the metacarpus
The metacarpus is the key region for the architecture of the hand.It is where the five digital rays form and then develop into the thumb and fingers. While in neutral position, the longitudinal axis of each finger is parallel, and the longitudinal axis of the abducted thumb and the spread fingers converge to an intersection point in the capitate (**a**). However, when bending the finger joints, the axes converge to an intersection point in the scaphoid (**b**). Only the knowledge of these defined basic anatomic positions allows for the diagnosis of mal-alignments caused by injury (most notably rotational mal-alignments of the fingers, meaning "twisting" of phalanges as a result of a fracture [see inset]). The five digital rays are connected by three functionally significant arches (**c**): a longitudinal arch along the third ray, a metacarpal arch, and a transverse carpal arch.

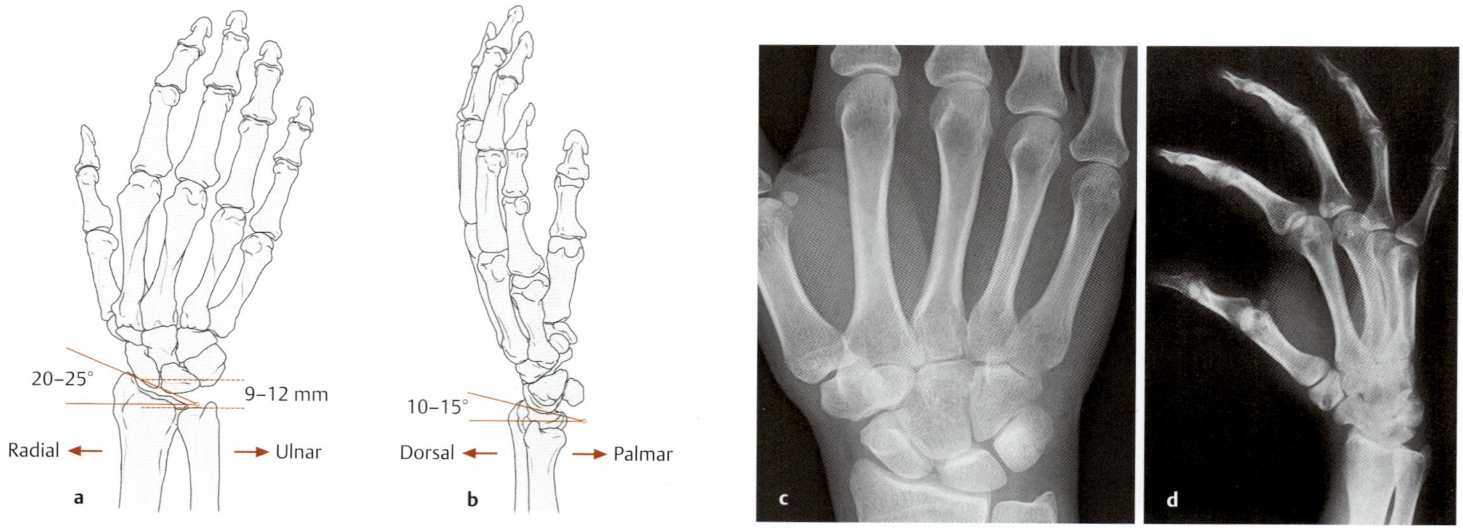

B Inclination angle of the articulating surfaces along the distal radius

a Radioulnar inclination angle, right hand, dorsal view.
b Dorsopalmar inclination angle, right hand, ulnar view.
c X-ray image of carpal region, dorsopalmar beam path and
d radioulnar beam path (**c** from: Böhni U, Lauper M, Locher H, eds. Manuelle Medizin 2. 2nd ed. Stuttgart: Thieme; 2020; **d** from: Möller T, editor. Röntgennormalbefunde. 6th ed. Stuttgart: Thieme; 2019).

The distal radius helps form the radiocarpal and distal radioulnar joints. In addition, it supports the ulnocarpal disk and the strong dorsal and palmar extrinsic ligaments (see p. 298). It is primarily responsible for carpal load transmission within the longitudinally arranged columns (scaphoid column, lunate column, and triquetrum, see p. 266) and thus is prone to injury (see **C**). For a harmonious relationship between the parts forming the radiocarpal joint in terms of optimal mobility of the hand, the position of the part of the socket formed by the radius is of importance. The carpal articulating surface of the ulna is not vertical to the longitudinal axis of the forearm, but in a radioulnar inclination angle of 20 to 25° (*ulnar inclination*, Böhler I angle) and a dorsopalmar inclination angle of 10 to 15° (*palmar inclination*, Böhler II angle). The distal radial length relative to the ulna (tip of the styloid process of the radius to the carpal articulating surface of the ulna) is approximately 9 to 12 mm (important for optimal mobility of the hand).

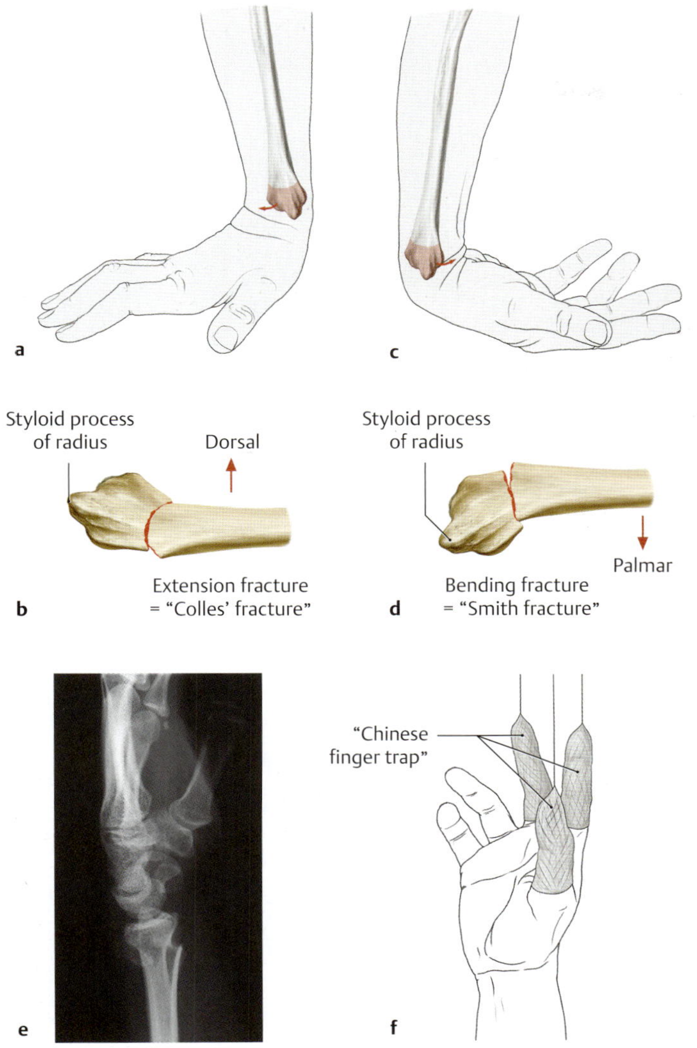

C Distal radius fractures

Accounting for 20 to 25% of all fractures, the distal (or near the wrist) radius fracture as a result of falling on the hand is the most common fracture in humans. Affected are almost 80% of women over the age of 50 (main cause: postmenopausal osteoporosis). Depending on the position of the carpus relative to the distal radius at the moment of impact, 90% of falls result in extension fractures (Colles' fracture; see **a** and **b**), and 10%, in flexion fractures (see **e**, from Henne-Bruns D, Dürig M, Kremer B. Chirurgie. 2nd ed. Stuttgart: Thieme; 2003).

In distal radius fractures, one generally differentiates between extra- and intra-artciular fractures, with the extra-articular fractures typically located 3 to 4 cm proximal to the radiocarpal joint. The standard criteria for such a diagnosis are determined by taking conventional X-ray images of the wrists in two planes (see **e** lateral view). The therapeutic procedure (conservatively using casts or surgically through osetosynthesis) depends on the angle and direction of dislocation (fracture stability), the course of the fracture line (intra-/extra-articular), and the severity of accompanying injuries (e.g., involvement of the ulna, most notably the styloid process). Less complicated (nondislocated, easy-to-reposition) and primarily stable fractures are conservatively treated using "Chinese finger traps" (**f**) with the help of an image intensifier. Axial alignment, most notably of the original radial length and angle (ulnar and palmar inclination angles, see **Ba** and **Bb**), are being restored through vertical extension and joint immobilization with the help of a dorsopalmar splint. With intra-artciular fractures involving large articular fragments, osteosynthetic stabilization is generally recommended.

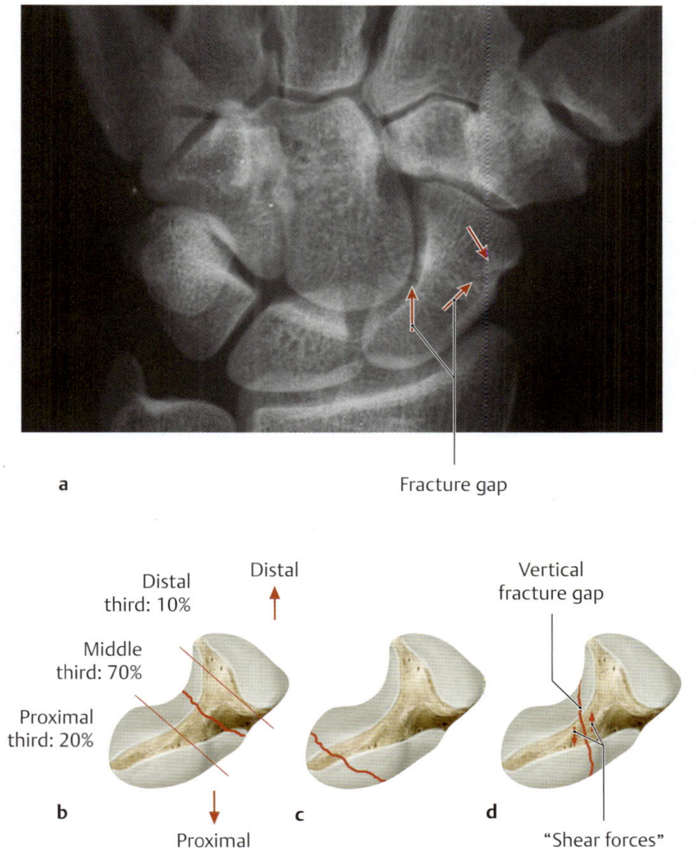

D Scaphoid fractures

a Scaphoid fracture as seen on X-ray, dorsopalmar view (from: Stäbler A, Ertl-Wagner B, editors. Radiologie-Trainer Bewegungsapparat, 3rd ed. Stuttgart: Thieme; 2015).

b Frequency and distribution of scaphoid fractures.

Carpal fractures, most notably scaphoid fractures (accounting for two thirds of all cases), are another injury caused by falling on the outstretched hand extended in the dorsal direction. Unlike distal radial fractures (see **C**), young people are predominantly affected by scaphoid fractures (typical sports injury). During physical examination, the symptoms can be relatively discrete. Usually the symptoms include tenderness on palpation of the area around the anatomic snuff box with simultaneous radial or ulnar abduction and compressive pain around the thumb and index finger. In the case of a suspected scaphoid fracture, conventional radiographs of the wrist in four different planes (so-called scaphoid quartet series) should be taken (see **a**, white arrow), in order to determine the direction of the fracture gap. If the X-ray image fails to confirm the suspected diagnosis 10 to 14 days after initially stabilizing the fracture, a follow-up X-ray must be taken (if needed using CT scans). At this point, the process of resorption in the fracture hematoma is usually completed so that the fracture gap is wider and thus more clearly visible. Depending on the location, scaphoid fractures are classified into fractures of the proximal third, central third, or distal third (see **b**). The healing process is especially lengthy in fractures of the proximal third (up to 3 months in an upper arm cast for stabilization with inclusion of the thumb basal joint, see **c**) because this part of the bone has few blood vessels (the scaphoid receives the majority of its blood supply via distal vessels). The healing process is just as lengthy in slanted or vertical fractures due to the resulting shearing forces pushing in opposite directions (see **d**).

Note: The scaphoid bone is involved in all movements of the hand, which makes long-term stabilization rather difficult. Thus, pseudo-osteoarthritis is a typical complication of scaphoid fracture (= false joint after failed fracture healing, see p. 42).

1.11 The Joints of the Shoulder: Overview and Clavicular Joints

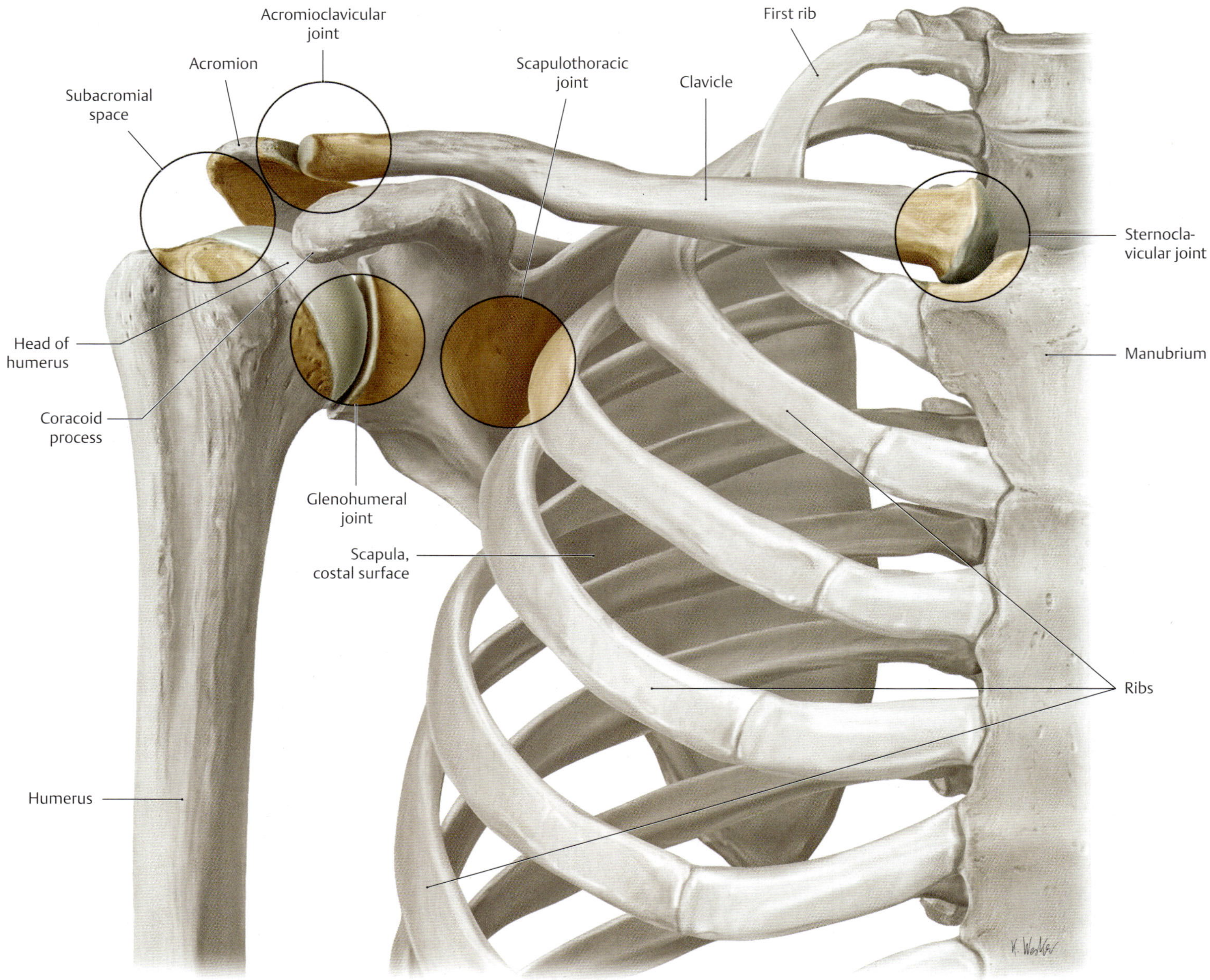

A The five joints of the shoulder
Right shoulder, anterior view. A total of five joints contribute to the wide range of arm motions at the shoulder joint. There are three true shoulder joints and two functional articulations:

- **True joints:**
  1. Sternoclavicular joint
  2. Acromioclavicular joint
  3. Glenohumeral joint
- **Functional articulations:**
  4. Subacromial space: a space lined with bursae (subacromial and subdeltoid bursae) that allows gliding between the acromion and the rotator cuff (= muscular cuff of the glenohumeral joint, consisting of the supraspinatus, infraspinatus, subscapularis, and teres minor muscles, which press the head of the humerus into the glenoid cavity; see p. 317).
  5. Scapulothoracic joint: loose connective tissue between the subscapularis and serratus anterior muscles that allows gliding of the scapula on the chest wall.

Besides the true joints and functional articulations, the two ligamentous attachments between the clavicle and first rib (costoclavicular ligament) and between the clavicle and coracoid process (coracoclavicular ligament) contribute to the mobility of the upper limb. All of these structures together comprise a functional unit, and free mobility in all the joints is necessary to achieve a full range of motion. This expansive mobility is gained at the cost of stability, however. Since the shoulder has a loose capsule and weak reinforcing ligaments, it must rely on the stabilizing effect of the rotator cuff tendons. As the upper limb changed in mammalian evolution from an organ of support to one of manipulation, the soft tissues and their pathology assumed increasing importance. As a result, a large percentage of shoulder disorders involve the soft tissues.

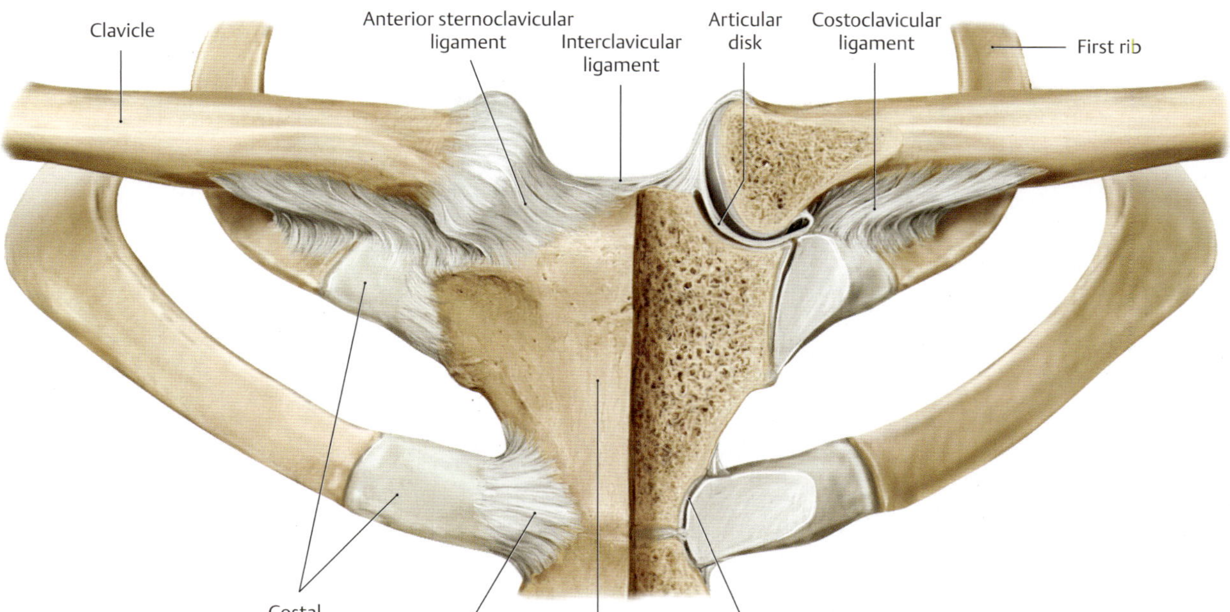

B The sternoclavicular joint and its ligaments
Anterior view. The sternoclavicular joint (also called the medial clavicular joint) and the acromioclavicular joint (the lateral clavicular joint, see below) together make up the true joints within the shoulder girdle itself. In the figure, a coronal section has been made through the sternum and adjacent clavicle to demonstrate the inside of the left sternoclavicular joint. A fibrocartilaginous articular disk compensates for the mismatch of surfaces between the two saddle-shaped articular faces of the clavicle and manubrium.

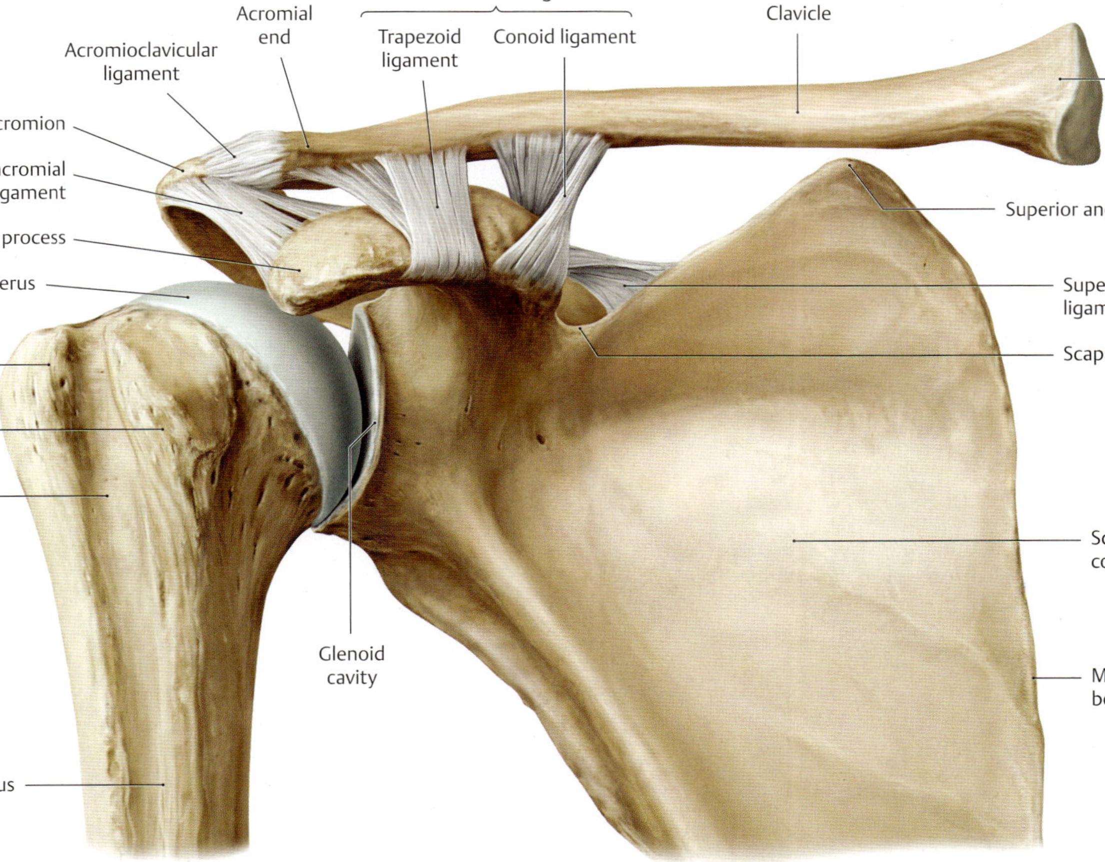

C The acromioclavicular joint and its ligaments
Right shoulder, anterior view. The acromioclavicular joint (the lateral clavicular joint) has the form of a plane joint. Because the articulating surfaces are flat, they must be held in place by strong ligaments (acromioclavicular, coracoacromial, and coracoclavicular ligaments). This greatly limits the mobility of the acromioclavicular joint. In some individuals the acromioclavicular joint has a variably shaped articular disk that gives the joint greater mobility.

1.12 The Joints of the Shoulder: Ligaments of the Clavicular and Scapulothoracic Joints

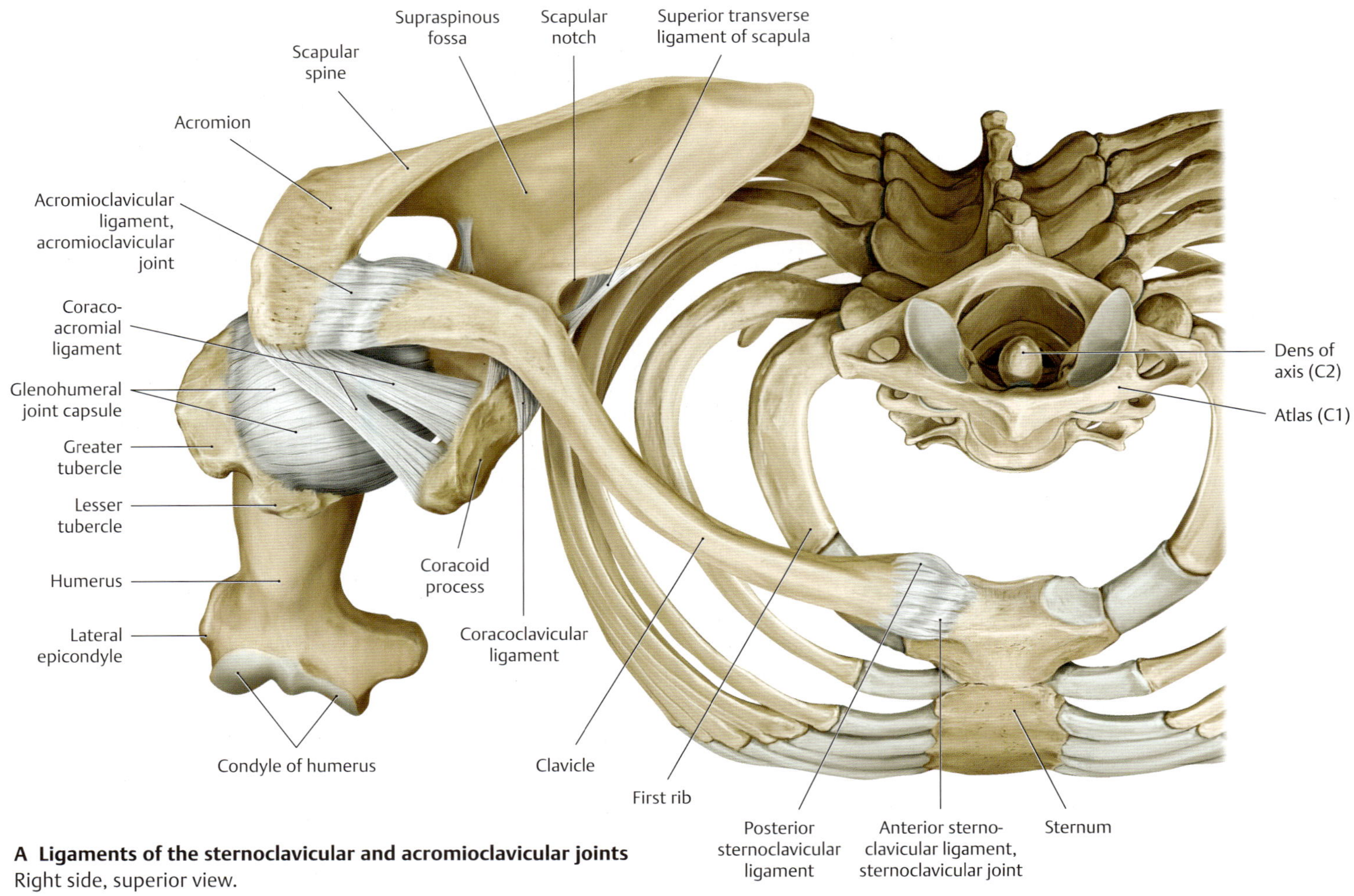

A Ligaments of the sternoclavicular and acromioclavicular joints
Right side, superior view.

B Injuries to the acromioclavicular ligaments
These injuries often result from falling on a shoulder or an outstretched arm. According to Tossy, they can be classified into three types:

- **Tossy I:** The acromioclavicular and coracoclavicular ligaments are stretched but still intact.
- **Tossy II:** The acromioclavicular ligament is ruptured, with subluxation of the joint.
- **Tossy III:** Ligaments are all disrupted, with complete dislocation of the acromioclavicular joint.

Rockwood added three more types that occur less frequently:

- **Rockwood IV:** Dislocation of the clavicle shifts dorsally, due to the clavicular part of the deltoid being pulled off the clavicle.
- **Rockwood V:** Dislocation of lateral end of the clavicle is increased in the cranial direction, due to the deltoid and trapezius being pulled off the clavicle.
- **Rockwood VI:** The lateral end of the clavicle underneath the acromion or the coracoid is dislocated (very rare).

Depending on the extent of the injury, the so-called piano-key phenomenon can be triggered through palpation (caution: painful!). The lateral end of the clavicle, which is elevated due to the injury, can be reduced by applying pressure from the cranial direction, but it pops back up when pressure is released. Radiographs in different planes will show widening of the space in the acromioclavicular joint. Comparative-stress radiographs with the patient holding approximately 10-kg weights in each hand will reveal the extent of upward displacement of the lateral end of the clavicle on the affected side (is not performed in cases of visible partial rupture of ligaments in order to avoid further damage).

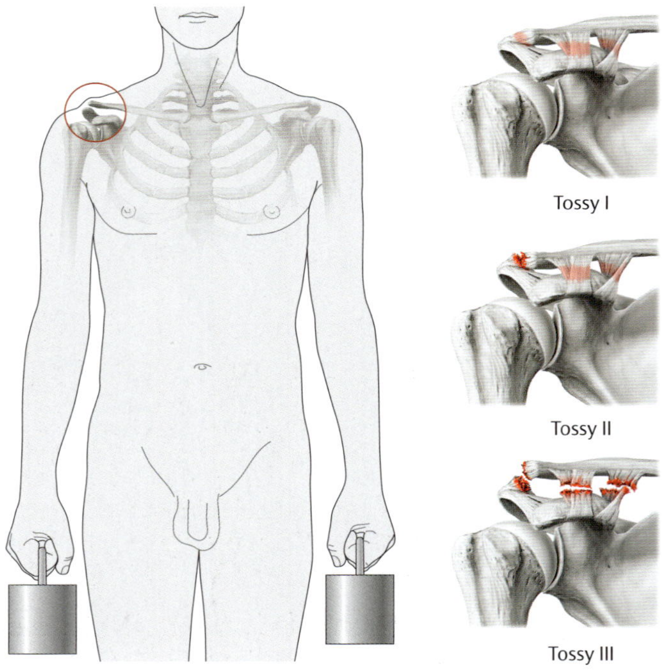

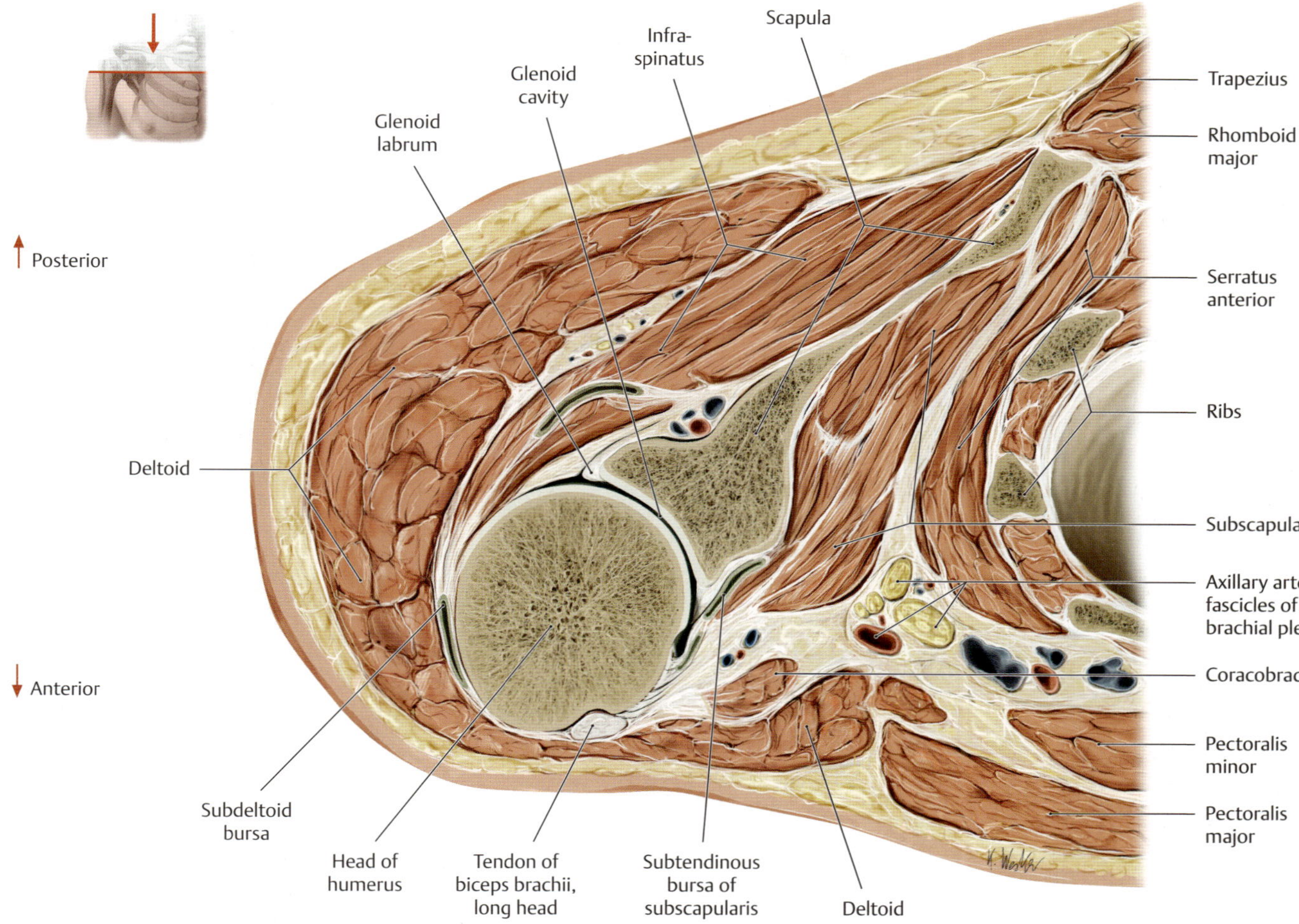

C Transverse section through the right shoulder joint

Superior view. In all movements of the shoulder girdle, the scapula glides on a curved surface of loose connective tissue between the serratus anterior and subscapularis muscles (see **D**). This surface can be considered a "scapulothoracic" joint that allows the scapula not only to change the position of the shoulder (translational motion), but also to pivot, with the glenohumeral joint maintained in a relatively stable position on the thorax (rotational motion) (see p. 286) (drawing based on a specimen from the Anatomical Collection at the University of Kiel).

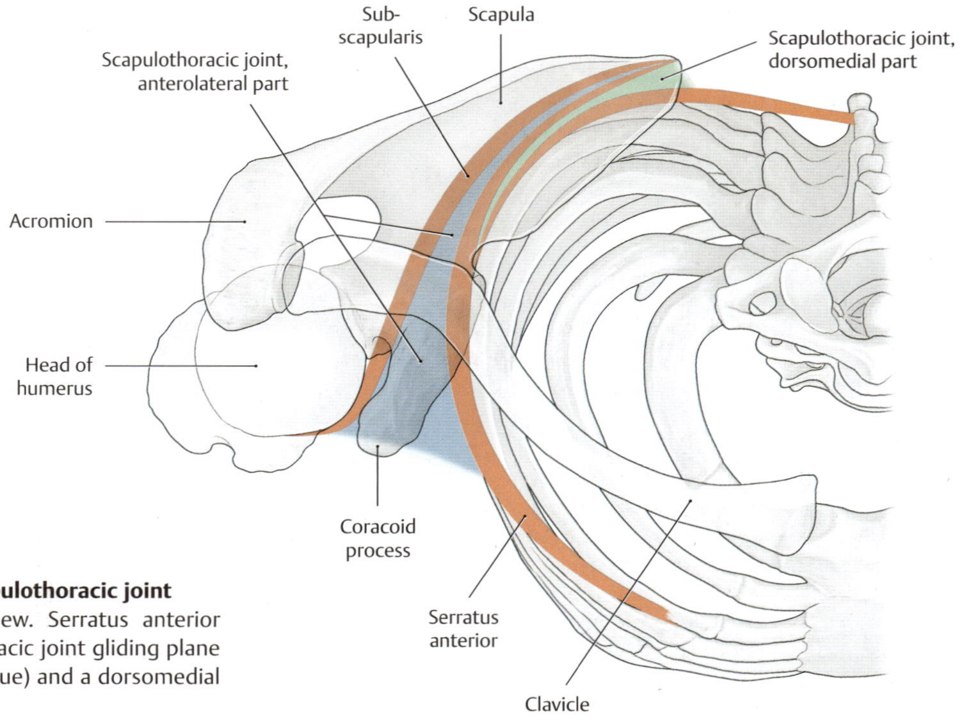

D Location of the scapulothoracic joint

Right side, superior view. Serratus anterior divides the scapulothoracic joint gliding plane into an anterolateral (blue) and a dorsomedial (green) part.

1.13 The Joints of the Shoulder: Articular Surfaces, Articular Capsule, and Articular Cavity of Glenohumeral Joint

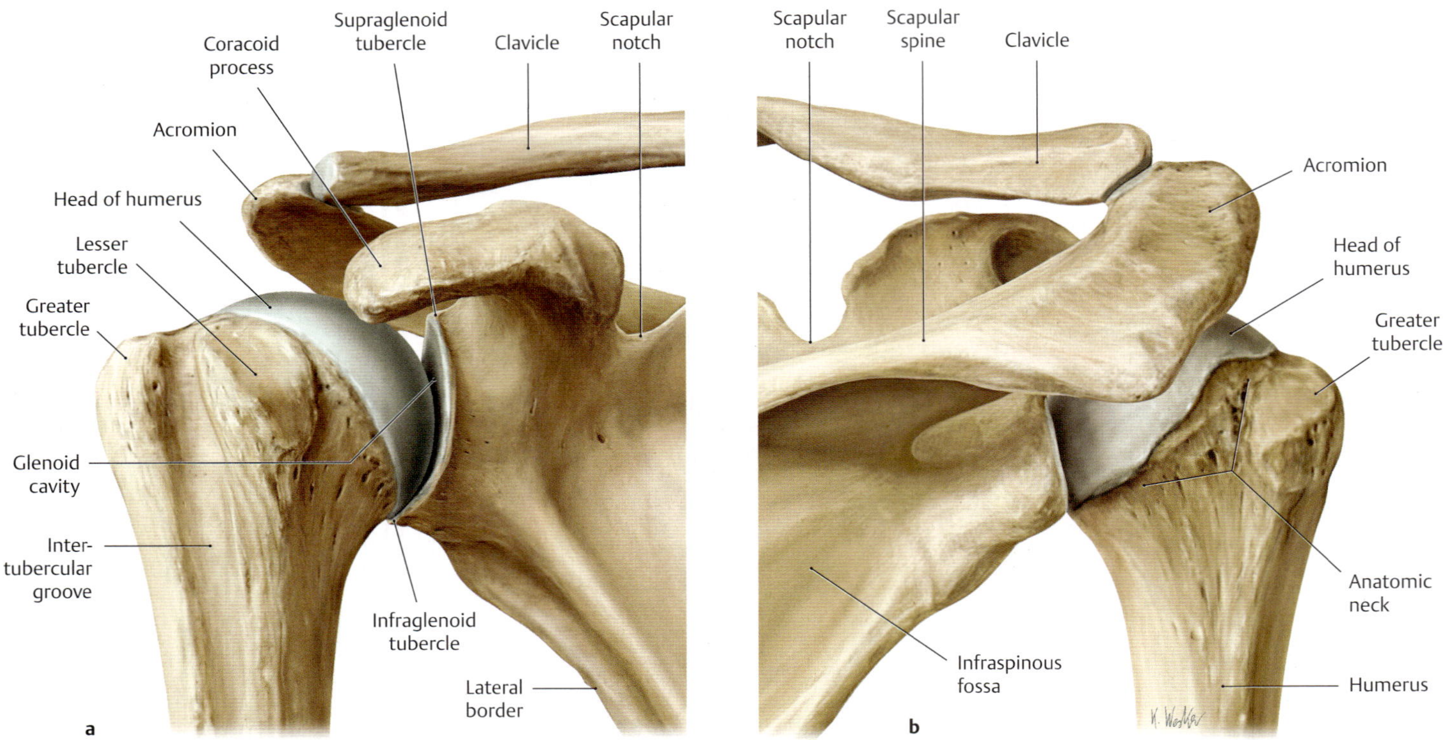

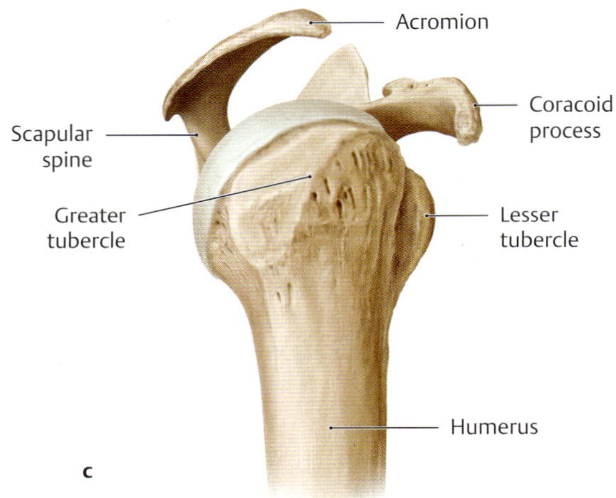

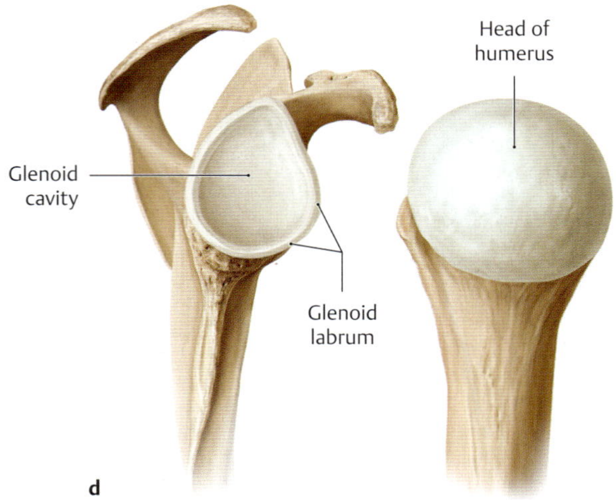

A Articulating bony elements of the right shoulder joint (glenohumeral joint) and shoulder dislocation

a Anterior view.
b Posterior view.
c Lateral view.
d Articulating surfaces.

In the shoulder joint—the most mobile joint in the human body but also the most susceptible to injury—the head of the humerus articulates with the glenoid cavity of the scapula to form a spheroidal type of joint. The articular surface of the scapula, which is three to four times smaller than the head of the humerus, is enlarged by a rim of fibrocartilage, the glenoid labrum, which is approximately 5 mm wide at its base (see **d**). While this size discrepancy of the articulating surfaces serves to increase the range of shoulder motion, it compromises the stability of the joint. Since the joint capsule and ligaments are weak, the rotator cuff tendons are the primary stabilizers of the glenohumeral joint (see p. 318).

Dislocations of the shoulder joint are notoriously common. Approximately 45% of all dislocations involve the shoulder joint. In typical cases the head of the humerus dislocates anteriorly or anteroinferiorly in response to forcible external rotation of the raised arm. Whereas considerable trauma is generally needed to cause the initial dislocation, certain movements of the shoulder (e.g., excessive arm rotation during sleep) may be sufficient to redislocate the humeral head from the glenoid cavity (recurrent shoulder dislocation).

In addition to physical examination (based on postion of the arm, occurring pain, and palpation of shoulder contour), diagnostic procedures for a dislocated shoulder include X-rays in two planes. In injuries caused by dislocation, especially anterior luxation, which account for most cases, mainly the rim of the socket (torn-off labrum, so-called *Bankart lesion*) and humeral head (impression fracture caused by rim of socket, so-called *Hill-Sachs lesion*) are affected. Additional important complications include simultaneous damage to the axillary nerve (test sensitivity in the autonomous zone), the axillary artery (test capillary refill time of the radial artery), and, particularly in older patients, the rotator cuff (rupture).

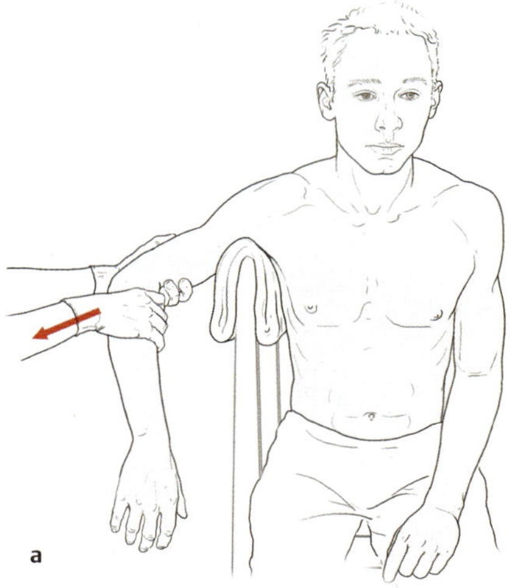

a

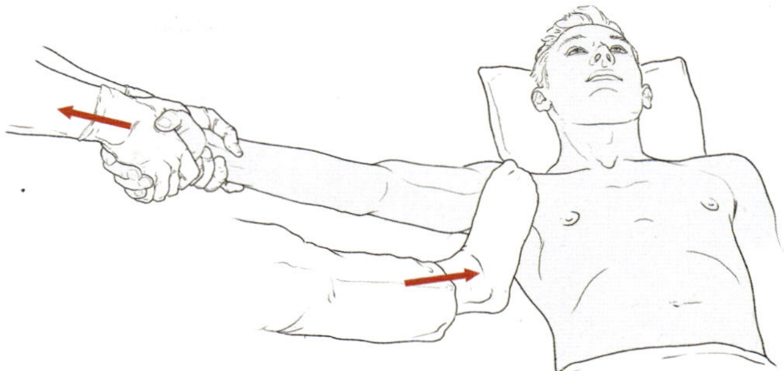

b

B Repositioning of shoulder dislocation
a Repositioning according to Arlt;
b repositioning according to Hippocrates

There are several techniques to reduce a dislocated shoulder. In all cases, the repositioning should be performed with the patient under sedation or analgesia, possibly even anesthesia. According to the Arlt method, the patient sits in a chair with the affected arm hanging downward over the upholstered armrest. Reposition is achieved through longitudinal traction, with the armrest serving as a support. The method as described by Hippocrates requires the patient to lie on the floor. Here, too, reduction is performed through longitudinal traction of the arm, with the doctor's foot placed in the axilla, thereby serving as a counter bearing.

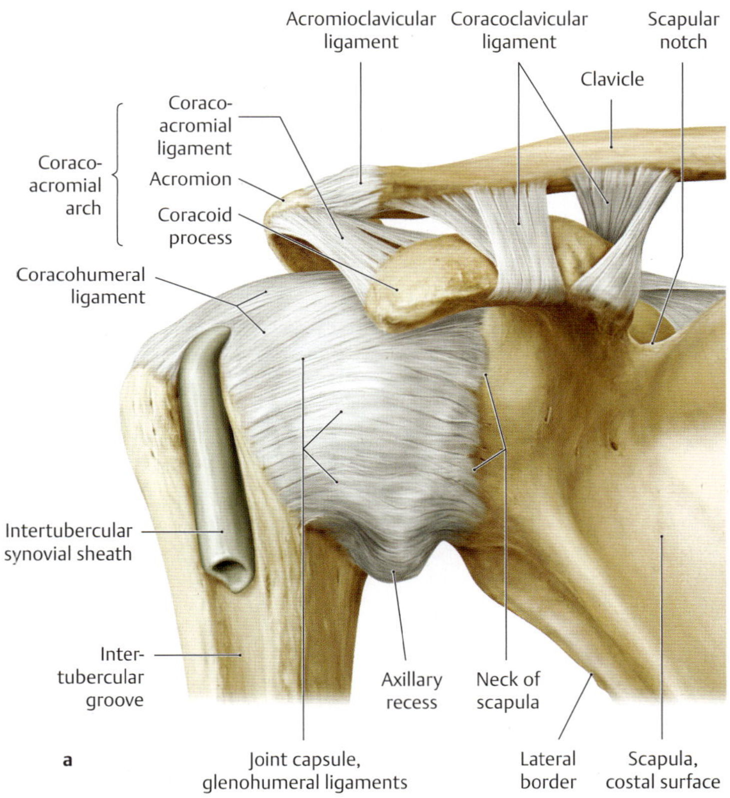

a

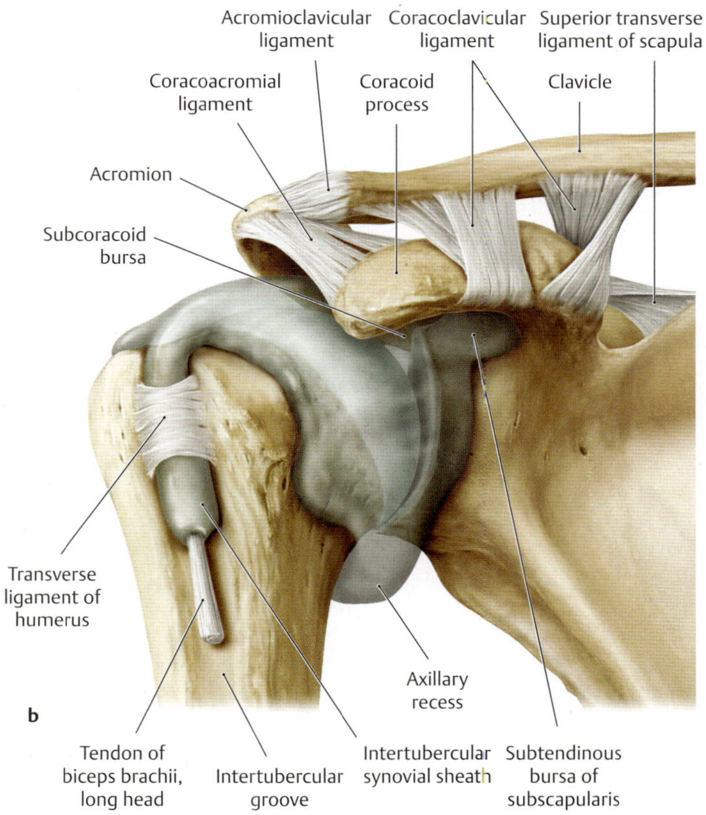

b

C Capsule, ligaments, and joint cavity of the right shoulder
a Anterior view.
b Joint cavity with capsule removed, anterior view.

The capsule of the shoulder joint is broad and very thin posteriorly, where it is not reinforced by ligaments. But it is strengthened anteriorly by three ligaments (the superior, medial, and inferior glenohumeral ligaments; see p. 276) and superiorly by the coracohumeral ligament. The coracoacromial ligament, acromion, and coracoid process together form the coracoacromial arch, which helps stabilize the humeral head in the glenoid cavity but also limits upward movement of the humerus. When the arm is hanging at the side, the lower part of the joint capsule, which is not reinforced by muscle, sags to form the axillary recess. This redundant fold provides a reserve capacity that is particularly useful during abduction movements of the arm. With prolonged disuse of the arm, the axillary recess may become atrophic or obliterated by adhesions, causing significant limitation of arm motion. The cavity of the shoulder joint is connected to the adjacent bursae. The subtendinous bursa of the subscapularis muscle and the subcoracoid bursa consistently communicate with the joint cavity. The tendon sheath of the long head of the biceps brachii also communicates with the joint cavity in its passage through the intertubercular groove.

Note: The glenohumeral ligaments are usually clearly visible only at the inner side of the capsule.

1.14 The Joints of the Shoulder: Ligaments, Reinforcing Capsule, and Rotator Interval of the Glenohumeral Joint

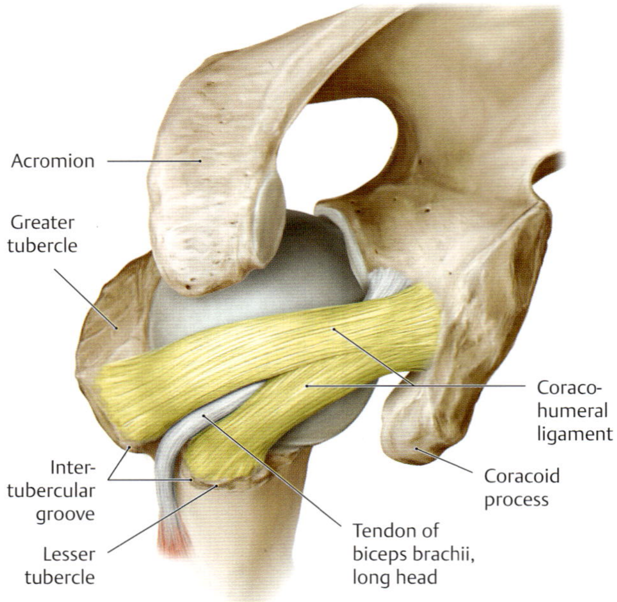

A Coracohumeral ligament
Right shoulder joint, superior view
The coracohumeral ligament arises from the base of the coracoid process as a strong, broad band. Its two parts pass from there to the greater and lesser tubercles of the humerus. The coracohumeral ligament stabilizes the tendon of the long head of the biceps brachii as it passes posterior to the coracohumeral ligament and before it passes through the intertubercular groove.

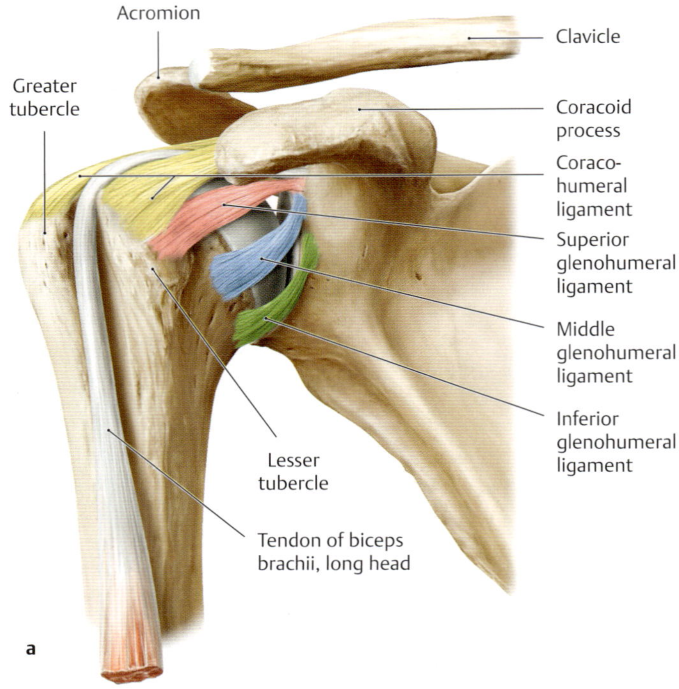

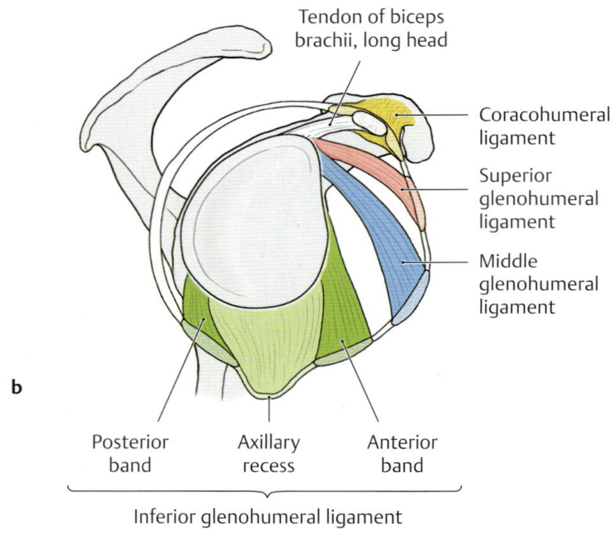

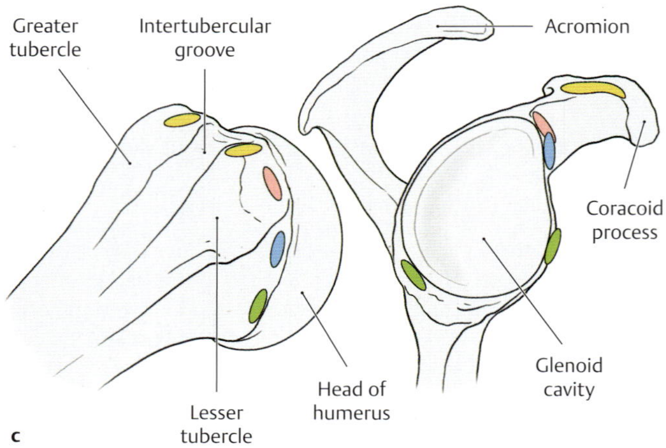

B Ligaments reinforcing capsules
a Right shoulder joint, anterior view; **b** schematic representation of the ligaments reinforcing the capsule after removal of the humeral head and cutting through both the capsule and the tendons of insertion of the rotator cuff muscles, lateral view; **c** origin and insertion of ligaments reinforcing capsule, lateral view.
The capsule of the shoulder joint is relatively loose and thin posteriorly. Anteriorly, it is reinforced by the glenohumeral ligaments, which differ significantly in shape and are usually clearly visible only when viewed arthroscopically (see p. 283).

- **Superior glenohumeral ligament:** It passes from the upper margin of the glenoid cavity to the intertubercular groove and to the lesser tubercle, forming, together with the coracohumeral ligament, the rotator interval (see **D**).
- **Middle glenohumeral ligament:** It passes at almost a right angle to the tendon of insertion of the subscapularis, from the upper margin of the glenoid cavity to the anatomic neck of the humerus.
- **Inferior glenohumeral ligament:** It consists of three parts, an anterior and posterior band and the axillary recess located between the bands. All three parts extend from the inferior margin of the glenoid cavity to the middle of the anatomic neck of the humerus, with the middle part (axillary recess) extending down to the surgical neck. The inferior glenohumeral ligament forms a hammock, especially during abduction, and is of particular significance to the anterior-inferior stability of the shoulder.

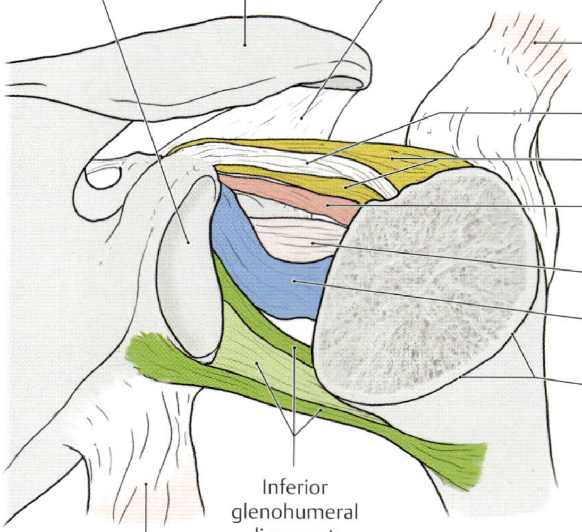

C Ligaments reinforcing the capsule
Right shoulder joint, posterior view.
View from behind the anterior articular capsule after removal of the humeral head.

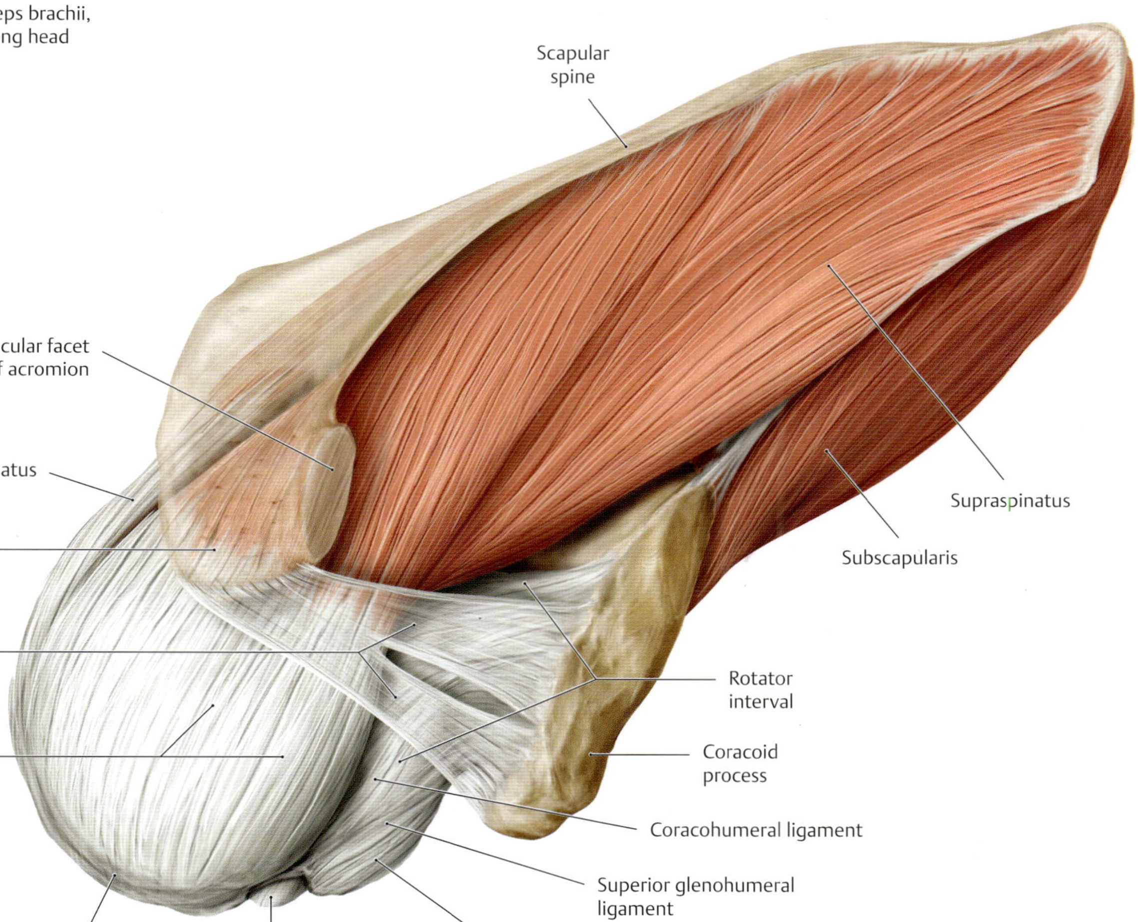

D Rotator interval
Right shoulder, superior view; clavicle and deltoid have been removed. The rotator interval is the capsular area or gap between the superior edge of the subscapularis and the anterior edge of the supraspinatus. The glenohumeral and coracohumeral ligaments are primarily responsible for reinforcing the articular capsule. The ligaments join together in the rotator interval to form the so-called biceps pulley, which surrounds the tendon of the long head of the biceps brachii and prevents it from being dislocated in an anterior-medial direction. The superior glenohumeral ligament forms the bottom and the coracohumeral ligament forms the top of this "sling" or "pulley" (see **Bb**). In addition, fibers from the tendons of insertion of the subscapularis and supraspinatus are interwoven in the rotator interval.

1.15 The Joints of the Shoulder: The Subacromial Space

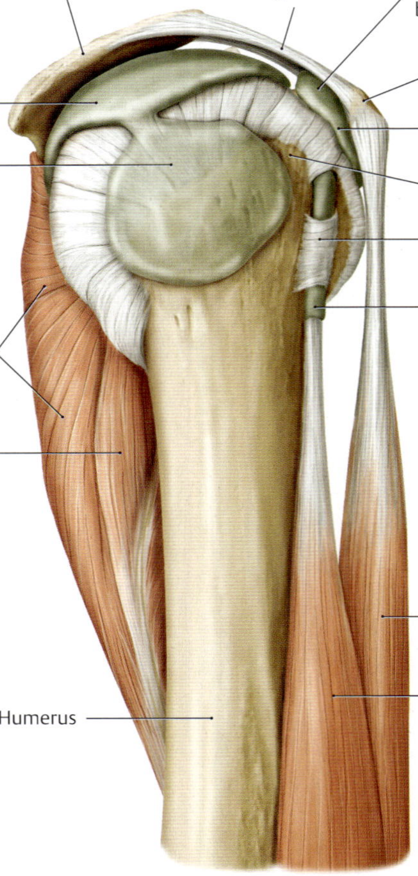

A Subacromial space, right shoulder
Lateral view. The deltoid muscle has been removed to demonstrate the following structures:

- The attachments of the rotator cuff muscles (supraspinatus, infraspinatus, teres minor, and subscapularis) to the proximal humerus (see also **B**).
- The tendons of origin of the biceps brachii muscle.
- The subacromial space with the subacromial bursa, which consistently communicates with the subdeltoid bursa.

The two bursae allow frictionless gliding between the humeral head and the rotator cuff tendons (especially the supraspinatus and the upper part of the infraspinatus) beneath the coracoacromial arch during abduction and elevation of the arm (see p. 287).

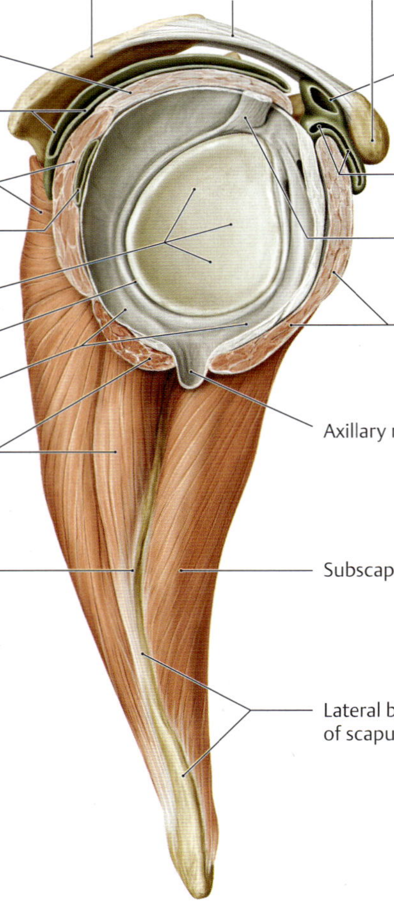

B Subacromial bursa and glenoid cavity of the right shoulder joint
Lateral view. The humeral head has been removed and the rotator cuff tendons of insertion have been divided to demonstrate the glenoid cavity of the shoulder joint. The *glenoid labrum* enlarges and deepens the glenoid only very slightly. Just before inserting into the humeral head, the muscles of the rotator cuff send tendinous expansions to the joint capsule that help to press the humeral head into the glenoid cavity. The subacromial bursa is located between the coracoacromial arch and the tendons that insert on the humeral head (see **D**).

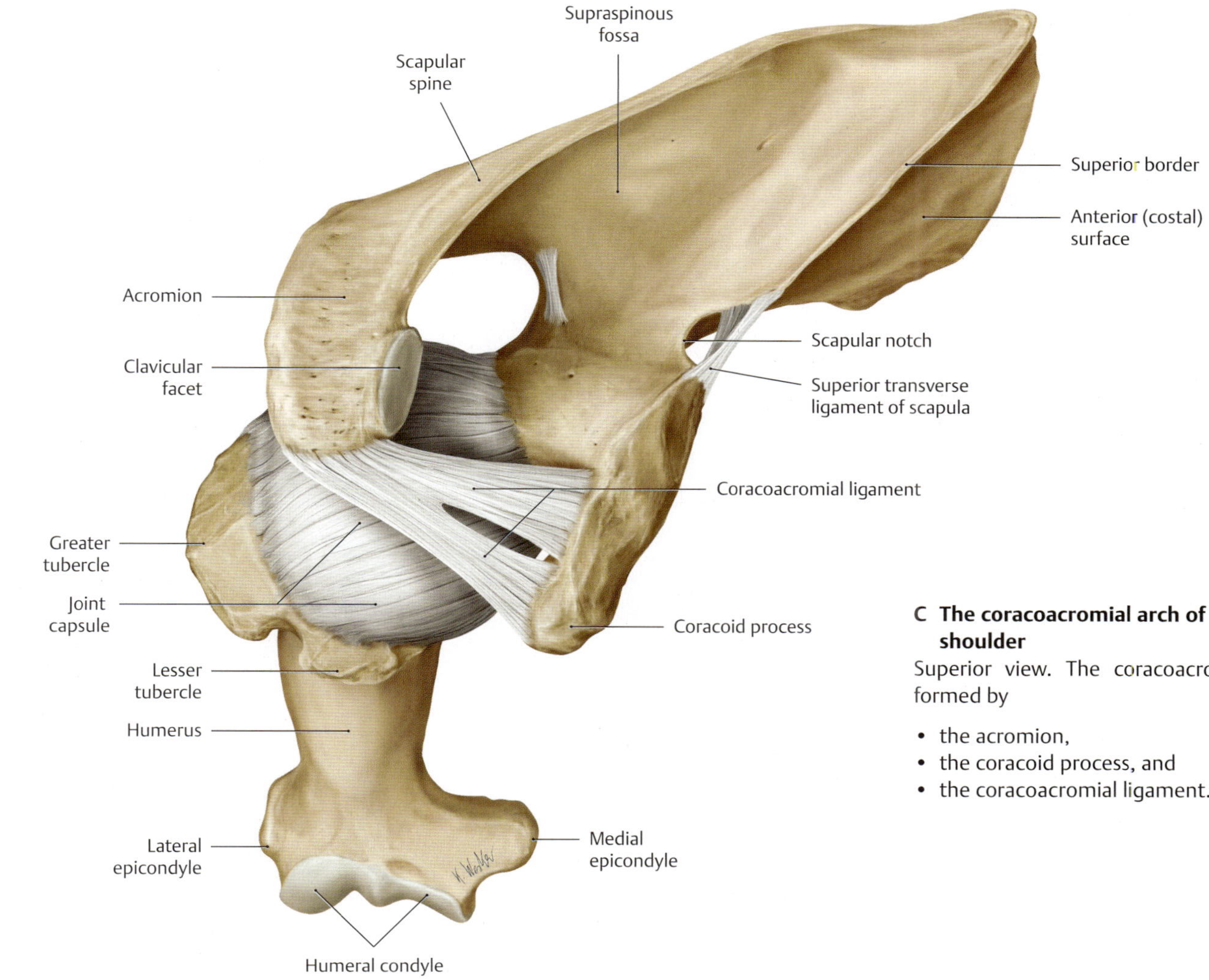

C The coracoacromial arch of the right shoulder

Superior view. The coracoacromial arch is formed by

- the acromion,
- the coracoid process, and
- the coracoacromial ligament.

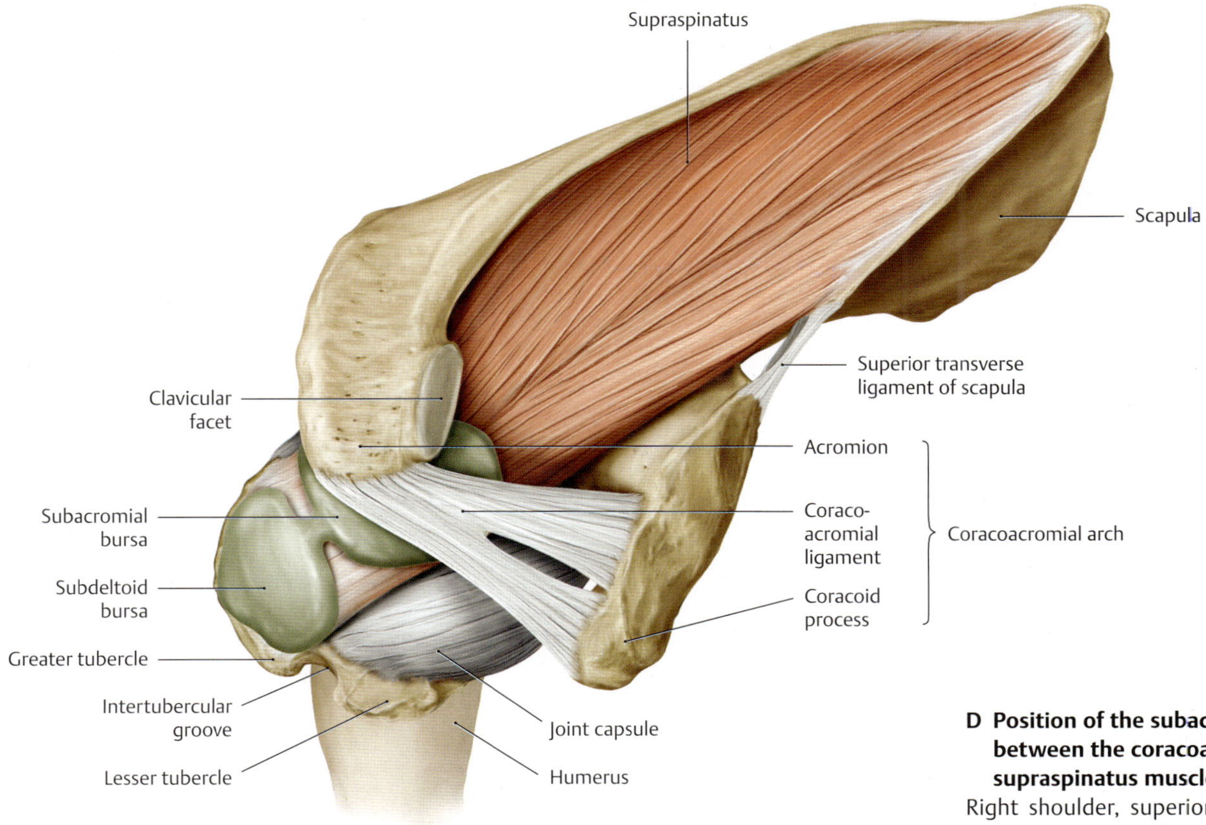

D Position of the subacromial bursa between the coracoacromial arch and supraspinatus muscle

Right shoulder, superior view. See also the "supraspinatus syndrome," p. 280.

1.16 The Subacromial and Subdeltoid Bursae

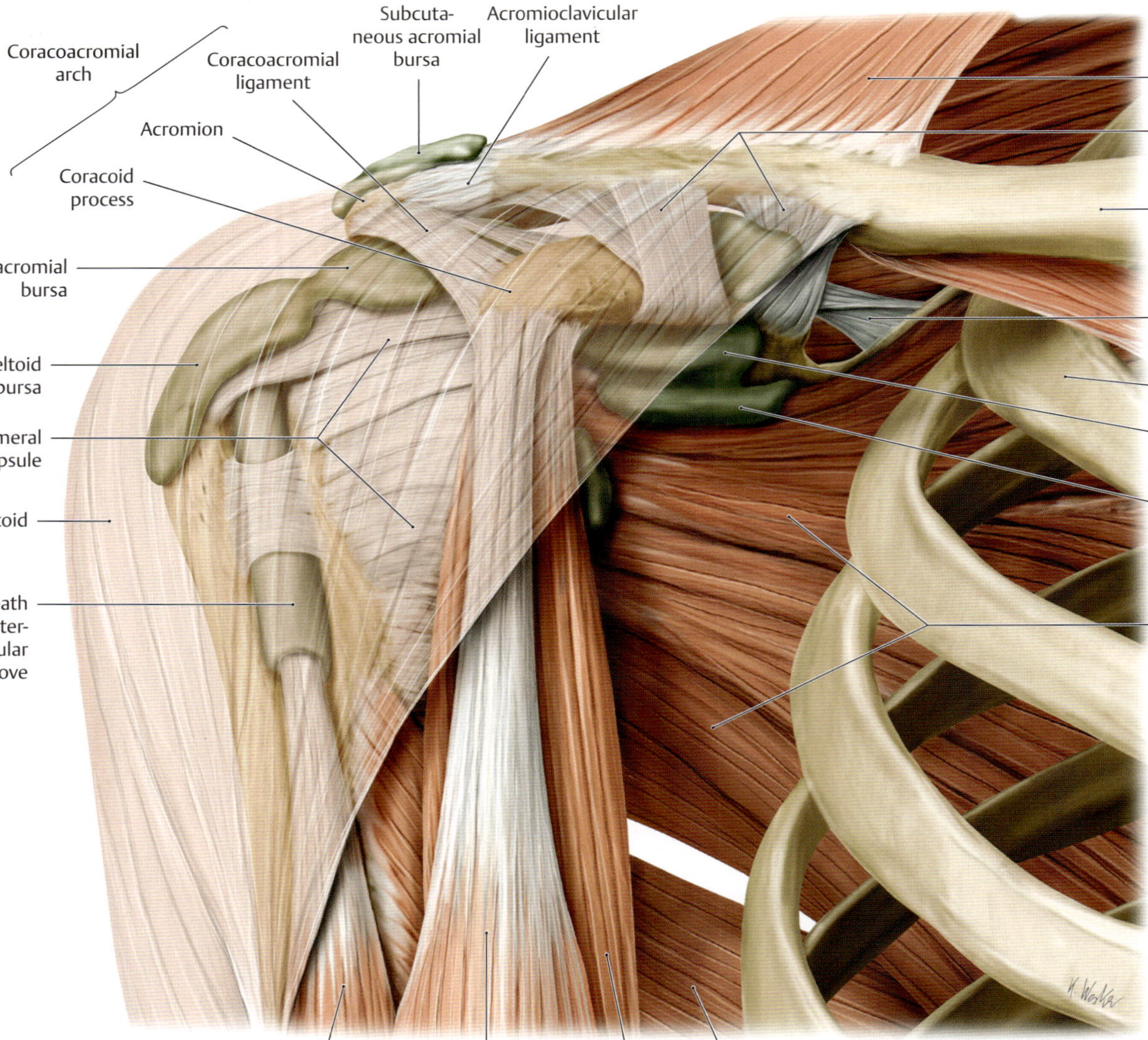

A Location of bursae in the right shoulder
Anterior view. The pectoralis major and minor and serratus anterior muscles have been removed. The location of the bursae can be seen through the deltoid muscle, which is lightly shaded.
Note in particular the coracoacromial arch and the underlying subacromial bursa.

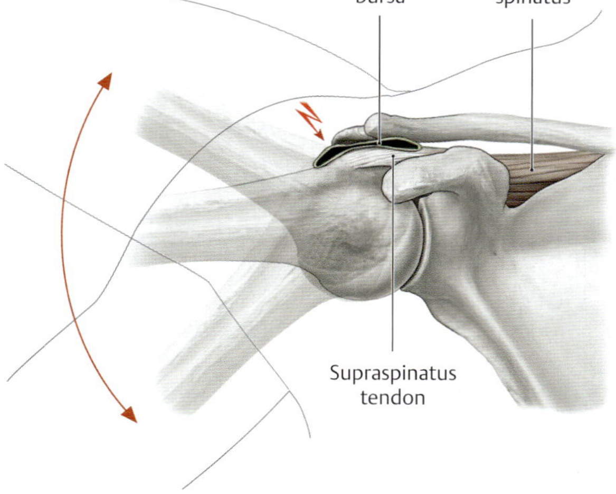

B Supraspinatus (impingement) syndrome
A supraspinatus tendon that has been thickened by calcification or other degenerative changes can be caught underneath the acromion, impinging upon the subacromial bursa, when the arm is abducted (see **D**). Pain from this supraspinatus, or impingement, syndrome occurs in an arc of motion between 60 and 120° of abduction

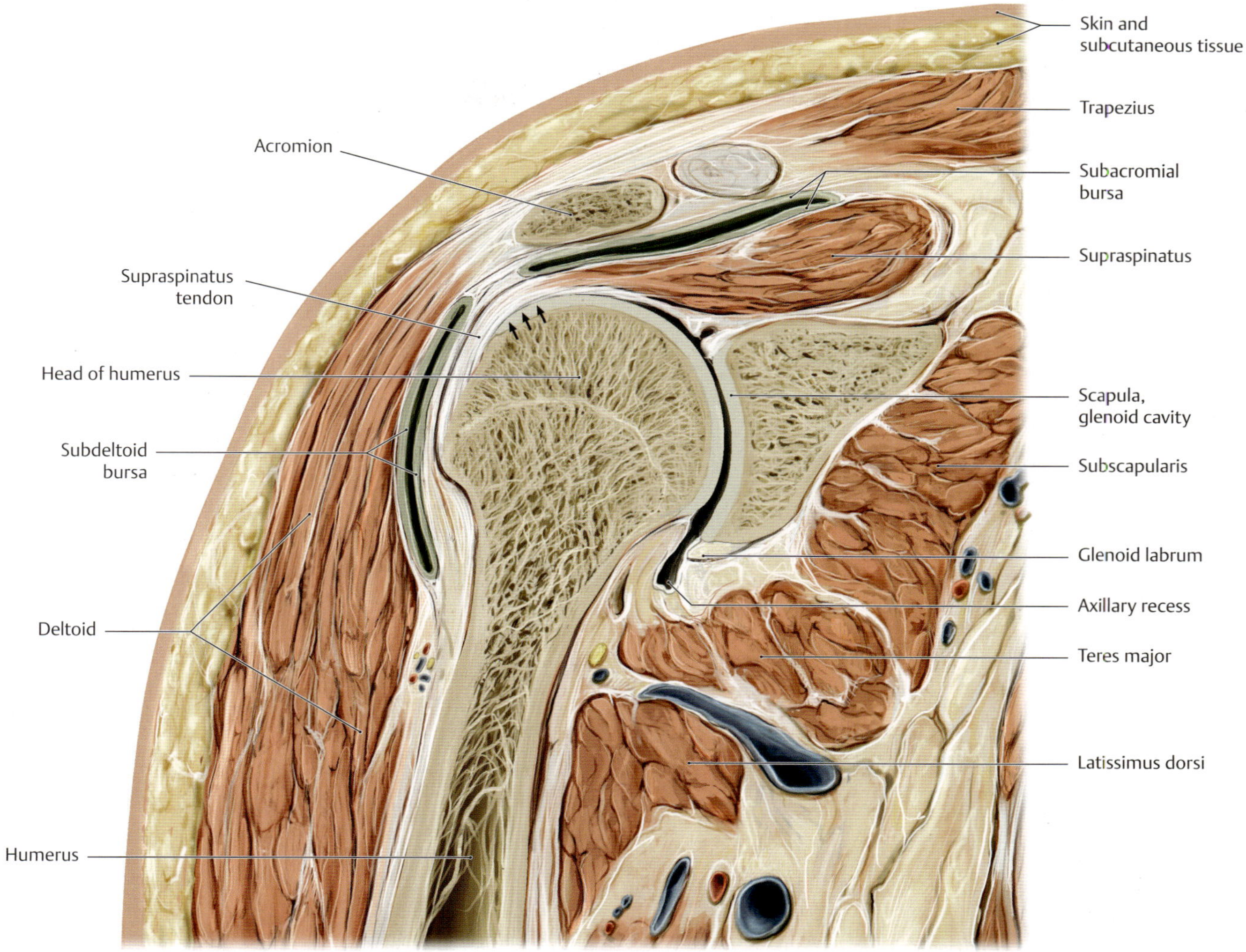

C Coronal section through the right shoulder joint
Anterior view. The tendon of insertion of the supraspinatus muscle differs structurally from ordinary traction tendons. Its distal course gives it the function of a gliding tendon, which passes over the fulcrum of the humeral head (arrows). In that area, located approximately 1 to 2 cm proximal to its insertion on the greater tubercle, the tendon tissue in contact with the humeral head is composed of fibrocartilage. This fibrocartilage zone is avascular, representing an adaptation of the tendon to the pressure loads imposed by the bony fulcrum (drawing based on a specimen from the Anatomical Collection at the University of Kiel).

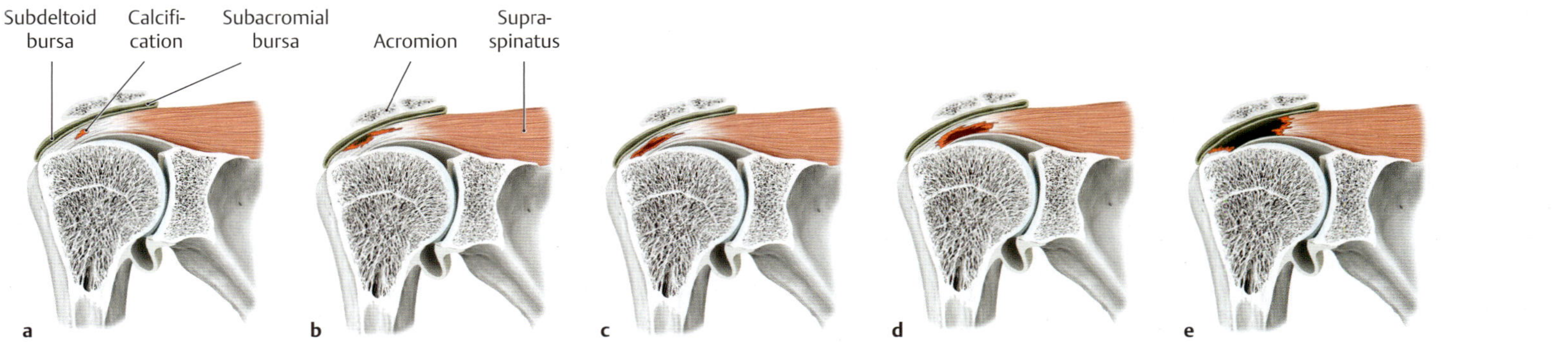

D Damage to the supraspinatus tendon
a Calcification of the tendon of insertion of the supraspinatus, resulting from degenerative changes in the area surrounding the fibrocartilaginous zone (see also **C**); **b–d** partial defects in the supraspinatus tendon (**b** bursal-sided, **c** intratendinous, and **d** articular-sided); **e** complete defect (tear) of the supraspinatus tendon (also called rotator cuff rupture).
Degenerative changes of the rotator cuff, primarily the supraspinatus tendon, together with the clinical effects on the tendon of the long head of the biceps brachii, as well as the subacromial and subdeltoid bursa, generally lead to a painful functional disorder in the shoulder area. The cause of this so-called subacromial syndrome (or subacromial pain syndrome) is the pathologic contact of the calcified and thickened supraspinatus tendon with the acromion, or subacromial impingement syndrome. After a complete rotator cuff rupture (in 95% of all cases, the supraspinatus tendon is involved), the subacromial and subdeltoid bursae communciate with the articular cavity. The supraspinatus normally contributes significantly to the first 10° of abduction (the "starter function"; see p. 316).

1.17 Shoulder Arthroscopy

A Overview of arthroscopy of the shoulder joint

Arthroscopy of the shoulder joint has taken on an increasingly important role in the field of arthroscopic surgery. Initially, it was predominantly a diagnostic procedure. Over time, as a result of technical advances, it has become a surgical technique for resection and reconstruction of the shoulder joint, increasingly replacing open shoulder surgeries. Arthroscopy is now considered primarily a therapeutic surgical procedure, which also allows a dynamic functional evaluation of the capsule ligament structures (e.g., in case of instability). Currently, all surgeries are possible as part of arthroscopy of the shoulder joint, with the exception of endoprosthesis. **Indications** for arthroscopy of the shoulder joint can be classified into three groups:

- Shoulder instability, e.g., labral lesion (Bankart lesion), osteochondral lesion (Hill-Sachs lesion);
- Degenerative changes, e.g., partial or complete rotator cuff rupture, subacromial impingement syndrome (tendonitis calcarea), biceps brachii tendon; and
- Inflammatory changes, e.g., synovialitis, adhesive capsulitis.

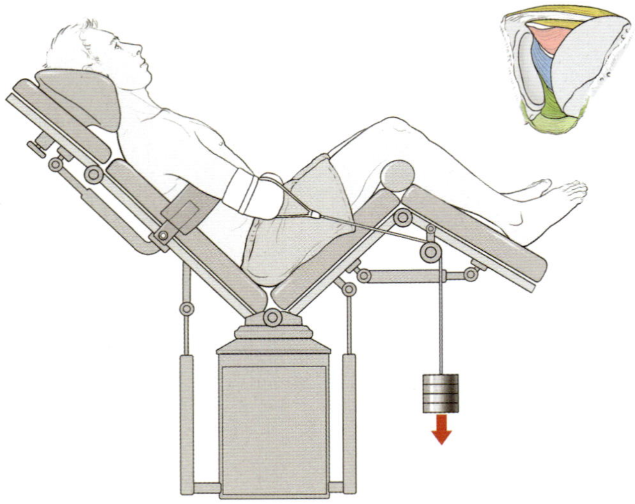

B Patient positioning

Two standard positions are available:

- Lateral position
- Half-sitting position (so-called beach-chair position)

In the half-sitting position, the patient sits upright with the back flexed at 60°. The advantage of this position is that it allows full rotation and abduction of the arm during surgery. In addition, in case the operation needs to be converted to an open procedure, there is no need to repostion the patient, nor does it require additional washing or surgical drapes. Furthermore, the arm can be extended using traction weights in order to get a better view inside the joint.

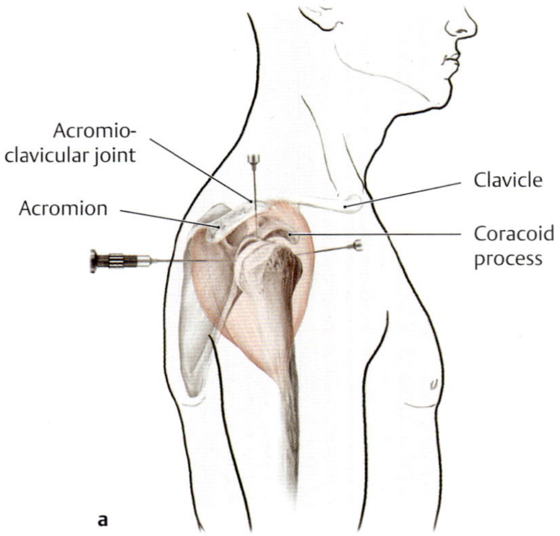

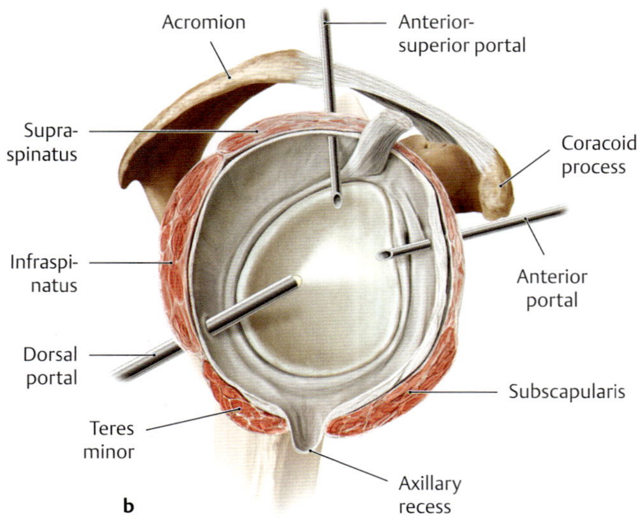

C Shoulder arthroscopy portals

a Right shoulder joint, lateral view; for better orientation, the most important anatomic landmarks have been marked: coracoid process, lateral end of the clavicle, acromion, and acromioclavicular joint; **b** arthroscopic portals, right shoulder joint (humeral head removed), lateral view. With the arthroscopic portals, one differentiates between

- Instrument portals (through which instruments are inserted) and
- Optical portals.

The standard portal for the 30°-angled arthroscope is created by making a stab incision 1 cm inferior and 1.5 cm medial to the posterolateral acromial margin (dorsal portal). Afterward, a blunt trocar is used to perforate first the subcutaneous tissue followed by the deltoid and the dorsal capsule in the direction of the coracoid process. Then, the trocar is replaced by a camera, and the articular cavity is washed. The anterior and anterior-superior portals are commonly used to insert the instruments. Depending on the type of surgery, additional portals can be created (e.g., the lateral portal for access to the subacromial space). Whereas the anterior portal is created by incising the skin lateral to the tip of the coracoid process and accessing the joint above the subscapularis tendon, the anterior-superior portal is made just in front of the anterolateral angle of the acromion with the joint being accessed just behind the tendon of the long head of the biceps brachii. Moving the arthroscope to different portals allows for a complete overview of the articular space.

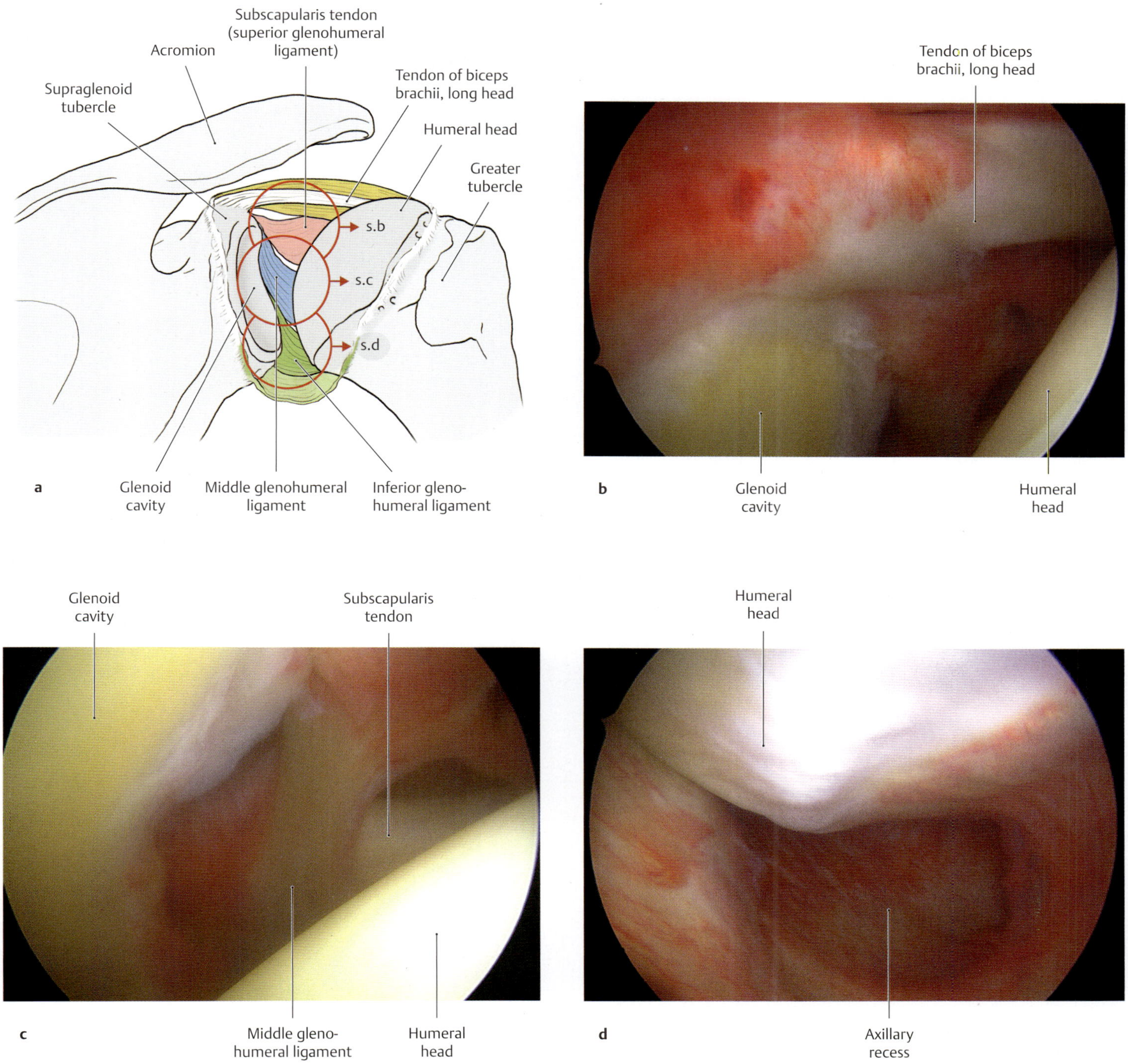

D Arthroscopic anatomy of the shoulder joint

a Schematic representation of the arthroscopically visible fields of sight; right shoulder, dorsal view; **b–d** corresponding arthroscopy images (from Merk H, Jerosch J, Hrsg. Arthroskopie des Schultergelenks. Stuttgart: Thieme; 2000).

For the arthroscopic evaluation of the intra-articular structures, a standardized test using examining hooks is recommended. The goal of this examination is the inspection of the entire articular space. In the beach-chair position (see **B**) with dorsal view, the glenoid cavity is oriented vertically. Over the opposite humeral head, the tendon of the long head of the biceps brachii passes horizontally through the superior articular space to the supraglenoid tubercle and is clearly visible as an anatomic landmark. The tendon of the long head of the biceps brachii together with the glenoid cavity and head of the humerus form the anterior triangle with the subscapularis tendon and the transverse humeral ligament (**b**, **c**). By tilting the lens upward, the insertion of the tendon of the long head of the biceps brachii into the intertubercular groove in the rotator interval becomes visible at the anterior edge of the supraspinatus tendon. The superior glenohumeral ligament and the anterior edge of the supraspinatus tendon form a U-shaped sling, which crosses underneath the tendon of the long head of the biceps brachii. Furthermore, the undersurface of the supraspinatus and infraspinatus tendons can be made visible. The inferior glenohumeral ligament can be best evaluated from the anterior portal looking toward the axillary recess. It extends—with its anterior and posterior fibers as well as the axillary recess located between them—to the inferior edge of the glenoid cavity.

1.18 X-ray and Cross-sectional Anatomy of the Shoulder

A Conventional diagnostic radiology of the right shoulder joint

In primary diagnosis, clinical assessments and sonographic examinations (see **B**) are followed by X-ray radiography. Additionally, complicated injuries can be examined with the help of computed tomography and magnetic resonance imaging (see **C**). Similar to the X-ray radiography in evaluating bones and joints, it is recommended to take two images of two perpendicular planes:

- Anteroposterior image (**a** and **b**) and
- Axial (transverse) image (**c** and **d**).

In order to avoid overlap in X-ray images, the glenoid cavity needs to be placed at a 30° angle when taking an anteroposterior image. At slight external rotation, the head of the humerus and the glenoid cavity are shown without overlap; the greater tubercle forms the lateral contour. In the axial image, the patient lies on his or her back with the arm slightly rotated outward and abducted. The X-ray cassette is positioned superior to the shoulder with the beam hitting the axilla inferiorly. As a result, the head of the humerus and glenoid cavity are shown at a right angle to the anteroposterior image (enhancing the detection of fractures) (**a** and **c** from: Echtermeyer V, Bartsch S, eds. Praxisbuch Schulter. 2nd ed. Stuttgart: Thieme; 2004).

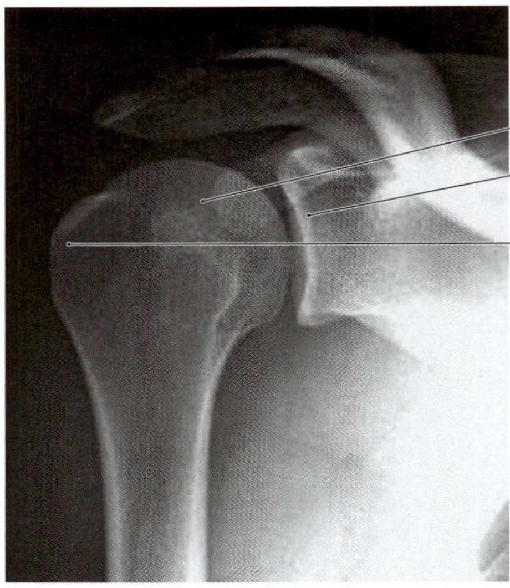

a

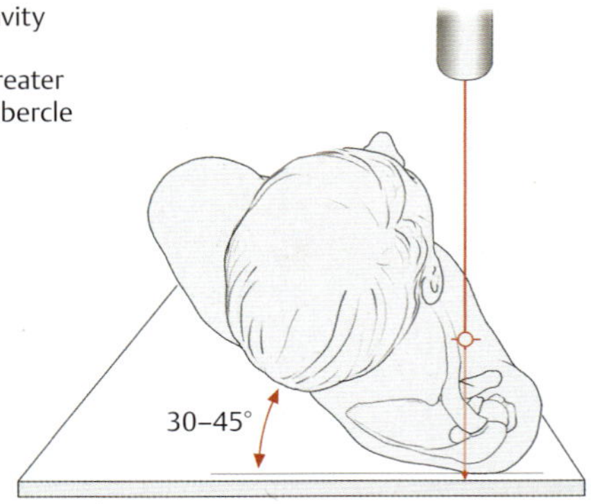

b

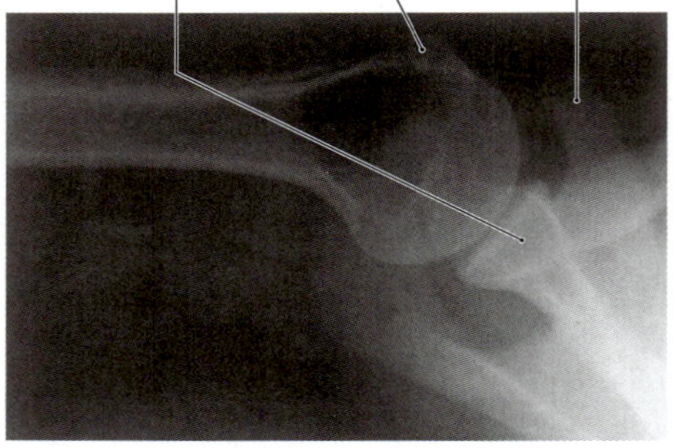

c

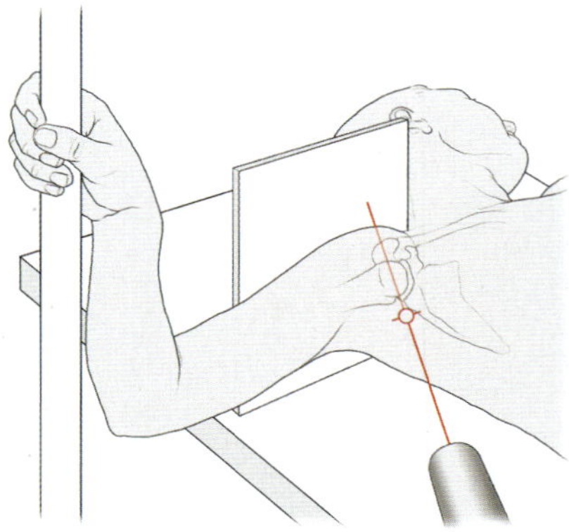

d

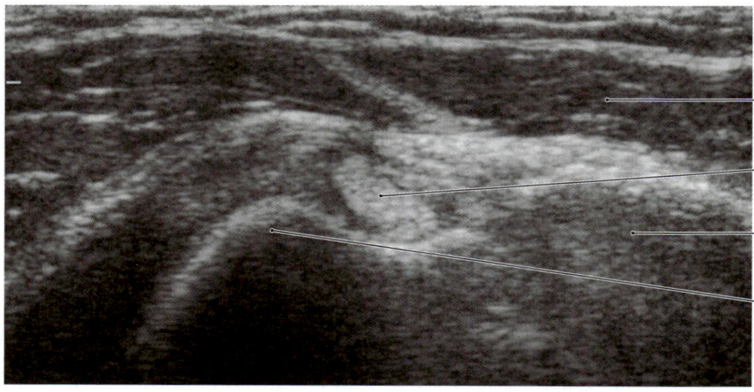

a

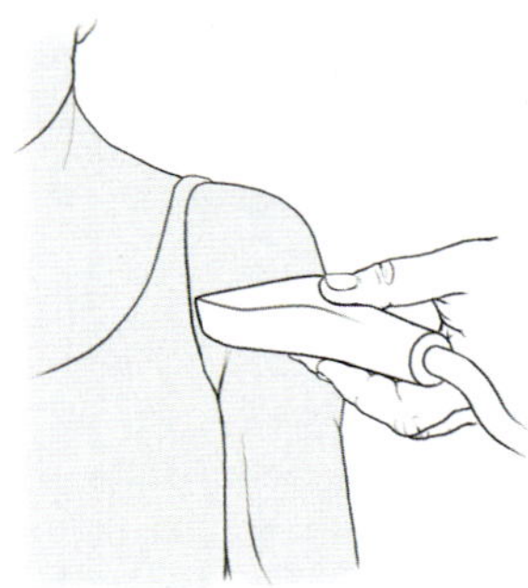

b

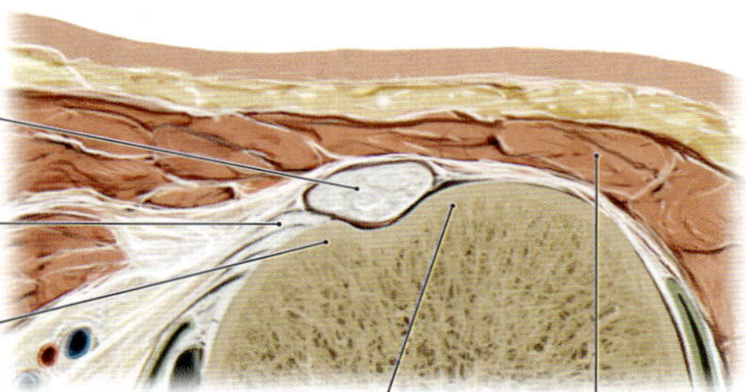

c

B Diagnostic ultrasound of the anterior region of the left shoulder

a Sonogram; **b** position of the transducer at the level of the intertubercular groove of the left shoulder joint; **c** schematic representation of the sonogram (caudal view of the transverse section) (**a** from: Konermann W, Gruber G. Ultraschalldiagnostik der Bewegungsorgane, 3rd ed. Thieme: Stuttgart; 2011).

Like diagnostic X-rays, in standardized sonographic examination of joints, two almost perpendicular planes (transverse and longitudinal) are used. Rotation of the arm, combined with different positions of the transducer, allows for an all-encompassing examination of the shoulder joint.

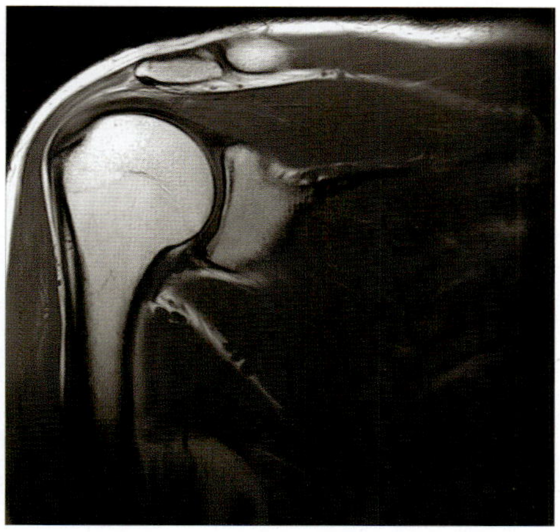

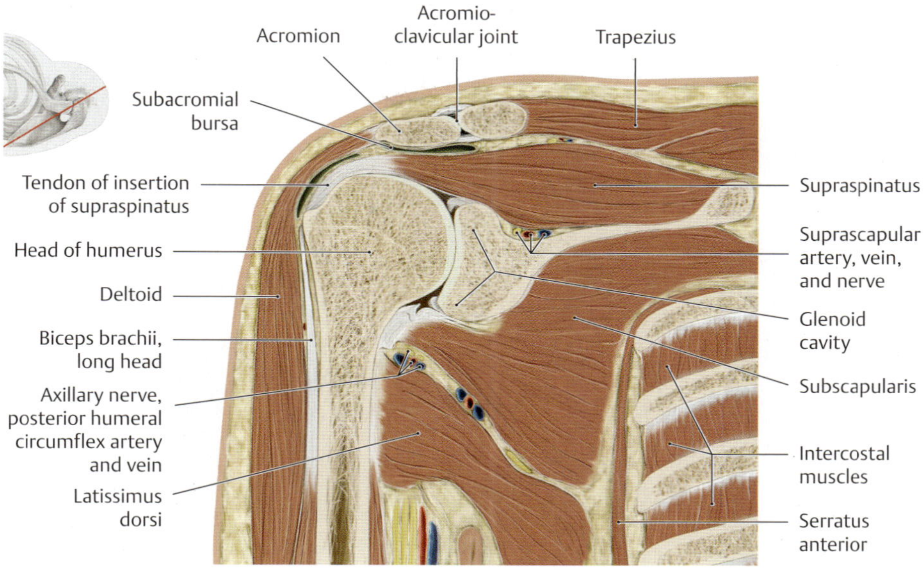

a Coronal, T1-weighted image (cross section parallel to the supraspinatus and perpendicular to the glenoid cavity)

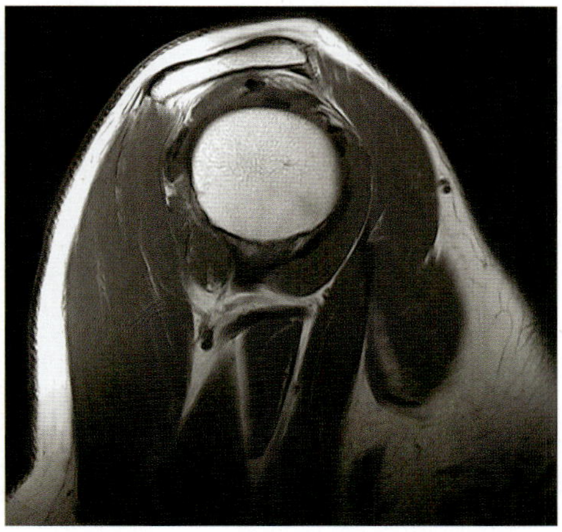

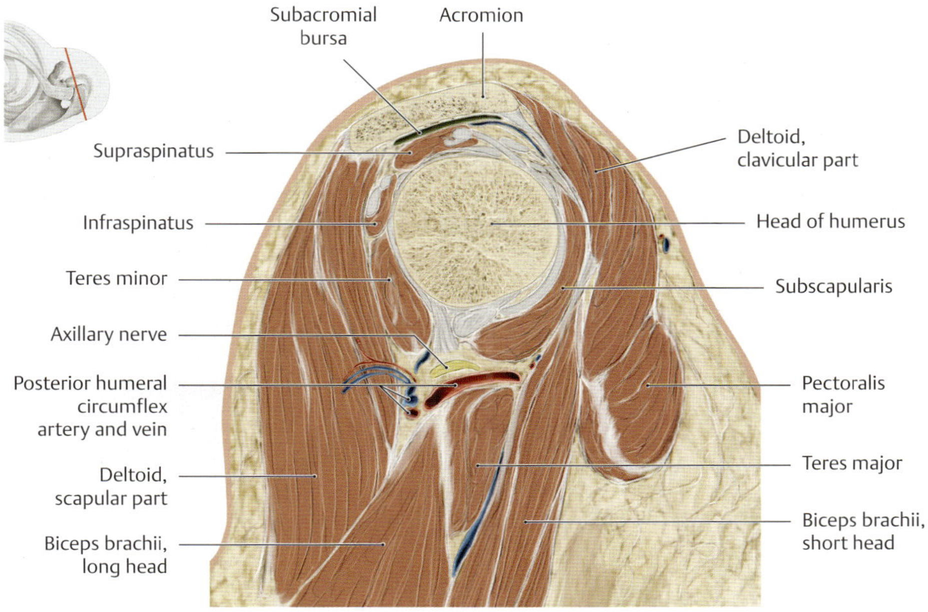

b Sagittal, T1-weighted image (cross section parallel to the glenoid cavity)

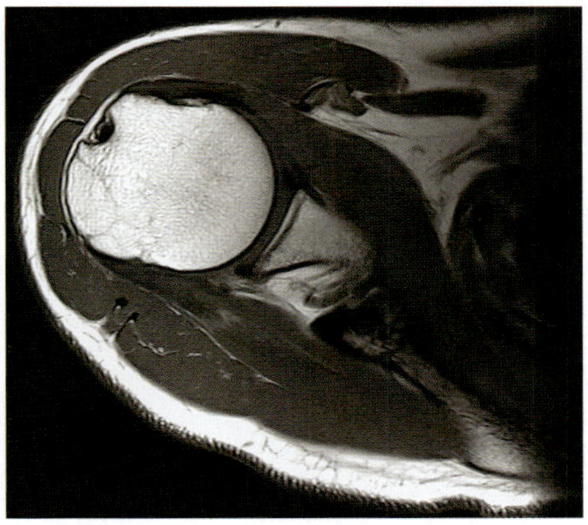

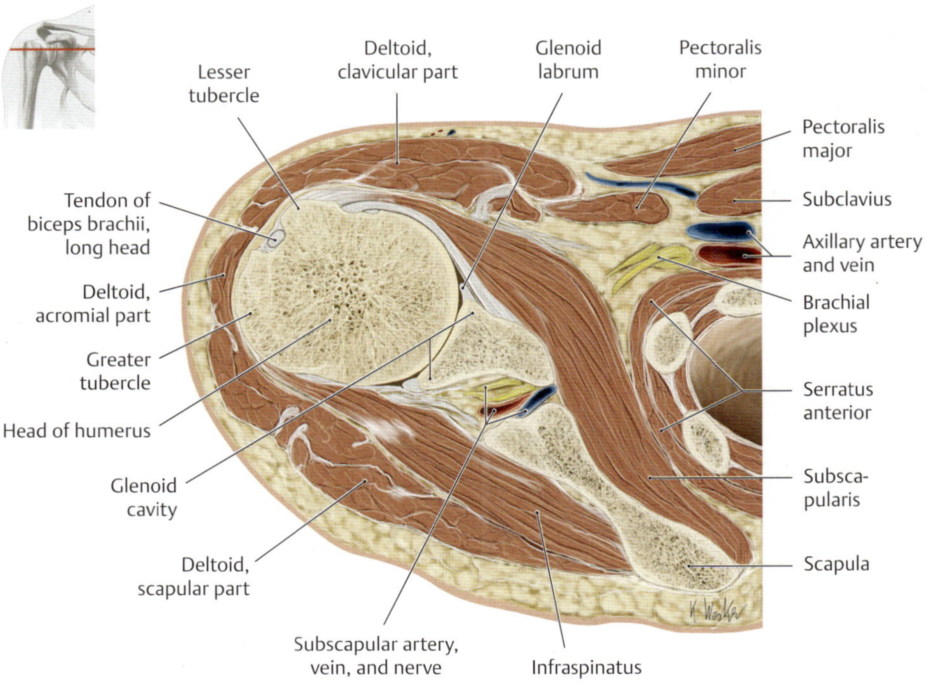

c Axial (transverse), T1-weighted image

C MRI examination of the right shoulder joint in three planes
(From Möller TB, Reif E. Taschenatlas der Schnittbildanatomie, Band III. Thieme: 2nd ed. Stuttgart; 2019.)
Note: Axial images are always viewed inferiorly.

1.19 Movements of the Shoulder Girdle and Shoulder Joint

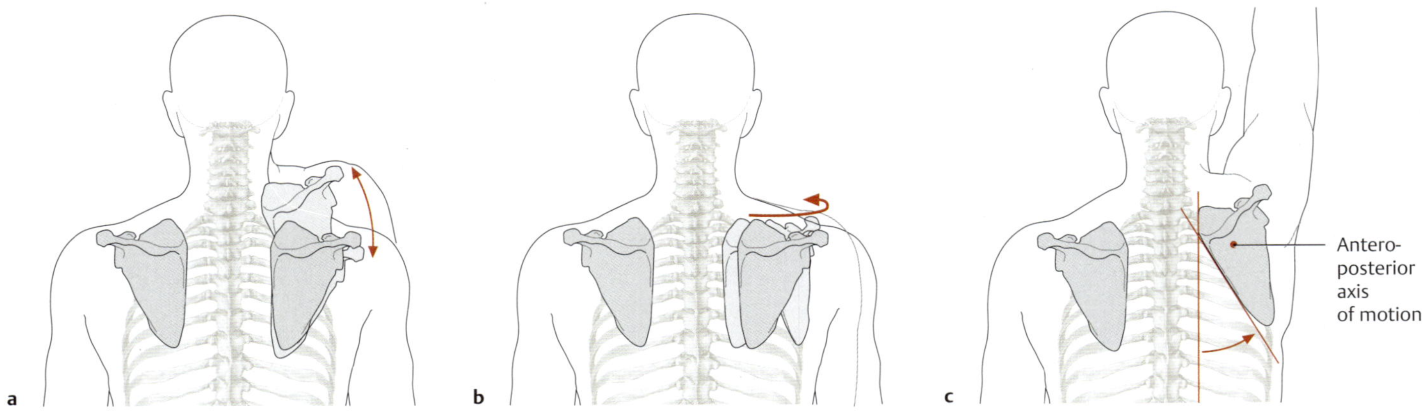

A Movements of the scapula
The sternoclavicular and acromioclavicular joints are mechanically linked in such a way that all movements of the clavicle are accompanied by movements of the scapula. The scapula moves by gliding on the chest wall in the scapulothoracic joint. Both its movement and its fixation are effected by muscular slings. The following types of scapular movement are distinguished:

- **a** Elevation and depression (during elevation and depression of the shoulder girdle): translation of the scapula in the craniocaudal direction.
- **b** Abduction and adduction (during protraction and retraction of the shoulder girdle): horizontal translation of the scapula in the posteromedial-to-anterolateral direction.
- **c** Lateral rotation of the inferior angle (during abduction or elevation of the arm): rotation of the scapula about an anteroposterior axis through the center of the scapula.With an approximately 60° range of rotation, the inferior angle of the scapula moves about 10 cm laterally, while the superior angle moves about 2 to 3 cm inferomedially.

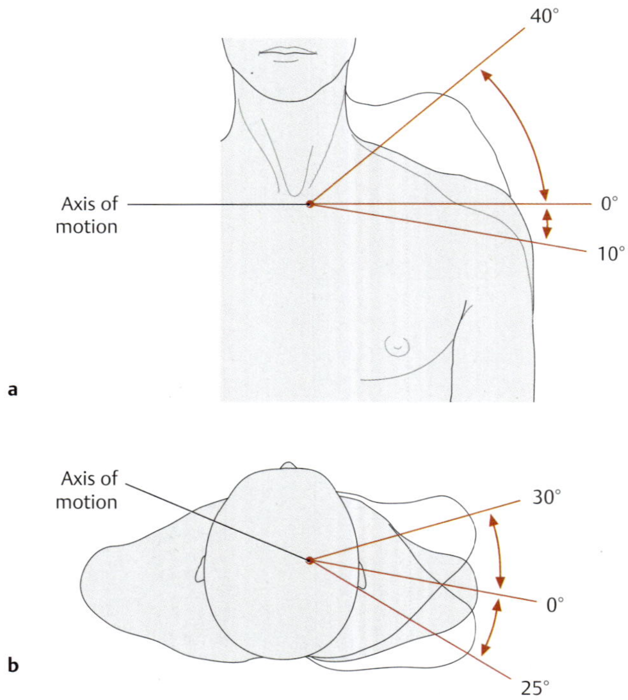

B Movements (and range of motion) in the sternoclavicular joint
- **a** Elevation and depression of the shoulder about a parasagittal axis.
- **b** Protraction and retraction of the shoulder about a longitudinal (vertical) axis.

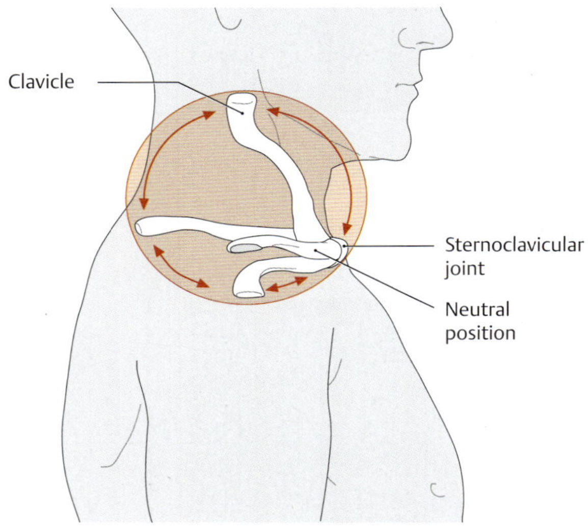

C Range of motion of the clavicle
Right clavicle, lateral view. Viewing the range of motion of the clavicle in the sternoclavicular joint from the lateral view, we find that the clavicle moves roughly within a conical shell whose apex is directed toward the sternum, and with a slightly oval base approximately 10 to 13 cm in diameter. The clavicle also rotates about its own axis, particularly during elevation of the shoulder girdle, where its S shape significantly increases the range of shoulder elevation. The range of this rotation is approximately 45° and creates a third degree of freedom that gives the sternoclavicular joint the *function* of a *spheroidal joint*.

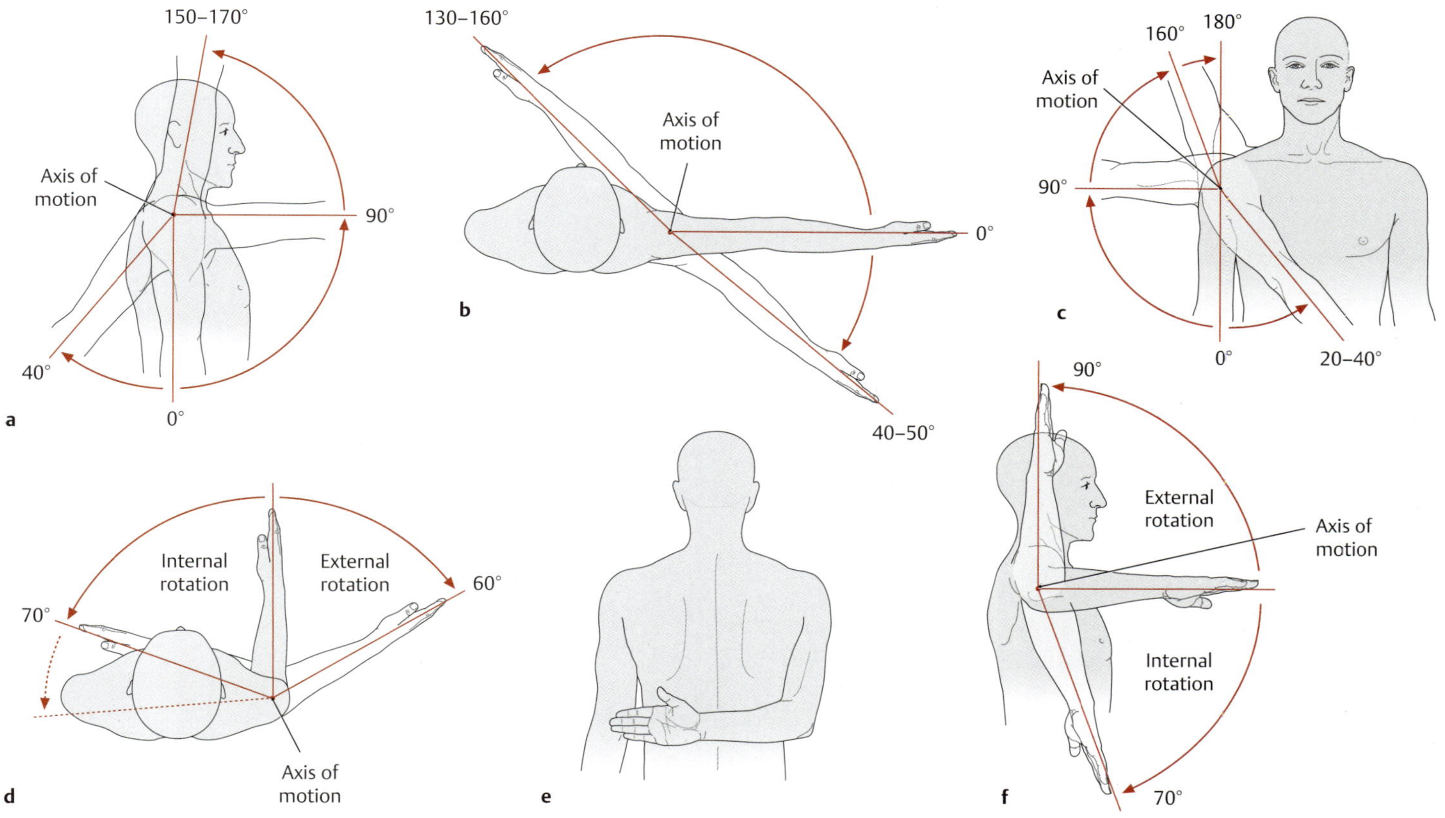

D Movements in the shoulder joint

As a typical spheroidal joint, the shoulder joint allows movements about **three** mutually perpendicular **cardinal axes** with three degrees of freedom and a total of six main directions of movement:

- **Movements about a horizontal (= transverse) axis:** anteversion and retroversion movements (flexion and extension), either from the neutral (0°) position (= arm hanging down, see **a**) or with the arm abducted 90° (**b**, then also referred to as *horizontal movement*).
- **Movements about a sagittal axis:** lateral abduction and medial abduction (see **c**). Past 80 to 90° of abduction, an automatic external rotation occurs. This keeps the greater tubercle from impinging on the coracoacromial arch. When the arm is internally rotated, the range of abduction is decreased to approximately 60°.
  Note: Abduction past 90° is often referred to as "elevation." In clinical parlance, however, the term "elevation" is generally applied to all vertical movements. Whereas abduction or elevation of the arm up to 160° also requires rotation of the scapula about a sagittal axis (past about 60°), elevation beyond this up to 180° in the coronal plane can only be achieved with simultaneous external rotation and inclination of the spine toward to the contralateral side.
- **Movements about a vertical axis (= longitudinal or shaft axis of the humerus):** internal and external rotation (see **d–f**). When the forearm is flexed at the elbow during these movements, the forearm can be used as a pointer. When the arm is hanging at the side, the maximum range of internal rotation is limited by the trunk. Placing the arm behind the back is equivalent to 95° of internal rotation (**e**). When the arm is abducted 90°, the range of external rotation is increased, while the maximum range of internal rotation is slightly reduced (**f**). The maximum range of motion in any case is achieved by moving the shoulder girdle as well.

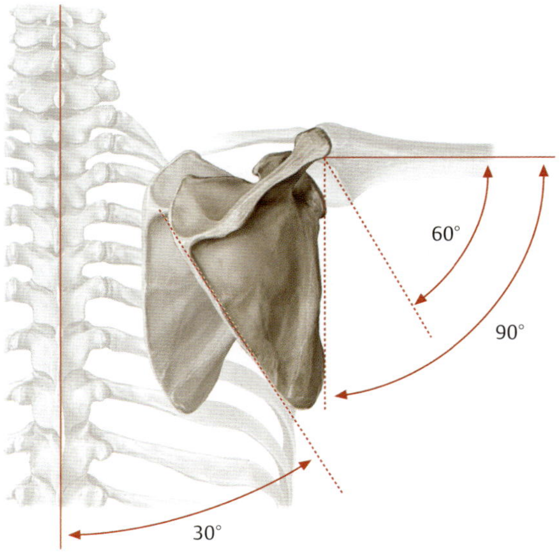

E Humeroscapular rhythm

The arm and scapula move in a 2:1 ratio during abduction. This means that when the arm is abducted 90°, for example, 60° of that movement occurs in the glenohumeral joint, while 30° is accomplished by concomitant movement of the shoulder girdle. This "humeroscapular rhythm" is dependent upon freedom of movement of the scapula during abduction. Diseases of the shoulder joint can alter this rhythm, often causing the scapula to begin its rotation considerably earlier. This is best illustrated by cases involving ankylosis or arthrodesis of the glenohumeral joint (pathological or intentional operative stiffening or fixation), in which movements in the shoulder girdle alone still permit the arm to be abducted 40 to 60° and still allow for one third the normal range of flexion/extension.

1.20 The Elbow Joint as a Whole

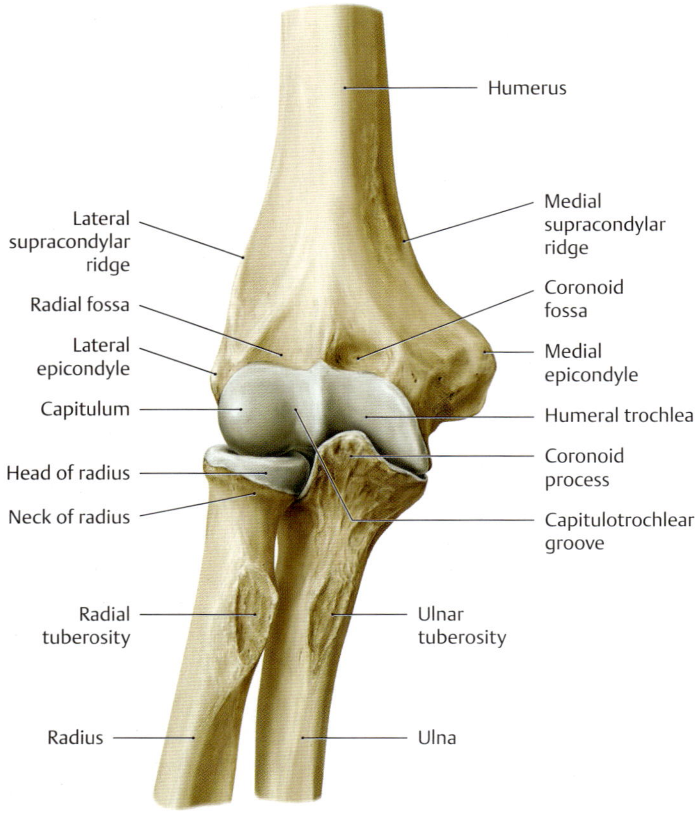

a Anterior view

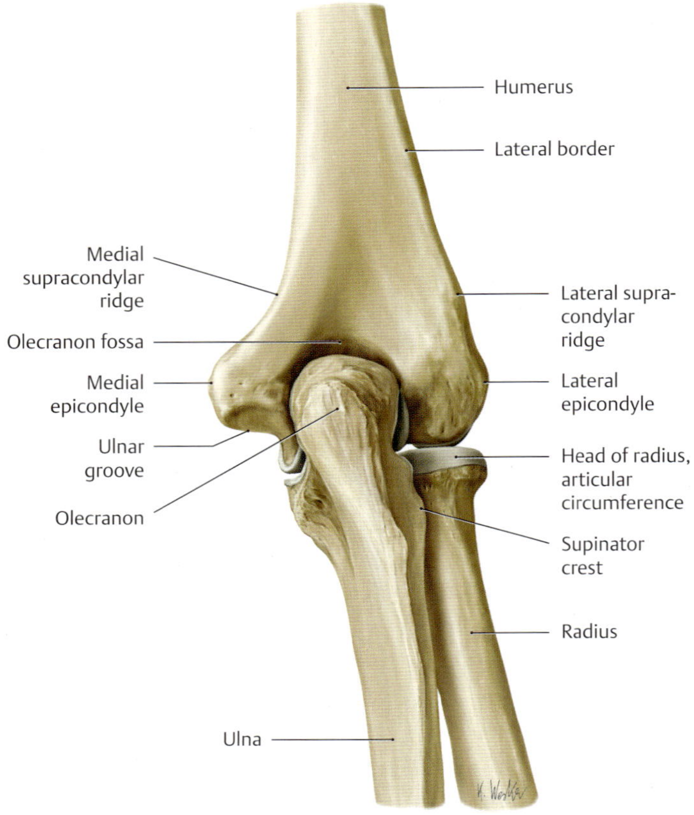

b Posterior view

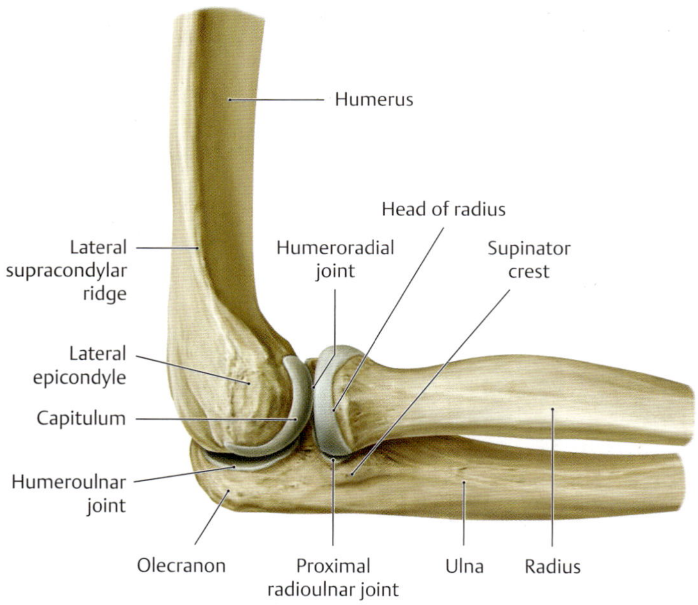

c Lateral view

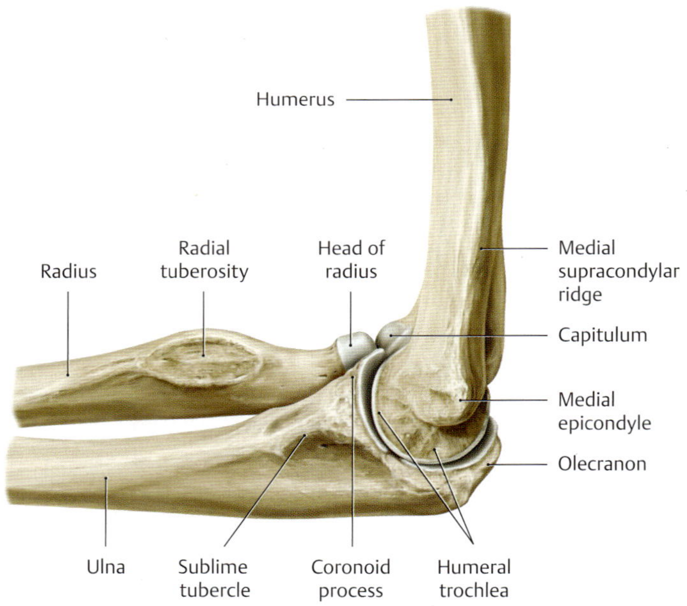

d Medial view

A The articulating skeletal elements of the right elbow joint

The *humerus*, *radius*, and *ulna* articulate with one another at the elbow (cubital) joint. The elbow consists of three articulations:

- The humeroulnar joint between the humerus and ulna.
- The humeroradial joint between the humerus and radius.
- The proximal radioulnar joint between the proximal ends of the ulna and radius.

Note: The term "sublime tubercle" is a clinical term that is not part of standard anatomic nomenclature.

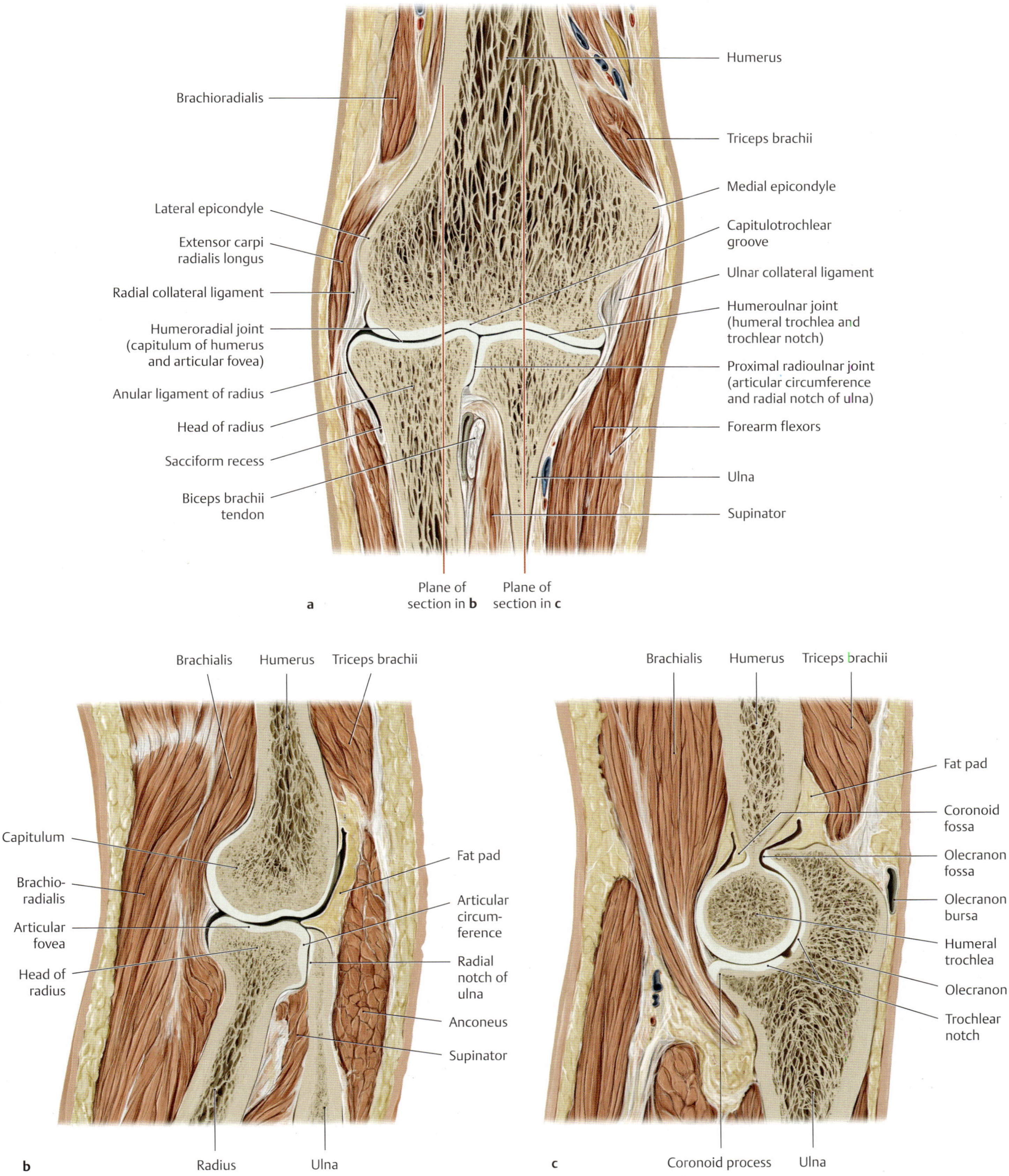

B Skeletal and soft-tissue elements of the right elbow joint

a Coronal section viewed from the front (*note* the planes of section shown in **b** and **c**).

b Sagittal section through the humeroradial joint and proximal radioulnar joint, medial view.

c Sagittal section through the humeroulnar joint, medial view.

(Drawings based on specimens from the Anatomical Collection at the University of Kiel.)

1.21 The Elbow Joint: Capsule and Ligaments

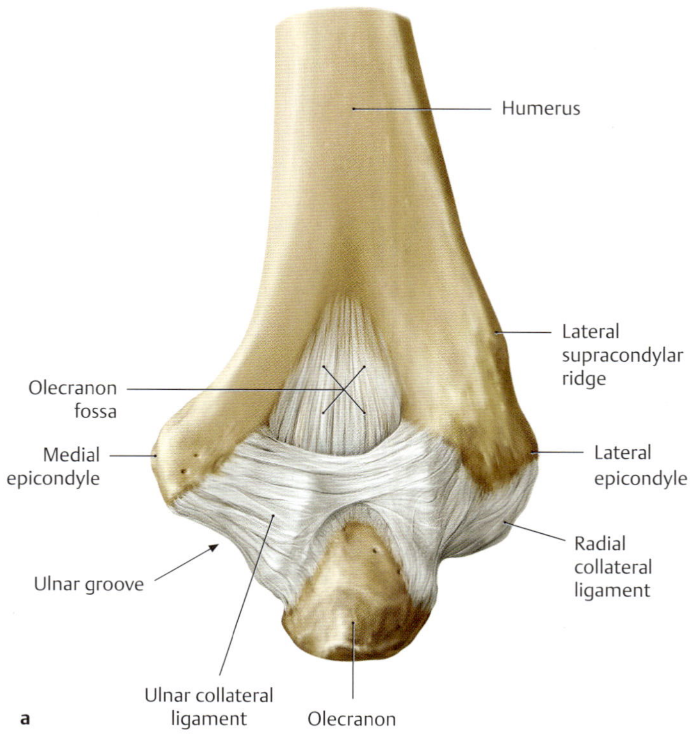

A The capsule and ligaments of the right elbow joint in 90° flexion
a Posterior view, **b** medial view, **c** lateral view with forearm pronated.
Both the humeroradial and humeroulnar joints are stabilized by powerful collateral ligaments that strengthen the lateral portions of the joint capsule. These structures, called the *ulnar collateral ligament* and *radial collateral ligament*, have a fan-shaped arrangement that can give lateral stability to the joint in any position. The anular ligament of the radius (see also **D**) stabilizes the proximal radioulnar joint.

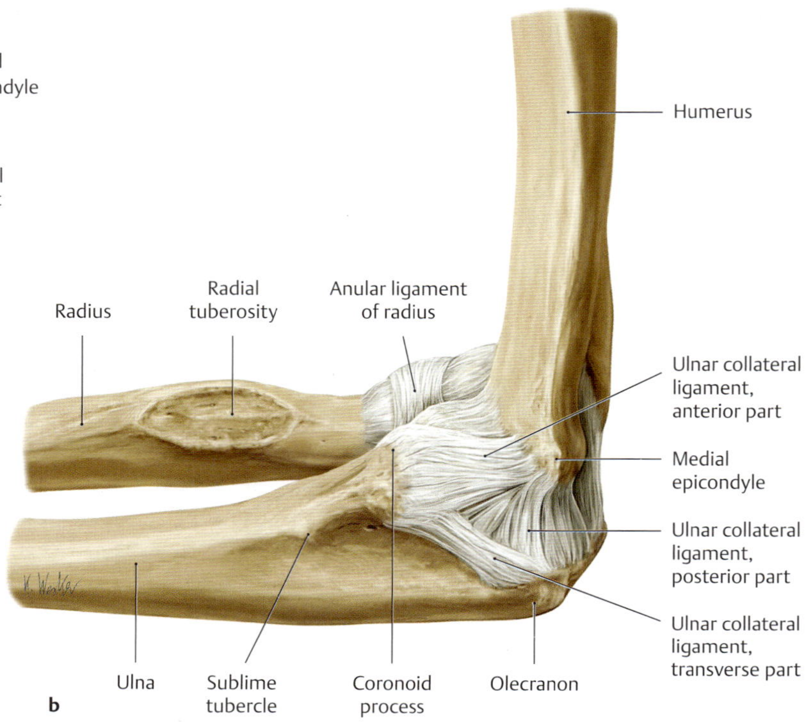

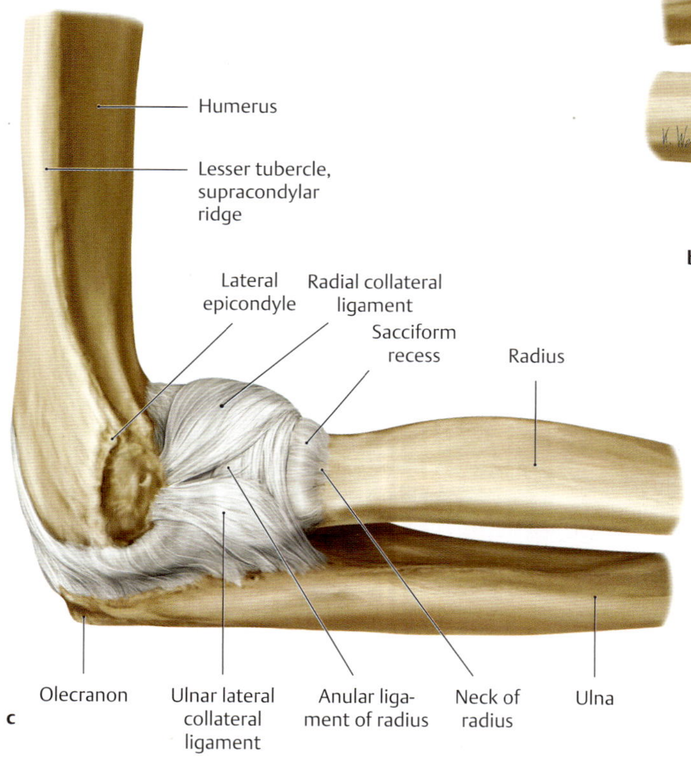

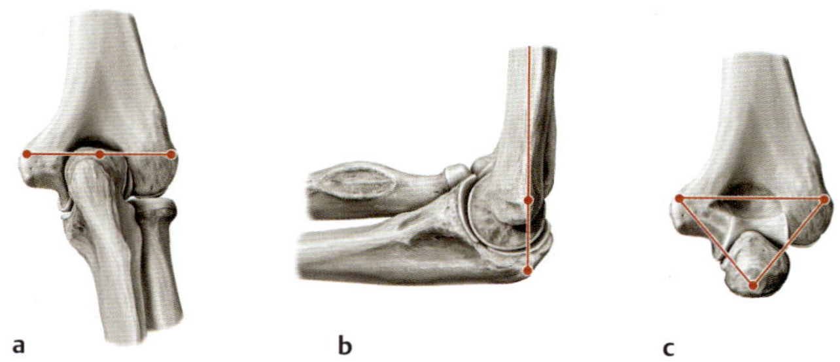

B The Hueter line and triangle
a Extension, posterior view; **b** flexion, medial view; **c** flexion, posterior view.
The epicondyles and the olecranon lie on a straight line when viewed from the posterior view in the extended elbow. They also lie on a straight line when viewed from the lateral view in the flexed elbow. But when the flexed elbow is viewed from behind, the two epicondyles and the tip of the olecranon form an equilateral triangle. Fractures and dislocations alter the shape of the triangle.

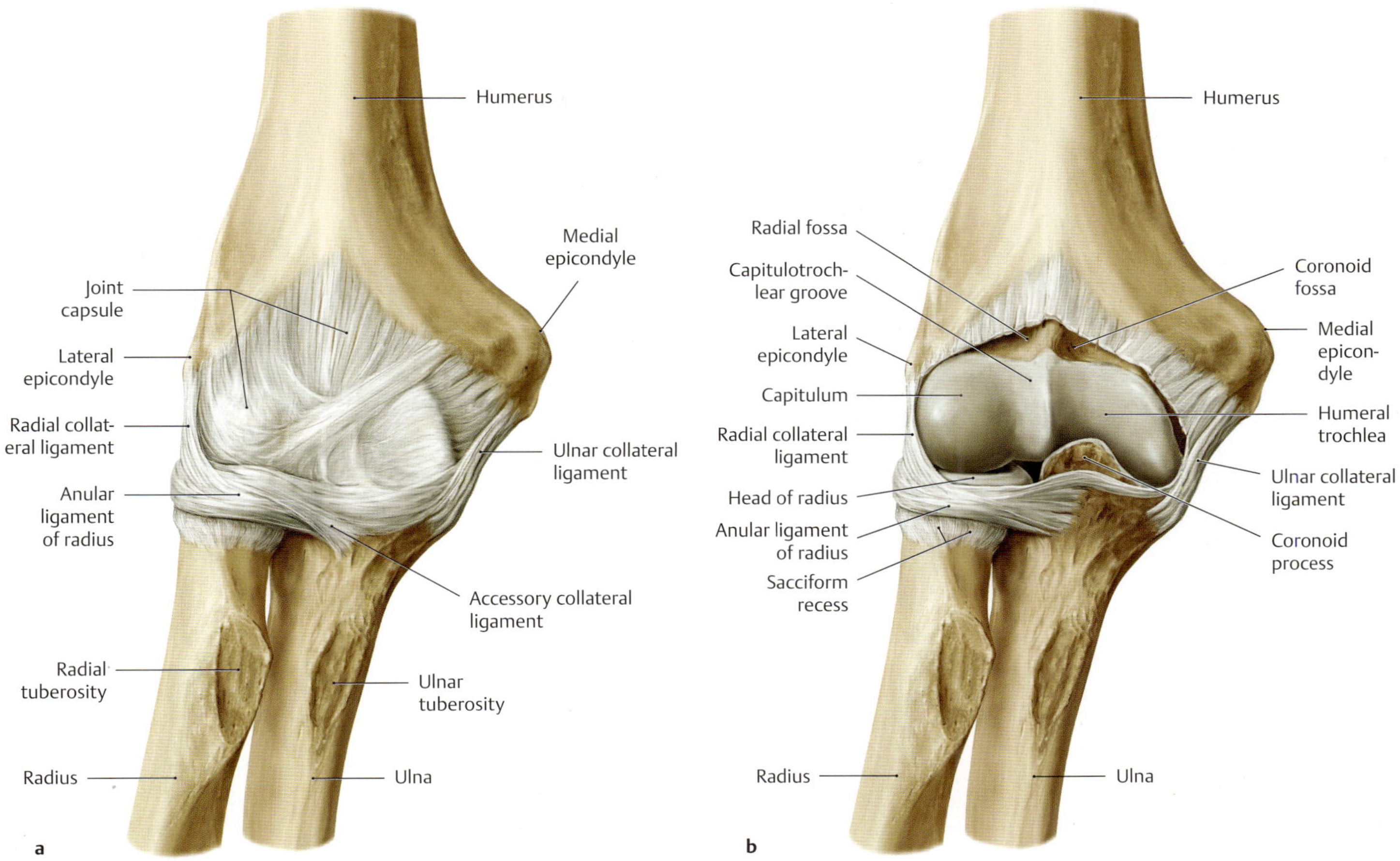

C The capsule and ligaments of the right elbow joint in extension

a Anterior view, **b** Anterior view with the ventral portions of the capsule removed.

The joint capsule of the elbow encloses all three articulations in the elbow joint complex. While the capsule is thin anteriorly and posteriorly, it is reinforced on each side by the collateral ligaments (see **A**). Over the end of the radius, the joint capsule is expanded below the annular ligament to form the sacciform recess—a redundant tissue fold that provides a reserve capacity during pronation and supination of the forearm. During flexion and extension, the brachialis and anconeus muscles tighten the joint capsule to prevent entrapment of the capsule between the articular surfaces (see p. 326).

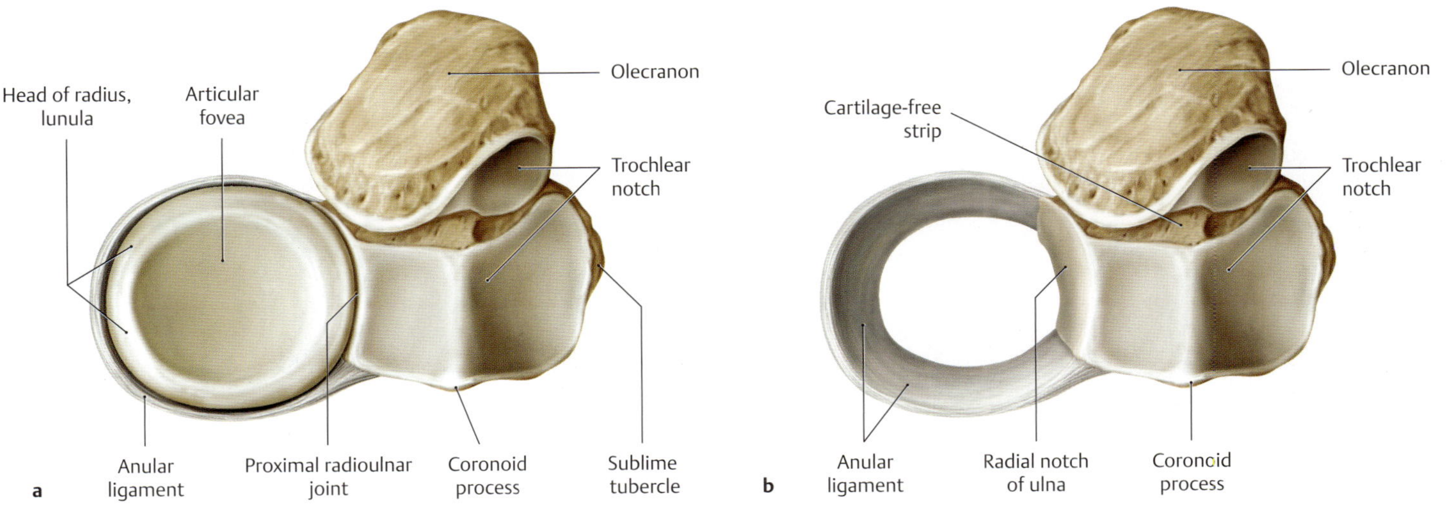

D Course of the anular ligament in the right proximal radioulnar joint

a View of the proximal articular surfaces of the radius and ulna after removal of the humerus **b** Same view as in **a** with the radius also removed

The anular ligament is of key importance in stabilizing the proximal radioulnar joint. It runs from the anterior to the posterior border of the radial notch of the ulna (= cartilage-covered articular surface on the ulna), wrapping around the radial head and pressing it into the ulnar articular surface. Histologically, the inner surface of the anular ligament has the fibrocartilaginous structure of a gliding tendon, enabling it to withstand the compressive loads that are transmitted to the ligament.

1.22 Imaging Studies of the Elbow

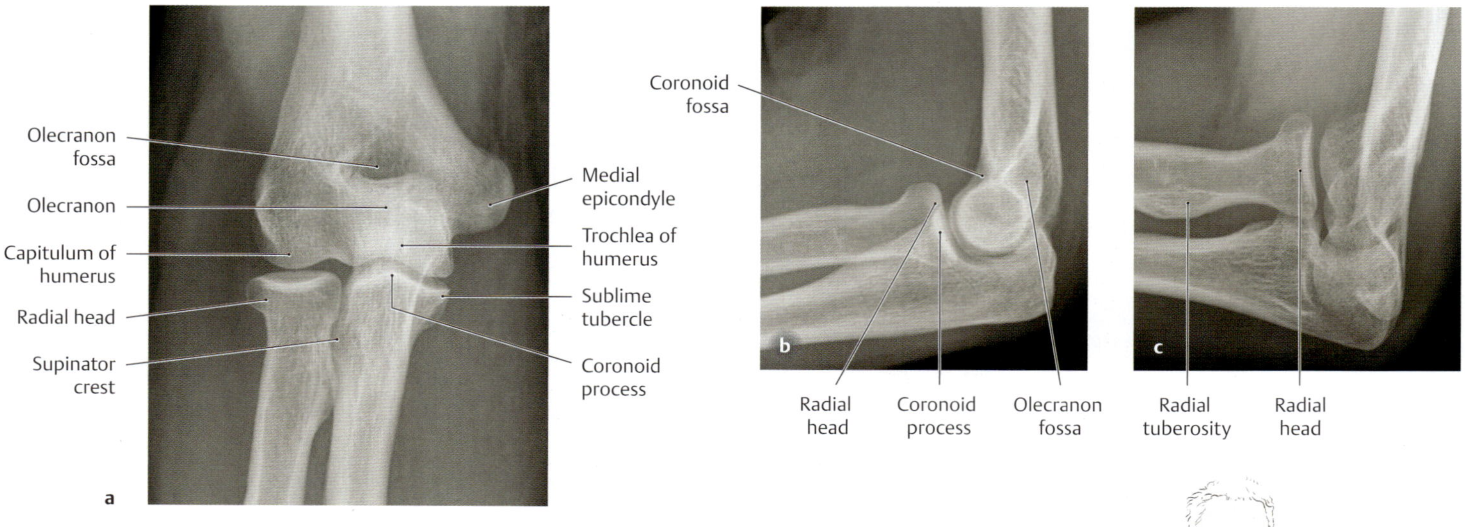

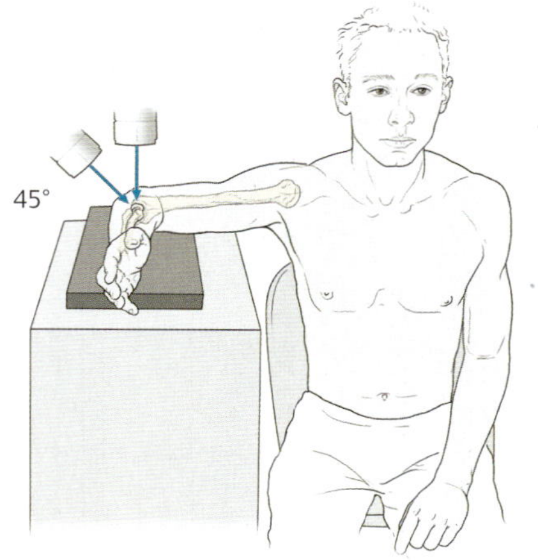

A Conventional radiographs of the right elbow

a Anterior-posterior (AP) image; **b** and **c** lateral images; **c** represents what is known as a spot view of the radial head (see below); **d** imaging technique: The lateral films (like **b** and **c** here) are obtained with the arm in 90° abduction, the elbow flexed 90°, and the hand partially pronated (in **b** the beam is vertical from above). Where a fracture of the radial head is suspected, a spot view of the radial head (see **c**) should also be obtained (also lateral but in a 45° projection). It is indicated especially in non-displaced fractures of the radial head (radiographs: Hans-Peter Sobotta, M.D., Stiftung HEH Braunschweig).

Conventional radiographs in two planes are also indicated in these cases:

- Suspected acute dislocation of the elbow (second most common dislocation with 5–6 cases per 100,000 adults per year, see **Ba**),
- Associated fractures (such as coronoid fracture, i.e., fracture of the coronoid process [see **Bb**], radial head fracture [see **C**], and the risk of chronic instability) as a posttraumatic complication,
- After reduction to evaluate joint congruity, an indicator of possible joint instability,
- To demonstrate classic radiologic signs of primary and posttraumatic arthrosis of the elbow (such as osteophytes, joint space narrowing, subchondral sclerosis, etc.).

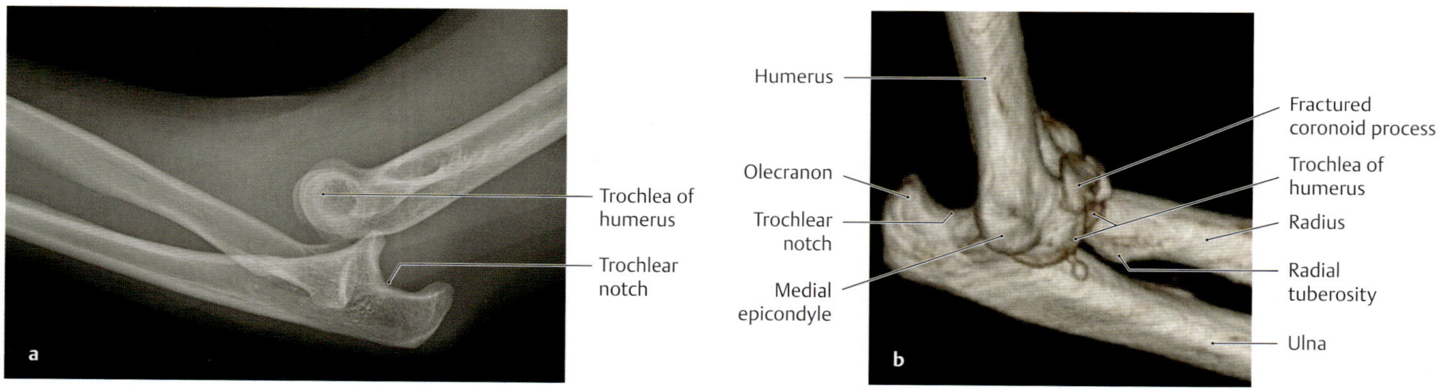

B Acute elbow dislocation

a Dislocation without associated injury (original image: Prof. S. Müller-Hülsbeck, M.D., Klinik für Diagnostische Radiologie. DIAKO Krankenhaus gGmbH, Flensburg); **b** medial view of a 3D CT reconstruction of a "terrible triad" injury: complex elbow dislocation with fractures of the radial head (Mason type II fracture, see **Cd**) and coronoid process and ruptures of the LCL and/or MCL (from: Müller L, Hollinger B, Burkhart K, editors. Expertise Ellenbogen. Stuttgart: Thieme; 2016).

Note: CT 3D reconstruction (**b**) permits analysis of fracture lines, displaced fragments, and the extent of areas of comminution and is therefore indicated with complex bony and intra-articular injuries. MRI is the imaging modality of choice for more detailed evaluation of the capsular ligaments.

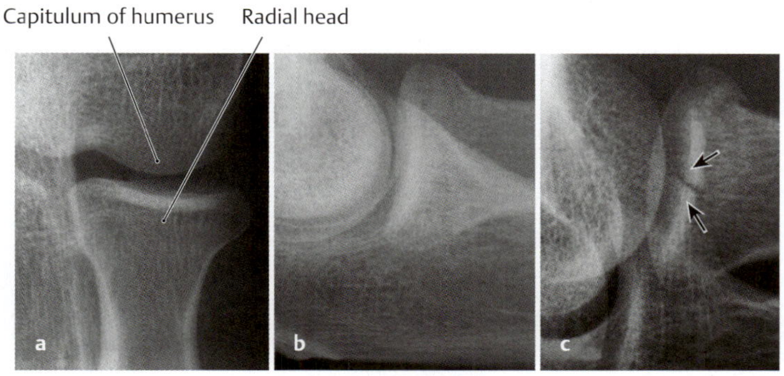

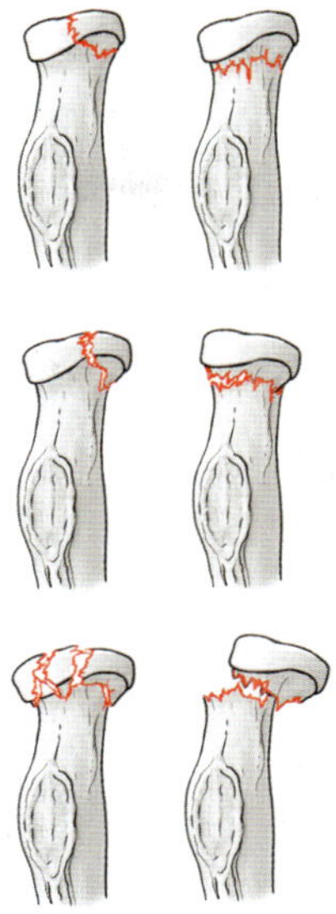

Type I

- Nondisplaced or minimally displaced (chisel fractures, marginal fractures)
- Single fragment
- Intra-articular or extra-articular

Type II

- Displaced (> 2 mm)
- Single fragment
- Intra-articular or extra-articular

Type III

- Comminuted fractures
- Fully displaced radial head

d

C Nondisplaced fracture of the radial head (Mason type I, chisel fracture) with joint involvement (left elbow)

a Anterior-posterior (AP) radiograph; **b** lateral radiograph; **c** 45° projection, spot view of the radial head, see **Ad** (from: Bohndorf K, Imhof H, Fischer W, eds. Radiologische Diagnostik der Knochen und Gelenke. 2nd ed. Stuttgart: Thieme; 2006); **d** Mason classification for determining the severity of a radial head fracture.

Note: Fractures of the radial head and neck are the most common elbow injuries in adults, accounting for about 50% of all cases. They result from indirect trauma from a fall on the outstretched arm with compression of the radial head against the capitulum of the humerus. About half of these fractures are nondisplaced (Mason type I fractures) and are therefore difficult to detect on conventional radiographs. They are often directly visible only on a 45° projection (fracture line = black arrow).

E Common clinical abbreviations for the elbow

MCL	Medial collateral ligament	LCL	Lateral collateral ligament
AMCL	Anterior bundle of MCL	LUCL	Lateral ulnar collateral ligament
PMCL	Posterior bundle of MCL	RCL	Radial collateral ligament
Cooper ligament		AL	Annular ligament
		ALCL	Accessory lateral collateral ligament
HRJ	Humeroradial joint	PRUJ	Proximal radioulnar joint
HUJ	Humeroulnar joint	DRUJ	Distal radioulnar joint

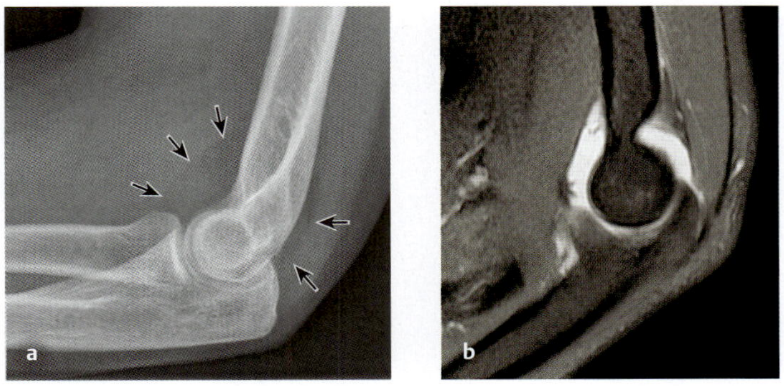

D Fat pad sign

a Lateral radiograph (original image: Hans-Peter Sobotta, M.D., Stiftung HEH Braunschweig); **b** on MR image with hyperintensity indicative of joint effusion (source: MVZ blikk. Braunschweig GmbH). The elbow joint contains small anterior and posterior fat pads in the subintima, residing between the synovial membrane and the fibrous capsule. In a healthy elbow only the *anterior* fat pads are detectable on radiographs. They appear as a fine radiolucent line close to the bone. In contrast, the *posterior* fat pads are hidden within the olecranon fossa (see p. 289, **Bc**). When superfluous fluid (capsule effusion or hemarthrosis) accumulates in the joint secondary to a fracture (or other clinical pictures such as rheumatoid arthritis or infection), the subintimal anterior fat pad is lifted off the bone and the posterior fat pad is also detectable and appears to be detached from the bone (black arrows). The "fat pad sign" in the elbow is thus a radiologic sign that aids in diagnosing fractures and other clinical pictures. It is evaluated on conventional lateral radiographs of the elbow in 90° flexion.

Note: Given a corresponding history and clinical symptoms, the presence of a positive fat pad sign is sufficient to diagnose a fracture. With proper imaging technique (strict lateral projection with the elbow in 90° flexion) it has high sensitivity and specificity for detecting a fracture.

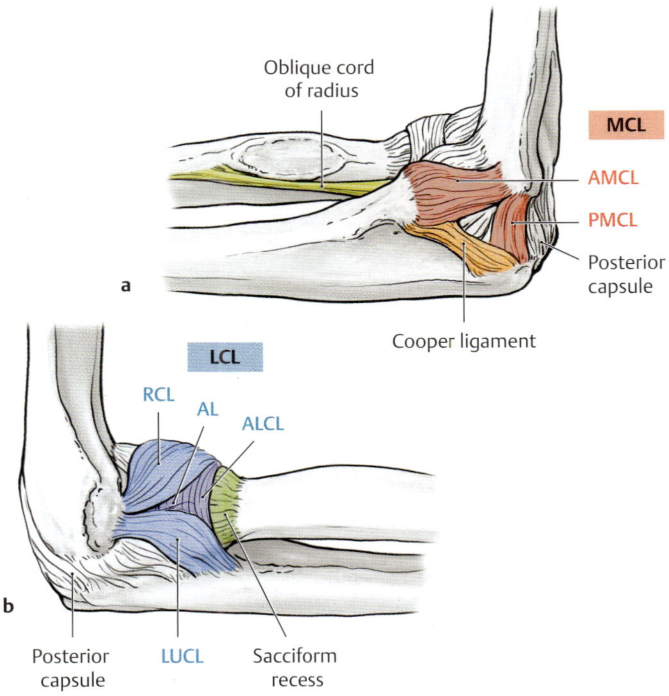

F Medial (a) and lateral (b) ligaments reinforcing the capsule in the right elbow (see Table **E** for abbreviations)

1.23 The Forearm: Proximal and Distal Radioulnar Joints

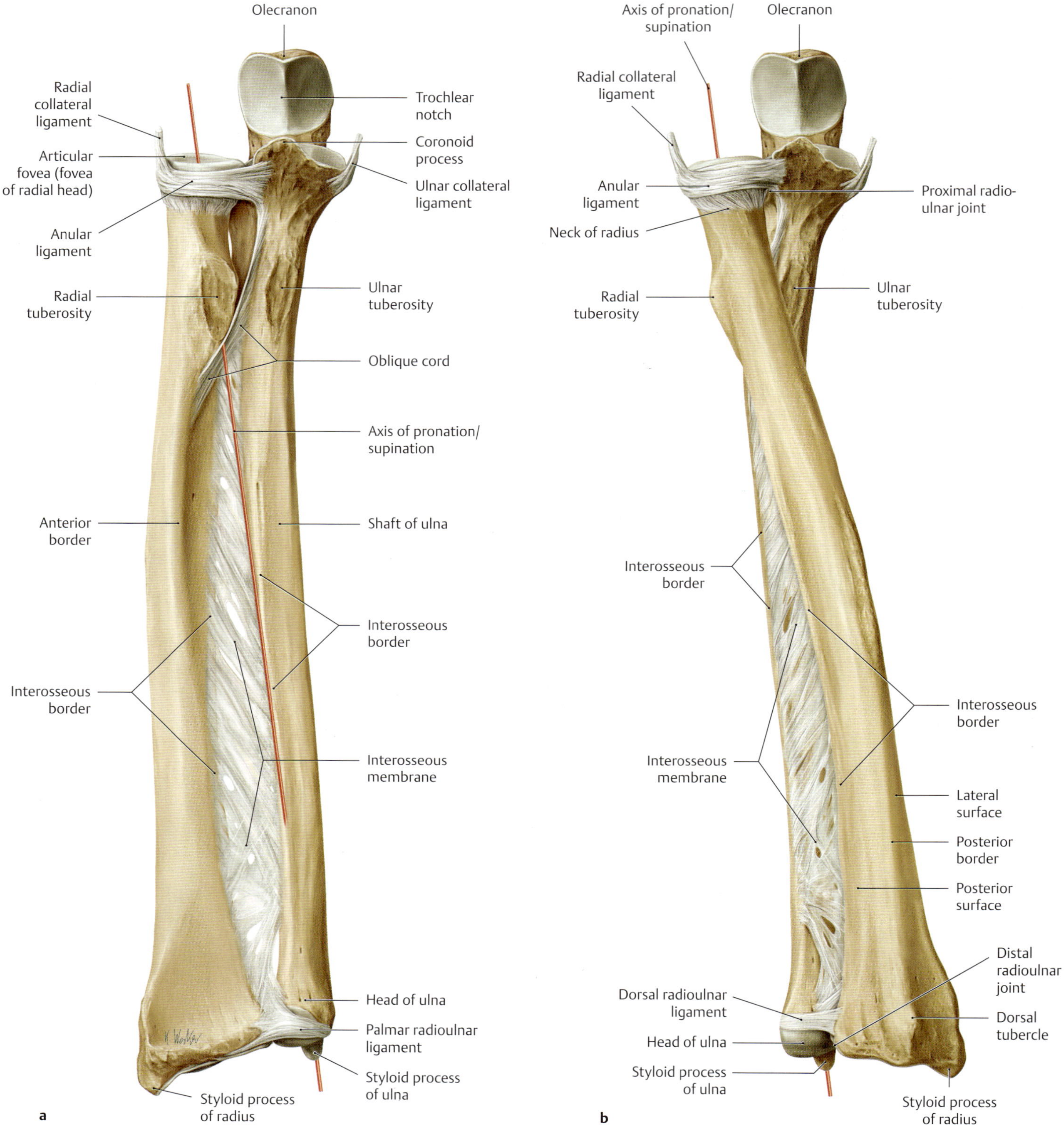

A Ligaments and axes for pronation and supination in the proximal and distal radioulnar joints
Right forearm, anterior view.

a Supination (the radius and ulna are parallel to each other).
b Pronation (the radius crosses over the ulna).

The proximal radioulnar joint functions together with the distal radioulnar joint to enable pronation and supination movements of the hand. The movements of both joints are functionally interlinked by the interosseous membrane, so that the movement of one is necessarily associated with movement of the other. The axis for pronation and supination runs obliquely from the center of the humeral capitulum (not shown) through the center of the radial articular fovea down to the styloid process of the ulna.

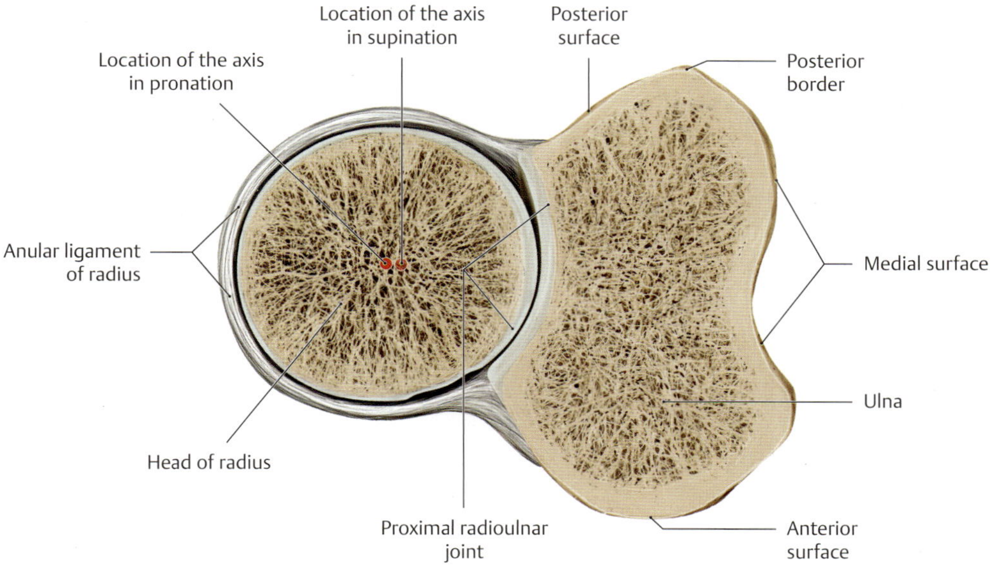

B Cross section through the right proximal radioulnar joint in pronation

Distal view. Owing to the slightly oval shape of the radial head, the pronation/supination axis that runs through the radial head moves approximately 2 mm radially during pronation (the long diameter of the radial head is transverse when pronation is reached). This ensures that when the hand is pronated, there will be sufficient space for the radial tuberosity within the interosseous space (= the space between the radial tuberosity and oblique cord; see **Aa**, for example).

Note the thicker articular cartilage of the radial articular circumference on the pronation side. This thickening occurs as an adaptation to the greater articular pressure in the proximal radioulnar joint in the pronated position.

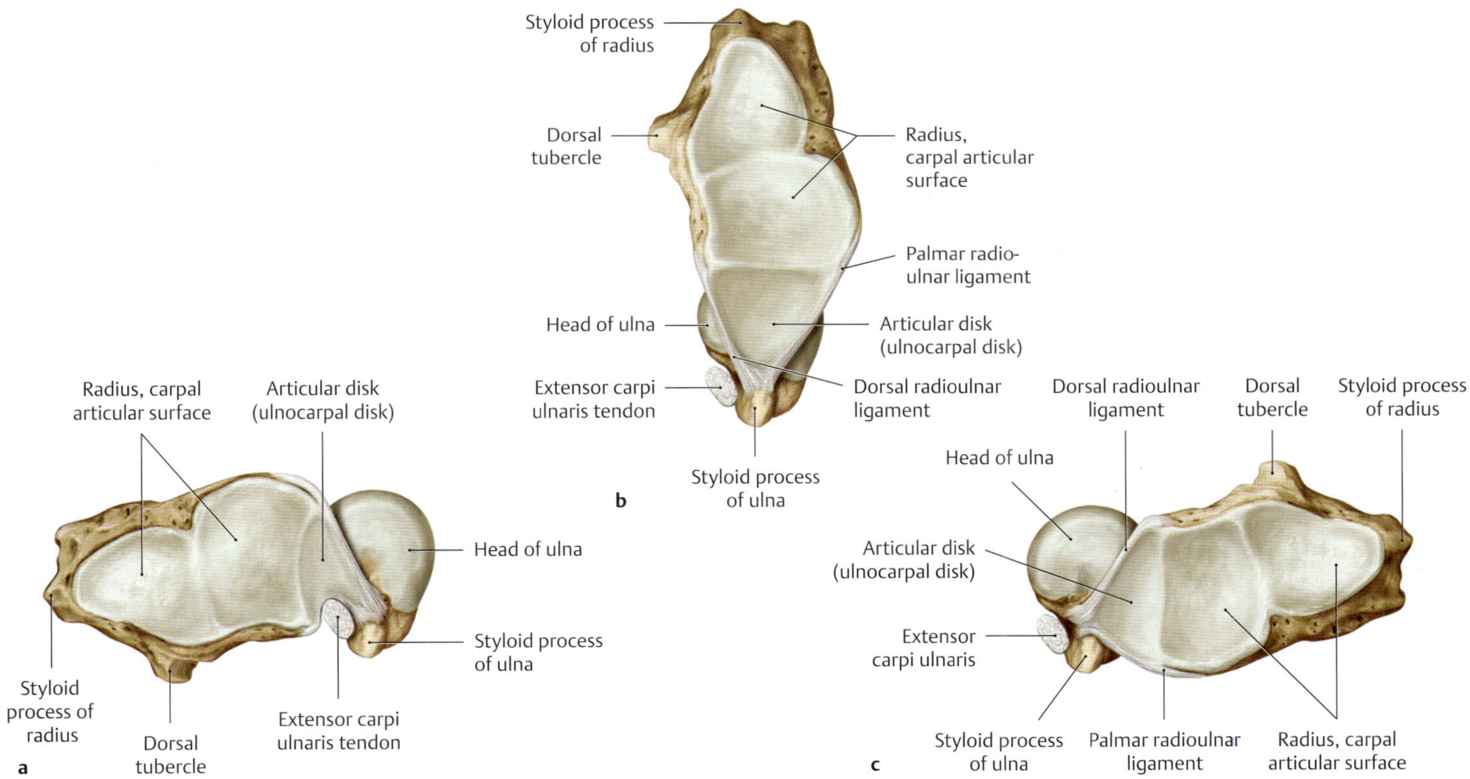

C Rotation of the radius and ulna during pronation and supination

View of the distal articular surfaces of the radius and ulna of the right forearm. For clarity, the ulnocarpal disk is not shown.

a Supination
b Semipronation
c Pronation

The dorsal and palmar radioulnar ligaments are part of the "ulnocarpal complex," which serves to stabilize the distal radioulnar joint. The mode of contact between the two distal articular surfaces varies with the position of the radius and ulna. They are in close apposition only in an intermediate (semipronated or neutral) position.

1.24 Movements of the Elbow and Radioulnar Joints

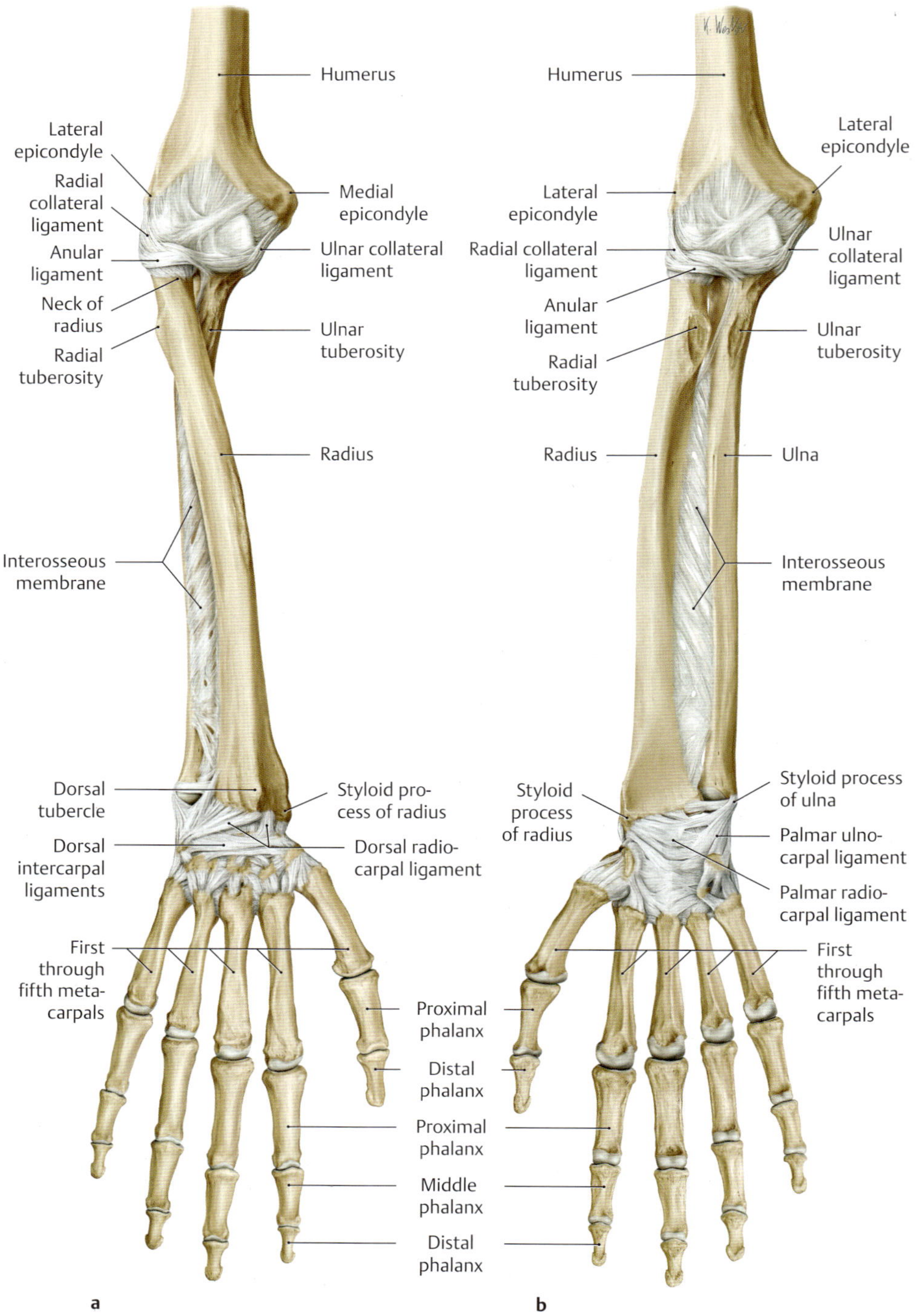

A Pronation and supination of the right hand
Anterior view.
a Pronation, **b** supination.
Pronation and supination of the hand make it possible to raise an object to the mouth for eating and to touch any area of the body for protection or cleaning. Pronation and supination are also essential for the working hand in actions such as turning a screwdriver, screwing in a light bulb, emptying a bucket, and unlocking a door. The range of hand movements can be further increased by adding movements of the shoulder girdle and trunk. This may be done, for example, to enable a full 360° twisting movement of the hand.

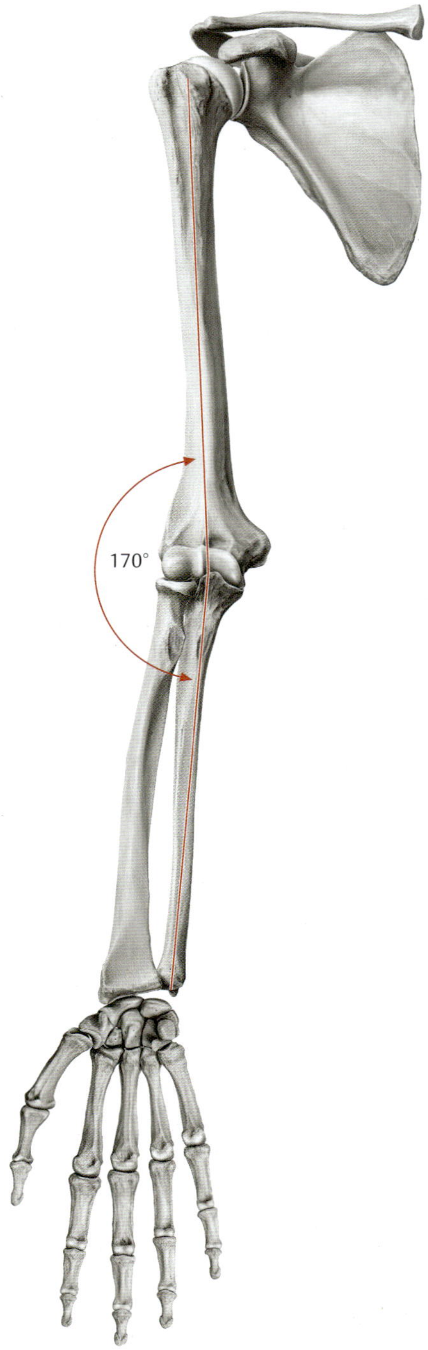

B Normal valgus of the elbow joint
Skeleton of the right upper limb with the forearm supinated. Anterior view.
The shape of the humeral trochlea (see p. 288) results in a normal valgus angulation between the humeral shaft and ulna (cubitus valgus) [valgus = bend or twist outward, away from the body axis]. This applies particularly during extension and supination. This "cubital angle" equals approximately 170°.

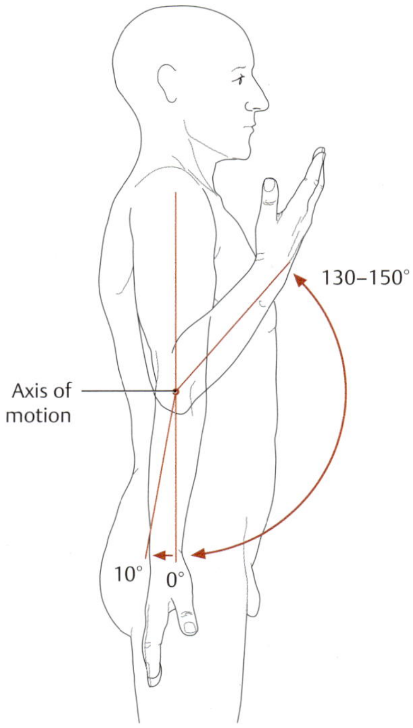

C Range of motion in the humeroradial and humeroulnar joints of the elbow

The flexion/extension axis of the forearm runs below the epicondyles through the capitulum and trochlea of the humerus. Starting from the neutral (0°) position, both joints have a maximum range of 150° in flexion and approximately 10° in extension. Both movements are constrained either by soft tissues (muscles, fat, etc. = soft-tissue restraint) or by bone (olecranon = bony restraint).

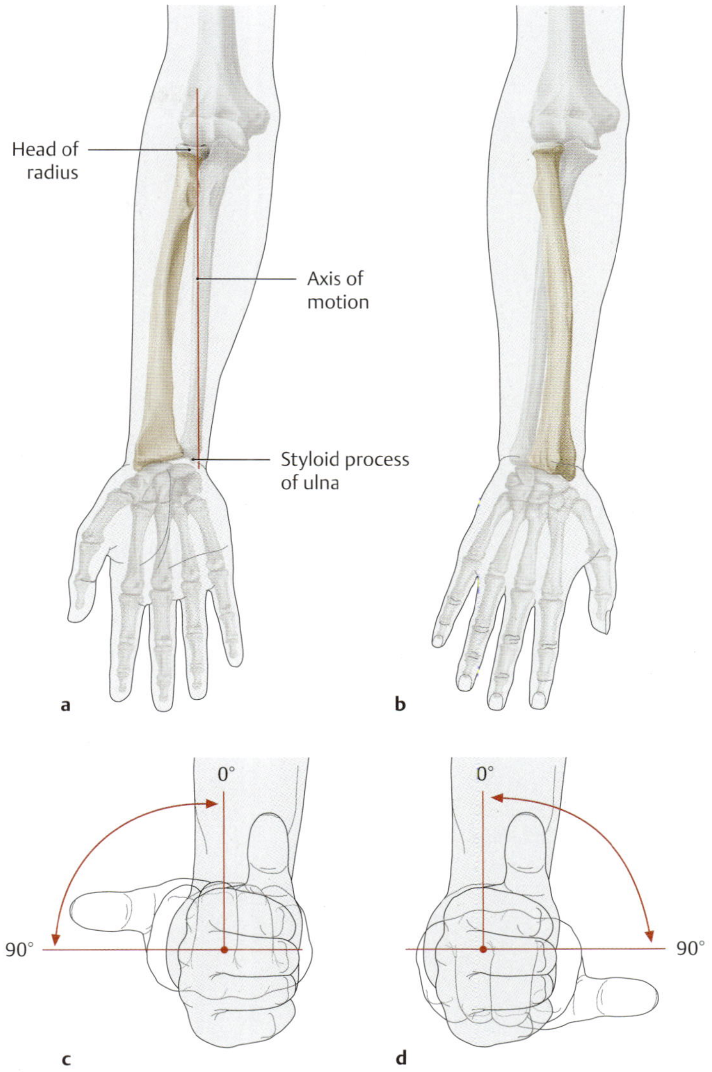

D Range and axis of pronation/supination of the right hand

The neutral (0°) position of the hand and forearm is also called semipronation. The axis of pronation/supination extends through the head of the radius and the styloid process of the ulna.

- **a** Supination (the radius and ulna are parallel to each other).
- **b** Pronation (the radius crosses over the ulna).
- **c** Supination of the hand with the elbow flexed, viewed from the front (the palmar surface of the hand is up).
- **d** Pronation of the hand with the elbow flexed, viewed from the front (the palmar surface of the hand is down).

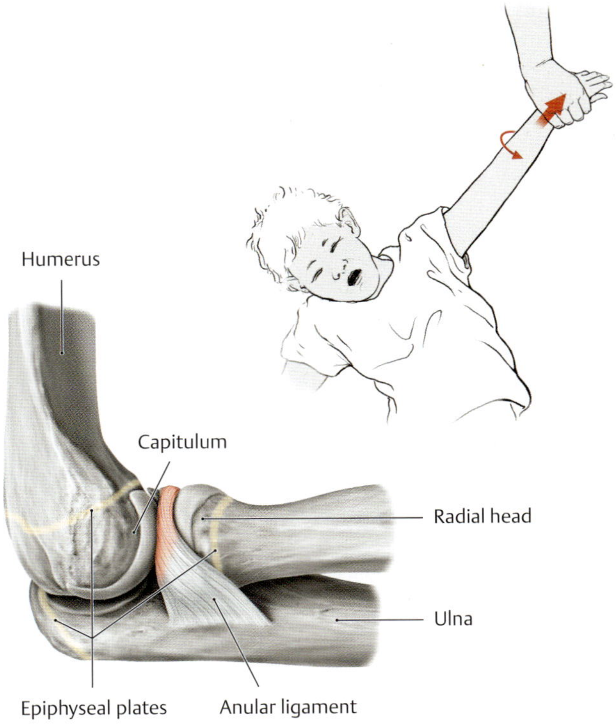

E Pulled elbow, or nursemaid's elbow

Pulled elbow is a very common injury in children (5- to 7-year-olds; with increasing age, ligaments become stronger, reducing the risk of injuries). The head of the radius slips under the anular ligament (radial head subluxation), which results from a sudden pull on a child's arm. As a result, the annular ligament gets stuck between the radius and the capitulum, the elbow joint is locked in a slighly bent position, the arm remains turned inward (pronated position). Due to subluxation and accompanying pain, the child keeps the arm immobile and allows it to hang downward, which makes the arm appear to be paralyzed (pseudoparesis; so-called Chassaignac's paralysis). The clinical findings and the required X-ray examination in two planes to rule out any bone injuries (epiphyseal fracture of the head of the radius) confirm the diagnosis. After reduction, during which the bent elbow joint is extended while firmly supinating the wrist, the child is symptom-free within minutes.

1.25 Overview of the Ligaments of the Hand

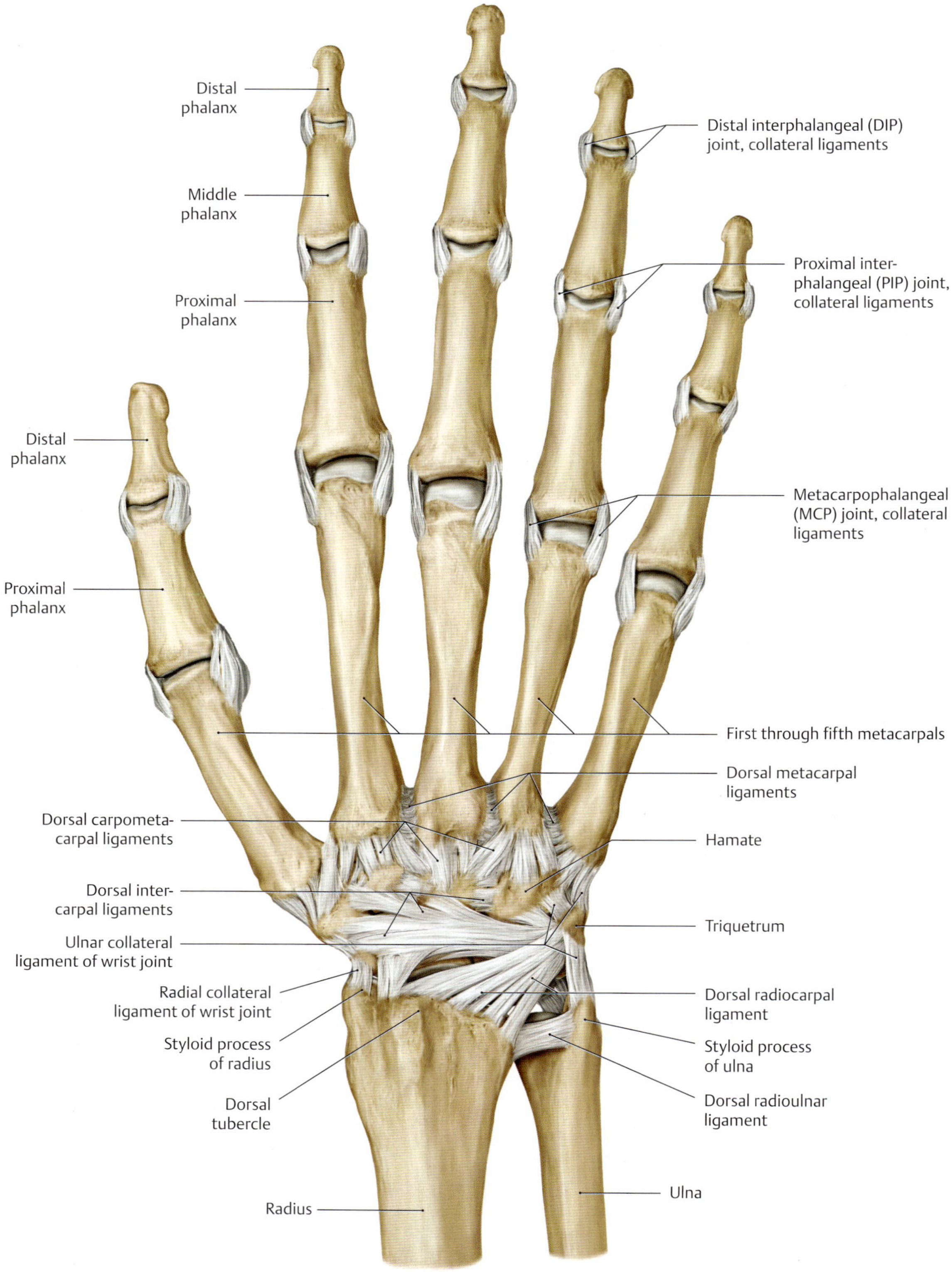

a Dorsal view

A The ligaments of the right hand

Carpal ligaments connect adjacent bones, limiting excess movement and providing stability to the wrist. Their courses vary. They are closely interwoven, and so they are difficult to dissect. The so-called extrinsic carpal ligaments, which run closer to the surface, are interwoven with the articular capsule and primarily act to stabilize it. Deeper, the so-called intrinsic carpal ligaments (see p. 300), divide the joint space into individual compartments.

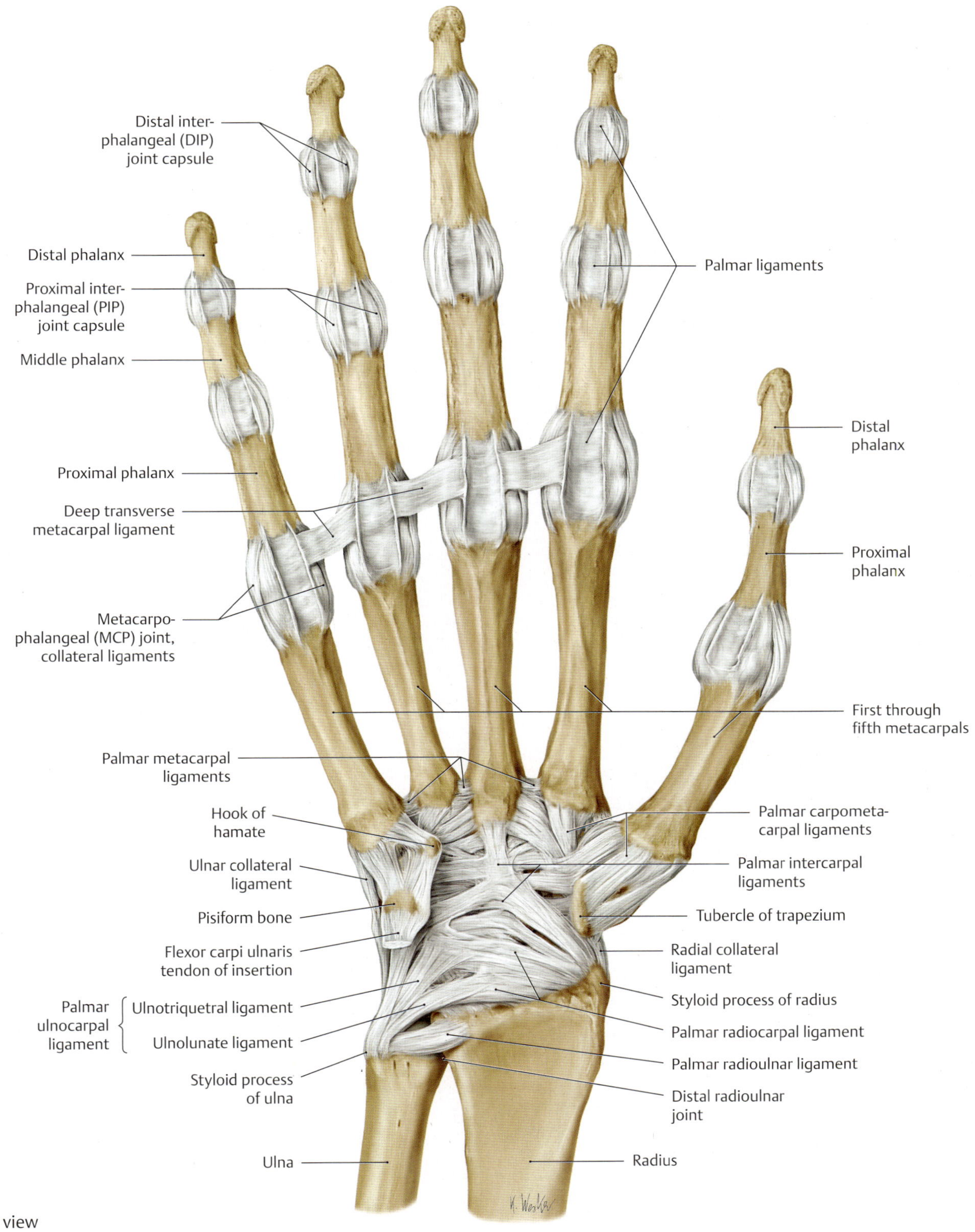

b Palmar view

Aside from the common classification as extrinsic and intrinsic, the carpal ligaments are also classified according to their location and arrangement:

- The ligaments between the forearm and carpal bones (radiocarpal and ulnocarpal ligaments, collateral ligaments).
- The ligaments between individual carpal bones (intercarpal ligaments).
- The ligaments between the carpal and metacarpal bones (carpometacarpal ligaments).
- The ligaments between the bases of the metacarpals (metacarpal ligaments).

1.26 Intrinsic Ligaments of the Hand, Compartments of the Joint, and Ulnocarpal Complex

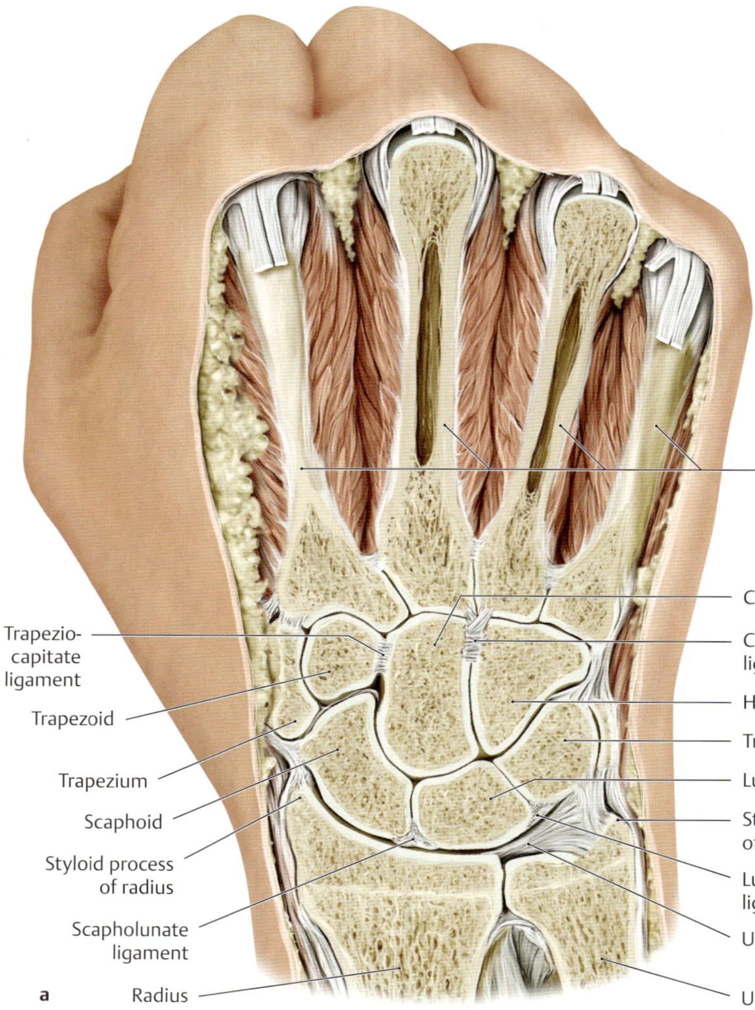

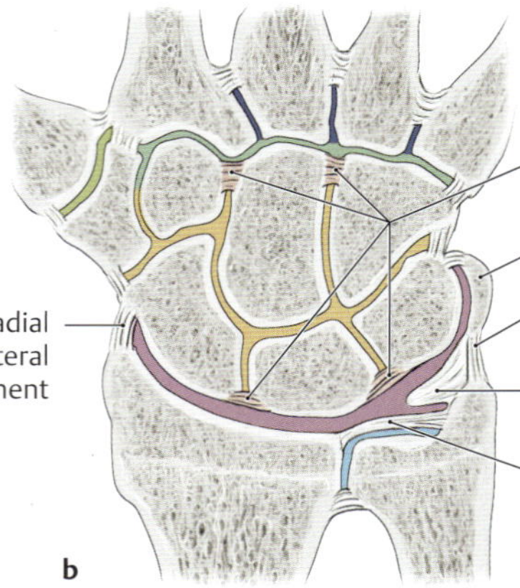

c

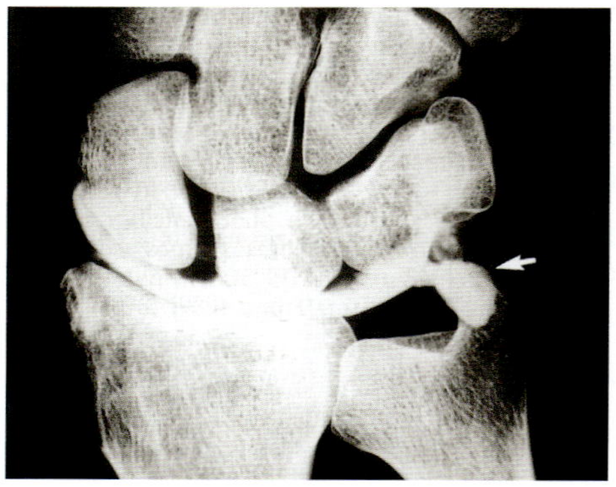

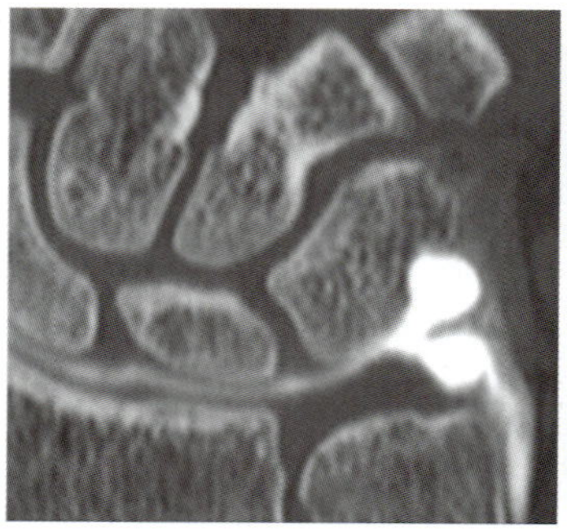

d

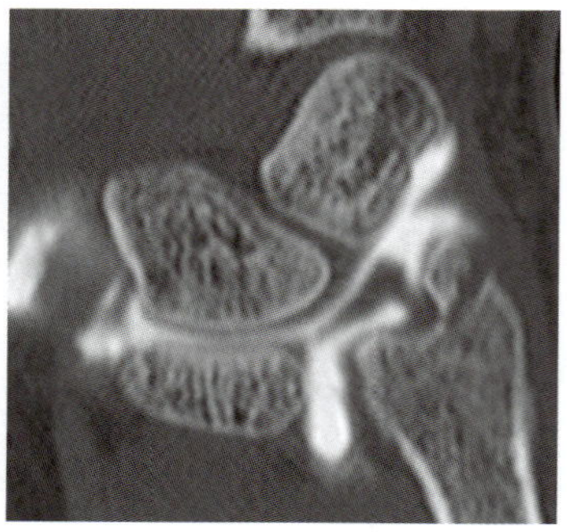

e

A Interosseous ligaments and compartments within the carpus

a Coronal section of the right carpus, dorsal view (drawing based on a specimen from the Anatomical Collection of the University of Kiel); **b** schematic representation of compartments (dorsal view of right hand); **c** arthrography of the radiocarpal joint (courtesy of Dr. J. Koebke, Anatomical Institute, University of Cologne); **d** and e CT arthrography of the radiocarpal joint (from Bohndorf K, Imhof H, Fischer W. Radiologische Diagnostik der Knochen und Gelenke. 2nd ed. Stuttgart: Thieme; 2006): **d** intact compartment; **e** lesion of the ulnocarpal disk with contrast agent crossing over into the radioulnar joint.

In addition to the extrinsic ligaments of the carpus (see p. 298), which reinforce the articular capsule, there are intrinsic ligaments. These interosseous ligaments, along with the ulnocarpal disk (triangular disk), divide the intra-articular space into compartments, some of which are completely enclosed. These compartments include (see **b**)

- distal radioulnar joint,
- radiocarpal joint,
- medial carpal compartment,
- carpometacarpal joint,
- intermetacarpal compartment, and
- thumb saddle joint.

Knowledge of these compartments is clinically significant for conducting and interpreting arthrographies (**c–e**).

The main intrinsic interosseous ligaments are those of the distal carpal row (capitohamate and trapeziocapitate ligaments), the proximal carpal row (lunotriquetral and scapholunate ligaments) (see **a**), and the ulnocapral disk, the most important structure of the ulnocarpal complex (see **b**). These ligament structures often succumb to degeneration or are affected in cases of carpal injuries. Degenerative changes in the ulnocarpal disk can already be detected in people in their 20s, but degeneration of the interosseous ligaments of the proximal carpal row are more commonly seen in older patients (in 30% of all cases).

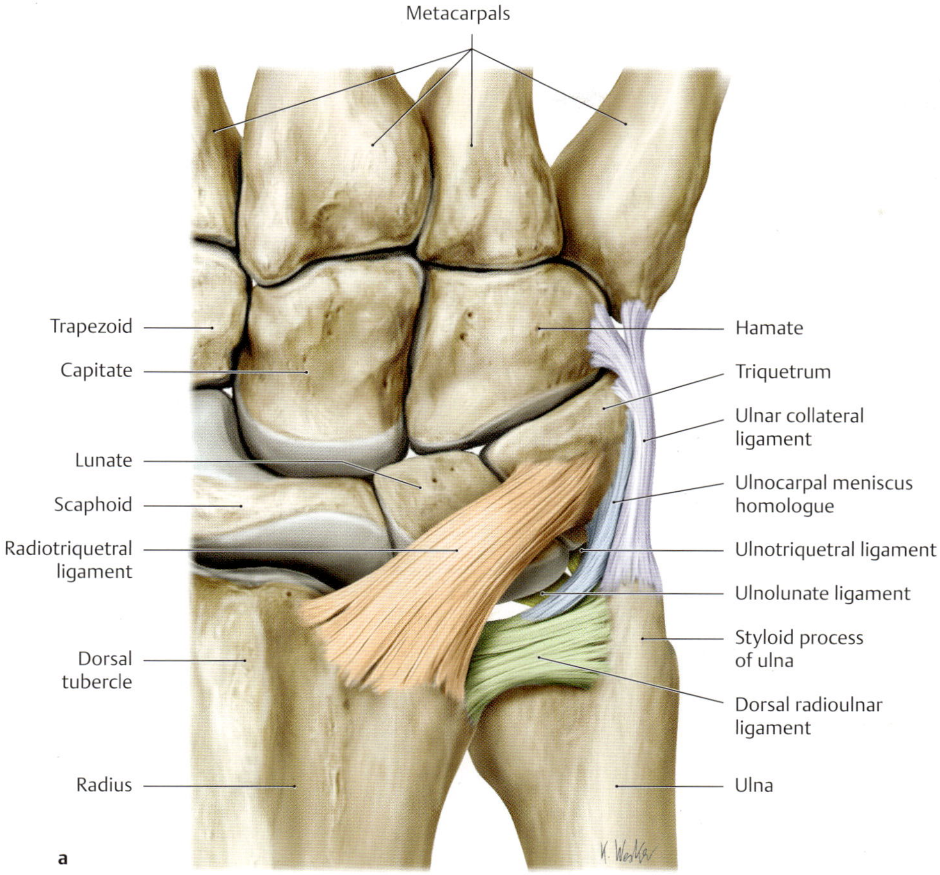

a

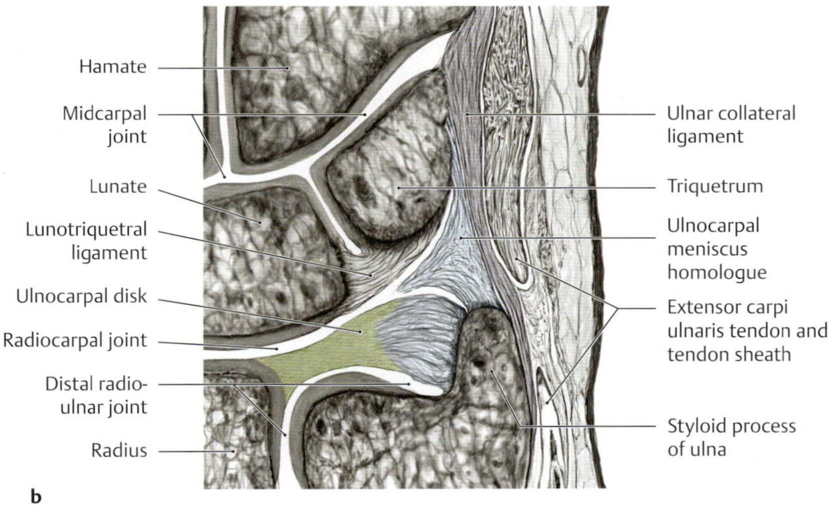

b

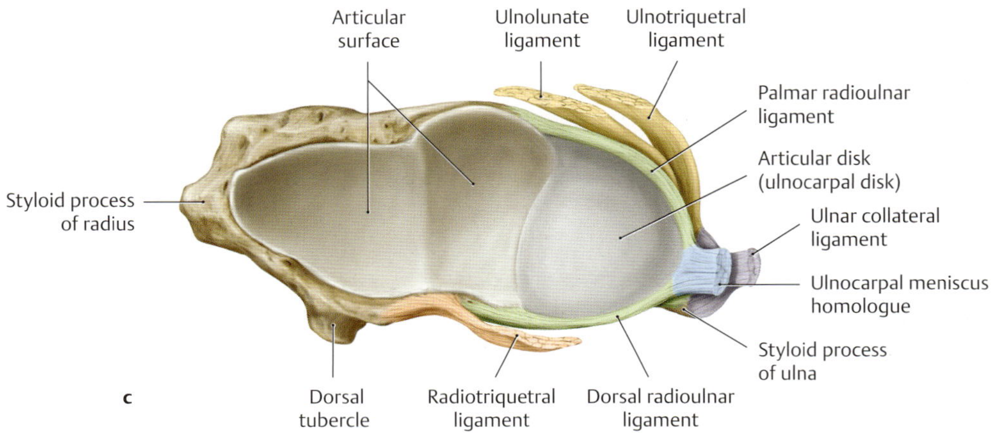

c

B The ulnocarpal complex

a Ulnocarpal complex of the right hand, dorsal view; **b** schematic representation of a histologic specimen of an ulnocarpal complex (from Schmidt HM, Lanz U. Chirurgische Anatomie der Hand. 2nd ed. Stuttgart: Thieme; 2003); **c** ulnocarpal complex of the right hand, distal view.

The ulnocarpal complex, which appears triangular in shape (synonym: triangular fibrocartilage complex, TFCC), is a combination of ligaments and disks and serves as a connection between the distal ulna, the distal radioulnar joint, and the proximal carpal row. In the case of injuries to the ulnocarpal complex, patients usually report ulnar-sided wrist pain. Functionally, the ulnocarpal complex is divided into

- Ulnocarpal disk (triangular disk)
- Dorsal and palmar radioulnar ligaments
- Ulnolunate and ulnotriquetral ligaments
- Ulnocarpal meniscus
- Ulnar collateral ligament
- Radiotriquetral ligament (a component of the dorsal radiocarpal ligament)

The ulnocarpal disk, which consists of fibrocartilage, extends transversely and is located between the distal ulna and the triquetrum or the lunate. It arises from the distal edge of the radius of the ulnar notch in the hyaline cartilage, and two of its fibrous bands (often) extend both to the styloid process of the ulna and the base of the distal ulna. The outer margins of the disk are attached to the dorsal and palmar radioulnar ligaments. The central and radial parts of the fibrocartilaginous disk are slightly vascular and thus take considerably longer to heal following injuries. Degenerative changes, too, are a common occurrence in this area. The ulnocarpal disk is not to be confused with the ulnocarpal meniscus, the collagen fibers of which extend from the dorsal and ulnar edges of the ulnocarpal disk to the palmar side of the triquetrum. The ulnocarpal meniscus bridges the ulnar intra-articular space of the proximal wrist, which is wide in humans and especially in ulnar abduction helps to increase the force-transmitting surface.

1.27 The Carpal Tunnel

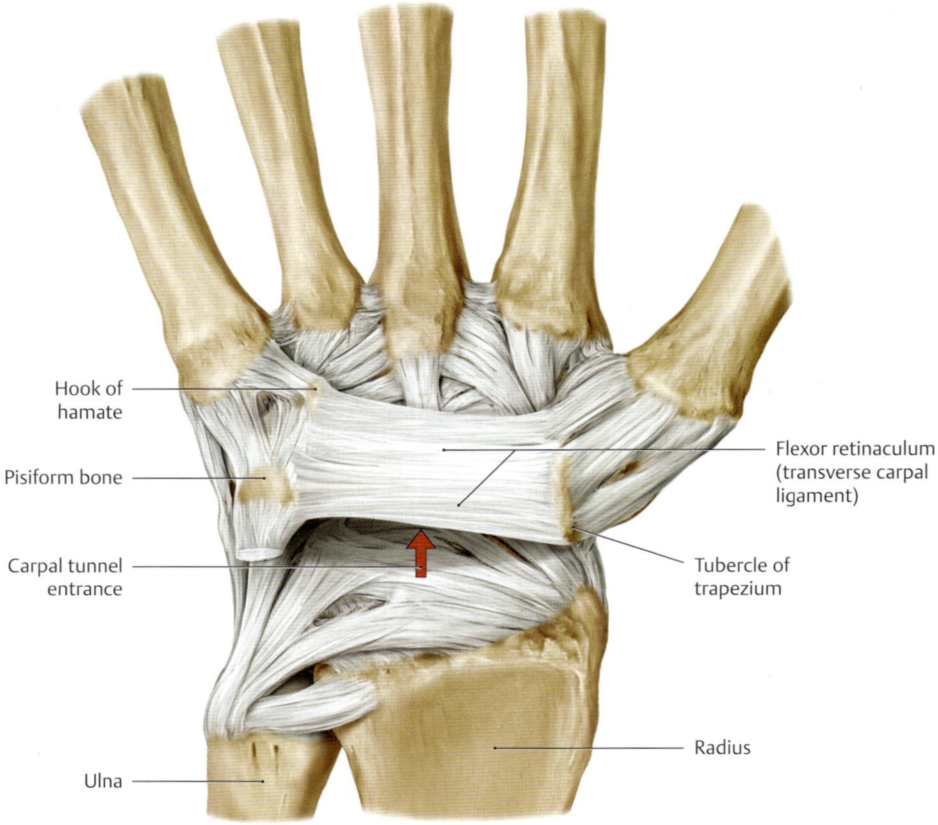

A Flexor retinaculum (transverse carpal ligament) and carpal tunnel, right hand

Anterior view. The bony elements of the wrist form a concave groove on the palmar side (see also **C**), which is closed by the flexor retinaculum (referred to clinically as the transverse carpal ligament) to form a fibro-osseous tunnel called the carpal tunnel or carpal canal. The narrowest part of this canal is located approximately 1 cm beyond the midline of the distal row of carpal bones (see **D**). The cross-sectional area of the tunnel at that site measures only about 1.6 cm^2. The carpal tunnel is traversed by a total of *nine flexor tendons* (enclosed in tendon sheaths and embedded in connective tissue) and the *median nerve* (see p. 386). The tight fit of sensitive neurovascular structures with closely apposed, frequently moving tendons in this narrow space often causes problems when any of the structures swell or degenerate, leading to carpal tunnel syndrome. Narrowing of the tunnel can entrap or compress the median nerve, altering its function both by direct mechanical action and by restricting the blood flow within the nerve sheath. With chronic compression, the median nerve itself begins to degenerate beyond the site of entrapment, causing progressive pain and paresthesia and, ultimately, denervation and wasting of the muscles it serves, particularly the abductor pollicis brevis (see p. 386).

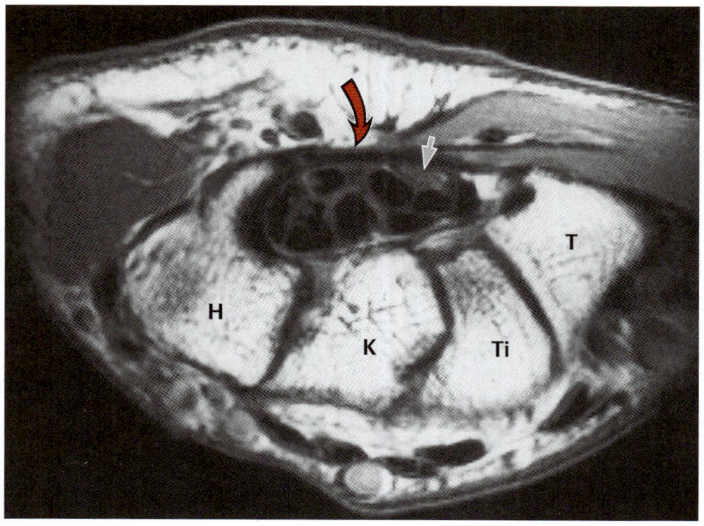

B Axial magnetic resonance image (T1-weighted) of the right hand at the level of the carpal tunnel

Proximal view. The flexor retinaculum (transverse carpal ligament) can be recognized as a band of low signal intensity (red arrow). Just below it toward the radial side is the median nerve (small arrow), whose water and lipid contents cause it to display a higher signal intensity than the superficial and deep flexor tendons. The primary diagnosis of carpal tunnel syndrome is based on clinical signs and electrophysiological measurements such as nerve conduction velocity. While conventional radiographs and CT scans can detect bony causes of the syndrome, magnetic resonance imaging can also demonstrate soft-tissue causes (e.g., edema or swelling of the median nerve, fibrosis, neuroma).

H = Hamate
K = Capitate
T = Trapezium
Ti = Trapezoid

(From: Vahlensieck M, Reiser M. MRT des Bewegungsapparates. 4th ed. Stuttgart: Thieme; 2014)

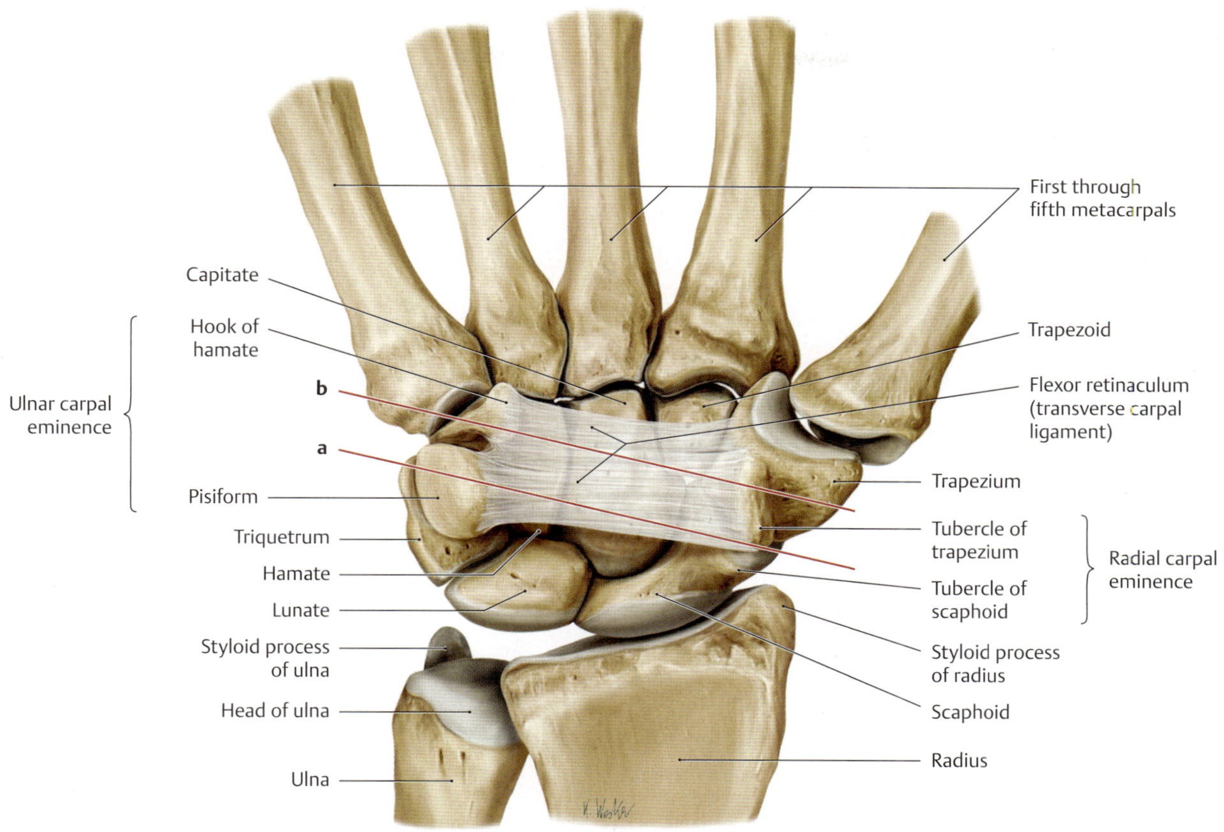

C Bony boundaries of the carpal tunnel of the right hand
Anterior view. The carpal bones form a convex arch on the dorsal side of the wrist and a concave arch on the palmar side. This creates a carpal tunnel on the palmar side, which is bounded by bony elevations on the radial and ulnar sides (the radial and ulnar carpal eminences). The tubercles of the trapezium and scaphoid form the palpable eminence on the radial side, while the hook of the hamate and the pisiform bone form the eminence on the ulnar side. Stretched between them is the flexor retinaculum (transverse carpal ligament), which closes the carpal tunnel on the palmar side (the planes of section marked **a** and **b** correspond to the cross sections in **D**).

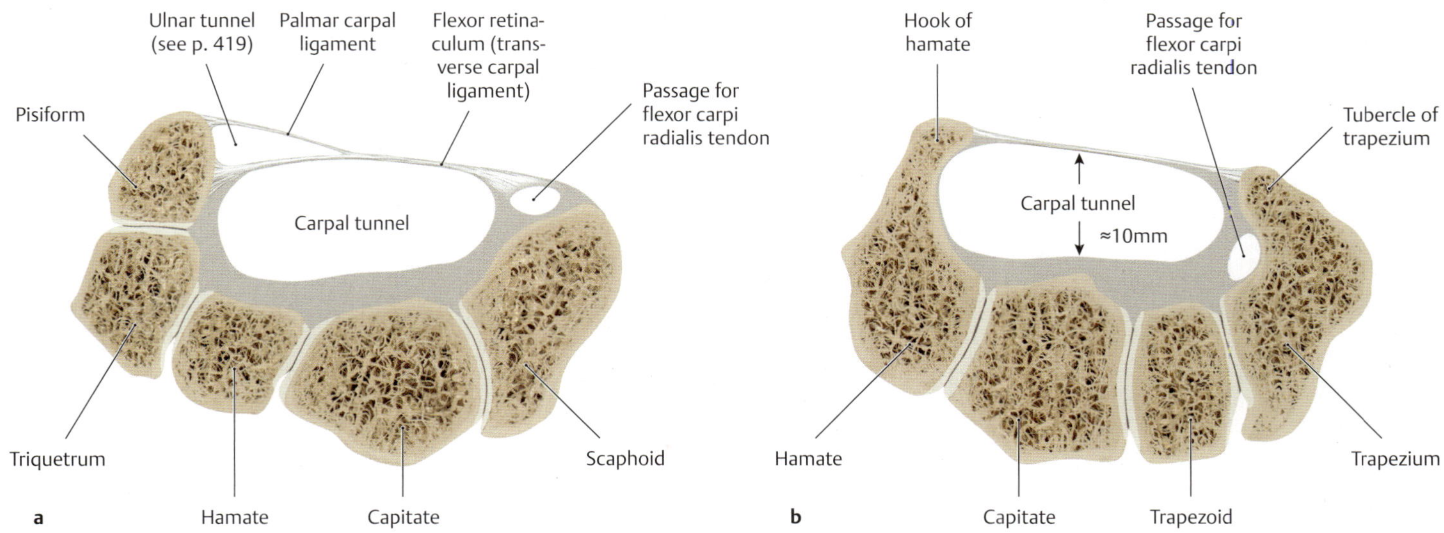

D Cross sections through the carpal tunnel
a Cross section through the proximal part of the carpal tunnel (plane **a** in **C**).
b Cross section through the distal part of the carpal tunnel (plane **b** in **C**).

Note: The carpal tunnel is narrowest over the center of the distal row (approximately 10 mm) of carpal bones (**b**) (after Schmidt and Lanz). In all, the carpal tunnel has a median cross-sectional area of approximately 1.6 to 1.7 cm^2.

1.28 The Ligaments of the Fingers

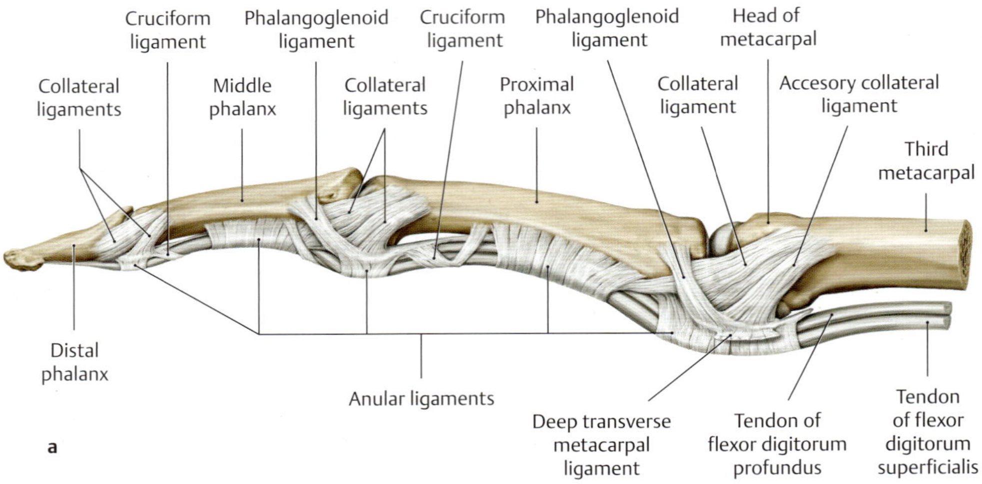

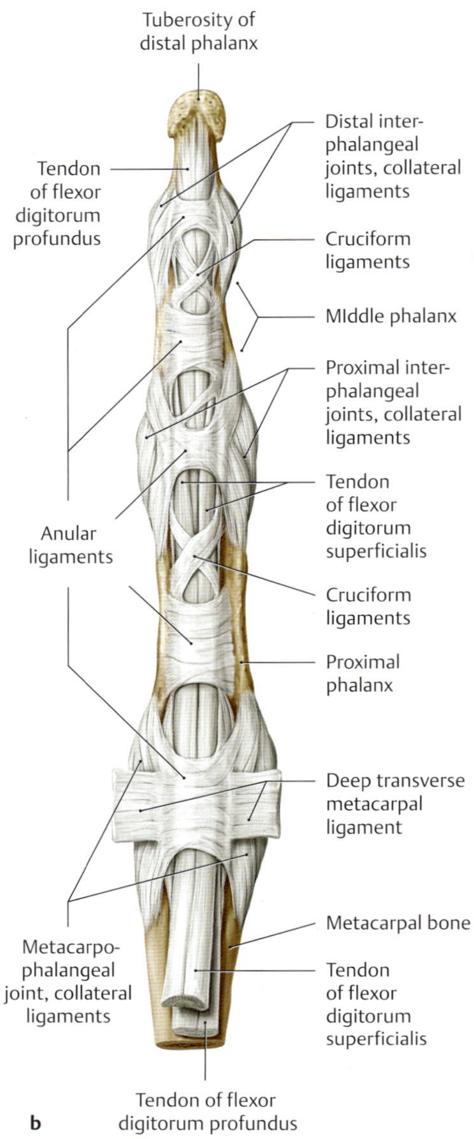

A The joint capsules, ligaments, and digital tendon sheath of the right middle finger
a Lateral view, **b** Palmar view.
The long flexor tendons (flexor digitorum superficialis and profundus) run in a strong, common synovial tendon sheath (not shown here) on the palmar side of the fingers. The *tendon sheaths* are guide mechanisms that allow for frictionless gliding of the long flexor tendons. The outer fibrous layer of the tendon sheaths, the stratum fibrosum, is strengthened by *anular ligaments* and *cruciform ligaments* (see **B**), which also bind the sheaths to the palmar surface of the phalanx and prevent palmar deviation of the sheaths during flexion. The gaps between the anular and cruciform ligaments are necessary to allow flexion of the fingers (see also p. 358, Musculature: Topographical Anatomy).

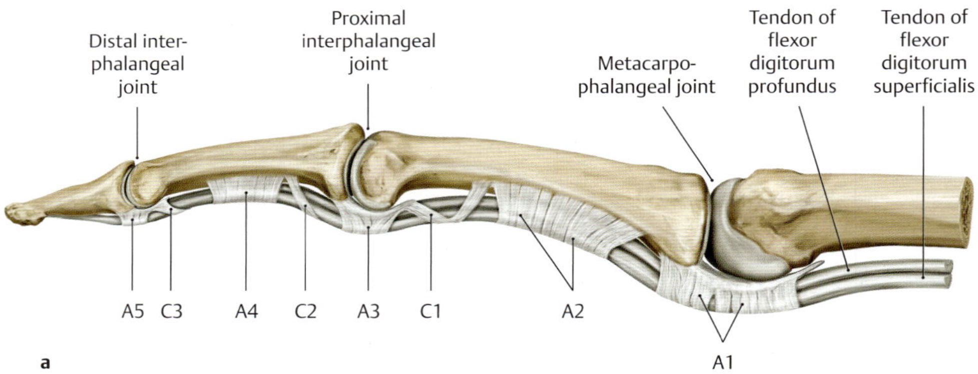

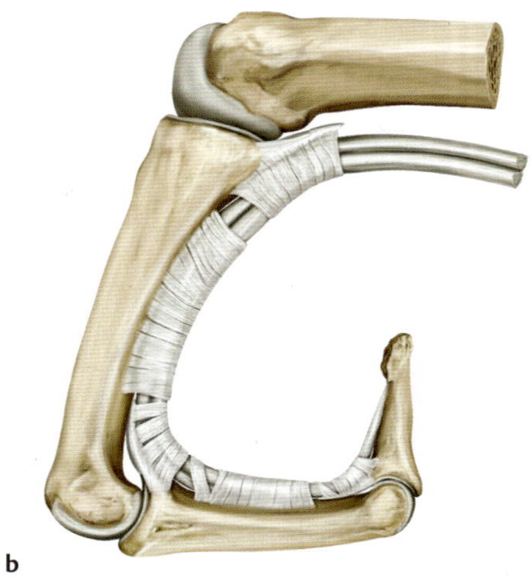

B Ligaments reinforcing the digital tendon sheath
a Lateral view in extension, **b** lateral view in flexion.
A1–5 = anular ligaments, C1–3 = cruciate ligaments.

- First anular ligament (A1): at the level of the metacarpophalangeal joint
- Second anular ligament (A2): on the shaft of the proximal phalanx
- Third anular ligament (A3): at the level of the proximal interphalangeal joint
- Fourth anular ligament (A4): on the shaft of the middle phalanx
- Fifth anular ligament (A5): at the level of the distal interphalangeal joint

The cruciform ligaments are highly variable in their course.

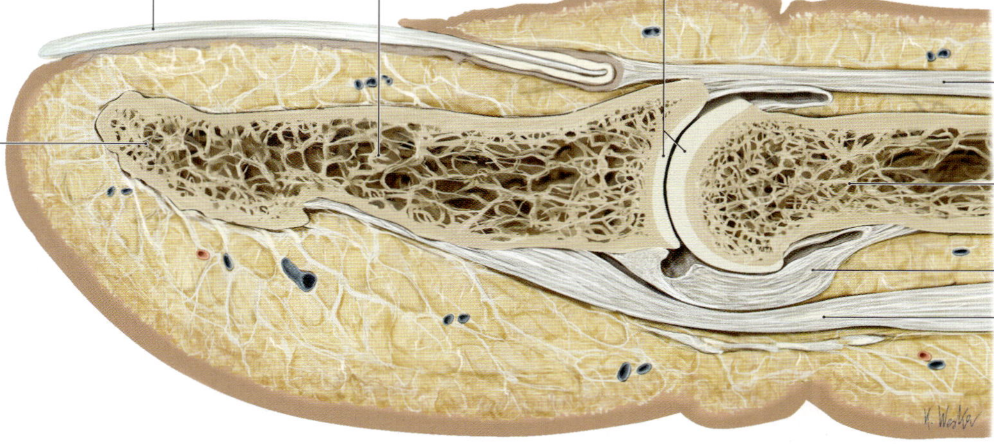

C Longitudinal section through the distal part of a finger

In the metacarpophalangeal joint, as well as the proximal and distal interphalangeal joints, the palmar articular surfaces of the phalanges are enlarged proximally by a fibrocartilaginous plate called the palmar ligament (volar plate). The palmar ligaments also form the floor of the digital tendon sheaths at these locations.

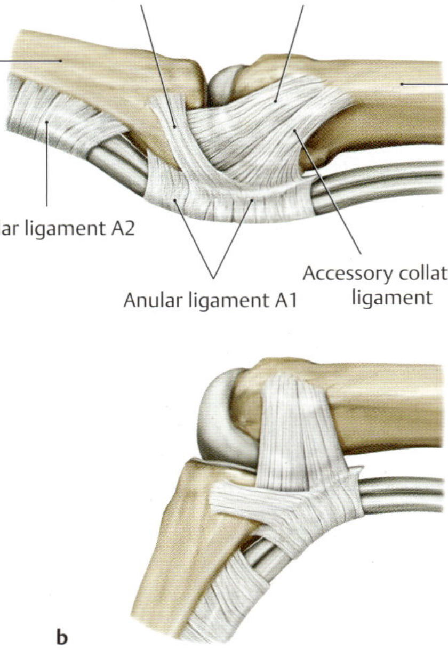

D The capsule and ligaments of the metacarpophalangeal joint

a Extension, **b** flexion. Lateral view.

Note: The collateral ligament is lax in extension and taut in flexion. For this reason, the finger joints should always be placed in a "functional position" (e.g., with the metacarpophalangeal joints flexed approximately 50 to 60°, see p. 309) if the hand is to be immobilized (e.g., in a cast) for a long period of time. If this is not done and the finger joints remain extended for a prolonged period, the collateral ligaments will shorten and create an extension deformity after the cast is removed. The accessory collateral ligament and phalangoglenoid ligament are taut in both flexion and extension and act mainly as restraints to limit extension.

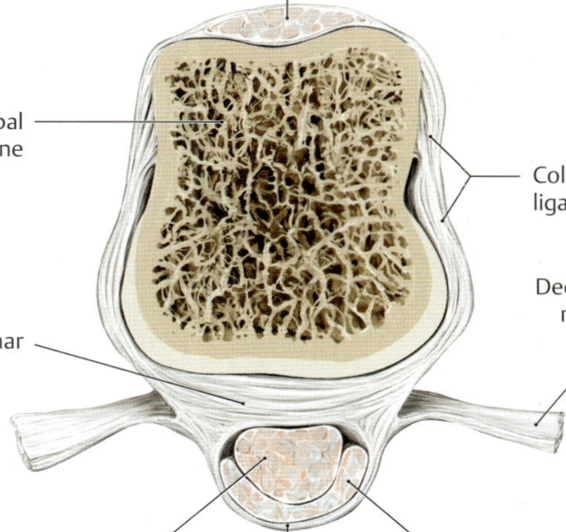

E Cross section through the head of the third metacarpal of the right hand

Proximal view. At the level of the second through fifth metacarpal heads, the volar fibrocartilage plates (palmar ligaments) are interconnected by transverse bands, the deep transverse metacarpal ligaments. By binding the palmar ligaments to the A1 anular ligaments (see **B**) of the flexor tendon sheaths, they also strengthen the distal metacarpus and stabilize the transverse metacarpal arch.

1.29 The Carpometacarpal Joint of the Thumb

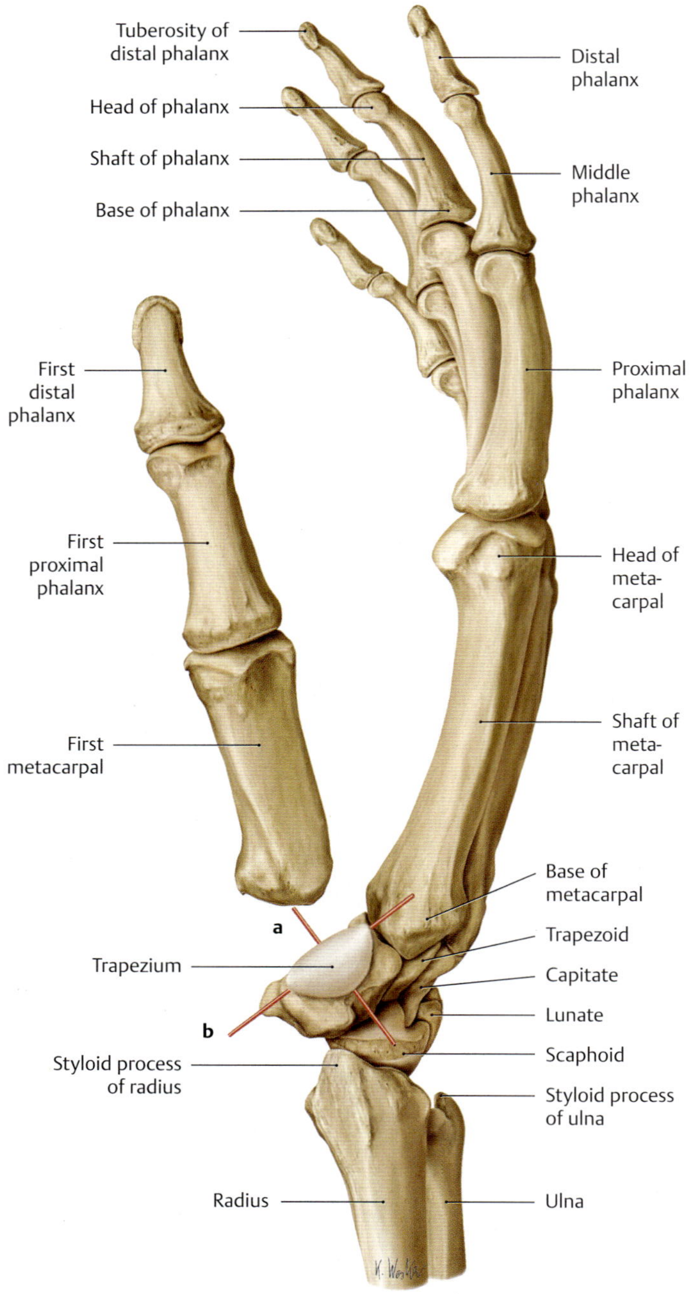

A Axes of motion of the carpometacarpal joint of the thumb
Skeleton of the right hand, radial view. The first metacarpal bone has been moved slightly distally to facilitate orientation. The saddle-shaped articular surfaces of the trapezium and first metacarpal allow movements about two cardinal axes:

- An abduction/adduction axis (**a**)
- A flexion/extension axis (**b**)

While the axis for abduction/adduction runs approximately on a dorsopalmar line, the axis for flexion/extension runs transversely through the sellar limb of the trapezium. When the thumb is moved toward the small finger (opposition), a rotary movement takes place about a longitudinal axis through the first metacarpal bone (third degree of freedom). This oppositional movement of the thumb—essential for precision grasping movements of the hand—is made possible by the natural mismatch of the articular surfaces (see **F**).

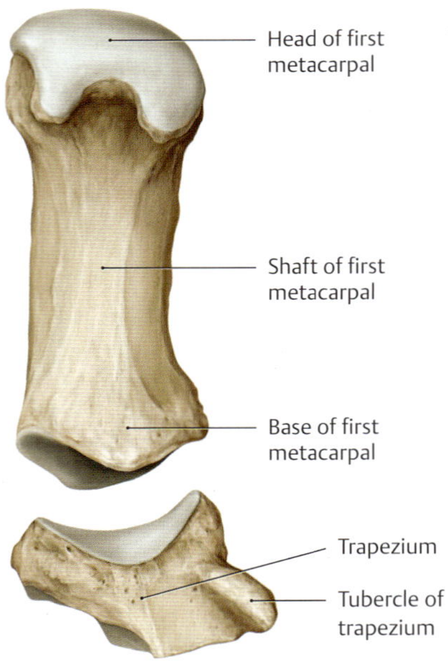

B Articulating surfaces of the carpometacarpal joint of the thumb
Palmar-ulnar view. The articular surface of the trapezium is convex in the dorsopalmar direction and concave in the radioulnar direction. This is opposite to the curvatures found in the corresponding articular surface of the first metacarpal bone.

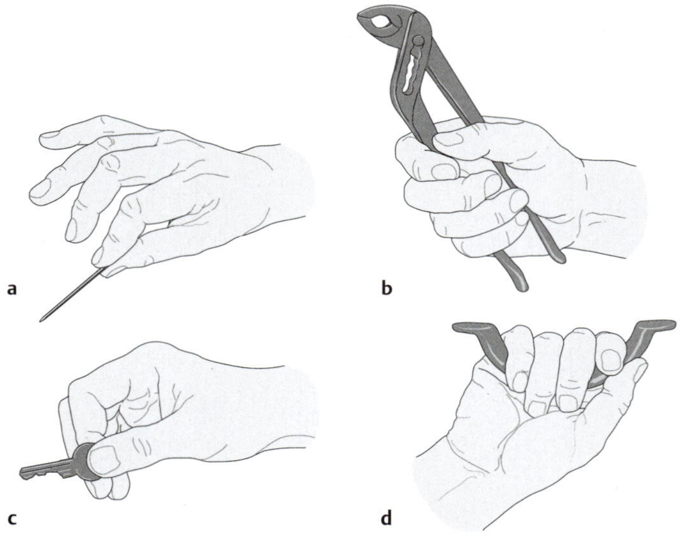

C Types of grip
Normal hand actions can be reduced to four primary types of grip:
a Pinch or precision grip, **b** power grip, **c** key grip, **d** hook grip.
The clinical examination should include function testing of the hand, giving particular attention to disturbances of fine motor skills and gross strength. It is important, for example, to evaluate the pinch grip and key grip, the pinch grip between the thumb and index finger having fundamental importance for the function of the hand. This is why, in issues pertaining to workers' compensation, loss of the thumb or index finger is considered to have a more serious impact on occupational capacity than the loss of the other long fingers.

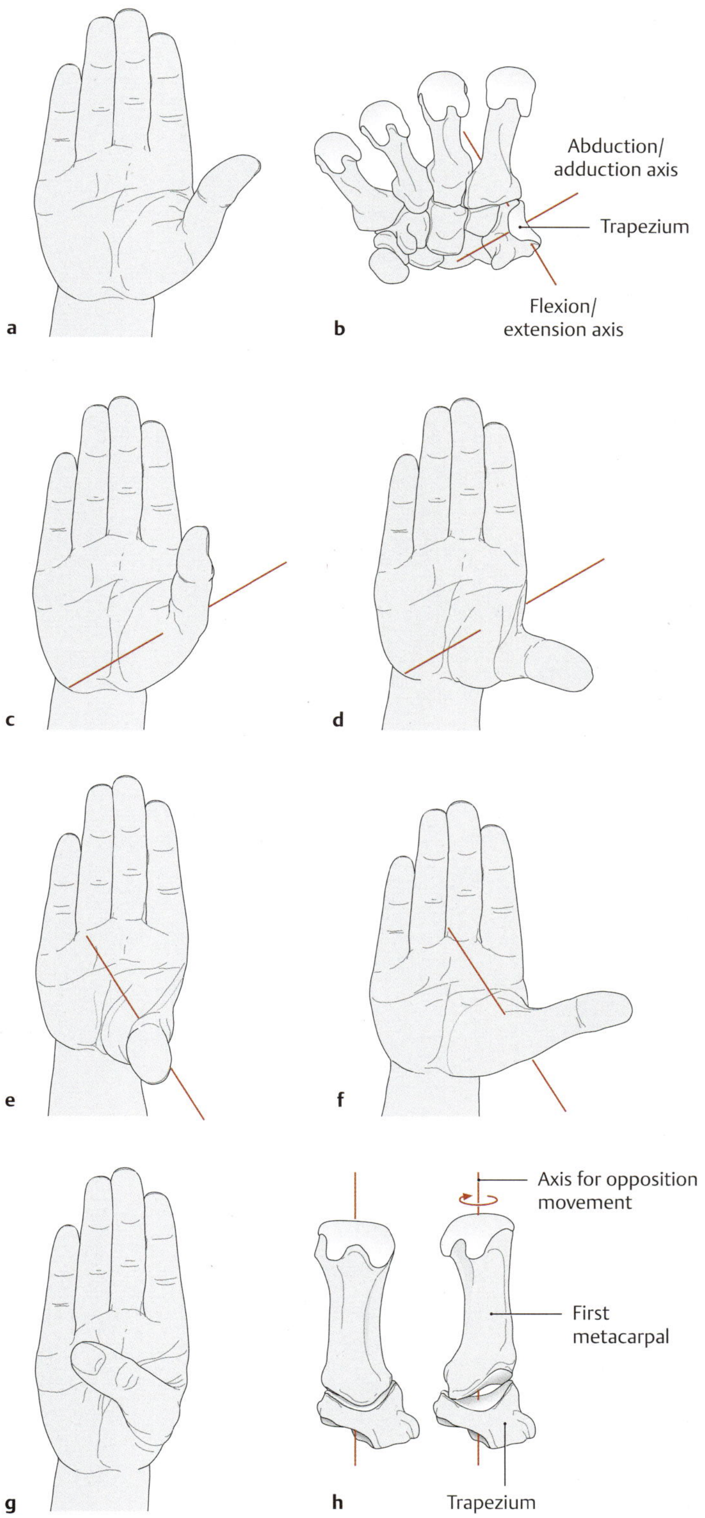

D Movements in the carpometacarpal joint of the thumb
Right hand. Palmar view.

- **a** The neutral (0°) position.
- **b** Axes of motion in the carpometacarpal joint of the thumb.
- **c** Adduction.
- **d** Abduction.
- **e** Flexion.
- **f** Extension.
- **g** Opposition.
- **h** Axis for opposition of the thumb. As the first metacarpal rotates, its area of contact with the articular surface of the trapezium is greatly diminished (see **F**).

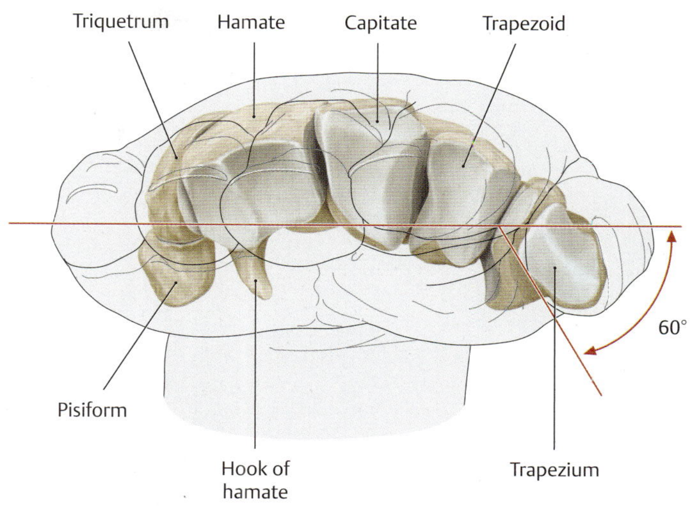

E Relationship of the thumb to the fingers in the neutral (0°) position
Right hand, distal view. Owing to the concave arch of the carpal bones, the scaphoid and trapezium have a markedly radiopalmar orientation. As a result, the metacarpus of the thumb is not placed in line with the other fingers but is rotated approximately 60° toward the palm.

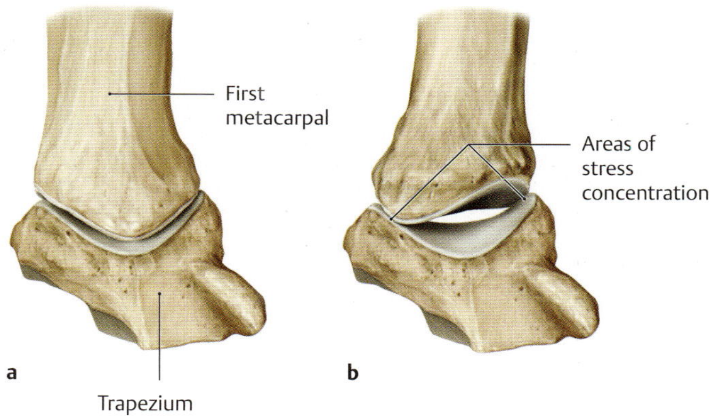

F Rotation-induced incongruity of the carpometacarpal joint during opposition of the thumb
a Neutral (0°) position, **b** the thumb in opposition.
As a sellar (saddle-shaped) joint, the carpometacarpal joint of the thumb is subjected to functional stresses that may promote osteoarthritis. The potentially harmful stresses are created by rotation of the first metacarpal bone during opposition of the thumb. When the thumb is maximally opposed, this rotation greatly reduces the surface area available for stress transfer across the joint (contrast this with the large area available in **a**). This concentration of stresses in a localized area predisposes to degenerative changes in the ascending sellar limb of the first metacarpal bone and the articular surface of the trapezium (first carpometacarpal osteoarthritis).

1.30 Movements of the Hand and Finger Joints

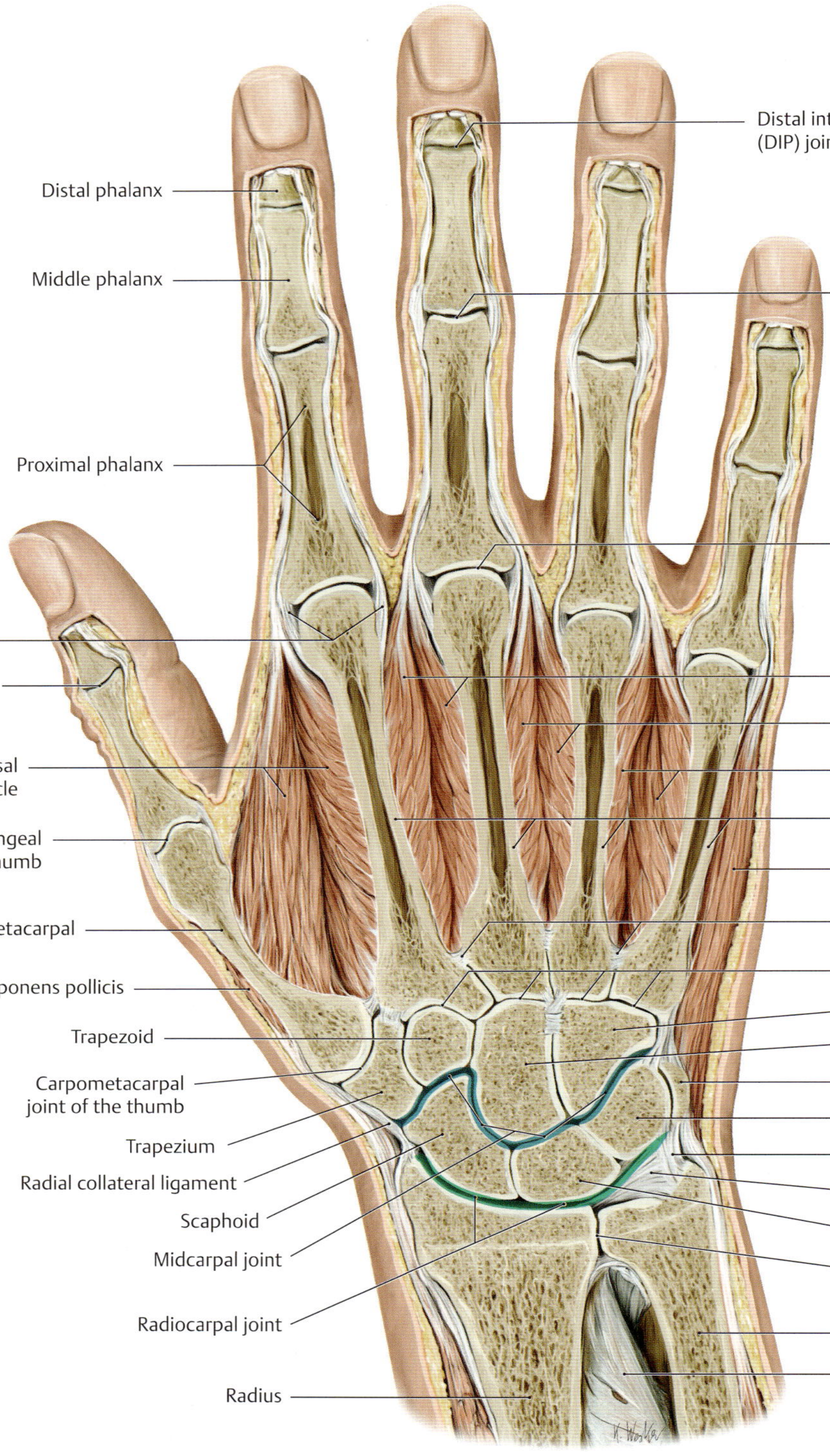

A Transverse section through the right hand
Posterior view. The hand and forearm are connected at the **radiocarpal and midcarpal joints** (both indicated by blue lines in the drawing). Morphologically, the *radiocarpal joint* is an ovoid or ellipsoidal joint, while the *midcarpal joint* is an interdigitating hinge joint (with an approximately S-shaped joint space between the proximal and distal rows of carpal bones). Except for the carpometacarpal joint of the thumb, the joints between the distal row of carpal bones and the bases of the metacarpals (the carpometacarpal joints) are *amphiarthroses* (joined by fibrocartilage) that permit very little motion.

The **finger joints** are classified as follows:

- *Metacarpophalangeal joints* between the metacarpal bones and the proximal phalanges (MCP joints, spheroidal type).
- *Proximal interphalangeal joints* between the proximal and middle phalanges (PIP joints, hinge type).
- *Distal interphalangeal joints* between the middle and distal phalanges (DIP joints, hinge type).

Since the *thumb* lacks a middle phalanx, it has only two joints: a metacarpophalangeal joint and an interphalangeal joint (drawing based on a specimen from the Anatomical Collection at the University of Kiel).

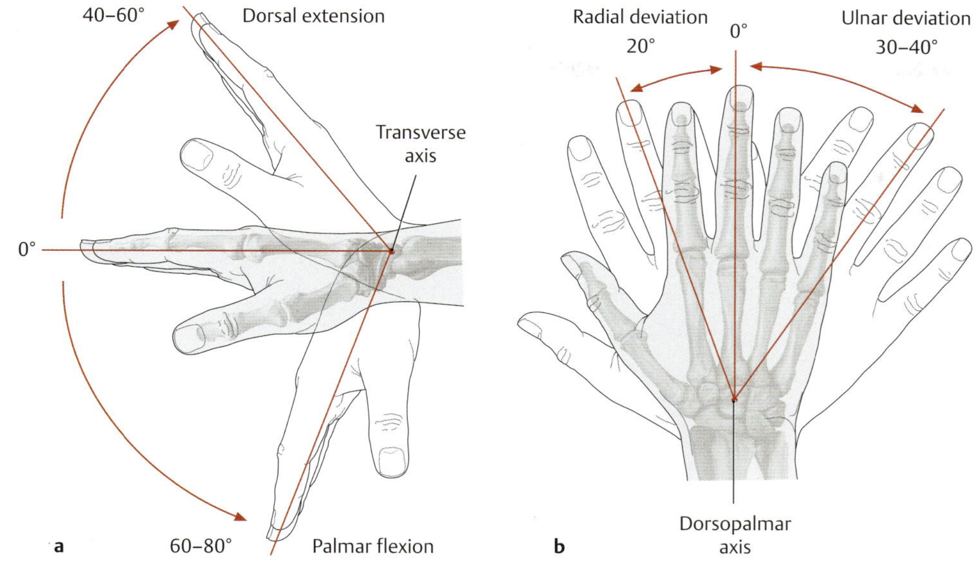

B Movements of the radiocarpal and midcarpal joints

Starting from the neutral (0°) position, palmar flexion and dorsal extension are performed about a transverse axis (**a**), while radial and ulnar deviation occurs about a dorsopalmar axis (**b**). The *transverse* axis runs through the lunate bone for the radiocarpal joint and through the capitate bone for the midcarpal joint. The *dorsopalmar* axis runs through the capitate bone. Thus, while palmar flexion and dorsal extension can occur in both the radiocarpal and midcarpal joints, radial and ulnar deviation can occur only in the radiocarpal joint.

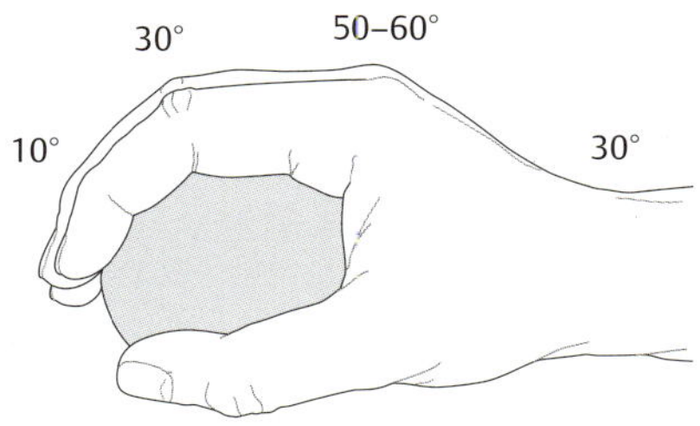

C Functional position of the hand

For postoperative immobilization of the hand, the desired position of the wrist and fingers should be considered when the cast, splint, or other device is applied. Otherwise the ligaments may shorten, and the hand can no longer assume a normal resting position.

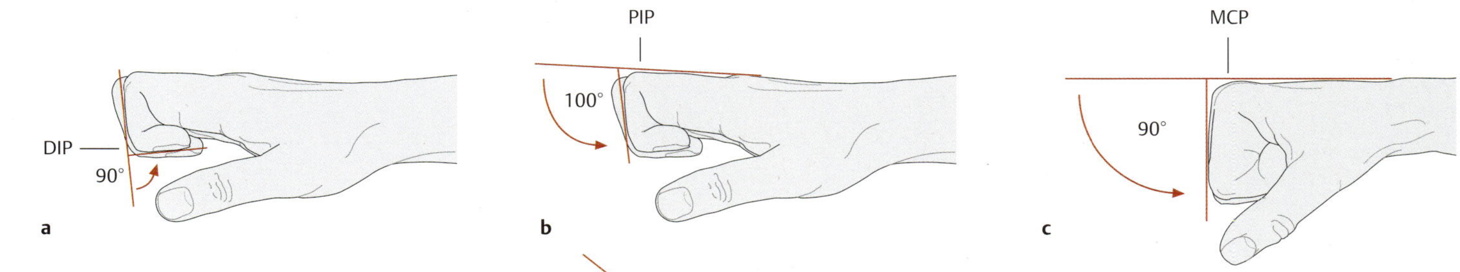

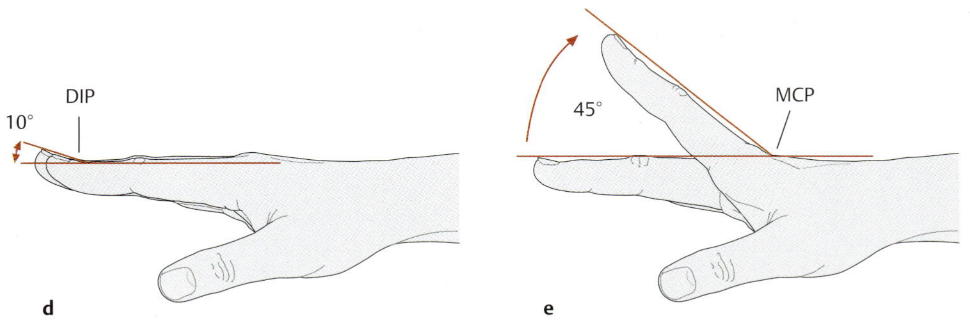

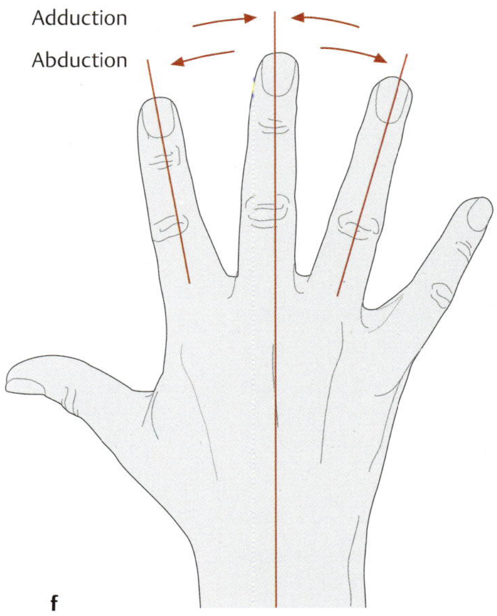

D Range of motion of the finger joints

The proximal interphalangeal (PIP) and distal interphalangeal (DIP) joints are pure hinge joints with only one degree of freedom (flexion extension). The metacarpophalangeal (MCP) joints of the second through fifth fingers are shaped like spheroidal joints with three theretcal degrees of freedom, but rotation is so limited by the collateral ligaments that only two degrees of freedom exist: flexion/extension and abduction/adduction.The following specific movements of the finger joints are distinguished:

- **a** Flexion in the DIP joint.
- **b** Flexion in the PIP joint.
- **c** Flexion in the MCP joint.
- **d** Extension in the DIP joint.
- **e** Extension in the MCP joint.
- **f** Abduction and adduction in the MCP joints (spreading the fingers apart and bringing them together about a dorsopalmar axis through the heads of the metacarpals).

Abduction/adduction movements are described in relation to the middle finger: all movements away from the middle finger are classified as abduction, all movements toward the middle finger as adduction.

2.1 Functional Muscle Groups

A Principles used in classification of the muscles of the upper limb

The muscles of the upper limb can be classified according to various criteria. An optimum system for classification should be logical and clear. The following criteria are suitable for classifying muscles:

- Origin
- Topography
- Function
- Innervation

While function and topography in the upper limb are often interrelated (muscles with the same action on a joint are often located close together), muscles that have similar actions in the shoulder region (e.g., muscles of the shoulder joint and shoulder girdle) vary considerably in their location. The following classification (**B**), then, is a compromise between topographical and functional considerations. In section **C**, a different muscle classification system, based on innervation, is presented. The grouping of muscles by the pattern of their innervation reveals features of their embryological and phylogenetic origin and provides clinical insights into the clusters of consequences from damage to particular nerves.

B Functional-topographical classification of the muscles of the upper limb

Muscles of the shoulder girdle

Shoulder girdle muscles that have migrated from the head
- Trapezius
- Sternocleidomastoid
- Omohyoid

Posterior muscles of the trunk and shoulder girdle
- Rhomboid major
- Rhomboid minor
- Levator scapulae

Anterior muscles of the trunk and shoulder girdle
- Subclavius
- Pectoralis minor
- Serratus anterior

Muscles of the shoulder joint

Posterior shoulder muscles
- Supraspinatus
- Infraspinatus
- Teres minor
- Subscapularis
- Deltoid
- Latissimus dorsi
- Teres major

Anterior shoulder muscles
- Pectoralis major
- Coracobrachialis

Muscles of the arm

Posterior arm muscles
- Triceps brachii
- Anconeus

Anterior arm muscles
- Brachialis
- Biceps brachii

Muscles of the forearm

Posterior forearm muscles
- Superficial extensors
  - Extensor digitorum
  - Extensor digiti minimi
  - Extensor carpi ulnaris
- Deep extensors
  - Supinator
  - Abductor pollicis longus
  - Extensor pollicis brevis
  - Extensor pollicis longus
  - Extensor indicis

Anterior forearm muscles
- Superficial flexors
  - Pronator teres
  - Flexor digitorum superficialis
  - Flexor carpi radialis
  - Flexor carpi ulnaris
  - Palmaris longus
- Deep flexors
  - Flexor digitorum profundus
  - Flexor pollicis longus
  - Pronator quadratus

Radial forearm muscles
- Radialis group
  - Brachioradialis
  - Extensor carpi radialis longus
  - Extensor carpi radialis brevis

Muscles of the hand

Metacarpal muscles
- First through fourth lumbricals
- First through fourth dorsal interossei
- First through third palmar interossei

Thenar muscles
- Abductor pollicis brevis
- Adductor pollicis
- Flexor pollicis brevis
- Opponens pollicis

Hypothenar muscles
- Abductor digiti minimi
- Flexor digiti minimi brevis
- Opponens digiti minimi
- Palmaris brevis

C Classification of the muscles of the upper limb by their innervation

Almost all the muscles of the upper limb are innervated by the brachial plexus arising from spinal cord segments C 5–T 1. Exceptions are the trapezius, sternocleidomastoid, and omohyoid; originating in vertebrate evolution as muscles of the head, they are supplied by cranial nerve XI (accessory nerve) and the cervical plexus (ansa cervicalis).

Nerve	Innervated muscles
Accessory nerve	Trapezius Sternocleidomastoid
Ansa cervicalis	Omohyoid
Dorsal scapular nerve	Levator scapulae Rhomboid major Rhomboid minor
Suprascapular nerve	Supraspinatus Infraspinatus
Long thoracic nerve	Serratus anterior
Nerve to the subclavius	Subclavius
Subscapular nerves	Subscapularis (upper and lower) Teres major (upper)
Thoracodorsal nerve	Latissimus dorsi
Medial and lateral pectoral nerves	Pectoralis major Pectoralis minor
Musculocutaneous nerve	Coracobrachialis Biceps brachii Brachialis
Axillary nerve	Deltoid Teres minor
Radial nerve	Triceps brachii Anconeus Supinator Brachioradialis Extensor carpi radialis longus Extensor carpi radialis brevis Extensor digitorum Extensor digiti minimi Extensor carpi ulnaris Extensor pollicis longus Extensor pollicis brevis Extensor indicis Abductor pollicis longus
Median nerve	Pronator teres Pronator quadratus Palmaris longus Flexor carpi radialis Flexor pollicis longus Flexor digitorum profundus (half) Flexor digitorum superficialis Abductor pollicis brevis Opponens pollicis Flexor pollicis brevis (superficial head) First and second lumbricals
Ulnar nerve	Flexor carpi ulnaris Flexor digitorum profundus (half) Palmaris brevis Flexor digiti minimi brevis Abductor digiti minimi Opponens digiti minimi Adductor pollicis Flexor pollicis brevis (deep head) Palmar and dorsal interossei Third and fourth lumbricals

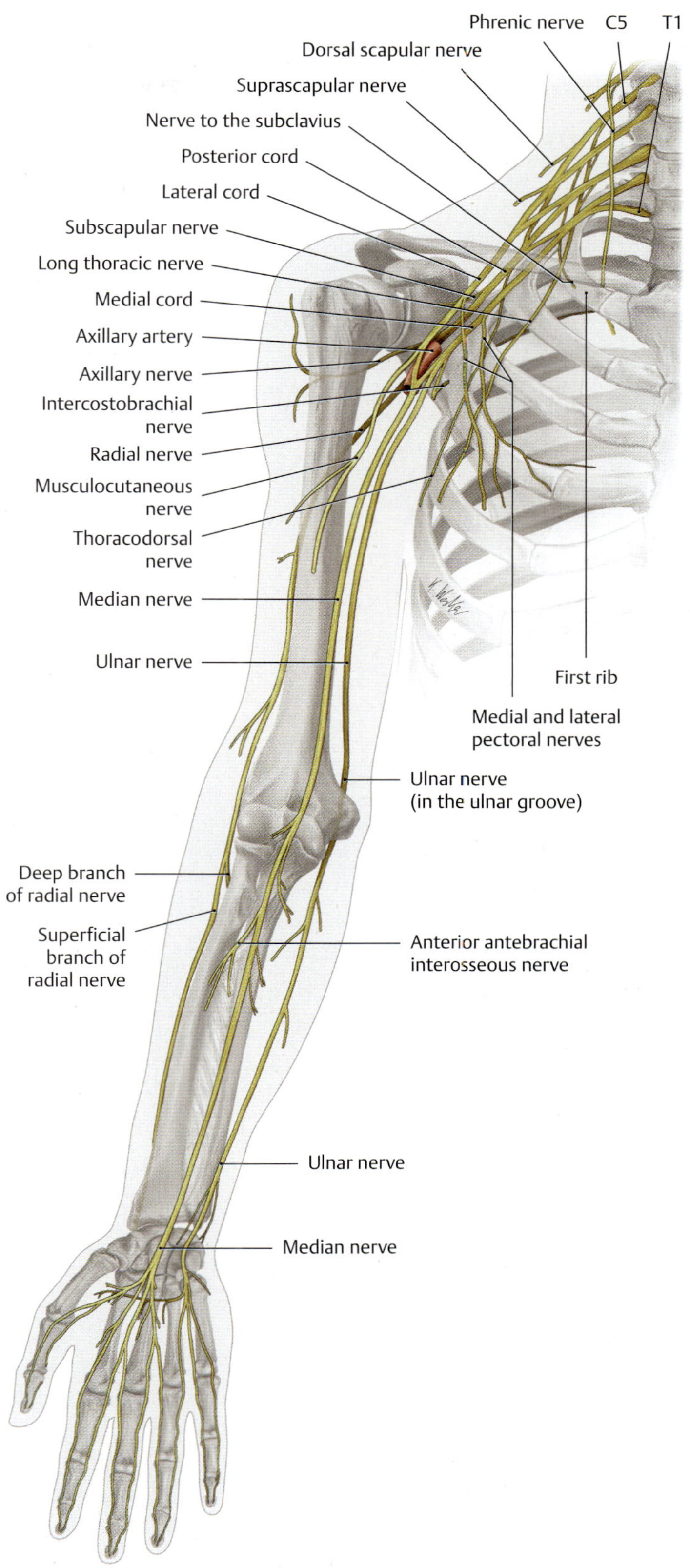

D Overview of the motor branches of the brachial plexus that supply the muscles of the upper limb

With the outgrowth of the limb buds from the trunk during embryonic development, the branches of the brachial plexus follow the genetically determined posterior extensor muscles and anterior flexor muscles. The nerves for the extensors (radial and axillary nerves) arise from the three posterior divisions of the brachial plexus, while the nerves for the flexors (musculocutaneous nerve, ulnar nerve, median nerve) arise from the three anterior divisions of the plexus (see p. 408, Neurovascular Systems: Topographical Anatomy).

2.2 The Muscles of the Shoulder Girdle: Trapezius, Sternocleidomastoid, and Omohyoid

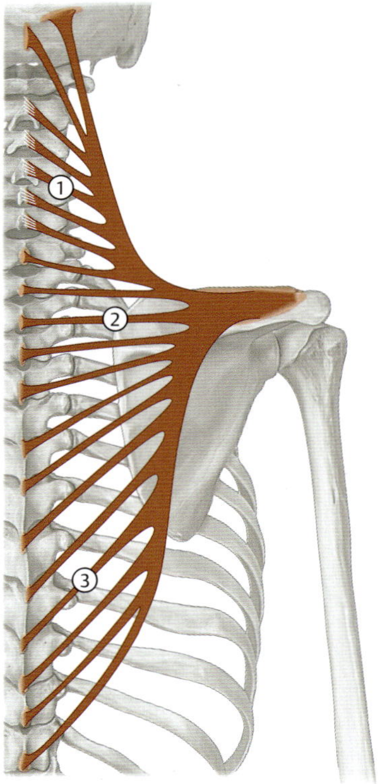

A **Schematic of the trapezius**

Origin:	① Descending part[1]: • Occipital bone (superior nuchal line and external occipital protuberance) • The spinous processes of all cervical vertebrae via the nuchal ligament ② Transverse part: Broad aponeurosis at the level of the T 1–T 4 spinous processes ③ Ascending part: Spinous processes of T 5–T 12
Insertion:	• Lateral third of the clavicle (descending part) • Acromion (transverse part) • Scapular spine (ascending part)
Actions:	• Descending part: – Draws the scapula obliquely upward and rotates the glenoid cavity inferiorly (acting with the inferior part of the serratus anterior) – Tilts the head to the same side and rotates it to the opposite side (with the shoulder girdle fixed) • Transverse part: draws the scapula medially • Ascending part: draws the scapula medially downward (supports the rotating action of the descending part) • Entire muscle: steadies the scapula on the thorax
Innervation:	Accessory nerve (CN XI) and cervical plexus (C 2–C 4)

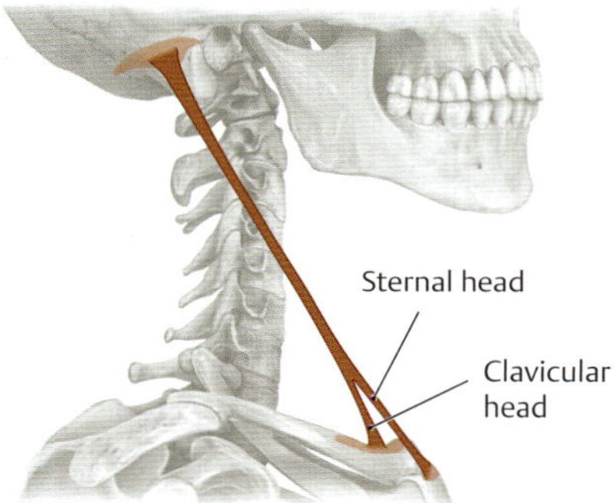

B **Schematic of the sternocleidomastoid**

Origin:	• Sternal head: manubrium • Clavicular head: medial third of the clavicle
Insertion:	Mastoid process and superior nuchal line
Actions:	• Unilateral: – Tilts the head to the same side – Rotates the head to the opposite side • Bilateral: – Extends the head – Assists in respiration when the head is fixed
Innervation:	Accessory nerve (CN XI) and the cervical plexus (C 1–C 2)

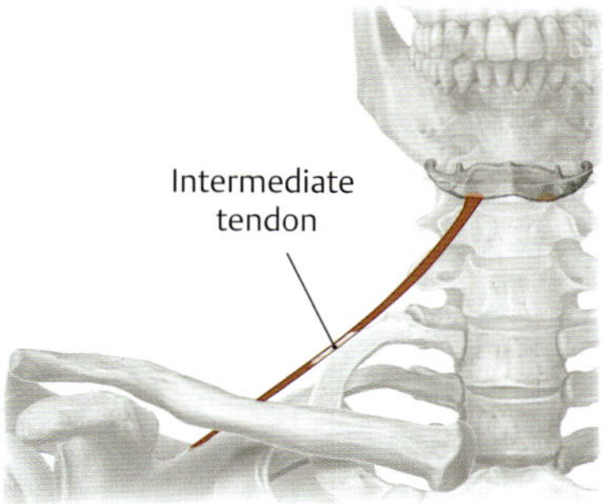

C **Schematic of the omohyoid**

Omohyoid	
Origin:	Superior border of the scapula
Insertion:	Body of the hyoid bone
Actions:	• Depresses (fixes) the hyoid bone • Moves the larynx and hyoid bone downward (for phonation and the final phase of swallowing) • Tenses the cervical fascia with its intermediate tendon and maintains patency of the internal jugular vein
Innervation:	Ansa cervicalis (C 1–C 4) from the cervical plexus

[1] The tables and associated diagrams above are intended to give a systematic overview of the named muscles and their actions, while the drawings on p. 313 are intended to display the muscles as they would appear in a dissection. Not all structures listed in the tables are shown in the drawings on p. 313 because they are not all visible in those views.

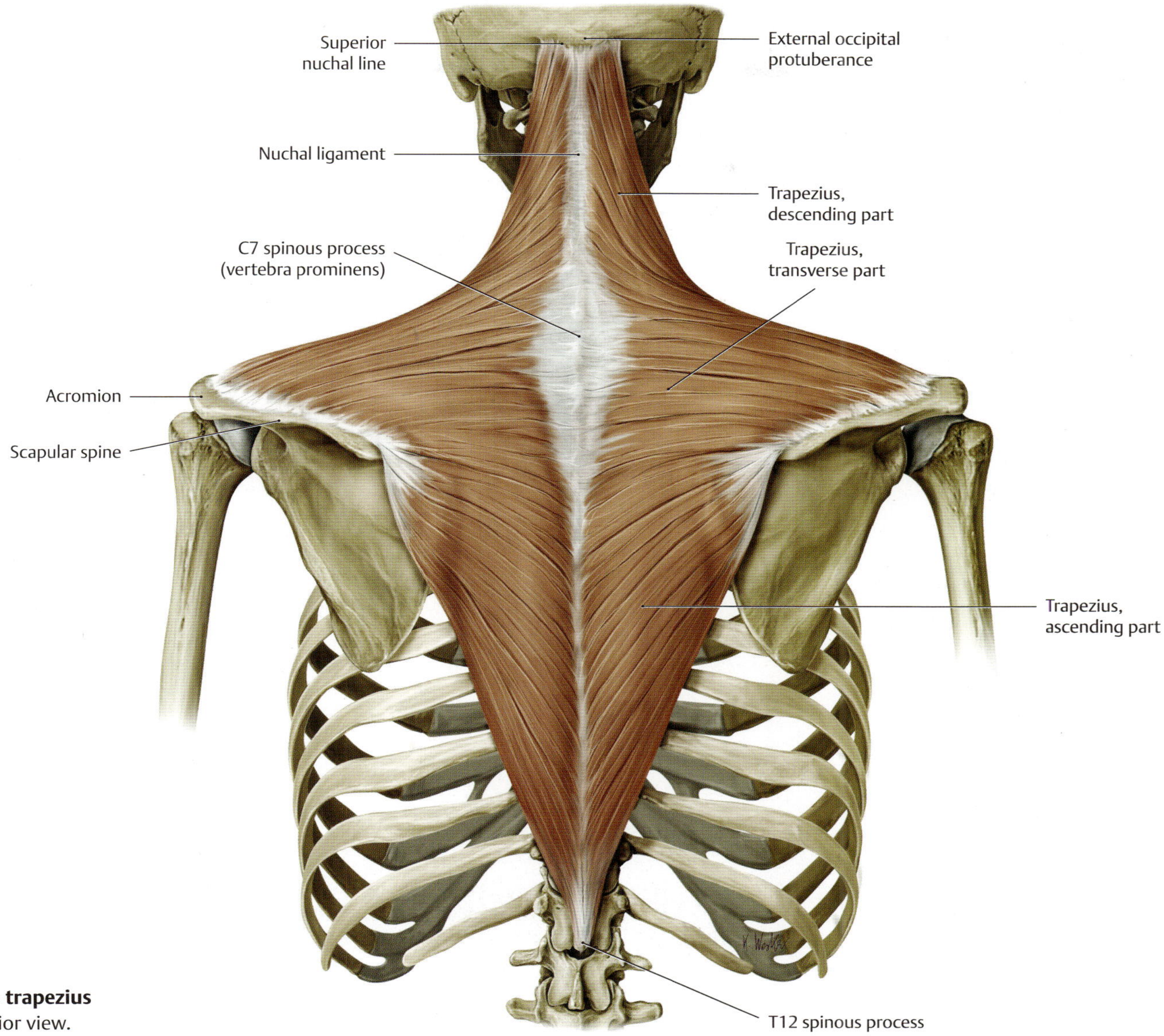

D The trapezius
Posterior view.

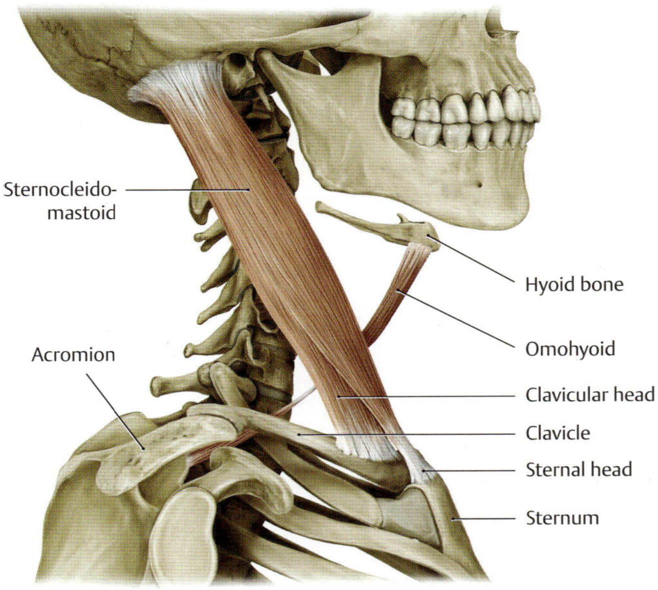

E The sternocleidomastoid and omohyoid
Right side, lateral view.

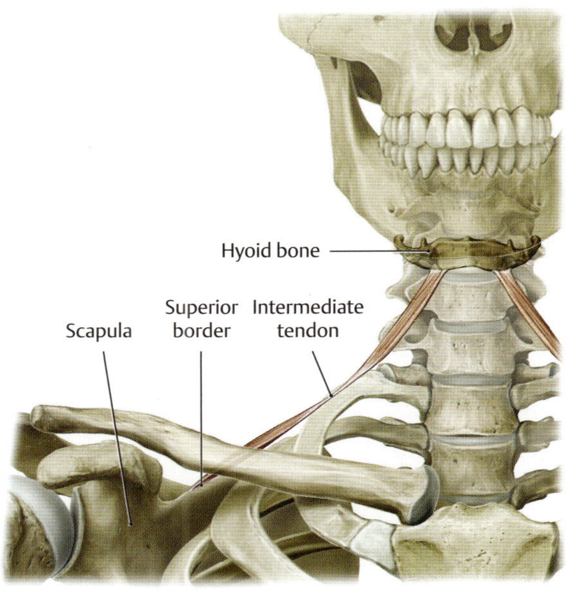

F The omohyoid
Right side, anterior view.

2.3 The Muscles of the Shoulder Girdle: Serratus Anterior, Subclavius, Pectoralis Minor, Levator Scapulae, and Rhomboids Major and Minor

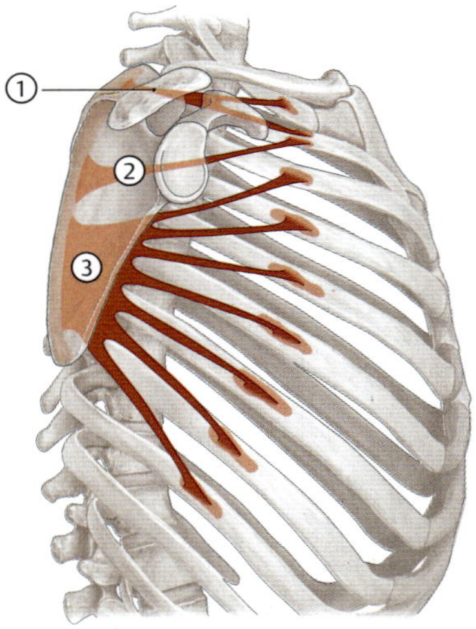

A Schematic of the serratus anterior

Serratus anterior

Origin: First through ninth ribs
Insertion: Scapula: ① Superior part (costal and dorsal surfaces of superior angle)
② Intermediate part (costal surface of medial border)
③ Inferior part (costal and dorsal surfaces of inferior angle and costal surface of medial border)
Actions:
- Entire muscle: draws the scapula laterally forward, elevates the ribs when the shoulder girdle is fixed (assists in respiration)
- Inferior part: rotates the scapula and draws its inferior angle laterally forward (rotates glenoid cavity superiorly), allowing for arm to be elevated above 90°
- Superior part: lowers the raised arm (antagonist to the inferior part)

Innervation: Long thoracic nerve (C5–C7)

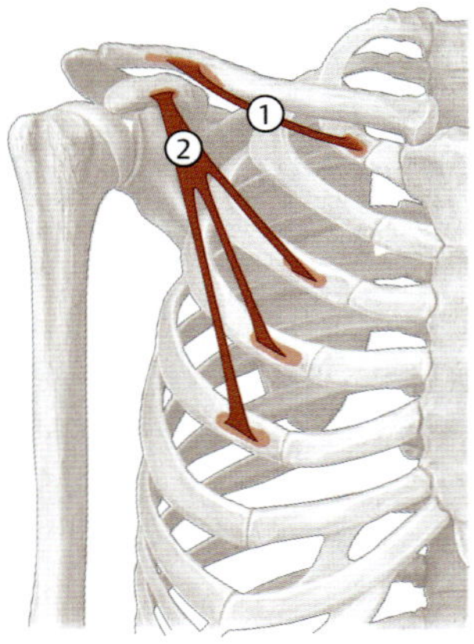

B Schematic of the subclavius and pectoralis minor

① **M. subclavius**

Origin: First rib (chondro-osseous junction)
Insertion: Inferior surface of the clavicle (lateral third)
Action: Steadies the clavicle in the sternoclavicular joint
Innervation: Nerve to the subclavius (C5, C6)

② **Pectoralis minor**

Origin: Third through fifth ribs
Insertion: Coracoid process of the scapula
Actions:
- Draws the scapula downward, causing its inferior angle to move posteromedially (lowers the raised arm), rotates glenoid inferiorly
- Assists in respiration

Innervation: Medial and lateral pectoral nerves (C8, T1)

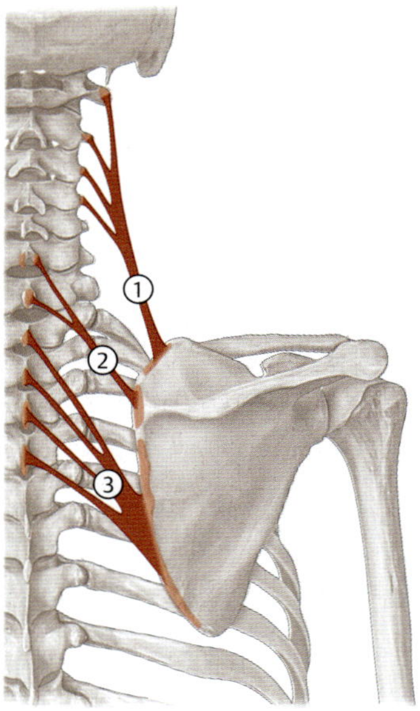

C Schematic of the levator scapulae and rhomboids minor and major

① **Levator scapulae**

Origin: Transverse processes of the C1–C4 vertebrae
Insertion: Superior angle of the scapula
Actions:
- Draws the scapula medially upward while moving the inferior angle medially (returns the raised arm to the neutral [0°] position)
- Inclines the neck toward the same side (when the scapula is fixed)

Innervation: Dorsal scapular nerve and cervical spinal nerves (C4–C5)

② **Rhomboid minor**

Origin: Spinous processes of the C6 and C7 vertebrae
Insertion: Medial border of the scapula (above the scapular spine)
Actions:
- Steadies the scapula
- Draws the scapula medially upward (returns the raised arm to the neutral [0°] position)

Innervation: Dorsal scapular nerve (C4–C5)

③ **Rhomboid major**

Origin: Spinous processes of the T 1–T 4 vertebrae
Insertion: Medial border of the scapula (below the scapular spine)
Actions:
- Steadies the scapula
- Draws the scapula medially upward (returns the raised arm to the neutral [0°] position)

Innervation: Dorsal scapular nerve (C4–C5)

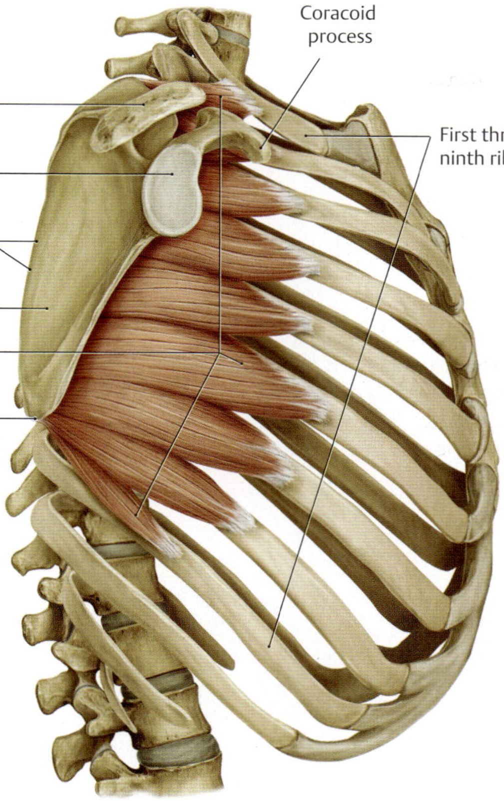

D The serratus anterior
Right side, lateral view.

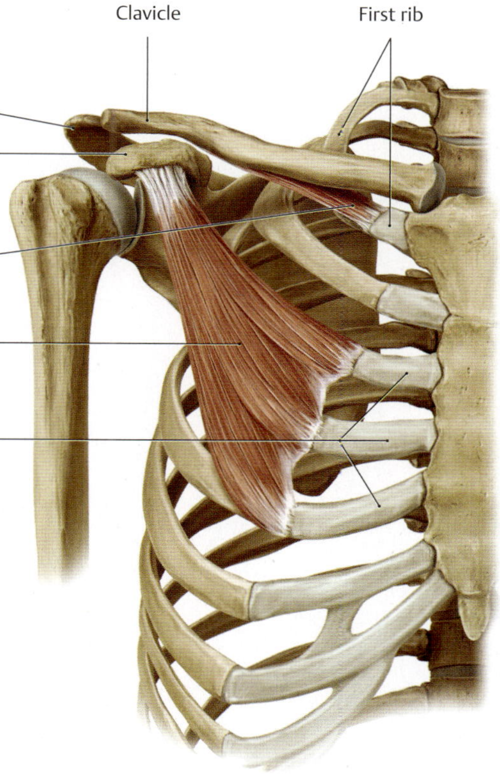

E The pectoralis minor and subclavius
Right side, anterior view.

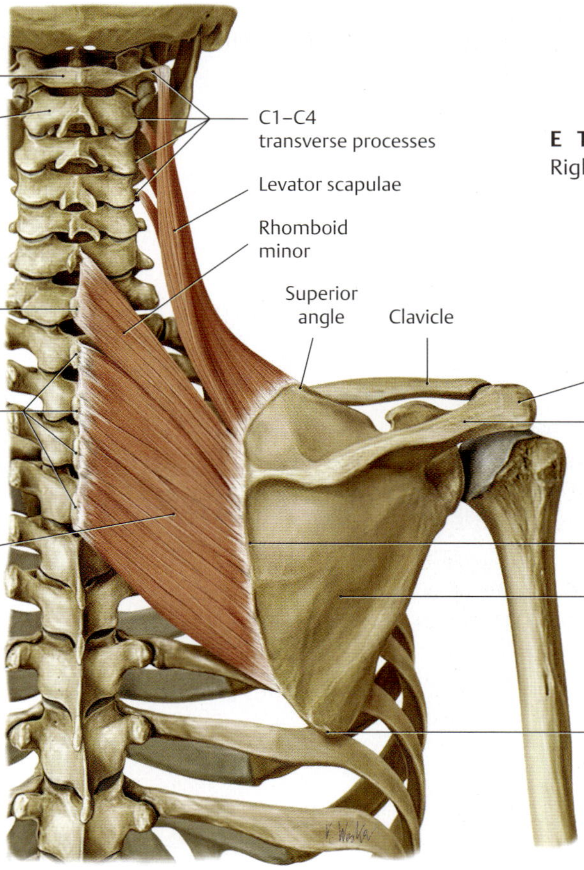

F The levator scapulae, rhomboid major, and rhomboid minor
Right side, posterior view.

2.4 The Posterior Muscles of the Shoulder Joint: The Rotator Cuff

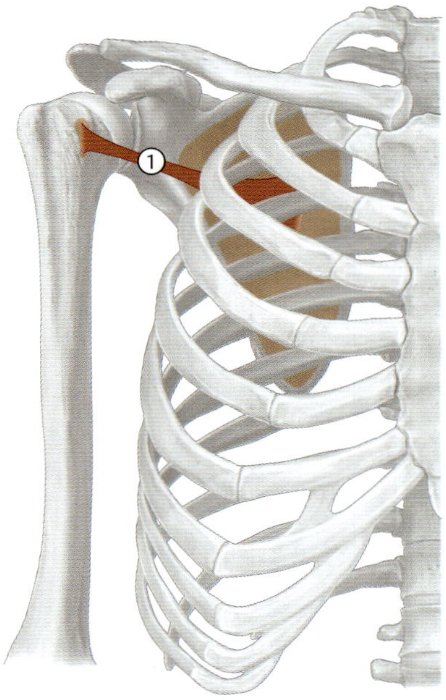

A Schematic of the subscapularis

① **Subscapularis**

Origin: Subscapular fossa of the scapula
Insertion: Lesser tubercle of the humerus
Action: Internal rotation
Innervation: Upper and lower subscapular nerves (C5–C8)

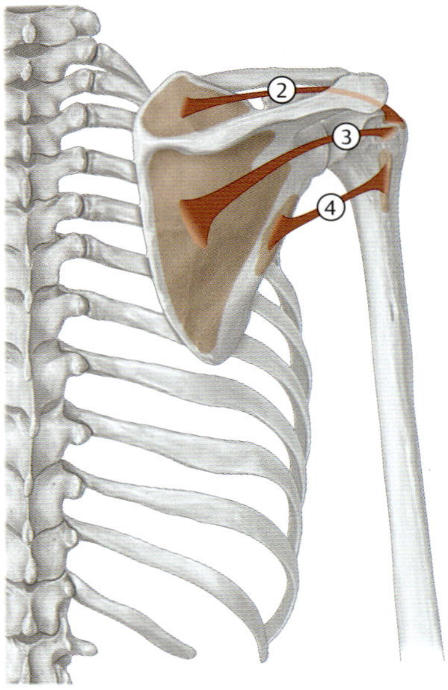

B Schematic of the supraspinatus, infraspinatus, and teres minor

② **Supraspinatus**

Origin: Supraspinous fossa of the scapula
Insertion: Greater tubercle of the humerus
Action: Abduction
Innervation: Suprascapular nerve (C4–C6)

③ **Infraspinatus**

Origin: Infraspinous fossa of the scapula
Insertion: Greater tubercle of the humerus
Funktion: External rotation
Innervation: Suprascapular nerve (C4–C6)

④ **Teres minor**

Origin: Lateral border of the scapula
Insertion: Greater tubercle of the humerus
Action: External rotation, weak adduction
Innervation: Axillary nerve (C5, C6)

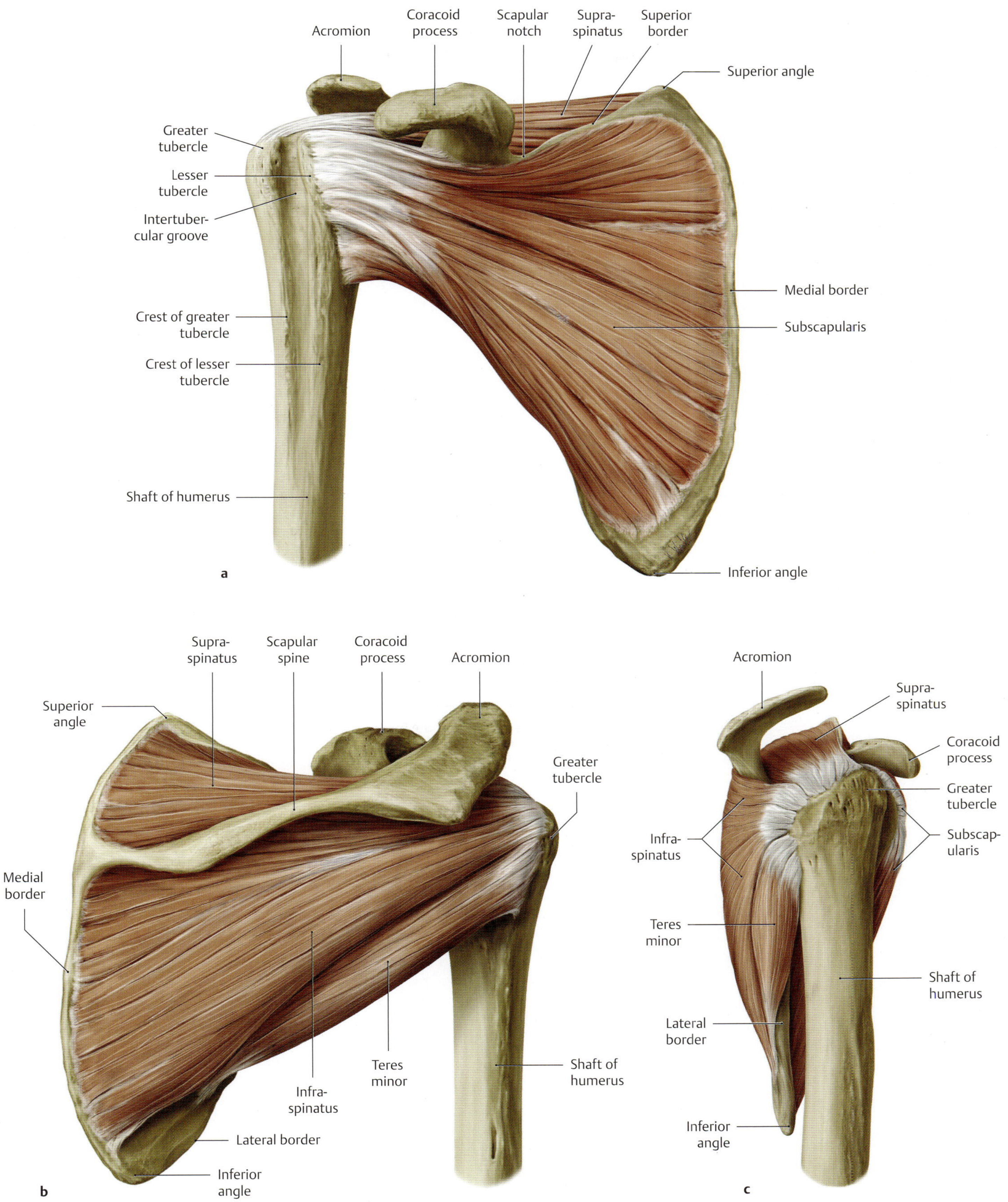

C Muscles of the rotator cuff: supraspinatus, infraspinatus, teres minor, and subscapularis
Right shoulder joint.

a Anterior view.
b Posterior view.
c Lateral view.

2.5 The Posterior Muscles of the Shoulder Joint: The Deltoid

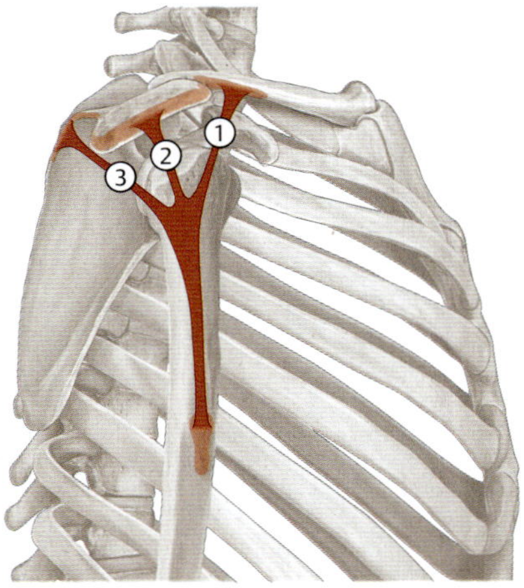

Origin:	① Clavicular part: lateral third of the clavicle ② Acromial part: acromion ③ Spinal part: scapular spine
Insertion:	Deltoid tuberosity on the humerus
Actions:	• Clavicular part: flexion (moves the arm and shoulder forward), internal rotation, adduction • Acromial part: abduction • Spinal part: extension (moves the arm and shoulder backward), external rotation, adduction
Innervation:	Between 60° and 90° of abduction, the clavicular and spinal parts of the deltoid assist the acromial part of the muscle with abduction. Axillary nerve (C5, C6)

A Schematic of the deltoid

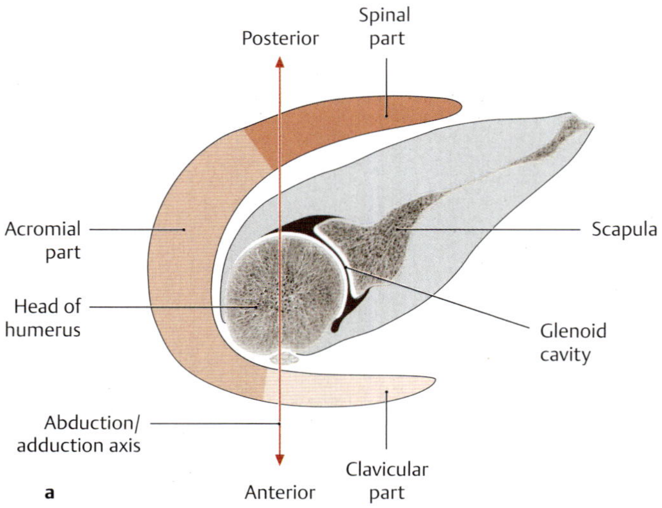

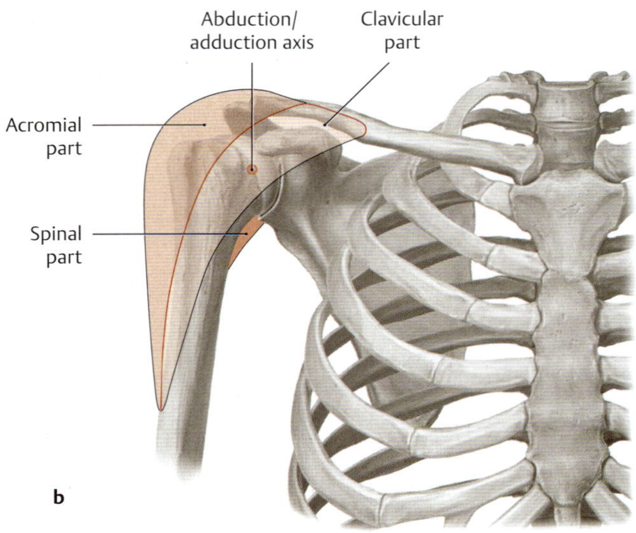

B The variable actions of the deltoid components
a Cross section through the right shoulder joint.
b Right shoulder joint in the neutral (0°) position, anterior view.
c Right shoulder joint in 60° of abduction, anterior view.

The actions of the three parts of the deltoid muscle (clavicular, acromial, and spinal) depend on their relationship to the position of the humerus and its axis of motion. As a result, the parts of the deltoid can act antagonistically as well as synergistically. At *less than 60°* abduction, the clavicular and spinal parts of the deltoid act as antagonists to the acromial part, but at *more than 60°* abduction, the acromial part assists in the abduction. Starting from the neutral (0°) position, the acromial part of the deltoid abducts the arm and steadies it in any position it assumes. When the arm is abducted past approximately 60°, the clavicular and spinal parts also become active as they move past the sagittal motion axis (abduction/adduction axis, **c**). This alters the action of these parts: they act as adductors below 60°, but when the arm passes 60°, they become abductors.

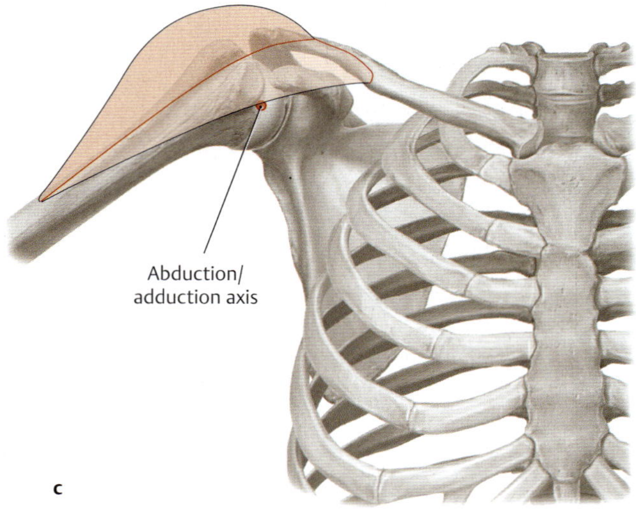

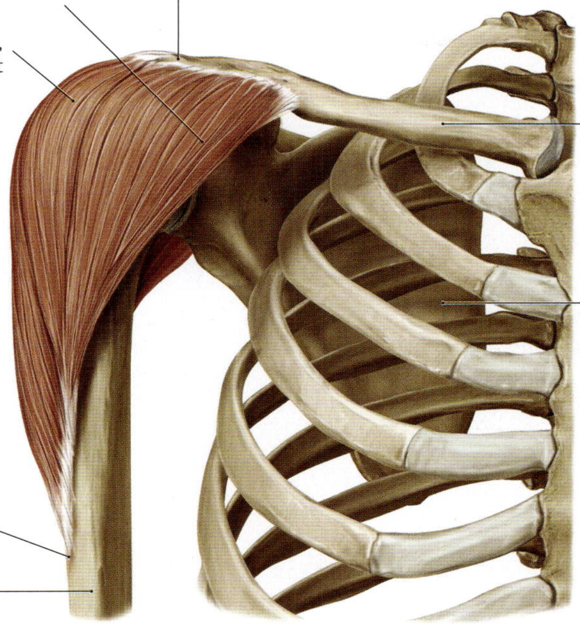

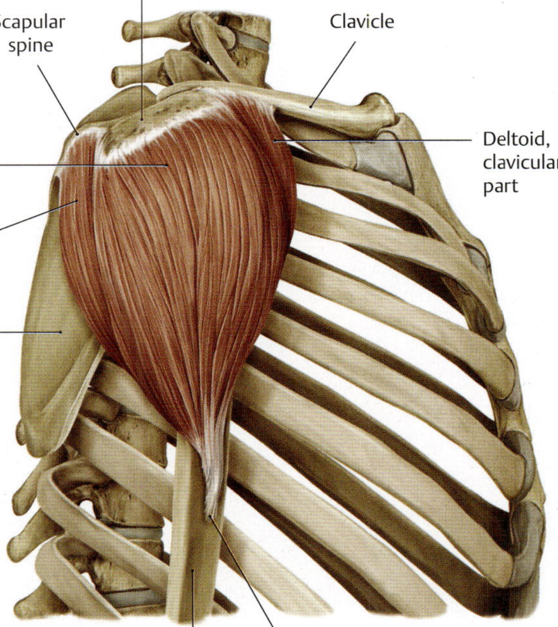

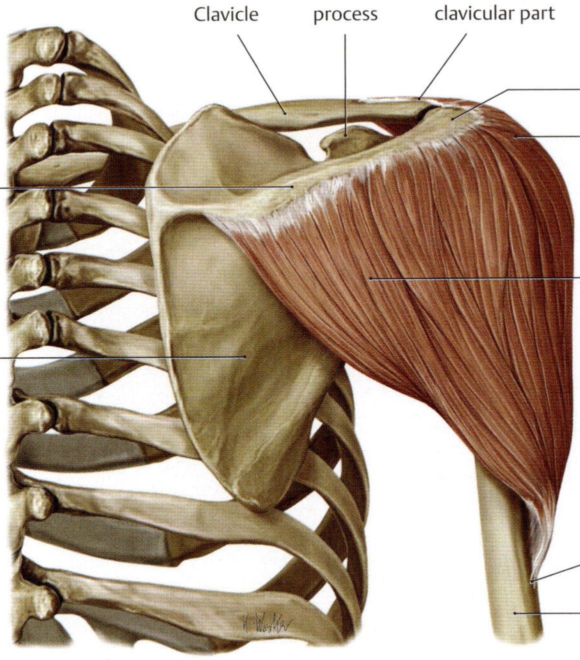

C The deltoid
Right shoulder joint.

a Anterior view.
b Posterior view.
c Lateral view.

2.6 The Posterior Muscles of the Shoulder Joint: Latissimus Dorsi and Teres Major

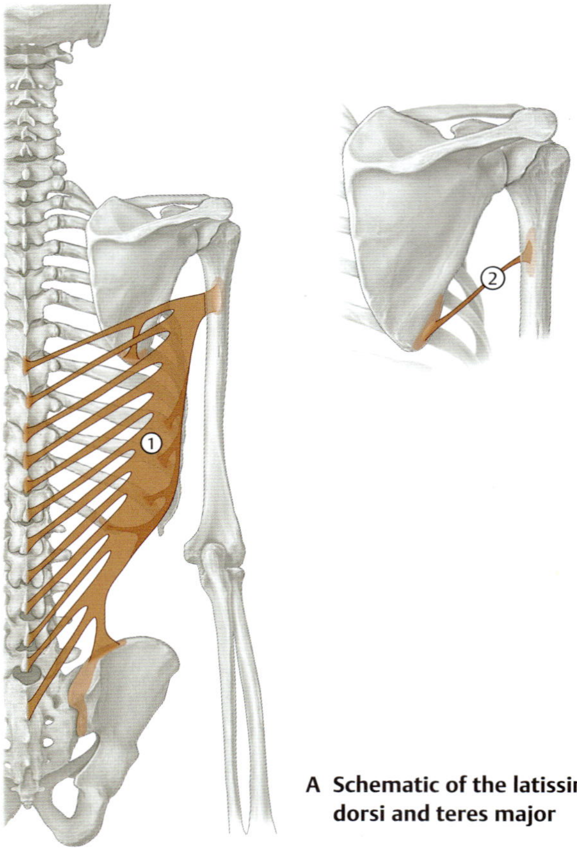

A Schematic of the latissimus dorsi and teres major

① **Latissimus dorsi**	
Origin:	• Vertebral part: – Spinous processes of the T7–T12 vertebrae – Thoracolumbar fascia of the spinous processes of all lumbar vertebrae and the sacrum • Iliac part: posterior third of the iliac crest • Costal part: Ninth through twelfth ribs • Scapular part: inferior angle of the scapula
Insertion:	Floor of the intertubercular groove of the humerus
Actions:	Internal rotation, adduction, extension (moves the arm backward), respiration (expiration, “cough muscle”)
Innervation:	Thoracodorsal nerve (C6–C8)

② **Teres major**	
Origin:	Inferior angle of the scapula
Insertion:	Crest of the lesser tubercle of the humerus
Actions:	Internal rotation, adduction, extension
Innervation:	Lower subscapular nerve (C5–C8; approx. 85% of cases) or thoracodorsal nerve (C6–C8, approx. 15% of cases)

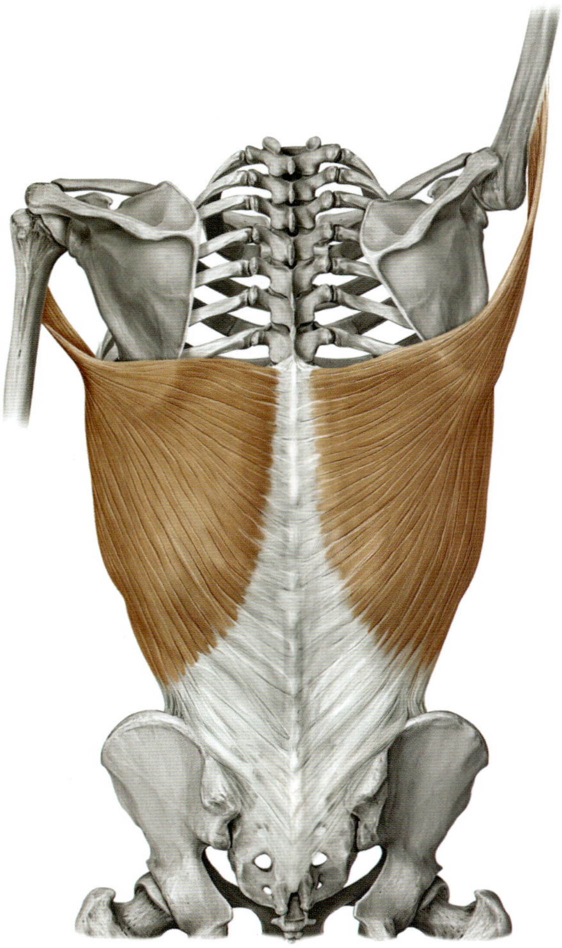

B Course of the tendon of insertion of the latissimus dorsi in the neutral position and in elevation

Posterior view. The latissimus dorsi muscle is most active in the abducted or elevated arm. Raising the arm untwists the muscle fibers in the area of insertion, increasing the muscle's stretch and maximizing the force it can exert. When the position of the arm is fixed, the latissimus dorsi can pull the body upward, as in the act of climbing, or can depress the arm against a resistance. This makes the latissimus dorsi an important muscle for paraplegics, for example, who can use the muscle to raise themselves from wheelchairs. In elevation (i.e., abduction or flexion over 90°), the humerus is automatically in maximum external rotation (known as the paradox of Codman). This forced external rotation is a more or less involuntary motion that prevents compression of the greater tuberosity of the humerus against the coracoacromial arch. Usually up to 50° of external rotation is possible. When the range of motion in abduction is a maximum of 180° as shown here, the humerus is in maximum external rotation, causing the crest of the lesser tuberosity to migrate laterally. The latissimus dorsi can perform its various functions, especially adduction, particularly effectively from this position.

The latissimus dorsi also aids in respiration. When the arm is immobilized (fixed point on the humerus), the lateral parts of the muscle in particular (costal part) support the inferior ribs (ninth through twelfth ribs) and constrict the inferior thoracic aperture. This aids in expiration. Where expiration is impaired, as it is in bronchial asthma or chronic bronchitis, the latissimus dorsi often becomes hypertrophied. As the costal part of the muscle also acts as a fixed point for the portion of the diaphragm arising from the ribs during coughing (contraction of the diaphragm), the latissimus dorsi is often referred to as the *cough muscle*.

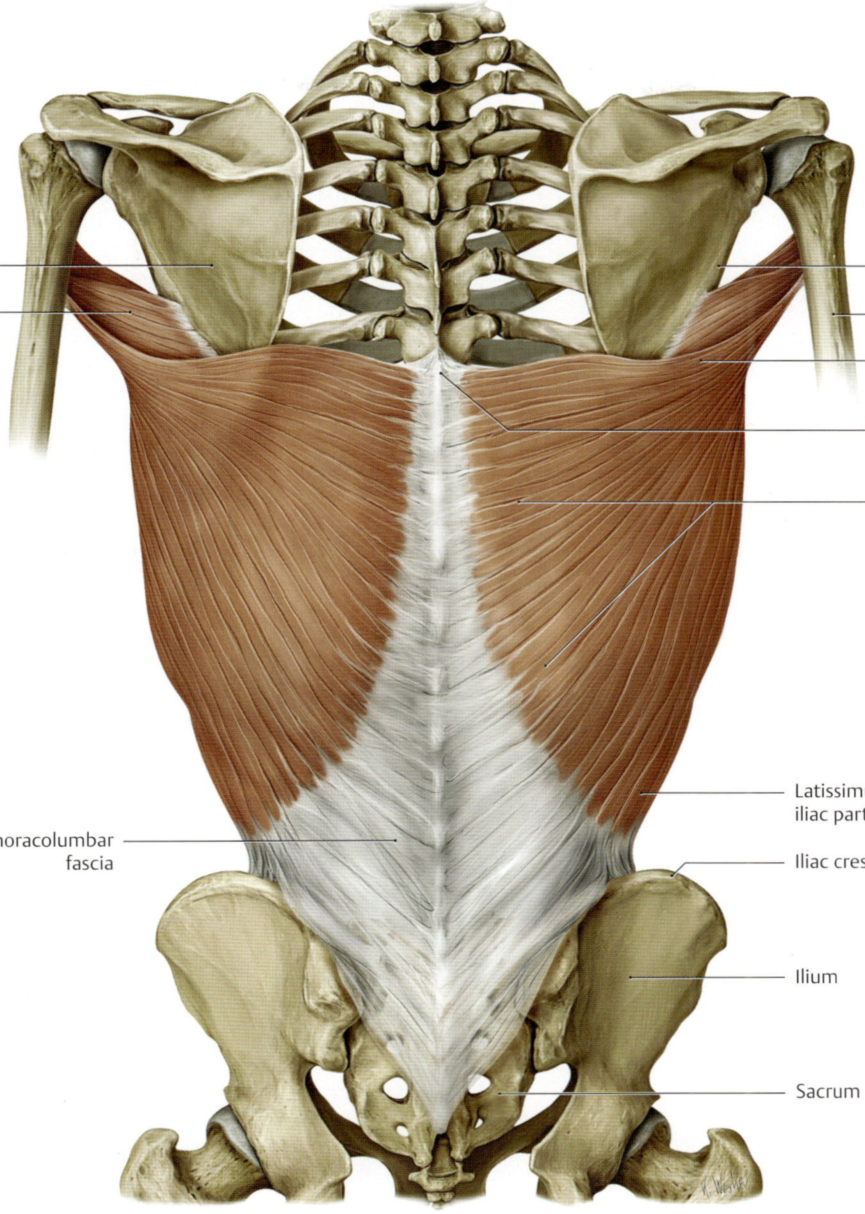

C The latissimus dorsi and teres major
Posterior view.

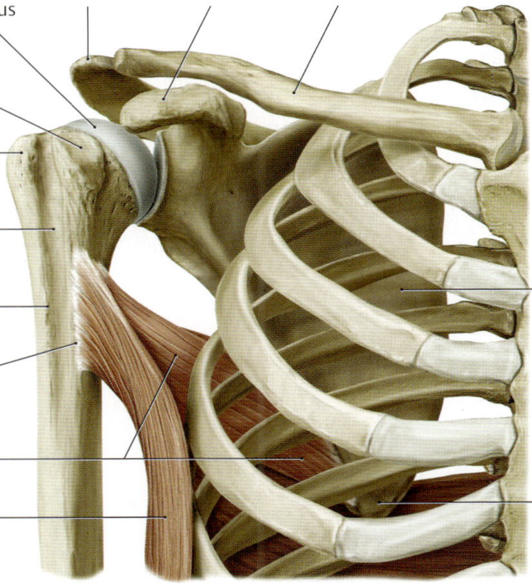

D Insertion of the latissimus dorsi on the floor of the intertubercular groove and the teres major on the crest of the lesser tubercle
Right side, anterior view.

2.7 The Anterior Muscles of the Shoulder Joint: Pectoralis Major and Coracobrachialis

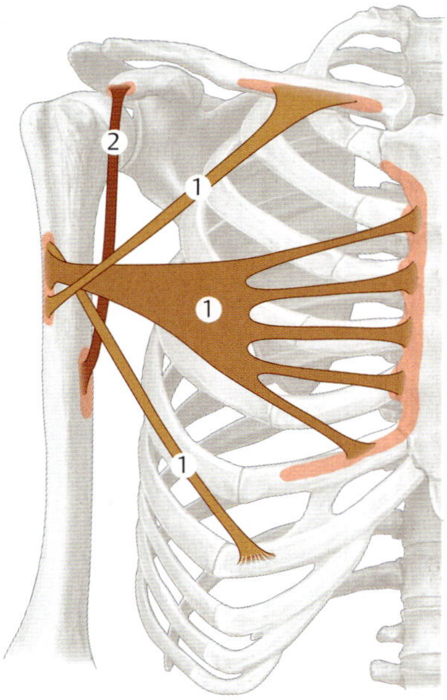

A Schematic of the pectoralis major and coracobrachialis

① **Pectoralis major**	
Origin:	• Clavicular part: medial half of the clavicle • Sternocostal part: sternum and the second through sixth costal cartilages • Abdominal part: anterior layer of the rectus sheath
Insertion:	Crest of the greater tubercle of the humerus
Actions:	• Adduction and internal rotation (entire muscle) • Flexion (clavicular and sternocostal parts) • Assists respiration when the shoulder girdle is fixed
Innervation:	Medial and lateral pectoral nerves (C5–T1)

② **Coracobrachialis**	
Origin:	Coracoid process of the scapula
Insertion:	Humerus (in line with the crest of the lesser tubercle)
Actions:	Flexion, adduction, internal rotation
Innervation:	Musculocutaneous nerve (C5, C6)

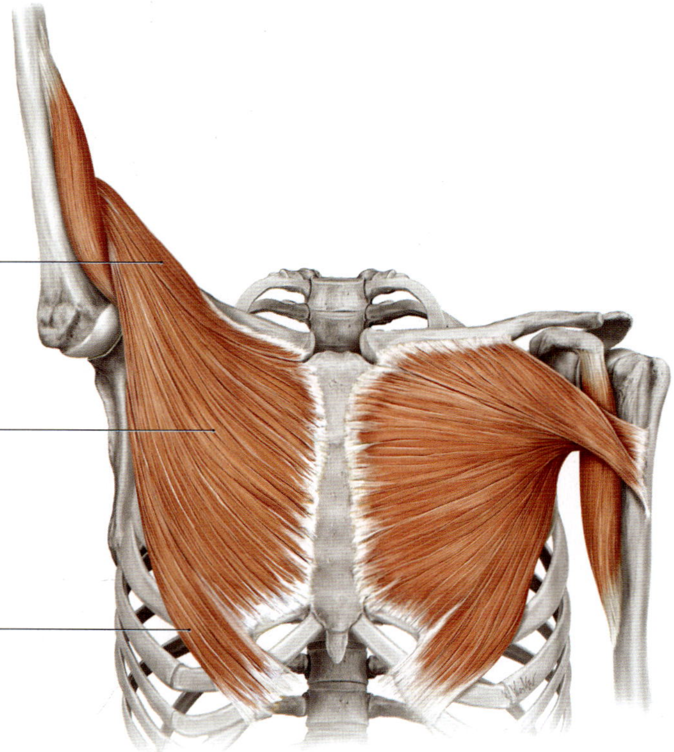

B Twisting of the tendon of insertion of the pectoralis major

Anterior view. The three parts of the pectoralis major (clavicular, sternocostal, and abdominal) converge laterally and insert on the crest of the greater tubercle by a broad tendon that has a horseshoe-shaped cross section. The tendon fiber bundles are twisted on themselves in such a way that the clavicular part inserts lower on the humerus than the sternocostal part, which inserts lower than the abdominal part. As with the latissimus dorsi, the muscle fibers of the pectoralis major become untwisted and stretched with increasing elevation of the arm, increasing the force that the muscle can exert.

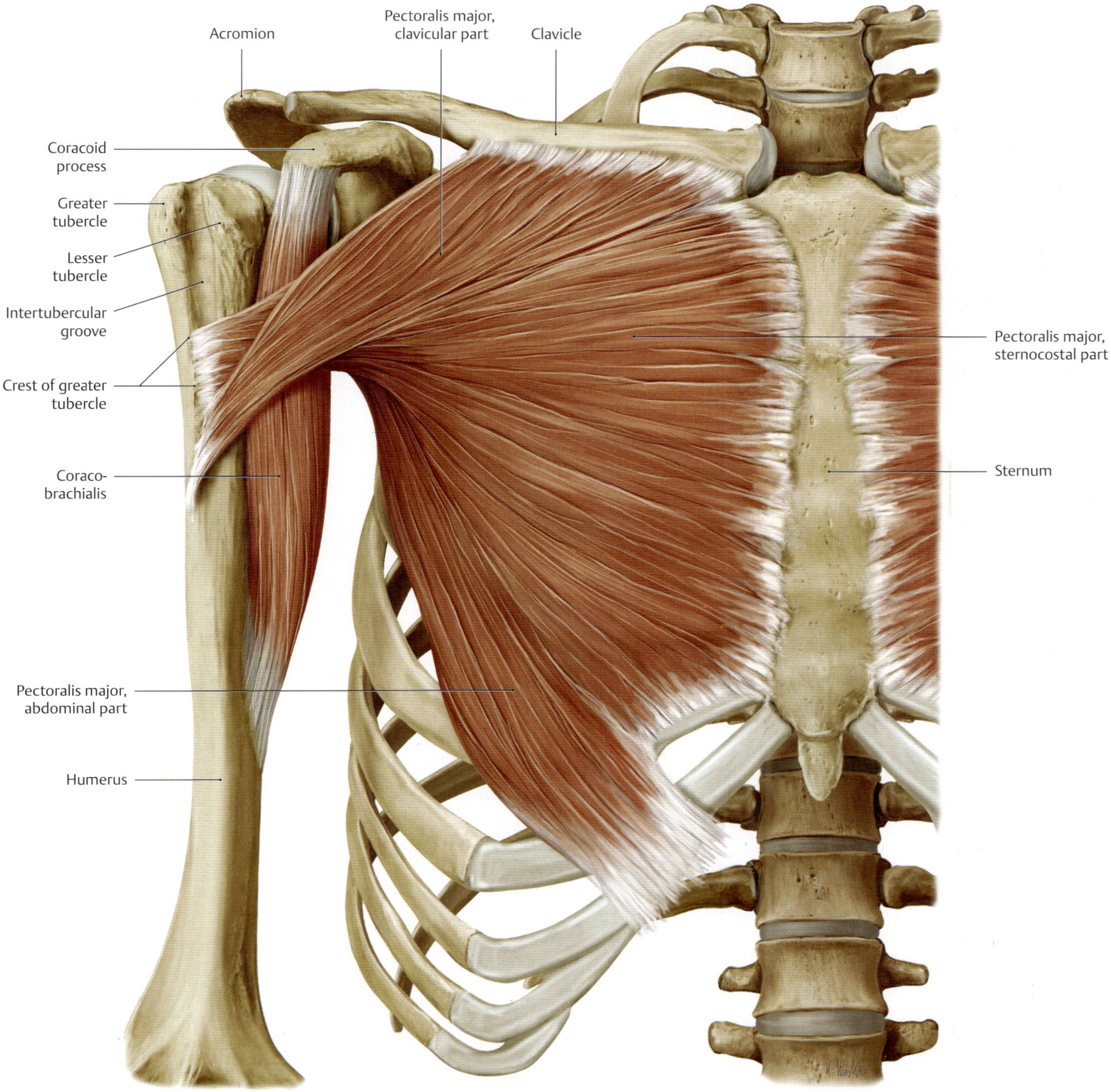

C The pectoralis major and coracobrachialis
Right side, anterior view.

2.8 The Anterior Muscles of the Arm: Biceps Brachii and Brachialis

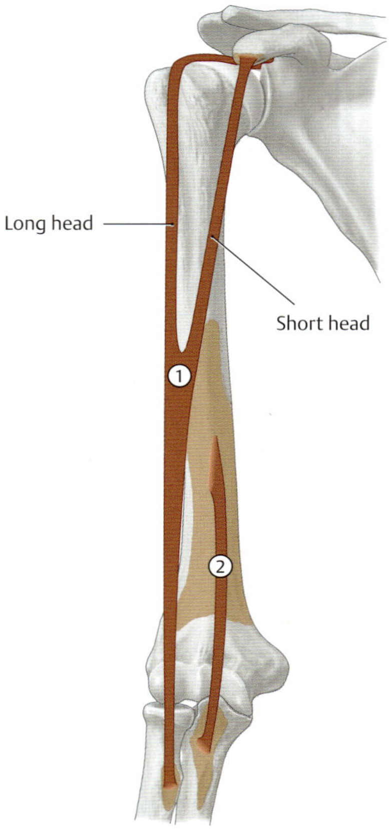

A Schematic of the biceps brachii and brachialis

① Biceps brachii

Origin:
- Long head: supraglenoid tubercle of the scapula
- Short head: coracoid process of the scapula

Insertion: Radial tuberosity and antebrachial fascia via bicipital aponeurosis

Actions:
- Elbow joint:
  - Flexion, supination (with the elbow flexed)
- Shoulder joint:
  - Flexion (forward motion of humerus)
  - Stabilization of humeral head during deltoid contraction
  - Abduction and internal (medial) rotation of the humerus (long head)

Innervation: Musculocutaneous nerve (C5–C7)

② Brachialis

Origin: Distal half of the anterior surface of the humerus, also the medial and lateral intermuscular septa

Insertion: Ulnar tuberosity

Actions: Flexion at the elbow joint

Innervation: Musculocutaneous nerve (C5–C7), radial nerve (C5–C6), variable branches that arise from the radial nerve in the radial tunnel as muscular branches

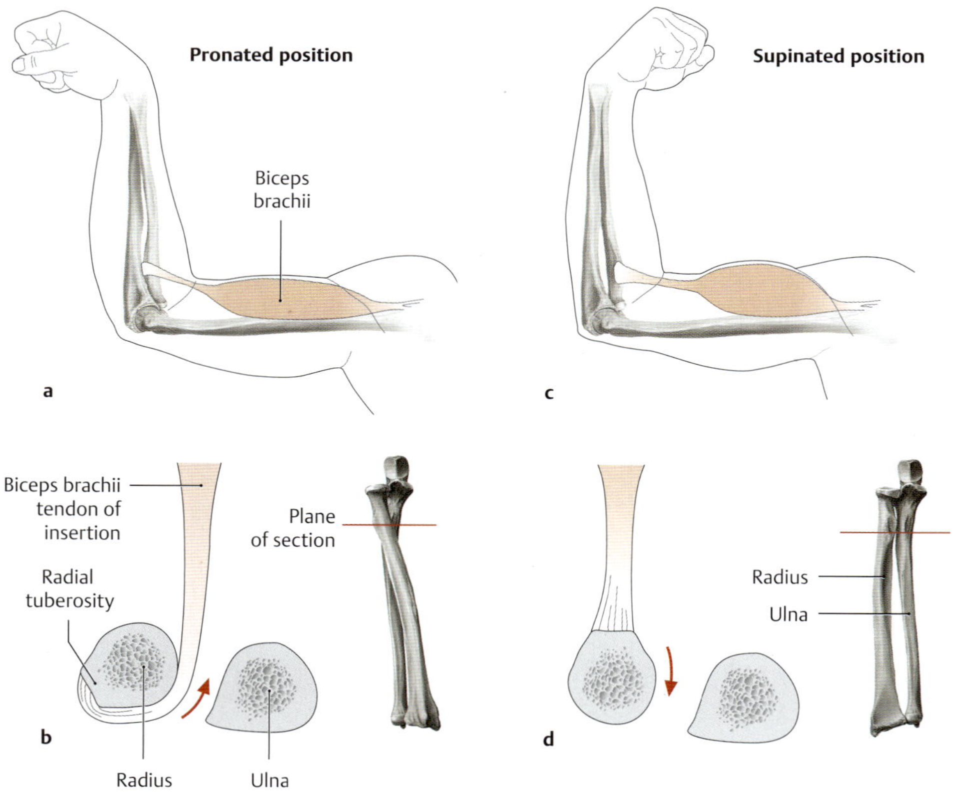

B Supinating action of the biceps brachii with the elbow flexed

a The forearm is pronated with the elbow flexed (right arm, medial view).
b Cross section at the level of the radial tuberosity with the forearm pronated (distal view).
c The forearm is supinated with the elbow flexed (right arm, medial view).
d Cross section at the level of the radial tuberosity with the forearm supinated (distal view).

When the elbow is flexed, the biceps brachii acts as a powerful supinator in addition to its role as a flexor, because the lever arm in that position is almost perpendicular to the axis of pronation/supination (see p. 294). This is why *supination movements* are particularly effective when the elbow is flexed. When the forearm is pronated (**a**), the tendon of insertion of the biceps brachii is wrapped around the radius (**b**). When the muscle then contracts to supinate (**c**) or flex the elbow, the tendon unwraps like a rope coiled around a crank (**b** to **d**).

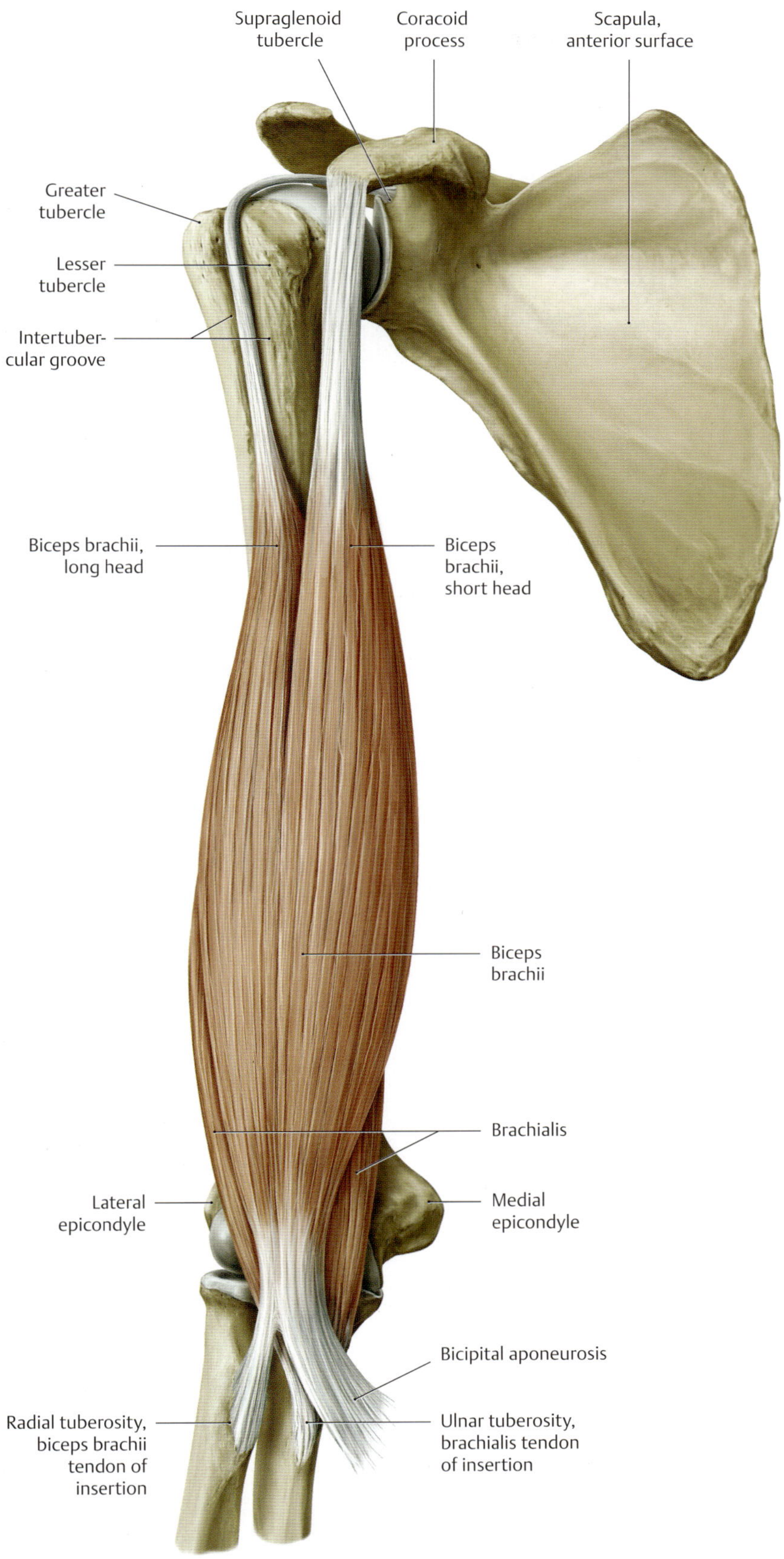

C The biceps brachii and brachialis
Right arm, anterior (ventral) view.

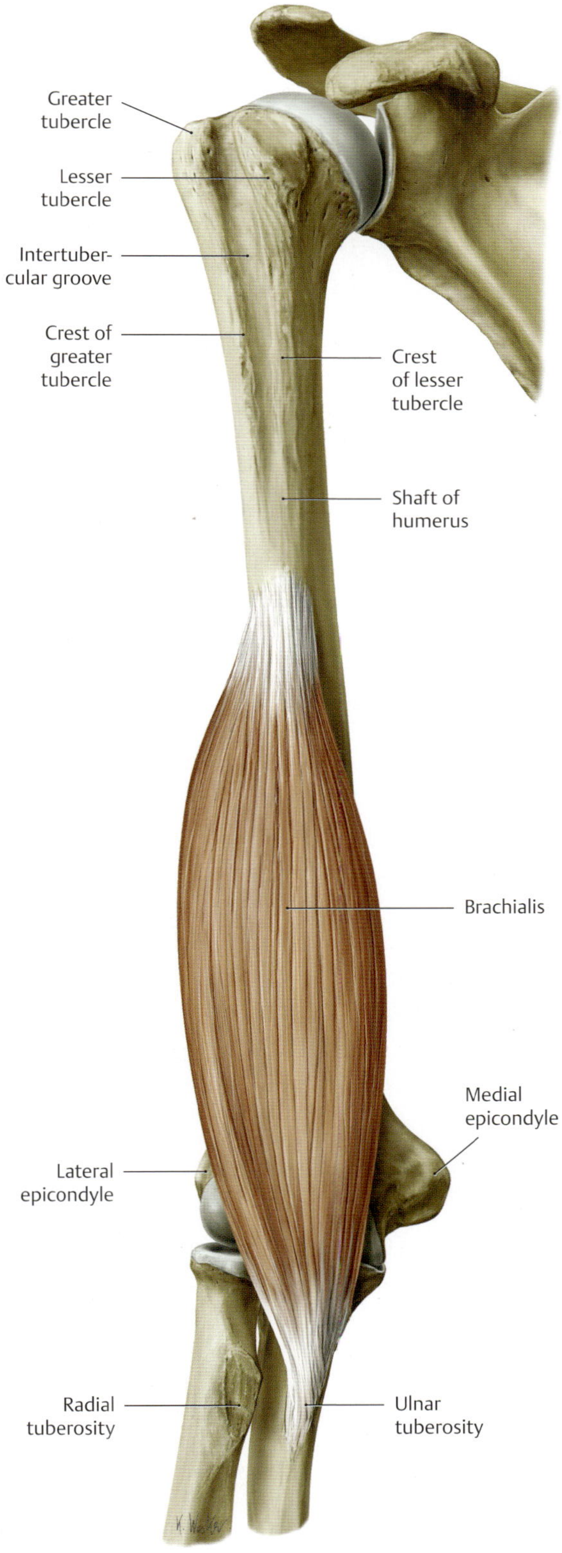

D The brachialis
Right arm, anterior (ventral) view.

2.9 The Posterior Muscles of the Arm: Triceps Brachii and Anconeus

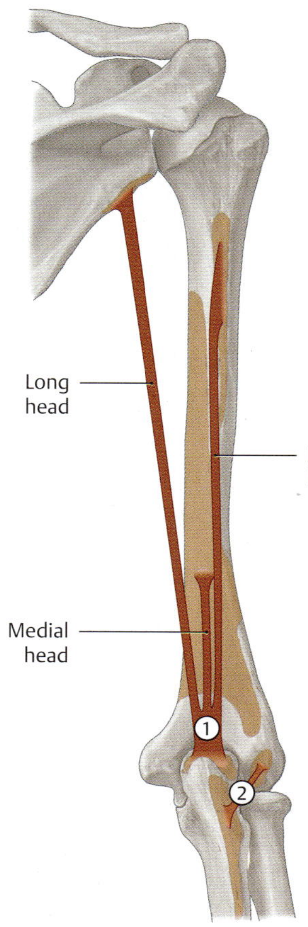

① **Triceps brachii**	
Origin:	• Long head: infraglenoid tubercle of the scapula • Medial head: posterior surface of the humerus, distal to the radial groove, and the medial intermuscular septum • Lateral head: posterior surface of the humerus, proximal to the radial groove, and the lateral intermuscular septum
Insertion:	Olecranon of the ulna
Actions:	• Elbow joint: extension • Shoulder joint, long head: backward movement and adduction of the arm
Innervation:	Radial nerve (C6–C8)
② **Anconeus**	
Origin:	Lateral epicondyle of the humerus (and the posterior joint capsule in some cases)
Insertion:	Olecranon of the ulna (radial surface)
Actions:	Extends the elbow and tightens its joint capsule
Innervation:	Radial nerve (C6–C8)

A Schematic of the triceps brachii and anconeus

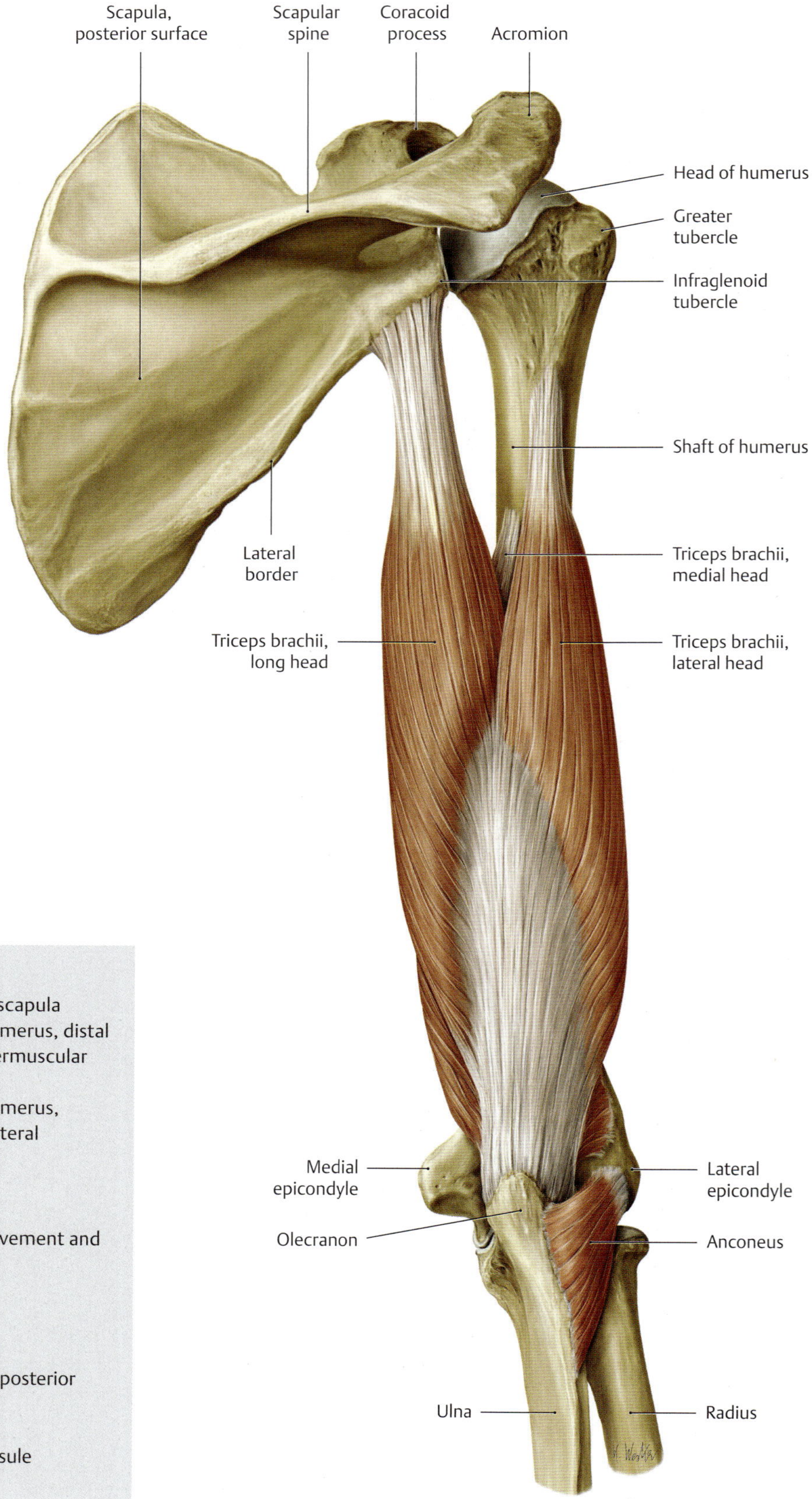

B The triceps brachii and anconeus
Right arm, posterior (dorsal) view.

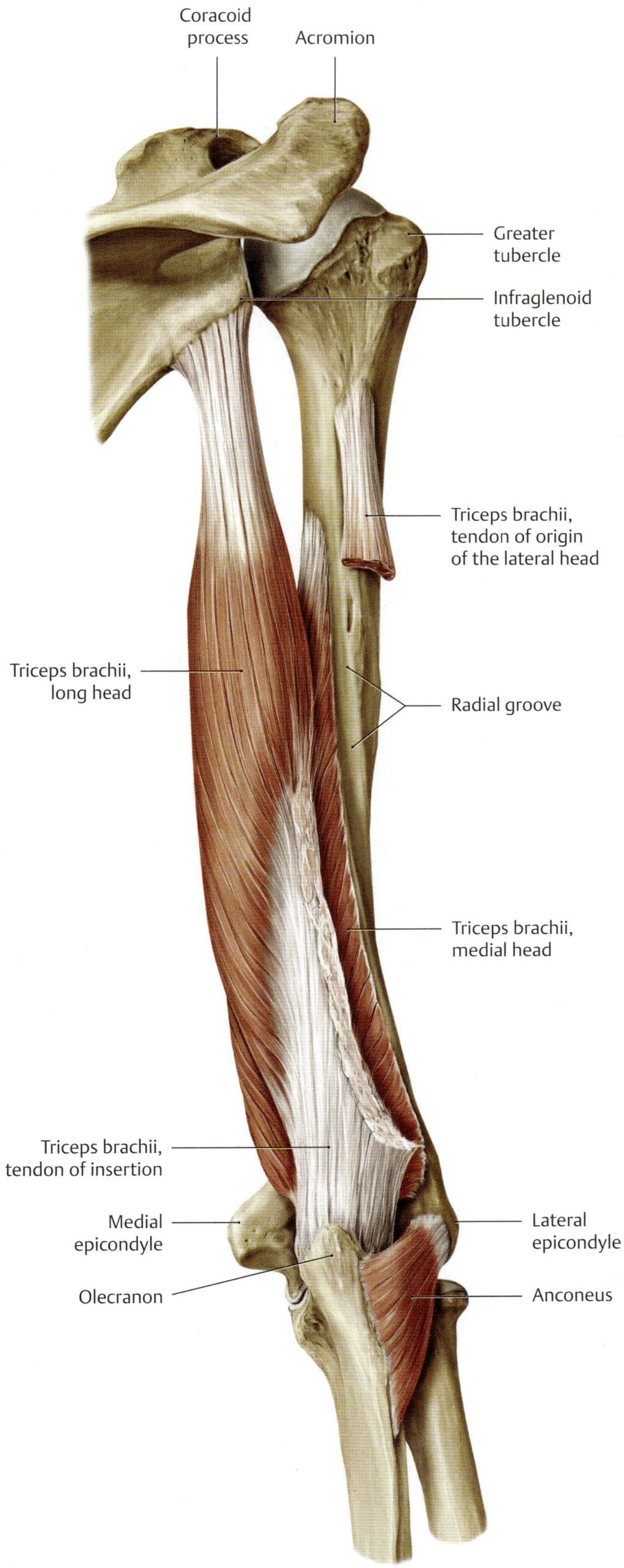

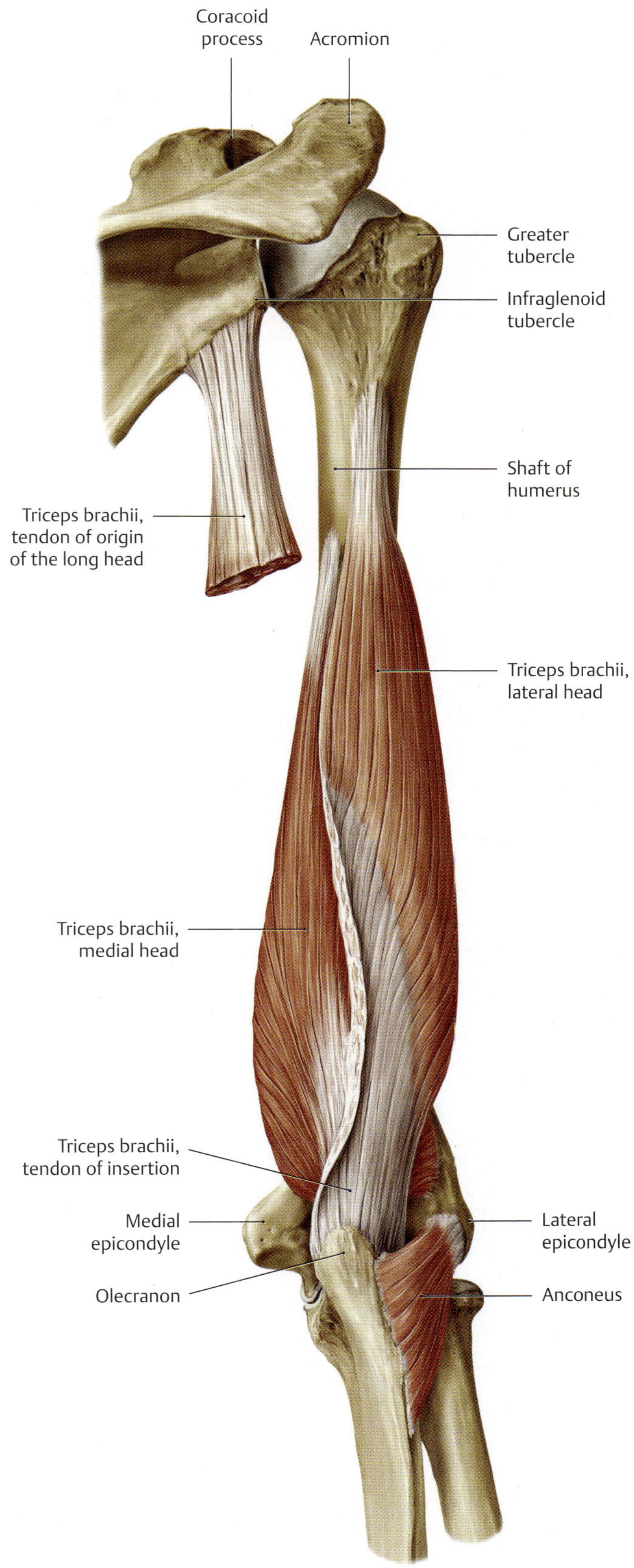

C The triceps brachii and anconeus
Right arm, posterior (dorsal) view. The lateral head of the triceps brachii has been partially removed.

D The triceps brachii and anconeus
Right arm, posterior (dorsal) view. The long head of the triceps brachii has been partially removed.

2.10 The Anterior Muscles of the Forearm: The Superficial and Deep Flexors

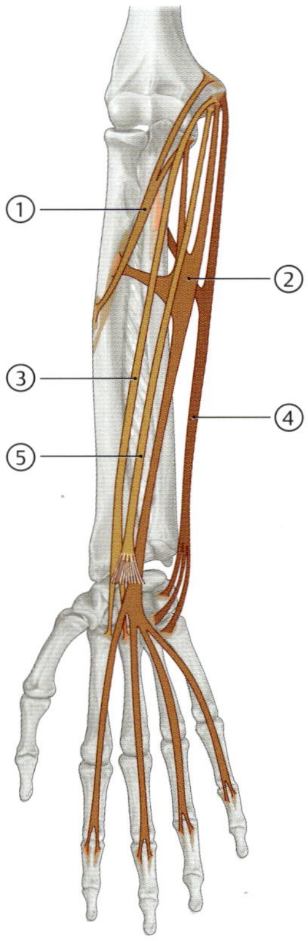

A Schematic of the superficial flexors

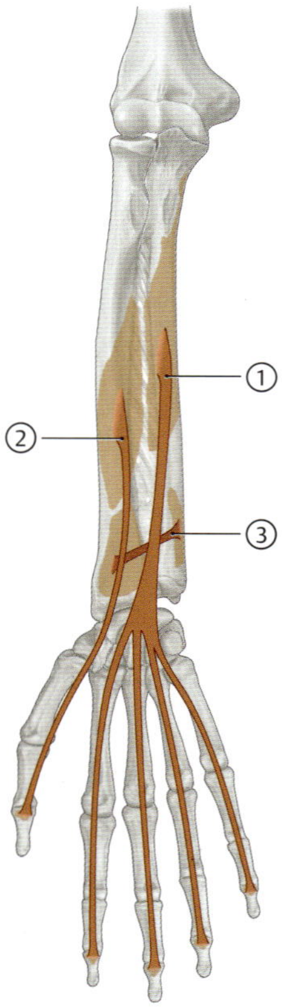

B Schematic of the deep flexors

① Pronator teres

Origin:
- Humeral head: medial epicondyle of the humerus
- Ulnar head: coronoid process of the ulna

Insertion: Lateral surface of the radius (distal to the supinator insertion)

Actions:
- Elbow: weak flexor
- Forearm joints: pronation

Innervation: Median nerve (C6)

② Flexor digitorum superficialis

Origin:
- Humeral head: medial epicondyle of the humerus (common head)
- Ulnar head: coronoid process of the ulna
- Radial head: distal to the radial tuberosity

Insertion: The sides of the middle phalanges of the second through fifth digits

Actions:
- Elbow: weak flexor
- Wrist joints, metacarpophalangeal and proximal interphalangeal joints of the second through fifth digits: flexion

Innervation: Median nerve (C7–T1)

③ Flexor carpi radialis

Origin: Medial epicondyle of the humerus (common head)

Insertion: Base of the second metacarpal and sometimes of the third as well

Actions:
- Wrist joints: flexion, abduction (radial deviation)
- Elbow: weak pronation

Innervation: Median nerve (C6–C8)

④ Flexor carpi ulnaris

Origin:
- Humeral head: medial epicondyle of the humerus (common head)
- Ulnar head: olecranon

Insertion: Hook of the hamate, base of the fifth metacarpal, pisiform (as a sesamoid)

Actions: Wrist joints: flexion, adduction (ulnar deviation)

Innervation: Ulnar nerve (C8–T1)

⑤ Palmaris longus

Origin: Medial epicondyle of the humerus (common head)

Insertion: Palmar aponeurosis

Actions:
- Elbow: weak flexor
- Wrist joint: palmar flexion, tightens the palmar aponeurosis

Innervation: Median nerve (C8–T1)

① Flexor digitorum profundus

Origin: Proximal two thirds of the flexor surface of the ulna and the adjacent interosseous membrane

Insertion: Palmar surface of the distal phalanges of the second through fifth digits

Actions: Wrists and base, middle, and end joints of second through fifth digits: flexion

Innervation:
- Median nerve (radial part, second and third digits), C8, T1
- Ulnar nerve (ulnar part, fourth and fifth digits), C7–T1

② Flexor pollicis longus

Origin: Mid-anterior surface of the radius and the adjacent interosseous membrane

Insertion: Palmar surface of the distal phalanx of the thumb

Actions:
- Wrist joints: flexion and radial abduction of the hand
- Carpometacarpal joint of the thumb: opposition
- MCP and IP joints of the thumb: flexion

Innervation: Median nerve (C6–C8)

③ Pronator quadratus

Origin: Distal one fourth of the anterior surface of the ulna

Insertion: Distal one fourth of the anterior surface of the radius

Actions: Pronates the hand, stabilizes the distal radioulnar joint

Innervation: Median nerve (C8, T1)

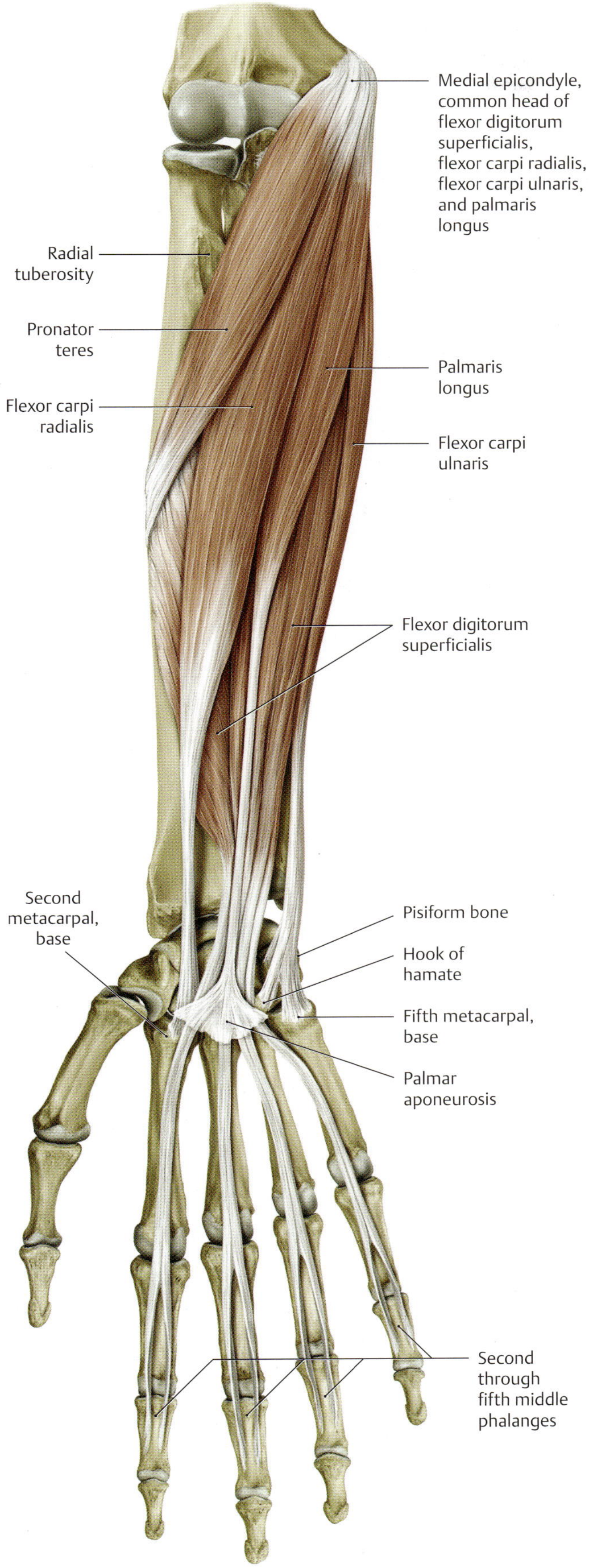

C The superficial flexors (pronator teres, flexor digitorum superficialis, flexor carpi radialis, flexor carpi ulnaris, and palmaris longus)
Right forearm, anterior view.

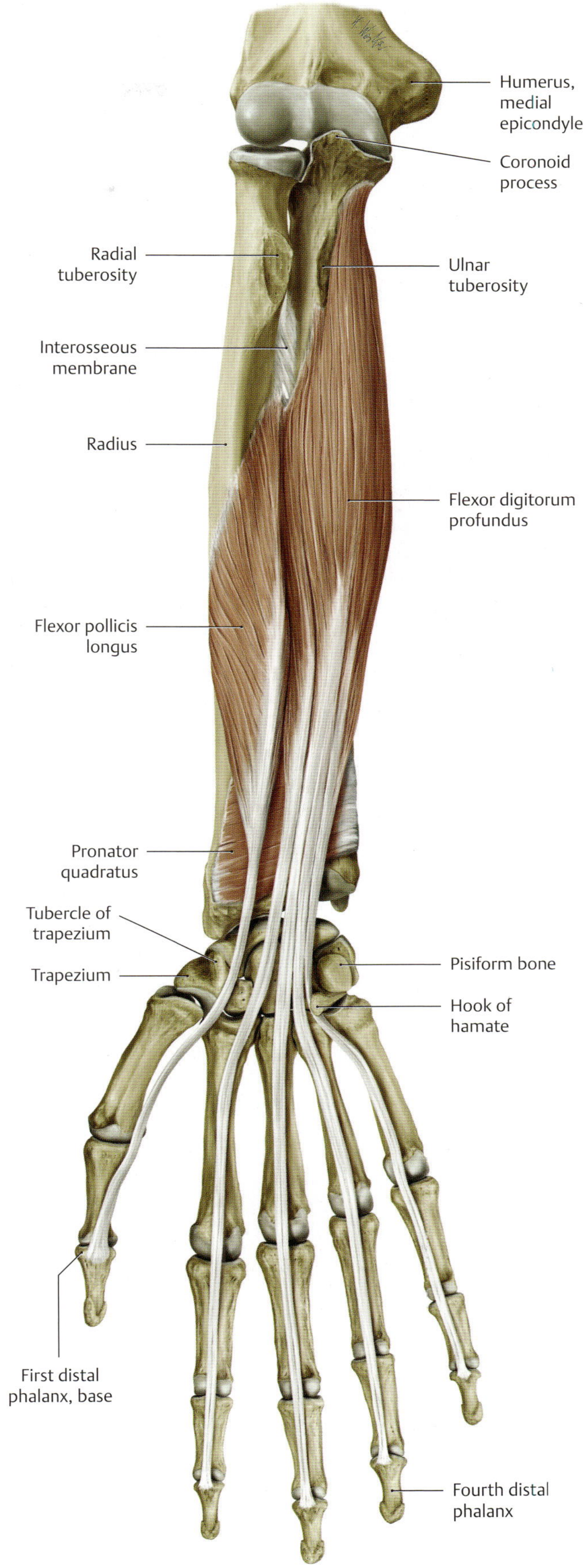

D The deep flexors (flexor digitorum profundus, flexor pollicis longus, and pronator quadratus)
Right forearm, anterior view.

2.11 The Muscles of the Forearm: The Radialis Muscles

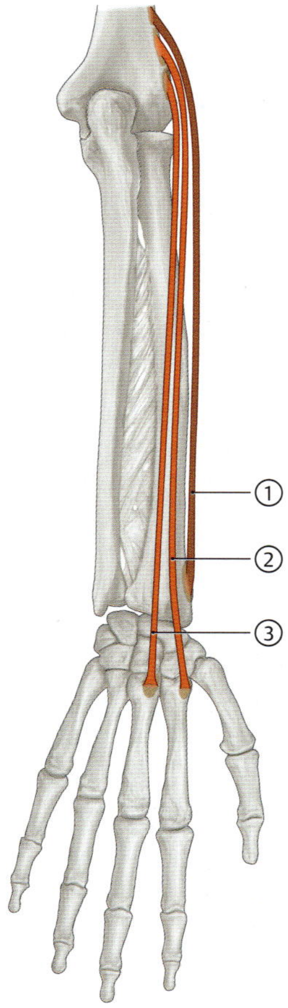

A Schematic of the radialis muscles

① **Brachioradialis**

Origin: Lateral surface of the distal humerus, lateral intermuscular septum
Insertion: Styloid process of the radius
Actions:
- Elbow joint: flexion
- Forearm joints: semipronation

Innervation: Radial nerve (C5–C7)

② **Extensor carpi radialis longus**

Origin: Lateral surface of the distal humerus (lateral supracondylar ridge), lateral intermuscular septum
Insertion: Dorsal base of the second metacarpal
Actions: Wrist joints: dorsal extension (assists in fist closure), abduction (radial deviation) of the hand
Innervation: Radial nerve (C5–C7)

③ **Extensor carpi radialis brevis**

Origin: Lateral epicondyle of the humerus
Insertion: Dorsal base of the third metacarpal
Actions: Wrist joints: dorsal extension (assists in fist closure), abduction (radial deviation) of the hand
Innervation: Radial nerve (C5–C7)

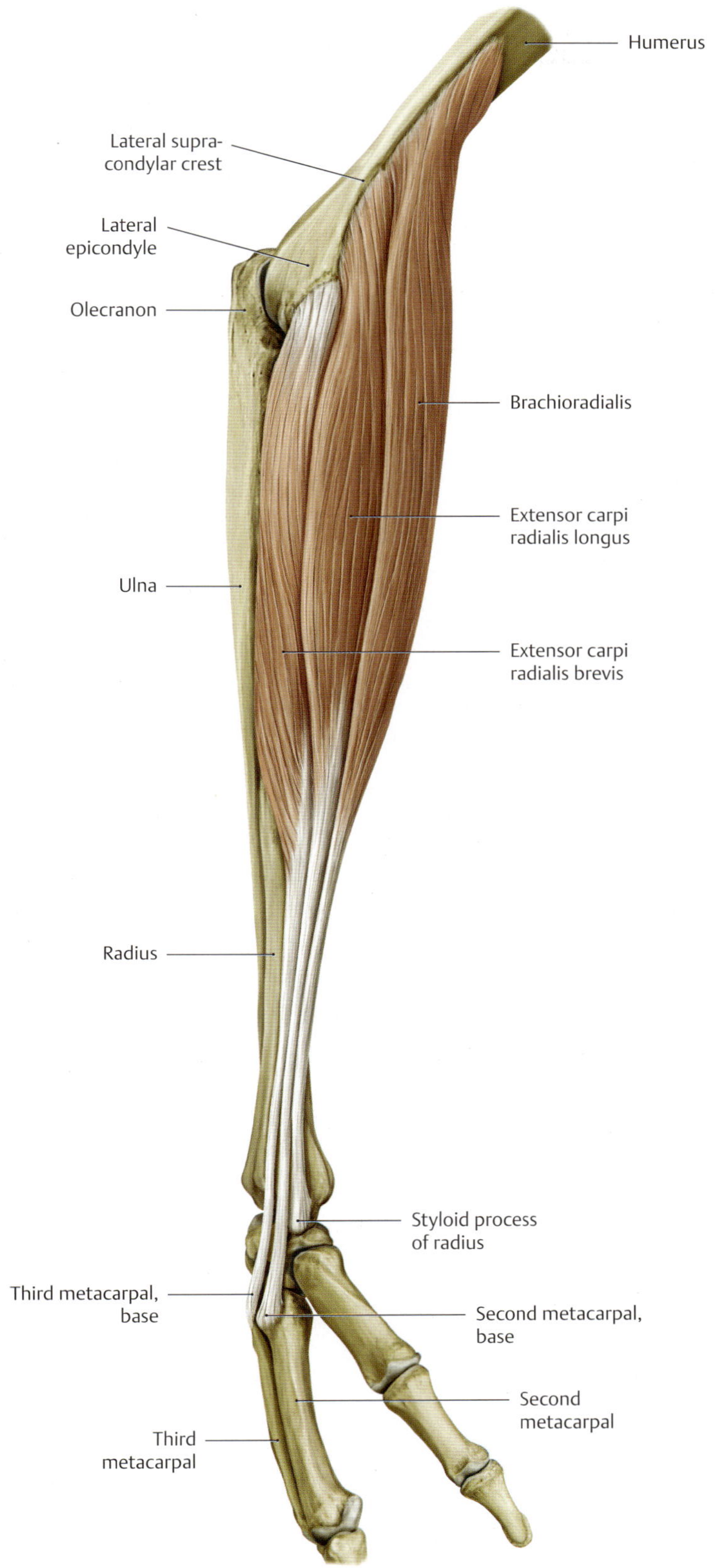

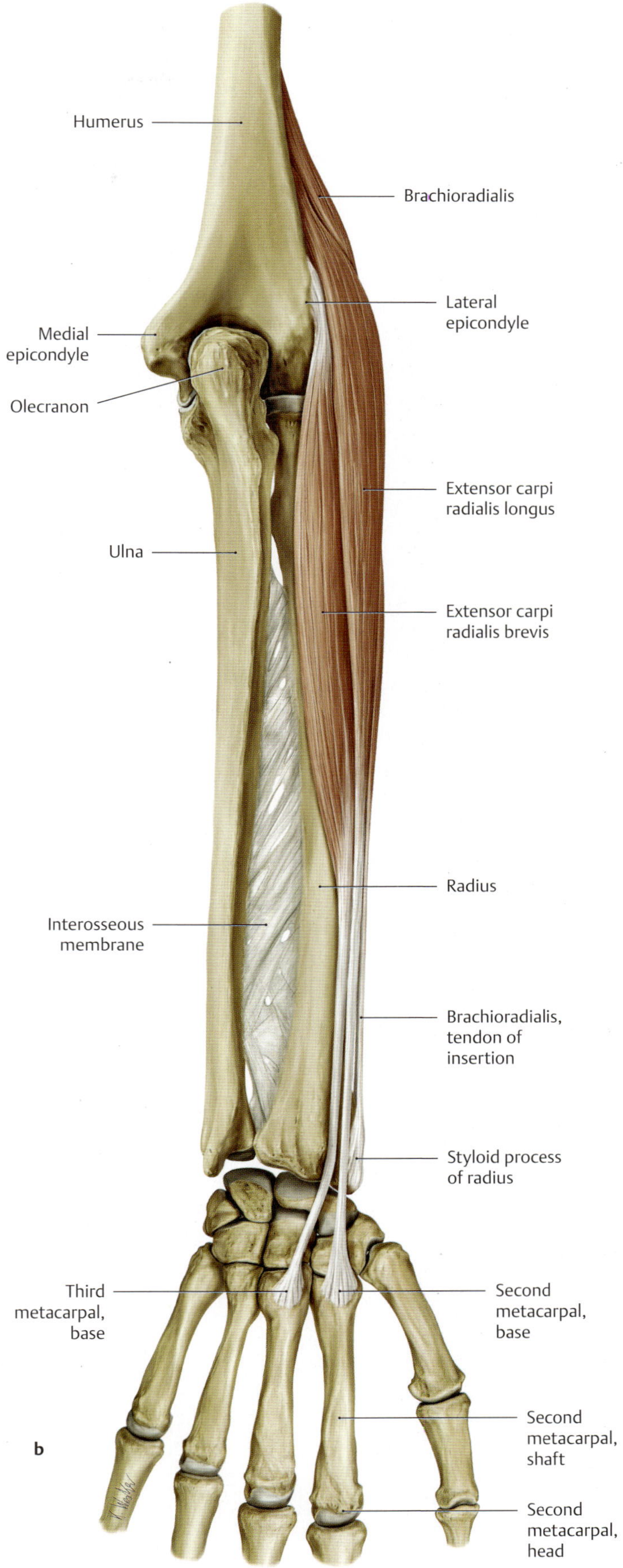

B The radialis muscles (brachioradialis, extensor carpi radialis longus, and extensor carpi radialis brevis)
Right forearm.

a Lateral (radial) view.
b Posterior (dorsal) view.

2.12 The Posterior Muscles of the Forearm: The Superficial and Deep Extensors

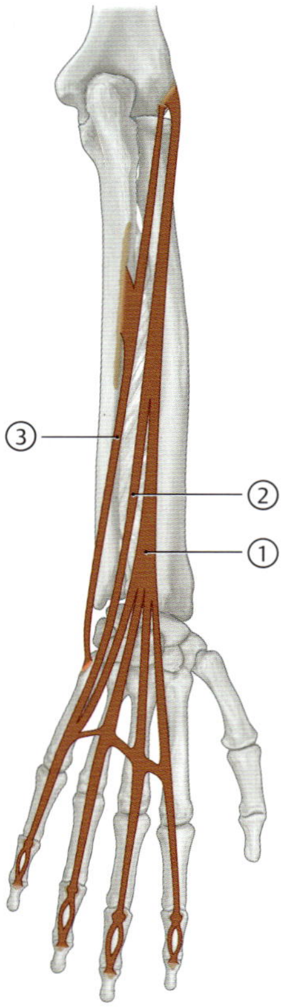

A Schematic of the superficial extensors

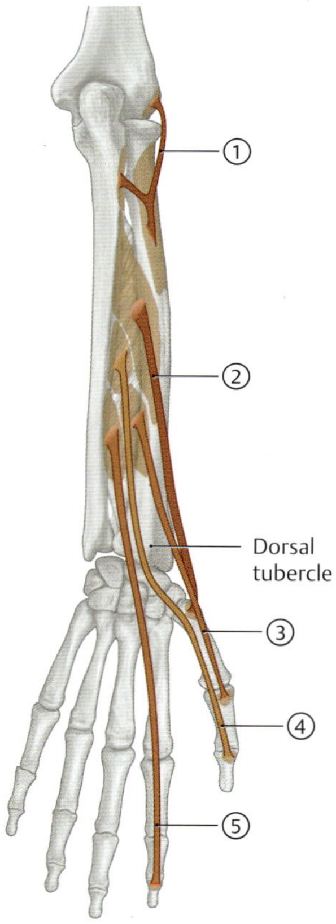

B Schematic of the deep extensors

① **Extensor digitorum**

Origin: Common head (lateral epicondyle of the humerus)
Insertion: Dorsal digital expansion of the second through fifth digits
Actions:
- Wrist joints: extension
- MCP, PIP, and DIP joints of the second through fifth digits: extension and abduction of the fingers

Innervation: Radial nerve (C6–C8)

② **Extensor digiti minimi**

Origin: Common head (lateral epicondyle of the humerus)
Insertion: Dorsal digital expansion of the fifth digit
Actions:
- Wrist joints: extension, adduction (ulnar deviation)
- MCP, PIP, and DIP joints of the fifth digit: extension and abduction of the fifth digit

Innervation: Radial nerve (C6–C8)

③ **Extensor carpi ulnaris**

Origin: Common head (lateral epicondyle of the humerus), ulnar head (dorsal surface of the ulna)
Insertion: Base of the fifth metacarpal
Actions: Wrist joints: extension, adduction (ulnar deviation) of the hand
Innervation: Radial nerve (C6–C8)

① **Supinator**

Origin: Olecranon of the ulna (supinator crest of the ulna), lateral epicondyle of the humerus, radial collateral ligament, anular ligament of the radius
Insertion: Radius (between the radial tuberosity and the insertion of pronator teres)
Actions: Radioulnar joint: supination
Innervation: Radial nerve (C5, C6)

② **Abductor pollicis longus***

Origin: Dorsal surfaces of the radius and ulna, also the interosseous membrane
Insertion: Base of the first metacarpal
Actions:
- Radiocarpal joint: abduction (radial deviation) of the hand
- Carpometacarpal joint of the thumb: abduction

Innervation: Radial nerve (C6–C8)

③ **Extensor pollicis brevis***

Origin: Posterior surface of the radius and the interosseous membrane (distal to abductor pollicis longus)
Insertion: Base of the proximal phalanx of the thumb
Actions:
- Radiocarpal joint: abduction (radial deviation) of the hand
- Carpometacarpal and MCP joints of the thumb: extension

Innervation: Radial nerve (C6–C8)

④ **Extensor pollicis longus***

Origin: Posterior surface of the ulna and the interosseous membrane
Insertion: Base of the distal phalanx of the thumb
Actions:
- Wrist joints: extension and abduction (radial deviation) of the hand
- Carpometacarpal joint of the thumb: adduction
- MCP and interphalangeal joints of the thumb: extension

Innervation: Radial nerve (C6–C8)

⑤ **Extensor indicis***

Origin: Posterior surface of the ulna and the interosseous membrane
Insertion: Posterior digital expansion of the second digit
Actions:
- Wrist joint: extension
- MCP, PIP, and DIP joints of the second digit: extension

Innervation: Radial nerve (C6–C8)

* Help with supination

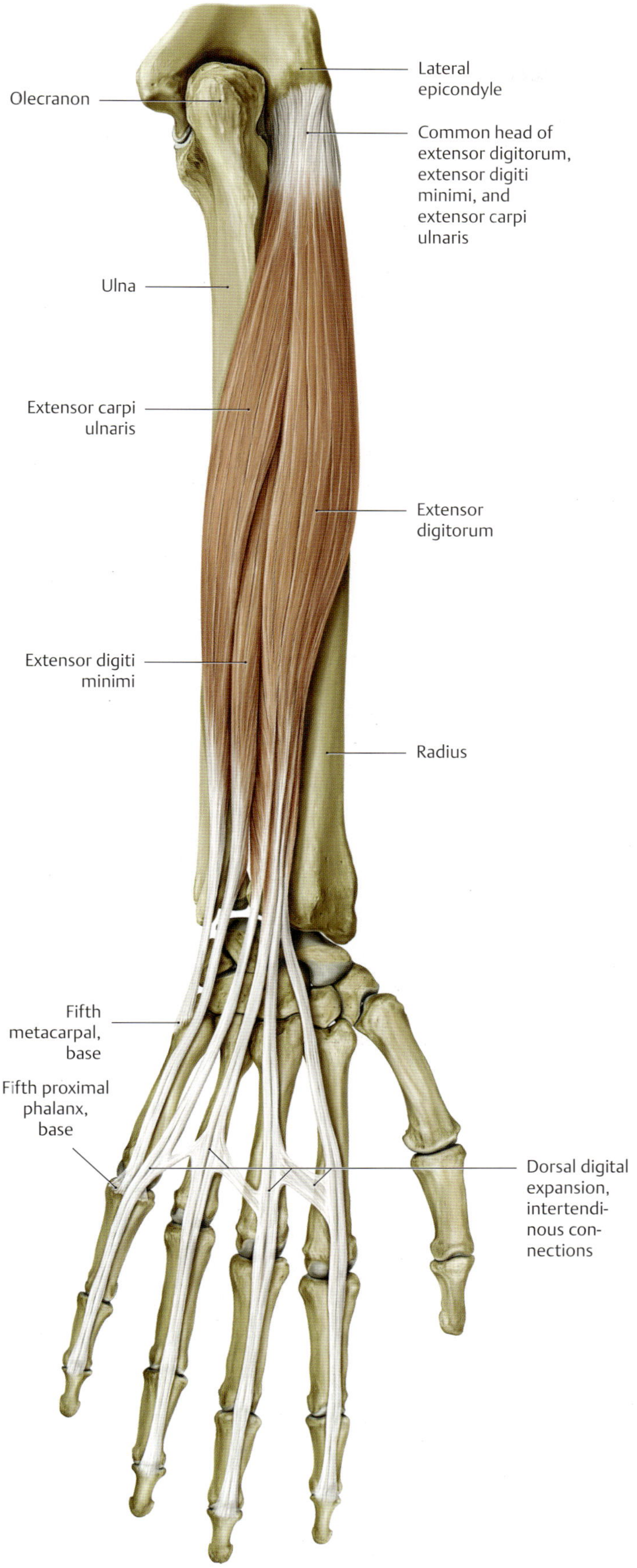

C The superficial extensors (extensor digitorum, extensor digiti minimi, and extensor carpi ulnaris)
Right forearm, posterior (dorsal) view. For the structure of the dorsal digital expansion, see p. 361, Fig. **C**.

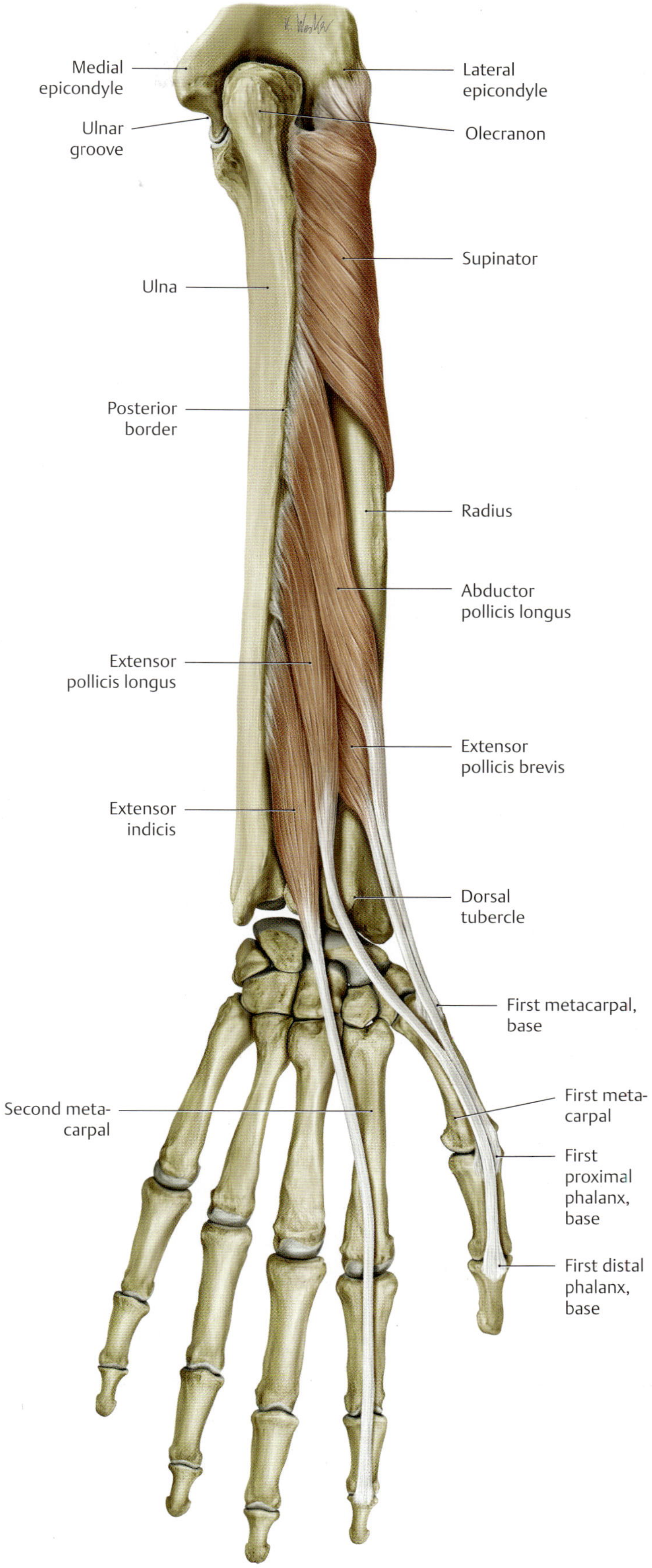

D The deep extensors (supinator, abductor pollicis longus, extensor pollicis brevis, extensor pollicis longus, and extensor indicis)
Right forearm, posterior (dorsal) view.

2.13 The Intrinsic Muscles of the Hand: The Thenar and Hypothenar Muscles

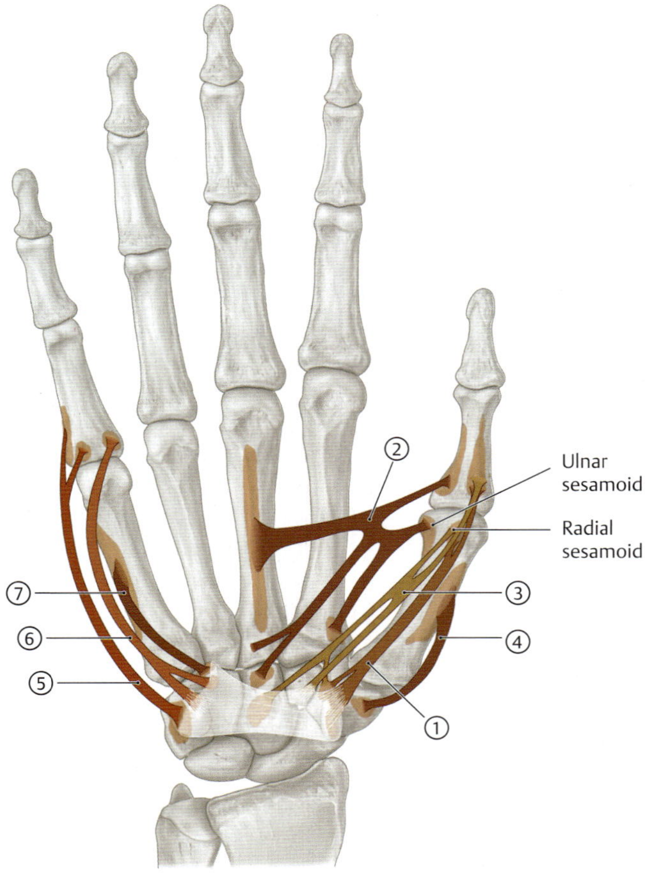

A Schematic of the thenar (①–④) and hypothenar (⑤–⑦) muscles

① Abductor pollicis brevis

Origin: Scaphoid and trapezium, flexor retinaculum
Insertion: Base of the proximal phalanx of the thumb (via the radial sesamoid)
Actions: Abduction of the thumb
Innervation: Median nerve (C6, C7)

② Adductor pollicis

Origin:
- Transverse head: palmar surface of the third metacarpal
- Oblique head: capitate, base of second and third metacarpals

Insertion: Base of the proximal phalanx of the thumb (via the ulnar sesamoid)
Actions:
- CMC joint of the thumb: opposition
- MCP joint of the thumb: flexion

Innervation: Ulnar nerve (C8, T1)

③ Flexor pollicis brevis

Origin:
- Superficial head: flexor retinaculum
- Deep head: capitate, trapezium

Insertion: Base of the proximal phalanx of the thumb (via the radial sesamoid)
Actions:
- CMC joint of the thumb: flexion, opposition
- MCP joint of the thumb: flexion

Innervation:
- Superficial head: Median nerve (C6–T1)
- Deep head: Ulnar nerve (C8, T1)

④ Opponens pollicis

Origin: Trapezium
Insertion: Radial border of the first metacarpal
Actions: CMC joint of the thumb: opposition
Innervation: Median nerve (C6, C7)

⑤ Abductor digiti minimi

Origin: Pisiform
Insertion: Ulnar base of the proximal phalanx and the dorsal digital expansion of the fifth digit
Actions:
- MCP joint of the little finger: flexion and abduction of the little finger
- PIP and DIP joints of the little finger: extension

Innervation: Ulnar nerve (C8, T1)

⑥ Flexor digiti minimi brevis

Origin: Hook of the hamate, flexor retinaculum
Insertion: Base of the proximal phalanx of the fifth digit
Actions: MCP joint of the little finger: flexion
Innervation: Ulnar nerve (C8, T1)

⑦ Opponens digiti minimi

Origin: Hook of the hamate and flexor retinaculum
Insertion: Ulnar border of the fifth metacarpal
Actions: Draws the metacarpal in the palmar direction (opposition)
Innervation: Ulnar nerve (C8, T1)

Palmaris brevis (not shown, see also pp. 357 and 362)

Origin: Ulnar border of the palmar aponeurosis
Insertion: Skin of the hypothenar eminence
Actions: Tightens the palmar aponeurosis (protective function)
Innervation: Ulnar nerve (C8, T1)

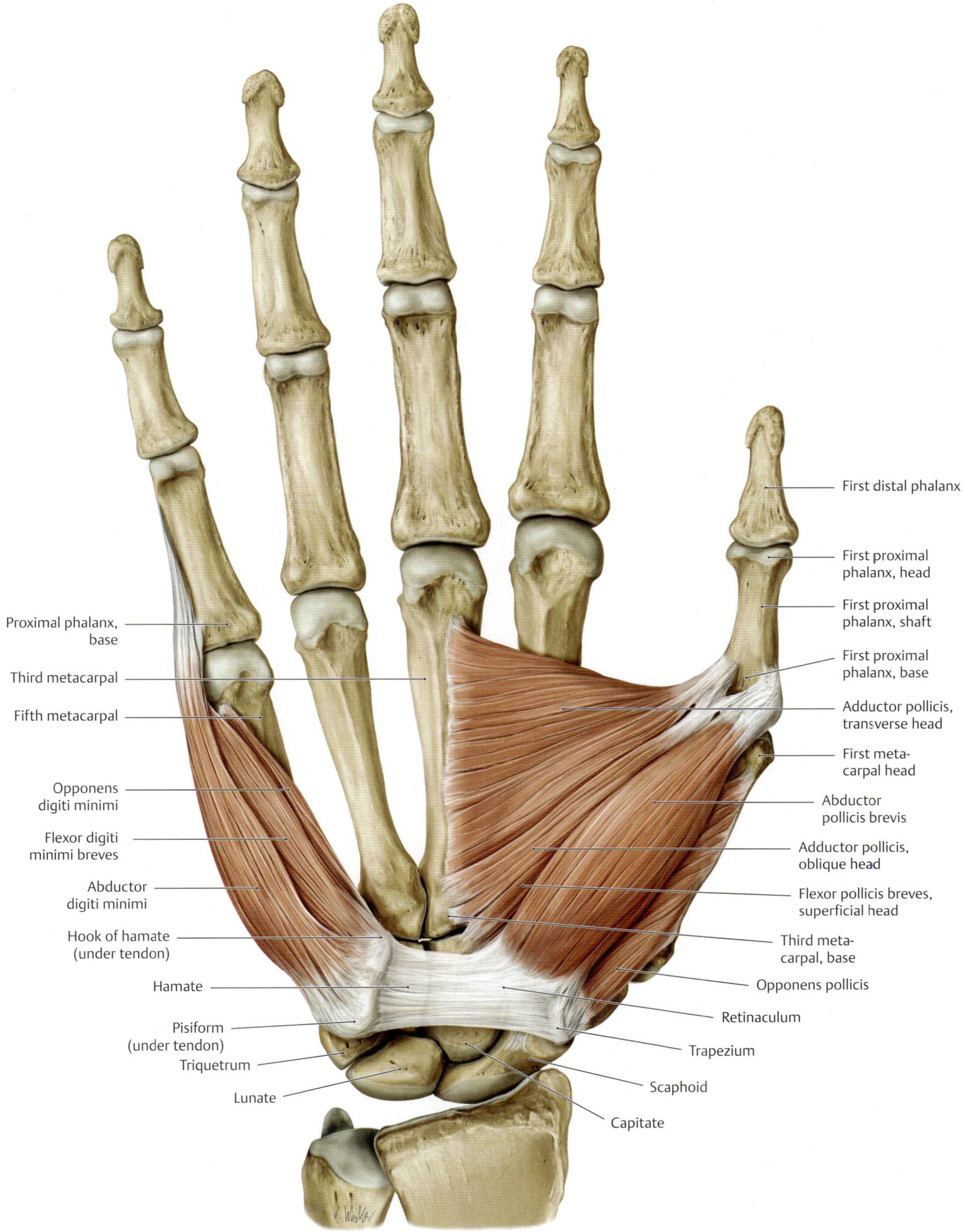

B The thenar muscles (abductor pollicis brevis, adductor pollicis, flexor pollicis brevis, and opponens pollicis) and the hypothenar muscles (abductor digiti minimi, flexor digiti minimi, and opponens digiti minimi)

Right hand, anterior (palmar) view.

2.14 The Intrinsic Muscles of the Hand: Lumbricals and Interossei (Metacarpal Muscles)

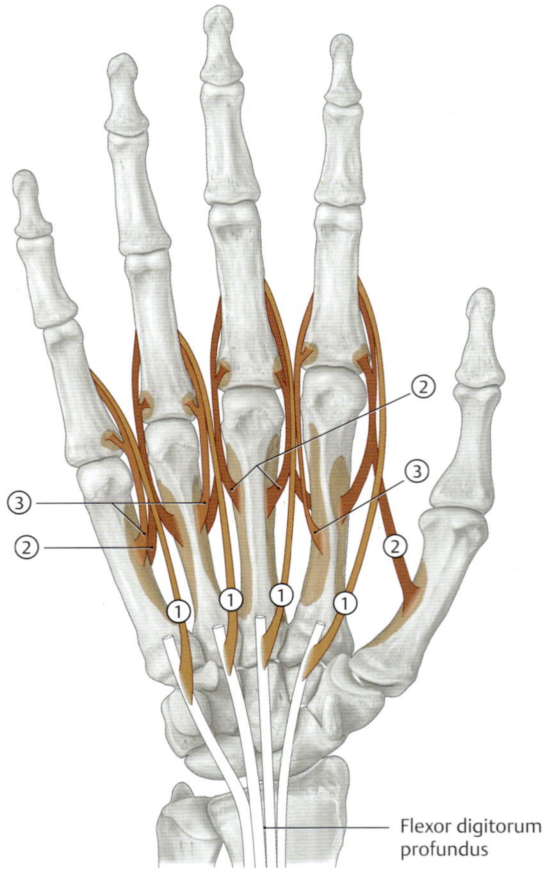

A Schematic of the metacarpal muscles

① **First through fourth lumbrical muscles**

Origin: Radial sides of the tendons of flexor digitorum profundus (variable)

Insertion:
- First lumbrical: dorsal digital expansion of the second digit (index finger)
- Second lumbrical: dorsal digital expansion of the third digit (middle finger)
- Third lumbrical: dorsal digital expansion of the fourth digit (ring finger)
- Fourth lumbrical: dorsal digital expansion of the fifth digit (little finger)

Actions:
- MCP joints of the second through fifth digits: flexion
- PIP and DIP joints of the second through fifth digits: extension

Innervation:
- Median nerve, C8, T1 (first and second lumbricals)
- Ulnar nerve, C8, T1 (third and fourth lumbricals)

② **First through fourth dorsal interossei**

Origin: By two heads from adjacent sides of the first through fifth metacarpals

Insertion:
- Dorsal digital expansion of the second through fourth digits, base of the proximal phalanx
- First interosseus: radial side of the second proximal phalanx (index finger)
- Second interosseus: radial side of the third proximal phalanx (middle finger)
- Third interosseus: ulnar side of the third proximal phalanx (middle finger)
- Fourth interosseus: ulnar side of the fourth proximal phalanx (ring finger)

Actions:
- MCP joints of the second through fourth digits: flexion
- PIP and DIP joints of the second through fourth digits: extension and abduction of the fingers (abduction of the index and ring fingers from the middle finger)

Innervation: Ulnar nerve (C8, T1)

③ **First through third palmar interossei**

Origin:
- First interosseus: ulnar side of the second metacarpal (index finger)
- Second interosseus: radial side of the fourth metacarpal (ring finger)
- Third interosseus: radial side of the fifth metacarpal (little finger)

Insertion: Dorsal digital expansion and base of the proximal phalanx of the associated finger

Actions:
- MCP joints of the second, fourth, and fifth digits: flexion
- PIP and DIP joints of the second, fourth, and fifth digits: extension and adduction of the fingers (adduction of the second, fourth, and fifth digits toward the middle finger)

Innervation: Ulnar nerve (C8, T1)

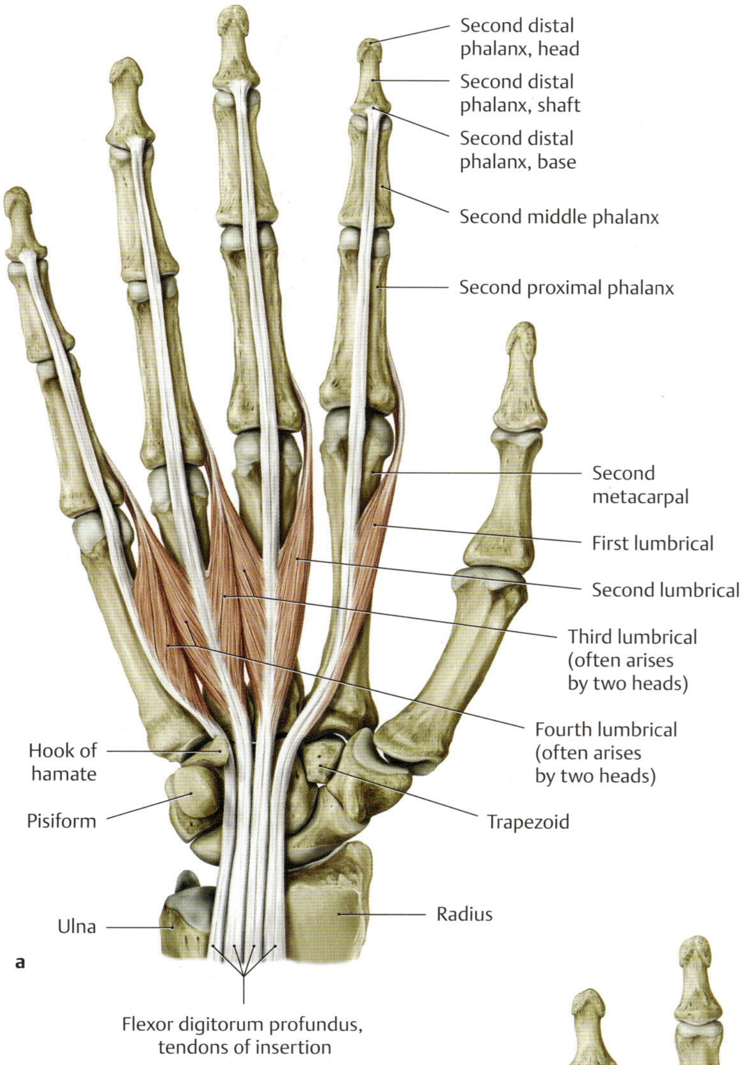

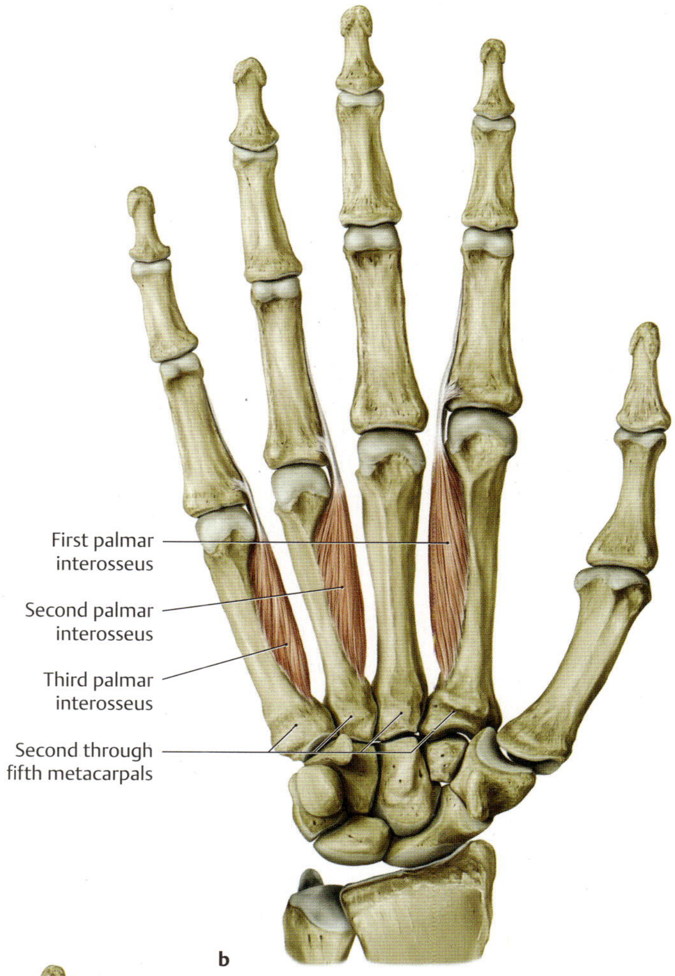

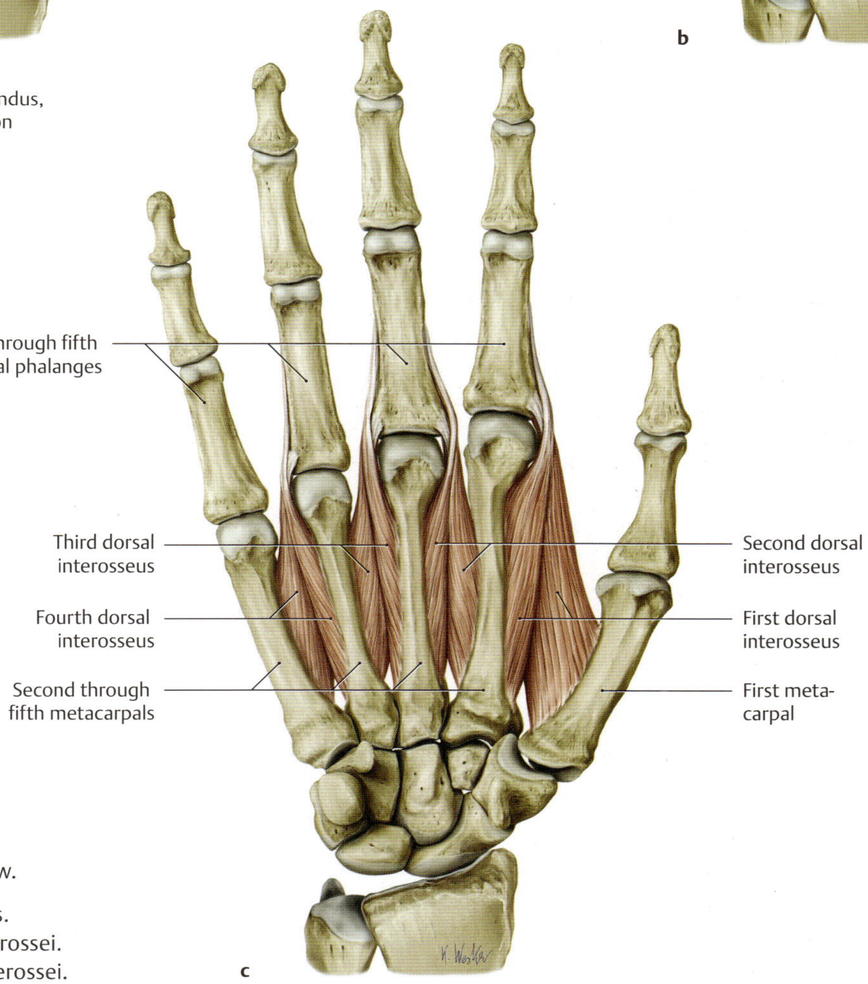

B The muscles of the hand
Right hand, anterior (palmar) view.

- **a** First through fourth lumbricals.
- **b** First through third palmar interossei.
- **c** First through fourth dorsal interossei.

2.15 Overview of Muscle Functions: Shoulder (Glenohumeral) Joint

A Movements of the shoulder (glenohumeral) joint*

Type of movement	Extent of movement	Muscle	Innervation	Responsible nerve segment
Flexion	90° (>90° = elevation)	• Deltoid (clavicular part) • Biceps brachii • Pectoralis major (clavicular and sternocostal parts) • Coracobrachialis	• Axillary nerve • Musculocutaneous nerve • Medial and lateral pectoral nerves • Musculocutaneous nerve	• C5–C6 • C5–C7 • C5–T1 • C5–C7
Extension	40°	• Latissimus dorsi • Teres major • Triceps brachii, (long head) • Deltoid (spinal part)	• Thoracodorsal nerve • Lower subscapular nerve • Radial nerve • Axillary nerve	• C6–C8 • C5–C8 • C6–C8 • C5–C6
Abduction	90° (>90° = elevation)	• Deltoid (acromial part), above 60°, the entire muscle • Supraspinatus • Biceps brachii (long head)	• Axillary nerve • Suprascapular nerve • Musculocutaneous nerve	• C5–C6 • C4–C6 • C5–C7
Adduction	20–40°	• Pectoralis major • Latissimus dorsi • Triceps brachii (long head) • Teres major • Deltoid (clavicular and spinal parts) • Coracobrachialis	• Medial and lateral pectoral nerves • Thoracodorsal nerve • Radial nerve • Lower subscapular nerve • Axillary nerve • Musculocutaneous nerve • Musculocutaneous nerve	• C5–T1 • C6–C8 • C6–C8 • C5–C8 • C5–C6 • C5–C7 • C6–C7
Internal rotation	50–95°	• Subscapularis • Pectoralis major • Biceps brachii, (long head) • Deltoid (clavicular part) • Teres major • Latissimus dorsi	• Upper and lower subscapular nerve • Medial and lateral pectoral nerves • Musculocutaneous nerve • Axillary nerve • Lower subscapular nerve • Thoracodorsal nerve	• C5–C8 • C5–T1 • C5–C7 • C5–C6 • C5–C8 • C6–C8
External rotation	60–90° (depending on extent of flexion)	• Infraspinatus • Teres minor • Deltoid (spinal part)	• Suprascapular nerve • Axillary nerve • Axillary nerve	• C4–C6 • C5–C6 • C5–C6

*The starting point of the functional approach is movement within the joint. The muscles, which are involved in all movements, are listed in the order of their strength. Remember, however, that although individual muscles may be involved in particular movements, their functional significance may be negligible. In order to describe the many types of movement in a joint, each joint (respective of its degree of freedom) is assigned clearly defined basic movements along defined axes. Only the combination of these basic movements allows for a range of movements depending on the individual joint.

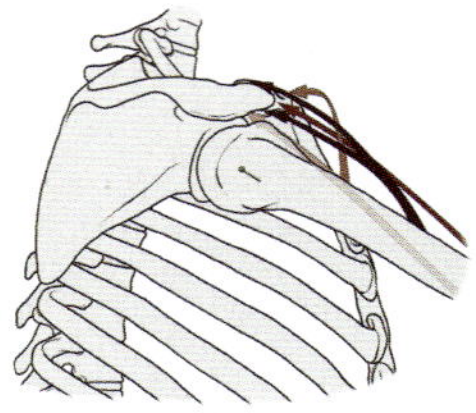
a Flexion

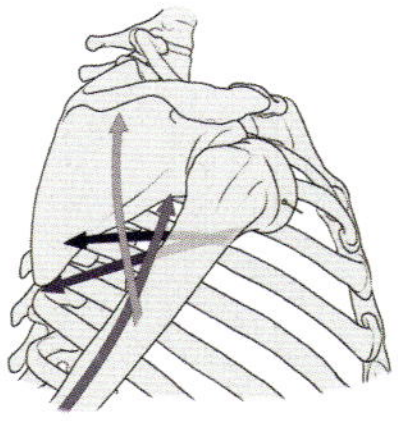
b Extension

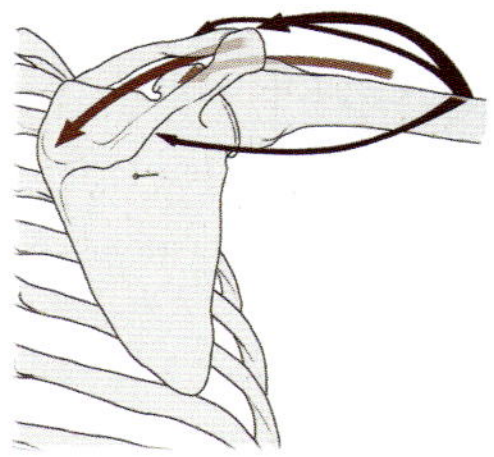
c Abduction

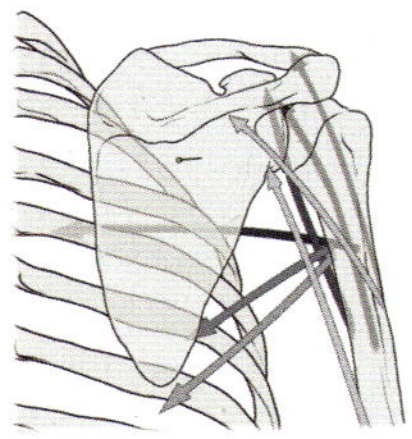
d Adduction

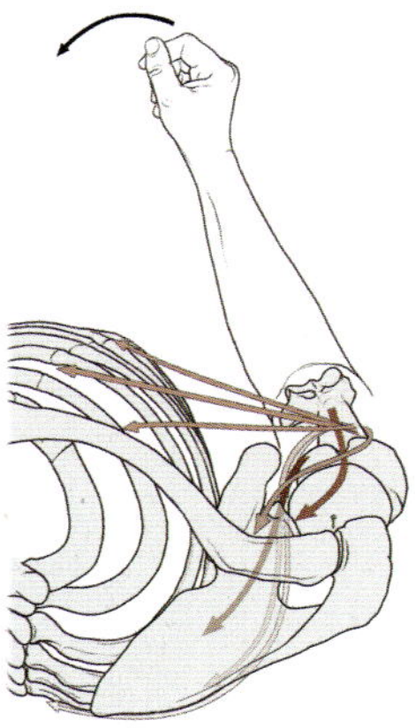
e Internal rotation

f External rotation

B Movements of the shoulder joint

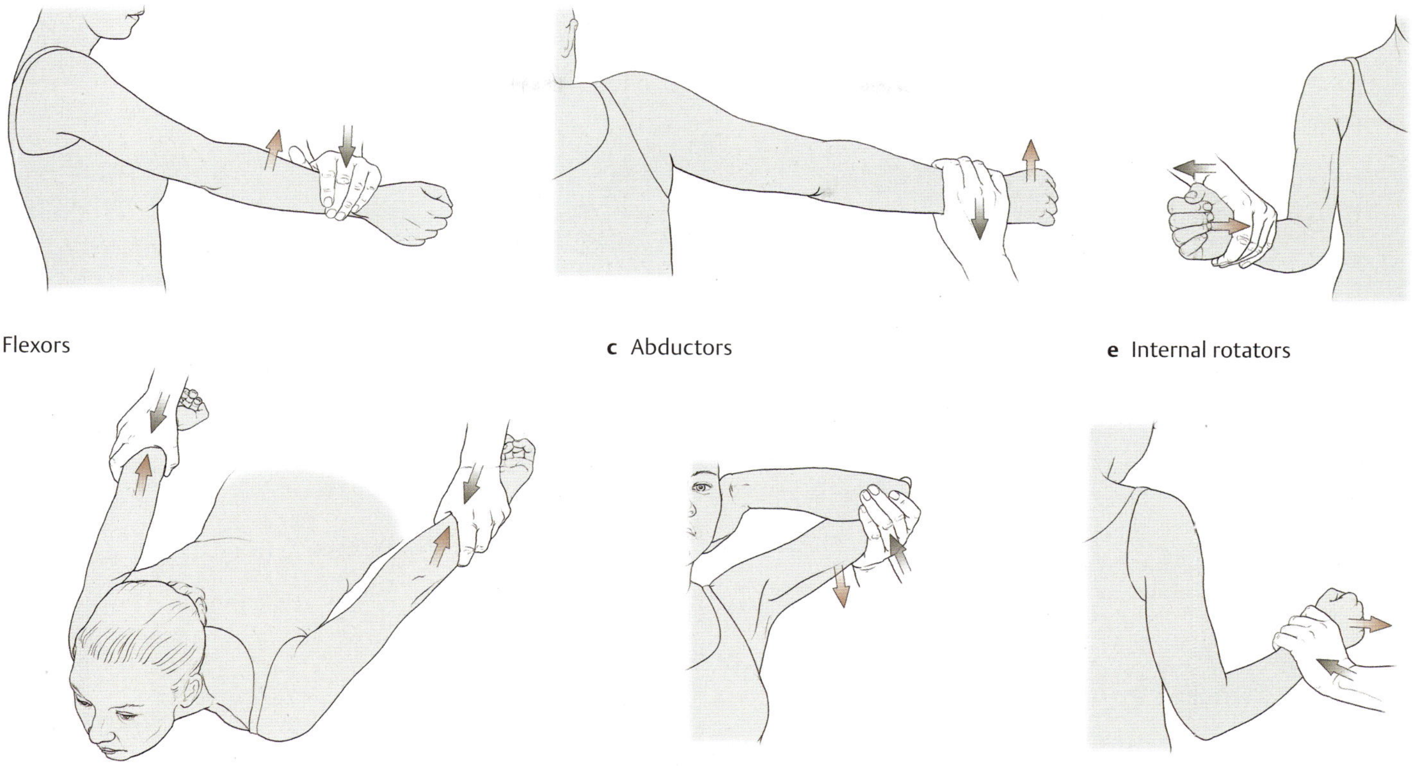

a Flexors

b Extensors

c Abductors

d Adductors

e Internal rotators

f External rotators

C Functional muscle testing in the shoulder joint

In order to test muscle strength, the patient is asked to actively move the muscle group against some resistance from the examiner. The force generated by the muscle action is measured using a scale from 0 to 5.

D Clinical symptoms of shoulder joint muscle shortening or weakness

Muscles	Symptoms of muscle shortening	Symptoms of muscle weakness
Flexors	Shoulder is commonly protracted.	In order to compensate for muscle weakness, the acromion is lifted, and the upper body, due to increased lumbar lordosis, moves posteriorly. In case of isolated weakness of the deltoid, the shoulder contours are flattened with the acromion protuding.
Extensors	Significantly impaired movements in elevation and external rotation.	Weakness in the extensors becomes apparent when extreme loads are placed on the shoulder, e.g., propping oneself up using both arms.
Abductors	Impaired movement as a result of shortened muscles (primarily supraspinatus and biceps brachii), especially in extension, adduction, and external rotation of the shoulder joint.	The arm cannot maintain the arm in abduction against gravity. Flattening of the contours of the shoulder with typical indentation below the acromion, often subluxation of the shoulder joint with the arm hanging down.
Adductors	Protracted shoulder and increased thoracic kyphosis, along with compensatory cervical lordosis. Significantly impaired arm movements in flexion and abduction above the head.	Dysfunction in sequence of abdominal muscles of the opposite side, meaning mainly hitting and chopping motions with the affected limb are impeded. Difficulty holding large or heavy objects at waist level.
Internal rotators	Significantly impaired movement when lifting the arm above the head with simultaneous external rotation.	An isolated weakness of the internal rotators is rare and is compensated for by the flexors (most of the daily chores are performed in flexion and internal rotation).
External rotators	Shortening of the external rotators is rare (mostly following long periods of immobilization) and leads to impaired internal rotation.	Insufficent inferior glide of the head of the humerus during elevation, resulting in compensatory activation of abductors (e.g., supraspinatus), pulling up of the shoulder, and increased lateral trunk flexion.

2.16 Overview of Muscle Functions: Elbow Joint

A Movements of the elbow joint*

Type of movement	Extent of movement	Muscle	Innervation	Responsible nerve segment
Flexion	130–150°	• Biceps brachii • Brachialis • Brachioradialis • Extensors carpi radialis longus and brevis • Pronator teres • Flexor digitorum superficialis • Palmaris longus	• Musculocutaneous nerve • Musculocutaneous nerve • Radial nerve • Radial nerve • Median nerve • Median nerve • Median nerve	• C5–C7 • C5–C7 • C5–C7 • C5–C7 • C6 • C7–T1 • C8–T1
Extension	10°	• Triceps brachii • Anconeus	• Radial nerve • Radial nerve	• C6–C8 • C6–C8
*Supination**	90°	• Biceps brachii • Supinator • Abductor pollicis longus • Extensor pollicis brevis • Extensor pollicis longus • Extensor indicis • Brachioradialis (pronated)	• Musculocutaneous nerve • Radial nerve • Radial nerve • Radial nerve • Radial nerve • Radial nerve • Radial nerve	• C5–C7 • C5–C6 • C6–C8 • C6–C8 • C6–C8 • C6–C8 • C5–C7
*Pronation**	90°	• Pronator quadratus • Pronator teres • Flexor carpi radialis • Brachioradialis (supinated)	• Median nerve • Median nerve • Median nerve • Radial nerve	• C8–T1 • C6 • C6–C8 • C5–C7

*In addition to the elbow joint, the distal radioulnar joint is involved in turning the forearm.

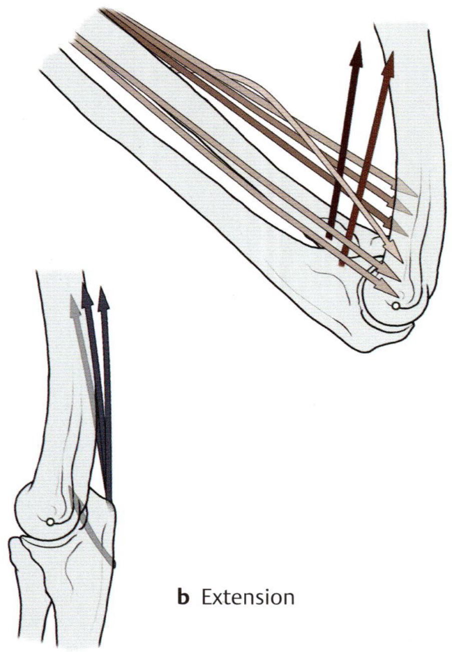

a Flexion

b Extension

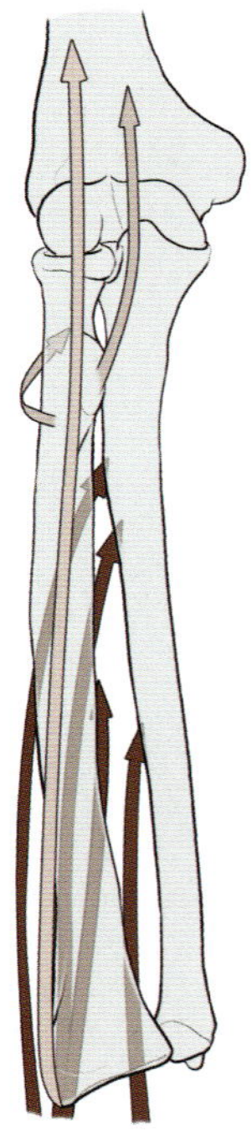

c Supination

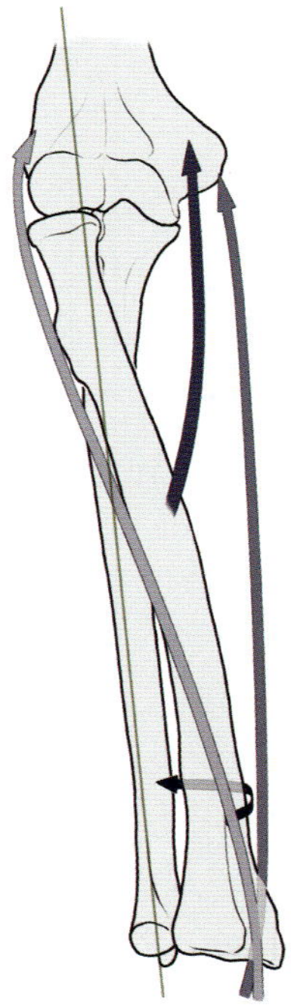

d Pronation

B Movements of the elbow joint

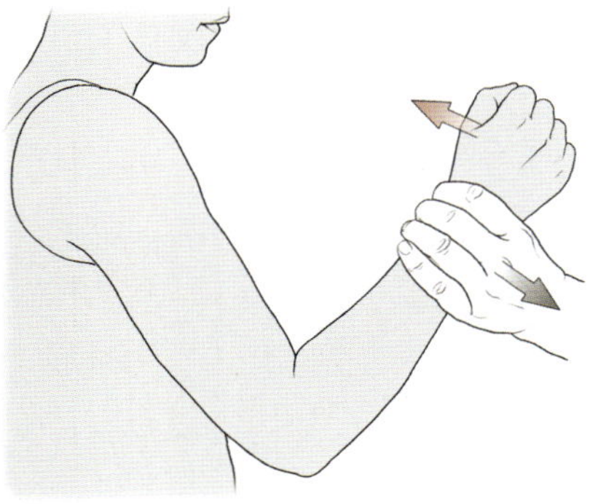

a Flexors

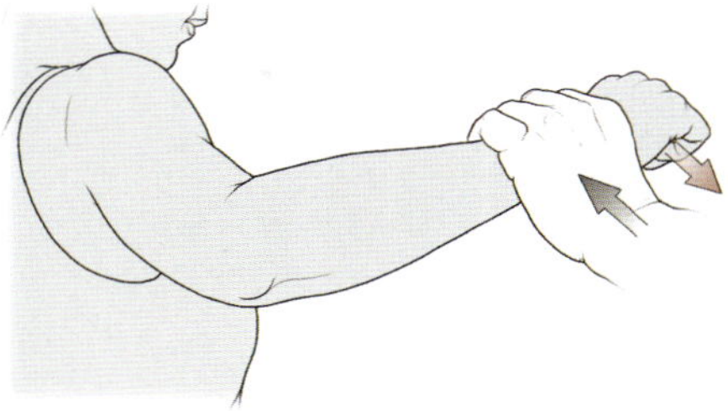

b Extensors

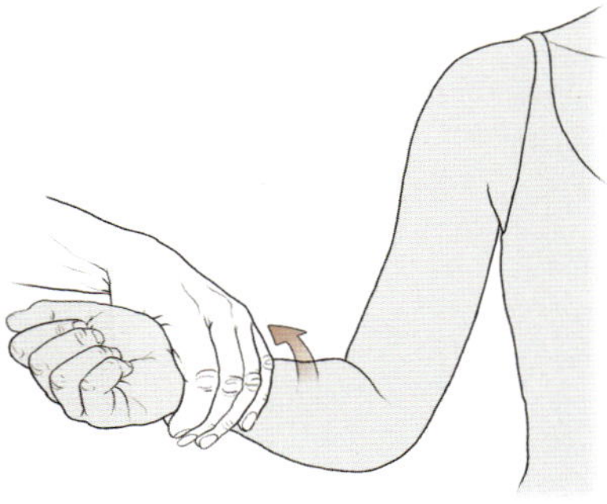

c Supinators (elbow joint is flexed)

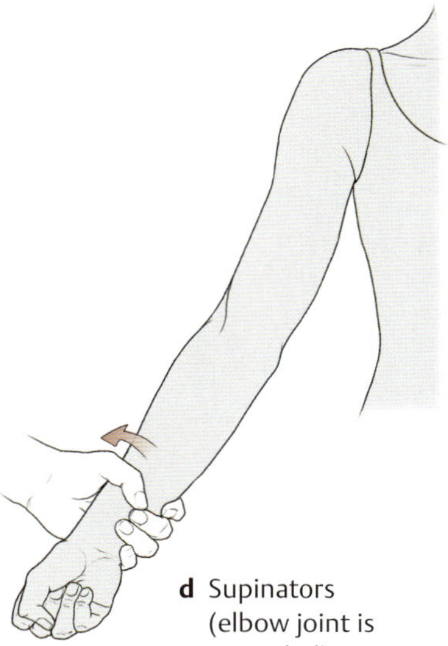

d Supinators (elbow joint is extended)

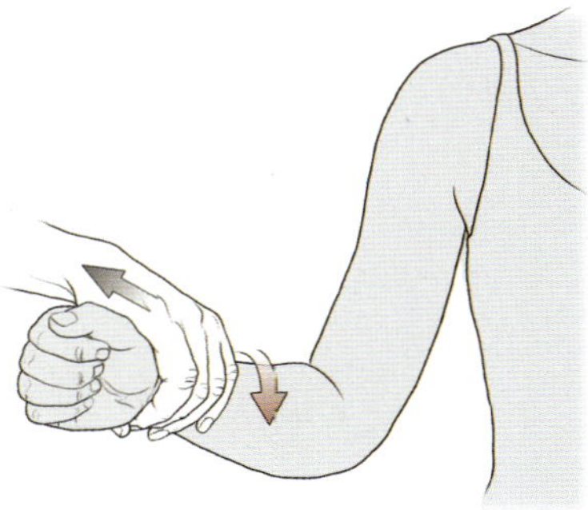

e Pronators

C Functional testing of muscles of the elbow joint

D Clinical symptoms of elbow joint muscle shortening and weakness

Muscles	Symptoms of muscle shortening	Symptoms of muscle weakness
Flexors	Flexion contracture of the elbow with involvement of the biceps brachii, resulting in significant functional impairment, requiring a shift from supination to pronation.	Muscle weakness, e.g., of the biceps brachii, leads to preponderance of pronation; thus, eating, for instance (moving a spoon to the mouth), is significantly impaired.
Extensors	Extension contracture of the elbow resulting in significant functional impairment in routine daily movements.	It is not possible to throw an object. The affected patient is not able to use a walking device since he or she cannot apply any weight on his or her hands due to inability to extend the elbow.
Supinators	Impaired movement when forearm is pronated. Compensated for by increased internal rotation and abduction of the shoulder joint.	Weakness in supinators becomes apparent in numerous routine daily movements: e.g., unlocking a door, turning off a faucet, or tightening a screw.
Pronators	Impaired movement when forearm is supinated. Compensated for by increased external rotation and adduction of the shoulder joint.	Routine daily movements are hampered (see above), such as unlocking a door or turning off a faucet.

2.17 Overview of Muscle Functions: Wrist

A Movement of the proximal and distal wrist (radiocarpal and mediocarpal joints)

Type of movement	Extent of movement	Muscle	Innervation	Responsible nerve segment
Palmar flexion	60–80°	• Flexor digitorum superficialis • Flexor digitorum profundus • Flexor carpi ulnaris • Flexor pollicis longus • Flexor carpi radialis • Palmaris longus	• Median nerve • Median nerve • Ulnar nerve • Ulnar nerve • Median nerve • Median nerve • Median nerve	• C7–T1 • C7–T1 • C8–T1 • C8–T1 • C6–C8 • C6–C8 • C8–T1
Dorsal extension	40–60°	• Extensor digitorum • Extensors carpi radialis longus and brevis • Extensor carpi ulnaris • Extensor indicis • Extensor pollicis longus • Extensor digiti minimi	• Radial nerve • Radial nerve • Radial nerve • Radial nerve • Radial nerve • Radial nerve	• C6–C8 • C5–C7 • C6–C8 • C6–C8 • C6–C8 • C6–C8
Abduction (radial deviation)	20°	• Extensors carpi radialis longus and brevis • Abductor pollicis longus • Extensor pollicis brevis • Extensor pollicis longus • Flexor pollicis longus • Flexor carpi radialis	• Radial nerve • Radial nerve • Radial nerve • Radial nerve • Median nerve • Median nerve	• C5–C7 • C6–C8 • C6–C8 • C6–C8 • C6–C8 • C6–C8
Abduction (ulnar deviation)	30–40°	• Extensor carpi ulnaris • Flexor carpi ulnaris • Extensor digiti minimi	• Radial nerve • Ulnar nerve • Radial nerve	• C6–C8 • C8–T1 • C6–C8

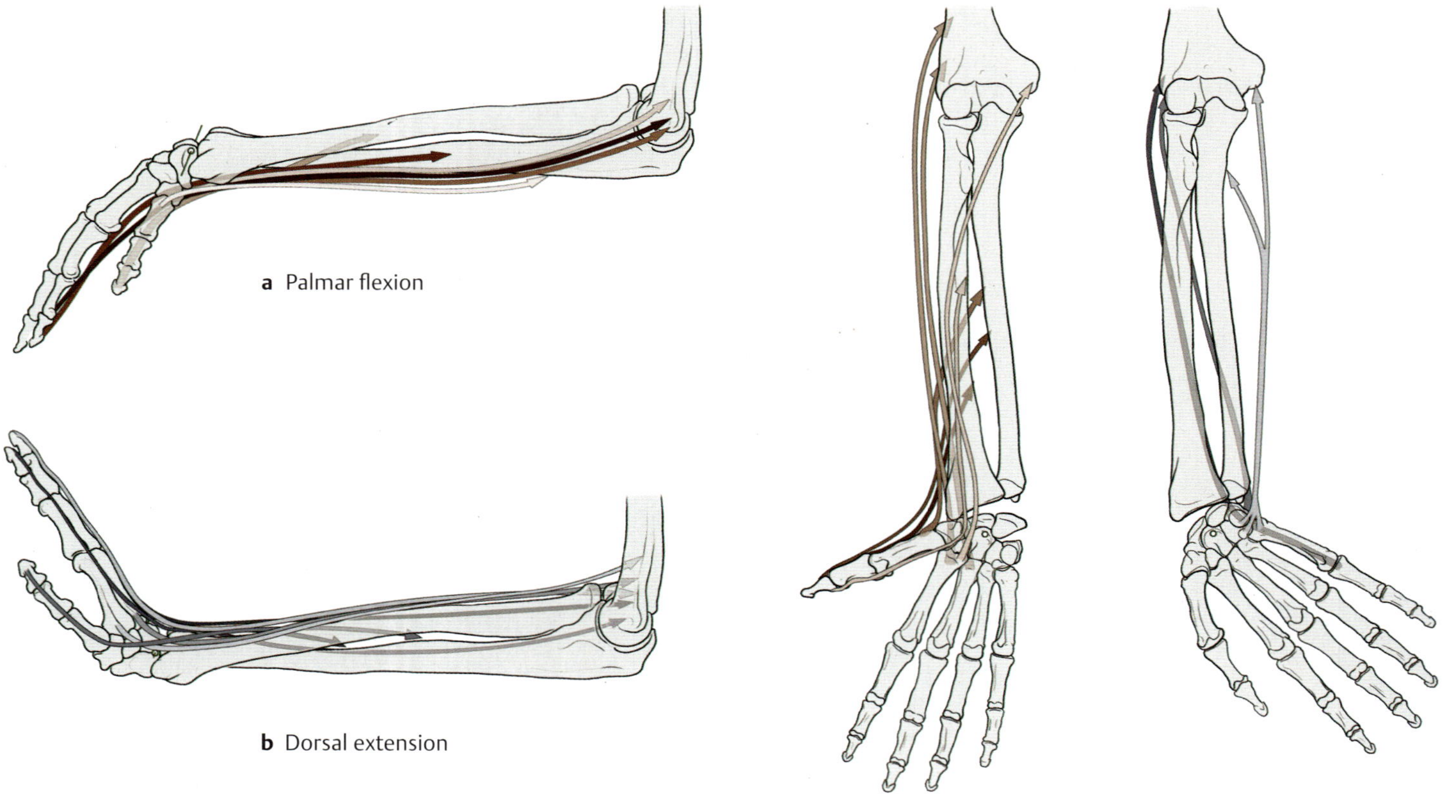

a Palmar flexion

b Dorsal extension

c Radial abduction

d Ulnar abduction

B Movements of the proximal and distal wrist

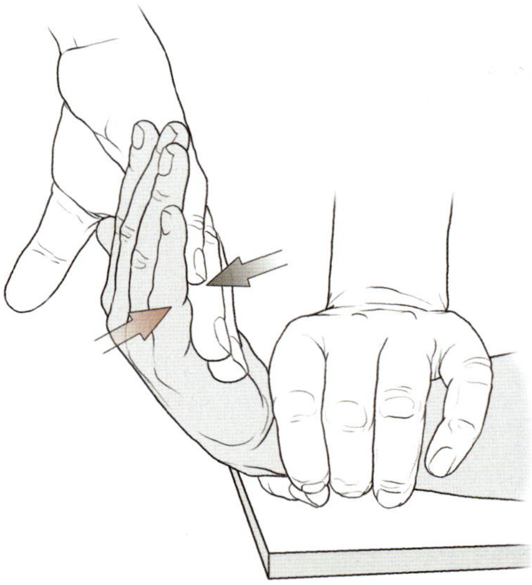

a Palmar flexion

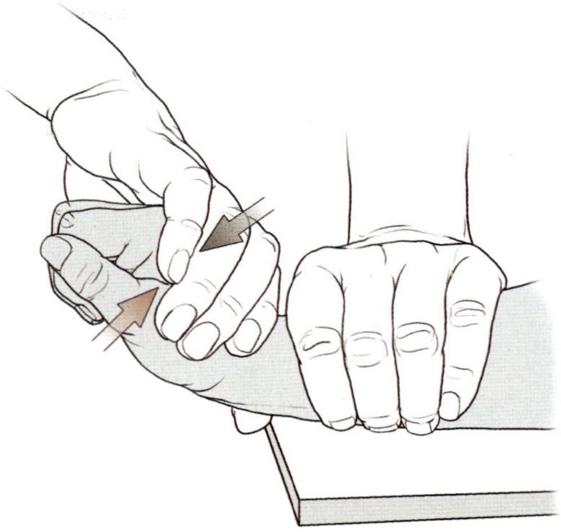

b Dorsal extension

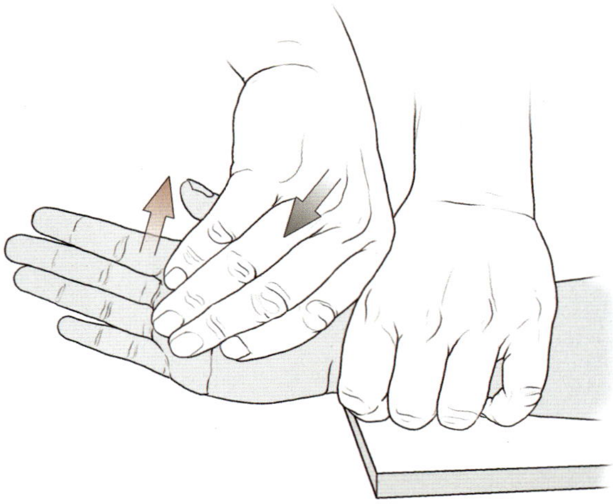

c Radial abduction

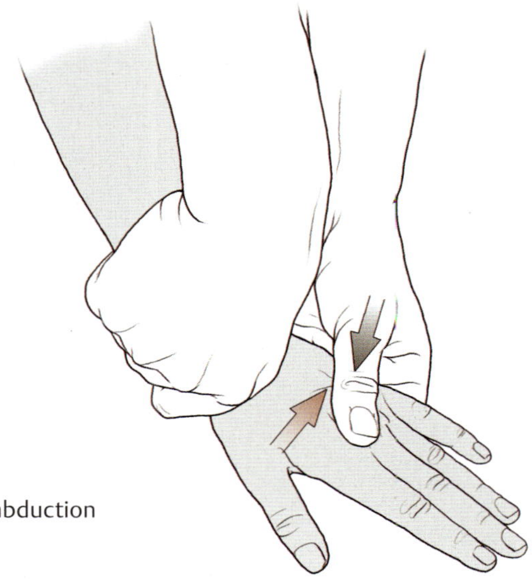

d Ulnar abduction

C Functional muscle testing of the proximal and distal wrist

D Clinical symptoms of muscle shortening and weakness of the proximal and distal wrist

Muscles	Symptoms of muscle shortening	Symptoms of muscle weakness
Palmar flexors	Impaired movement in dorsal extension of the wrist and finger joints, combined with constant overloading to palmar flexors, which leads to epicondylitis medialis (golfer's elbow).	When lifting heavy objects, the affected patient is not able to sufficiently stabilize the wrist with supinated forearm; thus, the wrist tilts in a dorsal direction. In the long run, this leads to injuries resulting from overload of the tendons of origin of the finger and hand flexors.
Dorsal extensors	Impaired movement in palmar flexion of wrist and finger joints, combined/accompanied by constant overloading to dorsal extensors, which leads to epicondylitis lateralis (tennis elbow).	When lifting heavy objects, the affected patient is not able to sufficiently stabilize the wrist with pronated forearm; thus, the wrist tilts in a palmar direction. In the long run, this leads to injuries resulting from overload of the tendons of origin of the finger and hand extensors.
Radial abductors	Weakening of ulnar abduction.	Always combined with weakened dorsal extension and palmar flexion.
Ulnar abductors	Weakening of radial abduction.	Loss of strength functionally less visible.

3.1 The Posterior Muscles of the Shoulder Girdle and Shoulder Joint

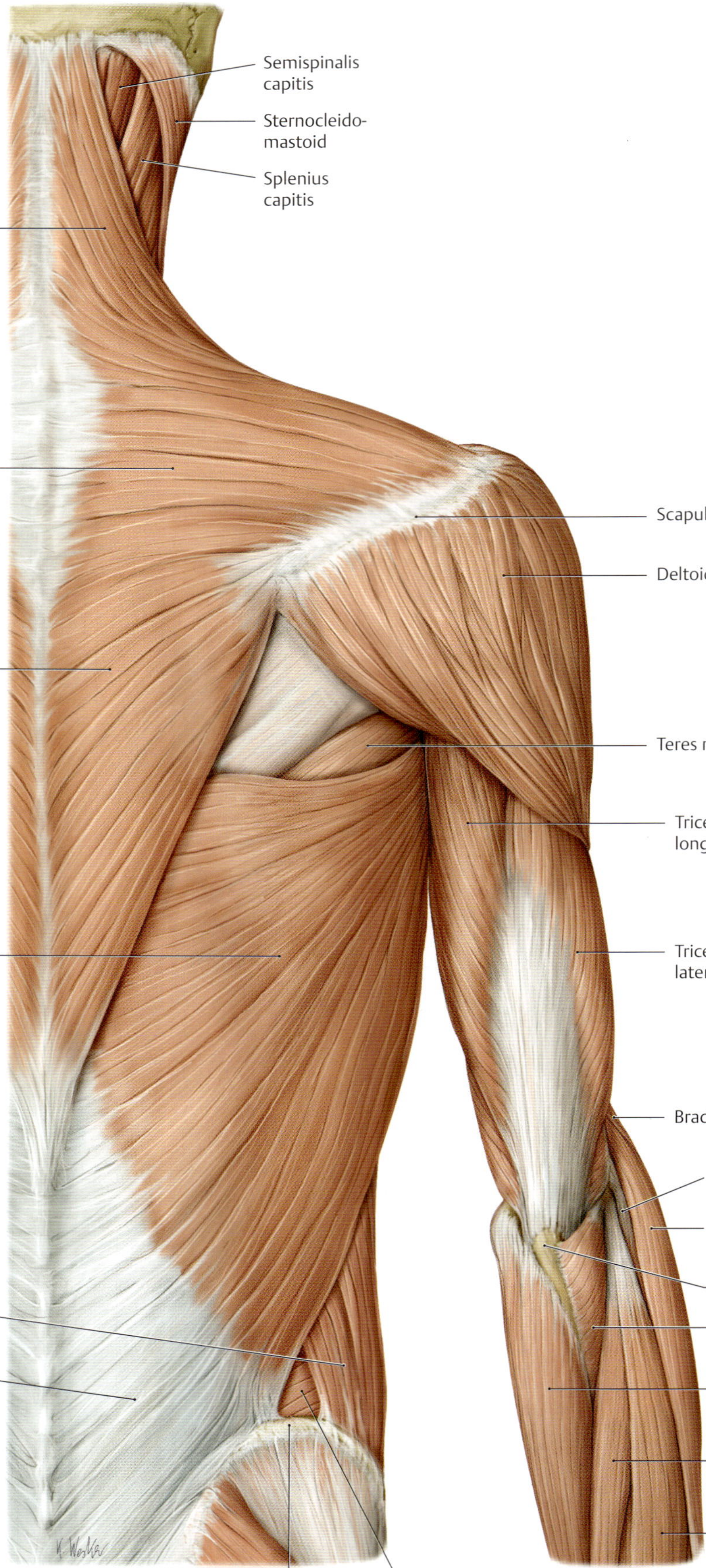

A Muscles of the shoulder girdle and shoulder joint
Right half of the body, posterior view. Superficial layer.

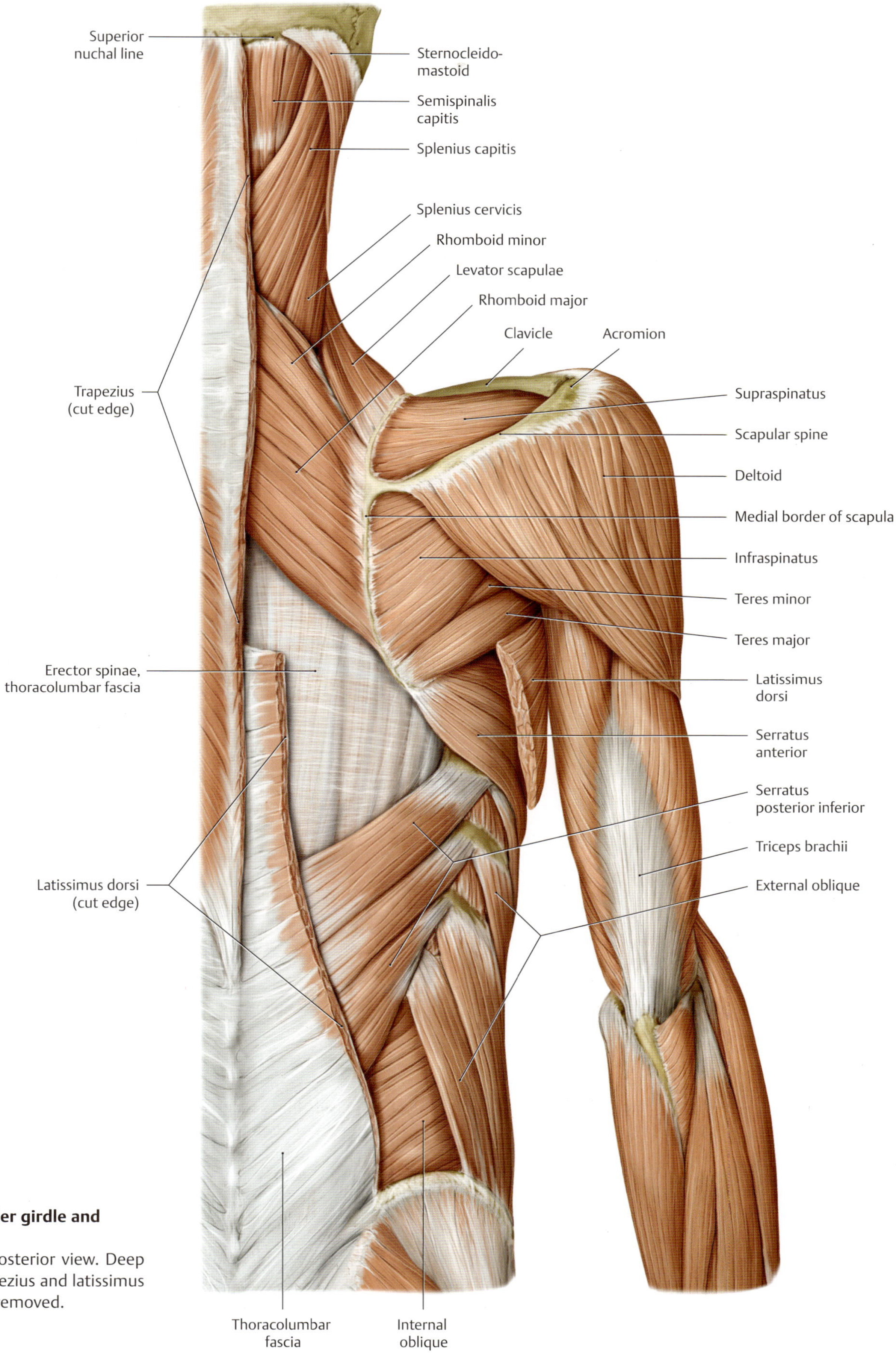

B Muscles of the shoulder girdle and shoulder joint
Right half of the body, posterior view. Deep layer. Portions of the trapezius and latissimus dorsi muscles have been removed.

3.2 The Posterior Muscles of the Shoulder Joint and Arm

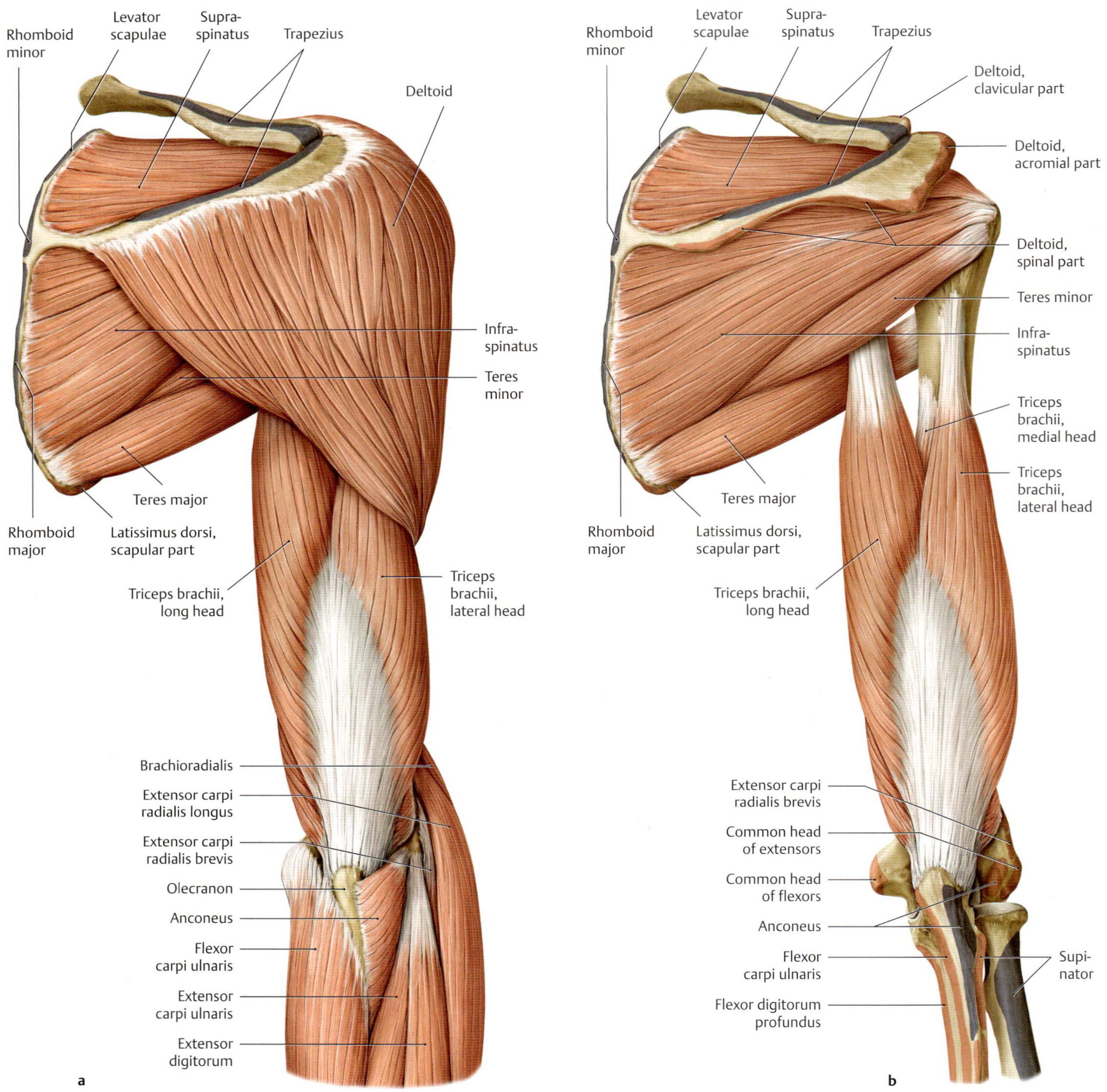

A Muscles of the right shoulder and upper arm, posterior view
Origin shown in red, insertion in blue.

a After removal of the trapezius.
b After removal of the deltoid and forearm muscles.

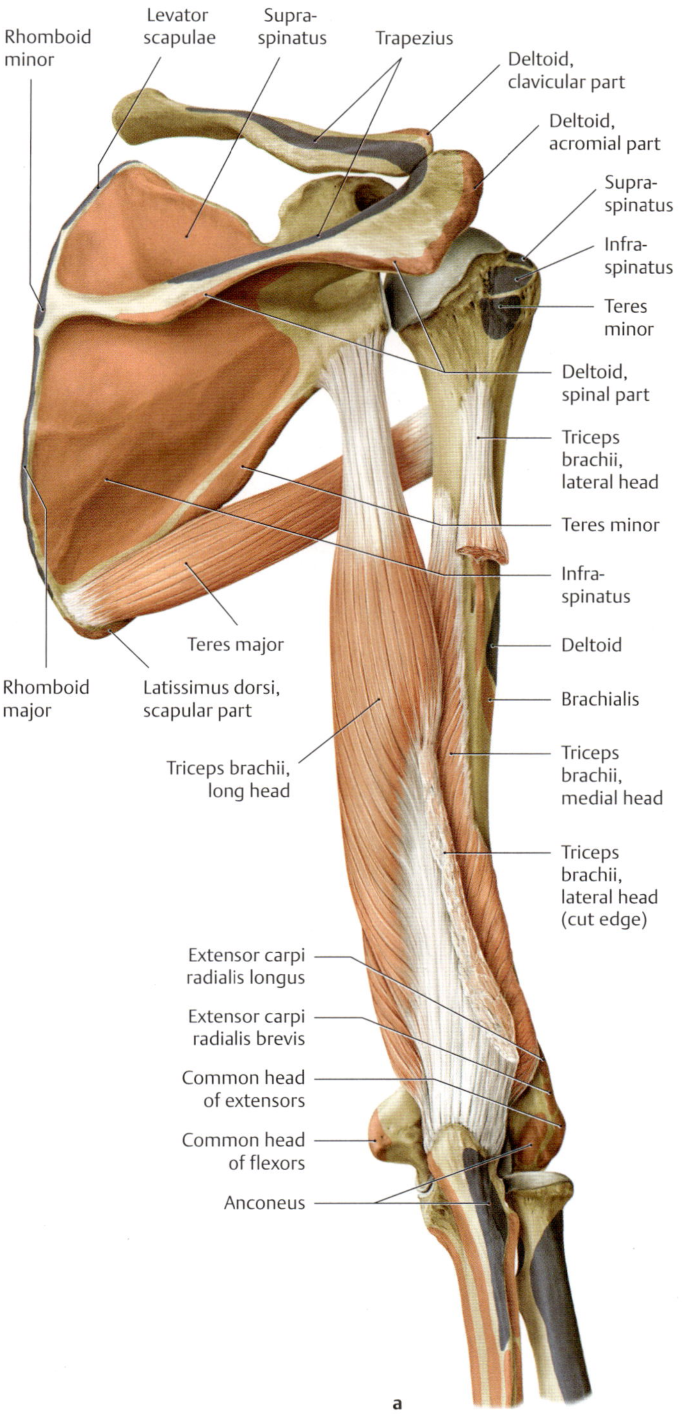

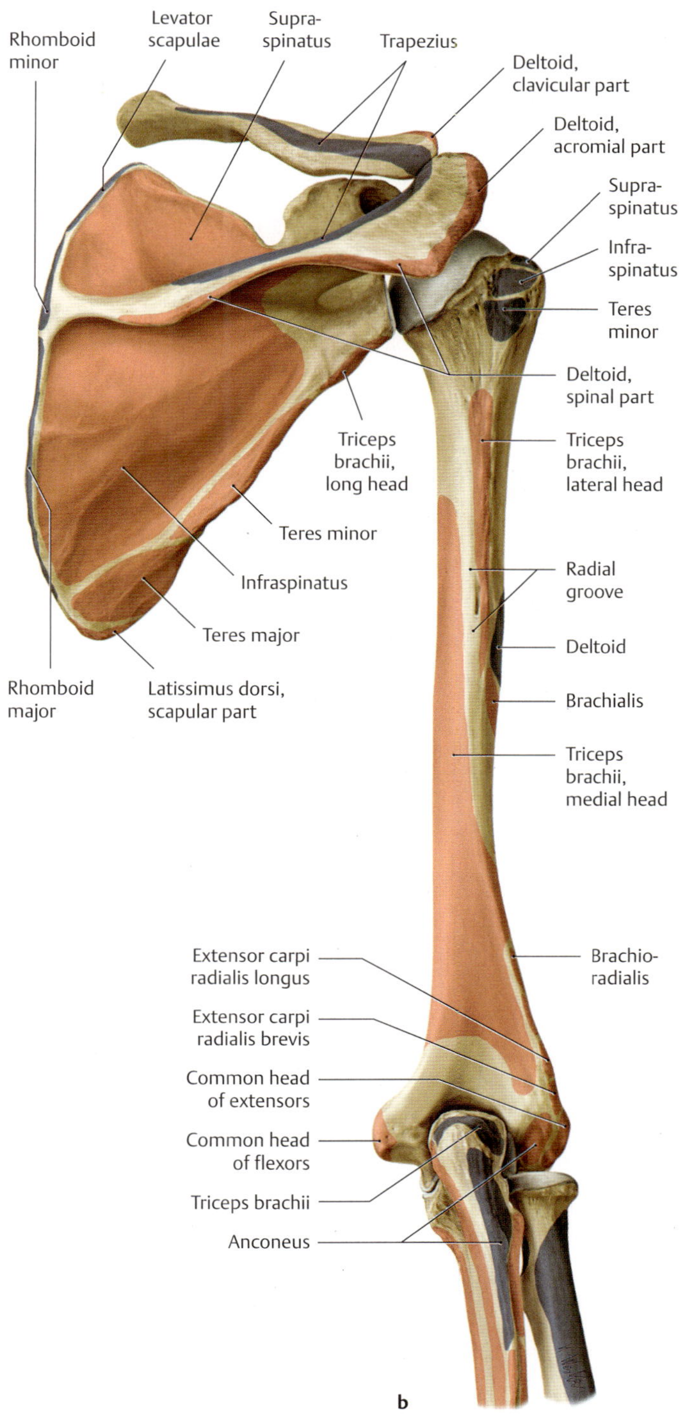

B Muscles of the right shoulder and upper arm, posterior view
Origin shown in red, insertion in blue.

a The supraspinatus, infraspinatus, and teres minor have been removed. The lateral head of the triceps brachii has been partially removed.
b All the muscles have been removed.

3.3 The Anterior Muscles of the Shoulder Girdle and Shoulder Joint

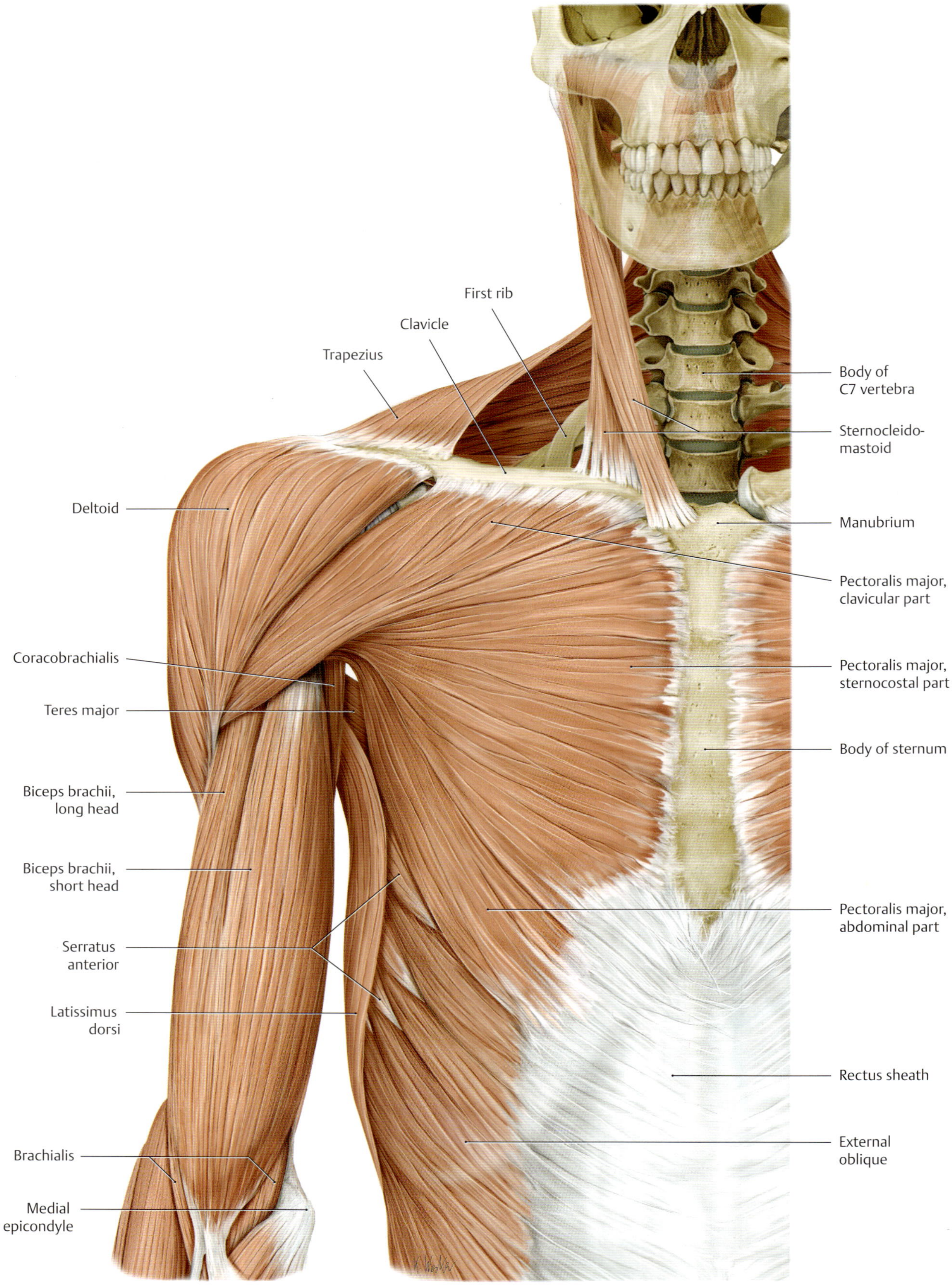

A Muscles of the shoulder and arm
Right side, anterior view.

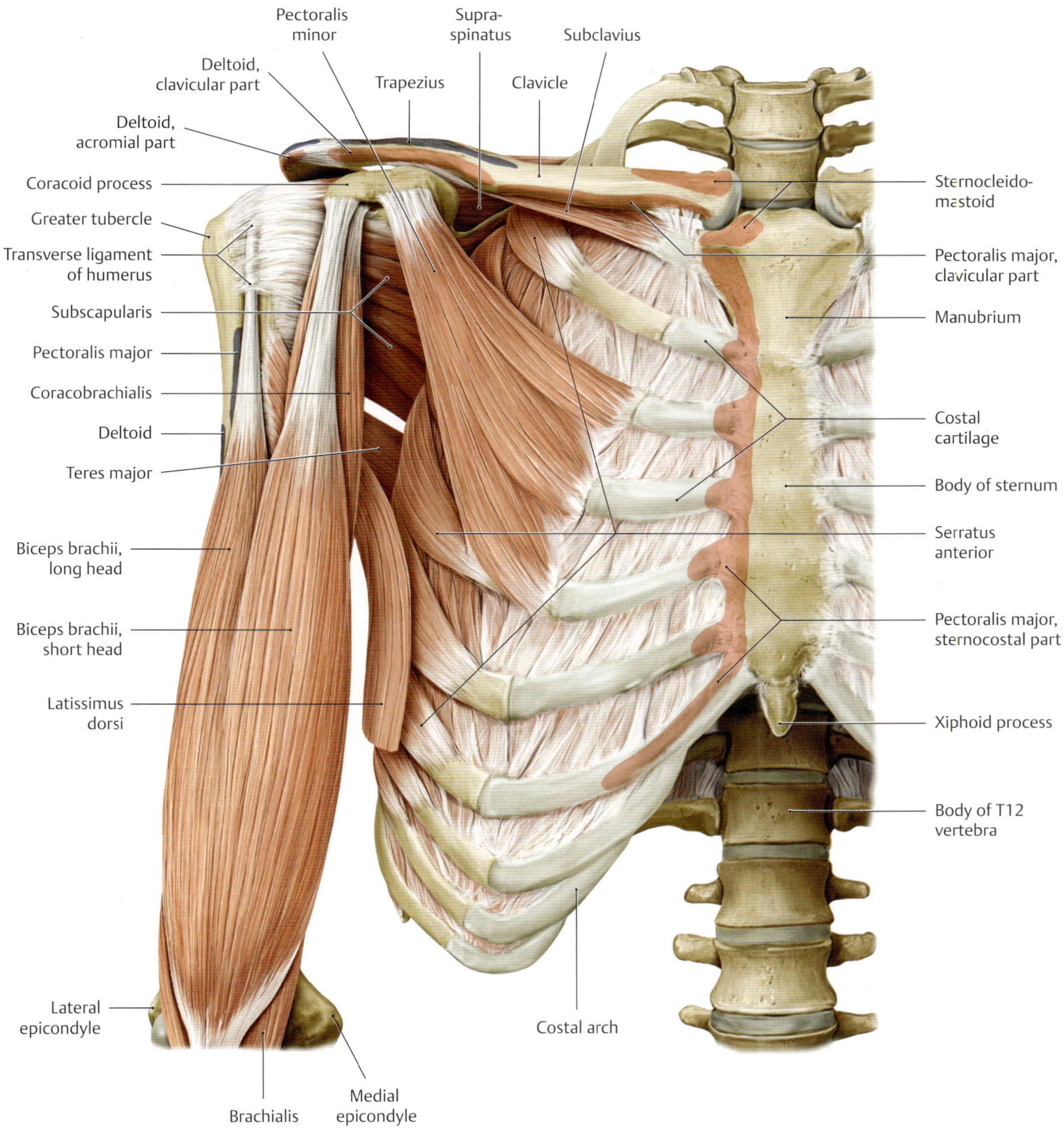

B Muscles of the right shoulder and upper arm, anterior view
Origin shown in red, insertion in blue. The sternocleidomastoid, trapezius, pectoralis major, deltoid, and external oblique muscles have been removed completely and the latissimus dorsi partially.

* Indicates irregularly present ligament structures that cross the Intertubercular sulcus.

3.4 The Muscles of the Shoulder Joint and Anterior Arm

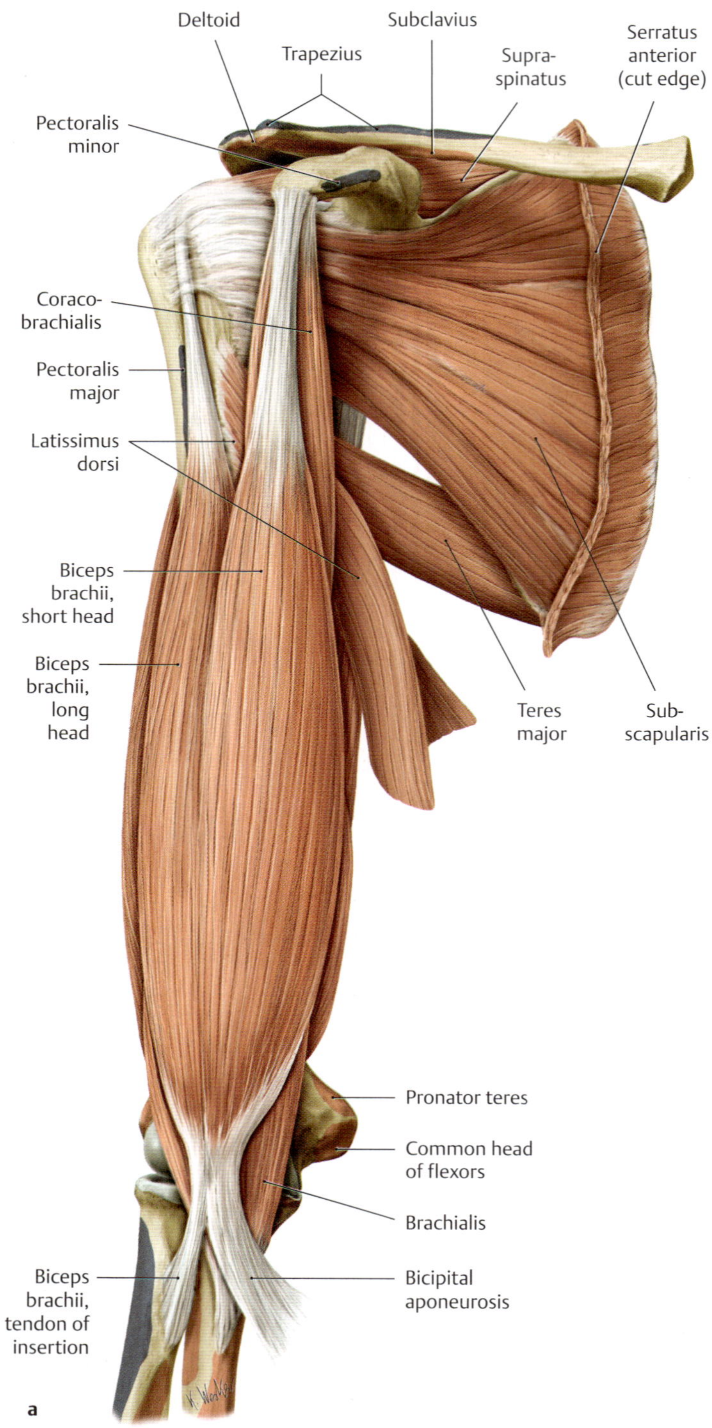

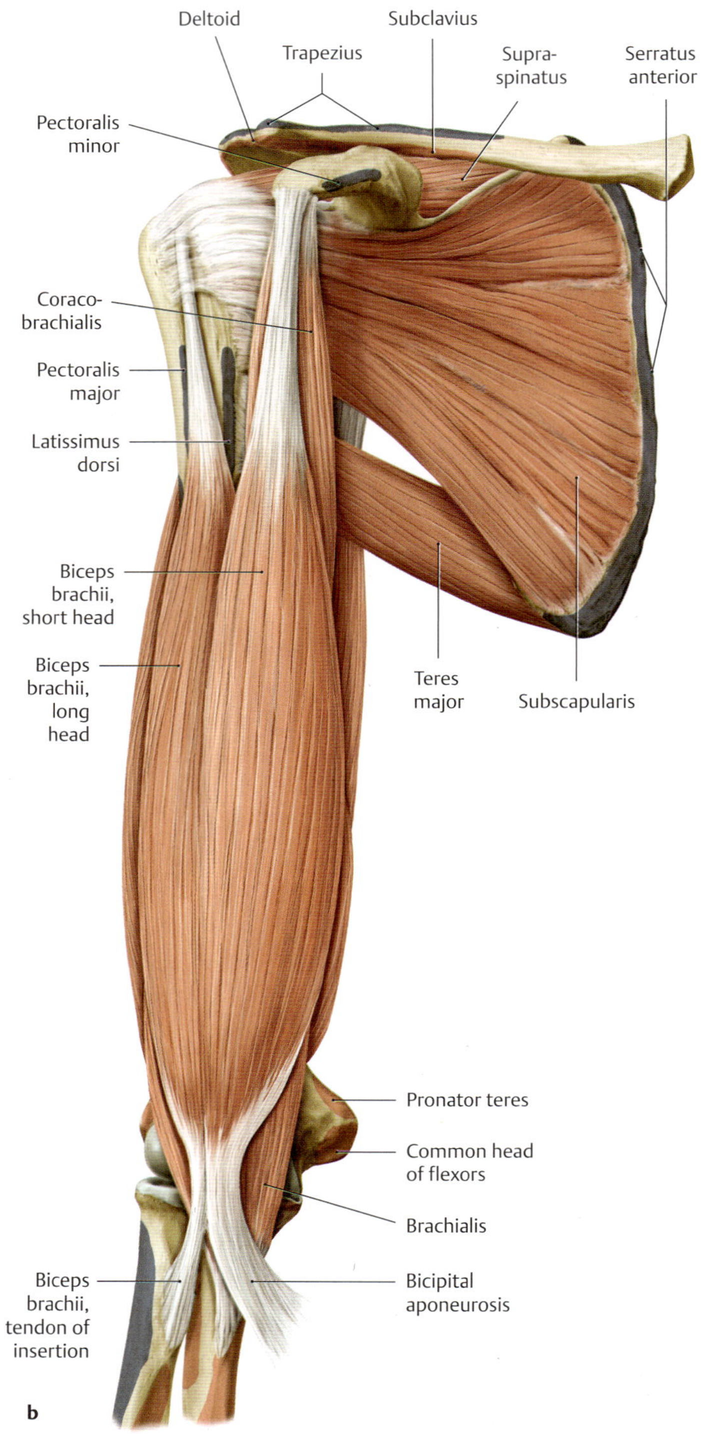

A Muscles of the right shoulder and upper arm, anterior view
Origin shown in red, insertion in blue.

a After removal of the thoracic skeleton. Latissimus dorsi and serratus anterior have been removed to their insertions.

b Latissimus dorsi and serratus anterior have been completely removed.

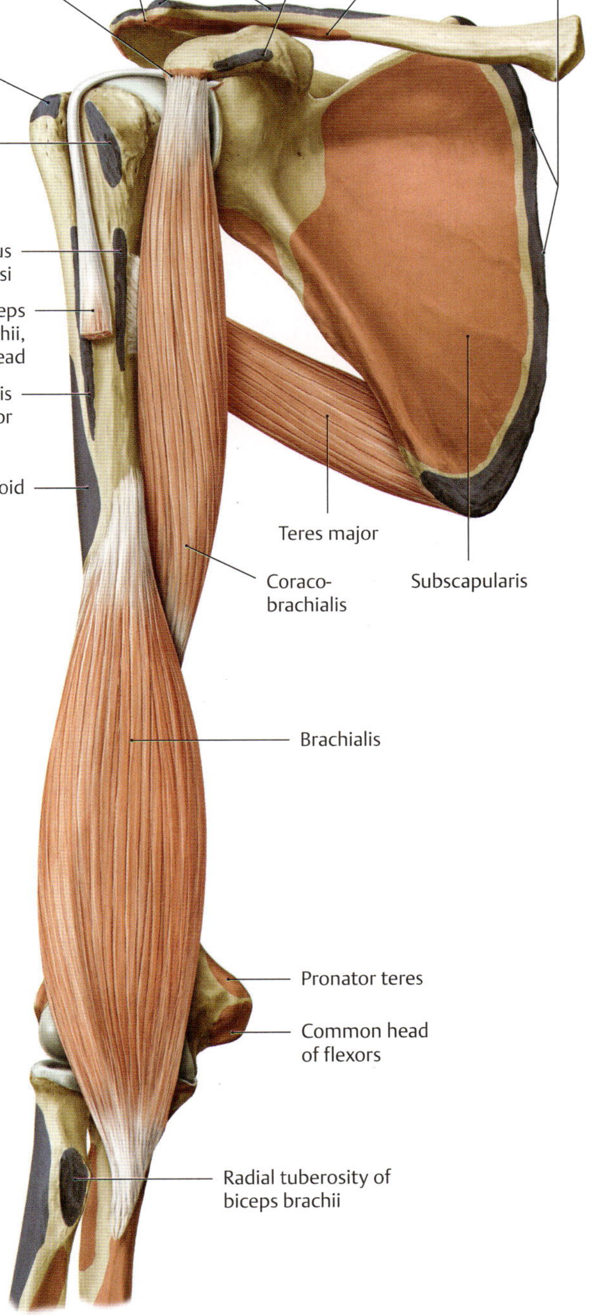

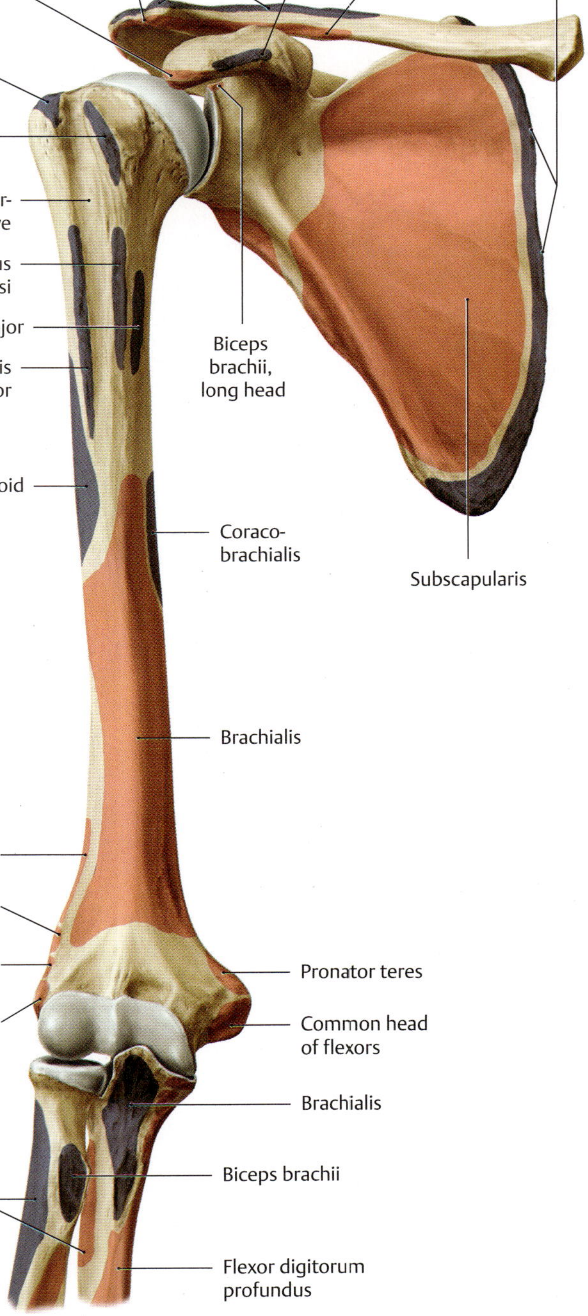

B Muscles of the right shoulder and upper arm, anterior view
Origin shown in red, insertion in blue.

a After removal of the thoracic skeleton and the subscapularis and supraspinatus muscles. The biceps brachii has been removed to the tendon of origin of its long head (note its course through the intertubercular groove).

b All the muscles have been removed.

3.5 The Muscles of the Anterior Forearm

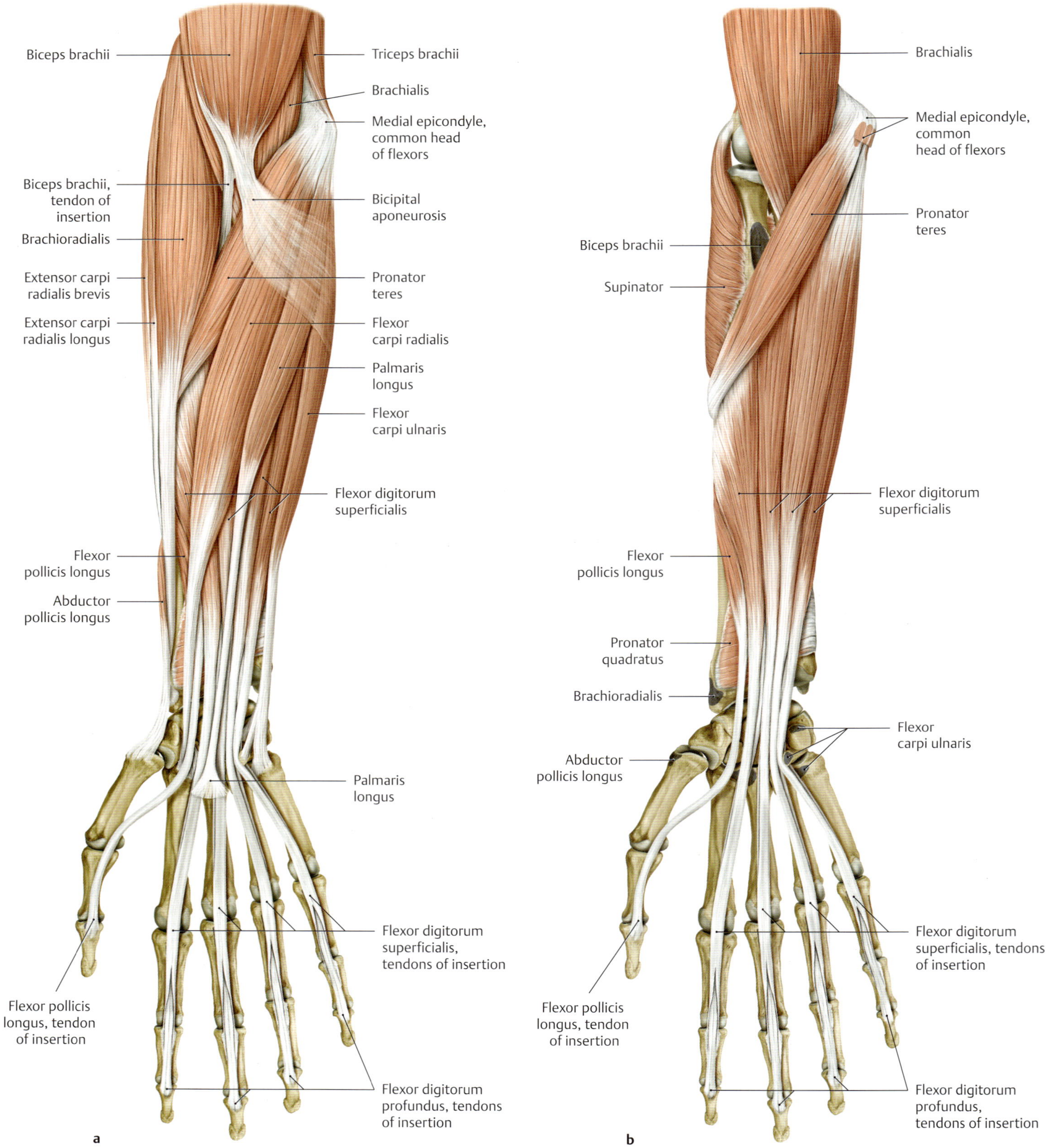

A Muscles of the right forearm, anterior view
Origin shown in red, insertion in blue.

a The superficial flexors and the radialis group are shown.

b The radialis group (brachioradialis, extensor carpi radialis longus, extensor carpi radialis brevis) has been completely removed, along with the flexor carpi radialis, flexor carpi ulnaris, abductor pollicis longus, palmaris longus, and biceps brachii.

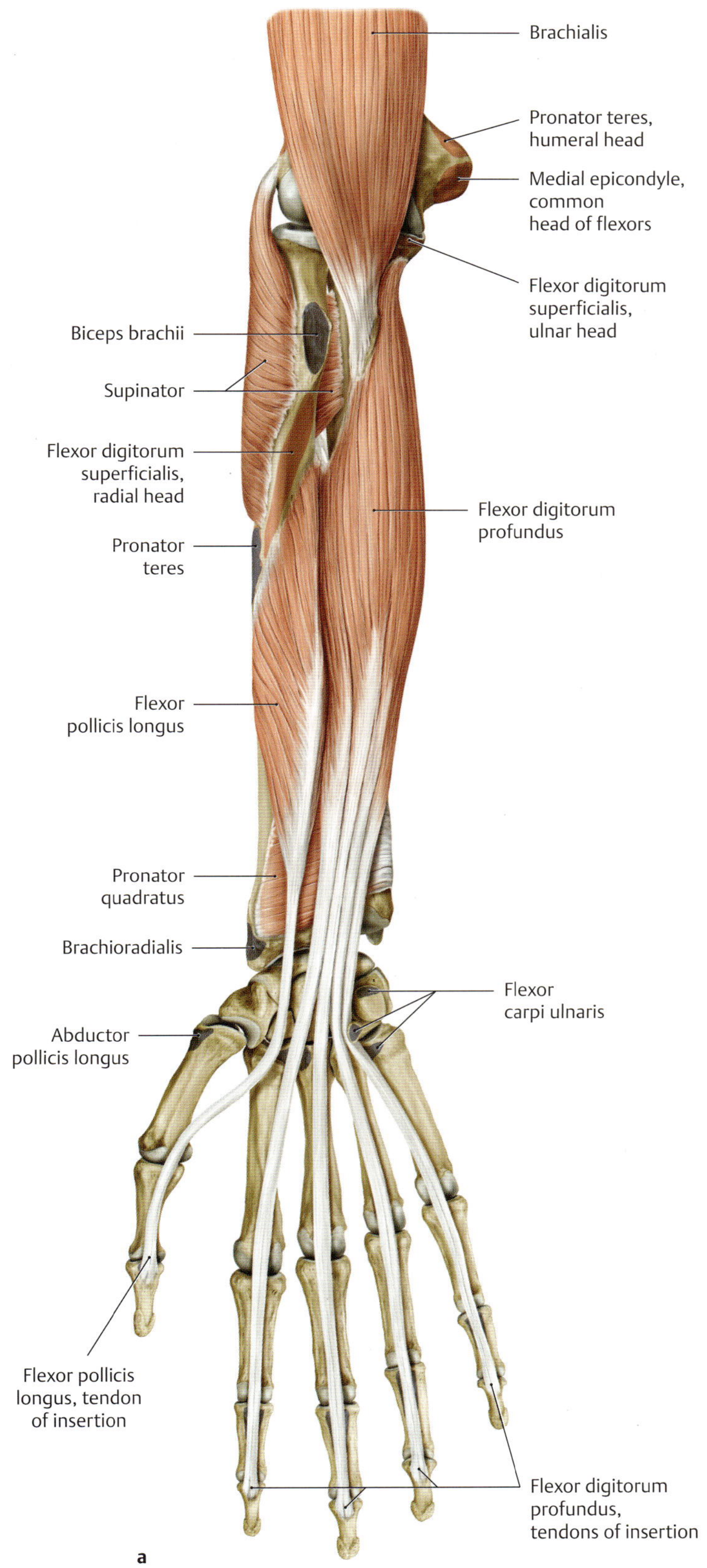

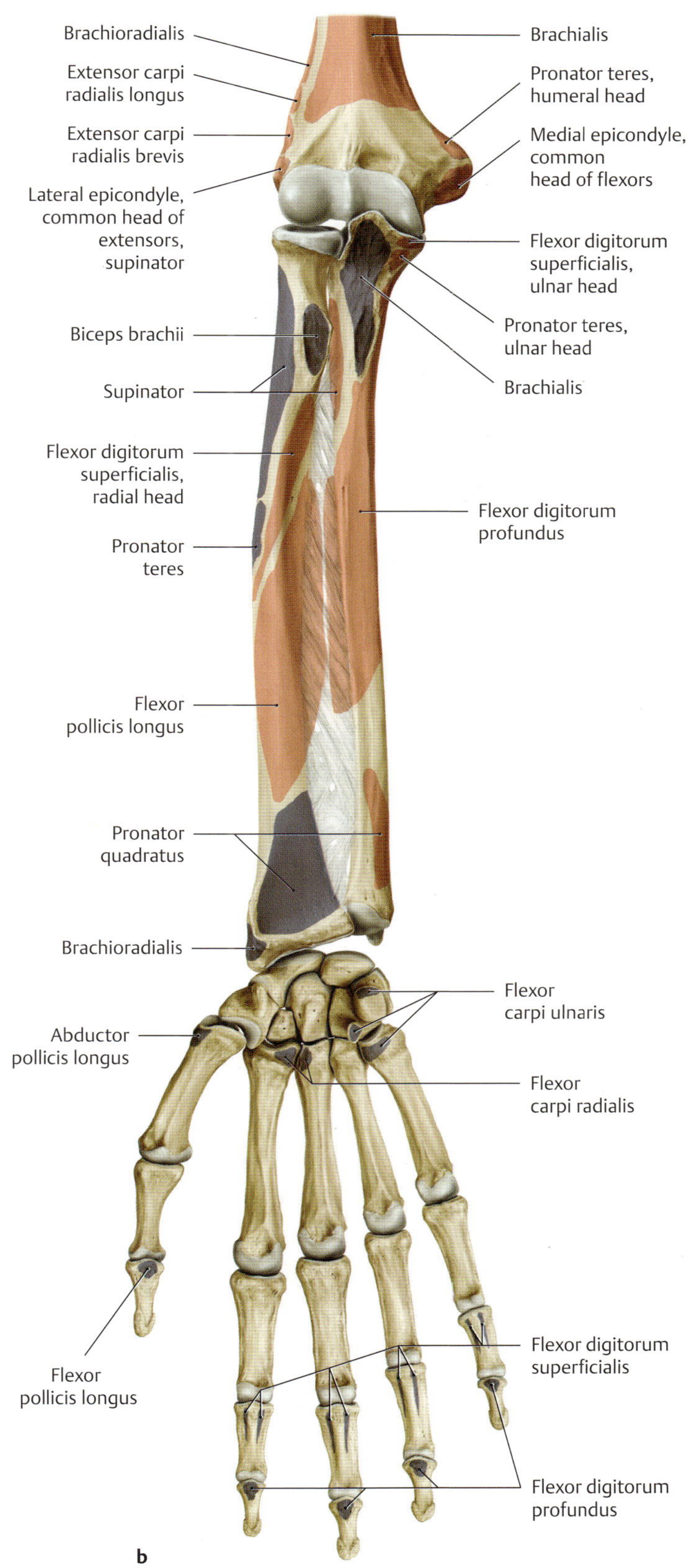

B Muscles of the right forearm, anterior view
Origin shown in red, insertion in blue.

a Pronator teres and flexor digitorum superficialis have been removed.

b All the muscles have been removed.

3.6 The Muscles of the Posterior Forearm

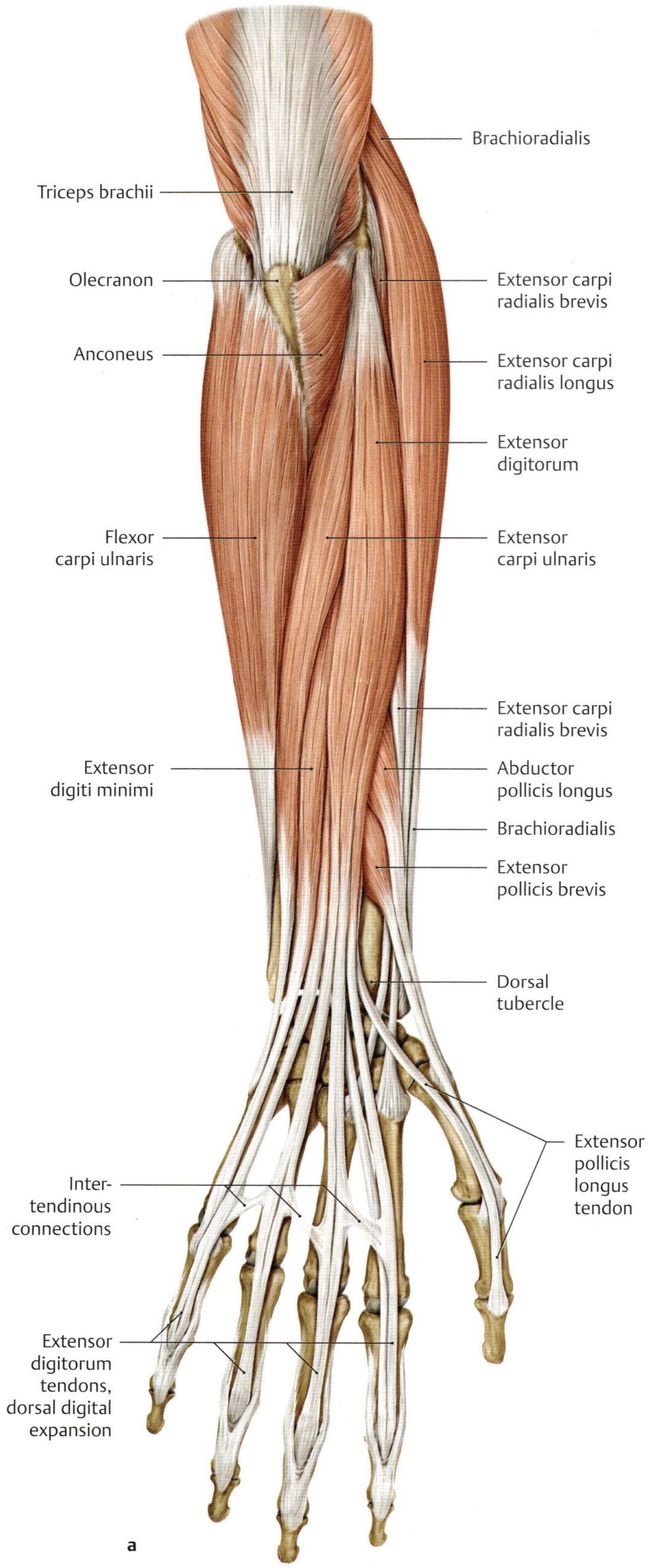

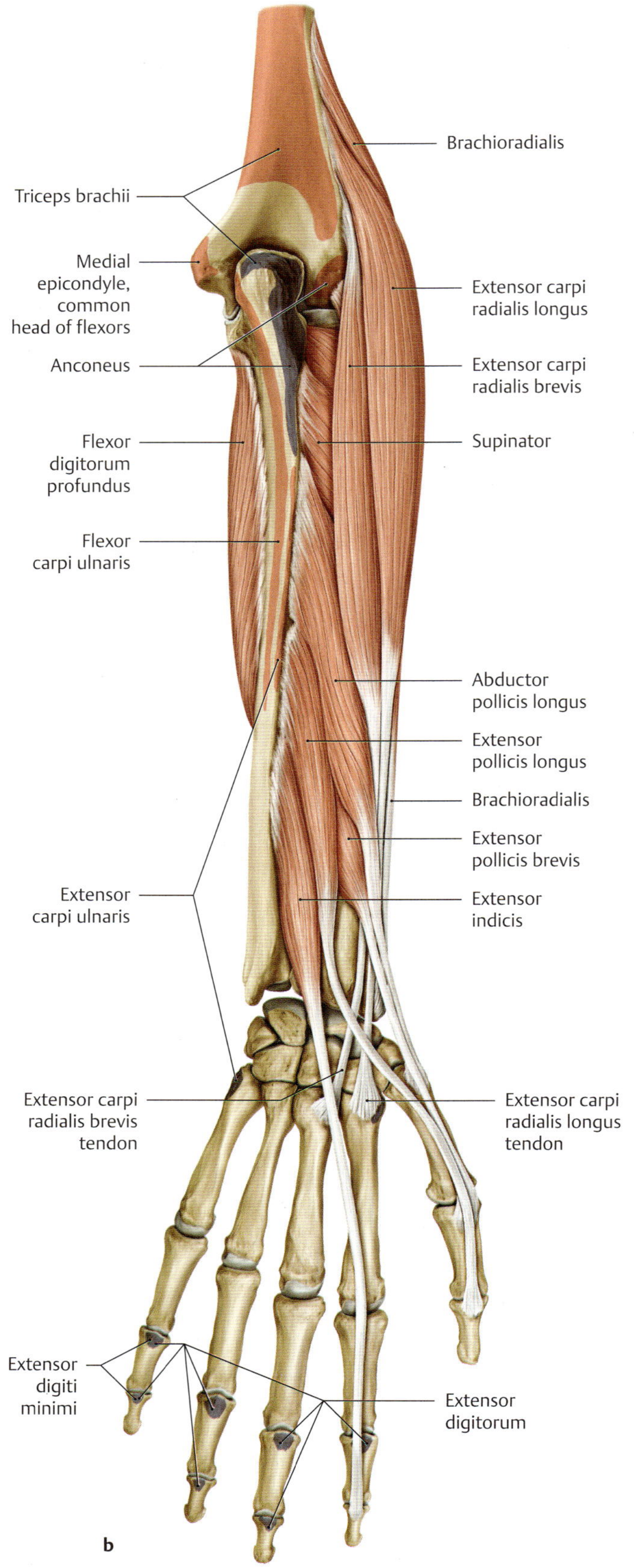

A Muscles of the right forearm, posterior view
Origin shown in red, insertion in blue.

a The superficial extensors and the radialis group are shown.

b Triceps brachii, anconeus, flexor carpi ulnaris, extensor carpi ulnaris, and extensor digitorum have been removed.

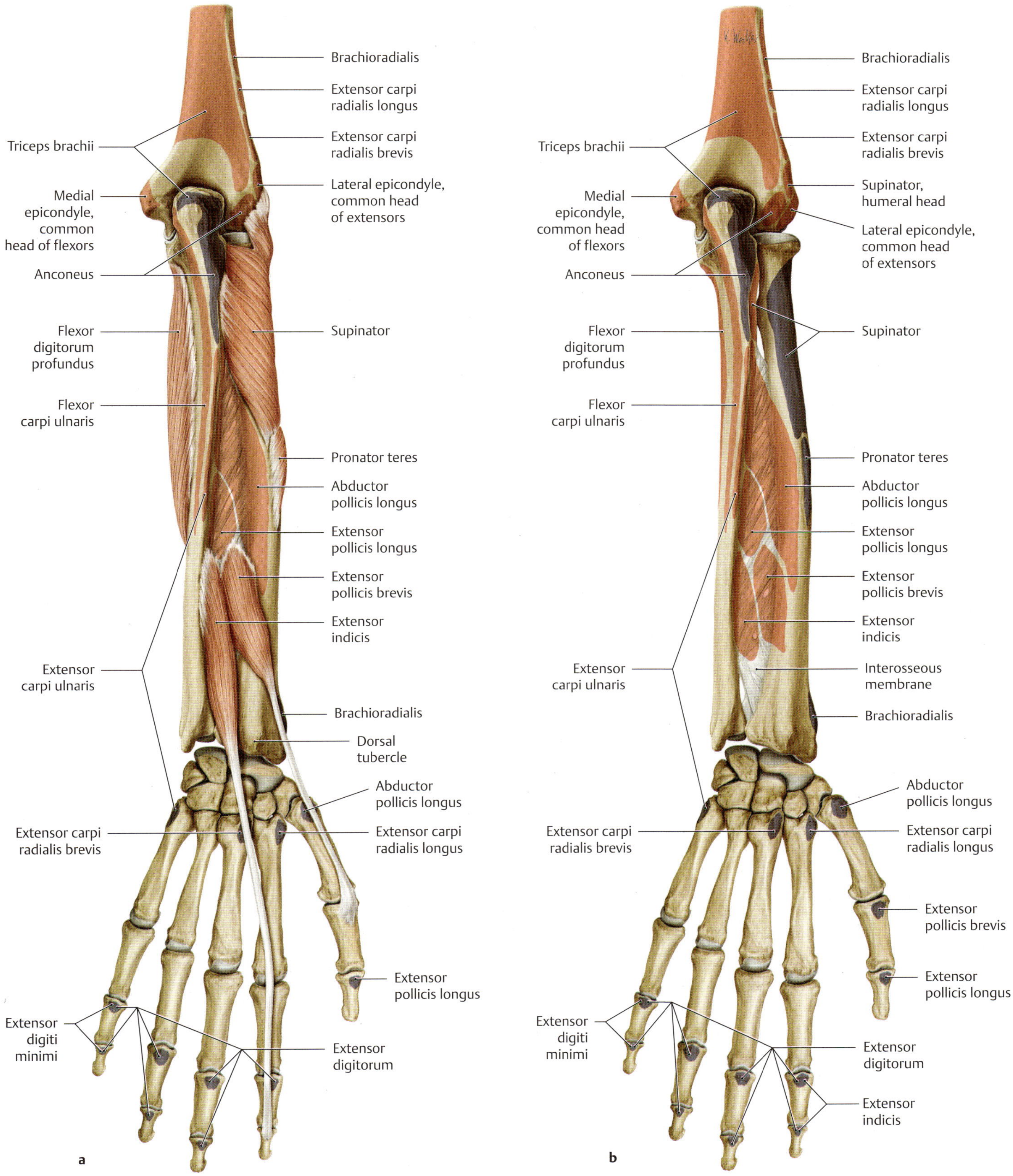

B Muscles of the right forearm, posterior view
Origin shown in red, insertion in blue.
Note the interosseous membrane, which contributes to the origin of several muscles in the forearm.

a Abductor pollicis longus, extensor pollicis longus, and the radialis group have been removed.
b All the muscles have been removed.

3.7 Cross-sectional Anatomy of the Arm and Forearm

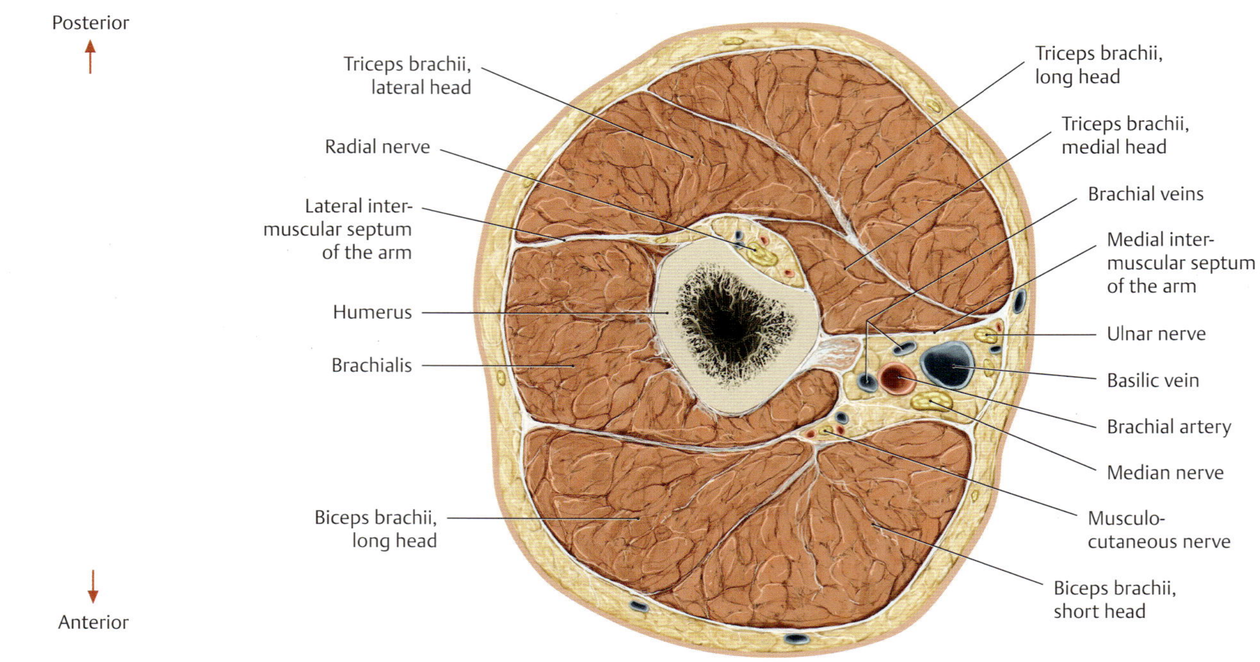

A Cross section through the right arm
Proximal view. The location of the sectional plane is shown in **C**.

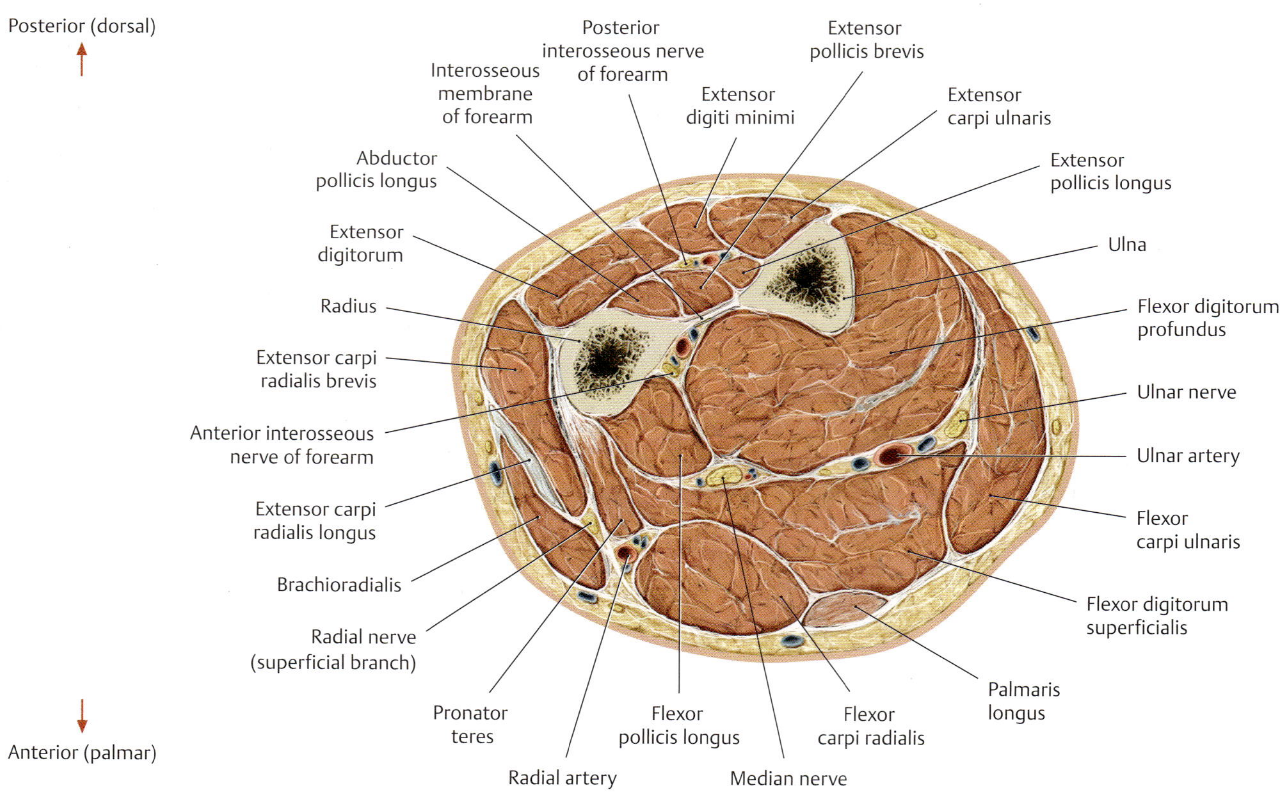

B Cross section through the right forearm
Proximal view. The location of the plane of section is shown in **D**.

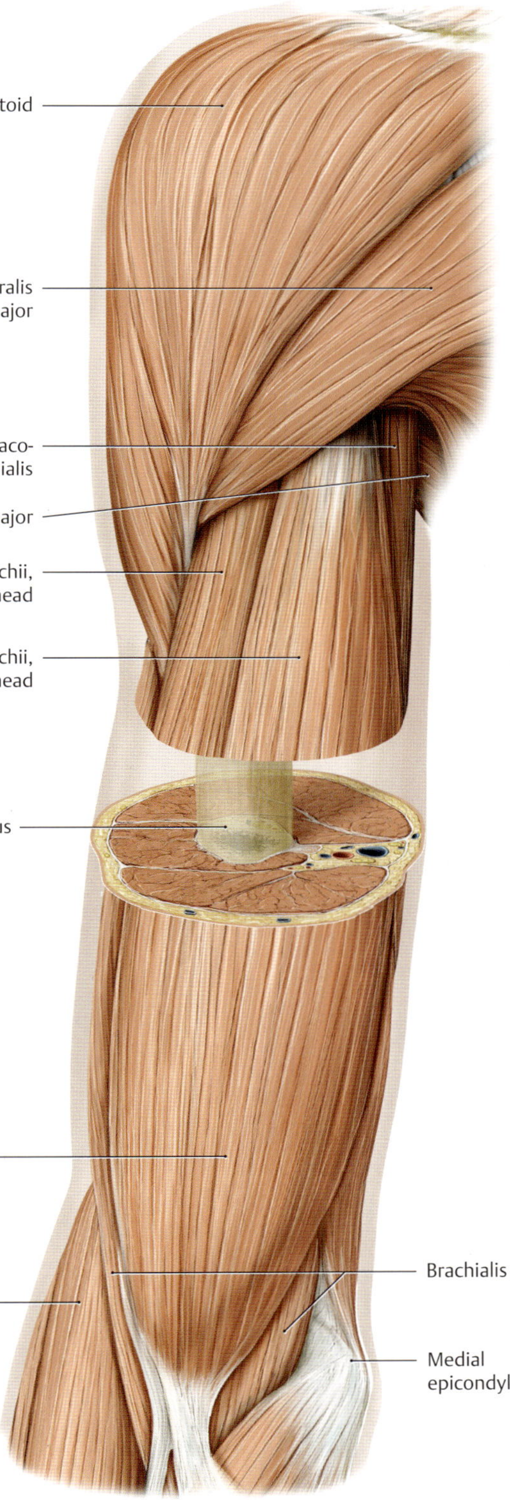

C "Windowed" dissection of the right arm
Anterior view.

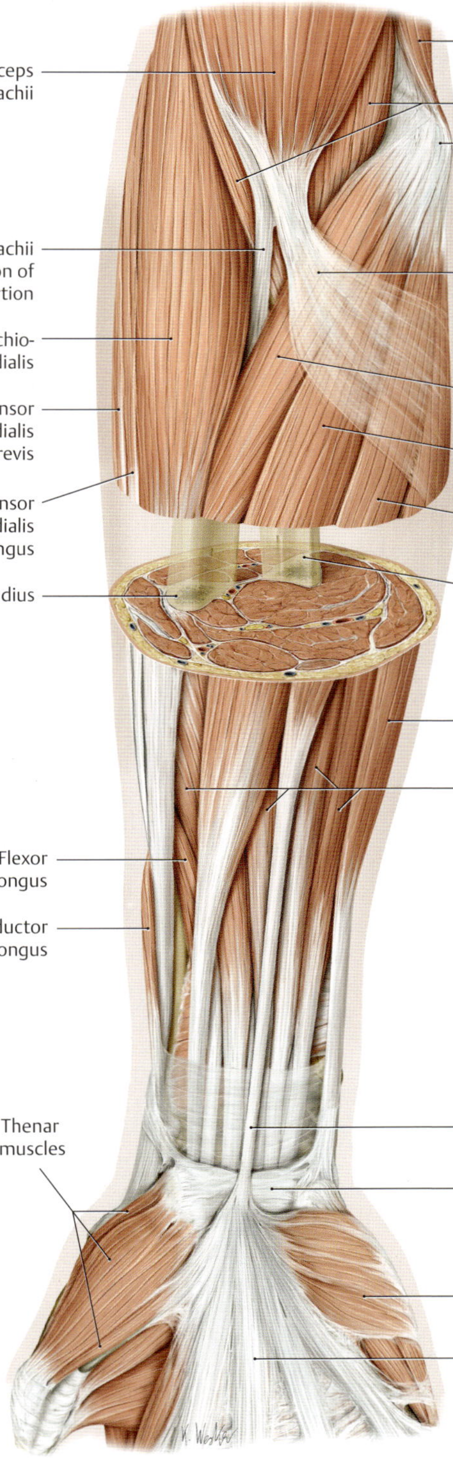

D "Windowed" dissection of the right forearm
Anterior view.

3.8 The Tendon Sheaths of the Hand

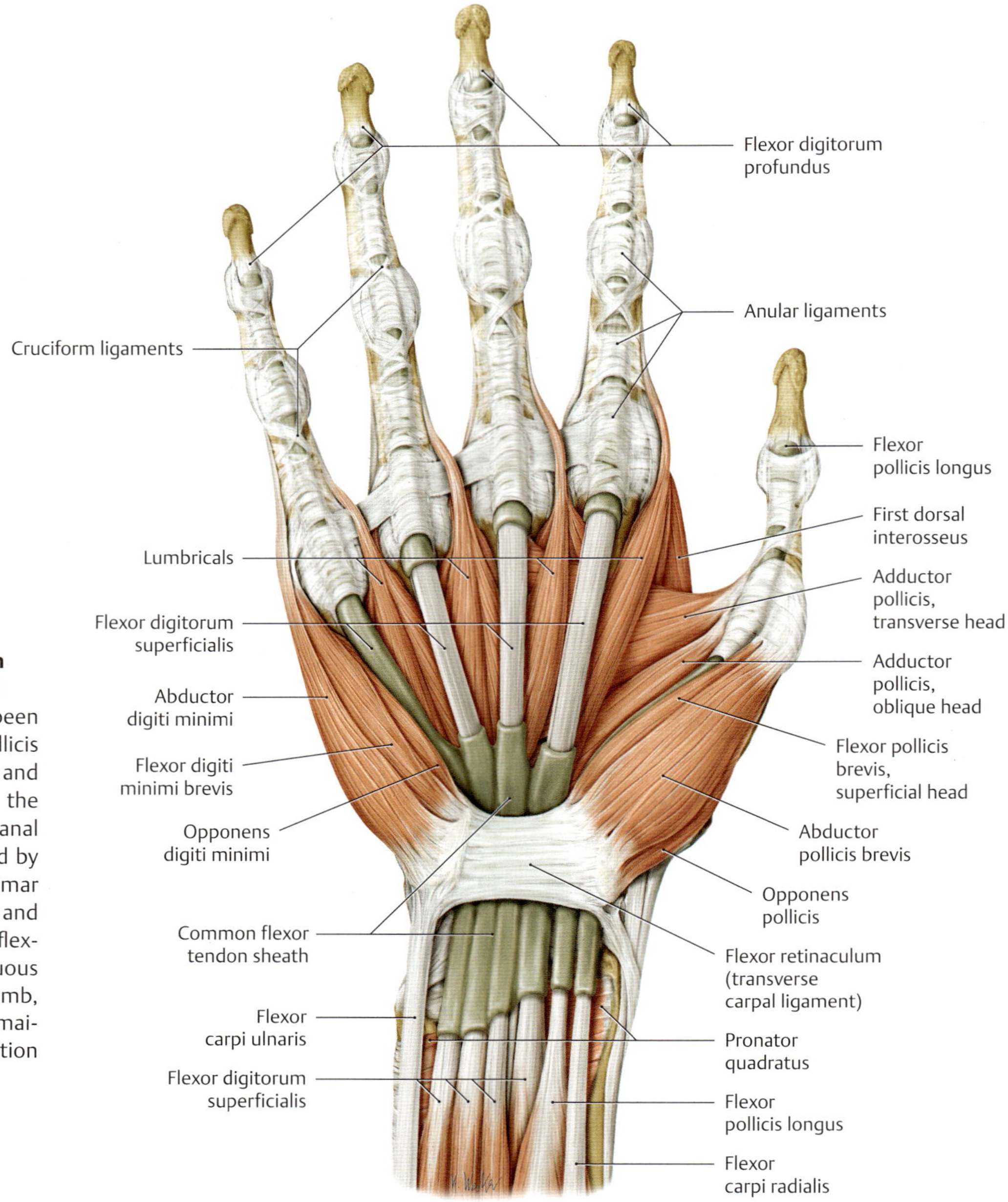

A Carpal and digital tendon sheaths on the palmar surface of the right hand
The palmar aponeurosis (see p. 362) has been removed. The tendons of the flexor pollicis longus, flexor digitorum superficialis, and flexor digitorum profundus run from the distal forearm through a fibro-osseous canal (carpal tunnel) to the palm, accompanied by the median nerve and protected by the palmar carpal tendon sheaths (see also p. 302 and p. 416). The carpal tendon sheath of the flexor pollicis longus is consistently continuous with the digital tendon sheath of the thumb, while the digital tendon sheaths of the remaining fingers show variable communication with the carpal tendon sheaths (see **B**).

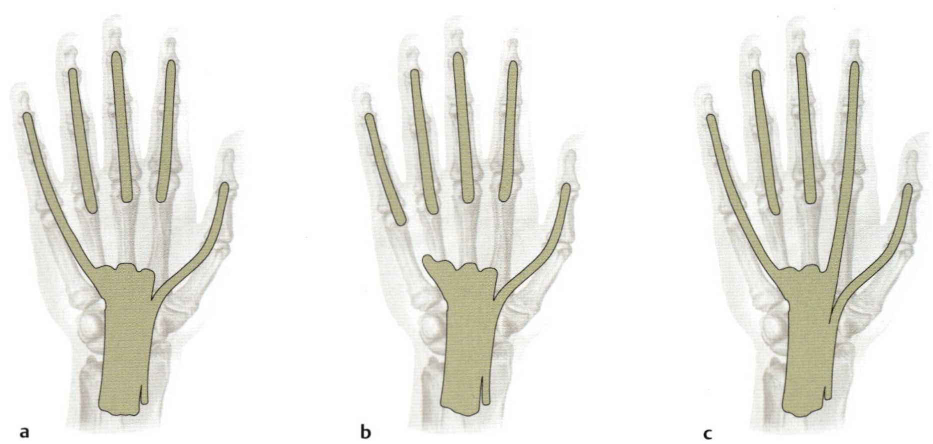

B Communication between the digital and carpal tendon sheaths
Right hand, anterior view.

- **a** In 71.4% of cases (Scheldrup 1951) the digital tendon sheath of the little finger communicates directly with the carpal tendon sheath, while the other tendon sheaths of the second through fourth digits extend only from the metacarpophalangeal joint to the distal interphalangeal joint.
- **b** In 17.4% of cases the carpal tendon sheath does not communicate with the digital tendon sheath of the little finger.
- **c** Besides the tendon sheath of the little finger, the carpal tendon sheath may occasionally be continuous with the digital tendon sheath of the index finger (3.5%) or of the ring finger (3%).

C Dorsal tendon compartments for the extensor tendons

First tendon compartment:	Abductor pollicis longus Extensor pollicis brevis
Second tendon compartment:	Extensor carpi radialis longus and brevis
Third tendon compartment:	Extensor pollicis longus
Fourth tendon compartment:	Extensor digitorum Extensor indicis
Fifth tendon compartment:	Extensor digiti minimi
Sixth tendon compartment:	Extensor carpi ulnaris

The location of the tendon compartments is shown in **D**.

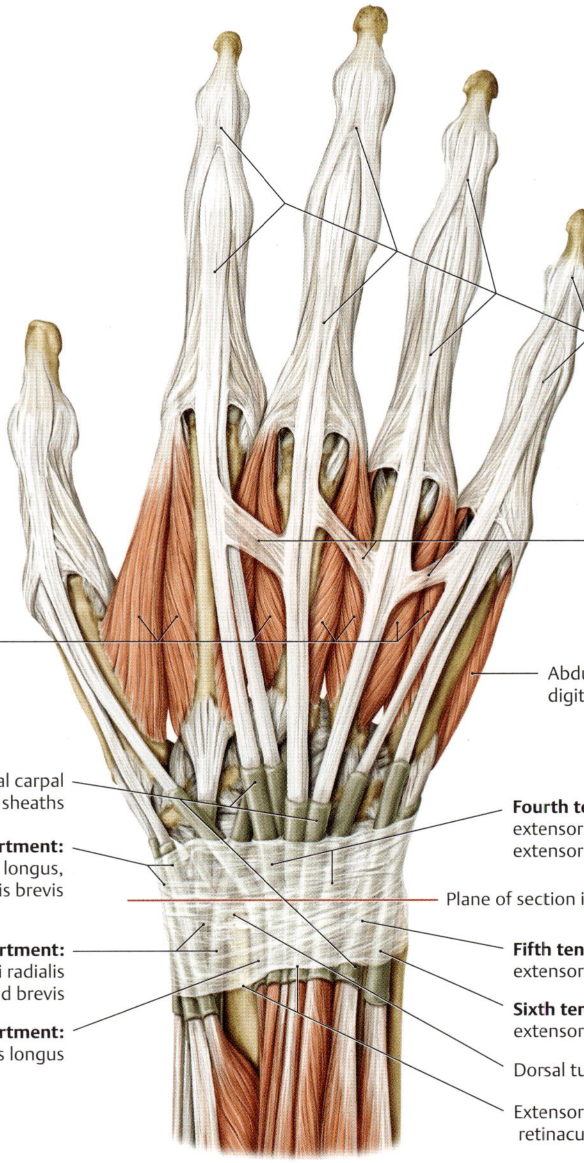

D Extensor retinaculum and dorsal carpal tendon sheaths

Right hand, posterior view. The extensor retinaculum is part of the antebrachial (forearm) fascia. Its transverse fibers strengthen the fibrous layer of the tendon sheaths and fix them to the dorsum of the hand. Deep to the extensor retinaculum are *tendon sheath compartments*, which transmit the long extensor tendons singly or in groups. There are a total of six of these compartments, numbered 1 to 6 from the radial side to the ulnar side of the wrist (their contents are shown in **C**).

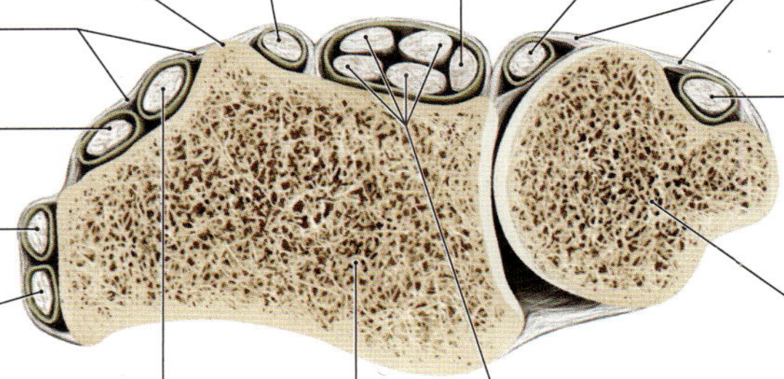

E Schematic cross section through the forearm at the level of the distal radioulnar joint

Right forearm, proximal view. (Location of the sectional plane is shown in **D**.) Vertical connective tissue septa extend anteriorly from the deep surface of the extensor retinaculum to the bone or joint capsule and form six fibro-osseous canals, the tendon sheath compartments of the extensor tendons (extensor tendon compartments).
Note also the dorsal tubercle, which redirects the tendon of insertion of the extensor pollicis longus to the thumb (see also **D**).

3.9 The Dorsal Digital Expansion

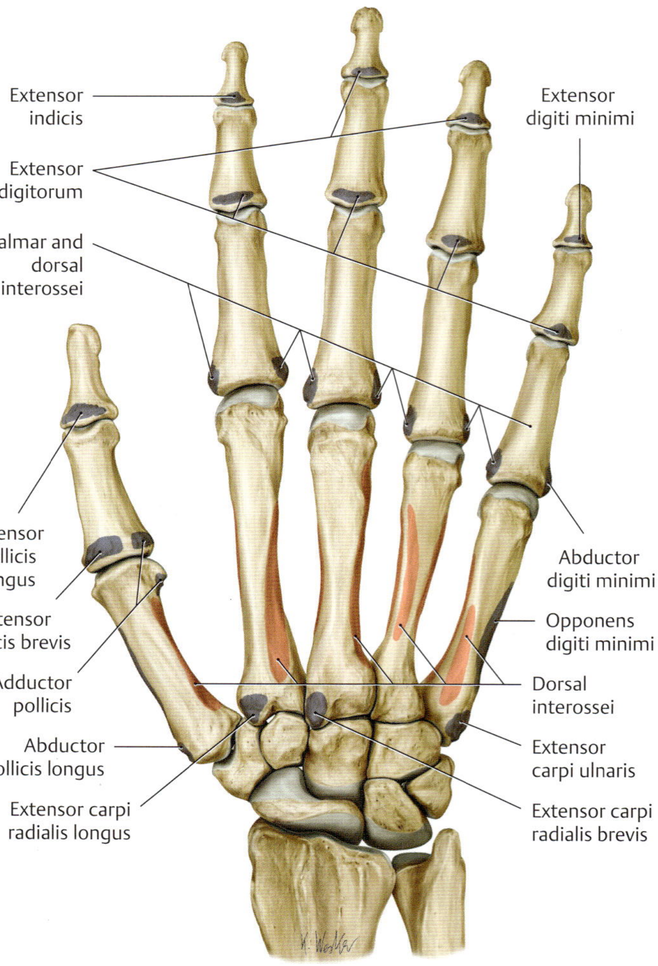

A Origins and insertions of the dorsal muscles of the right hand
Origin shown in red, insertion in blue.

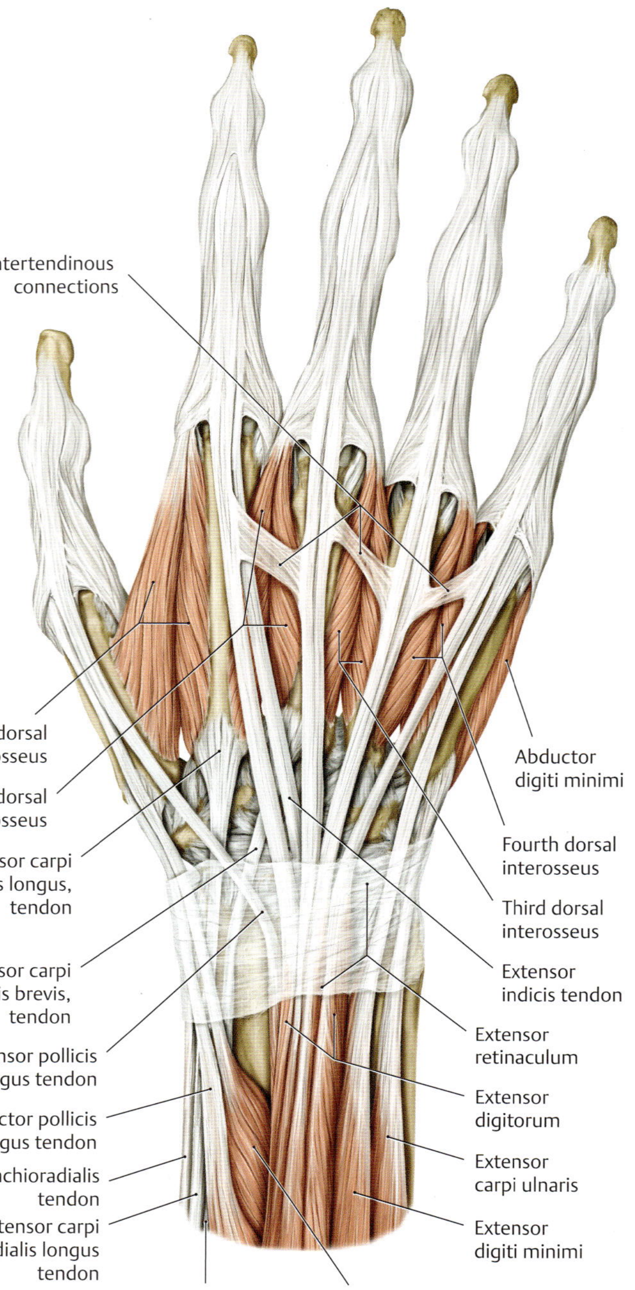

B Extensor tendons and intertendinous connections on the dorsum of the right hand

The tendons of insertion of the extensor digitorum are interlinked by variable oblique bands called *intertendinous connections*. The most proximal of the intertendinous connections are those extending between the index and middle fingers. No such connection is present on the tendon of the extensor indicis. The extensor digitorum inserts by a variable number of tendons. Generally, all of the fingers have at least two extensor tendon elements. In addition, the *index finger* and *little finger* have their own extensor muscles (extensor indicis and extensor digiti minimi) whose tendons always run on the ulnar side of the tendons of the common extensor digitorum. Because the index finger and little finger have their own extensors, they can more easily be moved independently of the other fingers.

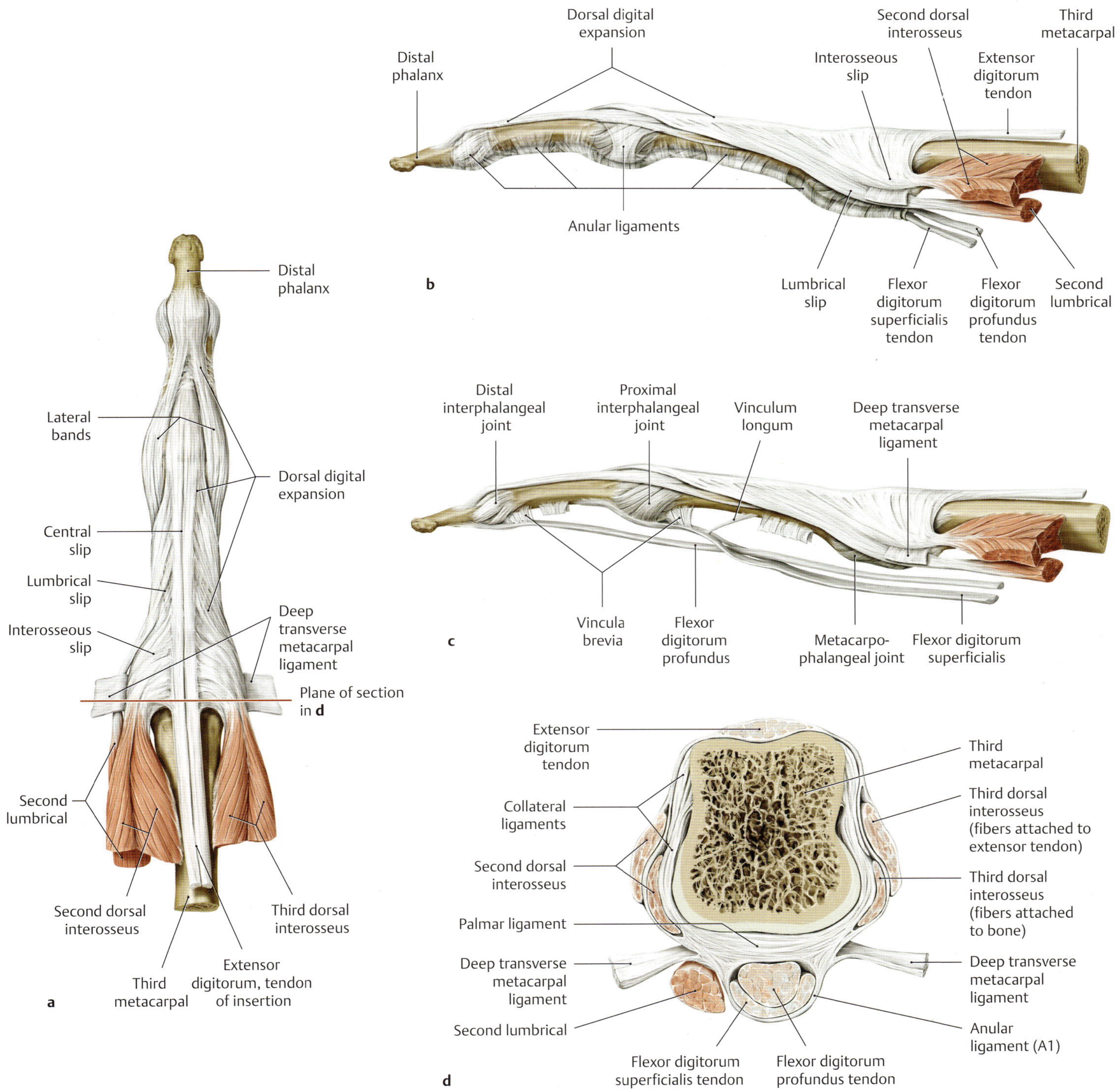

C Dorsal digital expansion
Dorsal digital expansion of the middle finger of the right hand.

a Posterior view.
b Radial view.
c After opening the common tendon sheath of the flexor digitorum superficialis and profundus muscles.
d Cross section at the level of the metacarpal head.

The dorsal digital expansion is more than an aponeurosis that incorporates slips from the extensor digitorum, lumbrical, and interosseous tendons. It is a complex system of interwoven fiber bands, joined by loose connective tissue to the periosteum of the phalanges. The dorsal digital expansion consists of a *central slip* and *lateral bands*, each of which has a *lateral part* and a *medial part*. The lateral part of the expansion receives slips from the tendons of the lumbrical and interosseous muscles (see **a**). This complex arrangement makes it possible for the long digital flexors and the short hand muscles to act on all three of the finger joints.

3.10 The Intrinsic Muscles of the Hand: Superficial Layer

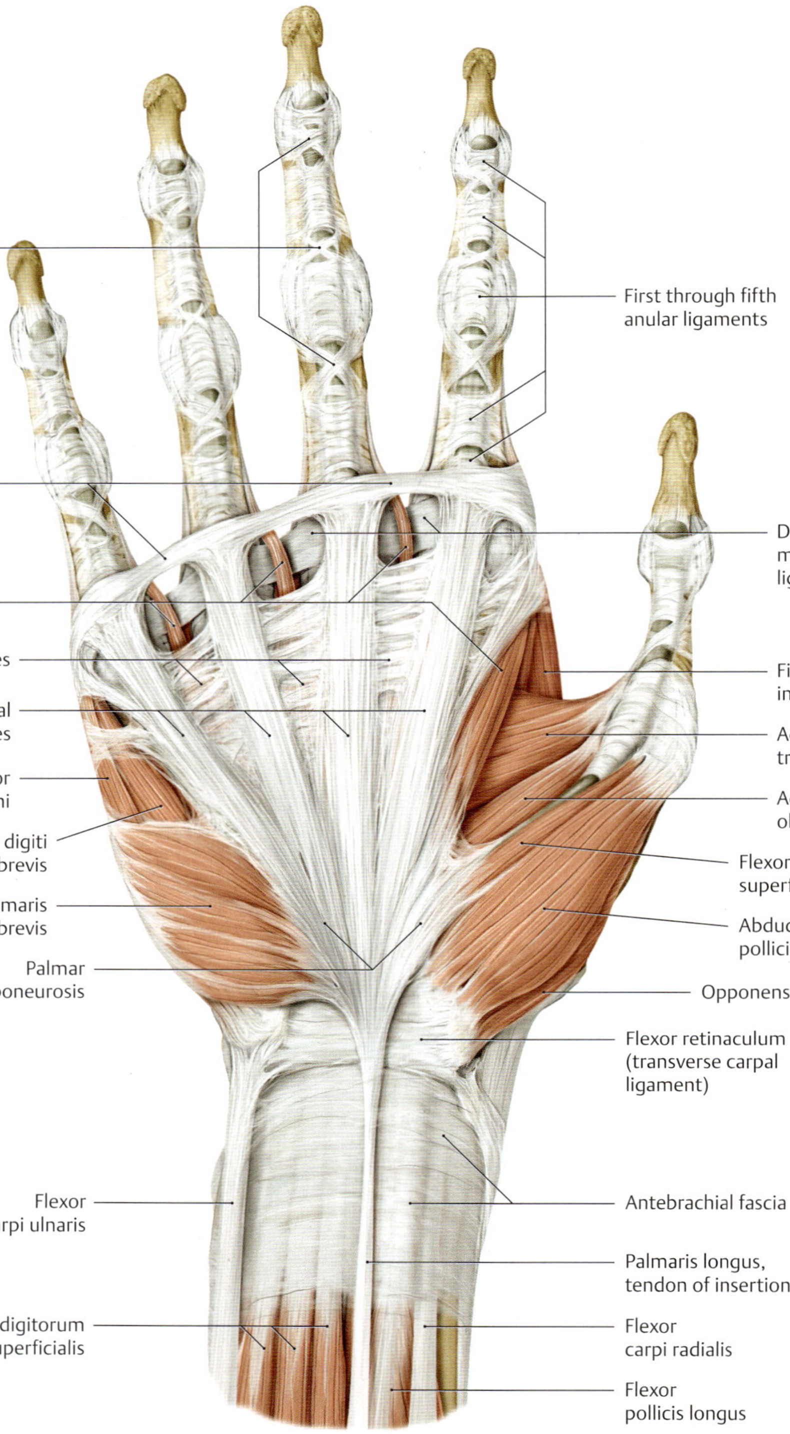

A The palmar aponeurosis and Dupuytren's contracture

Right hand, anterior view. The muscular fascia of the palm is thickened by firm connective tissue to form the *palmar aponeurosis*, which separates the palm from the subcutaneous fat to protect the soft tissues. It is composed mainly of longitudinal fiber bundles (*longitudinal fascicles*), which give it a fan-shaped arrangement. The longitudinal fascicles are held together by transverse fiber bundles (*transverse fascicles*) at the level of the metacarpal bones and by the superficial transverse metacarpal ligament at the level of the metacarpophalangeal joints. Two muscles, the palmaris brevis and palmaris longus, keep the palmar aponeurosis tense and prevent it from contracting, especially when the hand is clenched into a fist. Gradual atrophy or contracture of the palmar aponeurosis leads to progressive shortening of the palmar fascia that chiefly affects the little finger and ring finger (*Dupuytren's contracture*). Over a period of years, the contracture may become so severe that the fingers assume a fixed flexed position with the fingertips touching the palm; this seriously compromises the grasping ability of the hand. The causes of Dupuytren's contracture are poorly understood, but it is a relatively common condition that is most prevalent in men over 40 years of age and is associated with chronic liver disease (e.g., cirrhosis). Treatment generally consists of complete surgical removal of the palmar aponeurosis.

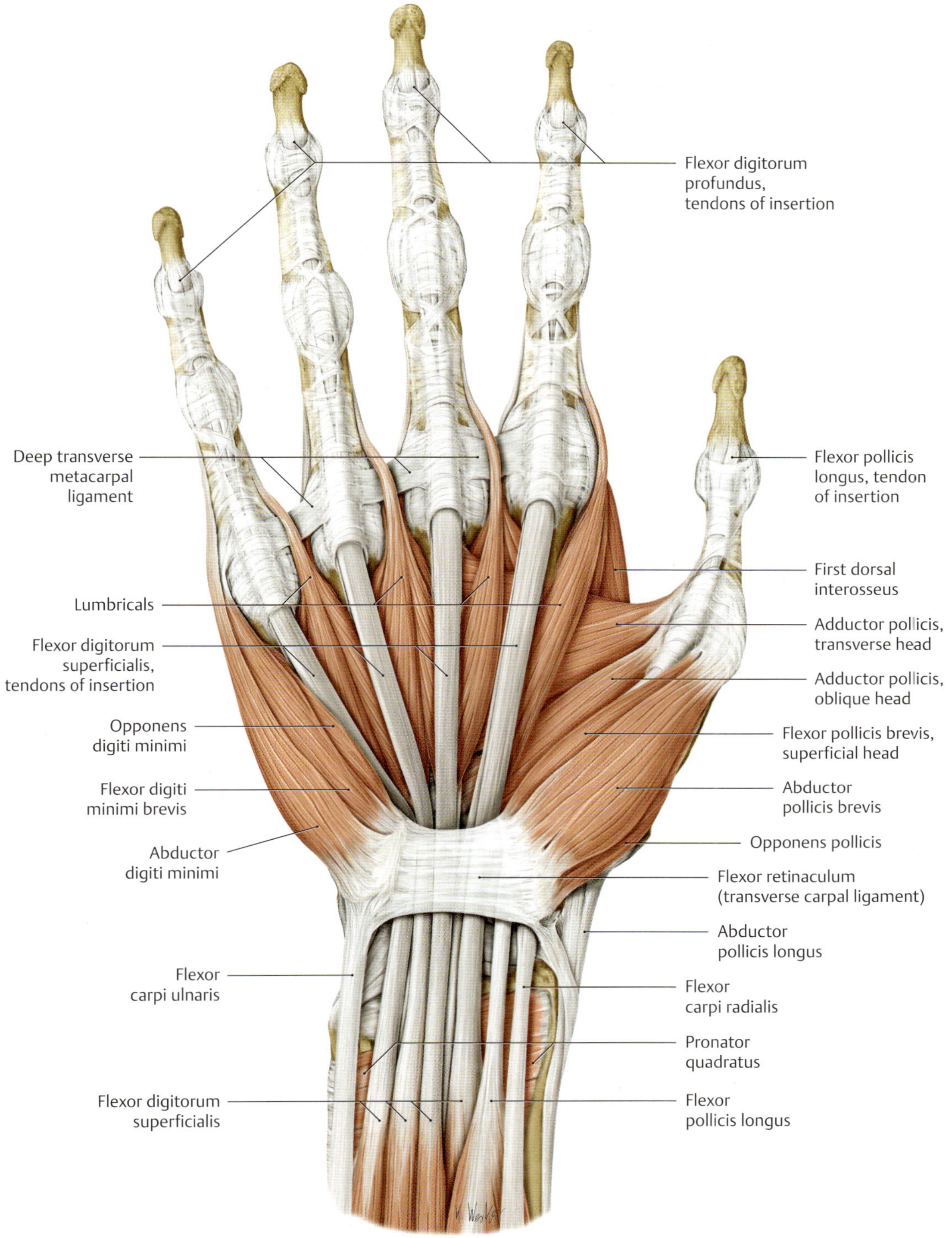

B Superficial layer of the muscles of the right hand after removal of the palmar aponeurosis

Anterior view. The palmar aponeurosis, antebrachial (forearm) fascia, and palmaris brevis and longus muscles have been removed, along with the palmar and carpal tendon sheaths.

3.11 The Intrinsic Muscles of the Hand: Middle Layer

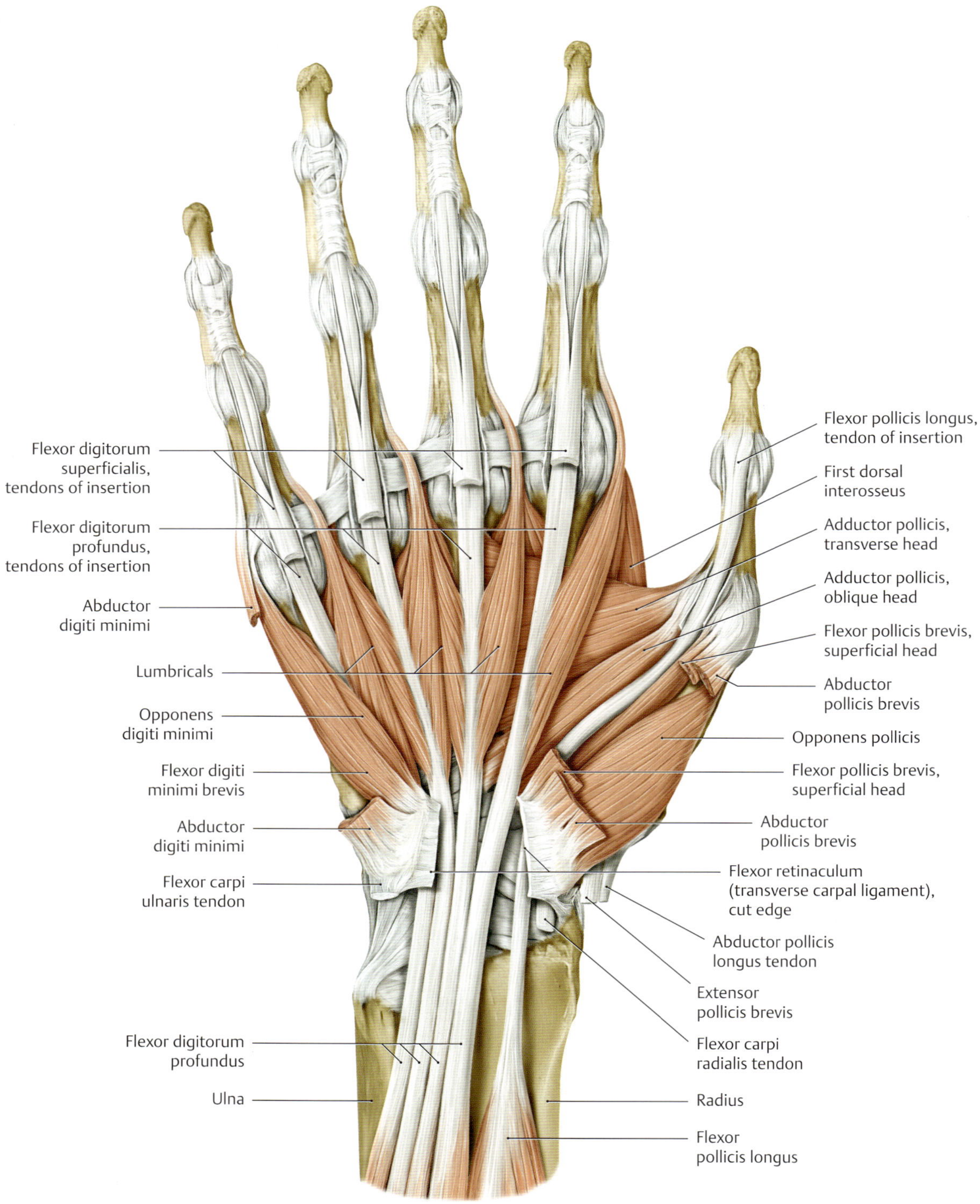

A Middle layer of the muscles of the right hand
Anterior view. The flexor digitorum superficialis muscle has been removed, and its four tendons of insertion have been divided at the level of the metacarpophalangeal joints. The first through third anular ligaments have been cut open to reveal the flexor tendons on the fingers. The flexor retinaculum (transverse carpal ligament) has been partially removed to open the carpal tunnel. Of the thenar muscles, portions of the abductor pollicis brevis and flexor pollicis brevis (superficial head) have been removed. Part of the abductor digiti minimi has been resected on the hypothenar side.

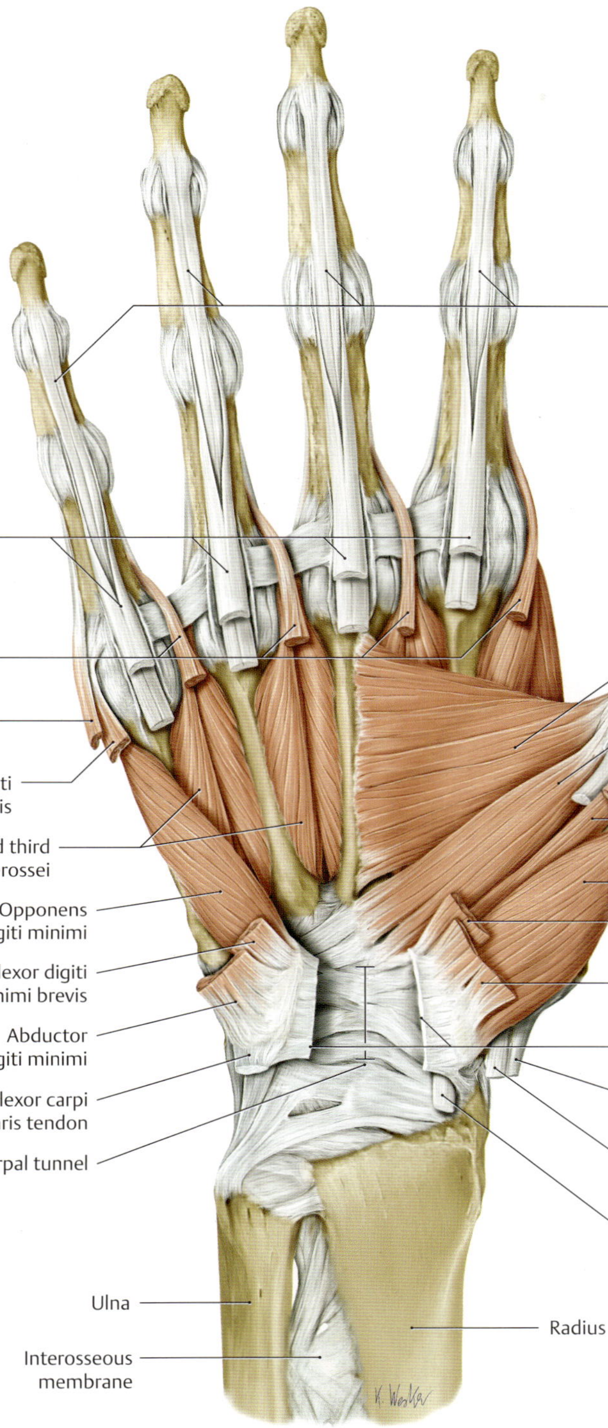

B Middle layer of the muscles of the right hand
Anterior view. The flexor digitorum profundus muscle has been removed, and its four tendons of insertion and the lumbricals arising from them have been divided. The flexor pollicis longus and flexor digiti minimi brevis muscles have also been removed.

3.12 The Intrinsic Muscles of the Hand: Deep Layer

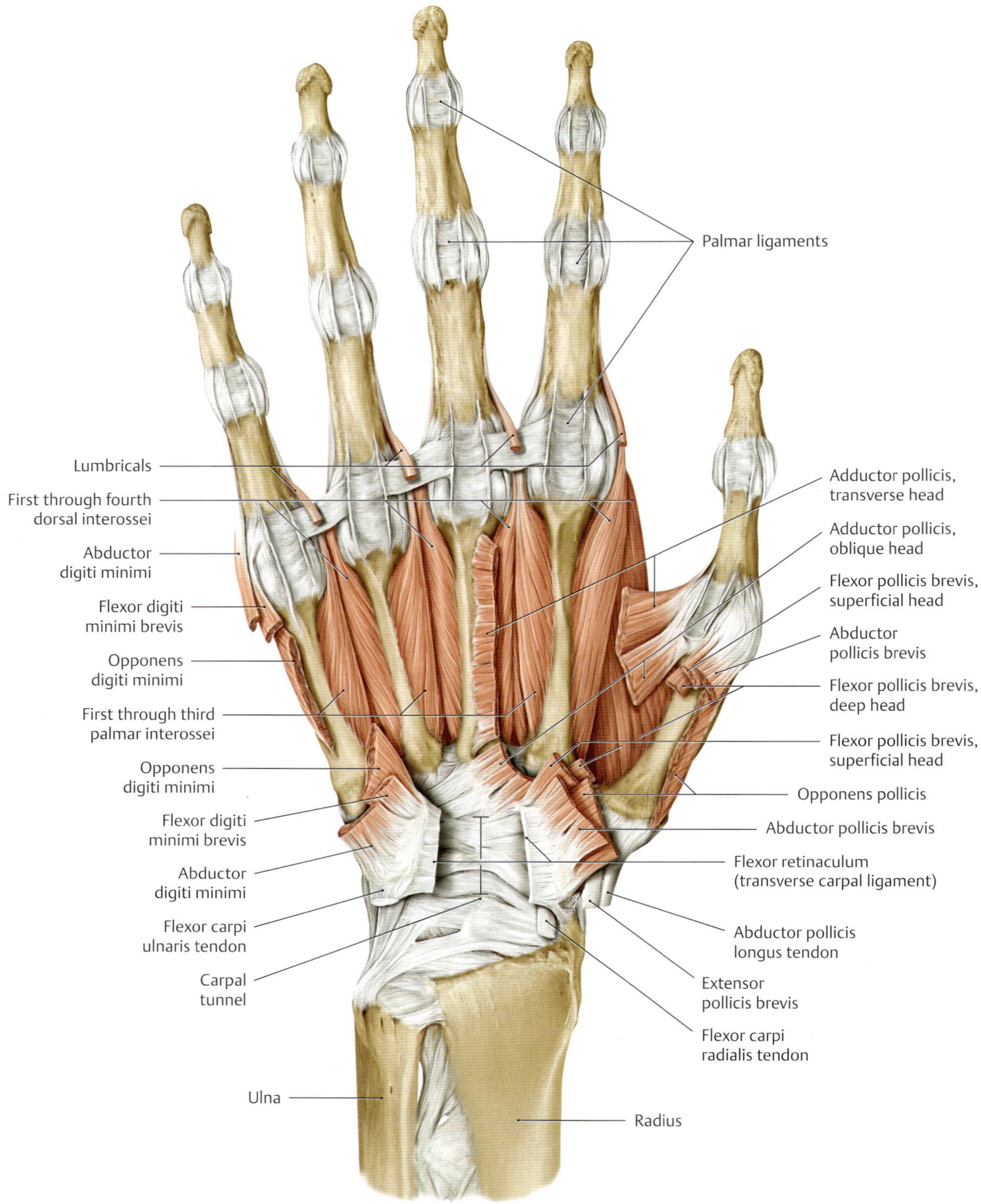

A Deep layer of the muscles of the right hand
Anterior view. The tendons of insertion, tendon sheaths, and anular ligaments of the long digital flexors have been completely removed. *Note* the exposed palmar ligaments, which combine with the tendon sheaths to form a trough that directs the long flexor tendons (see p. 305). The first dorsal interosseus and first palmar interosseus muscles have been almost completely exposed by removal of the adductor pollicis. Both the opponens pollicis and the opponens digiti minimi have been partially removed.

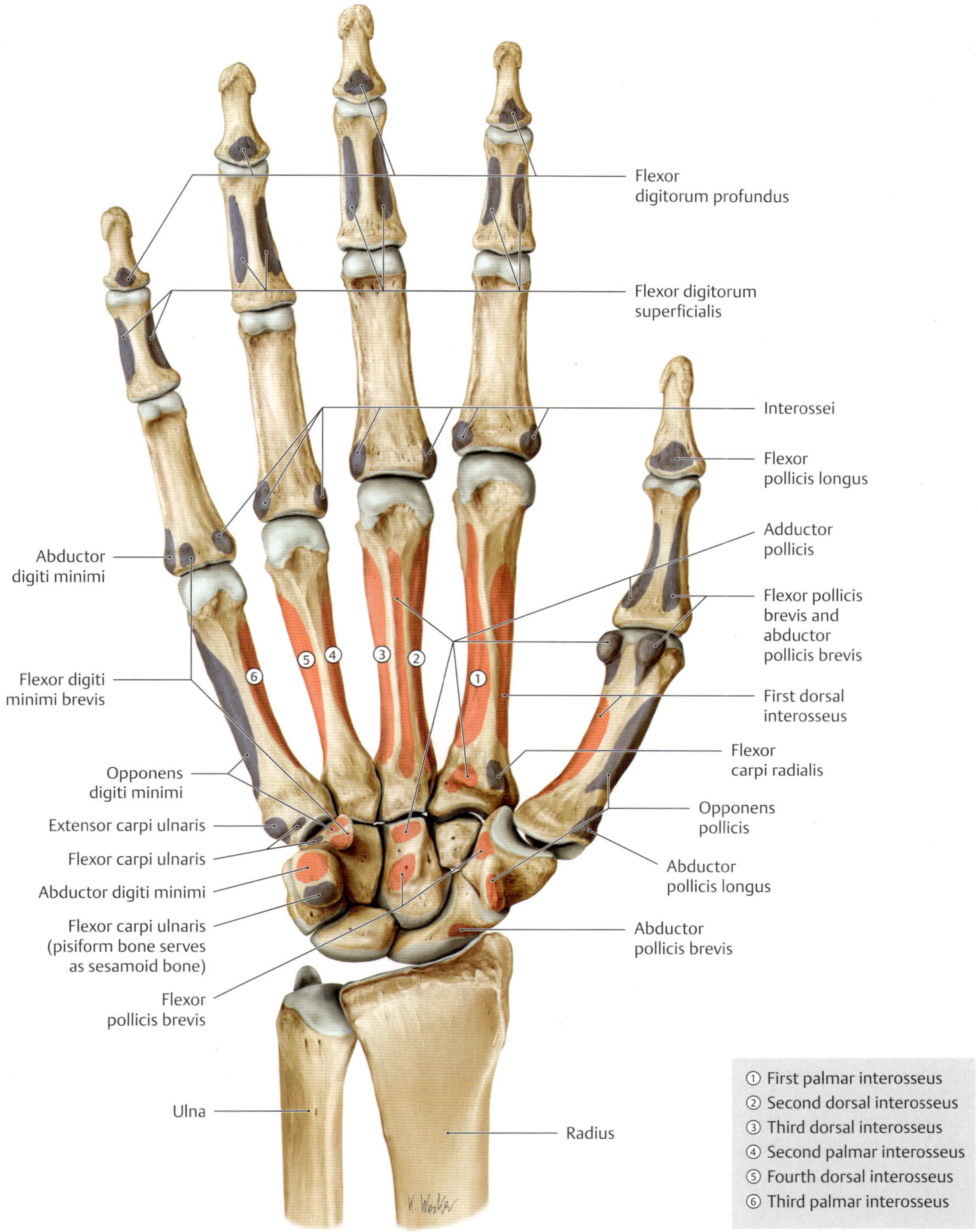

B Origins and insertions of the palmar muscles of the right hand
Origin shown in red, insertion in blue.

4.1 The Arteries

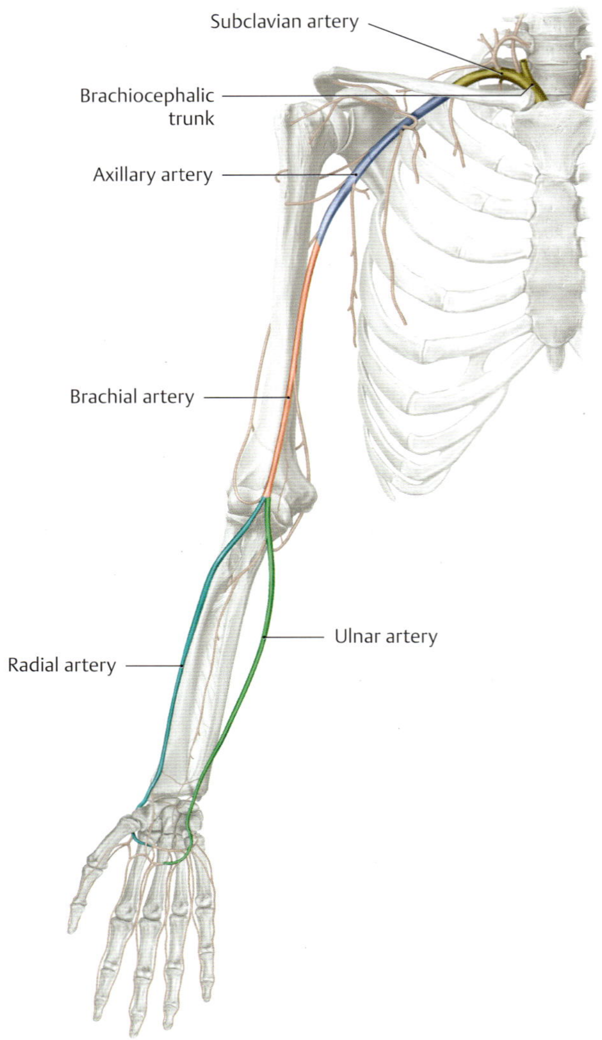

A Course of the different segments of the arteries supplying the shoulder and arm

Subclavian artery: The right subclavian artery arises from the brachiocephalic trunk (as shown here), and the left arises directly from the aortic arch. The vessel runs over the first rib between the anterior and middle scalene (interscalene space, scalene interval) and continues as the axillary artery (see below) on reaching the lateral border of the rib. Unlike the other arteries pictured here, the subclavian artery supplies blood not only to the upper limb (i.e., the shoulder girdle and arm) but also to

- a portion of the neck,
- the cerebral circulation, and
- the anterior chest wall.

Axillary artery: The continuation of the subclavian, the axillary artery runs from the lateral border of the first rib to the inferior border of the teres major muscle.
Brachial artery: The brachial artery is the continuation of the axillary artery. It ends at the elbow joint by dividing into the radial and ulnar arteries.
Radial artery: The radial artery runs distally on the radial side of the forearm from the division of the brachial artery, passing between the brachioradialis and flexor carpi radialis muscles on its way to the wrist. It terminates in the deep palmar arch.
Ulnar artery: This second division of the brachial artery runs below the pronator teres on the ulnar side of the forearm, under cover of the flexor carpi ulnaris, to the superficial palmar arch.

B Overview of the arteries of the shoulder and arm
The arteries of the shoulder and arm vary considerably in their origins and branching patterns (the principal variants are reviewed in Chapter 5, Neurovascular Systems: Topographical Anatomy). The branches are listed below in the order in which they arise from the parent vessels.

Branches of the subclavian artery
- Branches of the subclavian artery
- Vertebral artery
- Internal thoracic artery (internal mammary artery)
- Thyrocervical trunk
  - Inferior thyroid artery
  - Ascending cervical artery
  - Suprascapular artery
  - Transverse cervical artery
- Costocervical trunk
  - Deep cervical artery
  - Supreme intercostal artery

Branches of the axillary artery
- Branches of the axillary artery
- Superior thoracic artery
- Thoracoacromial artery
  - Acromial branch
  - Clavicular branch
  - Deltoid branch
  - Pectoral branch
- Lateral thoracic artery
- Subscapular artery
  - Thoracodorsal artery
  - Circumflex scapular artery
- Anterior circumflex humeral artery
- Posterior circumflex humeral artery

Branches of the brachial artery
- Deep artery of the arm (deep brachial artery)
  - Medial collateral artery
  - Radial collateral artery
- Superior ulnar collateral artery (arterial network of the elbow)
- Inferior ulnar collateral artery (arterial network of the elbow)

Branches of the radial artery
- Radial recurrent artery (arterial network of the elbow)
- Palmar carpal branch (palmar carpal network)
- Superficial palmar branch (superficial palmar arch)
- Dorsal carpal branch (dorsal carpal network)
  - Dorsal metacarpal arteries
  - Dorsal digital arteries
- Princeps pollicis artery
- Radialis indicis artery
- Deep palmar arch
  - Palmar metacarpal arteries
  - Perforating branches

Branches of the ulnar artery
- Ulnar recurrent artery (arterial network of the elbow)
- Common interosseous artery
  - Posterior interosseous artery
  - Recurrent interosseous artery
  - Anterior interosseous artery
- Palmar carpal branch (palmar carpal network)
- Dorsal carpal branch (dorsal carpal network)
- Deep palmar branch (deep palmar arch)
- Superficial palmar arch
  - Common palmar digital arteries
  - Proper palmar digital arteries

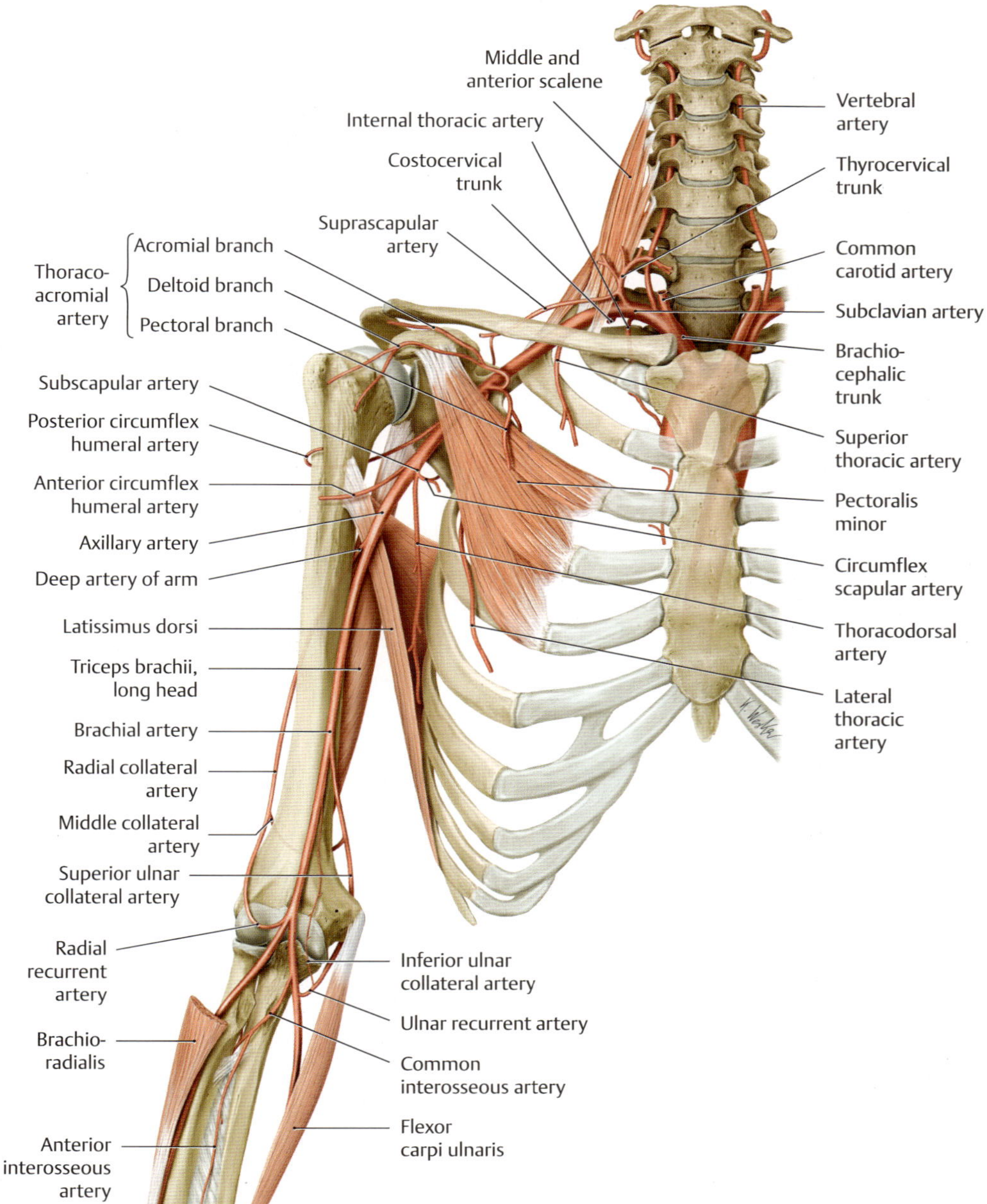

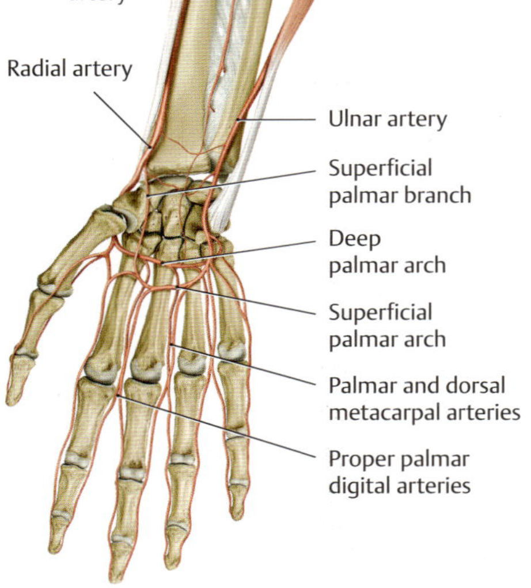

C Arteries of the right upper limb
Anterior view with the forearm supinated. For clarity, some of the arteries listed in **B** are not illustrated.

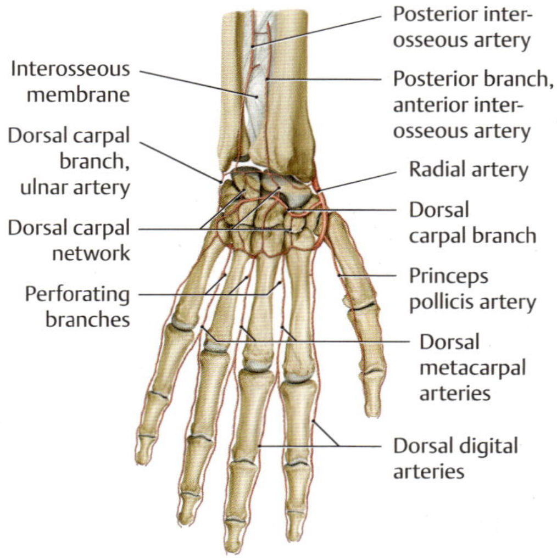

D Arteries of the right hand
Posterior view.

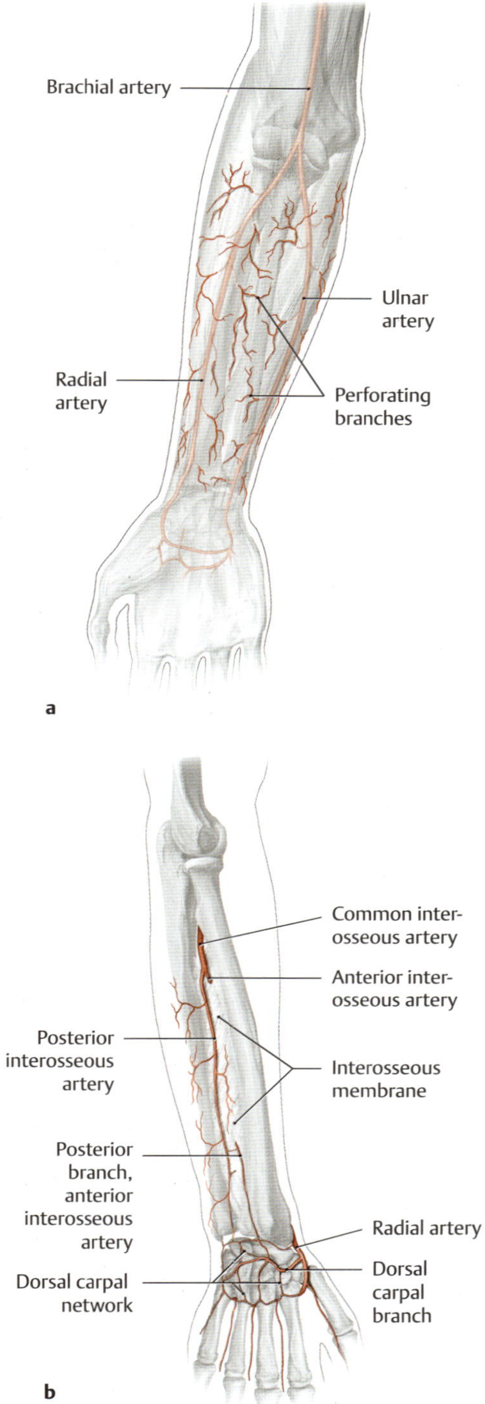

E Main arterial branches in the right forearm (perforating branches)

a From the radial artery and ulnar artery (forearm supinated, anterior view).

b From the posterior interosseous artery (forearm pronated, posterior view).

Fasciocutaneous flaps with an excellent vascular pedicle can be harvested from the thin skin of the forearm. Composed of skin, subcutaneous tissue, and fascia, these flaps are supplied by branches of the major arteries and their accompanying veins. The skin flaps carry this vascular supply with them when they are transferred to the recipient site.

4.2 The Veins

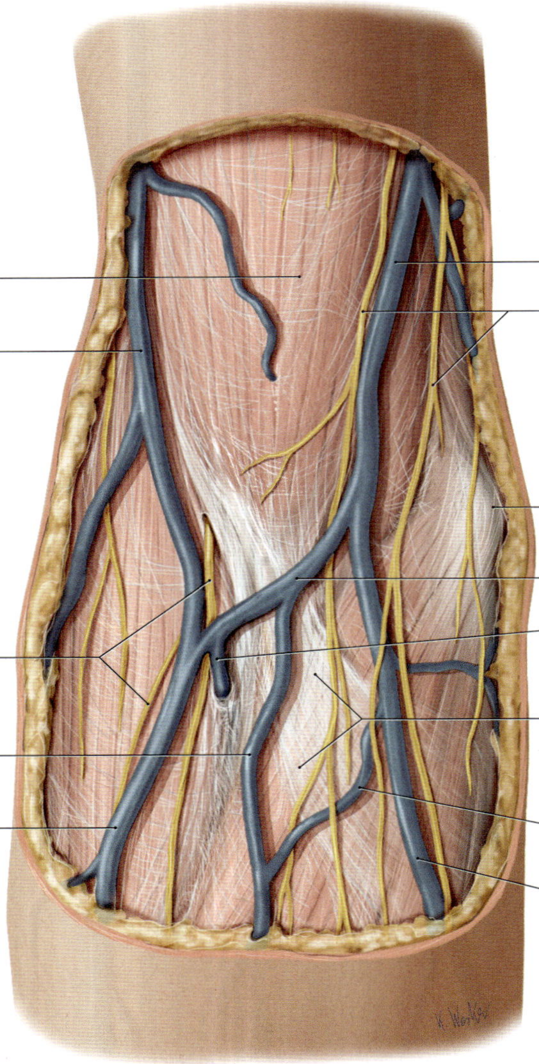

A Cutaneous veins and cutaneous nerves of the right elbow

Anterior view. The subcutaneous veins of the elbow are excellent sites for administering intravenous injections and drawing blood owing to their size and accessibility and the relatively thin skin in that region. But given their close relationship to the cutaneous nerves, as illustrated by the proximity of the basilic vein to the medial antebrachial cutaneous nerve, injections into these veins may cause severe transient pain, as in cases where an accidental "paravascular" injection irritates the surrounding connective tissue. "Rolling veins" refers to a condition in which the subcutaneous veins are exceptionally mobile within the subcutaneous fat. In approximately 3% of cases the ulnar artery may pass over the surface of the flexor muscles (superficial ulnar artery, see also p. 409). An unintended intra*arterial* injection can have devastating consequences with certain medications. This complication can be avoided by palpating the vessel and confirming arterial-type pulsations before giving the injection and always drawing a small amount of blood back into the syringe (dark red = venous blood, bright red = arterial blood) before depressing the plunger.

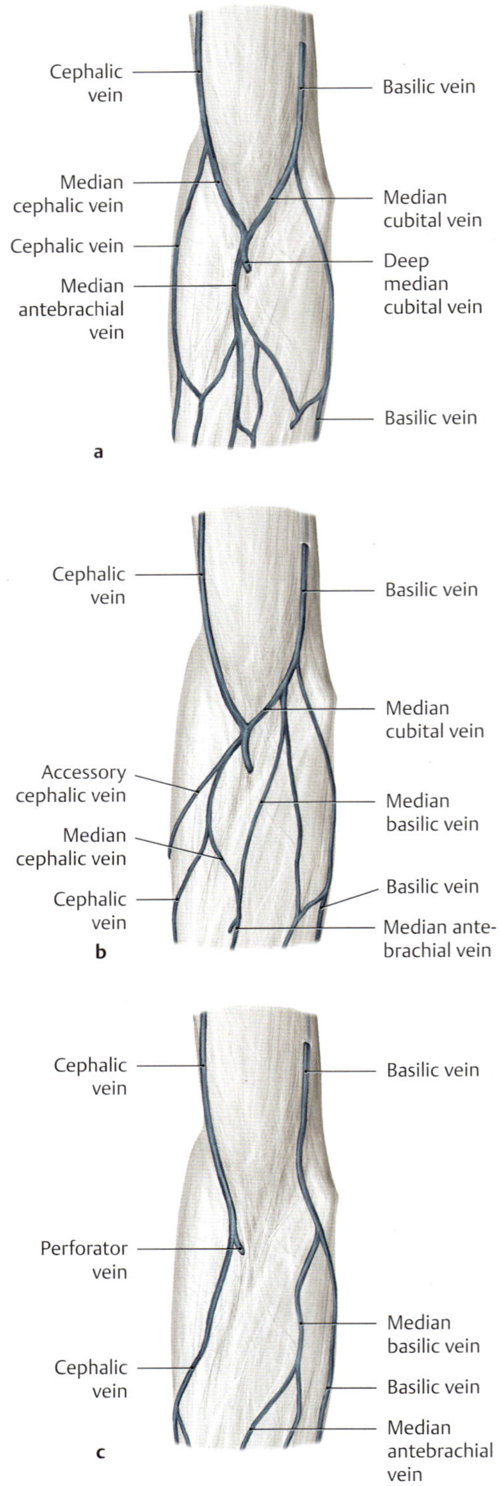

B Cubital fossa of the right arm: variable course of the subcutaneous veins

a M-shaped venous pattern above the median antebrachial vein.

b Presence of an accessory cephalic vein from the venous plexuses on the extensor side of the forearm.

c Absence of the median cubital vein.

All of the illustrated variants are common.

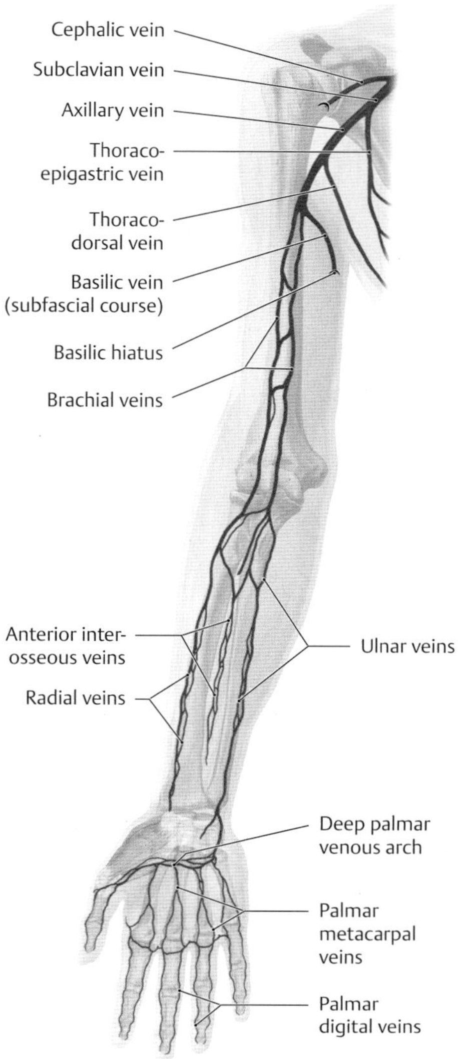

C Deep veins of the right upper limb
Anterior view.

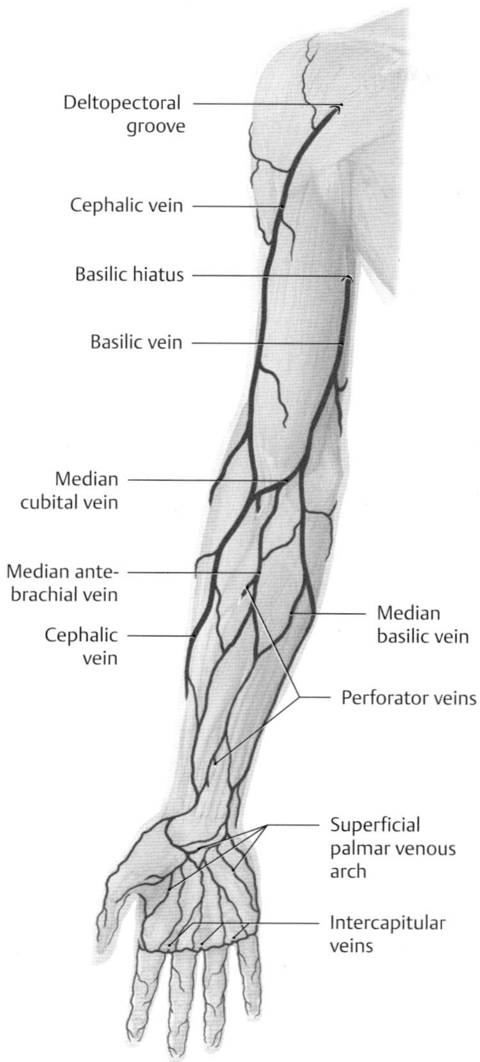

D Superficial veins of the right upper limb
Anterior view. The main longitudinal trunks of the subcutaneous venous network of the arm are the median antebrachial vein, the basilic vein, and the cephalic vein.
Median antebrachial vein: This vein, unlike the cephalic and basilic, receives blood mainly from the cutaneous veins on the dorsum of the hand, draining the *flexor side of the forearm*. The *variable* median antebrachial vein opens into the corresponding longitudinal veins at the elbow, usually by way of the median cephalic vein and median basilic vein.
Basilic vein: This vein begins at the elbow, first ascending in the *epifascial* plane in the medial bicipital groove to the basilic hiatus, where it pierces the fascia in the middle of the arm. It terminates in a *subfascial* plane at the ulnar brachial vein.
Cephalic vein: In the *arm* the cephalic vein first ascends on the lateral side of the biceps brachii, then enters a groove between the deltoid and pectoralis major muscles (the deltopectoral groove). It finally opens into the axillary vein in the clavipectoral triangle (see p. 394).

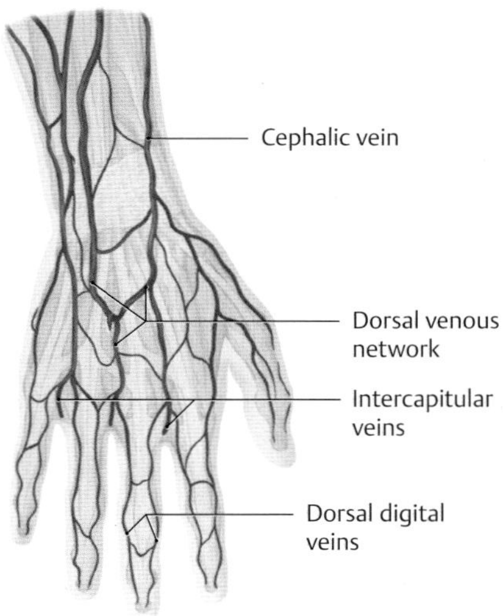

E Superficial veins of the dorsum of the right hand

F Overview of the main superficial and deep veins of the upper limb
Numerous connections exist between the deep and superficial veins of the arm—the perforator veins. Valves are incorporated into the veins at regular intervals, increasing the efficiency of venous return (see p. 73).

Deep veins of the upper limb
- Subclavian vein
- Axillary vein
- Brachial veins
- Ulnar veins
- Radial veins
- Anterior interosseous veins
- Posterior interosseous veins
- Deep palmar venous arch
- Palmar metacarpal veins

Superficial veins of the upper limb
- Cephalic vein
- Accessory cephalic vein
- Basilic vein
- Median cubital vein
- Median antebrachial vein
- Median cephalic vein
- Median basilic vein
- Dorsal venous network of the hand
- Superficial palmar venous arch

4.3 The Lymphatic Vessels and Lymph Nodes

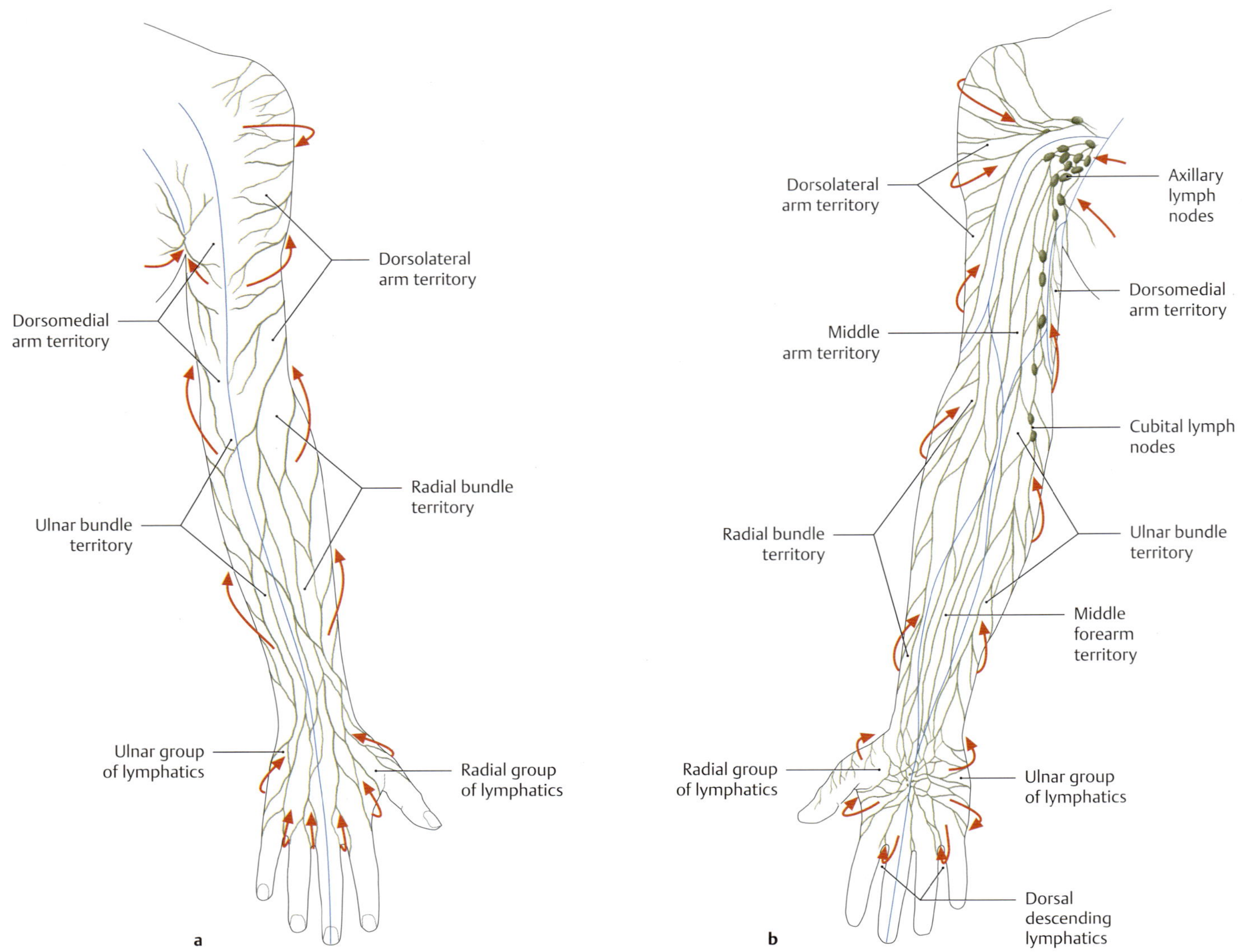

A Lymph vessels of the upper limb (after Schmidt and Lanz)
a Posterior view, **b** anterior view. The lymph vessels (lymphatics) in the upper limb are of two types:

- Superficial (epifascial) lymphatics
- Deep lymphatics

While the deep lymphatics of the upper limb accompany the arteries and deep veins, the superficial lymphatics lie in the subcutaneous tissue. In the forearm, they are most closely related to the cephalic and basilic veins. Numerous anastomoses exist between the deep and superficial systems. The arrows in the diagrams indicate the main directions of lymphatic drainage. Inflammations and infections of the hand generally incite a painful swelling of the axillary lymph nodes. When the lymph vessels are also involved, they are visible as red streaks beneath the skin (lymphangitis).

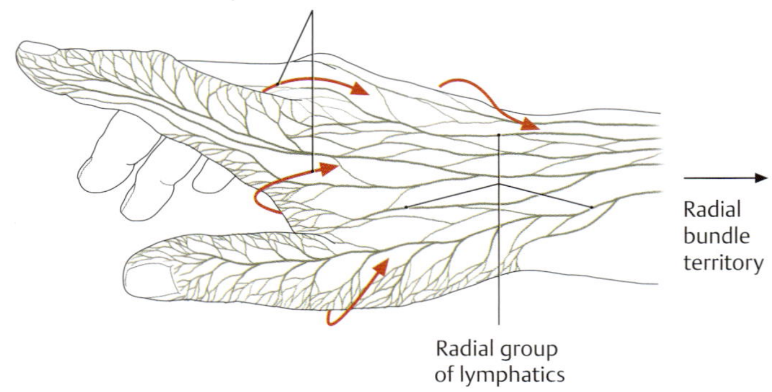

B Lymphatic drainage of the thumb, index finger, and dorsum of the hand (after Schmidt and Lanz)
The thumb, index finger, and part of the middle finger are drained by a radial group of lymph vessels that pass directly to the axillary lymph nodes. The other fingers are drained by an ulnar group of lymphatics (not shown here) that end at the cubital lymph nodes.

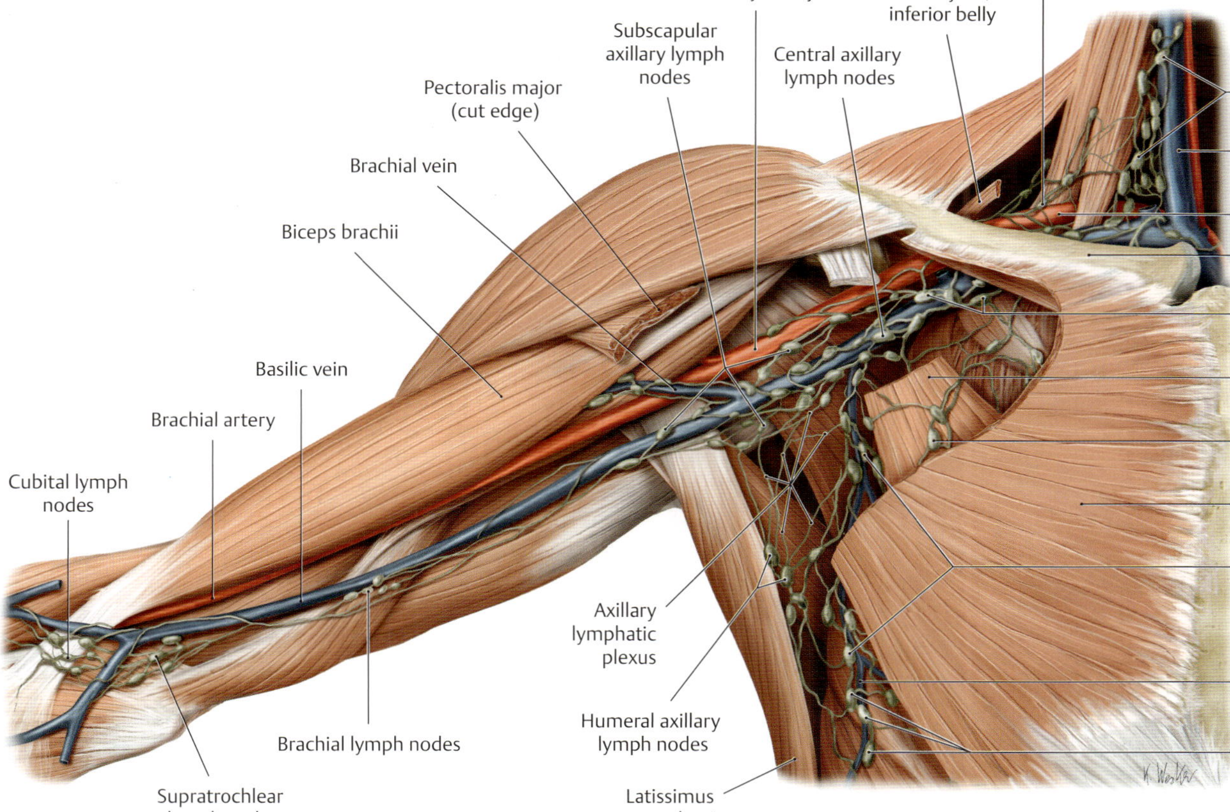

C Regional lymph nodes of the right upper limb
Anterior view. The lymph nodes of the axilla (*axillary lymph nodes*) are important collecting stations for the arm, shoulder girdle, and anterior chest wall. The 30 to 60 lymph nodes of the axilla are divided into several groups or levels, numbered I to III (see **E**), which are interconnected by lymph vessels. Taken together, the lymphatics in this region form an *axillary lymphatic plexus* lying within the fatty tissue. Lymphatic drainage from the axilla is collected in the subclavian trunk (not shown here). On the right side, the lymph is conveyed by the right jugular trunk and right bronchomediastinal trunk to the right lymphatic duct, which opens into the junction of the right subclavian and internal jugular veins (see p. 202).

D The axillary lymph nodes, grouped by levels
(after Henne-Bruns, Dürig, and Kremer)

Level I: lower axillary group
(lateral to pectoralis minor)
- Pectoral axillary lymph nodes
- Subscapular axillary lymph nodes
- Humeral axillary lymph nodes
- Paramammary lymph nodes

Level II: middle axillary group
(along the pectoralis minor)
- Interpectoral axillary lymph nodes
- Central axillary lymph nodes

Level III: upper, infraclavicular group
(medial to pectoralis minor)
- Apical axillary lymph nodes

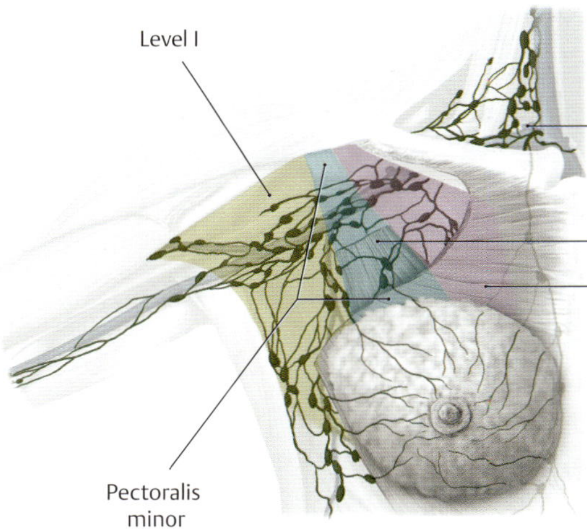

E Classification of axillary lymph nodes by level
The axillary lymph nodes have major clinical importance in breast cancer. A malignant breast tumor will metastasize (seed tumor cells) to the axillary nodes as it grows. As a guide for surgical removal, the axillary lymph nodes can be segregated into groups arranged in three levels, based on their relationship to the pectoralis minor muscle (see p. 217).

- Level I: all the lymph nodes lateral to the pectoralis minor.
- Level II: all the lymph nodes along the pectoralis minor.
- Level III: all the lymph nodes medial to the pectoralis minor (see p. 314).

4.4 The Brachial Plexus: Structure

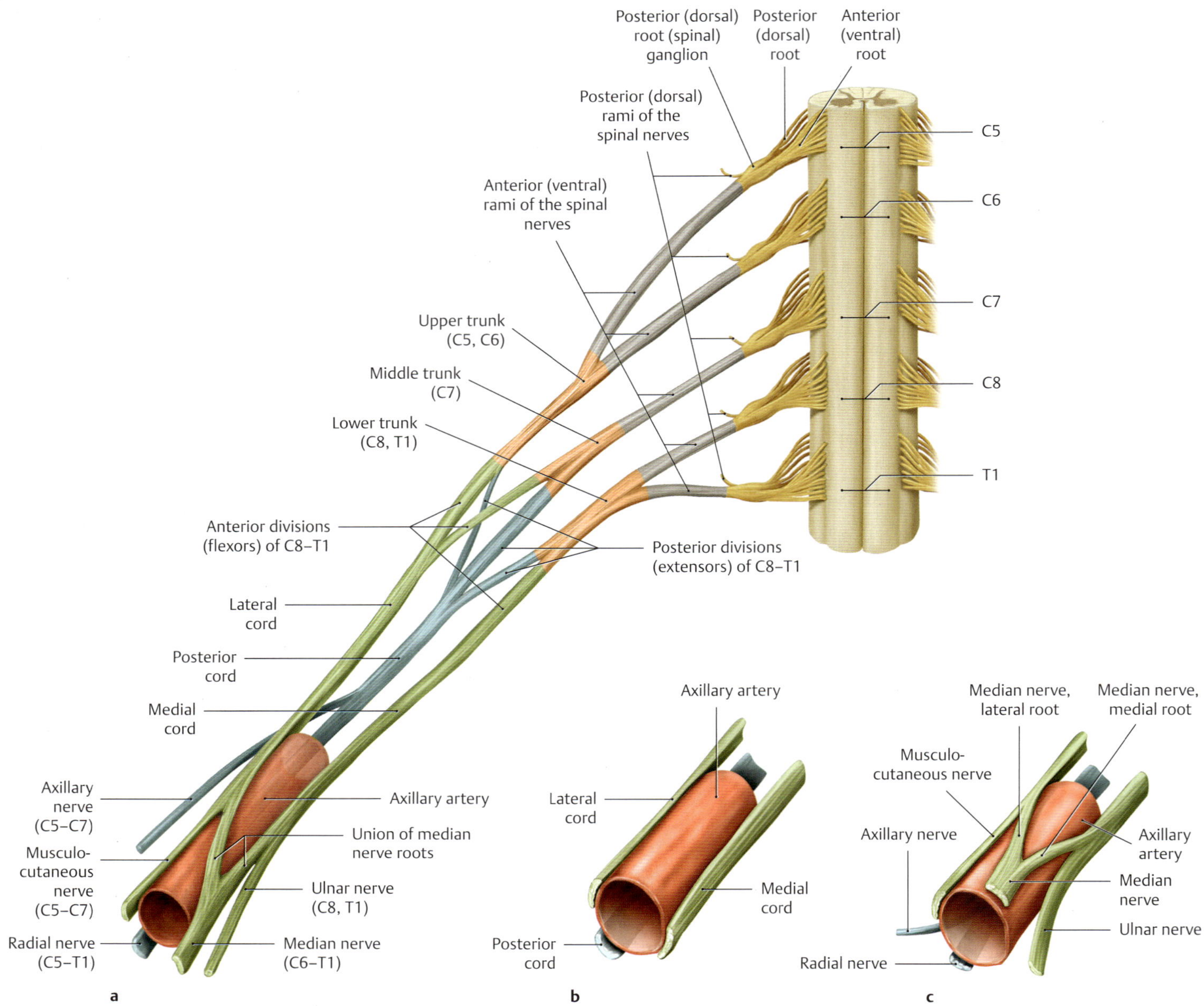

A Schematic representation of the structure of the brachial plexus

a Names and sequence of the various components of the brachial plexus.

b Relationship of the lateral, medial, and posterior cords of the brachial plexus to the axillary artery.

c Subdivision of the brachial plexus cords into their main branches.

B Number and location of the main components of the brachial plexus

Components	Number	Location
1. Plexus roots (anterior rami of the spinal nerves from cord segments C5–T1)	5	Between anterior and middle scalene (interscalene space)
2. The primary trunks: upper, middle, and lower	3	Lateral to the interscalene space and above the clavicle
3. The three anterior and three posterior divisions	6	Posterior to the clavicle
4. The lateral, medial, and posterior cords	3	In the axilla, posterior to the pectoralis minor

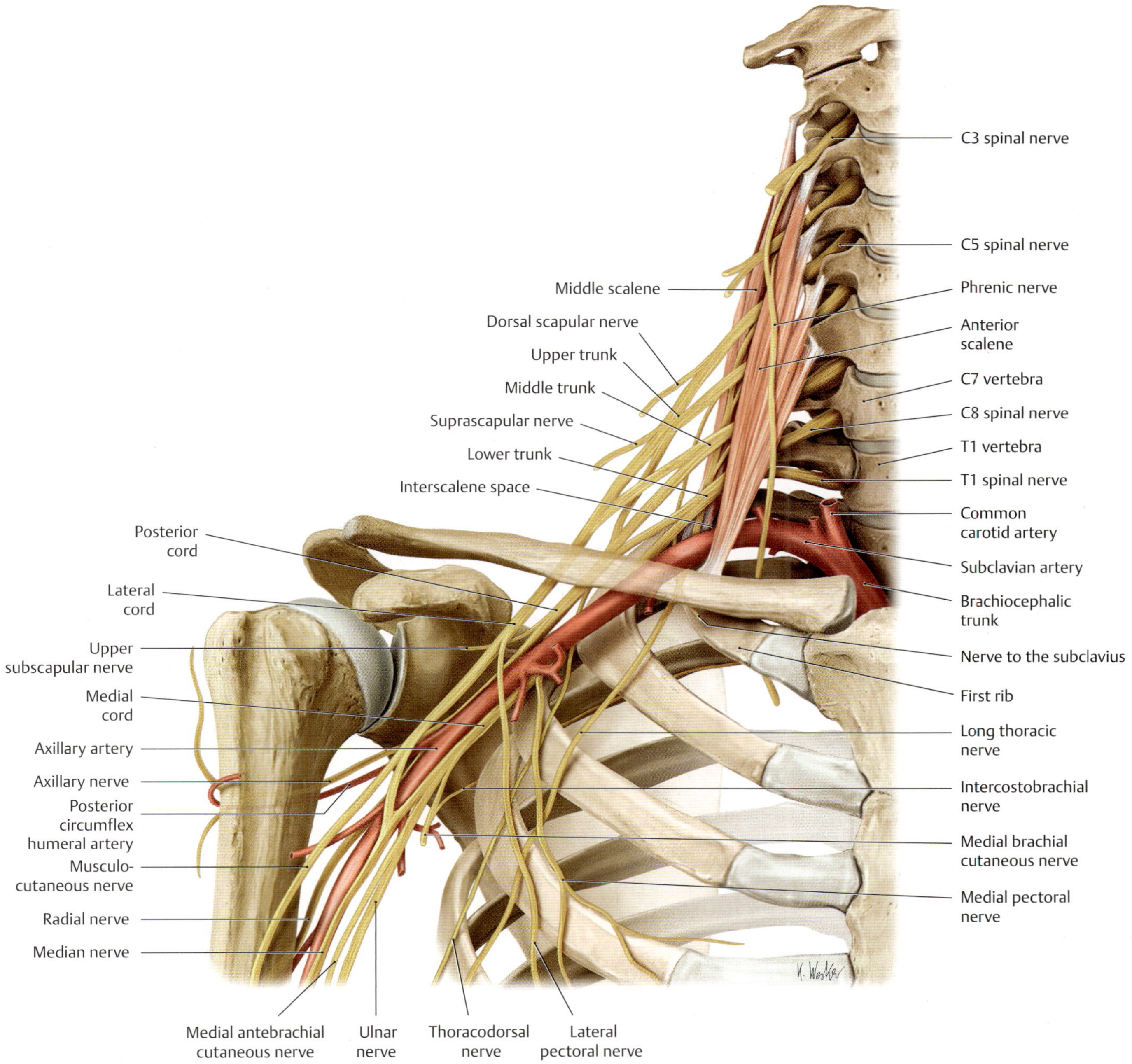

C Course of the brachial plexus and its relation to the thorax after passing through the interscalene space
Right side, anterior view.

D Spinal cord segments and nerves of the brachial plexus

Brachial plexus trunks and associated spinal cord segments
- Upper trunk C5 + C6
- Middle trunk C7
- Lower trunk C8 + T1

Brachial plexus cords and associated spinal cord segments
- Lateral cord C5–C7
- Medial cord C8–T1
- Posterior cord C5–T1

Nerves of the supraclavicular part of the brachial plexus (direct branches from the anterior rami or plexus trunks)
- Dorsal scapular nerve
- Long thoracic nerve
- Suprascapular nerve
- Nerve to the subclavius

Nerves of the infraclavicular part of the brachial plexus (short and long branches from the plexus cords)
- Lateral cord
  - Musculocutaneous nerve
  - Lateral pectoral nerve
  - Median nerve (lateral root)
- Medial cord
  - Median nerve (medial root)
  - Ulnar nerve
  - Medial pectoral nerve
  - Medial brachial cutaneous nerve
  - Medial antebrachial cutaneous nerve
- Posterior cord
  - Radial nerve
  - Axillary nerve
  - Upper subscapular nerve
  - Lower subscapular nerve
  - Thoracodorsal nerve

4.5 The Brachial Plexus: Supraclavicular Part

A Supraclavicular part of the brachial plexus
The supraclavicular part of the brachial plexus includes all nerves that arise directly from the plexus roots (anterior rami of the spinal nerves) or from the plexus trunks in the lateral cervical triangle between the anterior and middle scalene muscles. The different nerves of the supraclavicular plexus are predisposed to paralysis and/or compression in varying degrees based on their location and course (see **B–D**).

Nerve	Segment	Innervated muscle
Dorsal scapular nerve	C4, 5	• Levator scapulae • Rhomboid major • Rhomboid minor
Suprascapular nerve	C4–C6	• Supraspinatus • Infraspinatus
Long thoracic nerve	C5–C7	• Serratus anterior
Nerve to the subclavius	C5, C6	• Subclavius

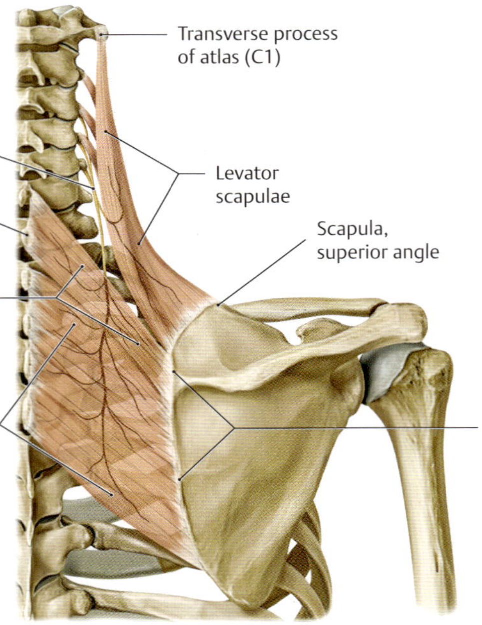

B The dorsal scapular nerve
Isolated paralysis of the dorsal scapular nerve is extremely rare owing to the protected location of the nerve between the deep nuchal muscles, the levator scapulae, and the rhomboids.

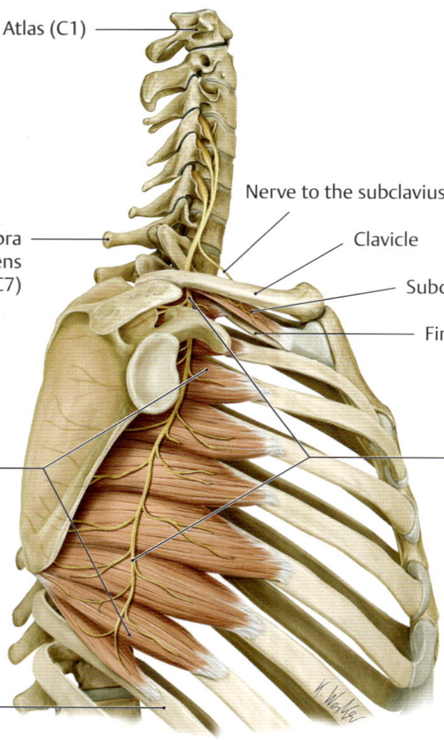

C The long thoracic nerve and nerve to the subclavius
Its long, superficial course on the serratus anterior along the lateral chest wall makes the long thoracic nerve susceptible to mechanical injury. The prolonged wearing of a heavy backpack is a common mechanism for this type of lesion. In iatrogenic cases, the nerve may be damaged by an axillary lymphadenectomy performed for a metastatic breast tumor. Clinically, loss of the serratus anterior muscle causes the medial scapular border to become elevated from the chest wall. This "winging" of the scapula is most conspicuous when the arm is raised forward, and generally the arm cannot be elevated past 90°.

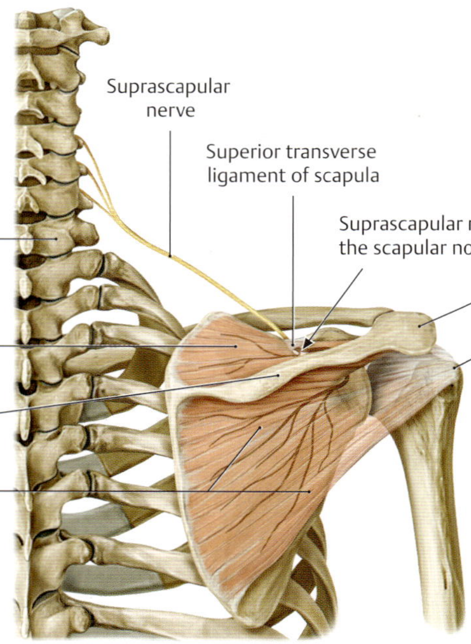

D The suprascapular nerve
Injuries and chronic compression of the suprascapular nerve are uncommon conditions that lead to atrophy of the supraspinatus and infraspinatus muscles with weakness of arm abduction (especially during the initial phase owing to the "starter" function of the supraspinatus) and external rotation of the arm. Besides an isolated injury, the nerve may become compressed in the fibro-osseous canal between the scapular notch and the superior transverse scapular ligament (which is occasionally ossified to form a bony canal). The resulting symptoms are known collectively as "scapular notch syndrome."

E Brachial plexus compression syndrome based on narrow anatomic passages in the shoulder region

In its course from the intervertebral foramina to the nerves of the upper limb, the brachial plexus must negotiate several narrow passages in which it may become compressed by surrounding structures. There are also extrinsic factors, such as carrying heavy loads, that may exert direct pressure on the brachial plexus. Several types of compression syndrome are distinguished:

1. Scalene syndrome or cervical rib syndrome: neurovascular compression in the interscalene space caused by a cervical rib or ligamentous structure (see **F**)
2. Costoclavicular syndrome: narrowing of the space between the first rib and clavicle (see **G**)
3. Hyperabduction syndrome: compression of the brachial plexus by the pectoralis minor muscle and coracoid process when the upper arm is raised above the head (see **H**)
4. Chronic heavy load on the shoulder girdle (e.g., "backpack paralysis")

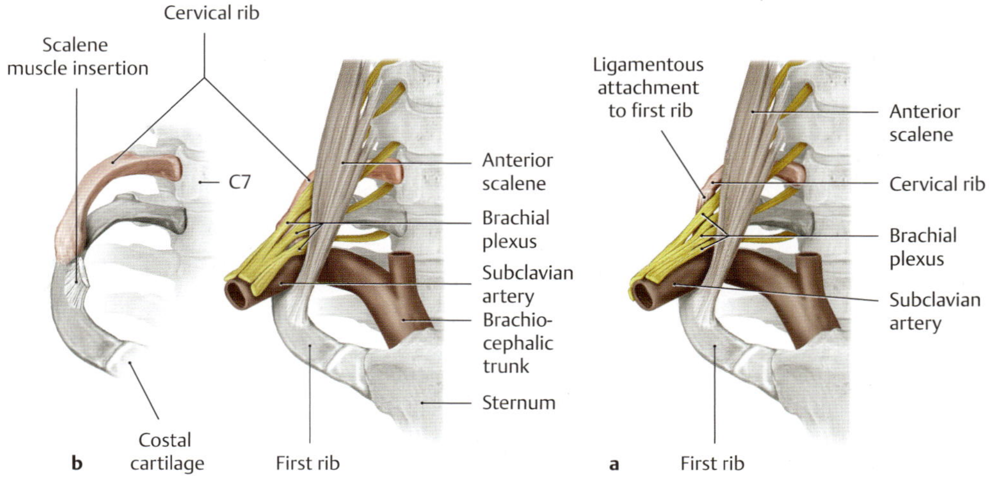

F Scalene syndrome due to narrowing of the interscalene space by a cervical rib

In approximately 1% of the population, cervical ribs may narrow the interscalene space bounded by the anterior and middle scalene muscles and the first rib. In this condition the trunks of the brachial plexus that pass through the interscalene space along with the subclavian artery are compressed from behind and below, placing varying degrees of tension on the neurovascular bundle. If there is no bony contact between a short cervical rib and the first rib (**b**), that site is often occupied by a ligamentous structure that can also cause neurovascular compression. The main clinical manifestations are pain radiating down the arm, chiefly to the ulnar side of the hand, and circulatory impairment caused by mechanical irritation of the periarterial sympathetic plexus of the subclavian artery.

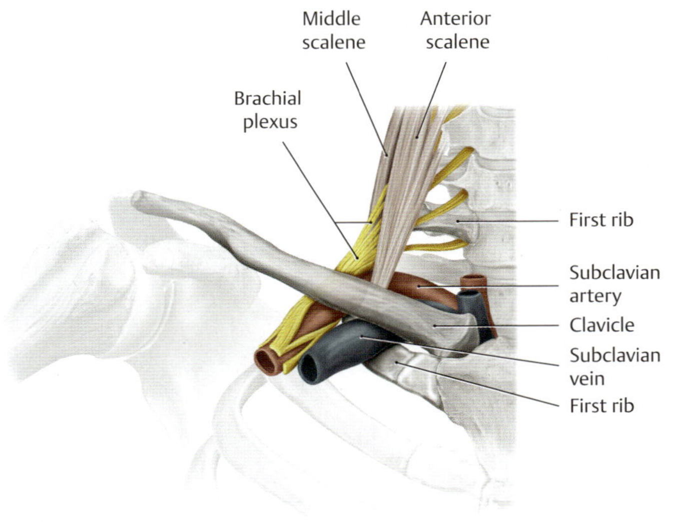

G Costoclavicular syndrome due to compression of the neurovascular bundle between the first rib and clavicle

Narrowing of the costoclavicular space is a rare condition that is most common in persons with drooping shoulders, a flat back, retracted shoulders (from carrying heavy loads), a deformed first rib, or a previous clavicular fracture. Any narrowing of the costoclavicular space can be aggravated by lowering and retracting the shoulder girdle. The complaints are similar to those in scalene syndrome and may be accompanied by signs of venous stasis caused by impaired return through the subclavian vein.

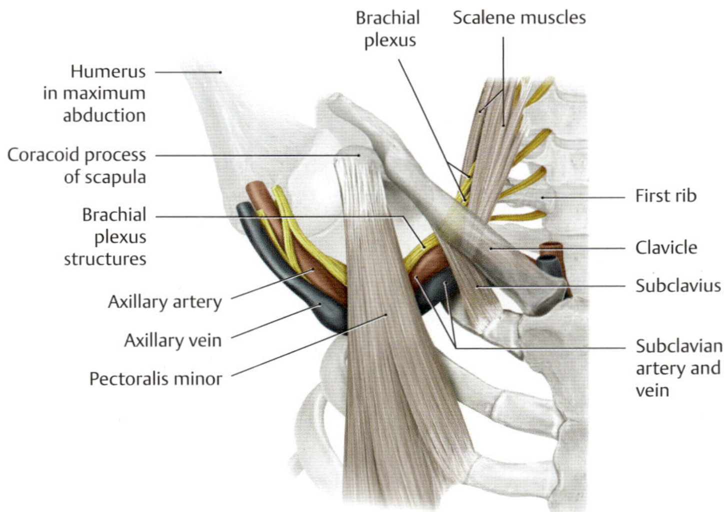

H Hyperabduction syndrome due to compression of the neurovascular bundle below the pectoralis minor and coracoid process

This rare syndrome is caused by neurovascular compression beneath the tendon of the pectoralis minor under the coracoid process. It is precipitated by maximum abduction or elevation of the arm on the affected side. A simple clinical test consists of pulling the arm upward and backward and holding it there. In normal cases, a definite radial artery pulse should still be palpable after 1 to 2 minutes, and the patient should not complain of radiating pain.

4.6 Infraclavicular Part of the Brachial Plexus: Overview and Short Branches

A Infraclavicular part of the brachial plexus
The infraclavicular part of the brachial plexus includes all the nerves that leave the plexus at the level of the plexus cords—the *short branches*—and those that continue down the arm as terminal branches of the individual plexus cords—the *long branches*. These nerves are reviewed below, beginning with the short branches.

Nerve	Segment	Innervated muscle	Innervated muscle
Part I: Short branches			
• Upper and lower subscapular nerves	C5–C8	• Subscapularis • Teres major (upper only)	–
• Thoracodorsal nerve	C6–C8	• Latissimus dorsi	–
• Medial and lateral pectoral nerves	C5–T1	• Pectoralis major • Pectoralis minor	–
• Medial brachial cutaneous nerve	T1	–	• Medial brachial cutaneous nerve
• Medial antebrachial cutaneous nerve	C8–T1	–	• Medial antebrachial cutaneous nerve
• Intercostobrachial nerves*	T2–T3	–	• Lateral cutaneous branches
Part II: Long branches			
• Musculocutaneous nerve (see p. 380)	C5–C7	• Coracobrachialis • Biceps brachii • Brachialis	• Lateral antebrachial cutaneous nerve
• Axillary nerve (see p. 381)	C5–C6	• Deltoid • Teres minor	• Superior lateral brachial cutaneous nerve
• Radial nerve (see p. 382)	C5–T1	• Brachialis (contribution) • Triceps brachii • Anconeus • Supinator • Brachioradialis • Extensor carpi radialis longus • Extensor carpi radialis brevis • Extensor digitorum • Extensor digiti minimi • Extensor carpi ulnaris • Extensor pollicis longus • Extensor pollicis brevis • Extensor indicis • Abductor pollicis longus	• Inferior lateral brachial cutaneous nerve • Posterior brachial cutaneous nerve • Posterior antebrachial cutaneous nerve • Superficial branch of radial nerve
• Median nerve (see p. 386)	C6–T1	• Pronator teres • Pronator quadratus • Palmaris longus • Flexor carpi radialis • Flexor pollicis longus • Flexor digitorum profundus (half) • Flexor digitorum superficialis • Abductor pollicis brevis • Opponens pollicis • Flexor pollicis brevis (superficial head) • First and second lumbricals	• Palmar branch of median nerve • Common and proper palmar digital nerves
• Ulnar nerve (see p. 384)	C8–T1	• Flexor carpi ulnaris • Flexor digitorum profundus (half) • Palmaris brevis • Flexor digiti minimi brevis • Abductor digiti minimi • Opponens digiti minimi • Adductor pollicis • Flexor pollicis brevis (deep head) • Palmar and dorsal interosseous muscles • Third and fourth lumbricals	• Palmar branch of ulnar nerve • Dorsal branch of ulnar nerve • Dorsal digital nerves • Common and proper palmar digital nerves

* These are the cutaneous branches of the intercostal nerves 2 and 3, which accompany the medial brachial cutaneous nerve.

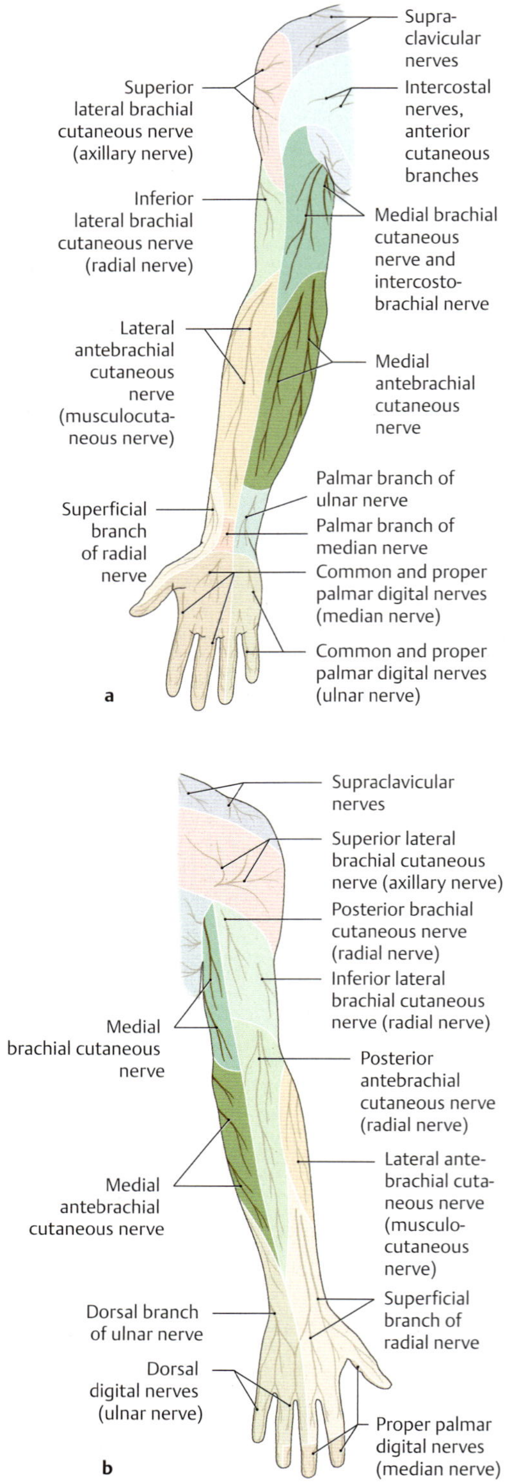

B Sensory distribution of the medial brachial cutaneous nerve and medial antebrachial cutaneous nerve of the right arm
a Anterior view, **b** posterior view.

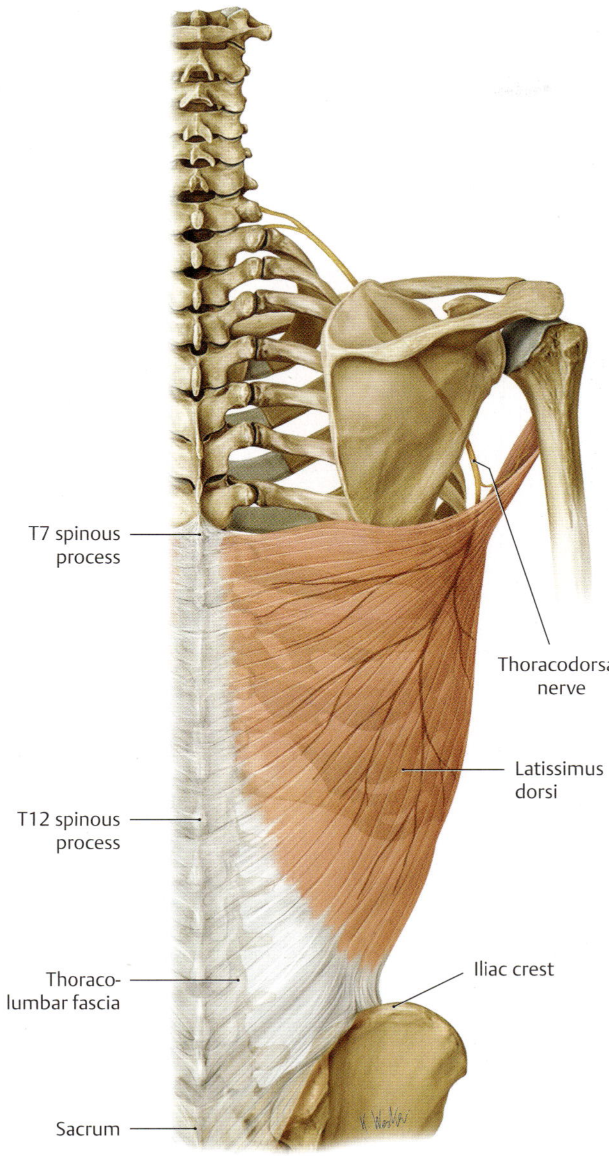

C The thoracodorsal nerve
Right side, posterior view.

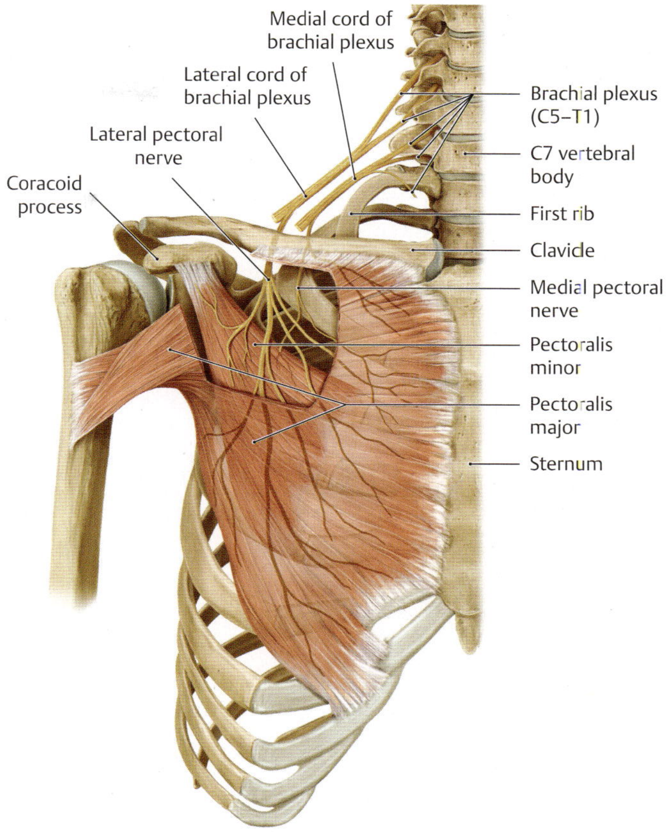

D The medial and lateral pectoral nerves
Right side, anterior view.

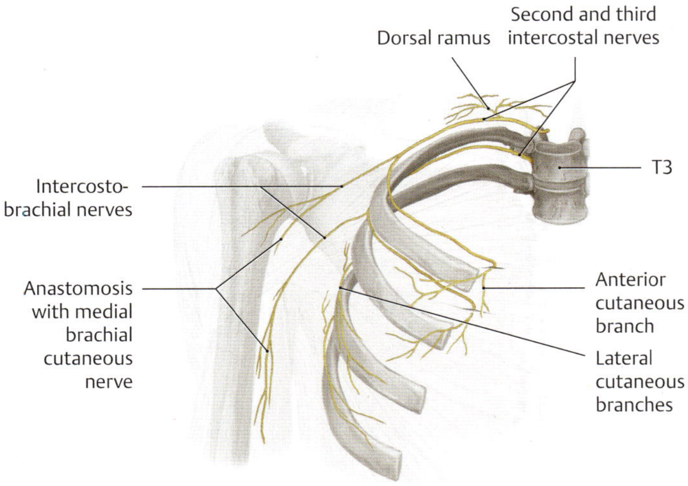

E Origin and cutaneous distribution of the intercostobrachial nerves of the right arm
Anterior view.

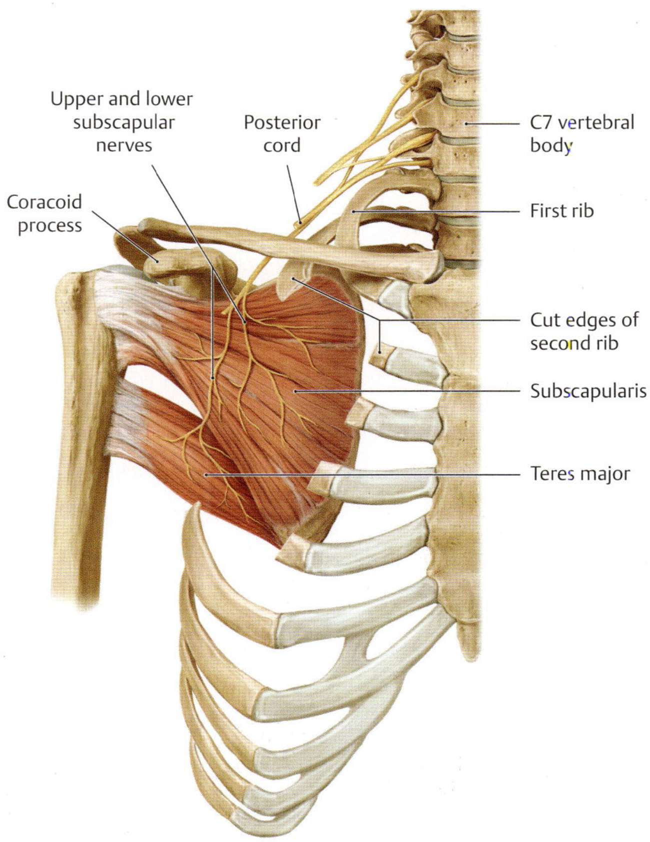

F The upper and lower subscapular nerves
Right side, anterior view. The ribs have been partially removed.

4.7 Infraclavicular Part of the Brachial Plexus: The Musculocutaneous Nerve and Axillary Nerve

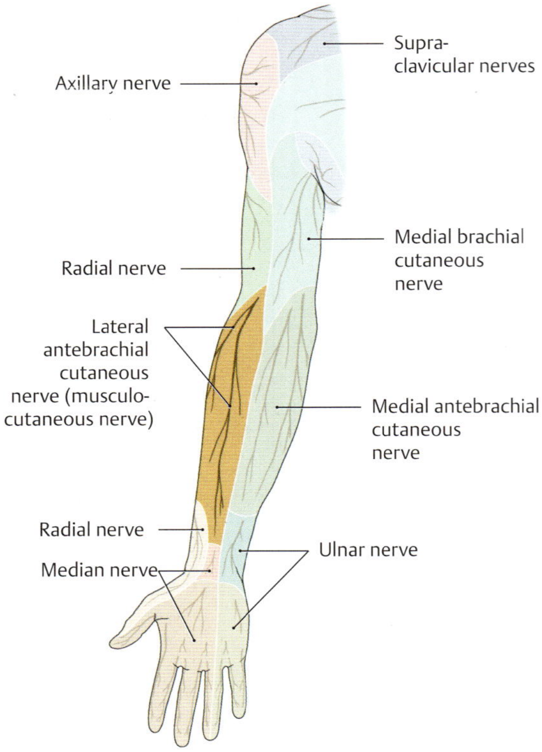

A Sensory distribution of the lateral antebrachial cutaneous nerve
Anterior view.

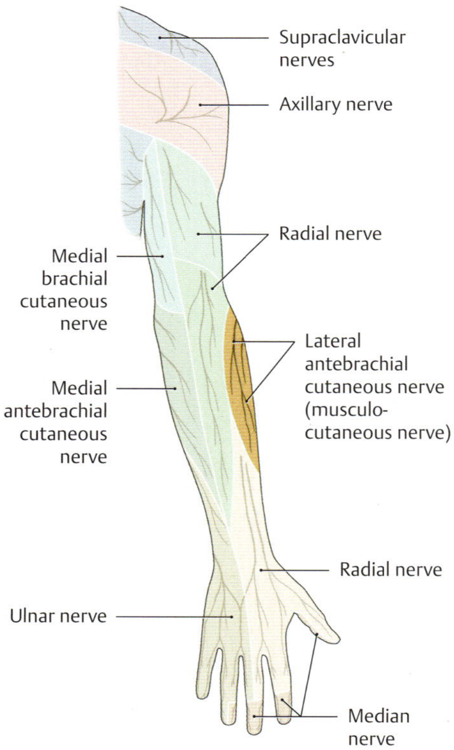

B Sensory distribution of the lateral antebrachial cutaneous nerve
Posterior view.

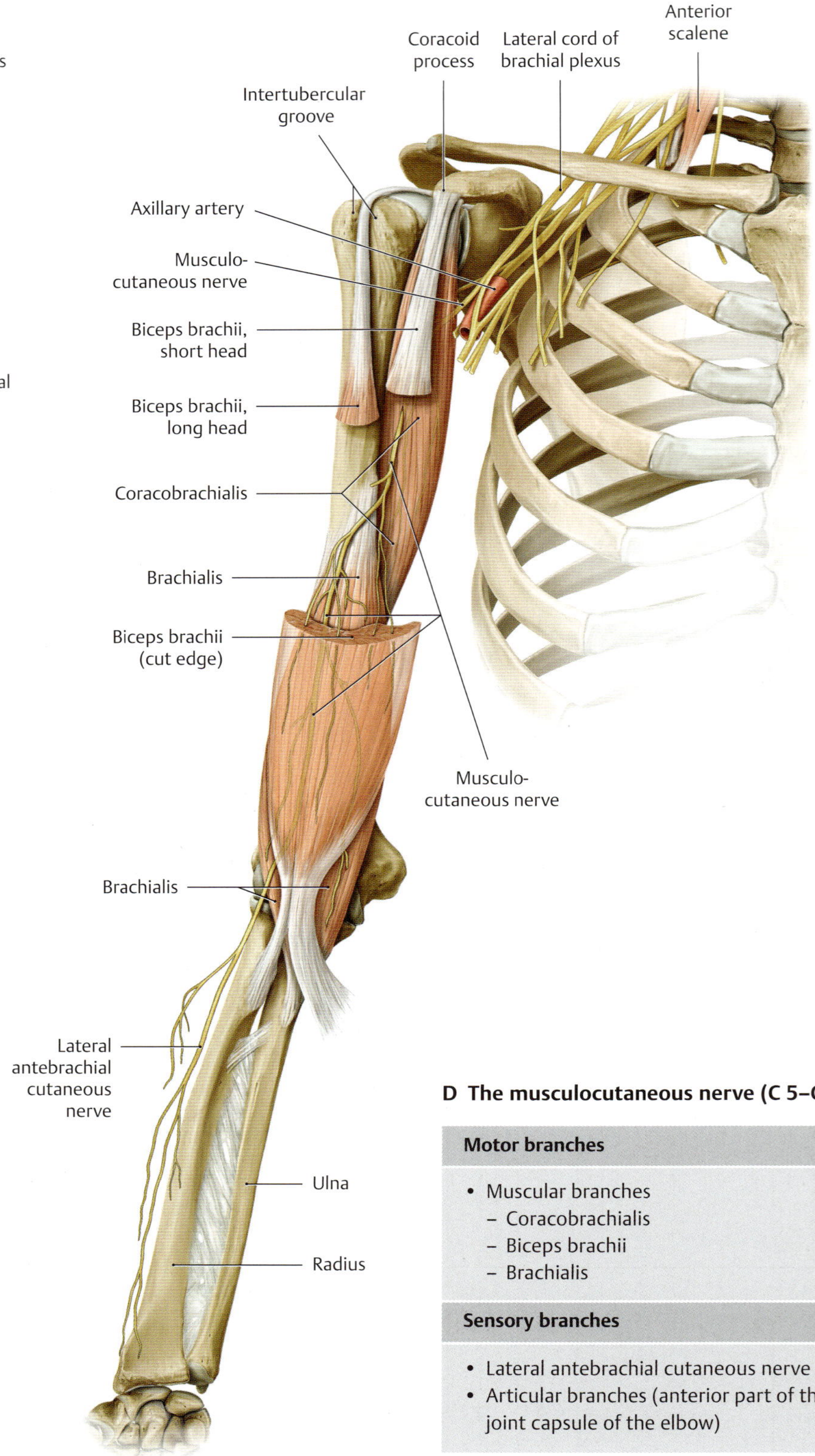

C Course of the musculocutaneous nerve after leaving the lateral cord of the brachial plexus
Right upper limb, anterior view. The musculocutaneous nerve leaves the lateral cord of the brachial plexus as a mixed nerve (one with motor and sensory branches) at the level of the lateral border of the pectoralis minor (not shown here) and runs a short course before piercing the coracobrachialis. It then runs between the biceps brachii and brachialis muscles to the elbow, where its terminal sensory branch supplies the skin on the radial side of the forearm.

D The musculocutaneous nerve (C 5–C 7)

Motor branches
• Muscular branches – Coracobrachialis – Biceps brachii – Brachialis
Sensory branches
• Lateral antebrachial cutaneous nerve • Articular branches (anterior part of the joint capsule of the elbow)

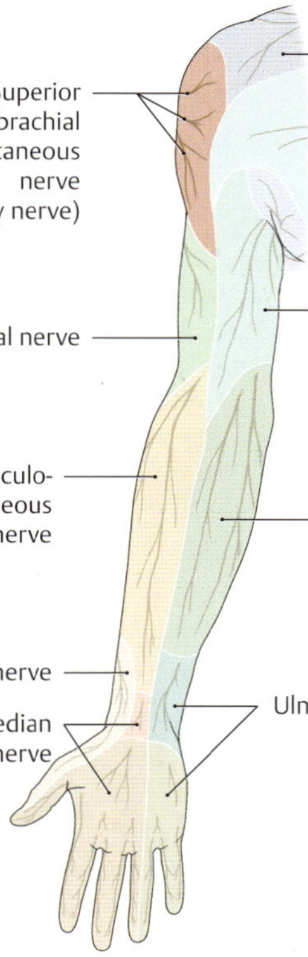

E Sensory distribution of the superior lateral brachial cutaneous nerve
Anterior view.

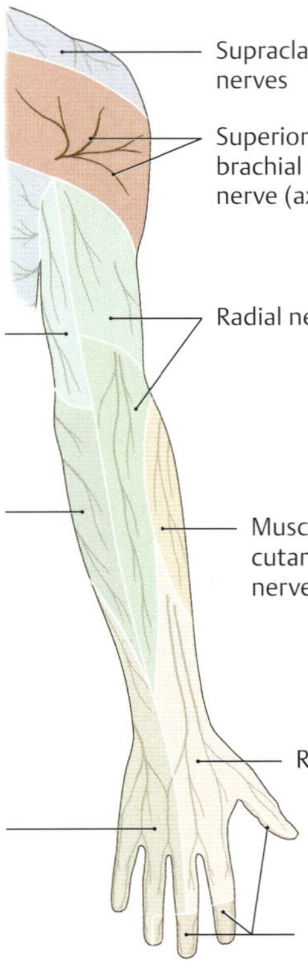

F Sensory distribution of the superior lateral brachial cutaneous nerve
Posterior view.

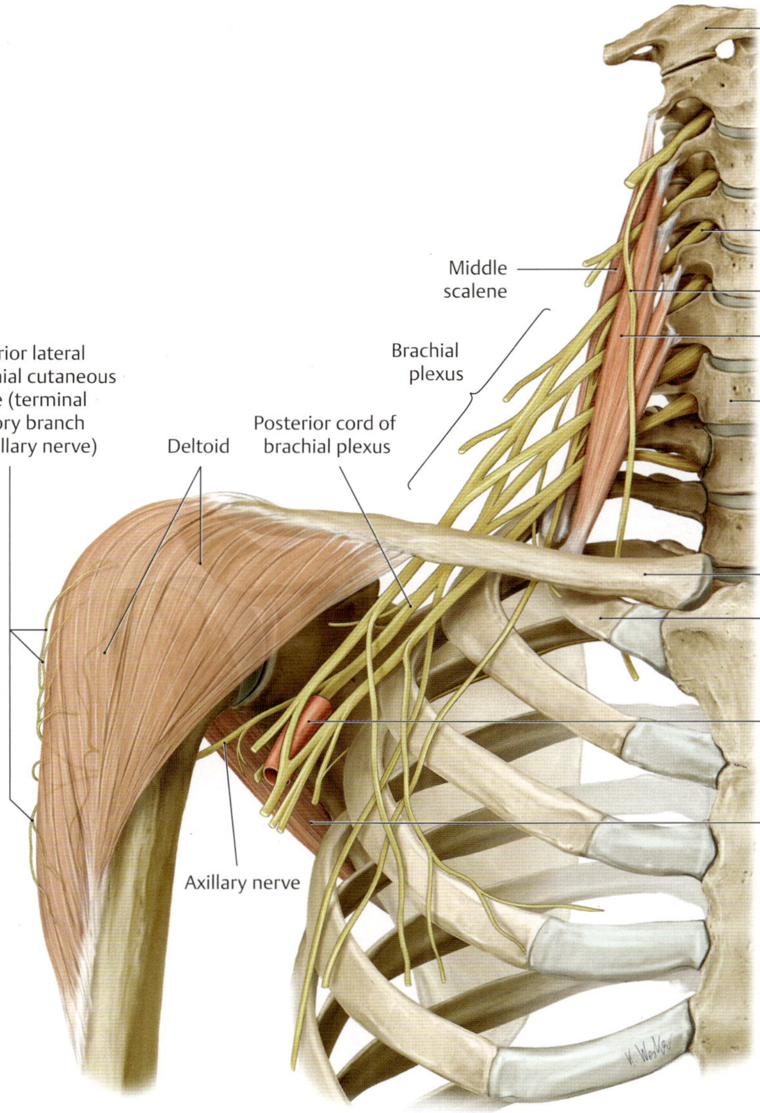

G Course of the axillary nerve after leaving the posterior cord of the brachial plexus
Right upper limb, anterior view. The axillary nerve leaves the posterior cord of the brachial plexus as a mixed nerve and runs backward through the deep part of the axilla, passing directly below the shoulder joint. It courses through the quadrangular space of the axilla (with the posterior circumflex humeral artery) and along the surgical neck to the posterior side of the proximal humerus. Its terminal sensory branch supplies the skin over the deltoid muscle. Isolated *axillary nerve palsy* may occur following an antero-inferior shoulder dislocation (or a traumatic reduction attempt), a humeral fracture at the level of the surgical neck, or prolonged pressure from an improperly adjusted crutch in the axilla. Often the axillary nerve innervates the long head of the triceps brachii in addition to the radial nerve.

H The axillary nerve (C 5 and C 6)

Motor branches
• Muscular branches – Deltoid – Teres minor
Sensory branch
• Superior lateral brachial cutaneous nerve

4.8 Infraclavicular Part of the Brachial Plexus: The Radial Nerve

A The radial nerve (C 5–T 1)

Motor branches
• Muscular branches (from the radial nerve) – Brachialis (contribution) – Triceps brachii – Anconeus – Brachioradialis – Extensor carpi radialis longus – Extensor carpi radialis brevis • Deep branch (terminal branch: posterior interosseous nerve) – Supinator – Extensor digitorum – Extensor digiti minimi – Extensor carpi ulnaris – Extensor pollicis longus – Extensor pollicis brevis – Extensor indicis – Abductor pollicis longus
Sensory branches
• Articular branches (from the radial nerve) – Capsule of the shoulder joint • Articular branches (from the posterior interosseous nerve) – Joint capsule of the wrist and the four radial metacarpophalangeal joints • Posterior brachial cutaneous nerve • Inferior lateral brachial cutaneous nerve • Posterior antebrachial cutaneous nerve • Superficial branches – Dorsal digital nerves – Ulnar communicating branch

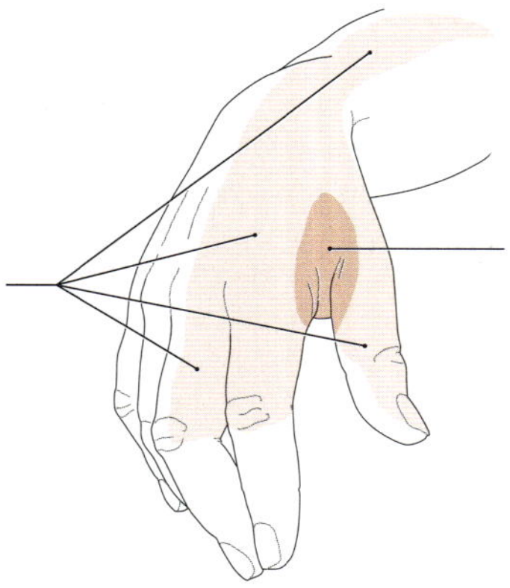

C Wrist drop due to proximal and midlevel radial nerve lesions
When the radial nerve is damaged, the patient can no longer actively extend the hand at the wrist, and *wrist drop* (drop hand) is said to be present. Besides the dropped position of the wrist, clinical examination reveals areas of sensory loss on the radial surface of the dorsum and on the extensor surface of the thumb, index finger, and the radial half of the middle finger extending to the proximal interphalangeal joint. The sensory deficits are often confined to the area of the hand that receives sensory innervation exclusively from the radial nerve (the interosseous space between the thumb and index finger).

B Traumatic lesions and compression syndromes involving the radial nerve
The radial nerve may be damaged anywhere in its course as a result of injury or chronic compression. The clinical features depend critically on the *site of the lesion*. As a general rule, the more proximal the site of the lesion, the greater the number of extensor muscles that are affected. The characteristic feature of a proximal ("high") radial nerve lesion is *wrist drop* (see **C**), in which the patient is unable to extend the wrist or the metacarpophalangeal joints. Lesions at some sites may additionally cause sensory disturbances (pain, paresthesia, numbness), particularly in the exclusive sensory territory of the superficial branch on the radial side of the dorsum (first interosseous space between the thumb and index finger).

Proximal radial nerve lesion
• Chronic pressure in the axilla (e.g., due to prolonged crutch use). **Clinical features:** typical dropped wrist with loss of the triceps brachii (and sensory disturbances). • Traumatic lesion due to a humeral shaft fracture at the level of the radial groove (spiral canal). **Clinical features:** usually a typical dropped wrist *without involvement of the triceps brachii*, since the muscular branches that supply the triceps brachii leave the radial nerve just before it enters the radial groove (sensory disturbances are present, however). • Chronic compression of the radial nerve against the bony floor of the radial groove (e.g., during sleep or due to improper positioning of the patient during general anesthesia, exuberant callus formation after a fracture, or a tendon expansion from the lateral head of the triceps brachii). "Park bench palsy" is a common form caused by draping the arm over the back of a park bench. **Clinical features:** dropped wrist *without involvement of the triceps brachii*. Sensory disturbances are present. The prognosis is usually favorable, and the palsy should resolve in a few days.
Midlevel radial nerve lesion
• Chronic compression of the radial nerve in its passage through the lateral intermuscular septum and in the radial tunnel (e.g., by bridging vessels and connective-tissue septa). **Clinical features:** dropped wrist with sensory disturbances.
Distal radial nerve lesion
• Compression of the deep branch of the radial nerve at its entry into the supinator canal by a sharp-edged tendon of the superficial part of the supinator muscle: supinator syndrome or distal radial nerve compression syndrome. **Clinical features:** no typical wrist drop and no sensory disturbances involving the hand (before entering the supinator canal, the deep branch gives off the purely sensory superficial branch and muscular branches for the supinator, brachioradialis, and extensor carpi radialis longus and brevis). There are palsies involving the extensor pollicis brevis and longus, abductor pollicis longus, extensor digitorum, extensor indicis, and extensor carpi ulnaris. • Trauma to the deep radial nerve branch caused by a fracture or dislocation of the radius. **Clinical features:** no wrist drop and no sensory disturbances.

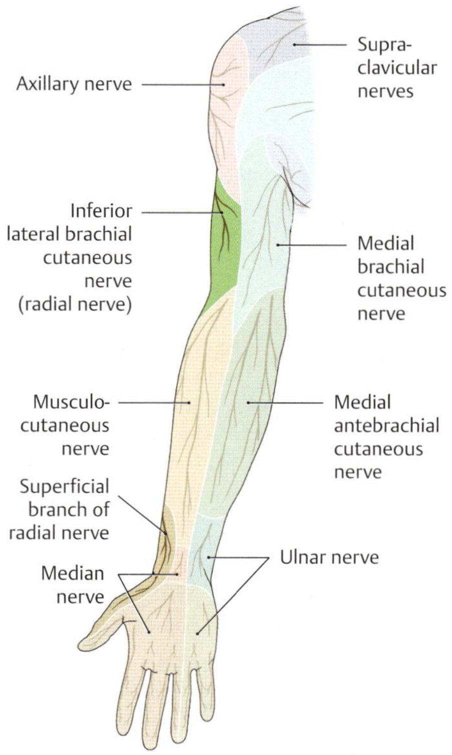

D Sensory distribution of the radial nerve
Anterior view.

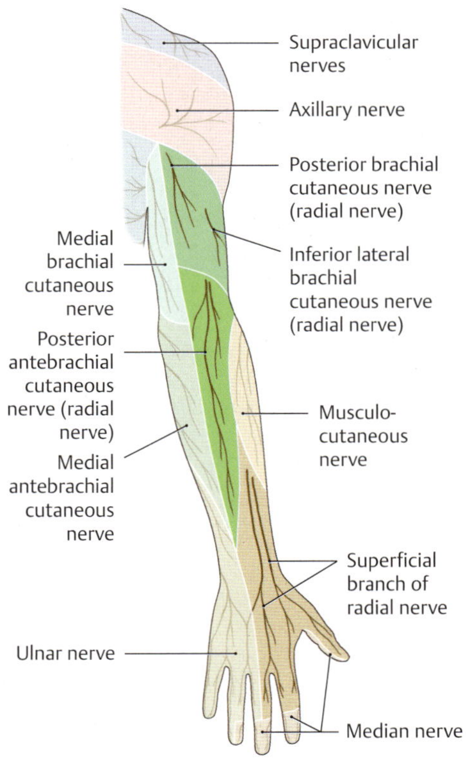

E Sensory distribution of the radial nerve
Posterior view.

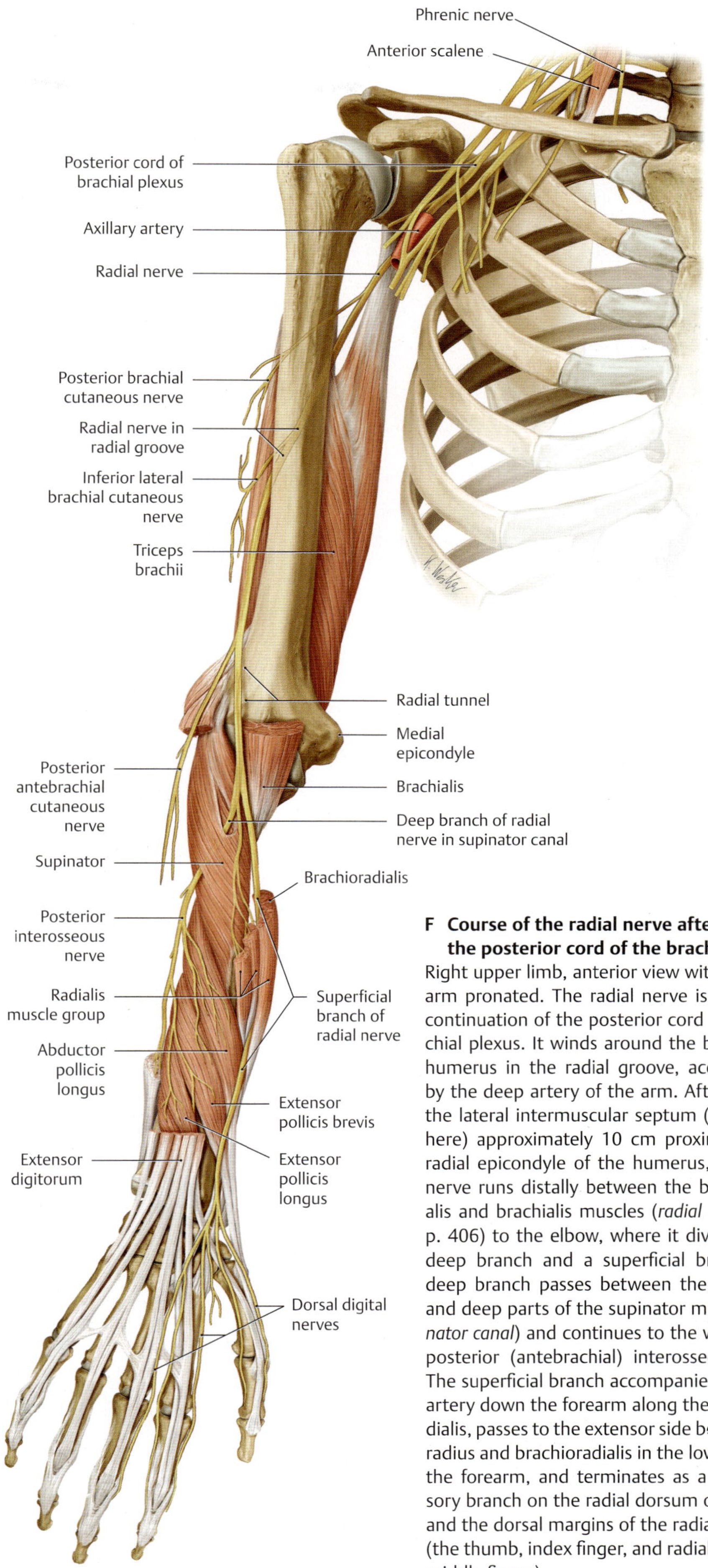

F Course of the radial nerve after leaving the posterior cord of the brachial plexus
Right upper limb, anterior view with the forearm pronated. The radial nerve is the direct continuation of the posterior cord of the brachial plexus. It winds around the back of the humerus in the radial groove, accompanied by the deep artery of the arm. After piercing the lateral intermuscular septum (not shown here) approximately 10 cm proximal to the radial epicondyle of the humerus, the radial nerve runs distally between the brachioradialis and brachialis muscles (*radial tunnel*, see p. 406) to the elbow, where it divides into a deep branch and a superficial branch. The deep branch passes between the superficial and deep parts of the supinator muscle (*supinator canal*) and continues to the wrist as the posterior (antebrachial) interosseous nerve. The superficial branch accompanies the radial artery down the forearm along the brachioradialis, passes to the extensor side between the radius and brachioradialis in the lower third of the forearm, and terminates as a main sensory branch on the radial dorsum of the hand and the dorsal margins of the radial 2½ digits (the thumb, index finger, and radial half of the middle finger).

4.9 Infraclavicular Part of the Brachial Plexus: The Ulnar Nerve

A The ulnar nerve (C8, T1)

Motor branches
• Muscular branches (directly from the ulnar half) – Flexor carpi ulnaris – Flexor digitorum profundus (ulnar half) • Muscular branch (from the superficial ulnar nerve branch) – Palmaris brevis • Muscular branches (from the deep ulnar nerve branch) – Abductor digiti minimi – Flexor digiti minimi brevis – Opponens digiti minimi – Third and fourth lumbricals – Palmar and dorsal interosseous muscles – Adductor pollicis – Flexor pollicis brevis (deep head)
Sensory branches
• Articular branches – Capsule of the elbow joint and of the carpal and metacarpophalangeal joints • Dorsal branch of the ulnar nerve (terminal branches: dorsal digital nerves) • Palmar branch of the ulnar nerve • Proper palmar digital nerve (from the superficial branch) • Common palmar digital nerve (from the superficial branch; terminal branches: proper palmar digital nerves)

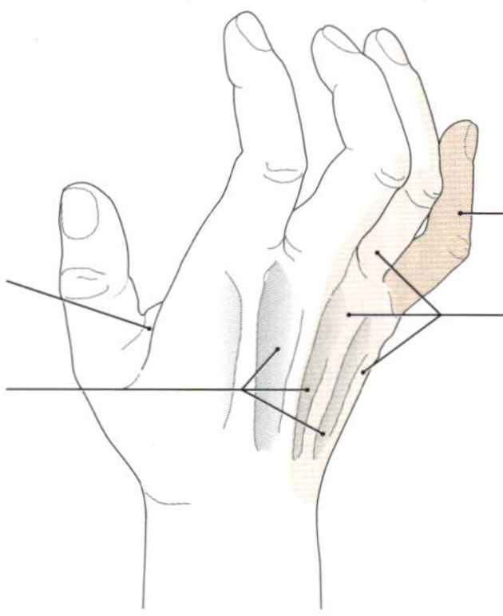

C Claw hand due to an ulnar nerve lesion
Besides the typical clawlike appearance of the hand, atrophy of the interossei leads to hollowing of the interosseous spaces in the metacarpus. Sensory abnormalities are frequently confined to the little finger (exclusive sensory territory of the ulnar nerve).

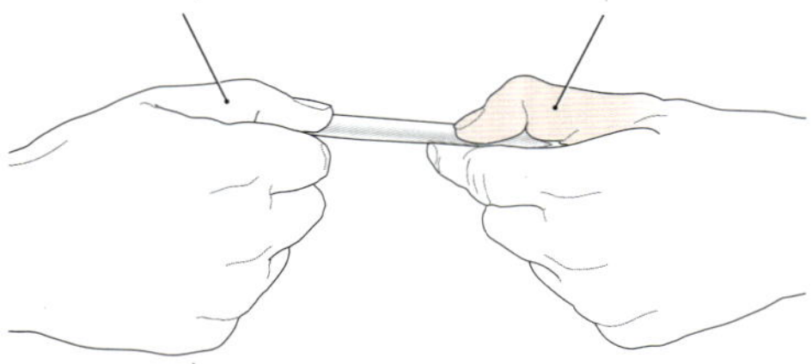

B Traumatic lesions and compression syndromes involving the ulnar nerve
Ulnar nerve palsy is the *most common peripheral nerve paralysis*. The characteristic feature of an ulnar nerve lesion is a "*claw hand*" deformity (see **C**), in which *loss of the interosseous muscles* causes the fingers to be hyperextended at the metacarpophalangeal joints and slightly flexed at the proximal and distal interphalangeal joints. The deformity is least pronounced in the index and middle fingers because the first and second lumbrical muscles, which are innervated by the median nerve, can partially compensate for the clawing of those fingers. The thumb is markedly hyperextended due to the loss of the adductor pollicis and the dominance of the extensor pollicis longus and abductor pollicis. The interossei muscles atrophy in 2 to 3 months; this is most conspicuous in the first interosseous space and is accompanied by hypothenar atrophy. Sensory disturbances affect the ulnar portion of the hand, the ulnar half of the ring finger, and the entire little finger.

Proximal ulnar nerve lesion
• Traumatic lesions, usually occurring at the elbow joint due to the exposed position of the nerve in the ulnar groove (e.g., pressure from resting on the arm), displacement of the nerve from its groove, or articular injuries due to fractures. • Chronic pressure on the nerve in the ulnar groove due to degenerative or inflammatory changes in the elbow joint, or chronic traction on the nerve caused by repetitive flexion and extension at the elbow joint (sulcus ulnaris syndrome). • Possible compression between the tendons of origin of the flexor carpi ulnaris (cubital tunnel syndrome). **Clinical features:** claw hand and sensory disturbances.
Midlevel ulnar nerve lesion
• Traumatic lesions at the wrist (e.g., lacerations). • Chronic compression of the nerve in the ulnar tunnel, a fibro-osseous canal between the palmar carpal ligament, pisiform bone, and flexor retinaculum (ulnar tunnel syndrome, see p. 419). **Clinical features:** claw hand and sensory disturbances that spare the hypothenar region (palmar branch is intact).
Distal ulnar nerve lesion
• Compression of the deep branch of the ulnar nerve in the palm due to chronic pressure (e.g., from an air hammer or other tools). **Clinical features:** claw hand with no sensory disturbances (superficial branch is intact).

D Positive "Froment sign" in the left hand
A positive Froment sign indicates *palsy of the adductor pollicis muscle*. When the patient is told to hold a piece of paper firmly between the thumb and index finger, he or she must use the flexor pollicis longus, which is innervated by the *median nerve*, rather than the paralyzed adductor pollicis, which is innervated by the *ulnar nerve*. Flexing the thumb at the interphalangeal joint signifies a positive test.

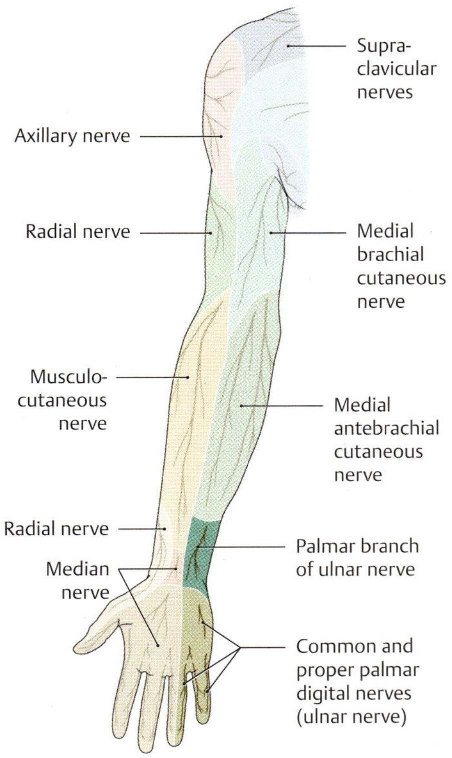

E Sensory distribution of the ulnar nerve
Anterior view.

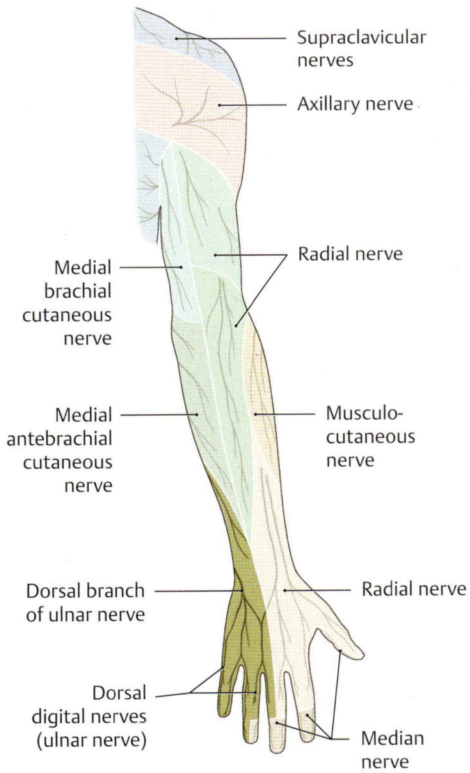

F Sensory distribution of the ulnar nerve
Posterior view.

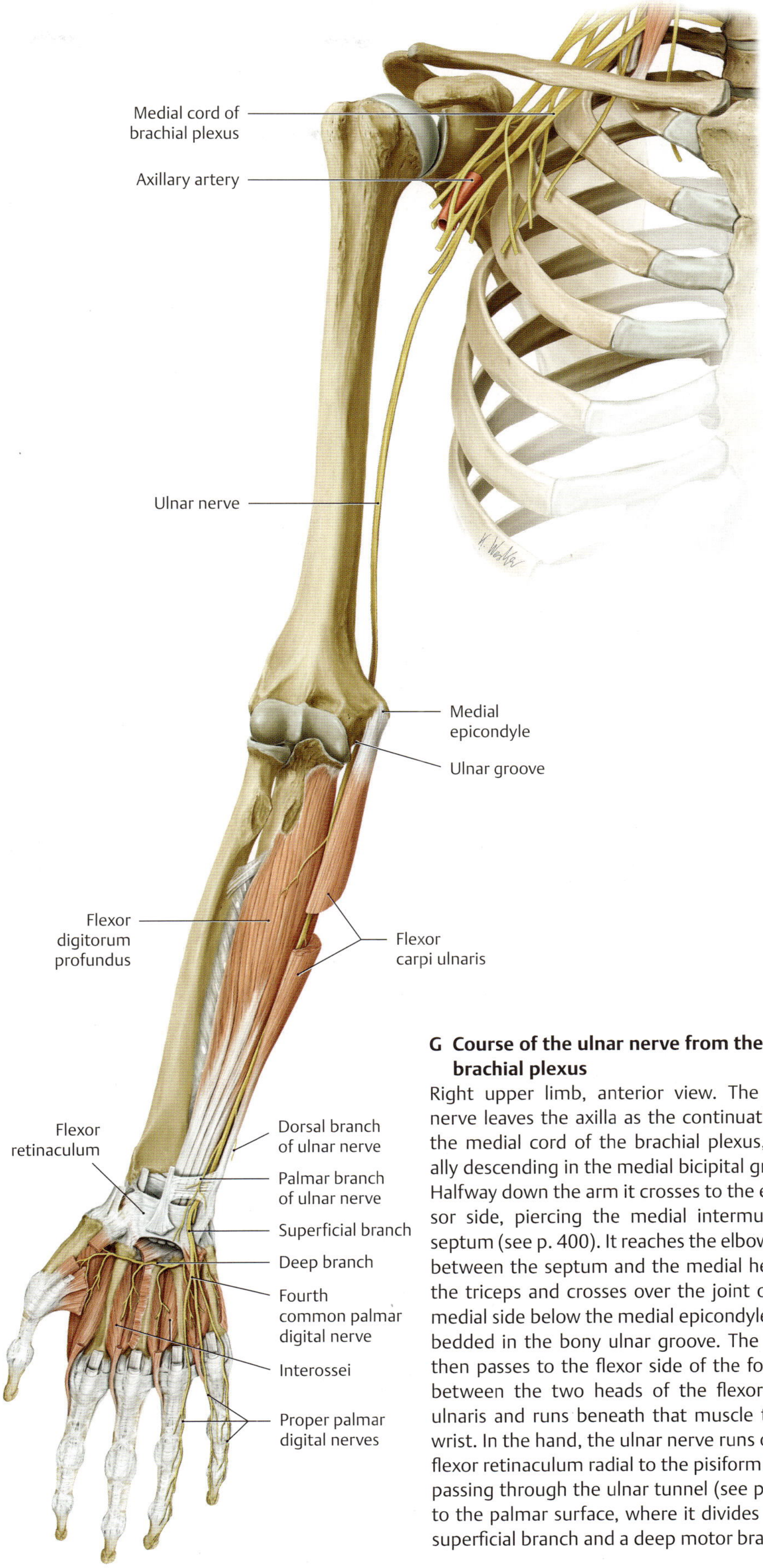

G Course of the ulnar nerve from the brachial plexus

Right upper limb, anterior view. The ulnar nerve leaves the axilla as the continuation of the medial cord of the brachial plexus, initially descending in the medial bicipital groove. Halfway down the arm it crosses to the extensor side, piercing the medial intermuscular septum (see p. 400). It reaches the elbow joint between the septum and the medial head of the triceps and crosses over the joint on the medial side below the medial epicondyle, embedded in the bony ulnar groove. The nerve then passes to the flexor side of the forearm between the two heads of the flexor carpi ulnaris and runs beneath that muscle to the wrist. In the hand, the ulnar nerve runs on the flexor retinaculum radial to the pisiform bone, passing through the ulnar tunnel (see p. 419) to the palmar surface, where it divides into a superficial branch and a deep motor branch.

4.10 Infraclavicular Part of the Brachial Plexus: The Median Nerve

A The median nerve (C 6–T 1)

Motor branches
• Muscular branches (directly from the median nerve) – Pronator teres – Flexor carpi radialis – Palmaris longus – Flexor digitorum superficialis • Muscular branches (from the anterior antebrachial interosseous nerve) – Pronator quadratus – Flexor pollicis longus – Flexor digitorum profundus (radial half) • Thenar muscular branch ("thenar branch") – Abductor pollicis brevis – Flexor pollicis brevis (superficial head) – Opponens pollicis • Muscular branches (from the common palmar digital nerves) – First and second lumbricals
Sensory branches
• Articular branches – Capsules of the elbow joint and wrist joints • Palmar branch of median nerve (thenar eminence) • Communicating branch to ulnar nerve • Common palmar digital nerves • Proper palmar digital nerves (fingers)

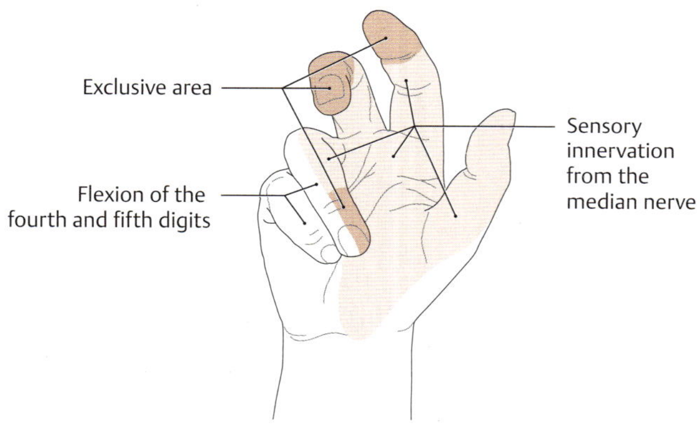

C "Hand of benediction" following a proximal median nerve lesion

When patients try to make a fist, they can flex only the ulnar fingers. This is the "hand of benediction" deformity. There may be associated sensory disturbances, particularly in the autonomous area of the nerve (tips of the radial 3½ digits).

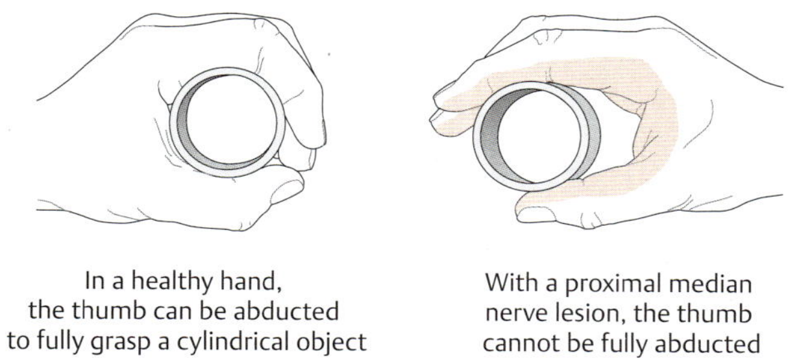

B Traumatic lesions and compression syndromes involving the median nerve

Median nerve lesions in the arm caused by an acute injury or chronic pressure are among the most common peripheral nerve lesions. The clinical manifestations depend on the *site of the lesion*. The two main categories are proximal and distal nerve lesions, as illustrated by the *pronator teres syndrome* and the *carpal tunnel syndrome*. The hallmark of a proximal median nerve lesion is the "*hand of benediction*," which occurs when the patient tries to clench the hand into a fist (loss of the long digital flexors except for the part of the flexor digitorum profundus supplied by the ulnar nerve). This contrasts with distal median nerve lesions, which present selectively with thenar atrophy and sensory disturbances, as in carpal tunnel syndrome.

Proximal median nerve lesion
• Traumatic injury caused by a fracture or dislocation of the elbow joint. • Chronic pressure injury from an anomalous supracondylar process connected to the medial epicondyle by a ligament ("Struthers' ligament," see p. 407), pressure from a tight bicipital aponeurosis, or a *pronator teres syndrome* in which the nerve is squeezed between the two heads of the pronator teres. **Clinical features:** Typical "hand of benediction" when fist closure is attempted, with incomplete pronation, loss of thumb opposition, impaired grasping ability, atrophy of the thenar muscles, and sensory disturbances affecting the radial part of the palm and radial 3½ digits (also autonomic trophic disturbances, such as decreased sweat secretion and increased cutaneous blood flow). The patient also has a positive "*bottle sign*" in which the fingers and thumb cannot fully close around a cylindrical object due to weakness of the abductor pollicis brevis.
Distal median nerve lesion
• The superficial location of the nerve in the distal forearm makes it vulnerable to cuts and lacerations (e.g., in attempted suicide). • Chronic compression of the median nerve in the carpal tunnel (most common compression syndrome affecting the median nerve: carpal tunnel syndrome). Compression or entrapment of the nerve within the carpal tunnel can have various causes, such as fractures and dislocations of the carpal bones, inflammatory changes in the tendon sheaths, muscle variants (e.g., lumbricals passing through the carpal tunnel), and connective = tissue proliferation due to endocrine hormonal changes (diabetes mellitus, pregnancy, menopause). **Clinical features:** oath hand is not present. Initial signs consist of sensory disturbances (paresthesias and dysesthesias), chiefly affecting the tips of the index and middle fingers and thumb due to increased carpal tunnel pressure resulting from prolonged flexion or extension of the wrist during sleep ("brachialgia paraesthetica nocturna"). Chronic or severe damage leads to motor deficits involving the thenar muscles (thenar atrophy) with preservation of thenar sensation (intact palmar branch of the median nerve) and a positive bottle sign (see **D**).

D Positive "bottle sign" in the right hand

When proximal and distal median nerve lesions are present, the thumb and fingers cannot completely encircle a cylindrical vessel with the affected hand due to weakness or loss of the abductor pollicis brevis.

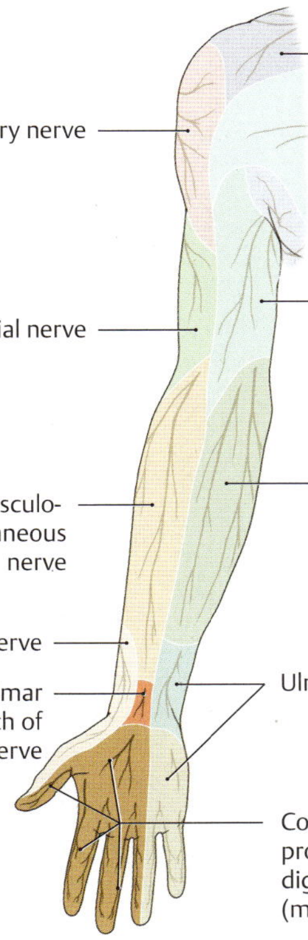

E Sensory distribution of the median nerve
Anterior view.

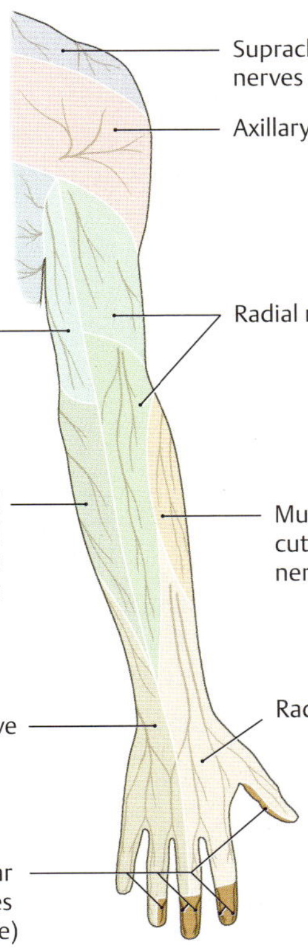

F Sensory distribution of the median nerve
Posterior view.

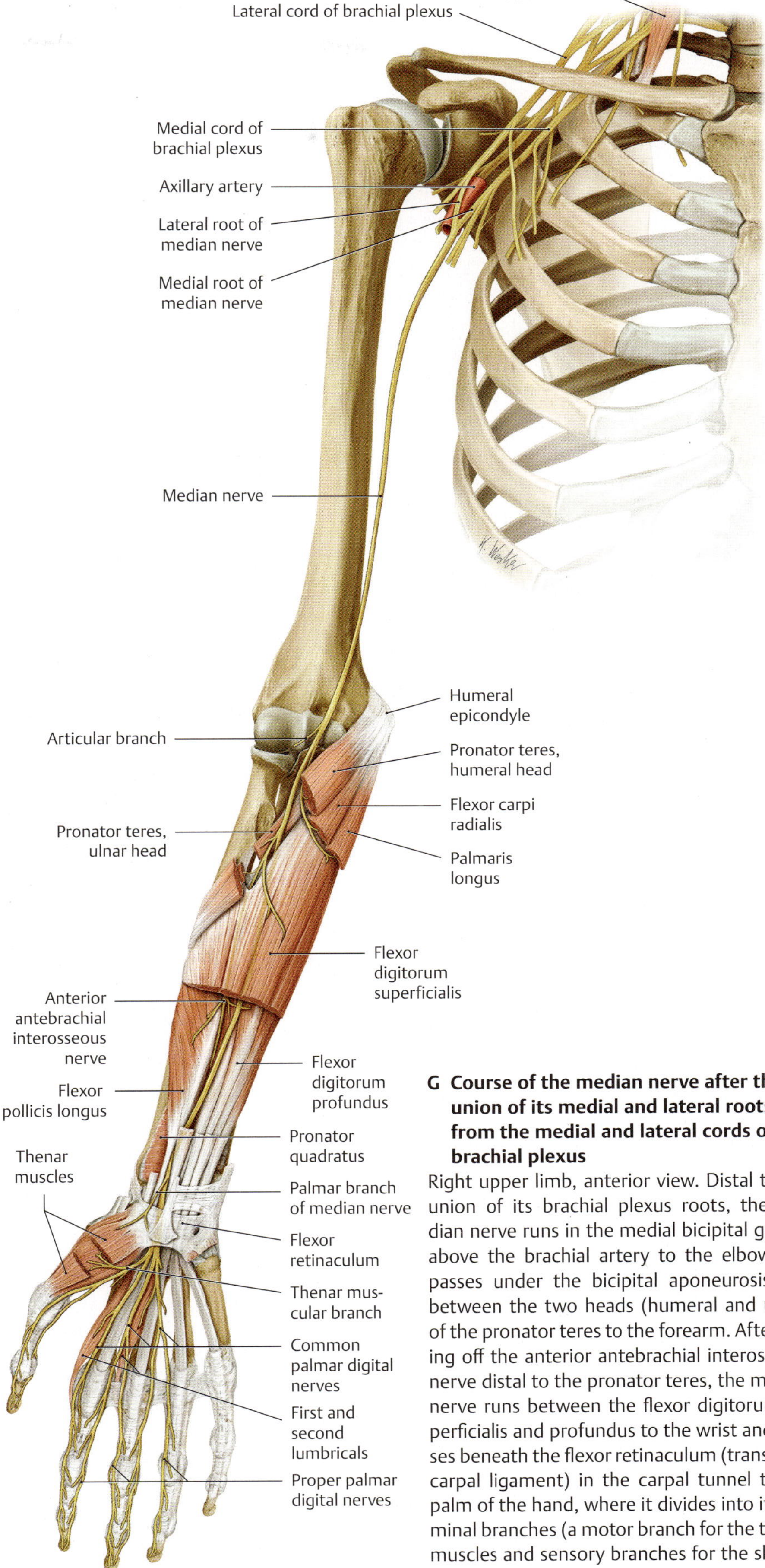

G Course of the median nerve after the union of its medial and lateral roots from the medial and lateral cords of the brachial plexus

Right upper limb, anterior view. Distal to the union of its brachial plexus roots, the median nerve runs in the medial bicipital groove above the brachial artery to the elbow and passes under the bicipital aponeurosis and between the two heads (humeral and ulnar) of the pronator teres to the forearm. After giving off the anterior antebrachial interosseous nerve distal to the pronator teres, the median nerve runs between the flexor digitorum superficialis and profundus to the wrist and passes beneath the flexor retinaculum (transverse carpal ligament) in the carpal tunnel to the palm of the hand, where it divides into its terminal branches (a motor branch for the thenar muscles and sensory branches for the skin on the palmar side of the radial 3½ digits).

5.1 Surface Anatomy and Superficial Nerves and Vessels: Anterior View

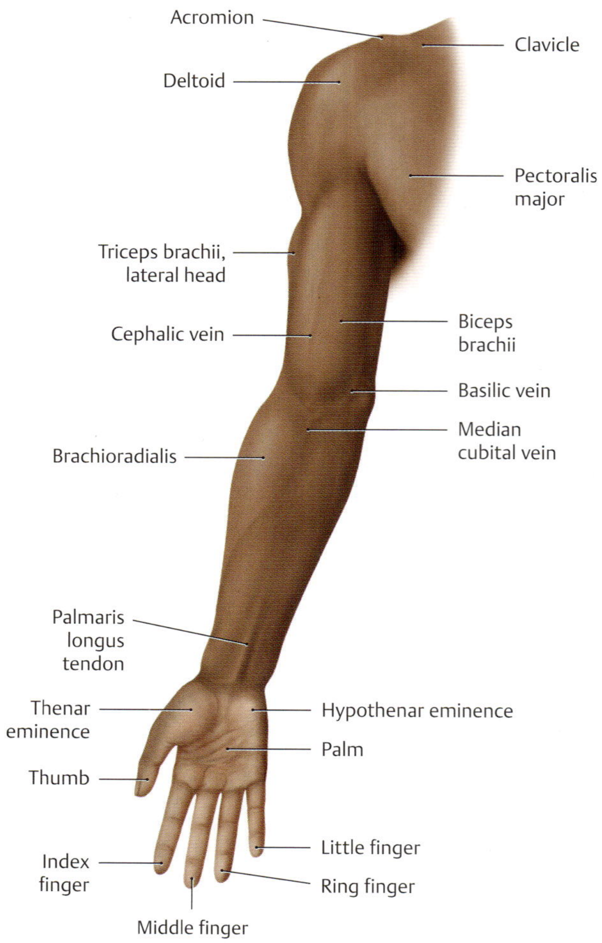

A Surface anatomy of the right upper limb
Anterior view. The palpable bony landmarks of the upper limb are reviewed on p. 251.

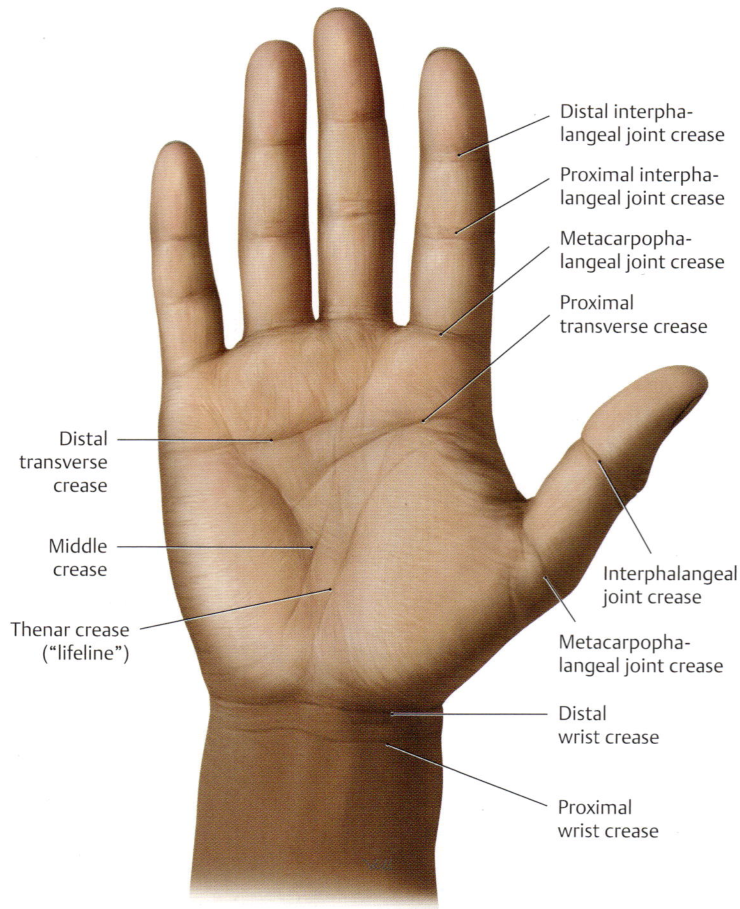

B Hand lines and flexion creases on the right palm with the wrist in slight flexion
The *proximal wrist crease*, located approximately one finger width from the palm, coincides with the distal epiphyseal lines of the radius and ulna. The *distal wrist crease* usually overlies the midcarpal joint.

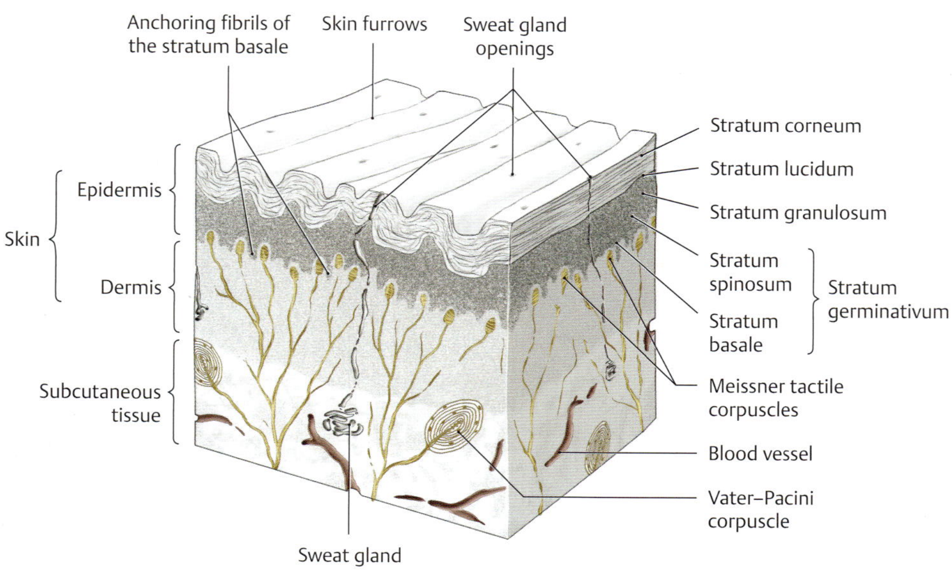

C Schematic structure of the ridged skin on the palm of the hand
The smooth, thin skin of the forearm gives way to the thicker, ridged skin on the palm of the hand. The papillary ridges are particularly high on the palmar skin of the fingers and, at 0.1 to 0.4 mm, are distinctly visible. The ridge pattern (dermatoglyphs) found on the bulbs of the fingers is unique for each individual. The tactile sensitivity of the fingertips is closely linked to the spatial distribution of tactile corpuscles and free nerve endings (e.g., 75–80 Vater–Pacini corpuscles per finger and approximately 100 free nerve endings per square millimeter).

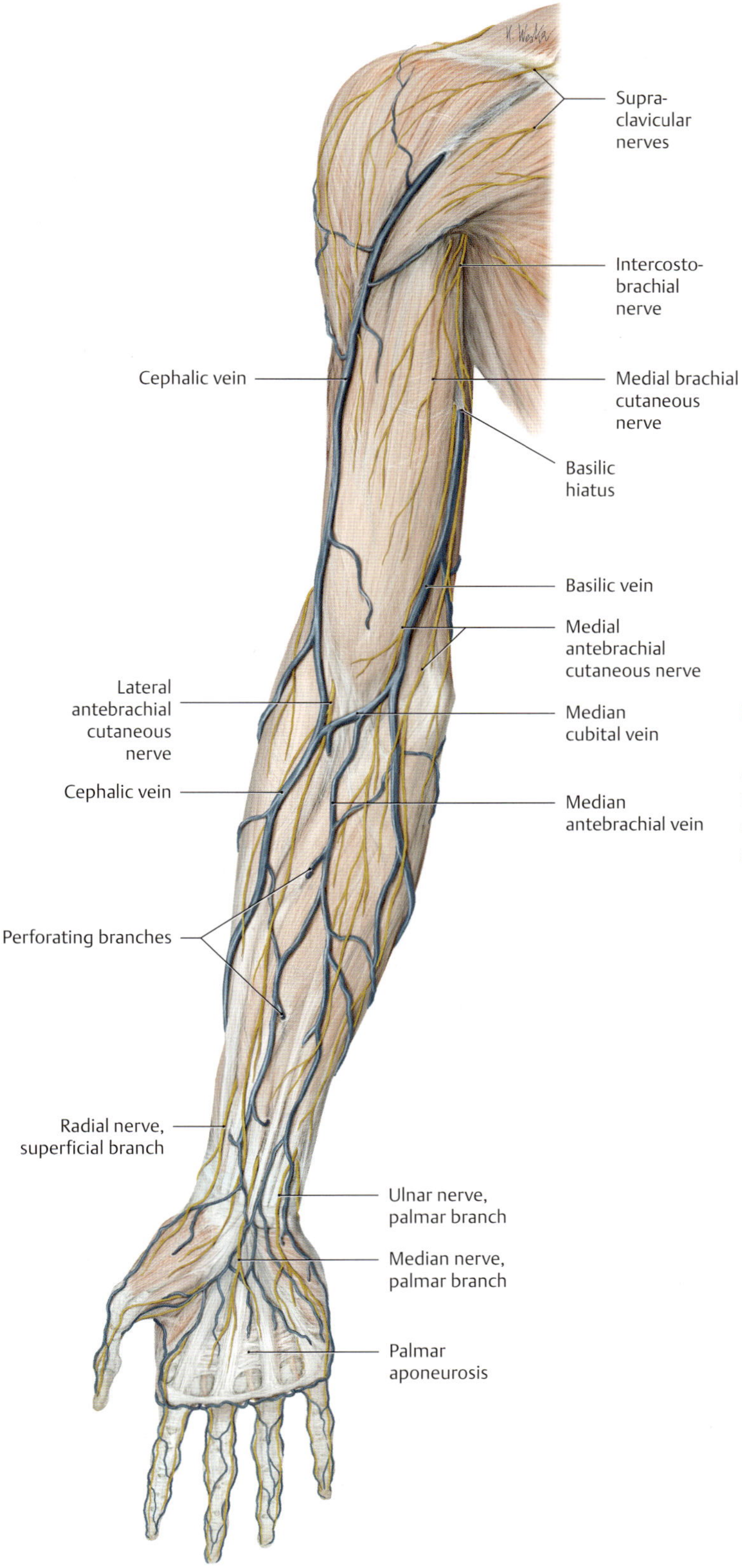

D Superficial cutaneous veins and nerves of the right upper limb
Anterior view. The arrangement of cutaneous veins about the elbow can vary considerably (see p. 370). This dissection does not show the cutaneous arteries that perforate the antebrachial fascia (particularly those arising from the radial artery, see also p. 369).

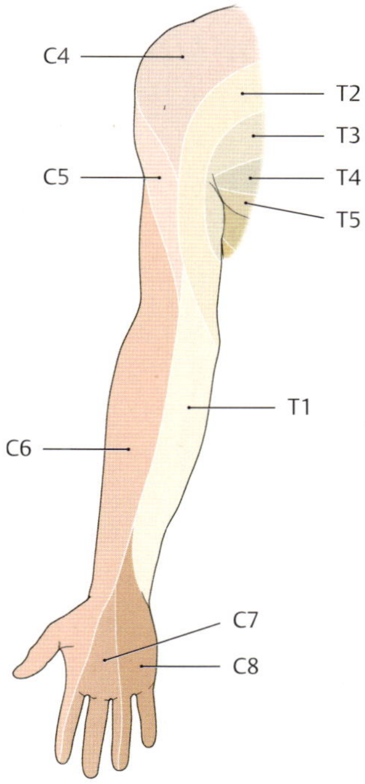

E Radicular (segmental) cutaneous innervation pattern (dermatomes) in the right upper limb
Anterior view. With the outgrowth of the upper limb during development, the sensory cutaneous segments become elongated in varying degrees to form narrow bands. In the process, segments C5–C7 become separated from the body wall.

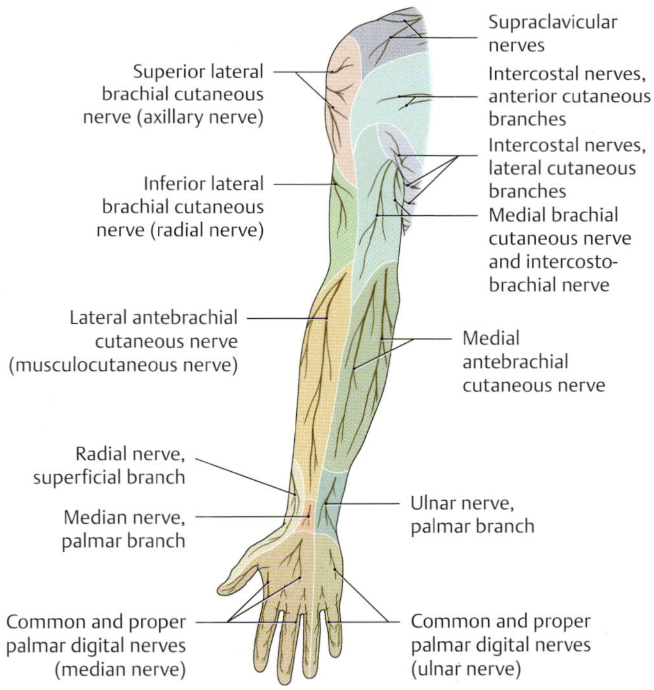

F Pattern of peripheral sensory cutaneous innervation in the right upper limb
Anterior view. The territories supplied by the peripheral cutaneous nerves (cutaneous branches) correspond to the areas of cutaneous nerve branching in the subcutaneous connective tissue that are demonstrable by dissection. The area served exclusively by a single nerve and thus rendered completely anesthetic by a lesion is much smaller because the individual sensory territories overlap extensively.
Note: The sensory loss following damage to a *peripheral nerve* shows a completely different pattern than that of a damaged *nerve* root.

5.2 Surface Anatomy and Superficial Nerves and Vessels: Posterior View

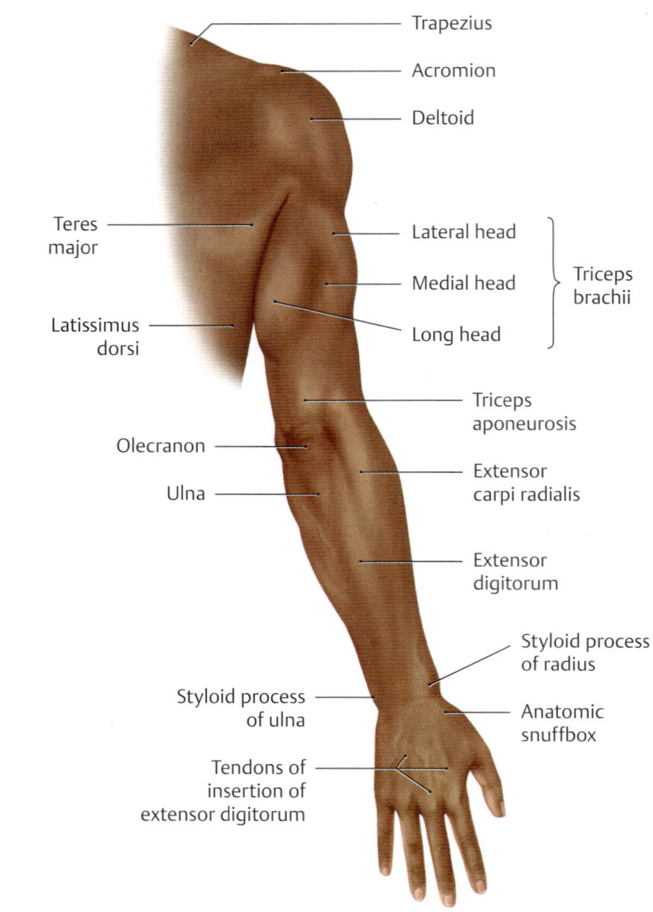

A Surface anatomy of the right upper limb
Posterior view. The palpable bony landmarks of the upper limb are reviewed on p. 251.

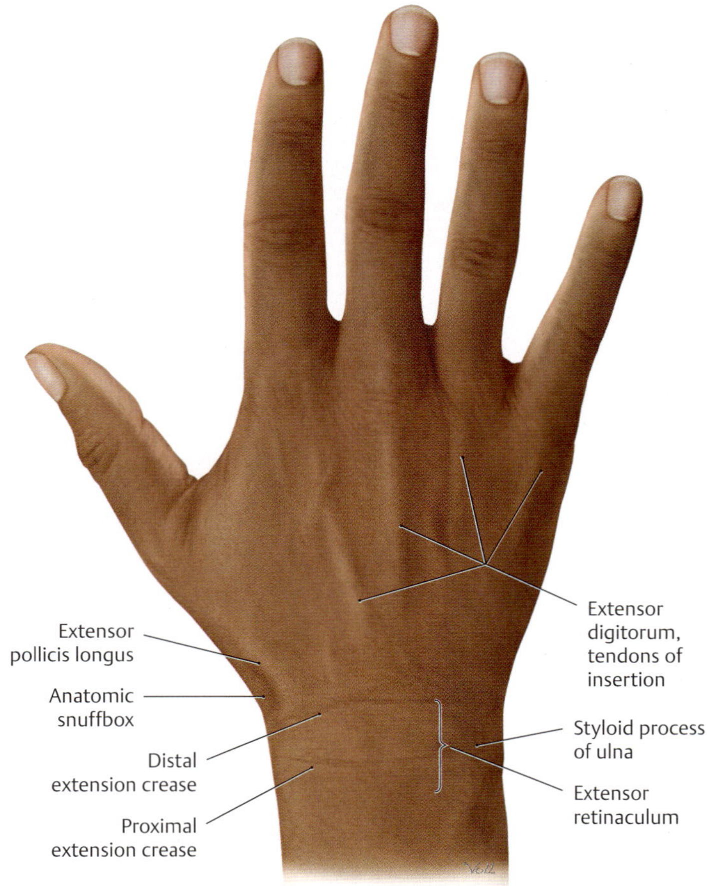

B Extension creases on the dorsum of the right hand
In contrast to the palm, the dorsal surfaces of the hand and fingers bear indistinct extension creases that deepen with maximum dorsiflexion of the hand. The most proximal crease overlies the styloid process of the ulna, while the most distal crease approximately overlies the distal margin of the extensor retinaculum. Unlike the hairless ridged skin of the palm, the dorsum of the hand is covered by smooth, thin, hair-bearing skin.

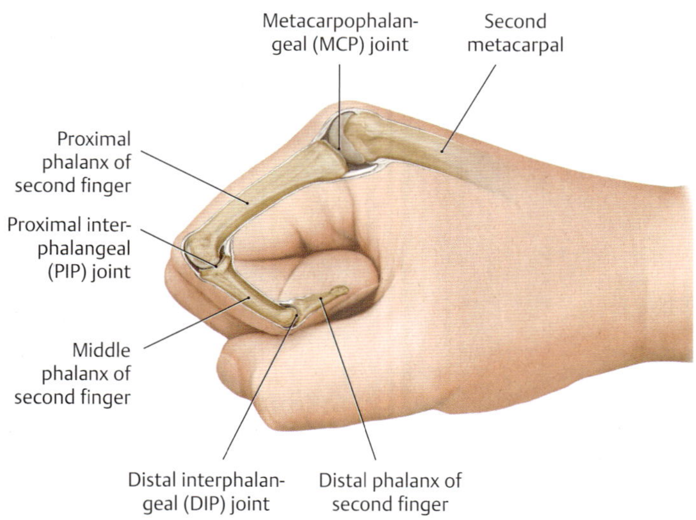

C Location of the MCP, PIP, and DIP joint spaces
Right hand closed into a fist, radial view.

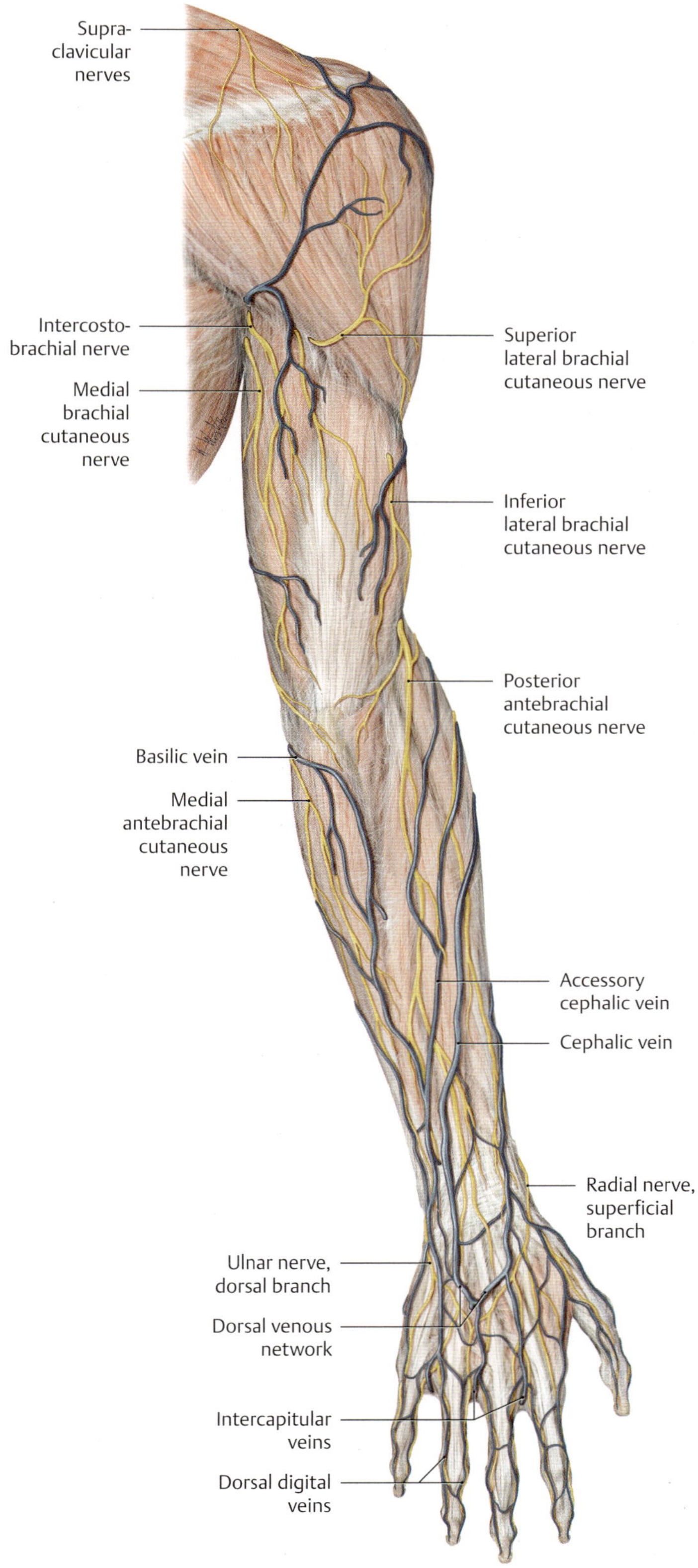

D Superficial cutaneous veins (subcutaneous veins) and cutaneous nerves of the right upper limb

Posterior view. The epifascial veins of the dorsum of the hand (dorsal venous network) display a highly variable branching pattern. Generally, the epifascial veins are clearly visible beneath the skin, receiving tributaries that include perforating veins from the palmar side of the hand. The cephalic vein on the radial side of the hand provides for most of the dorsal venous drainage, while the basilic vein provides for a lesser degree on the ulnar side. This dissection does not show the main branches of the posterior interosseous artery that perforate the antebrachial fascia on the back of the forearm (see also p. 369).

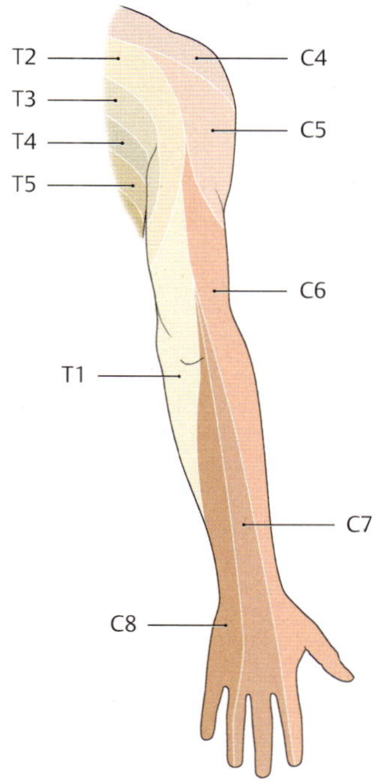

E Radicular (segmental) cutaneous innervation pattern (dermatomes) in the right upper limb

Posterior view. With the outgrowth of the limb during development, the sensory cutaneous segments become elongated in varying degrees to form narrow bands. As this occurs, segments C5–C7 become separated from the body wall.

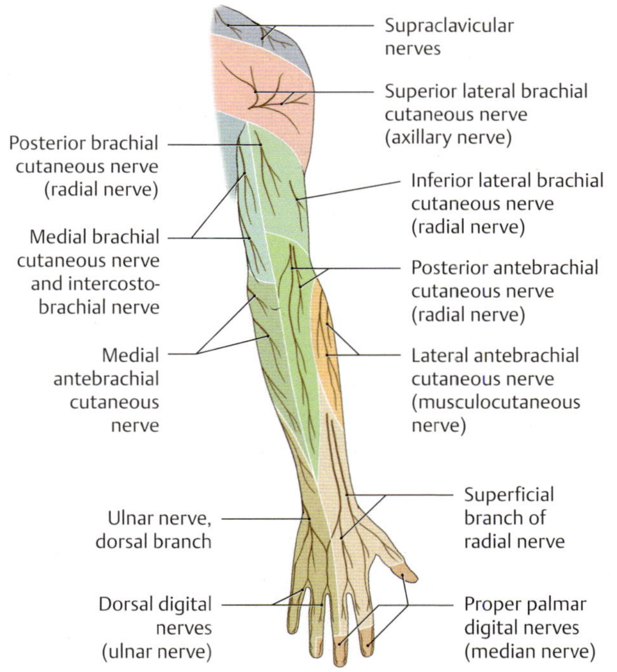

F Peripheral sensory cutaneous innervation pattern in the right upper limb

Posterior view. The color-coded areas that are supplied by the peripheral cutaneous nerves (cutaneous branches) correspond to the areas of cutaneous nerve branching in the subcutaneous connective tissue that are demonstrable by dissection. The areas of exclusive non-overlapping innervation of a specific nerve is much smaller.

Note that the sensory loss following damage to a *peripheral nerve* shows a completely different pattern from that caused by damage to a *nerve root* (see **E**).

5.3 The Shoulder Region: Anterior View

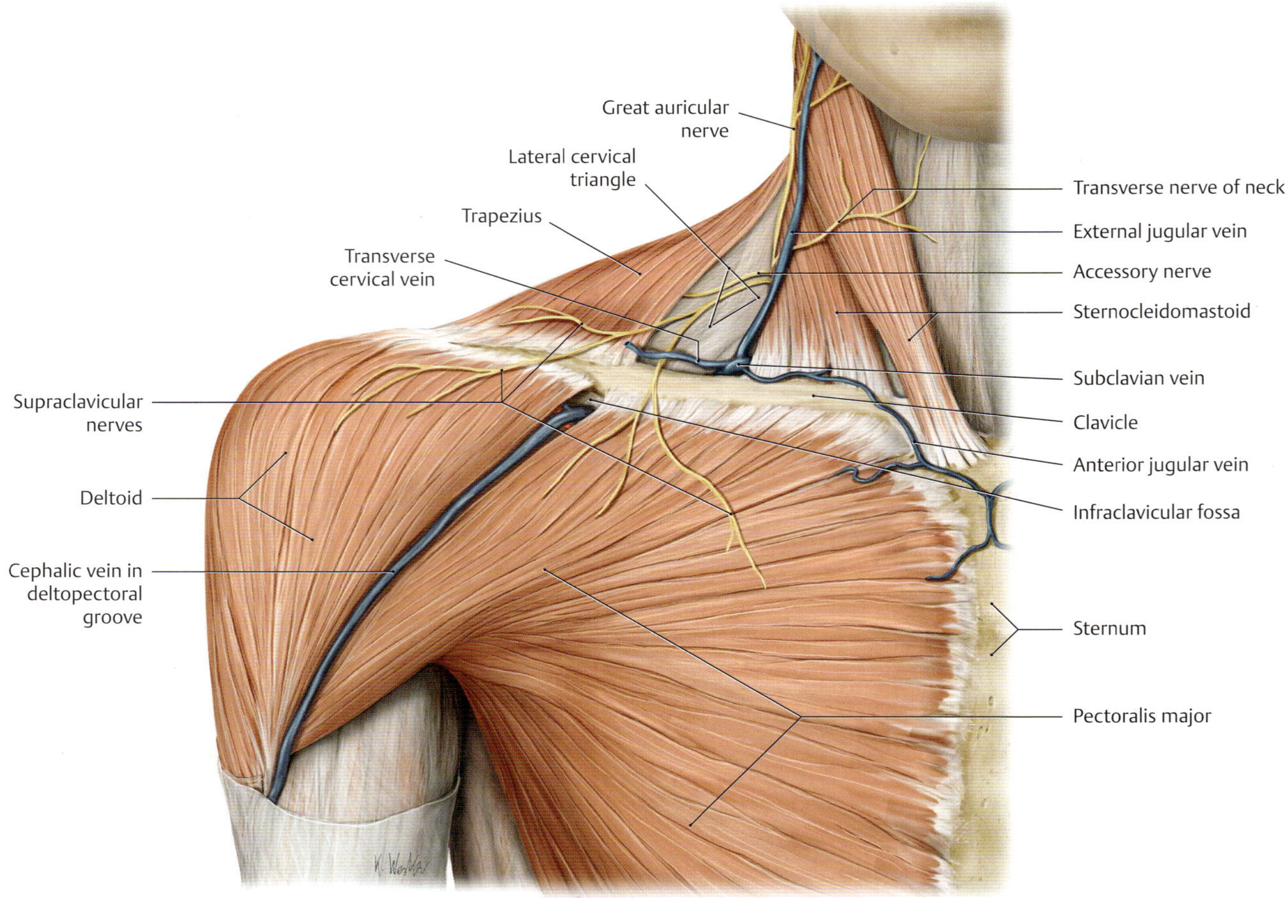

A Superficial veins and nerves of the right shoulder and neck region
Anterior view. The skin, platysma, muscle fasciae, and superficial layer of the cervical fascia have been removed in this dissection to demonstrate the branches of the cervical plexus (e.g., the great auricular nerve) and the superficial veins of the lateral and anterior neck. The *external jugular vein* and *anterior jugular vein* (not shown) are visible through the skin when the patient is lying supine and the veins are well filled. When *right-sided heart failure* is present, these veins may be engorged due to the damming back of venous blood and may be visible even when the patient is sitting upright. The *cephalic vein* crosses the shoulder in the groove between the pectoralis major and deltoid muscles (*deltopectoral groove*) and empties into the axillary or subclavian vein. This site of entry into the axillary vein, and thus into the deep veins, is visible and palpable on the skin as the infraclavicular fossa. At the level of the lateral border of the first rib, the axillary vein becomes the subclavian vein.

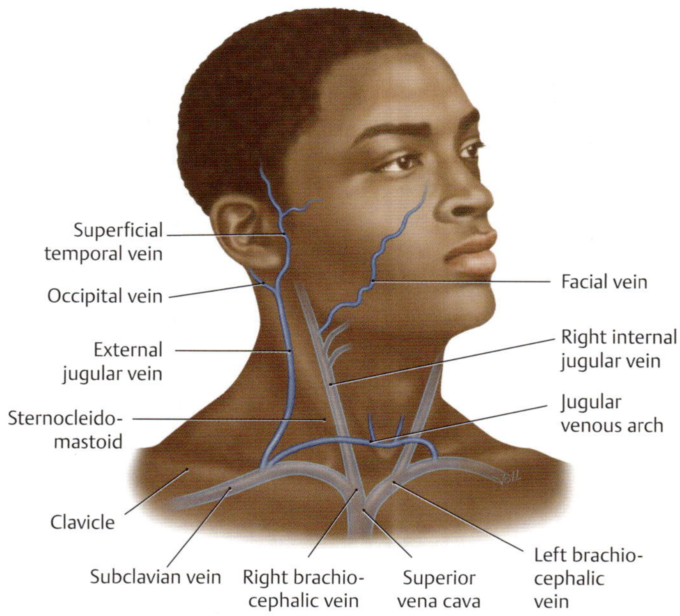

B Relationship of major superficial and deep veins in the neck to the sternocleidomastoid muscle
Anterior view. The internal jugular vein runs almost straight downward from the jugular foramen and unites with the subclavian vein just lateral to the sternoclavicular joint to form the brachiocephalic vein. When its course is projected onto the side of the neck, it follows a line drawn from the earlobe to the medial end of the clavicle. The internal jugular vein is crossed obliquely in its lower third by the sternocleidomastoid muscle, while the external jugular vein runs obliquely downward on the muscle and opens into the subclavian vein.

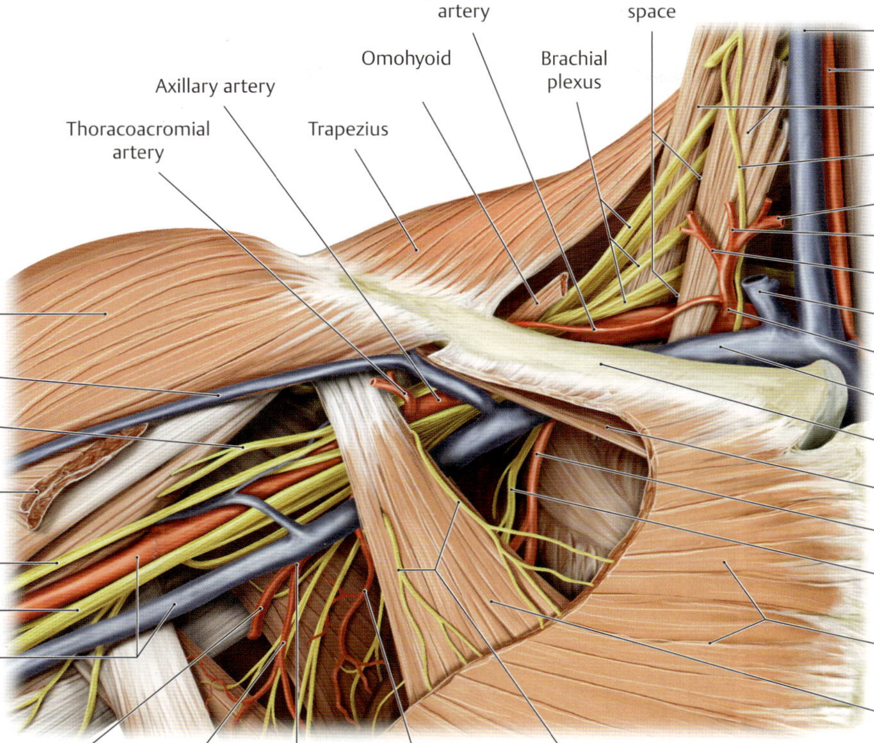

C Course of the right subclavian artery in the lateral neck region
Anterior view. The sternocleidomastoid and omohyoid muscles and all layers of the cervical fasciae have been removed to demonstrate the deep lateral cervical triangle and the passage of the subclavian artery and brachial plexus through the space between the anterior and middle scalene muscles (interscalene space). At the level of the first rib the subclavian artery becomes the axillary artery, which enters the axilla posterior to the tendon of insertion of the pectoralis minor.

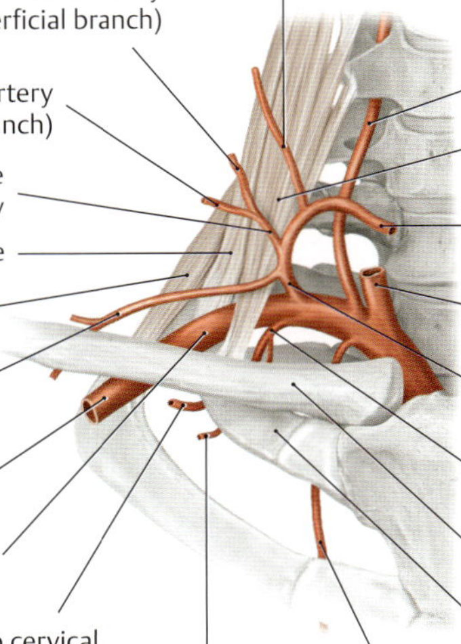

D Origin and branches of the right subclavian artery
Anterior view.

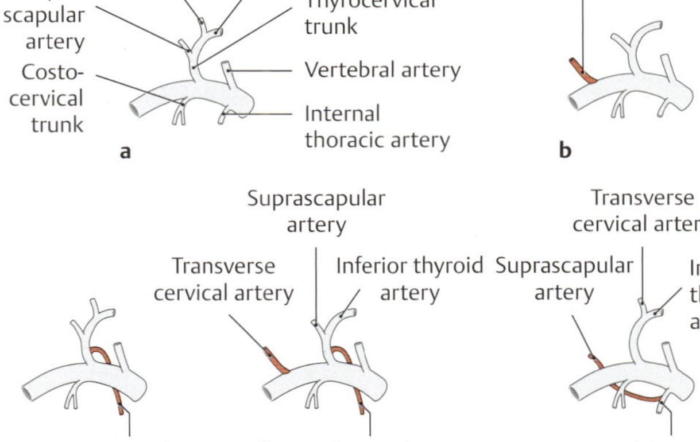

E Branches of the subclavian artery: normal anatomy and variants
(after Lippert and Pabst)

a Normally (30% of cases), the subclavian artery gives off the following branches:
- Thyrocervical trunk with the inferior thyroid artery, suprascapular artery, and transverse cervical artery
- Vertebral artery
- Internal thoracic artery
- Costocervical trunk

b–e Variants:

b The transverse cervical artery arises separately from the subclavian artery (30%).

c The internal thoracic artery arises from the thyrocervical trunk (10%).

d The thyrocervical trunk is made up of the inferior thyroid artery, suprascapular artery, and internal thoracic artery (8%).

e The subclavian artery gives off two main branches:
1. One with the inferior thyroid and transverse cervical arteries.
2. One with the internal thoracic and suprascapular arteries (4%).

5.4 The Axilla: Anterior Wall

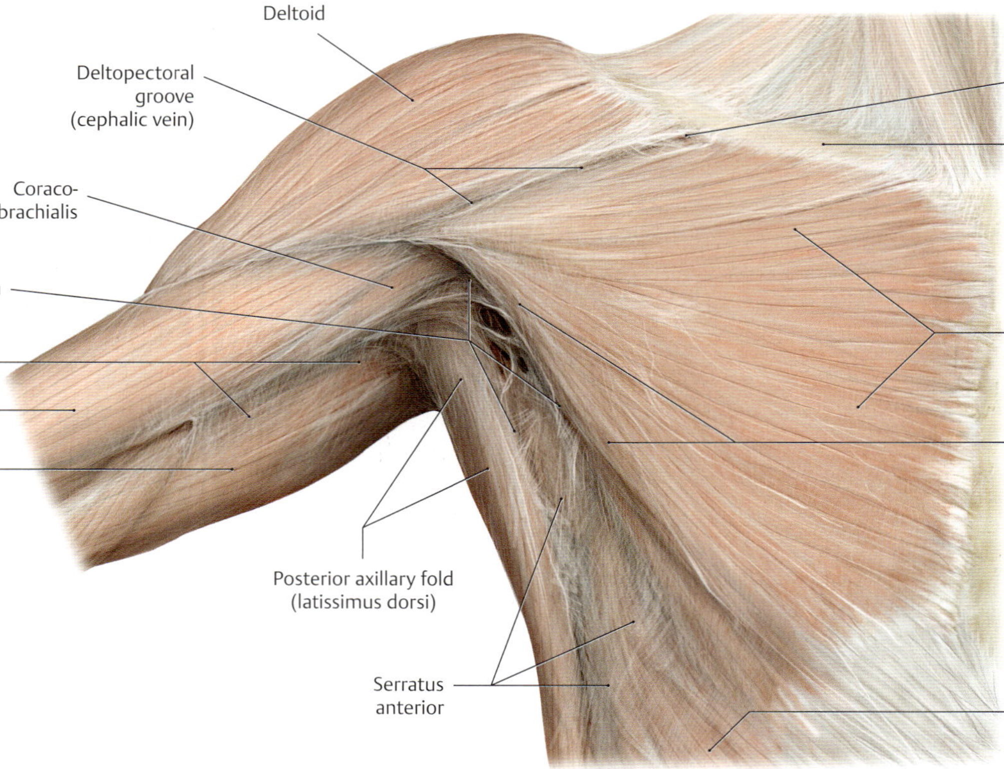

A The walls and fasciae of the right axilla
Anterior view. With the arm abducted, the axilla (axillary fossa) resembles a four-sided pyramid whose apex is approximately at the center of the clavicle and whose base is represented by the axillary *fascia*. The walls of the axilla are formed by various muscles and their fasciae:

Anterior wall: The anterior wall of the axilla consists of the pectoralis major and minor and the clavipectoral fascia (the pectoralis minor is not shown here; see **C** and **b**).
Posterior wall: This consists of the subscapularis, teres major (not shown here, see p. 320), and latissimus dorsi.

Lateral wall: This is narrow and formed by the intertubercular groove of the humerus.
Medial wall: This is formed by the lateral thoracic wall (ribs 1–4 and associated intercostal muscles) and the serratus anterior.

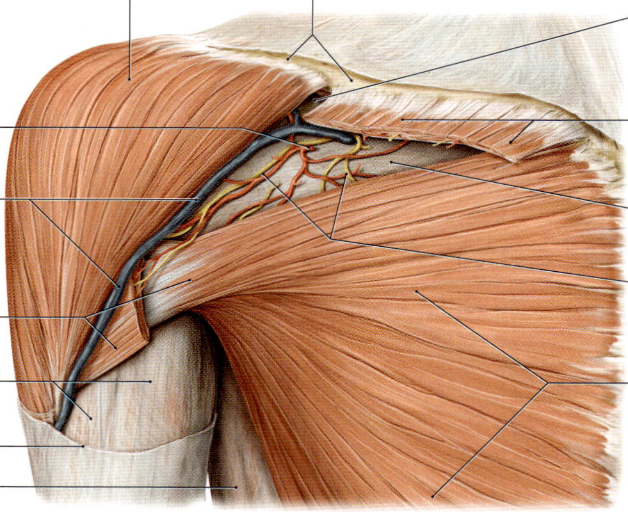

B The clavipectoral triangle and clavipectoral fascia
Right shoulder, anterior view. The clavicular part of the pectoralis major has been removed. In the clavipectoral triangle bounded by the deltoid, pectoralis major, and clavicle, the cephalic vein runs upward in the delto-pectoral groove, pierces the clavipectoral fascia, and drains into the axillary vein at the level of the infraclavicular fossa.

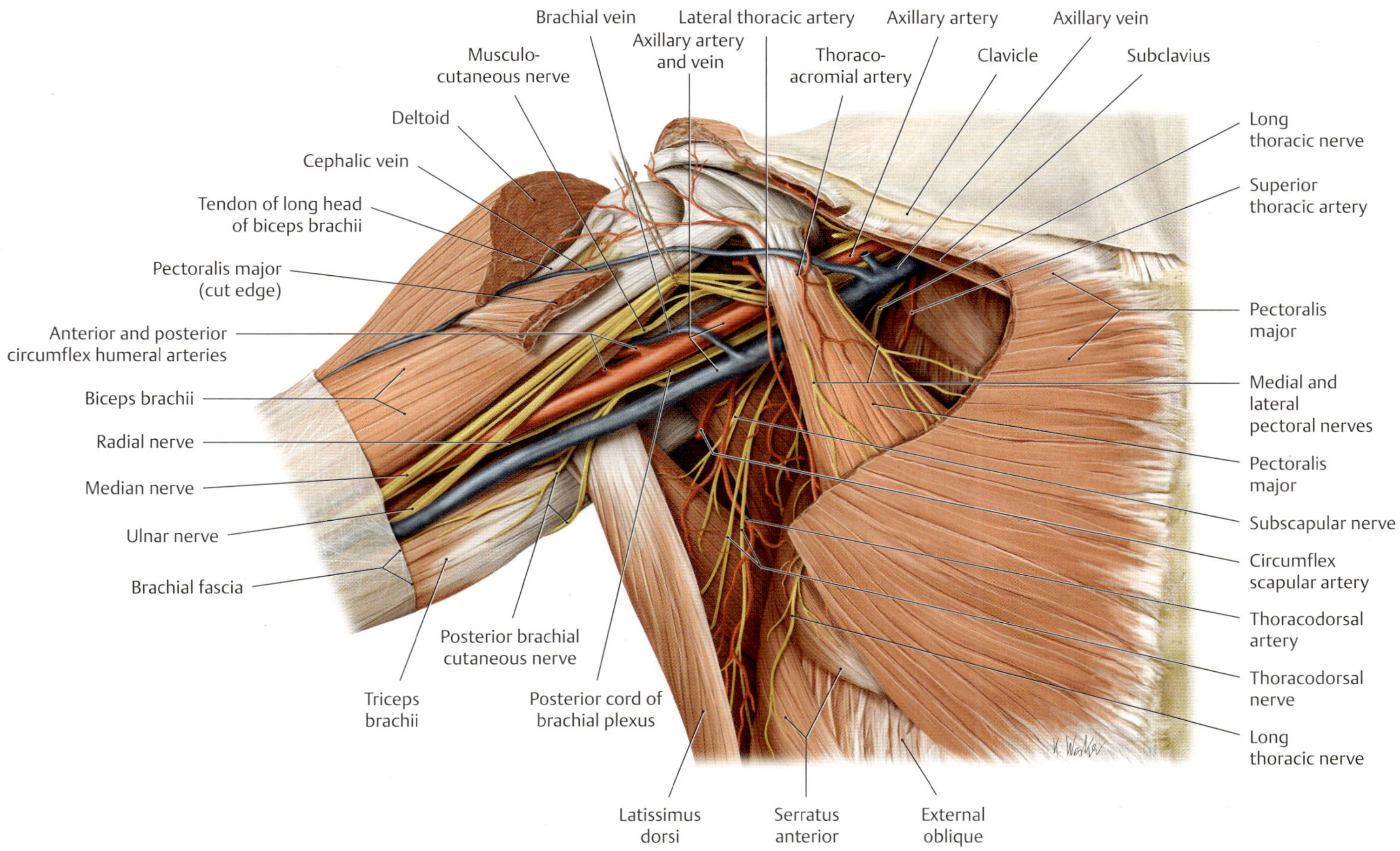

C The axilla after removal of the pectoralis major and clavipectoral fascia

Right shoulder, anterior view. The axillary artery runs approximately 2 cm below the coracoid process and posterior to the pectoralis minor. It relates laterally to the lateral cord of the brachial plexus and medially to the medial cord (both are retracted slightly upward in the drawing). The posterior cord of the brachial plexus, which runs behind the axillary artery, is just visible.

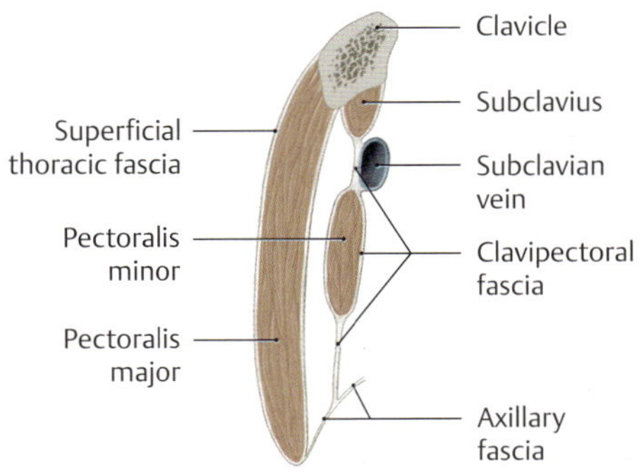

D Location of the superficial and deep thoracic fasciae

Sagittal section through the anterior wall of the right axilla. The clavipectoral fascia, known also as the "deep" thoracic fascia, encloses the pectoralis minor and subclavius muscles and covers the subclavian vein while being fused to its wall. The fascia is made tense by the pectoralis minor. The clavipectoral fascia exerts traction on the vein wall that can keep its lumen patent, thus facilitating venous return to the superior vena cava.

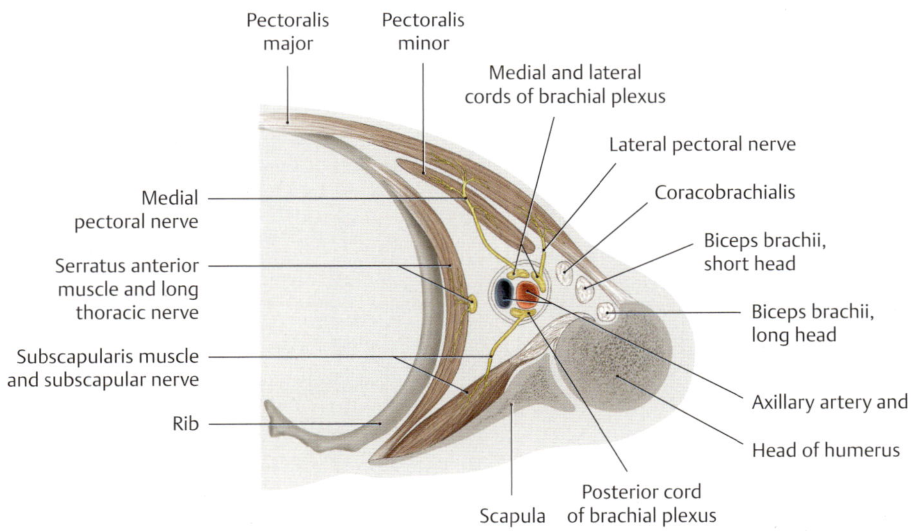

E Schematic transverse section through the right axilla

Superior view. The three muscular walls and the bony lateral wall of the axilla are clearly delineated in this view. Neurovascular structures (axillary artery and vein plus the medial, lateral, and posterior cords of the brachial plexus) traverse the axilla, invested by a fibrous sheath and embedded in the axillary fat.

5.5 The Axilla: Posterior Wall

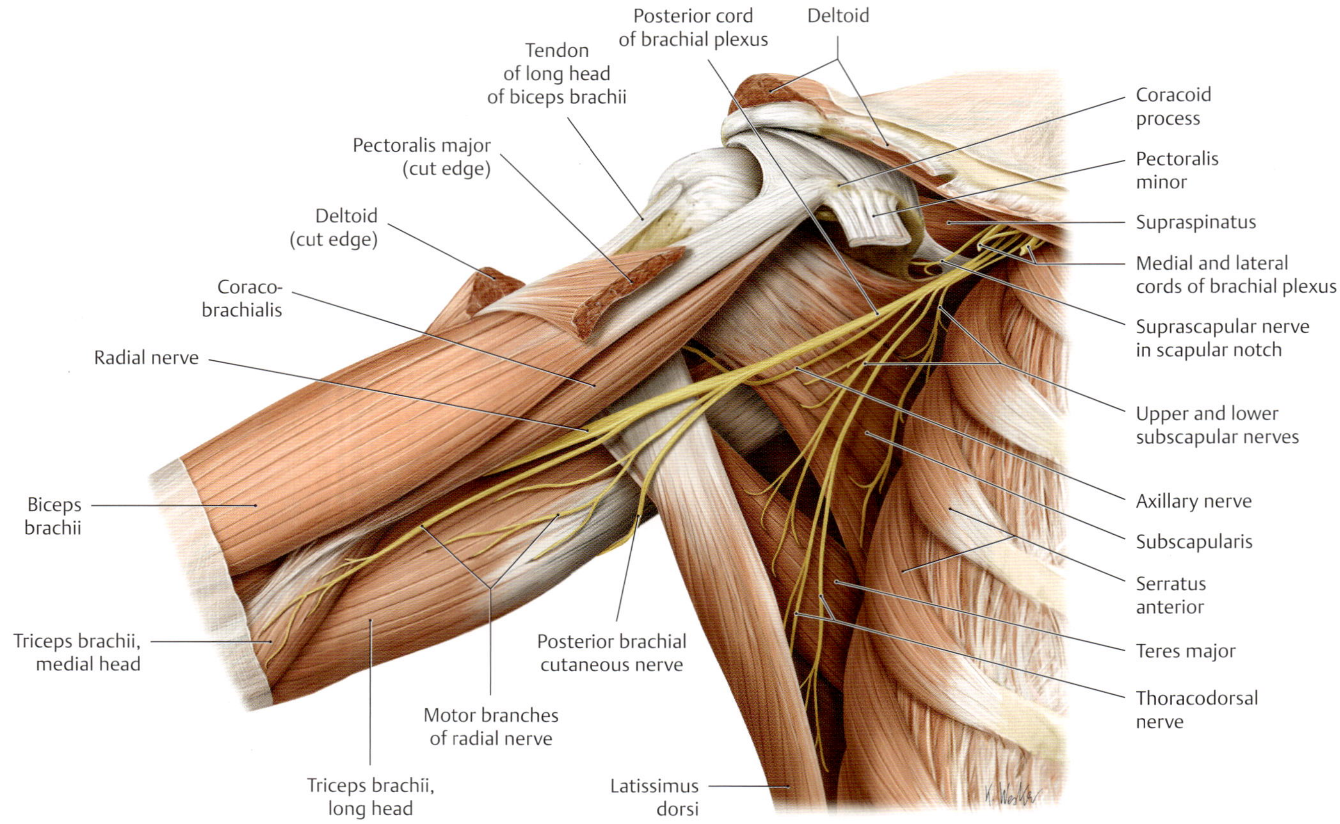

A Posterior wall of the axilla with the posterior cord and its branches

Right shoulder, anterior view. The medial and lateral cords of the brachial plexus and the axillary vessels have been removed to demonstrate the course of the *posterior* cord and its branches in the posterior axilla.

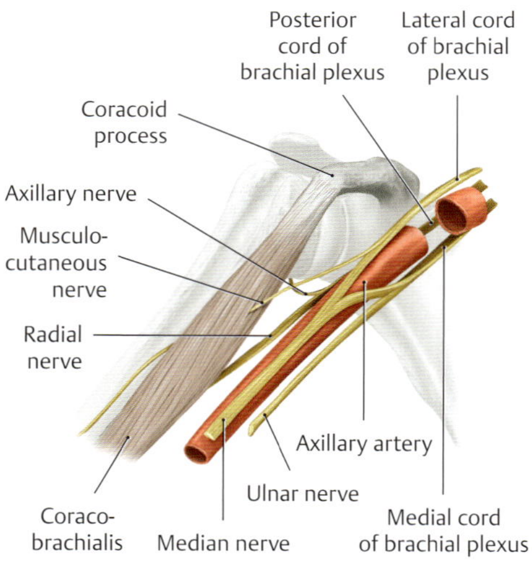

B Relationship of the medial, lateral, and posterior cords of the brachial plexus to the axillary artery

Note that the musculocutaneous nerve passes through the coracobrachialis, which aids in locating the nerve. Very rarely, this nerve may be compressed as it pierces the muscle.

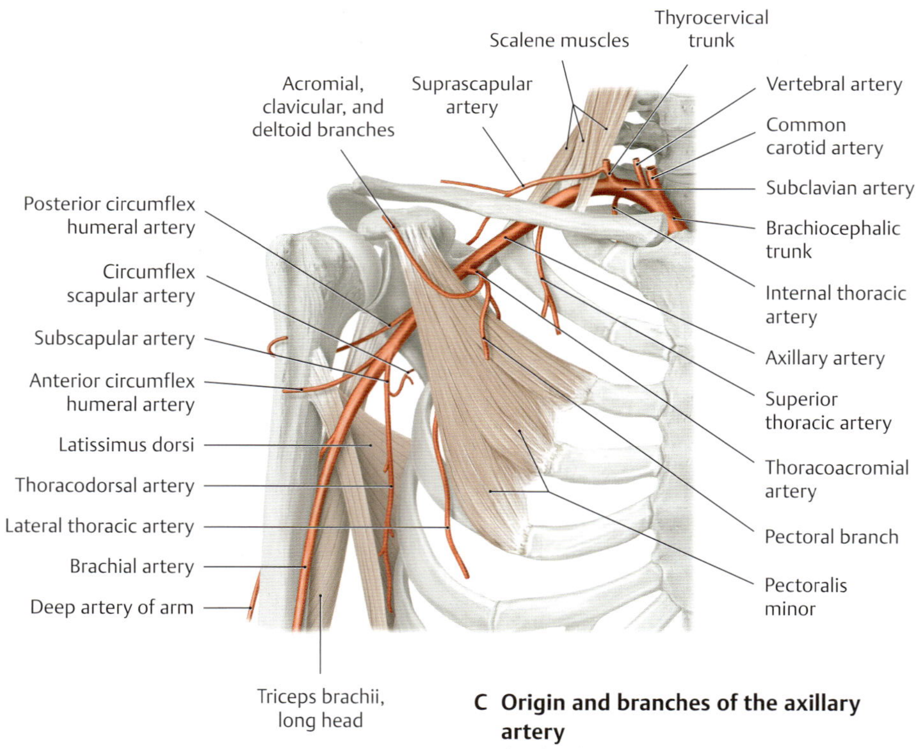

C Origin and branches of the axillary artery

Right shoulder, anterior view.

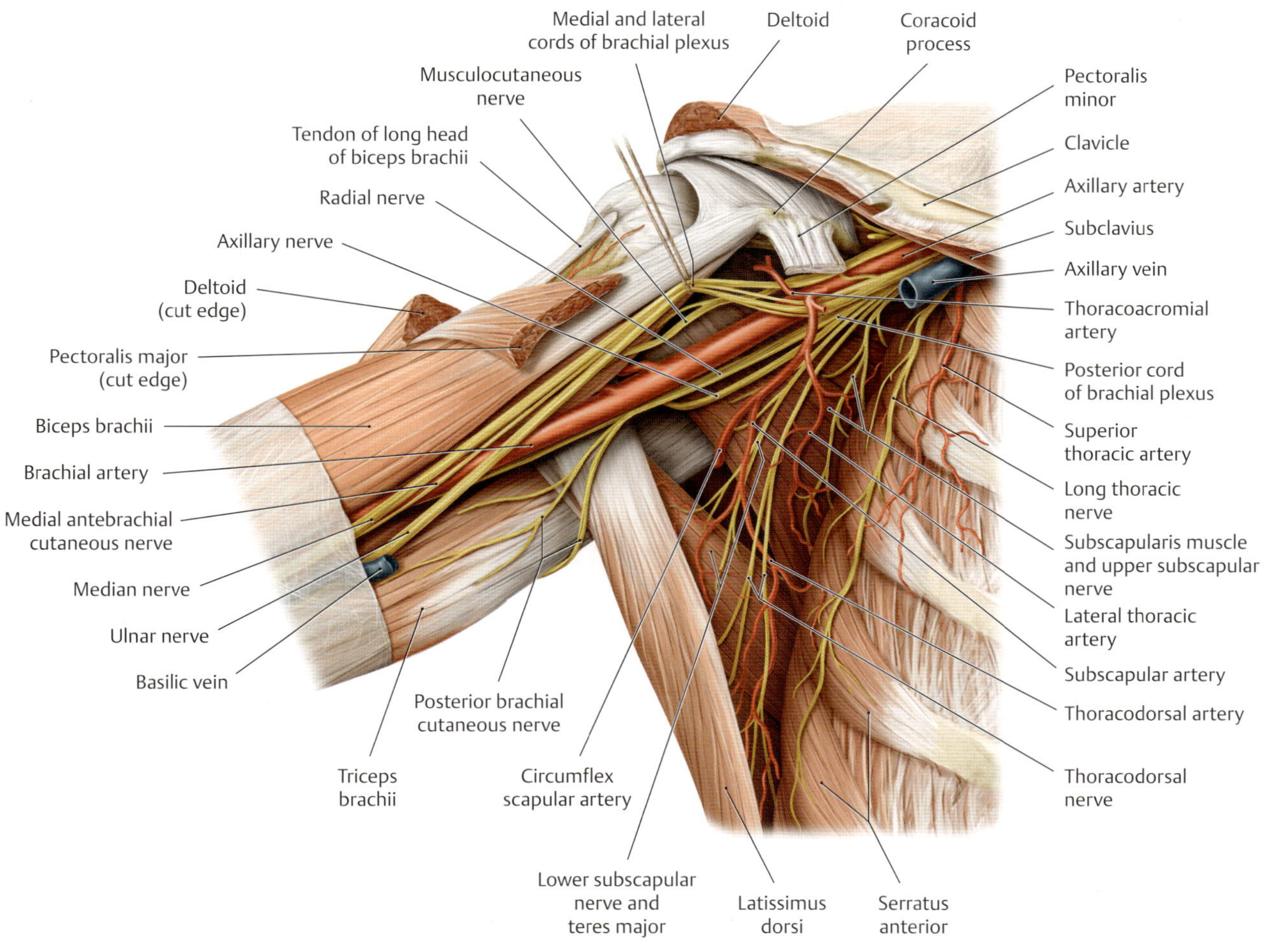

D Axilla with the entire anterior wall removed
Right shoulder, anterior view. The axillary vein has been removed, and the medial and lateral cords of the brachial plexus have been retracted upward to show more clearly the location and course of the posterior cord and its terminal branches, the radial nerve and axillary nerve.

Note the superficial course of the long thoracic nerve on the serratus anterior.

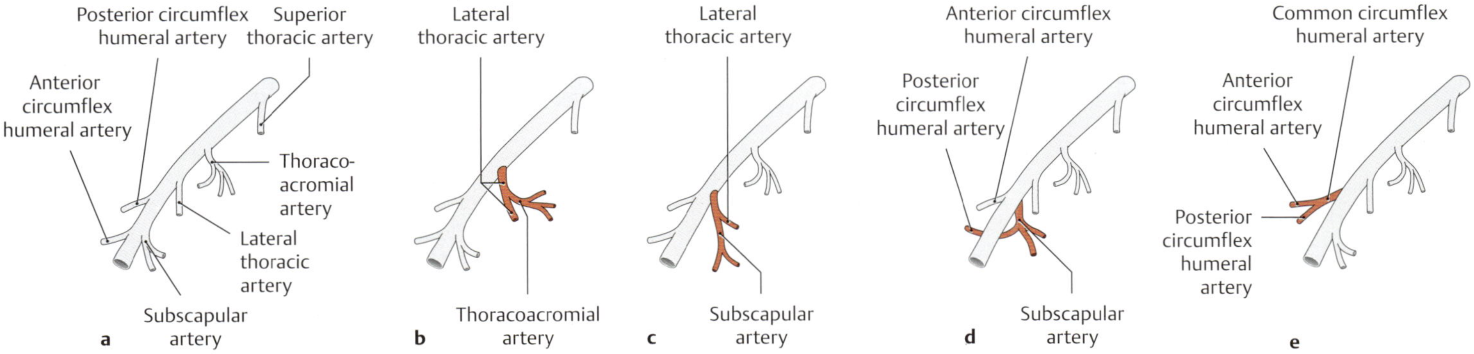

E Branches of the axillary artery: normal anatomy and variants (after Lippert and Pabst)

a Normally (40% of cases), the axillary artery gives off the following branches:
Superior thoracic artery, thoracoacromial artery, lateral thoracic artery, subscapular artery, anterior circumflex humeral artery, and posterior circumflex humeral artery.

b–e Variants:

b The thoracoacromial artery arises from the lateral thoracic artery (10% of cases).

c Common origin of the lateral thoracic artery and subscapular artery (10% of cases).

d The posterior circumflex humeral artery arises from the subscapular artery (20% of cases).

e Common origin of the anterior and posterior circumflex humeral arteries (20% of cases). The common segment formed by both arteries is termed the common circumflex humeral artery.

5.6 Conduction Anesthesia of the Brachial Plexus: Principle, Pathways, and Administering the Block

A Principle of peripheral conduction anesthesia

Peripheral conduction anesthesia is a *regional anesthetic technique* that blocks conduction of action potentials. The area under anesthesia is thus *distal* to the puncture site. It is possible to anesthetize both individual peripheral nerves and the entire plexus.

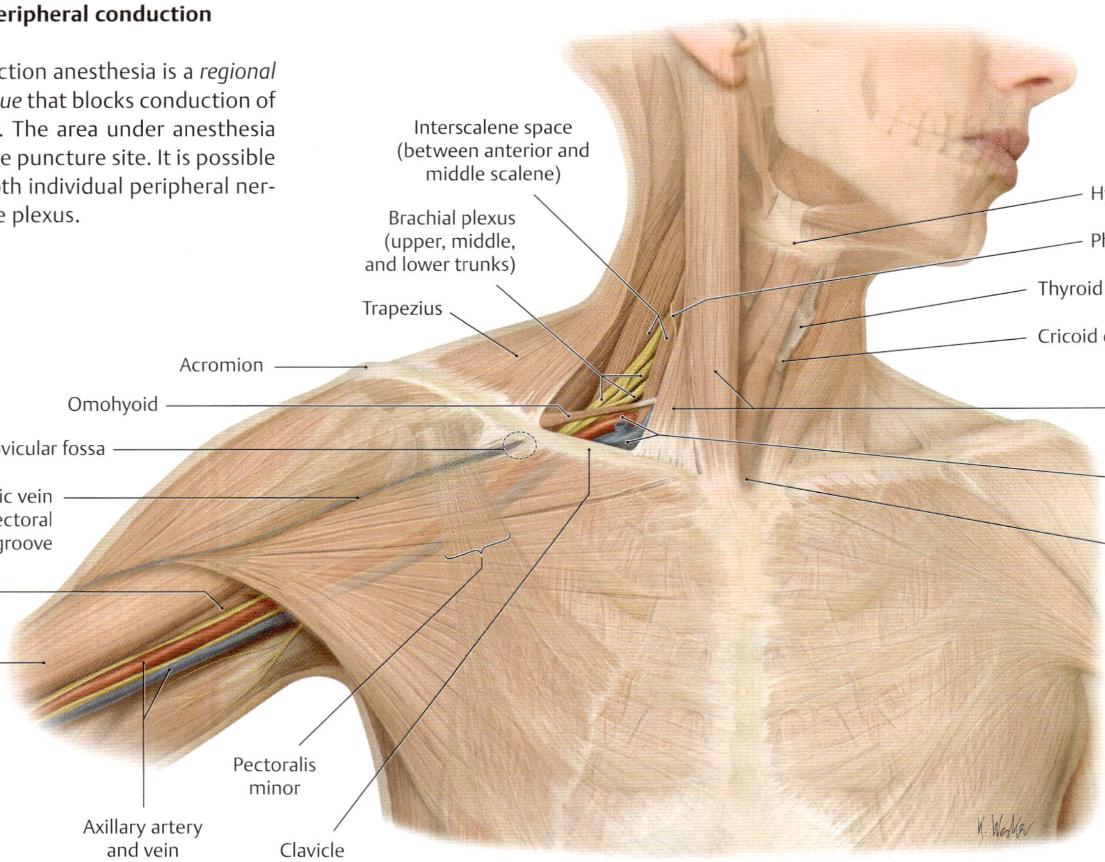

B Topography of the brachial plexus and anatomic landmarks

The brachial plexus is responsible for the motor and sensory nerve supply of the upper limbs. It is formed from the anterior rami of the spinal nerves C5–T1 (see p. 374). In the course of the plexus, first the trunks are formed, followed by the divisions, and then the cords. The trunks are located at the level of the interscalene space. The divisions are above and behind the clavicle. The cords run infraclavicular and start cranial or lateral to the axillary artery and at the level of the axilla, posterior (posterior cord), lateral (lateral cord), and medial (medial cord) to the axillary artery.

Note the important anatomic landmarks for the individual pathways: sternocleidomastoid, cricoid cartilage, thyroid, anterior and middle scalenes (interscalene space), clavicle, acromion, jugular fossa, infraclavicular fossa (Mohrenheim's fossa), coracobrachialis, and axillary artery. In addition, it is recommended to memorize the topographic anatomy in relation to the following structures, which could potentially get damaged: phrenic nerve, recurrent laryngeal nerve, cervical or thoracic and cervical sympathetic ganglia (e.g., stellate ganglion), vertebral artery, cervical epidural and subarachnoid space, and pleural dome.

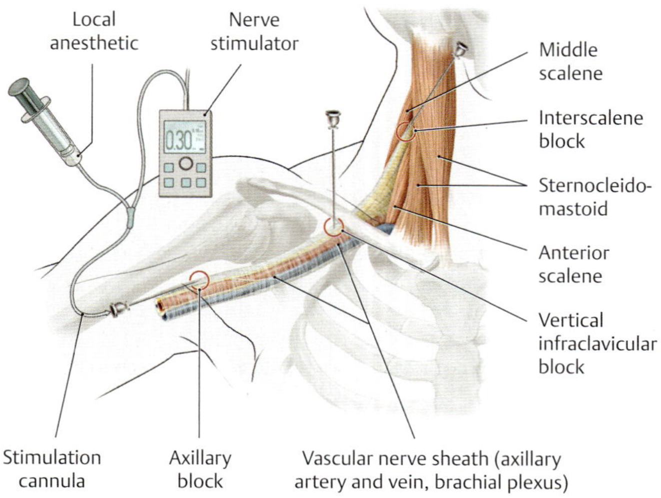

C Brachial plexus sheath and electrical nerve stimulation

From passage through the interscalene space to the axillary region, the entire brachial plexus, together with the accompanying axillary artery and vein, is enclosed by a connective-tissue sheath. Within this sheath, local anesthetic can dissipate more or less evenly, anesthetizing all nerves in this area. In order to target and effectively block one particular nerve, electrical nerve stimulation is performed. A stimulation cannula, only the tip of which is not isolated, emits a defined current pulse. This current pulse can trigger action potentials at the respective motor axons (for more details about the possible responses, see **E**). With correct positioning of the cannula, ensuing injection of 1 to 2 mL of a suitable local anesthetic should lead to immediate suppression of muscle function (extinction phenomenon).

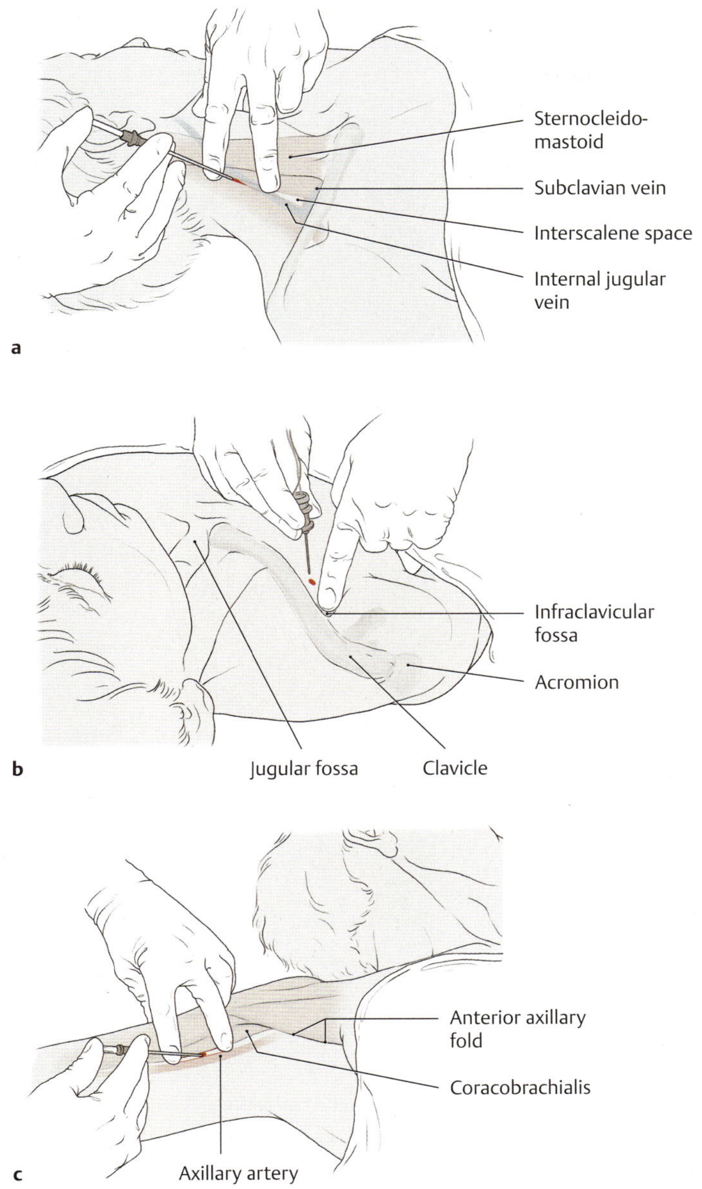

D Clinically important pathways for brachial plexus block

In order to avoid mechanical lesions, usually the use of an electrical nerve stimulation device (see **C**) and atraumatic needles is recommended. Among common contraindications for all procedures mentioned here are infections of the puncture site and manifest coagulation disorders. Among more specific contraindications are contralateral phrenic nerve paresis and contralateral recurrent nerve paresis.

a Interscalene approach according to Meier: Allows for the most proximal brachial plexus block and thus procedures involving the neck and shoulder. Anatomic landmark is the posterior edge of the sternocleidomastoid at the level of the thyroid notch 2 cm above the cricoid cartilage. Puncture is made from the cranial direction down toward the interscalene space with the needle at 30° to the skin.

b Vertical-infraclavicular block according to Kilka, Geiger, and Mehrkens: Advantage over the axillary block (see below): in addition to blocking the cords, it safely blocks the musculocutaneous nerve. The anatomic landmarks are the anterior edge of the acromion and the center of the jugular fossa. The midpoint of the connecting line between those two points marks the puncture site and the lower edge of the clavicle. For better orientation, the index finger is placed in infraclavicular fossa (Mohrenheim's fossa). Puncture is made from medial of the index finger and strictly vertical to the exam table cover. A medial direction of puncture and puncturing too deep is to be avoided in any case due to the risk of pneumothorax.

c Axillary brachial plexus anesthesia: Most established, technically easiest, and least risky pathway to the brachial plexus. Among the indications are all procedures involving the hand, forearm, and distal upper arm. Anatomic landmark is the axillary artery medial to the coracobrachialis, which is usually easily palpable. Use two fingers to palpate the gap between the axillary artery and coracobrachialis, which is located directly distal to the anterior axillary fold (lateral edge of the pectoralis major). The puncture is made in this gap with the needle being held at a 30 to 45° angle parallel to the artery. After penetrating the vessel nerve sheath, which is felt as a sturdy, elastic resistance, the needle is lowered and advanced tangentially in the proximal direction all the way. By using a nerve stimulation device, the position of the tip of the needle can be optimized. "Problematic nerves" (= those difficult to anesthetize) include the radial nerve, which runs behind the axillary artery, and the musculocutaneous nerve, which has left the vessel nerve sheath significantly more proximal.

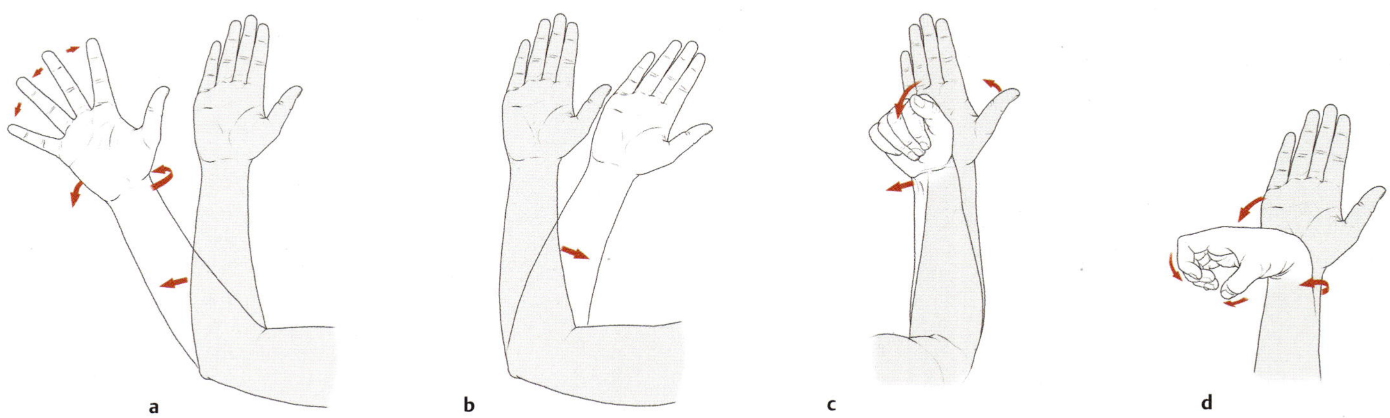

E Motor response of individual nerves of the upper limbs following electrical nerve stimulation

a Radial nerve: elbow joint extension (triceps brachii), extension and radial abduction of the wrist, supination of forearm, and finger extension.

b Musculocutaneous nerve: elbow joint flexion (biceps brachii).

c Ulnar nerve: ulnar flexion of the wrist, flexion of the metacarpophalangeal joints II–V.

d Median nerve: flexion and pronation of the wrist, flexion of the phalanges, including the thumb.

5.7 The Anterior Brachial Region

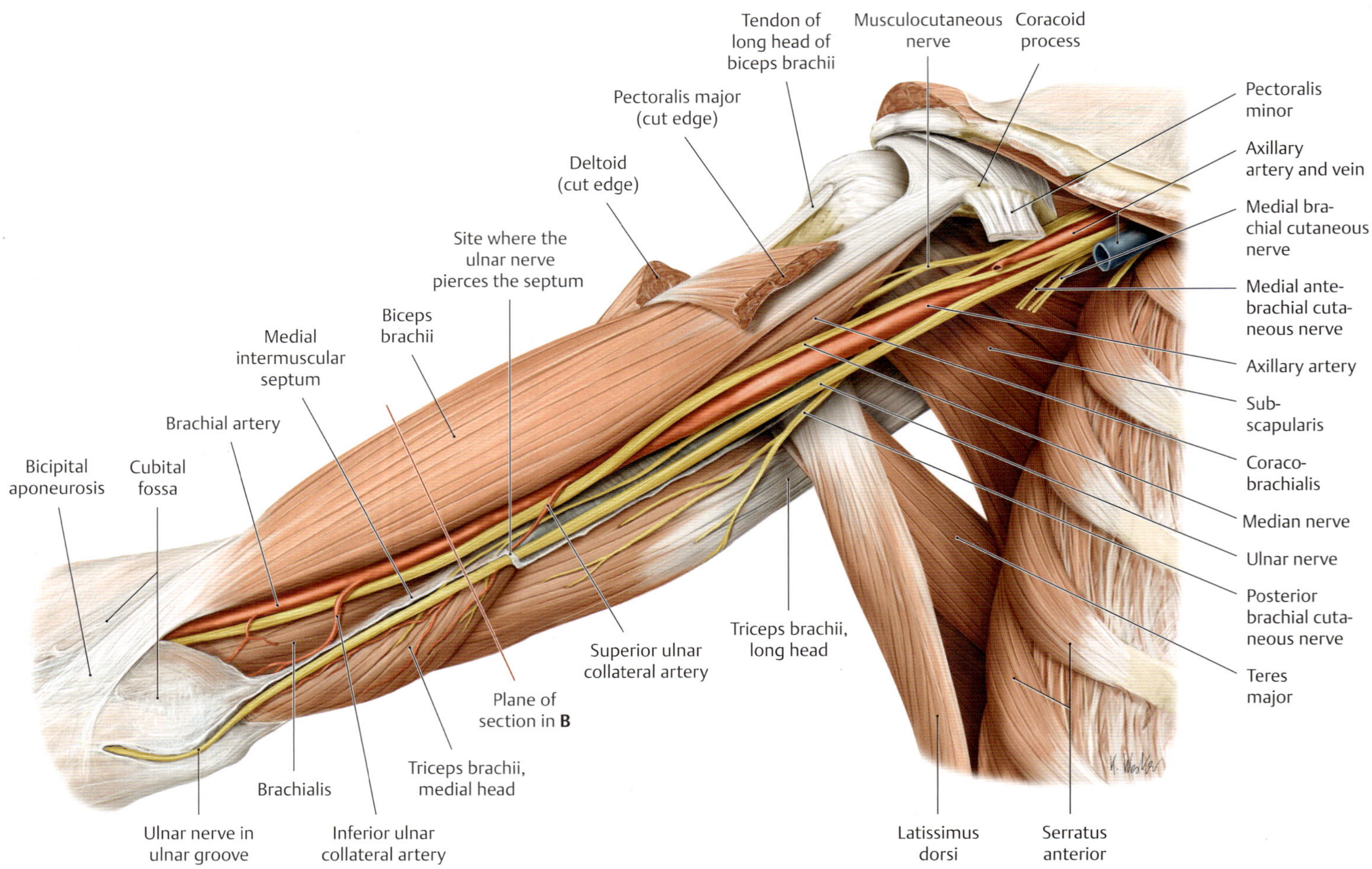

A Main neurovascular tract of the arm: the medial bicipital groove
The right arm has been abducted and slightly rotated externally and is viewed from the anterior view. The deltoid, pectoralis major, and pectoralis minor muscles have been removed. The medial bicipital groove is a subcutaneous longitudinal groove on the medial side of the arm that is bounded deeply by the biceps brachii and brachialis muscles and the medial intermuscular septum of the arm. It marks the location of the main neurovascular tract of the arm extending from the axilla to the cubital fossa. The *most superficial* structure of the tract is the *medial antebrachial cutaneous nerve*, which leaves the medial bicipital groove at the basilic hiatus in company with the basilic vein (see p. 389). The *most medial* structure is the *ulnar nerve*, which initially courses on the medial intermuscular septum. In the lower third of the arm, the ulnar nerve pierces the intermuscular septum and passes to the back of the septum, entering the ulnar groove on the medial epicondyle of the humerus. The *deep part of the medial bicipital groove* transmits the principal artery of the arm, the *brachial artery*, which extends from the axilla to the elbow accompanied by the *median nerve*.

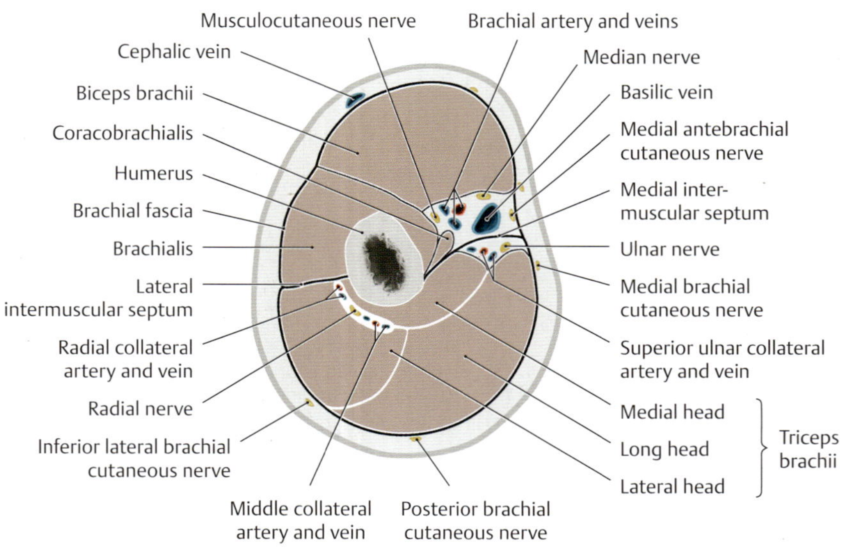

B Cross section through the middle third of the right arm
Distal view.
Note for orientation that the basilic hiatus (where the basilic vein perforates the deep fascia medial to the biceps) is distal to (below) the level of this section. Thus, the basilic vein and medial antebrachial cutaneous nerve are subfascial. The ulnar nerve and ulnar collateral artery have left the medial bicipital groove and have pierced the medial intermuscular septum, and thus lie posterior to it, in this section. Proximal to (above) this level, the artery deep of the arm (deep brachial artery) has split into its two terminal branches, the radial collateral and middle collateral arteries, which are seen here posterior to the humerus.

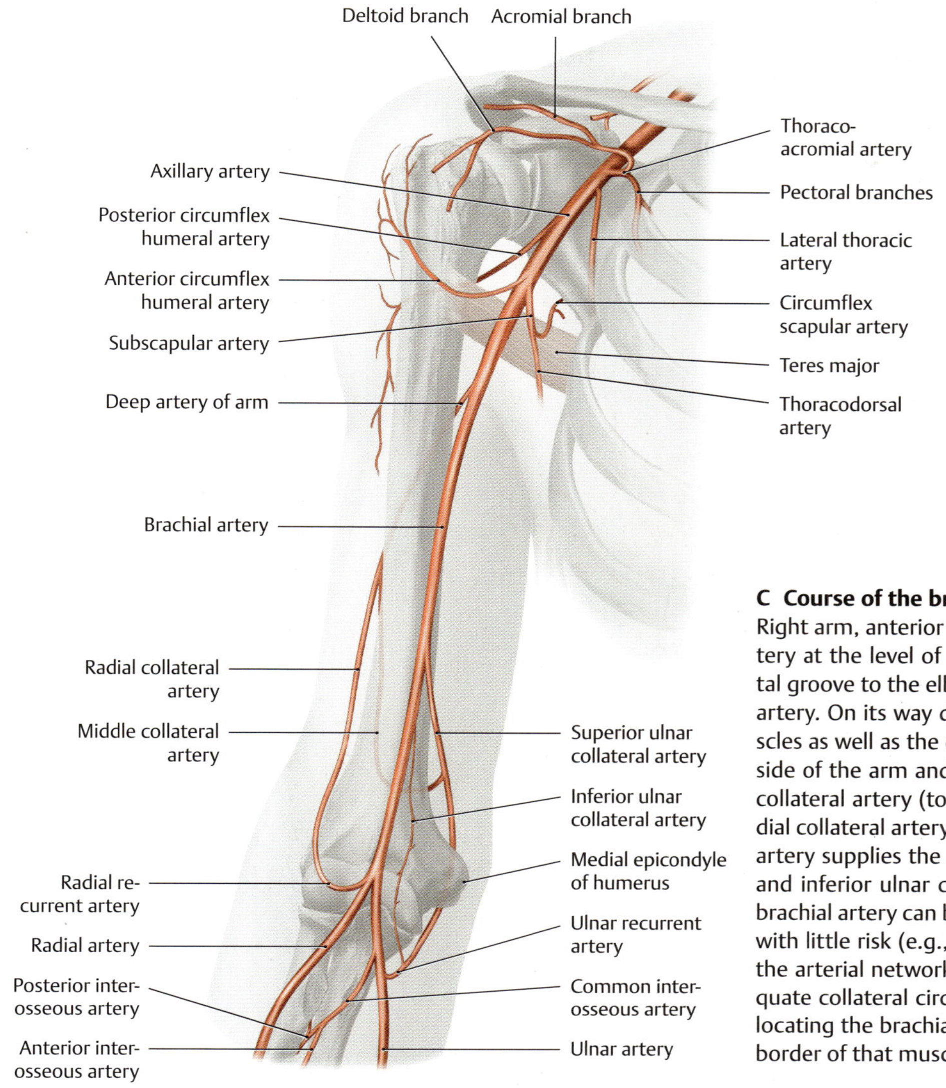

C Course of the brachial artery in the arm
Right arm, anterior view. The brachial artery arises from the axillary artery at the level of the teres major and descends in the medial bicipital groove to the elbow. There it divides into the radial artery and ulnar artery. On its way down the arm, it gives off branches to the arm muscles as well as the deep artery of the arm, which runs on the extensor side of the arm and divides distal to the radial groove into the middle collateral artery (to the medial head of the triceps brachii) and the radial collateral artery (to the arterial network of the elbow). The brachial artery supplies the arterial network of the elbow through the superior and inferior ulnar collateral arteries. It is significant clinically that the brachial artery can be ligated distal to the origin of the profunda brachii with little risk (e.g., to control heavy posttraumatic bleeding) because the arterial network of the elbow (see **C**, p. 407) can establish an adequate collateral circulation. The biceps brachii is a useful landmark for locating the brachial artery, whose pulse is palpable all along the ulnar border of that muscle.

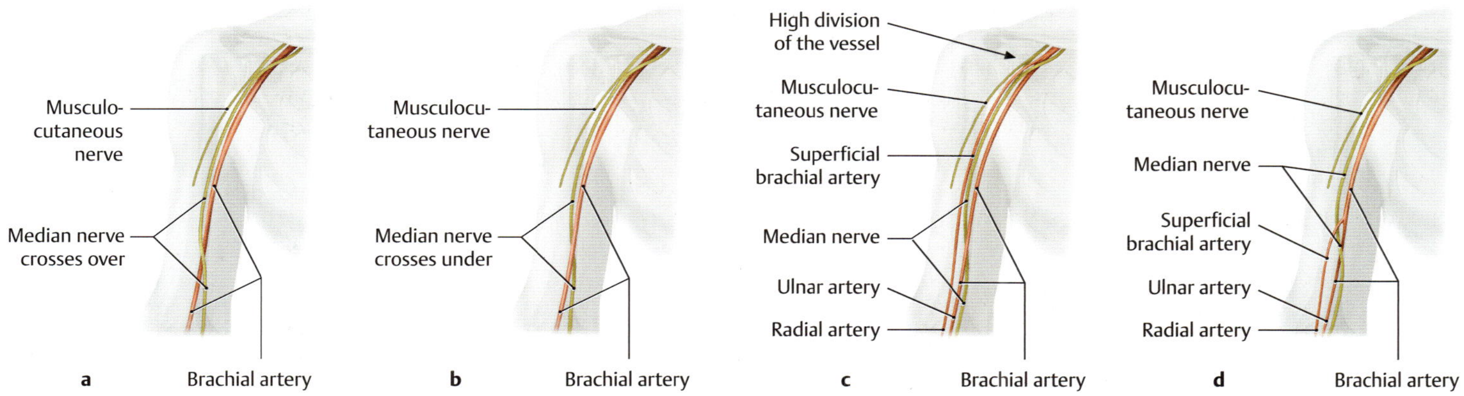

D Course of the brachial artery in the arm: normal anatomy and variants (after von Lanz and Wachsmuth)
Right shoulder, anterior view.

a Usually (74% of cases) the median nerve *crosses over* the brachial artery in the lower third of the arm.

b–d Variants:

b The median nerve *crosses under* the brachial artery (very rare, 1% of cases).

c, d The brachial artery divides into a *superficial* brachial artery and a brachial artery while still in the arm ("high division" pattern, 25% of cases). Both of these arteries may be well developed and may flank the union of the median nerve roots and the median nerve itself. In this case the radial artery arises from the superficial brachial artery ("high origin of the radial artery"), while the ulnar artery is the continuation of the brachial artery (see p. 409).

5.8 The Shoulder Region: Posterior and Superior Views

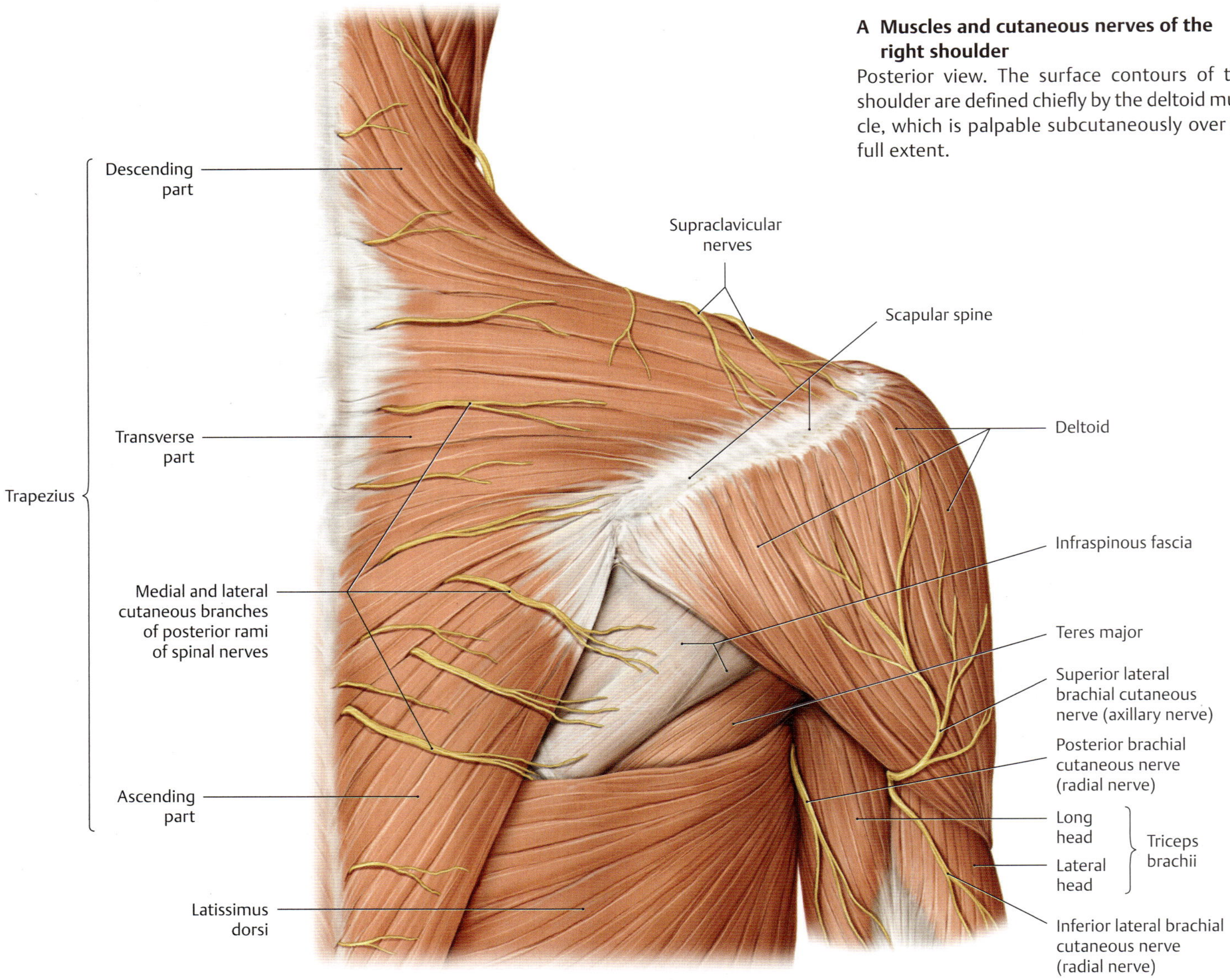

A Muscles and cutaneous nerves of the right shoulder

Posterior view. The surface contours of the shoulder are defined chiefly by the deltoid muscle, which is palpable subcutaneously over its full extent.

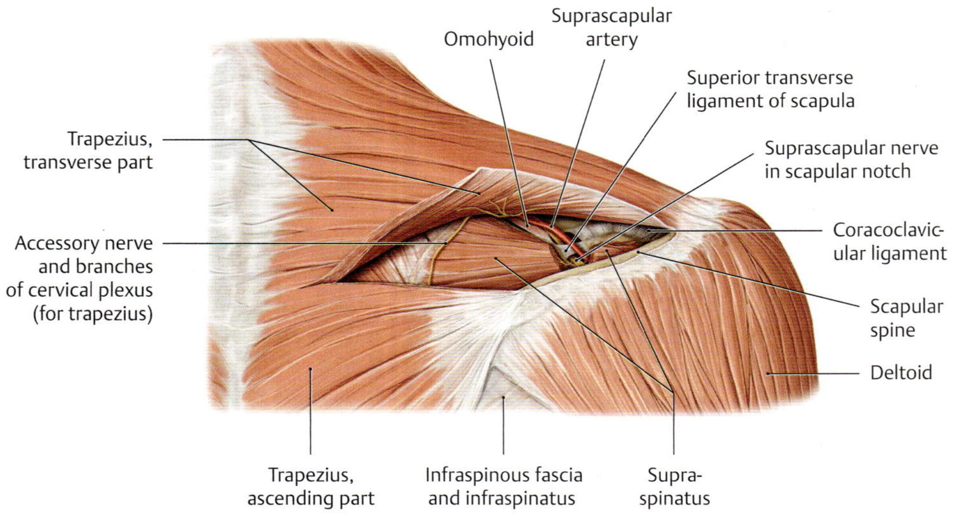

B Suprascapular region of the right shoulder from the posterior view

A flap from the transverse part of the trapezius muscle has been raised to reveal the suprascapular region, and the central portion of the supraspinatus has been removed.

Note the course of the suprascapular nerve in its fibro-osseous canal *below* the superior transverse ligament of the scapula in the scapular notch. Compression of this nerve in its canal, especially on extreme external rotation of the shoulder, can lead to paralysis of the supra- and infraspinatus muscles (*scapular notch syndrome*). Ossification of the superior transverse ligament of the scapula creates a scapular foramen that can also cause suprascapular nerve compression (see p. 255).

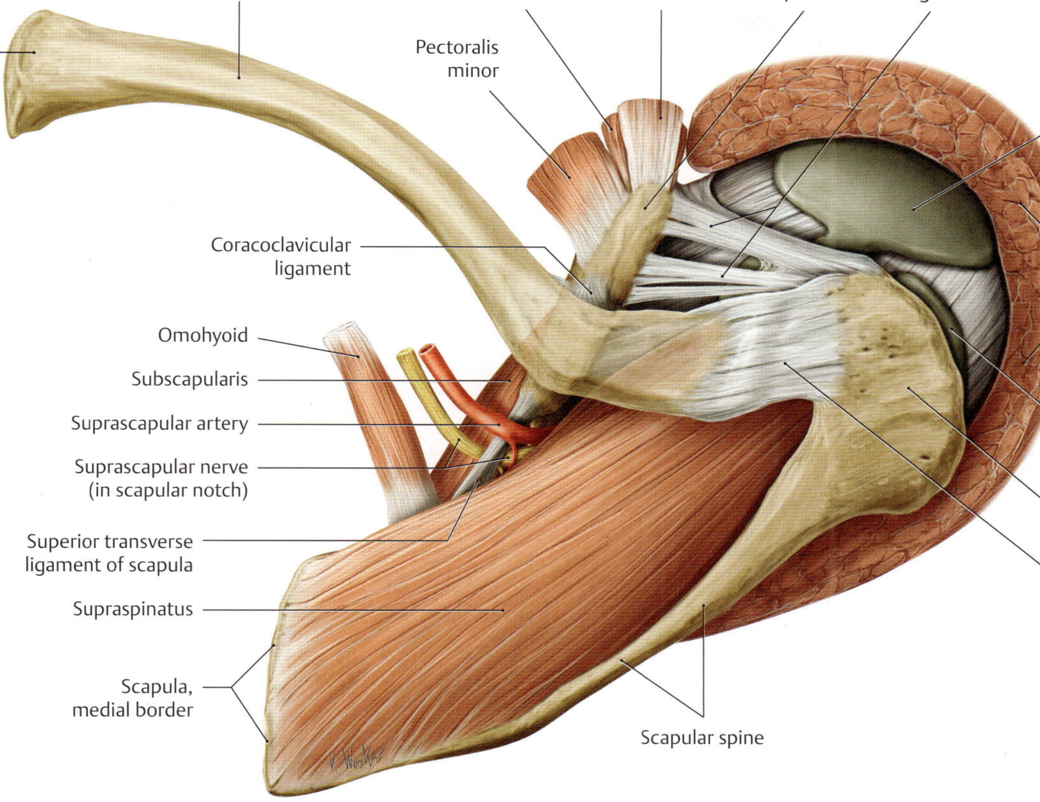

C Suprascapular region of the right shoulder, viewed from above
The trapezius and deltoid muscles have been removed to demonstrate the *supraspinatus*, which originates in the supraspinous fossa and passes laterally beneath the subacromial bursa in the subacromial space to insert on the greater tubercle. The suprascapular artery and nerve run along the anterior border of the supraspinatus at the level of the superior transverse ligament of the scapula, just lateral to the omohyoid insertion—the artery above the ligament and the nerve below it (see **B** and **D**).

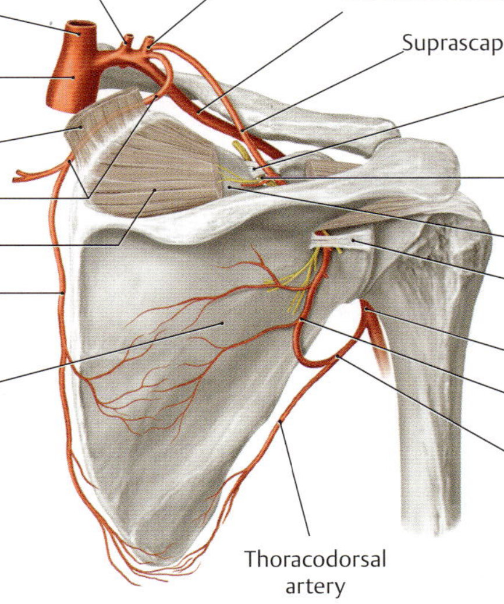

D Scapular arcade
Right scapula, posterior view. The suprascapular artery arises from the thyrocervical trunk and passes over the superior transverse ligament of the scapula to enter the supraspinous fossa. From there it runs past the neck of the scapula, passing *under* the *inferior* transverse ligament of the scapula (often absent), and enters the infraspinous fossa, where it communicates with the circumflex scapular artery (from the subscapular artery) and the deep branch (dorsal scapular artery) of the transverse cervical artery.
Note the anastomosis between the suprascapular artery and the circumflex scapular artery (*scapular arcade*). It is important clinically because it can provide a collateral circulation in response to ligation or occlusion of the axillary artery (see also p. 404).

5.9 The Posterior Brachial Region

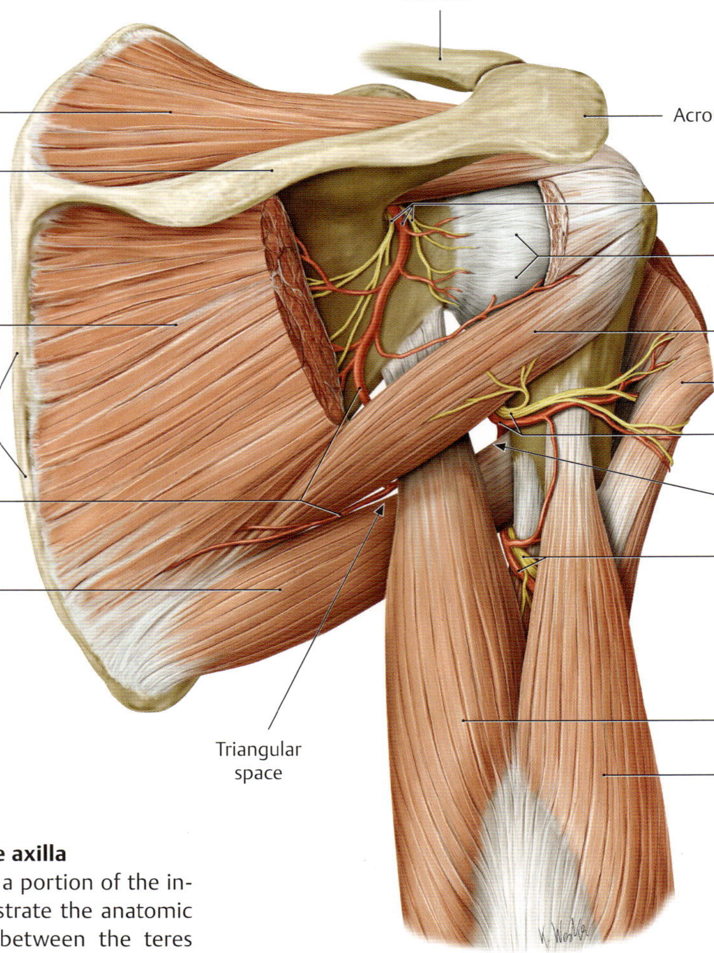

A The triangular and quadrangular spaces of the axilla
Right side, posterior view. Most of the deltoid and a portion of the infraspinatus muscle have been removed to demonstrate the anatomic relationships more clearly. The slitlike opening between the teres minor, teres major, and humerus is subdivided by the long head of the triceps brachii into a *quadrangular space* (lateral axillary foramen) and a *triangular space* (medial axillary foramen).

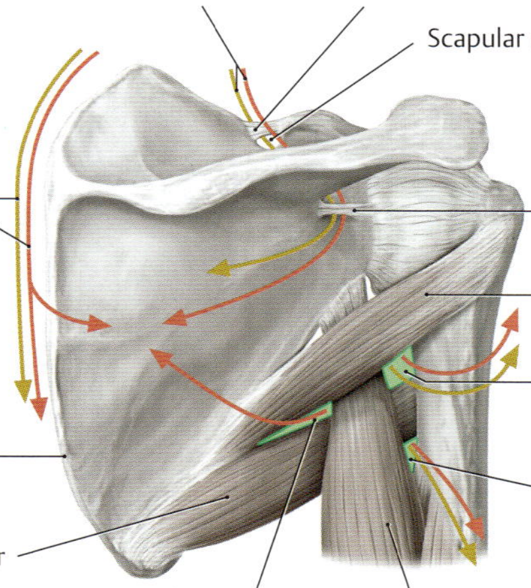

B Neurovascular tracts associated with the scapula
Right side, posterior view. The triangular and quadrangular spaces of the axilla and the triceps hiatus provide important passageways that transmit neurovascular structures from the anterior to the posterior scapular region.

Passageways	Structures transmitted
• Triangular space	Circumflex scapular artery
• Quadrangular space	Posterior circumflex humeral artery and axillary nerve
• Triceps hiatus	Deep artery of arm and radial nerve

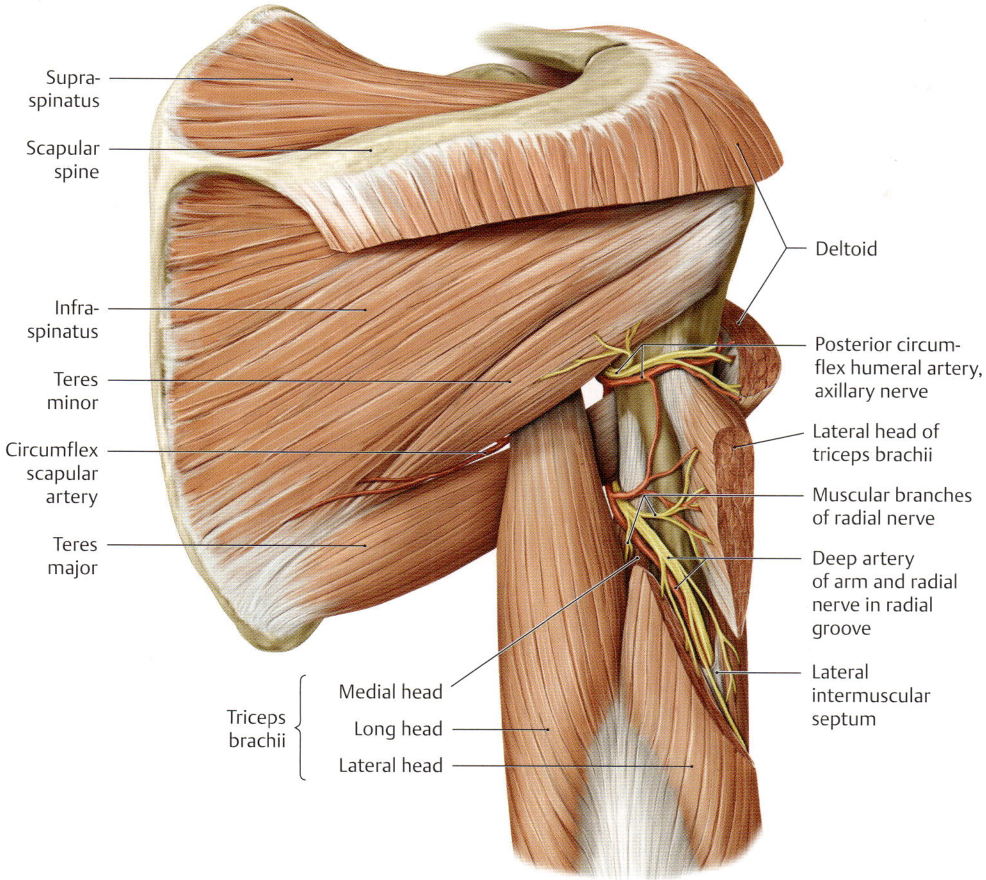

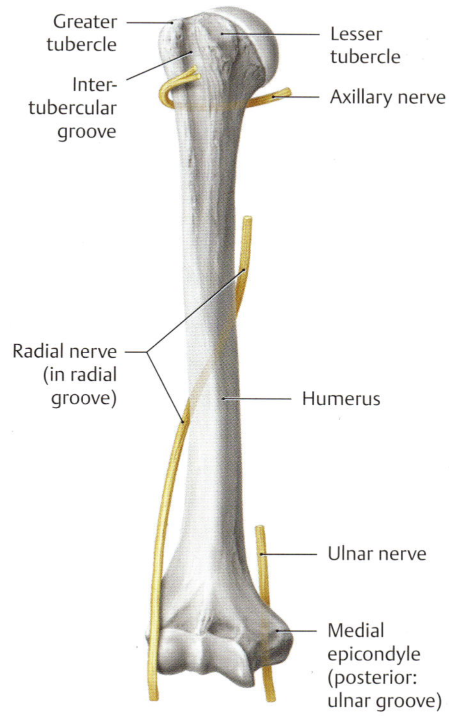

E Nerves that are closely related to the humerus
Right humerus, anterior view.

C Course of the radial nerve in the radial groove
Right shoulder and upper arm, posterior view. The lateral head of the triceps brachii has been divided to show how the radial nerve spirals around the humerus. The dissection shows the bony radial groove between the origins of the medial and lateral heads of the triceps brachii. At the distal end of the groove, the radial nerve passes through the lateral intermuscular septum to the front of the humerus and continues in the radial tunnel to the cubital fossa (not shown here; see also p. 406). *Note* that the radial nerve branches for the triceps brachii arise proximal to the radial groove. Thus, the triceps brachii may still be functional after a humeral shaft fracture at the level of the radial groove, even though the radial nerve has been damaged, because the muscular branches to the triceps arise proximal to the site of the lesion.

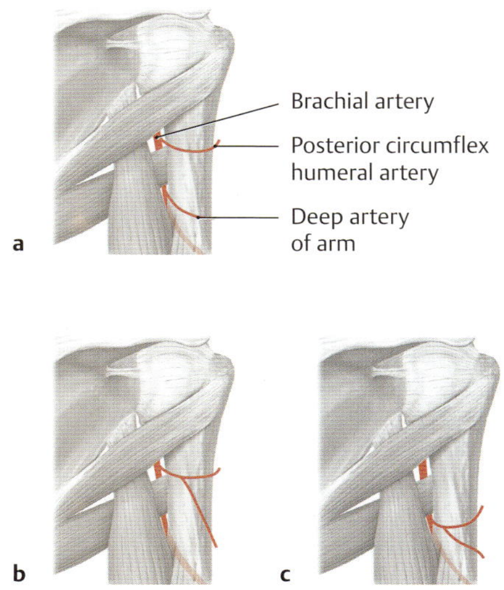

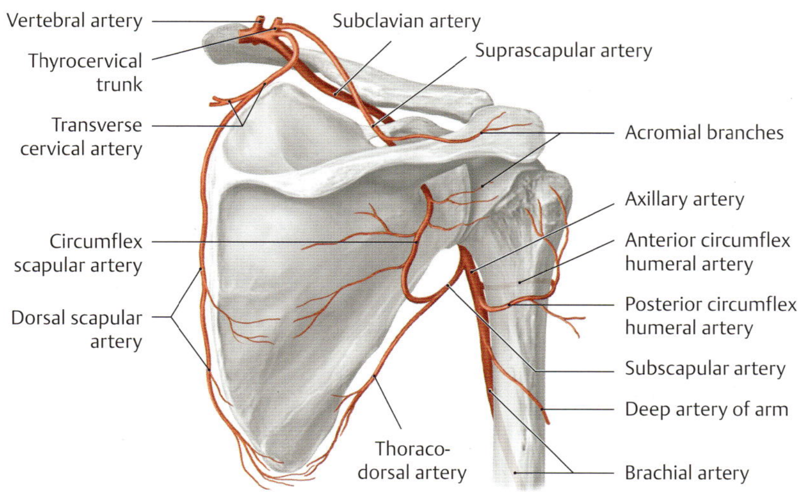

D Arterial supply to the scapular region
Right shoulder, posterior view.

F Branches of the brachial artery: normal anatomy and variants (after von Lanz and Wachsmuth)

a Typically (77% of cases), the deep artery of the arm and the posterior circumflex humeral artery arise from the brachial artery.

b, c Variants:

b The deep artery of the arm arises from the posterior circumflex humeral artery (7%).

c As in **b**, but the posterior circumflex humeral artery runs through the triceps hiatus rather than the quadrangular space (16%).

5.10 The Elbow (Cubital Region)

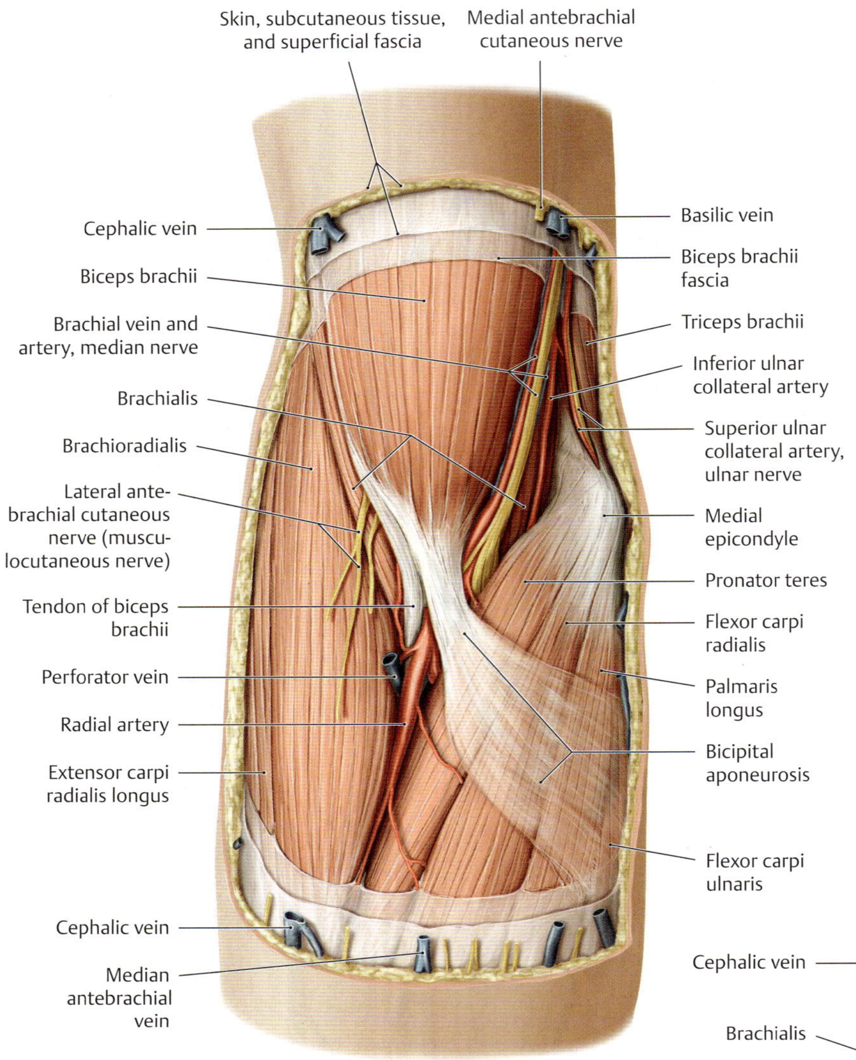

A Cubital fossa after removal of the fasciae and epifascial neurovascular structures
Right arm, anterior view. The *median nerve* and the *musculocutaneous nerve* with its main branches are clearly visible (variants in the course of the median nerve are shown in **D**). The brachioradialis muscle must be retracted (as in **B**) before the radial nerve can be seen.

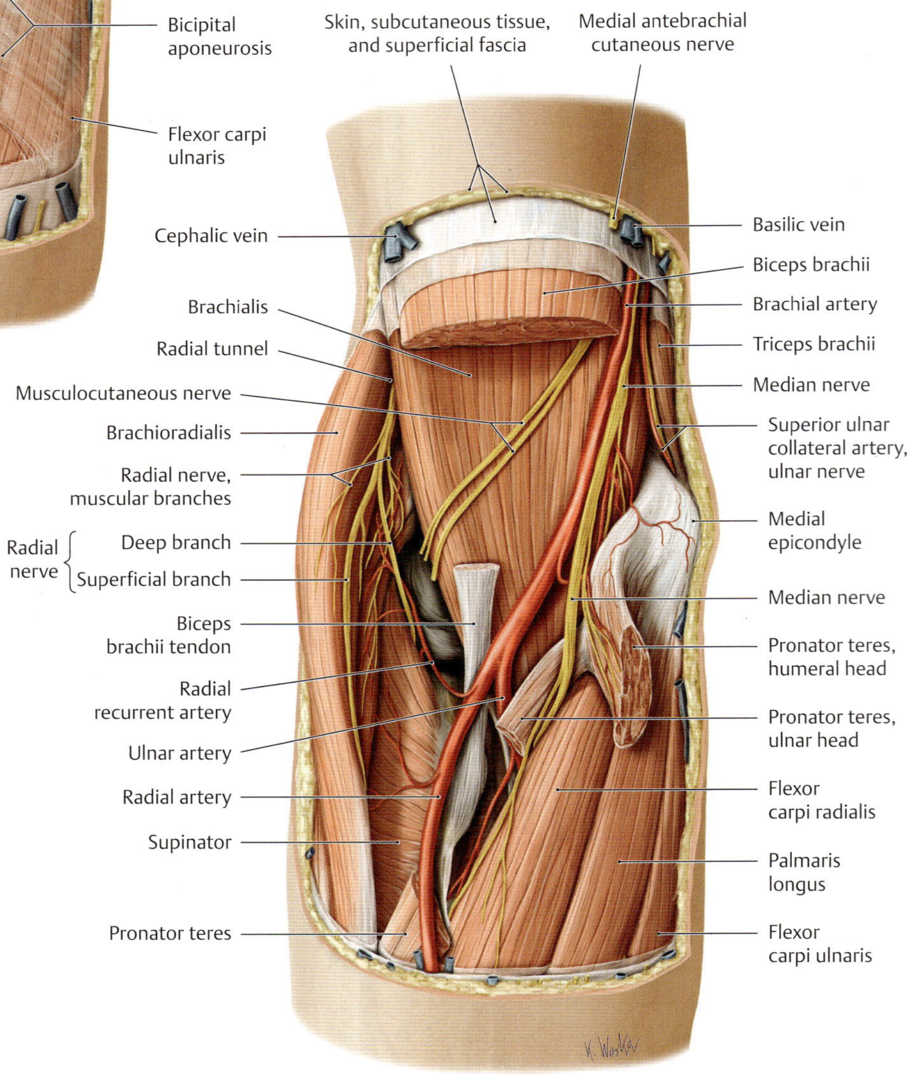

B Deep dissection of the cubital fossa
Right arm, anterior view. The distal muscle belly of the biceps brachii has been removed, and the muscles of the radialis group (brachioradialis, extensor carpi radialis longus, and extensor carpi radialis brevis) have been pulled aside to better demonstrate the course of the *radial nerve*. After passing through the radial tunnel, the nerve gives off its sensory superficial branch and its muscular branches to the muscles of the radialis group before entering the supinator (see p. 407). Portions of the pronator teres have been retracted medially to demonstrate the course of the median nerve between the two heads of that muscle.

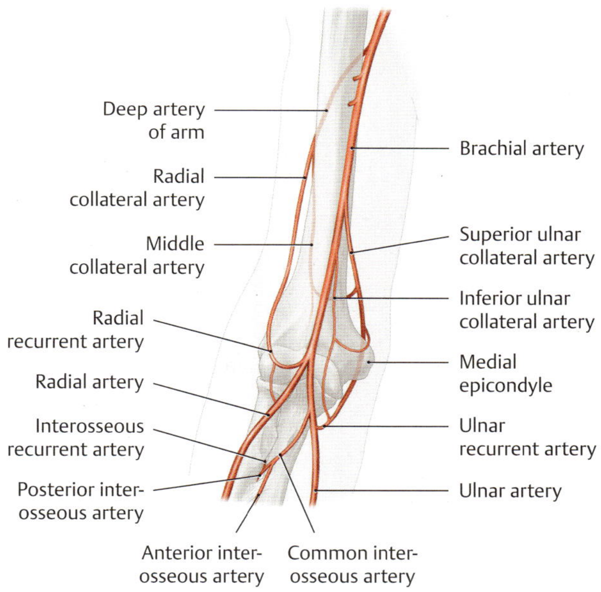

C Arterial anastomoses around the elbow joint: the arterial network of the elbow
Right arm, anterior view. The arterial anastomoses in the elbow region form a vascular network that is fed by several arteries:

- The middle collateral artery and radial collateral artery from the deep artery of the arm (communicate with the radial artery via the radial recurrent artery and interosseous recurrent artery).
- The superior ulnar collateral artery and inferior ulnar collateral artery from the brachial artery (communicate with the ulnar artery via the ulnar recurrent artery).

Because of this arterial network, the brachial artery can be ligated distal to the origin of the deep artery of the arm without compromising the blood supply to the elbow region.

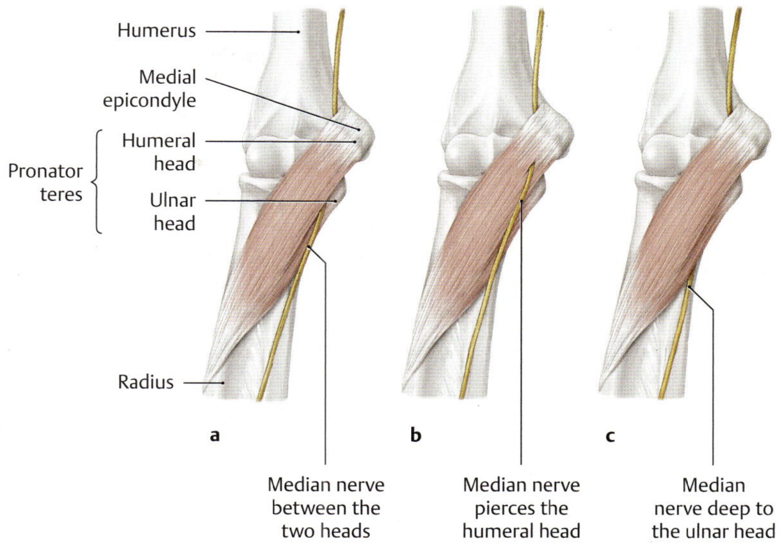

D Relationship of the median nerve to the pronator teres: normal anatomy and variants (after von Lanz and Wachsmuth)
Right arm, anterior view.

a In the great majority of cases (95%) the median nerve runs between the two heads of the pronator teres.
b, c Variants:
b The median nerve pierces the humeral head of the pronator teres (2% of cases).
c The median nerve runs on the bone beneath the ulnar head of the pronator teres (3% of cases).

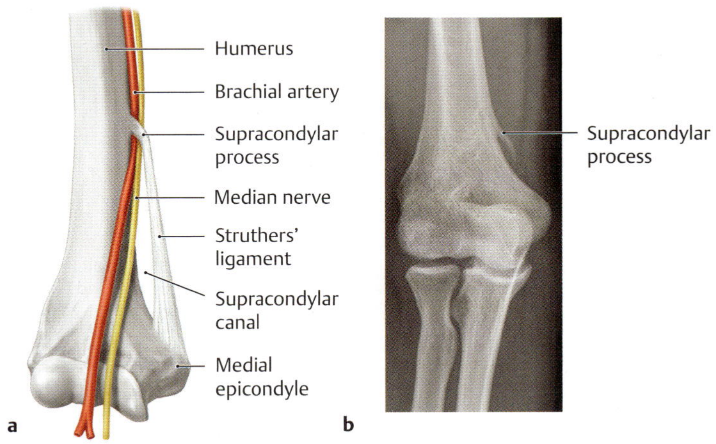

E Supracondylar process on the humerus
a Right distal humerus, anterior view;
b AP radiograph. The *supracondylar process* is an unusual anomaly (0.7% of the population), a bony outgrowth above the medial epicondyle. When present, it can serve as an attachment for a connective tissue band referred to as Struthers' ligament, which ends on the medial epicondyle. The resulting fibro-osseous *supracondylar canal* can entrap and compress the brachial artery and median nerve. (Radiograph: Hans-Peter Sobotta, M.D., Stiftung HEH Braunschweig.)

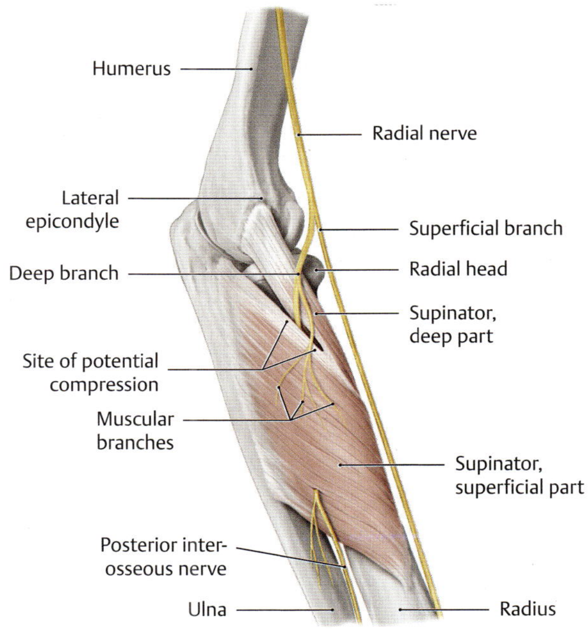

F Relationship of the radial nerve to the supinator
Right elbow region, radial view. Just proximal to the supinator muscle, the radial nerve divides into its deep motor branch and superficial sensory branch. This arrangement can lead to entrapment and compression of the deep motor branch (see p. 382), with resulting selective palsy of the extensor muscles (and abductor pollicis longus) served by this nerve.

5.11 The Anterior Forearm Region

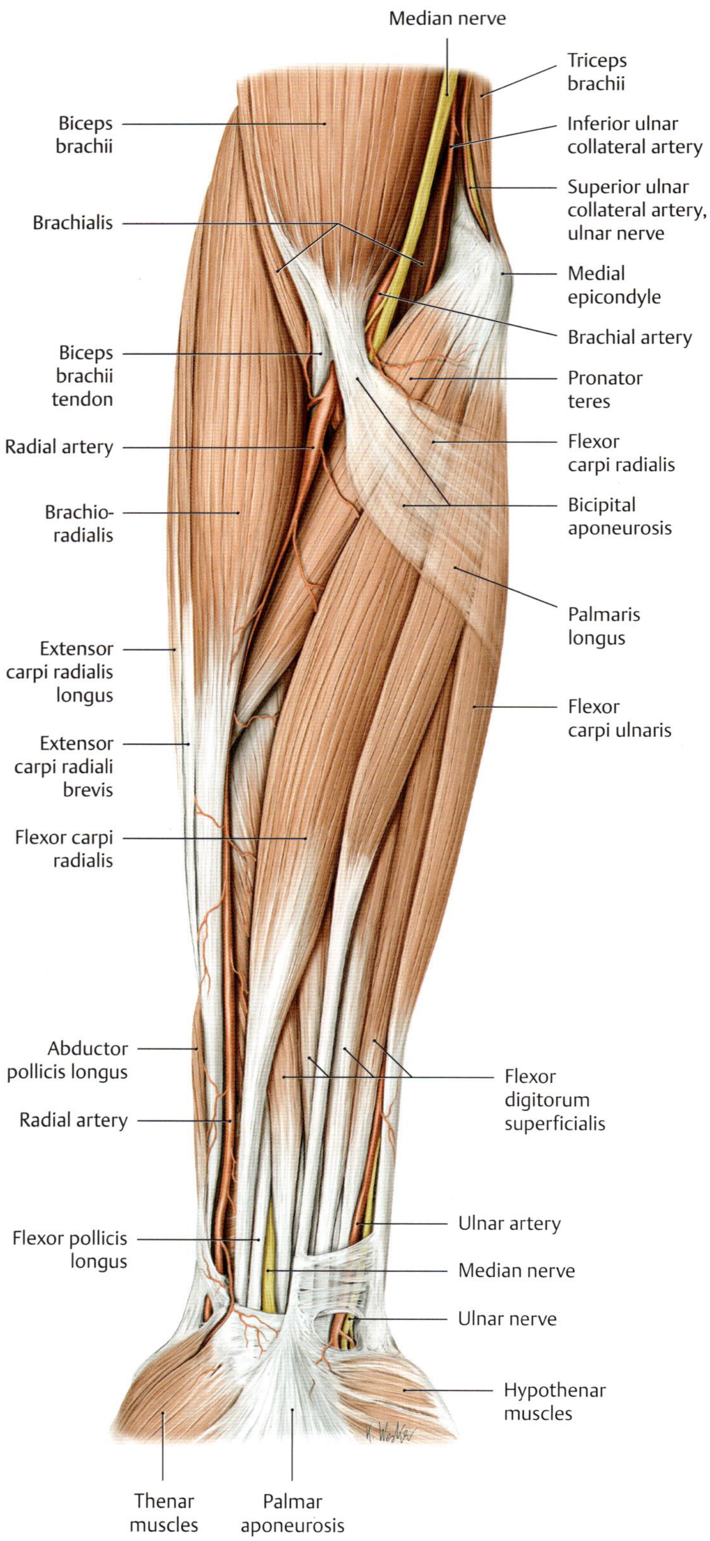

A Right forearm, anterior view, superficial layer
The fasciae and superficial neurovascular structures have been removed. Most of the forearm's neurovascular structures are obscured in this view. (The superficial veins are shown in **D**, p. 387.)

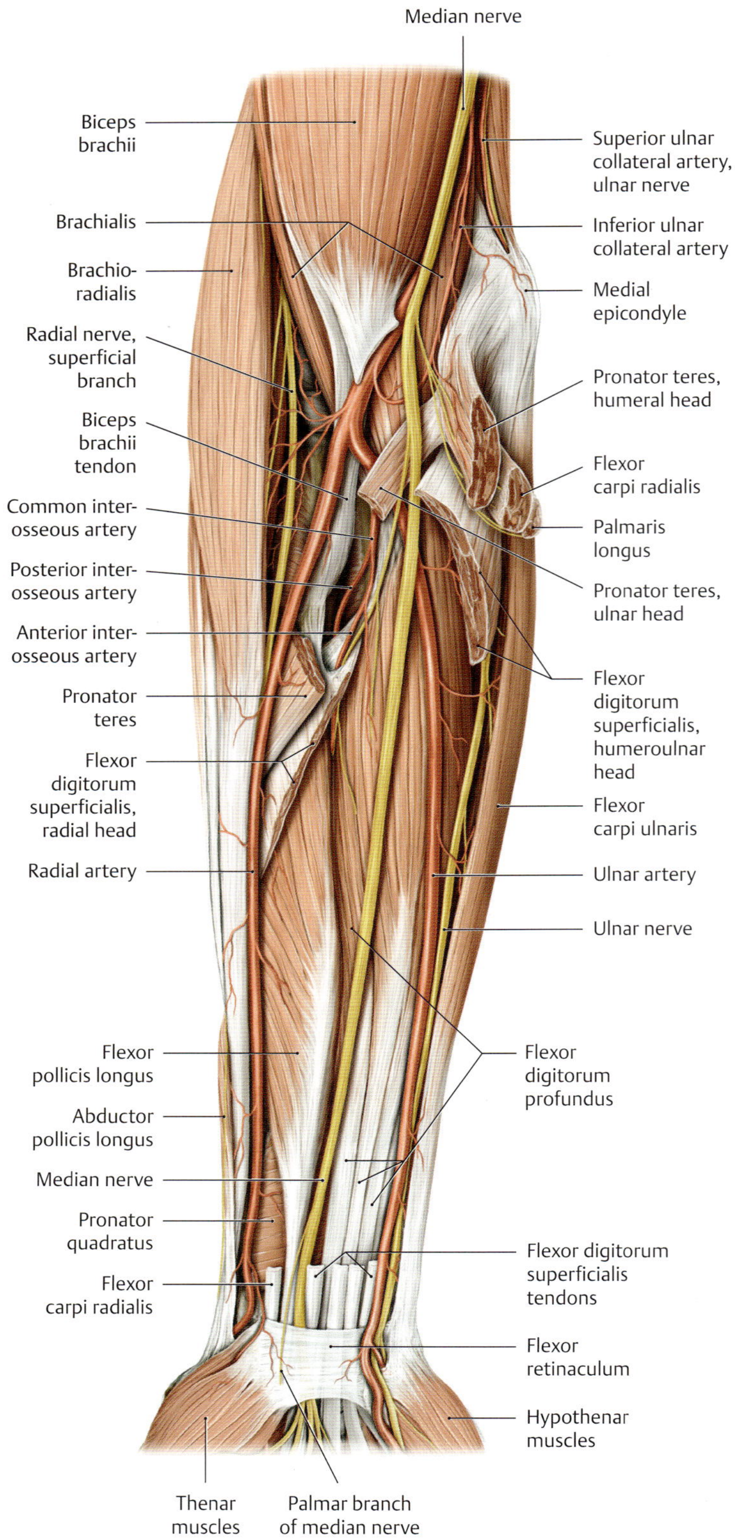

B Right forearm, anterior view, deep layer
The pronator teres, flexor digitorum superficialis, palmaris longus, and flexor carpi radialis have been partially removed to demonstrate the median nerve, the superficial branch of the radial nerve, and the radial and ulnar arteries (variants in the course of the arteries are shown in **D**).

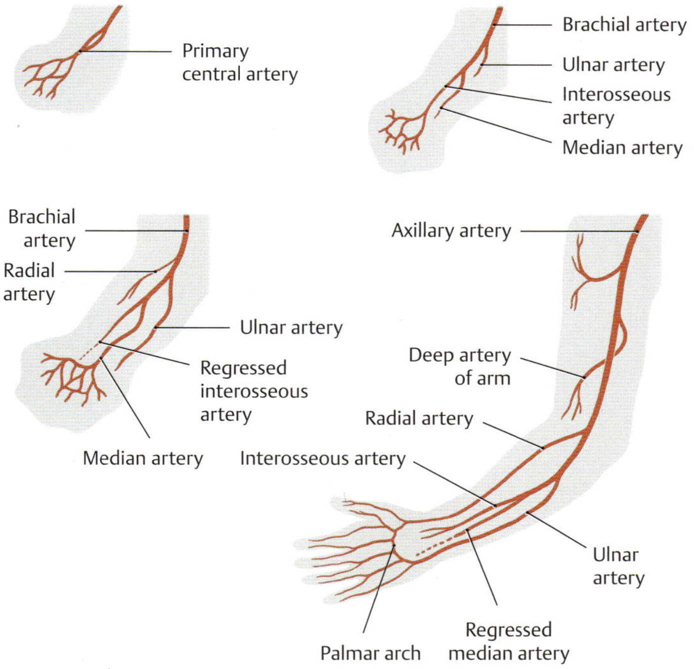

C Development of the arteries of the upper limb

The vascular system in the embryonic arm does not develop directly into the mature anatomy but undergoes several basic changes. The early limb bud is supplied by a central vascular trunk that develops distally into the *common interosseous artery*. As development proceeds, a second longitudinal trunk, called the *median artery*, is formed parallel to the median nerve. That vessel provides most of the blood supply to the forearm and hand, while the interosseous artery regresses. Finally, the initially small muscular branches enlarge on the ulnar and radial sides to form the *ulnar artery* and the *radial artery*, which replace the median artery in primates and assume its functions. The interosseous artery remains the principal vessel of the arm in nonmammals, while the median artery is the dominant vessel in lower mammals. The interosseous and median arteries may persist in humans as well-developed atavistic anomalies (see **D**), providing most of the blood supply to the palm of the hand.

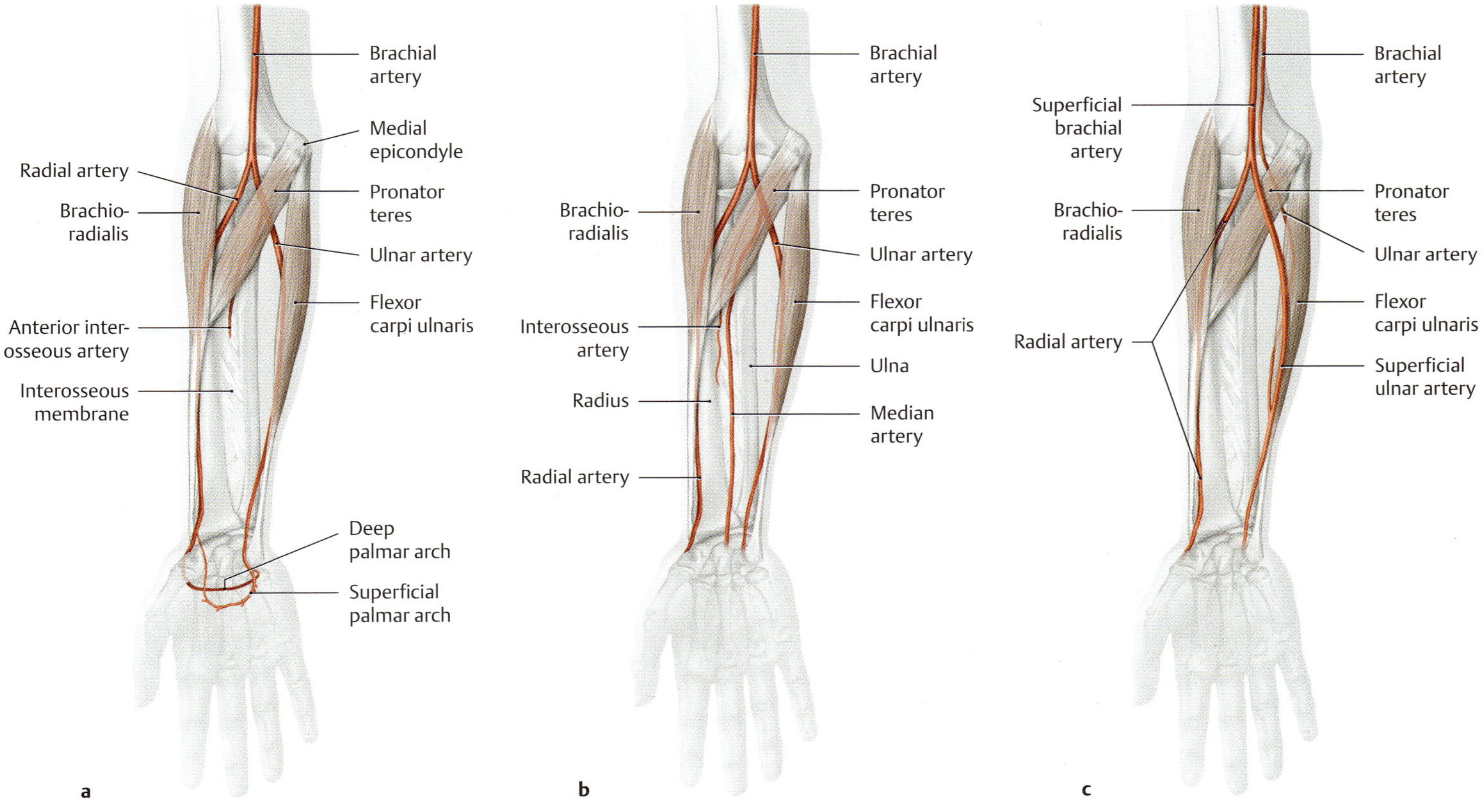

D Arteries of the forearm: normal anatomy and variants
(after Lippert and Pabst) Right forearm, anterior view.

a Typical arterial anatomy in the forearm (84% of cases).
b, c Variants:
b A persistent median artery is present, often arising from the ulnar artery distal to the origin of the common interosseous artery (8% of cases).
c Accessory superficial arteries are present in the forearm (superficial antebrachial arteries, 8% of cases), such as a superficial ulnar artery arising from a superficial brachial artery, which runs over the surface of the flexor muscles and may unite distally with the ulnar artery. The presence of this vessel is a potential hazard during intravenous injections in the cubital area (see p. 370). Accessory superficial arteries most commonly develop in cases where the brachial artery divides in the arm into a superficial brachial artery (which becomes the radial artery in the forearm) and a brachial artery (which becomes the ulnar artery in the forearm; see "high division" pattern, p. 389).

5.12 The Posterior Forearm Region and the Dorsum of the Hand

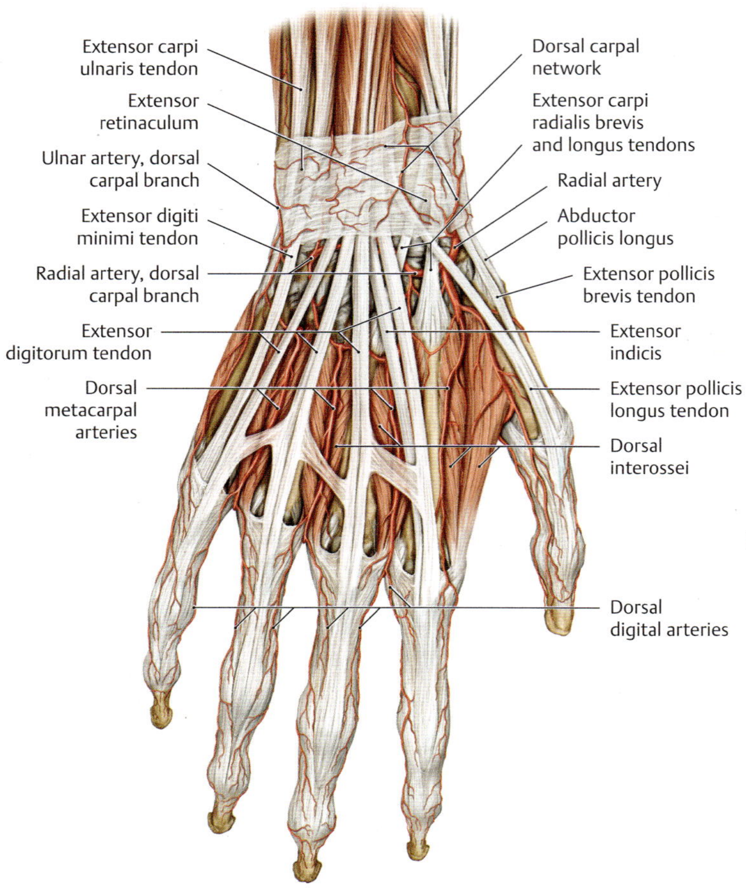

A Arteries of the dorsum of the hand and the extensor sides of the fingers in the right hand

The skin, subcutaneous tissue, and dorsal fascia of the hand have been removed to demonstrate the dorsal arteries (for clarity, the veins and nerves have also been removed). The dorsum of the hand receives most of its blood supply from the radial artery, while the ulnar artery contributes only one small vessel (the dorsal carpal branch). The perforating branches, however, create numerous connections between the palmar and dorsal arteries of the hand. In the fingers, these connections are provided by lateral anastomoses between the dorsal digital arteries and the proper palmar digital arteries (not seen here).

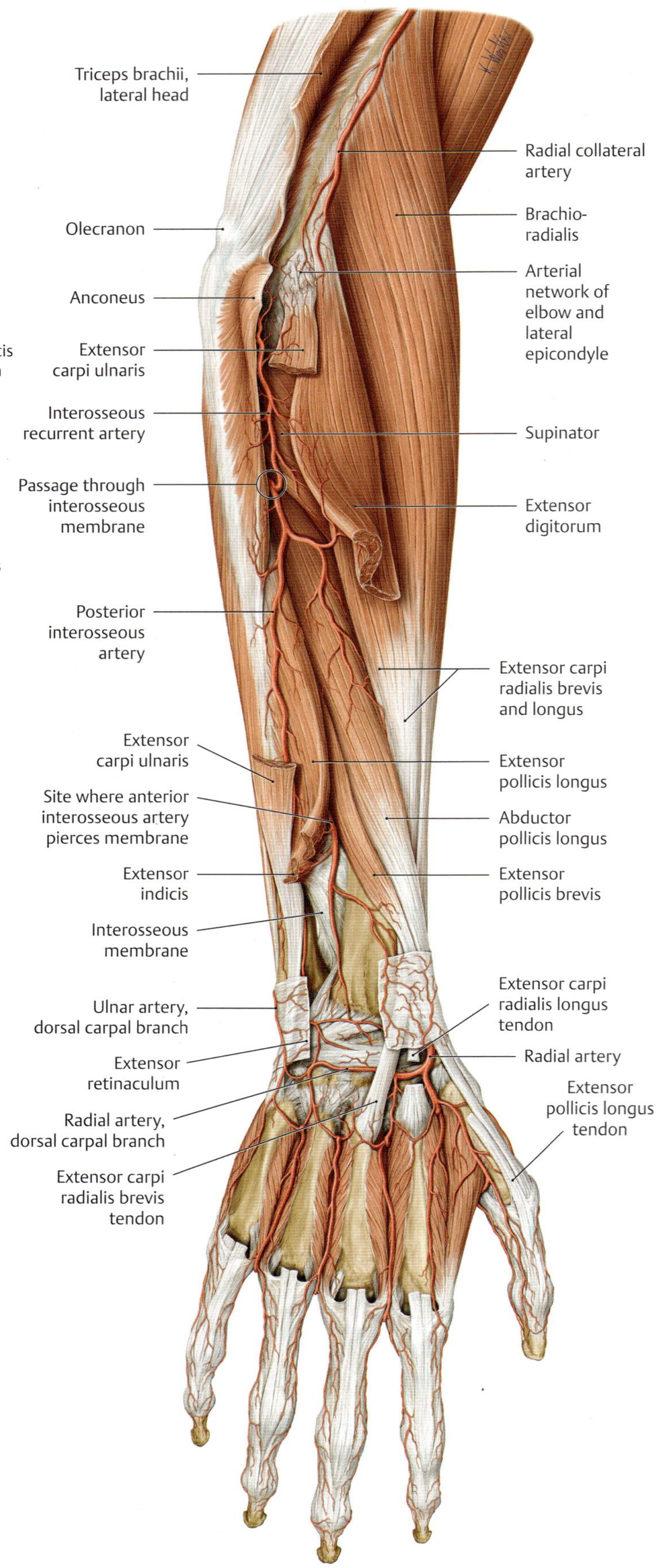

B Deep dissection of the arteries on the extensor side of the forearm and the dorsum of the hand in the right arm

In the elbow region the anconeus muscle has been released from its origin and reflected to the side. The triceps brachii has also been released from its origin at a more proximal level. On the extensor side of the forearm, the extensor carpi ulnaris and extensor digitorum have been partially resected.

Note how the posterior interosseous artery pierces the interosseous membrane just below the lower border of the supinator and enters the extensor compartment of the forearm. In the distal forearm, parts of the extensor pollicis longus and extensor indicis have been removed to demonstrate the site where the anterior interosseous artery pierces the interosseous membrane to reach the back of the forearm. Both arteries are important sources of blood to the extensor compartment.

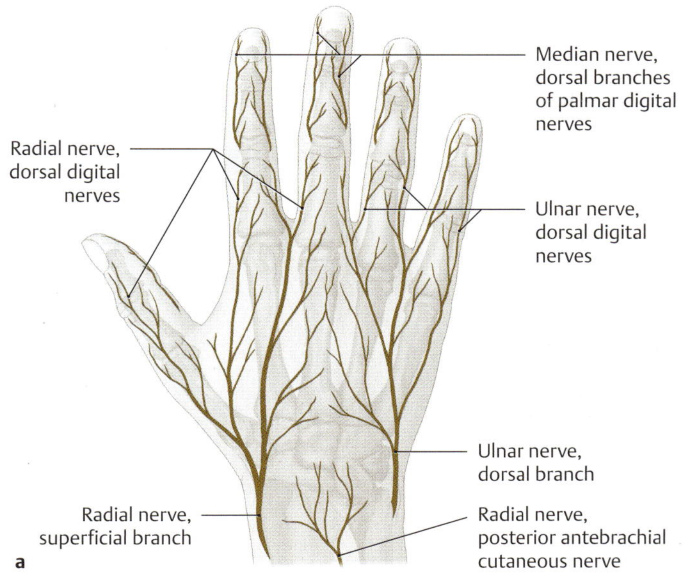

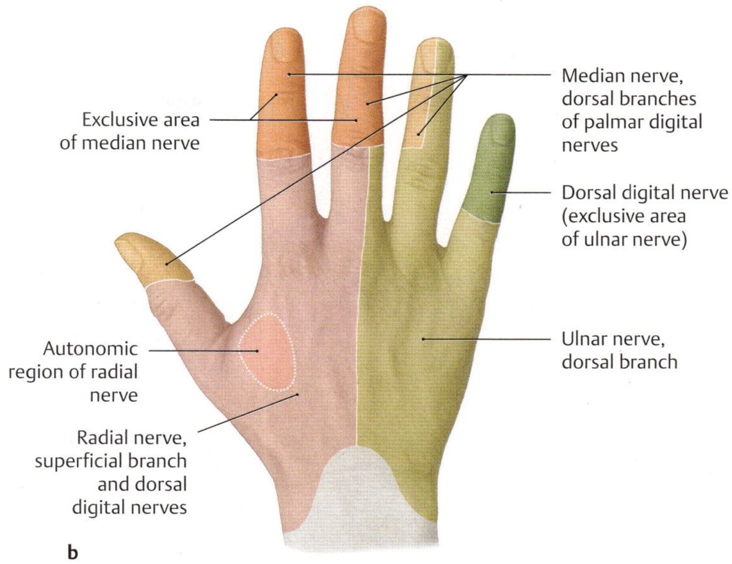

C Nerve supply to the dorsum of the hand
Right hand, posterior view.

a The cutaneous nerves on the dorsum of the hand. *Note* that the index and middle fingers and the radial part of the ring finger are supplied by different nerves in their proximal and distal portions.

- *Distal:* by the dorsal branches of the palmar digital nerves from the *median nerve*.
- *Proximal:* by the dorsal digital nerves from the *radial nerve* (to about the proximal interphalangeal joints of the index and middle fingers) and from the *ulnar nerve* (also to the level of the proximal interphalangeal joints of the middle and ring fingers).

b Exclusive and overlapping areas of sensory innervation on the dorsum of the hand, for ulnar, median, and radial nerves. In lighter shading, areas are indicated that receive sensory innervation mostly from the corresponding nerve. Because the sensory distribution of each nerve in actuality overlaps extensively with the adjacent sensory territories of the other nerves, an isolated nerve lesion does not render its entire territory anesthetic. Instead, extensive or complete loss of sensation is restricted to areas where there is little or no overlap, depicted in heavier shading.

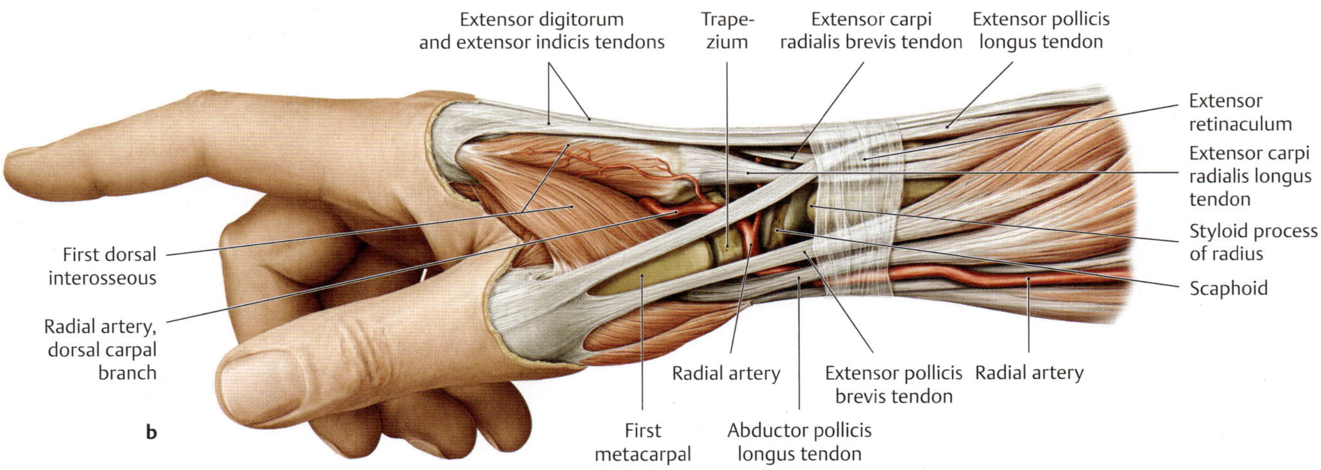

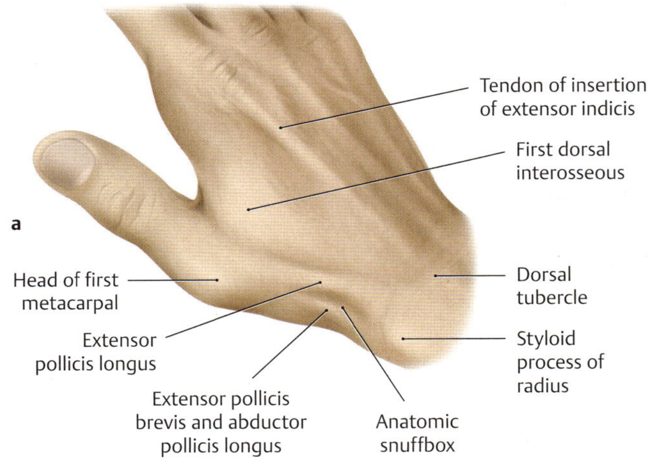

D Boundaries of the anatomic snuffbox

a Surface anatomy of the dorsum of the right hand, posterolateral view.

b Muscles and tendons of the dorsum of the right hand, radial view. The three-sided "anatomic snuffbox" is bounded on the palmar side by the tendons of insertion of the abductor pollicis longus and extensor pollicis brevis and dorsally by the tendon of insertion of the extensor pollicis longus. The floor is formed mostly by the scaphoid and trapezium carpal bones. Fractures of the scaphoid bone are thus often associated with deep tenderness in the snuffbox. The snuffbox is bounded proximally by the extensor retinaculum.

Note that the radial artery runs deep in the snuffbox between the trapezium and scaphoid, providing a landmark for dissection.

5.13 The Palm of the Hand: Epifascial Nerves and Vessels

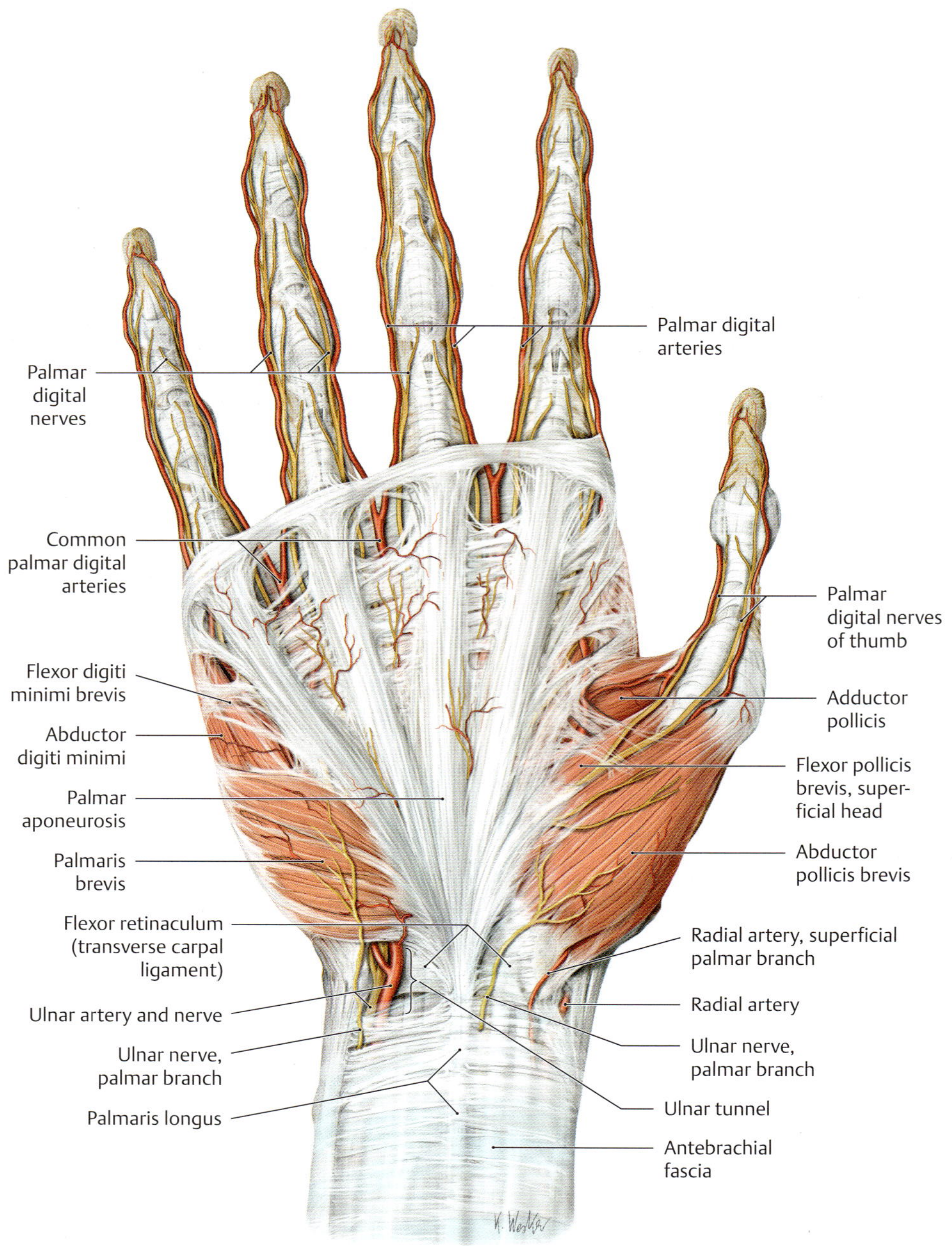

A Superficial arteries and nerves of the palm

Right hand, anterior view. All of the fasciae except the palmar aponeurosis have been removed to demonstrate the superficial neurovascular structures of the palm. The palmar carpal ligament has also been removed to demonstrate the neurovascular structures that pass through the ulnar tunnel (the ulnar artery and nerve; see p. 419).

Note the superficial palmar branch of the radial artery, which is highly variable in its course. In the case shown, it runs between the origins of the abductor and flexor pollicis brevis to the palm of the hand. In approximately 30% of cases it combines with the ulnar artery to form the superficial palmar arch (not shown here; see also p. 414).

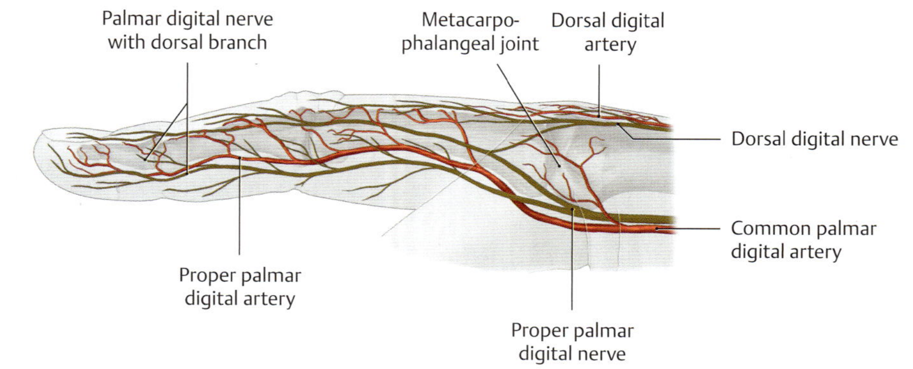

B Nerves and vessels of the right middle finger

Lateral view. The arteries of the palm are anterior to the nerves, but they are dorsal to the nerves in the fingers (generally crossing at the level of the metacarpophalangeal joint). The lateral and distal dorsal surfaces of the fingers are supplied by branches of the proper palmar digital nerves (from the median nerve).

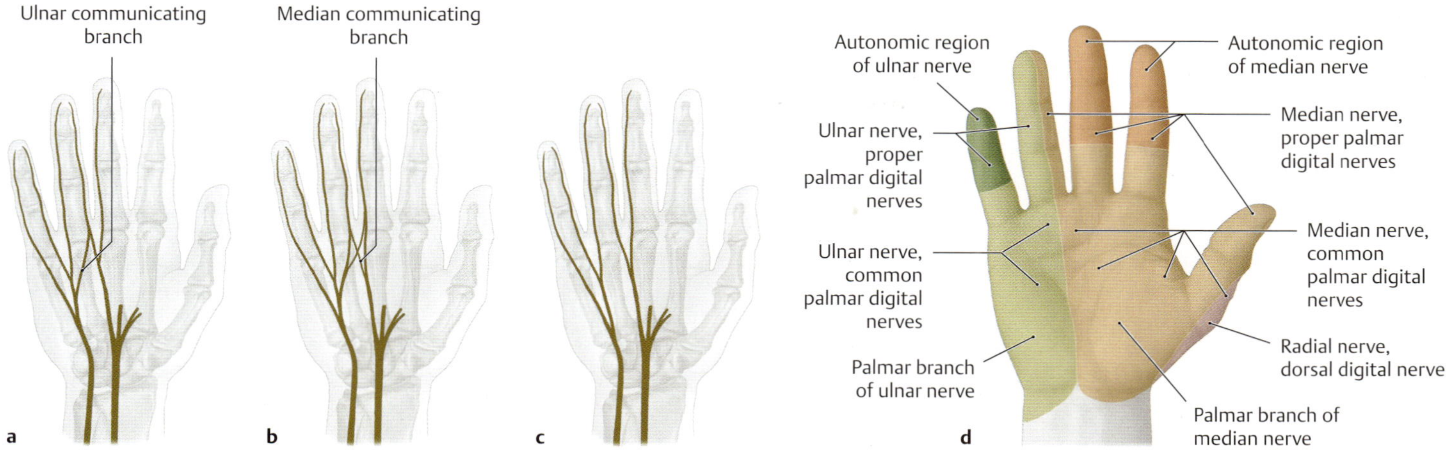

C Nerve supply to the palm of the hand
Right hand, anterior view.

a–c Innervation patterns in the palm of the hand (after Schmidt and Lanz). The sensory innervation pattern is marked by the presence of connecting branches between the median nerve and ulnar nerve. The following innervation patterns are most frequently encountered:

a Most commonly (46% of cases) the median nerve and ulnar nerve are interconnected by an ulnar communicating branch.

b Variant 1 (20% of cases): The median nerve and ulnar nerve are cross-connected by an ulnar communicating branch and a median communicating branch.

c Variant 2 (20% of cases): There are no communicating branches between the median and ulnar nerves.

d Exclusive and overlapping areas of sensory innervation of the hand, palmar view. Non-overlapping areas of exclusive innervation are indicated by darker shading. Compare with the dorsal aspect (**C**, p. 411).

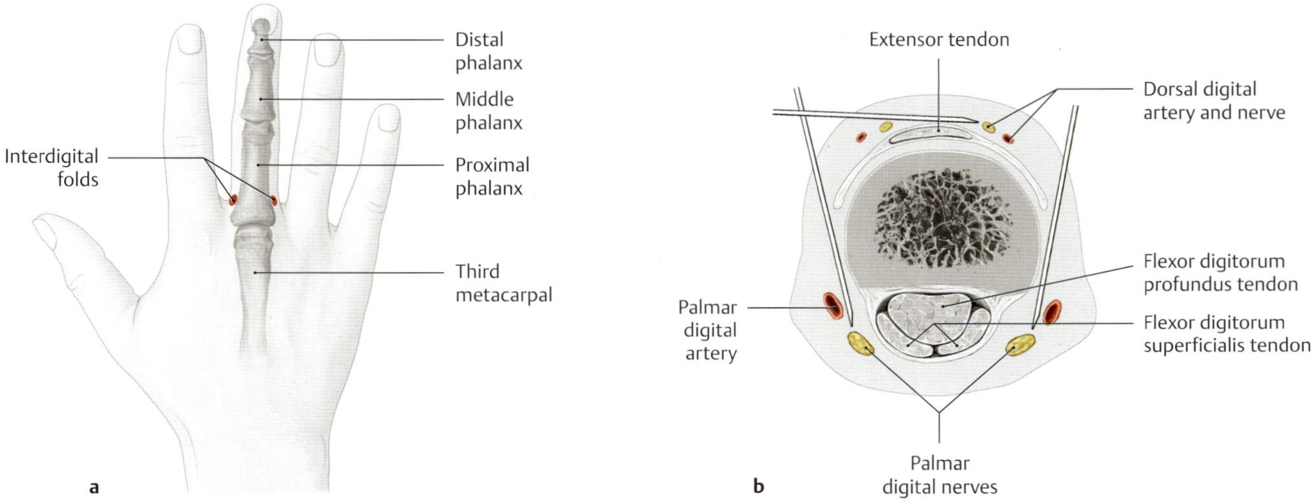

D The Oberst nerve block
Right hand, posterior view. This type of local anesthesia is clinically useful for injuries of the fingers, specifically for wounds that require suturing.

a The injection sites are located in the interdigital folds.

b After the dorsal nerve branches have been numbed, the needle is advanced on both the radial and ulnar sides toward the palmar nerves, and a subcutaneous bolus of 1 to 2 mL of local anesthetic is injected at each site.

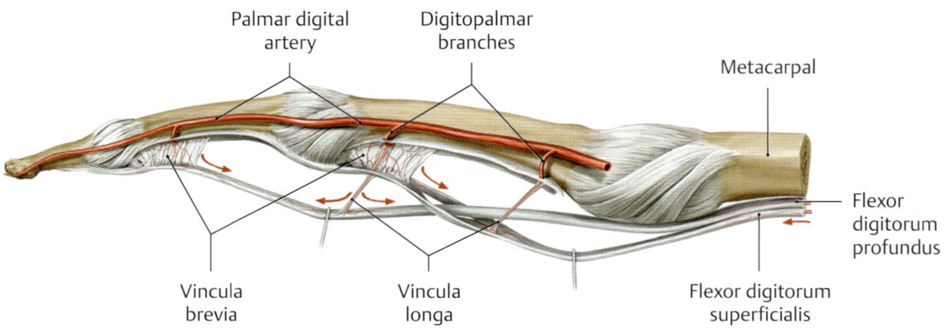

E Blood supply to the flexor tendons of the finger within the tendon sheath
Right middle finger, lateral view. The flexor tendons are supplied within their sheath by branches of the palmar digital arteries that are transmitted to the tendons through the mesotendons (vinculum longum and vinculum breve).

5.14 The Palm of the Hand: Vascular Supply

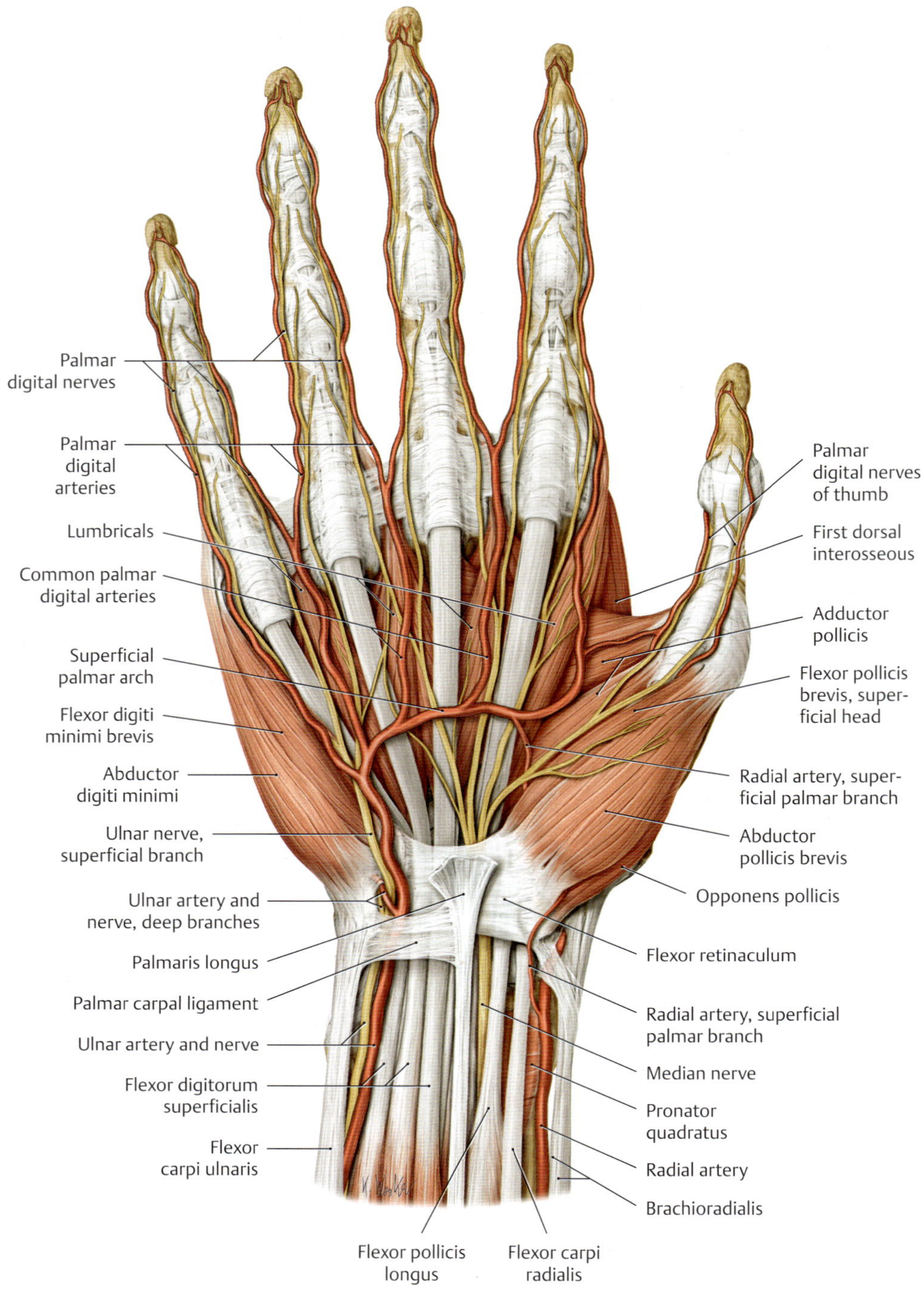

A The superficial palmar arch and its branches
Right hand, anterior (palmar) view. The palmar aponeurosis and other fasciae have been removed to demonstrate the *superficial palmar arch* (variants are shown in **B**).

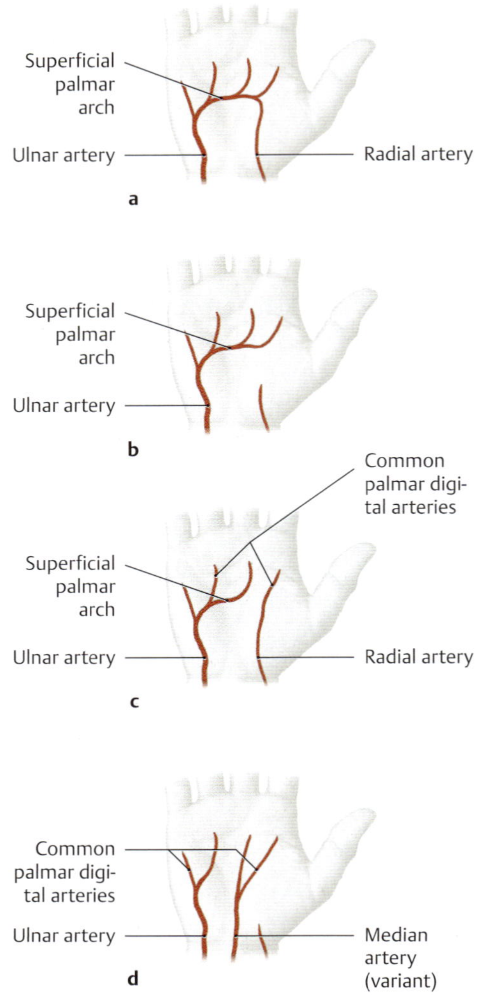

B The superficial palmar arch: normal anatomy and variants (after Lippert and Pabst)
Right hand, palmar view (no studies have been done on possible variants of the deep palmar arch).

a Normally (37% of cases), the radial artery and ulnar artery contribute equally to the superficial palmar arch.

b–d Variants:

b The palmar arch arises entirely from the ulnar artery (37% of cases).

c All of the common palmar digital arteries arise from the ulnar artery except for the first, which arises from the radial artery (13%).

d The ulnar artery and a median artery give off the common palmar digital arteries (very rare).

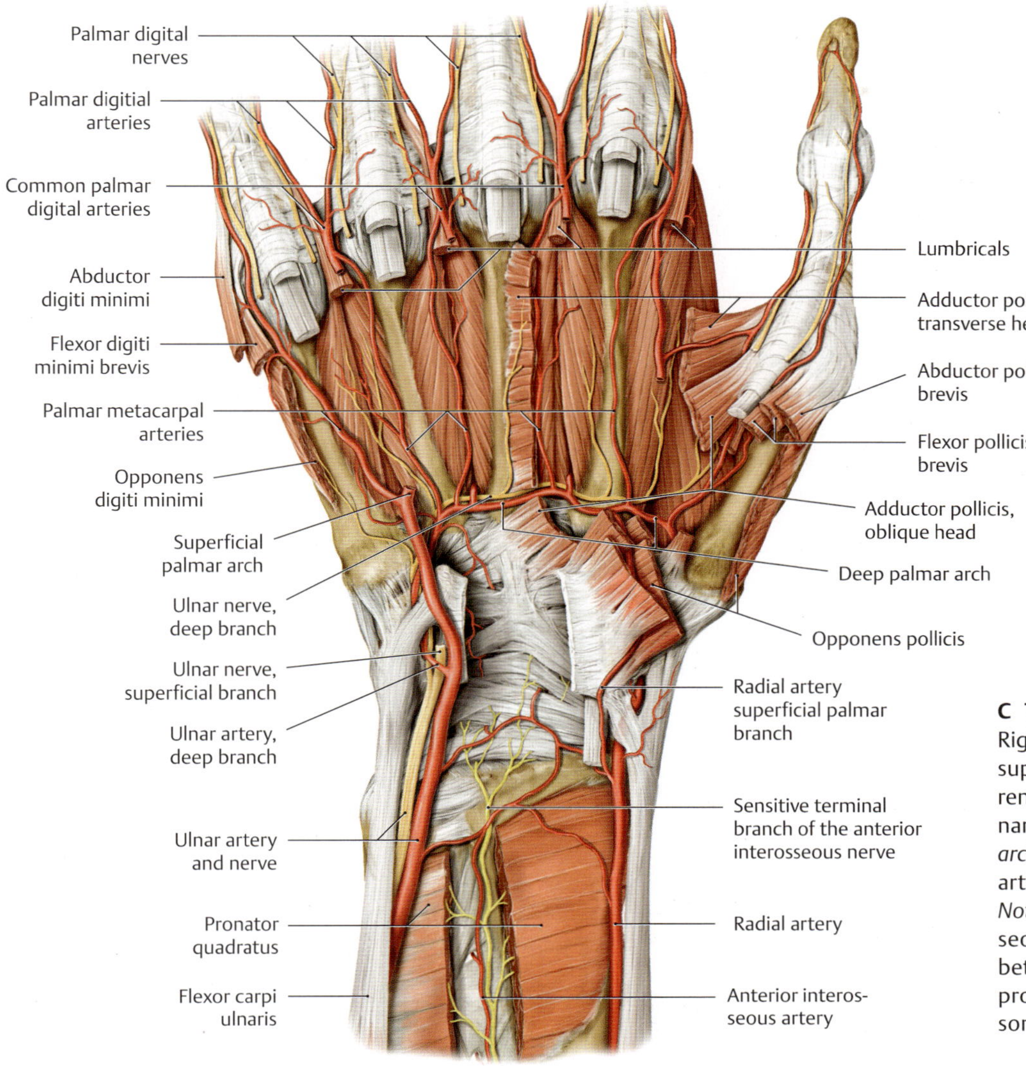

C The deep palmar arch and its branches
Right hand, anterior (palmar) view. The superficial and deep flexor tendons have been removed along with the thenar and hypothenar muscles to demonstrate the *deep palmar arch* as the terminal branch of the radial artery.
Note the end branch of the anterior interosseous antebrachial nerve, which extends between the interosseous membrane and pronator quadratus distally and supplies sensory innervation to the capsule of the wrists.

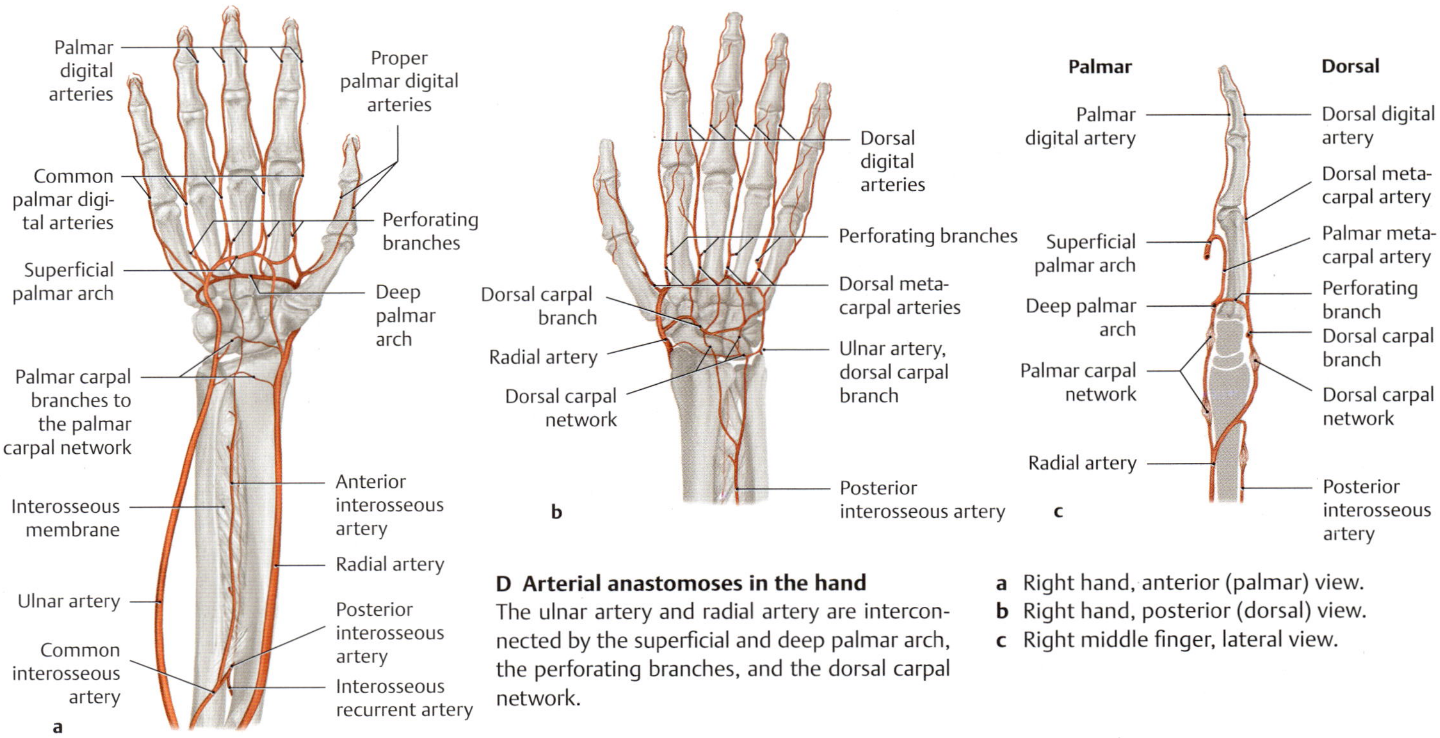

D Arterial anastomoses in the hand
The ulnar artery and radial artery are interconnected by the superficial and deep palmar arch, the perforating branches, and the dorsal carpal network.

a Right hand, anterior (palmar) view.
b Right hand, posterior (dorsal) view.
c Right middle finger, lateral view.

5.15 The Carpal Tunnel

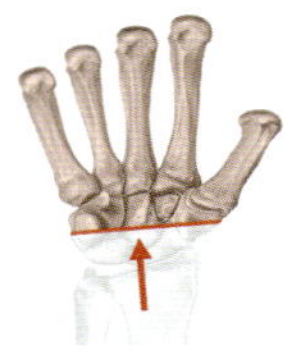

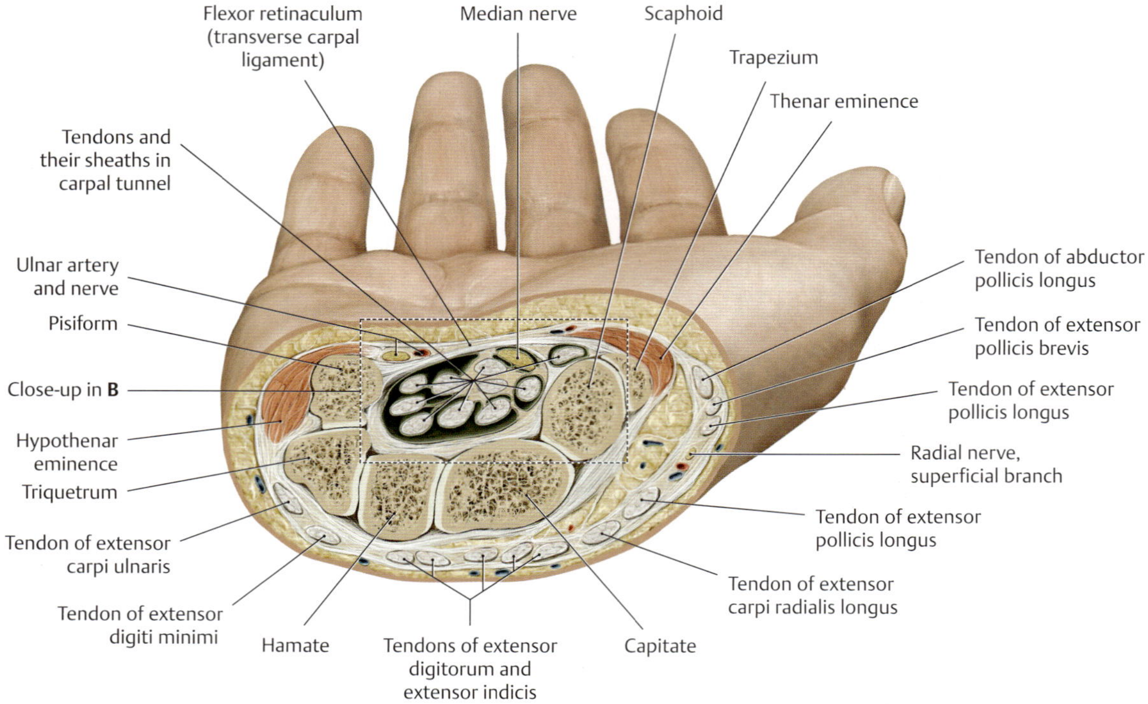

A Cross section through the right wrist (see also **B**)
Distal view. The *carpal tunnel* is a fibro-osseous canal (see p. 302) through which pass the median nerve and the tendons of insertion of the flexor digitorum superficialis, and flexor digitorum profundus, flexor pollicis longus. Its dorsal boundary is formed by the carpal sulcus on the anterior surface of the carpal bones, and its palmar boundary is the flexor retinaculum (also known clinically as the transverse carpal ligament). The ulnar artery and ulnar nerve pass through the *ulnar tunnel* on the palmar side of the flexor retinaculum (see p. 419).

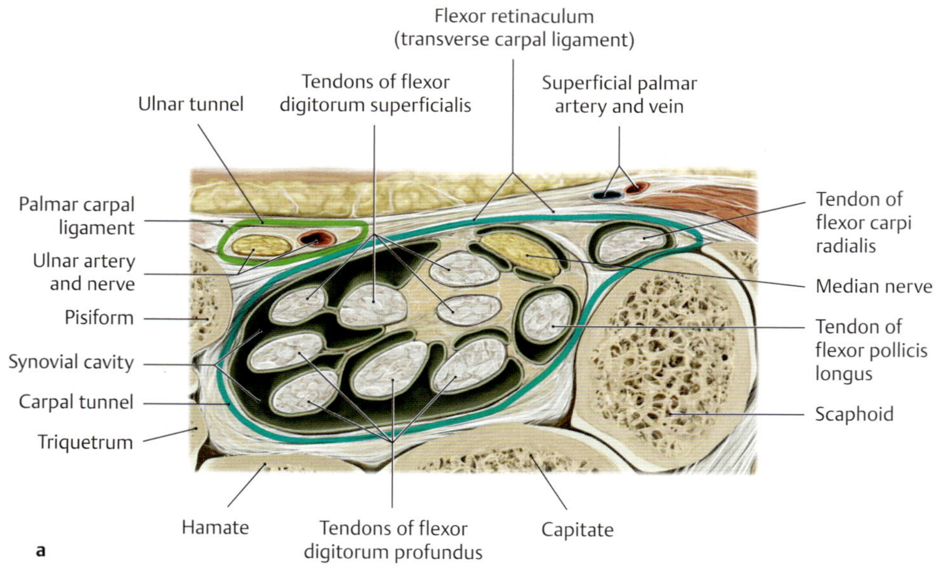

a

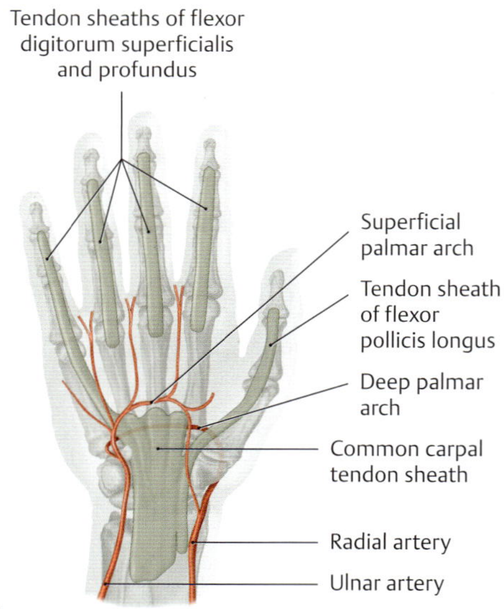

b

B Relationship of the palmar arches to the carpal and digital tendon sheaths

a Tendon sheaths in the carpal tunnel (detail from **A**). The long flexor tendons pass through the carpal tunnel, encased in their palmar tendon sheaths. The tendons of the flexor digitorum superficialis and profundus are contained in their own ulnar synovial sheath. Radial to this sheath is the tendon of the flexor pollicis longus. The common mesotendon of all the digital flexor tendons is attached to the radial and palmar walls of the carpal tunnel. The median nerve generally occupies a separate space just deep to the flexor retinaculum (the variable course of the tendon sheaths is described on p. 358).

b Relationship of the carpal and digital tendon sheaths to the palmar arches.

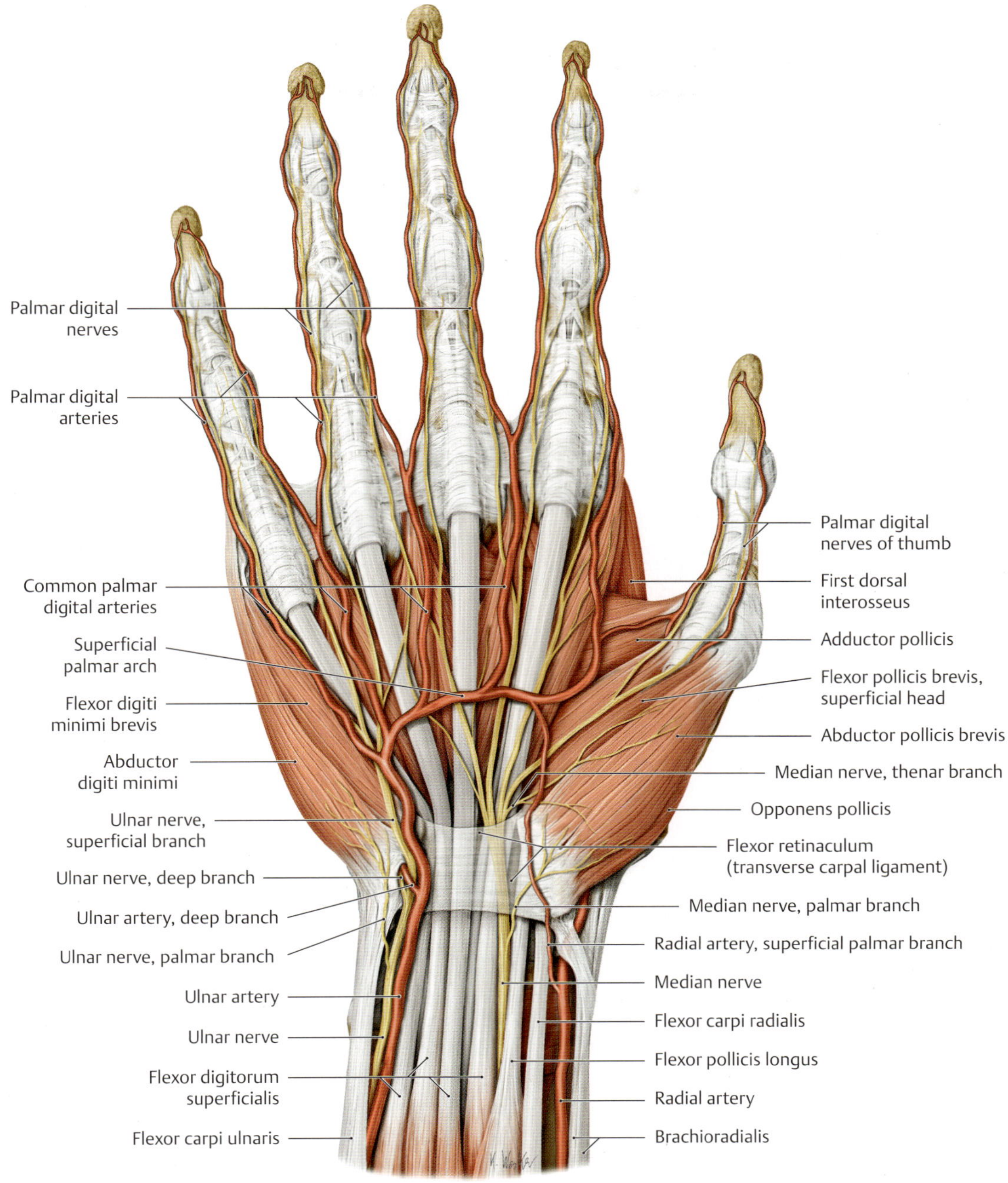

C View into the carpal tunnel of the right hand

Anterior (palmar) view. The flexor retinaculum is transparent to reveal the contents of the carpal tunnel. The ulnar tunnel has been opened to expose the ulnar artery and nerve.

Note the superficial course of the median nerve in the carpal tunnel and the origin of its thenar (motor) branch just distal to the flexor retinaculum (variants are shown in **D**). During surgical division of the flexor retinaculum for carpal tunnel syndrome, the hand surgeon must be aware of its variable course to avoid cutting the thenar branch.

The superficial palmar branch of the radial artery runs on the flexor retinaculum in the case shown here, but frequently it passes through the thenar muscles (see p. 334).

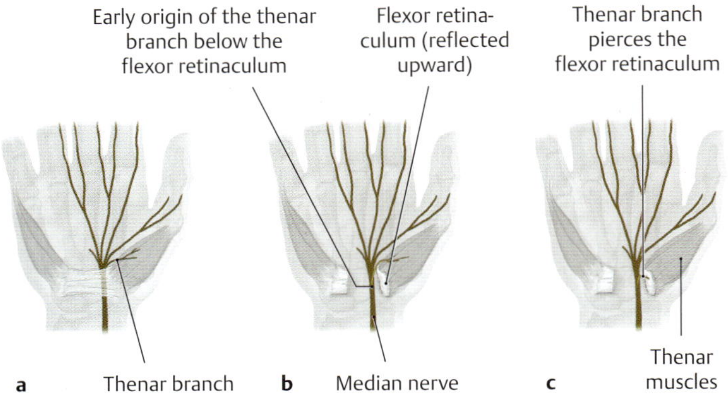

D Origin of the thenar (motor) branch of the median nerve: normal anatomy and variants (after Schmidt and Lanz)

- **a** Normally (46 % of cases) the median nerve gives off its thenar branch distal to the flexor retinaculum (transverse carpal ligament).
- **b, c Variants:**
- **b** The thenar branch has a subligamentatous origin and course (31% of cases).
- **c** The thenar branch pierces the flexor retinaculum (transverse carpal ligament; approximately 23% of all cases), making it vulnerable during surgical division of the retinaculum.

5.16 The Ulnar Tunnel and Anterior Carpal Region

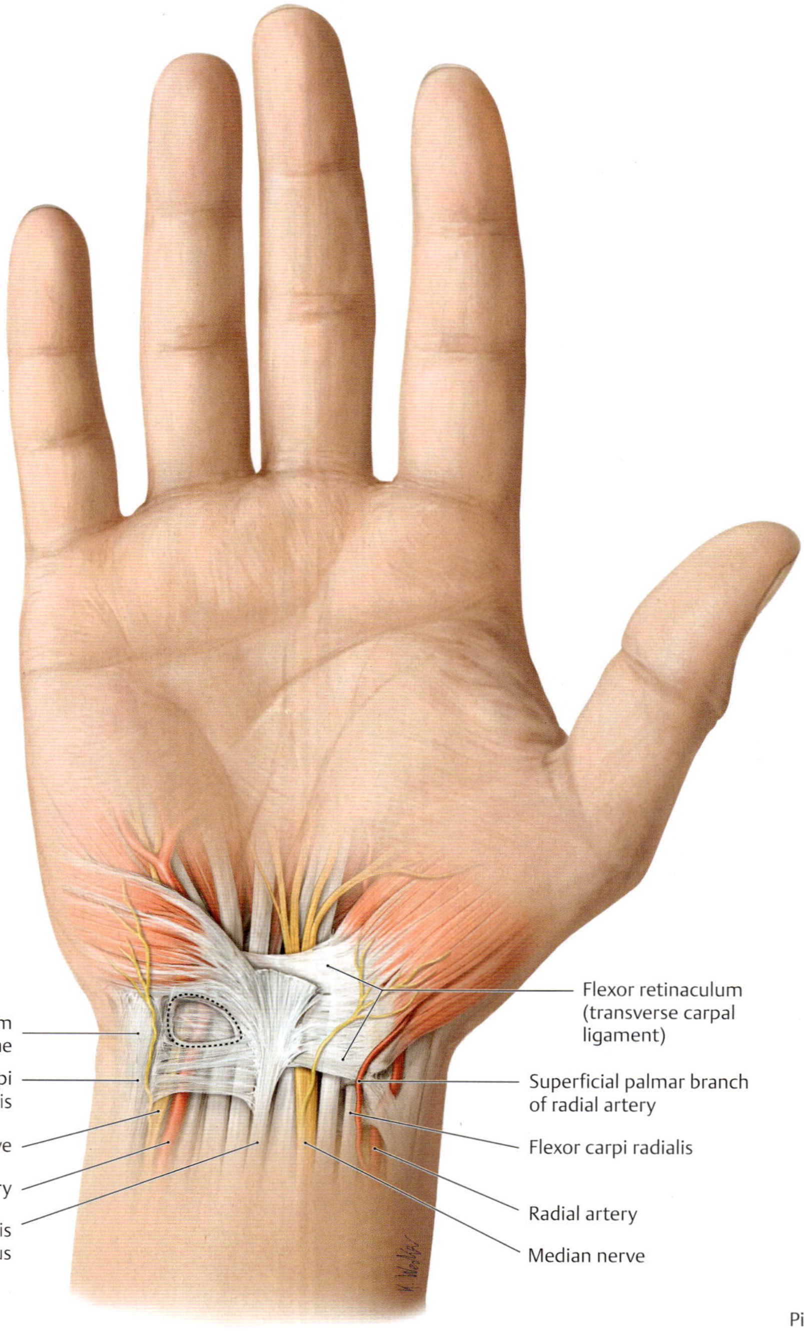

A Superficial structures of the anterior carpal region

Right hand, anterior view, demonstrating the superficial structures of the anterior carpal region. The distal border of the anterior carpal region, which is bounded distally by the flexor retinaculum (transverse carpal ligament). The tendons of insertion of the flexor carpi ulnaris, palmaris longus, and flexor carpi radialis in particular are clearly visible and palpable beneath the skin, especially when the fist is tightly clenched and the wrist is in slight flexion (see **B**). The tendon of the flexor carpi radialis is a useful landmark for locating the radial artery pulse. Almost all European specialist associations now recommend the radial artery as the puncture site of first choice for percutaneous coronary intervention. The radial artery has thus replaced the femoral artery as the classic access site for catheter systems although catheter insertion into this artery is less straightforward from a technical standpoint. The great advantage of the radial artery for catheterization is its superficial location. The risk of injuring important structures such as veins and nerves is slight, a consideration that particularly benefits obese patients and patients on anticoagulants. The flexor carpi ulnaris tendon is palpable proximally over the pisiform bone.

Note: Due to their superficial course, the median and ulnar nerves and the radial and ulnar arteries are particularly susceptible to injury from lacerations of the wrist.

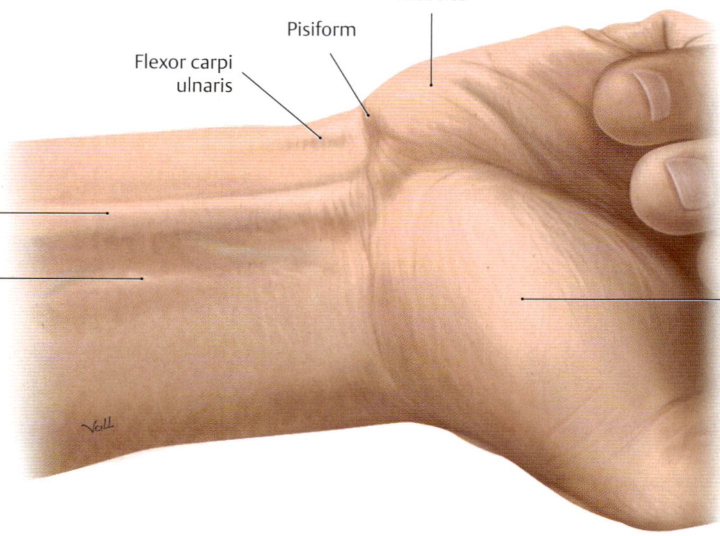

B Surface anatomy of the right wrist

Anterior view.

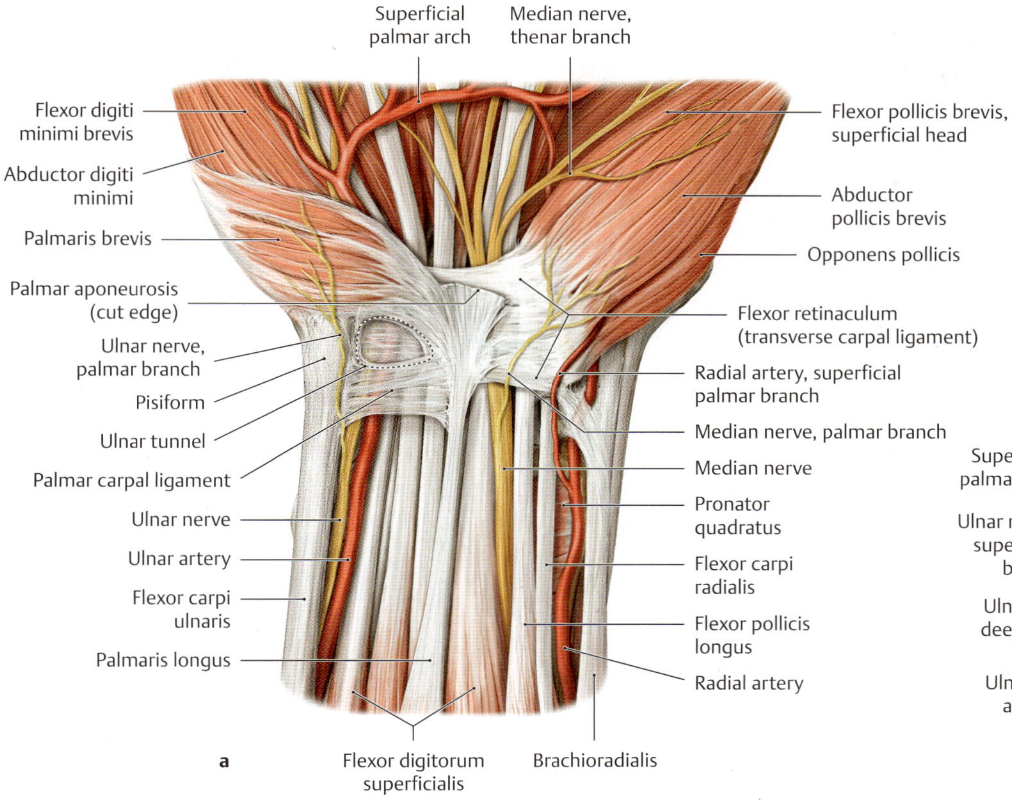

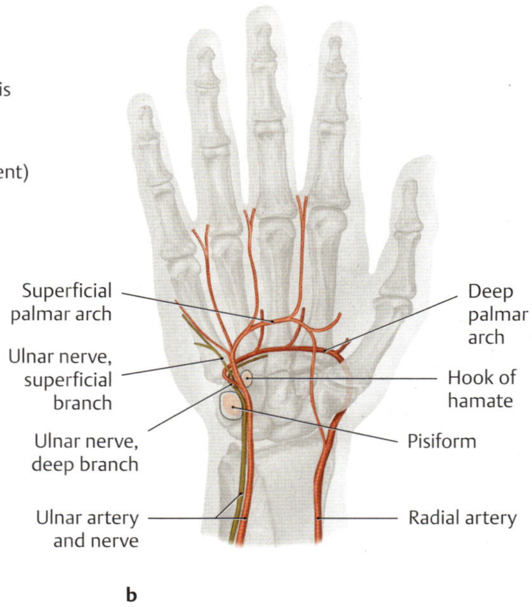

C Course of the ulnar artery and nerve in the ulnar tunnel and deep palm

Right hand, anterior view.

a The palmar aponeurosis and antebrachial fascia have been removed to demonstrate the course of the ulnar artery and nerve through the ulnar tunnel.

b Bony landmarks within the ulnar tunnel. The pisiform bone on the ulnar side of the wrist and the hook of the hamate bone, which lies more distally and radially, provide the bony landmarks between which the ulnar artery and nerve pass through the ulnar tunnel.

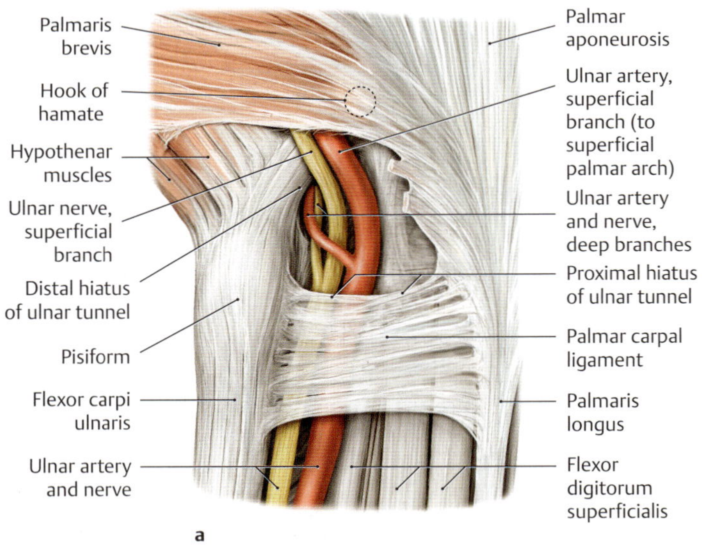

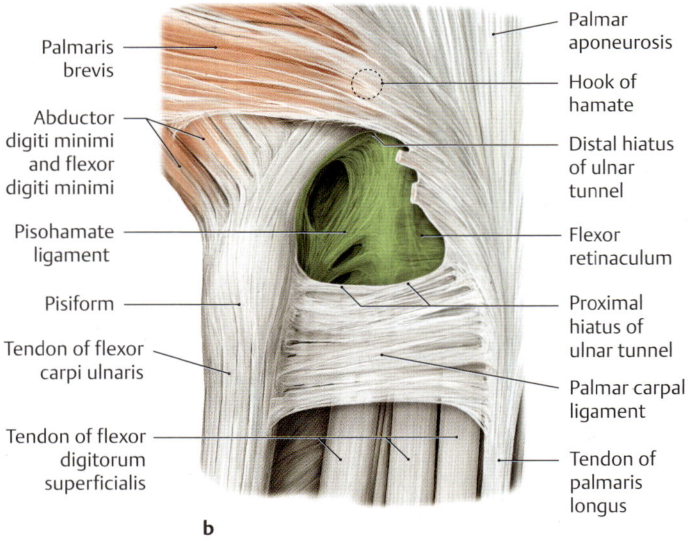

D Apertures and walls of the ulnar tunnel with (a) and without (b) the nerves and vessels

Right hand, anterior view. The *palmar* roof of the ulnar tunnel is formed by skin and subcutaneous fat, the palmar carpal ligament (proximal), and the palmaris brevis muscle (distal). The ulnar tunnel is bounded *dorsally* by the flexor retinaculum (transverse carpal ligament) and the pisohamate ligament. The *entrance* to the tunnel begins at the level of the pisiform bone below the palmar carpal ligament (proximal hiatus). The *outlet* is at the level of the hook of the hamate, marked by a taut, transverse, crescent-shaped tendinous arch between the pisiform and the hook of the hamate (distal hiatus), the latter giving attachment to the flexor digiti minimi brevis. The *deep branches* of the ulnar artery and nerve reach the central compartment of the palm on the pisohamate ligament, passing deep to the tendinous arch. The *superficial branches* of the artery and nerve run distally above the tendinous arch, passing deep to the palmaris brevis.

D Lower Limb

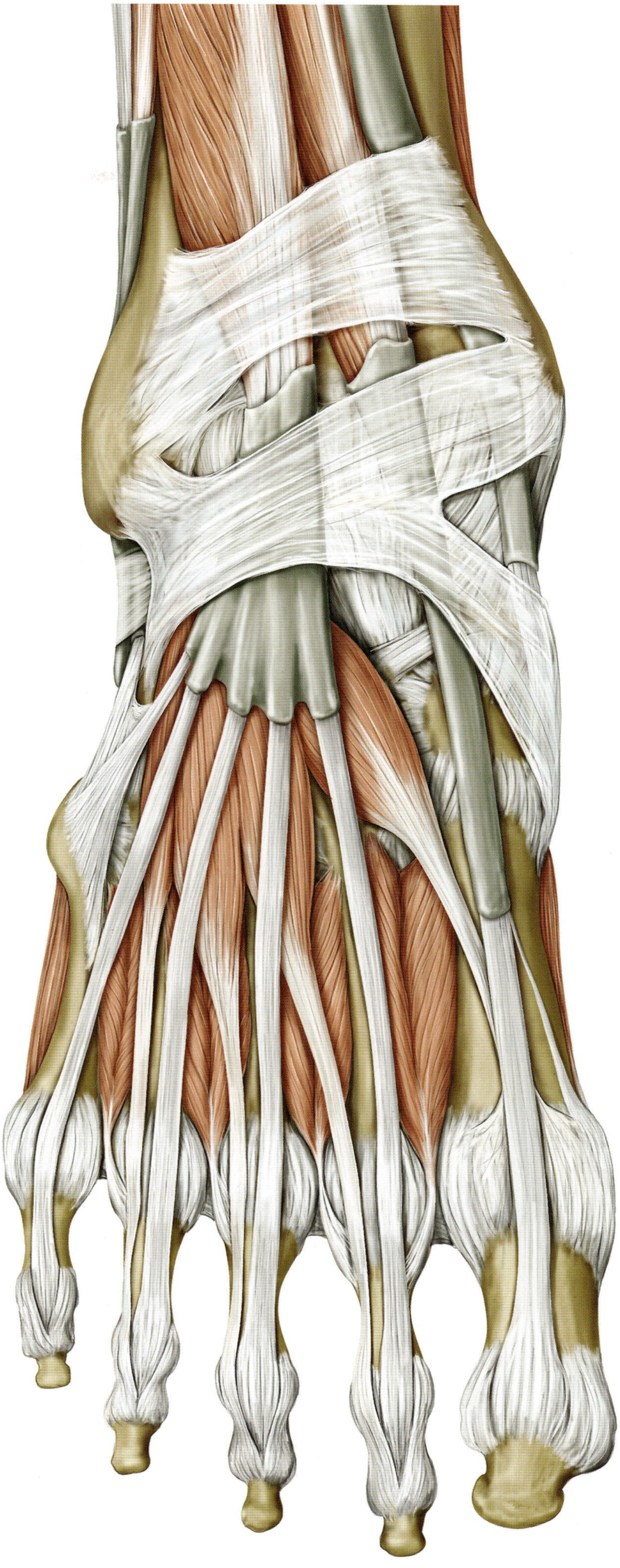

1.1 The Lower Limb: General Aspects

A Unique features and specialized function of the human lower limb

The evolution of the lower limb into a mechanism specifically adapted for bipedal locomotion, along with the specialization of the upper limb for visually guided manipulation, is a distinctive feature of the anatomy of the human primate. The uniquely human conformation of shapes and proportions is the end result of a process that rearranged the primate center of gravity and the positions of internal organs, dramatically altering the form and biomechanics of the trunk to produce a progressively more efficient bipedal gait. Other primates have the capacity to assume an erect body posture and to walk upright, but only for short periods and at a much greater relative expenditure of energy. The habitual upright gait of humans has been achieved through a series of anatomic adaptations of the musculoskeletal system. The most critical of these adaptations occurred in the vertebral column and pelvis. The design of the human vertebral column differs markedly from that of other primates: the simple "arch and cord" construction of the chimpanzee spine has been abandoned in favor of the human double-S-shaped curve, which allows the human axial skeleton to act as a shock-absorbing spring (see p. 111), while shifting the entire weight of the trunk over the load-bearing surface of the feet. This shift to an upright posture has imposed the full weight of the abdominal viscera upon the pelvis. Concomitantly, the iliac wings of the pelvis have spread farther apart, and the sacrum has broadened, to generate a structure in humans that is now specialized for bearing the load of the viscera. The efficiency of upright gait has been improved further by stabilization of the pelvis and secure anchoring to the spine via the sacrum. The unique proportions of the human lower limb provide a dramatic demonstration of the extent of this specialization. Because their function is more exclusively directed toward support and locomotion, the legs are exceptionally long and powerful in humans. While the leg length is 111% of trunk length in orangutans and 128% in chimpanzees, it measures 171% of trunk length in humans. The specialization of the human lower limb for bipedal gait is also reflected in the substantial changes in function of certain muscles, particularly the gluteal muscles, the knee-joint extensors, and the muscles of the calf.

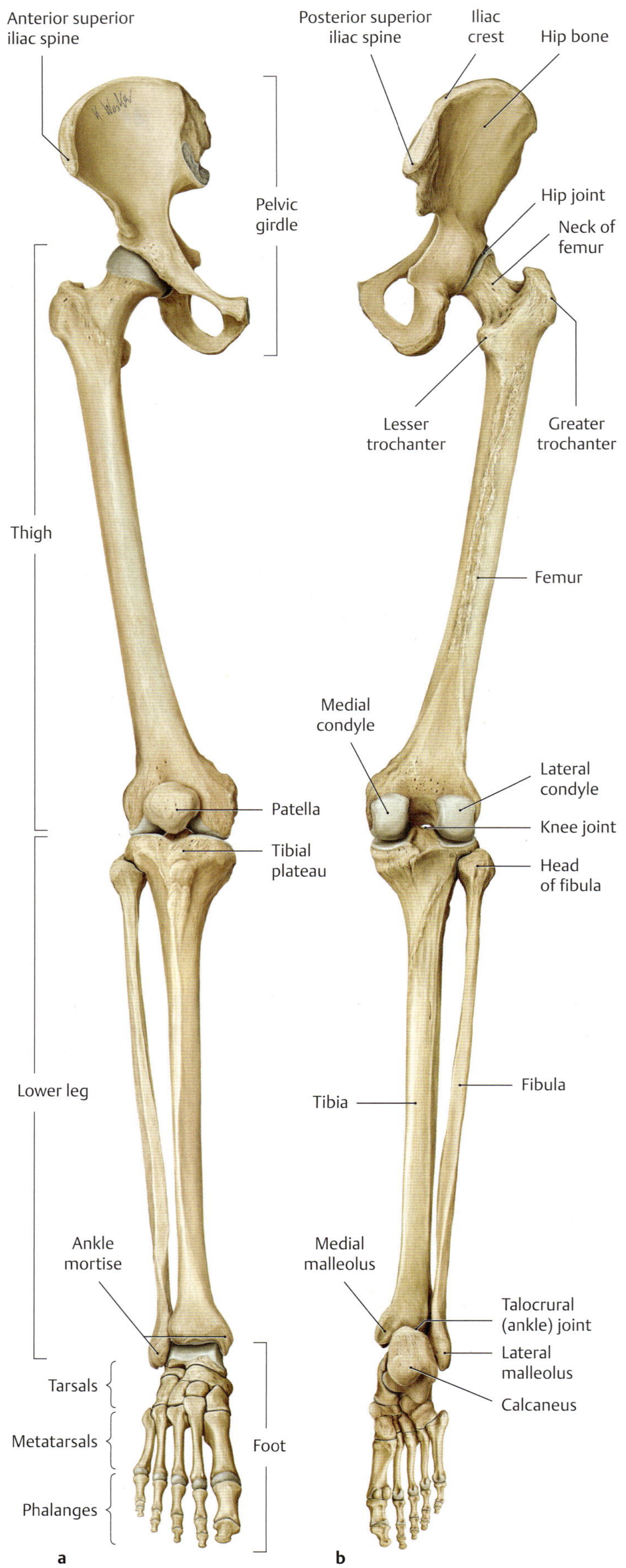

B Overview of the skeleton of the lower limb

a Right lower limb, anterior view.

b Right lower limb, posterior view (the foot is in maximum plantar flexion in both views).

As in the upper limb, the skeleton of the lower limb consists of a limb girdle and the attached free limb.

- The pelvic girdle in adults is formed by the paired hip bones (ossa coxae, innominate bones). They differ from the shoulder girdle in that they are firmly integrated into the axial skeleton through the sacroiliac joints (see p. 150). The two hip bones combine with the sacrum and pubic symphysis to form the *pelvic ring* (see p. 427).
- The free limb consists of the thigh (femur), the leg (tibia and fibula), and the foot. It is connected to the pelvic girdle by the hip joint.

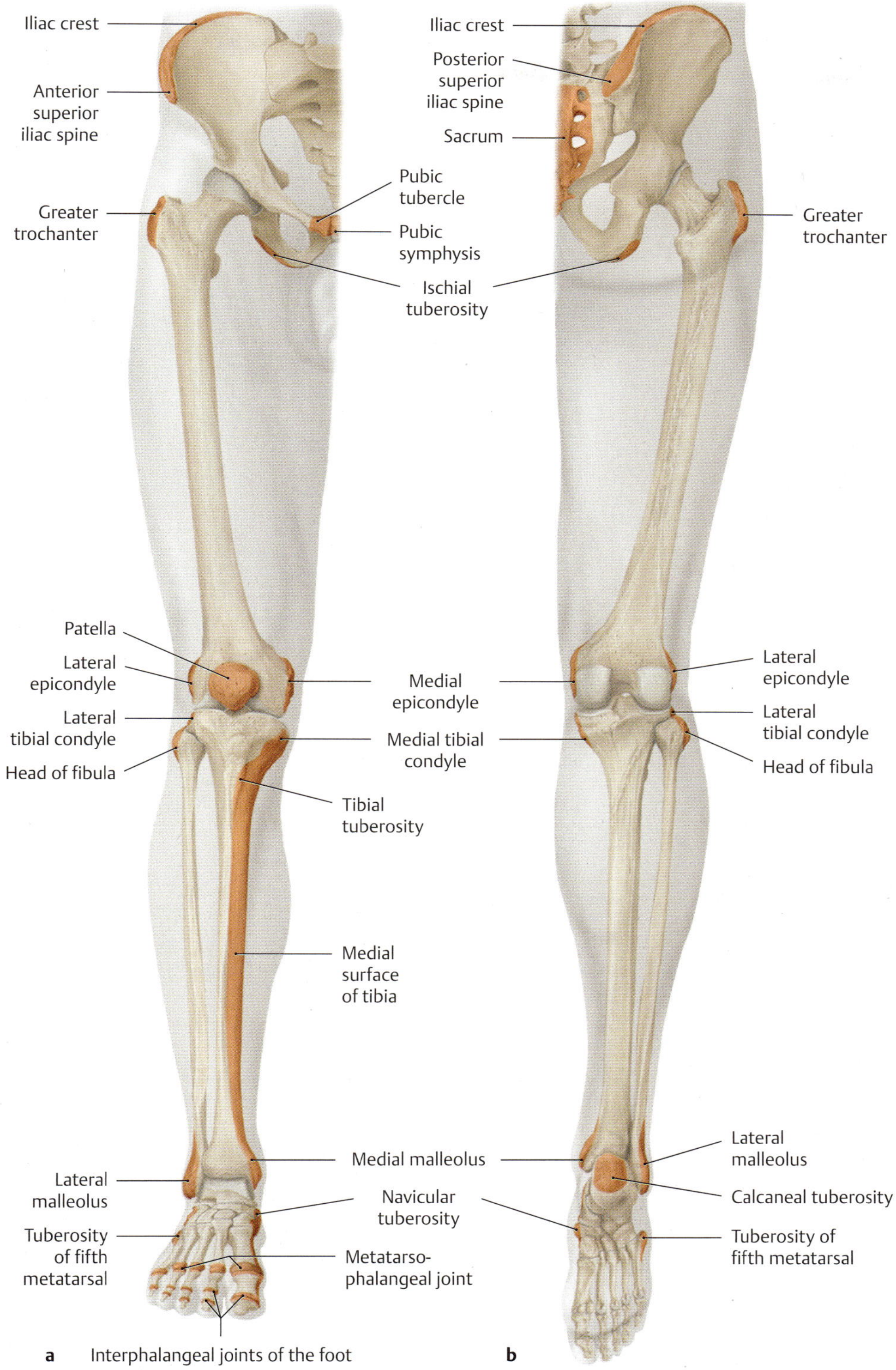

C Palpable bony prominences of the right lower limb

a Anterior view, **b** posterior view.

Almost all the skeletal elements of the lower limb have bony prominences, margins, or surfaces (e.g., the medial tibial surface) that can be palpated through the skin and soft tissues. The only exceptions are structures that are largely covered by muscle, such as the hip joint, the neck and shaft of the femur, and large portions of the fibular shaft. Several standard anatomic landmarks have been defined in the lower limb for use in measuring the length of the leg and certain skeletal elements. They are the anterior superior iliac spine, the greater trochanter of the femur, the medial joint space of the knee (superior margin of the medial tibial condyle), and the medial malleolus. The clinical evaluation of leg length discrepancy is important because "true" shortening of the leg (a disparity of anatomic leg lengths), as well as functional leg shortening (e.g., due to muscle contractures), can lead to *pelvic tilt* and scoliotic deformity of the spine (see p. 141).

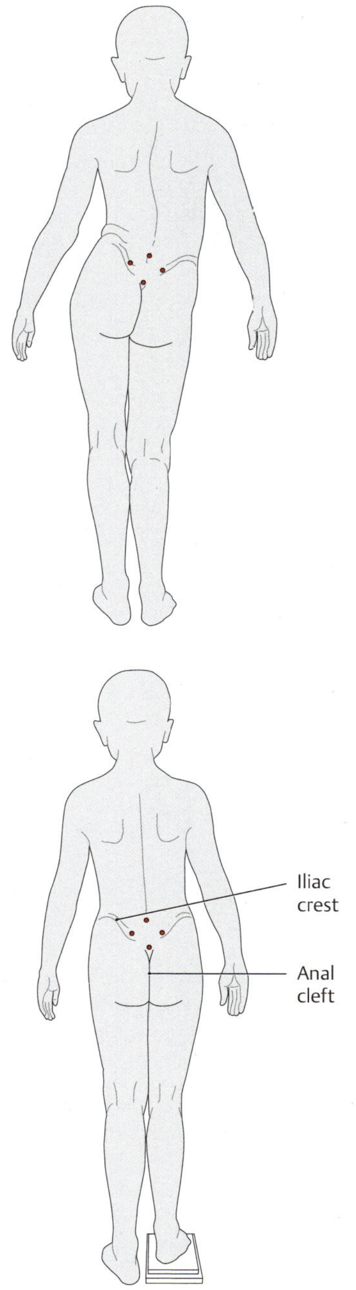

D Leg length measurement in the standing position

Leg length discrepancy can be measured with reasonable accuracy in the standing patient by placing wooden blocks of known thickness (0.5 cm, 1 cm, 2 cm) beneath the foot of the shorter leg until the pelvis is horizontal. Horizontal position is confirmed when noting that both iliac crests are at the same level when palpated from behind and the anal fissure is vertical. If the pelvis cannot be leveled by placing blocks under the apparently shorter limb, then a "functional" leg length discrepancy is present rather than a "true" discrepancy. Most cases of this kind are caused by a fixed pelvic tilt secondary to a hip joint contracture or scoliosis. The measured leg lengths in these cases may actually be equal, and the pelvic tilt only mimics a length discrepancy.

1.2 The Anatomic and Mechanical Axes of the Lower Limb

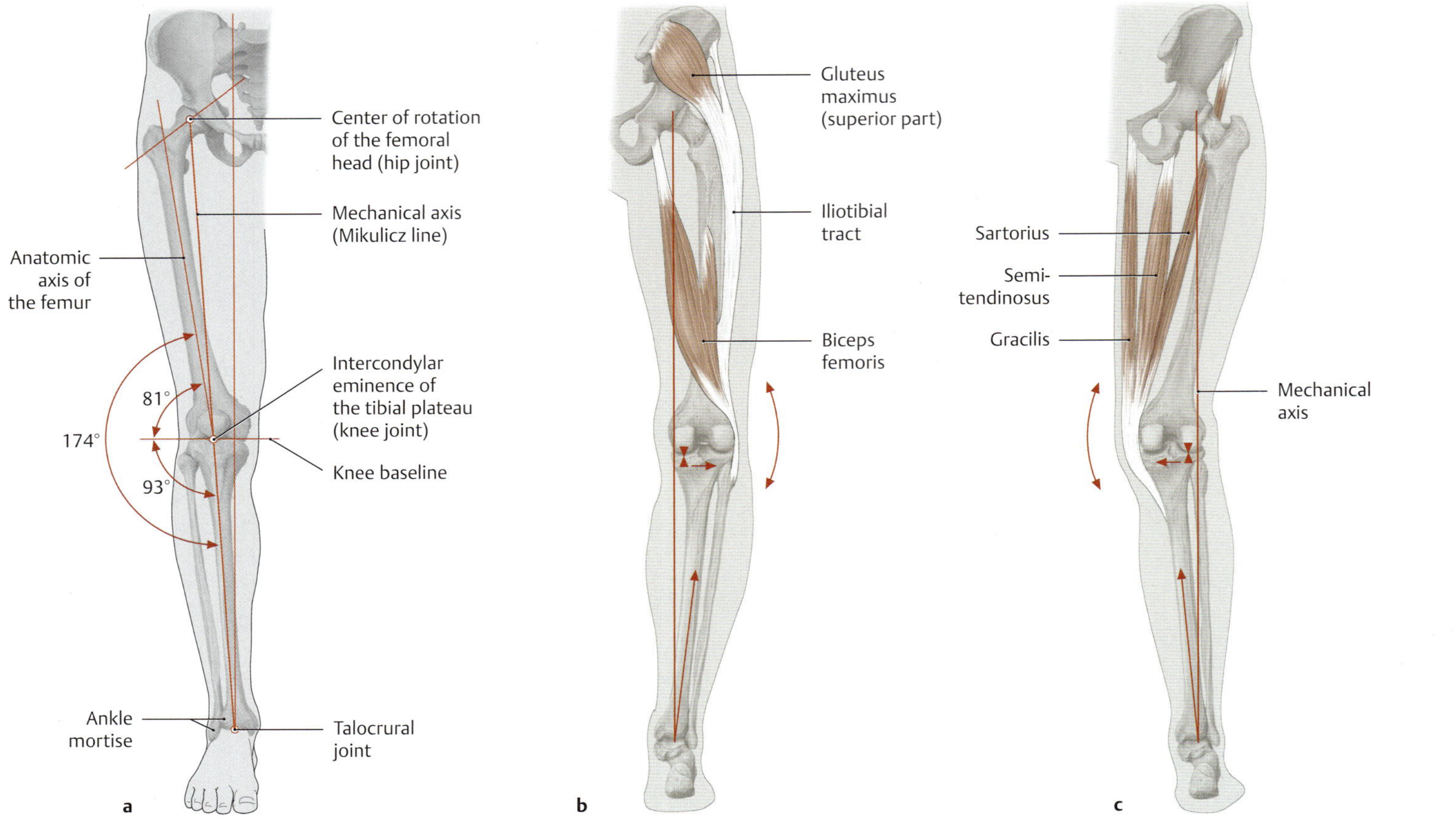

A The mechanical axis of the leg (the Mikulicz line)

a Normal mechanical axis, anterior view.
b Mechanical axis in genu varum, posterior view.
c Mechanical axis in genu valgum, posterior view.

In an individual with normal axial alignment, the large joints of the lower limb (the hip, knee, and ankle) lie on a straight line that represents the mechanical longitudinal axis of the leg (the *Mikulicz line*). This mechanical axis extends from the center of rotation of the femoral head through the intercondylar eminence of the tibial plateau and down through the center of the ankle *mortise* (the pocket created by the fibula and tibia for the talus in the ankle joint, from Arabic, *murtazz*, fastened). While the mechanical axis and anatomic axis coincide in the *tibial shaft*, the anatomic and mechanical axes of the *femoral shaft* diverge at a 6° angle. Thus, the longitudinal anatomic axes of the femur and tibia do not lie on a straight line but form a laterally open angle of 174° at the level of the knee joint in the coronal plane (the *femorotibial angle*). In genu varum (**b**) the center of the knee joint is lateral to the mechanical axis, and in genu valgum (**c**) it is medial to the mechanical axis. Both conditions impose abnormal, unbalanced loads on the joints (see **B**) that gradually cause degenerative changes to develop in the bone and cartilage (osteoarthritis of the knee) accompanied by stretching of the associated joint capsule, ligaments, and muscles. In genu varum (**b**), for example, the medial joint complex of the knee is subjected to abnormal pressure, while the lateral joint structures (e.g., the lateral collateral ligament), iliotibial tract, and biceps femoris muscle are subjected to abnormal tension. Genu varum also places greater stress on the lateral border of the foot, resulting in a fallen pedal arch.

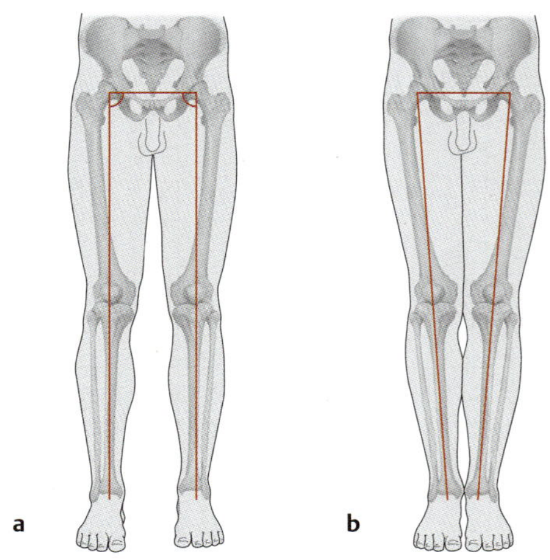

B Position of the mechanical axes with the feet slightly apart and together

Anterior view.

a In upright stance with the feet placed slightly apart, the mechanical axis runs almost vertically through the center of the three large joints.
b The legs are generally considered "straight" if, when the feet are together, the opposing medial malleoli and knees are touching. Accordingly, the intercondylar distance and the intermalleolar distance between the legs provide an index for the measurement of genu varum and genu valgum. When this stance is attempted, an intercondylar distance greater than 3 cm or an intermalleolar distance greater than 5 cm is considered abnormal (see **C**).

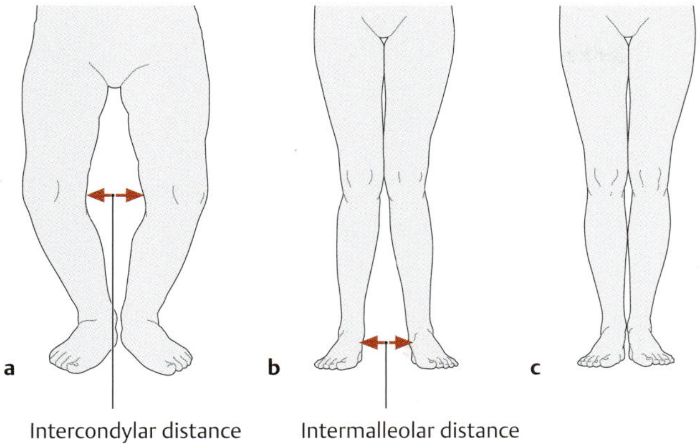

C The normal leg axes at different ages
a Infant, **b** small child, **c** school-age child.
Up to about 20° of genu varum is considered normal during the first year of life. Up to about 10° of genu valgum is also considered normal through 2 years of age. By the time the child enters school, the legs are essentially straight as a result of musculoskeletal growth.

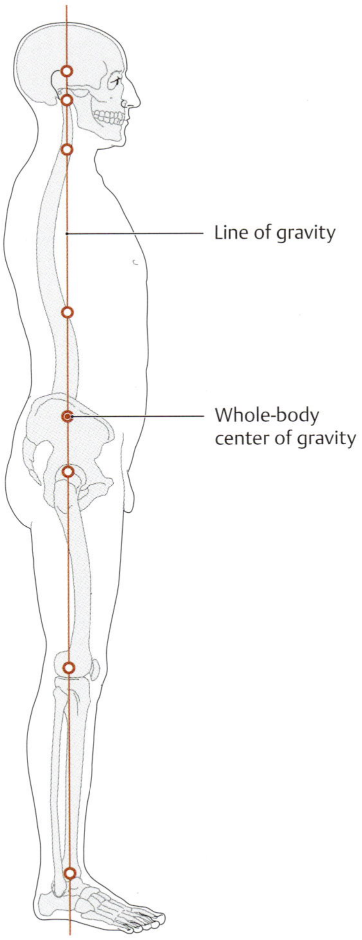

D Normal anatomic position in relation to the line of gravity
Right lateral view. The *line of gravity* runs vertically from the whole-body center of gravity to the ground. In normal upright humans, it intersects the external auditory canal, the dens of the axis (odontoid process of C2), the inflection points between the normal curves in the vertebral column (between the cervical and thoracic curves, and thoracic and lumbar curves), the whole-body center of gravity, and the hip, knee, and ankle joints. Chronic deviation of any reference point from this line imposes abnormal stresses on different clusters of musculoskeletal elements.

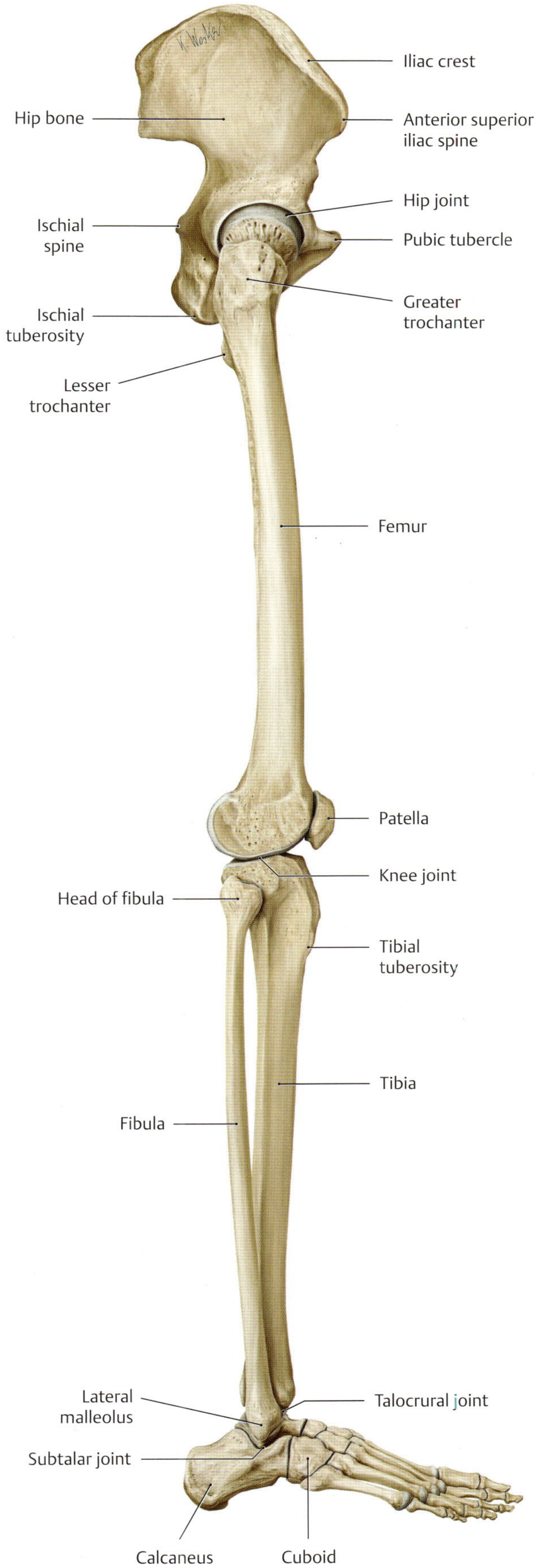

E Skeleton of the right lower limb
Right lateral view.

1.3 The Bones of the Pelvic Girdle

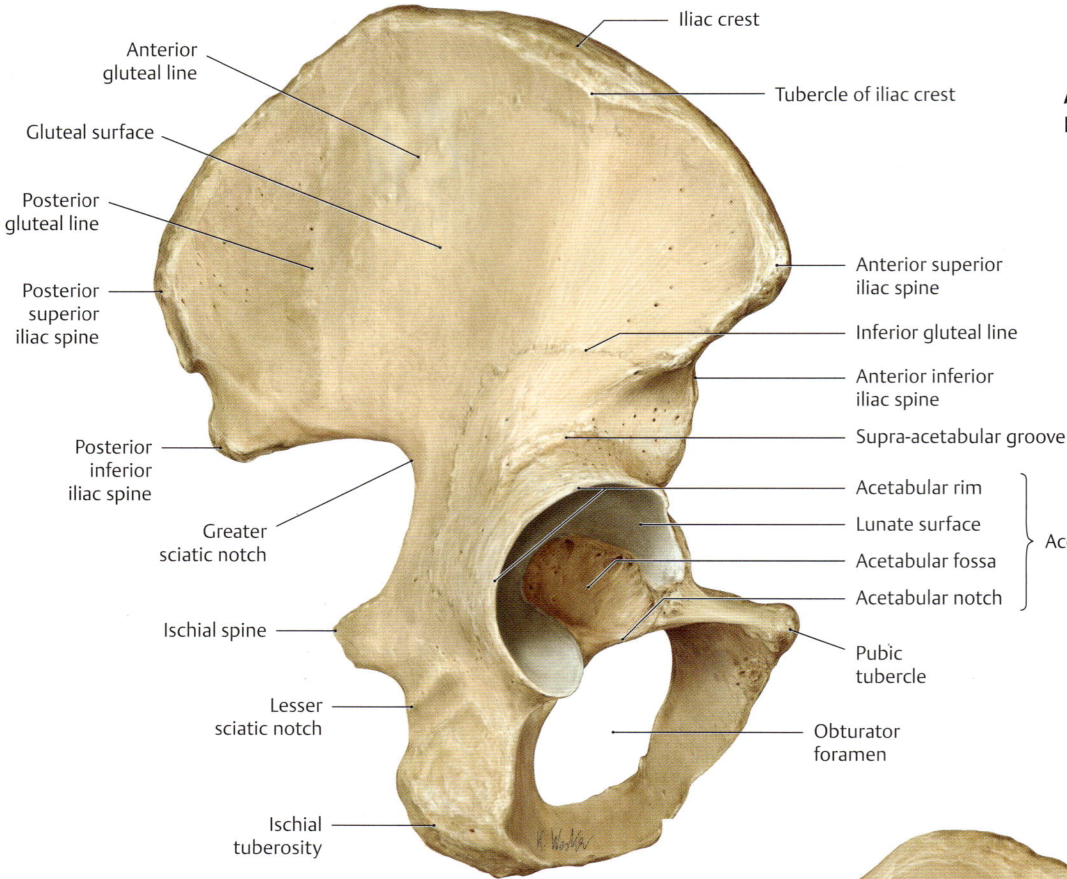

A The right hip bone
Lateral view.

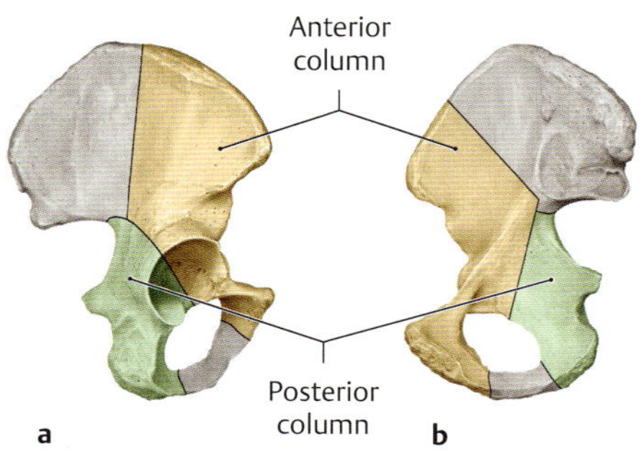

C Column principle of the hip bone
a Lateral view, **b** medial view.
Major external trauma (e.g., traffic accidents) often result in fractures of the pelvis or acetabulum. In order to classify such fractures, the hip bone or acetabulum (according to Letournel) is divided into a short posterior column and a significantly larger anterior column. The major lines of force of the pelvis run through these two columns. In case of major trauma, the impact is first transmitted through the femoral neck to the acetabulum and then to the columns. The type of fracture depends on the position of the femoral neck at the time of forceful impact (compare p. 447).

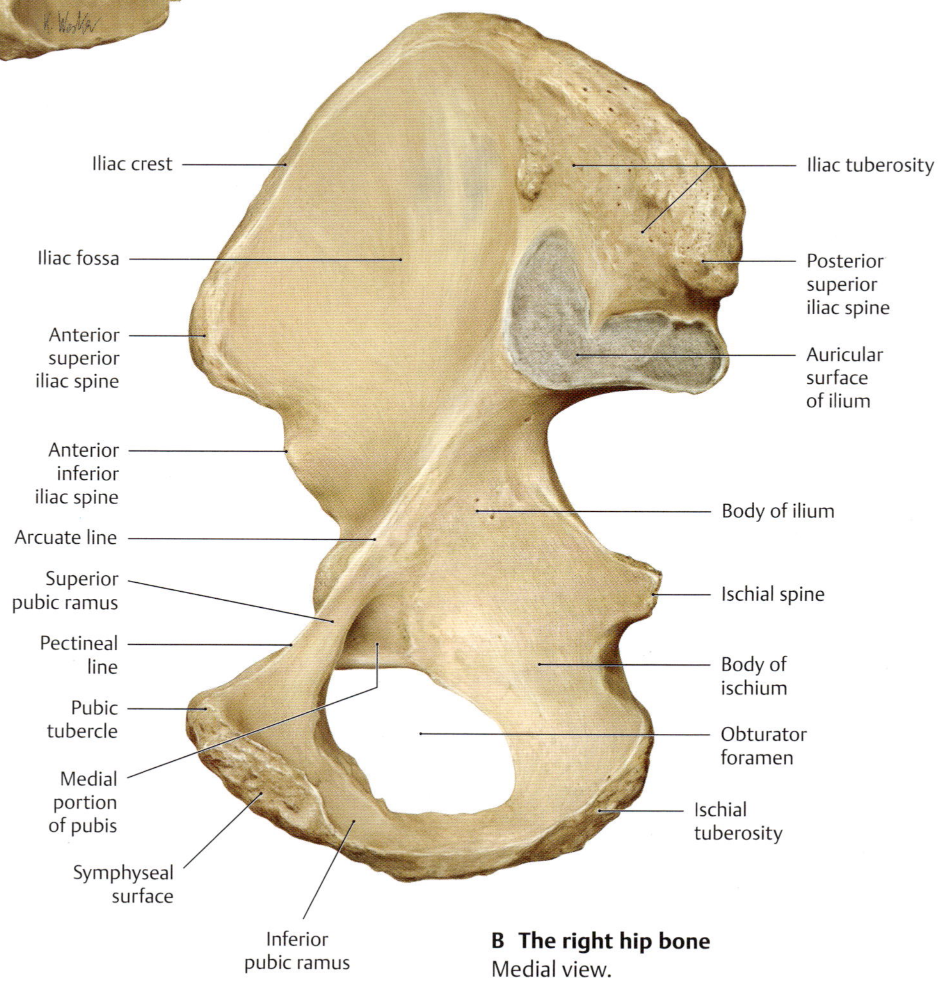

B The right hip bone
Medial view.

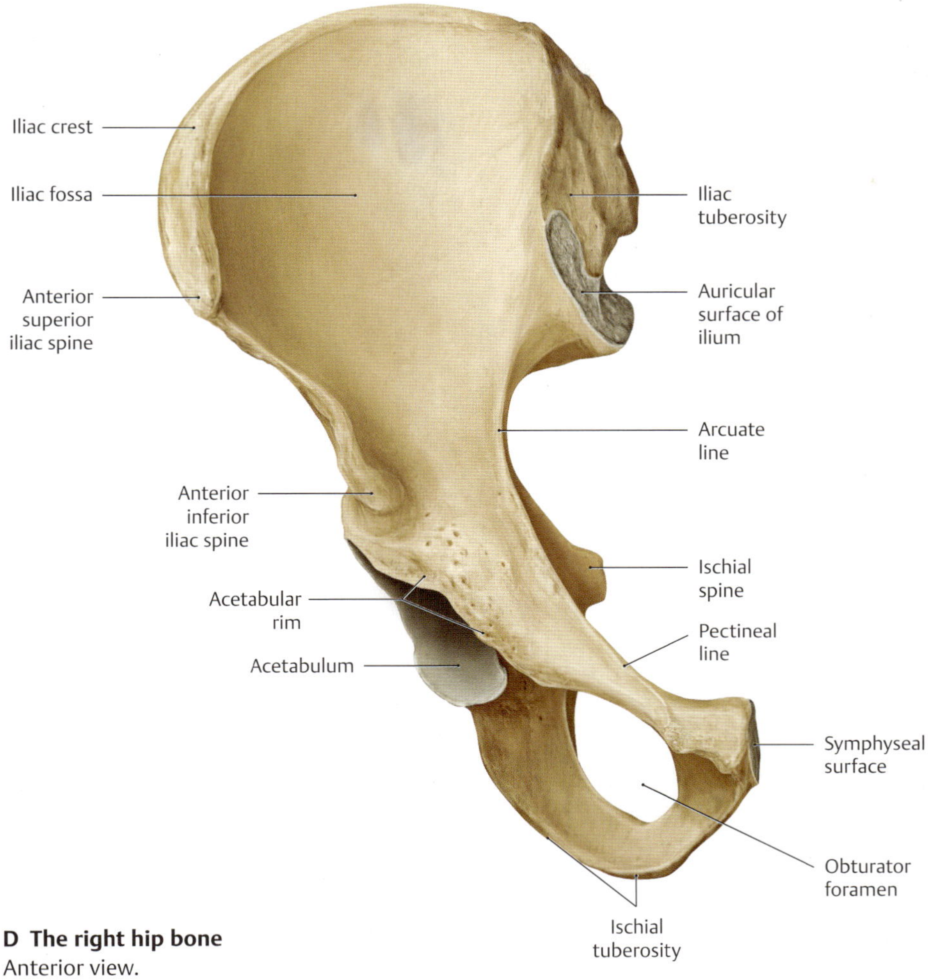

D The right hip bone
Anterior view.

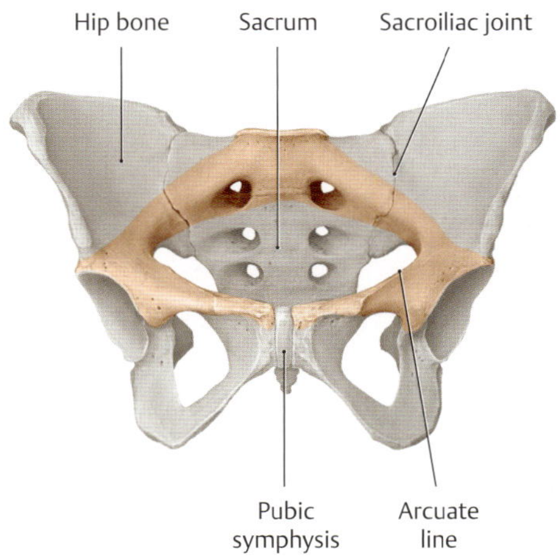

E The pelvic girdle and pelvic ring
Anterior view. The paired hip bones that make up the pelvic girdle are connected to each other at the cartilaginous pubic symphysis and to the sacrum at the sacroiliac joints (see p. 150). This creates a stable ring, the bony pelvic ring (shaded in red), that permits very little motion. This stability throughout the pelvic ring is an important prerequisite for the transfer of trunk loads to the lower limb necessary for normal gait.

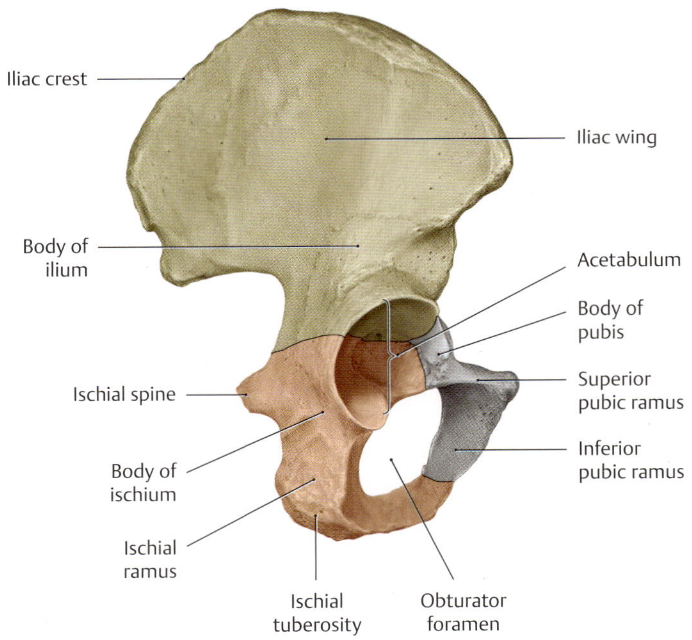

F The triradiate cartilage of a right hip bone: the junction of the ilium, ischium, and pubis
Lateral view.

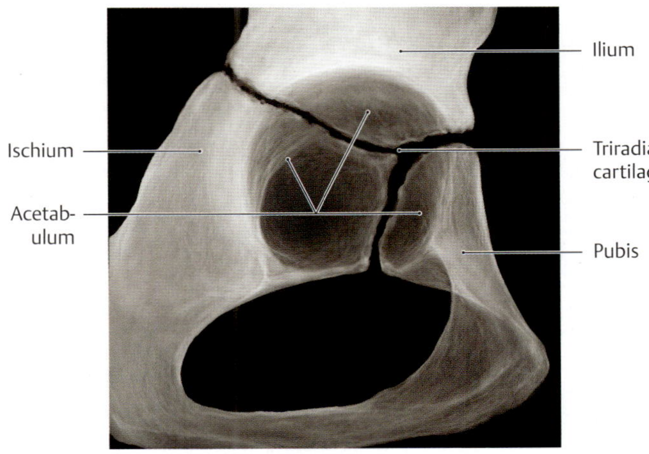

G Schematic radiograph of the right acetabulum of a child
Lateral view (lateral projection). The bony elements of the hip bone come together in the acetabulum, with the ilium and ischium each comprising two fifths of the acetabulum and the pubis one fifth. Definitive fusion of the Y-shaped growth plate (triradiate cartilage) occurs between the 14th and 16th years of life.

1.4 The Femur: Importance of the Femoral Neck Angle

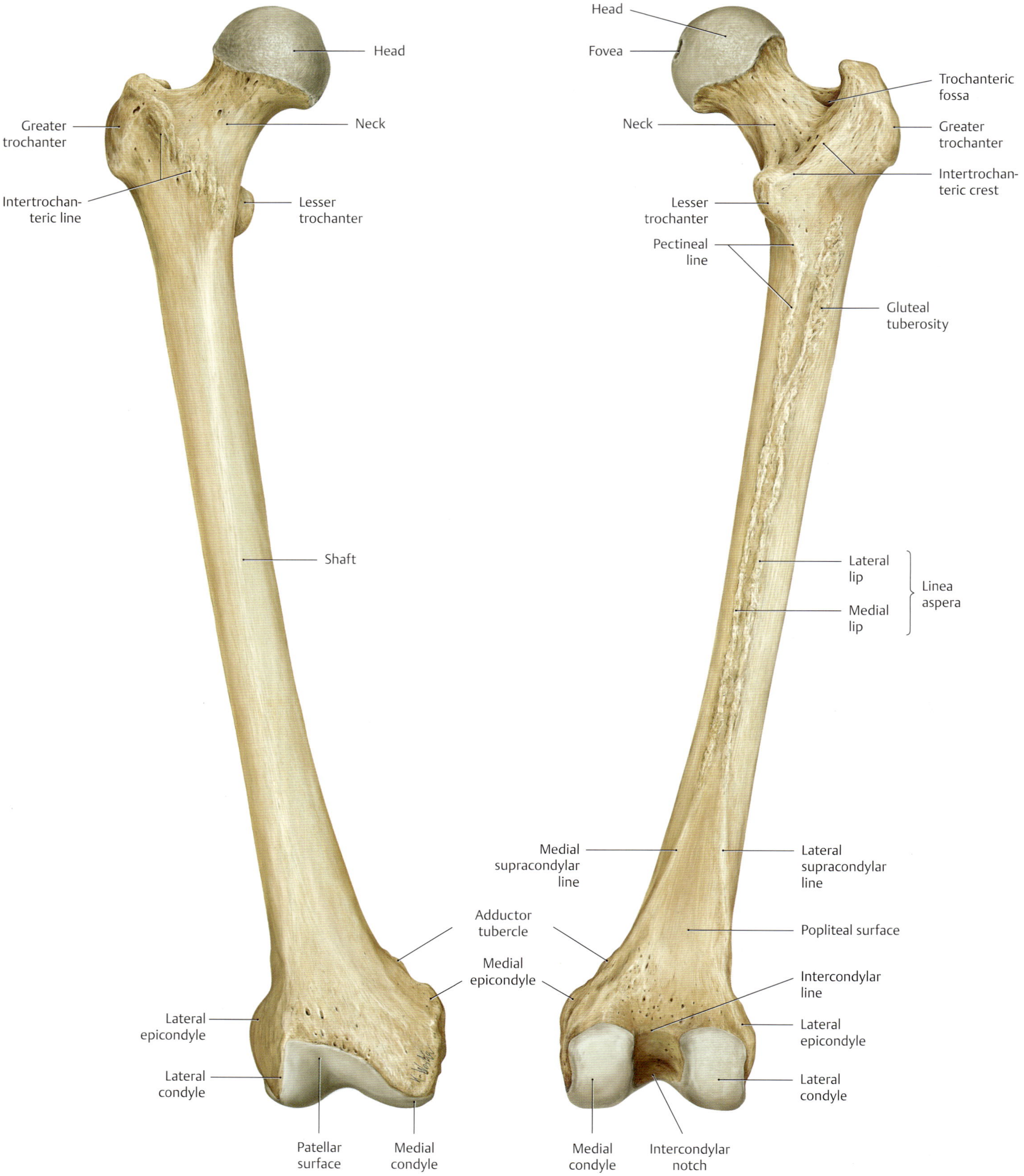

A The right femur
Anterior view.

B The right femur
Posterior view.

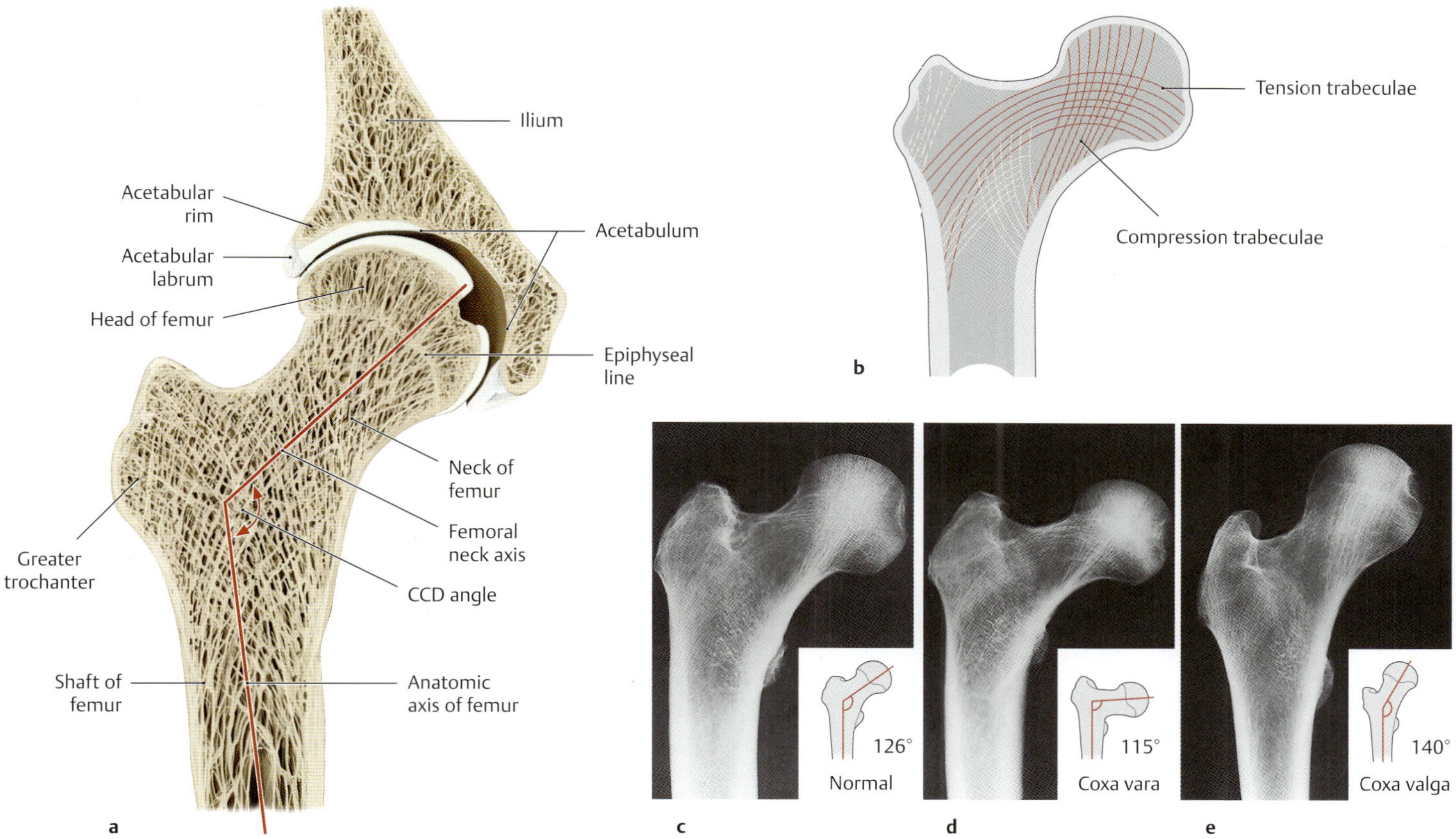

C The arrangement and prominence of tension trabeculae and compression trabeculae as a function of the femoral neck angle
Right femur, anterior view.

a Coronal section through the right hip joint at the level of the fovea on the femoral head. The angle between the longitudinal axis of the femoral neck and the axis of the femoral shaft is called the femoral neck angle or CCD angle (centrum-collum-diaphysis angle). This angle normally measures approximately 126° in adults and 150° in newborns. It decreases continually during growth due to the constant bone remodeling that occurs in response to the changing stress patterns across the hip.

b The trabecular pattern associated with a normal femoral neck angle.
c–e Radiographs in the sagittal projection.
c Normal femoral neck angle with a normal bending load.
d A *decreased* femoral neck angle (coxa vara) leads to a greater bending load with higher tensile stresses, thereby stimulating the formation of more tension trabeculae.
e An *increased* femoral neck angle (coxa valga) leads to a greater pressure load with higher compressive stresses, stimulating the formation of more compression trabeculae.

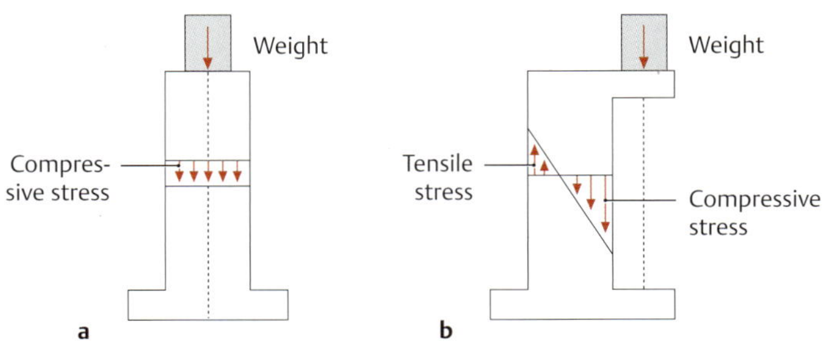

D Compressive and tensile stresses in a bone model

a An *axial* (centered) weight placed atop a Plexiglas model of a pillar creates a uniform pressure load that is evenly distributed over the cross section of the pillar and whose sum is equal to the applied weight.
b A *nonaxial* (eccentric) weight placed on an overhang creates a bending load that generates both tensile and compressive stresses in the pillar.

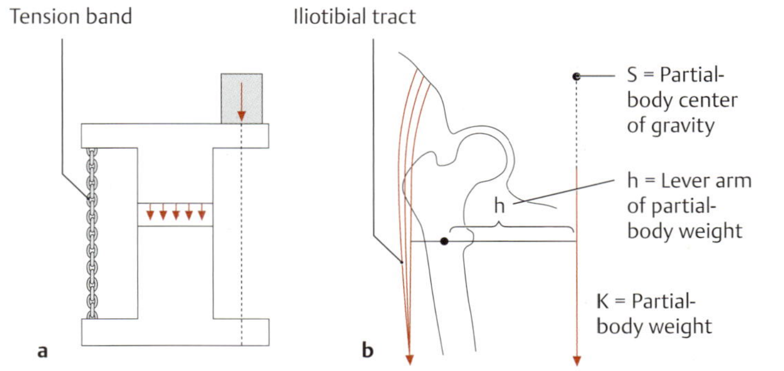

E The principle of the tension band (after Pauwels)

a The bending load acting on an I-beam model can be reduced by placing a high-tensile-strength member (chain) on the side opposite the bending force. This added member transforms the bending load into a pure compressive load.
b In the leg, the fascia lata on the lateral side of the thigh is thickened to form the iliotibial tract (see p. 497). By functioning as a tension band, the iliotibial tract reduces the bending loads on the proximal femur.

1.5 The Femoral Head and Deformities of the Femoral Neck

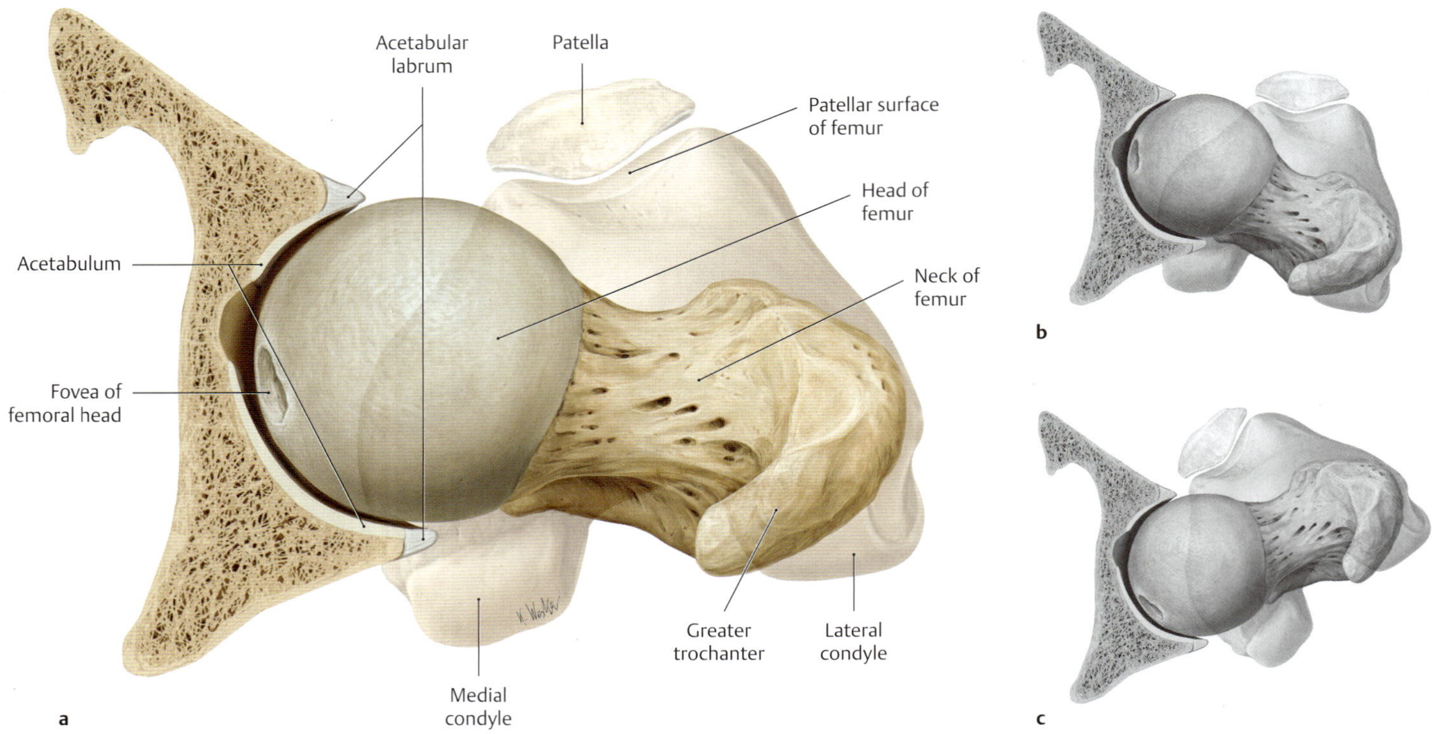

A The right femur
Proximal view. For clarity, the acetabulum has been sectioned in the horizontal plane. The distal end of the femur (with patella) has been added in light shading.
a Hip joint with centered femoral head, **b** hip joint externally rotated, **c** hip joint internally rotated.

Note the orientation of the acetabulum, which is angled forward by approximately 17°. This anterior angle affects the stability and "seating" of the femoral head in the hip joint (see p. 441). With the femoral head centered in the acetabulum and the femur slightly medially rotated (**a**), the distal femur and thus the knee joint point slightly inward (physiologic internal rotation of the knee). The position of the foot is additionally influenced by the external rotation of the tibia (see p. 435) (see **D**).

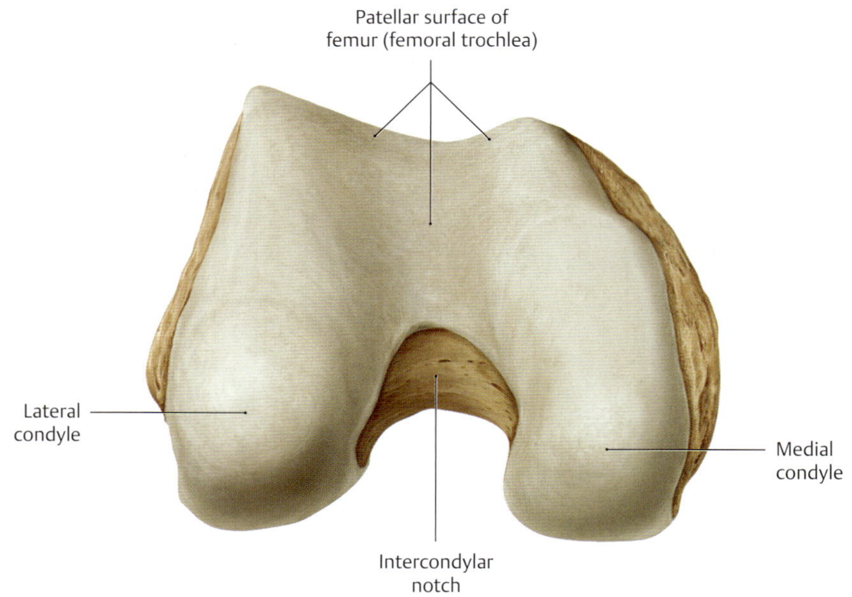

B The right femur
Distal view.

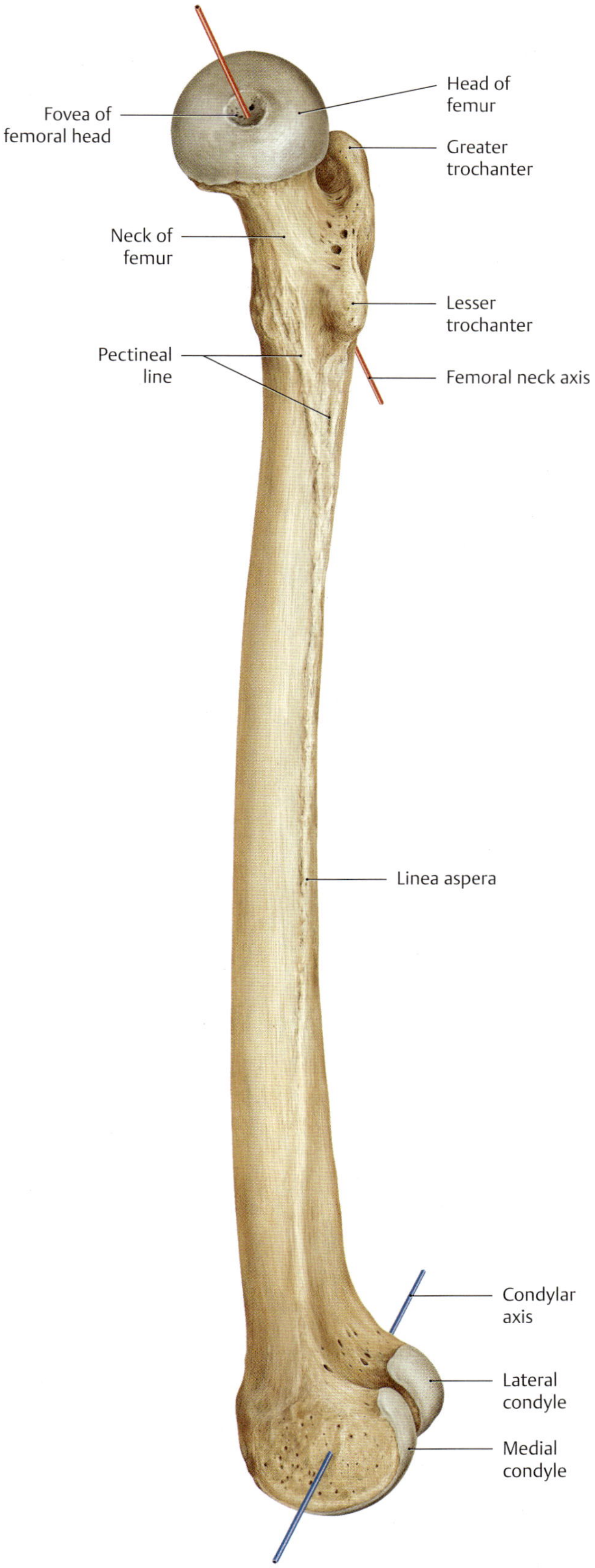

C The right femur
Medial view.
Note the transverse condylar axis and the femoral neck axis. When the axes are superimposed, the two lines intersect each other at a 12° angle in adults (anteversion angle, see also **D** and **A**). This angle is considerably larger at birth, measuring 30° to 40°, but decreases to the normal adult value by the end of the 2nd decade.

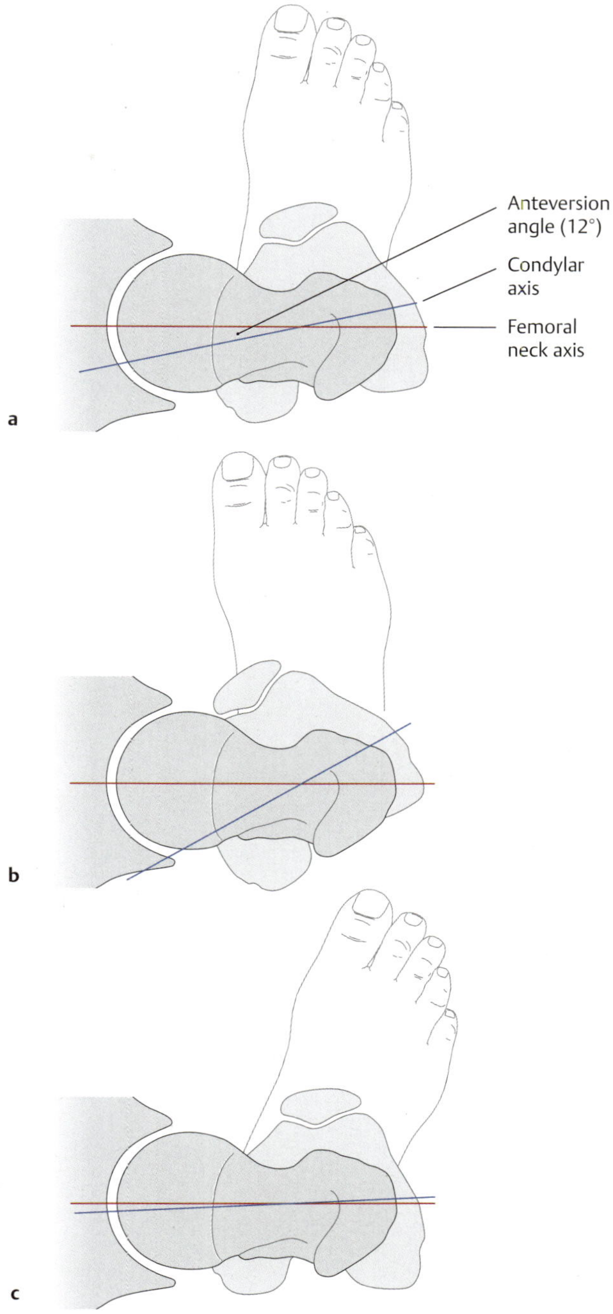

D Rotational deformities of the femoral neck
Right hip joint, superior view. Increased or decreased torsion of the femoral shaft results in torsion angles of varying size. When the hip is centered, this leads to increased internal or external rotation of the leg with a corresponding change in gait (a "toeing-in" or "toeing-out" gait). When the condylar axis is taken as the reference point, femoral torsion may be described as normal (**a**), increased (**b**), or decreased (**c**).

- **a** A normal anteversion angle of approximately 12° with the foot directed forward (taking into account tibial torsion of 23°, see p. 435).
- **b** An increased anteversion angle (*coxa anteverta*) typically leads to a toeing-in gait accompanied by a pronounced limitation of external rotation.
- **c** The femoral neck is retroverted (points backward in relation to the condylar axis). The result is *coxa retroverta* with a toeing-out gait.

1.6 The Patella

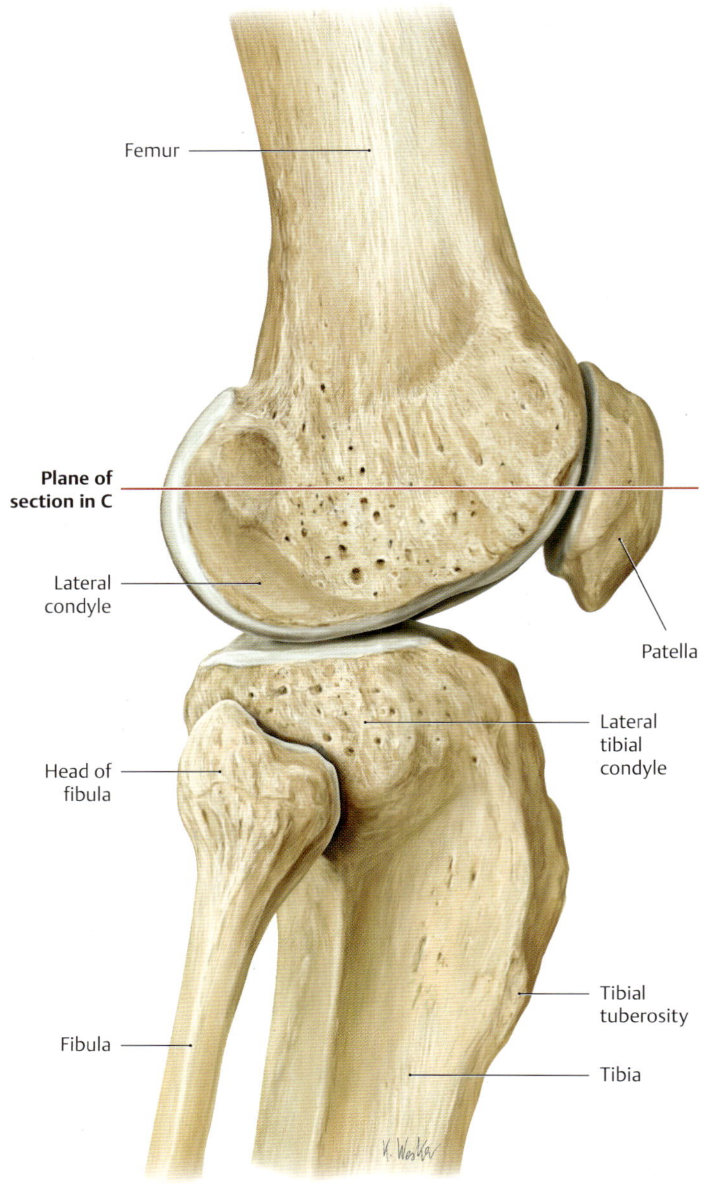

A Location of the patella
Right knee joint, lateral view. The red line indicates the plane of section in **C**.

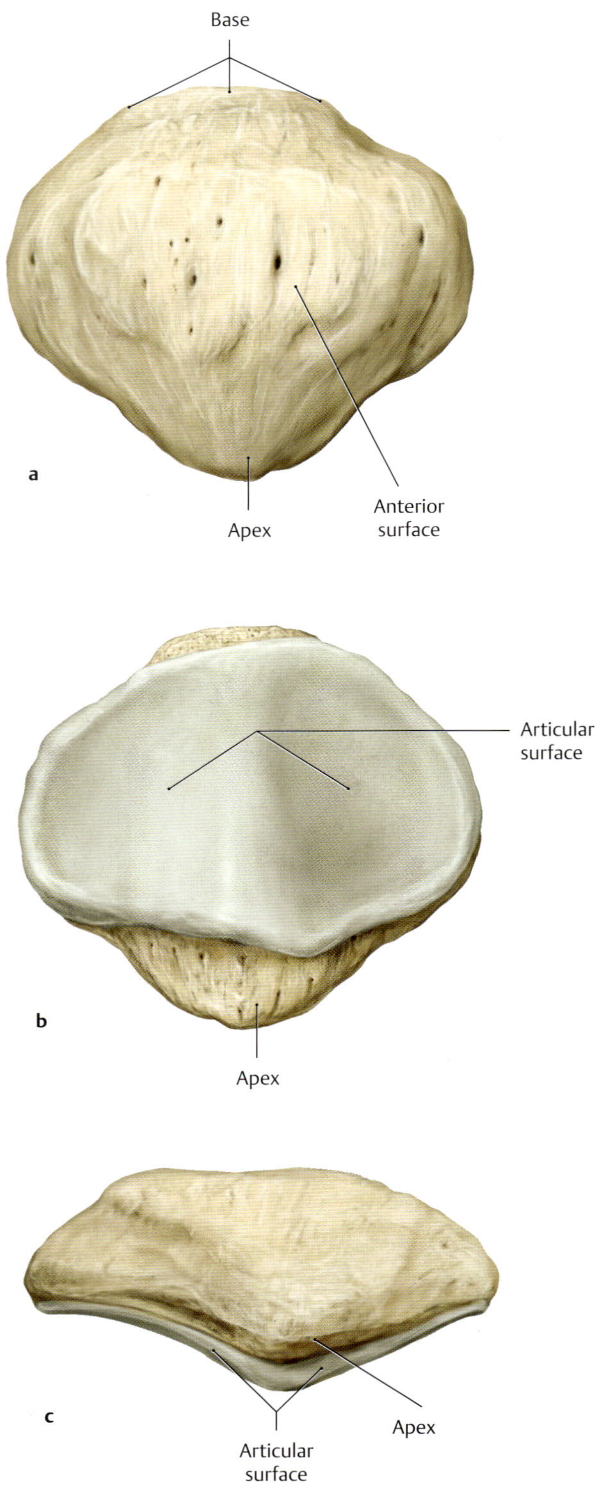

B Right patella
a Anterior view, **b** posterior view, **c** distal view.

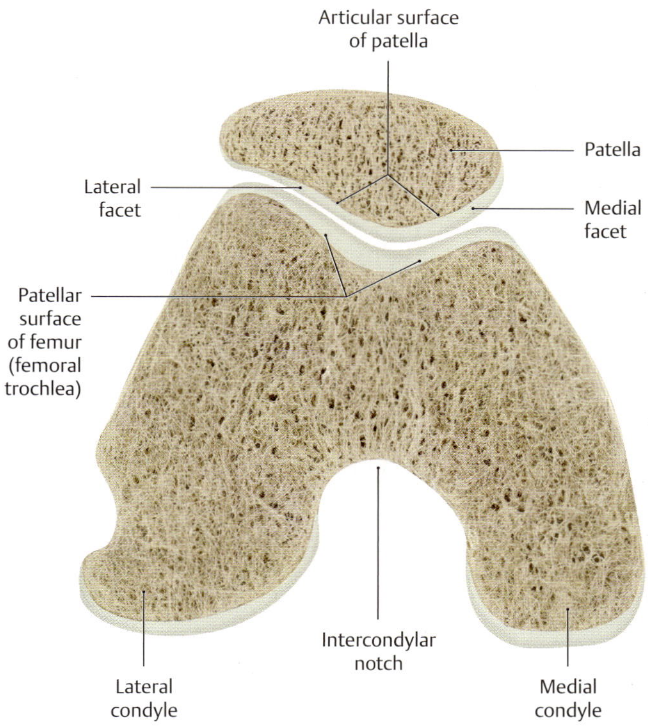

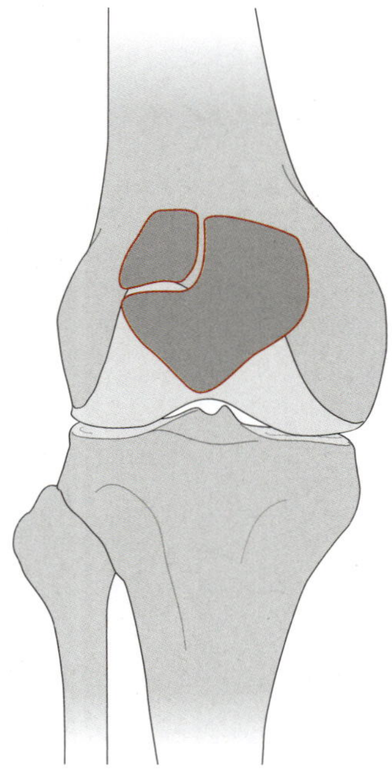

C Cross section through the femoropatellar joint
Right knee, distal view. The level of the cross section is shown in **A**. The femoropatellar joint is the site where the patellar surface of the femur, often called the femoral trochlea (by analogy with the distal humerus), articulates with the posterior articular surface of the patella. The patella is a sesamoid bone (the largest sesamoid) embedded in the quadriceps tendon. The patella is well centered when the ridge on the undersurface of the patella is seated within the groove of the femoral trochlea. The main functional role of the patella is to lengthen the effective lever arm of the quadriceps femoris muscle (the only extensor muscle of the knee), thereby reducing the force required to extend the knee joint (see also p. 500).

D Bipartite patella
Because the patella develops from multiple ossification centers, the failure of an ossification center to fuse results in a two-part (bipartite) patella. The upper lateral quadrant of the patella is most commonly affected. A fracture should always be considered in the radiographic differential diagnosis of a bipartite patella.

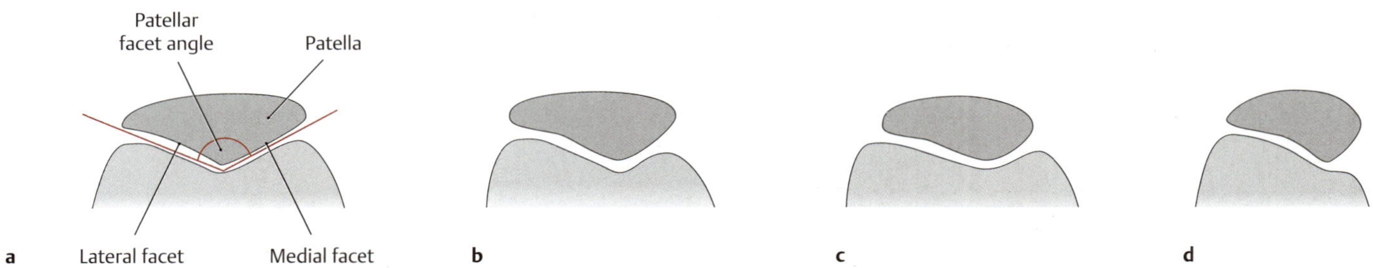

E The evaluation of patellar shape
Diagrams of tangential radiographs of the patella ("sunrise" view: supine position, knee flexed 60°, caudocranial beam directed parallel to the posterior patellar surface). Each diagram shows the relation of the patella to the femoral trochlea in a horizontal plane through the right knee joint. The posterior articular surface of the patella bears a vertical ridge dividing it into a lateral facet and a medial facet. Generally, the lateral facet is slightly concave, while the medial facet is slightly convex. The angle between the lateral and medial facets, called the patellar facet angle, is normally 130° ± 10°. Wiberg, Baumgart, and Ficat devised the following scheme for the classification of patellar shape based on the facet angle:

- **a** Patella with medial and lateral facets of approximately equal size and a facet angle within the normal range.
- **b** Most common patellar shape with a slightly smaller medial facet.
- **c** A distinctly smaller medial facet ("medial hypoplasia").
- **d** Patellar dysplasia with a very steep medial facet ("hunter's hat" configuration).

Besides the various patellar shapes, the patellar surface of the femur (the femoral trochlea) has a variable morphology (described in the Hepp classification system). Developmental dysplasias of the patella and femoral trochlea lead to patellar instability marked by recurrent lateral or medial subluxation or dislocation of the patella.

1.7 The Tibia and Fibula

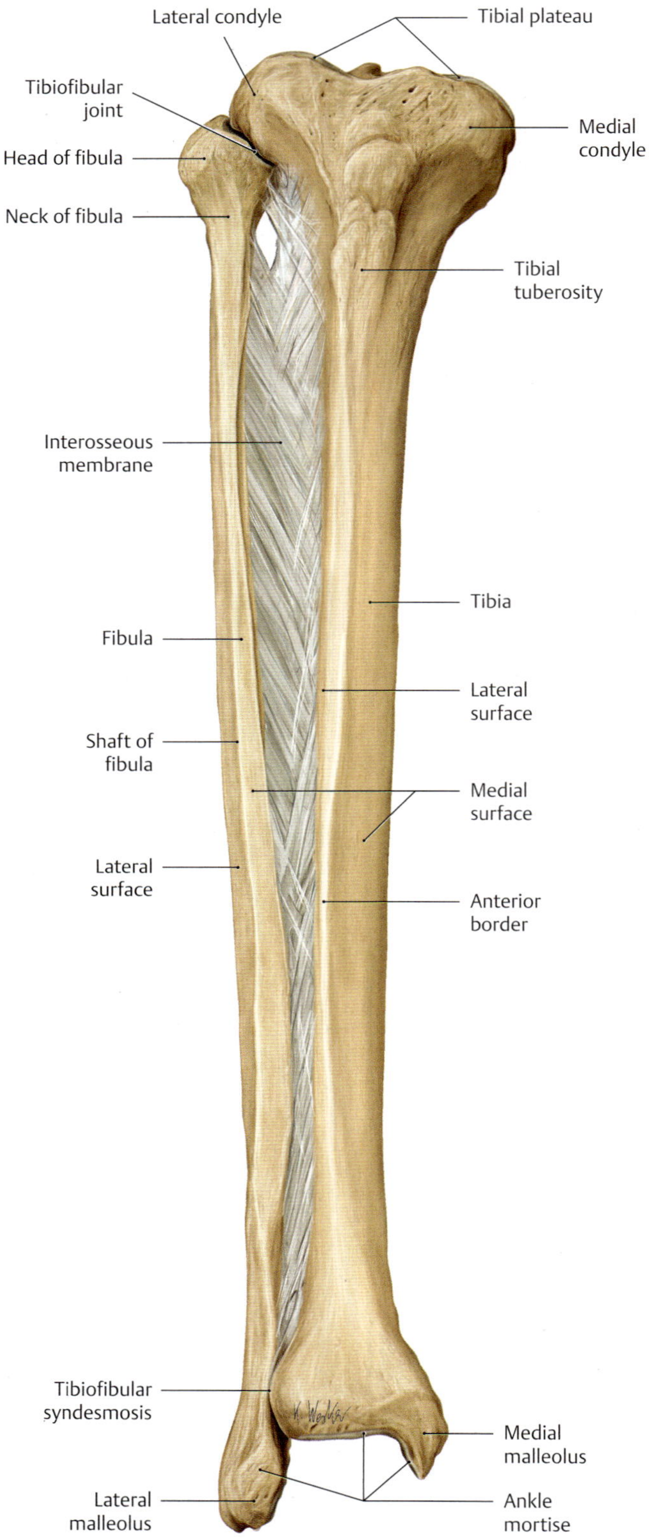

A The tibia, fibula, and crural interosseous membrane
Right leg, anterior view. The tibia and fibula articulate at two joints that allow only very limited motion (rotation). Proximally, near the knee, is the synovial tibiofibular joint; distally, at the ankle, is the tibiofibular *syndesmosis* (fibrous joint with bony elements united by ligaments). The crural interosseous membrane (see also **F**) is a sheet of tough connective tissue that serves as an origin for several muscles in the leg. Additionally, it acts with the tibiofibular syndesmosis to stabilize the ankle mortise.

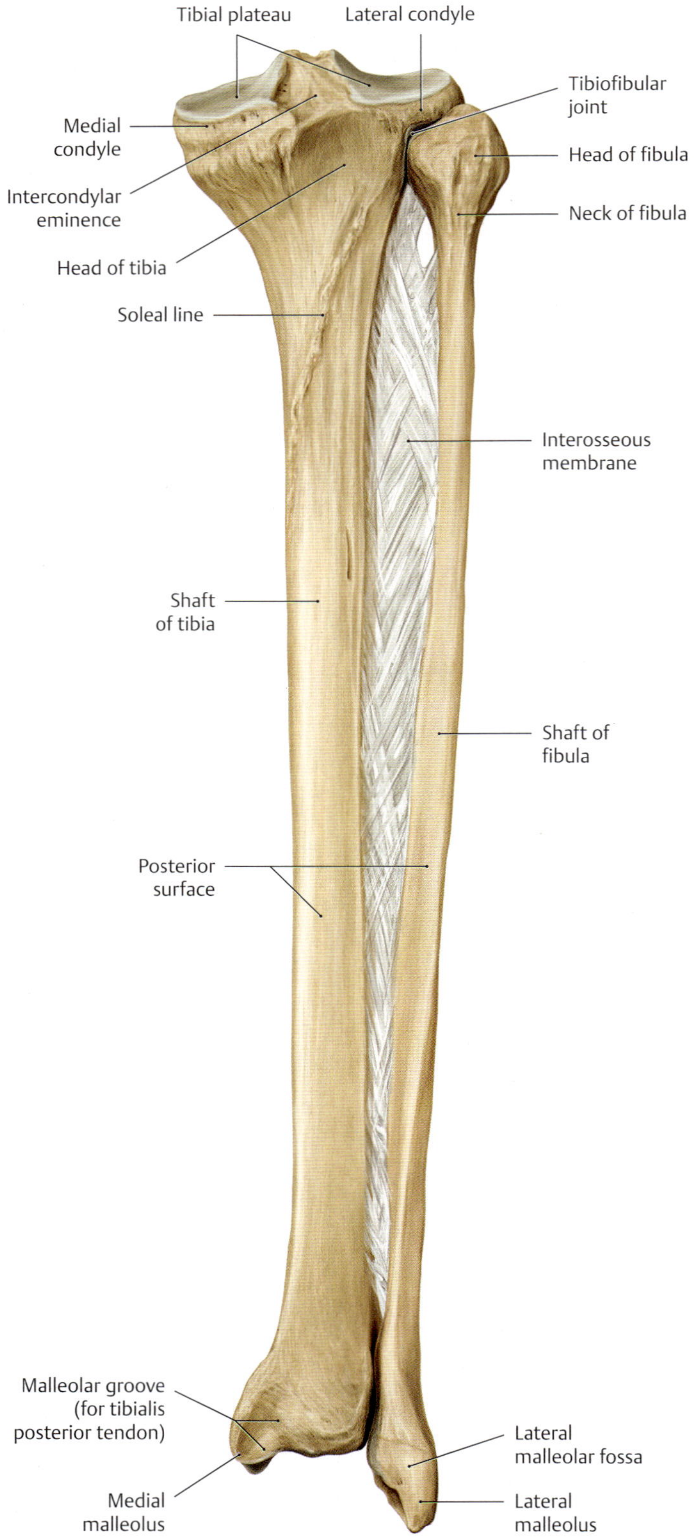

B The tibia, fibula, and crural interosseous membrane
Right leg, posterior view. In adults the tibial plateau has a 5–7 degree posterior to anterior inclination (known as the "tibial slope").

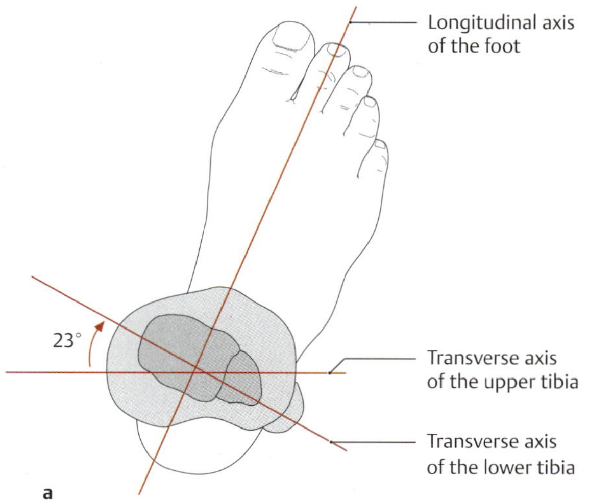

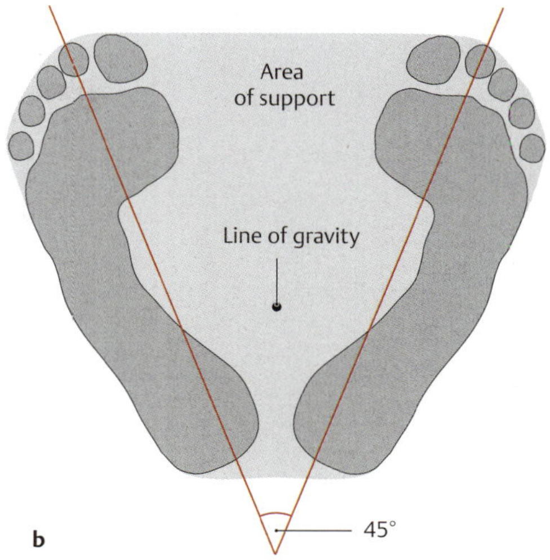

C Normal orientation of the tibia and its role in stability
When the transverse axes of the upper tibia (tibial plateau) and lower tibia (ankle mortise) are superimposed, they form an angle of approximately 23°; i. e., the transverse axis of the ankle joint is rotated 23° laterally relative to the transverse axis of the tibial plateau (*normal tibial orientation*, **a**). As a result of this, the longitudinal anatomic axis of the foot does not lie in the sagittal plane, and the toes point outward when the upper tibia is directed forward (**b**). This significantly improves the stability of bipedal stance by placing the line of gravity close to the center of the area of support.

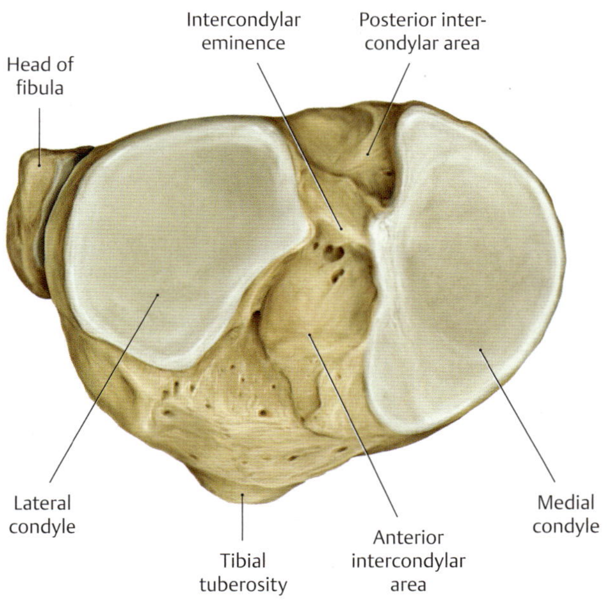

D The right tibial plateau
Proximal view.

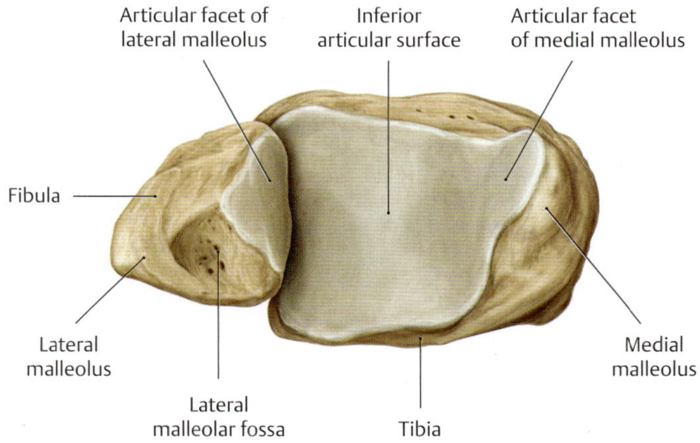

E The right ankle mortise
Distal view.

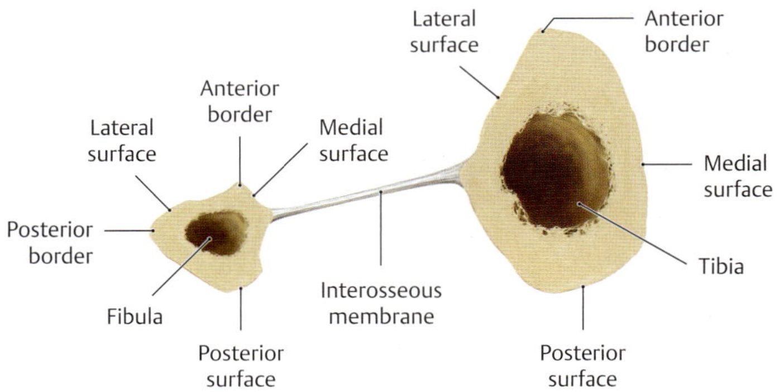

F Cross section through the middle third of the right leg
Distal view.

1.8 The Bones of the Foot from the Dorsal and Plantar Views

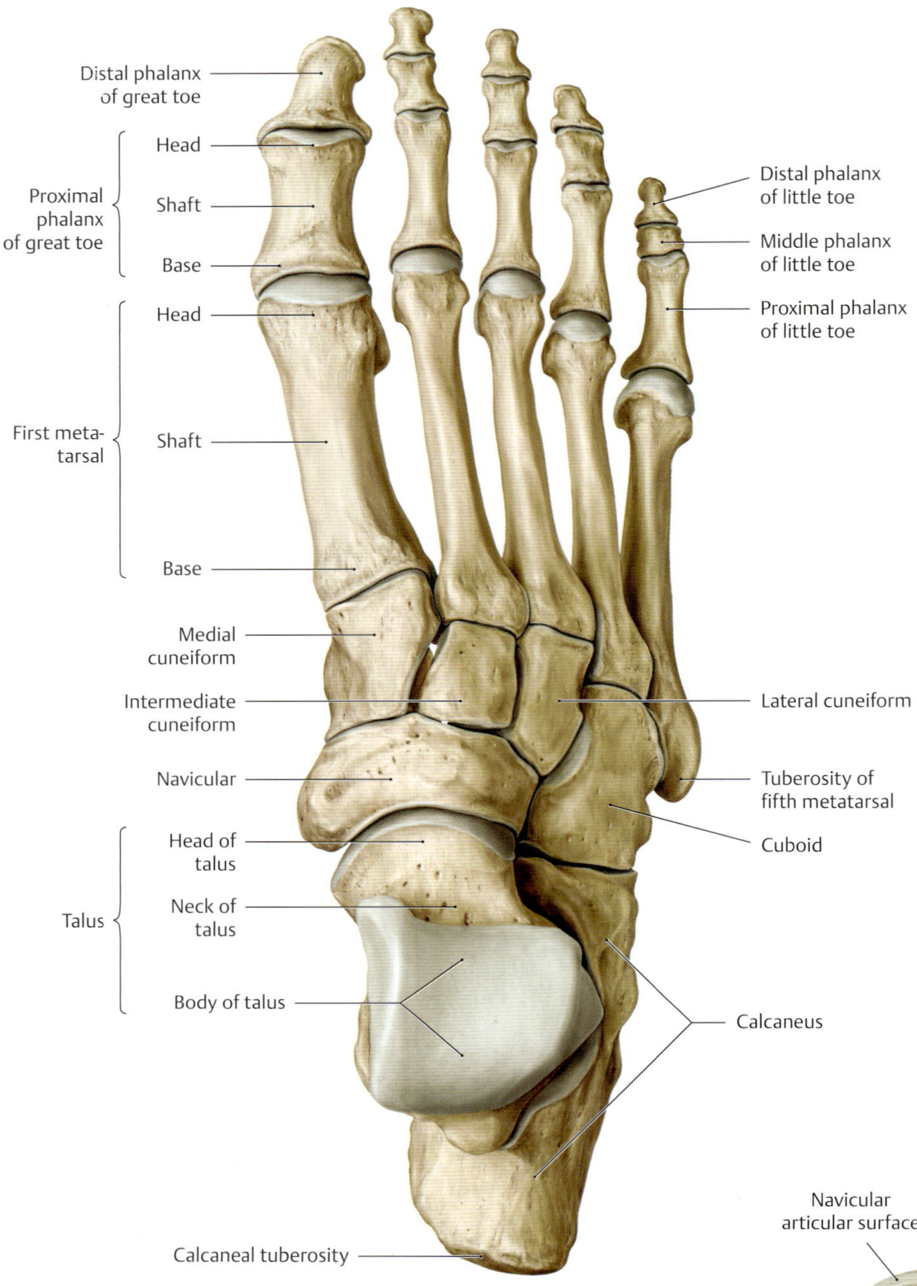

A The bones of the right foot
Dorsal view.

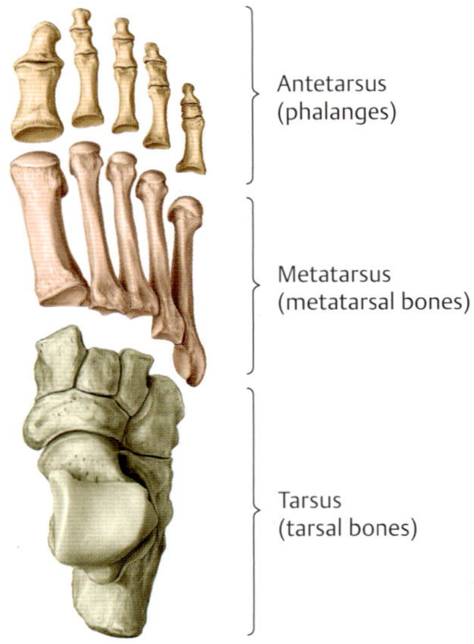

C Anatomic subdivisions of the pedal skeleton
Right foot, dorsal view. In the nomenclature of *descriptive* anatomy, the skeletal elements of the foot are subdivided into three adjoining segments:

- The tarsus, composed of seven bones.
- The metatarsus, composed of five bones.
- The antetarsus, composed of 14 phalanges.

Compare this diagram with the *functional* subdivisions in **D**.

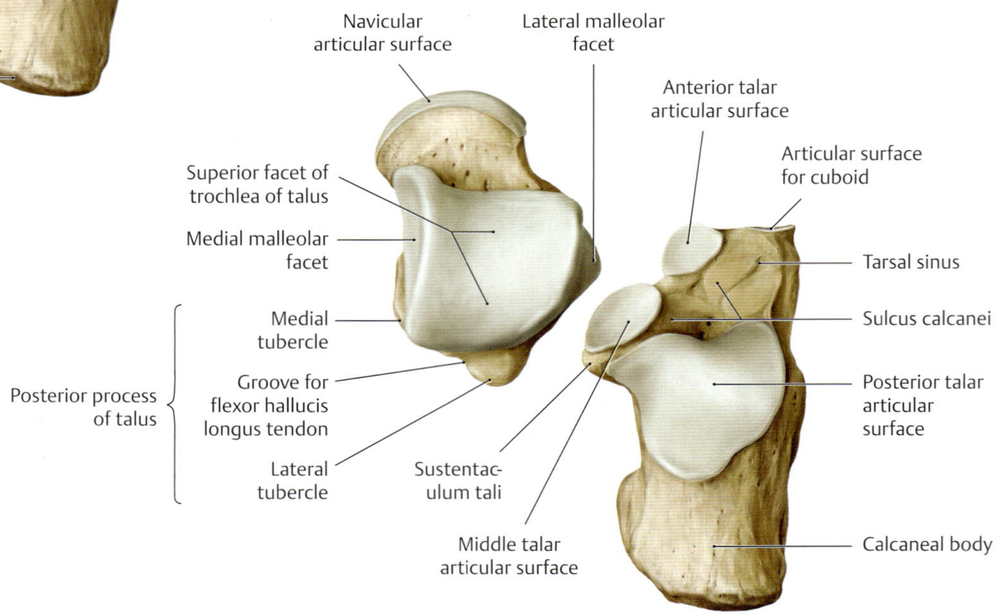

B The right talus and calcaneus
Dorsal view. The two tarsal bones have been separated at the subtalar joint to demonstrate their articular surfaces.

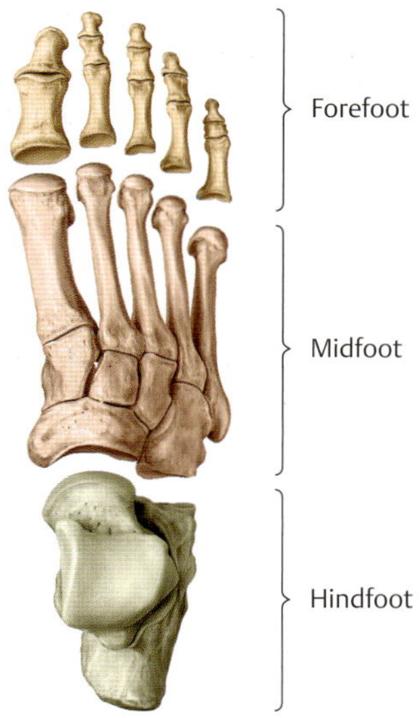

D Functional subdivisions of the pedal skeleton
Right foot, dorsal view. The skeleton of the foot is often subdivided as follows based on functional and clinical criteria:

- The hindfoot (calcaneus and talus)
- The midfoot (cuboid, navicular, cuneiforms and metatarsals)
- The forefoot (the phalanges)

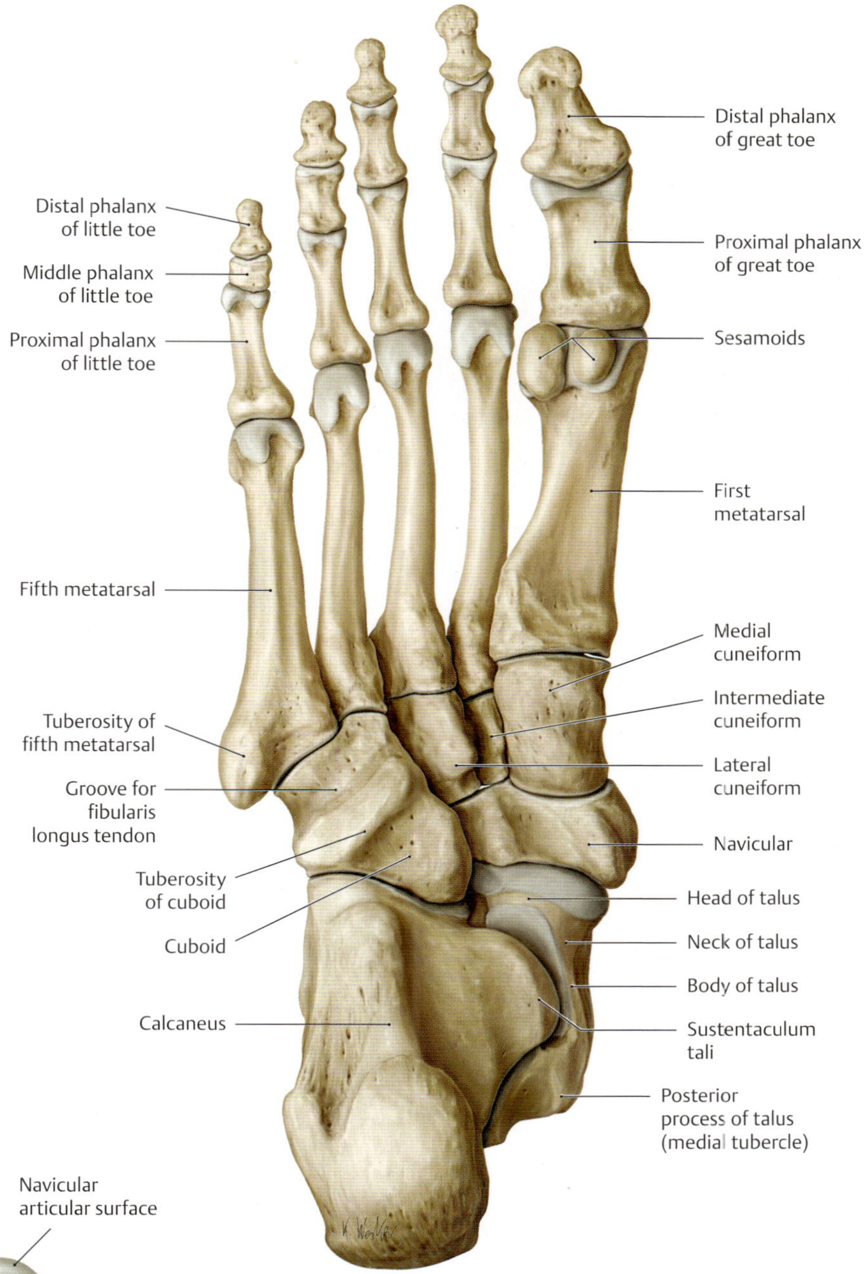

E The bones of the right foot
Plantar view.

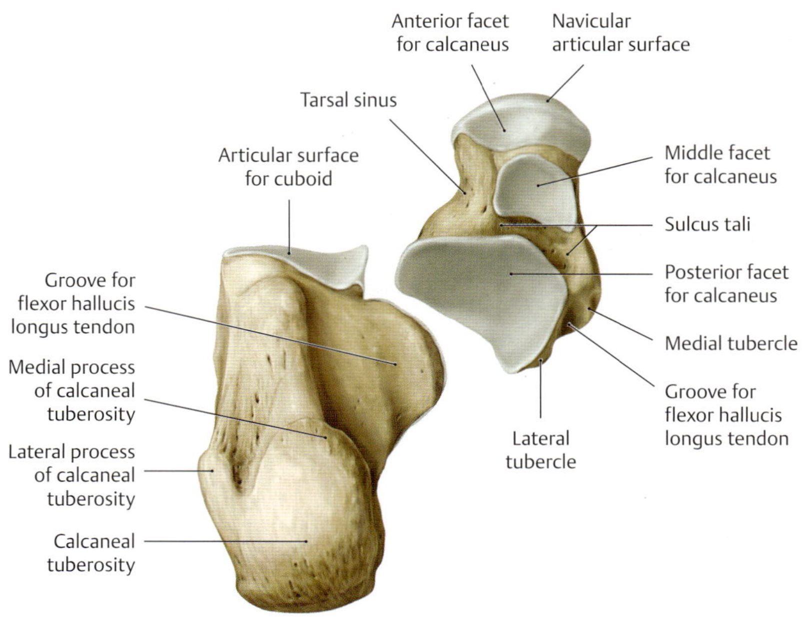

F The right talus and calcaneus
Plantar view. The two tarsal bones have been separated at the subtalar joint to demonstrate their articular surfaces.

1.9 The Bones of the Foot from the Lateral and Medial Views; Accessory Tarsal Bones

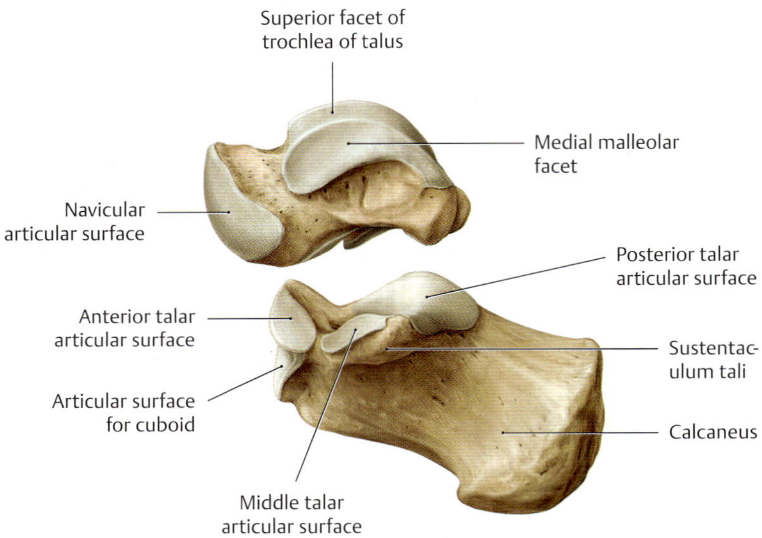

A The right talus and calcaneus
Medial view. The two tarsal bones have been separated at the subtalar joint to demonstrate their articular surfaces.

C Sustentaculum tali ("supports talus")

- The sustentaculum tali is a bony protuberance of the medial calcaneus.
- It is palpable approximately 1.5 cm below the tip of the medial malleolus.
- It supports the talus on its balcony-like protrusion.
- It marks the end of the tarsal canal, which divides the two chambers of the lower ankle joint (see p. 475).
- It serves as a support for the tendon of the flexor hallucis longus (thereby supporting the calcaneus in upright position, see p. 482).
- It has a longitudinal groove for the flexor digitorum longus (see p. 482).
- Two ligaments insert at the sustentaculum tali: the spring ligament (calcaneonavicular plantar ligament) and part of the deltoid ligament (see p. 476).
- It is a common location for fractures associated with snowboarding (see also p. 439).

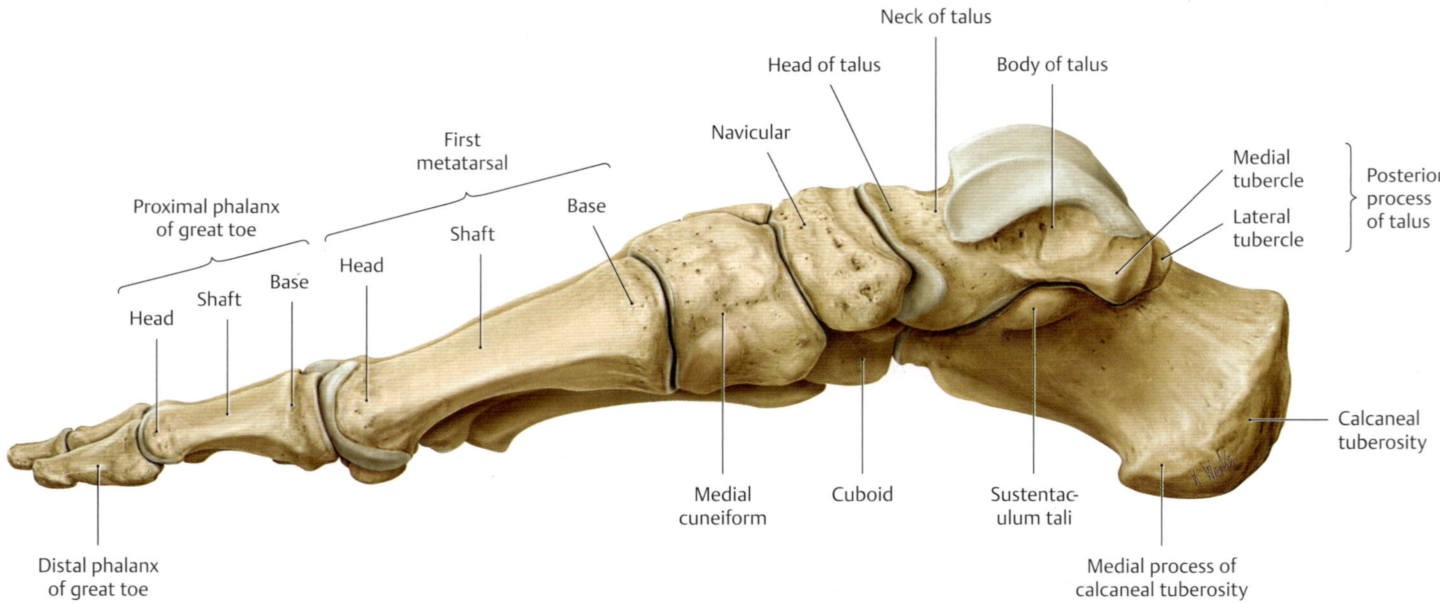

B The bones of the right foot
Medial view.

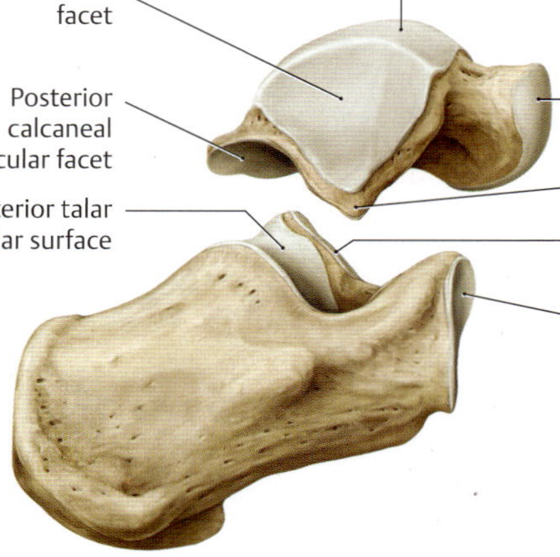

D The right talus and calcaneus
Lateral view. The two tarsal bones have been separated at the subtalar joint to demonstrate their articular surfaces.
Note: Fractures of the lateral process of the talus belong to the peripheral talar fractures and are usually indicative of a subtalar dislocation. Such an avulsion fracture of the lateral process of the talus is known as a "snowboarder's fracture" as it is a typical injury of snowboarders (see also **C**, p. 438).

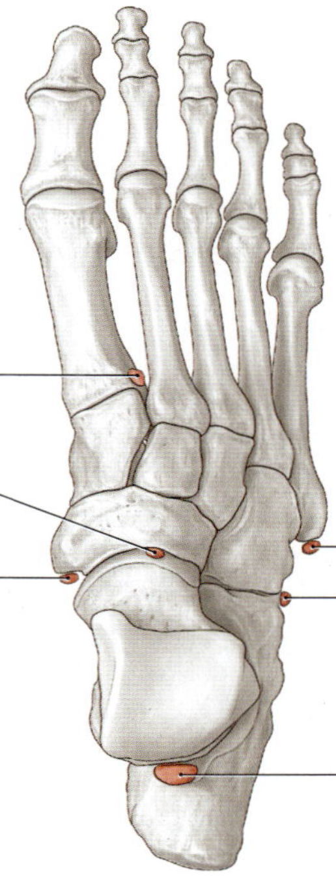

F Accessory tarsal bones
Right foot, dorsal view. A number of accessory (inconstant) ossicles are sometimes found in the foot. While they rarely cause complaints, they do require differentiation from fractures (see **E**). A clinically important accessory bone is the external tibial bone, which can be a source of discomfort when tight shoes are worn.

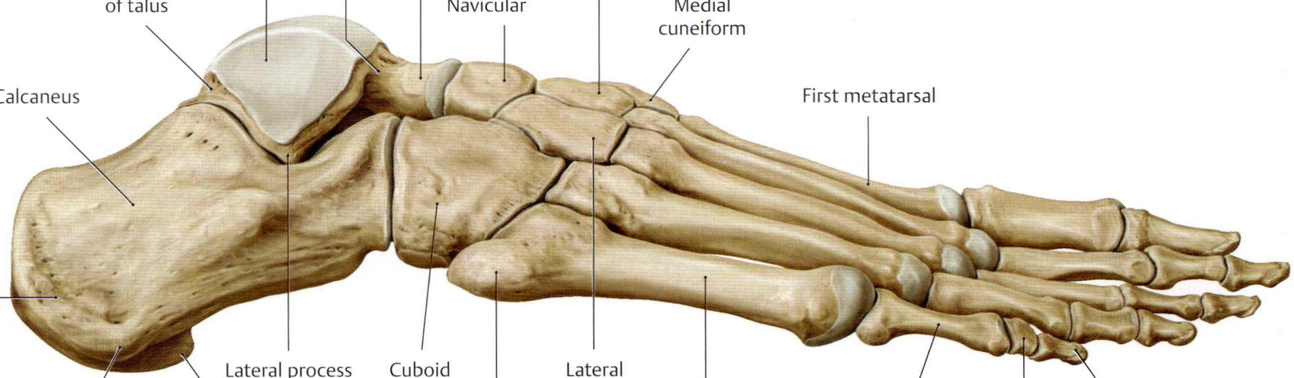

E The bones of the right foot
Lateral view. With an incidence of about 6.7 per 100,000 people, metatarsal fractures are among the most common injuries to the foot. The base of the fifth metatarsal, especially its tuberosity, is most often involved (30% of all metatarsal fractures).

1.10 The Hip Joint: Articulating Bones

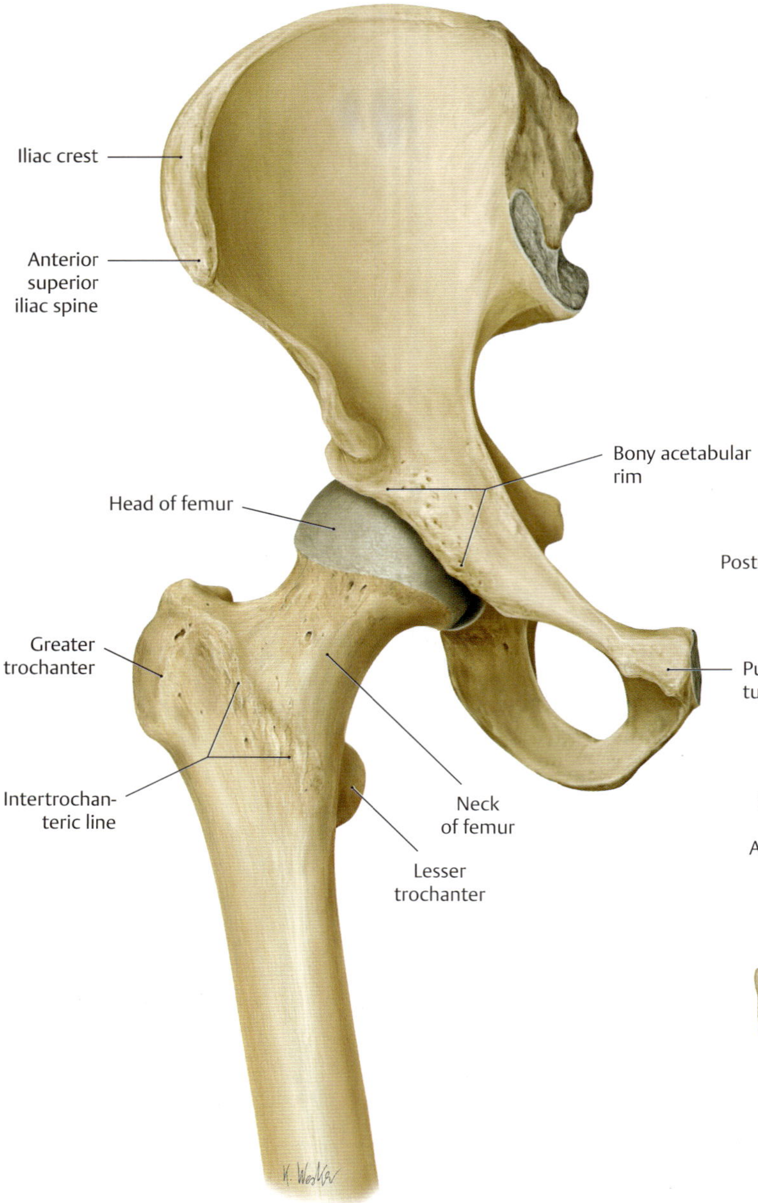

A The right hip joint from the anterior view
In the hip joint, the head of the femur articulates with the acetabulum of the pelvis. Owing to the shape of the two articulating bones, the joint is a special type of spheroidal (ball-and-socket) joint. The roughly spherical femoral head, which has an average radius of curvature of approximately 2.5 cm, is largely contained within the acetabulum (see also **C**).

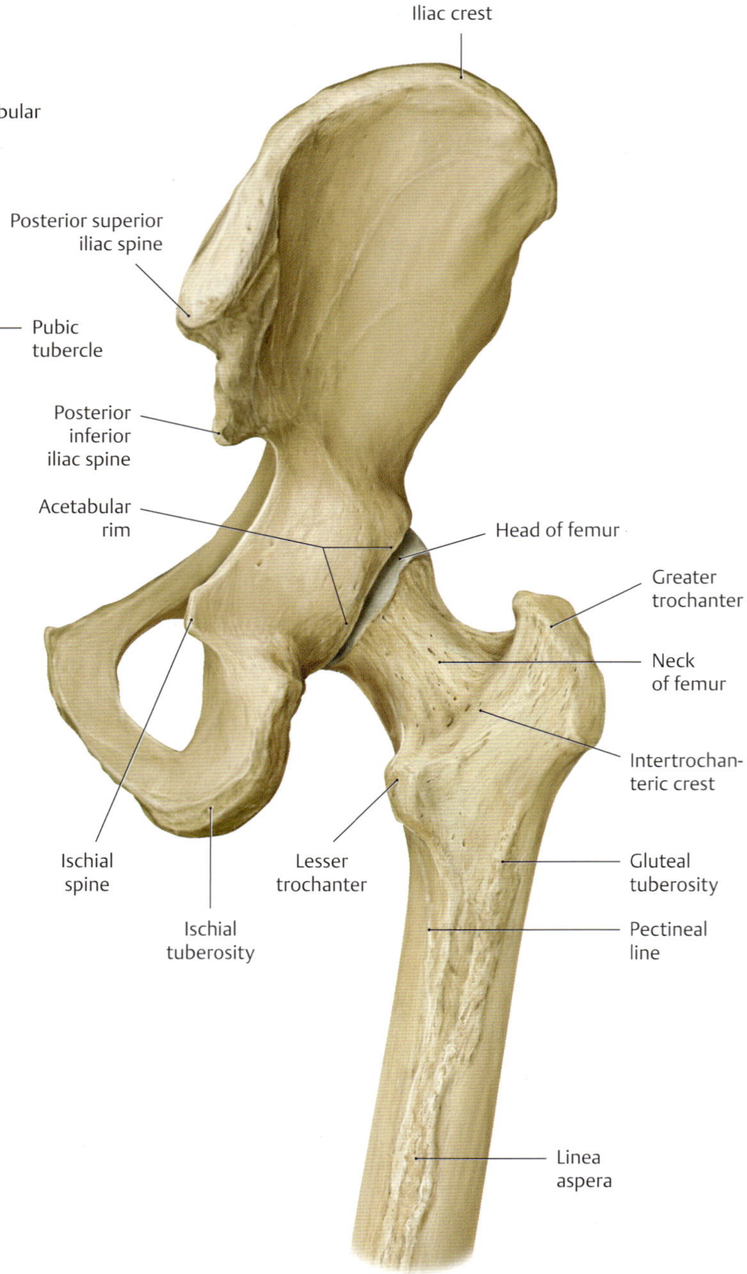

B The right hip joint from the posterior view

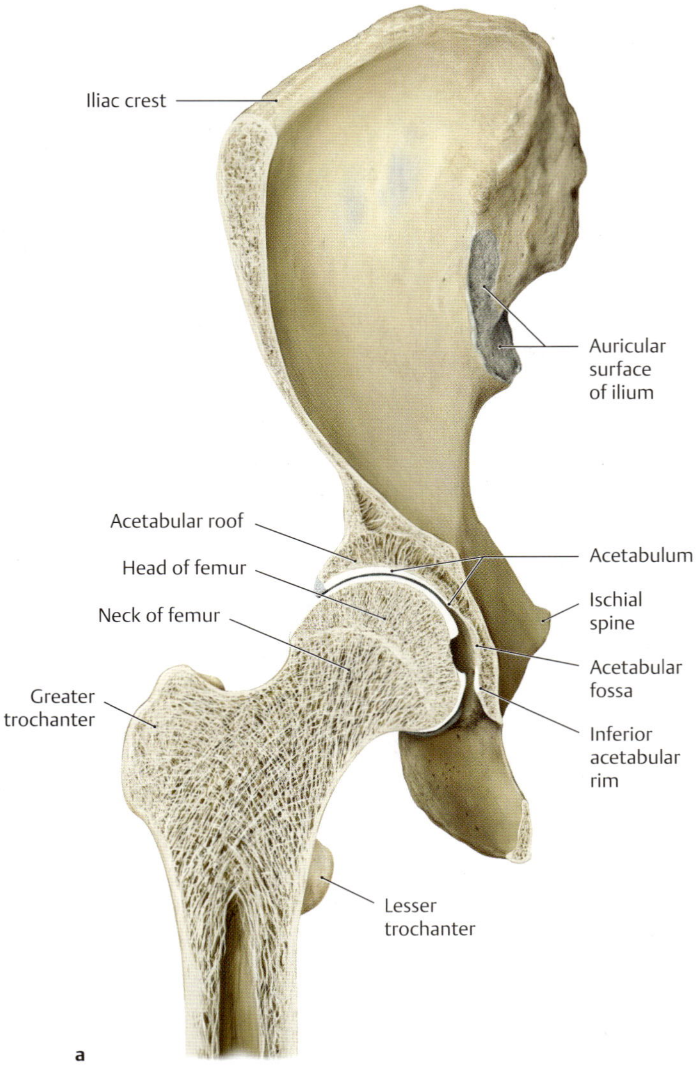

a

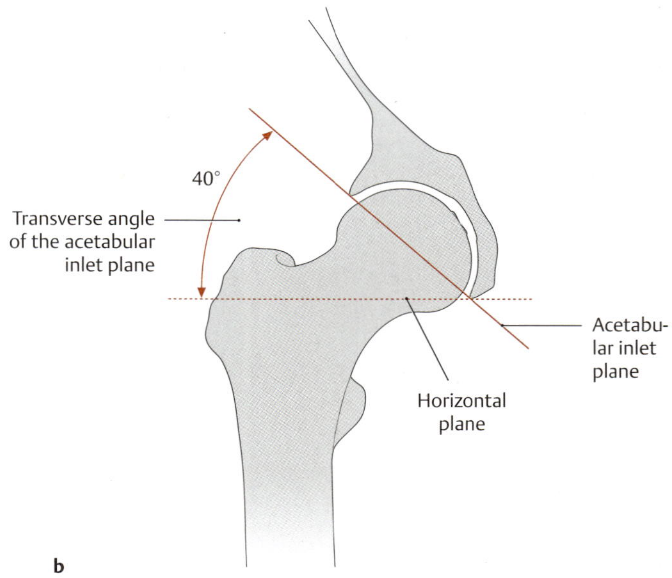

b

C Transverse angle of the acetabular inlet plane in the adult
Right hip joint, anterior view. Coronal section at the level of the acetabular fossa. The acetabular inlet plane, or bony acetabular rim, faces inferolaterally (*transverse angle*) and also anteroinferiorly (*sagittal angle*; see **D**). The inferolateral tilt of the acetabulum can be determined by drawing a line from the superior acetabular rim to the inferior acetabular rim (lowest point of the acetabular notch) and measuring the angle between that line and the true horizontal. This transverse angle normally measures approximately 51° at birth, 45° at 10 years of age, and 40° in adults (after Ullmann and Sharp). The value of the transverse angle affects several parameters, including the degree of lateral coverage of the femoral head by the acetabulum (the *center-edge angle* of Wiberg, see p. 453).

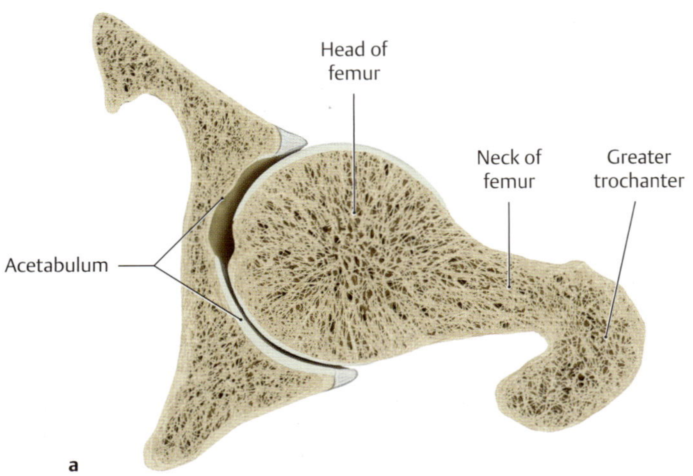

a

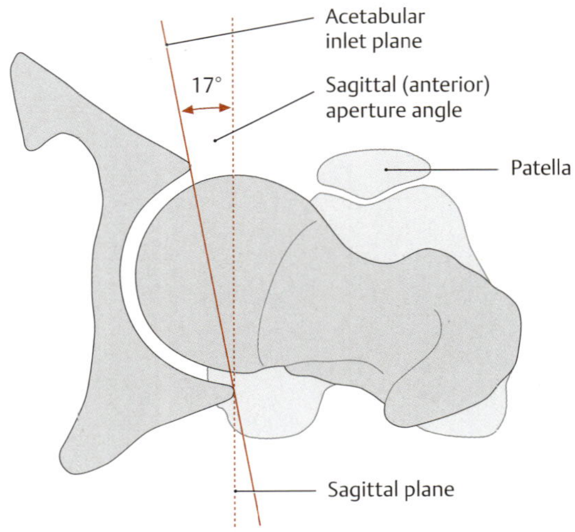

b

D Sagittal angle of the acetabular inlet plane in the adult
Right hip joint, superior view. Horizontal section through the center of the femoral head.

The bony acetabular rim is angled anteroinferiorly relative to the sagittal plane (compare this with the horizontal plane in **C**). This aperture angle measures approximately 7° at birth and increases to 17° by adulthood (after Chassard and Lapiné).

1.11 The Ligaments of the Hip Joint: Stabilization of the Femoral Head

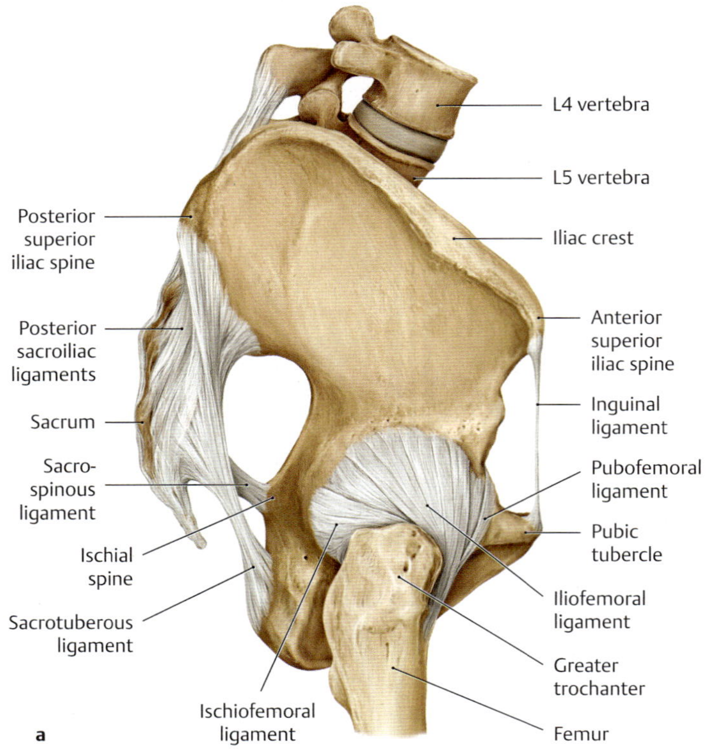

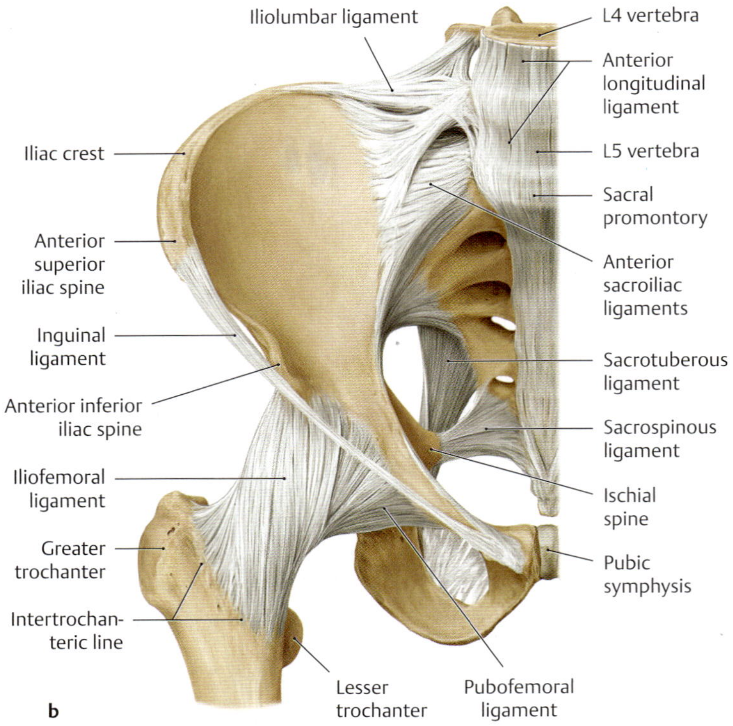

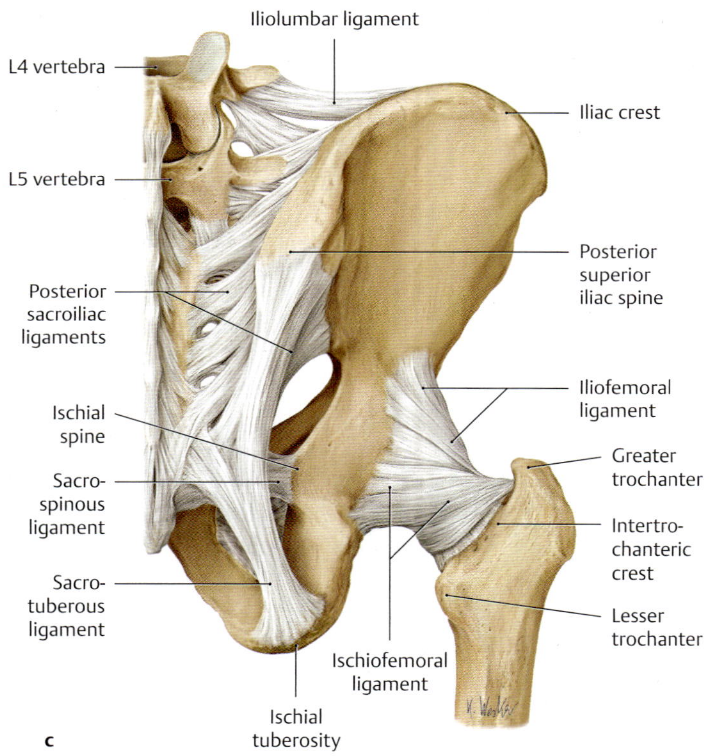

A The ligaments of the right hip joint

a Lateral view, **b** anterior view, **c** posterior view.

The strongest of the three ligaments, the iliofemoral ligament, arises from the anterior inferior iliac spine and fans out at the front of the hip, attaching along the intertrochanteric line (see **b**). With a tensile strength greater than 350 N, it is the most powerful ligament in the human body and provides an important constraint for the hip joint: it keeps the pelvis from tilting posteriorly in the upright stance, without the need for muscular effort. It also limits adduction of the extended limb (particularly the lateral elements of the ligament) and stabilizes the pelvis on the stance side during gait; i.e., it acts with the small gluteal muscles to keep the pelvis from tilting toward the swing side.

B The ligaments of the hip joint

- Iliofemoral ligament
- Pubofemoral ligament
- Ischiofemoral ligament
- Zona orbicularis (anular ligament)*
- Ligament of head of femur**

* Not visible externally, it encircles the femoral neck like a buttonhole (see p. 445, **C**).

** Has no mechanical function, but transmits vessels that supply the femoral head (see also p. 445).

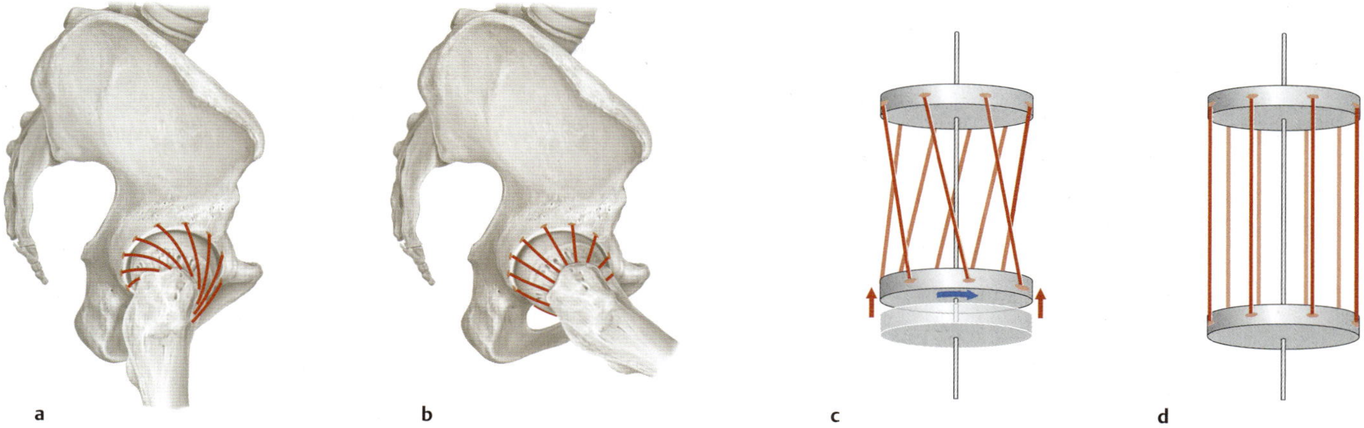

C Actions of the ligaments as a function of joint position

a Right hip joint in extension, lateral view. The capsular ligaments of the hip joint (see facing page) form a ringlike collar that encircles the femoral neck. When the hip is extended, these ligaments become twisted upon themselves (as shown here), pushing the femoral head more firmly into the acetabulum (joint-stabilizing function of the ligaments).

b Right hip joint in flexion, lateral view. During flexion (anteversion), the ligament fibers are lax and press the femoral head less firmly into the acetabulum, allowing a greater degree of femoral mobility.

c, d The twisting mechanism of the capsular ligaments can be represented by a model consisting of two disks interconnected by parallel bands. The situation in **c** represents the position of the ligaments when the hip joint is extended. When one of the two disks rotates (blue arrow), the bands become twisted and draw the two disks closer together (red arrows). Panel **d** models the situation in the flexed hip. The ligaments are no longer twisted, so the distance between the two disks is increased.

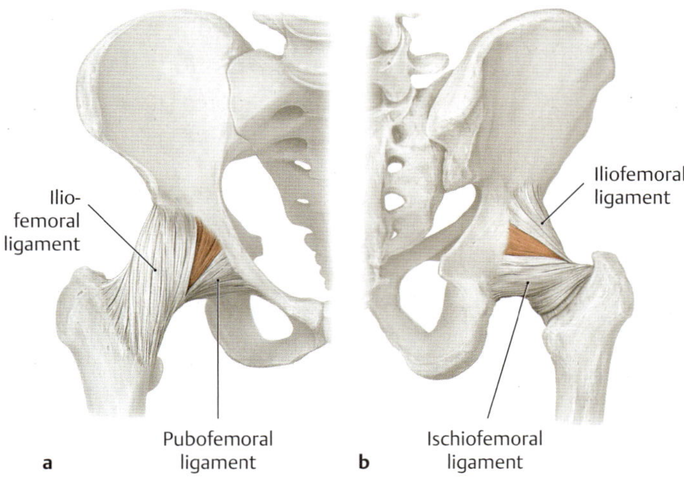

D Weak spots in the capsule of the right hip joint

a Anterior view, **b** posterior view.

There are weak spots in the joint capsule (color-shaded areas) located between the ligaments that strengthen the fibrous membrane (see **A**). External trauma may cause the femoral head to dislocate from the acetabulum at these sites (see **E**).

The combination of great ligament strength and the close congruity of the femoral head in the acetabulum makes the hip joint very stable and dislocations relatively rare. The situation is different, however, following a hip replacement arthroplasty. The hip joint ligaments must be at least partially divided to implant the prosthesis, and the risk of dislocation is markedly increased.

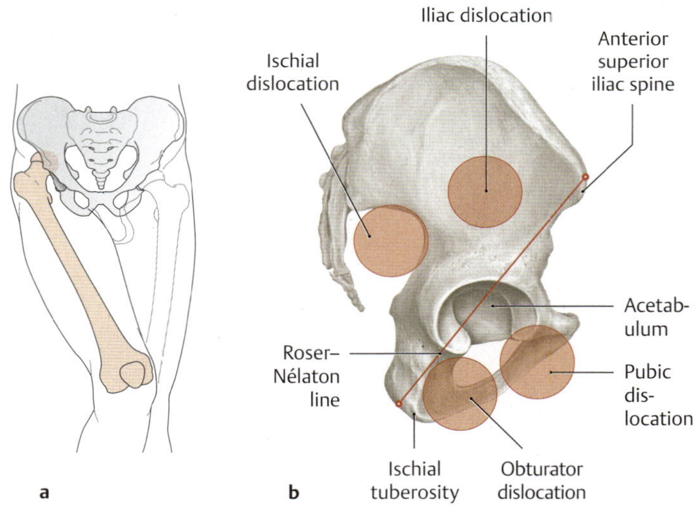

E Traumatic dislocation of the hip

a It is most common for the femoral head to dislocate upward and backward from the acetabulum (iliac dislocation) between the iliofemoral ligament and ischiofemoral ligament. Typically, this is caused by a fall from a great height, a motor vehicle accident (front-end collision), etc. In this type of dislocation, the leg assumes a position of adduction and slight internal rotation.

b Lateral view. Position of the femoral head in various types of dislocation. The greater trochanter may be above or below the Roser–Nélaton line (line between the ischial tuberosity and the anterior superior iliac spine). In an intact hip with the thigh flexed 45°, the greater trochanter is projected precisely onto this line.

1.12 The Ligaments of the Hip Joint: Nutrition of the Femoral Head

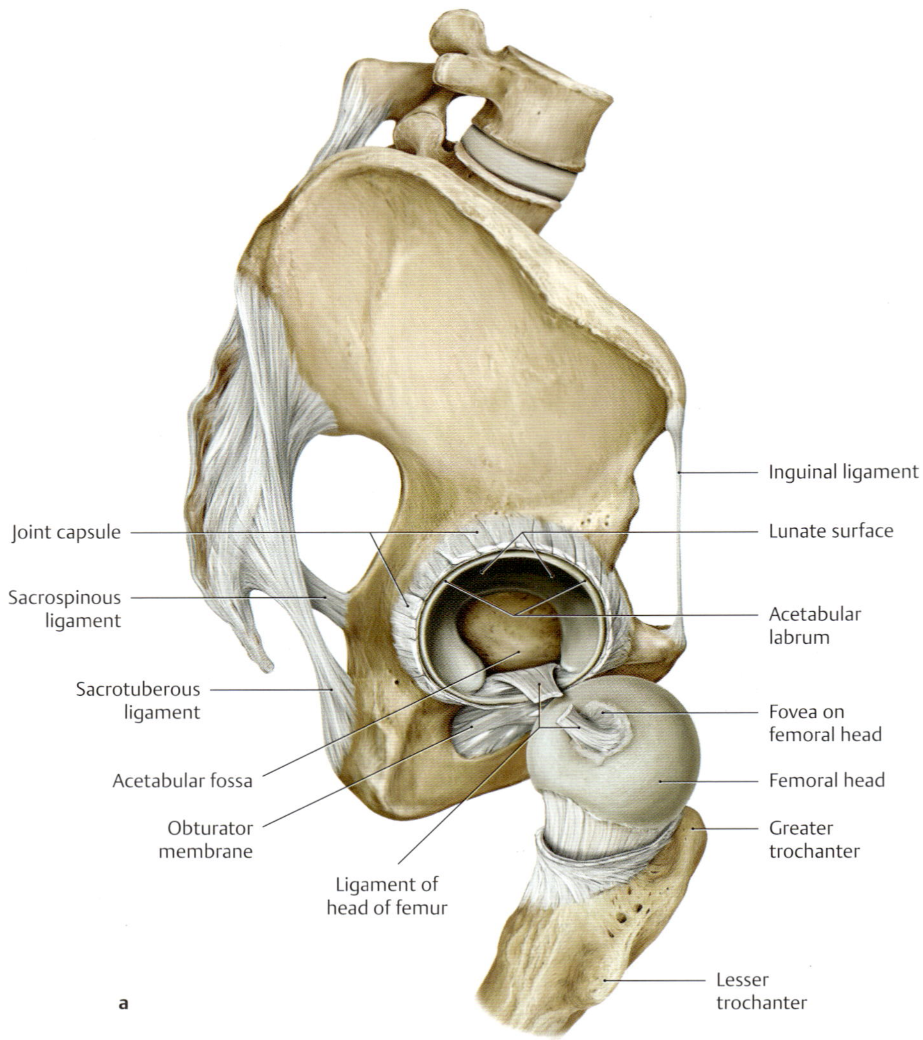

a

A The ligaments of the right hip joint

a Lateral view. The joint capsule has been divided at the level of the acetabular labrum, and the femoral head has been dislocated to expose the divided ligament of the head of the femur. This ligament transmits important nutrient blood vessels for the femoral head.

b Anterior view. The fibrous membrane of the joint capsule has been removed at the level of the femoral neck to show the conformation of the synovial membrane. This membrane extends laterally from the acetabular rim, and about 1 cm proximal to the attachment of the fibrous membrane, it is reflected onto the femoral neck within the joint cavity. It continues up the femoral neck to the chondro-osseous junction of the femoral head (see also the coronal section in **C**).

c Posterior view.

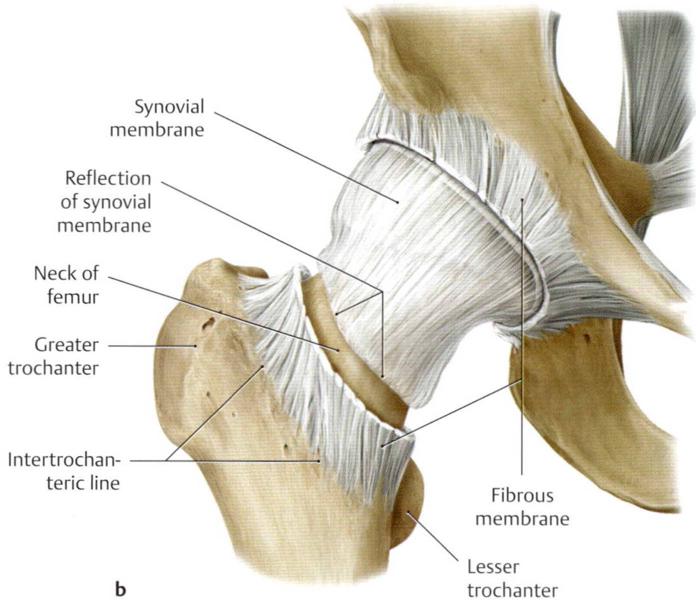

b

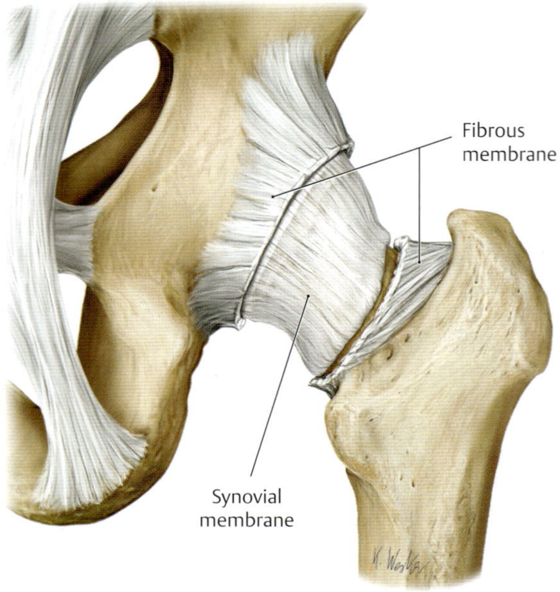

c

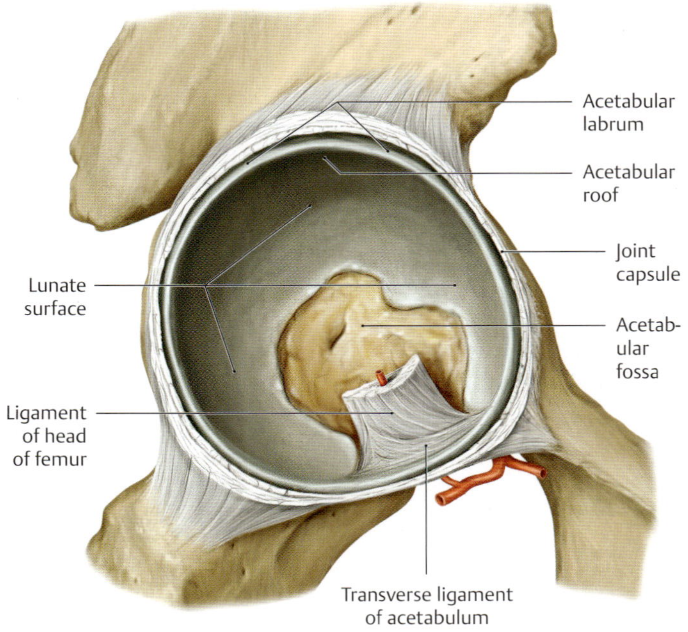

B The acetabulum of the right hip joint with the femoral head removed

Lateral view. The cartilage-covered articular surface of the acetabulum is crescent-shaped (lunate surface) and is broadest and thickest over the acetabular roof. The lunate surface is bounded externally by the slightly protruding bony rim of the acetabulum, which is extended by a lip (the acetabular labrum) composed of tough connective tissue and fibrocartilage. The cartilaginous articular surface lines much of the acetabular fossa, which is occupied by loose, fibrofatty tissue and is bounded inferiorly by the transverse acetabular ligament in the area of the acetabular notch (not visible here). The ligament of the head of the femur, which has been sectioned in the drawing, transmits blood vessels that nourish the femoral head (see **C**).

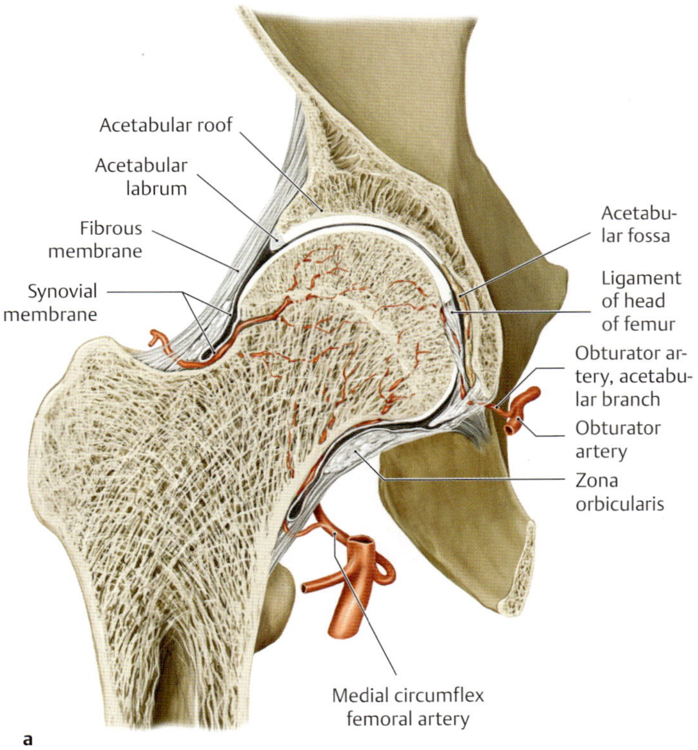

a

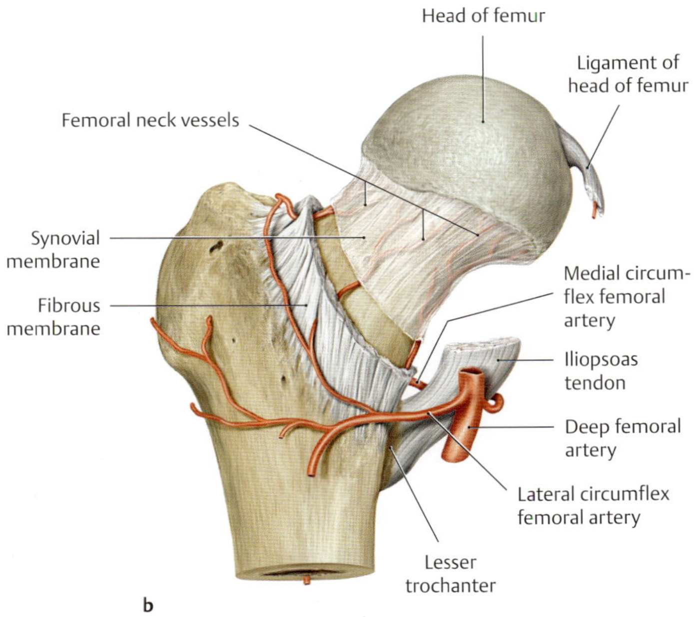

b

C The blood supply to the femoral head

a Coronal section through the right hip joint, anterior view.

b Course of the femoral neck vessels in relation to the joint capsule (right femur, anterior view).

The femoral head derives its blood supply from the lateral and medial circumflex femoral arteries and the artery of the ligament of the head of the femur, which branches from the obturator artery (see p. 568).

If the anastomoses between the vessels of the ligament of the head of the femur and the femoral neck vessels are absent or deficient due to the avulsion of blood vessels caused by a dislocation or femoral neck fracture, the bony tissue in the head of the femur may become necrotic (avascular necrosis of the femoral head).

1.13 Cross-sectional and X-ray Anatomy of the Hip Joint. Typical Medical Condition of the Elderly: Femoral Neck Fractures

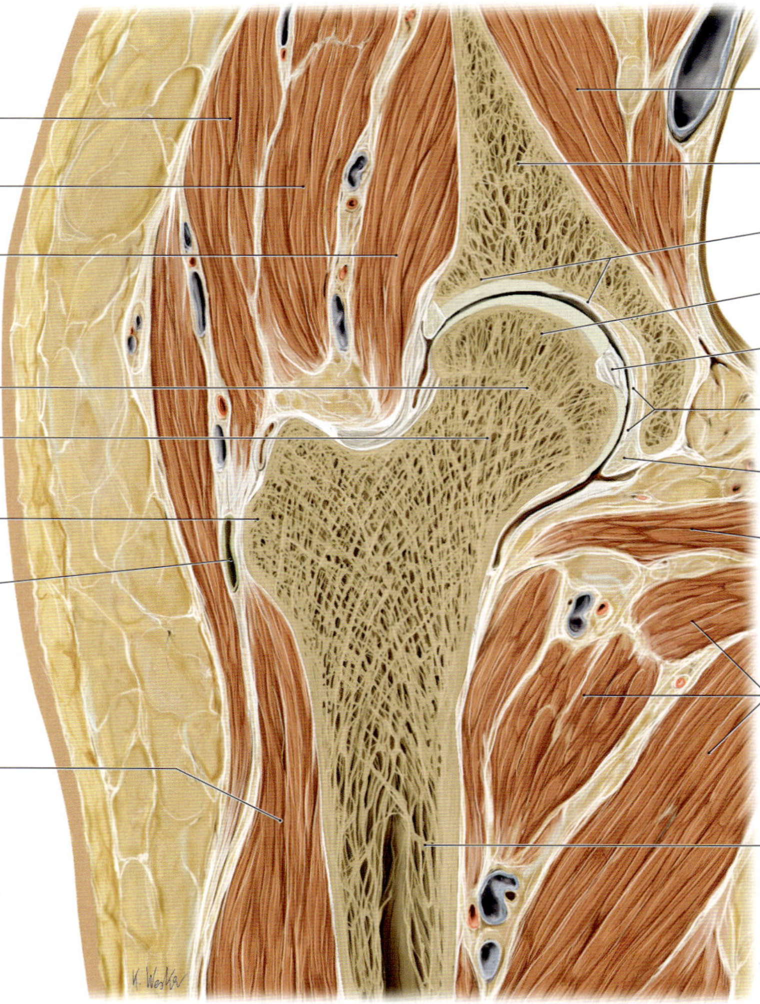

A Coronal section through the right hip joint
Anterior view (drawing based on a specimen from the Anatomical Collection of Kiel University).

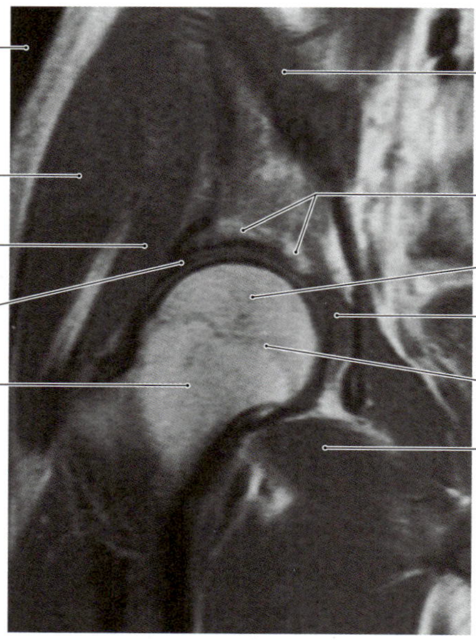

B MRI of the hip region: coronal T1-weighted spin-echo (SE) image at the level of the acetabular fossa (from Vahlensieck M, Reiser M. MRT des Bewegungsapparates. 4th ed. Stuttgart: Thieme 2014).

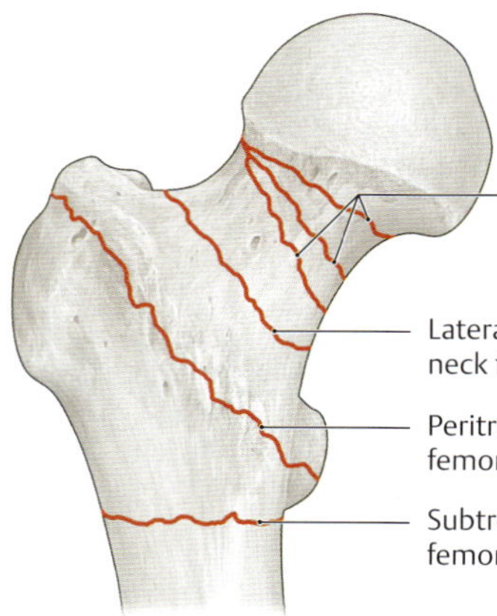

C Classification of proximal femoral fractures
Among the femoral fractures close to the hip joint, the medial femoral fractures (see **F**) are typical injuries of the osteoporotic bones of the elderly. Often, the cause is minimal trauma, such as falling on the greater trochanter or the outstretched leg.

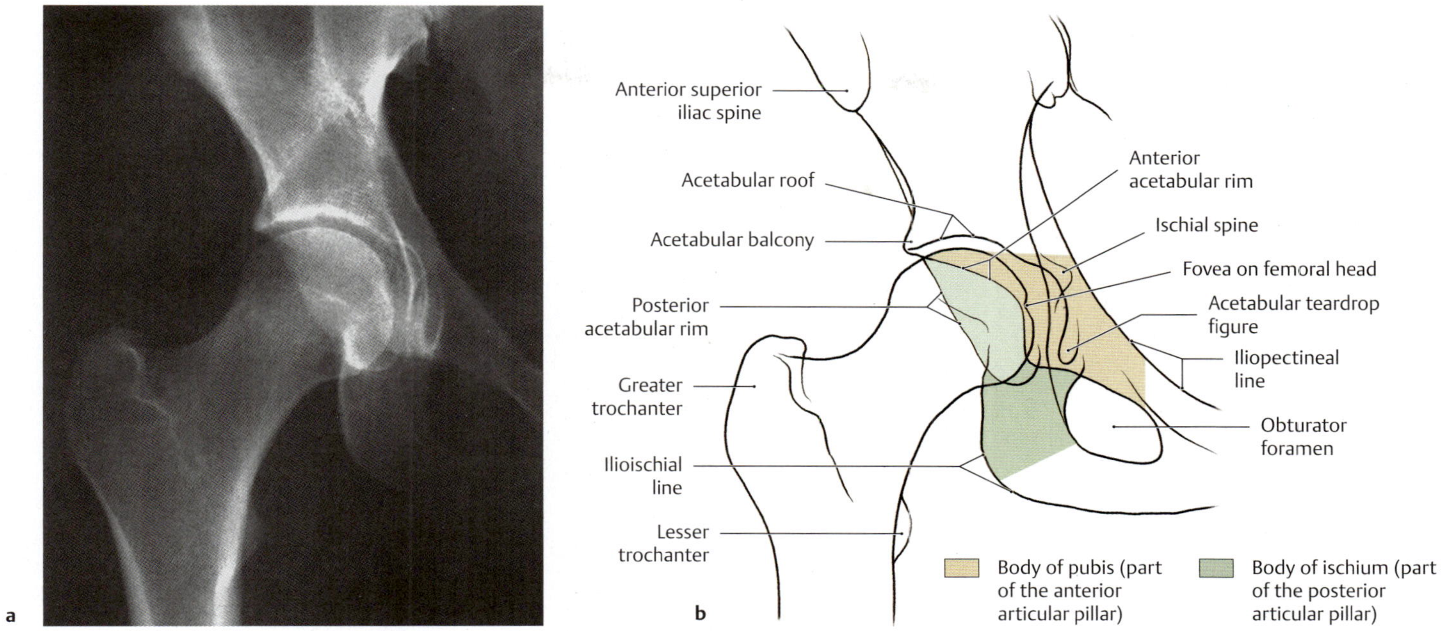

D X-ray of the hip joint and diagnostic reference levels in case of acetabular fractures

a X-ray of the hip joint in a sagittal beam path (section of a pelvis survey radiograph; from Möller TB, Reif E. Taschenatlas der Röntgenanatomie. 6th ed. Stuttgart: Thieme; 2016); **b** reference lines important for diagnostic radiology, mainly X-rays of the hip socket.

The pelvis survey radiograph is supplemented only when needed (i.e., when it is not sufficent for making a diagnosis, e.g., in undisplaced femoral fractures) with the help of particular positioning (e.g., ala and obturator view in which the healthy or injured side, respectively, is lifted at 45°) or MRI (see **E**) or CT. The CT is required as soon as an acetabular fracture has been diagnosed.

Particular reference lines in the anteroposterior (AP) view of the pelvis are important for diagnosis or for therapeutic surgical procedures in cases of acetabular fractures: anterior and posterior socket rim, acetabular roof, Köhler's teardrop (equivalent to the depth of the acetabular fossa), and iliopectineal or ilioischial line.

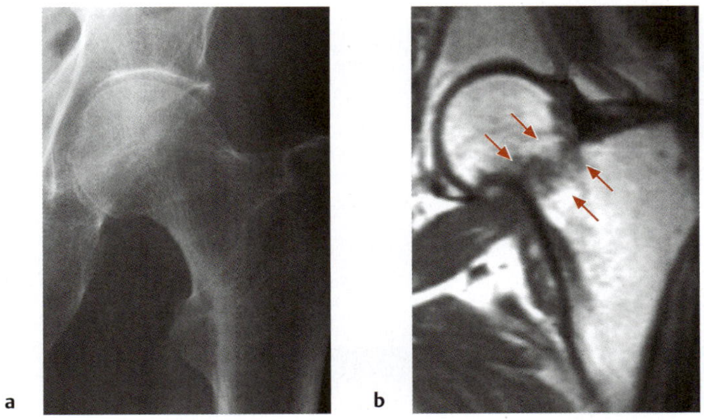

E Radiologic diagnosis of proximal femoral fractures
(from Bohndorf K, Imhof H, Fischer W.Radiologische Diagnostik der Knochen und Gelenke. 2nd ed. Stuttgart: Thieme; 2006)

a Normal X-ray of a medial, undisplaced femoral fracture in a sagittal beam path; **b** MRI identifies fracture: T1-weighted coronal cross section of the same patient with fracture-related edema (red arrows).

Whereas diagnosing displaced factures usually is not a problem since they are visible in AP views of the hip (see **Da**), both undisplaced femoral fractures and stress fractures often appear in conventional X-rays only as slight abnormalities of the trabecular bone structure, so an MRI is needed to identify a fracture (because fracture-related edema can be identified through the lesser signals it emits).

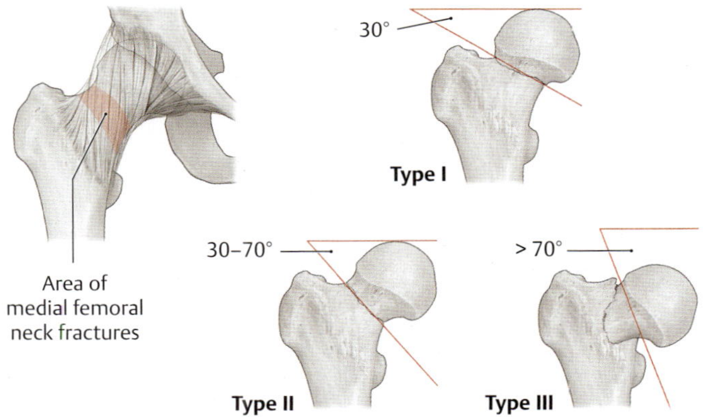

F Medial femoral fractures (classification according to Pauwels)

Medial femoral fractures are much more common than lateral femoral fractures (95% of cases compared to 5% of cases). Medial femoral fractures are always intracapsular and due to common complications associated with them (e.g., ischemic femoral head necrosis, delayed fracture healing, and development of pseudarthrosis) are of particular clinical relevance. In particular, the damage to epiphyseal vessels (see p. 445) as a result of intracapsular femoral head fractures leads to impaired blood flow to the femoral head. Medial femoral fractures are classified according to Pauwels by the inclination angle of the fracture line relative to the horizontal line (type I: 0–30°, type II: 30–70°, and type III: > 70°). The steeper the angle, meaning the steeper the course of the fracture line, the greater the risk of the femoral head sliding off and the greater the danger of pseudarthrosis.

1.14 Cross-sectional Anatomy of the Hip Joint: Sonographic Representation of Hip Joint Effusion

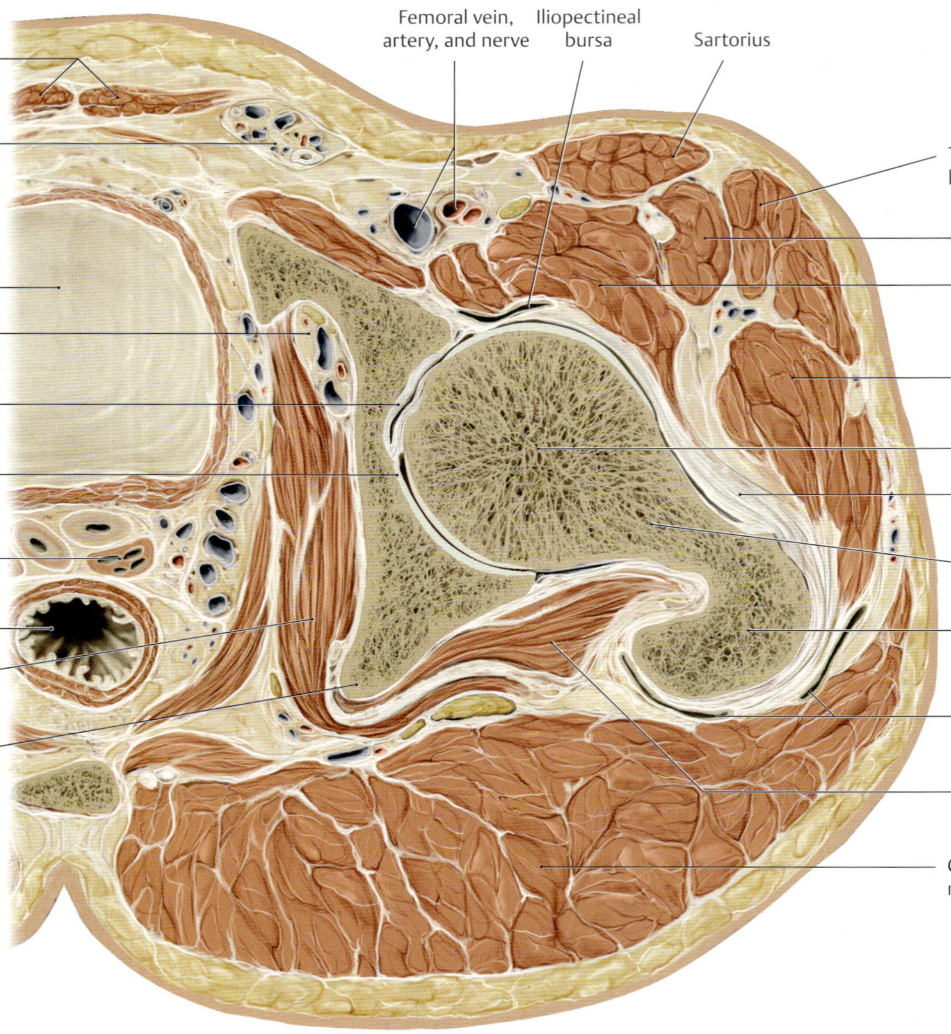

A Transverse section through the right hip joint
Superior view (drawing based on a specimen from the Anatomical Collection of Kiel University).

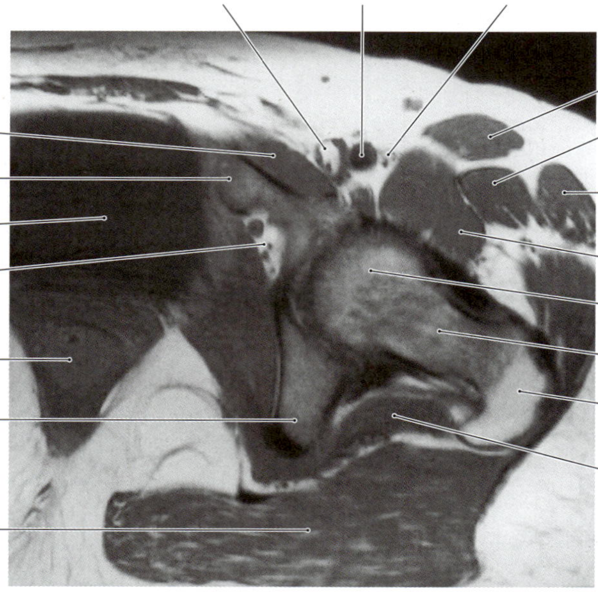

B MRI of the hip region: axial (transverse) T1-weighted SE image at the level of the femoral neck (from: Vahlensieck M, Reiser M. MRT des Bewegungsapparates. 4th ed. Stuttgart: Thieme; 2017)
The synovial bursae shown in **A** are not visible because in T1-weighted MRI, they always appear as low signal intensity and are barely distinguishable from muscles, which are also low signal intensity.
Note: With MRI, axial (transverse) cross sections are always inferior views. See p. 496 for the course of the obturator internus.

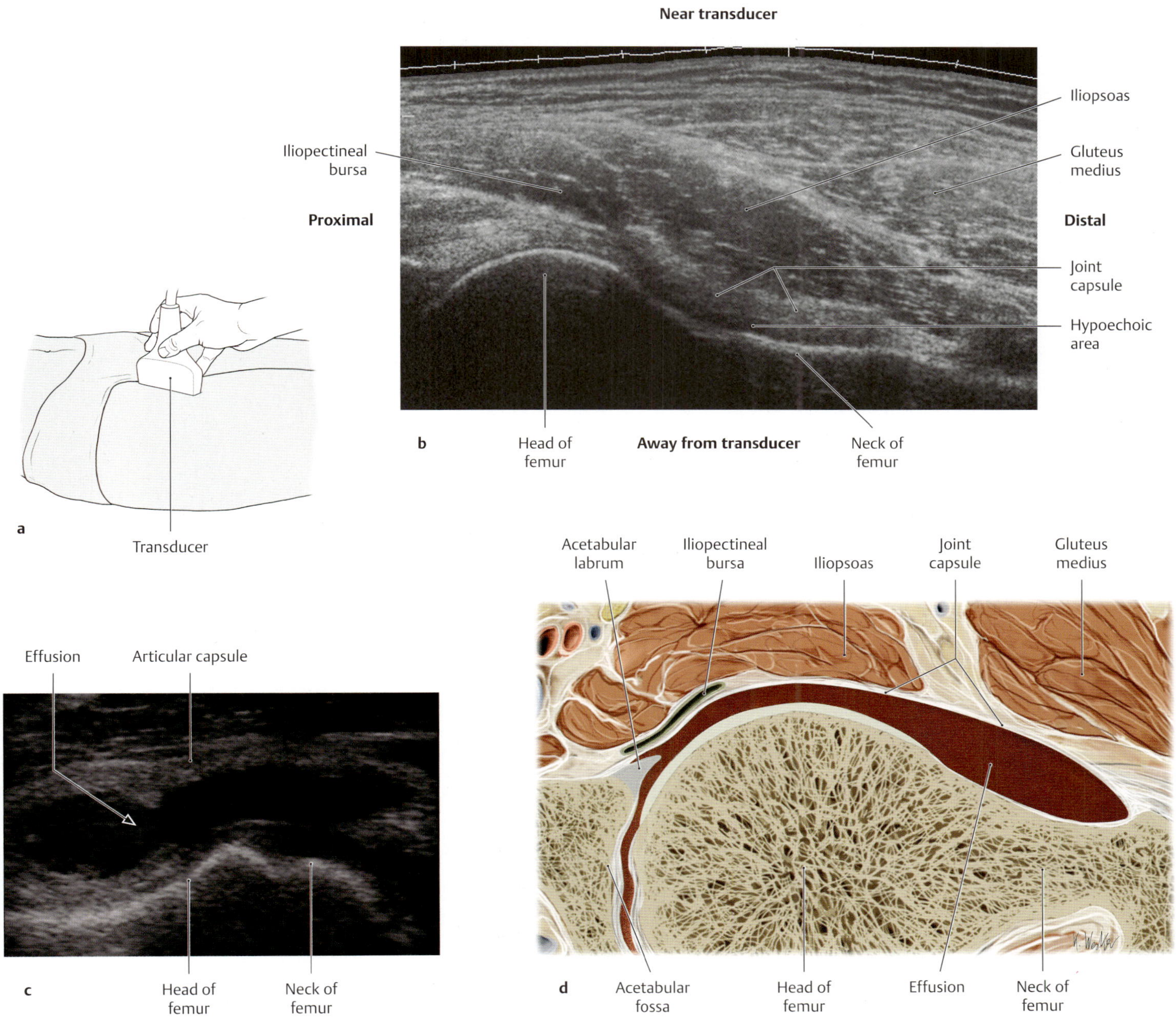

C Diagostic ultrasonography of a longitudinal section of the hip: normal and showing hip joint effusion

a Position of the transducer from the front and in the longitudinal direction to the femoral neck; **b** sonogram of a normal finding (from Konermann W, Gruber G. Ultraschalldiagnostik der Bewegungsorgane. 3rd ed. Stuttgart: Thieme; 2011); **c** sonogram of hip joint effusion (ultrasound image: Hans-Peter Sobotta, M.D., Stiftung HEH Braunschweig); **d** schematic view of the sonogram. Ultrasonography delivers real-time pictures. Structures at the top of the screen are close to the transducer; those at the bottom are farther away; those on the left side are proximal, those on the right side are distal.

In addition to the diagnostic ultrasound of the embryonic hip joint, the standardized diagnostic ultrasound of the infant hip joint (see p. 452) is of great significance as it is cost-effective and quickly done. Just as in diagnostic X-rays, two almost perpendicular sectional planes (transverse and longitudinal relative to the femoral neck) are recorded. The sonographic examination of the hip joint is performed with the patient lying down. The hip and knee joints are in neutral position (**a**). The longitudinal section allows for a very good evaluation of the anterior parts of the articular capsule and the osseous and periarticular structures of the hip joint. The surface contours of the anterior margin, the crescent-shaped femoral head, and the femoral neck appear as echoic structures. The articular capsule runs parallel to the femoral head and neck and is generally separated from the femoral neck by a narrow low echoic area (**b**). Hip disorders, which are accompanied by increased intra-articular volume (e.g., effusion as part of synovialitis or bacterial coxitis), can be very well shown with the help of an anterior longitudinal section because the effusion appears as a capsular distention at the anterior side of the femoral neck (see **c**). A lateral difference of more than 2 mm between the healthy and affected area (between the articular capsule and the femoral neck) is viewed as significant and implies increased intra-articular volume.

Note: Diagnostic ultrasound helps to show periarticular fluid buildup, e.g., in the case of trochanteric bursitis (= above the greater trochanter).

1.15 The Movements and Biomechanics of the Hip Joint

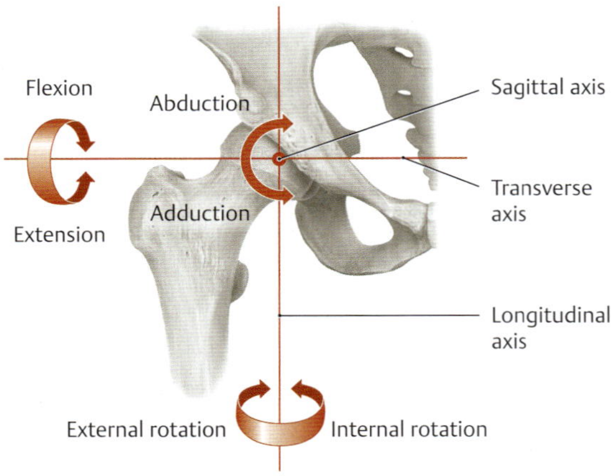

A The axes of motion in the hip joint
Right hip joint, anterior view. As a spheroidal joint, the hip has three principal axes of motion, all of which pass through the center of the femoral head (the rotational center of the hip) and are mutually perpendicular. Accordingly, the joint has three degrees of freedom, allowing movement in six principal directions:

1. Transverse axis: flexion and extension.
2. Sagittal axis: abduction and adduction.
3. Longitudinal axis: internal rotation and external rotation.

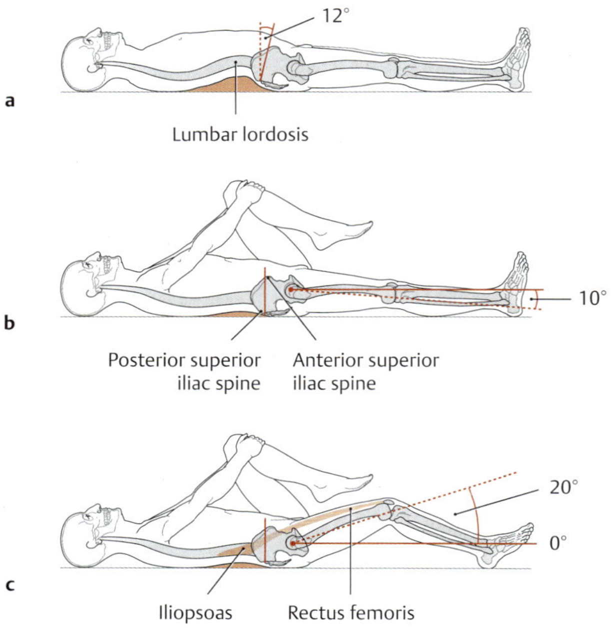

B Range of extension of the right hip joint determined with the Thomas maneuver
The Thomas maneuver is used to measure the range of hip extension while the patient lies supine on a hard surface.

a Starting position with a slight degree of anterior pelvic tilt (approximately 12°). In this position it cannot be determined whether or not there is a flexion contraction of the hip joint. This is because the patient can compensate for any limitation of extension by increasing lumbar lordosis (arching the lower back to hyperlordosis) and increasing the degree of anterior pelvic tilt.
b Pelvic tilt can be temporarily eliminated by drawing the opposite hip joint (in this case, the left hip) into a position of maximum flexion. If the right thigh remains flat on the table, the right hip joint will be in approximately 10–20° of extension (*normal extension*).
c If *hip extension is limited* (e.g., due to a shortened rectus femoris or iliopsoas muscle), and the opposite (left) hip is placed in maximum flexion, the femur of the affected leg will be raised from the table by an amount equal to the loss of extension. Increased lumbar lordosis is generally present when hip extension is limited and is easily detected clinically by palpating the lower back.

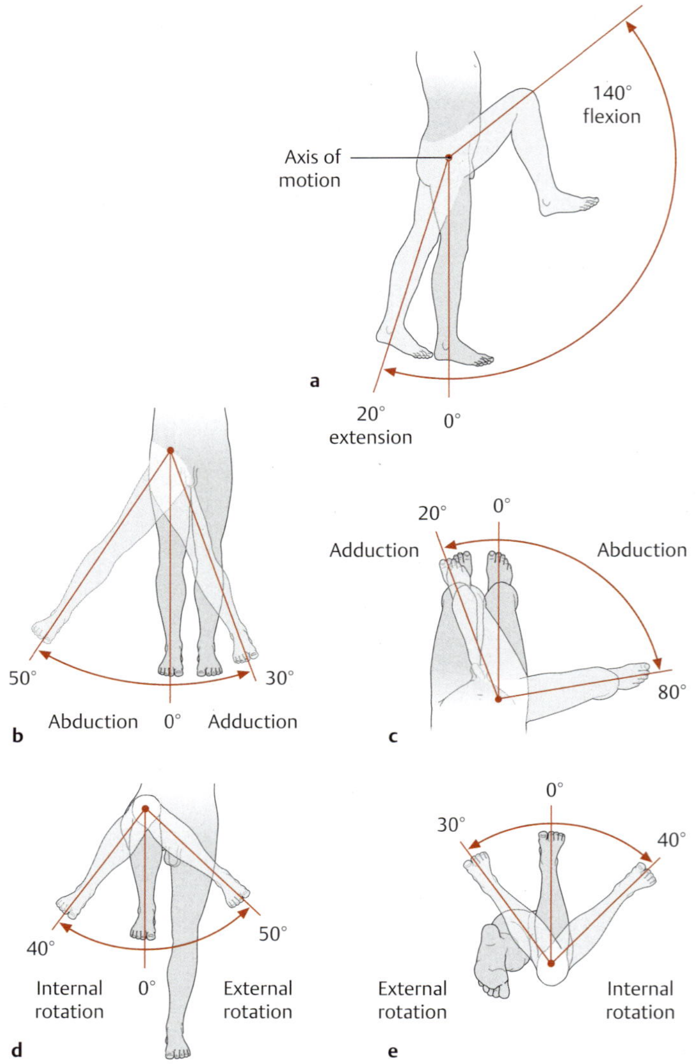

C Range of motion of the hip joint from the neutral (0°) position
The range of motion of the hip joint is measured using the neutral zero method (see p. 52).

a Range of flexion/extension.
b Range of abduction/adduction with the hip extended.
c Range of abduction/adduction with the hip flexed 90°.
d Range of internal rotation/external rotation with the hip flexed 90°.
e Range of internal rotation/external rotation in the prone position with the hip extended (when measuring rotation, the examiner uses the leg, flexed 90°, as a pointer to determine the range of motion).

D Hip joint load depending on the femoral neck angle

Each illustration shows a one-legged stance, standing on the right leg, anterior view. **a** Normal femoral neck angle, around 126°, see p. 429 (the femoral neck angle is also known as the CCD angle—centrum-collum-diaphysis angle, with centrum standing for the center of the femoral head); **b** increased femoral neck angle (coxa valga); **c** decreased femoral neck angle (coxa vara).

The resultant force vector *R* determines the hip joint load when walking or standing on one leg. It is calculated using the partial body weight *K*, muscle force *M*, and length of the lever arms (see p. 53). With a normal CCD angle, the lever arm of the body weight is around 3 times more than the one of the muscle force, meaning in a one-legged stance, the hip joint load is 4 times (*R* = 4) larger than the body weight *K*. Since the lever arm of muscle force in coxa valga (coxa vara) decreases (increases), the hip joint load increases in coxa valga (*R* = 7) and decreases in coxa vara (*R* = 3). S is the partial-body center of gravity.

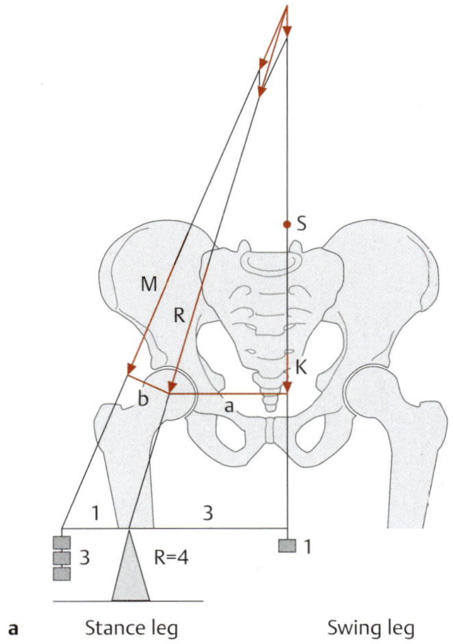

Note: Due to this relationship, surgically changing the femoral neck angle (osteotomy) influences the hip joint load.

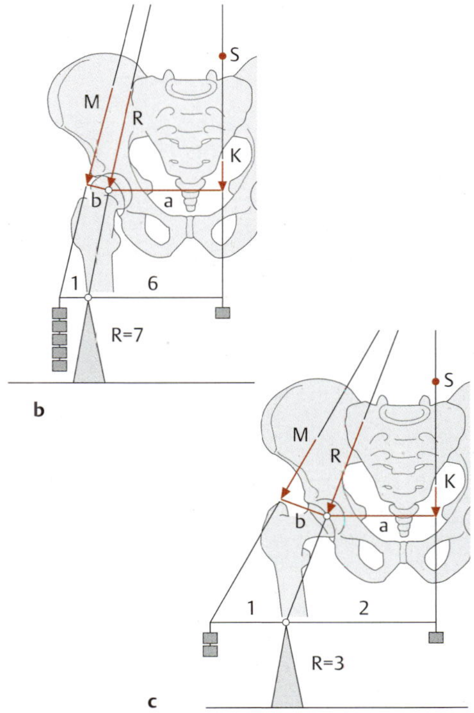

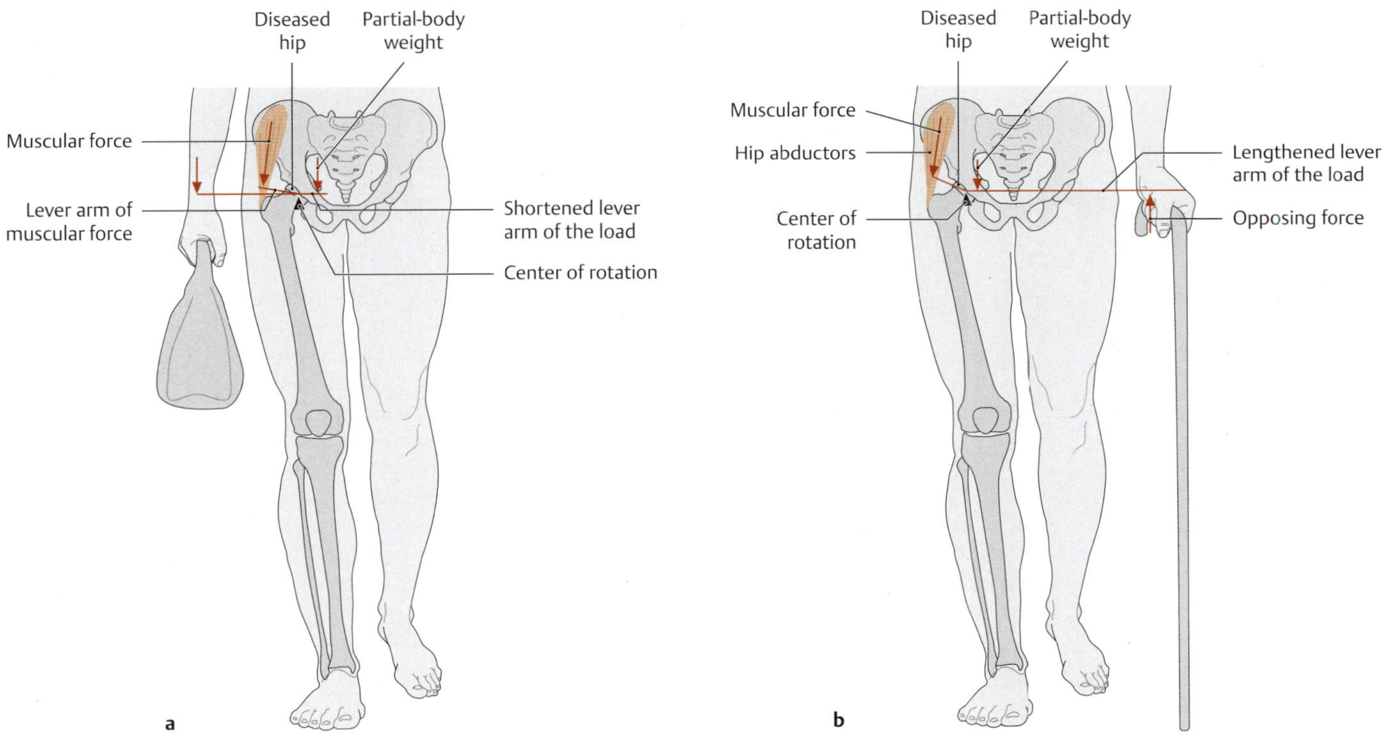

E Reducing the stresses on an osteoarthritic right hip

Anterior view. In patients with advanced osteoarthritis of the hip, various measures can be taken to alleviate the stresses, and thus the pain, on the affected side.

- **a** Shift the partial-body center of gravity (see above) toward the affected side. One way to do this is by carrying a shopping bag *on the affected right side*, as shown here. This moves the partial-body center of gravity closer to the center of the femoral head, thereby *shortening* the lever arm of the load (in this case, the partial-body weight) and also reducing the torque generated by the partial-body weight. The same effect is produced by adopting a *Duchenne limp*—an unconscious response in which the patient leans over the affected side with the upper body during the stance phase of gait (see also p. 554).
- **b** Use a cane *on the unaffected (left) side*. While this *lengthens* the lever arm of the load (the partial-body weight), it also provides a force (the cane) that counteracts the body load at the end of that lever arm. This reduces the torque generated by the load (as in **a**).

1.16 The Development of the Hip Joint

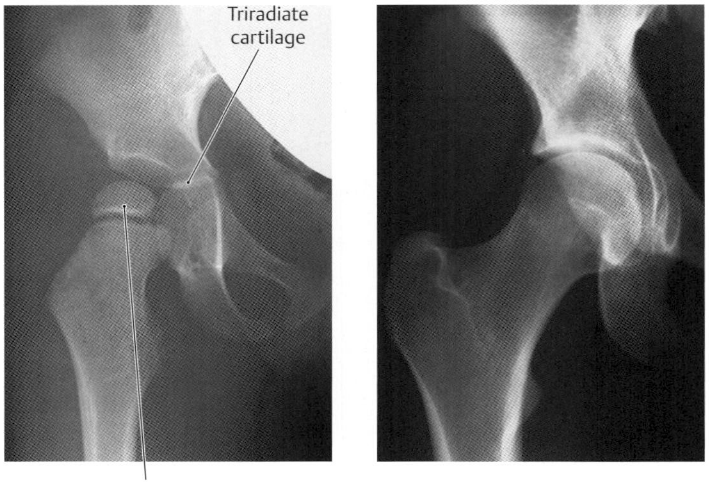

A Radiographic appearance of the right hip joint
Anteroposterior projection.

a Boy, 2 years of age (original film from: Prof. S. Müller-Hülsbeck, M.D., Institut für Diagnostische und Interventionelle Radiologie/Neuroradiologie, DIAKO Krankenhaus gGmbH Flensburg).
Note: The ossification center for the femoral head is already visible (see **B**).

b Man, 25 years of age (from Möller TB, Reif E. Taschenatlas der Röntgenanatomie. 7th ed. Stuttgart: Thieme; 2020).

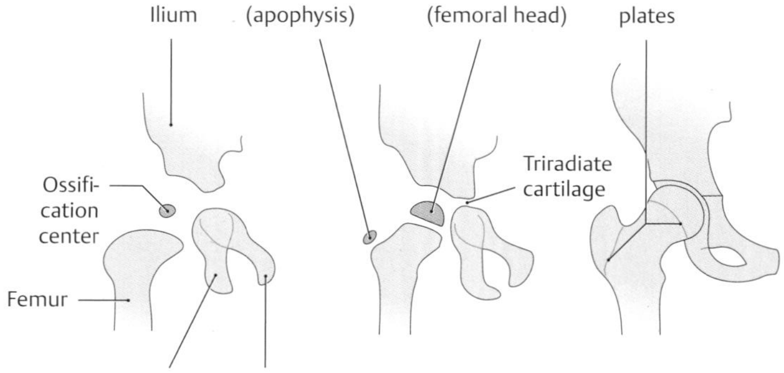

B Stages in the development of the hip joint
Schematic representations of AP radiographs of the right hip joint. The ossification centers are indicated by dark shading.

a Six-month-old child with ossification center for the femoral head;
b Four-year-old child with ossification centers for the femoral head and greater trochanter;
c Fifteen-year-old adolescent; the epiphyseal plates have not yet fused.

The anatomic differentiation of all structures that comprise the hip joint is largely complete by the 12th week of development (crown–rump length 80 mm). Whereas ossification of the acetabulum begins between the 3rd and 6th months of fetal development, the ossification center for the capital femoral epiphysis (femoral head) does not appear until about 5 to 6 months after birth. The ossification center for the apophysis of the greater trochanter appears during the 4th year of life. Fusion of the epiphyseal plates takes place between 16 and 18 years of age in the proximal femur and at about 15 years of age in the triradiate cartilage.

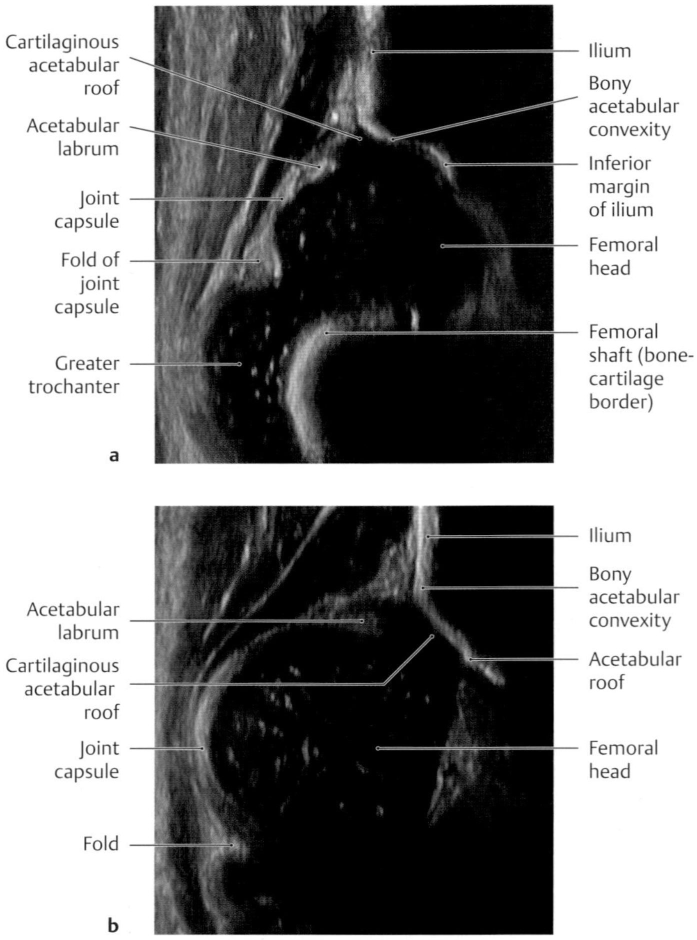

C Diagnostic ultrasound studies of the infant hip after Graf (1984)
a Mature infant hip type Ia, **b** immature decentered infant hip type III (= dislocated femoral head); both images show the right hip (original images: Jörg Detlev Moritz, M.D., Kiel).

Ultrasound examination of the hip after Graf is the most important screening method for early postnatal detection of **delayed maturation of the hip** (dysplasia) and for diagnosing **subluxation** and **dislocation** of the hip to allow prompt treatment (goal: abduction treatment or similar methods in mild forms, joint reduction in severe forms). It was introduced as a standard procedure in Germany in 1996 and is generally performed as part of the examination at the age of 3 to 5 weeks. Its great advantage is that it can be performed without subjecting the patient to ionizing radiation. The child is placed in the lateral decubitus position for the examination; the transducer is placed longitudinally above the hip perpendicular to the skin (taking care to avoid errors from tilting the transducer). The ultrasound images must be obtained in the **standard planes of section** defined by Graf (= middle of the acetabular roof, inferior margin of the ilium, and acetabular labrum). The shape of the bony and cartilaginous acetabular roof is assessed to determine the alpha angle for the degree of bony coverage and the beta angle for the degree of cartilaginous coverage of the femoral head. The measured angles, the bony shape, and the formation of the bony and cartilaginous acetabular convexity form the basis of the sonographic classification of hip dysplasia after Graf. A total of eight types of hips are differentiated (Ia/b, IIa/b, IIc, D, III, and IV). The classification of hip type involves the identification of eight anatomic structures (see **a**).

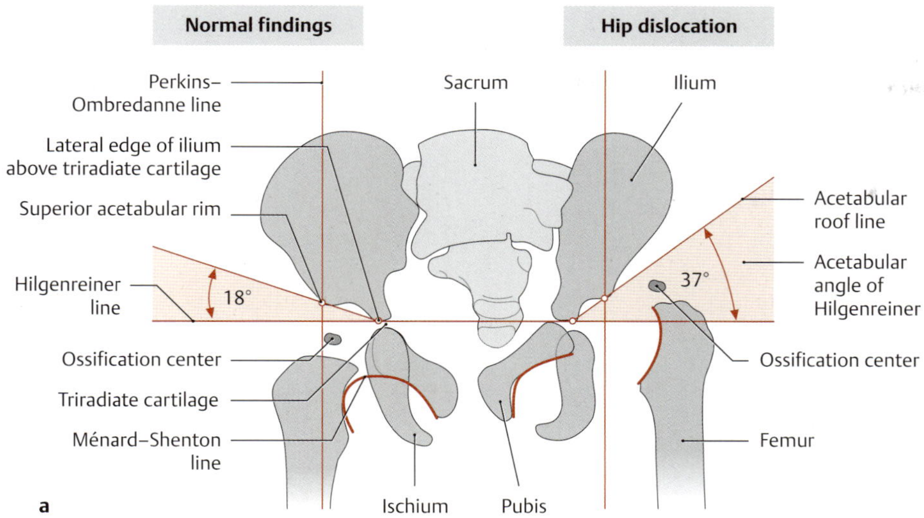

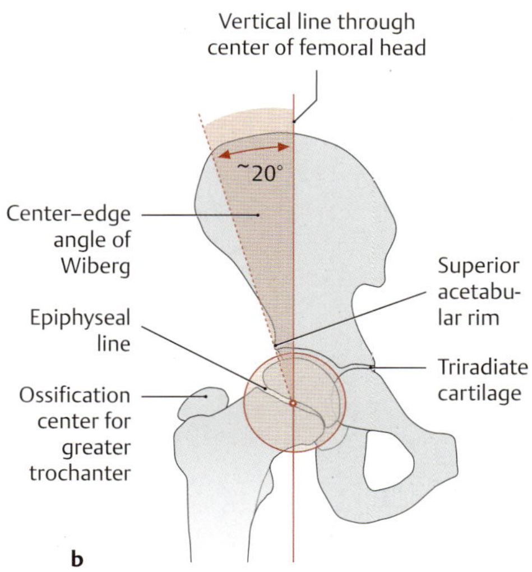

D Radiographic evaluation of the pediatric hip

Schematic representations of AP pelvic radiographs. Radiographic evaluation of the infant hip is feasible after 3 months of age, by which time there has been sufficient ossification of the joint. Both hips should always be imaged on the same radiograph.

a Normal findings (left half of figure) contrasted with findings in congenital hip dislocation (right half of figure) in a 2-year-old child. The following standard reference lines are used in the radiographic analysis of the infant hip:

- *Hilgenreiner line:* connects the inferolateral edge of the ilium above the triradiate cartilage on both sides.
- *Perkins–Ombredanne line:* drawn from the most lateral edge of the acetabular roof, perpendicular to the Hilgenreiner line.
- *Menard–Shenton line:* curved line drawn from the superior border of the obturator foramen along the medial border of the femoral neck.
- *Acetabular angle of Hilgenreiner (AC angle):* angle formed by the intersection of the Hilgenreiner line and a line connecting the superior acetabular rim with the lowest part of the ilium at the triradiate cartilage (see p. 427). This angle measures approximately 35° at birth, approximately 25° at 1 year of age, and should be less than 10° by 15 years of age.

Typically, the acetabular angle of Hilgenreiner is increased on the affected (left) side, while the center–edge angle of Wiberg (see below) is decreased. Additionally, there is a discontinuity in the Ménard–Shenton line, and the Perkins–Ombredanne line runs medial to the femoral shaft.

b Evaluation of lateral femoral head coverage based on the center–edge angle of Wiberg (drawing of a radiograph of the right hip in a 5-year-old child). The angle is formed by a vertical line through the center of the femoral head (within the future epiphyseal line) and a line drawn from the center of the femoral head to the superior acetabular rim. The center–edge angle should not be less than 10° between 1 and 4 years of age, and it should be in the range of 15 to 20° at 5 years of age.

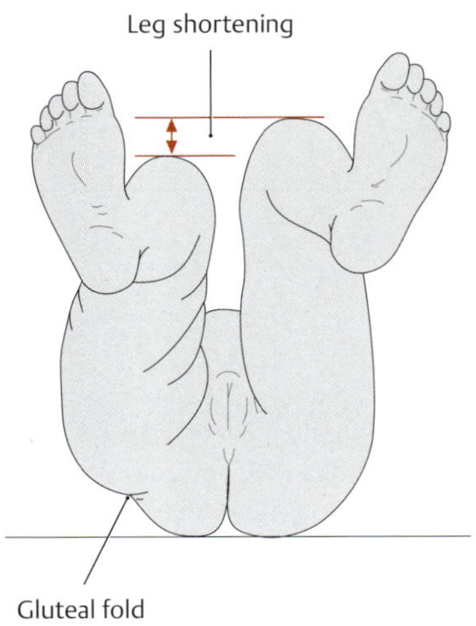

E Clinical examination of congenital dysplasia and dislocation of the hip

Hip dysplasia is characterized by an abnormal development of the acetabulum (acetabular dysplasia) in which the steep, flattened acetabular roof provides insufficient coverage for the femoral head (see also **D**). The principal complication is dislocation of the hip, since the femoral head is poorly contained in the dysplastic acetabulum and may be displaced upward and backward as a result of muscular traction or external loads. The *etiology* of hip dysplasia and dislocation is related to endogenous factors (familial disposition, maternal hormone status), as well as exogenous factors. The following *clinical signs* may direct attention to a dysplastic or dislocated hip:

- Instability of the hip joint: paucity of kicking activity or a positive Ortolani click caused by subluxation of the femoral head. The Ortolani test requires a very experienced examiner. While still considered part of the clinical examination, it is performed less often today owing to the availability of ultrasound.
- Leg shortening with asymmetry of the posterior leg folds and gluteal folds.
- Limitation of abduction due to increased reflex tension from the hip adductors.

1.17 The Knee Joint: Articulating Bones

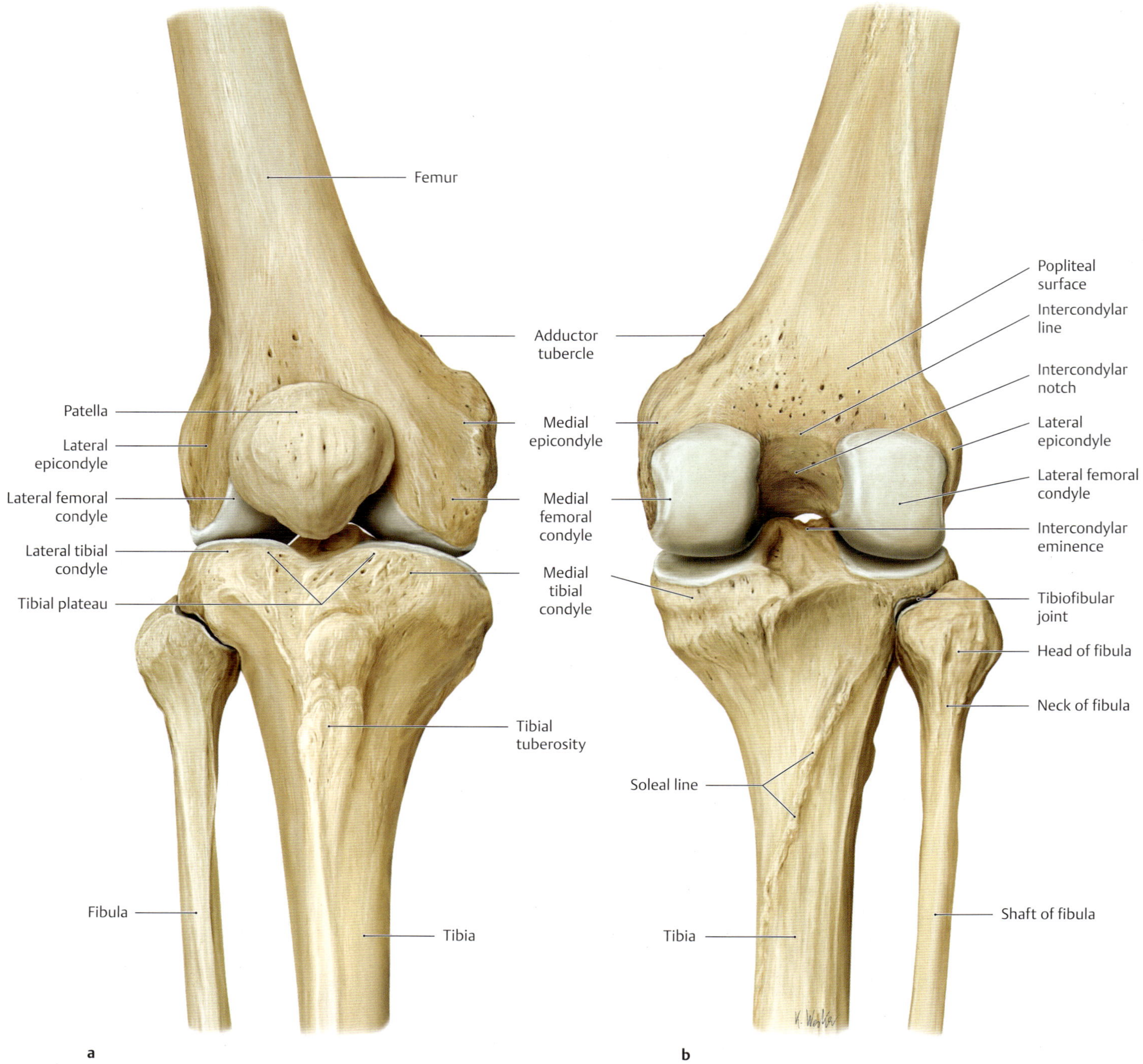

A The right knee joint from the anterior view (a) and posterior view (b)

Three bones articulate at the knee joint: the femur, tibia, and patella. The femur and tibia form the femorotibial joint, while the femur and patella form the femoropatellar joint. Both joints are contained within a common capsule and have communicating articular cavities (see p. 462). Contrasting with the elbow joint, where the forearm bones articulate with the humerus, the fibula is not included in the knee joint. It forms a separate, rigid articulation with the tibia called the tibiofibular joint.

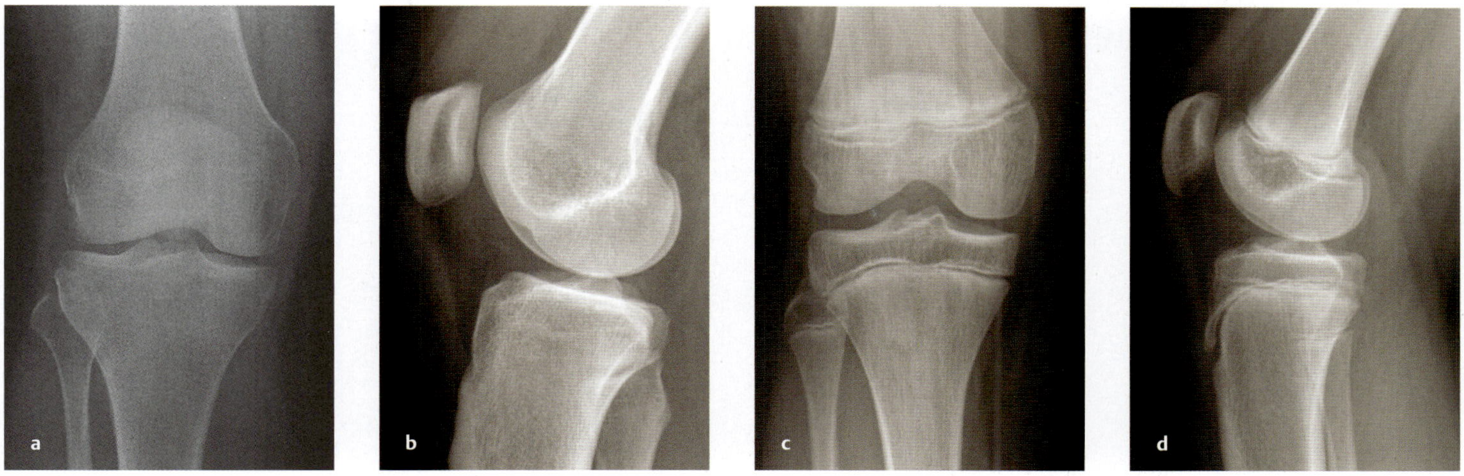

B Knee joint

Right knee joint of an adult (**a** and **b**) and a child (**c** and **d**); **a** and **c** anteroposterior projection; **b** and **d** lateral projection (**a** original film from: Prof. S. Müller-Hülsbeck, M.D., Institut für Diagnostische und Interventionelle Radiologie/Neuroradiologie, DIAKO Krankenhaus gGmbH Flensburg; **b–d** original film from: Hans-Peter Sobotta, M.D., Stiftung HEH Braunschweig).

The radiographic examination of the knee joint involves three standard films, that is to say the knee is visualized in three planes:

- Anteroposterior, to evaluate the width of the joint space and the contour of the tibial head;
- Lateral, to evaluate the shape of the femoral condyles and determine the height of the patella;
- Tangential, to evaluate the femoropatellar articulation and the position of the patella in the femoral trochlea (see **C**).

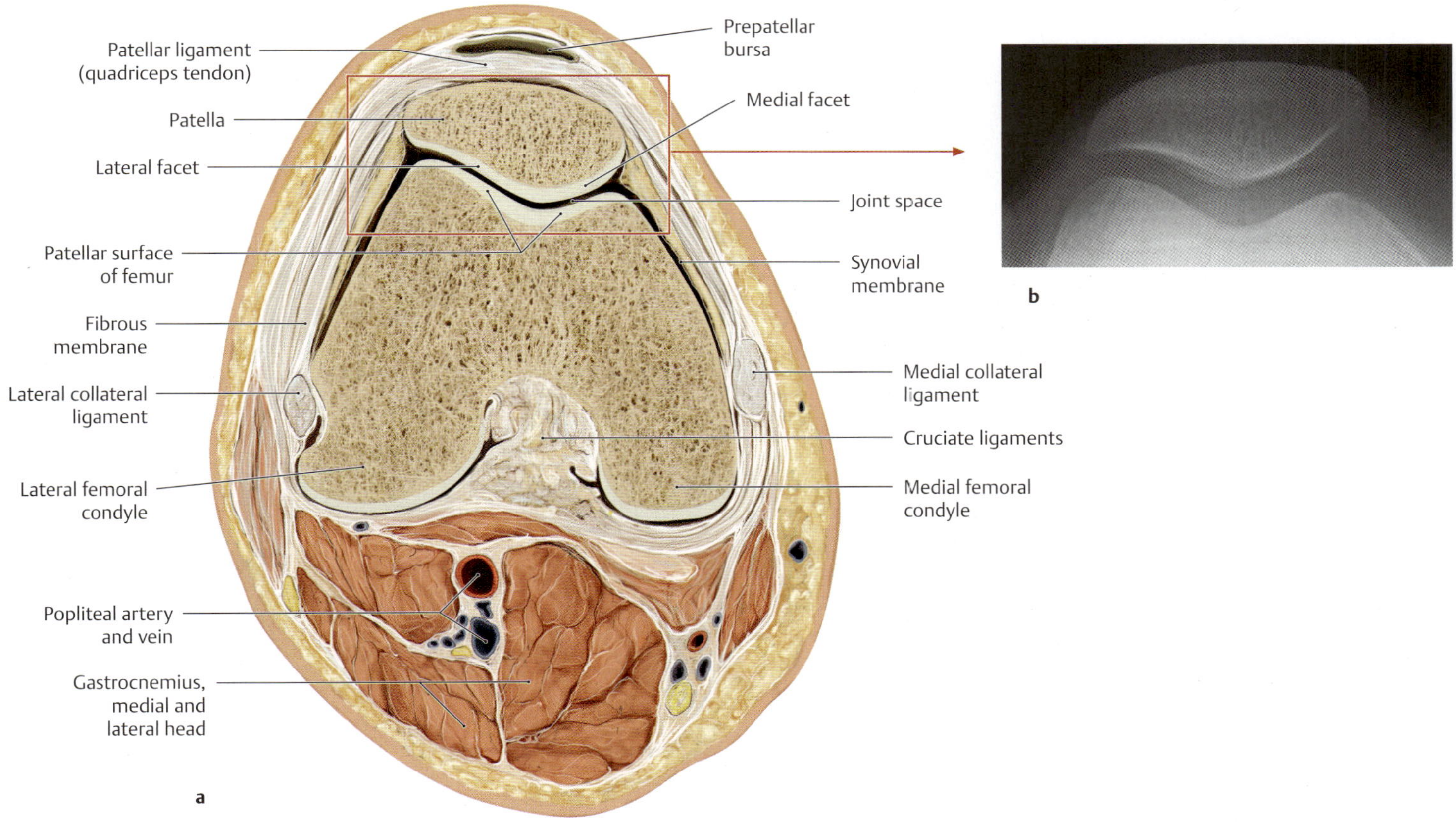

C The femoropatellar joint

a Transverse section at the level of the femoropatellar joint. Right knee joint in slight flexion, distal view (drawn from a specimen in the Anatomical Collection of Kiel University).

b Tangential radiographic view of the patella and femoral trochlea ("sunrise" view of the right knee joint in 60° flexion with the beam parallel to the posterior patellar surface). This view is excellent for evaluating the articular surface of the patella and the femoral trochlea. The radiographic "joint space" appears particularly wide owing to the relatively thick articular cartilage in this region (articular cartilage is not visible on radiographs). (Original film from: Prof. S. Müller-Hülsbeck, M.D., Institut für Diagnostische und Interventionelle Radiologie/Neuroradiologie, DIAKO Krankenhaus gGmbH Flensburg.)

1.18 The Ligaments of the Knee Joint: An Overview

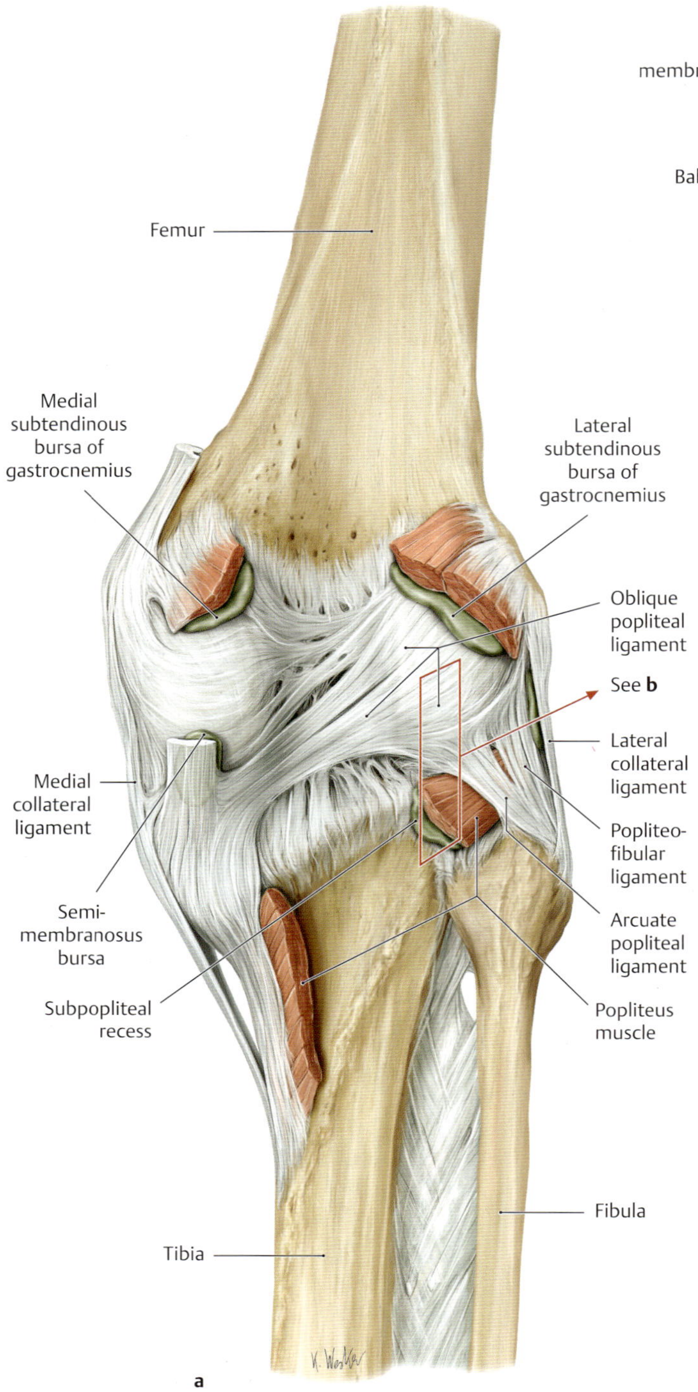

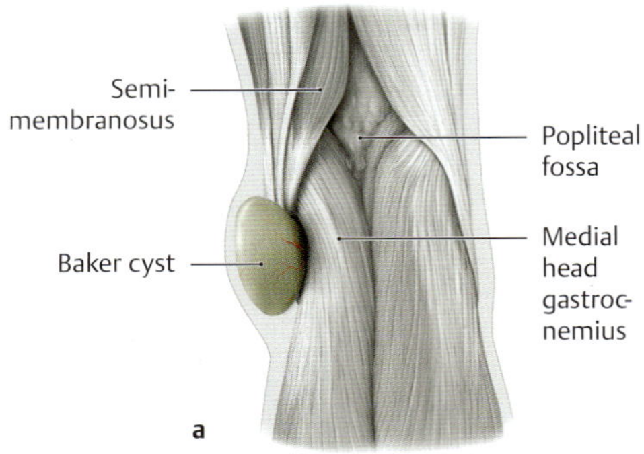

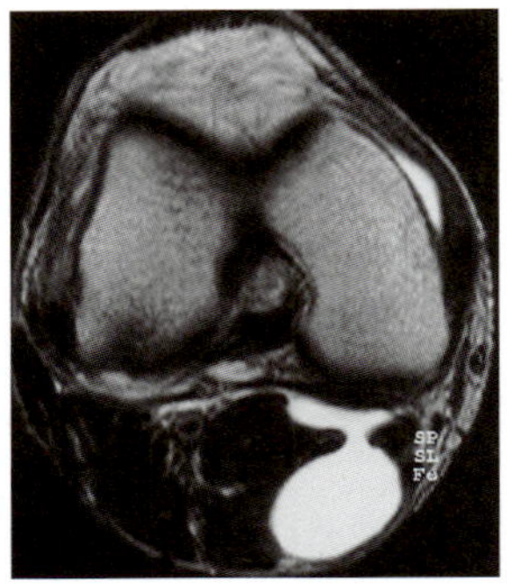

B Baker cyst in the popliteal fossa

a Right knee with Baker cyst. Posterior view.

b Axial T2-weighted MRI of a knee joint (from Vahlensieck M, Reiser M. MRT des Bewegungsapparates. 3rd edition. Stuttgart, Thieme 2006): The MR image clearly demonstrates a Baker cyst communicating with the joint space as a bright area of high signal intensity.

A Baker cyst is an enlarged, synovia-filled sac formed by the fusion of the semimembranosus bursa with the subtendinous bursa deep to the gastrocnemius. It lies in the medial popliteal fossa at the level of the medial posterior femoral condyle. Such a popliteal synovial cyst often develops secondary to a chronic inflammation such as rheumatoid arthritis that leads to increased production of synovial fluid with correspondingly increased intra-articular pressure in the knee capsule. The excess fluid flows through a sort of channel from the joint into the fused gastrocnemius-semimembranosus bursa where it accumulates (see **b**). This leads to painful swelling in the knee.

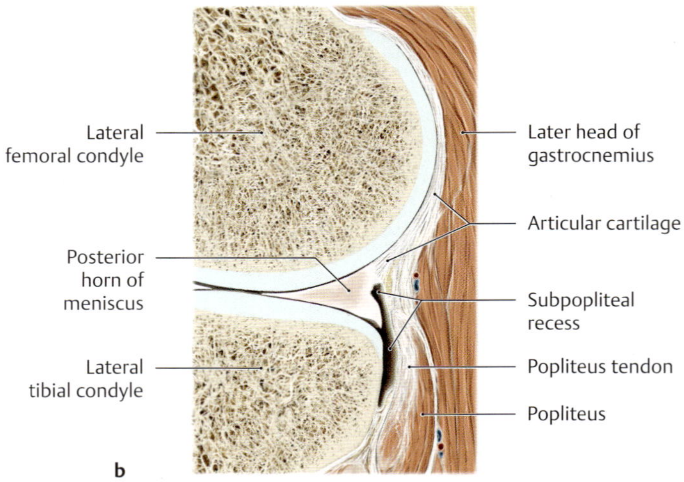

A The capsule, ligaments, and periarticular bursae of the popliteal fossa

a Right knee, posterior view.

b Sagittal section through the posterolateral corner of the capsule of a human knee at the level of the subpopliteal recess. The lateral head of the gastrocnemius has been shortened for clarity. Medial view (drawing of a specimen of the anatomic collection of the University of Kiel), see **a**.

Besides the ligaments that reinforce the joint capsule on the posterior aspect of the knee (the oblique popliteal ligament and arcuate popliteal ligament), the capsule is strengthened posteriorly by the tendinous attachments of the muscles in the popliteal region. There are several sites where the joint cavity communicates with periarticular bursae—these include the subpopliteal recess, the semimembranosus bursa, and the medial subtendinous bursa of the gastrocnemius.

Note: The popliteofibular ligament extends from the fibular head to the tendon of the popliteus and inserts together with it into the femur. Clinically, the popliteofibular ligament, the tendon of insertion of the popliteus, the lateral collateral ligament (fibular collateral ligament), and the iliotibial tract are referred to as the "posterior ligament complex." This acts as a passive stabilizer especially in external rotation and posterior tibial translation (Petersen and Zantop 2009).

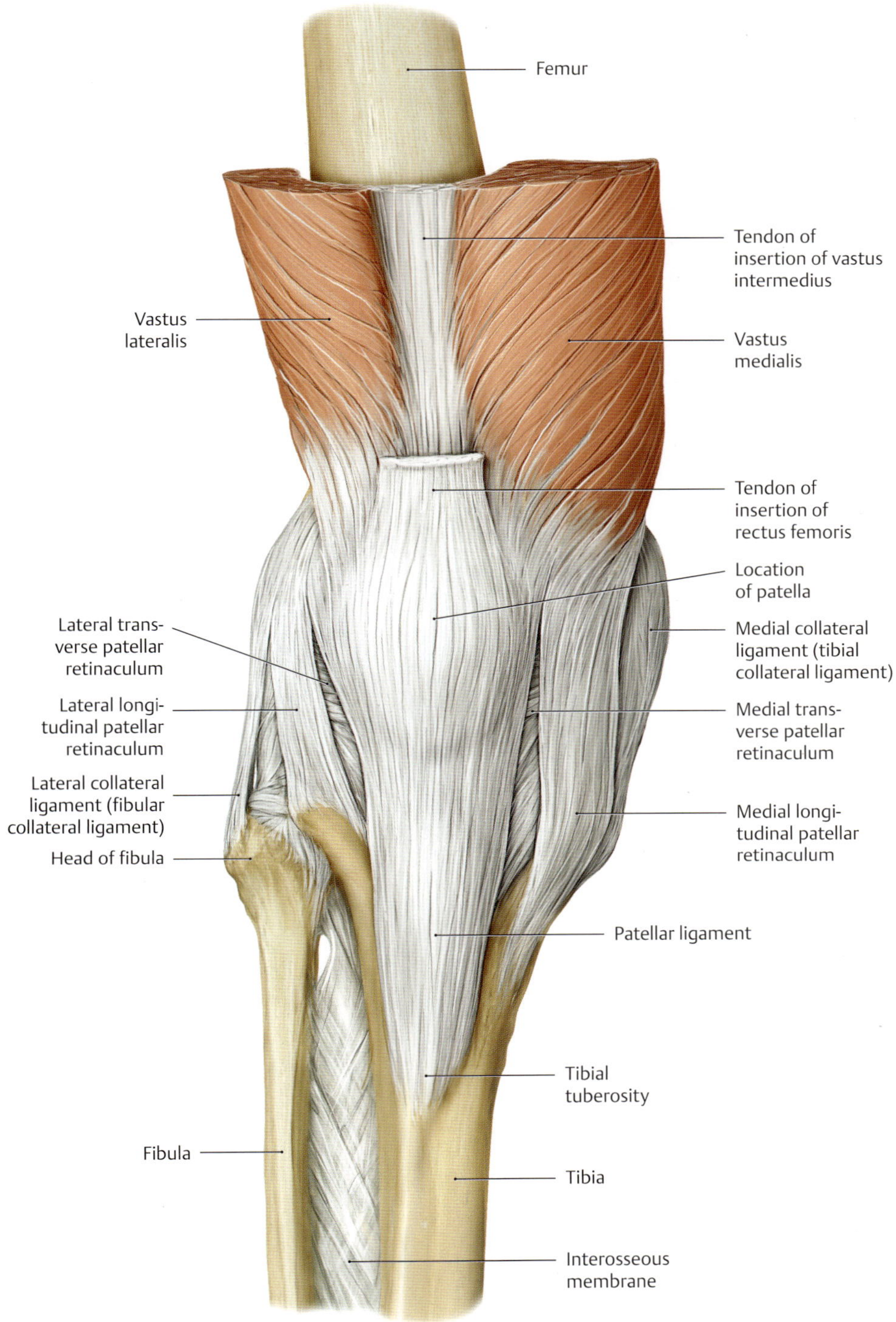

C The anterior and lateral capsule and ligaments of the right knee joint

Anterior view. The capsule and ligaments at the front of the knee serve mainly to stabilize the patella. The key stabilizers for this purpose are the tendons of insertion of the rectus femoris and vastus medialis and lateralis, the longitudinal and transverse patellar retinacula, and, at a deeper level, the meniscopatellar ligaments.

Note: The capsule and ligaments, in particular the medial and lateral collateral ligaments (see **E** and p. 458-459), in combination with other structures form the posteromedial and posterolateral ligament complex (from the posterolateral ligament complex, see **A**).

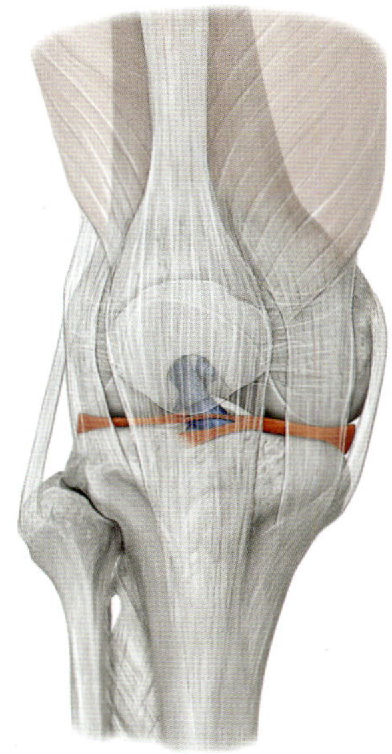

D Location of the cruciate ligaments and menisci in the knee joint

Anterior view of the right knee, in which the capsule and patella are shown in light shading. The cruciate ligaments are colored dark blue, the menisci red.

E Overview of the ligaments of the knee joint

Because its articulating bony surfaces are not closely apposed over a large area, the knee must rely upon a group of strong and extensive ligaments for stability. These ligaments of the knee joint can be segregated into two groups, extrinsic and intrinsic.

Extrinsic ligaments
• Anterior side – Patellar ligament – Medial longitudinal patellar retinaculum – Lateral longitudinal patellar retinaculum – Medial transverse patellar retinaculum – Lateral transverse patellar retinaculum • Medial and lateral sides – Medial collateral ligament (tibial collateral ligament) – Lateral collateral ligament (fibular collateral ligament) • Posterior side – Oblique popliteal ligament – Arcuate popliteal ligament
Intrinsic ligaments
– Anterior cruciate ligament – Posterior cruciate ligament – Transverse ligament of the knee – Posterior meniscofemoral ligament

1.19 The Knee Joint: The Cruciate and Collateral Ligaments

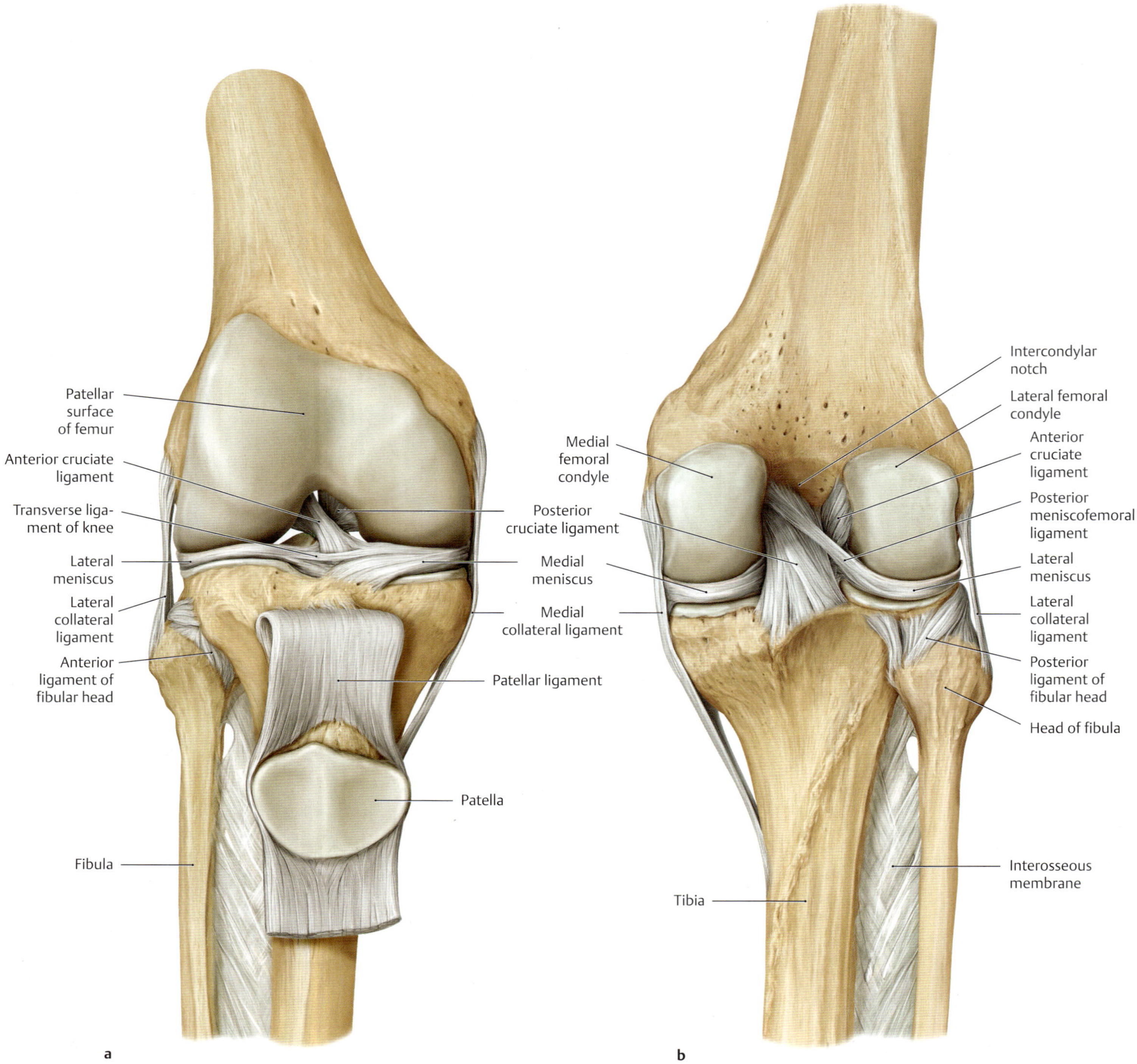

A The cruciate ligaments of the right knee joint

a Anterior view. The patellar ligament has been reflected downward with the attached patella.

b Posterior view. The cruciate ligaments of the knee stretch between the anterior and posterior intercondylar areas of the tibia (not visible here; see p. 460) and the intercondylar notch of the femur.

- The anterior cruciate ligament runs from the anterior intercondylar area of the tibia to the medial surface of the lateral femoral condyle. From an anatomic and especially a functional standpoint, two separate, steeply coursing fiber bundles are differentiated: an anteromedial (AM) and a posteromedial (PM) bundle. See p. 466: Rupture of the anterior cruciate ligament.
- The posterior cruciate ligament is thicker than the anterior ligament and is oriented at a right angle to it (cruciate = cross), passing from the posterior intercondylar area to the lateral surface of the medial femoral condyle. The posterior cruciate ligament also exhibits two fibrous bands: a stout anterolateral band and a more slender posteromedial one.

The cruciate ligaments keep the articular surfaces of the femur and tibia in contact while stabilizing the knee joint primarily in the sagittal plane. Some portions of the cruciate ligaments are taut in every position of the joint.

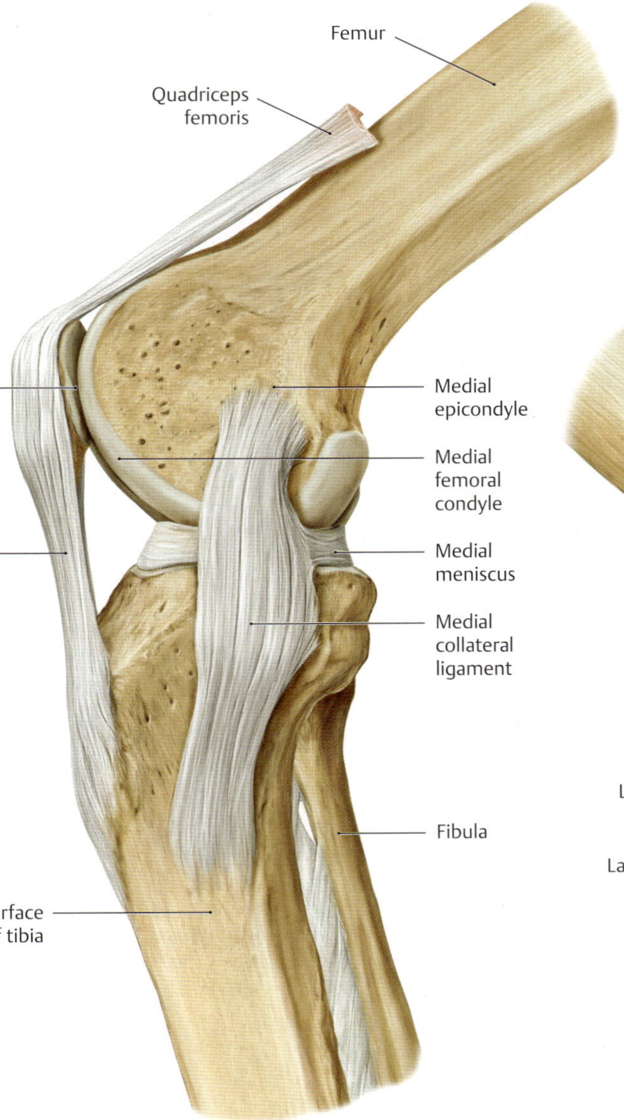

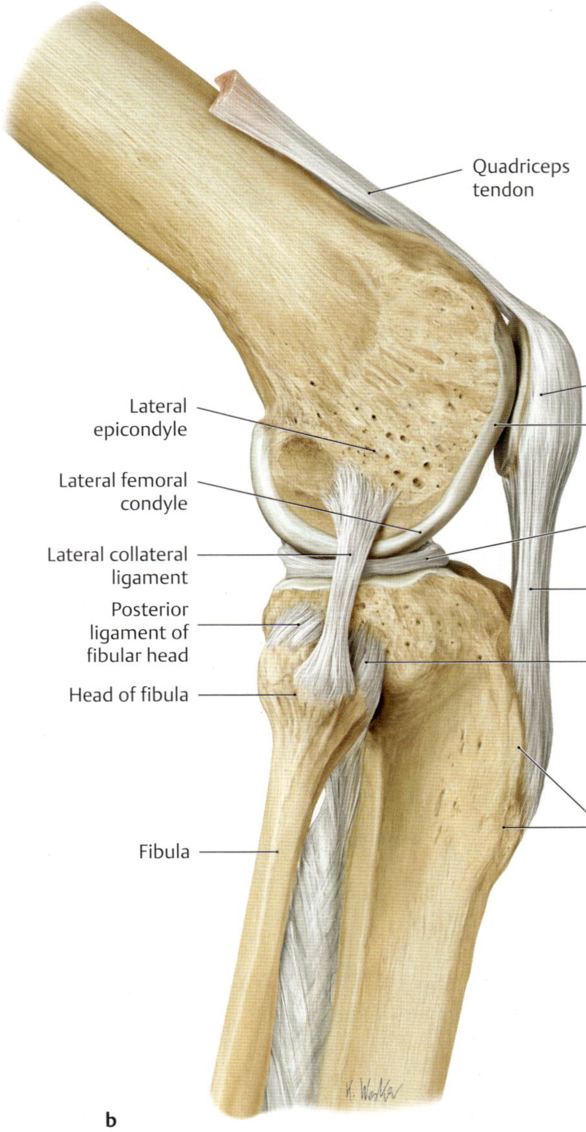

B The collateral ligaments and patellar ligament of the right knee joint

a Medial view, **b** lateral view.

The knee joint has two collateral ligaments:

- The medial collateral ligament on the medial view of the knee.
- The lateral collateral ligament on the lateral view of the knee.

The *medial collateral ligament* (tibial collateral ligament) is the broader of the two ligaments. It runs obliquely downward and forward from the medial epicondyle of the femur to the medial surface of the upper tibia approximately 7 to 8 cm below the tibial plateau. Superficial and deep portions are often differentiated. The two portions are separated by loose connective tissue. The deep portion of the medial collateral ligament is closely connected to the medial meniscus and usually ends directly inferior to the tibial plateau. The *lateral collateral ligament* (fibular collateral ligament) is a round cord that runs obliquely downward and backward from the lateral epicondyle of the femur to the head of the fibula. Both of the collateral ligaments are taut when the knee is in *extension* (see **A**). When the knee is in *flexion*, the radius of curvature is decreased, and the origins and insertions of the collateral ligaments move closer together, causing the ligaments to become lax. Both collateral ligaments stabilize the knee joint in the coronal plane. Thus, damage or rupture of these ligaments can be diagnosed by examining the mediolateral stability of the knee and the extent of medial and lateral opening of the joint space with manipulation.

Note the different relationship of each collateral ligament to the joint capsule and associated meniscus. The *medial* collateral ligament is firmly attached both to the capsule and the medial meniscus, whereas the *lateral* collateral ligament has no direct contact with the capsule or the lateral meniscus. As a result, the medial meniscus is less mobile than the lateral meniscus and is thus far more susceptible to injury (see also p. 461).

1.20 The Knee Joint: The Menisci

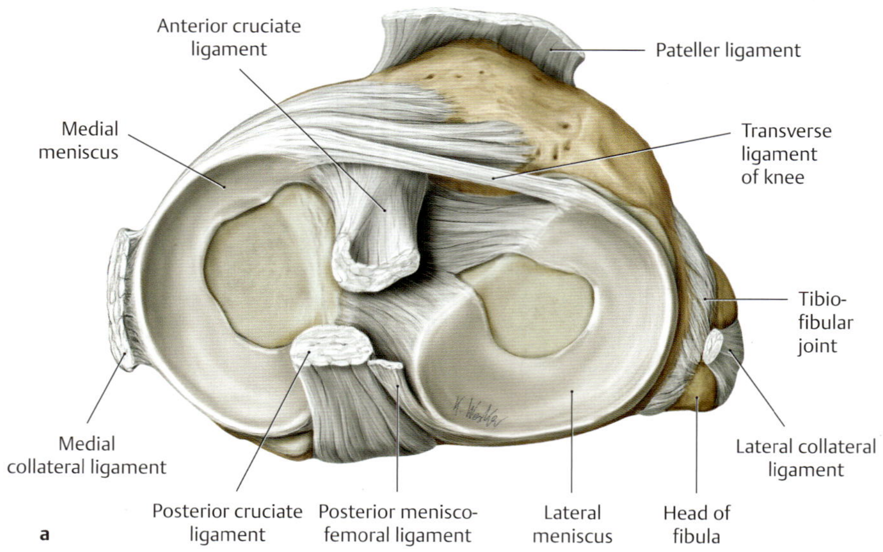

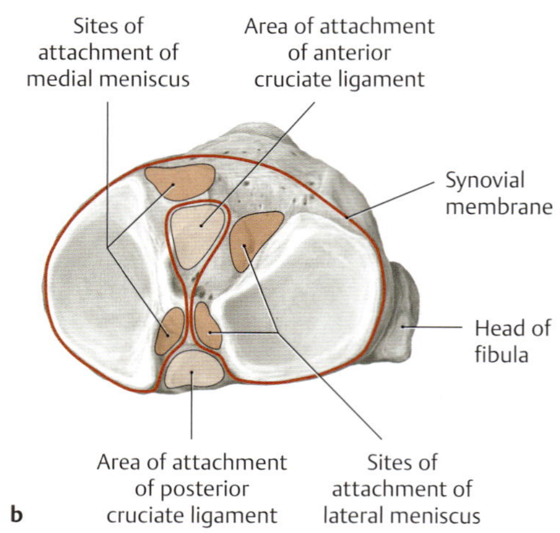

A The tibial plateau with the medial and lateral menisci and the sites of attachment of the menisci and cruciate ligaments

Right tibial plateau, proximal view with the cruciate and collateral ligaments divided and the femur removed.

a Shape and attachments of the menisci: The medial and lateral menisci are both crescent-shaped when viewed from above (L. *meniscus* = crescent). Their ends (the anterior and posterior horns) are attached by short ligaments to the bone of the anterior and posterior intercondylar areas of the tibia. The *lateral meniscus* forms almost a complete ring, while the *medial meniscus* has a more semicircular shape. On the whole, the medial meniscus is less mobile than the lateral meniscus because its points of attachment to the bone are spaced farther apart (see **b**), and it is also firmly attached peripherally to the medial collateral ligament. The lateral meniscus, by contrast, has no attachment to the lateral collateral ligament (see **E**).

b Sites of attachment of the medial and lateral menisci and the cruciate ligaments: The red line indicates the tibial attachment of the synovial membrane, which covers the cruciate ligaments anteriorly and at the sides. The cruciate ligaments lie in the subsynovial connective tissue of the joint capsule and are covered posteriorly by the heavy fibrous membrane. Because the cruciate ligaments migrate forward into the knee joint during development, they are extracapsular but intra-articular in their location (see also p. 462), and they derive their blood supply from the popliteal fossa (middle genicular artery, see p. 579).

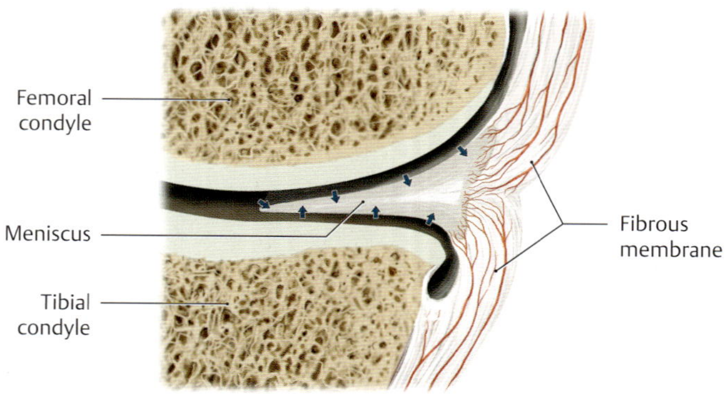

B Blood supply to the menisci

Schematic coronal section through the femorotibial joint. The fibrous portions of the menisci located adjacent to the capsule have a rich blood supply (the medial and lateral inferior articular arteries from the popliteal artery, see p. 579). But the more central portions of the menisci, composed of fibrocartilage, are avascular and are nourished entirely by the synovial fluid (arrows).

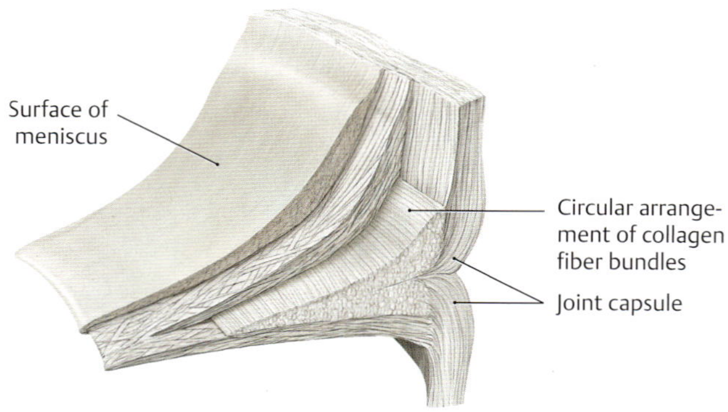

C Structure of the meniscus

The meniscus has a wedge-shaped cross section, the base of the wedge being directed toward the periphery and attached to the joint capsule. The surface facing the tibial plateau is flat, while the upper surface facing the femoral condyles is concave. The central, inner two thirds of the menisci are composed of fibrocartilage and the outer third of tough connective tissue. The bundles of collagen fibers in both the fibrocartilage and the connective tissue have a predominantly circular arrangement, reflecting the high tensile stresses that develop in the menisci. The ability of the meniscal tissue to move outward in response to loading is similar to that found in intervertebral disks (converting pressure to tensile forces).

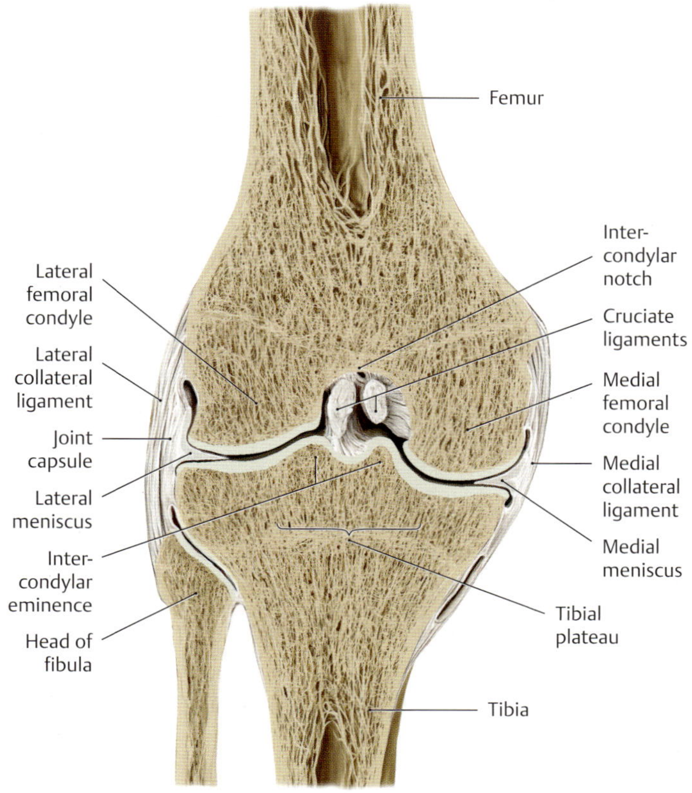

D Coronal section through the femorotibial joint
Right knee, anterior view. An essential task of the menisci is to increase the surface area available for load transfer across the knee joint. With their different curvatures, the menisci compensate for the mismatch in the articulating surfaces of the femur and tibia. They absorb approximately one third of the loads imposed on the knee, and they help to distribute the pressures more evenly within the femorotibial joint.

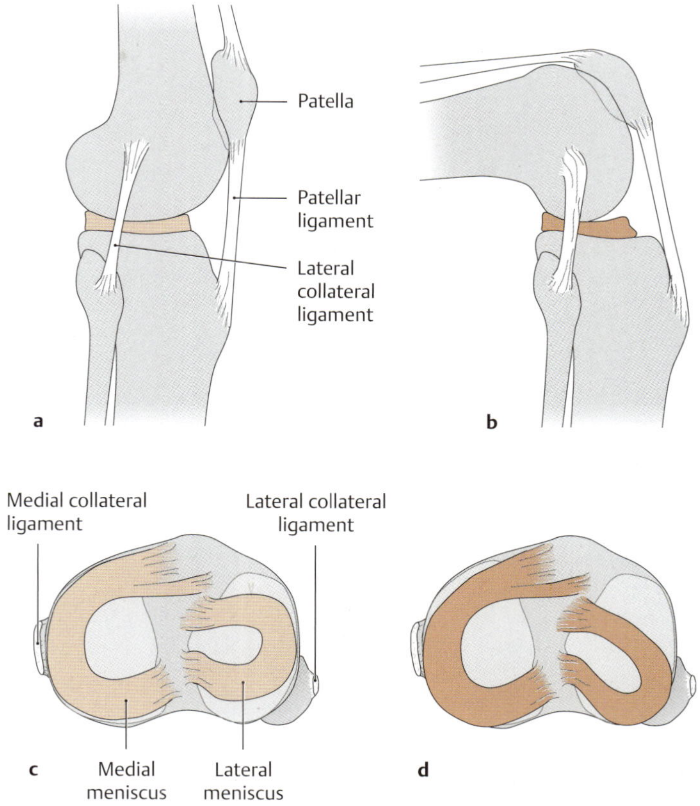

E Movements of the menisci during flexion of the knee
The drawings show a right knee joint from the lateral view in extension (**a**) and flexion (**b**) and the associated tibial plateau viewed from above in extension (**c**) and flexion (**d**).
Note that the medial meniscus, which is anchored more securely than the lateral meniscus, undergoes considerably less displacement during knee flexion.

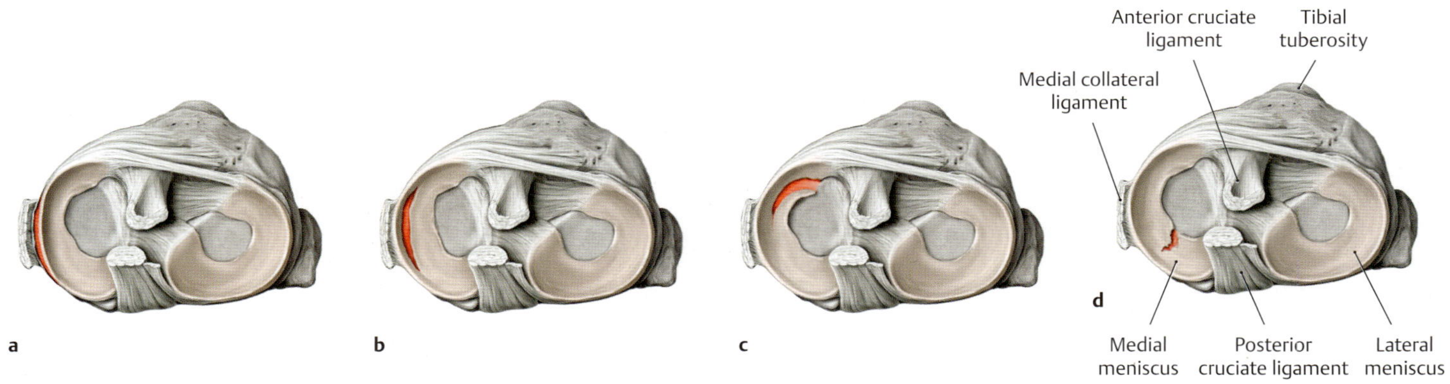

F Different patterns of meniscal tears
Right tibial plateau, proximal view.

a Peripheral tear.
b Bucket-handle tear.
c Longitudinal or flap tear of the anterior horn.
d Radial tear of the posterior horn.

The medial meniscus, being less mobile, is injured far more often than the lateral meniscus. Meniscal injuries most commonly result from sudden extension or rotational movements of the flexed knee (external and internal rotation) while the leg is fixed, as may occur while skiing or playing soccer. The resultant shearing forces can tear the substance of the meniscus or avulse it from its peripheral attachment. The cardinal feature of a fresh meniscal injury is a painful limitation of active and passive knee extension immediately after the trauma, while the patient favors the knee by keeping it slightly flexed. Degenerative changes in the menisci occur with aging and are exacerbated by excessive loads and angular deformities of the knee (genu varum or valgum, see p. 424).

1.21 The Knee Joint: Capsule and Joint Cavity

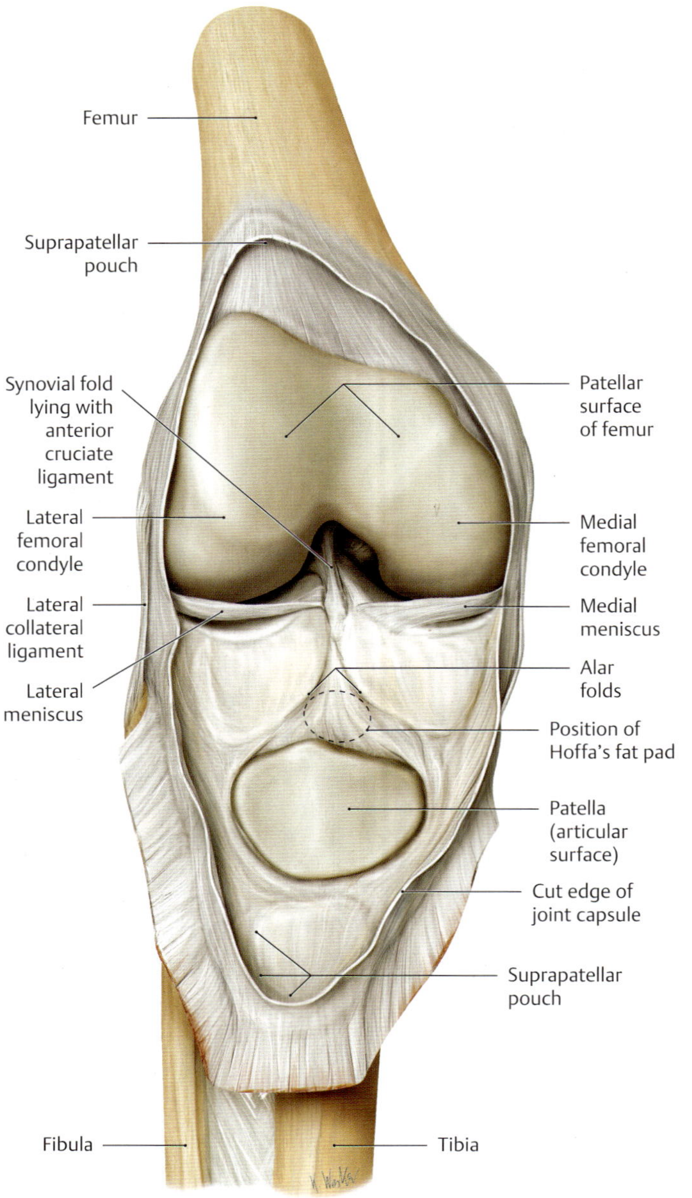

A The right knee with the joint capsule opened
The patella has been reflected downward. In the anterior menisco-femoral portion of the joint capsule, variable folds of the capsule project into the joint cavity (alar folds on both sides of the infrapatellar (Hoffa's) fat pad), increasing its capacity.

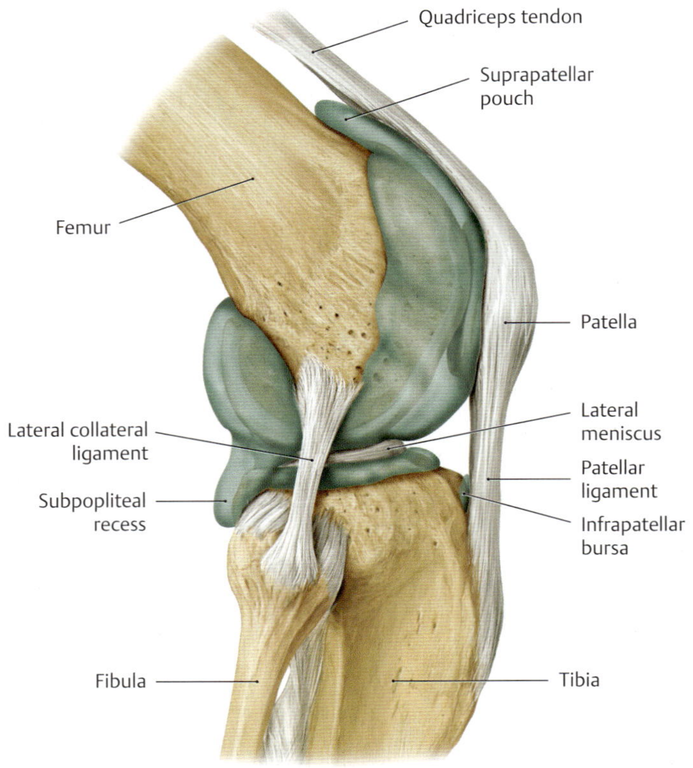

B Extent of the joint cavity
Right knee joint, lateral view. The joint cavity was demonstrated by injecting a liquid plastic into the knee joint and later removing the capsule after the plastic had cured.

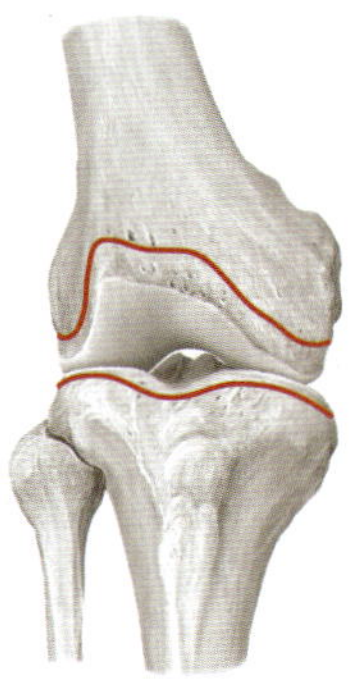

C Anterior femoral and tibial attachments of the joint capsule
Right knee joint, anterior view.

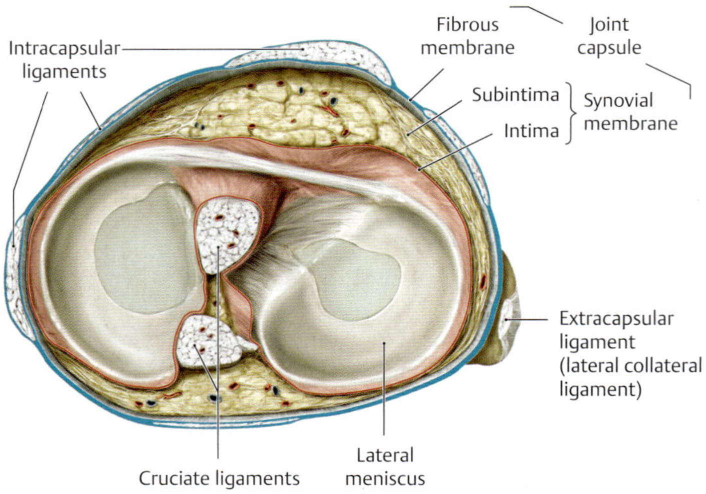

D Structure of the articular capsule
Right knee after removal of femur (articular capsule and ligaments severed), superior view.
Note: The cruciate ligaments run in the subintima of the synovial membrane and are thus intracapsular but extra-articular (compare p. 45). In contrast, the menisci are intra-articular because they are not covered by a synovial intima and have direct contact with synovial fluid (see p. 44).

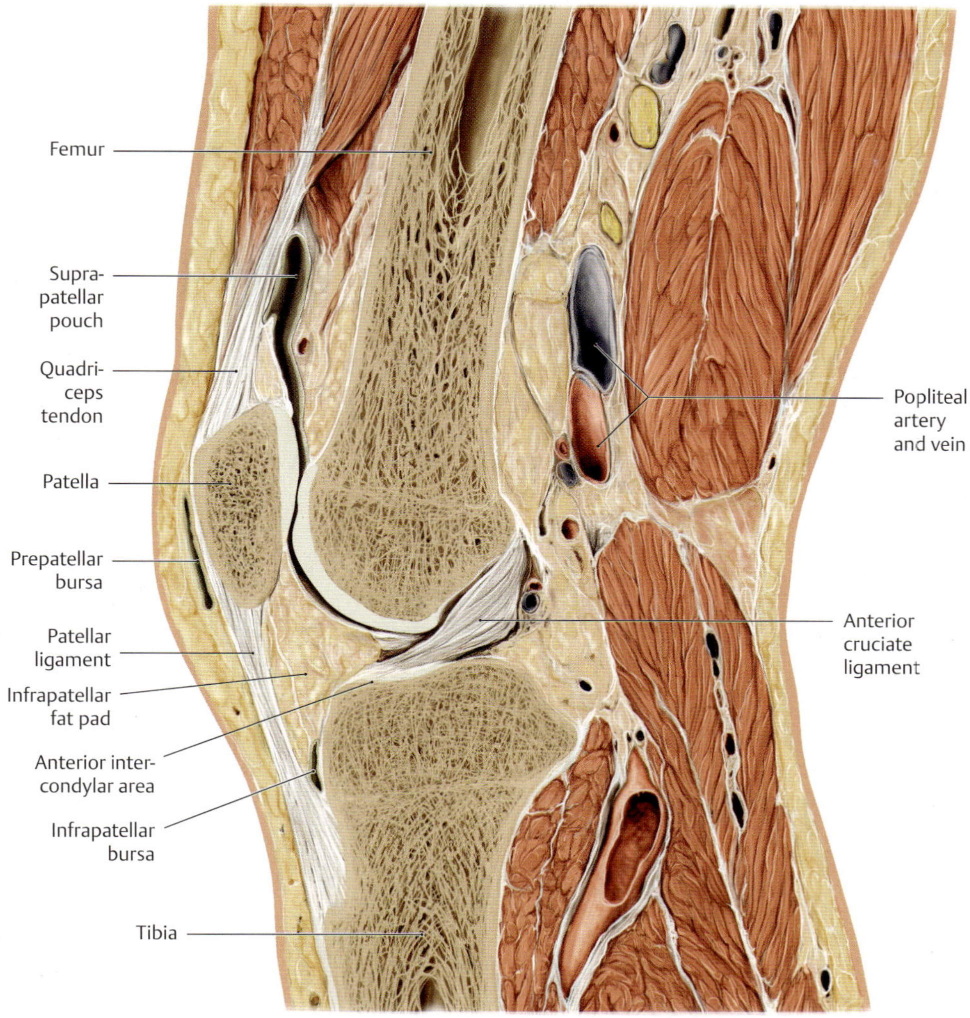

E Mid-sagittal section through the right knee joint

Note the extent of the suprapatellar pouch (also called the suprapatellar *bursa*) and compare it with **F**. *Note* also the placement of the infrapatellar fat pad between the anterior intercondylar area and the deep surface of the patellar ligament. A fall onto the knee or chronic mechanical irritation due to frequent kneeling can cause pain and inflammation of the bursae about the patella: infrapatellar bursitis ("clergyman's knee") and prepatellar bursitis (drawn from a specimen in the Anatomical Collection of Kiel University).

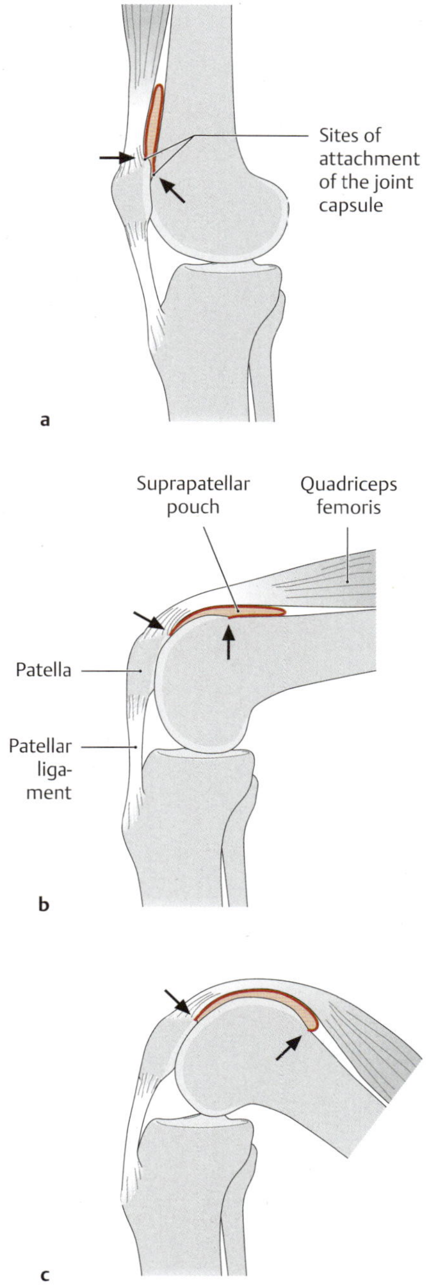

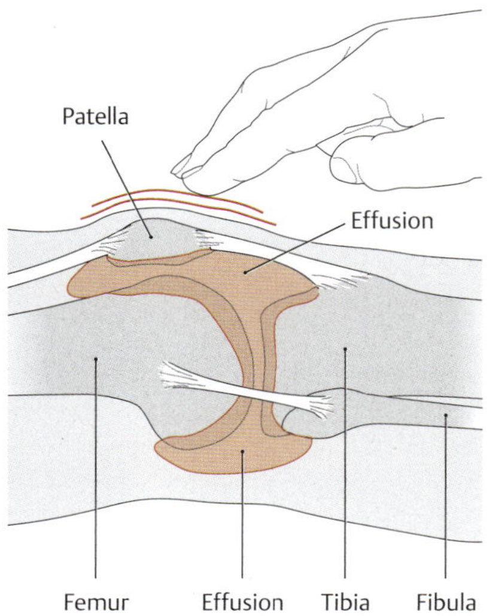

F The "ballottable patella sign" of knee effusion

When an effusion develops in the knee joint due to inflammatory changes or injury, various degrees of joint swelling may be seen. To differentiate an intra-articular effusion from swelling of the joint capsule itself, the leg is placed in a position of maximum extension. This will force the (potentially increased) intra-articular fluid out of the suprapatellar pouch and into the space between the patella and femur. The examiner then pushes the patella downward with the index finger. If there is excessive fluid in the joint, the patella will rebound when released, signifying a positive test.

G Unfolding of the suprapatellar pouch during flexion

Right knee joint, medial view.

- **a** Neutral (0°) position.
- **b** 80° of flexion.
- **c** 130° of flexion.

The suprapatellar pouch extends proximally from the superior pole of the patella, turns back distally, and inserts at the chondro-osseous junction on the patellar surface of the femur. This redundant fold provides a reserve capacity when the knee is flexed, opening up completely past about 130° of flexion.

1.22 Knee: Measuring Range of Motion and Evaluating Function of the Capsule and Ligaments

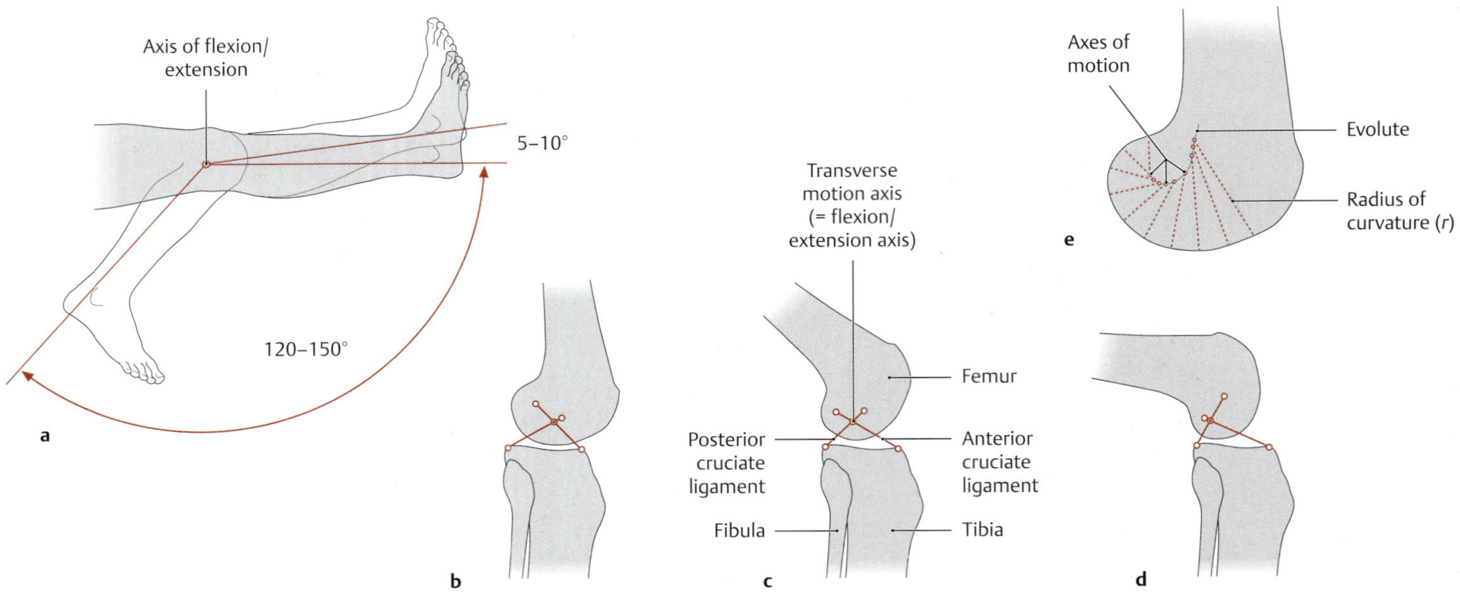

A Flexion and extension of the knee joint
Right knee joint, lateral view.
Flexion and extension of the knee joint take place about a transverse axis (**a**) that passes through the dynamic center of rotation in any joint position. That center is located at the point where both the collateral ligaments and the cruciate ligaments intersect (**b**). With increasing flexion of the knee (**c**, **d**), the dynamic flexion axis moves upward and backward along a curved line (the evolute, **e**). The momentary distance from that curve to the articular surface of the femur is equal to the changing radius of curvature (*r*) of the femoral condyle. The total range of motion, especially in flexion, depends on various parameters (soft-tissue restraints, active insufficiency or hamstring tightness, see p. 503).

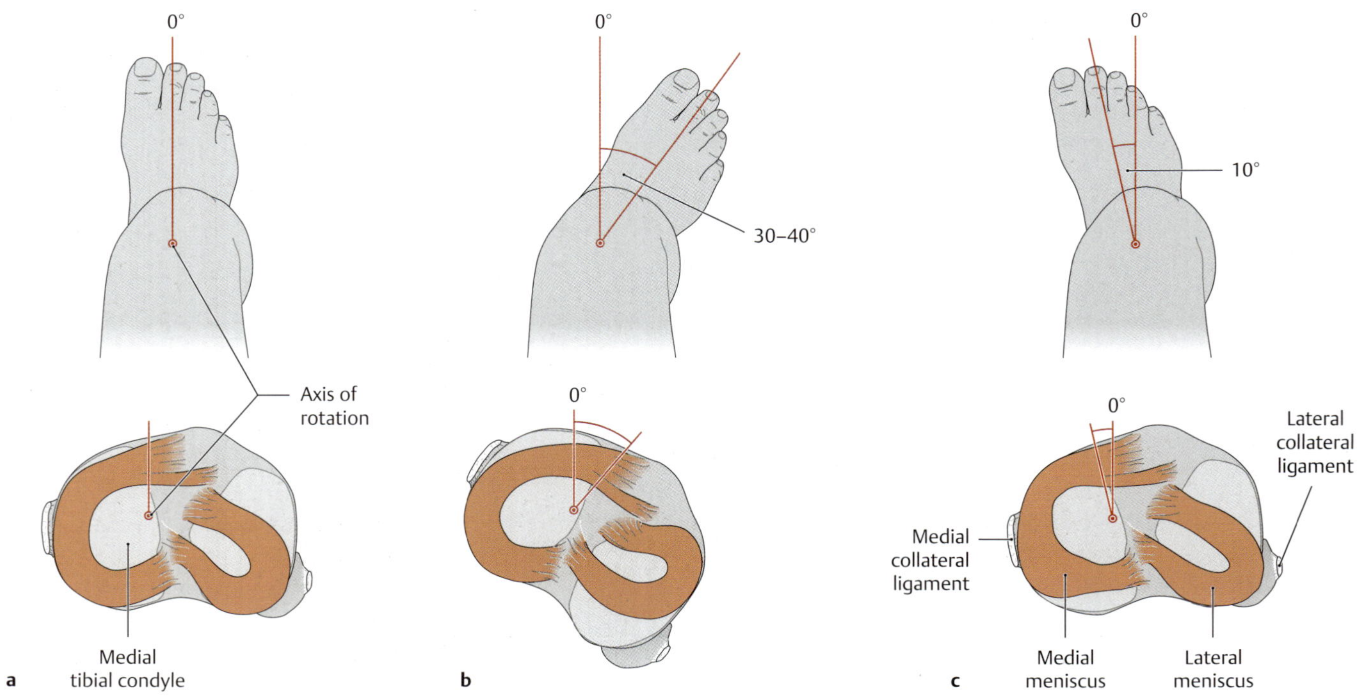

B Rotational movements of the tibia relative to the femur with the knee flexed 90°
Right knee joint, proximal view of the flexed knee and corresponding tibial plateau.

a Neutral (0°) position.
b External rotation.
c Internal rotation.

The axis of tibial rotation runs vertically through the medial part of the medial tibial condyle. Because the cruciate ligaments (not shown here) twist around each other during internal rotation, the range of internal rotation in the knee (approximately 10°) is considerably smaller than the range of external rotation (30–40°). As a result, the majority of cruciate ligament tears occur during internal rotation and involve the anterior cruciate ligament.
Note the different degrees of displacement of the lateral and medial menisci.

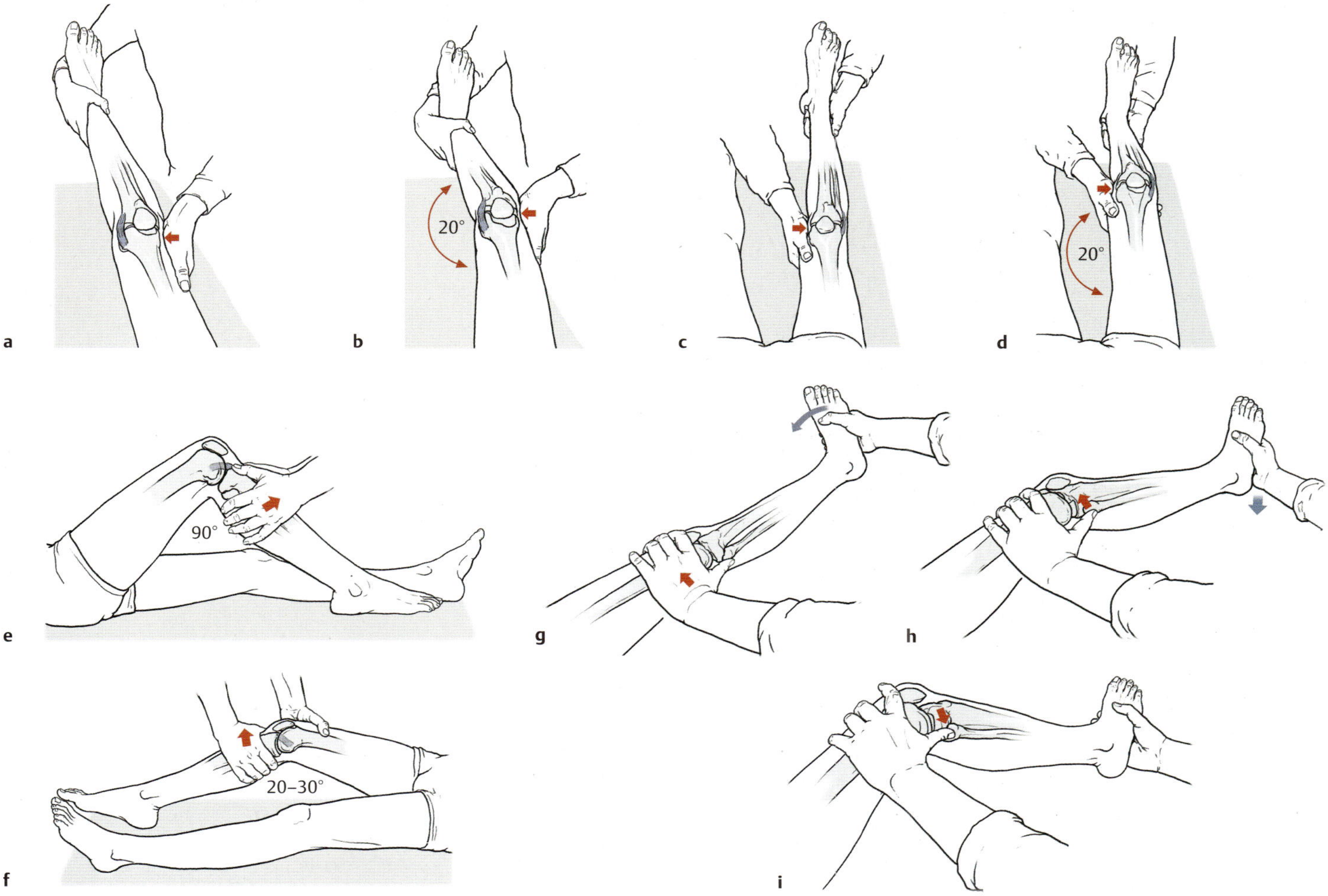

C Evaluating function of the capsule and ligaments (selection)
About 90% of all injuries to the capsule and ligaments can be diagnosed sufficiently reliably on the basis of a detailed history and clinical examination. Depending on the mechanism of injury they produce more or less complex instabilities of the knee.
Note: In order to evaluate stability, it is essential to know in which position the knee is stabilized by precisely those structures one wants to evaluate. Tests of stability can be classified as follows:

a–d Valgus and varus stress test (testing lateral opening of the knee)
The patient lies supine with the knee either extended or in 20° of flexion. The examiner grasps the patient's lower leg with one hand at the level of the ankle. With the other hand, the examiner produces a valgus or varus stress, pressing against the knee from the outside (**a** and **b**) or inside (**c** and **d**). Increased opening under valgus stress in extension suggests not only injury to the medial collateral ligament but also injury to the posteromedial capsule and cruciate ligaments. In contrast, valgus stress at 20° of knee flexion permits isolated evaluation of the medial collateral ligament. Increased opening under varus stress in extension suggests not only injury to the lateral collateral ligament but also injury to the posterior capsule and cruciate ligaments. Increased opening under varus stress at 20° of knee flexion suggests isolated injury to the lateral collateral ligament.

e and f Drawer test and Lachman test
The two tests are performed for a suspected injury to the anterior or posterior cruciate ligament (anterior or posterior drawer test and Lachman test), whereby the Lachman test is particularly sensitive. The illustrations show the respective anterior version as the anterior cruciate ligament (ACL) is injured more often than the posterior one (see p. 466). In the anterior drawer test (**e**) the patient lies supine. The hip is flexed 45°, the knee 90°, and the foot is immobilized with the buttocks. The examiner grasps the tibial head with both hands and attempts to shift the tibia anteriorly. In the anterior Lachman test (**f**) the patient also lies supine. However, the knee and hip are both flexed 20–30°. This is because in this rest position close to extension the ACL assumes an essential stabilizing function as the effective lever arm of the hamstrings is minimized. Therefore, damage to the anterior cruciate ligament will be particularly apparent in this position. The examiner grasps the thigh with one hand and pulls the tibia anteriorly with the other. In both the anterior drawer test and the Lachman test a soft endpoint suggests a compete cruciate tear, whereas a hard endpoint suggests an intact ligament or a partial tear.
"Soft endpoint" means that the tibia does not stop abruptly when one tries to push it anteriorly or posteriorly, whereas a "hard endpoint" means that it stops abruptly.

g–i Pivot shift test (test for dynamic anterior subluxation)
Here too, the illustration shows the version of the anterior subluxation test used to demonstrate an ACL injury. The patient lies supine. The examiner stands next to the extremity to be tested. The examiner holds the patient's extended leg above the heel in internal rotation and applies a valgus stress via the proximal tibia (**g**). When this stress is applied in the presence of an anterior cruciate rupture, increased anterior translation (subluxation) of the lateral tibial plateau will occur sometime between full extension of the knee and 30° flexion (**h**). Slowly flexing the knee further will usually cause the tibial plateau to snap back into place at 20–30° of flexion due to the pull of the iliotibial tract, whose direction of pull changes at this angle of flexion (**i**). The patient experiences this as an unpleasant feeling of instability.

1.23 Knee: Rupture of the Anterior Cruciate Ligament

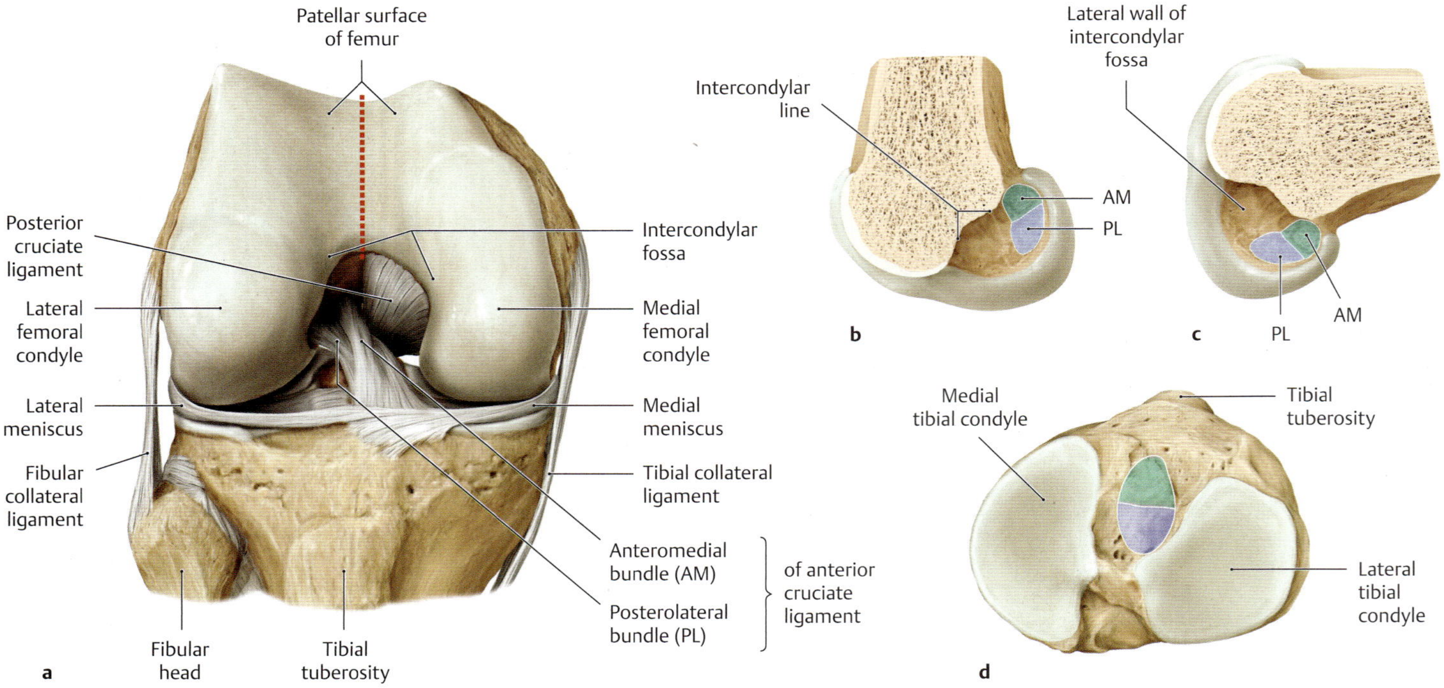

A Anatomy of the anterior cruciate ligament (ACL)

a The right knee in flexion. Anterior view with the patella and anterior capsule removed. **b** and **c** Position of the origins ("footprints") of the anteromedial and posterolateral fiber bundles in the lateral wall of the intercondylar fossa in extension (**b**) and flexion (**c**). **d** Insertions of the two fiber bundles on the tibial plateau.

The two fiber bundles of the anterior cruciate ligament are referred to as AM and PL bundles according to their tibial insertion (see **d**). The femoral origin of both bundles in the posterior lateral intercondylar fossa borders superiorly on the intercondylar line (AM bundle) and extends along the bone-cartilage border (PL bundle) (see **c**). Occasionally an additional bony ridge known as the "lateral bifurcate ridge" may be observed between the two origins. As they course through the intercondylar fossa, the two bundles lie behind each other so that the anterior cruciate ligament exhibits an oval cross section (see **d**). Because of their different lengths (AM bundle = approximately 38 mm and PL bundle = approximately 20 mm) and the different angle at which they course, different parts of the bundles are taut or relaxed in different positions of the knee (see **B**).

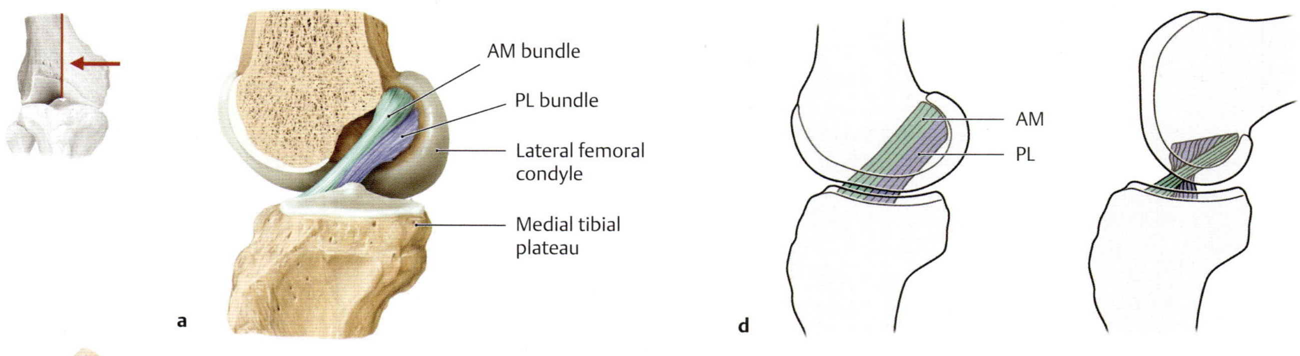

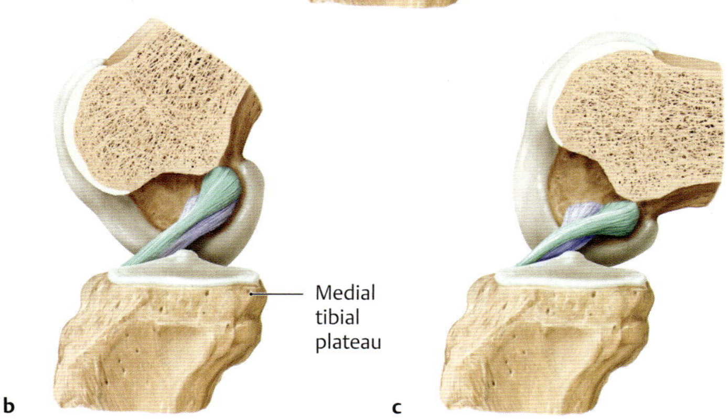

B Function of the anterior cruciate ligament

a Right knee in various degrees of flexion (**a** = 0°, **b** = 35°, and **c** = 90°). View of the lateral wall of the intercondylar fossa with the medial femoral condyle removed (see Navigator). **d** Schematic diagram of the fiber bundle under tension in various degree of flexion.

The two fiber bundles (AM and PL bundle) exhibit reciprocal degrees of tension, i.e., the AM bundle is under tension when the knee is in flexion, whereas the PL bundle is under tension primarily in extension. In this manner the anterior cruciate ligament ensures anterior stability as well as rotational stability in various degree of flexion.

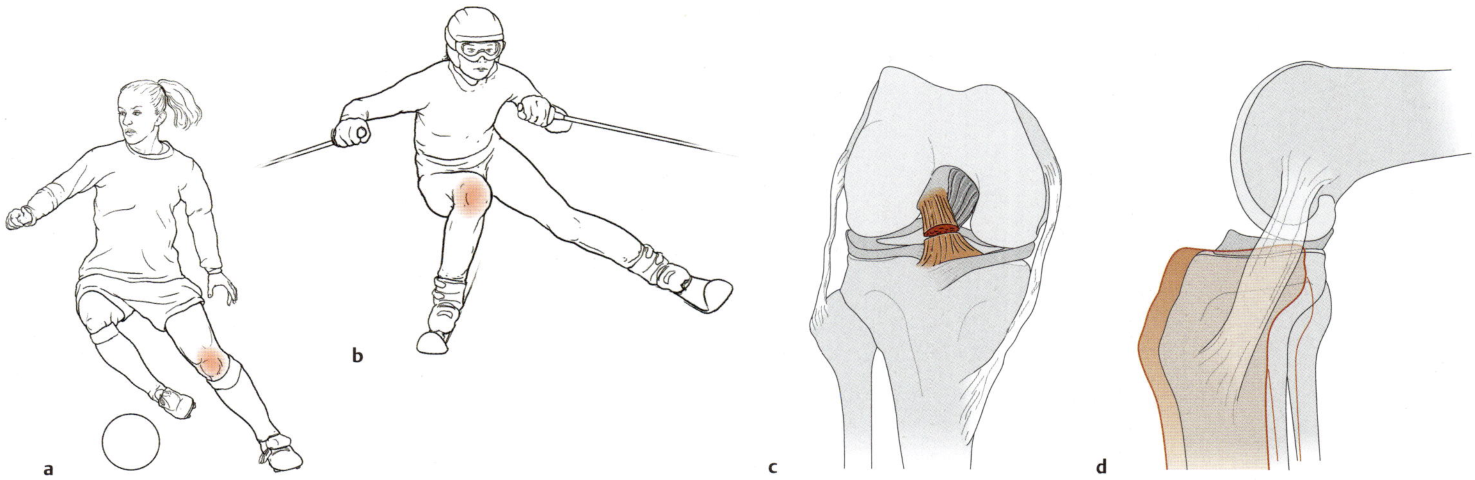

C Rupture of the anterior cruciate ligament: mechanisms of injury, epidemiology, and diagnostics

a Typical mechanism of injury with a sudden change of direction when running. The knee is flexed, in valgus, and externally rotated.

b Typical mechanism of injury with immobilized lower leg and greatly flexed knee (internal rotation trauma with the body's center of gravity behind the knee).

c and **d** Status post rupture of the anterior cruciate ligament. Anterior view (**c**) and medial view (**d**) of the right knee in flexion.

The rupture of the anterior cruciate ligament (ACL) is a typical sequela of sports injuries and occurs about 10 times as often as a rupture of the posterior cruciate ligament. Twenty percent of all knee injuries involve an ACL rupture. It is thus the most common knee injury of all (in Germany currently approximately 35,000 per year, in the USA 100,000 per year). Seventy percent of ACL ruptures occur between the ages of 15 and 45, whereby women are affected significantly more often than men. Common associated injuries include ruptures of the medial collateral ligament and the medial meniscus ("unhappy triad" injury). In addition to a precise history, careful inspection, imaging studies, and specific diagnostic tests are crucial. One of the classic tests for evaluating function of the capsule and ligaments is the drawer test (see p. 465). A positive anterior drawer suggests a rupture of the anterior cruciate ligament, a positive posterior drawer rupture of the posterior cruciate ligament.

Note: Clinical interpretation of drawer tests can be difficult. For example, an anterior drawer test can be negative in the presence of a ruptured anterior cruciate ligament due to the tension of the hamstrings. The pain of acute injury may make it impossible to bend the knee 90° which is necessary to perform the drawer test.

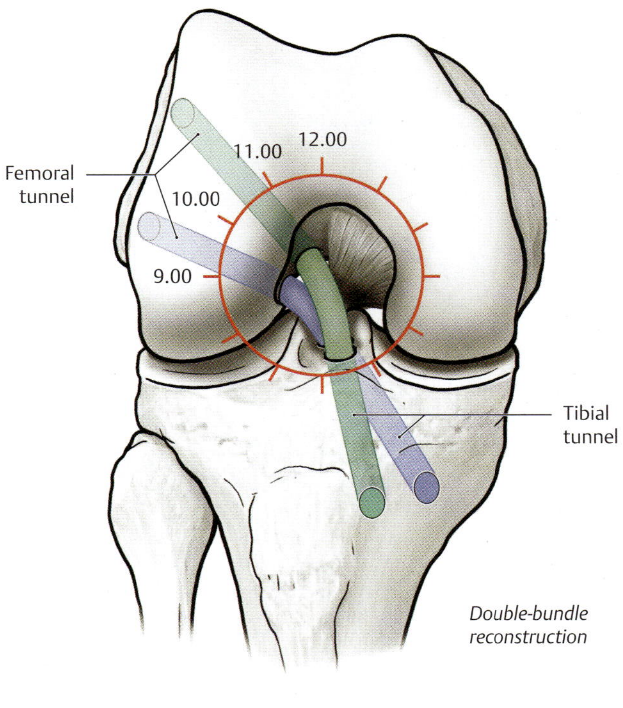

D Rupture of the anterior cruciate ligament: Reconstruction of the double bundle

Restoring the anterior-posterior stability and rotational stability of the anterior cruciate ligament (ACL) requires the replacement of both fiber bundles (AM and PL bundle) (see **B**). In a double-bundle reconstruction, 4 bony tunnels are created arthoscopically with the knee flexed 110° and two grafts are pulled through in the position of the AM and PL bundles. On transverse cross-sectional images of a right knee, the graft should exit the intercondylar fossa at 9:30 (PL bundle) and at 10:30 (AM bundle). The tibial insertion of the AM bundle lies at 30% of the maximum diameter of the tibia, the insertion of the PL bundle at 44%. Today autologous semimembranosus or gracilis tendons are used as grafts. The patellar tendon grafts that were used for many years exhibit a significantly higher donor site morbidity than the tendons of the other muscles. The double-bundle technique promises to improve the functional result of surgery with respect to anterior-posterior translation and rotational stability (see Petersen and Zantop 2009). However, as this is a very complex operation the single-bundle technique with one bony tunnel each and one graft is most often employed and is regarded as the gold standard.

Surgical reconstruction of the cruciate ligament is recommended especially for active athletes. The goal is to achieve an anatomical reconstruction of the compromised structures to prevent ligament, meniscus, and cartilage damage (this decreases the risk of osteoarthritis). Improper positioning of the reconstructed ACL can impair mobility and joint kinematics.

1.24 Cross-sectional Anatomy of the Knee

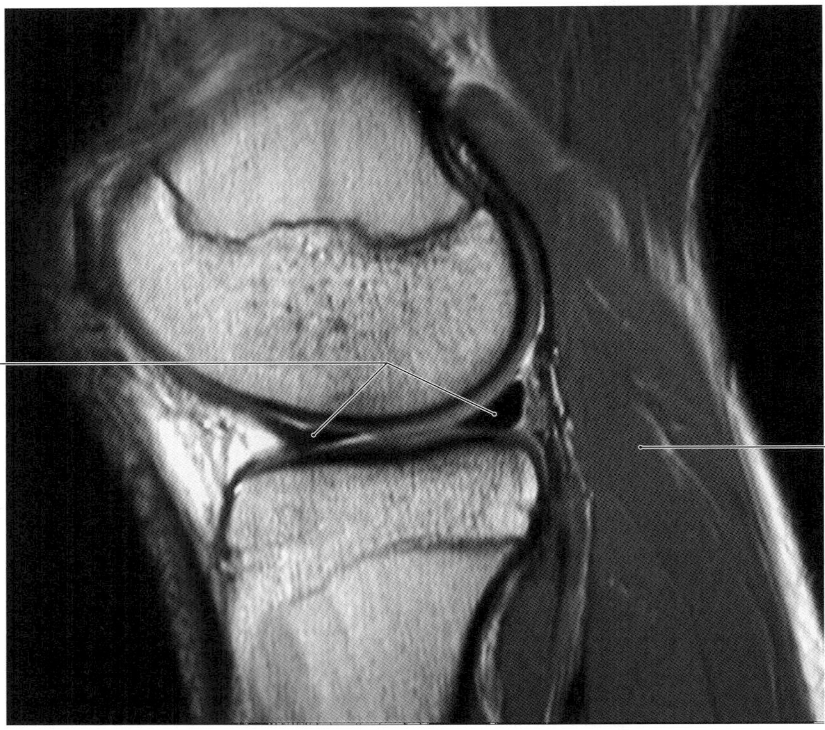

a

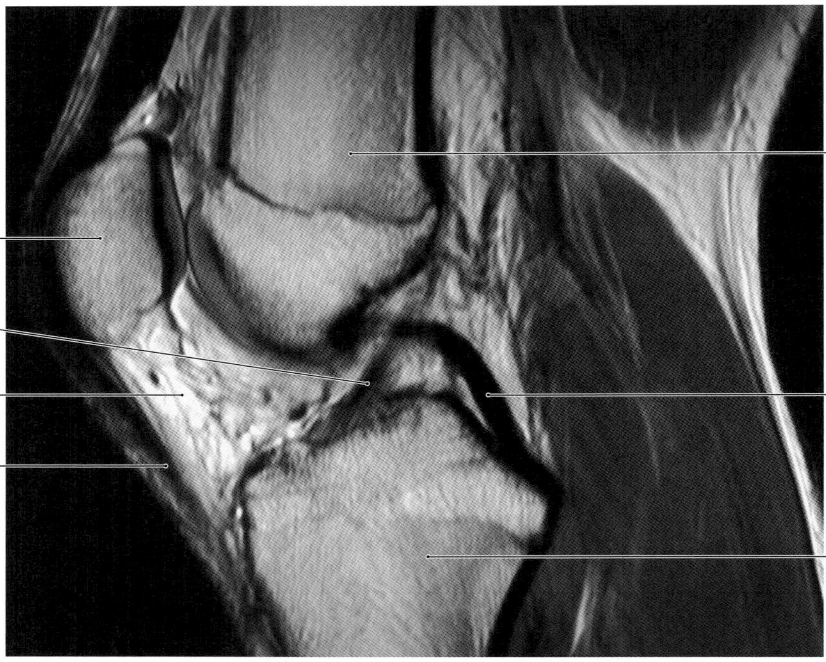

b

A Magnetic resonance imaging of the knee joint
(from Vahlensieck M, Reiser M. MRT des Bewegungsapparates. 4th ed. Stuttgart: Thieme; 2014)

T1-weighted sagittal sequences at the level of the outer meniscus (**a**) and the cruciate ligaments (**b**).

The strength of MRI when evaluating joints includes the differentiated representation of the joint's internal structure, the capsular ligament apparatus, the surrounding soft tissue, and the subchondral bone. In addition to the T1-weighted images, which are primarily suited for anatomic orientation and tissue characterization, the T2-weighted proton-density fat saturation sequences (PDFS) play an important role in the diagnosis of joint damage because they are very sensitive to edematous changes (e.g., fractures) and allow for the high signal intensity of the hyaline articular cartilage (see **B**). (In MRI, fat emits a high signal, which, however, is often adverse and thus needs to be suppressed.)

Note: In sagittal cross sections, the posterior cruciate ligament is clearly visible in its entirety, even its archlike course from posterior to superior. In contrast, the anterior cruciate ligament is only visible in its entirety when the knee joint, as shown here, is externally rotated at 15 to 20°.

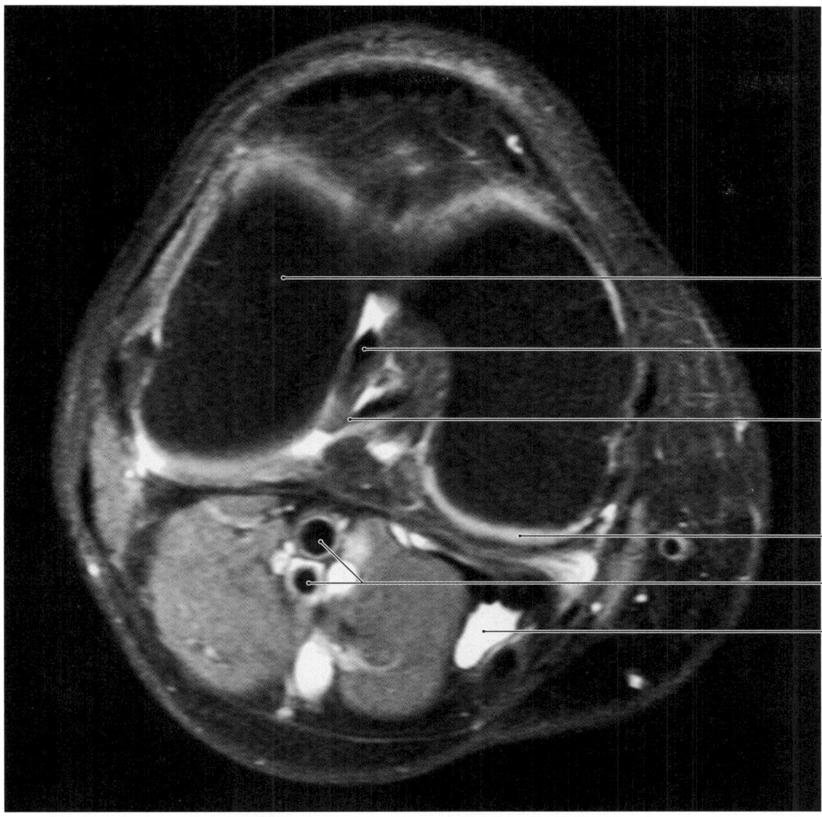

a

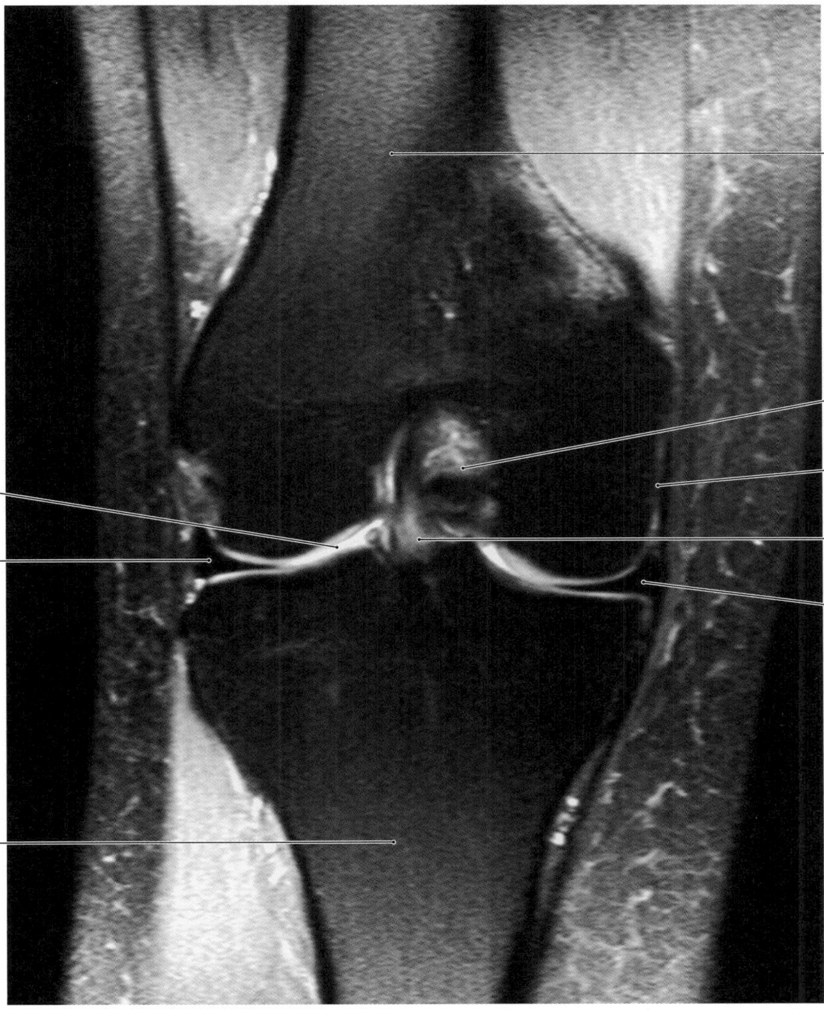

b

B Magnetic resonance imaging of the knee joint (from Vahlensieck M, Reiser M. MRT des Bewegungsapparates. 4th ed. Stuttgart: Thieme: 2014)

T2-weighted axial (**a**) and coronal (**b**) proton-density fat saturation (PDFS) sequences.

Note the high signal intensity of the hyaline articular cartilage, which allows for evaluation of cartilage damage.

1.25 The Joints of the Foot: Overview of the Articulating Bones and Joints

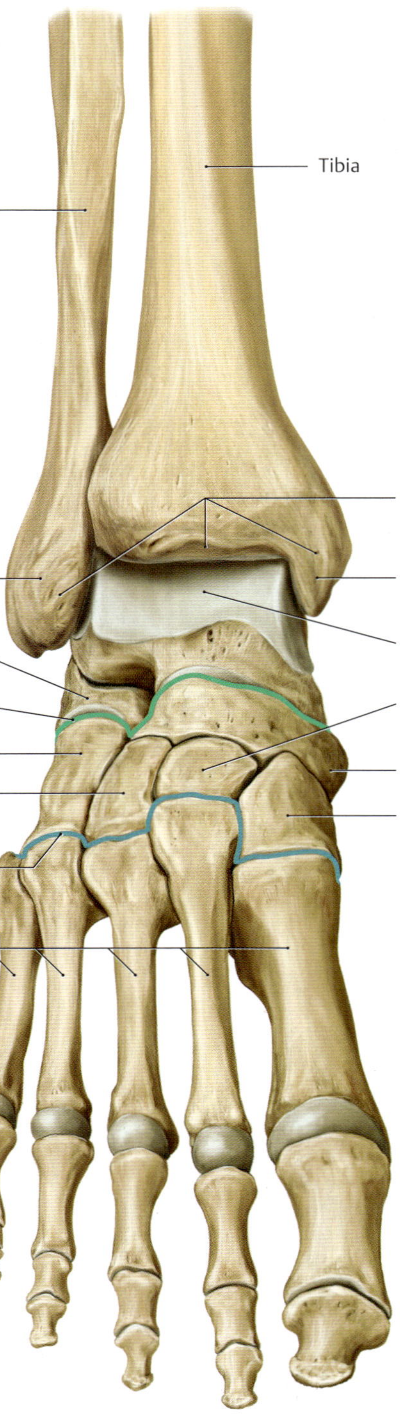

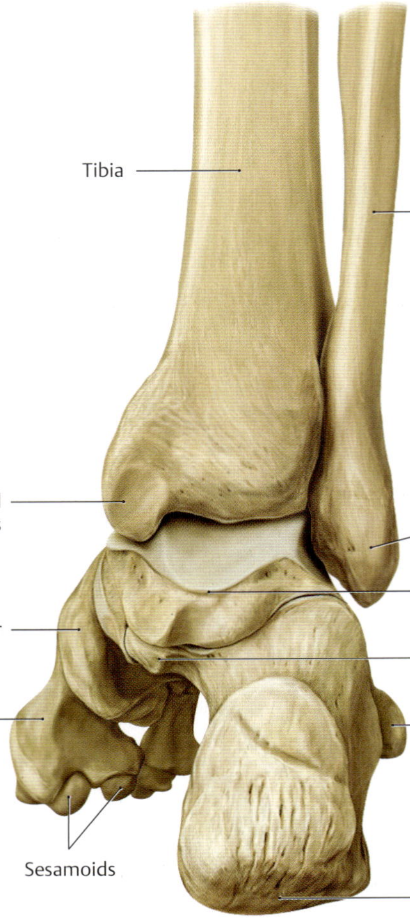

A The articulating bones in different joints of the right foot

a Anterior view with the talocrural joint in plantar flexion.
b Posterior view with the foot in the neutral (0°) position.

B Overview of the joints in the foot

- Talocrural joint (ankle joint)
- Subtalar joint (talocalcaneal joint and talocalcaneonavicular joint)*
- Calcaneocuboid joint (between the calcaneus and cuboid bone)
- Talonavicular joint (between the talus and navicular bone)
- Transverse tarsal joint**
- Cuneonavicular joint (between the cuneiform and navicular bones)
- Intercuneiform joints (between the cuneiform bones)
- Cuneocuboid joint (between the lateral cuneiform and cuboid bones)
- Tarsometatarsal joints
- Intermetatarsal joints (between the bases of the metatarsal bones)
- Metatarsophalangeal joints
- Proximal interphalangeal joints
- Distal interphalangeal joints

* In the subtalar joint, the talus articulates with the calcaneus and the navicular bone to form two separate articulations: the talocalcaneal joint posteriorly and the talocalcaneonavicular joint anteriorly. Both are often referred to collectively as the "subtalar joint."

** Consists of the calcaneocuboid joint and talonavicular joint.

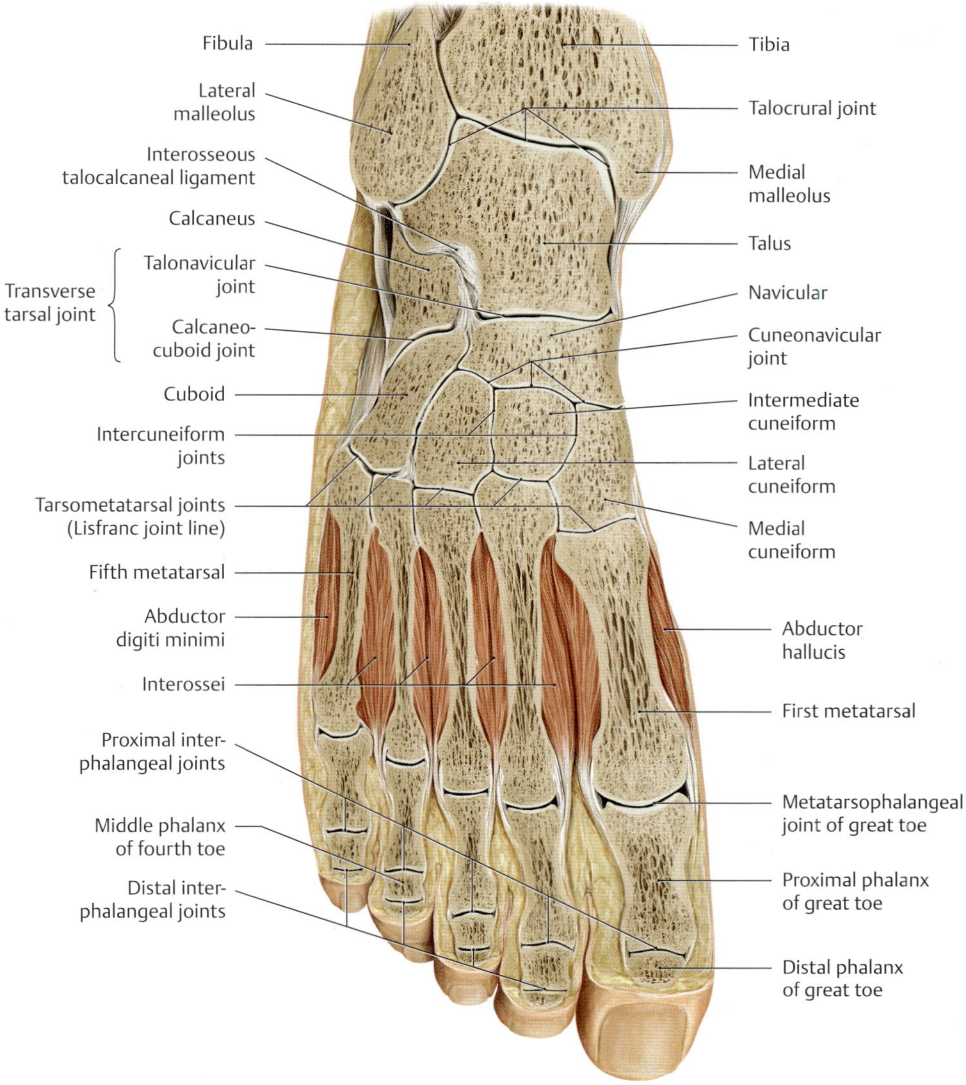

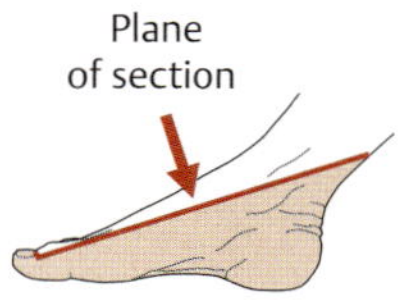

C Oblique transverse section through the foot
Right foot, superior view. The foot is plantar flexed at the talocrural (ankle) joint (drawn from a specimen in the Anatomical Collection of Kiel University).

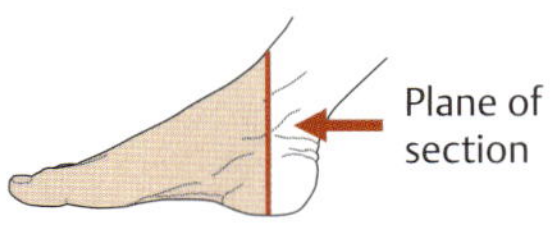

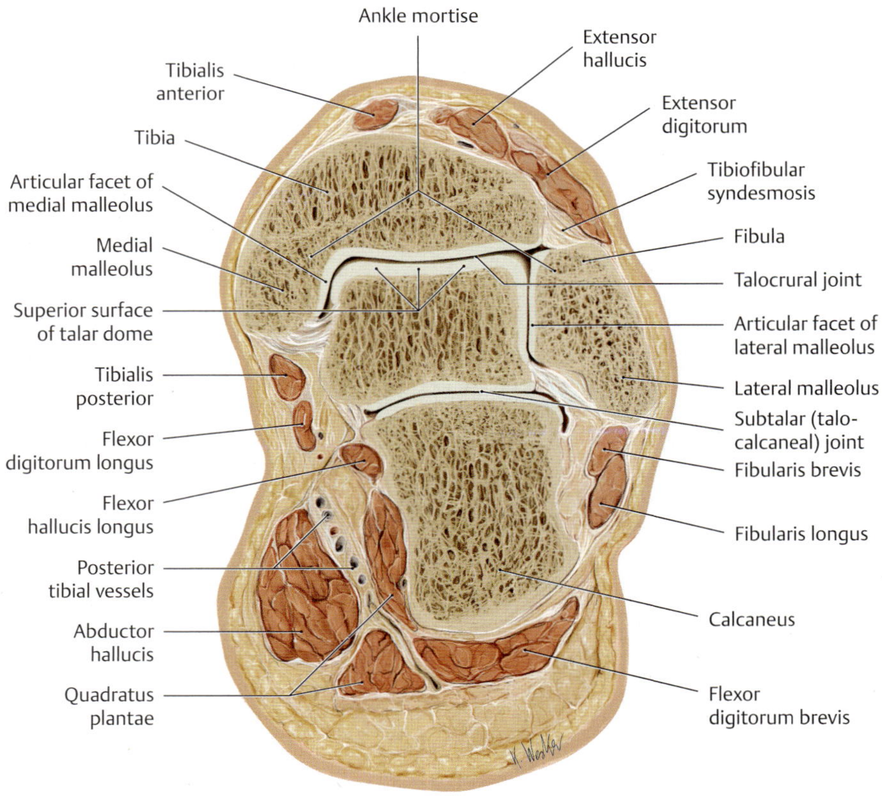

D Coronal section through the talocrural and subtalar joints
Right foot, proximal view. The talocrural joint is plantar flexed, and the subtalar joint is sectioned through its posterior compartment (drawn from a specimen in the Anatomical Collection of Kiel University).

1.26 The Joints of the Foot: Articular Surfaces

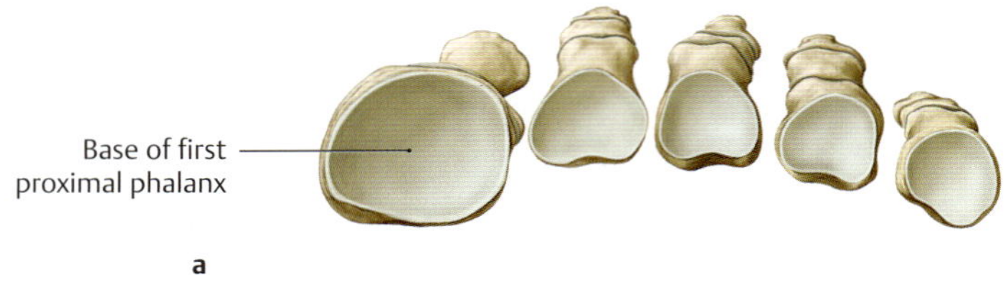

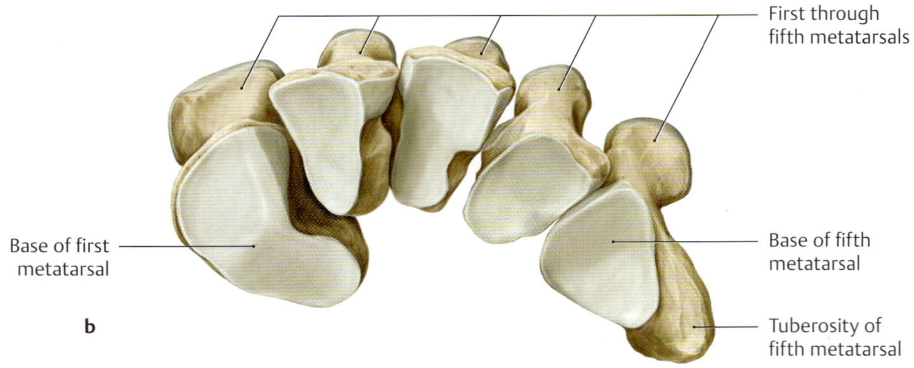

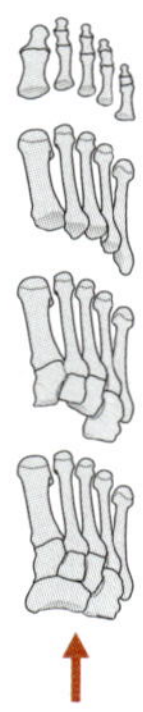

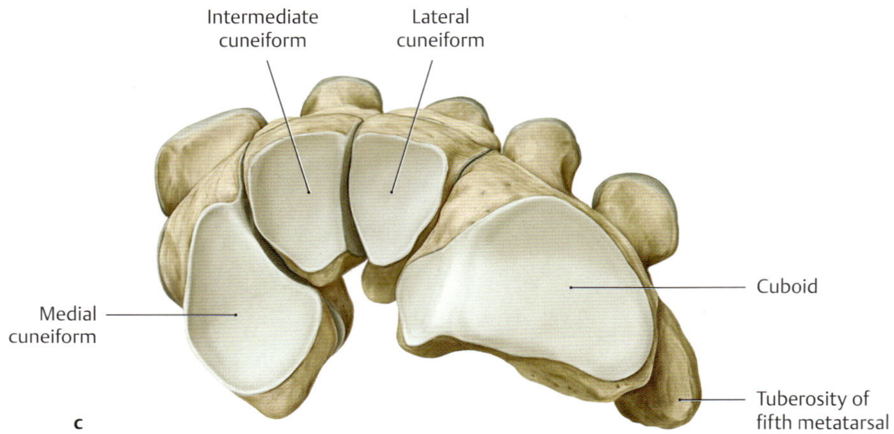

A Proximal articular surfaces
Right foot, proximal view.

- **a** Metatarsophalangeal joints: bases of the first through fifth proximal phalanges.
- **b** Tarsometatarsal joints: bases of the first through fifth metatarsals.
- **c** Cuneonavicular joint and calcaneocuboid joint: proximal articular surfaces of the medial, intermediate, and lateral cuneiform bones and the cuboid.
- **d** Talonavicular joint and calcaneocuboid joint: proximal articular surfaces of the navicular and cuboid bones.

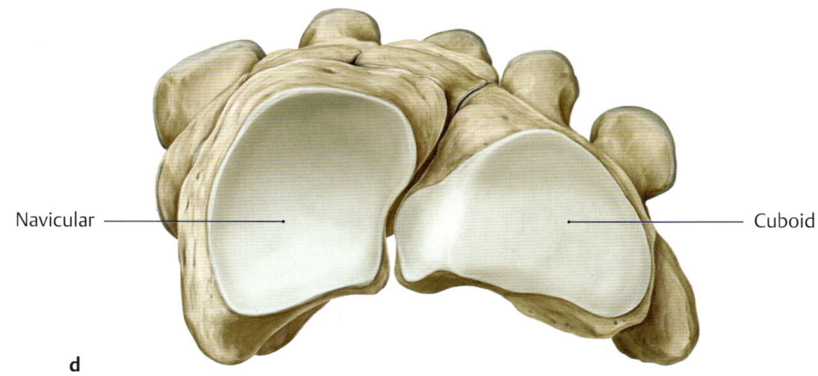

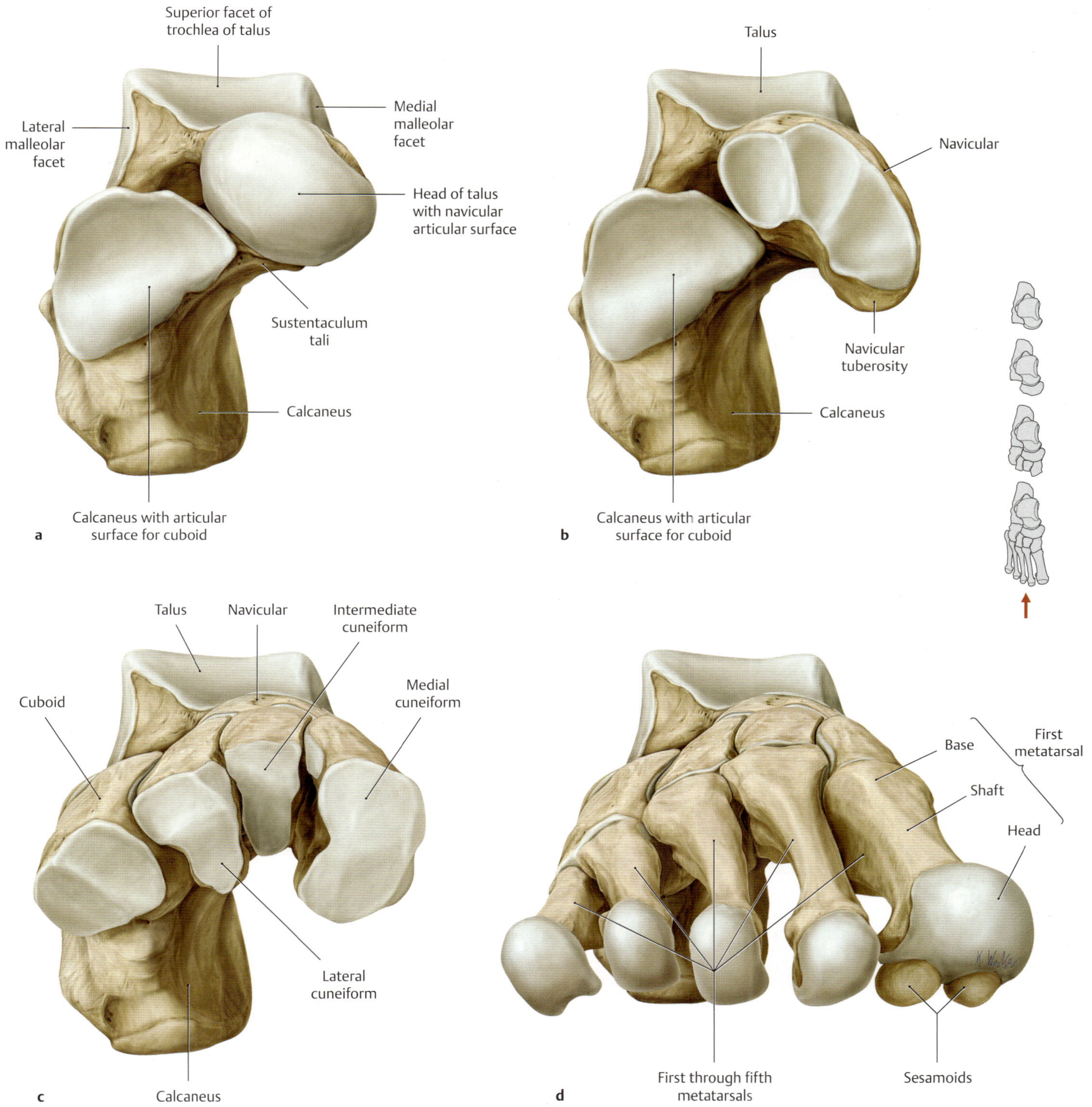

B Distal articular surfaces
Right foot, distal view.

a The talonavicular joint and calcaneocuboid joint: distal articular surfaces of the calcaneus and talus.

b The cuneonavicular joint and calcaneocuboid joint: distal articular surfaces of the navicular and calcaneus.

c The tarsometatarsal joints: distal articular surfaces of the medial, intermediate, and lateral cuneiform and the cuboid.

d Metatarsophalangeal joints: heads of the first through fifth metatarsals.

1.27 The Joints of the Foot: The Talocrural and Subtalar Joints

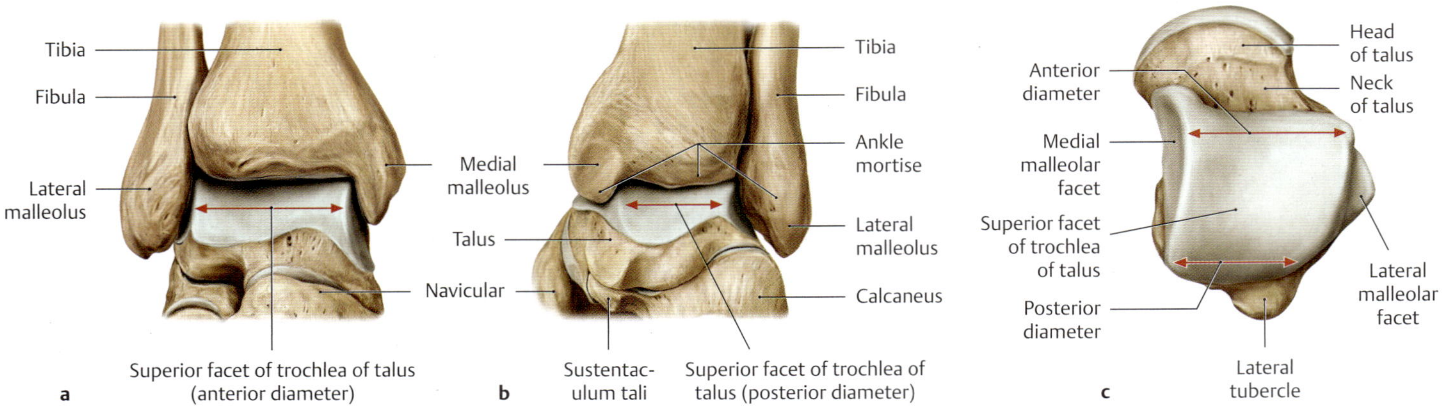

A The articulating skeletal elements of the talocrural joint
a Right foot, anterior view.
b Right foot, posterior view.
c Trochlea of the right talus, superior view.

The talocrural joint, called also the ankle joint, is formed by the distal ends of the tibia and fibula (the ankle mortise, see also **B**) articulating with the trochlea of the talus. This provides the talocrural joint with good bony and ligamentous stability and helps to stabilize the body in an erect posture. However, due to the shape of the talar trochlea (the anterior part, the superior surface, is approximately 5 to 6 mm broader than the posterior part), the bony stability of the talocrural joint differs in flexion and extension. When the broader anterior part of the trochlea articulates with the ankle mortise in *dorsiflexion* (where the foot moves closer to the leg, as in squatting), the syndesmotic ligaments (see p. 476) are tightly stretched, and there is excellent bony stability. But when the narrower posterior part of the trochlea comes in contact with the ankle mortise in *plantar flexion* (e.g., standing on the toes), the talus no longer provides a high degree of bony stability within the ankle mortise.

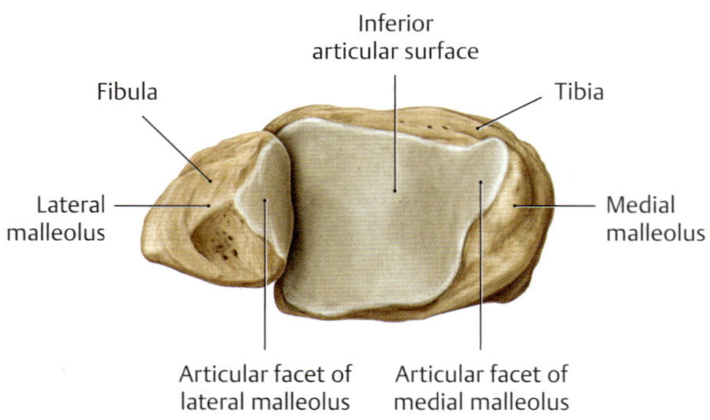

B The articular surfaces of the ankle mortise
Right foot, distal view.

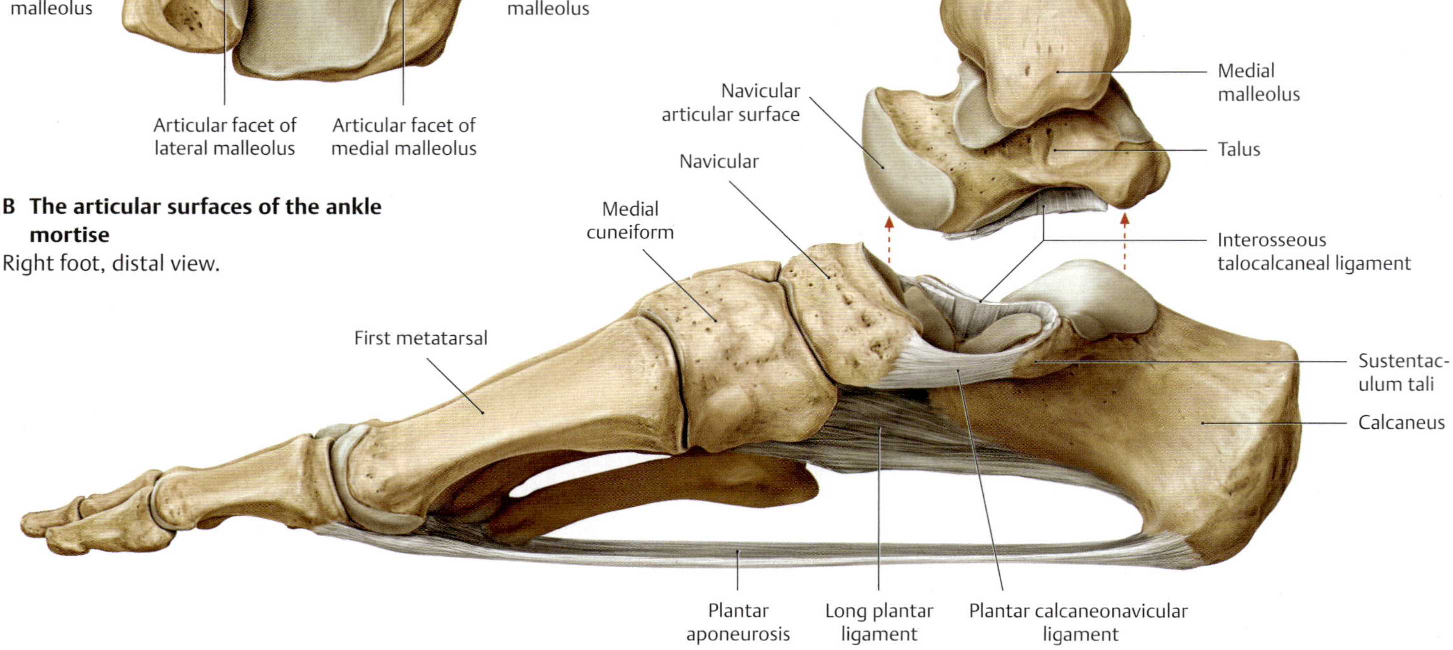

C Overview of an opened subtalar joint
Right foot, medial view. The interosseous talocalcaneal ligament has been divided, and the talus has been displaced upward to demonstrate the articular surfaces of the subtalar joint.

Note the course of the plantar calcaneonavicular ligament, which functions with the long plantar ligament and plantar aponeurosis to support the longitudinal arch of the foot (see also **D** and p. 483).

D Course of the plantar calcaneonavicular ligament and the long plantar ligament

Right foot, plantar view. The plantar calcaneonavicular (spring) ligament stretches between the sustentaculum tali and the navicular. It completes the bony socket of the talocalcaneal joint from the plantar side.

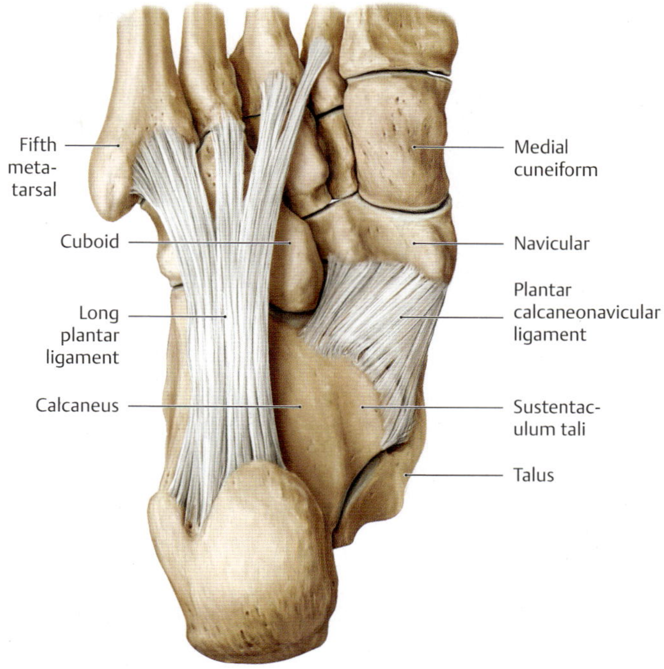

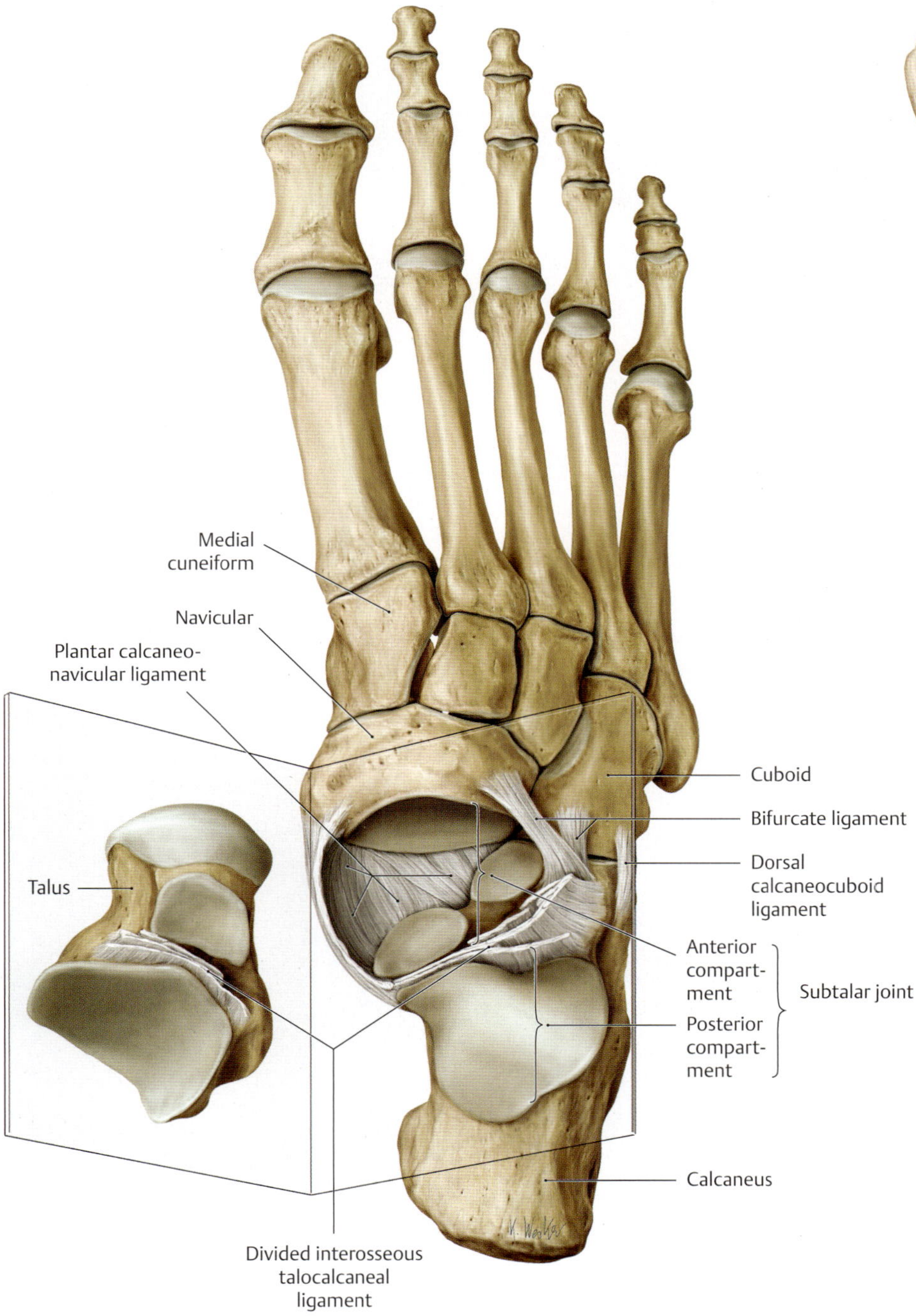

E The articular surfaces of an opened subtalar joint

Right foot, dorsal view (after separation of the talus). In the subtalar joint, the talus articulates with the calcaneus and the navicular. It consists of two completely separate articulations:

- posterior compartment (the talocalcaneal joint) and
- an anterior compartment (the talocalcaneonavicular joint).

The boundary between the two compartments is formed by the interosseous talocalcaneal ligament located in the tarsal canal (bony canal formed by the sulcus tali and sulcus calcanei; its entrance is the tarsal sinus). The plantar calcaneonavicular ligament, which has cartilage cells in its medial surface, loops like a tendon around the plantar head of the talus, which acts as a fulcrum. It stabilizes the position of the talus on the calcaneus and helps to support the apex of the longitudinal pedal arch (see p. 483). Overstretching of the plantar calcaneonavicular (spring) ligament due to flattening of the plantar vault promotes the development of flat foot.

1.28 The Ligaments of the Foot

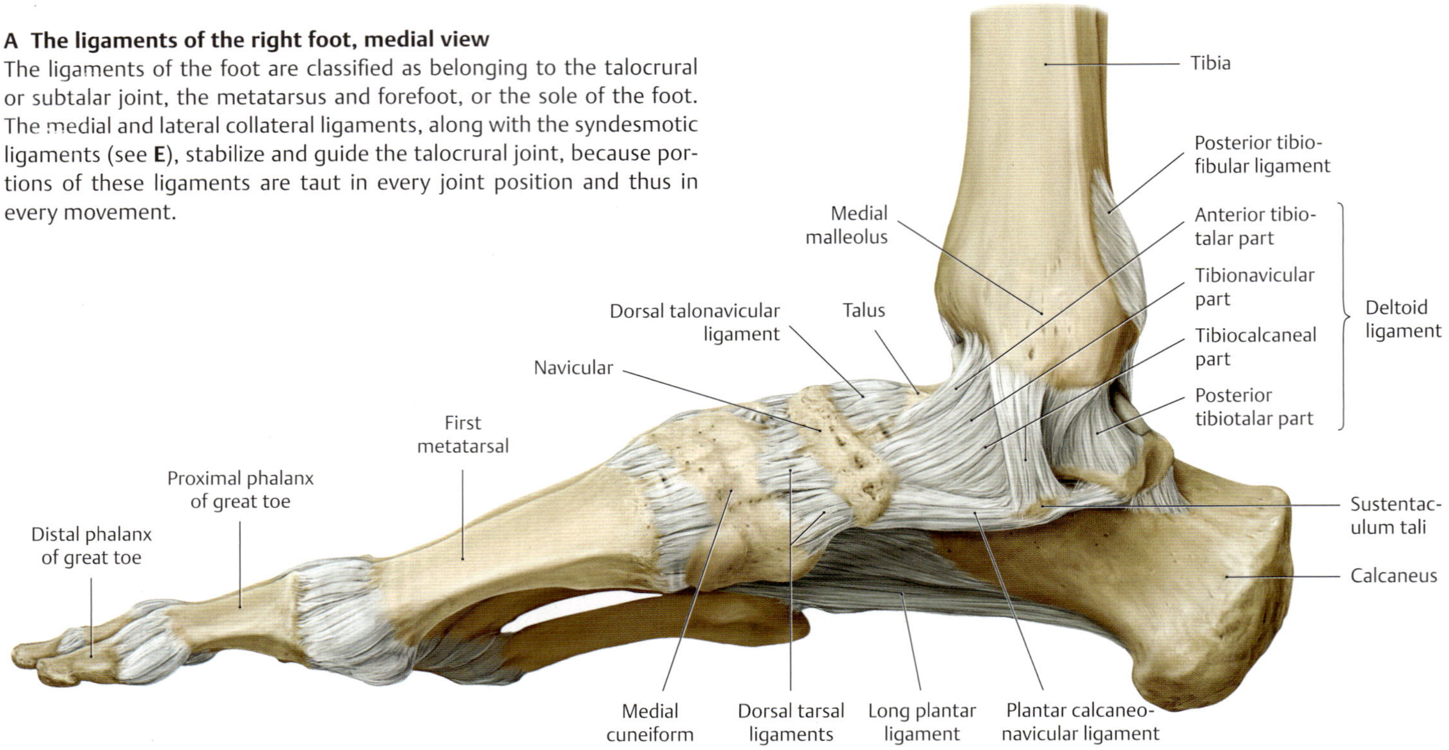

A The ligaments of the right foot, medial view
The ligaments of the foot are classified as belonging to the talocrural or subtalar joint, the metatarsus and forefoot, or the sole of the foot. The medial and lateral collateral ligaments, along with the syndesmotic ligaments (see **E**), stabilize and guide the talocrural joint, because portions of these ligaments are taut in every joint position and thus in every movement.

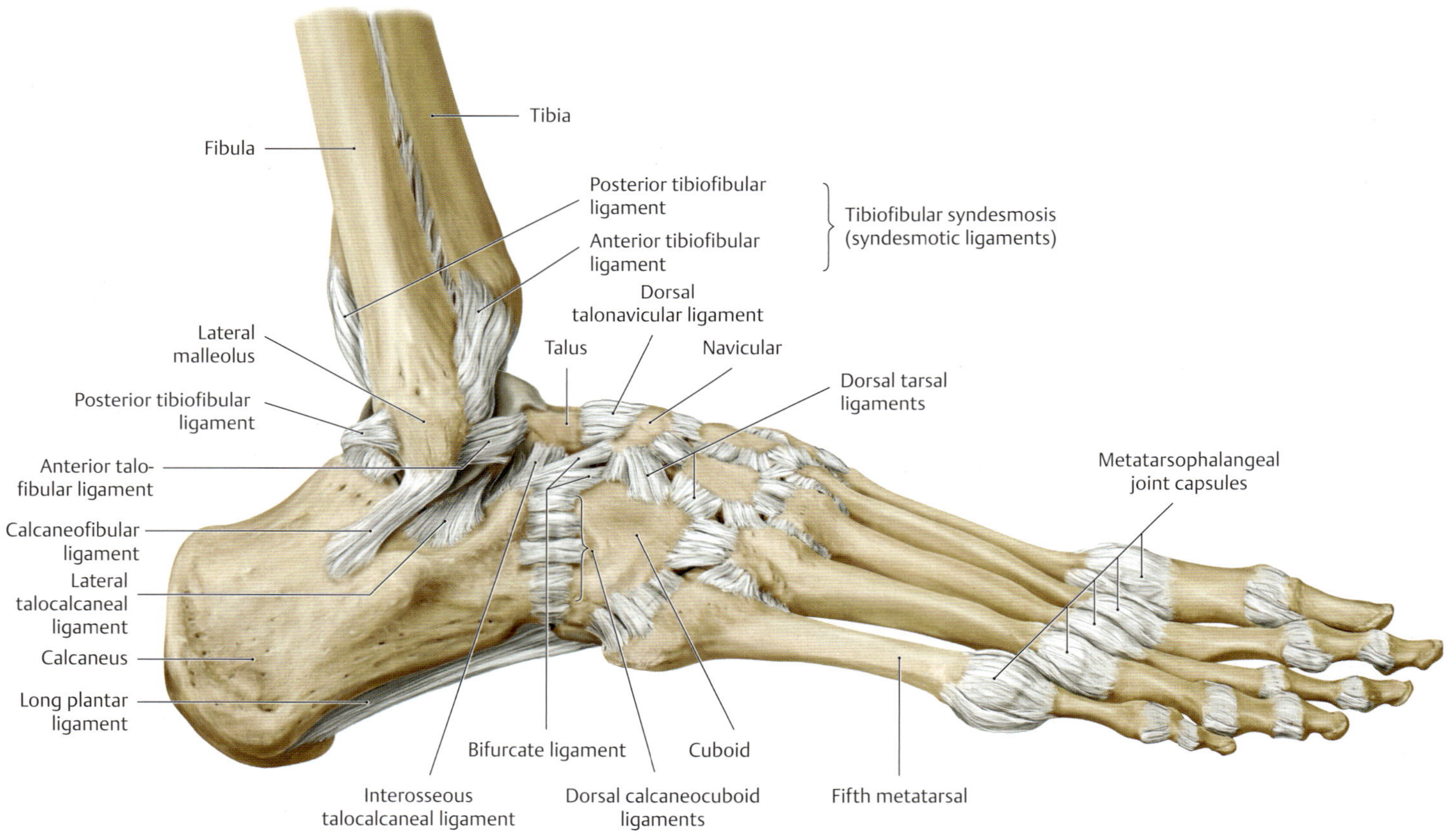

B The ligaments of the right foot, lateral view
Sprains of the ankle joint and especially of its lateral ligaments (usually supination trauma = buckling of the ankle in a supinated position) are extremely common injuries. They often occur during plantar flexion of the foot, a position that provides less bony stability to the talocrural joint (see p. 474). Most of these injuries occur during sports and other leisure activities when the ankle gives way on uneven ground. Typically, the trauma will cause stretching or tearing of the anterior talofibular ligament, the calcaneofibular ligament, or both. If the leg is twisted violently while the foot is fixed, there may also be separation of the ankle mortise with disruption of the tibiofibular syndesmosis (see **D**). The lateral ligaments often exhibit an anatomic variant of the anterior talofibular ligament: the ligament is doubled and its inferior strand is joined in an arc with the calcaneofibular ligament (Vega et al. 2018).

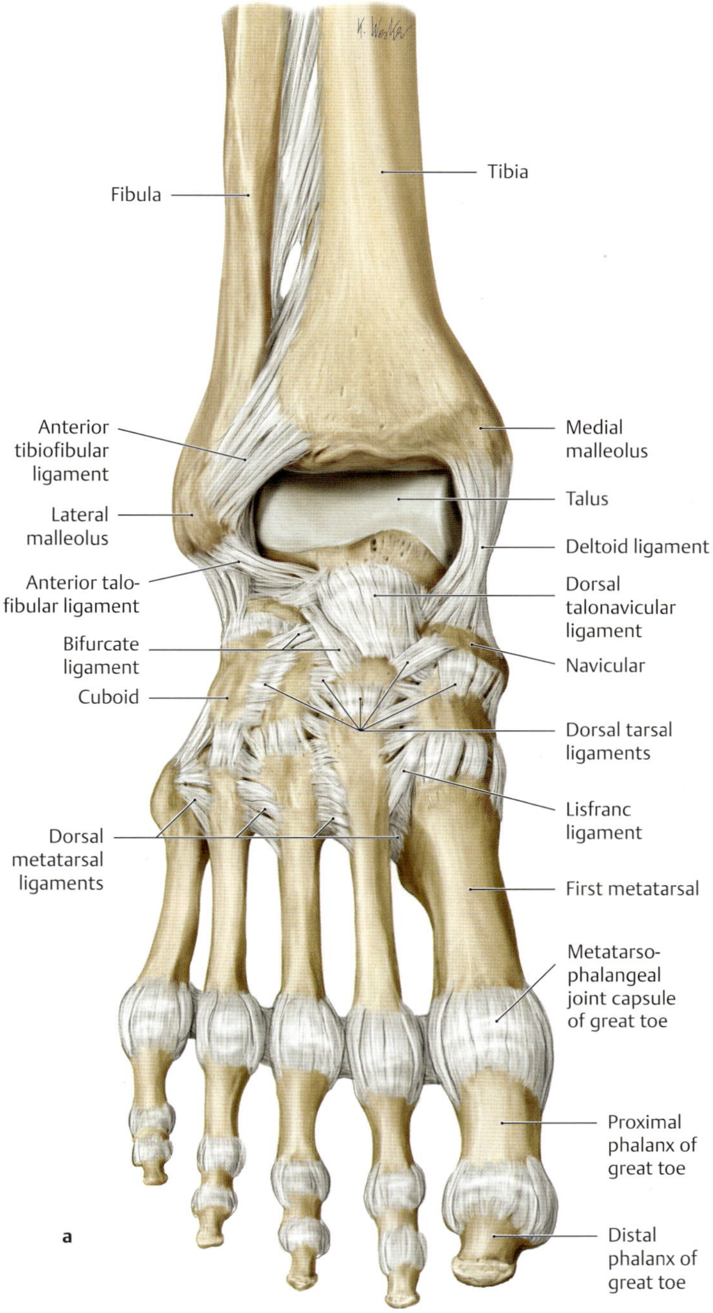

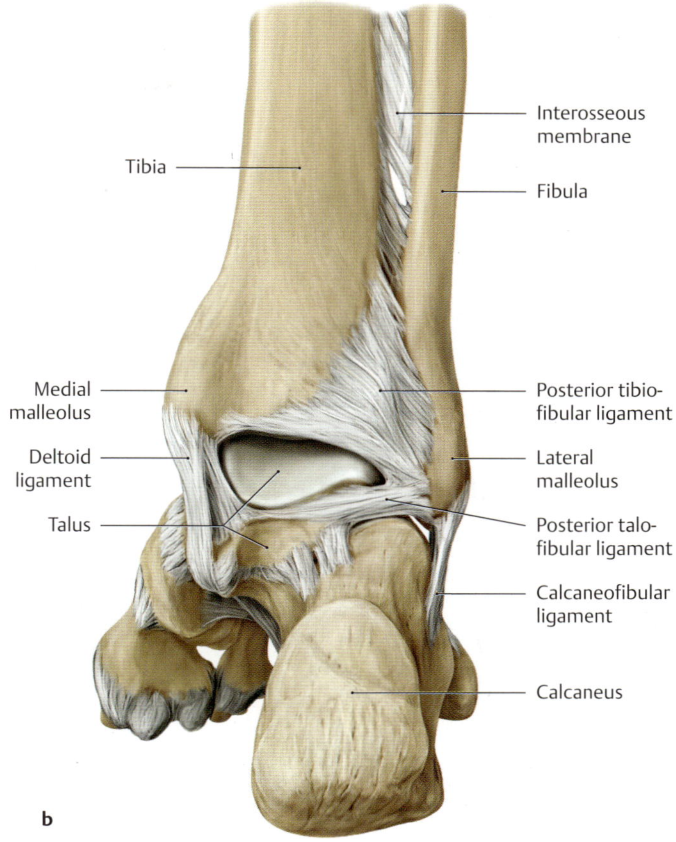

C The ligaments of the right foot

a Anterior view (talocrural joint in plantar flexion).

b Posterior view (plantigrade foot position).

The anterior and posterior portions of the talocrural joint capsule have been removed to demonstrate more clearly the placement of the ligaments.

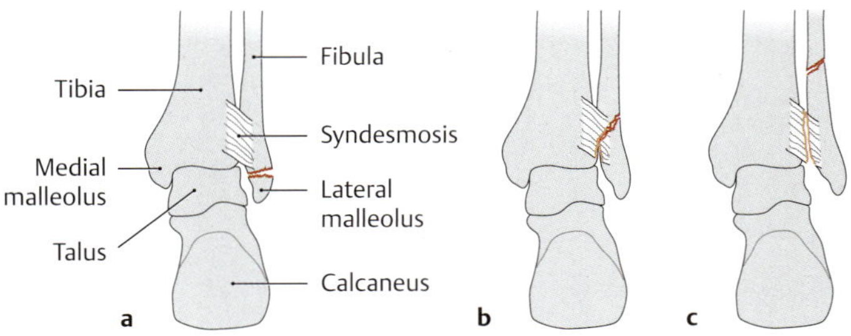

D Weber fractures

A Weber fracture is an avulsion fracture of the lateral malleolus of the fibula. Weber fractures are classified as type A, B, or C depending on whether the fibula is fractured below, level with, or above the syndesmosis. The syndesmosis *may* be torn in a Weber type B fracture (as shown here), but it is *always* torn in a Weber type C fracture.

E The ligaments of the talocrural joint (the ligaments of the subtalar joint are reviewed on p. 475).

Lateral ligaments*
- Anterior talofibular ligament
- Posterior talofibular ligament
- Calcaneofibular ligament

Medial ligaments*
- Deltoid ligament
  - Anterior tibiotalar part
  - Posterior tibiotalar part
  - Tibionavicular part
  - Tibiocalcaneal part

Syndesmotic ligaments of the ankle mortise
- Anterior tibiofibular ligament
- Posterior tibiofibular ligament

* The medial and lateral ligaments are also termed the medial and lateral collateral ligaments.

1.29 The Movements of the Foot

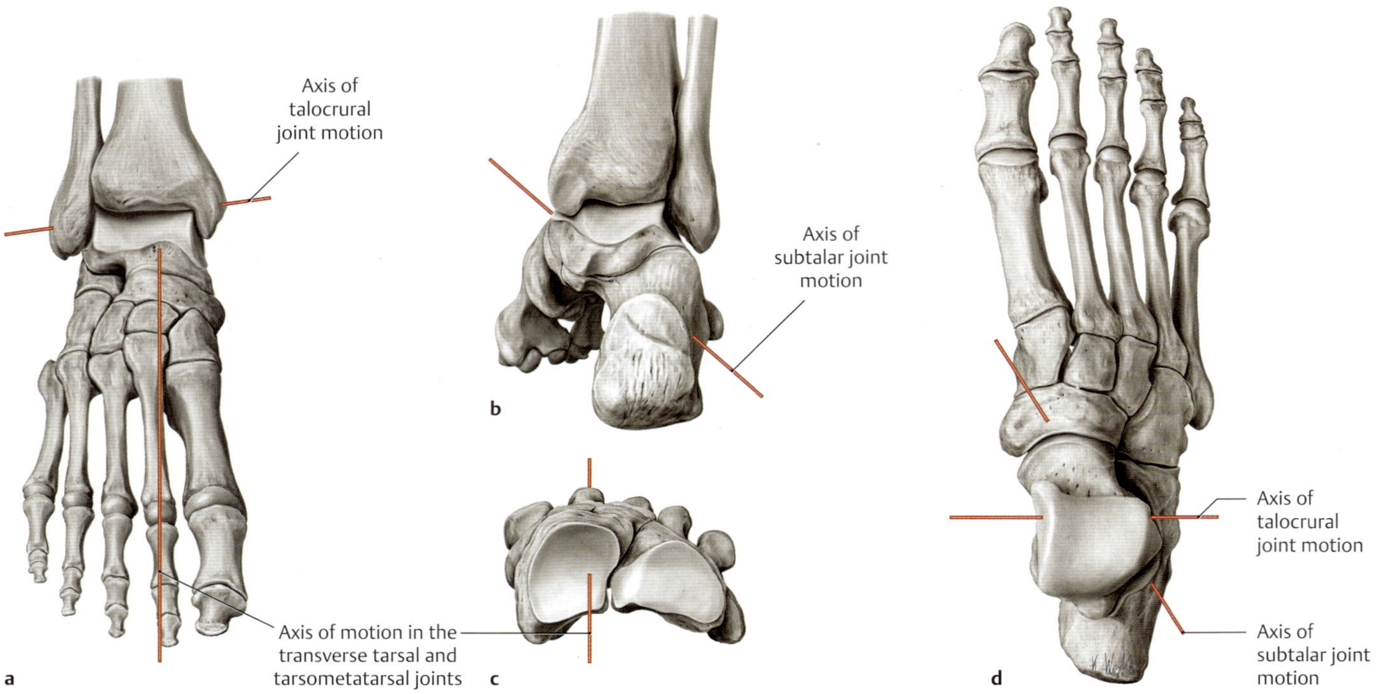

A The principal axes of motion in the right foot
a Anterior view with the talocrural joint in plantar flexion.
b Posterior view in the functional position (see **B**).
c Isolated right forefoot, proximal view.
d Superior view.

The axes of articular motion in the foot are complex, and the descriptions of movements in the pedal joints are often inconsistent and confusing. The following axes of motion are important in clinical parlance and for the testing of joint motion (compare with the facing page):

- **Axis of talocrural joint motion (plantar flexion/dorsiflexion):** This axis runs almost transversely through the lateral and medial malleoli. It forms an approximately 82° angle with the tibial shaft axis in the frontal plane, and it forms a 10° angle with the frontal plane on the medial side (**a**, **d**).
- **Axis of subtalar joint motion (inversion/eversion):** This axis runs obliquely upward through the foot in a posterolateral-to-anteromedial direction, i.e., from the lateral calcaneus through the medial portion of the tarsal canal to the center of the navicular bone. It forms an approximately 30° angle with the horizontal plane and a 20° angle with the sagittal plane (**b**, **d**).
- **Axis of forefoot motion in the transverse tarsal joint and tarsometatarsal joints (pronation/supination):** This axis lies approximately in the sagittal plane, running from the calcaneus through the navicular and along the second ray (**a**, **c**).

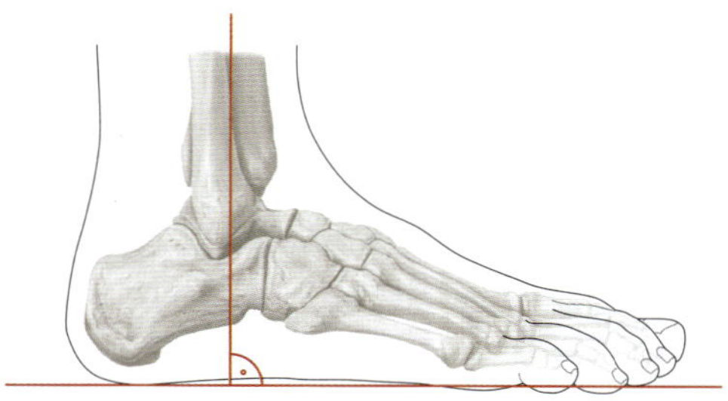

B The functional position of the foot
Right foot, lateral view. In the neutral (0°) position, the skeleton of the foot is angled approximately 90° relative to the skeleton of the leg. This *plantigrade foot position* is termed the "functional position" and is an important basis for normal standing and walking.

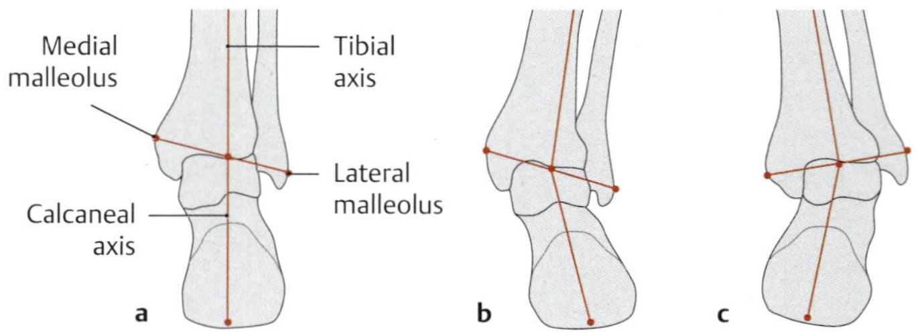

C Axis of the hindfoot
Distal right leg and hindfoot, posterior view.

a With normal axial alignment in the hindfoot, the tibial axis and calcaneal axis lie on a vertical line (pes rectus). The calcaneal axis bisects a line drawn between the two malleoli.
b Pes valgus: the foot is in an everted position.
c Pes varus: the foot is in an inverted position.

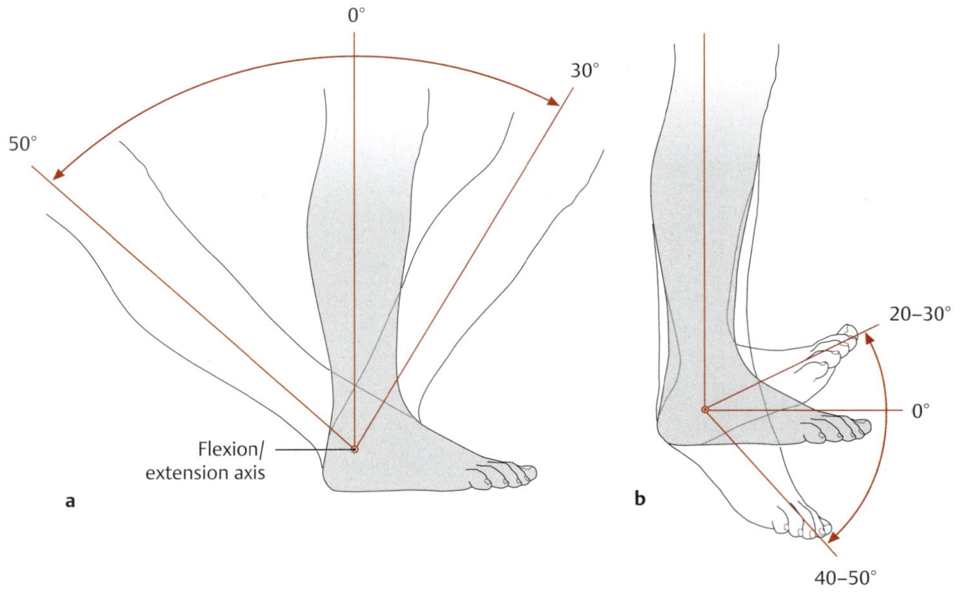

D Normal range of motion of the talocrural joint
Lateral view.

a Right foot on the ground (stance leg).
b Right foot off the ground (swing leg).

Starting from the neutral (0°) (plantigrade) position, the non-weight-bearing foot has an approximately 40 to 50° range of plantar flexion and an approximately 20 to 30° range of dorsiflexion (extension). When the foot is planted on the ground (in the stance phase of gait), the leg can be moved approximately 50° backward (plantar flexion) and 30° forward (dorsiflexion).

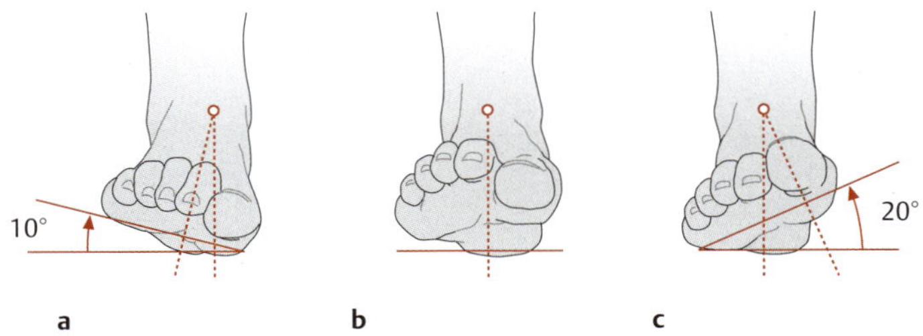

E Range of motion of the subtalar joint
Right foot, anterior view.

a Everted by 10°.
b Neutral (0°) position.
c Inverted by 20°.

Rotation of the calcaneus medially (inversion) and laterally (eversion) is measured from the neutral (0°) position. This is done clinically by holding the leg stationary and moving the calcaneus back and forth. Estimation of the range of inversion/eversion is based on the calcaneal axis.

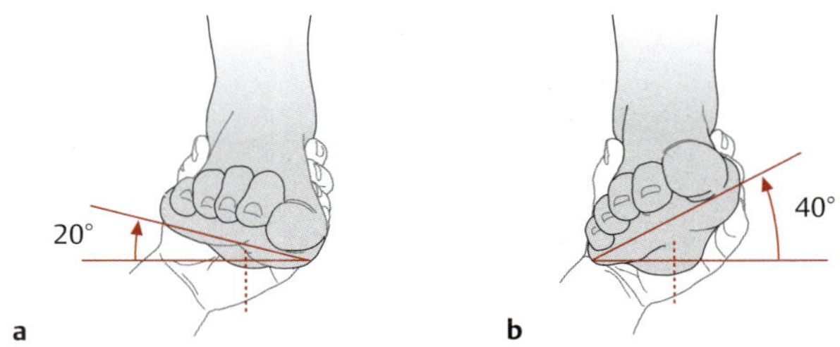

F Range of pronation/supination of the transverse tarsal and tarsometatarsal joints
Right foot, anterior view.

a Range of pronation of the forefoot: 20°.
b Range of supination of the forefoot: 40°.

Range of motion is tested with the hindfoot fixed. Pronation/supination of the forefoot is tested by rotating the forefoot outward relative to the hindfoot (raising the lateral border of the foot) or inward (raising the medial border of the foot).

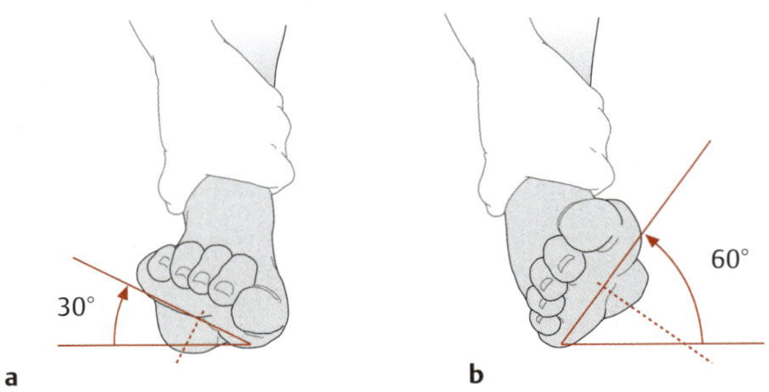

G Total range of motion of the forefoot and hindfoot
Right foot, anterior view.

a Eversion and pronation of the forefoot: 30°.
b Inversion and supination of the forefoot: 60°.

Because the movements in the joints are complex, and different joint movements are almost always mechanically coupled, the range of all joint movements can be assessed by holding the leg stationary and raising the entire foot in the medial and lateral directions.

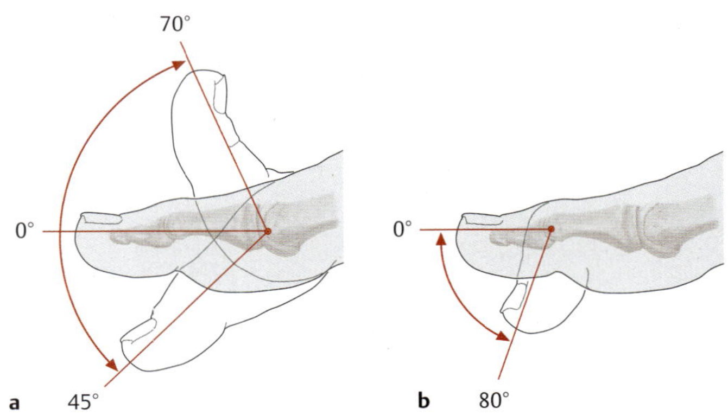

H Range of motion of the joints of the great toe
Lateral view.

a Flexion/extension of the first metatarsophalangeal joint.
b Flexion of the first interphalangeal joint.

The toes, especially the great toe, can be passively extended to approximately 90°. This is an important prerequisite for walking, especially during the phase between heel take-off and toe strike.

1.30 Overview of the Plantar Vault and the Transverse Arch

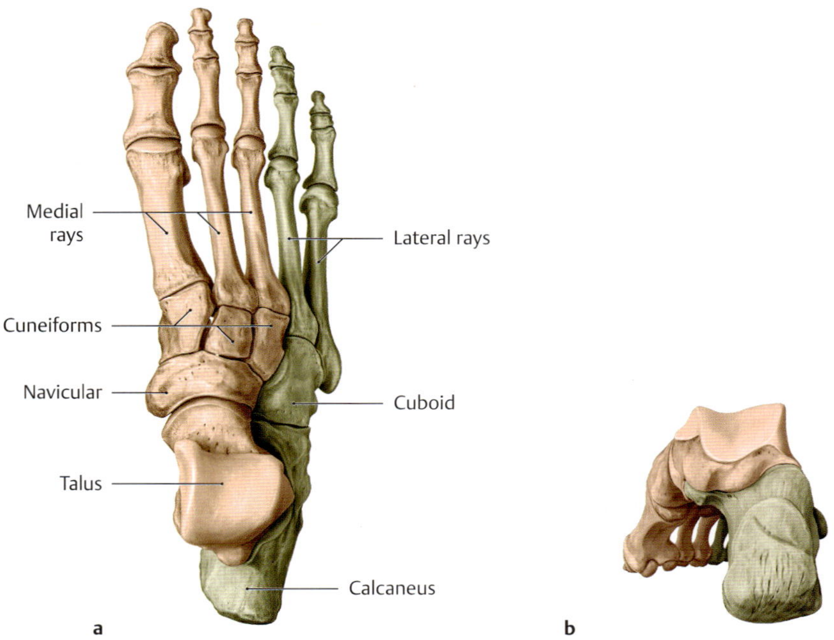

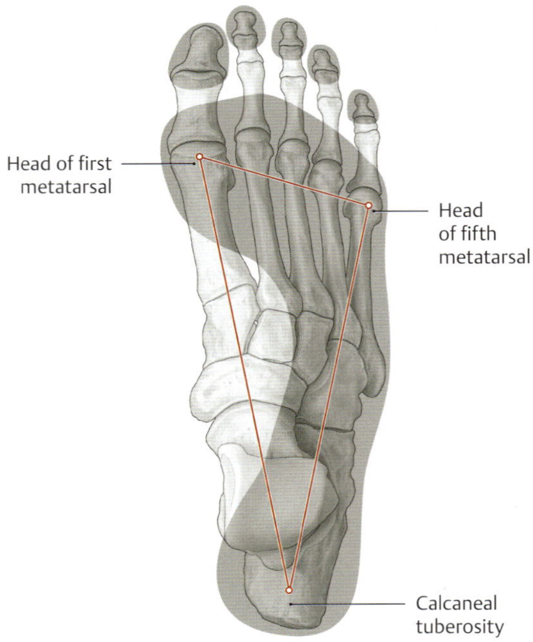

A The plantar vault

a Right foot, superior view.
b Right foot, posteromedial view.

From the perspective of structural engineering, the forces borne by the foot are distributed among two lateral (fibular) rays and three medial (tibial) rays. The lateral rays extend across the cuboid bone to the calcaneus, while the medial rays extend across the cuneiform and navicular bones to the talus. The arrangement of these rays—adjacent distally and overriding proximally—creates a longitudinal arch and a transverse arch in the sole of the foot. These plantar arches enable the foot to adapt optimally to uneven terrain, ensuring that the compressive forces can be transmitted under optimum mechanical conditions in any situation. The arches thus perform a kind of shock-absorber function, creating a springy flexibility that helps the foot absorb vertical loads. The deficient arches in a flatfoot or a splayfoot, for example, can lead to considerable pain during walking.

B The plantar architecture of the right foot

Superior view showing the bony points of support for the plantar vault and the associated footprint. The area that is outlined by interconnecting the bony supports (the calcaneal tuberosity and the heads of the first and fifth metatarsals) has the shape of a triangle. By contrast, the area of ground contact defined by the plantar soft tissues (the footprint or *podogram*) is considerably larger. The calluses typically found on the heel and the balls of the great and little toes confirm that these areas bear the brunt of the loads.

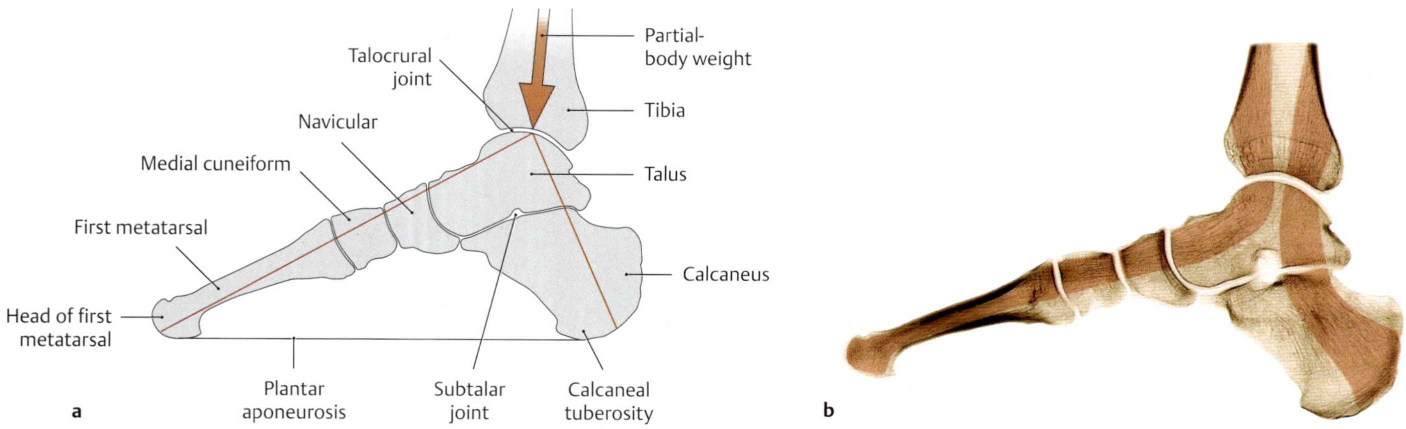

C Transfer of compressive stresses in the weight-bearing foot

Schematic sagittal section at the level of the first ray, medial view.

a During stance, the partial body weight on the talocrural joint is transferred across the talus to the forefoot and hindfoot.

b A schematic radiograph illustrates the parallel arrangement of the cancellous bony trabeculae. The pattern conforms to the compressive stresses (indicated by color shading) that result from the loads acting on the forefoot and hindfoot (**a**).

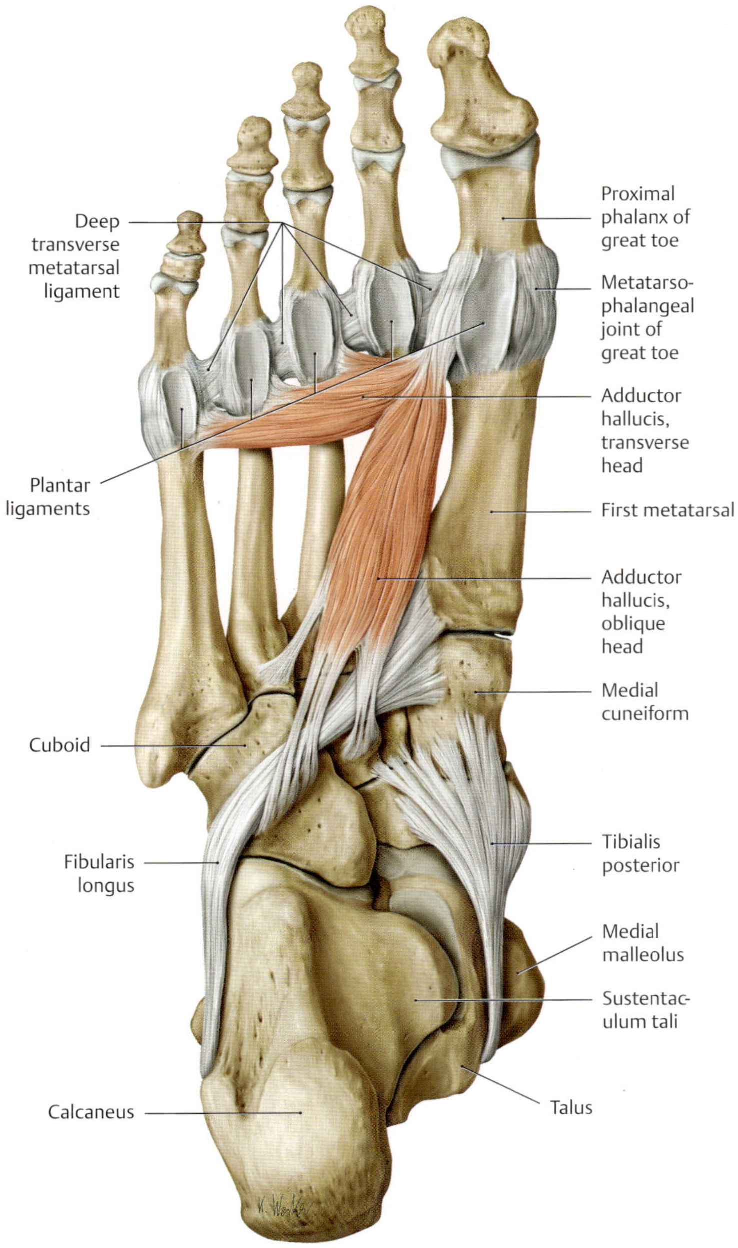

D Active and passive stabilizers of the transverse arch, plantar view

Right foot. Both active and passive stabilizing structures maintain the curvature of the transverse pedal arch. The passive stabilizers are ligaments, and the active stabilizers are muscles. In the foot, ligamentous structures are usually able to maintain the pedal arches without assistance from the muscles. But when the loads on the foot are increased, as during walking or running on uneven ground, active muscular forces are recruited to give additional support.

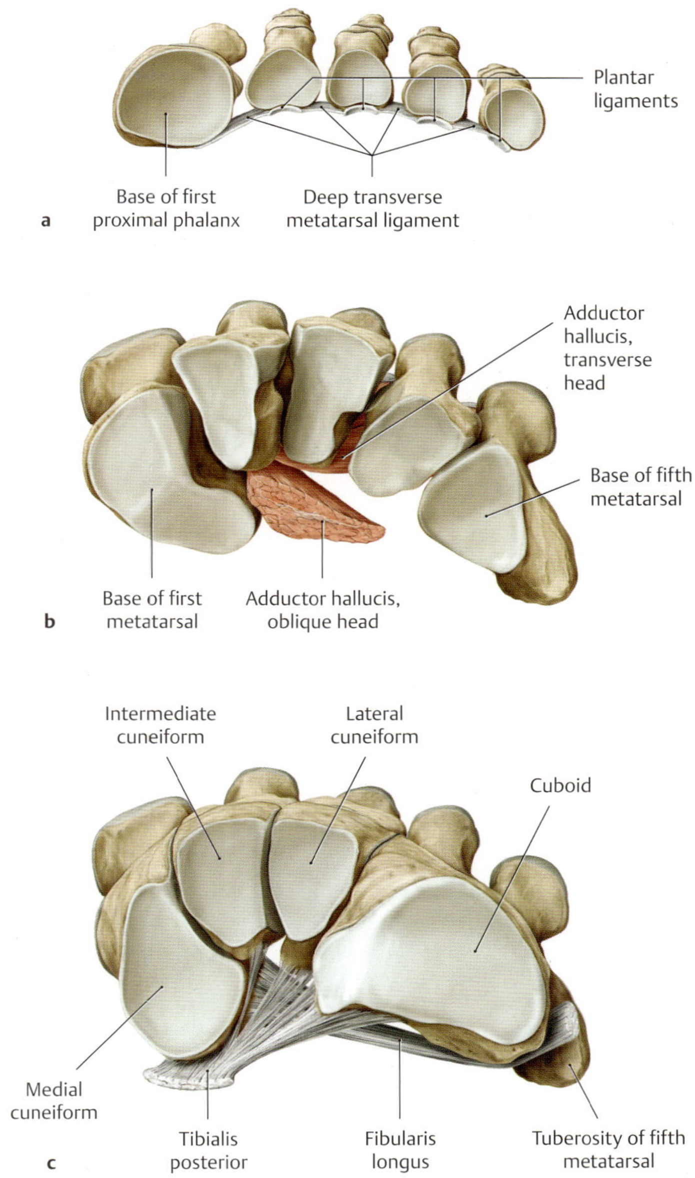

E Active and passive stabilizers of the transverse arch, proximal view

Right foot.

- **a** Arch stabilizers in the forefoot (anterior arch). The deep transverse metatarsal ligament stabilizes the anterior arch at the level of the metatarsal heads. The arch of the forefoot thus relies entirely on passive stabilizers, while the arches of the metatarsus and tarsus (**b**, **c**) have only active stabilizers.
- **b** Stabilizers of the metatarsal arch. The transverse head of the adductor hallucis is the primary muscular stabilizer of the metatarsal arch.
- **c** The principal arch-supporting muscle in the tarsal region is the fibularis longus. After winding around the cuboid, its tendon of insertion runs from the lateral border of the foot and across the sole to the medial cuneiform bone and the base of the first metatarsal. Another active stabilizer in this region is the tibialis posterior, whose tendon of insertion gives off expansions to the cuneiform bones. Like the fibularis longus, its oblique course enables it to support the longitudinal arch in addition to its transverse component.

1.31 The Longitudinal Arch of the Foot

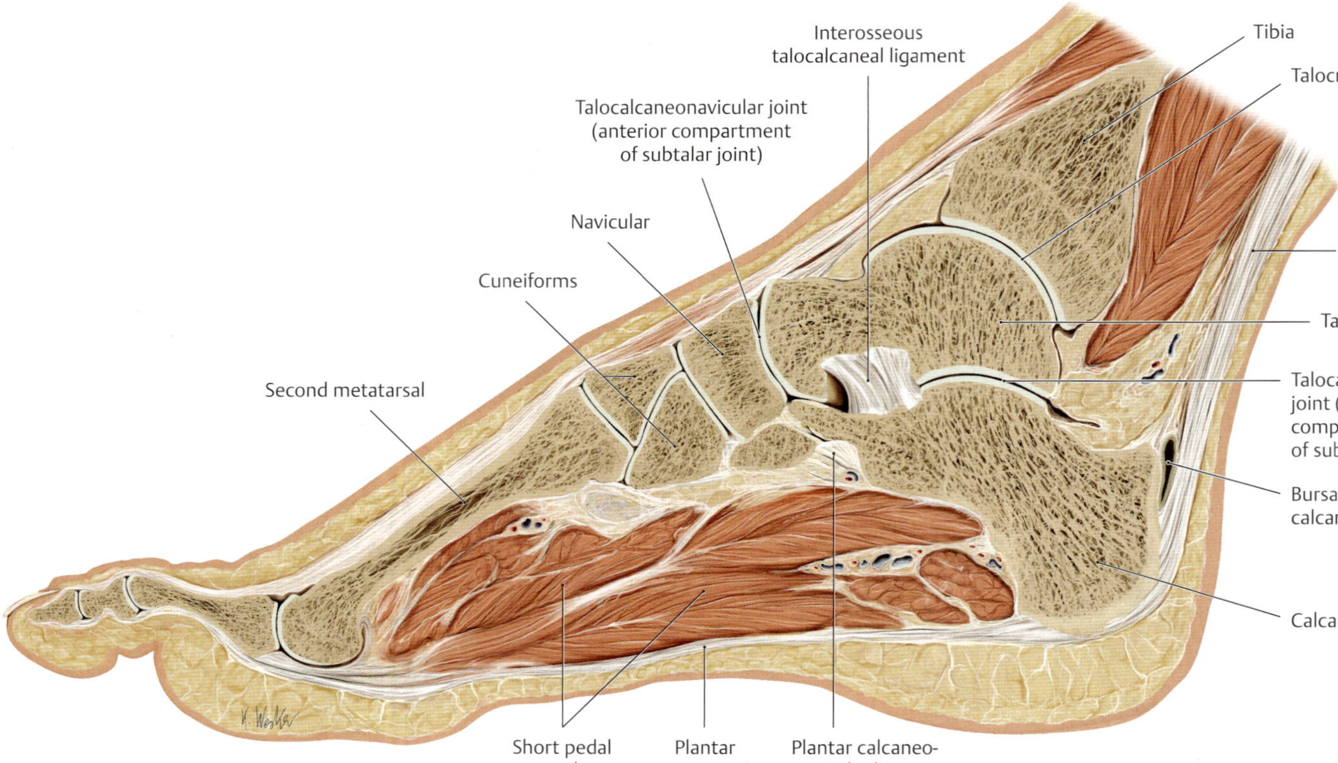

A Active stabilizers of the longitudinal arch

Sagittal section at the level of the second ray of the right foot, medial view. The second ray (consisting of the second toe, second metatarsal, intermediate cuneiform, navicular, and calcaneus) forms the highest arch within the overall longitudinal plantar arch, its height diminishing laterally. The main active stabilizers of the longitudinal arch are the *short muscles of the foot*: abductor hallucis, flexor hallucis brevis, flexor digitorum brevis, quadratus plantae, and abductor digiti minimi (drawn from a specimen in the Anatomical Collection of Kiel University).

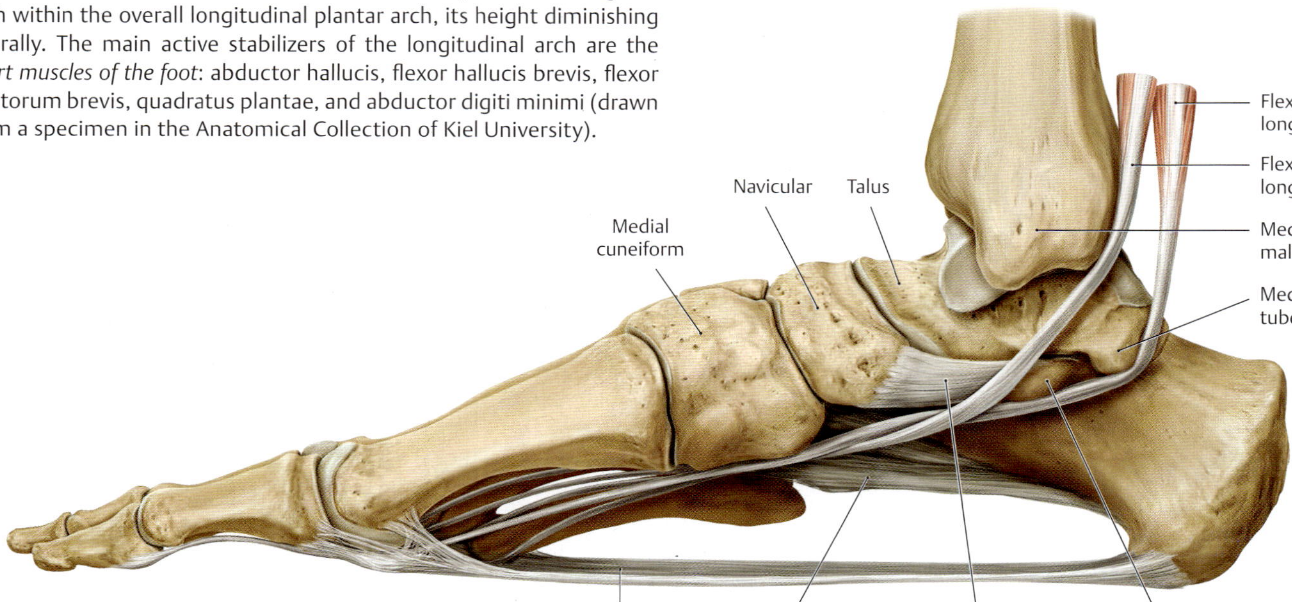

B Passive stabilizers of the longitudinal arch

Right foot, medial view. The main passive stabilizers of the longitudinal arch are the plantar aponeurosis, the long plantar ligament, and the plantar calcaneonavicular (spring) ligament. The plantar aponeurosis is particularly important owing to its long lever arm, while the plantar calcaneonavicular ligament is the weakest component (shortest distance from the apex of the longitudinal arch). *Note:* The tendons of insertion of the long flexors of the foot (flexor hallucis longus and flexor digitorum longus) also help to prevent sagging of the longitudinal arch, yet not passively but as an active tensioning structure. The flexor hallucis longus, which runs beneath the sustentaculum tali, is particularly effective in tightening the longitudinal arch like the chord of an arc.

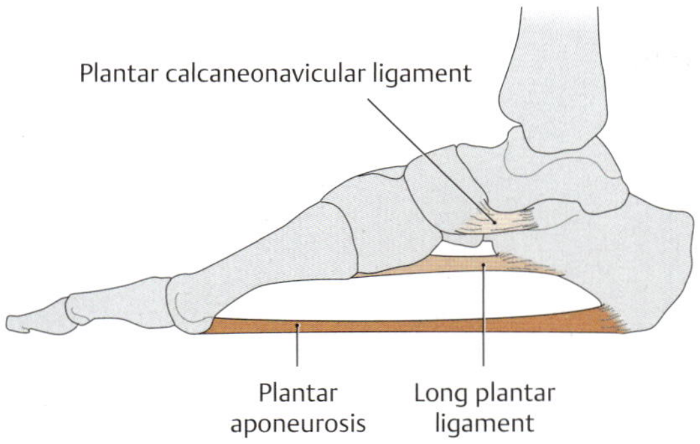

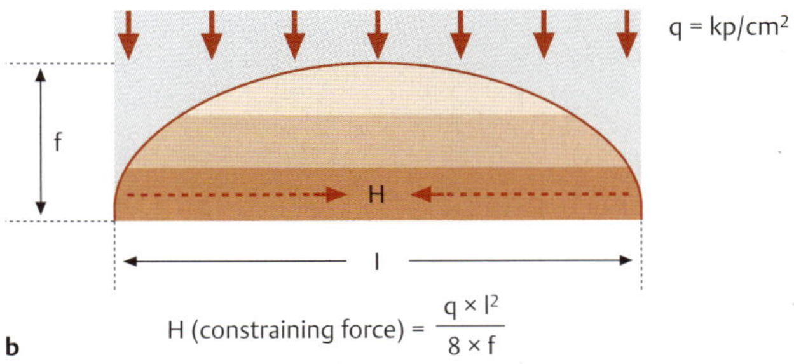

C Support of the longitudinal arch

a Ligamentous support of the longitudinal arch (right foot, medial view).

b Calculating the constraining force (*H*) needed to maintain the arch (after Rauber and Kopsch).

Comparing the longitudinal arch of the foot to a theoretical parabolic arch, we see that a constraining force (*H*) must be applied to maintain the arch curvature. The magnitude of this force depends on the load (*q*), the chord length of the arch (*l*), and the height of the arch (*f*). As a result, the structures that are most effective in maintaining the pedal arch are those closest to the ground, since the long lever arm of those structures requires the least expenditure of force. The formula also dictates that the constraining force must increase as the distance *l* between the points of support increases, or as the arch becomes flatter (smaller *f*).

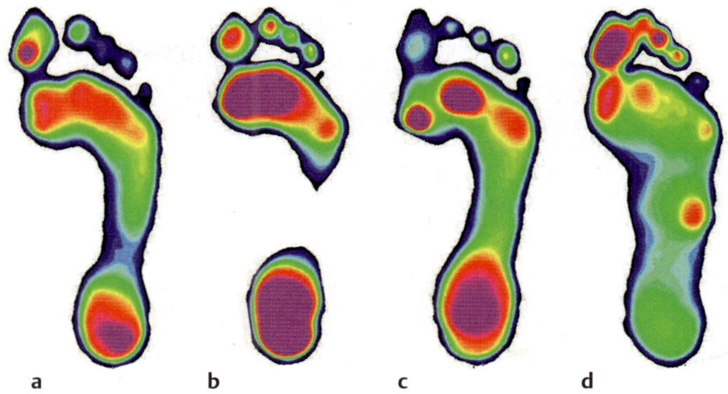

D Footprints (podograms) of right feet (dynamic measuring of foot pressure)

a Normal plantar arches (pes rectus).

b Increased height of the longitudinal arch (pes cavus).

c Loss of the transverse arch (splayfoot = pes transversoplanus).

d Loss of the longitudinal arch (flatfoot = pes planus).

In dynamic measuring of foot pressure, a pressure-sensitive measuring device (with approximately 4,000 sensors) analyzes gait from heel-strike to toe-off, gait symmetry, and all acting forces. The pressure distribution is color coded.

Foot deformities—deviations from a normal, healthy foot shape—may be congenital, or acquired through paralysis or trauma. Other structural abnormalities (static deformities) are caused by chronic loads imposed on the foot by the body weight (original images from: Michael Kriwat, Kiel).

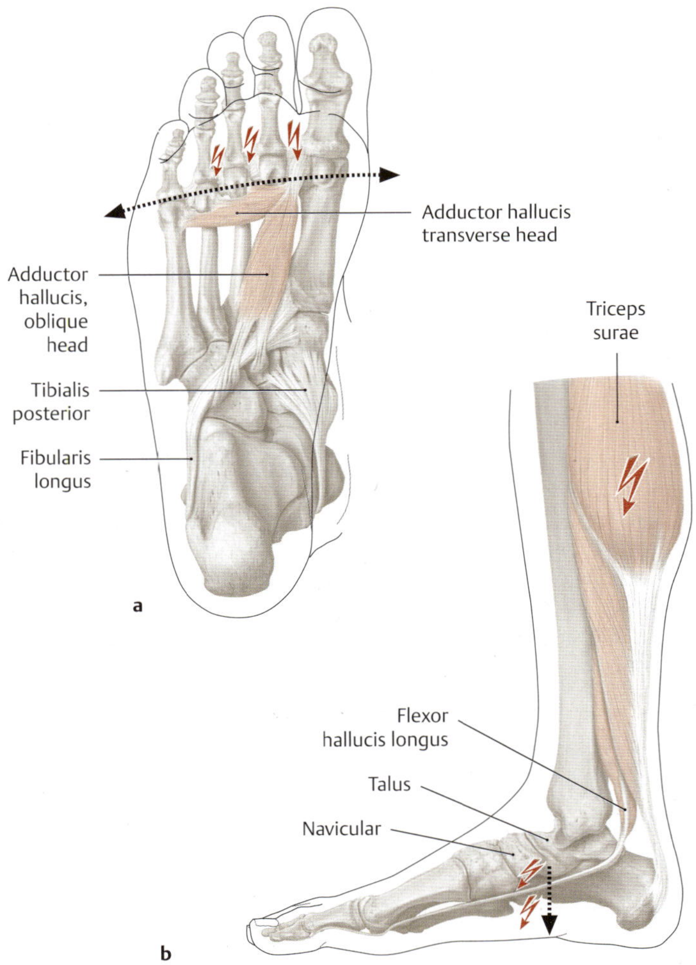

E Location of pain associated with splayfoot and flatfoot

a Right splayfoot viewed from the plantar aspect. The collapse of the transverse arch results in a broadened forefoot (arrows) with greater pressure acting on the heads of the second through fourth metatarsals and the associated metatarsophalangeal joints. Typically, very painful calluses will form between the balls of the great and little toes in this situation.

b Right flatfoot viewed from the medial aspect. With collapse of the longitudinal arch, marked by downward displacement of the talus and navicular (arrow), weight bearing often incites a diffuse foot pain that is most intense in the area of the stretched calcaneonavicular (spring) ligament. Calf pain may also develop as a result of sustained, increased tension on the calf muscles (and the pedal muscles as well, which must compensate for the deficient passive stabilizers).

1.32 The Sesamoid Bones and Plantar Plates of the Metatarsophalangeal Joints

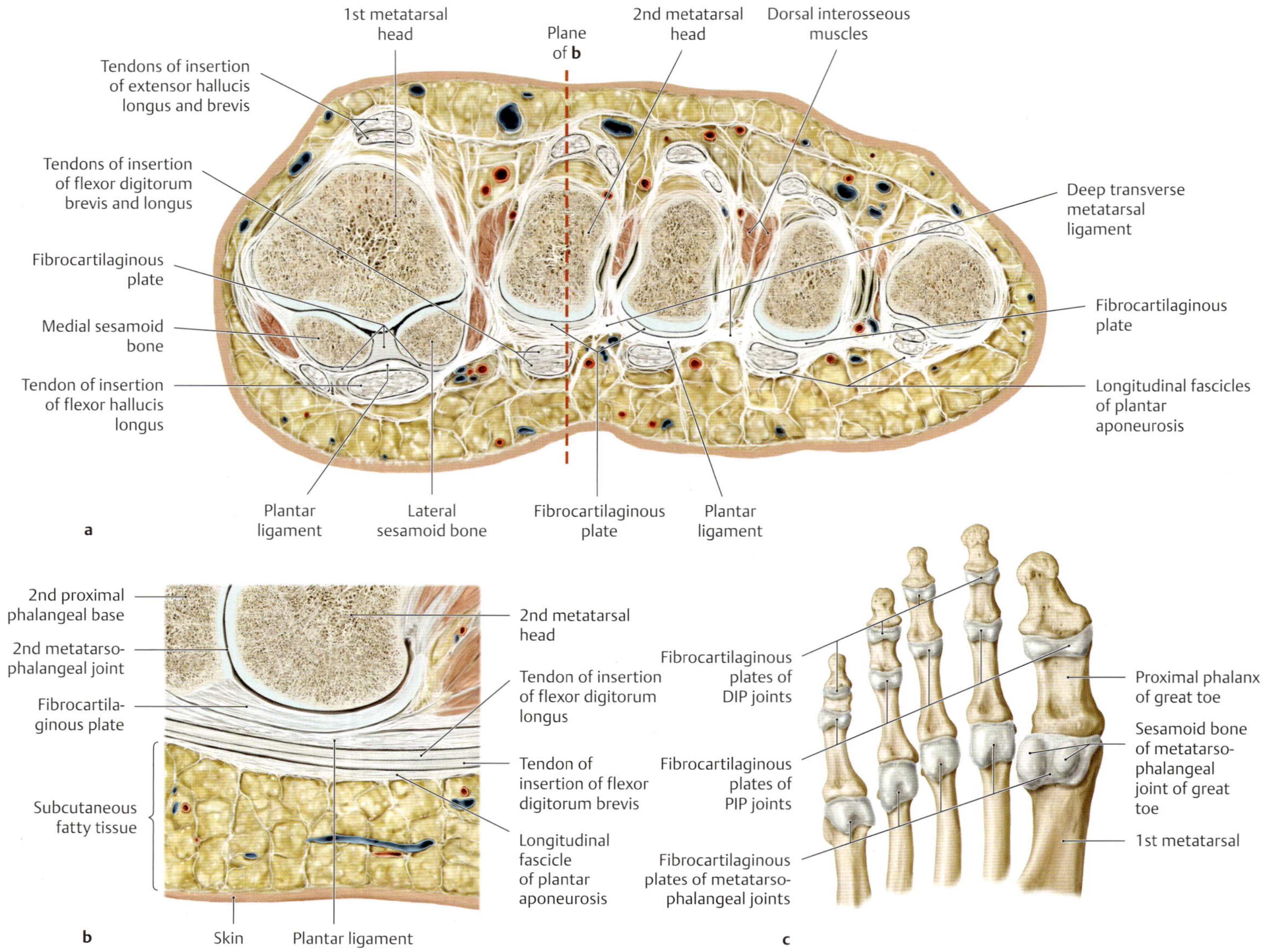

A Structure and function of the fibrocartilaginous plates in the joints of the toes

a Cross section through the right foot at the level of the metatarsal heads. Proximal view. Because the metatarsophalangeal joints are arranged in an arc, the section does not pass through the fibrocartilaginous plate of the fifth ray.

b Sagittal section at the level of the second ray through the metatarsophalangeal joint. Left view. The size difference between the articular cup and the condyloid head is clearly visible in this section. The fibrocartilaginous plate more than doubles the size of articular cup of the proximal phalanx. This allows it to distribute pressure on the joint uniformly over a large area. The principle is to enlarge the area on which the force acts (**a** and **b** drawn from a specimen of the anatomic collection of the University of Kiel).

c Isolated view of the fibrocartilaginous plates in the joints of the toes. Plantar view of the right foot

Notes: The sesamoid bones embedded in the fibrocartilaginous plate of the metatarsophalangeal joint of the great toe (see **B**). The plantar fibrocartilaginous plates in the metatarsophalangeal joints (see **a**) are about 1.5 cm long and 1 cm wide. Distally they consist of fibrocartilage, proximally of connective tissue. They reinforce the plantar joint capsule and are thickest at their narrow attachment immediately adjacent to the cartilage margin of the proximal phalanx. Their proximal ends extend as progressively thinning wedges to become a membranous portion of the joint capsule. Laterally, the fibrocartilaginous plates merge with the joint capsule, which is reinforced by collateral ligaments. Between the fibrocartilaginous portions of the plantar plate of the first through fifth rays lies the deep transverse metatarsal ligament. The plantar fibrocartilaginous plates act as a sort of articular labrum that extends the oval articular fossae of the proximal phalangeal bases and provides an extended supporting surface for the metatarsal heads. In this way the fibrocartilaginous plates distribute the pressure the forefoot absorbs while walking, thereby reducing the load on the metatarsal heads. The plantar ligaments lie plantar to the fibrocartilaginous plates. They form part of the fibrous portions of the grooves for the flexor tendon sheaths. Via their insertions into the plantar ligaments the interosseous muscles influence the position of the fibrocartilaginous plates. Chronically elevated pressure in the forefoot due to improper footwear (narrow shoes, high heels) permanently alters the position of the metatarsophalangeal joints, placing them in extreme dorsiflexion and promoting deformities such as hammer toe and claw toe. This is turn can cause stretching and tearing of the fibrocartilaginous plates in the metatarsophalangeal joints.

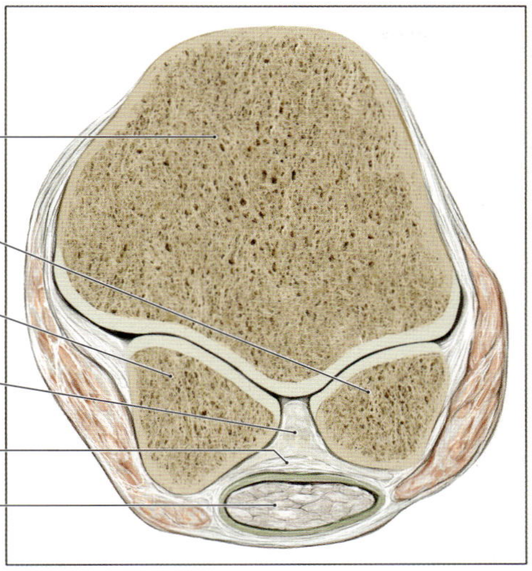

B Cross section through the head of the first metatarsal at the level of the sesamoids

Great toe of right foot, proximal view. The plane of section is indicated in **C**. The lateral and medial sesamoids are hemispheric bones, each presenting a slightly convex dorsal articular surface that articulates with the grooved plantar articular surfaces on the head of the first metatarsal. Sesamoids protect the tendons from excessive friction. They are important functionally for their ability to lengthen the effective lever arm of the muscle, so that muscular forces can be applied more efficiently. The development of sesamoids can be interpreted as a functional adaptation to the presence of pressure tendons.

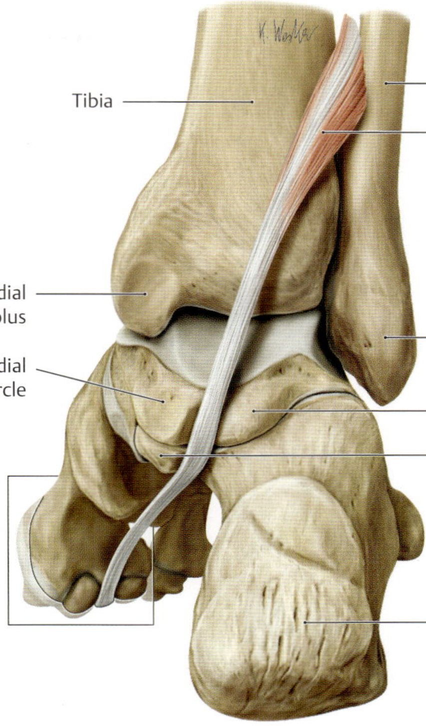

C The sesamoids of the great toe

Right foot, posteromedial view. The tendon of the flexor hallucis longus runs between the two sesamoids. The box around the sesamoids indicates the plane of section shown in **B**.

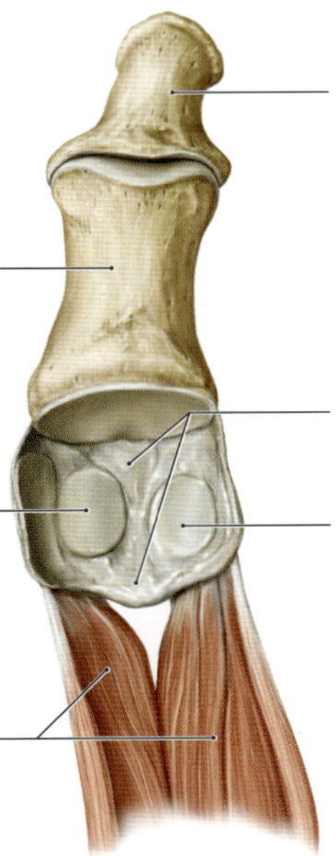

D Articular surfaces of the sesamoid bones

Dorsal view with the first metatarsal removed. The plantar fibrocartilaginous plate together with the capsule and ligaments of the sesamoid can tear when the metatarsophalangeal joint of the great toe is suddenly hyperextended, an injury known as "turf toe." This is seen in sports such as football, especially when played on artificial turf.

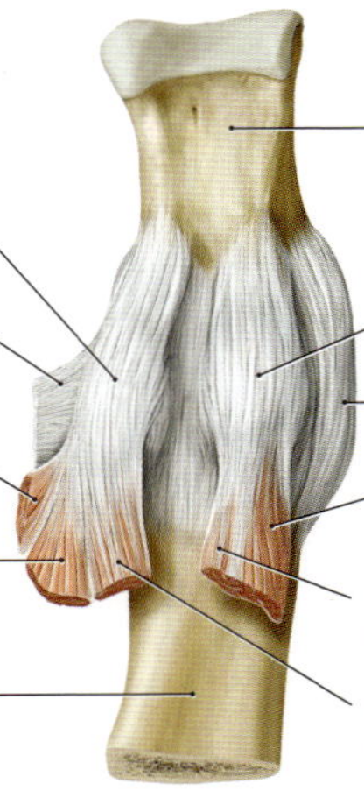

E The capsule and ligaments of the sesamoids and muscular attachments

First metatarsophalangeal joint of the right foot, plantar view. Both sesamoids are attached to the joint capsule and to the collateral ligaments of the metatarsophalangeal joint. They are embedded in the tendons of insertion of the following muscles:

- Medial sesamoid
  - Abductor hallucis
  - Medial head of flexor hallucis brevis
- Lateral sesamoid
  - Lateral head of flexor hallucis brevis
  - Transverse head of adductor hallucis
  - Oblique head of adductor hallucis

1.33 Degenerative Disorders of the First Ray: Hallux Valgus, Hallux Rigidus, and Hammer Toe

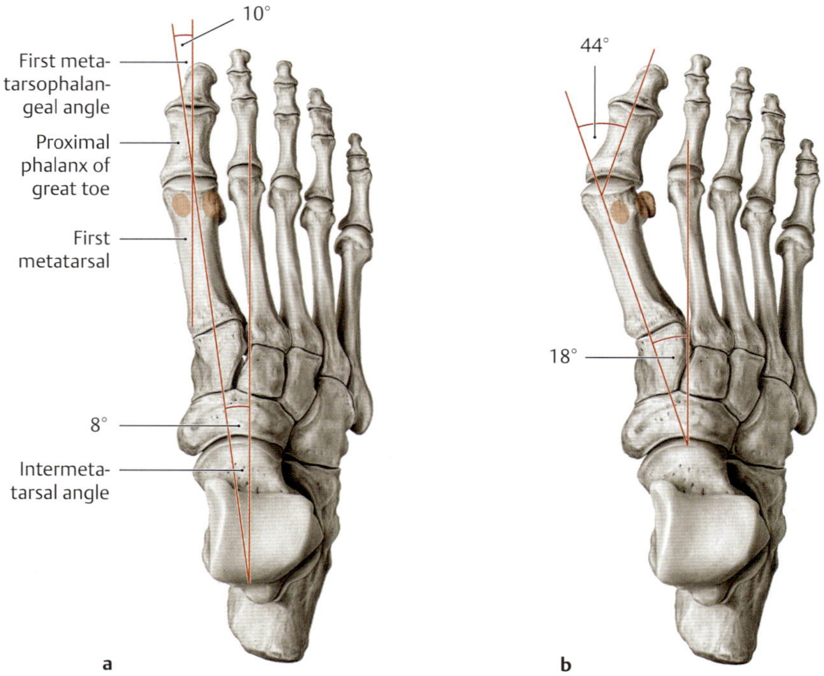

A Change in the first intermetatarsal angle and first metatarsophalangeal angle in hallux valgus
Right foot, superior view.

- **a** Skeleton of a normal right foot.
- **b** Lateral deviation of the first ray with subluxation of the metatarsophalangeal joint in hallux valgus.

In a normal foot, the *first intermetatarsal angle* (angle between the longitudinal axes of the first and second metatarsals) should not exceed 8°. The *first metatarsophalangeal angle* (angle between the longitudinal axes of the proximal phalanx of the great toe and the first metatarsal) should be less than 20°. In hallux valgus and also in splayfoot, which generally precedes it, both the intermetatarsal angle and metatarsophalangeal angle are significantly increased.

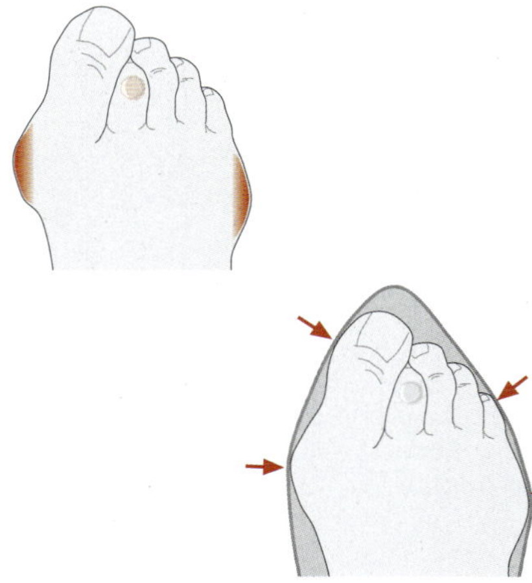

B Etiology of hallux valgus
Hallux valgus usually develops secondary to splayfoot. When a broadened forefoot is forced into a narrow, pointed shoe, the outer toes are crowded against the middle toes. This results in the pressure points and pain that are typical of hallux valgus and predominantly affect the medial side of the head of the first metatarsal, with chronic irritation of the first metatarsophalangeal joint and the overlying bursa (bursitis), in addition to reactive bone changes (exostosis). The middle toes are squeezed together anteriorly and become clawed (hammer toes, claw toes).

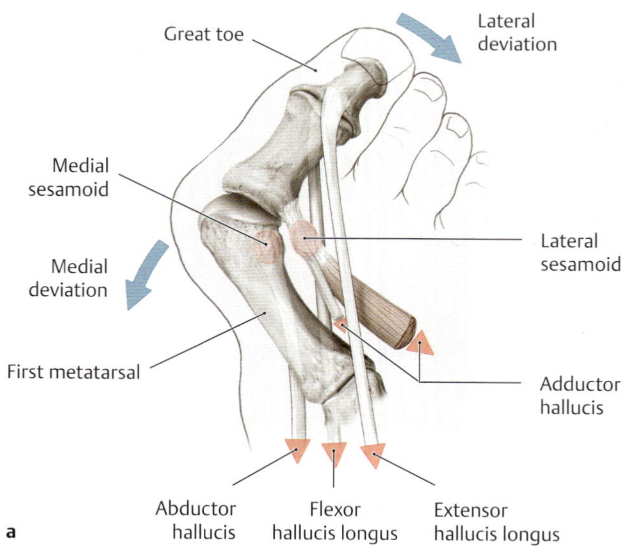

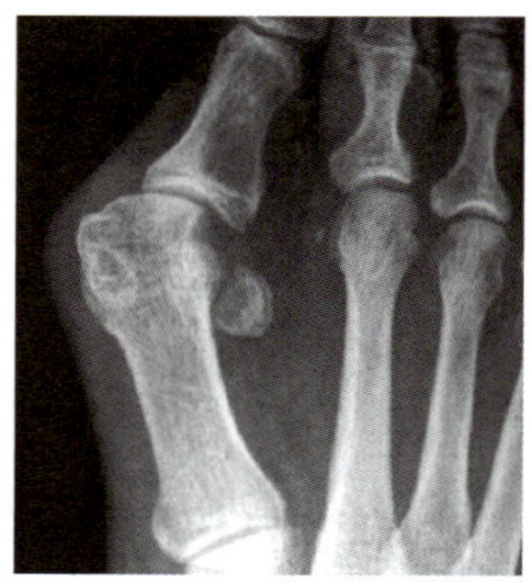

b

C Pathogenic mechanism of hallux valgus

- **a** Right forefoot, superior view. As the first metatarsal deviates medially and the great toe deviates laterally, a muscular imbalance develops marked by a change in the direction of tendon pull, which perpetuates and exacerbates the deformity. Most notably, the abductor hallucis moves laterally with the medial sesamoid, causing it to become an *adductor*. Meanwhile, the long flexor and extensor tendons move laterally, reinforcing the lateral angulation at the first metatarsophalangeal joint.
- **b** Hallux valgus on a radiograph (from Bohndorf K., Imhof H., Fischer W., Radiologische Diagnostik der Knochen und Gelenke. 2nd edition. Stuttgart, Thieme 2006).

Displacement of the lateral sesamoid bone is clearly demonstrated in addition to the severe hallux valgus deformity. The "secondary displacement" means that the sesamoid bone no longer protects the joint, making it easier for osteoarthritis to develop. Note the visible degenerative changes in the first metatarsophalangeal joint.

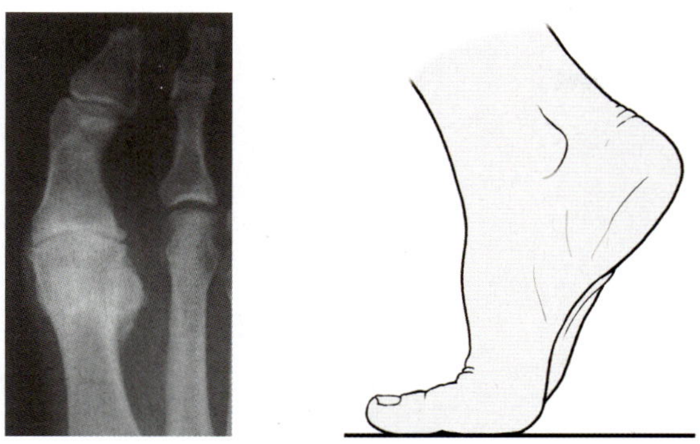

a b

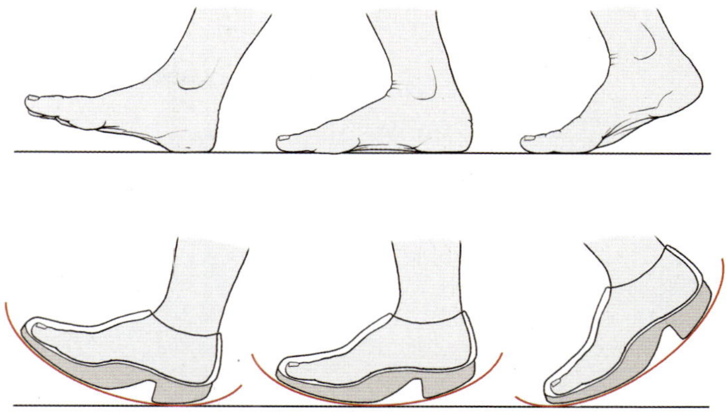

c

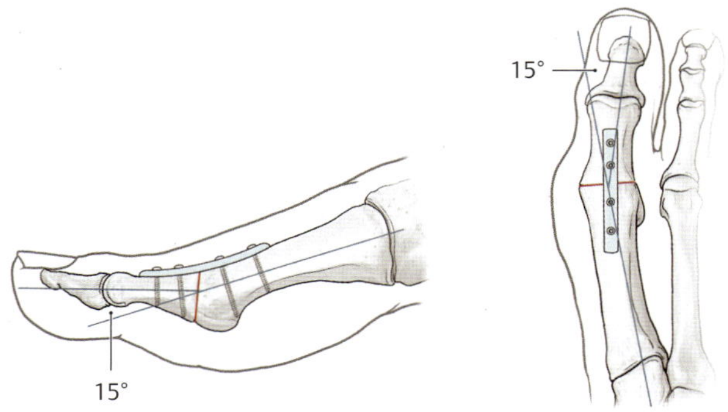

d

D Hallux rigidus: Pathology, clinical findings, and therapeutic options

a Hallux rigidus on a radiograph: The significantly narrowed joint space is clearly visible as are subchondral sclerosis, subchondral cysts, and osteophytes along the osteochondral junction (from: Niethard FU, Pfeil J. Orthopädie. Duale Reihe. 8th ed. Stuttgart: Thieme; 2017).
b Impaired toe off in hallux rigidus.
c Conservative therapy of hallux rigidus with a rocker sole on the shoe.
d Surgical therapy of hallux rigidus in the form of arthrodesis (surgical fusion of the metatarsophalangeal joint of the great toe).

Hallux rigidus is after hallux valgus (see p. 486) the second most common cause of symptoms in the metatarsophalangeal joint of the great toe (MTP joint). Why isolated osteoarthritis occurs only there is not known. Recurrent trauma, inflammation, and metabolic disorders such as gout have been discussed as possible causes. Painfully restricted mobility of the foot is typical of degenerative joint disease. Toe off in particular is painful as the great toe is pressed into dorsiflexion, which is hardly possible. Standing on the toes is also impossible. Affected patients avoid toeing off and bear weight only on the heel or lateral edge of the foot. The rocker sole (see **c**) on the shoe addresses this issue by making it possible to toe off with a stiff sole, which immobilizes the joints of the foot. The shoe with a stiff rocker sole moves on the ground like a rocking chair with shifting point of rotation. The foot itself remains in the same position throughout the stride (see **c**). In the presence of advanced active osteoarthritis (inflammation of the joint capsule due to abrasion of cartilage particles) and rheumatic symptoms, arthrodesis is the treatment of choice (see **d**). This surgical fusion immobilizes the metatarsophalangeal joint of the great toe in 15° of dorsiflexion and 15° valgus. This immobilization facilitates toeing off and allows the patient to wear stylish shoes.

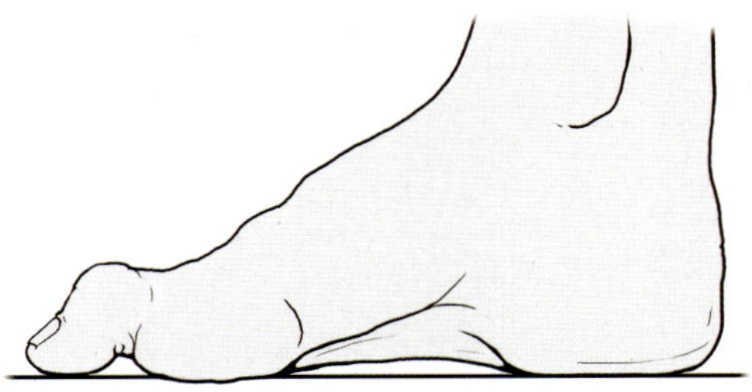

E Hammer toe

Hammer toe (hallux malleus) often accompanies pes cavus and is characterized by a flexion contracture of the interphalangeal joint. The lack of slight extension in the metatarsophalangeal joint also impairs toe off in hammer toe (see **Db**) and leads to painful pressure on the tip of the toe when toeing off.

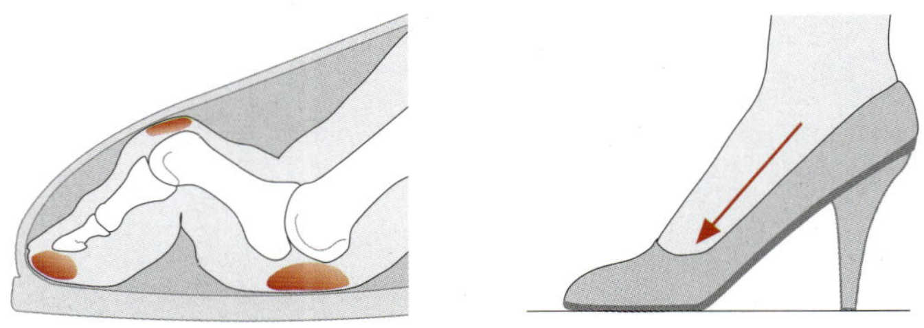

F Claw toes and hammer toes

Toe deformities are a very common associated feature of hallux valgus and splayfoot. When the foot is placed into a tight shoe with a high heel, it tends to slide forward and downward, and the resulting pressure leads to a typical cramped deformity with degenerative changes in the toe joints and painful callus formation. The claw toe deformity is characterized by marked hyperextensibility of the metatarsophalangeal joint with flexion of the proximal and distal interphalangeal joints. With hammer toes, there is less pronounced dorsiflexion of the metatarsophalangeal joint.

1.34 X-ray and Cross-sectional Anatomy of the Foot

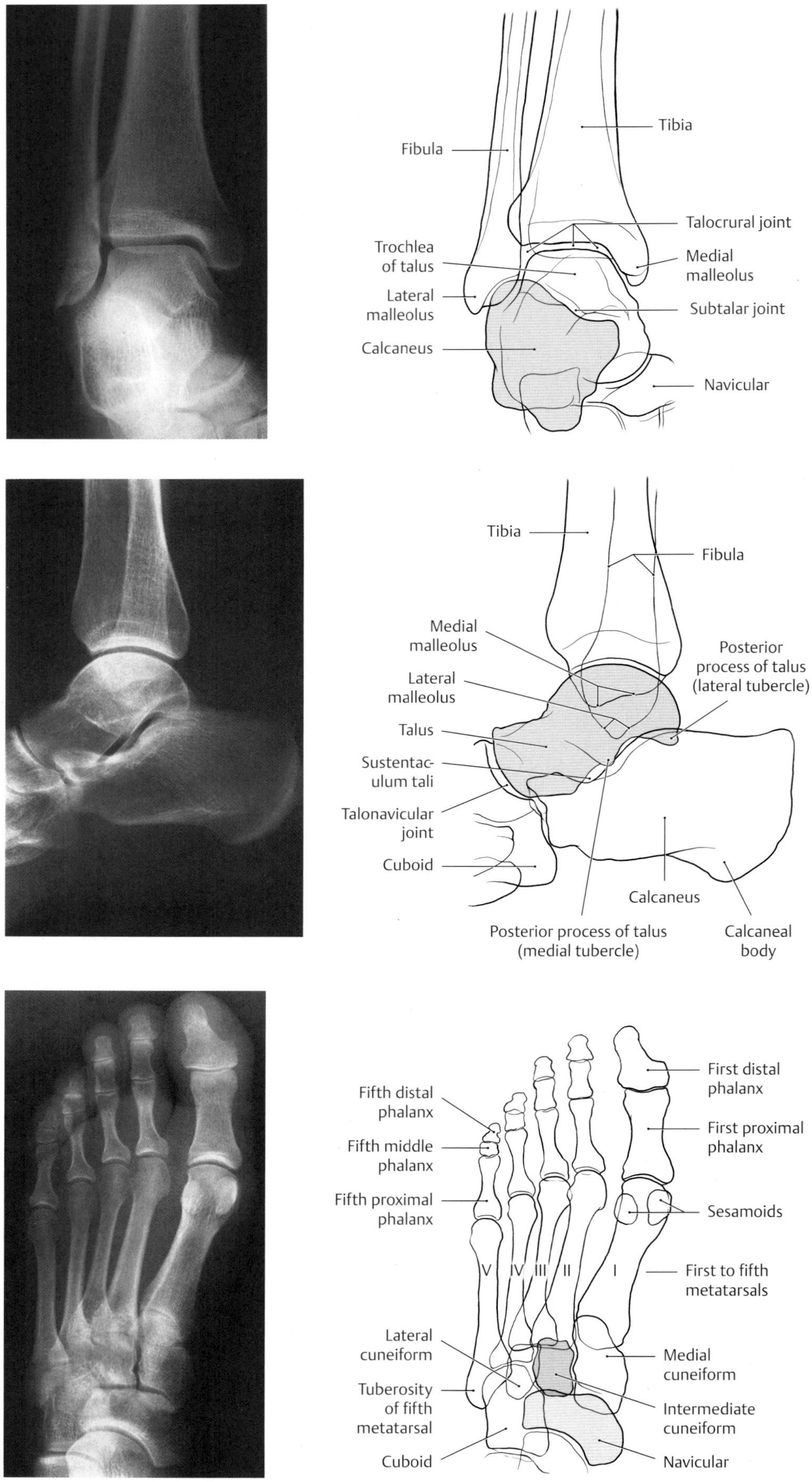

A X-ray anatomy of the foot
a Upper ankle joint in anterior-posterior beam path, **b** rearfoot in lateral beam path, **c** forefoot in dorsal plantar beam path (from Möller TB, Reif E. Taschenatlas der Röntgenanatomie. 7th ed. Stuttgart: Thieme; 2020).
The conventional X-ray is still the basis for medical imaging of the foot. It provides the highest spatial resolution and a good overview of the overall shape of the foot.

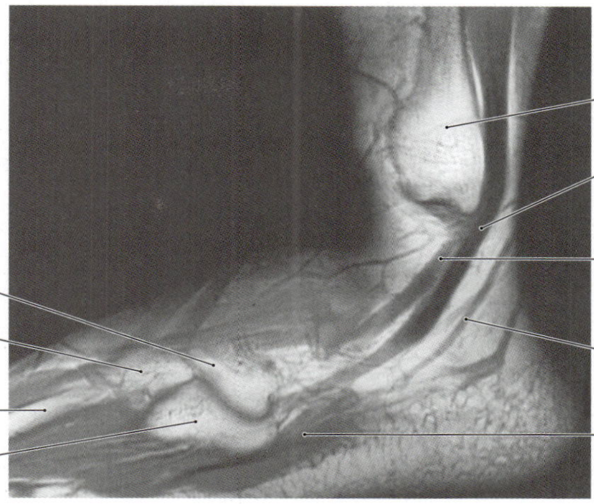

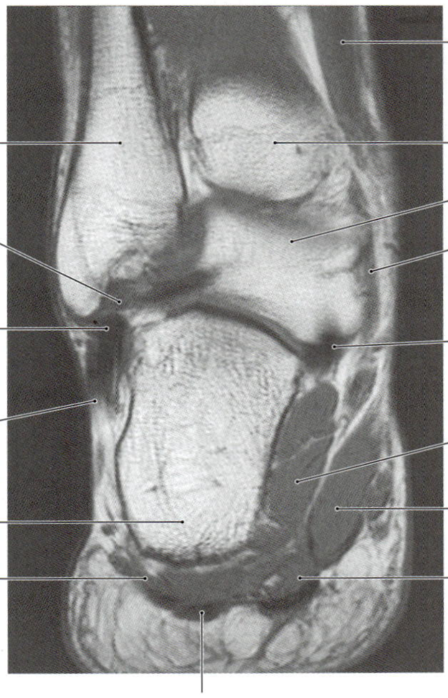

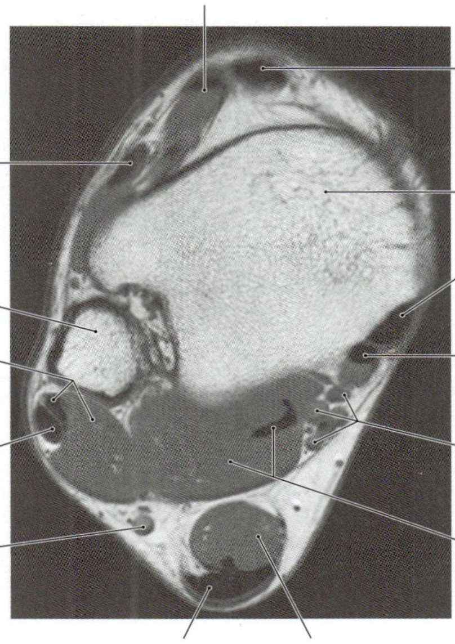

B Magnetic resonance imaging of the right foot (from: Vahlensieck M, Reiser M. MRT des Bewegungsapparates. 4th ed. Stuttgart: Thieme; 2014)

T1-weighted **a** sagittal, **b** coronal, and **c** axial (transverse) sequences.

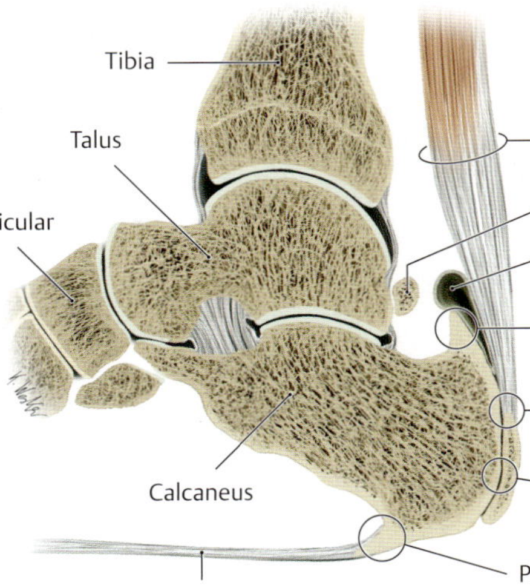

C Anatomic localization of clinically important disorders of the heel region

Heel pain is a very common symptom in patients of doctors specializing in foot surgery. In addition to the physical examination, only a few medical imaging techniques are necessary to make a diagnosis and treatment plan. Aside from typical bone diseases (osteoporosis and tumors), the following causes for heel pain need to be considered:

- Plantar heel spur: osseous exostosis at the calcaneal insertion of the plantar aponeurosis (often as a result of chronic plantar fascitis).
- Mechanical irritation of the "nerve of Baxter" (Baxter DE et al 1989; see also p. 544),
- Calcaneal apophysitis: defect in the bony apophysis of the calcaneal tuberosity during growth; pain results from overload of the cartilaginous growth plate.
- Posterior heel spur: prominent exostosis at the insertion on the calcaneus of the calcaneal (Achilles') tendon (insertional tendinopathy of the chondral apophyseal enthesis).
- Haglund's exostosis: proximally located dorsal calcaneal tuberosity (benign cartilaginous growth); pain results from inflammation of the adjacent bursa (bursa of the calcaneal tendon).
- Triangular bone: accessory bone of the foot at the posterior talus.
- Achillodynia: painful peridentinitis of the calcaneal (Achilles') tendon; symptoms include hyperthermia, swelling, and impaired function.

1.35 Human Gait

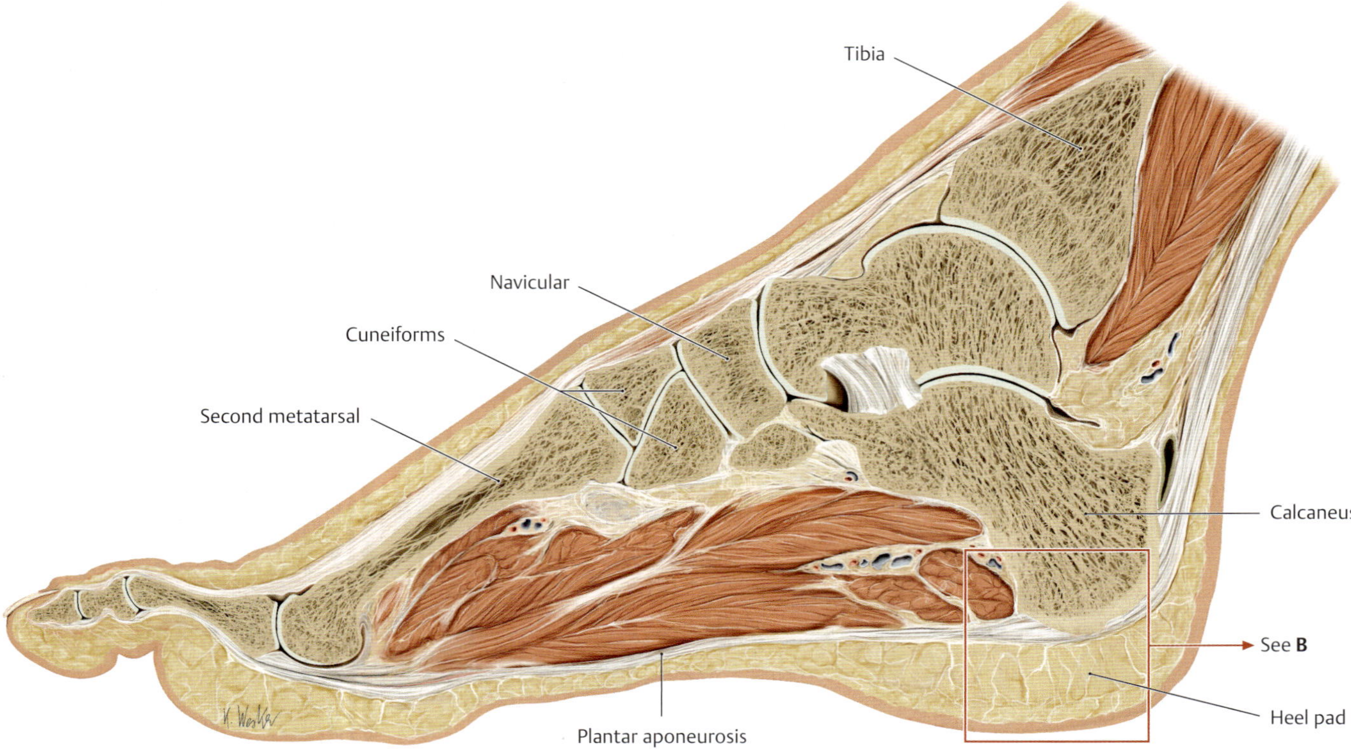

A The pressure chamber system in the sole of the foot
Sagittal section through the right foot at the level of the second ray, medial view (see detail in **B**).
During walking and particularly in the stance phase, large compressive forces are exerted on the heel pad and on the balls of the great and little toes. To distribute these concentrated forces more evenly over a larger area, the sole of the foot is covered by a layer of subcutaneous connective tissue up to 2 cm thick. As a functional adaptation to these demands, the tissue has a "pressure chamber" construction that acts as a shock absorber while also enhancing the mechanical stability of the sole. Without this pressure chamber construction, the loads on the foot would generate very high, localized stresses that would result in pressure necrosis (drawn from a specimen in the Anatomical Collection of Kiel University).

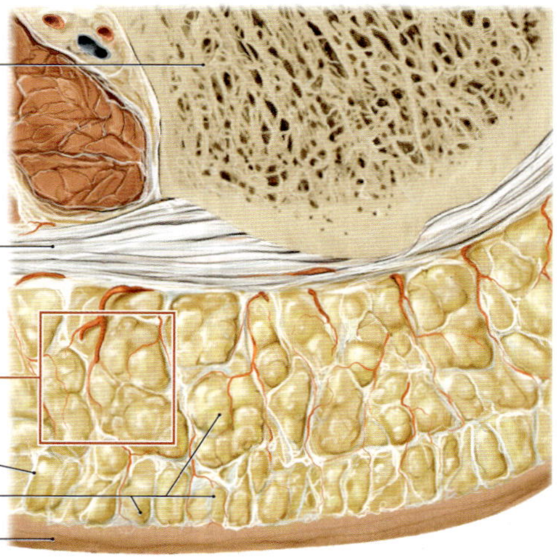

B The plantar pressure chambers
Detail from **A**.
Each of the pressure chambers contains an internal fibrofatty tissue covered externally by tough connective tissue made up of collagen fibers. These fibrous septa are firmly attached between the plantar aponeurosis and plantar skin and are supplied by an extensive network of blood vessels that further stabilize the walls of the pressure chambers (see close-up in **C**).

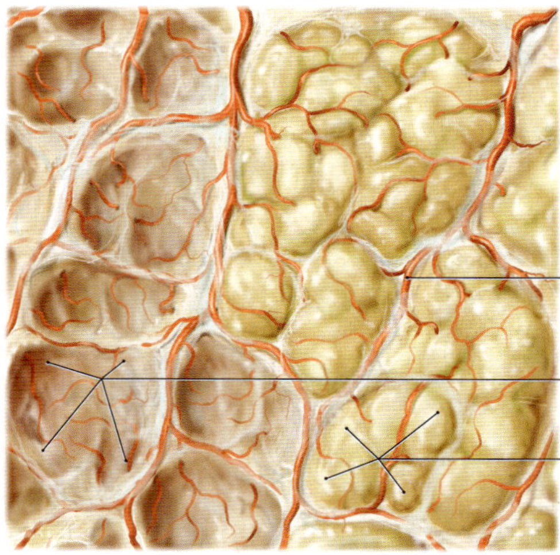

C Structure of the pressure chambers
Detail from **B**.
The fatty tissue has been removed from the chambers on the left side of the drawing to demonstrate the blood vessels that permeate the septa. (The sole of the foot is one of the most highly vascularized regions of the body surface.)

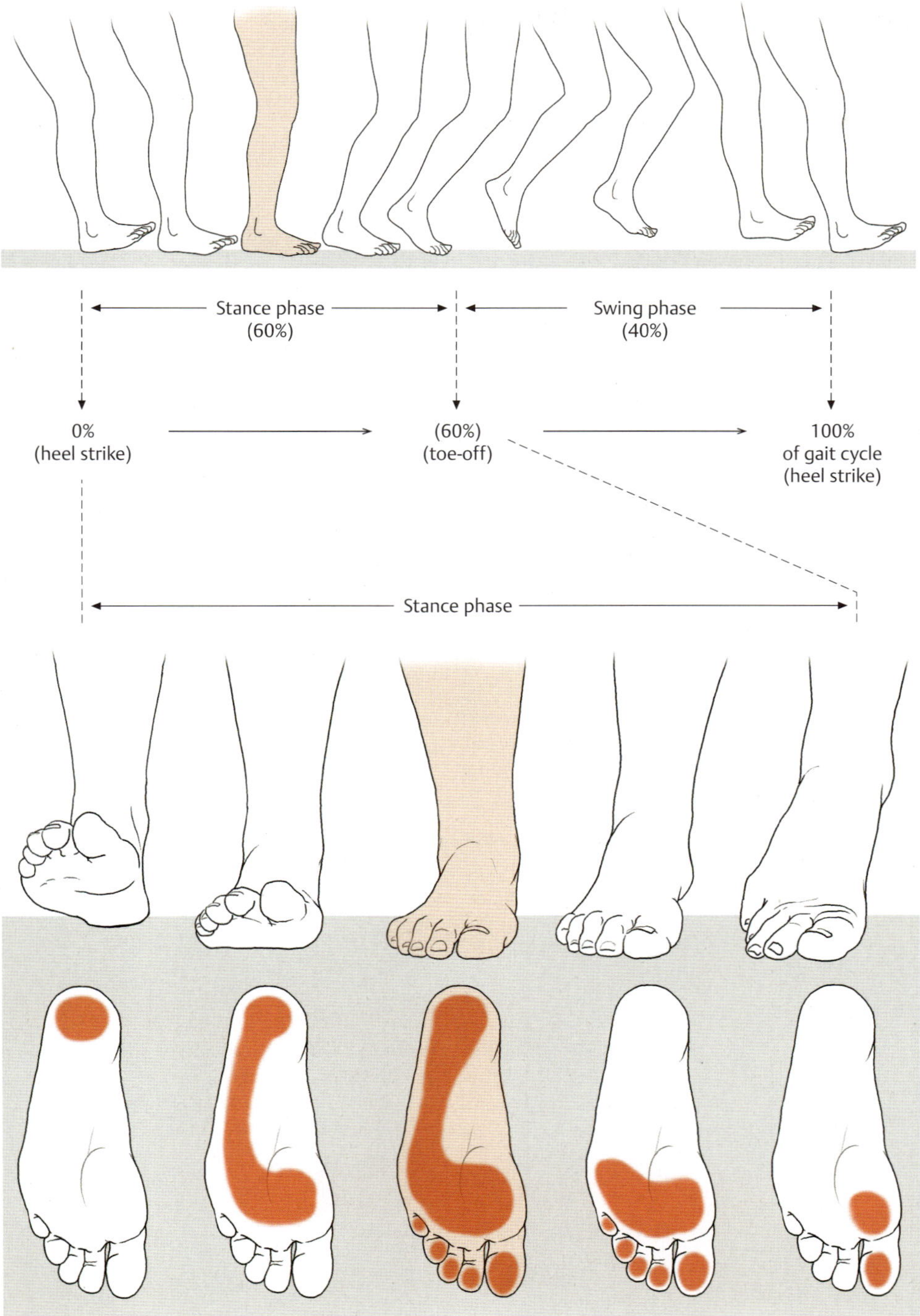

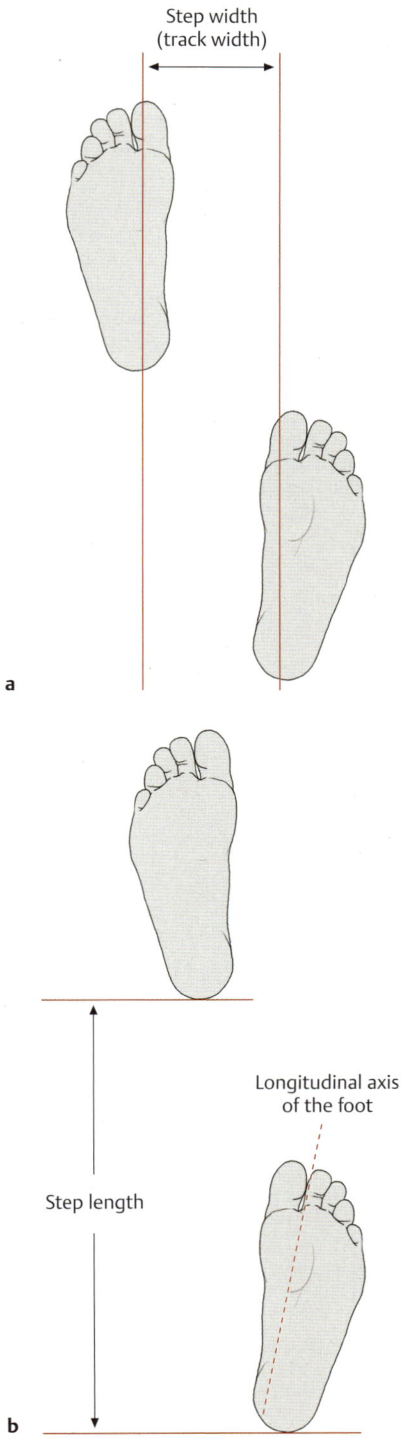

D Movements of the leg during one gait cycle

During normal walking, each leg functions alternately as the stance leg and swing leg. The stance phase begins when the heel contacts the ground (heel strike) and ends when the toes push off from the ground (toe-off). This phase makes up 60% of the gait cycle. The swing phase begins with the toe-off and ends with the heel strike. It makes up 40% of the gait cycle (100% of gait cycle = the period between two heel strikes of the same foot).

In the neutral position, the longitudinal axis of the foot relative to the line of direction shows an external rotation of 7°. This slight external rotation of the foot occurs also during walking.

E Step width (a) and step length (b)

The step width (track width) is evaluated from behind. Generally, it is narrower than the distance between the two hip joints. The step length (evaluated from the side) equals approximately 2 to 3 foot lengths.

The track width and step length define the area of support and thus play a critical role in stability. This is particularly important in hemiplegic patients, for example, in whom impaired proprioception can lead to instabilities of gait and stance.

2.1 The Muscles of the Lower Limb: Classification

In most mammals, the upper and lower limbs share many functions and have analogous functional groups of muscles. In humans, however, the specializations of the upper limb for manipulation and the lower limb for ambulation have imposed radically different requirements upon their respective muscle groups. For instance, the shoulder girdle has considerable freedom of motion on the trunk and is acted upon by an array of muscles, but the pelvic ring is firmly fixed to the vertebral column and changes position very little relative to the trunk, and has no comparable muscles to move it. In contrast, the hip and gluteal muscles have evolved into massive and powerful movers and stabilizers of the femur, counteracting the loads imposed by support of the whole body weight on two limbs and maintaining balance and stability during bipedal locomotion; these muscles are, in aggregate, larger than their counterparts that act upon the humerus, with a significantly different arrangement and orientation. As with the upper limb (see p. 310), the muscles of the lower limb can be classified on the basis of origin, topography, function, and innervation. Each such classification system has advantages and disadvantages, so several schemes are presented here. Segregation of muscles that act at the hip into specific functional groups is valid only for a particular joint position, because the axis of motion changes relative to the muscles as the joint is dynamically reoriented, causing abductors to become adductors, for example. Muscles surrounding the hip can be categorized topographically into an inner and outer group, relative to the pelvic girdle (see **A**). Muscles acting on the knee and foot can be grouped logically in an arrangement that uses both functional and topographical criteria, because these muscles tend to be clustered by functional groups into discrete compartments and act in a consistent way on joints with restricted ranges of motion. As with the upper limb, it is also instructive to categorize the lower limb's muscles by the pattern of their innervation (see **E**), a pattern that reveals the underlying logic of different clinical syndromes involving nerve damage.

A The hip and gluteal muscles

Inner hip muscles
- Psoas major
- Psoas minor
- Iliacus

} Act in unison as the "iliopsoas"

Outer hip muscles
- Gluteus maximus
- Gluteus medius
- Gluteus minimus
- Tensor fasciae latae
- Piriformis
- Obturator internus
- Gemelli
- Quadratus femoris

Muscles of the adductor group*
- Obturator externus
- Pectineus
- Adductor longus
- Adductor brevis
- Adductor magnus
- Adductor minimus
- Gracilis

* For functional reasons, the muscles of the adductor group, all of which are located on the medial side of the thigh, are classified as hip muscles because they act mainly on the hip joint.

B The thigh muscles

Anterior thigh muscles
- Sartorius
- Quadriceps femoris
  - Rectus femoris
  - Vastus medialis
  - Vastus lateralis
  - Vastus intermedius
  - (Articularis genus, the "fifth head" of quadriceps femoris, see p. 500)

Posterior thigh muscles*
- Biceps femoris
- Semimembranosus
- Semitendinosus
- Popliteus

} The hamstrings

* As a result of the altered position of the lower extremity (the genetically posterior portions have shifted to the anterior aspect and vice versa), the extensors lie on the anterior aspect of the thigh and the flexors on its posterior aspect (see p. 20).

C The leg muscles

Anterior compartment
- Tibialis anterior
- Extensor digitorum longus
- Extensor hallucis longus

Lateral compartment
- Fibularis* longus
- Fibularis brevis
- Fibularis tertius**

Posterior compartment

Superficial part
- Triceps surae
  - Soleus
  - Gastrocnemius (medial and lateral heads)
- Plantaris

Deep part
- Tibialis posterior
- Flexor digitorum longus
- Flexor hallucis longus

* Instead of "fibularis" the term "peroneus" can also be used.
** The fibularis tertius can also be classified as an extensor based on its function and innervation. Here it has been assigned to the fibularis group due to its name and its action on the subtalar joint (eversion).

D The short muscles of the foot

Dorsal muscles
- Extensor digitorum brevis
- Extensor hallucis brevis

Plantar muscles

Medial compartment
- Abductor hallucis
- Flexor hallucis brevis (medial and lateral heads)

Lateral compartment
- Abductor digiti minimi
- Flexor digiti minimi brevis
- Opponens digiti minimi

Central compartment
- Flexor digitorum brevis
- Adductor hallucis (transverse and oblique heads)
- Quadratus plantae
- First through fourth lumbricals
- First through third plantar interossei
- First through fourth dorsal interossei

E Classification of muscles based on their motor innervation

All the muscles of the lower limb are supplied by branches of the lumbar plexus (T12–L4) and the sacral plexus (L5–S3). They may be supplied by short, direct branches or by long nerves emanating from the corresponding plexus (see p. 548). See also **F**.

Nerve or plexus	Muscles supplied
Lumbar plexus	
Direct branches (muscular branches) (L1–L3)	Psoas major and minor
Nerves of the lower limb arising from the lumbar plexus	
Femoral nerve (L2–L4)	Iliacus, pectineus; sartorius, quadriceps femoris
Obturator nerve (L2–L4)	Obturator externus, pectineus, adductor longus, adductor brevis, adductor magnus (deep part), adductor minimus, gracilis
Sacral plexus	
Direct branches (muscular branches) (L5–S2)	Piriformis, obturator internus, gemelli, quadratus femoris
Nerves of the lower limb arising from the sacral plexus	
Superior gluteal nerve (L4–S1)	Tensor fasciae latae, gluteus medius and minimus
Inferior gluteal nerve (L5–S2)	Gluteus maximus
Sciatic nerve (L4–S3) (see **F**)	Adductor magnus (superficial part, tibial part), biceps femoris (long head, tibial part), biceps femoris (short head, fibular part), semimembranosus (tibial part), semitendinosus (tibial part)
• Common fibular* nerve (L4–S2)	
– Deep fibular nerve	Tibialis anterior, extensor digitorum longus and brevis, fibularis tertius, extensor hallucis longus and brevis
– Superficial fibular nerve	Fibularis longus and brevis
• Tibial nerve (L4–S3)	Popliteus, triceps surae, plantaris, tibialis posterior, flexor digitorum longus, flexor hallucis longus
– Medial plantar nerve	Abductor hallucis, flexor hallucis brevis (medial head), flexor digitorum brevis, first lumbrical
– Lateral plantar nerve	Flexor hallucis brevis (lateral head), adductor hallucis, abductor digiti minimi, flexor digiti minimi brevis, opponens digiti minimi, quadratus plantae, second to fourth lumbricals, first through third plantar interossei, first through fourth dorsal interossei

* The common fibular nerve and its divisions are also referred to as "peroneal nerves."

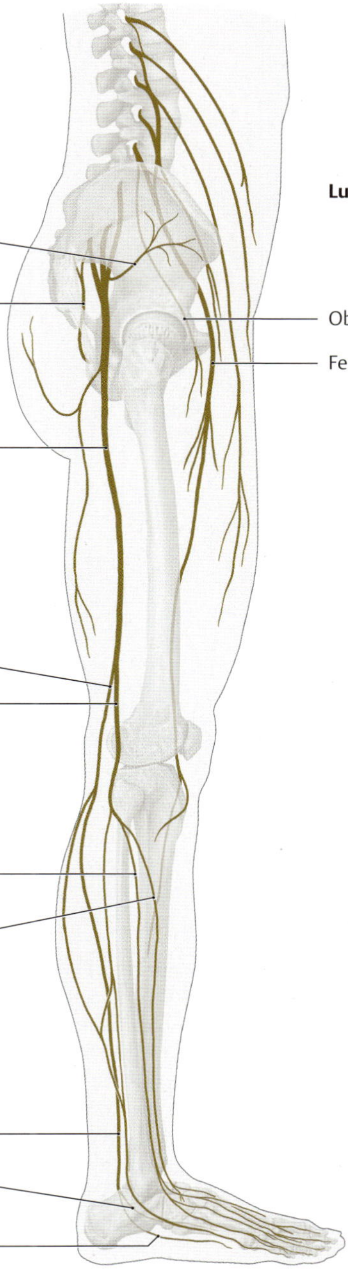

F The branches of the lumbosacral plexus that innervate the muscles of the lower limb

Right leg, lateral view. The anterior rami of the lumbar and sacral nerves, with contributions from the subcostal nerve and coccygeal nerve (not shown here), combine to form the lumbosacral plexus. While the branches arising from the lumbar plexus run *anterior* to the hip joint and mainly supply the muscles on the anterior and medial views of the thigh, the branches from the sacral plexus run *behind* the hip joint to supply the posterior thigh muscles and all the muscles of the leg and foot. The grossly visible division of the sciatic nerve into its two terminal branches (the tibial nerve and common fibular nerve) is generally located just above the knee joint, as pictured here (*low division*). But the nerve fibers that make up the two terminal branches become organized into bundles at a much more proximal level, where they already appear as separate nerve branches within their common fibrous sheath. In the *high division* pattern, the nerve divides into its terminal branches while still in the lesser pelvis (see p. 571).

2.2 The Hip and Gluteal Muscles: The Inner Hip Muscles

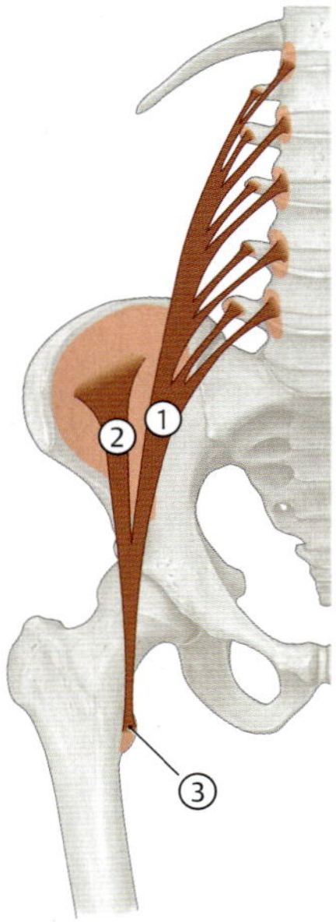

A Schematic of the inner hip muscles

Origin:	• ① Psoas major (superficial layer): lateral surfaces of the T12 vertebral body, the L1–L4 vertebral bodies, and the associated intervertebral disks • ① Psoas major (deep layer): costal processes of the L1–L5 vertebrae • ② Iliacus: the iliac fossa
Insertion:	Common insertion on the lesser trochanter of the femur as the iliopsoas ③ (psoas minor inserts into the iliopectineal arch, not depicted; see **B**)
Action:	• Hip joint: flexion and external rotation • Lumbar spine: unilateral contraction (with the femur fixed) bends the trunk laterally to the same side, bilateral contraction raises the trunk from the supine position
Innervation:	Femoral nerve (iliacus) and direct branches from the lumbar plexus (psoas) (L1–L3)

Properties and clinical aspects of the iliopsoas muscle

The iliopsoas is classified as a hip flexor, along with the rectus femoris, sartorius, and tensor fasciae latae. It is the most powerful flexor, its long vertical travel making it an important muscle for standing, walking, and running. As a typical postural muscle with a preponderance of slow-twitch red (type I) fibers, however, the iliopsoas is inherently susceptible to pathological shortening (particularly in older patients with a sedentary lifestyle or chronic immobilization conditions) and requires regular stretching to maintain normal tone (see pp. 60 and 517).

Shortening (contracture) of the hip flexors leads to

- increased anterior pelvic tilt,
- increased lumbar lordosis, and
- limitation of hip extension.

Unilateral shortening of the iliopsoas, in which the ilium on the affected side is tilted forward, can be diagnosed with the Thomas maneuver (see p. 450). This condition leads to pelvic torsion, in which the pelvis becomes twisted upon itself. This mainly alters the function of the sacroiliac joints but also compromises the intervertebral joints and the lumbosacral junction (increased lordosis of the lumbar spine with degenerative changes in the vertebral bodies, see p. 138). Patients with *bilateral* iliopsoas weakness or paralysis are unable to raise the trunk from the supine position, despite intact abdominal muscles, without using their arms and are greatly limited in their ability to walk and climb stairs without assistance.

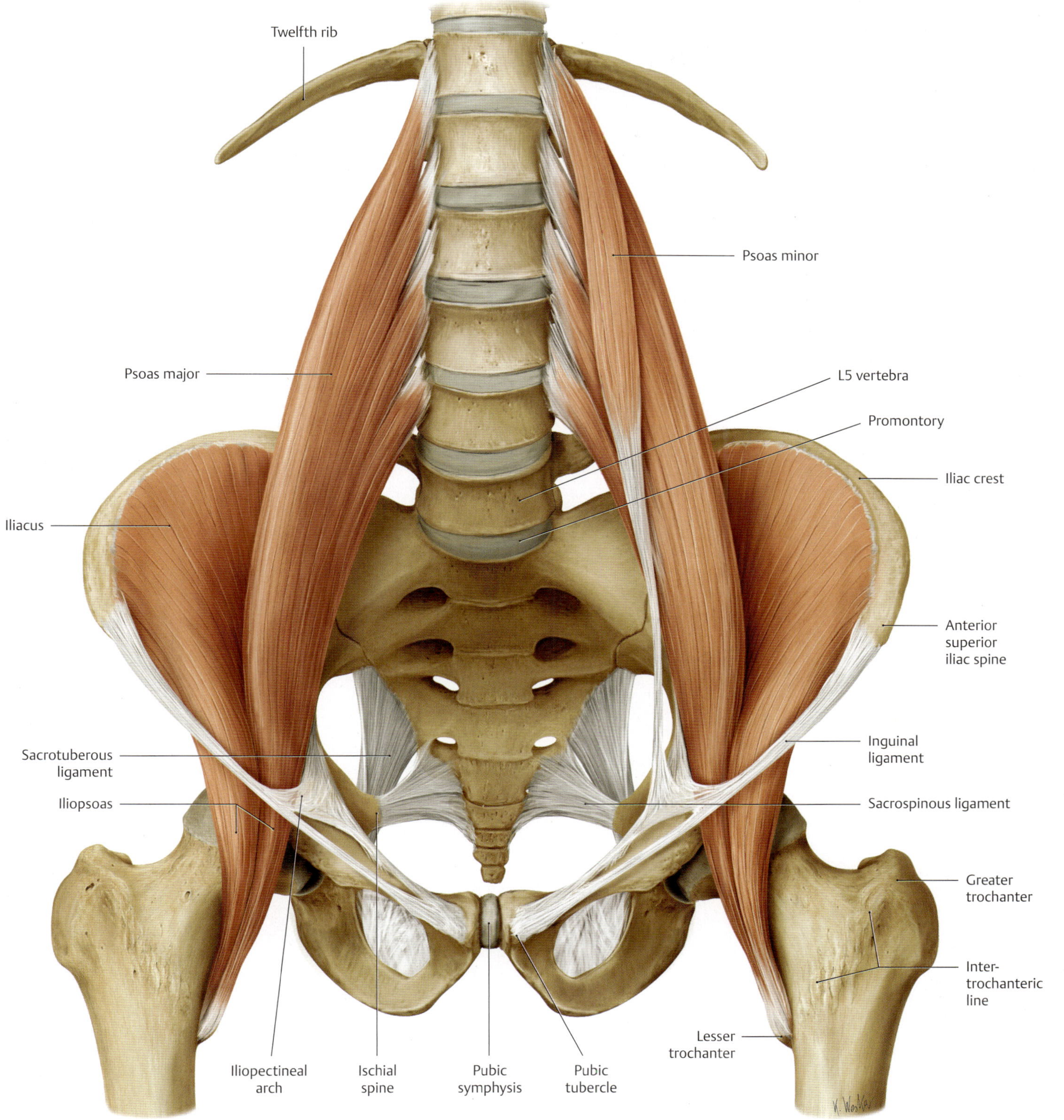

B The inner hip muscles
Anterior view.
Note: In the lower extremity, the directional terms "posterior" and "anterior" are better used than "dorsal" and "ventral"; for background, see rotation of the extremities on p. 20.

The psoas major unites with the iliacus at the level of the inguinal ligament to form a conjoined muscle, the iliopsoas. Approximately 50% of the population also has a psoas minor muscle (as shown here), which arises from the T12 and L1 vertebrae and inserts into the iliopectineal arch (iliac fascia).

2.3 The Hip and Gluteal Muscles: The Outer Hip Muscles

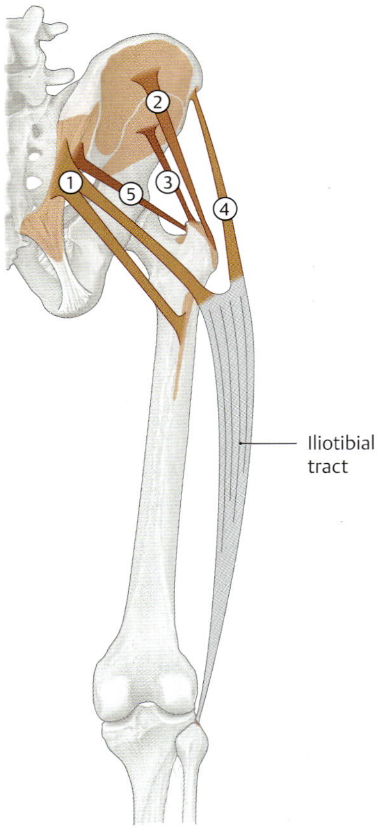

A Schematic of the vertically oriented outer hip muscles

① **Gluteus maximus**

Origin: Lateral part of the dorsal surface of the sacrum, posterior part of the gluteal surface of the ilium (behind the posterior gluteal line), also from the thoracolumbar fascia and sacrotuberous ligament

Insertion:
- Upper fibers: iliotibial tract
- Lower fibers: gluteal tuberosity

Action:
- Entire muscle: extends and externally rotates the hip, stabilizes the hip in both the sagittal and coronal planes
- Upper fibers: abduction
- Lower fibers: adduction

Innervation: Inferior gluteal nerve (L5–S2)

② **Gluteus medius**

Origin: Gluteal surface of the ilium (below the iliac crest between the anterior and posterior gluteal line)

Insertion: Lateral surface of the greater trochanter of the femur

Action:
- Entire muscle: abducts the hip, stabilizes the pelvis in the coronal plane
- Anterior part: flexion and internal rotation
- Posterior part: extension and external rotation

Innervation: Superior gluteal nerve (L4–S1)

③ **Gluteus minimus**

Origin: Gluteal surface of the ilium (below the origin of gluteus medius)

Insertion: Anterolateral surface of the greater trochanter of the femur

Action:
- Entire muscle: abducts the hip, stabilizes the pelvis in the coronal plane
- Anterior part: flexion and internal rotation
- Posterior part: extension and external rotation

Innervation: Superior gluteal nerve (L4–S1)

④ **Tensor fasciae latae**

Origin: Anterior superior iliac spine

Insertion: Iliotibial tract

Action:
- Tenses the fascia lata
- Abduction, flexion, and internal rotation of the hip joint

Innervation: Superior gluteal nerve (L4–S1)

⑤ **Piriformis**

Origin: Pelvic surface of the sacrum

Insertion: Apex of the greater trochanter of the femur

Action:
- External rotation, abduction, and extension of the hip joint
- Stabilizes the hip joint

Innervation: Direct branches from the sacral plexus (S1, S2)

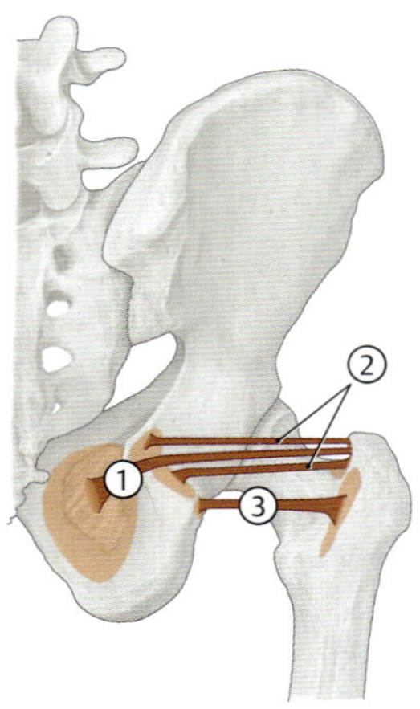

B Schematic of the horizontally oriented outer hip muscles

① **Obturator internus**

Origin: Inner surface of the obturator membrane and its bony boundaries

Insertion: Medial surface of the greater trochanter

Action: External rotation, adduction, and extension of the hip joint (also active in abduction, depending on the position of the joint)

Innervation: Direct branches from the sacral plexus (L5, S1)

② **Gemelli**

Origin:
- Superior gemellus: ischial spine
- Inferior gemellus: ischial tuberosity

Insertion: Jointly with obturator internus tendon (medial surface, greater trochanter)

Action: External rotation, adduction, and extension of the hip joint (also active in abduction, depending on the position of the joint)

Innervation: Direct branches from the sacral plexus (L5, S1)

③ **Quadratus femoris**

Origin: Lateral border of the ischial tuberosity

Insertion: Intertrochanteric crest of the femur

Action: External rotation and adduction of the hip joint

Innervation: Direct branches from the sacral plexus (L5–S2) and/or inferior gluteal nerve

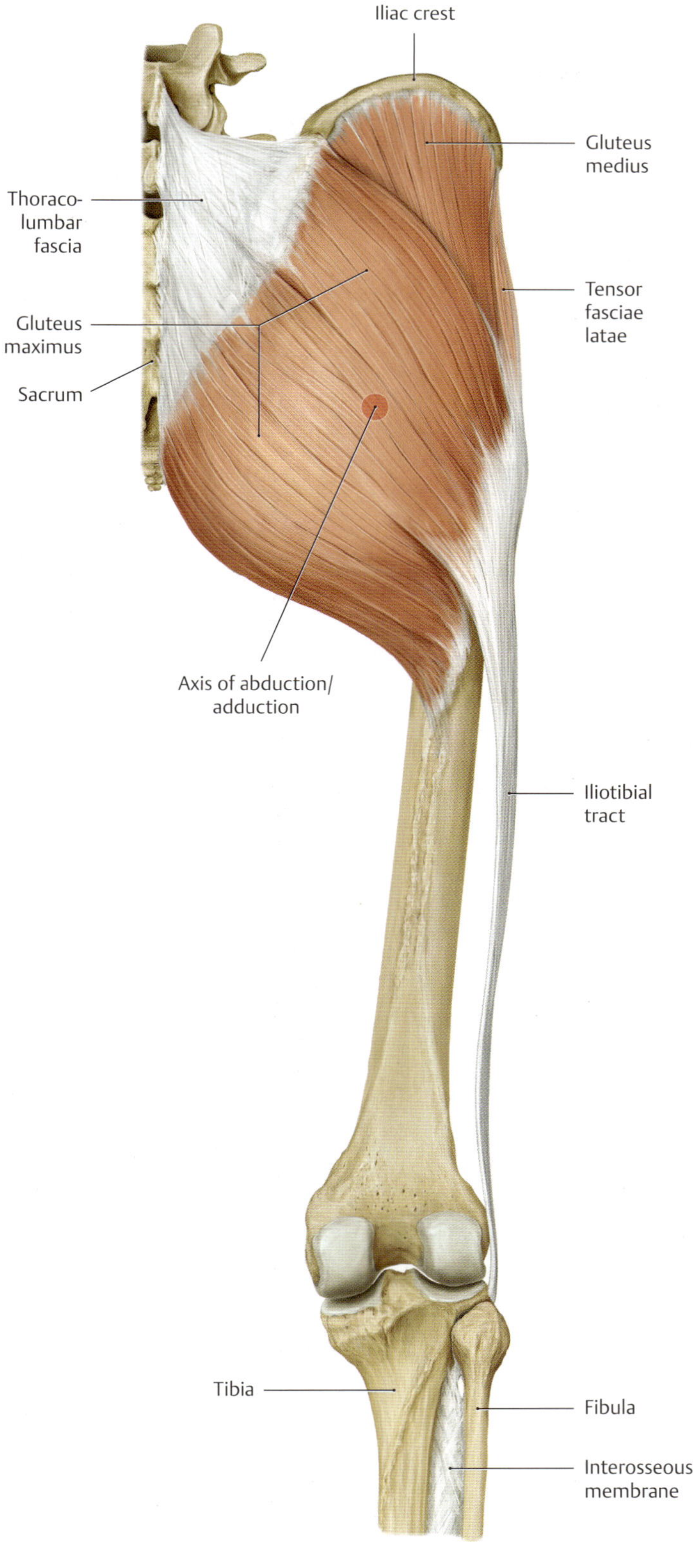

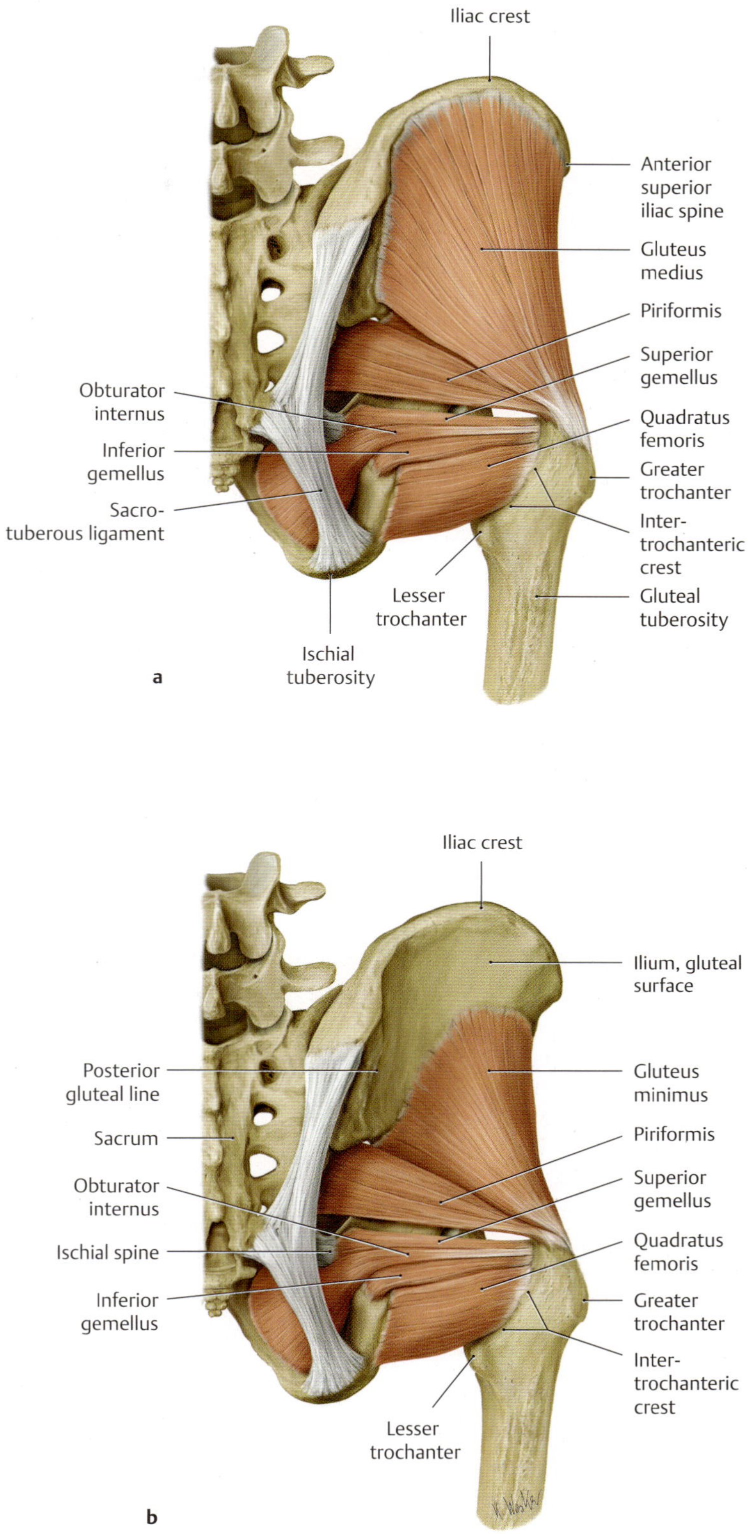

C The outer hip muscles: superficial layer
Right side, posterior view.
Note the position of the gluteus maximus muscle in relation to the axis of hip abduction and adduction. While the fibers of the gluteus maximus that run *above* the axis and insert on the tibia via the iliotibial tract are active in abducting the hip joint, the muscle fibers that run *below* the axis are active in adduction.

D The outer hip muscles: deep layer
Right side, posterior view.

a With the gluteus maximus removed.
b With the gluteus medius removed.

If there is weakness or paralysis of the small gluteal muscles (gluteus medius and minimus), the pelvis can no longer be stabilized in the coronal plane and will tilt toward the unaffected side (positive Trendelenburg sign, see also p. 554).

2.4 The Hip and Gluteal Muscles: The Adductor Group

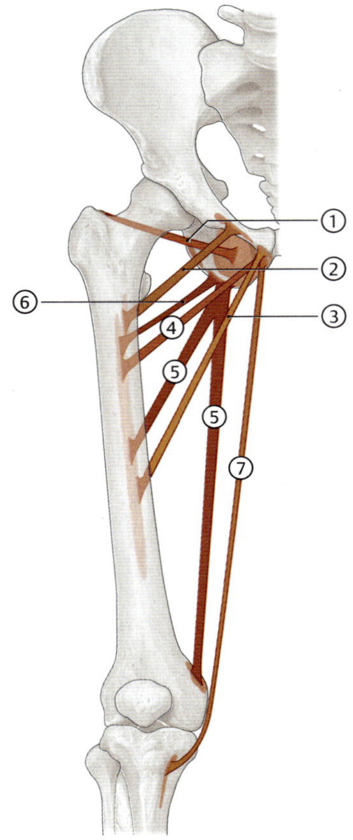

A Schematic of the adductors

① **Obturator externus**

Origin: Outer surface of the obturator membrane and its bony boundaries
Insertion: Trochanteric fossa of the femur
Action:
- Adduction and external rotation of the hip joint
- Stabilizes the pelvis in the sagittal plane

Innervation: Obturator nerve (L3, L4)

② **Pectineus**

Origin: Pecten pubis
Insertion: Pectineal line and the proximal linea aspera of the femur
Action:
- Adduction, external rotation, and slight flexion of the hip joint
- Stabilizes the pelvis in the coronal and sagittal planes

Innervation: Femoral nerve (L2–L4), obturator nerve (L2–L4)

③ **Adductor longus**

Origin: Superior pubic ramus and anterior side of the symphysis
Insertion: Linea aspera: medial lip in the middle third of the femur
Action:
- Adduction and flexion (up to 70°) of the hip joint (extends the hip past 80° of flexion)
- Stabilizes the pelvis in the coronal and sagittal planes

Innervation: Obturator nerve (L2–L4)

④ **Adductor brevis**

Origin: Inferior pubic ramus
Insertion: Linea aspera: medial lip in the upper third of the femur
Action:
- Adduction and flexion (up to 70°) of the hip joint (extends the hip past 80° of flexion)
- Stabilizes the pelvis in the coronal and sagittal planes

Innervation: Obturator nerve (L2, L3)

⑤ **Adductor magnus**

Origin: Inferior pubic ramus, ischial ramus, and ischial tuberosity
Insertion:
- Deep part ("fleshy insertion"): medial lip of linea aspera
- Superficial part ("tendinous insertion"): adductor tubercle of femur

Action:
- Adduction, extension, and slight flexion of the hip joint (the tendinous insertion is also active in internal rotation)
- Stabilizes the pelvis in the coronal and sagittal planes

Innervation:
- Deep part: obturator nerve (L2–L4)
- Superficial part: tibial nerve (L4)

⑥ **Adductor minimus (upper division of adductor magnus)**

Origin: Inferior pubic ramus
Insertion: Medial lip of the linea aspera
Action: Adduction, external rotation, and slight flexion of the hip joint
Innervation: Obturator nerve (L2–L4)

⑦ **Gracilis**

Origin: Inferior pubic ramus below the pubic symphysis
Insertion: Medial border of the tibial tuberosity (along with the tendons of sartorius and semitendinosus)
Action:
- Hip joint: adduction and flexion
- Knee joint: flexion and internal rotation

Innervation: Obturator nerve (L2, L3)

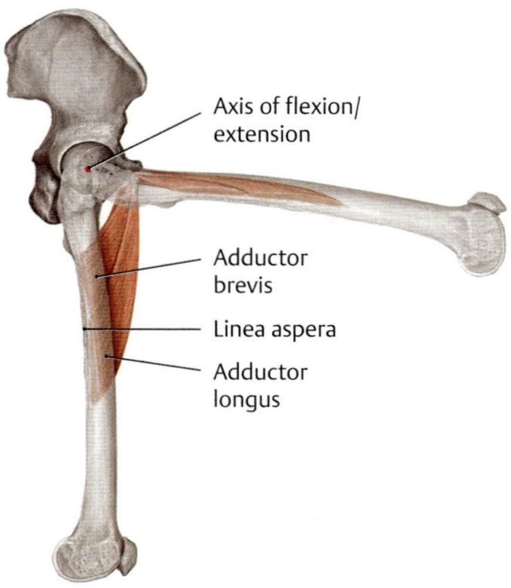

B Reversal of muscle actions, illustrated for the adductor brevis and longus
Right hip joint, lateral view. The femur in 80° of flexion is shown in lighter shading. In addition to their primary action as adductors, both muscles may be active in flexion and extension, depending on the joint position.

- They assist in flexion from the neutral (0°) position to approximately 70°.
- Their actions reverse past approximately 80° of flexion, and they become active in extension.

The flexor components of both muscles are transformed into extensor components as soon as their insertion (the linea aspera) moves higher than their origin (the inferior or superior pubic ramus).

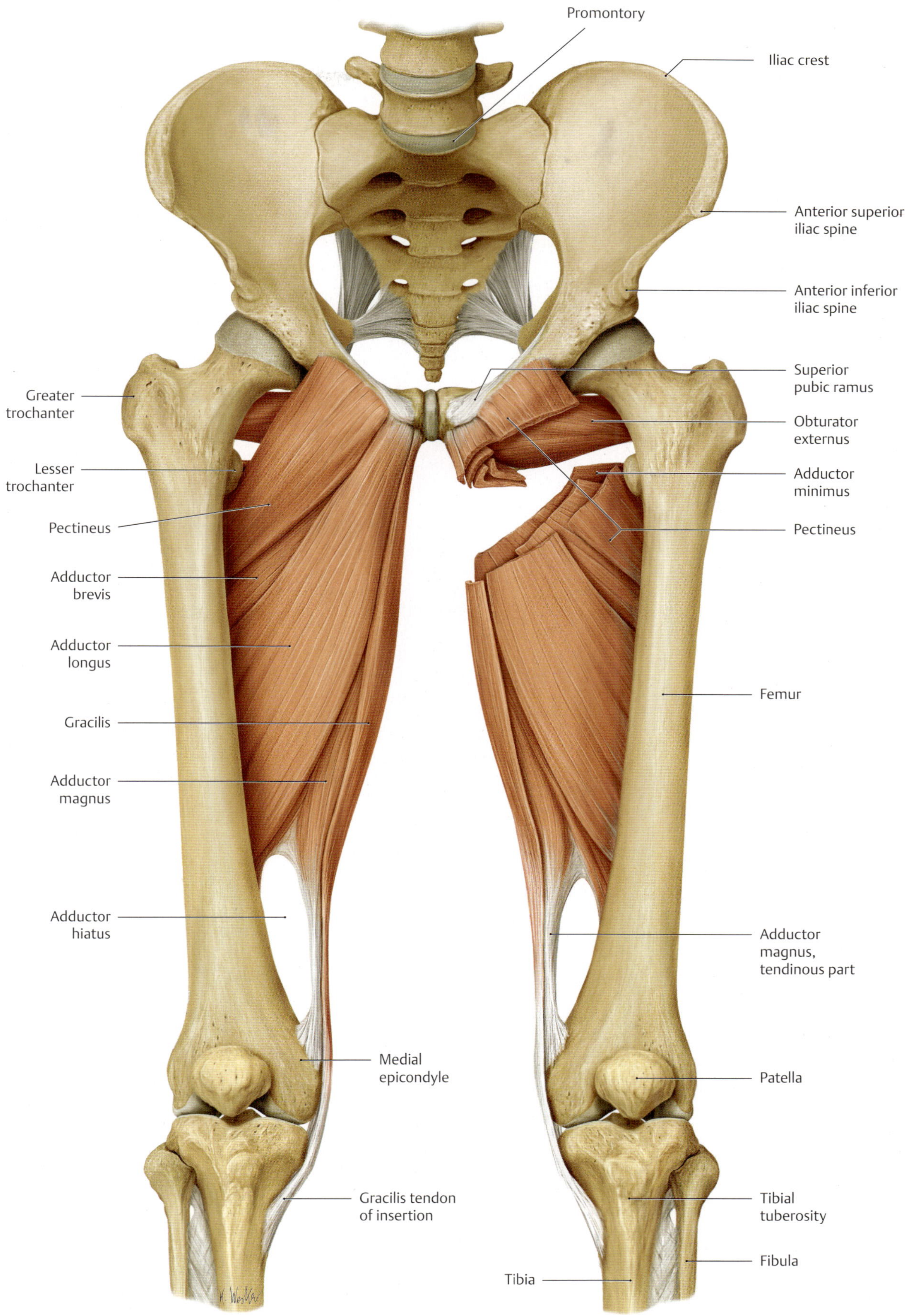

C The adductors (obturator externus; pectineus; adductors longus, brevis, magnus, and minimus; and gracilis)

Anterior view. A portion of the adductors, pectineus, and gracilis muscles on the left side have been removed just past their origins to demonstrate the course of the obturator externus more clearly.

Note: Unilateral shortening of the adductors leads to functional leg shortening on the affected side.

2.5 The Anterior Thigh Muscles: The Extensor Group

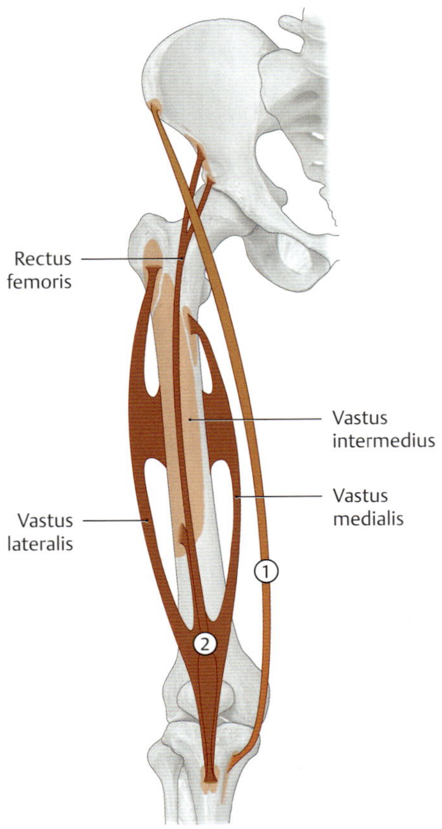

① Sartorius

Origin: Anterior superior iliac spine
Insertion: Medial to the tibial tuberosity (together with gracilis and semitendinosus)
Action:
- Hip joint: flexion, abduction, and external rotation
- Knee joint: flexion and internal rotation

Innervation: Femoral nerve (L2–L4)

② Quadriceps femoris

Origin:
- Rectus femoris: anterior inferior iliac spine (straight head), acetabular roof of the hip joint (reflected head)
- Vastus medialis: medial lip of the linea aspera, distal part of the intertrochanteric line
- Vastus lateralis: lateral lip of the linea aspera, lateral surface of the greater trochanter
- Vastus intermedius: anterior side of the femoral shaft
- Articularis genus (distal fibers of the vastus intermedius): anterior side of the femoral shaft at the level of the suprapatellar recess

Insertion:
- On the tibial tuberosity via the patellar ligament (entire muscle)
- Both sides of the tibial tuberosity on the medial and lateral condyles via the medial and lateral longitudinal patellar retinacula (vastus medialis and lateralis)
- The suprapatellar recess of the knee joint capsule (articularis genus)

Action:
- Hip joint: flexion (rectus femoris)
- Knee joint: extension (all parts), prevents entrapment of the capsule (articularis genus)

Innervation: Femoral nerve (L2–L4)

A Schematic of the extensors
At the level of the knee the sartorius has migrated posteriorly with the pes anserinus to a point posterior to the flexion and extension axis of the knee. It thus flexes the knee. However, because of its position it is included among the extensors of the thigh (once posterior musculature).

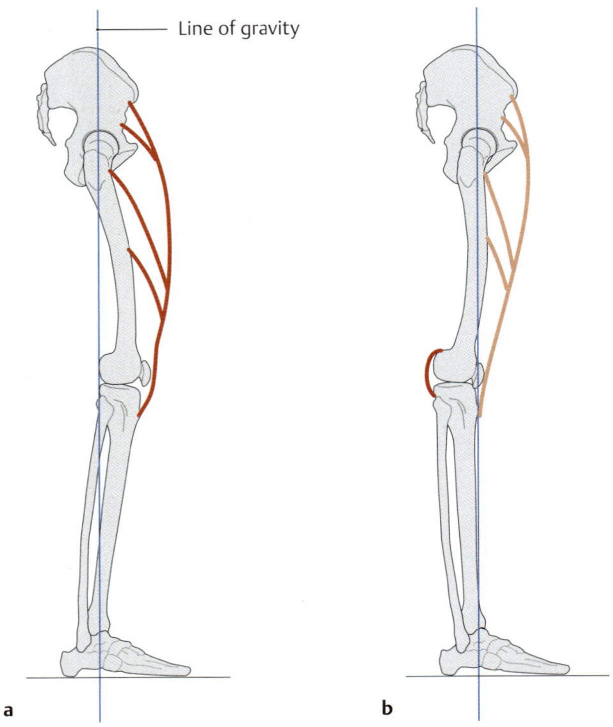

B Deficient stabilization of the knee joint due to weakness or paralysis of quadriceps femoris
Right lower limb, lateral view.

a When the quadriceps femoris is intact and the knee is in slight flexion, the line of gravity falls *behind* the transverse axis of knee motion. As the only extensor muscle of the knee joint, the quadriceps femoris keeps the body from tipping backward and ensures stability.

b With weakness or paralysis of the quadriceps femoris, the knee joint can no longer be actively extended. In order to stand upright, the patient must hyperextend the knee so that the line of gravity, and thus the whole-body center of gravity, is shifted forward, in front of the knee, to utilize gravity as the extending force. The joint is stabilized in this situation by the posterior capsule and ligaments of the knee.

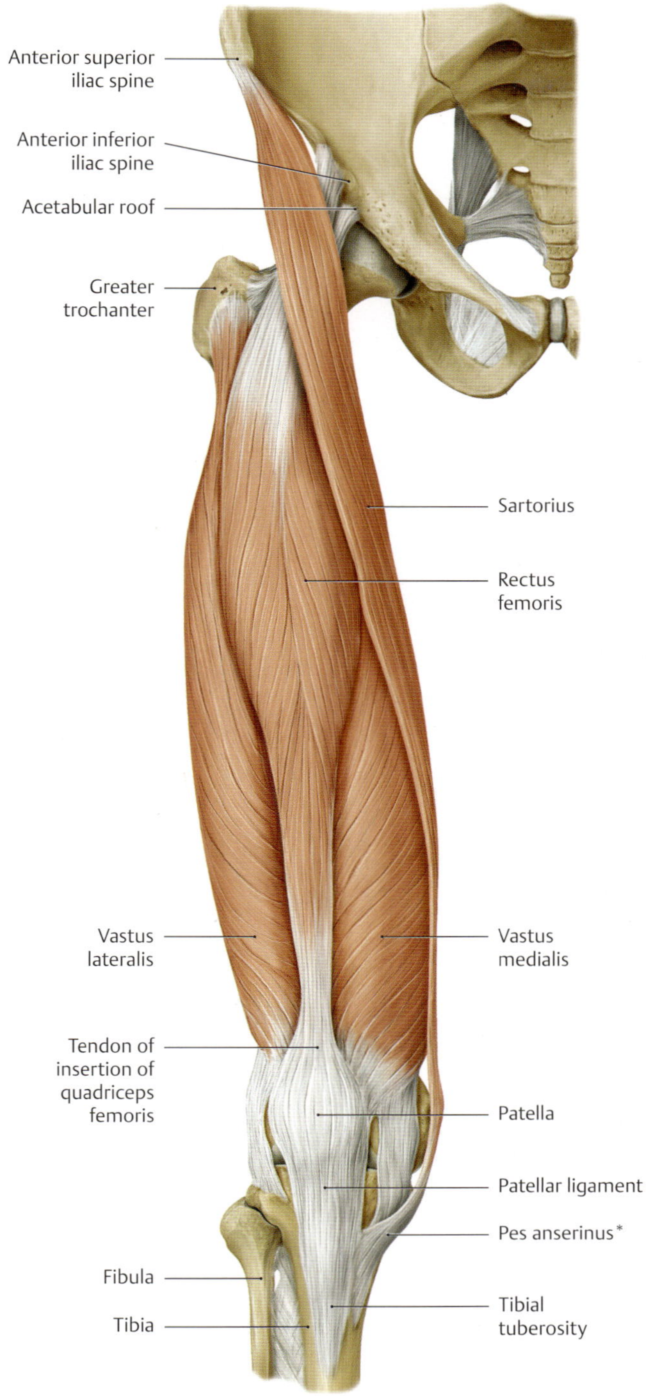

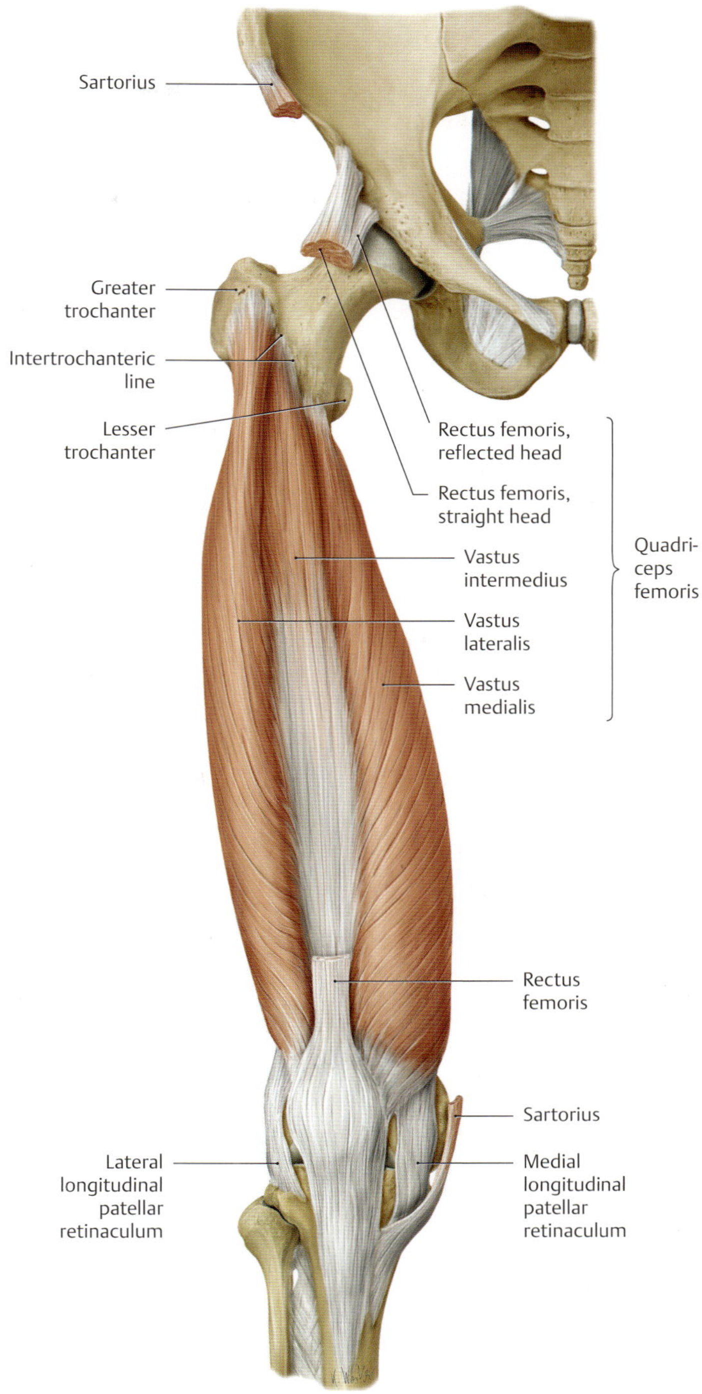

C The extensors (quadriceps femoris and sartorius)
Right side, anterior view. As its name implies, the quadriceps femoris is basically a four-headed muscle consisting of the rectus femoris and the vastus medialis, lateralis, and intermedius (the vastus intermedius, covered here by the rectus femoris, is visible in **D**). It may also be considered as having a fifth head, the articularis genus. The latter is composed of distal fibers of the vastus intermedius and so does not constitute a separate muscle. But because its fibers insert in the suprapatellar recess (not shown), unlike the other four parts, which all attach to the patellar ligament, the articularis genus is often regarded as the fifth head of the quadriceps femoris.
Note: The only biarticular part of the quadriceps femoris is the rectus femoris, which acts on both the hip and knee joints.

* The pes anserinus is the common tendinous expansion for the gracilis, sartorius, and semitendinosus.

D The extensors (deep portion of the quadriceps femoris)
Right side, anterior view. The sartorius and rectus femoris have been removed to their origins and insertions.
The area of origin of the rectus femoris is intimately related to the anterior aspect of the capsule of the hip joint, a relation with functional and clinical consequences. Pathological swelling of the joint capsule can cause pain that induces reflex reactions when the rectus is stretched during knee flexion; such reflex reactions are the basis of a useful test. With the patient lying prone, the examiner flexes the patient's knee. This causes passive stretching of the rectus femoris and adds significant pressure to a hip joint capsule already distended by effusion. The patient reflexively "escapes" the painful stimulus by raising the buttock, a positive "rectus sign."

2.6 The Posterior Thigh Muscles: The Flexor Group

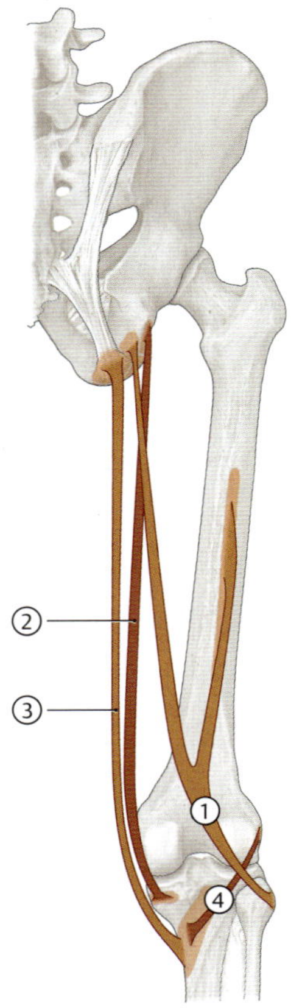

A Schematic of the flexors

① **Biceps femoris**

Origin:
- Long head: ischial tuberosity, sacrotuberous ligament (common head with semitendinosus)
- Short head: lateral lip of the linea aspera in the middle third of the femur

Insertion: Head of fibula

Action:
- Hip joint (long head): adduction, extends the hip, stabilizes the pelvis in the sagittal plane
- Knee joint (entire muscle): flexion and external rotation

Innervation:
- Tibial nerve, L5–S2 (long head)
- Common fibular nerve, L5–S2 (short head)

② **Semimembranosus**

Origin: Ischial tuberosity

Insertion: Medial tibial condyle, oblique popliteal ligament, popliteus fascia

Action:
- Hip joint: adduction, extends the hip, stabilizes the pelvis in the sagittal plane
- Knee joint: flexion and internal rotation

Innervation: Tibial nerve (L5–S2)

③ **Semitendinosus**

Origin: Ischial tuberosity and sacrotuberous ligament (common head with long head of biceps femoris)

Insertion: Medial to the tibial tuberosity in the pes anserinus (along with the tendons of gracilis and sartorius)

Action:
- Hip joint: adduction, extends the hip, stabilizes the pelvis in the sagittal plane
- Knee joint: flexion and internal rotation

Innervation: Tibial nerve (L5–S2)

④ **Popliteus**

Origin: Lateral femoral condyle, posterior horn of the lateral meniscus

Insertion: Posterior tibial surface (above the origin of soleus)

Action: Flexes and unlocks the knee by internally rotating the femur on the fixed head of the tibia 5°

Innervation: Tibial nerve (L4–S1)

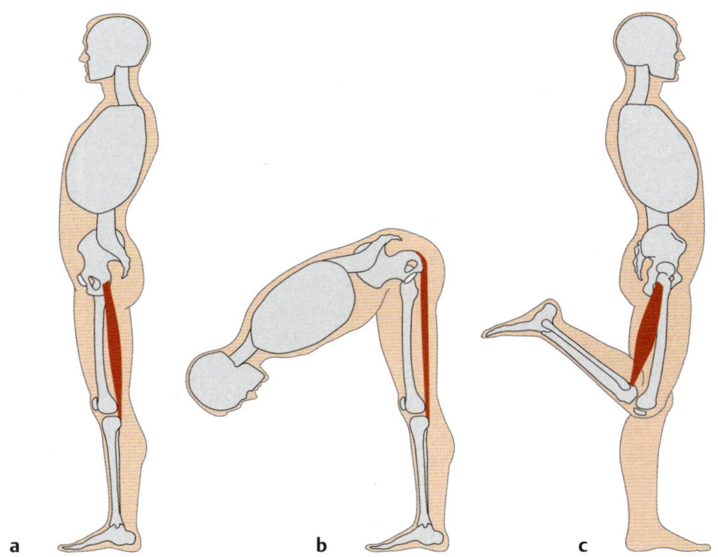

B Passive and active muscle insufficiency illustrated for ischiocrural muscles

a The ischiocrural hamstring muscles extend from the ischium over the hip joint and knee joint to the lower leg.

b Passive muscle insufficiency (lack of muscle length): With the knee outstretched, the ischiocrural muscles do not stretch enough to allow for maximal bending of the hip joint.

c Active muscle insufficiency (lack of muscle strength): With hip extended, the ischiocrural muscles cannot shorten enough to allow for maximal bending of the knee joint (see also muscle and soft tissue suspension, p. 52).

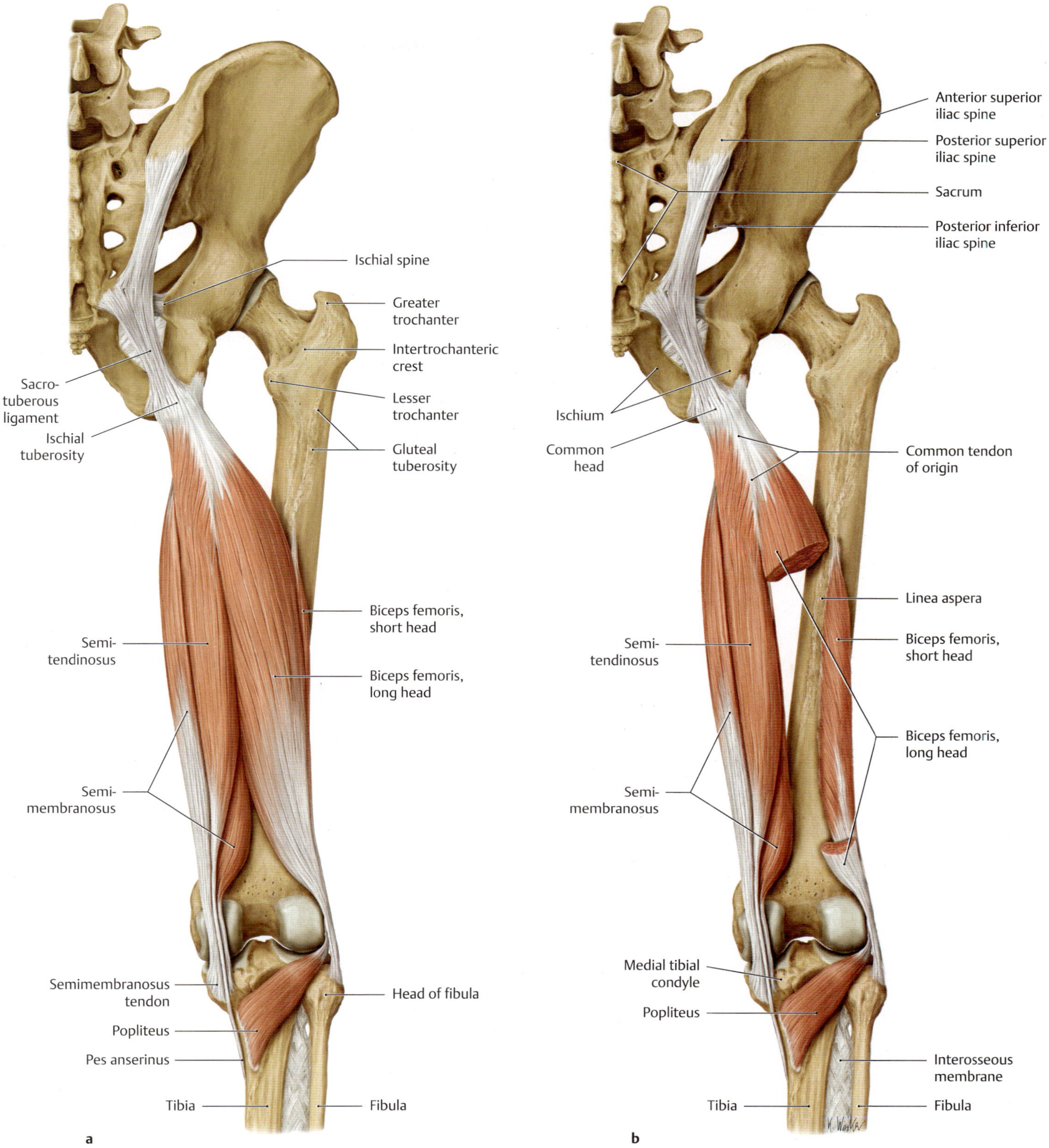

C The flexors (the hamstrings and popliteus)
Right side, posterior view.

a The *hamstrings* are the posterior thigh muscles that arise from the ischium and attach to the leg: biceps femoris, semimembranosus, and semitendinosus. They all cross two joints, except the short head of the biceps femoris.

Note: According to Battermann et al (2011) the greater portion of the semitendinosus arises from a common tendon of origin with the long head of the biceps femoris.

b A portion of the long head of the biceps femoris has been removed to display the short head and its origin from the lateral lip of the linea aspera.

2.7 The Leg Muscles: The Anterior and Lateral Compartments (Extensor and Fibularis Group)*

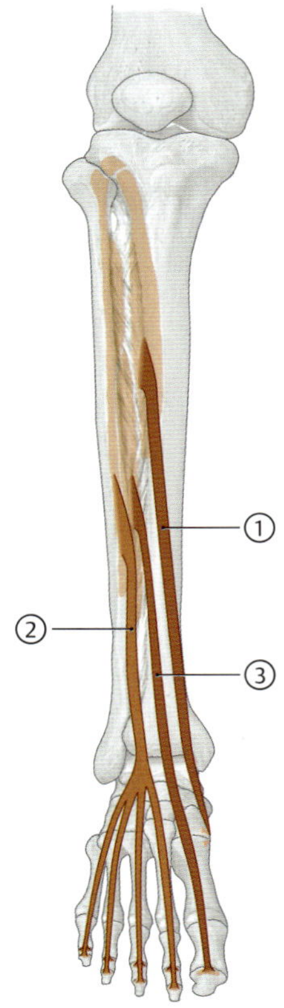

A Schematic of the anterior compartment

① **Tibialis anterior**

Origin: Upper two thirds of the lateral surface of the tibia, the crural interosseous membrane, and the highest part of the superficial crural fascia
Insertion: Medial and plantar surface of the medial cuneiform, the medial base of the first metatarsal
Action:
- Talocrural joint: dorsiflexion
- Subtalar joint: inversion (supination)

Innervation: Deep fibular nerve (L4, L5)

② **Extensor digitorum longus**

Origin: Lateral tibial condyle, head of the fibula, medial surface of the fibula, and the crural interosseous membrane
Insertion: By four slips to the dorsal aponeuroses of the second through fifth toes and the bases of the distal phalanges of the second through fifth toes
Action:
- Talocrural joint: dorsiflexion
- Subtalar joint: eversion (pronation)
- Extends the metatarsophalangeal and interphalangeal joints of the second through fifth toes

Innervation: Deep fibular nerve (L4, L5)

③ **Extensor hallucis longus**

Origin: Middle third of the medial surface of the fibula, the crural interosseous membrane
Insertion: Dorsal aponeurosis of the great toe and the base of its distal phalanx
Action:
- Talocrural joint: dorsiflexion
- Subtalar joint: active in both eversion and inversion (pronation/supination), depending on the initial position of the foot
- Extends the metatarsophalangeal and interphalangeal joints of the great toe

Innervation: Deep fibular nerve (L4, L5)

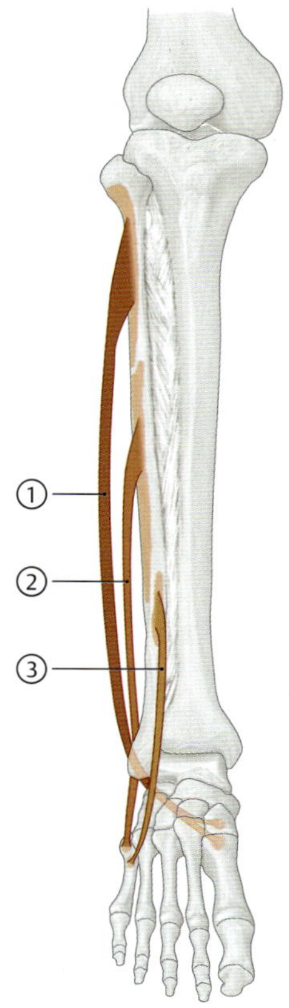

B Schematic of the lateral compartment

① **Fibularis longus**

Origin: Head of the fibula, proximal two thirds of the lateral surface of the fibula (arising partly from the intermuscular septa)
Insertion: Plantar side of the medial cuneiform, base of the first metatarsal
Action:
- Talocrural joint: plantar flexion
- Subtalar joint: eversion (pronation)
- Supports the transverse arch of the foot

Innervation: Superficial fibular nerve (L5, S1)

② **Fibularis brevis**

Origin: Distal half of the lateral surface of the fibula, intermuscular septa
Insertion: Tuberosity at the base of the fifth metatarsal (with an occasional division to the dorsal aponeurosis of the fifth toe)
Action:
- Talocrural joint: plantar flexion
- Subtalar joint: eversion (pronation)

Innervation: Superficial fibular nerve (L5, S1)

③ **Fibularis tertius** ** **(part of extensor digitorum longus)** (see also pp. 528 and 532)

Origin: Anterior border of the distal fibula
Insertion: Base of the fifth metatarsal
Action:
- Talocrural joint: dorsiflexion
- Subtalar joint: eversion (pronation)

Innervation: Deep fibular nerve (L4–S1)

** The fibularis tertius can also be classified as an extensor based on its function and innervation. Here it has been assigned to the fibularis group due to its name and its action on the subtalar joint (eversion).

* The term “peroneal” may also be used in place of the term “fibular.”

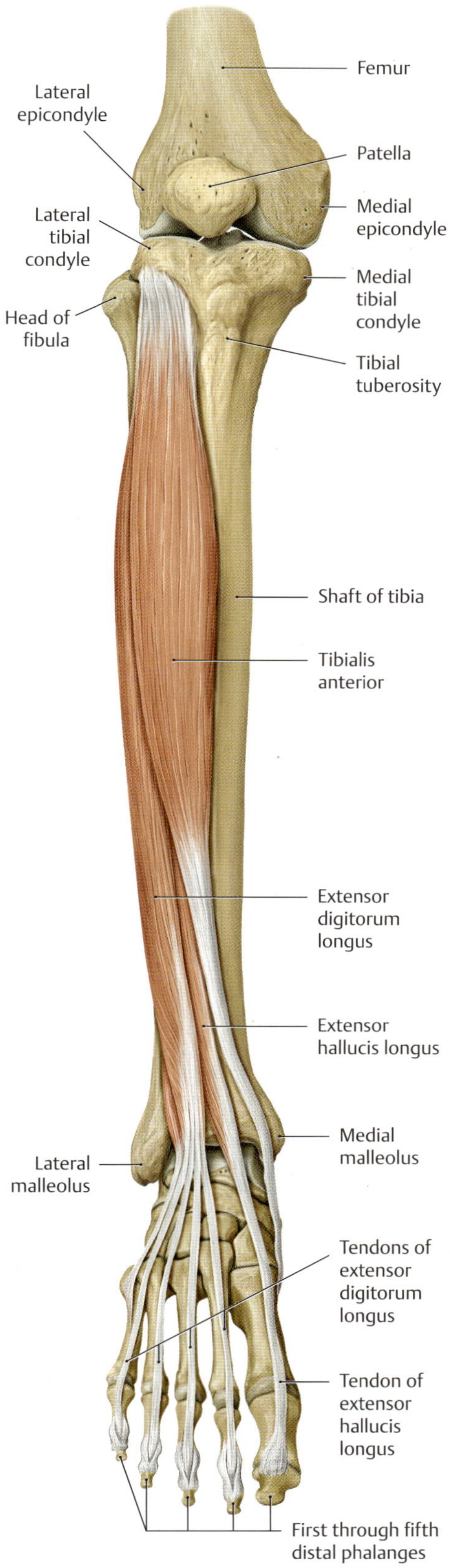

C The anterior compartment (tibialis anterior, extensor digitorum longus, and extensor hallucis longus)
Right leg, anterior view.

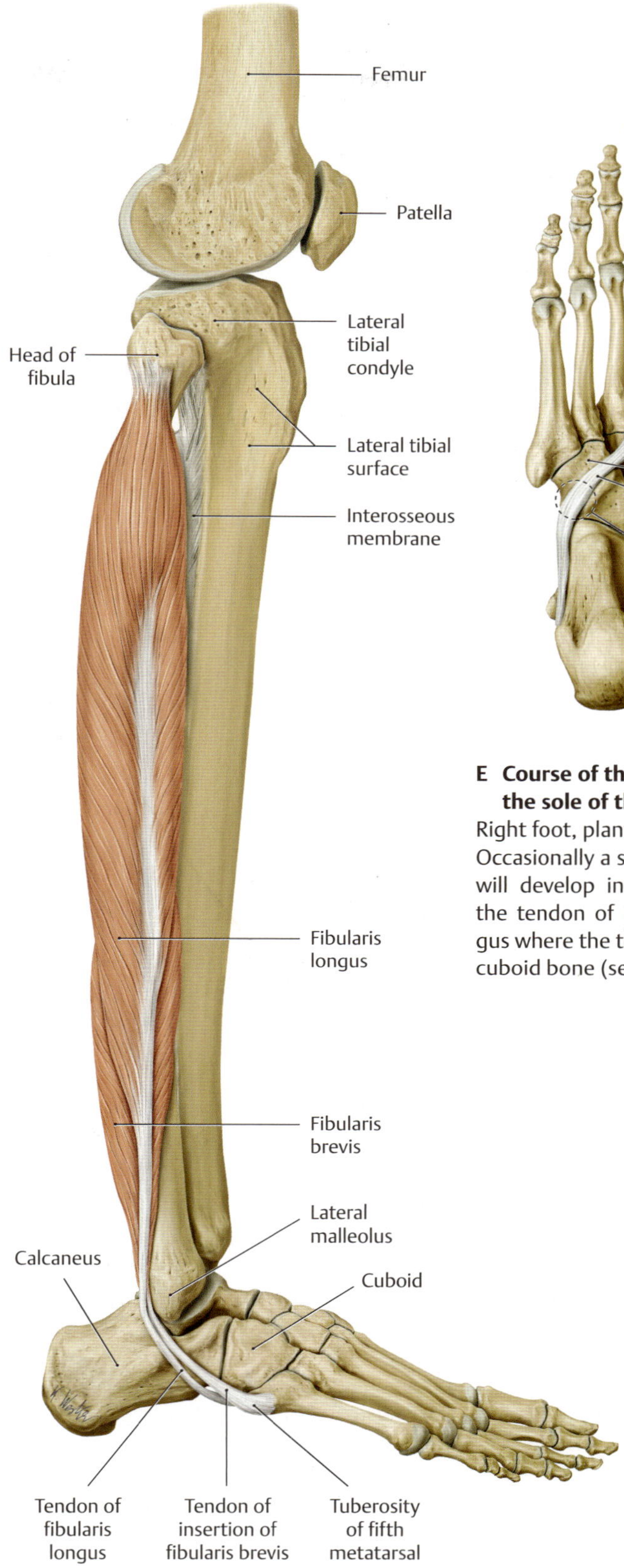

D The lateral compartment (fibularis longus and brevis)
Right leg, lateral view.

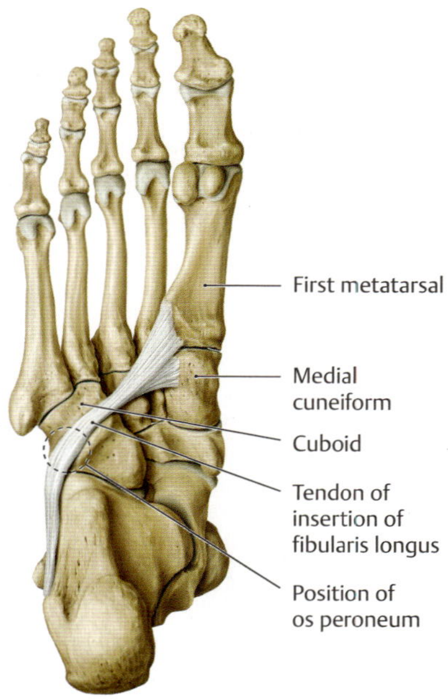

E Course of the fibularis longus tendon on the sole of the foot
Right foot, plantar view.
Occasionally a sesamoid bone (os peroneum) will develop in the compressed portion of the tendon of insertion of the fibularis longus where the tendon is deflected around the cuboid bone (see also p. 437).

2.8 The Leg Muscles: The Posterior Compartment (Superficial Flexor Group)

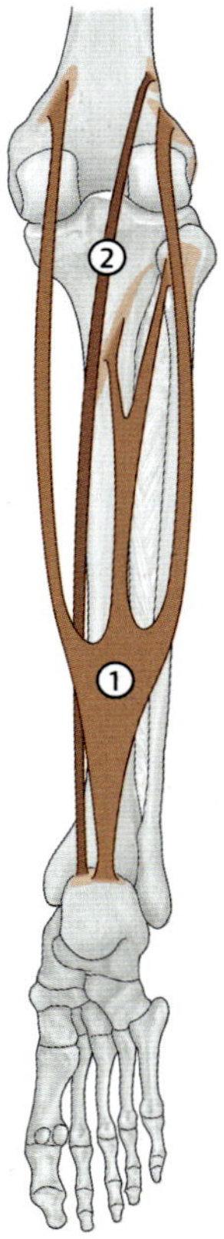

A Schematic of the superficial flexors

① Triceps surae

Origin:
- Soleus: posterior surface of the head and neck of the fibula; attached to the soleal line of the tibia via a tendinous arch
- Gastrocnemius, medial head: medial epicondyle of the femur
- Gastrocnemius, lateral head: lateral epicondyle of the femur

Insertion: The calcaneal tuberosity via the calcaneal (Achilles') tendon

Action:
- Talocrural joint: plantar flexion
- Knee joint: flexion (gastrocnemius)

Innervation: Tibial nerve (S1, S2)

② Plantaris

Origin: Lateral epicondyle proximal to the lateral head of gastrocnemius

Insertion: The calcaneal tuberosity via the calcaneal (Achilles') tendon

Action: Negligible due to its small cross section; may act to prevent compression of the posterior leg musculature during knee flexion

Innervation: Tibial nerve (S1, S2)

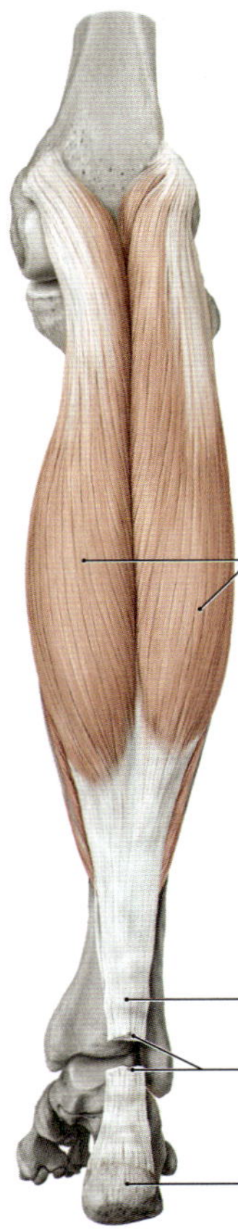

B Rupture of the calcaneal (Achilles') tendon

Right leg, posterior view. The calcaneal (Achilles') tendon is the common tendon of insertion of the muscles that comprise the triceps surae (the soleus and both heads of the gastrocnemius). The tendon has an average length of 20 to 25 cm, a mean cross-sectional area of approximately 70 to 80 mm^2, and a breaking strength of 60 to 100 N/mm^2. A healthy tendon can thus bear a load of nearly 1 ton. It is very unlikely, then, that the calcaneal tendon will rupture unless it has been subjected to chronic excessive loads (in high jumpers, for example). Repetitive microtrauma can compromise the blood supply to the tendon, causing it to degenerate and gradually lose its strength. This is particularly damaging in the area where the tendon already has the least blood flow: approximately 2 to 6 cm proximal to its insertion on the calcaneal tuberosity. This is the most common site of a degenerative calcaneal tendon rupture, which is eventually precipitated by a trivial injury. The rupture is accompanied by a whiplike snapping sound. Afterward the patient loses active plantar flexion and has only residual flexion from the deep flexor muscles.

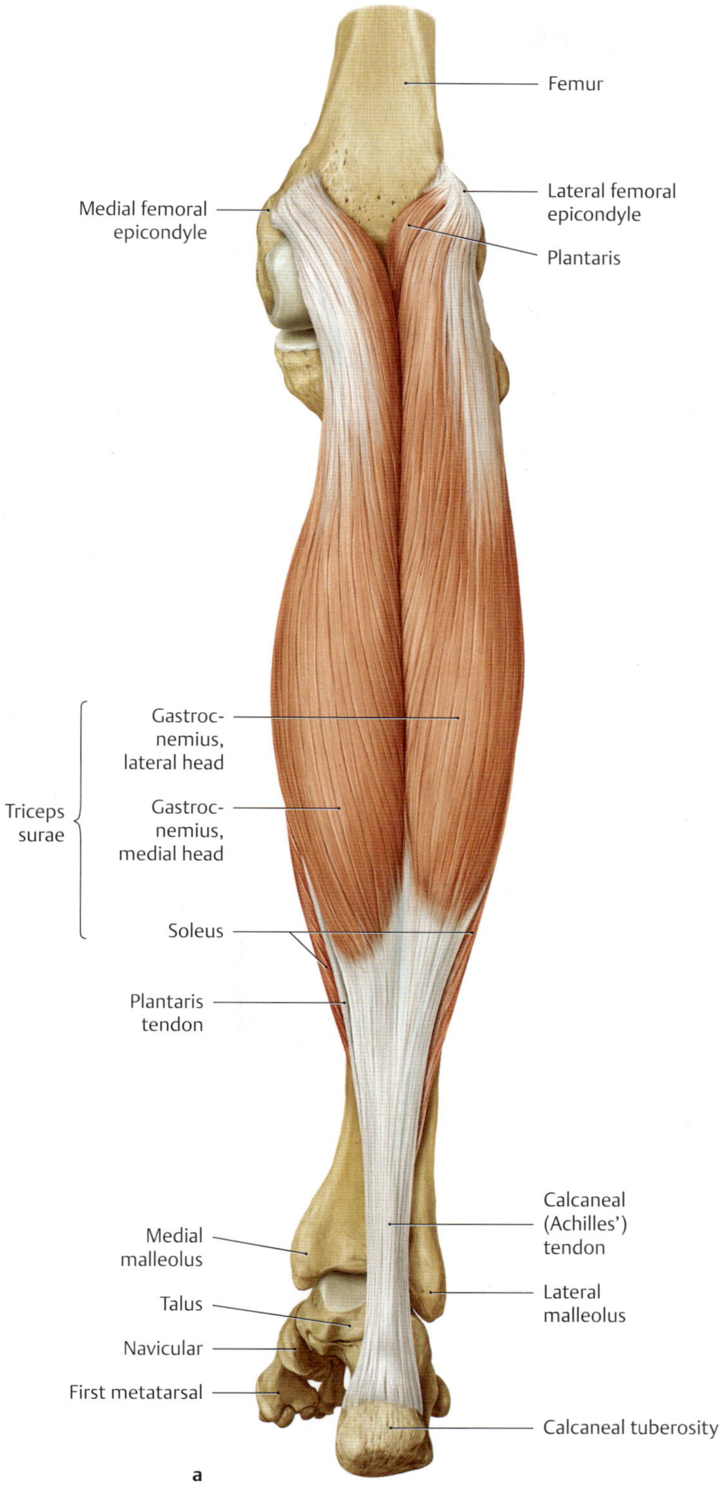

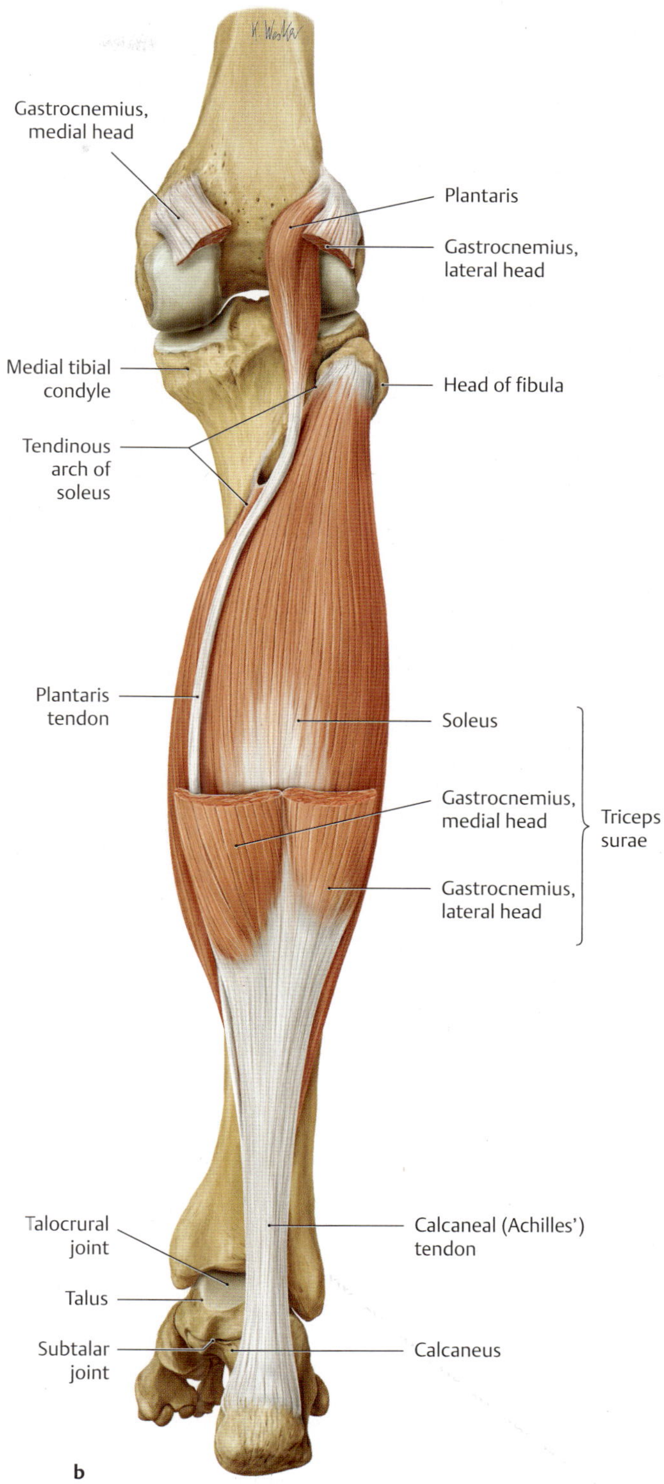

C The superficial flexors (triceps surae and plantaris)
Right leg, posterior view.

a The three heads of the triceps surae are clearly distinguishable: the medial and lateral heads of the gastrocnemius and the soleus. The plantaris, which arises proximal to the lateral head of the gastrocnemius, is often viewed as the fourth head of the triceps surae.

b Portions of the lateral and medial heads of the gastrocnemius have been removed to expose the soleus and the plantaris with its long, narrow tendon of insertion.

2.9 The Leg Muscles: The Posterior Compartment (Deep Flexor Group)

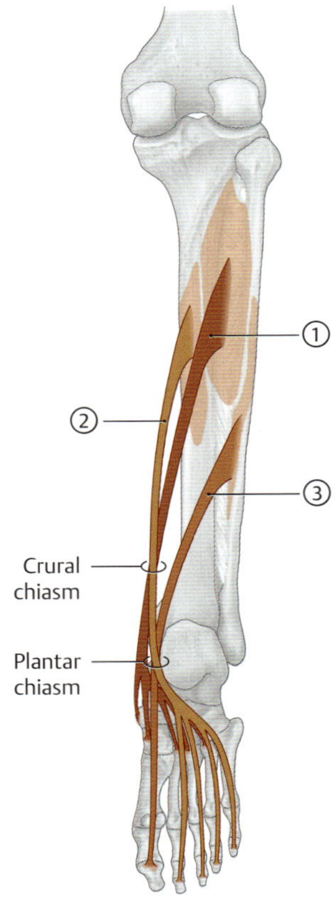

A Schematic of the deep flexors

① **Tibialis posterior**

Origin: Crural interosseous membrane and the adjacent borders of the tibia and fibula
Insertion: Tuberosity of the navicular; medial, intermediate, and lateral cuneiforms; bases of the second through fourth metatarsals
Action:
- Talocrural joint: plantar flexion
- Subtalar joint: inversion (supination)
- Supports the longitudinal and transverse arches of the foot

Innervation: Tibial nerve (L4, L5)

② **Flexor digitorum longus**

Origin: Middle third of the posterior surface of the tibia
Insertion: Bases of the second through fifth distal phalanges
Action:
- Talocrural joint: plantar flexion
- Subtalar joint: inversion (supination)
- Metatarsophalangeal and interphalangeal joints of the second through fifth toes: plantar flexion

Innervation: Tibial nerve (L5–S2)

③ **Flexor hallucis longus**

Origin: Distal two thirds of the posterior surface of the fibula, adjacent crural interosseous membrane
Insertion: Base of the distal phalanx of the great toe
Action:
- Talocrural joint: plantar flexion
- Subtalar joint: inversion (supination)
- Metatarsophalangeal and interphalangeal joints of the great toe: plantar flexion
- Supports the medial longitudinal arch of the foot

Innervation: Tibial nerve (L5–S2)

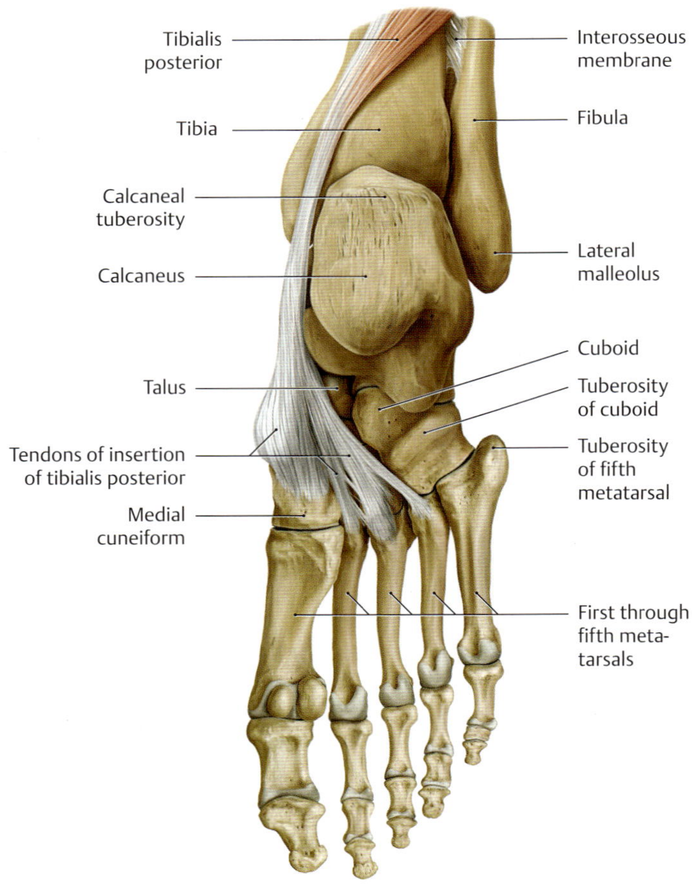

B Insertion of the tibialis posterior
Right foot in plantar flexion, plantar view. With its fan-shaped insertion, the tibialis posterior assists in stabilizing both the longitudinal and transverse arches of the foot.

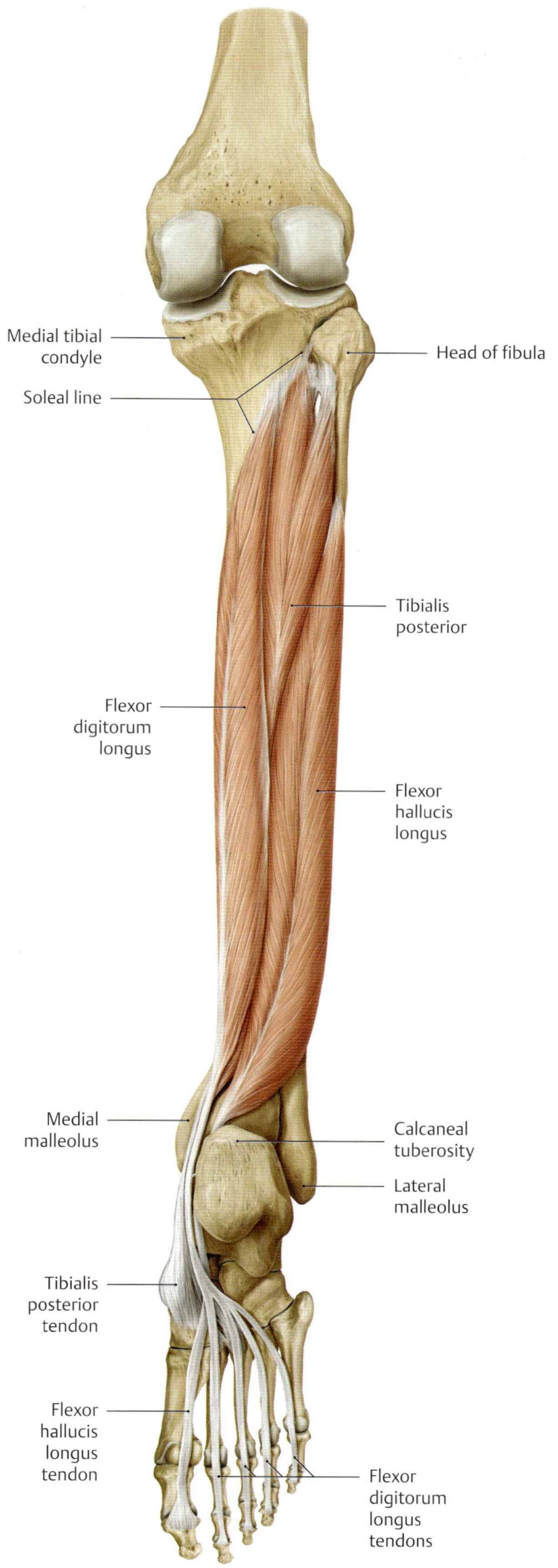

C The deep flexors (tibialis posterior, flexor digitorum longus, and flexor hallucis longus)
Right leg with the foot in plantar flexion, posterior view.

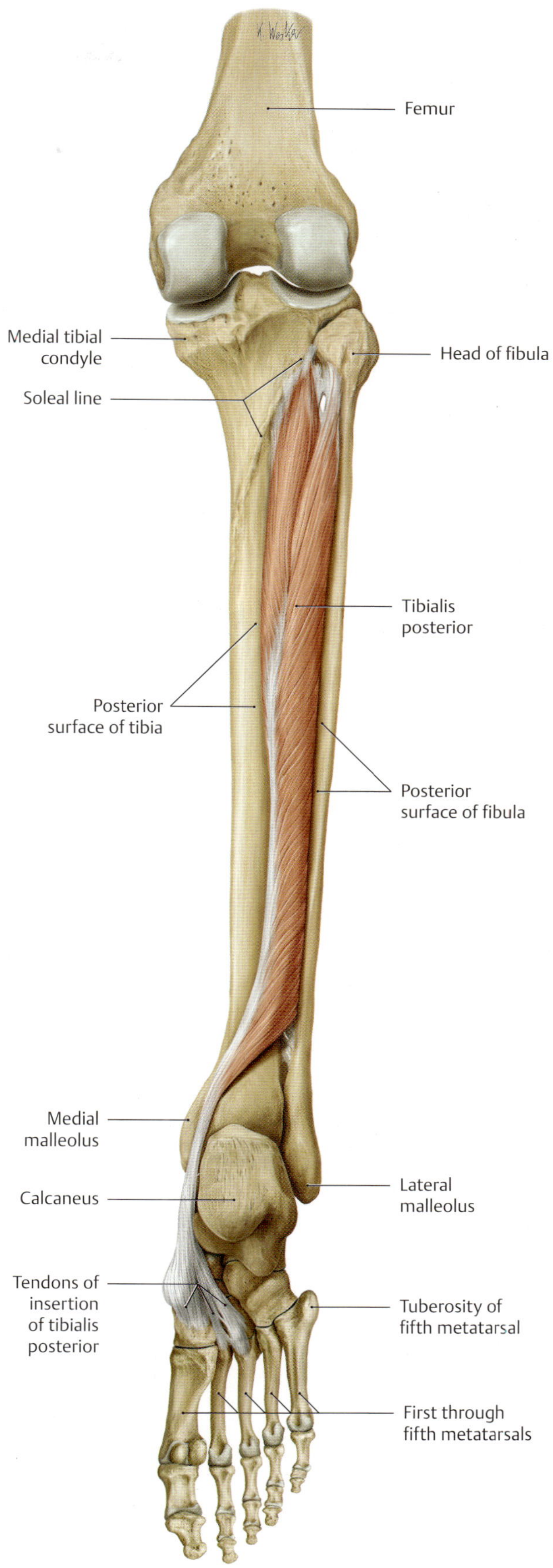

D The tibialis posterior
Right leg with the flexor digitorum longus and flexor hallucis longus removed. Foot in plantar flexion, posterior view.

2.10 The Short Muscles of the Foot: Dorsum and Medial and Lateral Compartments of the Plantar Surface

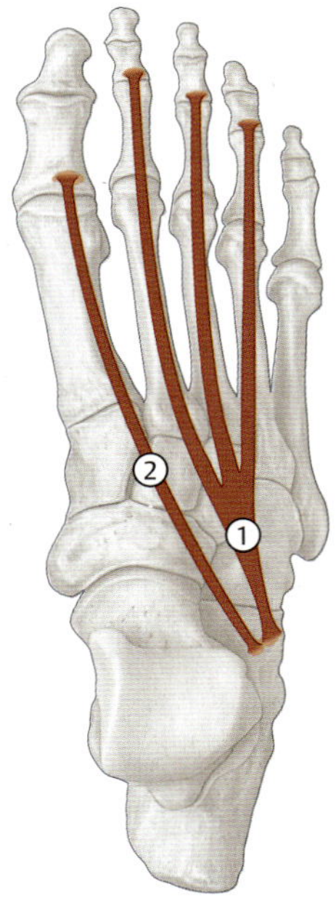

A Dorsal view

① **Extensor digitorum brevis**

Origin: Dorsal surface of the calcaneus
Insertion: Dorsal aponeurosis of the second through fourth toes, bases of the middle phalanges of these toes
Action: Extension of the metatarsophalangeal and proximal interphalangeal joints of the second through fourth toes
Innervation: Deep fibular* nerve (L5, S1)

② **Extensor hallucis brevis**

Origin: Dorsal surface of the calcaneus
Insertion: Dorsal aponeurosis of the great toe, base of the proximal phalanx of the great toe
Action: Extension of the metatarsophalangeal joint of the great toe
Innervation: Deep fibular nerve (L5, S1)

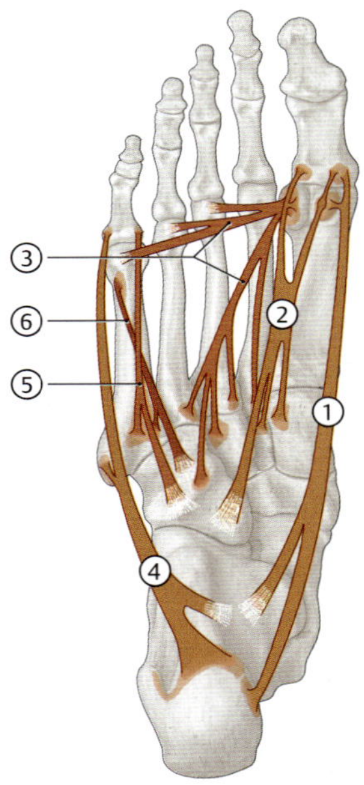

B Plantar view of the medial and lateral compartments

① **Abductor hallucis**

Origin: Medial process of the calcaneal tuberosity, plantar aponeurosis, flexor retinaculum
Insertion: Base of the proximal phalanx of the great toe via the medial sesamoid
Action: First metatarsophalangeal joint: flexion and medial abduction of the great toe; supports the longitudinal arch
Innervation: Medial plantar nerve (S1, S2)

② **Flexor hallucis brevis**

Origin: Cuboid, lateral cuneiform
Insertion:
- Medial head: base of the proximal phalanx of the great toe via the medial sesamoid
- Lateral head: base of the proximal phalanx of the great toe via the lateral sesamoid

Action: Flexes the first metatarsophalangeal joint, supports the longitudinal arch
Innervation: Medial plantar nerve (S1, S2)

③ **Adductor hallucis** (for clarity, the adductor hallucis is depicted here, although it is located in the central compartment)

Origin:
- Oblique head: bases of the second through fourth metatarsals, cuboid, lateral cuneiform
- Transverse head: metatarsophalangeal joints of the third through fifth toes, deep transverse metatarsal ligament

Insertion: Base of the first proximal phalanx by a common tendon via the lateral sesamoid
Action: Flexes the first metatarsophalangeal joint, adducts the great toe; transverse head supports the transverse arch, oblique head supports the longitudinal arch
Innervation: Lateral plantar nerve (S2, S3)

④ **Abductor digiti minimi**

Origin: Lateral process and inferior surface of the calcaneal tuberosity, plantar aponeurosis
Insertion: Base of the proximal phalanx of the little toe, tuberosity of the fifth metatarsal
Action: Flexes the metatarsophalangeal joint of the little toe, abducts the little toe, supports the longitudinal arch
Innervation: Lateral plantar nerve (S1–S3)

⑤ **Flexor digiti minimi brevis**

Origin: Base of the fifth metatarsal, long plantar ligament
Insertion: Base of the proximal phalanx of the little toe
Action: Flexes the metatarsophalangeal joint of the little toe
Innervation: Lateral plantar nerve (S2, S3)

⑥ **Opponens digiti minimi (often included with flexor digiti minimi brevis)**

Origin: Long plantar ligament, plantar tendon sheath of the fibularis longus
Insertion: Fifth metatarsal
Action: Pulls the fifth metatarsal slightly in the plantar and medial direction
Innervation: Lateral plantar nerve (S2, S3)

* The term "peroneal" may also be used in place of the term "fibular."

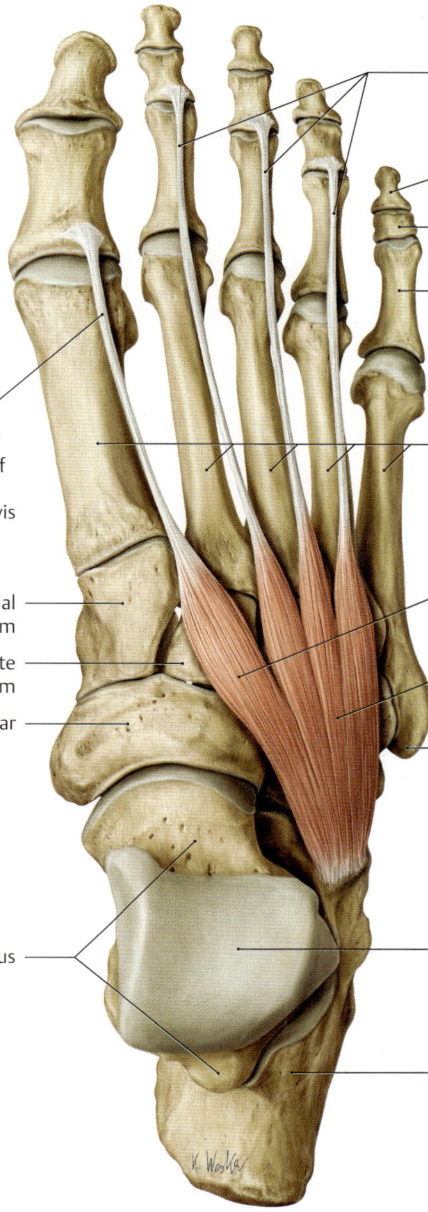

C The dorsal muscles of the foot (extensor digitorum brevis and extensor hallucis brevis)
Right foot, dorsal view.

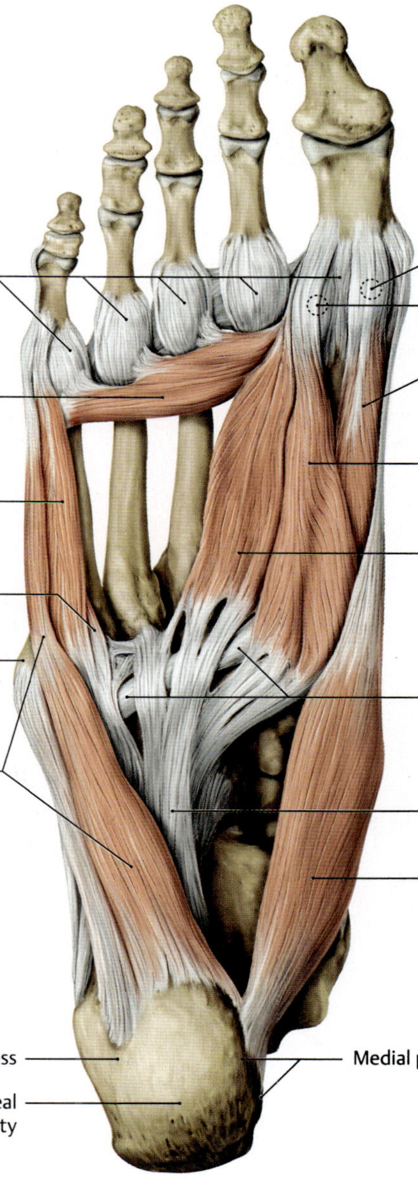

D The plantar muscles of the medial and lateral compartments (abductor hallucis, adductor hallucis,* flexor hallucis brevis, abductor digiti minimi, flexor digiti minimi brevis, and opponens digiti minimi)
Right foot, plantar view.

* The adductor hallucis is considered part of the central compartment (see p. 512).

2.11 The Short Muscles of the Foot: Central Compartment of the Plantar Surface*

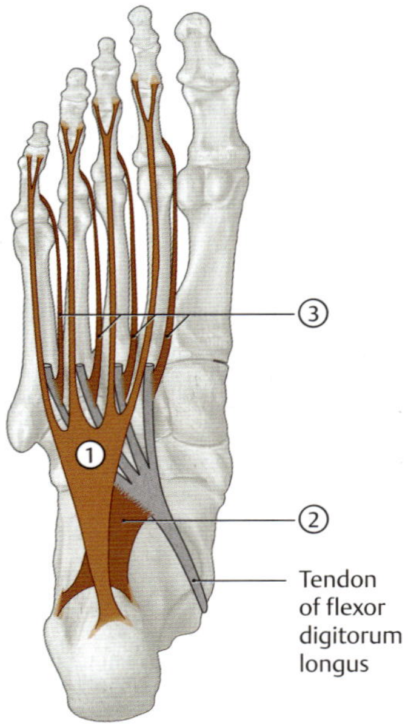

A Plantar view of flexor digitorum brevis, quadratus plantae, and lumbricals

① **Flexor digitorum brevis**

Origin: Medial tubercle of the calcaneal tuberosity, plantar aponeurosis
Insertion: The sides of the middle phalanges of the second through fifth toes
Action:
- Flexes the metatarsophalangeal and proximal interphalangeal joints of the second through fifth toes
- Supports the longitudinal arch of the foot

Innervation: Medial plantar nerve (S1, S2)

② **Quadratus plantae**

Origin: Medial and plantar borders on the plantar side of the calcaneal tuberosity
Insertion: Lateral border of the flexor digitorum longus tendon
Action: Redirects and augments the pull of the flexor digitorum longus
Innervation: Lateral plantar nerve (S1–S3)

③ **First through fourth lumbricals**

Origin: Medial borders of the flexor digitorum longus tendons
Insertion: Dorsal aponeuroses of the second through fifth toes
Action:
- Flexes the metatarsophalangeal joints of the second through fifth toes
- Extension of the interphalangeal joints of the second through fifth toes
- Moves the toes closer together (adducts the second through fifth toes toward the great toe)

Innervation:
- First lumbrical: medial plantar nerve (S2, S3)
- Second through fourth lumbricals: lateral plantar nerve (S2, S3)

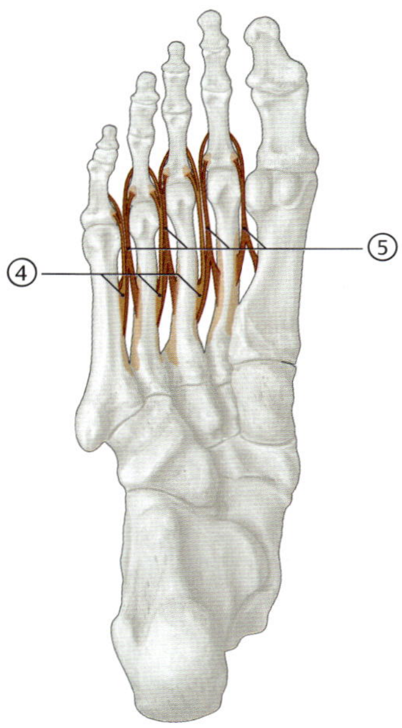

B Plantar view of plantar and dorsal interossei

④ **First through third plantar interossei**

Origin: Medial border of the third through fifth metatarsals
Insertion: Medial base of the proximal phalanx of the third through fifth toes
Action:
- Flexes the metatarsophalangeal joints of the third through fifth toes
- Extension of the interphalangeal joints of the third through fifth toes
- Moves the toes closer together (adducts the third through fifth toes toward the second toe)

Innervation: Lateral plantar nerve (S1, S2)

⑤ **First through fourth dorsal interossei**

Origin: By two heads from opposing sides of the first through fifth metatarsals
Insertion:
- First interosseus: medial base of the second proximal phalanx, dorsal aponeurosis of the second toe
- Second through fourth interossei: lateral base of the second through fourth proximal phalanges, dorsal aponeurosis of the second through fourth toes

Action:
- Flexes the metatarsophalangeal joints of the second through fourth toes
- Extension of the interphalangeal joints of the second through fourth toes
- Spreads the toes apart (abducts the third and fourth toes from the second toe)

Innervation: Lateral plantar nerve (S1, S2)

* The adductor hallucis, though part of the central compartment, is not pictured here (see p. 511).

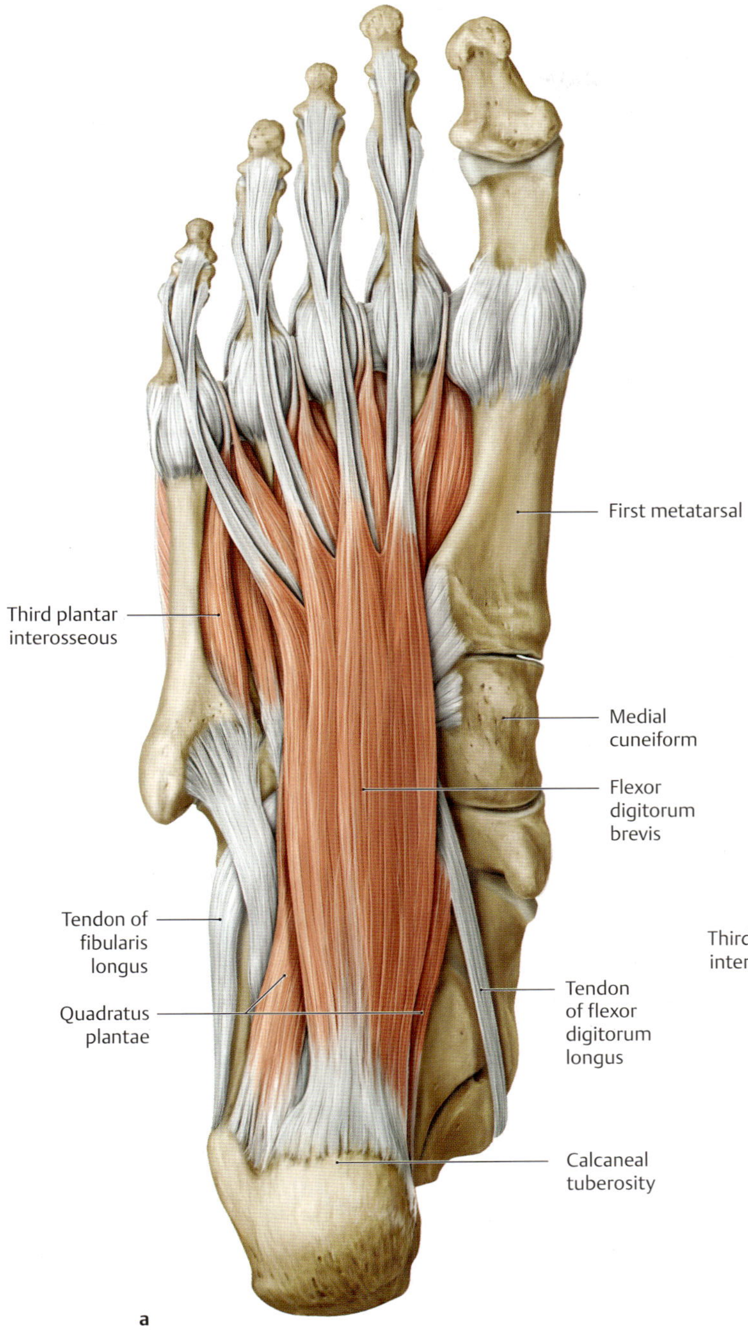

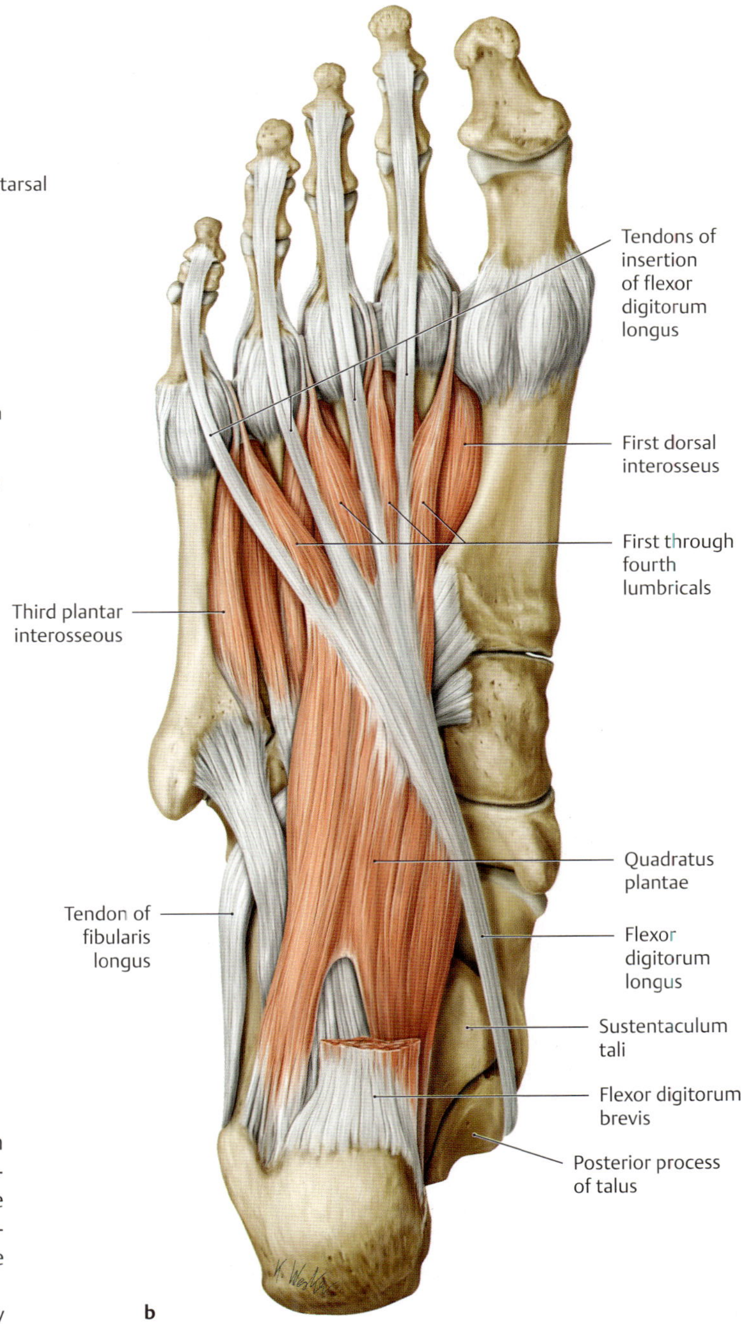

C The plantar muscles of the central compartment
Right foot, plantar view.

- **a** Flexor digitorum brevis, quadratus plantae, the first through fourth lumbricals (always numbered in the medial-to-lateral or tibial-to-fibular direction), the first through third plantar interossei, and the first through fourth dorsal interossei (for clarity, the adductor hallucis, though part of the central compartment, is not shown; see p. 511).
- **b** Flexor digitorum brevis has been removed to its origin to display more clearly the insertion of quadratus plantae on the lateral margin of the flexor digitorum longus tendon.

Note: The “movable origins” of the first through fourth lumbricals form the medial borders of the flexor digitorum longus tendons. When the flexor digitorum longus contracts and therefore shortens, the origins of the lumbricals move proximally. This “prestretching” of the lumbricals improves their ability to contract, enabling them to develop greater force.

2.12 Overview of Muscle Functions: Hip Joint

A Movements of the hip joint

Type of movement	Range of motion	Muscle	Innervation	Nerve segment
Flexion	120–140°	Iliopsoas (psoas major and iliacus)	Direct branches from lumbar plexus and femoral nerve	L1–L3
		Rectus femoris	Femoral nerve	L2–L4
		Tensor fasciae latae	Superior gluteal nerve	L4–S1
		Sartorius	Femoral nerve	L2–L4
		Pectineus	Femoral nerve, obturator nerve	L2–L4
		Adductor longus	Obturator nerve	L2–L4
		Adductor brevis	Obturator nerve	L2–L4
		Gracilis	Obturator nerve	L2–L4
		Gluteus medius and minimus, anterior parts	Superior gluteal nerve	L4–S1
Extension	20°	Gluteus maximus	Inferior gluteal nerve	L5–S2
		Semitendinosus	Tibial nerve	L5–S2
		Semimembranosus	Tibial nerve	L5–S2
		Biceps femoris, long head	Tibial nerve	L5–S2
		Gluteus medius and minimus (posterior parts)	Superior gluteal nerve	L4–S1
		Adductor magnus	Obturator nerve Tibial nerve	L2–L4
		Piriformis	Direct branches from sacral plexus	L4–L5
		Obturator internus	Direct branches from sacral plexus	
Abduction	50–80°	Gluteus medius	Superior gluteal nerve	L4–S1
		Tensor fasciae latae	Superior gluteal nerve	L4–L5
		Gluteus maximus (upper fibers)	Inferior gluteal nerve	L5–S2
		Gluteus minimus	Superior gluteal nerve	L4–L5
		Piriformis	Direct branches from sacral plexus	
		Sartorius	Femoral nerve	L1–L4
Adduction	20–30°	Adductor magnus	Obturator nerve Tibial nerve	L2–4 L4–5
		Adductor longus	Obturator nerve	L2–4
		Adductor brevis	Obturator nerve	L2–4
		Gluteus maximus (lower fibers)	Inferior gluteal nerve	L5–S2
		Pectineus	Femoral nerve, obturator nerve	L2–4
		Gracilis	Obturator nerve	L2–4
		Semitendinosus	Tibial nerve	L5–S2
		Semimembranosus	Tibial nerve	L5–S2
		Biceps femoris, long head	Tibial nerve	L5–S2
		Quadratus femoris	Direct branches from sacral plexus	L5–S2
		Obturator internus	Direct branches from sacral plexus	
		Obturator externus	Obturator nerve	L2–4
Internal rotation	40°	Gluteus medius and minimus (anterior parts)	Superior gluteal nerve	L4–S1
		Tensor fasciae latae	Superior gluteal nerve	L4–L5
		Adductor magnus (tendinoas insertion)	Tibial nerve	L4–L5
External rotation	30–50°	Gluteus maximus	Inferior gluteal nerve	L5–S1
		Obturator internus	Direct branches from the sacral plexus	
		Piriformis	Direct branches from the sacral plexus	
		Gemellus muscles	Direct branches from the sacral plexus	
		Quadratus femoris	Inferior gluteal nerve	L5–S2
		Obturator externus	Obturator nerve	L2–L4
		Gluteus medius and minimus (posterior parts)	Superior gluteal nerve	L4–S1
		Adductor magnus	Obturator nerve Tibial nerve	L2–L4 L4–L5
		Adductor longus	Obturator nerve	L2–L4
		Adductor brevis	Obturator nerve	L2–L4
		Pectineus	Obturator nerve	L2–L4
		Sartorius	Femoral nerve	L1–L4
		Iliopsoas (psoas major and iliacus)	Direct branches from lumbar plexus and femoral nerve	L1–L4

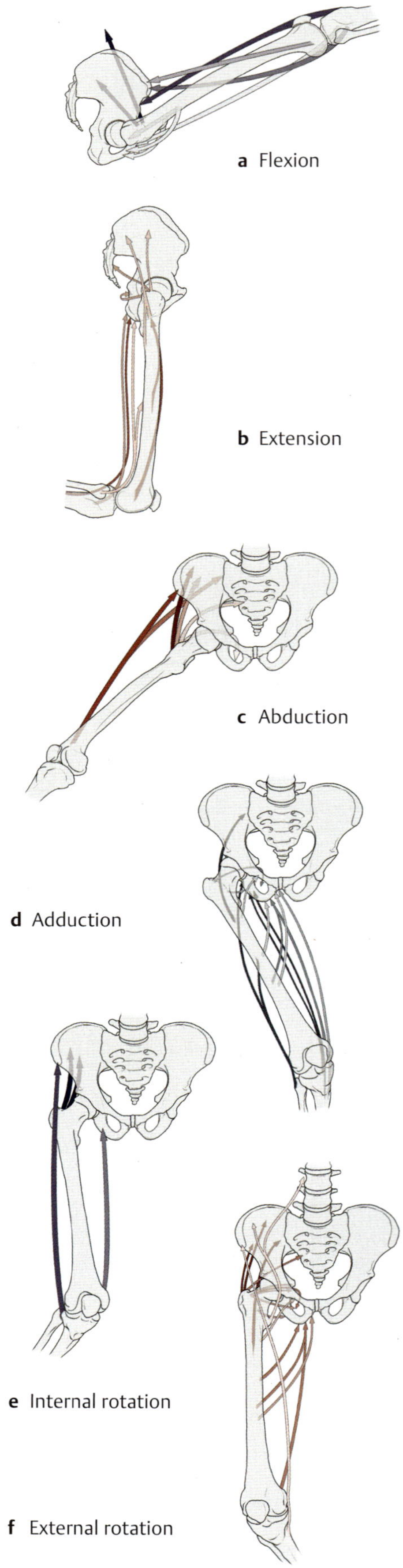

B Movements of the hip joint

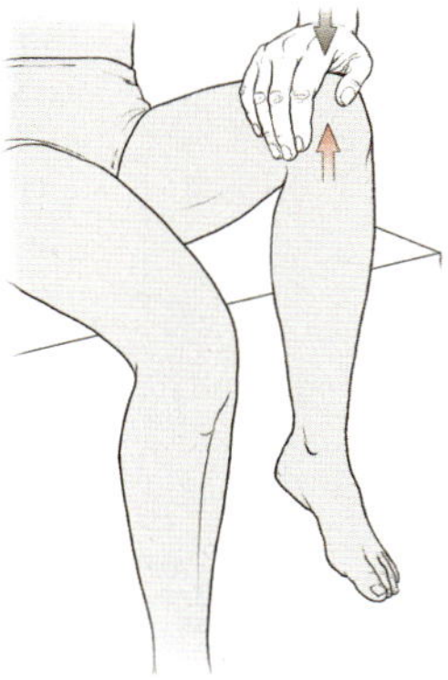

a Flexors

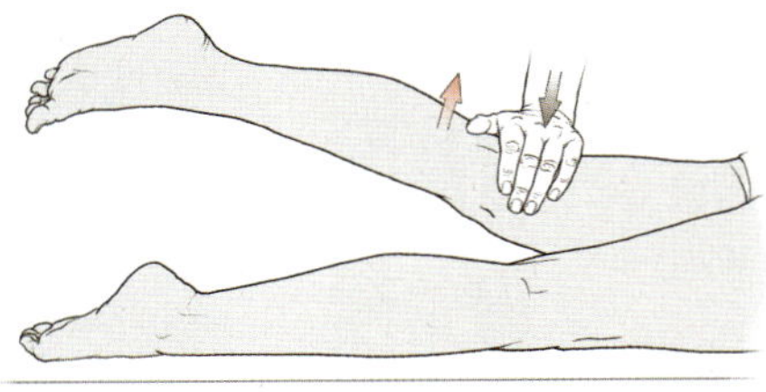

b Extensors

C Functional muscle testing of the hip joint
In order to test the strength of each muscle, the patient is asked to resist (push back against) the examiner. The force generated by the muscle action is measured using a scale from 0 to 5.

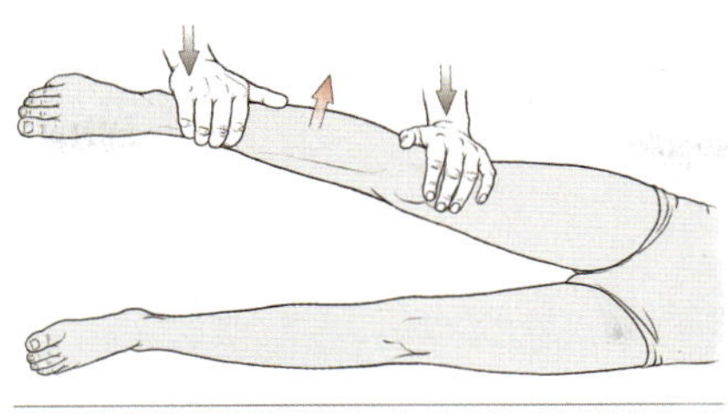

c Abductors

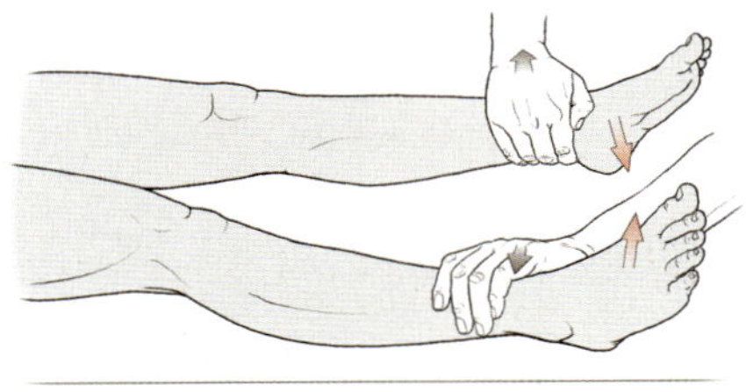

d Side-by-side comparison of adductors

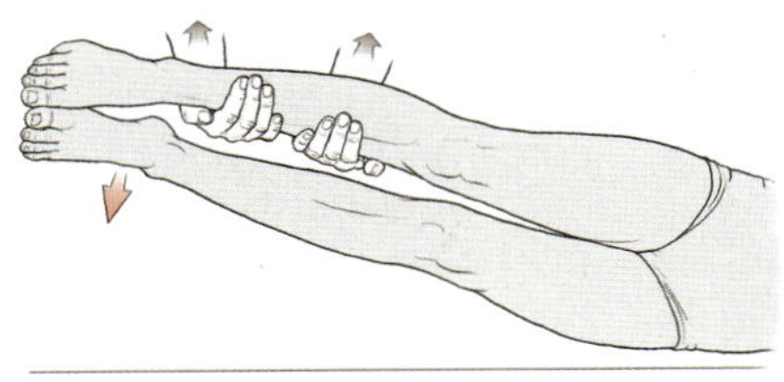

e Adductors in side-lying position with strength unimpaired

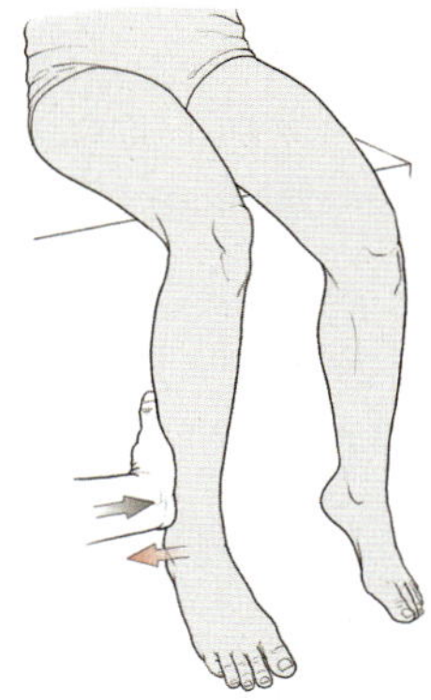

g Internal rotators

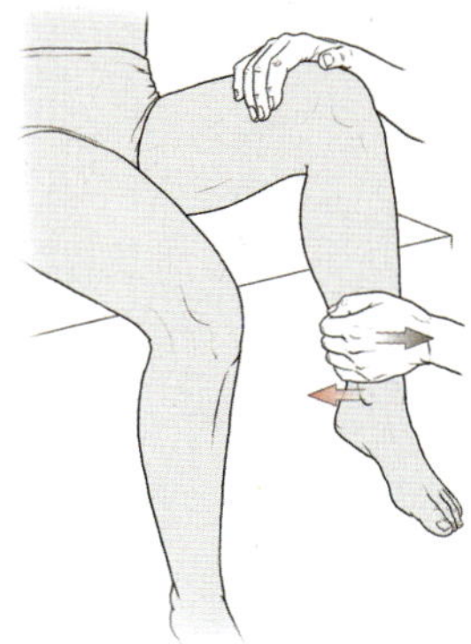

f External rotators

D Clinical symptoms of hip muscle shortening and weakness

Muscles	Symptoms of muscle shortening	Symptoms of muscle weakness
Flexors	Shortened hip flexors lead to the pelvis tilting forward with increased lumbar lordosis and impaired hip extension range. One-sided shortening leads to pelvic torsion with ensuing functional impairment mainly in the sacroiliac joint of the affected side.	Activities such as climbing stairs, walking uphill, shifting from supine position to sitting position, or moving the upper body forward when sitting up are considerably impaired with weak flexors. When walking, instead of hip joint flexion, the leg is moved forward by circumduction or through movements of the pelvis.
Extensors	Contractions of the gluteus maximus are rare. Shortened ischiocrural muscles lead to a characteristic posture with increased hip extension and flexion of the lumbar spine.	A lack of stability in stance phase is due to weak hip joint extensors. It is compensated for through shifting of the upper body (to achieve support from the iliofemoral ligament).
Abductors	Shortened abductors lead to pelvic shift in the frontal plane and functional leg elongation of the affected side. The body compensates for leg length discrepancy through increased flexion of the knee joint.	Due to the lack of muscular support, the pelvis cannot be kept in place in stance phase. It drops to the healthy side (positive Trendelenburg sign). In case of slight weakness, the upper body shifts to the affected side, thus preventing the pelvis from dropping (Duchenne sign).
Adductors	Shortened adductors also lead to pelvic shift in the frontal plane and functional leg elongation of the affected side.	Generally, weakness becomes noticeable during extreme movements, e.g., horse riding, skiing (e.g., rider cannot keep himself on horse).
Internal rotators	The hip joint can no longer rotate externally to the full extent, so sitting cross-legged is no longer possible.	Leads to a significant predominance of external rotation (= change of gait pattern: forward section of foot increasingly points outward when walking).
External rotators	The one-sided pull at the sacrum, especially with shortening of the short external rotators, leads to functional impairment of the sacroiliac joint of the affected side.	Increased internal rotation of the affected leg (forward section of foot increasingly points inward when walking).

2.13 Overview of Muscle Functions: Knee Joint

A Movements of the knee joint

Type of movement	Range of motion	Muscle	Innervation	Nerve segment
Flexion	120–150°	• Semimembranosus • Semitendinosus • Biceps femoris, long head • Biceps femoris, short head • Gracilis • Sartorius • Gastrocnemius, medial and lateral heads • Popliteus • Plantaris	• Tibial nerve • Tibial nerve • Tibial nerve • Common fibular nerve • Obturator nerve • Femoral nerve • Tibial nerve • Tibial nerve • Tibial nerve	• L5–S2 • L5–S2 • L5–S2 • L5–S2 • L2–L4 • L2–L4 • S1–S2 • L5–S2 • S1–S2
Extension	5–10°	• Quadriceps femoris – Rectus femoris – Vastus lateralis – Vastus medialis – Vastus intermedius	• Femoral nerve	• L2–L4
Internal rotation	10°	• Semimembranosus • Semitendinosus • Gracilis • Sartorius • Popliteus	• Tibial nerve • Tibial nerve • Obturator nerve • Femoral nerve • Tibial nerve	• L5–S2 • L5–S2 • L2–L4 • L2–L4 • L5–S2
External rotation	30–40°	• Biceps femoris – Long head – Short head	• Tibial nerve • Common fibular nerve	• L5–S2 • L5–S2

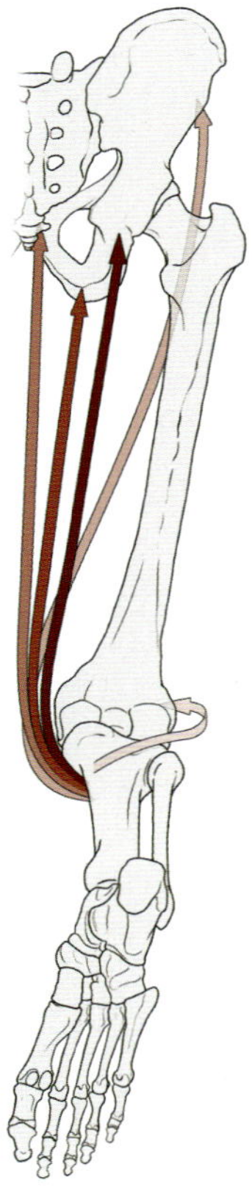

c Internal rotation with the knee joint flexed

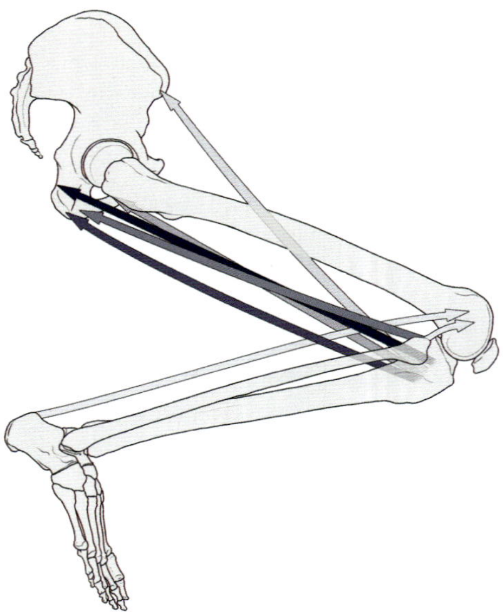

a Flexion

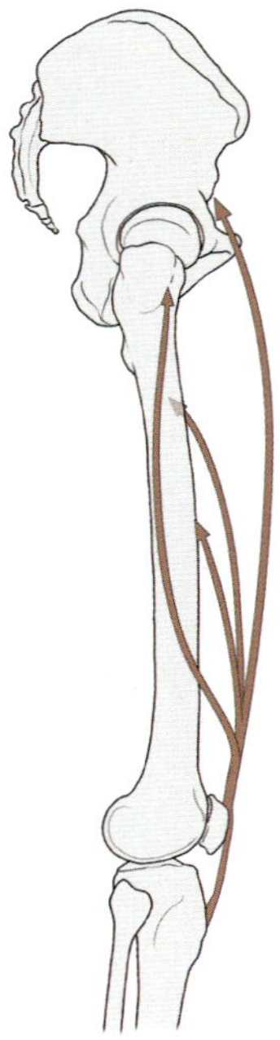

b Extension

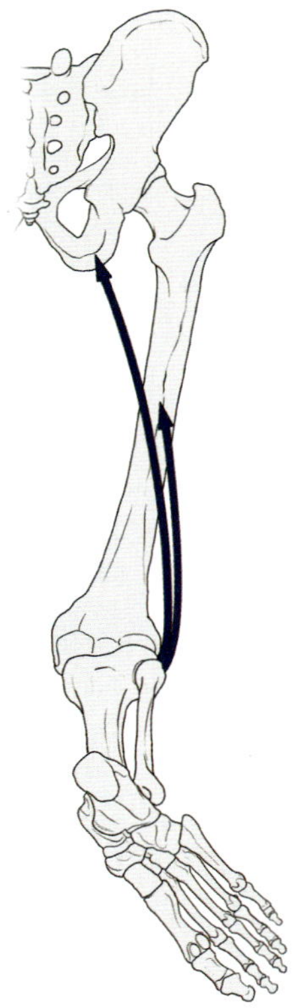

d External rotation with the knee joint extended

B Movements of the knee joint

C Clinical symptoms of knee muscle shortening and weakness

Muscles	Symptoms of muscle shortening	Symptoms of muscle weakness
Flexors	Impaired knee extension with bent hip or impaired hip flexion with hip outstretched. Shortened flexors, especially the ischiocrural (hamstring) muscles on both sides, lead to straightening of the pelvis and flattening of the lumbar spine. If the shortening affects only one side, the differing pull causes pelvic tilt and functional disorder of the sacroiliac joint.	Weakness of the ischiocrural muscles makes itself felt in both the knee and the hip joint. When standing, the result is increased pelvic tilt forward with simultaneous knee hyperextension. The clinical symptoms are similar to those of extensor weakness.
Extensors	In the case of muscle shortening, the rectus femoris is most commonly affected since it is the only part of the quadriceps femoris that crosses two joints. The result is impaired knee flexion or impaired hip extension with increased lumbar lordosis.	In case of muscle weakness, essential functions are significantly impaired, e.g., climbing stairs, walking uphill, standing up, and sitting down. Compensation occurs through shifting the center of gravity forward via knee hyperextension (genu recurvatum): "Gravity turns into stretching force."
Internal rotators	Limited external rotation	Increased external rotation of the lower leg
External rotators	Limited internal rotation	Increased internal rotation of the lower leg

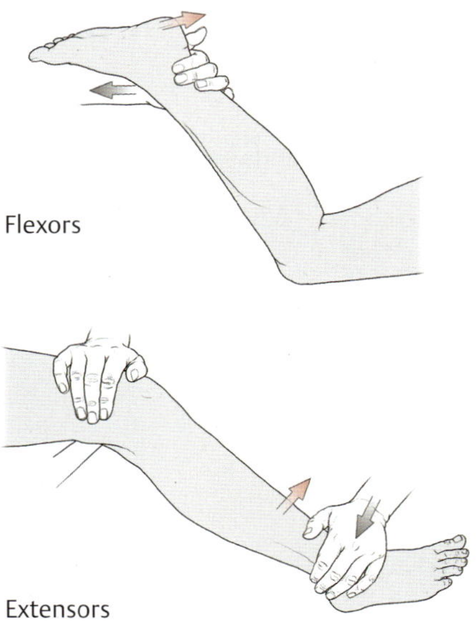

a Flexors

b Extensors

D Functional muscle testing of the knee joint

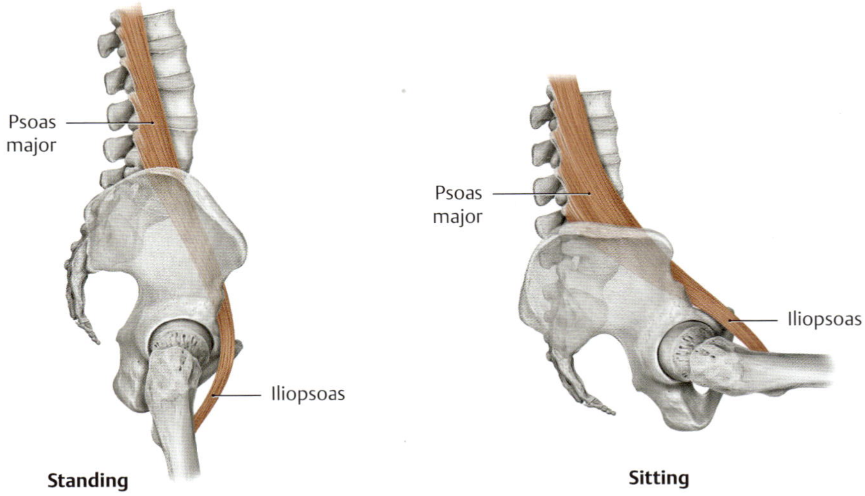

Standing **Sitting**

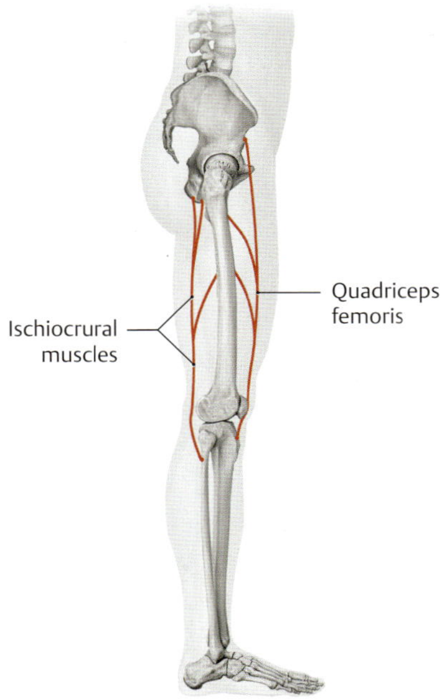

E Muscular dysbalance

Iliacus not shown. Special testing of every joint includes not only movement tests (neutral zero method, see p. 52) but also functional tests. The primary goal is not to test a muscle's force and ability to stretch but to detect muscular imbalances and coordination impairment. If there is an imbalance between a muscle and its antagonist, it is called a muscular dysbalance. What is commonly meant by this is an imbalance between tonic and phasic muscles.

Tonic muscles primarily consist of slow-twitch muscle fibers and in case of disorders tend to shorten (see p. 60). **Phasic muscles** consist of fast-twitch fibers and tend to weaken (atrophy) in case of disorders. Those disorders are the result of the insufficient adaptation of the musculoskeletal system to the civilized way of life, which includes prolonged sitting, inactivity, habitual movements, and abnormal postures in daily life and at work. Increased variety of motion and better quality of motion are easy ways to counteract the lack of stimuli to muscles.

Examples of the development of muscular dysbalances:

- Prolonged sitting leads to increased structural shortening of the iliopsoas muscles (these muscles reach their original length in standing position) and significant weakening of the hip extensors (mainly the gluteus maximus and ischiocrural muscles) and thus indirectly leads to increased susceptibility to disorders of the hyperlordotic lumbar spine.
- High heels lead to increased load on the quadriceps femoris with corresponding increased tonus; the ischiocrural muscles, as antagonists, however, shorten due to the constant lack of stimuli.

2.14 Overview of Muscle Functions: Ankle Joints

A Movements of the upper and lower ankle joint and the transverse tarsal joint

Type of movement	Range of motion	Muscle	Innervation	Nerve segment
Plantar flexion	40–50°	• Triceps surae • Fibularis longus • Fibularis brevis • Flexor hallucis longus • Flexor digitorum longus • Tibialis posterior • Plantaris	• Tibial nerve • Superficial fibular nerve • Superficial fibular nerve • Tibial nerve • Tibial nerve • Tibial nerve • Tibial nerve	• S1-S2 • L5–S1 • L5–S1 • L5–S2 • L5–S2 • L4–S1 • S1–S2
Dorsiflexion	20–30°	• Tibialis anterior • Extensor digitorum longus • Extensor hallucis longus • Fibularis tertius (variable)	• Deep fibular nerve • Deep fibular nerve • Deep fibular nerve • Deep fibular nerve	• L4–L5 • L4–S1 • L5–S1 • L4–S1
Inversion and supination	60°	• Triceps surae • Tibialis posterior • Flexor hallucis longus • Flexor digitorum longus • Tibialis anterior • (Extensor hallucis longus)	• Tibial nerve • Tibial nerve • Tibial nerve • Tibial nerve • Deep fibular nerve • Deep fibular nerve	• S1–S2 • L4–S1 • L5–S2 • L5–S2 • L4–L5 • L5–S1
Eversion and pronation	30°	• Fibularis longus • Fibularis brevis • Extensor digitorum longus • (Extensor hallucis longus)	• Superficial fibular nerve • Superficial fibular nerve • Deep fibular nerve • Deep fibular nerve	• L5–S1 • L5–S1 • L4–S1 • L5–S1

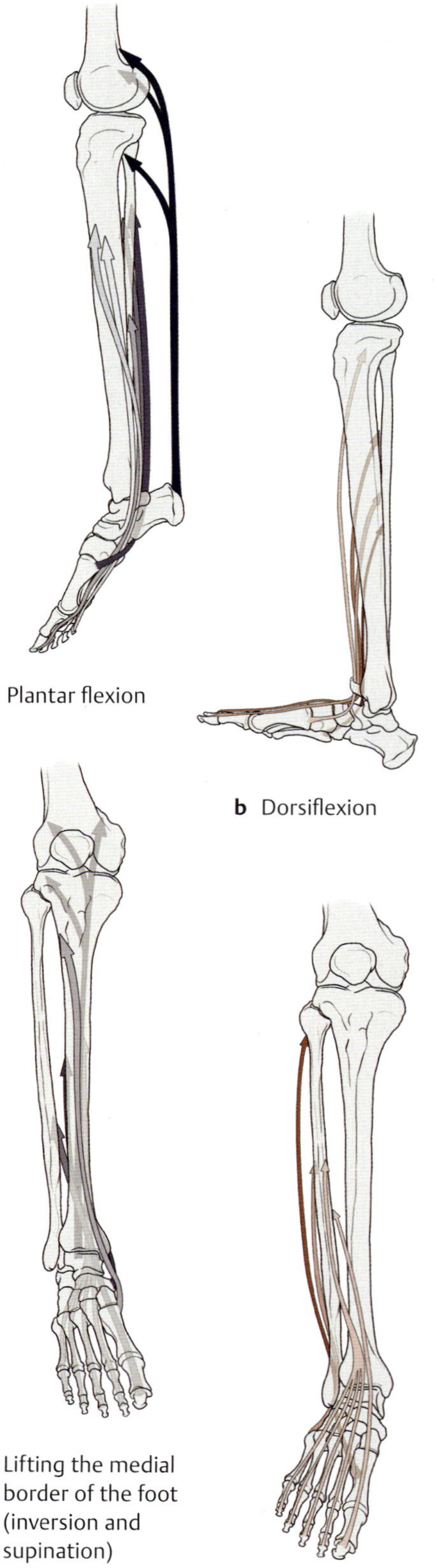

a Plantar flexion

b Dorsiflexion

c Lifting the medial border of the foot (inversion and supination)

d Lifting the lateral border of the foot (eversion and pronation)

B Movements of the upper and lower ankle joint and the transverse tarsal joint

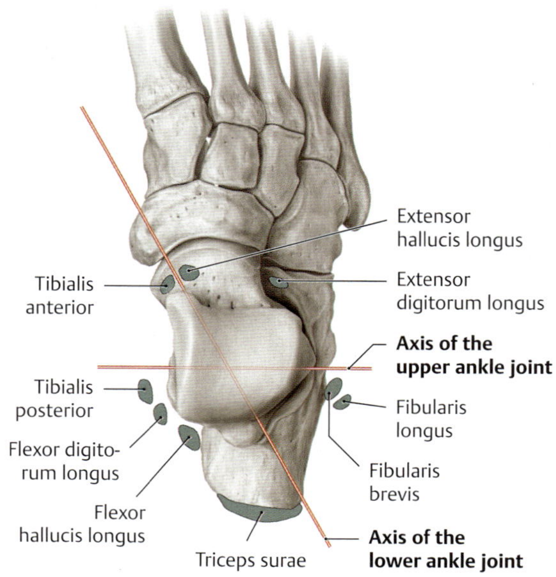

C Location of tendons of the long muscles of the foot relative to the axes of the upper and lower ankle joint

Superior view of both axes of the upper and lower ankle joint. Depending on the course of the tendons, the muscles can plantar flex and dorsal extend while at the same time pronating (everting) and supinating (inverting).

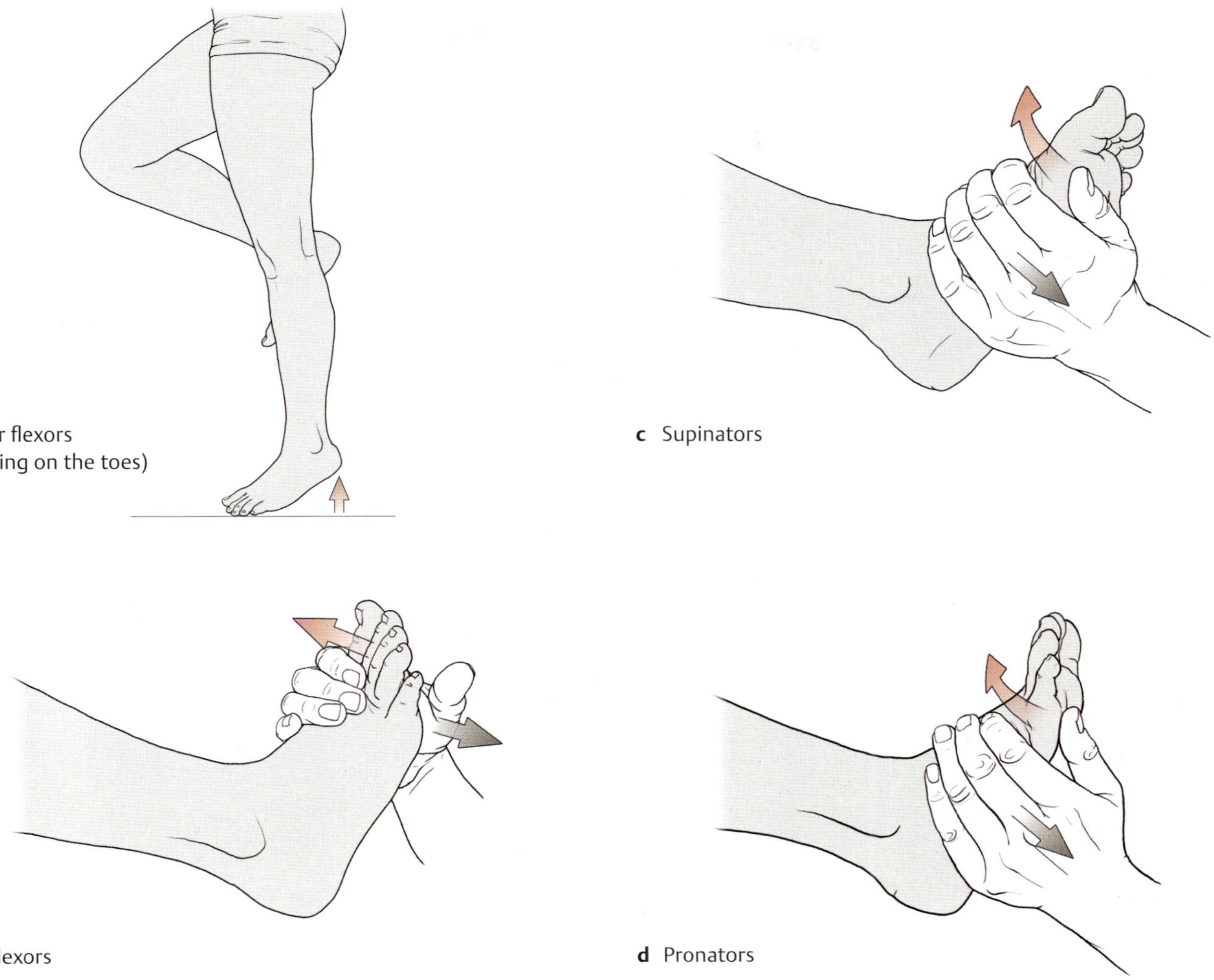

a Plantar flexors (standing on the toes)

b Dorsiflexors

c Supinators

d Pronators

D Functional muscle testing of the ankle joint

E Clinical symptoms of muscle shortening and weakness of the upper and lower ankle joint

Muscles	Symptoms of muscle shortening	Symptoms of muscle weakness
Plantar flexors	Shortening of plantar flexors, mainly the triceps surae, leads to club foot (pes equinovarus). The resulting functional leg elongation when standing is compensated for by increased knee flexion. When walking, during swing phase, increased hip flexon is required in order to allow for the functionally longer leg to swing forward.	Tiptoe position is usually not possible, and jumping power is considerably impaired. Also common is development of talipes calcaneus (no rollover during walking). In stance phase, there is increased knee extension.
Dorsiflexors	Shortening of dorsiflexors leads to impaired plantar flexion and impaired rollover during walking. Often development of club foot (no rollover during walking) develops.	The forward section of the foot cannot be sufficiently lifted during swing phase. To compensate for it, there is increased bending of the hip and knee. This characteristic gait pattern is also known as stork legs.
Supinators	Shortening leads to pes equinovarus. There is increased load on the lateral bottom of the foot during walking.	Weakened supinators result in valgus position (talipes valgus) with increased load on the inside edge of the foot.
Pronators	If the pronators (e.g., the fibularis muscles) are affected by shortening, the foot is plantar flexed and pronated (pes equinovalgus).	Weakened pronators often result in distortion of the upper and lower ankle joint (so-called supination trauma).

3.1 The Muscles of the Medial and Anterior Thigh, Hip, and Gluteal Region

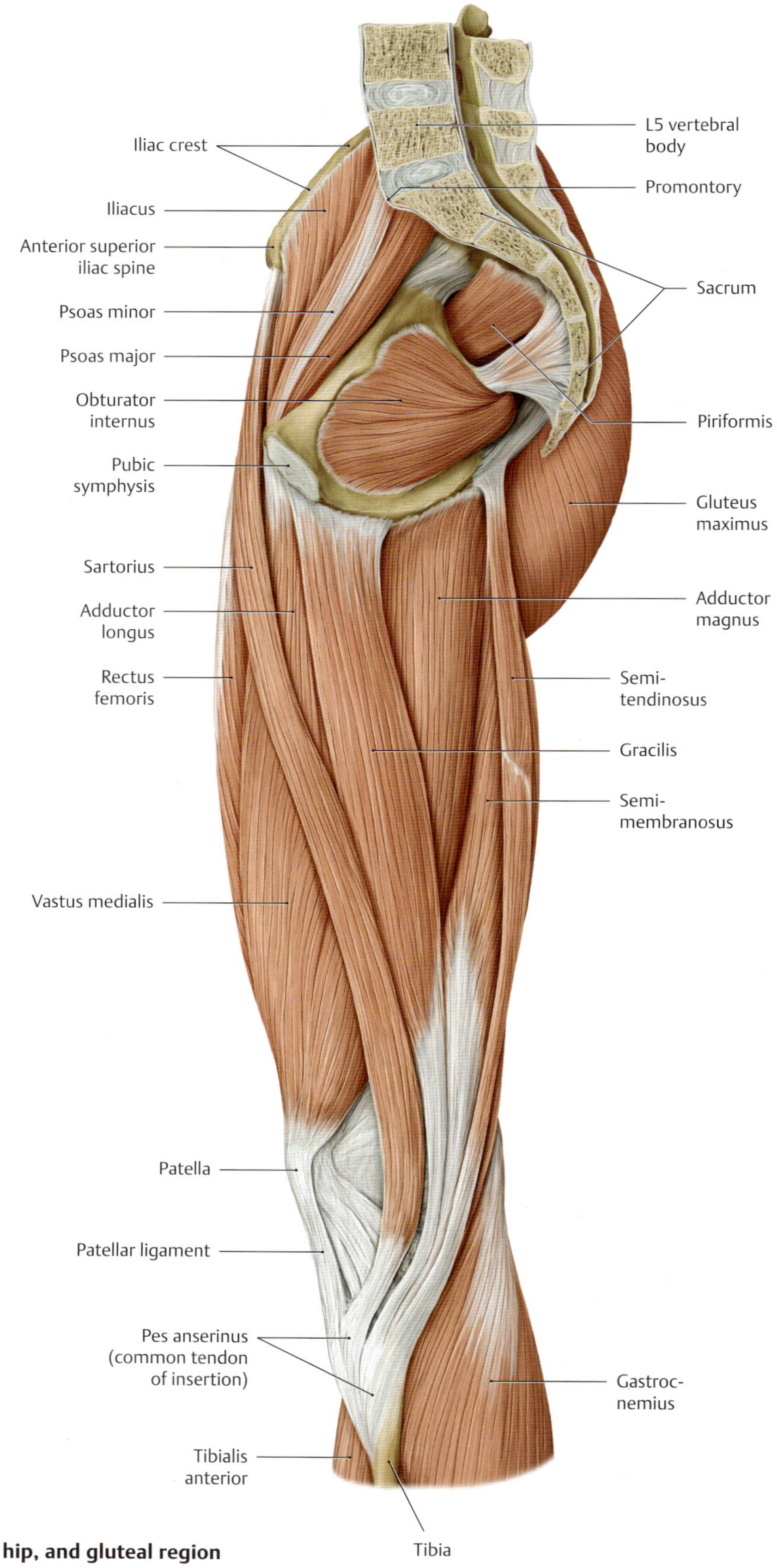

A The muscles of the thigh, hip, and gluteal region
Right side, medial view.

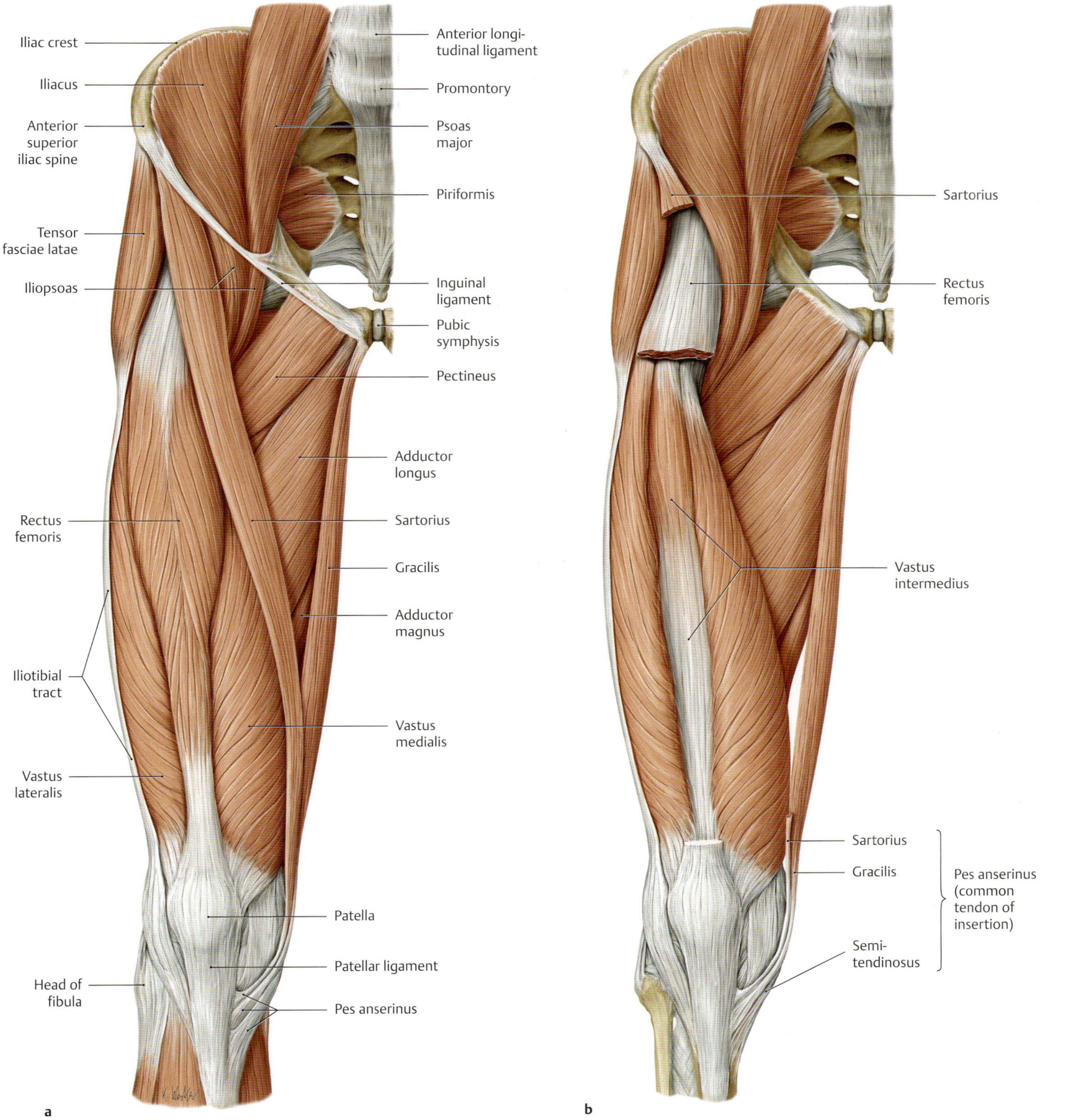

B The muscles of the thigh, hip, and gluteal region
Right side, anterior view.

a The fascia lata of the thigh (see p. 563) has been removed as far as the lateral iliotibial tract.

b Portions of the sartorius and rectus femoris have also been removed.

3.2 The Muscles of the Anterior Thigh, Hip, and Gluteal Region: Origins and Insertions

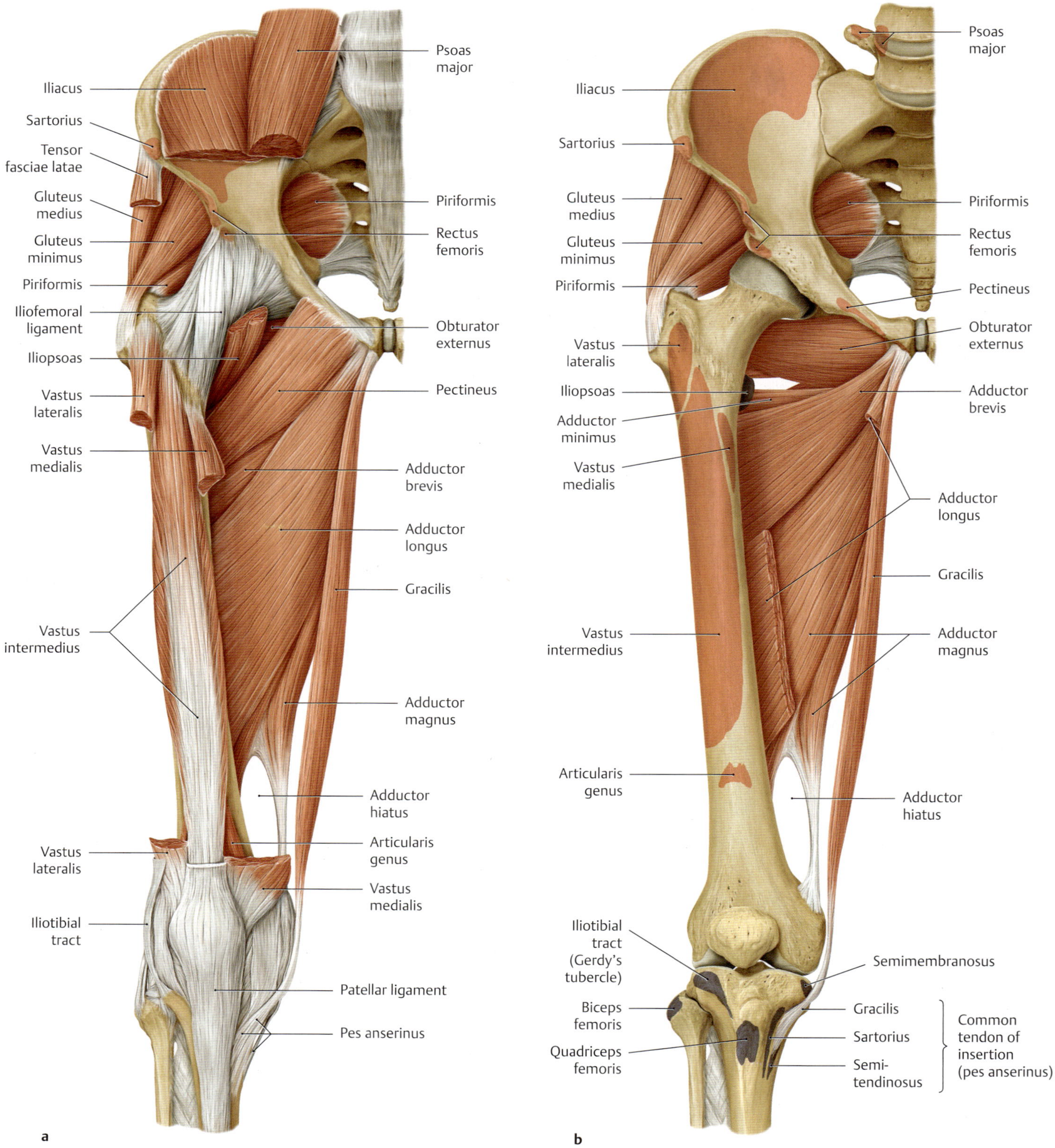

A The muscles of the right thigh, hip, and gluteal region, anterior view

Origin shown in red, insertion in blue.

a The iliopsoas and tensor fasciae latae have been partially removed. The sartorius, rectus femoris, vastus lateralis, and vastus medialis have been completely removed.

b The quadriceps femoris, iliopsoas, tensor fasciae latae, and pectineus have been completely removed. The midportion of the adductor longus has been removed.

The insertion of the iliotibial tract (Gerdy's tubercle) lies lateral and proximal to the tibial tuberosity and is identifiable as a palpable protuberance in the region of the lateral condyle of the tibia. Excessive stress on the iliotibial tract such as can occur with genu varum can lead to pain at the insertion. This is incorrectly referred to tenosynovitis at the muscular insertion; this is incorrect because the tubercle is not an apophysis.

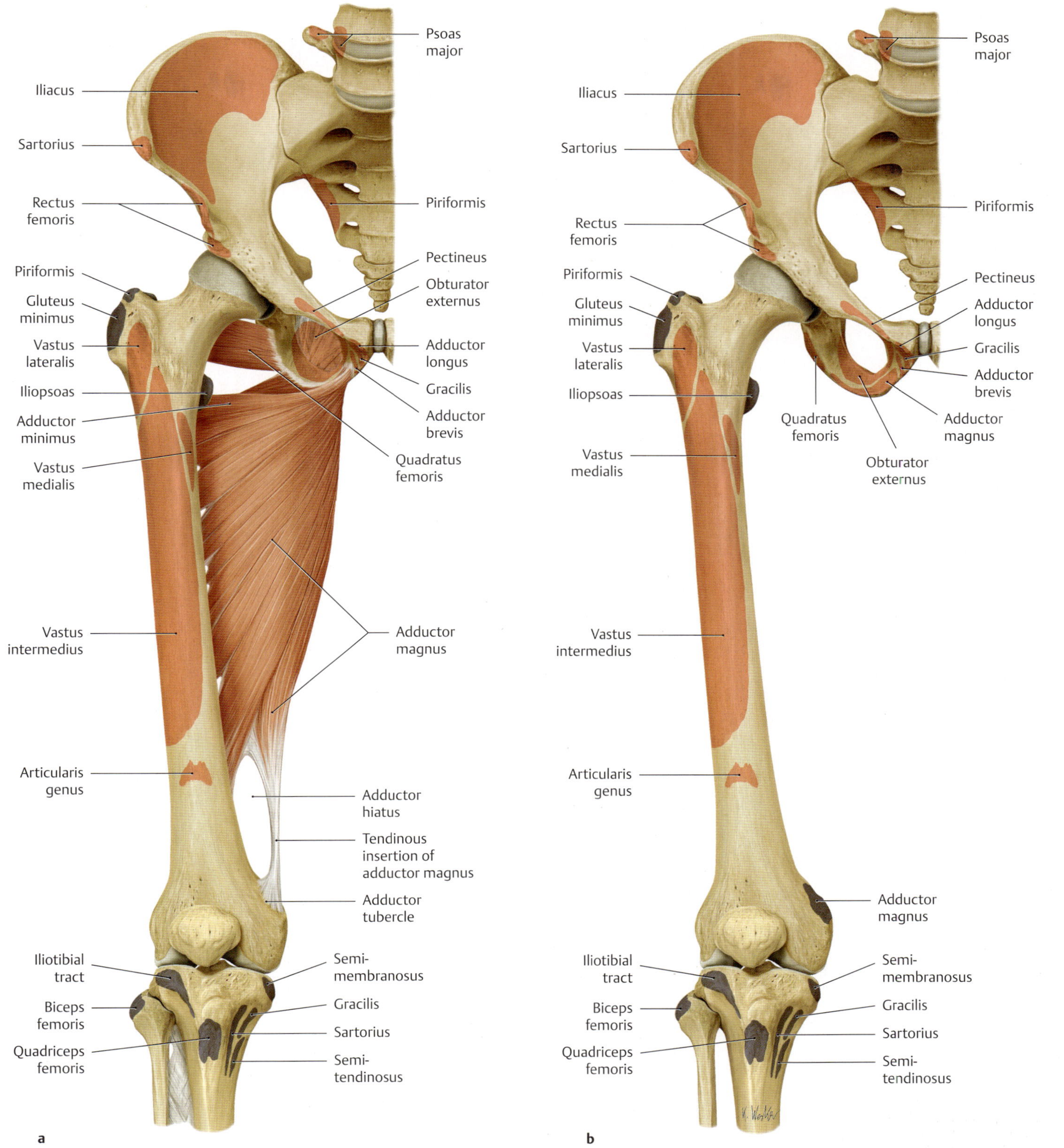

B The muscles of the right thigh, hip, and gluteal region, anterior view
Origin shown in red, insertion in blue.

a All of the muscles have been removed except for adductor magnus and quadratus femoris.

b All of the muscles have been removed.

Note the adductor hiatus, through which the femoral artery and vein enter the popliteal fossa of the leg.

3.3 The Muscles of the Lateral and Posterior Thigh, Hip, and Gluteal Region

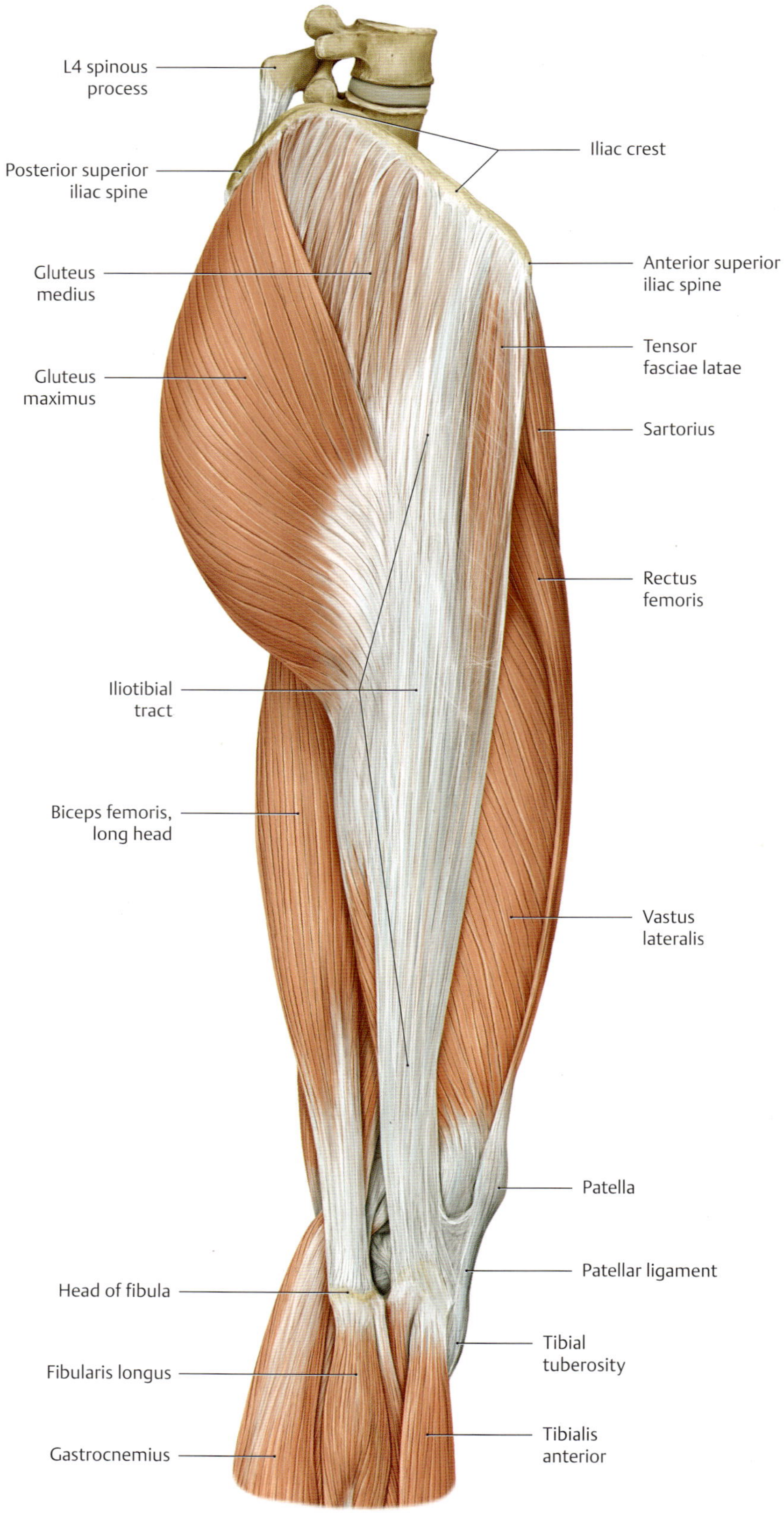

A The muscles of the thigh, hip, and gluteal region
Right side, lateral view. *Note* the tensor fasciae latae and gluteus maximus muscles, whose tendons of insertion strengthen and thicken the lateral part of the fascia lata. This thickened band, called the *iliotibial tract* because it runs between the *iliac* crest and the lateral side of the upper *tibia*, functions mechanically as a tension band to reduce the bending loads on the proximal femur (after Pauwels, see also p. 429).

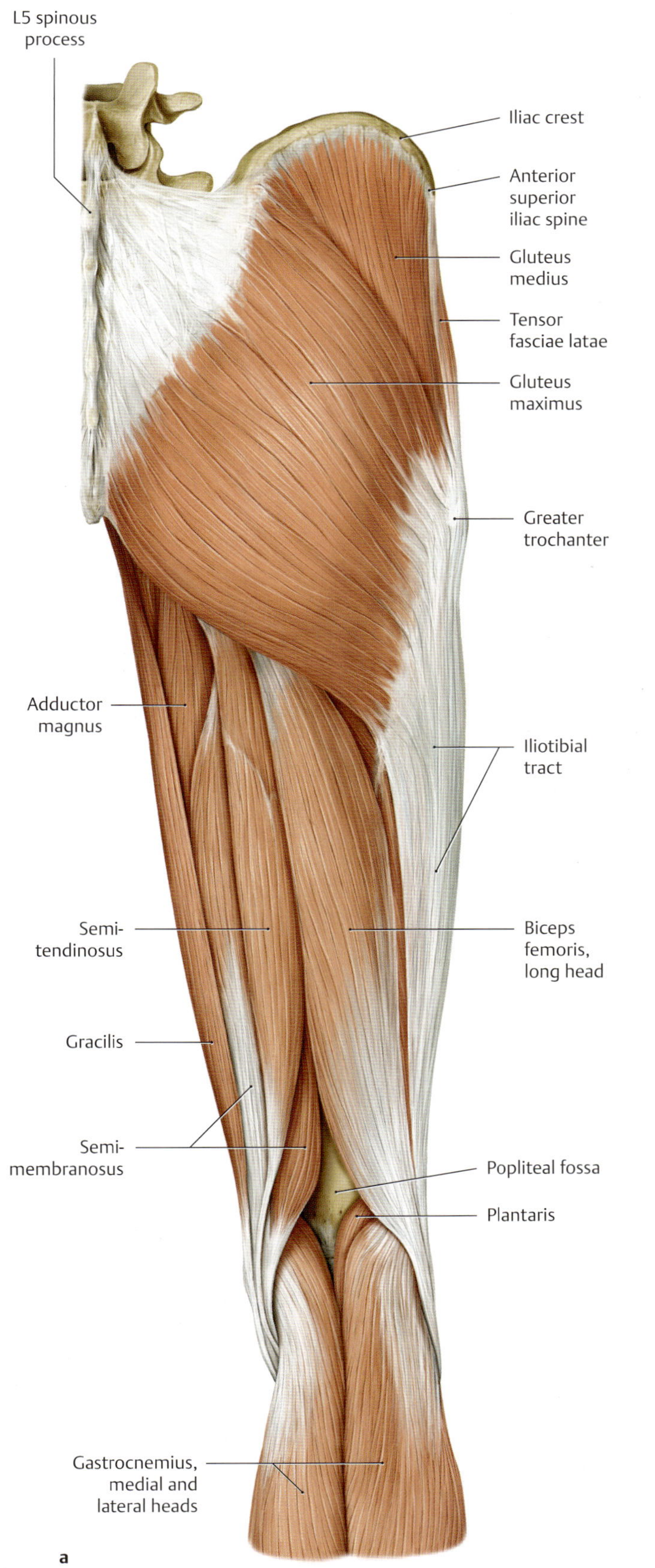

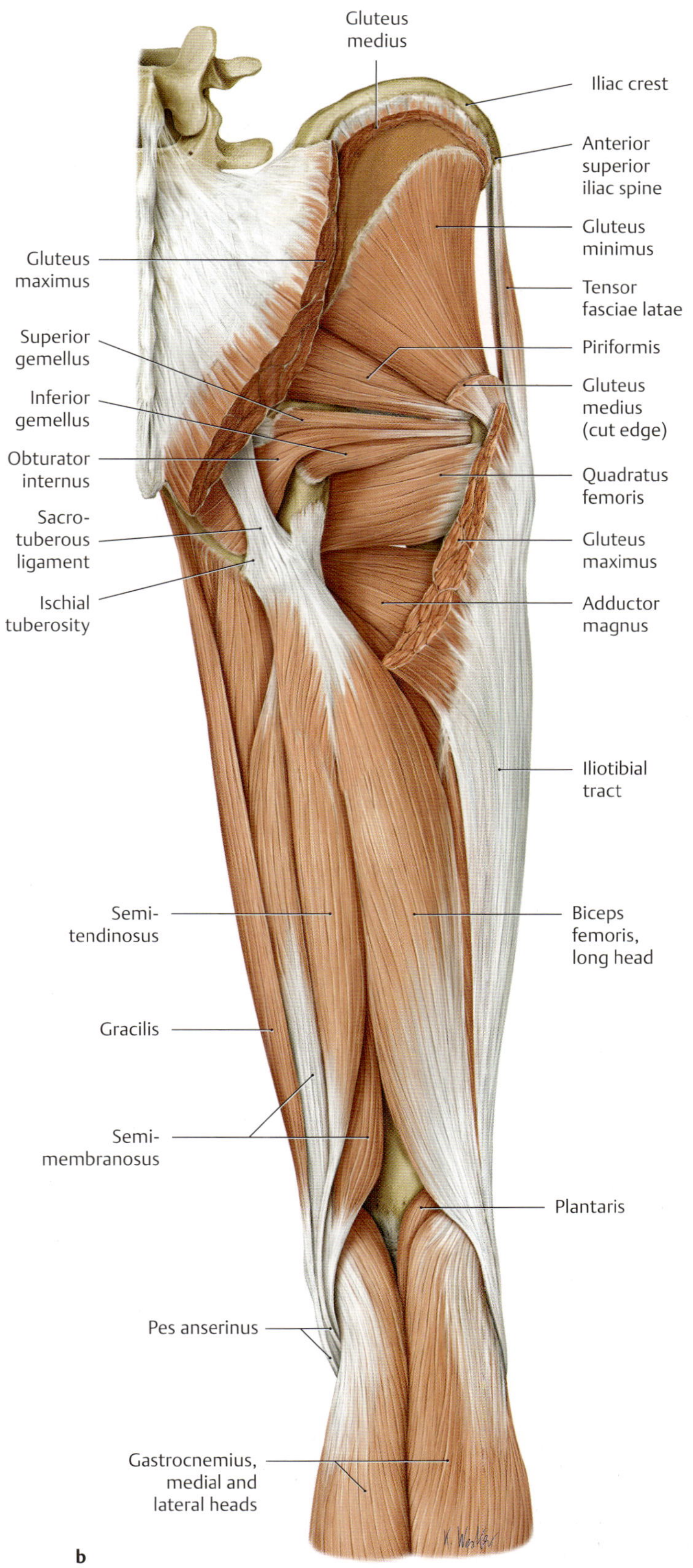

B The muscles of the thigh, hip, and gluteal region
Right side, posterior view.

a The fascia lata has been removed as far as the iliotibial tract (the portion over the buttock is called the gluteal fascia).

b The gluteus maximus and gluteus medius have been partially removed.

3.4 The Muscles of the Posterior Thigh, Hip, and Gluteal Region: Origins and Insertions

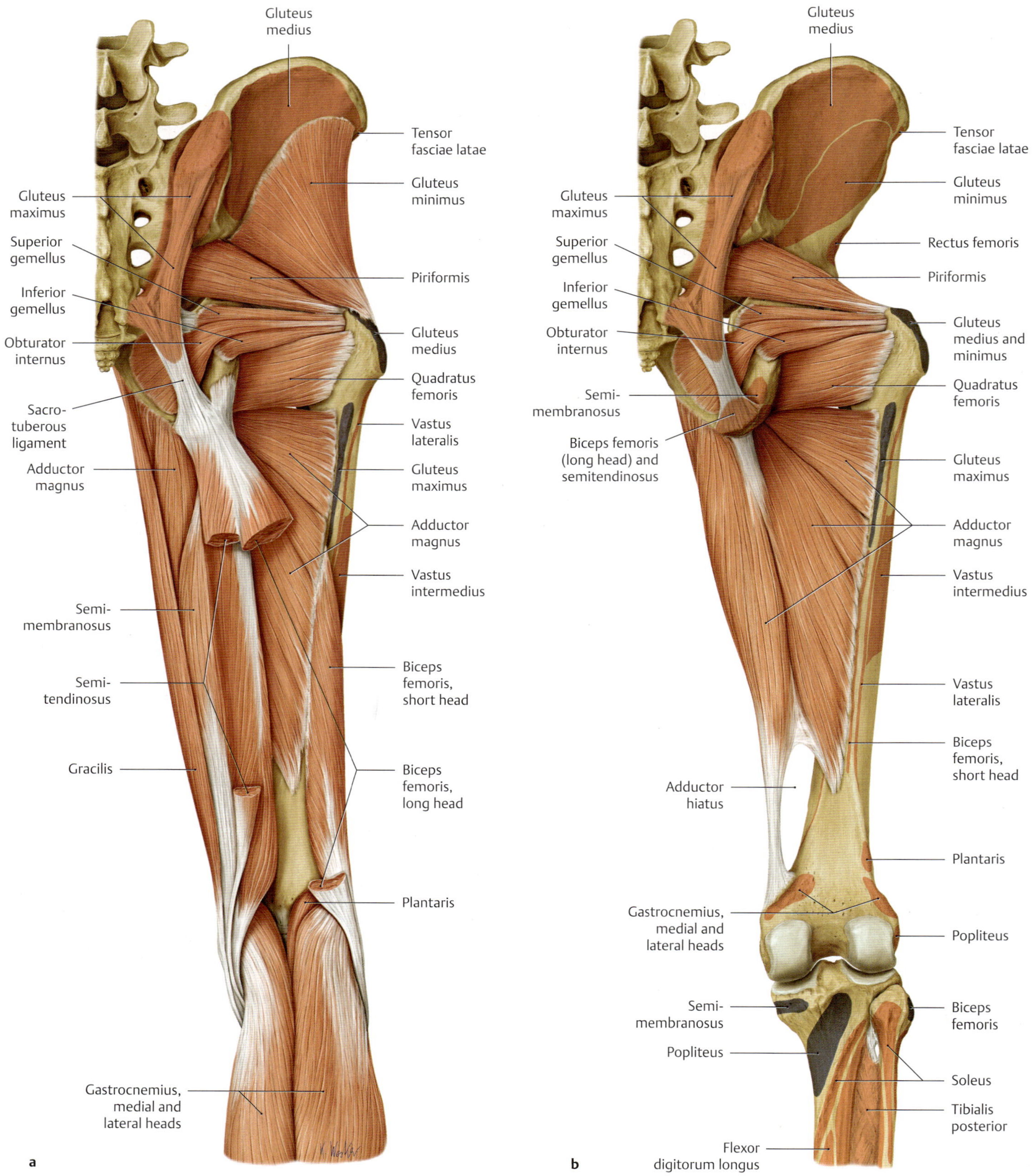

A The muscles of the right thigh, hip, and gluteal region, posterior view

Origin shown in red, insertion in blue.

a The semitendinosus and biceps femoris have been partially removed. The gluteus maximus and medius have been completely removed.

b The hamstrings (semitendinosus, semimembranosus, and biceps femoris) and gluteus minimus have been completely removed.

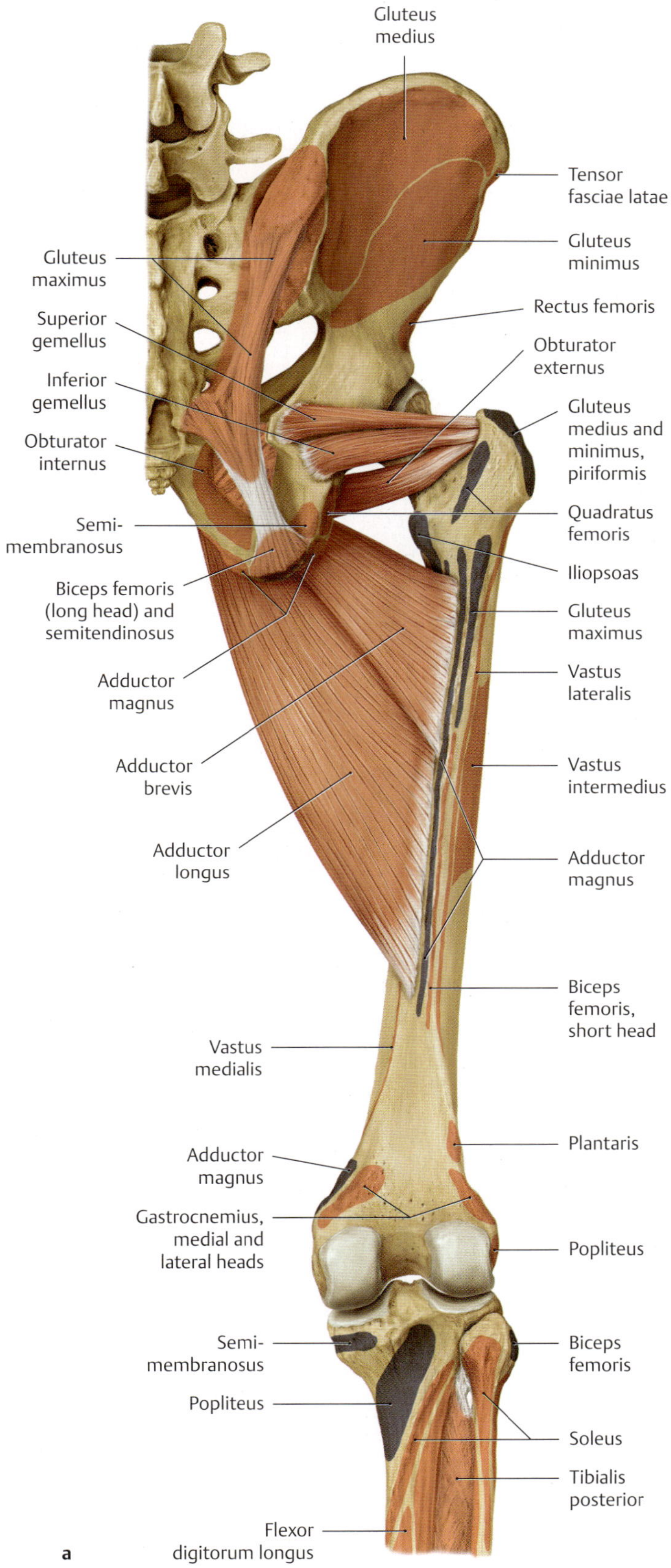

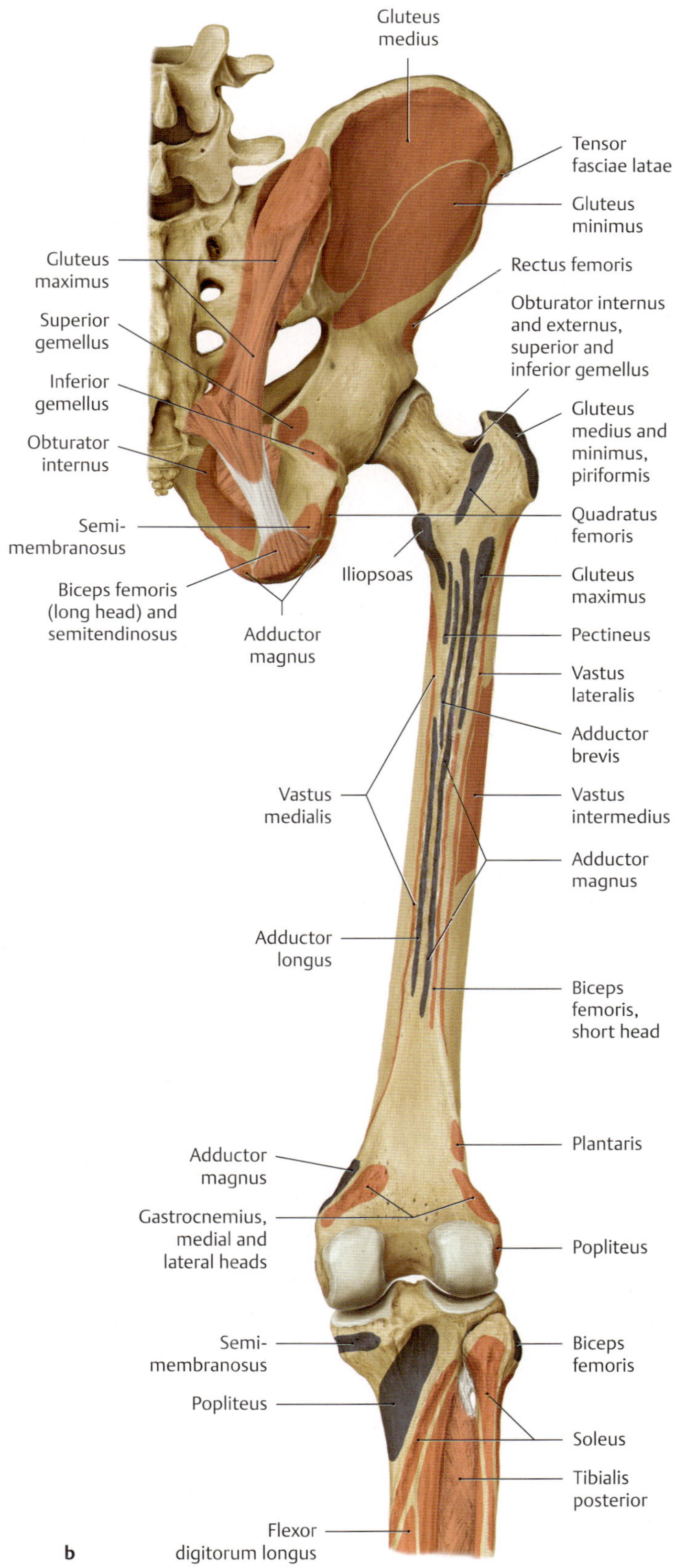

B The muscles of the right thigh, hip, and gluteal region, posterior view

Origin shown in red, insertion in blue.

a All muscles have been removed except for the adductor brevis, adductor longus, superior and inferior gemellus, and obturator externus.

b All of the muscles have been removed.

3.5 The Muscles of the Lateral and Anterior Leg: Origins and Insertions

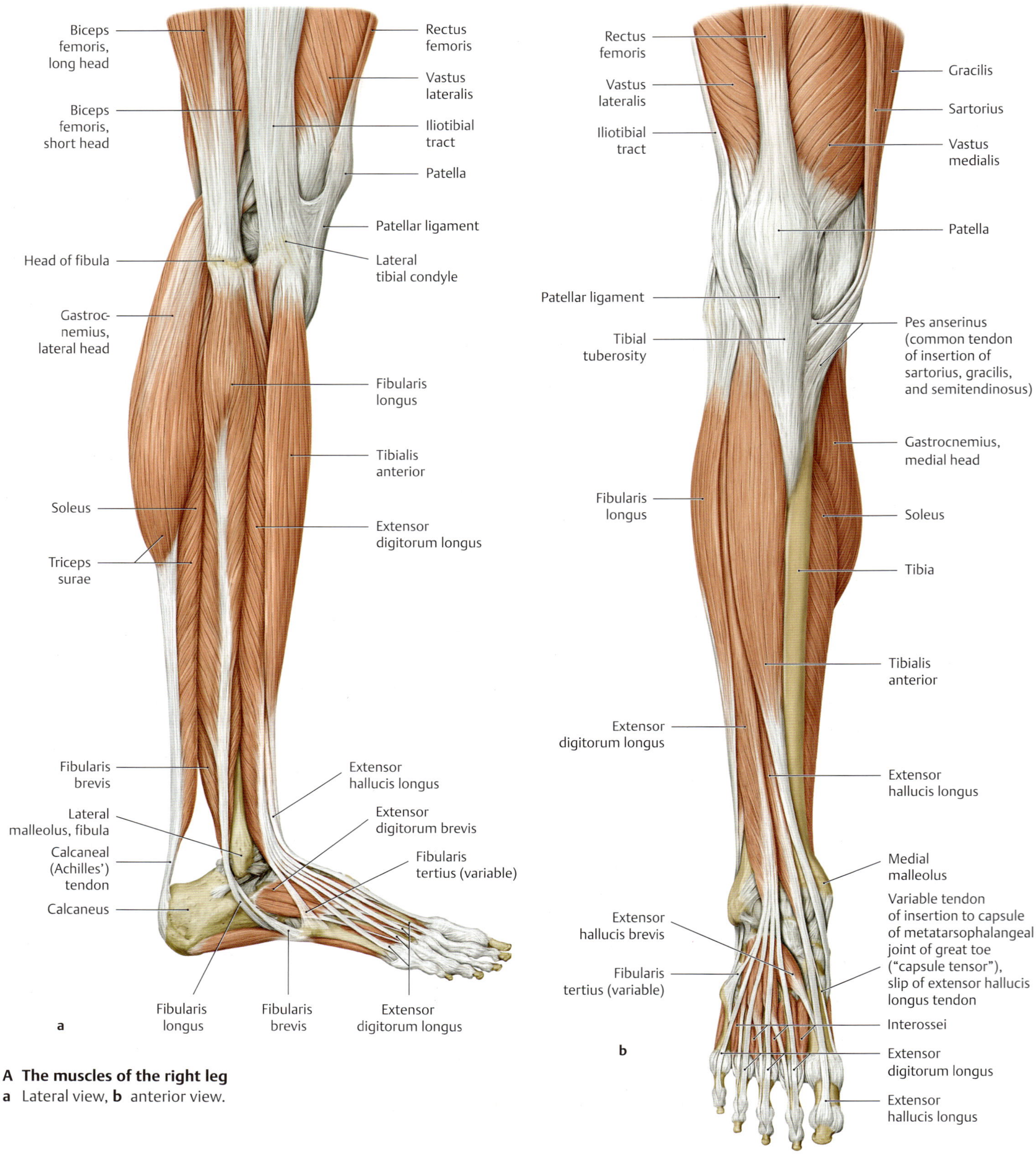

A The muscles of the right leg
a Lateral view, **b** anterior view.

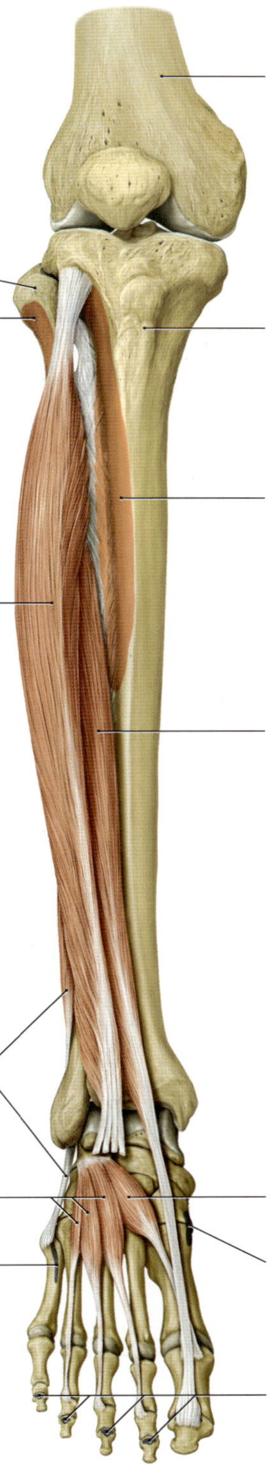

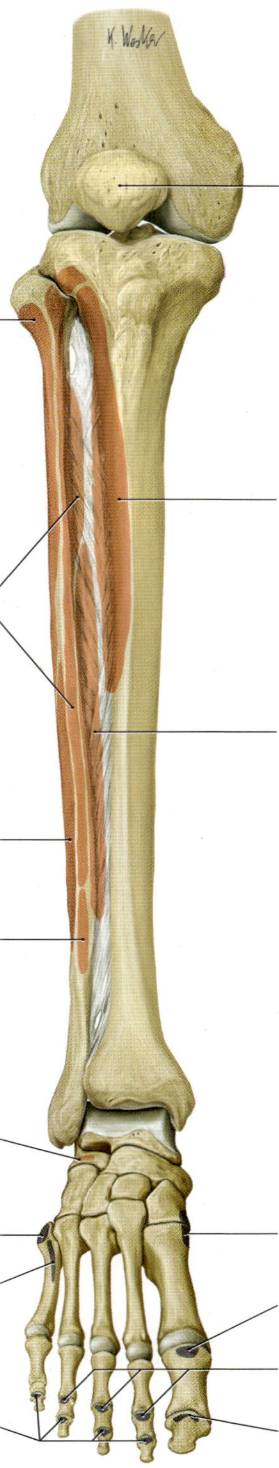

B Muscles of the right leg, anterior view
Origin shown in red, insertion in blue.

a The tibialis anterior and fibularis longus have been completely removed, as have the distal portions of the extensor digitorum longus tendons. The fibularis tertius is a division of the extensor digitorum longus.

b All of the muscles have been removed.

3.6 The Muscles of the Posterior Leg: Origins and Insertions

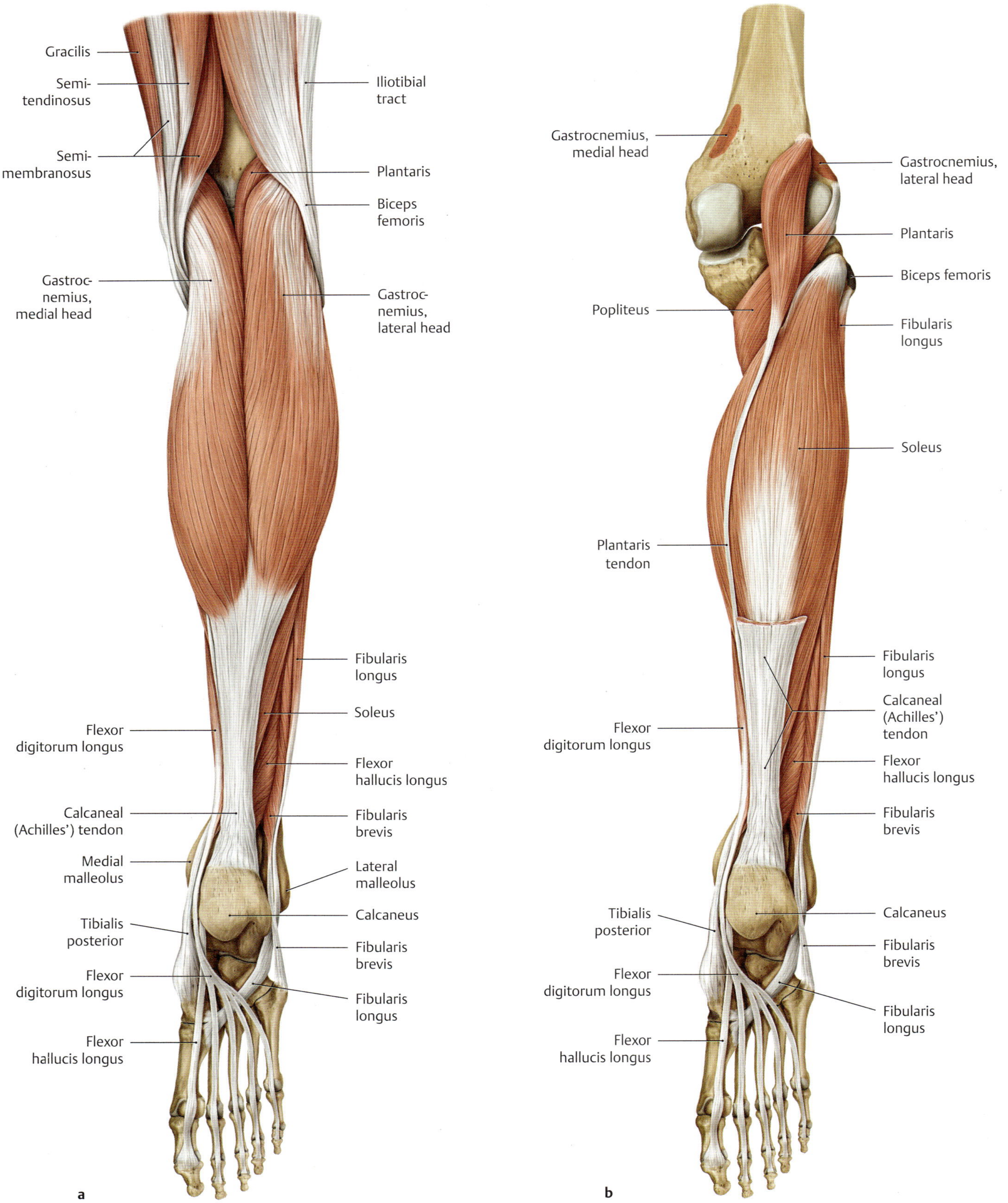

A Muscles of the right leg, posterior view
Origin shown in red, insertion in blue. The foot is shown in a plantar-flexed position to better demonstrate the plantar tendons.

a The bulge of the calf is produced mainly by the triceps surae muscle (= soleus plus the two heads of the gastrocnemius).
b Both heads of the gastrocnemius have been removed.

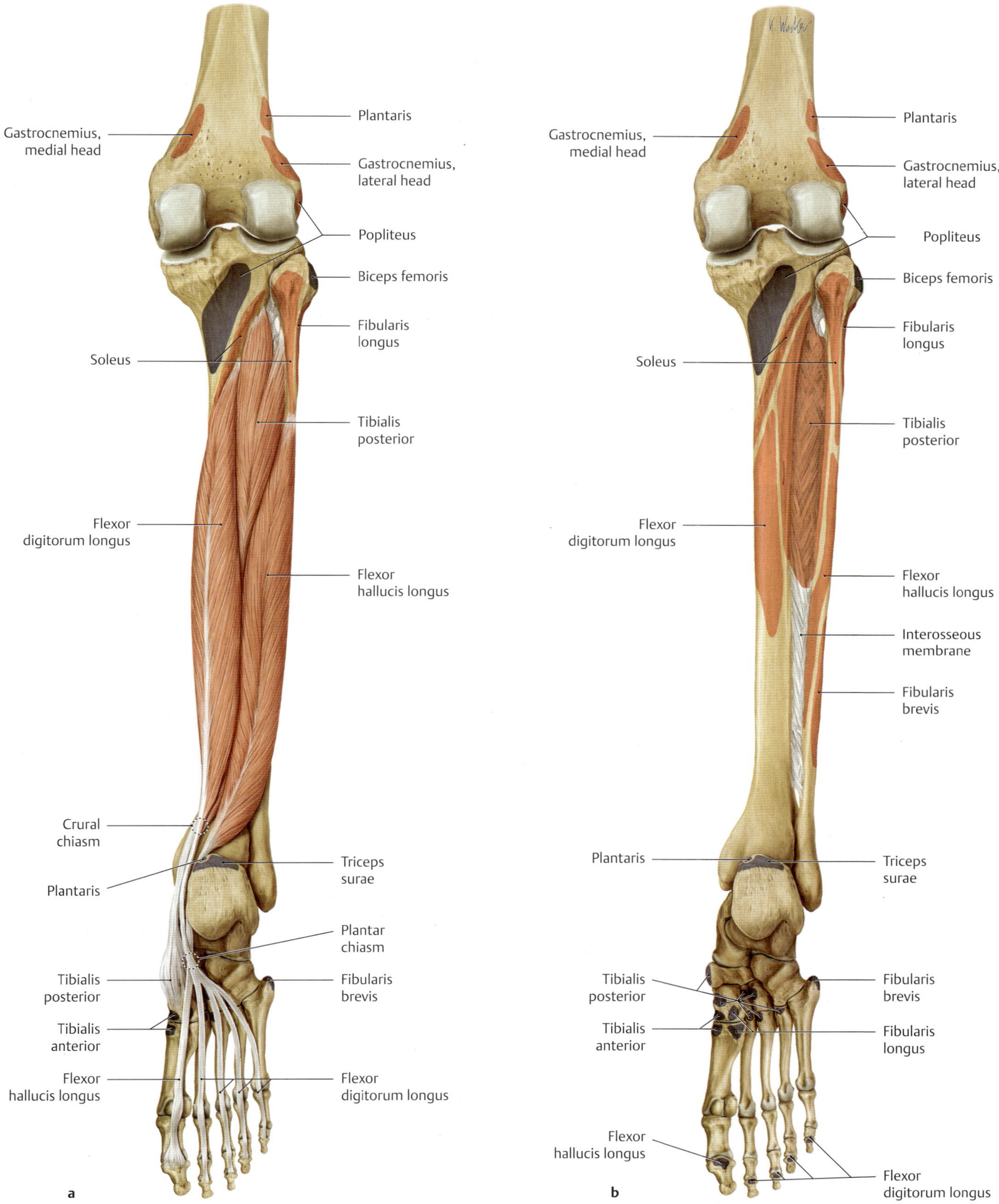

B Muscles of the right leg, posterior view
Origin shown in red, insertion in blue. The foot is shown in a plantar-flexed position to better demonstrate the plantar tendons.

a The triceps surae, plantaris, and popliteus muscles have been removed.

b All of the muscles have been removed.

3.7 The Tendon Sheaths and Retinacula of the Foot

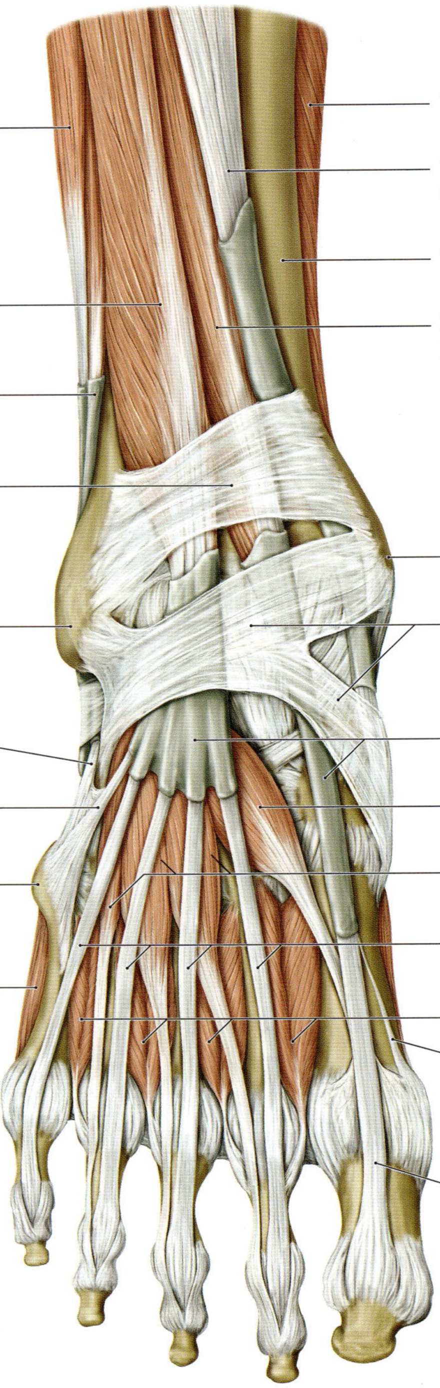

A The tendon sheaths and retinacula of the right foot
Anterior view. The foot is plantar-flexed, with superficial fascia removed, to display the deep fascial bands—*retinacula*—that hold in place the tendon sheaths of the long foot extensors and flexors. The superior and inferior extensor retinacula retain the long extensor tendons, allowing efficient redirection of the forces generated by their muscles (tibialis anterior, extensor digitorum longus, extensor hallucis longus, and fibularis tertius) while preventing the tendons from rising away from the bones of the ankle when the foot is dorsiflexed. Similarly, the fibular retinacula, laterally, hold the fibular muscle tendons in place posterior to the lateral malleolus (see **B**, part **b**), and the flexor retinaculum retains the long flexor tendons behind the medial malleolus (see **B**, part **a**), preventing displacement of these tendons while enabling them to operate smoothly regardless of the orientation of the ankle joint.

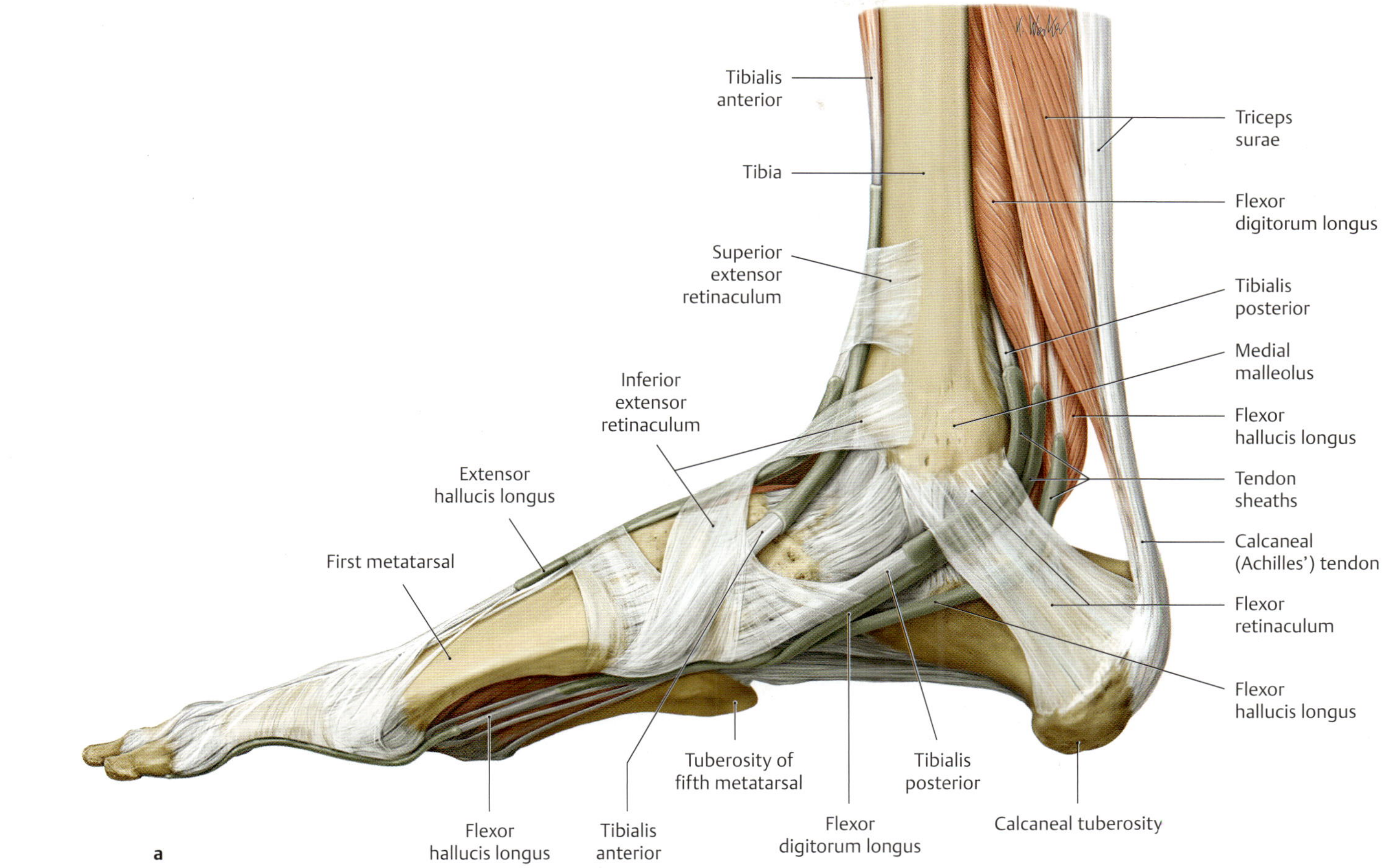

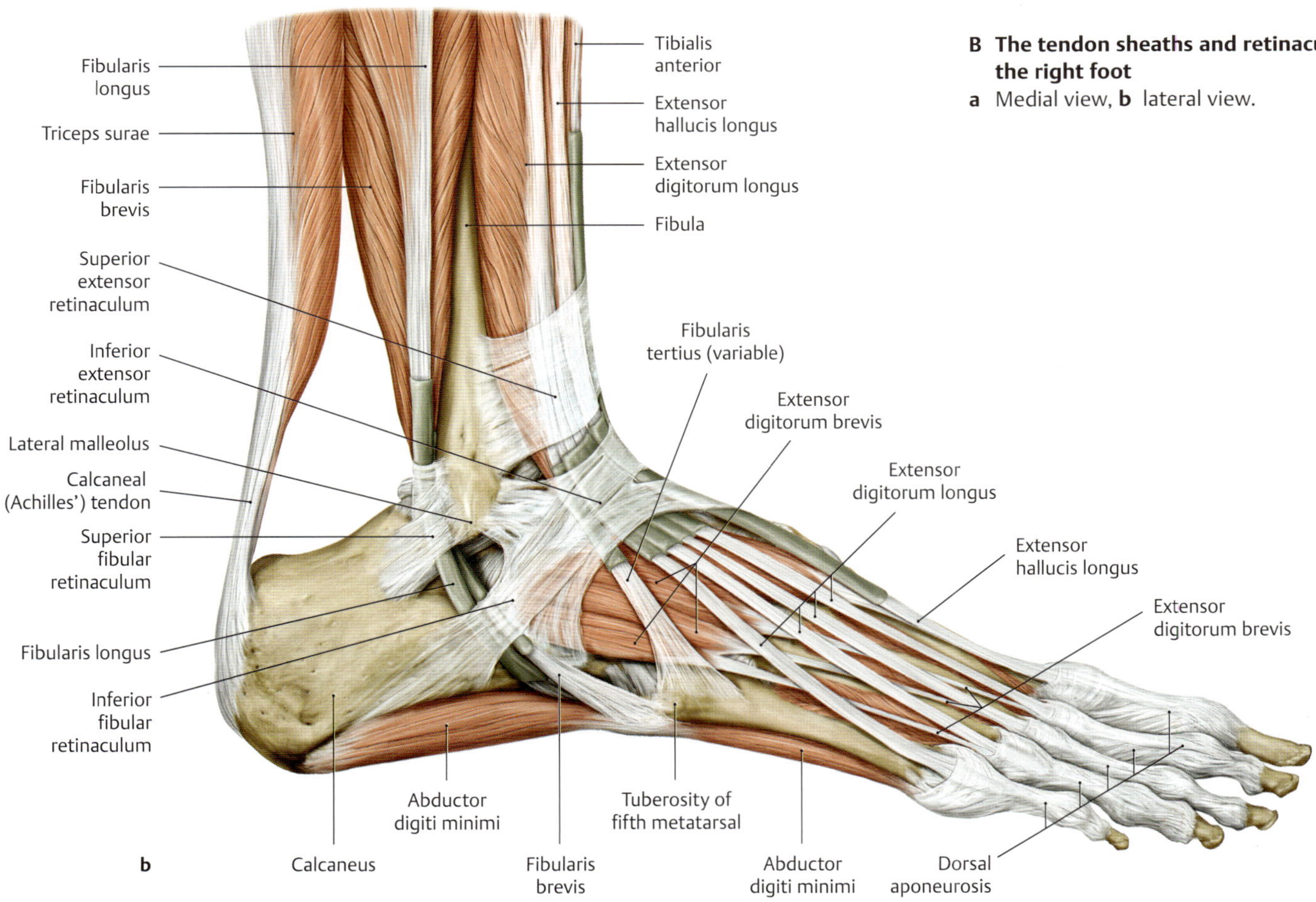

B The tendon sheaths and retinacula of the right foot

a Medial view, **b** lateral view.

3.8 The Intrinsic Foot Muscles from the Plantar View: The Plantar Aponeurosis and Superficial Layer

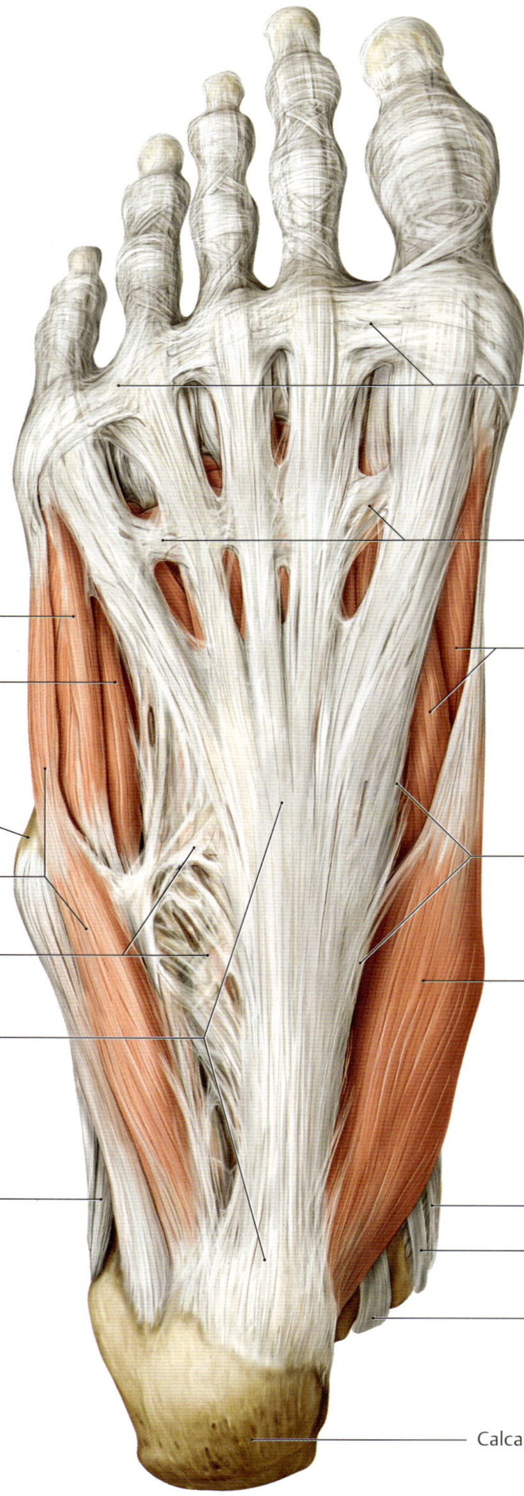

A The plantar aponeurosis of the right foot
Plantar view. The *plantar aponeurosis* is a tough aponeurotic sheet that is thicker centrally than medially and laterally and blends with the dorsal fascia (not shown here) at the borders of the foot. Two sagittal expansions of the thick *central* aponeurosis (the medial and lateral plantar septa) extend deep to the bones of the foot, defining the boundaries of three muscle compartments in the plantar region: the medial compartment, lateral compartment, and central compartment (not labeled here, see p. 512). The main function of the plantar aponeurosis is to give *passive* support to the longitudinal arch (see also p. 482).

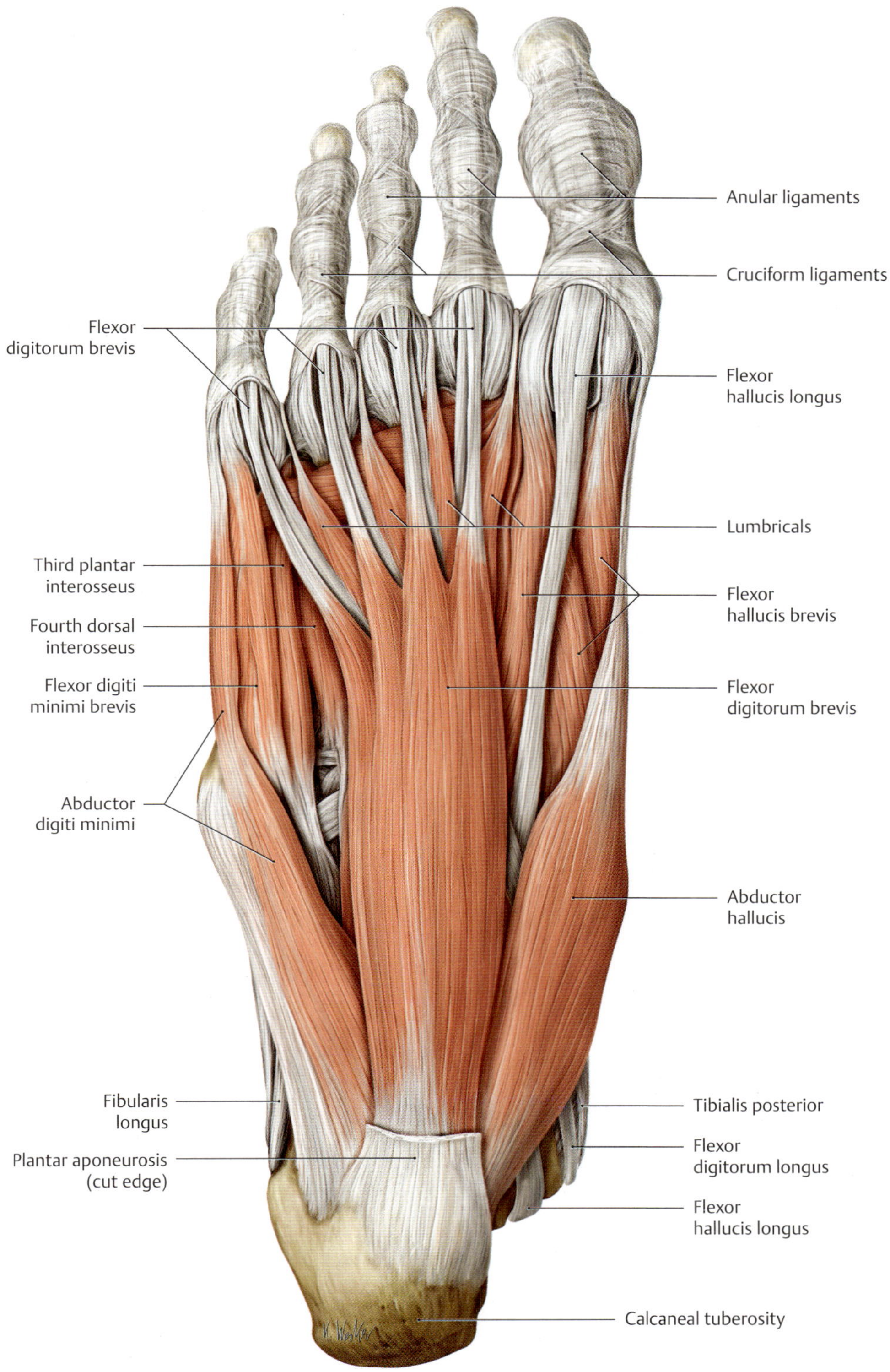

B The short muscles of the right foot
Plantar view. The entire plantar aponeurosis, including the superficial transverse metacarpal ligament, has been removed.

Note the anular ligaments on the plantar side of the toes. Together with the oblique cruciform ligaments, they strengthen the tendon sheaths and help to hold the tendons in position.

3.9 The Intrinsic Foot Muscles from the Plantar View: Middle Layer

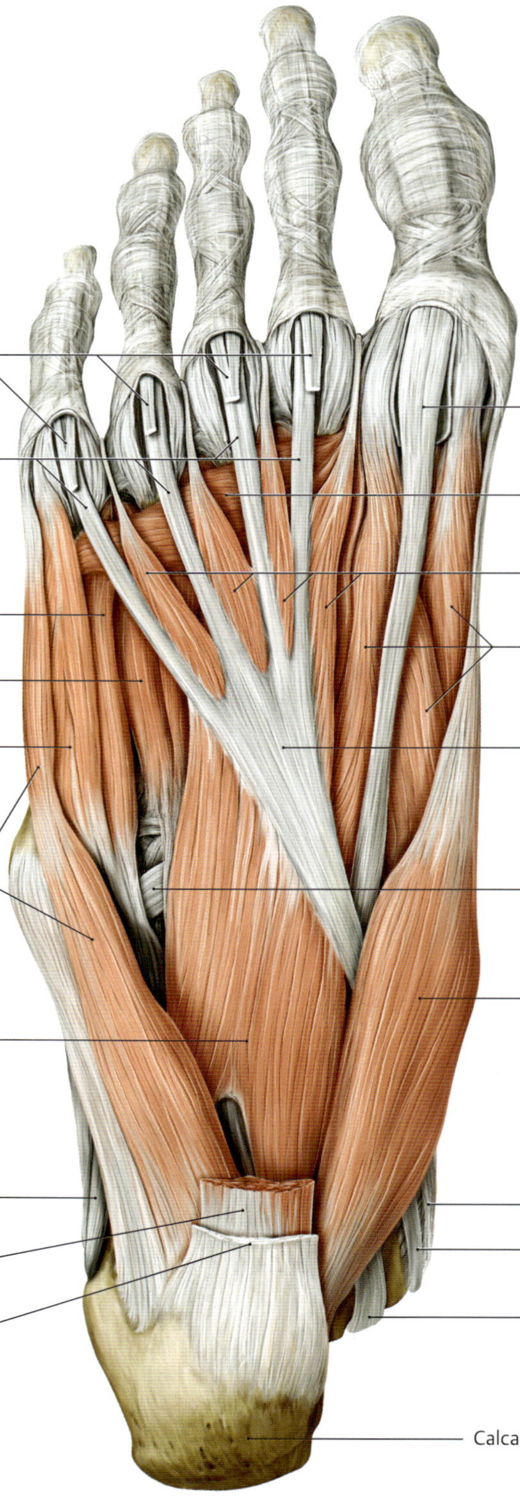

A The short muscles of the right foot
Plantar view. The plantar aponeurosis and flexor digitorum brevis have been removed.

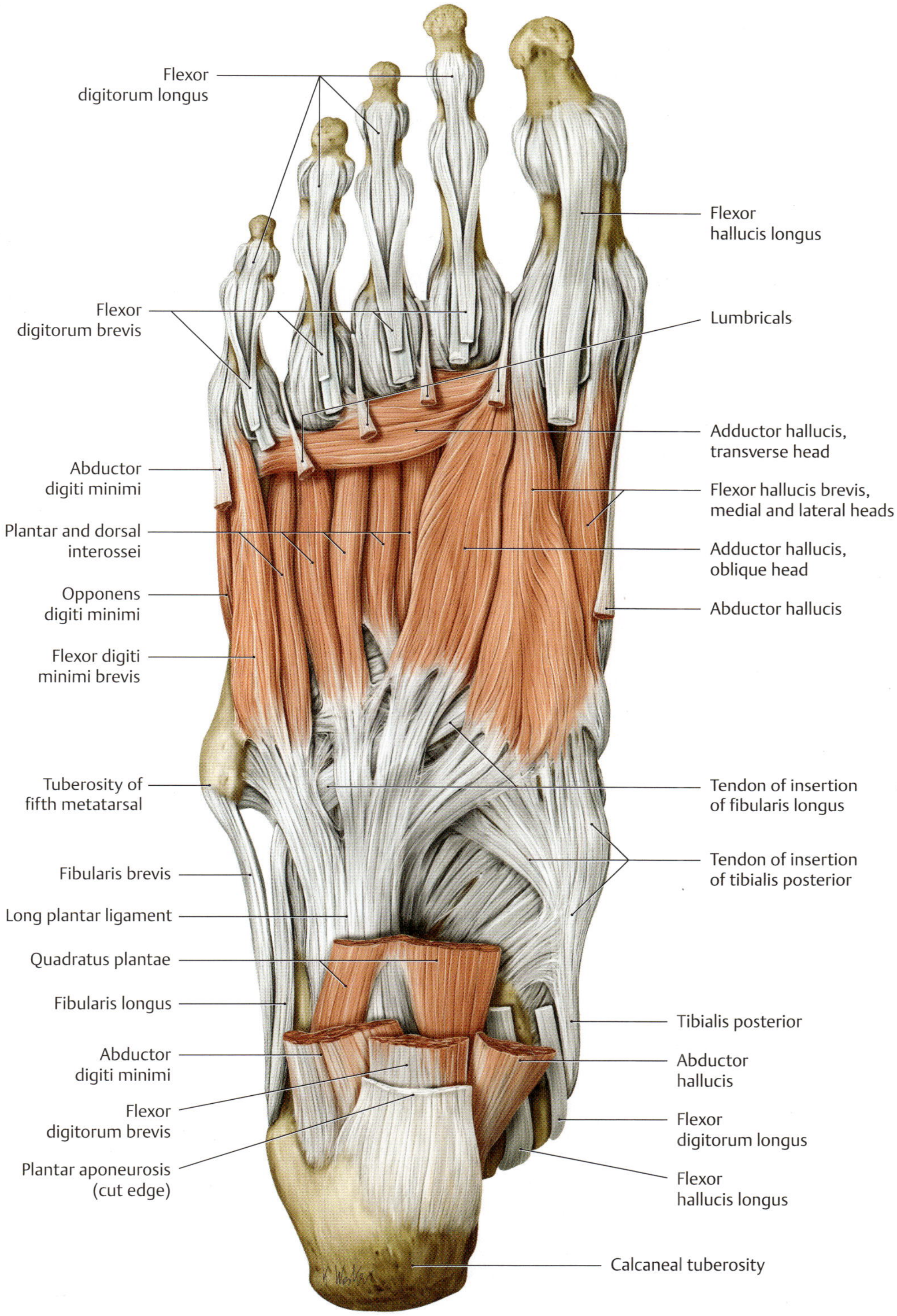

B The short muscles of the right foot
Plantar view. The plantar aponeurosis has been removed in addition to the following muscles: flexor digitorum brevis, abductor digiti minimi, abductor hallucis, quadratus plantae, the lumbricals, and the tendons of insertion of the flexor digitorum longus and flexor hallucis longus.

Note that each of the four tendons of insertion of the flexor digitorum brevis divides into two slips, and that the tendons of the flexor digitorum longus pass between these slips to insert on the distal phalanges.

3.10 The Intrinsic Foot Muscles from the Plantar View: Deep Layer and Origins and Insertions

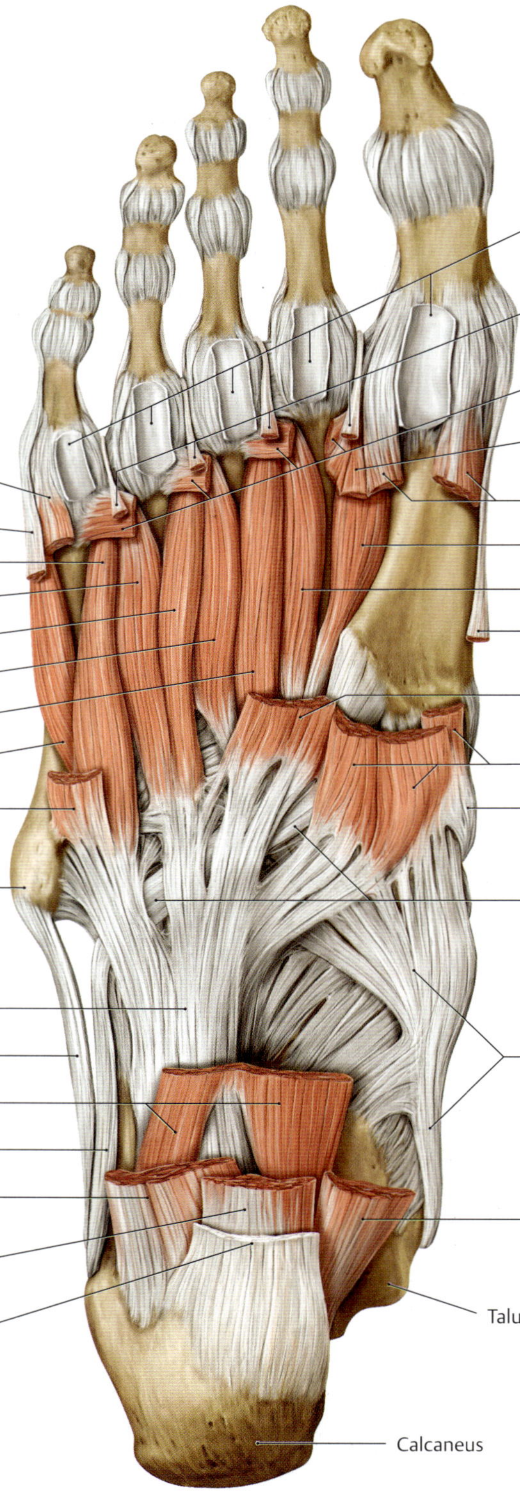

A The short muscles of the right foot
Plantar view. All of the short foot muscles except for the dorsal and plantar interossei have been removed, leaving behind their origins and insertions.

Note the course of the tibialis posterior and fibularis longus tendons of insertion, both of which help to support the transverse arch of the foot.

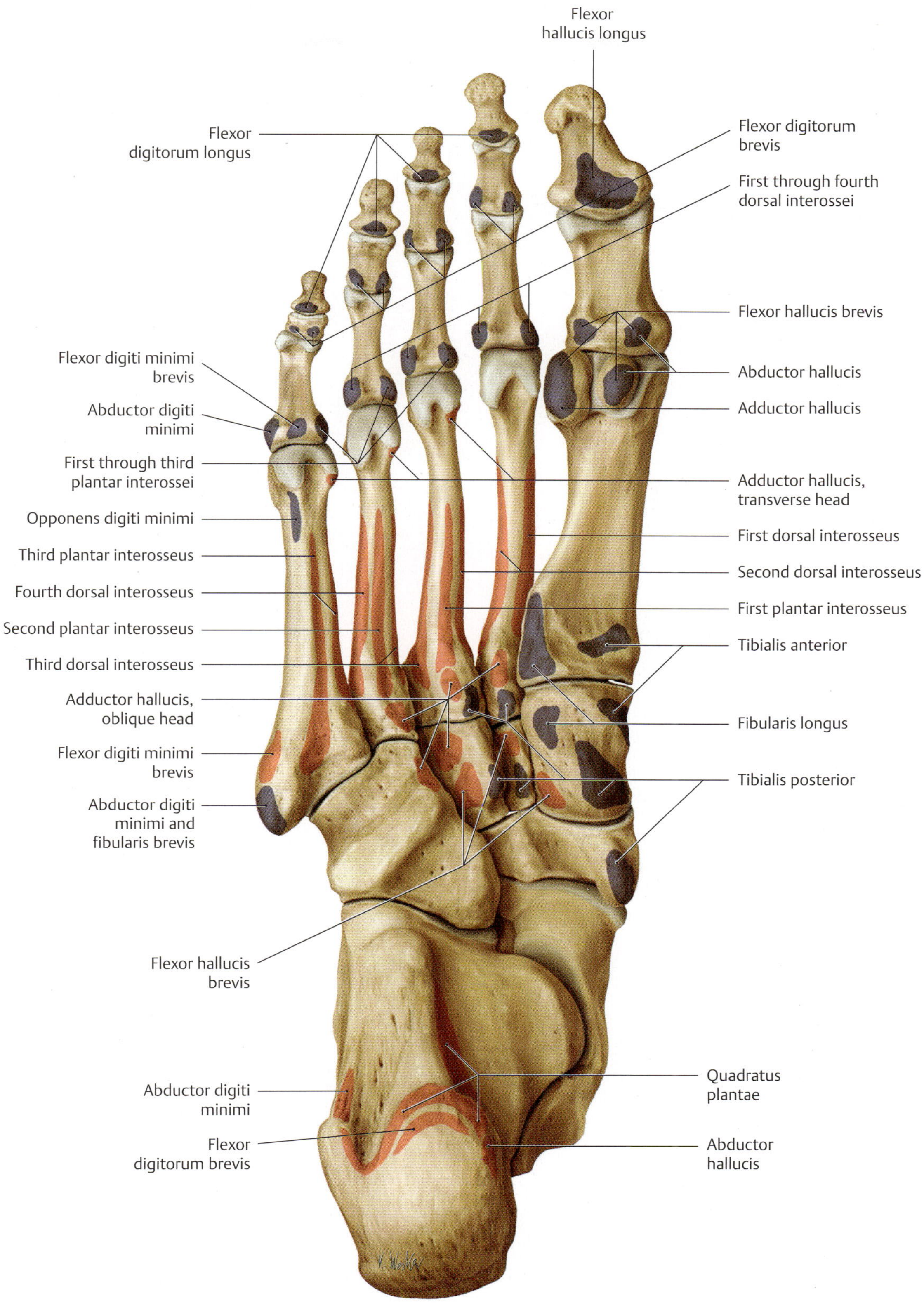

B Muscle origins and insertions of the right foot
Plantar view. Origin shown in red, insertion in blue.

3.11 Cross-sectional Anatomy of the Thigh, Leg, and Foot

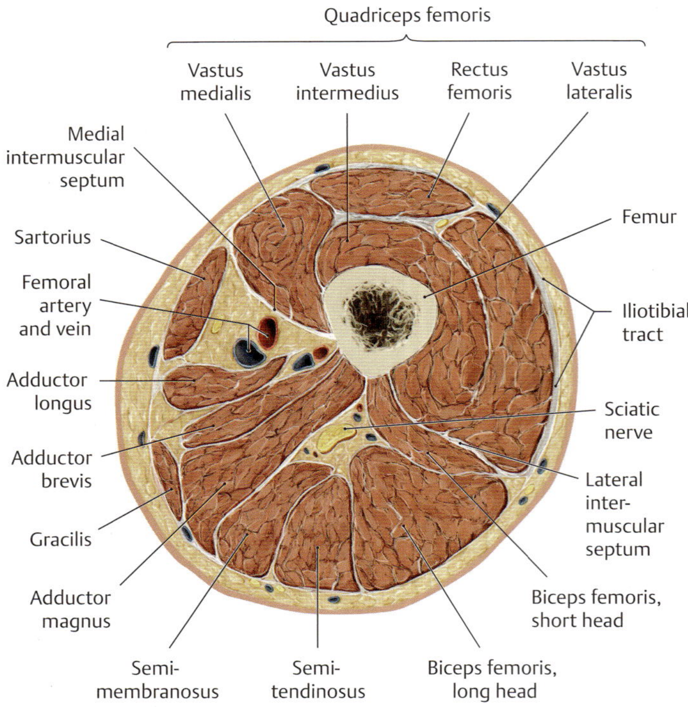

A Cross section through the right thigh
Proximal view. The level of the section is shown in **C**.

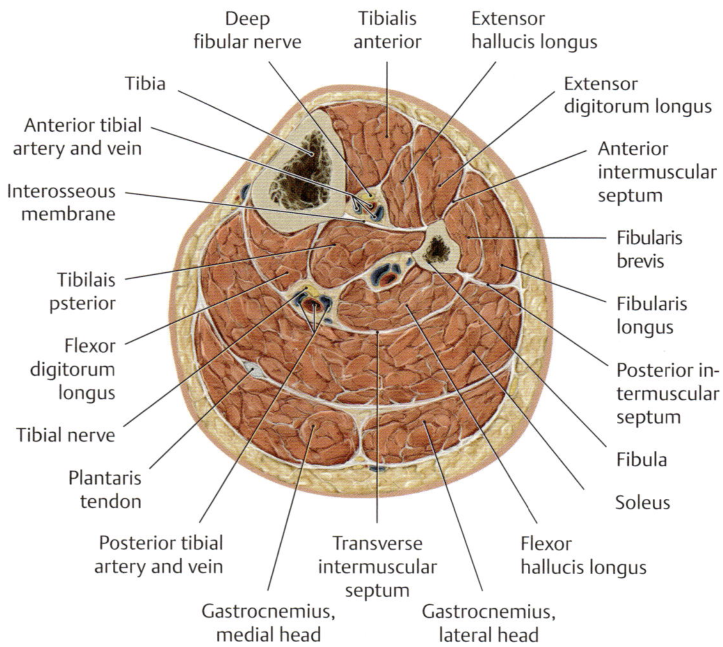

B Cross section through the right leg
Proximal view. The level of the section is shown in **C**.

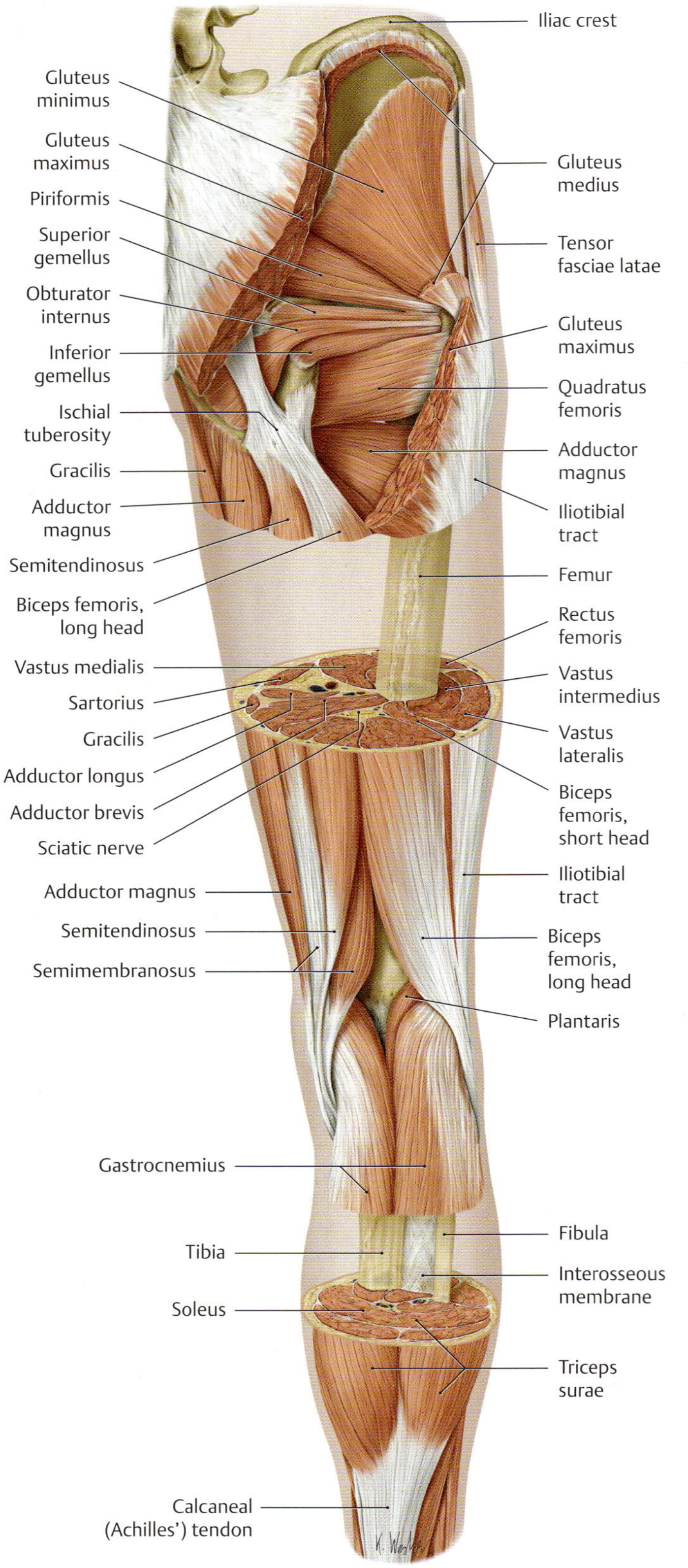

C "Windowed" dissection of the right lower limb
Posterior view. Portions of the gluteus maximus and medius have been removed (the removed cross sections are shown in **A** and **B**). The lower limb is one of the body regions most frequently examined by tomographic methods, and a knowledge of its cross-sectional anatomy is critically important in the identification of landmarks in both radiological and magnetic resonance (MR)-based images.

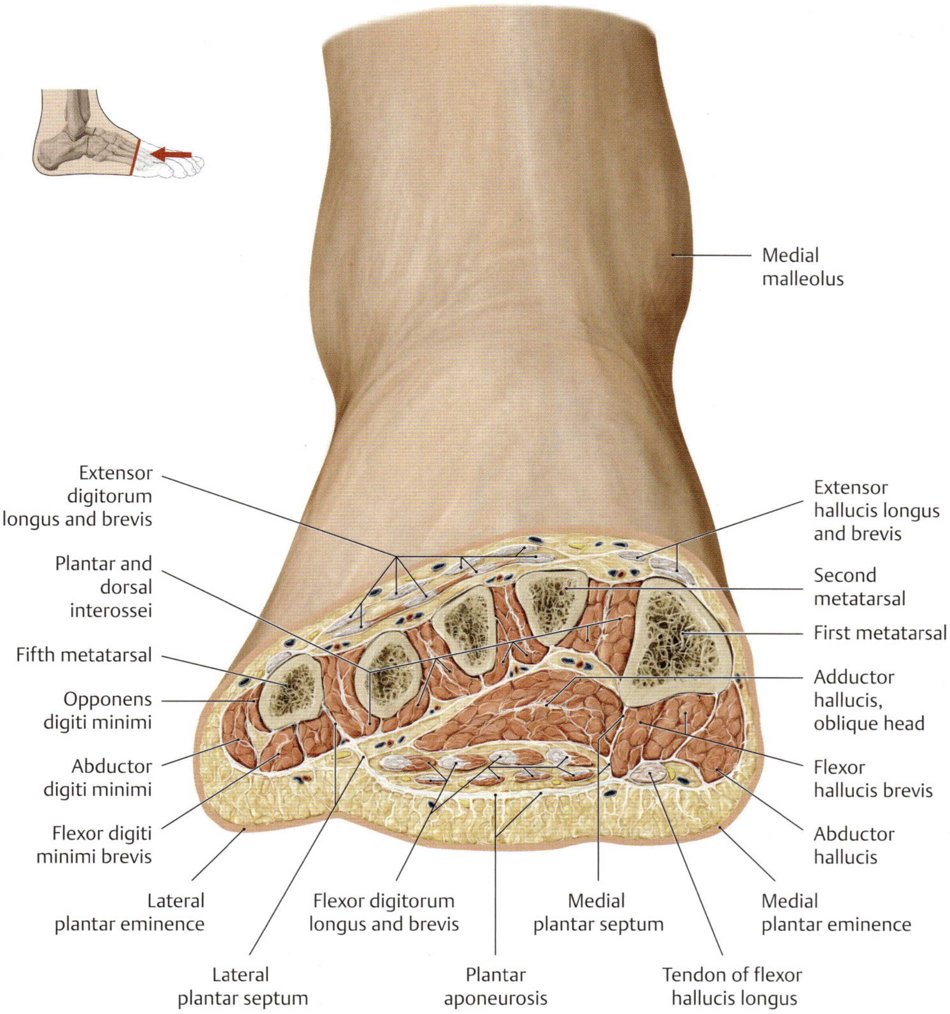

D Cross section through the right foot at the level of the metatarsals

View of the distal cut surface. The muscle compartments of the foot are formed mainly by the plantar aponeurosis, the medial and lateral plantar septa, and the deep plantar fascia (see also **E**). Foot injuries, such as fracture-dislocations of the tarsus and metatarsus, may lead to *compartment syndromes of the foot*. They are caused by increased tissue pressure in the affected compartment due to the local extravasation of blood. The raised pressure in the compartment leads to impaired venous drainage and diminished capillary perfusion, manifested clinically by swelling and pain. This leads in turn to neuromuscular dysfunction with circulatory compromise that may culminate in muscle necrosis (drawn from a specimen in the Anatomical Collection of Kiel University).

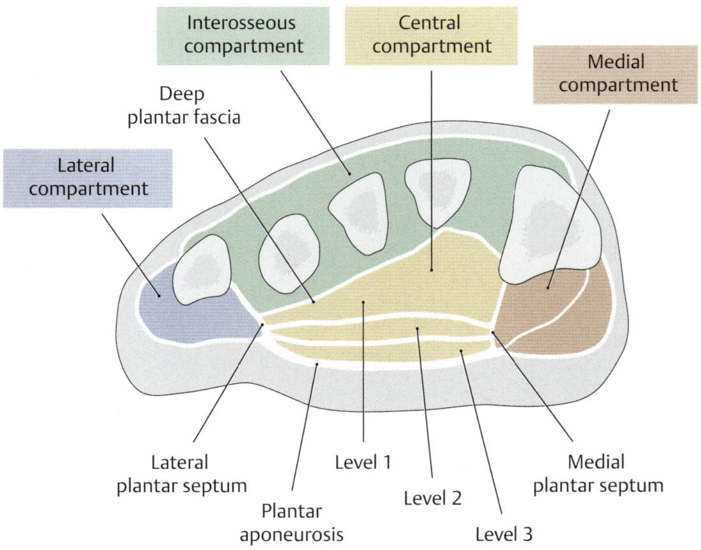

E Location of the compartments of the foot

Schematic cross section through a right foot, distal view. The different muscle compartments are indicated by color shading.

F The four compartments of the foot and their muscular contents (see also **E**)

Interosseous compartment

- Dorsal and plantar interossei

Medial compartment

- Abductor hallucis
- Flexor hallucis brevis
- Tendon of insertion of flexor hallucis longus

Lateral compartment

- Abductor digiti minimi
- Flexor digiti minimi brevis
- Opponens digiti minimi

Central compartment, consisting of three levels

- Level 1: adductor hallucis
- Level 2: quadratus plantae, lumbricals, and flexor digitorum longus tendons
- Level 3: flexor digitorum brevis

(after Mubarak and Hargens)

4.1 The Arteries

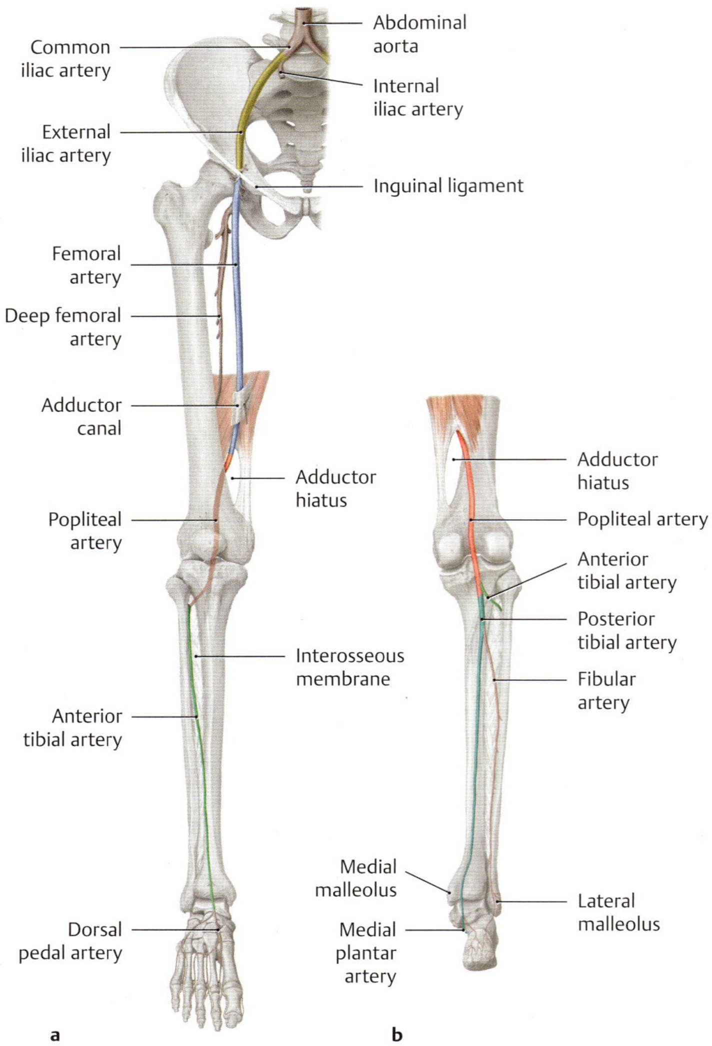

A Different segments of the arteries of the lower limb
a Right leg, anterior view; **b** right leg, posterior view. The different arterial segments are shown in different colors.

External iliac artery arises with the internal iliac artery from the common iliac artery and descends along the medial border of the psoas major through the vascular compartment (see p. 567). It becomes the femoral artery at the level of the inguinal ligament.
Femoral artery, the continuation of the external iliac artery, runs down the medial side of the thigh to the adductor canal, through which it passes from the anterior to the posterior side of the limb. On leaving the adductor hiatus, it becomes the popliteal artery.
Popliteal artery runs from the adductor hiatus through the popliteal fossa to the popliteus, dividing at the inferior border of that muscle into its terminal branches, the anterior and posterior tibial arteries.
Anterior tibial artery enters the extensor compartment of the leg at the upper border of the interosseous membrane and descends between the tibialis anterior and extensor hallucis longus. Distal to the extensor retinaculum, it continues onto the dorsum of the foot as the *dorsal pedal artery*.
Posterior tibial artery, the direct continuation of the popliteal artery, enters the flexor compartment of the leg and passes behind the medial malleolus. At that level it divides into its two terminal branches, the *medial* and *lateral plantar arteries* (the latter is shown in **D**), which continue onto the plantar side of the foot. The posterior tibial artery also gives rise to the fibular artery.

B Overview of the principal arteries of the lower limb
The arteries of the lower limb vary considerably in their origins and branching patterns (the main variants are reviewed in Chapter 5, Neurovascular Systems: Topographical Anatomy). The branches are listed in the order in which they arise from the parent vessels.

Branches of the external iliac artery
- Inferior epigastric artery
  - Cremasteric artery
  - Artery of the round ligament of the uterus
  - Pubic branch
- Deep circumflex iliac artery

Branches of the femoral* artery
- Superficial epigastric artery
- Superficial circumflex iliac artery
- Superficial external pudendal artery
- Deep external pudendal artery
- Deep femoral artery
  - Medial circumflex femoral artery
  - Lateral circumflex femoral artery
  - Perforating branches
- Descending genicular artery

Branches of the popliteal artery
- Posterior tibial recurrent artery (arterial network of the knee)
- Medial and lateral superior genicular arteries (arterial plexus, which runs mainly on the front of the knee)
- Sural arteries
- Middle genicular artery
- Medial and lateral inferior genicular arteries

Note that the paired superior and inferior genicular arteries form the *arterial anastomotic network* around the knee.

Branches of the anterior tibial artery
- Anterior tibial recurrent artery
- Anterior lateral malleolar artery
- Anterior medial malleolar artery
- Dorsal pedal artery
  - Lateral tarsal artery
  - Medial tarsal artery
  - Arcuate artery with the dorsal metatarsal arteries (→ dorsal digital arteries)

Branches of the posterior tibial artery
- Fibular artery
  - Perforating branch
  - Communicating branch
  - Lateral malleolar branches
  - Calcaneal branches
- Medial malleolar branch
- Calcaneal branches
- Medial plantar artery
  - Superficial branch
  - Deep branch (→ deep plantar arch)
- Lateral plantar artery (→ deep plantar arch)
- Plantar metatarsal arteries
- Common plantar digital arteries

* Often referred to clinically as the superficial femoral artery.
→ = is continuous with

Note: In addition, the lower limb is supplied by branches of the internal iliac artery (e.g., the obturator artery).

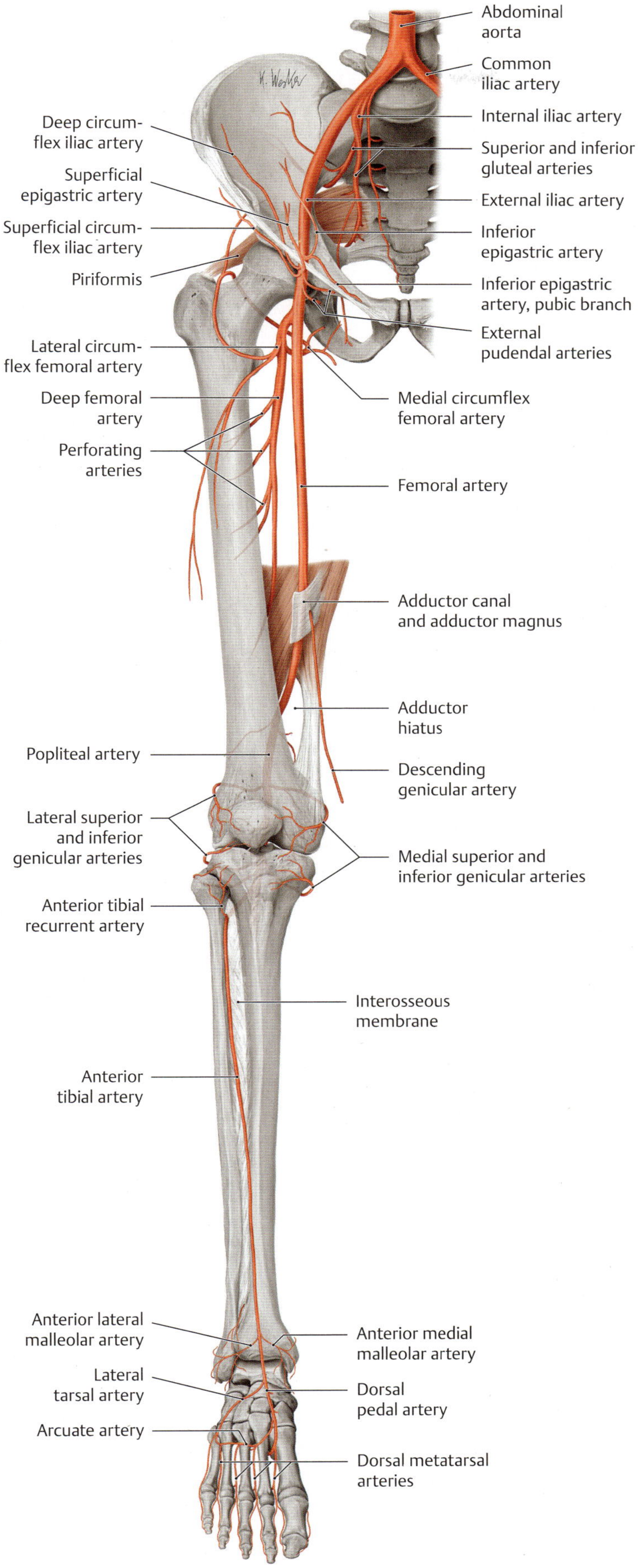

C The arteries of the lower limb
Right leg, anterior view with the foot in plantar flexion.

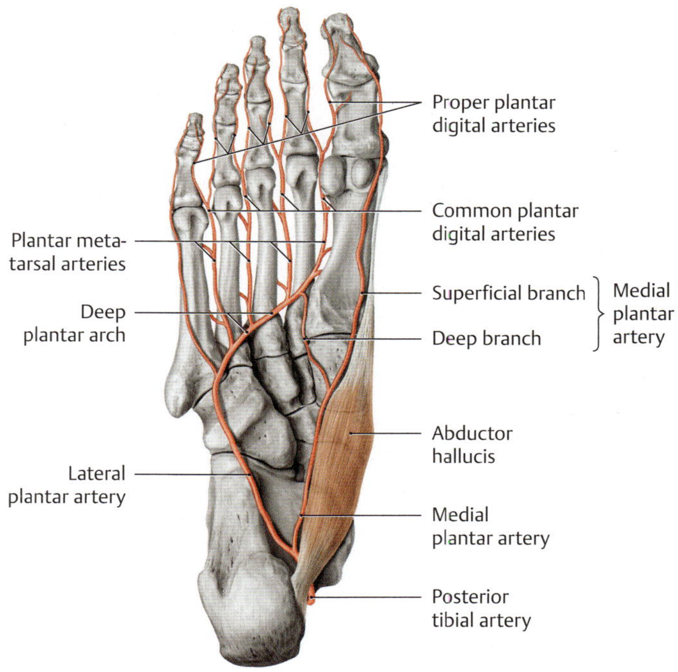

D The arteries of the sole of the foot
Right foot, plantar view.

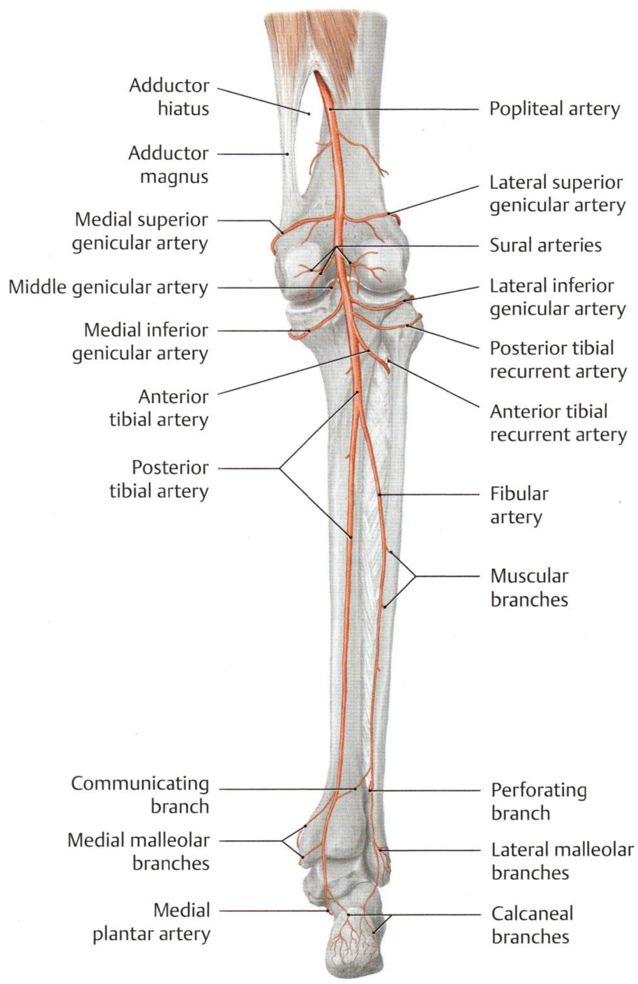

E The arteries of the popliteal fossa and leg
Right leg, posterior view.

4.2 The Veins

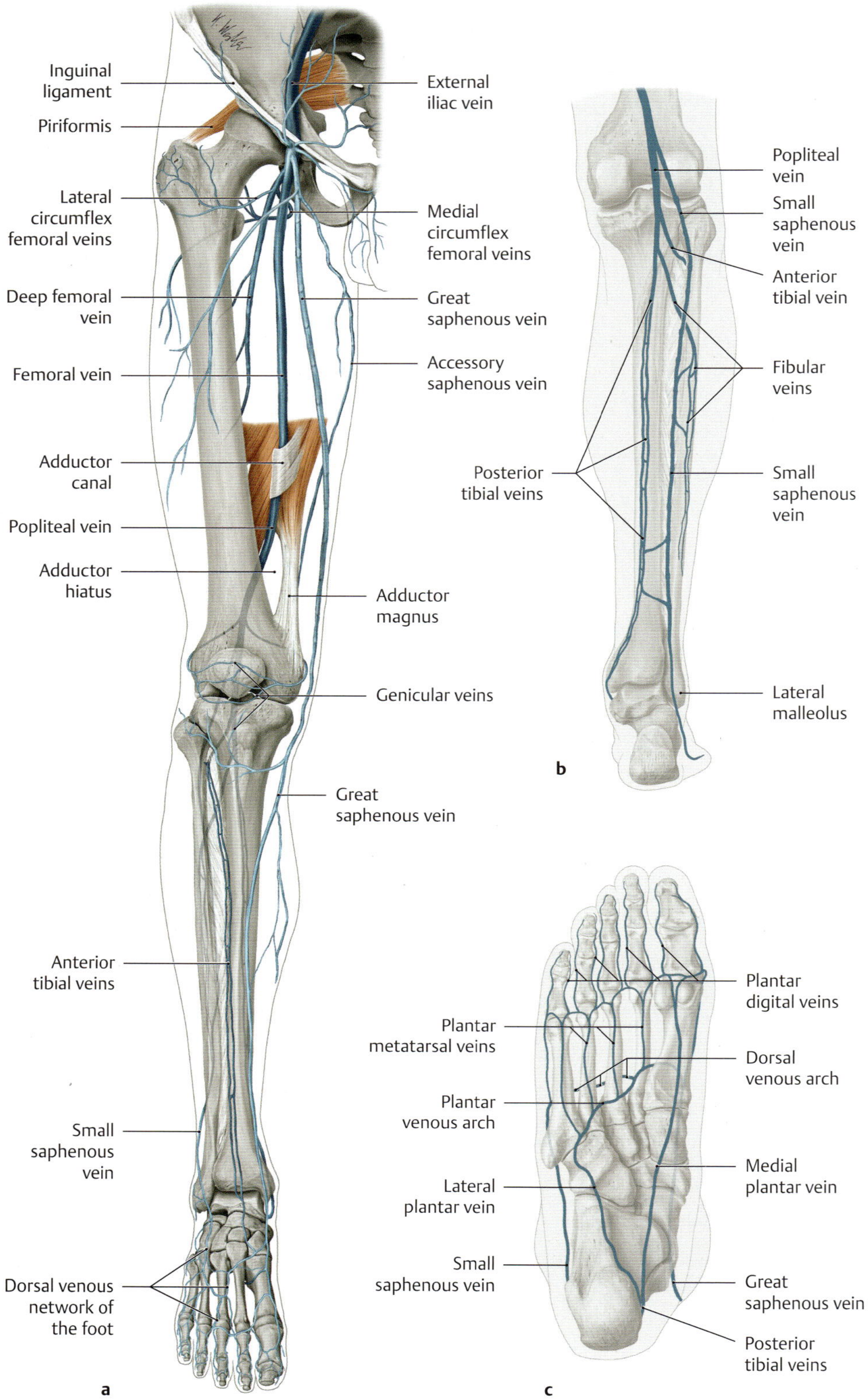

A The deep and superficial veins of the right lower limb

a Thigh, leg, and dorsum of the foot, anterior view.

b Leg, posterior view.

c Sole of the foot, plantar view.

For clarity, only the most important veins are demonstrated here.

B Overview of the principal veins of the lower limb

The veins of the lower limb are subdivided into three systems: a superficial (epifascial) system, a deep (intermuscular) system, and a *perforating* system that interconnects the superficial and deep veins. The upright human body posture places an exceptional load on the veins of the lower limb, which must act against the force of gravity in returning the blood to the heart (the deep venous system handles approximately 85% of the venous return, the superficial veins approximately 15%). A series of venous valves help to maintain the normal superficial-to-deep direction of blood flow (compare **E**). Note that, for the sake of clarity, not all of the veins in the table below have been depicted in these illustrations.

Deep veins of the lower limb

- Femoral vein
- Deep femoral vein
- Medial and lateral circumflex femoral veins
- Popliteal vein
- Sural veins
- Genicular veins
- Anterior and posterior tibial veins
- Fibular veins
- Dorsal and plantar metatarsal veins (see **Ac**)
- Plantar digital veins (see **Ac**)

Superficial lower limb veins

- Great saphenous vein
- External pudendal veins
- Superficial circumflex iliac vein
- Superficial epigastric vein
- Accessory saphenous vein
- Posterior arch vein
- Small saphenous vein (see **Cb**)
- Femoropopliteal vein (see **Cb**)
- Dorsal venous network (see **Ca**)
- Dorsal venous arch
- Plantar venous network
- Plantar venous arch

Perforating veins

Of the many perforating veins in the leg, three groups have the greatest clinical importance (see **E**):

- The Dodd group (medial side of the thigh, middle third)
- The Boyd group (medial side of the leg below the knee)
- The Cockett group (medial side of the distal leg)

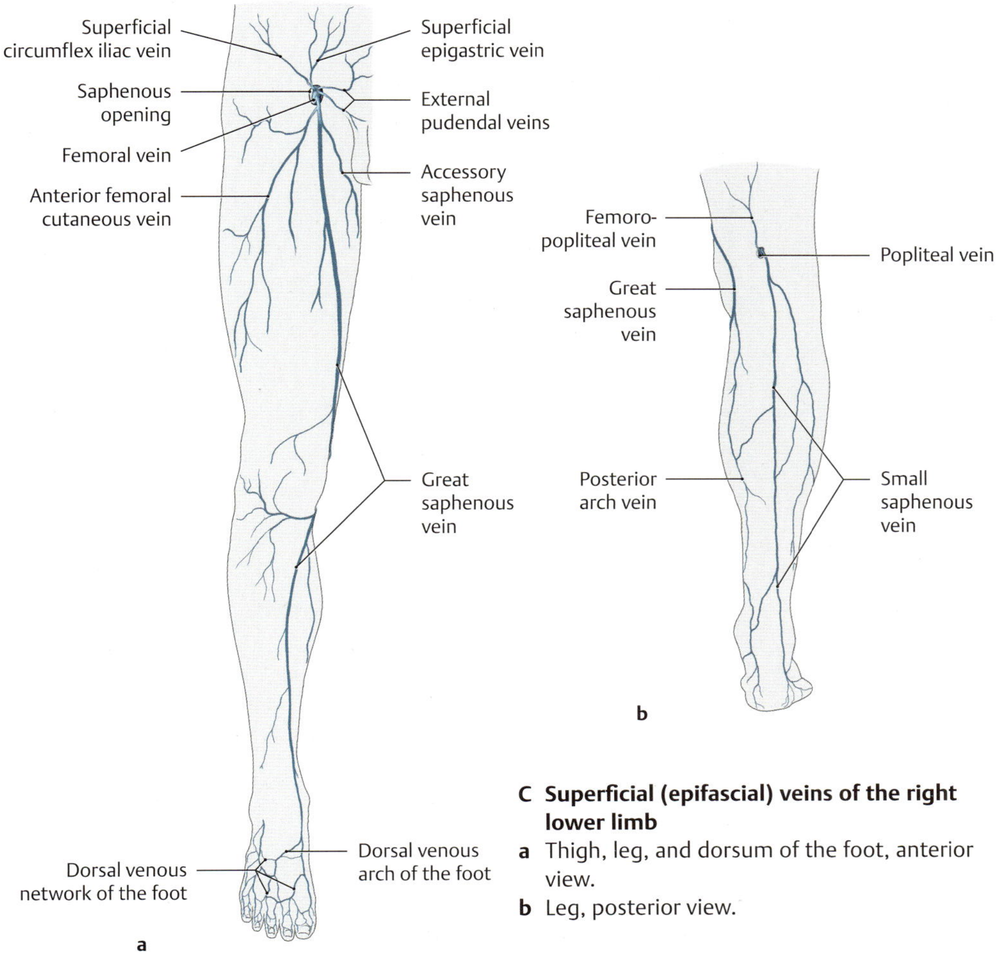

C Superficial (epifascial) veins of the right lower limb

a Thigh, leg, and dorsum of the foot, anterior view.

b Leg, posterior view.

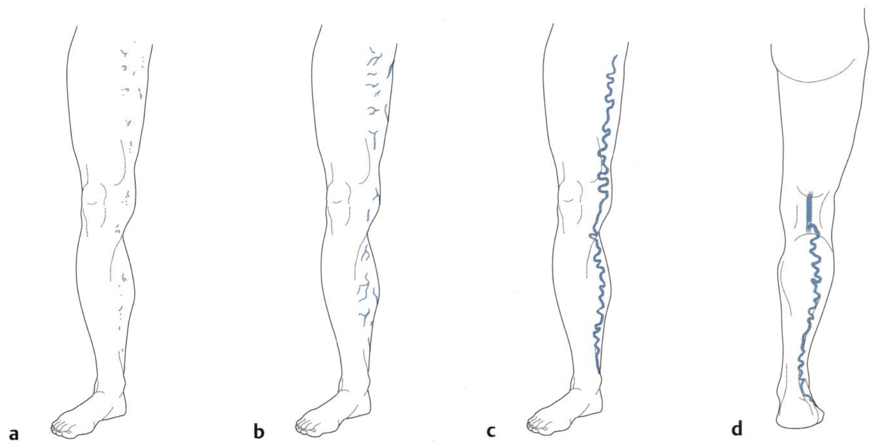

D Varices of the superficial leg veins

a Spider veins (tiny intradermal varices).

b Reticular varices (weblike dilations of small subcutaneous veins).

c Great saphenous varicosity.

d Small saphenous varicosity.

Varicose disease of the *superficial* leg veins is the most common *chronic* venous disease, affecting 15% of the adult population. Varicose veins can be classified as primary idiopathic varices (75%) or as secondary symptomatic varices. **Primary varices** generally result from degeneration of the vein wall leading to incompetence of the venous valves. **Secondary varices** result from chronic occlusion of the *deep* venous system with incompetence of the perforator veins and a reversal in the direction of venous flow. Besides chronic conditions, there are also important *acute* diseases that may affect the superficial venous system (e.g., thrombophlebitis) and deep venous system (e.g., venous thrombosis).

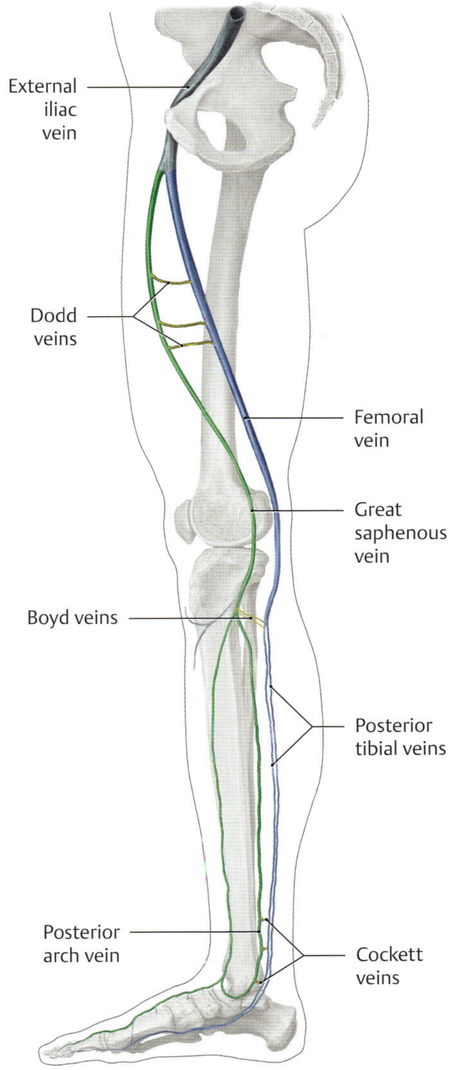

E Clinically important perforating veins

Right leg, medial view. Numerous perforating veins interconnect the deep and superficial venous systems of the leg. Their venous valves normally prevent blood flow from the deep veins to the superficial cutaneous veins. The clinically important members of this system are located between the deep veins and the tributary region of the great saphenous vein:

- **Dodd veins:** located between the great saphenous vein and femoral vein at the level of the adductor canal.
- **Boyd veins:** located between the great saphenous vein and posterior tibial veins on the medial side of the proximal leg.
- **Cockett veins (I–III):** located between a curved branch of the great saphenous vein behind the medial malleolus (the posterior arch vein) and the posterior tibial veins. The Cockett veins on the medial side of the distal leg are of special clinical importance because of this region's susceptibility to ulceration.

4.3 The Lymphatic Vessels and Lymph Nodes

A The superficial lymphatic system of the right lower limb

a Anterior view, **b** posterior view. (The arrows indicate the main directions of lymphatic drainage.)

The lymph in the lower limb is drained by a superficial (epifascial) system and a deep (subfascial) system, similar to the arrangement in the arm. The largest lymph vessels, called collectors, basically follow the course of the superficial veins (great and small saphenous veins) and deep veins (popliteal vein, femoral vein) and are interconnected by anastomoses located mostly in the popliteal and inguinal regions. While the superficial lymph vessels primarily drain the skin and subcutaneous tissue, the deep system drains lymph from the muscles, joints, and nerves. The superficial lymphatics consist of an anteromedial bundle and a posterolateral bundle. The **anteromedial bundle** runs along the great saphenous vein to the superficial *inguinal* lymph nodes. It drains all of the skin and subcutaneous tissue of the lower limb except for the lateral border of the foot and a narrow strip on the calf. Those areas are drained by the **posterolateral bundle** (see **b**), which thus receives drainage from a considerably smaller region. The lymph in the posterolateral bundle first passes along the small saphenous vein to the superficial *popliteal* lymph nodes and then drains through the deep popliteal lymph nodes to the deep inguinal lymph nodes.

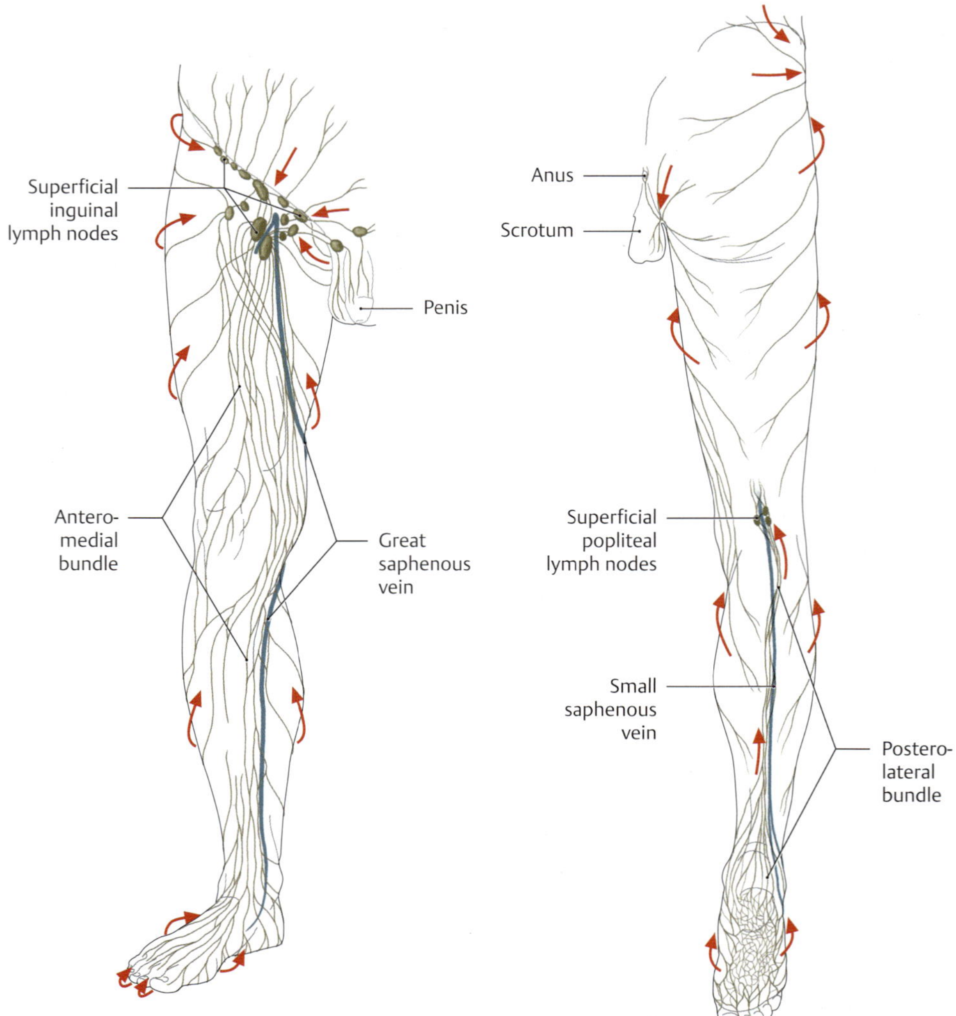

B The deep lymph nodes of the inguinal region

Right inguinal region after removal of the cribriform fascia about the saphenous opening, anterior view. The veins and lymphatic system above the inguinal ligament are shown in light shading. The deep inguinal lymph nodes are located near the termination of the great saphenous vein, medial to the femoral vein. They are important because *all the lymph from the limb* filters through them before reaching the iliac lymph nodes. The largest lymph node of this group (*Rosenmüller's lymph node*) is also the highest, placed at the level of the femoral canal. The group of pelvic lymph nodes, which includes the external iliac nodes, begins just above the inguinal ligament.

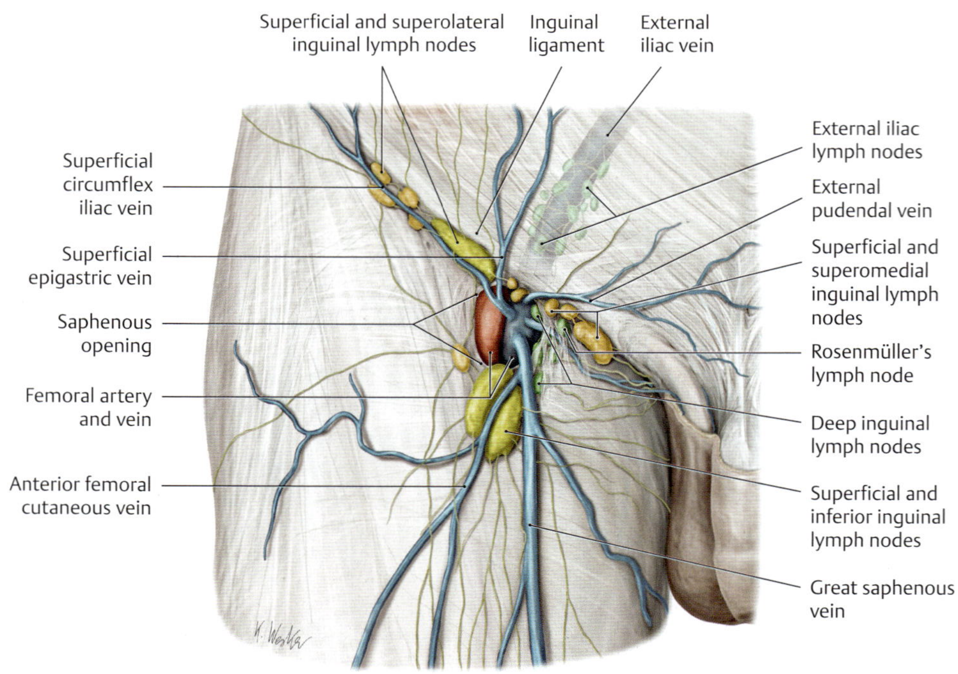

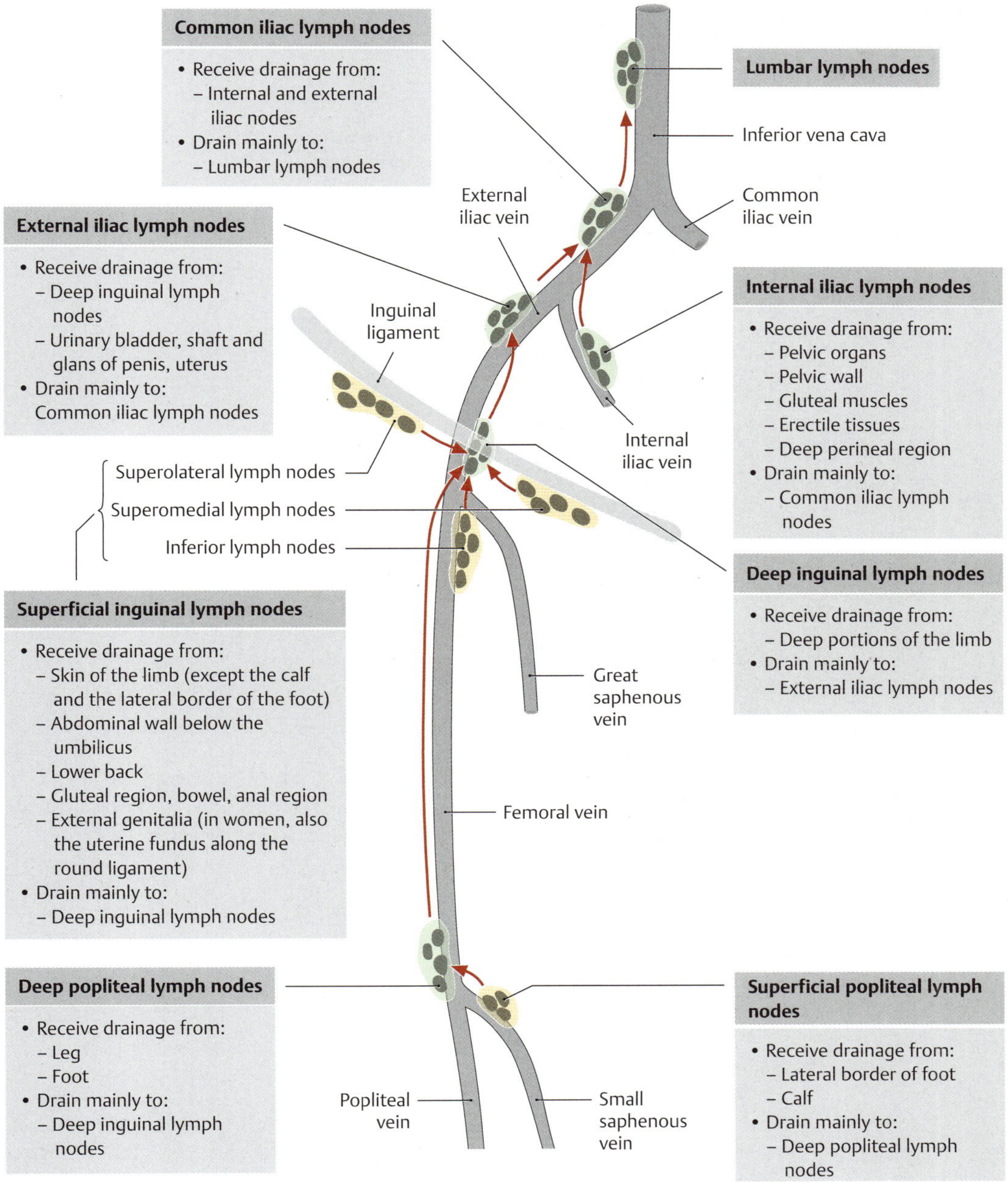

C The lymph node stations and drainage pathways in the lower limb

Right limb, anterior view. The arrows indicate the main directions of lymph flow in the superficial and deep lymphatic systems.

Note: Lymph from the skin and subcutaneous tissues of the calf and the lateral border of the foot passes through the superficial and deep popliteal lymph nodes along the deep system of lymphatics *directly* to the deep inguinal lymph nodes. By contrast, lymph from the rest of the skin of the lower limb first drains through the anteromedial bundle along the great saphenous vein to the superficial inguinal lymph nodes (see also **A**).

The *superficial inguinal lymph nodes*, located on the fascia lata, consist of the following:

- Lymph nodes arranged parallel to the inguinal ligament (the superomedial and superolateral inguinal lymph nodes)
- Lymph nodes distributed vertically along the terminal segment of the great saphenous vein (the inferior inguinal lymph nodes)

These nodes first drain into the deep inguinal lymph nodes (see **B**) and then along the external iliac vein to the external and common iliac lymph nodes, finally reaching the lumbar lymph nodes.

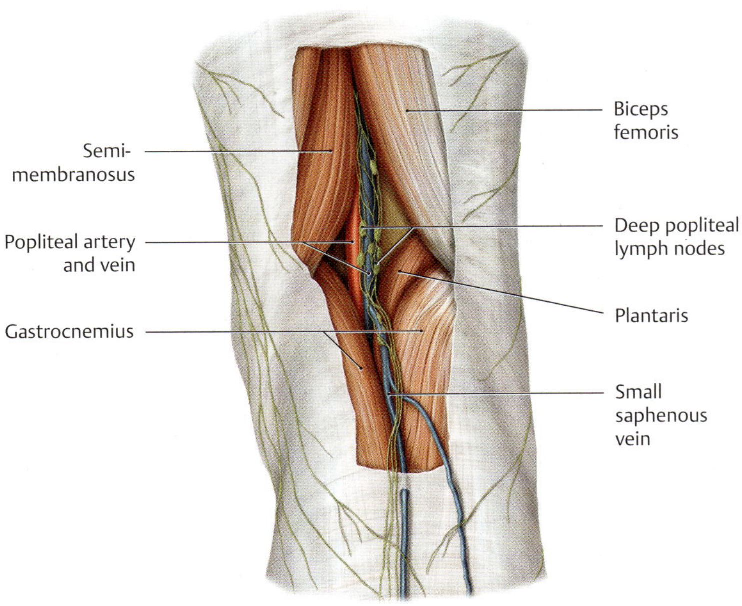

D The deep lymph nodes of the popliteal region

The popliteal fossa of the right leg, posterior view. Lymph from the deep lymphatics of the leg drains (through the deep popliteal nodes between the posterior knee joint capsule and popliteal vessels) along the femoral vein and then anteriorly through the adductor hiatus to the deep inguinal lymph nodes.

4.4 The Structure of the Lumbosacral Plexus

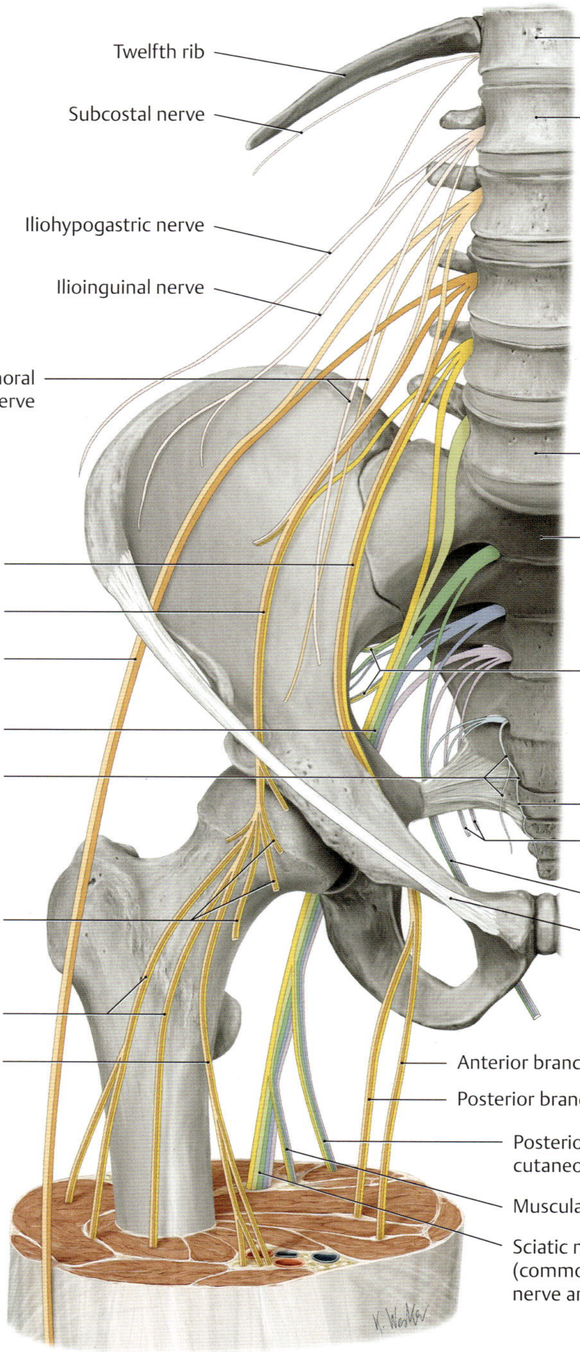

A The lumbosacral plexus and its branches

Right side, anterior view. For clarity, the muscles of the pelvis and lumbar spine have been removed. Lateral to the intervertebral foramina of the lumbar spine, the *ventral rami* of the first four *lumbar* spinal nerves (L1–L4) form the lumbar plexus and pass through the psoas major muscle. The smaller, muscular branches are distributed directly to the psoas major. The larger branches emerge from the muscle at various sites and pass sharply downward to reach the abdominal wall and thigh, except for the obturator nerve, which runs down the lateral wall of the lesser pelvis to the thigh. The ventral rami of the first four sacral spinal nerves (S1–S4) emerge from the anterior foramina of the sacrum and unite on the anterior surface of the piriformis with the anterior ramus of L5 to form the sacral plexus. The nerves from the sacral plexus are distributed to the back of the thigh, the leg, and the foot.

B Spinal cord segments and nerves of the lumbosacral plexus

The lumbosacral plexus supplies sensory and motor innervation to the lower limb. It is formed by the ventral rami of the lumbar and sacral spinal nerves, with contributions from the subcostal nerve (T12) and coccygeal nerve (Co1) (see **D**). The lumbosacral plexus is subdivided into the lumbar plexus and sacral plexus based on its distribution and topography.

Lumbar plexus (T12–L4)

- Iliohypogastric nerve (L1)
- Ilioinguinal nerve (L1)
- Genitofemoral nerve (L1, L2)
- Lateral femoral cutaneous nerve (L2, L3)
- Obturator nerve (L2–L4)
- Femoral nerve (L2–L4)
- Short, direct muscular branches to specific hip muscles

Sacral plexus (L4–S4)*

- Superior gluteal nerve (L4–S1)
- Inferior gluteal nerve (L5–S2)
- Posterior femoral cutaneous nerve (S1–S3)
- Sciatic nerve (L4–S3) with its two large branches:
  - Tibial nerve (L4–S3)
  - Common fibular nerve (L4–S2)
- Pudendal nerve (S2–S4)
- Short, direct muscular branches to specific hip muscles

* Often the sacral plexus is further subdivided into a sciatic plexus and a pudendal plexus. The main branch of the pudendal plexus, the pudendal nerve, supplies the skin and muscles of the pelvic floor, perineum, and external genitalia.

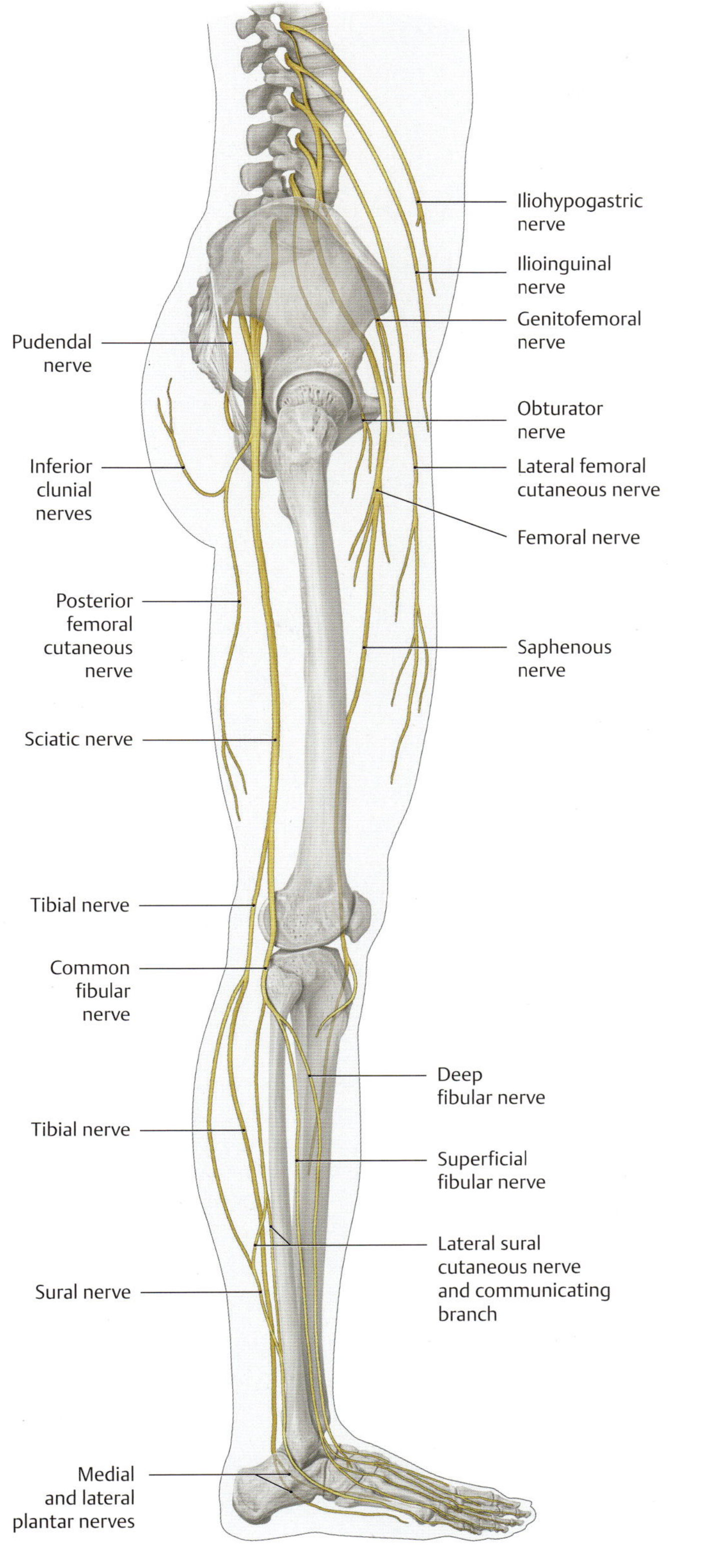

C Topography of the lumbosacral plexus

Right lower limb, lateral view. The nerves of the lumbar plexus reach the lower limb *in front of* the hip joint and mainly supply the *anterior side of the thigh*, while the nerves of the sacral plexus descend *behind* the hip joint and innervate the *posterior side of the thigh*, most of the *leg*, and the entire *foot*.

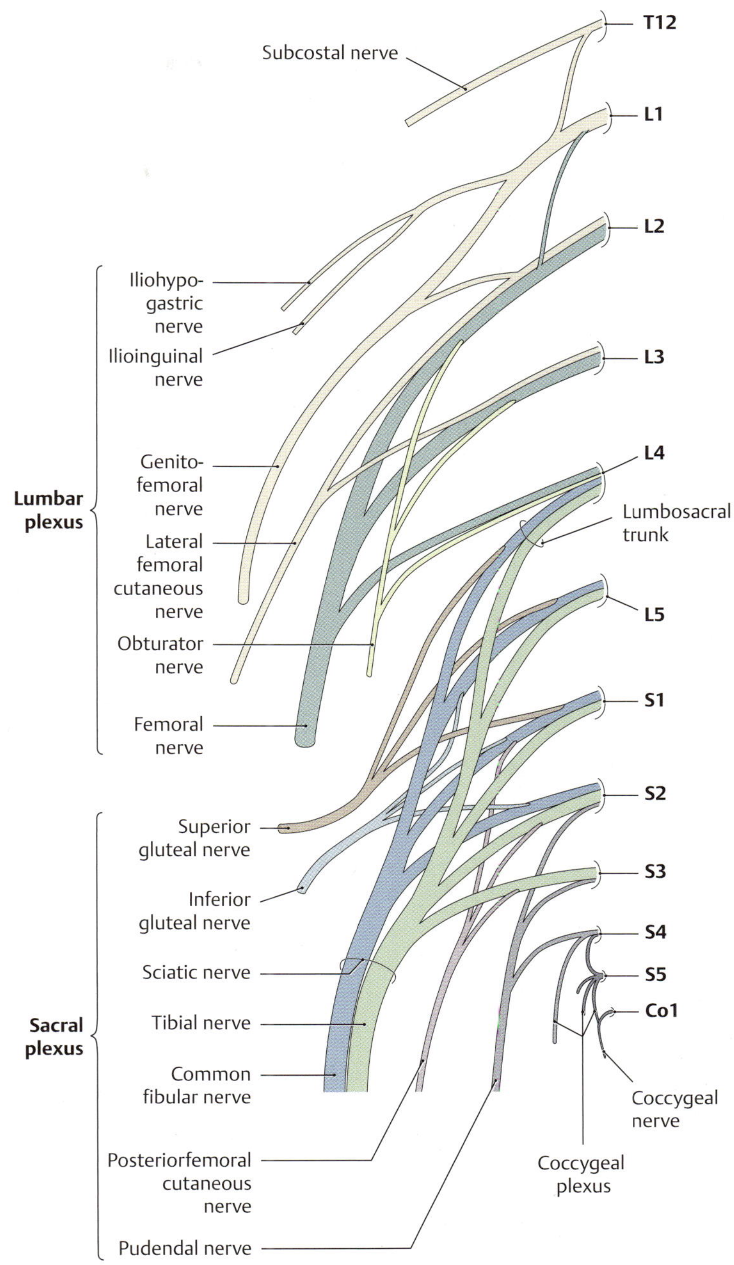

D Structure of the lumbosacral plexus

The lumbar nerves combine to form the lumbar plexus, while the sacral nerves form the sacral plexus. A portion of the L4 anterior ramus and all of the L5 anterior ramus combine to form the lumbosacral trunk. This trunk combines with the sacral plexus to form the sciatic nerve. The last spinal nerve, the coccygeal nerve, emerges from the sacral hiatus. It unites with the anterior rami of S4 and S5 to form the coccygeal plexus (see p. 560).

4.5 The Nerves of the Lumbar Plexus: The Iliohypogastric, Ilioinguinal, Genitofemoral, and Lateral Femoral Cutaneous Nerves

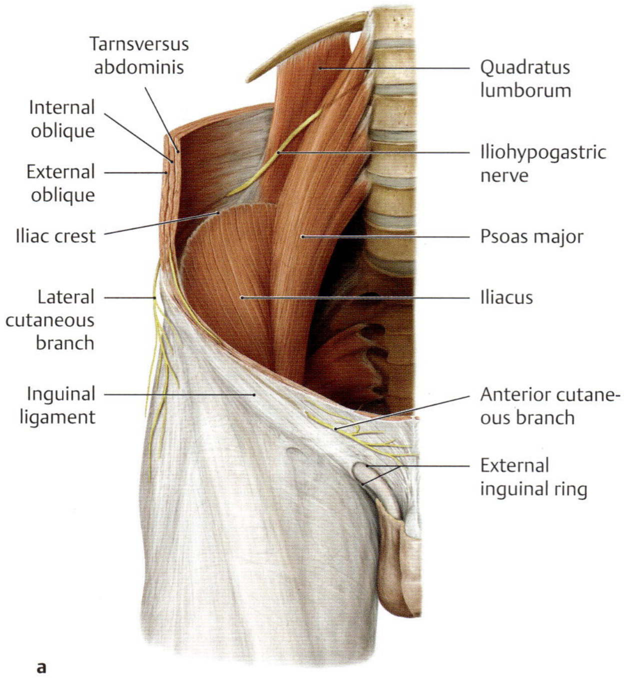

a

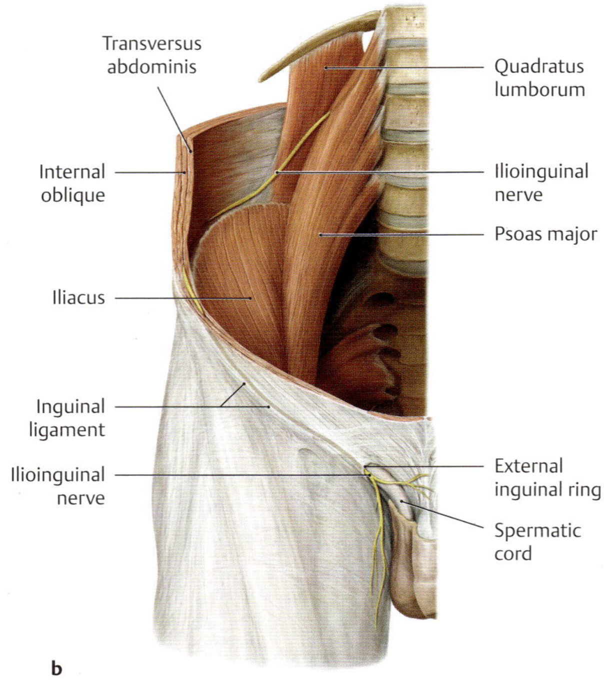

b

A Course of the iliohypogastric, ilioinguinal, genitofemoral, and lateral femoral cutaneous nerves after emerging from the lumbar plexus

Right lateral and posterior abdominal wall region, anterior view.

a The **iliohypogastric nerve** generally emerges with the ilioinguinal nerve (see **b**) at the lateral border of the psoas major and runs laterally and obliquely on the anterior surface of the quadratus lumborum. Approximately 3 to 4 cm past the lateral border of that muscle, it pierces the transversus abdominis and runs anteriorly above the iliac crest, passing between the transversus abdominis and internal oblique. After giving off several muscular branches to both of these muscles and a sensory lateral cutaneous branch to the skin of the lateral hip region, the terminal branch of the iliohypogastric nerve courses medially, running parallel to the inguinal ligament. Above the external inguinal ring it pierces the aponeurosis of the external oblique muscle and supplies a sensory anterior cutaneous branch to the skin above the inguinal ligament.

b The ilioinguinal nerve generally courses with the iliohypogastric nerve (see **a**) on the quadratus lumborum but soon separates from it and runs at the level of the iliac crest to the lateral abdominal wall, which it pierces at a variable location. It runs medially at the level of the inguinal ligament between the transversus abdominis and internal oblique, supplying twigs to both muscles, and it distributes sensory fibers through the external inguinal ring to the skin over the symphysis and to the lateral portion of the labia majora or scrotum.

B Overview of the nerves of the lumbar plexus

Nerve	Segment	Innervated muscles	**Cutaneous branches** (to the region receiving sensory innervation, see **C** and pp. 546 and 547)
• Iliohypogastric nerve	L1	• Transversus abdominis, internal oblique (the inferior portions of each)	• Anterior cutaneous branch • Lateral cutaneous branch
• Ilioinguinal nerve	L1	• Transversus abdominis, internal oblique (the inferior portions of each)	• Anterior scrotal nerves in males, anterior labial nerves in females
• Genitofemoral nerve	L1–L2	• Cremaster in males (genital branch)	• Genital branch, femoral branch
• Lateral femoral cutaneous nerve	L2–L3		• Lateral femoral cutaneous nerve
• Obturator nerve (see p. 552) – Anterior branch – Posterior branch	L2–L4	• Obturator externus • Adductor longus, adductor brevis, gracilis, pectineus • Adductor magnus	• Cutaneous branch
• Femoral nerve (see p. 553)	L2–L4	• Iliopsoas, pectineus, sartorius, quadriceps femoris	• Anterior cutaneous branches, saphenous nerve
• Short, direct muscular branches (see p. 552)	T12–L4	• Psoas major, quadratus lumborum, iliacus, intertransversarii lumborum	

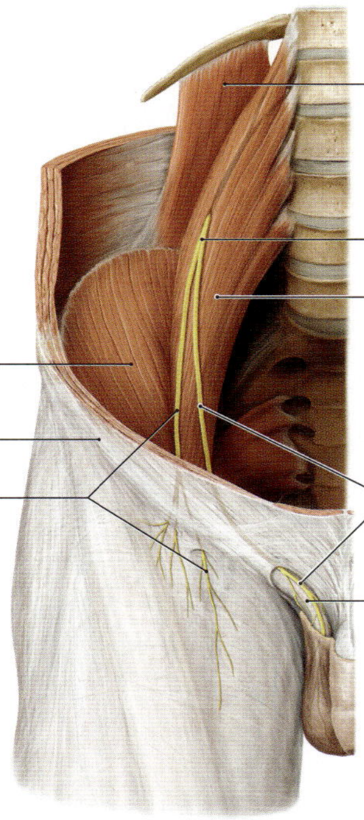

c

c The **genitofemoral nerve** pierces the psoas major and descends upon its anterior surface, dividing into its two terminal branches: the genital branch and femoral branch:

- The *purely sensory femoral branch* pierces the vascular compartment in the area of the saphenous opening (see p. 567) and becomes superficial, supplying the skin below the inguinal ligament in both sexes.
- The *mixed genital branch* runs in the spermatic cord in males. In females it initially passes through the inguinal canal accompanied by the round ligament of the uterus. In its further course it distributes sensory fibers to the scrotal skin in males and to the skin of the labia majora in females. It also supplies motor fibers to the cremaster muscle in males (see p. 184).

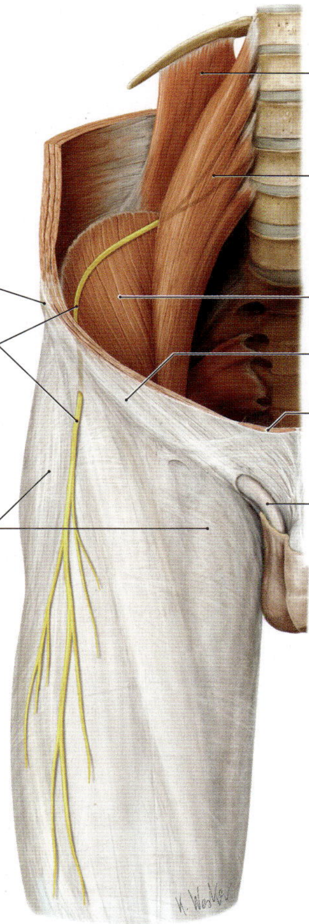

d

d The **lateral femoral cutaneous nerve** emerges from the lateral border of the psoas major and runs obliquely downward and laterally beneath the iliacus fascia toward the anterior superior iliac spine. Medial to the iliac spine, the nerve leaves the pelvis through the lateral muscular compartment (see p. 567) and first runs beneath the fascia lata and then upon it to the skin of the anterior thigh, piercing the fascia approximately 2 to 3 cm below the anterior superior iliac spine. The nerve is susceptible to occasional mechanical injury at its site of emergence from the pelvis below the inguinal ligament, as it makes an approximately 80° angle at that site and is susceptible to stretching, especially on extension of the hip. The nerve also has only scant coverage by fatty tissue at that location. Stretch injuries are manifested by sensory disturbances (paresthesias) or pain in the *lateral* part of the thigh.

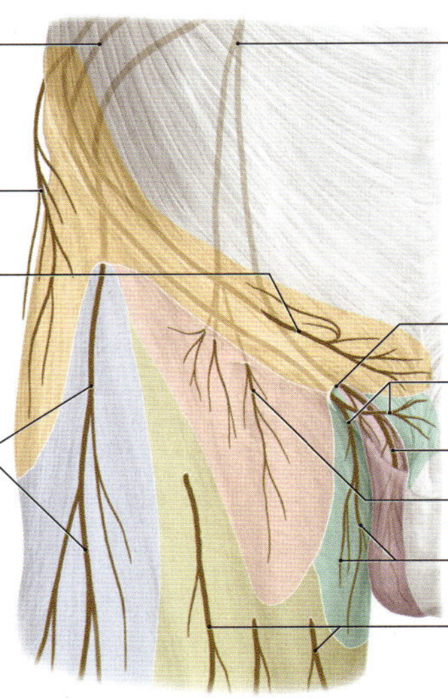

C Sensory innervation of the inguinal region and thigh

Right inguinal region of a male, anterior view. The territories of the various sensory nerves are indicated by different colors.

Note: Both the ilioinguinal nerve and the genital branch of the genitofemoral nerve pass through the external inguinal ring. The two nerves are frequently confused. The genital branch in males is located by first opening the spermatic cord. In females, the genital branch accompanies the uterine round ligament to supply the skin of the labia majora (see also **Ac**).

4.6 The Nerves of the Lumbar Plexus: The Obturator and Femoral Nerves

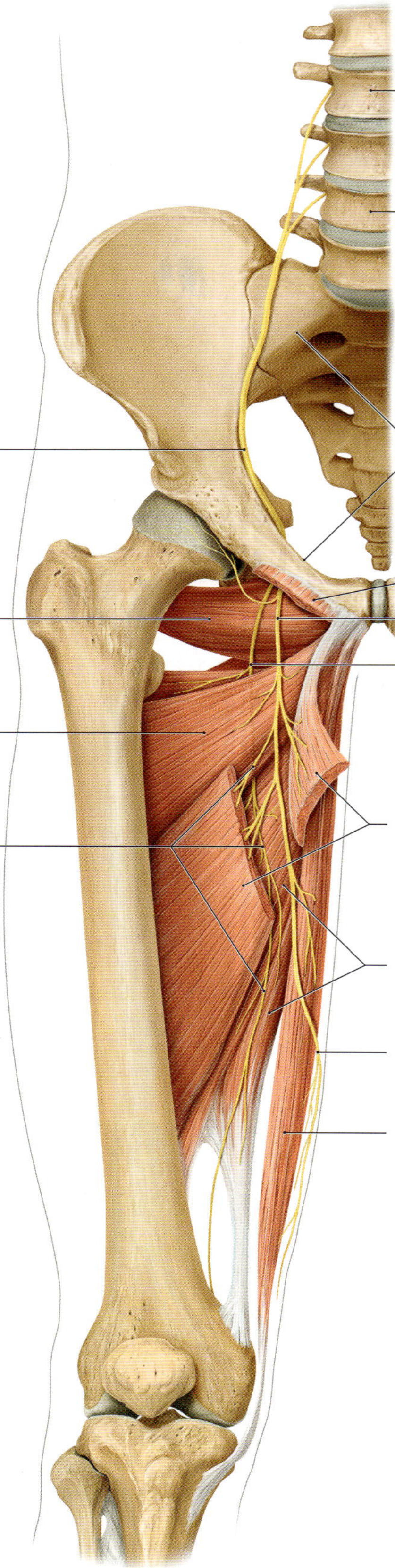

B Sensory distribution of the obturator nerve
Right leg, medial view.

A Course of the obturator nerve
The right inguinal region and thigh, anterior view. The obturator nerve receives fibers from the L 2–L 4 spinal segments. After leaving the lumbar plexus, it descends behind and medial to the psoas major (not shown here) toward the lesser pelvis and enters the obturator canal (not shown here) below the linea terminalis, accompanied by the obturator vessels. Farther distally it distributes muscular branches to the obturator externus and subsequently divides into an anterior branch and a posterior branch. These branches continue distally, passing respectively anterior and posterior to the adductor brevis, and supply motor innervation to the rest of the adductor muscles (pectineus, adductor longus, adductor brevis, adductor magnus, adductor minimus, and gracilis). The anterior branch gives off a terminal, sensory cutaneous branch at the anterior border of the gracilis, which pierces the fascia lata to supply a palm-sized area of skin on the medial view of the distal thigh. In evaluating motor deficits that are associated with obturator nerve injuries (e.g., intrapartum or due to pelvic fractures), it is important to know that the femoral nerve contributes to the supply of the pectineus muscle, while the sciatic nerve helps to supply the adductor magnus.

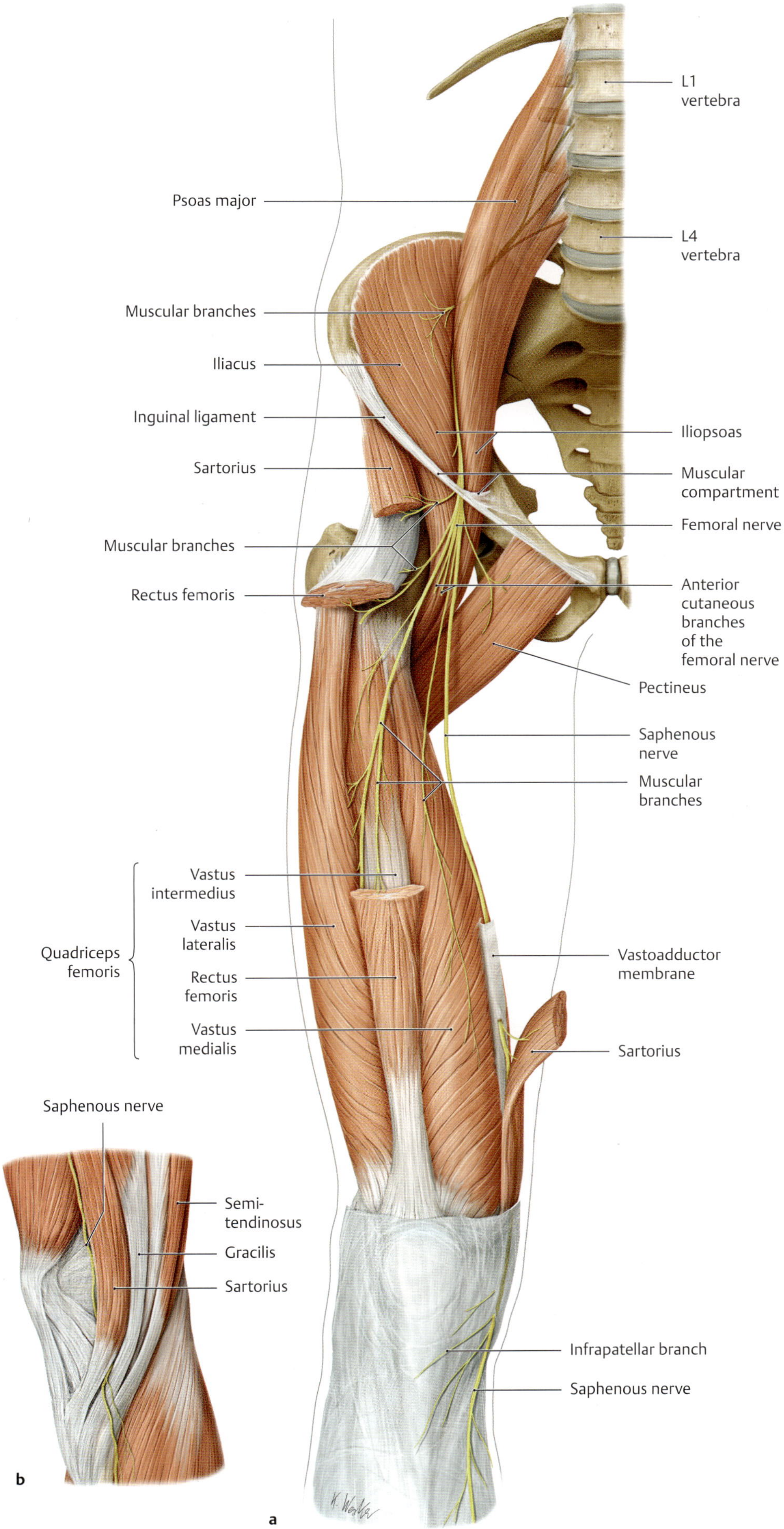

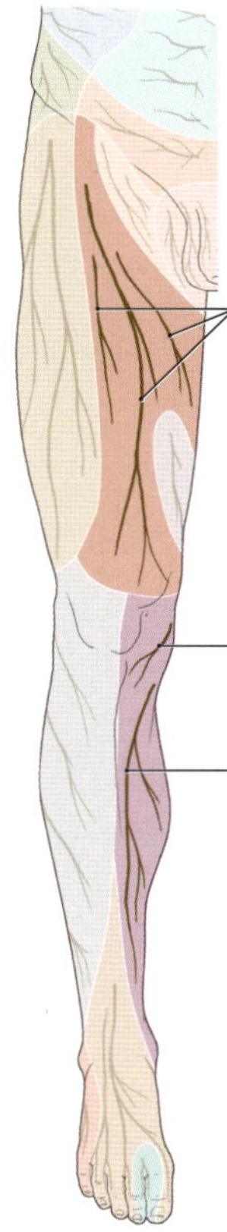

D Sensory distribution of the femoral nerve
Right lower limb, anterior view.

C Course of the femoral nerve

a Inguinal region and thigh of the right lower extremity. Anterior view.
b Course of the saphenous nerve deep to the sartorius. Medial view.

The right inguinal region and thigh, anterior view. As the largest and longest nerve of the lumbar plexus, the femoral nerve receives fibers from the second through fourth lumbar segments of the spinal cord. It supplies *motor* innervation to the iliopsoas, pectineus, sartorius, and quadriceps femoris muscles and *sensory* innervation to the skin of the anterior thigh, medial leg, and hindfoot (see **D**). The nerve, under cover of the psoas fascia, runs in a groove between the psoas major and iliacus to the medial muscular compartment while giving off branches to both muscles. Approximately 8 cm below the inguinal ligament, the femoral nerve divides into numerous cutaneous branches (anterior cutaneous branches) and muscular branches, as well as a long, terminal sensory branch that continues to the foot, the saphenous nerve. Initially, the saphenous nerve enters the adductor canal with the femoral vessels (below the vastoadductor membrane) but then leaves the canal through the vastoadductor membrane and passes with the sartorius toward the medial side of the knee (see **b**). After giving off a sensory infrapatellar branch to the skin of the medial knee, it follows the great saphenous vein to the skin of the medial leg and foot.

4.7 The Nerves of the Sacral Plexus: The Superior Gluteal, Inferior Gluteal, and Posterior Femoral Cutaneous Nerves

A Nerves of the sacral plexus (part I)
(For nerves of the sacral plexus, parts II and III, see pp. 556 and 560.)

Nerve	Segment	Innervated muscles	Cutaneous branches
• Superior gluteal nerve	L4–S1	• Gluteus medius • Gluteus minimus • Tensor fasciae latae	
• Inferior gluteal nerve	L5–S2	• Gluteus maximus	
• Posterior femoral cutaneous nerve	S1–S3		• Posterior femoral cutaneous nerve – Inferior clunial nerves – Perineal branches (see **F** for sensory distribution)
• *Direct branches from the plexus:* – Nerve of piriformis – Nerve of obturator internus	 S1–S2 L5–S2	 • Piriformis • Obturator internus • Gemelli	
– Nerve of quadratus femoris	L4–S1	• Quadratus femoris	

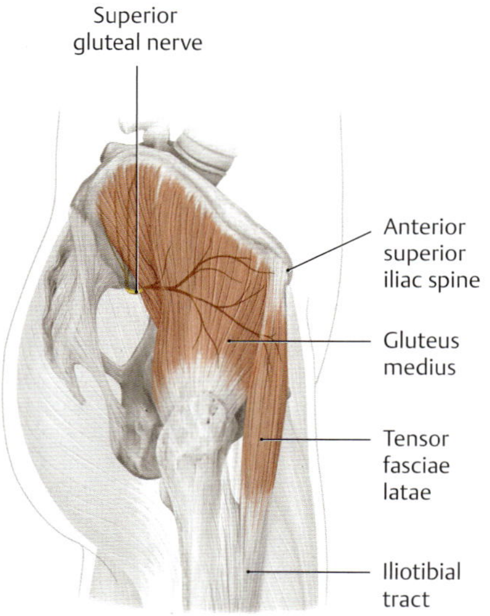

B Motor distribution of the superior gluteal nerve
Right hip region, lateral view. Accompanied by blood vessels with the same name, the superior gluteal nerve leaves the lesser pelvis through the greater sciatic foramen superior to the piriformis (see p. 572), runs in the intergluteal space, and supplies motor fibers to the small gluteal muscles (gluteus medius and minimus) and tensor fasciae latae.

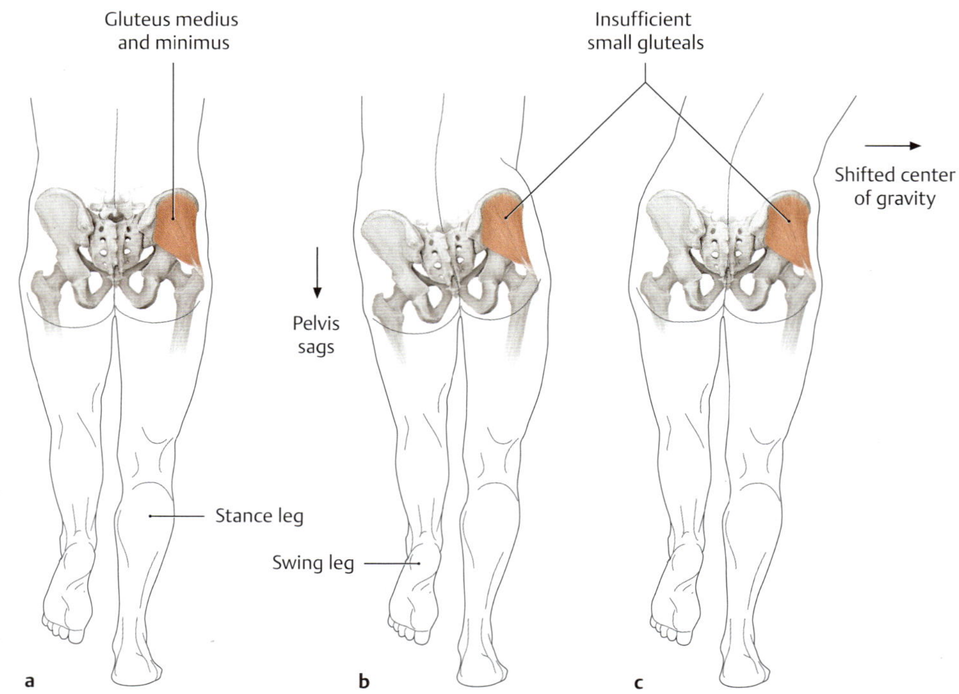

C Clinical indicators of small gluteal muscle weakness: the Trendelenburg sign and the Duchenne limp
Lower half of body, posterior view.

a In normal one-legged stance, the small gluteal muscles on the stance side can stabilize the pelvis in the coronal plane.

b Weakness or paralysis of the small gluteal muscles (e.g., due to a faulty intramuscular injection causing damage to the superior gluteal nerve) is manifested by weak abduction of the affected hip joint and an inability to stabilize the pelvis in the coronal plane. In a positive Trendelenburg test, the pelvis sags toward the *normal* unsupported side.

c Tilting the upper body toward the affected side shifts the center of gravity onto the stance side, thereby elevating the pelvis on the swing side (*Duchenne limp*). With bilateral loss of the small gluteals, the patient exhibits a typical waddling gait.

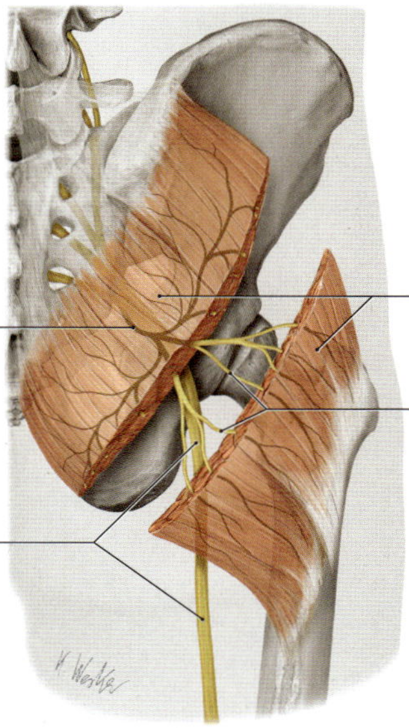

D Motor distribution of the inferior gluteal nerve

Right half of the pelvis, posterior view. The inferior gluteal nerve leaves the lesser pelvis with the sciatic nerve through the greater sciatic foramen inferior to the piriformis (see p. 572) and supplies numerous muscular branches to the gluteus maximus. Paralysis of the gluteus maximus causes little impairment of normal gait on even ground because the deficit is well compensated by the hamstrings (see p. 502). The affected patient is unable to run, jump, or climb stairs, however.

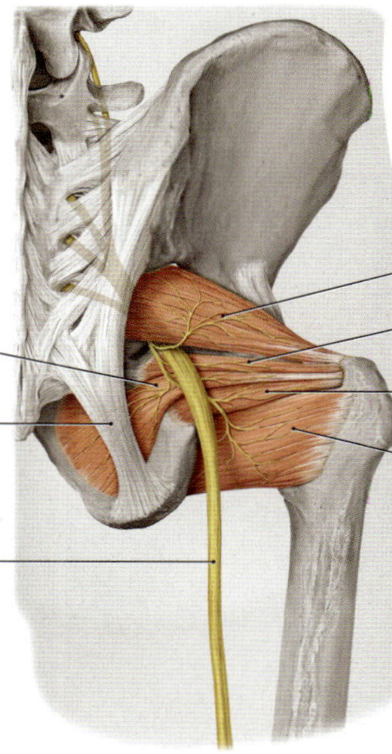

E Muscles that are supplied by direct branches from the sacral plexus

Right half of the pelvis, posterior view. The direct branches of the sacral plexus are listed in **A**.

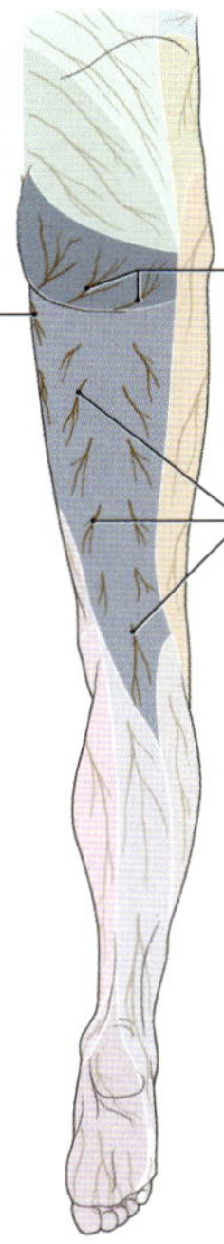

F Sensory distribution of the posterior femoral cutaneous nerve

Right lower limb, posterior view. Besides the skin of the posterior thigh, the posterior femoral cutaneous nerve distributes several branches to the skin of the gluteal sulcus (inferior clunial nerves), and its perineal branches supply the skin of the perineal region (darker shading indicates the exclusive area).

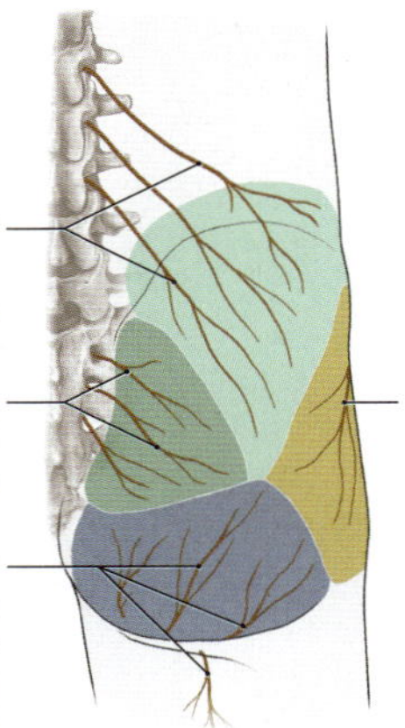

G Sensory innervation of the gluteal region

Right buttock, posterior view.

The gluteal region receives sensory innervation from portions of the sacral plexus and lumbar plexus (anterior rami of the spinal nerves) and also from posterior rami:

- From the sacral plexus: inferior clunial nerves (from the posterior femoral cutaneous nerve).
- From the lumbar plexus: the lateral branch of the iliohypogastric nerve.
- From posterior rami of the spinal nerves: the superior clunial nerves (posterior rami of L1–L3) and middle clunial nerves (posterior rami of S1–S3).

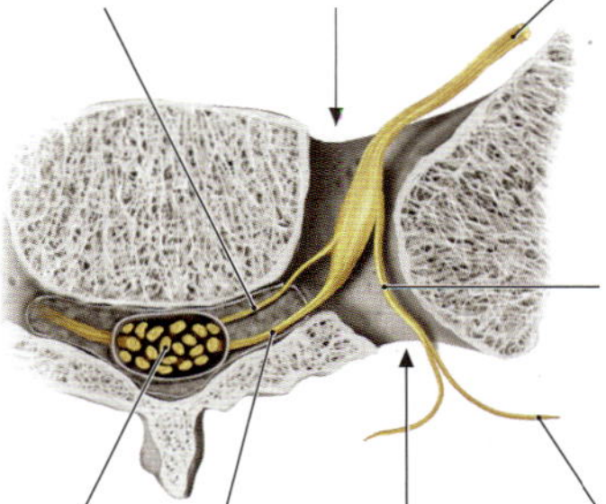

H An emerging sacral nerve

Horizontal section through the right half of the sacrum at the level of the sacral foramina. While the anterior ramus of a sacral nerve emerges from the sacrum through an anterior sacral foramen, the corresponding posterior ramus passes through the *posterior* sacral foramen to supply the skin of the buttock.

4.8 The Nerves of the Sacral Plexus: The Sciatic Nerve (Overview and Sensory Distribution)

A Nerves of the sacral plexus (part II)

The largest and longest of the peripheral nerves, the sciatic nerve, leaves the lesser pelvis through the greater sciatic foramen inferior to the piriformis and passes below the gluteus maximus to the back of the thigh. It divides into its two main branches, the tibial and common fibular nerves, at a variable level but generally before entering the popliteal fossa. The *muscular branches* of the sciatic nerve, however, can already be identified as consisting of a *fibular part* (Fib) and a *tibial part* (Tib) while still proximal to the bifurcation (see also p. 558). Injuries of the sciatic nerve may be caused by compression of the nerve at its emergence inferior to the piriformis (usually by extrinsic pressure, such as sitting). Other potential causes are misdirected intramuscular injections (in which the nerve is accidentally pricked), pelvic fractures, and surgical procedures (e.g., hip replacement).

Nerve	Segment	Innervated muscles	Cutaneous branches
Sciatic nerve	L4–S3	• Semitendinosus (Tib) • Semimembranosus (Tib) • Biceps femoris – Long head (Tib) – Short head (Fib) • Adductor magnus (Tib), medial part	
• Common fibular* nerve	L4–S2		• Lateral sural cutaneous nerve (→ sural nerve) • Fibular communicating branch
– Superficial fibular nerve		• Fibularis longus • Fibularis brevis	• Medial dorsal cutaneous nerve • Intermediate dorsal cutaneous nerve
– Deep fibular nerve		• Tibialis anterior • Extensor digitorum longus • Extensor digitorum brevis • Extensor hallucis longus • Extensor hallucis brevis • Fibularis tertius	• Lateral cutaneous nerve of the great toe • Medial cutaneous nerve of the second toe
• Tibial nerve	L4–S3	• Triceps surae • Plantaris • Popliteus • Tibialis posterior • Flexor digitorum longus • Flexor hallucis longus	• Medial sural cutaneous nerve (→ sural nerve) • Lateral calcaneal branches • Medial calcaneal branches • Lateral dorsal cutaneous nerve
– Medial plantar nerve		• Abductor hallucis • Flexor digitorum brevis • Flexor hallucis brevis, medial head • First lumbrical	• Proper plantar digital nerves
– Lateral plantar nerve		• Flexor hallucis brevis, lateral head • Quadratus plantae • Abductor digiti minimi (nerve of Baxter**) • Flexor digiti minimi brevis • Opponens digiti minimi • Second through fourth lumbricals • First through third plantar interossei • First through fourth dorsal interossei • Adductor hallucis	• Proper plantar digital nerves

* The term "peroneal" may also be used in place of the term "fibular."

** *Note:* The first mixed motor and sensory branch of the lateral plantar nerve to the abductor digiti minimi is known as the "nerve of Baxter" (after Donald Baxter who first described it). Because of its unusual course, this nerve can be the cause of pain in the heel. After arising at the level of the inlet of the posterior tarsal tunnel (see p. 581, Fig. **E**), the nerve of Baxter courses immediately beneath the malleolus medialis deep to the tendon of origin of the abductor hallucis (see p. 559, Fig. **B**) in the deep plane. In its further course, it approaches the insertion of the plantar aponeurosis, finally arriving at the lateral aspect of the foot to innervate the abductor digiti minimi (see p. 582, Fig. **B**)

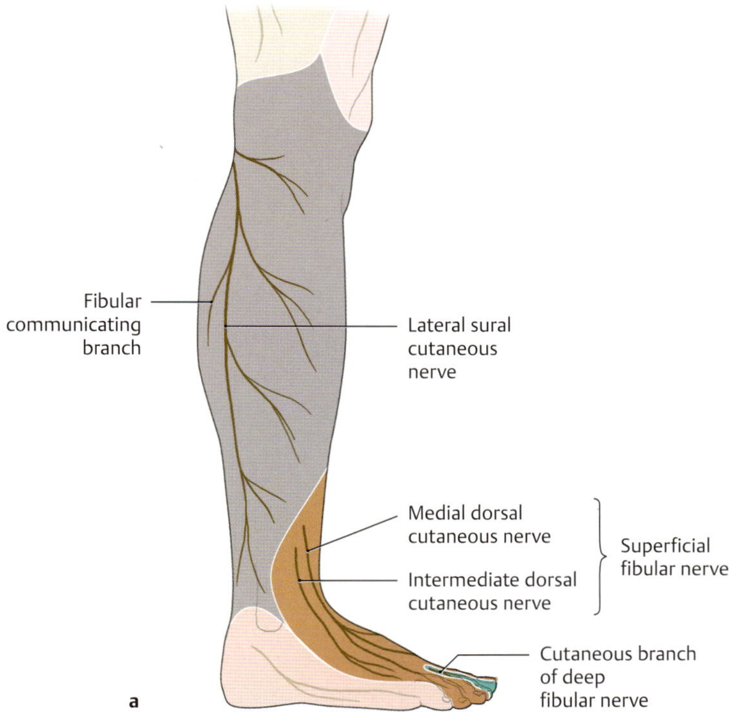

a

B Sensory distribution of the sciatic nerve
Right leg. **a** Lateral view, **b** anterior view, **c** posterior view.

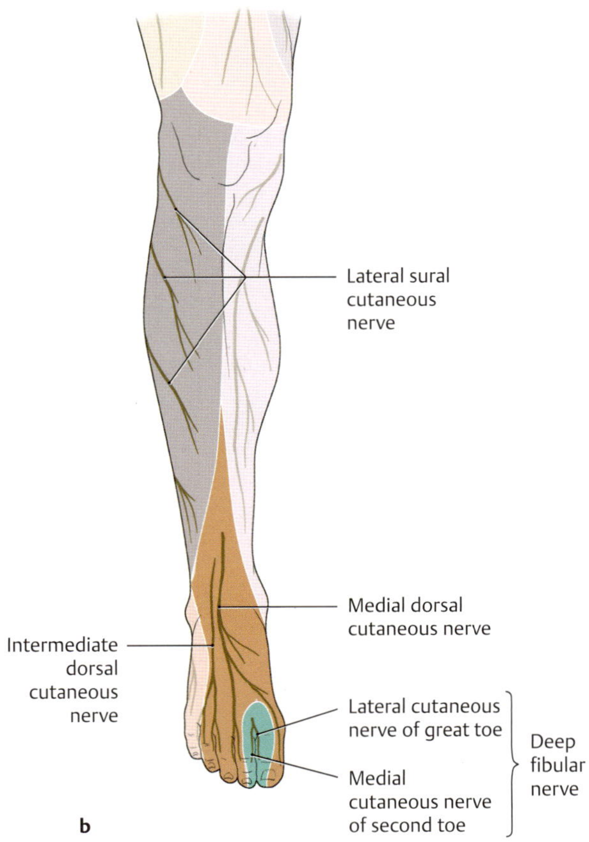

b

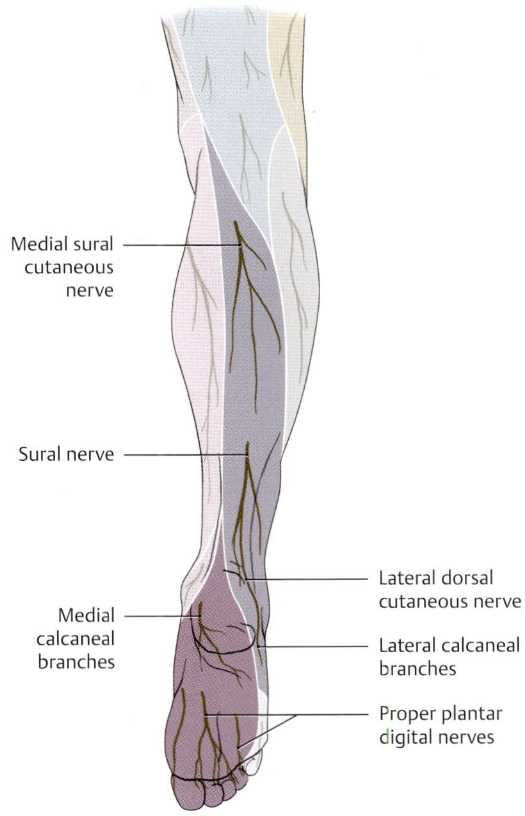

c

4.9 The Nerves of the Sacral Plexus: The Sciatic Nerve (Course and Motor Distribution)

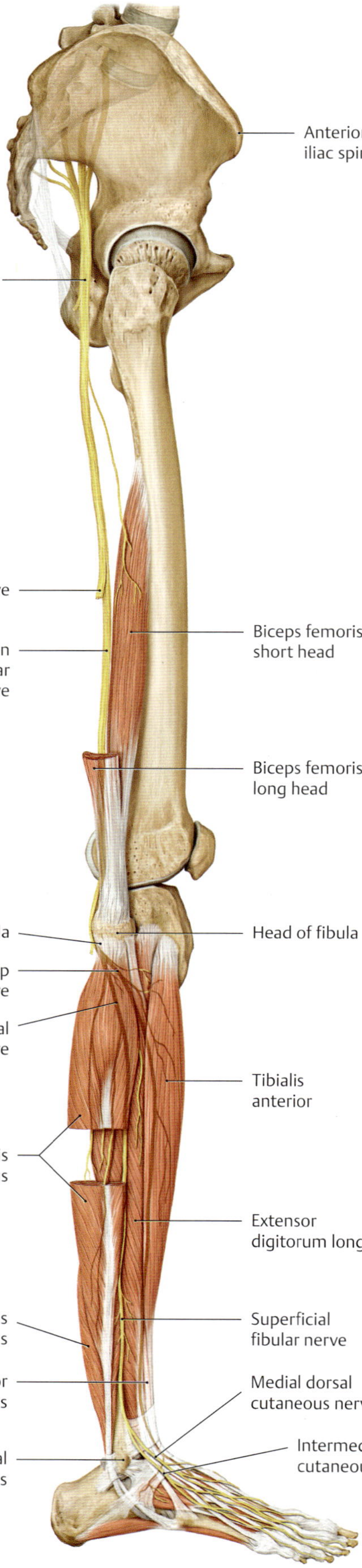

A Course and motor distribution of the sciatic nerve: the fibular part (common fibular nerve)

Right lower limb, lateral view. After giving off several muscular branches from its fibular part (to the short head of the biceps femoris), the sciatic nerve consistently divides in the lower third of the thigh into the tibial nerve and common fibular nerve. The common fibular nerve then follows the medial border of the biceps femoris to the head of the fibula and winds around the neck of the fibula to the front of the leg. Immediately after entering the fibularis longus, it divides into its two terminal branches, the *deep* fibular nerve and *superficial* fibular nerve. The superficial fibular nerve supplies the fibularis muscles and runs between the fibularis longus and fibula to the dorsum of the foot. The deep fibular nerve runs through the interosseous membrane to enter the extensor compartment. After supplying the tibialis anterior, extensor digitorum longus, and extensor hallucis longus, it runs in a groove between the tibialis anterior and extensor hallucis longus on the crural interosseous membrane, accompanied by the anterior tibial vessels, to the dorsum of the foot.

- If the nerve is damaged at the level of the fibular neck (a very exposed location!) before dividing into its two terminal branches, the result is weakness or paralysis of the anterior *and* lateral compartment muscles, resulting in foot drop with some inversion.
- If the nerve is damaged after dividing into its terminal branches, the result may be an *isolated* weakness or paralysis of the anterior compartment or the lateral muscles, depending on whether the *deep* fibular or the *superficial* fibular nerve is affected. Accordingly, the result may be weak dorsiflexion or weakness of eversion. An *isolated* lesion of the superficial fibular nerve generally affects only the *sensory* terminal branch, with pain involving the distal leg and dorsum of the foot. Gait disturbance will occur only with an isolated lesion of the deep fibular nerve (as in compartment syndrome caused by anterior compartment hemorrhage, see p. 585), resulting in foot drop and a "steppage gait." Increased flexion of the hip and knee joints is necessary to keep the toe from dragging the ground during the swing phase of gait.

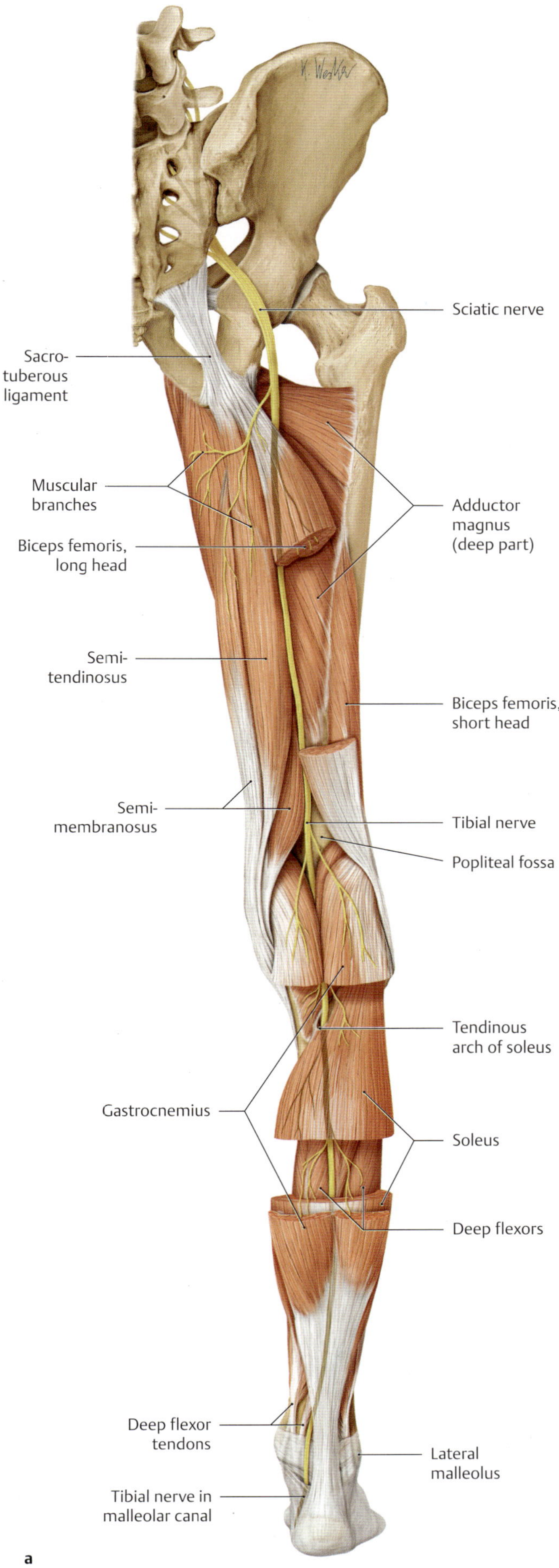

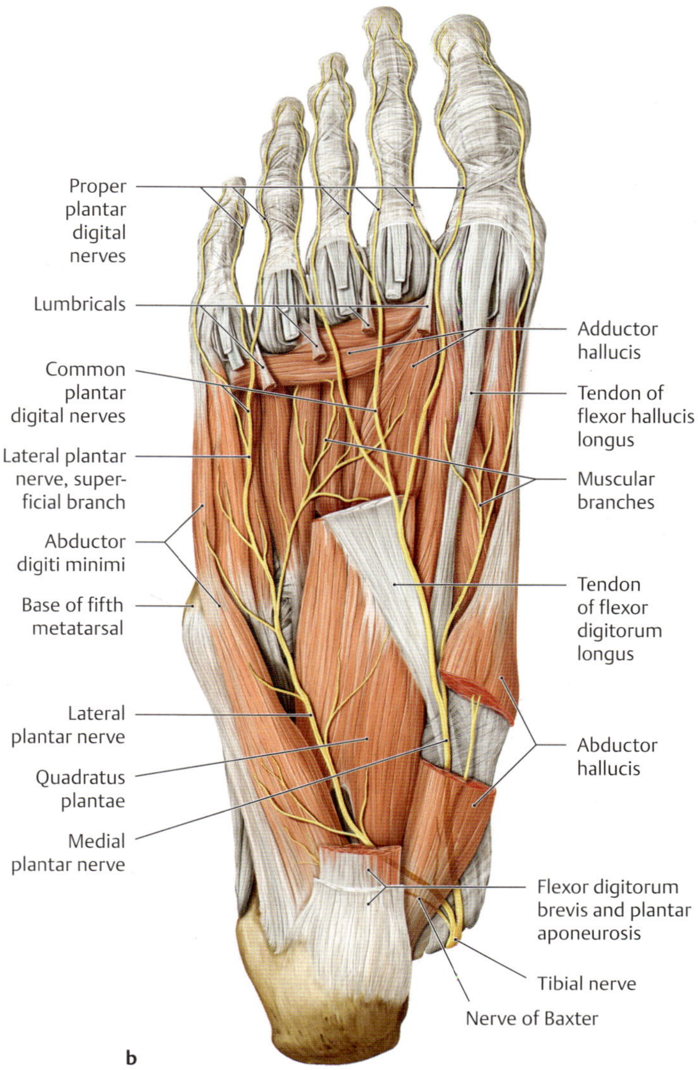

B Course and motor distribution of the sciatic nerve: the tibial part (tibial nerve)

a Right lower limb, posterior view; **b** right foot, plantar view. While still in the thigh, the tibial part of the sciatic nerve distributes several muscular branches to the semitendinosus, semimembranosus, biceps femoris (long head), and adductor magnus (medial part). After division of the sciatic nerve, the tibial nerve continues straight down through the center of the popliteal fossa and runs below the tendinous arch of the soleus muscle to the superficial and deep plantar flexors, which it supplies. In the deep posterior compartment, the tibial nerve continues distally in a neurovascular bundle with the posterior tibial vessels (not shown here) and passes through the tarsal tunnel, accompanied by the deep flexor tendons, to the plantar side of the foot (**b**). In passing through the tarsal tunnel, the tibial nerve divides into its two terminal branches (the lateral and medial plantar nerves), which supply all the muscles on the plantar side of the foot. Compression of the tibial nerve or its terminal branches at this site leads to an entrapment syndrome (= tarsal tunnel syndrome). This can result in pain and sensory disturbances affecting the sole of the foot or even palsies of the intrinsic foot muscles, particularly following severe nerve trauma in connection with a fracture of the tibial shaft or medial malleolus. Refer to p. 556 for the nerve of Baxter.

4.10 The Nerves of the Sacral Plexus: The Pudendal and Coccygeal Nerves

A Nerves of the sacral plexus (part III)
The pudendal nerve, the lowest branch of the sacral plexus, arises from a separate small plexus formed by the anterior rami of S1–S4; hence it is occasionally referred to as the *pudendal plexus*.

Nerve	Segment	Innervated muscles	Cutaneous branches
• Pudendal nerve (*pudendal plexus*)	S2–S4	• Pelvic floor muscles – Levator ani – Superficial transverse perineal – Deep transverse perineal – Bulbospongiosus – External anal sphincter – Urethral sphincter	• Inferior rectal nerves • Perineal nerves – Posterior labial nerves in females – Posterior scrotal nerves in males – Dorsal clitoral nerve in females – Dorsal penile nerve in males
• Coccygeal nerve (*coccygeal plexus*)	S5–Co2	• Coccygeus	• Anococcygeal nerve (anterior rami) • Dorsal rami

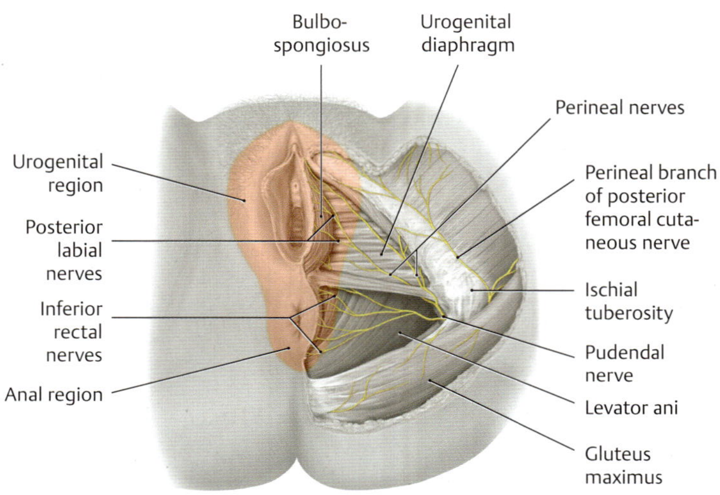

B The cutaneous branches of the pudendal nerve and its sensory distribution in the female
Lithotomy position, inferior view. The skin layers have been removed on the left side to demonstrate the terminal branches of the pudendal nerve in the ischioanal fossa (see p. 574). The area of cutaneous sensory innervation is color-shaded. Large portions of the urogenital region and anal region receive their sensory supply from the pudendal nerve. The skin area supplied by the pudendal nerve can be anesthetized during childbirth by infiltration anesthesia or a nerve block, enabling the obstetrician to perform and repair an episiotomy without pain (see p. 240). This can be done either by infiltrating the perineum between the anus and posterior fornix with a local anesthetic or by temporarily blocking the pudendal nerve with local anesthetic injected near the ischial spine (before the nerve has branched; see diagram in **C**).

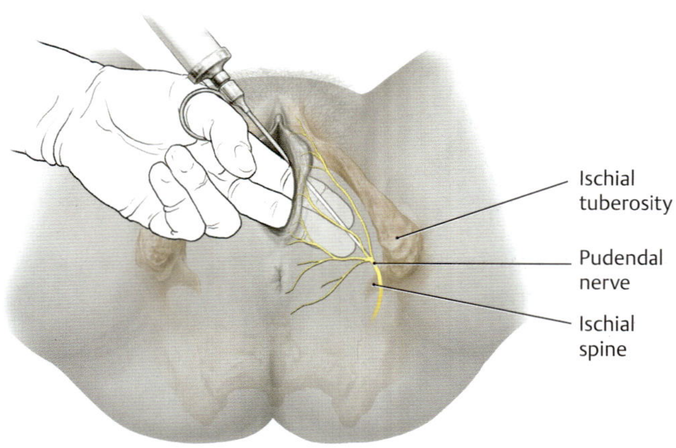

C Technique of a left-sided pudendal nerve block
Lithotomy position, inferior view. The most common type of conduction anesthesia used in vaginal deliveries is the pudendal nerve block, which renders the perineum, vulva, and lower third of the vagina insensitive to pain. In the *transvaginal approach*, a special guide cannula is introduced into the vagina, and 10 mL of a local anesthetic solution is injected approximately 1 cm above and 1 cm lateral to the palpable ischial spine on each side. An injection at this site will block the pudendal nerve *before* it has entered the pudendal canal (Alcock's canal) and *before* it has divided into its terminal branches. Often the nerve block is administered at the end of the expulsion stage to relieve stretch pain in the perineal region (see p. 240).

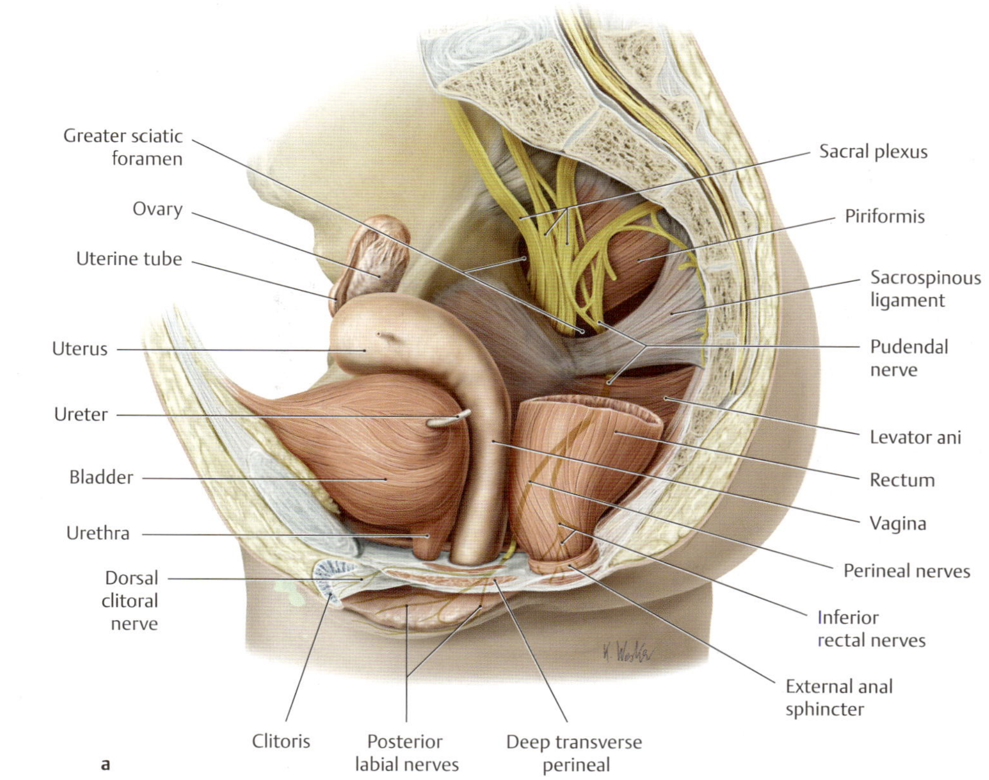

a

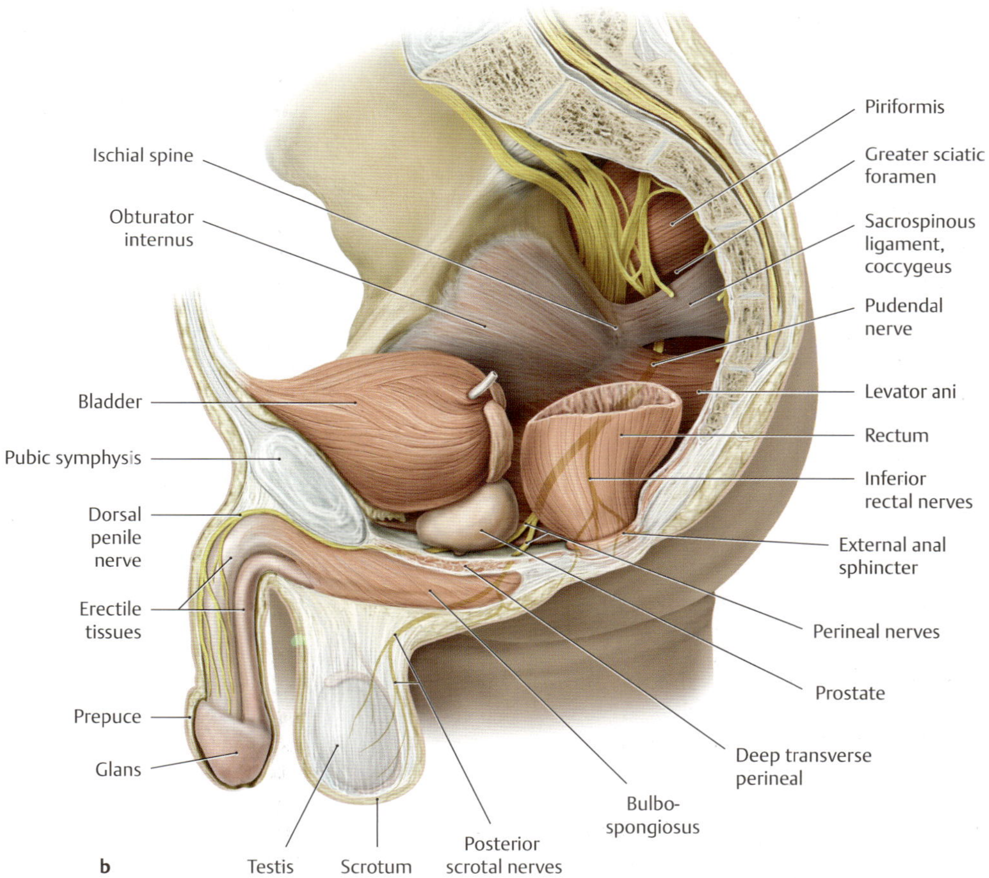

b

D Course of the pudendal nerve and coccygeal nerve in the female and male

a Sagittal section through the female pelvis, left lateral view. **b** Sagittal section through the male pelvis, left lateral view.

The **pudendal nerve** emerges from the lesser pelvis through the *greater* sciatic foramen. It then courses around the ischial spine and sacrospinous ligament and passes through the *lesser* sciatic foramen into the ischioanal (ischiorectal) fossa (see p. 574). It runs forward in the lateral wall of the fossa, embedded in a duplication of the obturator internus fascia (pudendal canal = Alcock's canal) and accompanied by the internal pudendal vessels (see p. 573). Below the symphysis it passes to the dorsum of the penis or the clitoris. The pudendal nerve gives off numerous *branches* from within the perineum:

- The *inferior rectal nerves* supply motor innervation to the external anal sphincter and sensory innervation to the skin around the anus.
- The *perineal nerves* distribute motor branches to the muscles of the perineum (see p. 170) and sensory branches to the skin of the posterior scrotum or labia majora and minora, to the skin of the penis or clitoris, and to the glans, prepuce, and erectile tissues.

Damage to the pudendal nerve (e.g., resulting from perineal injuries during childbirth) leads to loss of function of the muscles of the perineum, especially the sphincter muscles of the bladder and bowel, causing urinary and fecal incontinence. Pudendal nerve lesions can also lead to sexual dysfunction (e.g., male impotence).

The ventral rami of the fifth sacral nerve and the first or second coccygeal nerve form the **coccygeal nerve** (synonym: *coccygeal plexus*). This nerve and its terminal sensory branches, the anococcygeal nerves, pass along the anococcygeal ligament to supply the skin between the coccyx and anus.

5.1 Surface Anatomy and Superficial Nerves and Vessels: Anterior View

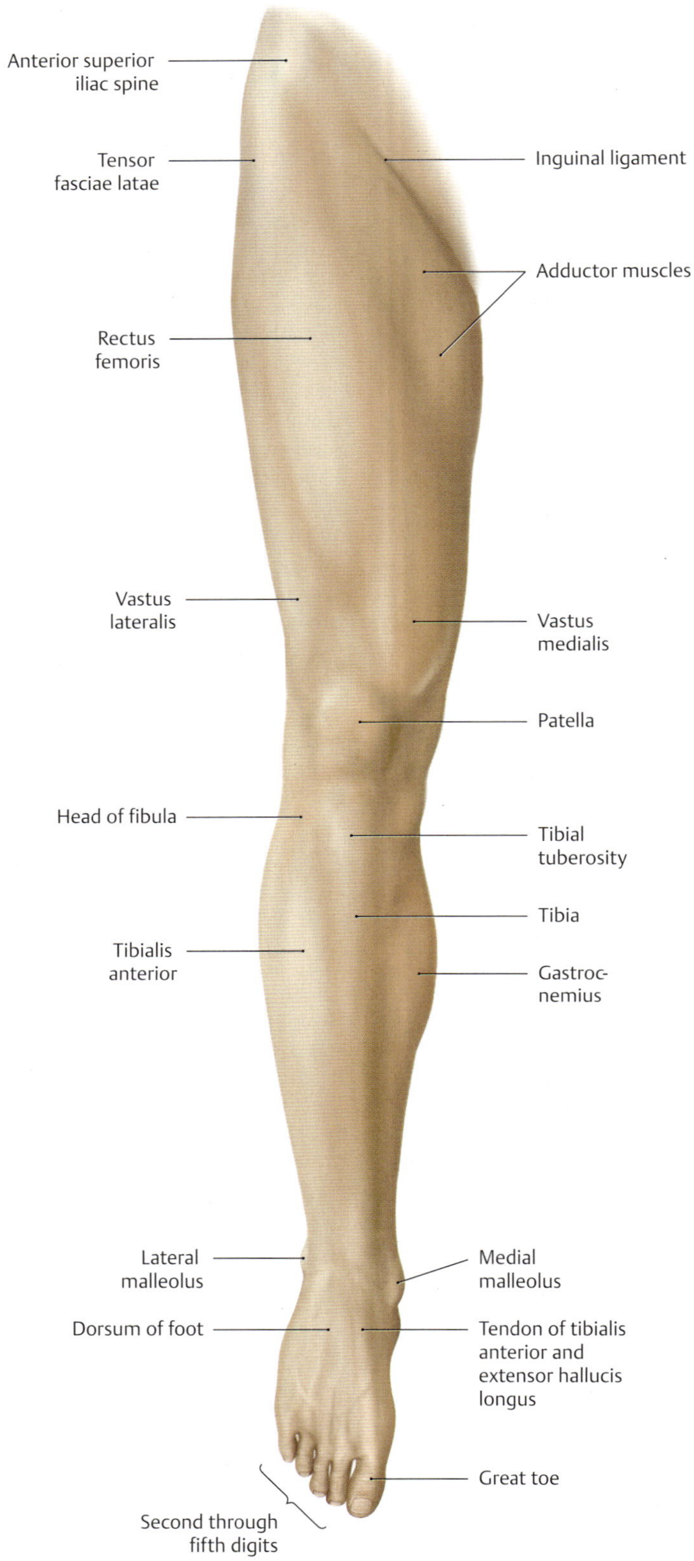

A Surface anatomy of the right lower limb
Palpable *bony prominences* on the lower limb are reviewed on p. 423.

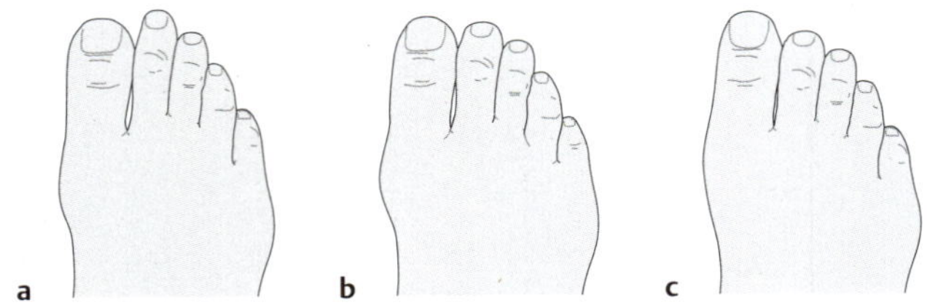

B The most common variants of the forefoot and toes
(after Debrunner and Lelievre)
Three types of foot shape are distinguished based on the relative lengths of the first and second toes:

- **a** The "Greek" type, in which the second toe is longer than the first.
- **b** The square type, in which the first and second toes are of equal length.
- **c** The "Egyptian" type, in which the first toe is longer than the second.

In the "Greek" type, the second metatarsal is generally longer than the metatarsal of the great toe. As a result, the head of the second metatarsal is often subject to painful overloading, especially when high heels are worn.

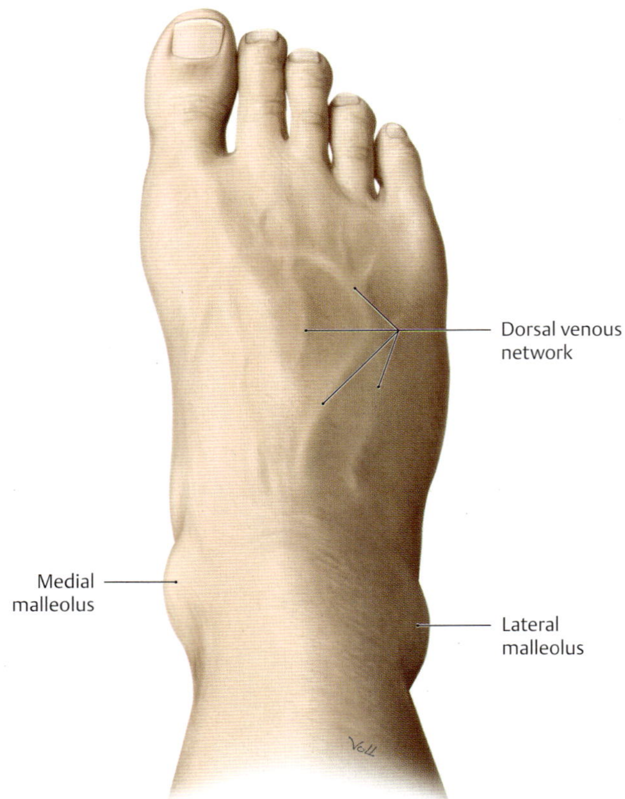

C The dorsal surface of the right foot
The superficial venous network is visible on the dorsum of the foot (compare with **D**).

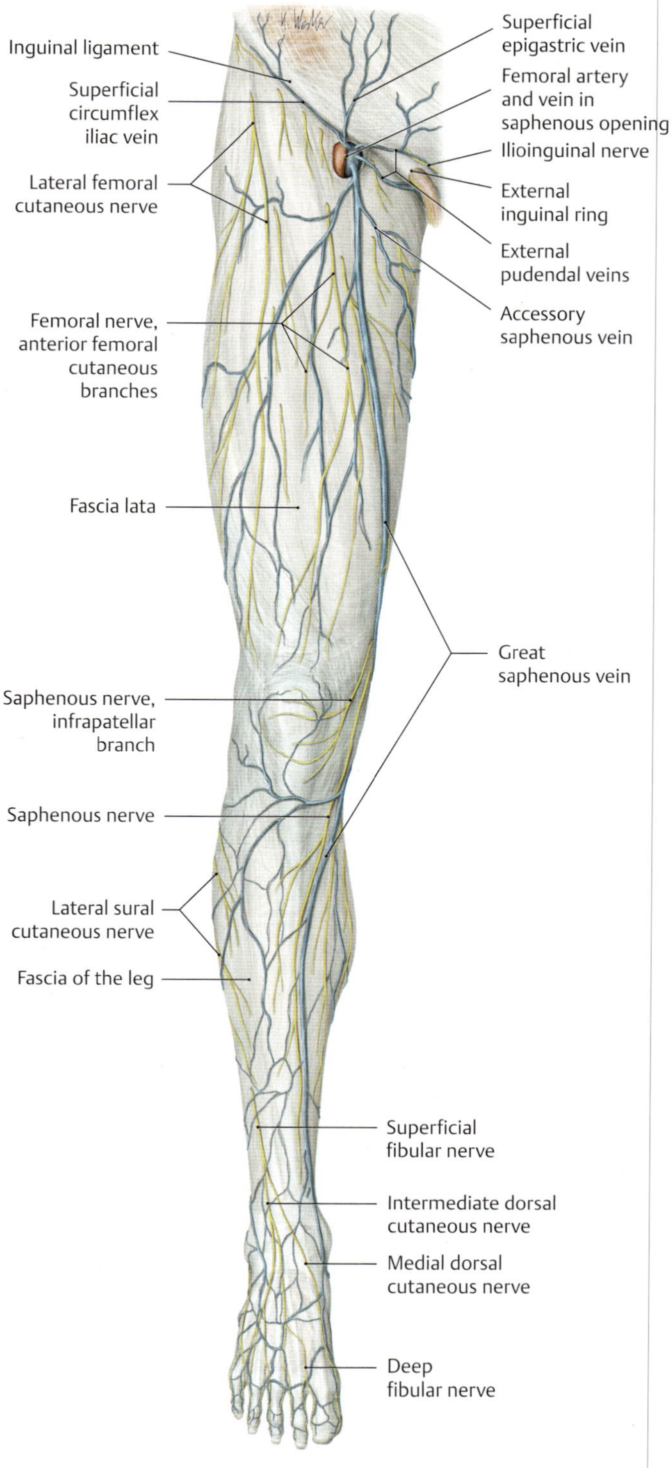

D Superficial cutaneous veins and nerves of the right lower limb

Anterior view. The dorsal venous network of the foot is drained by two large venous trunks (the great and small saphenous veins), which receive a variable pattern of cutaneous veins. While the small saphenous vein (see p. 565) enters the popliteal vein at the level of the popliteal fossa, the great saphenous vein extends up the medial side of the leg to a point just below the inguinal ligament, where it passes through the saphenous opening of the fascia lata to enter the femoral vein. The superficial veins of the lower limb are commonly affected by varicosity, causing them to become thickened, tortuous, and distinctly visible and palpable (see also p. 545).

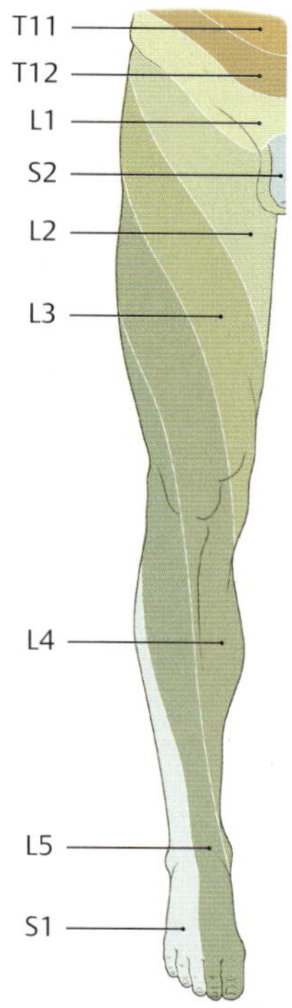

E Radicular (segmental) cutaneous innervation pattern (dermatomes) in the right lower limb

As in the arm, the outgrowth of the lower limb during development causes the sensory cutaneous segments to become elongated and drawn out into narrow bands. The L4, L5, and S1 segments in particular move so far peripherally that they no longer have any connection with the corresponding segments of the trunk.

Note that the dermatomes of the lumbar trunk segments lie mostly on the front of the leg, while those of the sacral segments are mostly on the back of the leg (see p. 94). This can be of diagnostic importance in patients with a herniated disk, for example, in order to determine the level of the herniation.

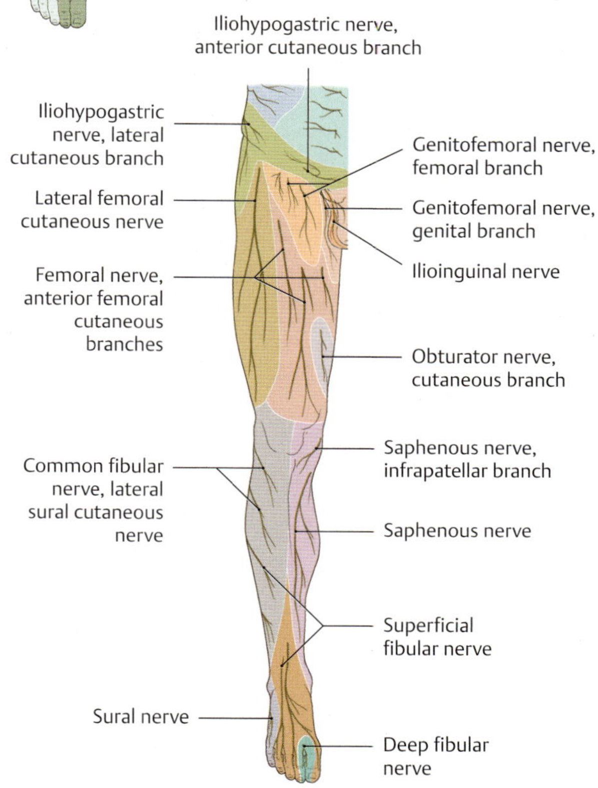

F Pattern of peripheral sensory cutaneous innervation in the right lower limb

As in the arm, the sensory distribution in the lower limb corresponds to the branching patterns of the peripheral cutaneous nerves in the subcutaneous connective tissue. The territories of the individual peripheral nerves overlap, especially at their margins. Hence the clinically determined *exclusive area* of a particular cutaneous nerve (the area supplied by that nerve alone) tends to be considerably smaller than the *maximum area* that can be demonstrated anatomically. For this reason, the traumatic disruption of a nerve causes a complete loss of sensation (anesthesia) in the exclusive area but often will cause only diminished sensation (hypoesthesia) at the perimeter of that area.

Note that the sensory loss resulting from a peripheral nerve injury presents a completely different pattern from that caused by injury to a nerve root (see p. 96).

5.2 Surface Anatomy and Superficial Nerves and Vessels: Posterior View

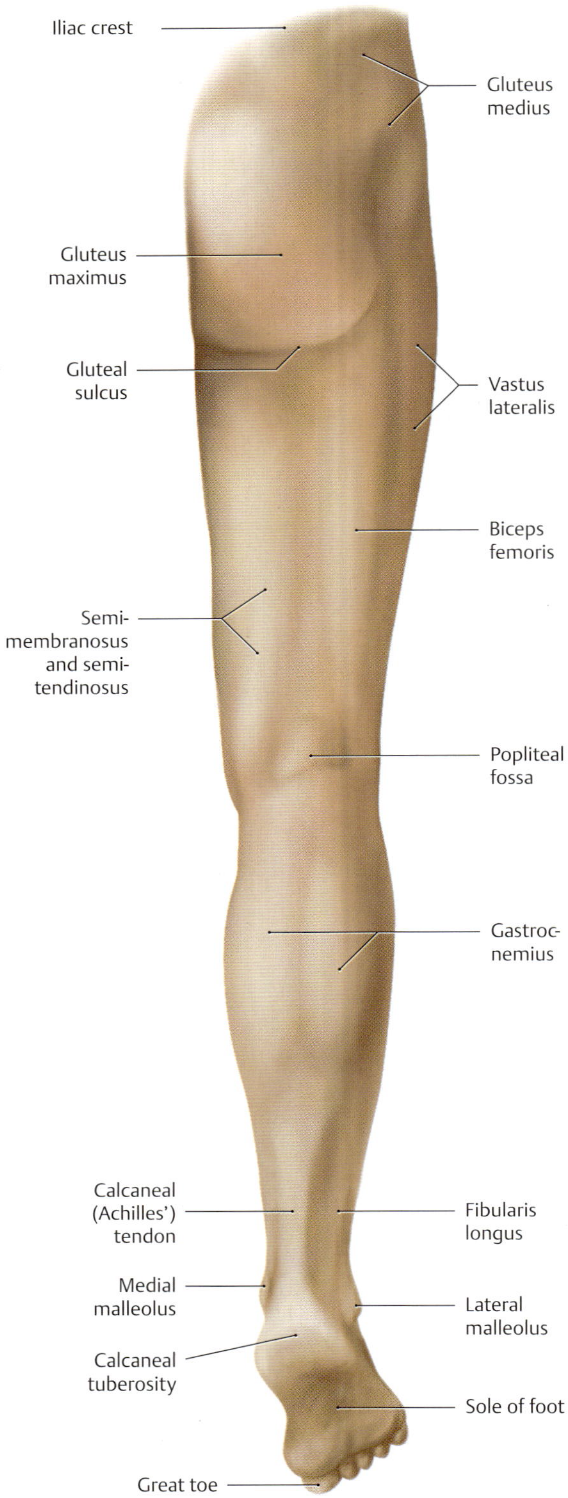

A Surface anatomy of the right lower limb
The foot is in plantar flexion. (Palpable *bony prominences* on the lower limb are reviewed on p. 423.)

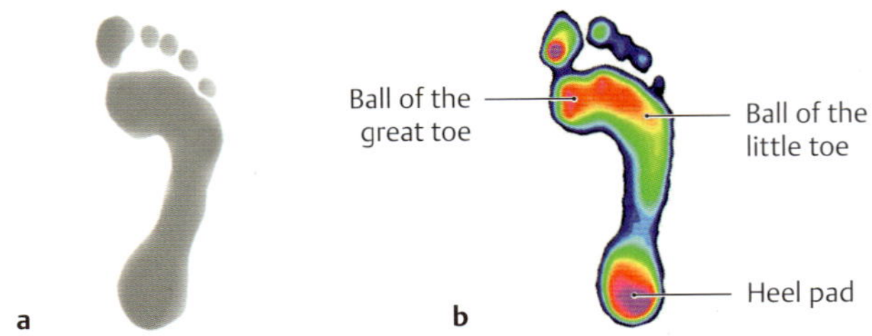

B Footprints (podograms) of the normal right foot in an adult
A podogram provides a graphic representation of the loads borne by the foot. Besides visual inspection of the sole of the foot, analysis of the podogram supplies the most useful information on the weight-bearing dynamics of the foot.

a Footprint created with an ink pad.
b Dynamic pressure podogram (see p. 483) with normal plantar arches (pes rectus). The pressure distribution is recorded with color coding; the image shows that the pressure is evenly distributed (red-violet) over all the major points of support (ball of the great toe and little toe and the heel) and that the intervening regions of the sole of the foot bear essentially no weight (original image **b**: Michael Kriwat, Kiel).

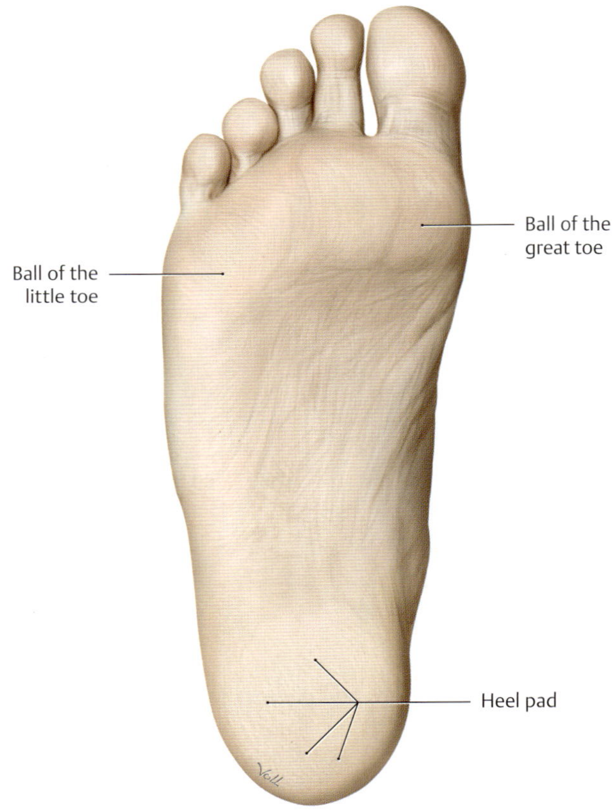

C The plantar surface of the right foot
The skin on the plantar surface of the foot serves as a sensory organ for contact with the ground, perceiving its consistency during stance and locomotion by means of receptors in the sole of the foot. Stresses acting on the heel pad and the balls of the great and little toes generate high local compressive forces at those sites, to which the subcutaneous connective tissue has adapted functionally by developing a system of *pressure chambers* (see p. 490).

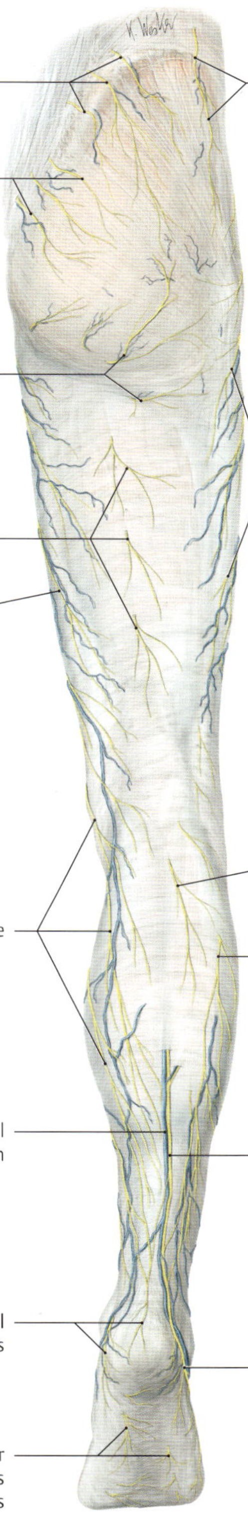

D The epifascial cutaneous veins and nerves of the right lower limb

Posterior view.

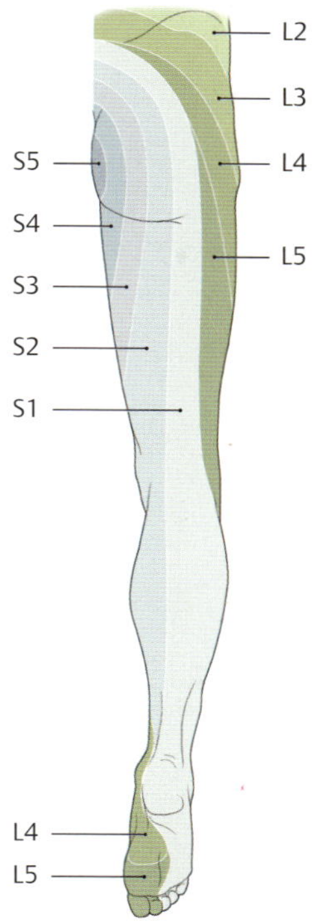

E Radicular (segmental) cutaneous innervation pattern (dermatomes) of the right lower limb

As in the arm, the outgrowth of the lower limb during development causes the sensory cutaneous segments to become elongated into narrow bands. The L4, L5, and S1 segments in particular move so far peripherally that they no longer have any connection with the corresponding segments of the trunk.

Note that the dermatomes of the lumbar trunk segments lie mostly on the front of the leg, while those of the sacral segments are mostly on the back of the leg (see p. 94). This can be of diagnostic importance in patients with a herniated disk, for example, in order to determine the level of the herniation.

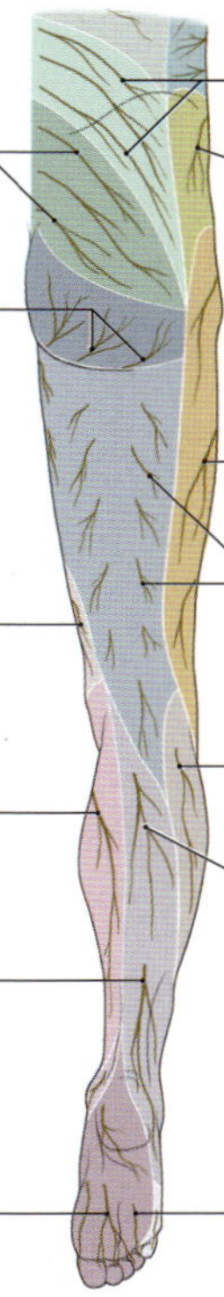

F Pattern of peripheral sensory cutaneous innervation in the right lower limb

As in the arm, the sensory distribution in the lower limb corresponds to the branching patterns of the peripheral cutaneous nerves in the subcutaneous connective tissue. The territories of the individual peripheral nerves overlap, especially at their margins. Hence the clinically determined *exclusive area* of a particular cutaneous nerve (the area supplied by that nerve alone) tends to be considerably smaller than the *maximum area* that can be demonstrated anatomically. For this reason, the traumatic disruption of a nerve causes a complete loss of sensation (anesthesia) in the exclusive area but often will cause only diminished sensation (hypoesthesia) at the perimeter of that area.

Note that the sensory loss resulting from a peripheral nerve injury presents a completely different pattern from that caused by injury to a nerve root (see p. 96).

5.3 The Anterior Femoral Region Including the Femoral Triangle

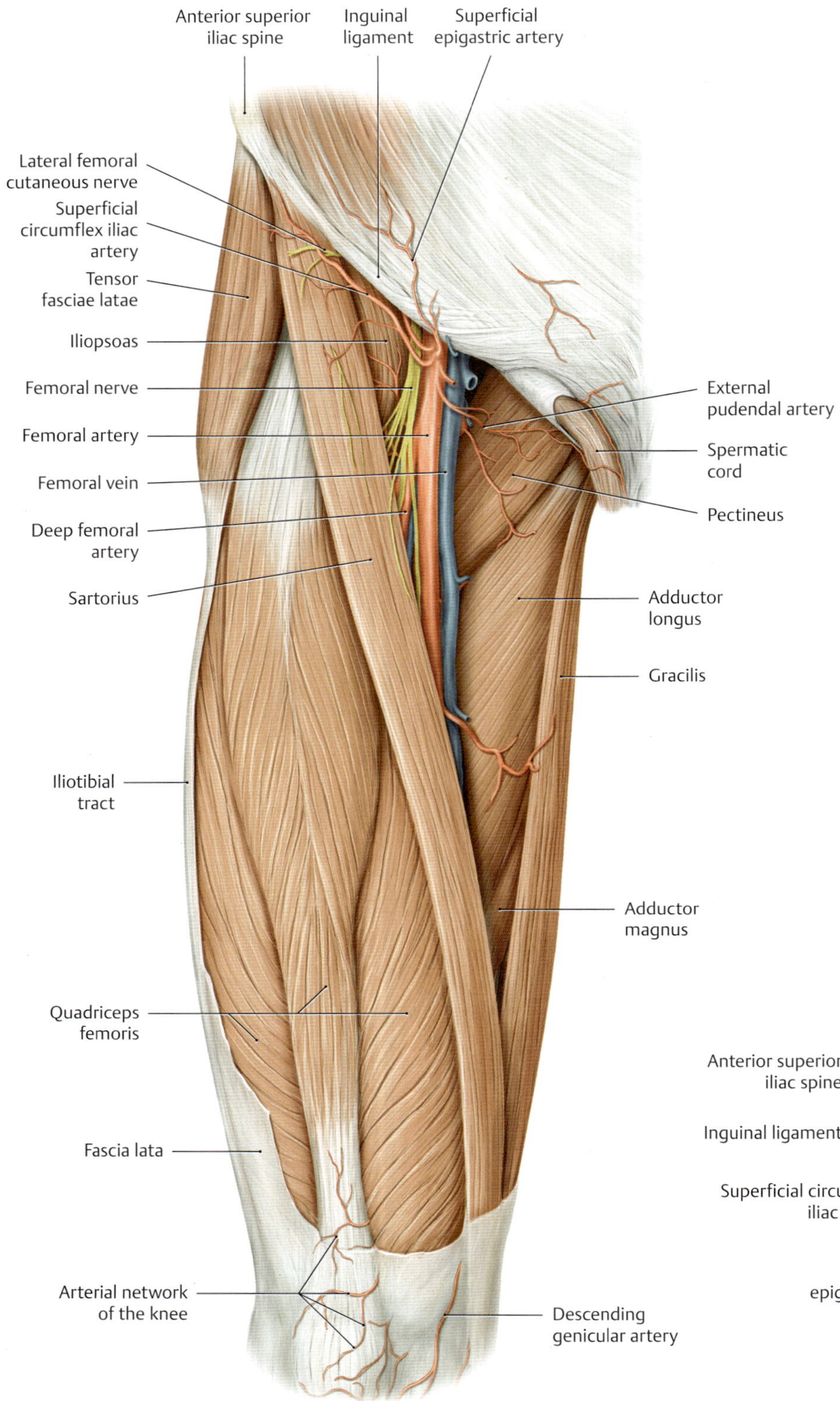

A The femoral triangle
Right thigh, anterior view. The skin, subcutaneous tissue, and fascia lata have been removed to demonstrate the neurovascular structures in the femoral triangle. The femoral triangle is bounded superiorly by the inguinal ligament, laterally by the sartorius muscle, and medially by the adductor longus. It contains the neurovascular structures that emerge from the pelvis and pass below the inguinal ligament to the anterior side of the thigh through the muscular compartment and vascular compartment (see also **C**). The posterior muscular wall of the femoral triangle is formed from lateral to medial by the iliopsoas and pectineus.

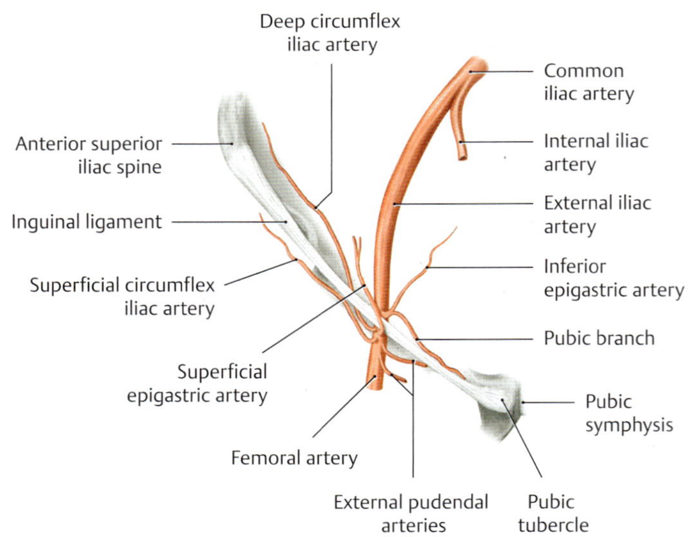

B The branches of the external iliac artery at its junction with the femoral artery in the region of the inguinal ligament
Note: The deep circumflex iliac artery is shown here as a branch of the external iliac artery at the level of the inguinal ligament. In fact, the origin of this artery is extremely variable and can also lie superior or inferior to the inguinal ligament.

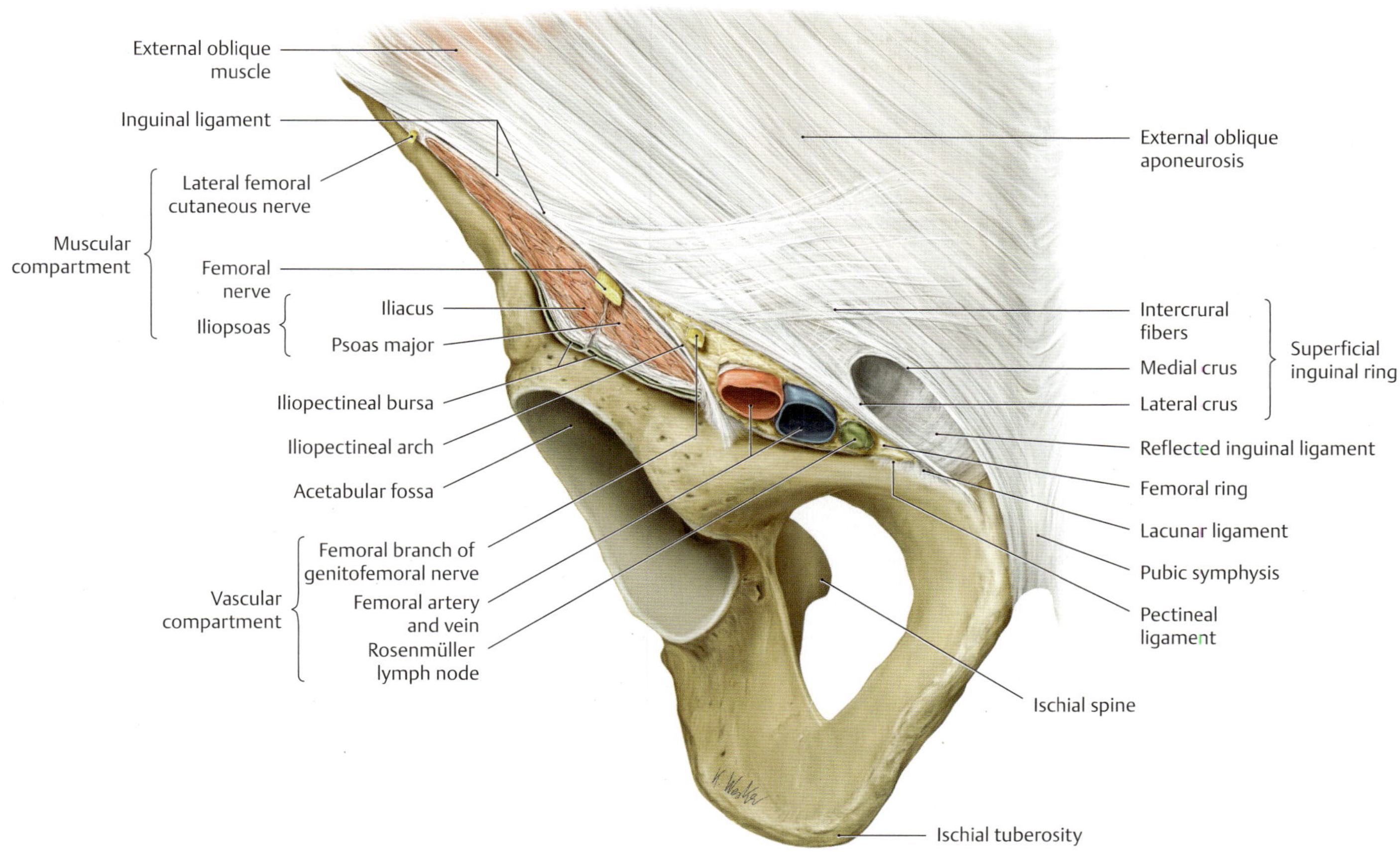

C Inguinal region and the contents of the muscular and vascular compartments

Anterior view. The drawing shows a portion of the right hip bone and the adjacent anterior inferior abdominal wall with the superficial inguinal ring and the contents of the muscular and vascular compartments below the inguinal ligament. The site of emergence of the muscles and vessels, bounded by the inguinal ligament and the superior pelvic rim, is subdivided by the fibrous iliopectineal arch.

The **vascular compartment** is located medial to the iliopectineal arch and is traversed from lateral to medial by the femoral branch of the genitofemoral nerve, the femoral artery and vein, and the deep inguinal lymphatic vessels (only one lymph node is shown here). The part of the vascular compartment that lies medial to the femoral vein is called the *femoral ring*. The lymph vessels from the thigh pass through that ring to enter the pelvis. The femoral ring is covered by a thin sheet of connective tissue called the femoral septum (not shown here), which usually contains a lymph node (the Rosenmüller node) belonging to the group of deep inguinal lymph nodes (see also p. 546).

The **muscular compartment** is lateral to the iliopectineal arch and is traversed by the iliopsoas muscle, femoral nerve, and lateral femoral cutaneous nerve.

Note the iliopectineal bursa located below the iliopsoas. It is the largest bursa of the hip region and communicates in 15% of cases with the joint cavity of the hip. For this reason, an inflammatory disease of the hip joint may incite inflammation of this bursa (bursitis). When inflamed, the iliopectineal bursa is frequently painful and swollen and may occasionally be mistaken for a neoplasm on MR images.

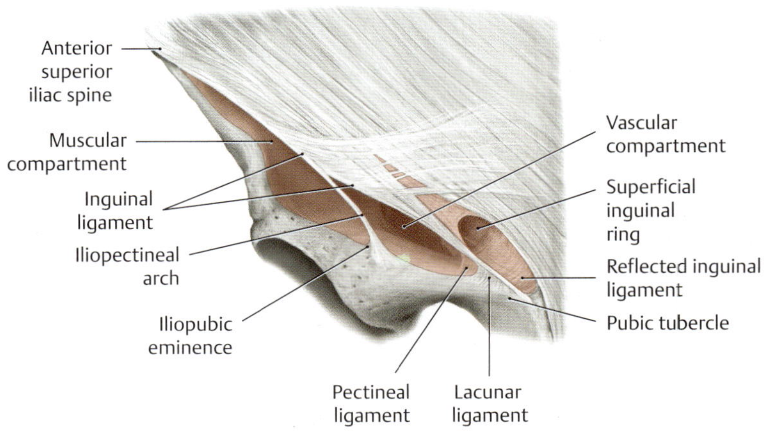

D The connective-tissue and bony boundaries of the muscular and vascular compartments

Diagram of the right inguinal region, anterior view. The connective-tissue boundary between the muscular and vascular compartments is formed by the iliopectineal arch, a thickened band in the medial portion of the iliacus fascia. It extends between the inguinal ligament and the iliopubic eminence. The fibrous band that curves downward from the medial attachment of the inguinal ligament is called the lacunar ligament, which extends further medially as the pectineal ligament (Cooper's ligament) and runs along the superior pubic ramus. This sharp-edged ligament defines the medial boundary of the vascular compartment (femoral ring) and may entrap the hernial sac in patients with a femoral hernia (see p. 222). Above the inguinal ligament is the external (superficial) inguinal ring, which is the external opening of the inguinal canal (see p. 218). The muscular compartment is bounded laterally by the anterior superior iliac spine.

5.4 Arterial Supply to the Thigh

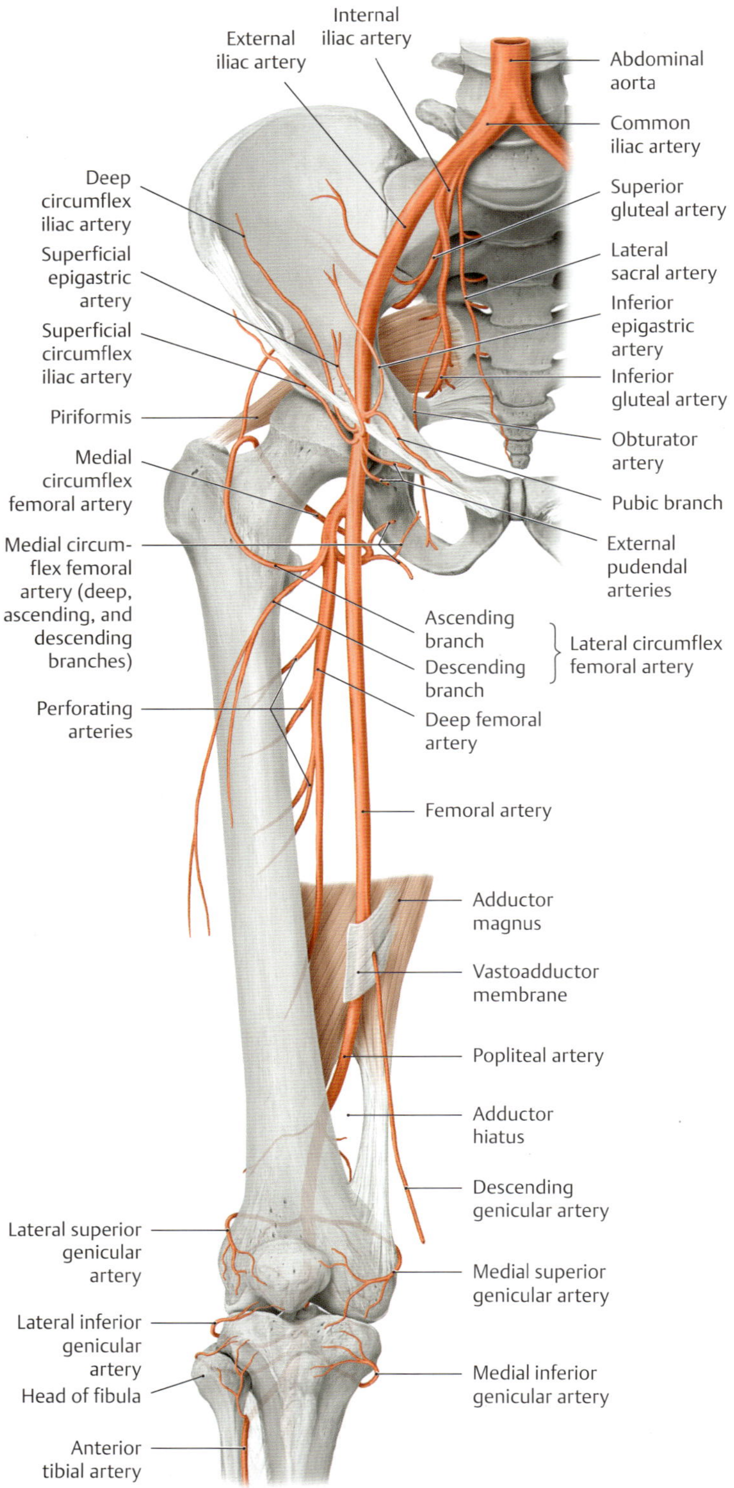

A Course and branches of the femoral artery

The femoral artery, the distal continuation of the external iliac artery, runs along the medial side of the thigh to the adductor canal, through which it passes to the back of the leg. After emerging from the adductor hiatus, it becomes the popliteal artery. In clinical parlance the femoral artery is often called the *superficial* femoral artery because of its superficial course down the front of the thigh, distinguishing it from the more deeply placed deep femoral artery that arises from it (see **D**).

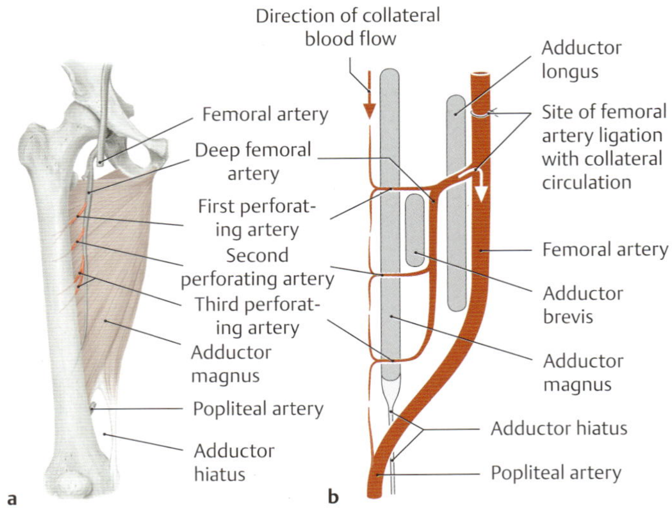

B Course of the deep femoral artery and sites where the perforating arteries pierce the adductor muscles

a Right thigh, anterior view; **b** schematic longitudinal section through the adductor muscles at the level of the perforating arteries. The deep femoral artery has approximately three to five terminal branches that pass from the front to the back of the thigh through the femoral insertions of the adductor muscles (= first through third perforating arteries) to supply the hamstring muscles (biceps femoris, semitendinosus, and semimembranosus). Generally, the arteries pierce the adductor muscles above and below the adductor brevis and just above the adductor hiatus. Ligation of the femoral artery proximal to the origin of the deep artery of the thigh is relatively well tolerated owing to a good collateral supply from branches of the internal iliac artery (superior gluteal artery and obturator artery).

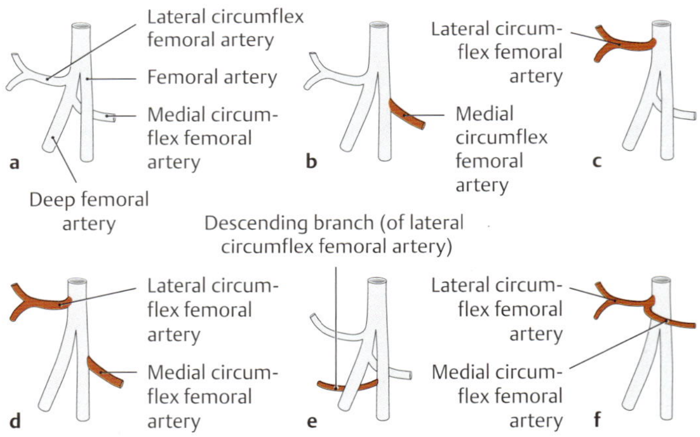

C Variants in the femoral artery branching pattern
(after Lippert and Pabst)

a Usually the deep femoral artery and medial and lateral circumflex femoral arteries arise from the femoral artery by a common trunk (58% of cases, also shown in the other figures on this page).
b The medial circumflex femoral artery arises directly from the femoral artery (18% of cases).
c The lateral circumflex femoral artery arises directly from the femoral artery (15% of cases).
d The circumflex arteries arise separately from the femoral artery (4% of cases).
e The descending branch of the lateral circumflex femoral artery springs directly from the femoral artery (3% of cases).
f The circumflex arteries arise by a common trunk (1% of cases).

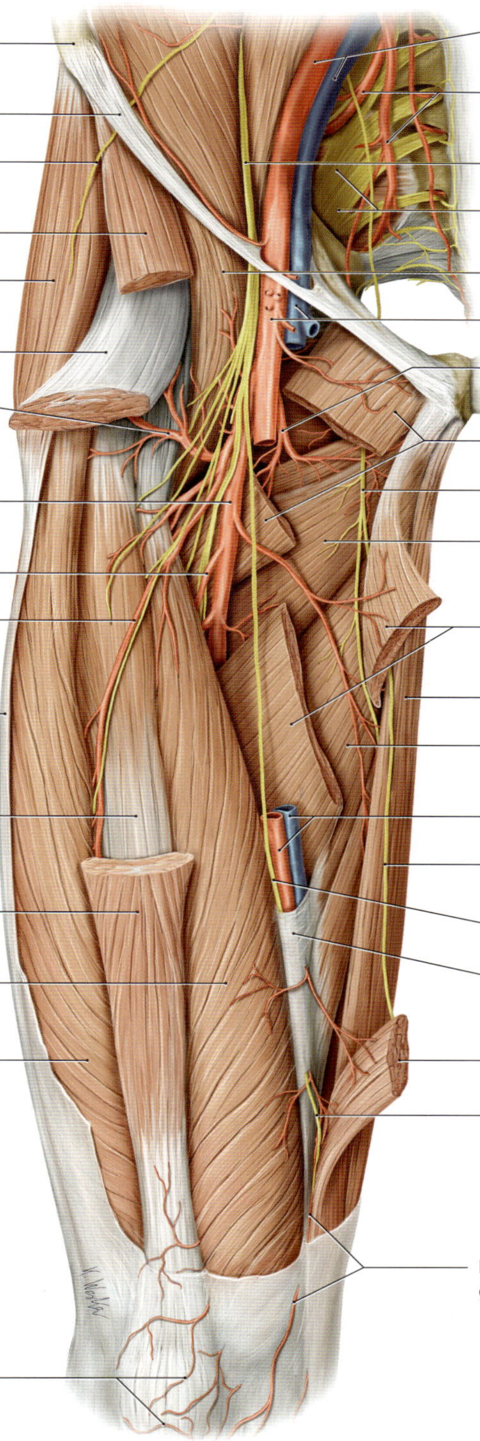

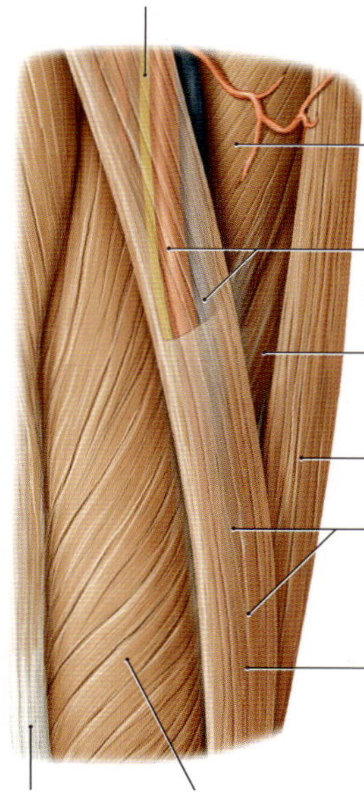

E The location of the adductor canal
Right thigh, anterior view. The saphenous nerve passes down the adductor canal on the anterior side of the thigh, accompanied by the femoral artery and vein. While both vessels continue toward the popliteal fossa through the adductor hiatus, the saphenous nerve pierces the vastoadductor membrane along with the descending genicular artery and passes to the medial side of the knee joint (see also **F**).

F The boundaries and contents of the adductor canal

Boundaries
- Adductor longus and magnus (posterior)
- Sartorius (medial)
- Vastoadductor membrane (anterior)
- Vastus medialis (lateral and anterior)

Contents
- Femoral artery
- Femoral vein
- Saphenous nerve } Pierce the vastoadductor membrane
- Descending genicular artery } Pierce the vastoadductor membrane

D The blood supply to the thigh from the deep femoral artery
Right thigh, anterior view. The sartorius, rectus femoris, adductor longus, and pectineus muscles have been partially removed, along with the central portion of the femoral artery, to demonstrate the course of the deep femoral artery. For clarity, the veins have also been removed to the level of the external iliac vein. This dissection does not show the anterior abdominal wall or the pelvic and abdominal organs above the level of the inguinal ligament. While the branches of the medial and lateral circumflex arteries mainly supply blood to the hip joint and extensors and adductors of the thigh, the terminal branches of the deep femoral artery (the first through third perforating arteries, see **B**) on the medial side of the femur pass to the back of the thigh through gaps in the insertions of the adductor muscles and supply the hamstrings (biceps femoris, semitendinosus, and semimembranosus).
Note: The vastoadductor membrane is pierced by the descending genicular artery and saphenous nerve (see **E** and **F**).

5.5 The Gluteal Region: Overview of Its Vessels and Nerves

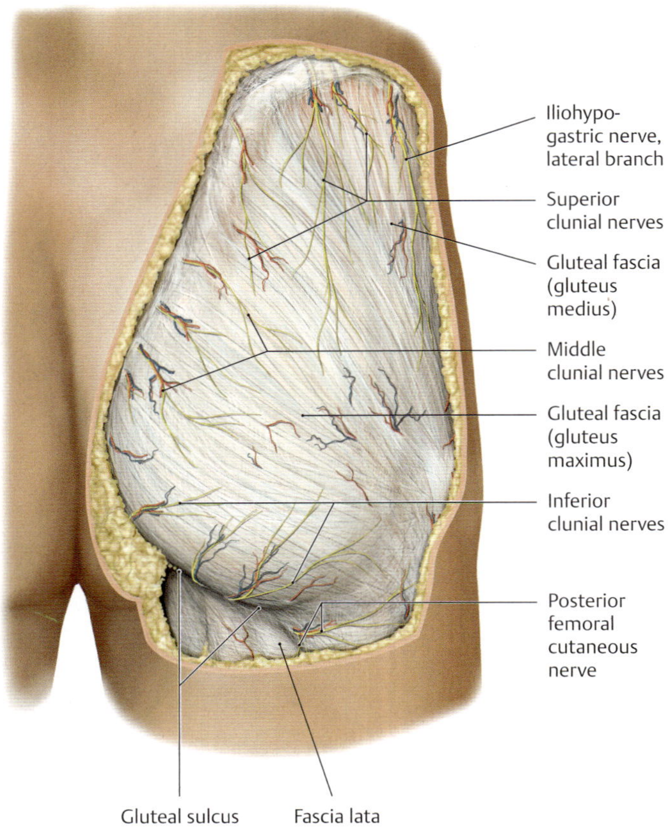

A The fasciae and cutaneous nerves of the superficial gluteal region
Right gluteal region, posterior view. The gluteal region is covered by the gluteal fascia, which is part of the fascia lata (although the term "fascia lata" strictly refers only to the part below the gluteus medius and maximus). The fascia covering the gluteus maximus forms septalike invaginations between the muscle bundles. At the junction of the gluteal region with the back of the upper thigh is the gently curved gluteal sulcus, in which thickened fiber tracts of the fascia lata run transversely across the thigh at the level of the ischial tuberosity.
Note on surface anatomy: The oblique inferior border of the gluteus maximus (see **B**) crosses the gluteal sulcus. Its course, then, is not identical to that of the gluteal sulcus.

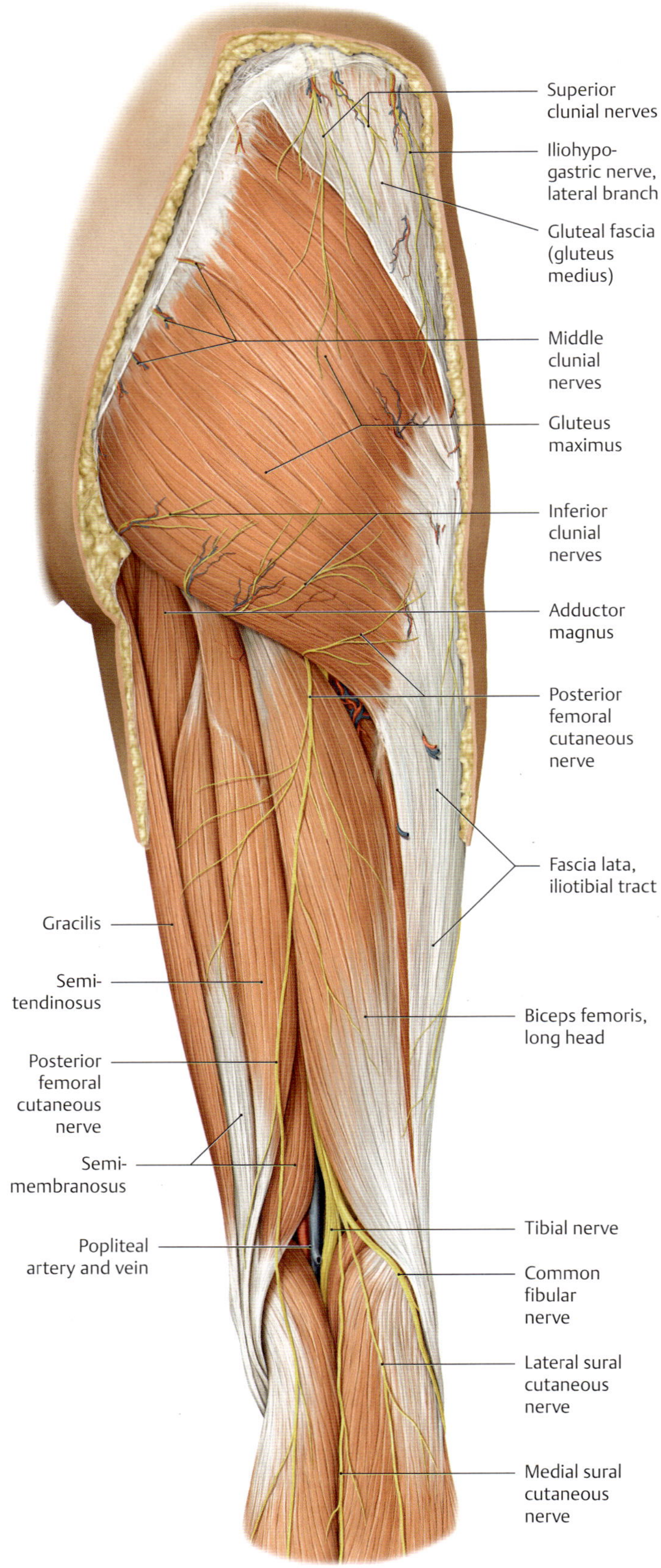

B The gluteal region and thigh with the fasciae removed
Right side, posterior view. With the fascia lata removed, the main trunk of the posterior femoral cutaneous nerve, which is subfascial over much of its course, can be traced into the popliteal fossa.

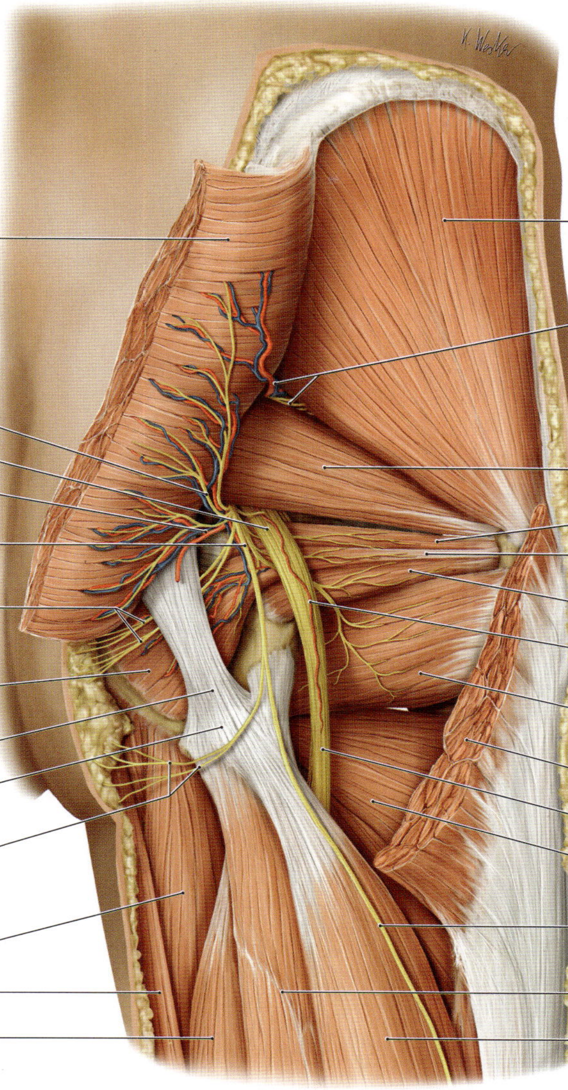

C The vessels and nerves of the deep gluteal region
Right side, posterior view, with the gluteus maximus partially removed. The neurovascular structures of the deep gluteal region traverse an extensive fatty and connective-tissue space below the gluteus maximus. The floor of this space is formed by the piriformis, obturator internus, superior gemellus, and quadratus femoris muscles. It communicates through the sciatic foramina with the connective-tissue spaces of the lesser pelvis and ischioanal fossa (not shown here). A useful topographic landmark is the piriformis, which extends from the pelvic surface of the sacrum through the greater sciatic foramen to the tip of the greater trochanter (see **A**, p. 572).

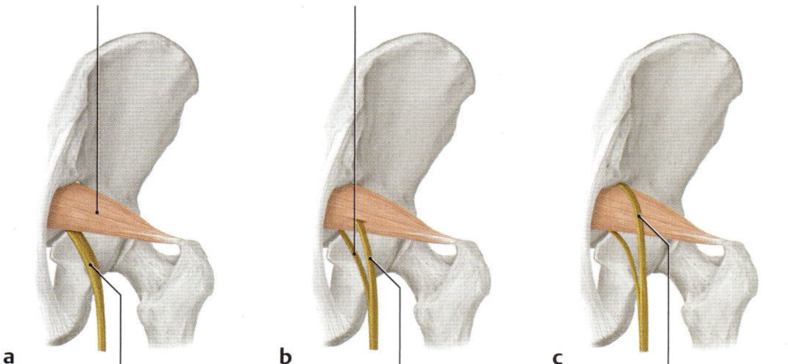

D Variable course of the sciatic nerve in relation to the piriformis

a The sciatic nerve leaves the lesser pelvis inferior to the piriformis (almost 85% of cases).

b This variant illustrates a *high division* of the sciatic nerve (approximately 15% of cases). In this pattern the fibular division (common fibular nerve) and sometimes the posterior femoral cutaneous nerve pass through the piriformis muscle and may become compressed at that location, causing a “piriformis syndrome.” Usually this term refers to the complaints that may develop following trauma to the gluteal region and are marked by severe gluteal pain. It is still uncertain whether these complaints are actually referable to the compression of sciatic nerve segments.

c In this variant the fibular part of the sciatic nerve leaves the lesser pelvis above the piriformis (rare, only about 0.5% of cases).

5.6 The Gluteal Region: The Sciatic Foramen and Sciatic Nerve

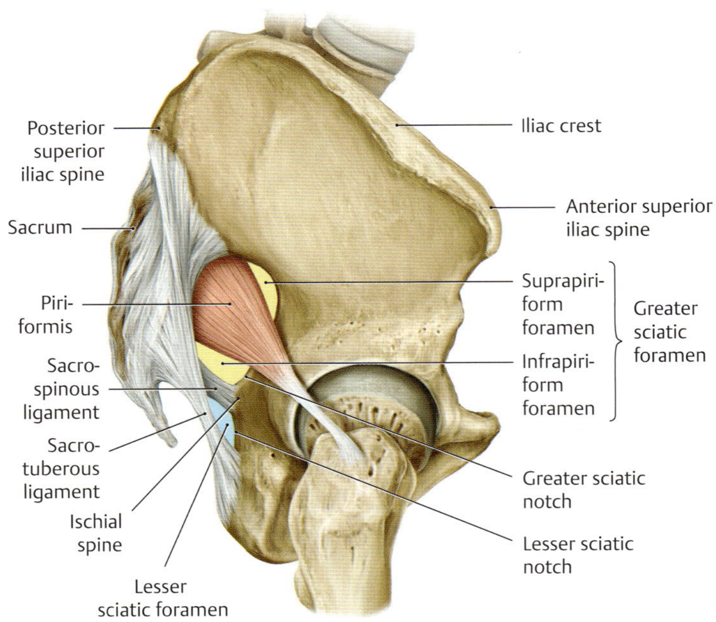

A Location of the greater and lesser sciatic foramina
Right hip bone, lateral view.

B The boundaries of the sciatic foramina and the structures that traverse them

The subgluteal connective-tissue space communicates through the sciatic foramina with the connective-tissue spaces of the lesser pelvis and the ischioanal fossa.

Foramen	Boundaries	Transmitted structures
• Greater sciatic foramen	• Greater sciatic notch • Sacrospinous ligament • Sacrum	• *Suprapiriform portion* – Superior gluteal artery and vein – Superior gluteal nerve • *Infrapiriform portion* – Inferior gluteal artery and vein – Inferior gluteal nerve – Internal pudendal artery and vein – Pudendal nerve – Sciatic nerve – Posterior femoral cutaneous nerve
• Lesser sciatic foramen	• Lesser sciatic notch • Sacrospinous ligament • Sacrotuberous ligament	– Internal pudendal artery and vein – Pudendal nerve – Obturator internus

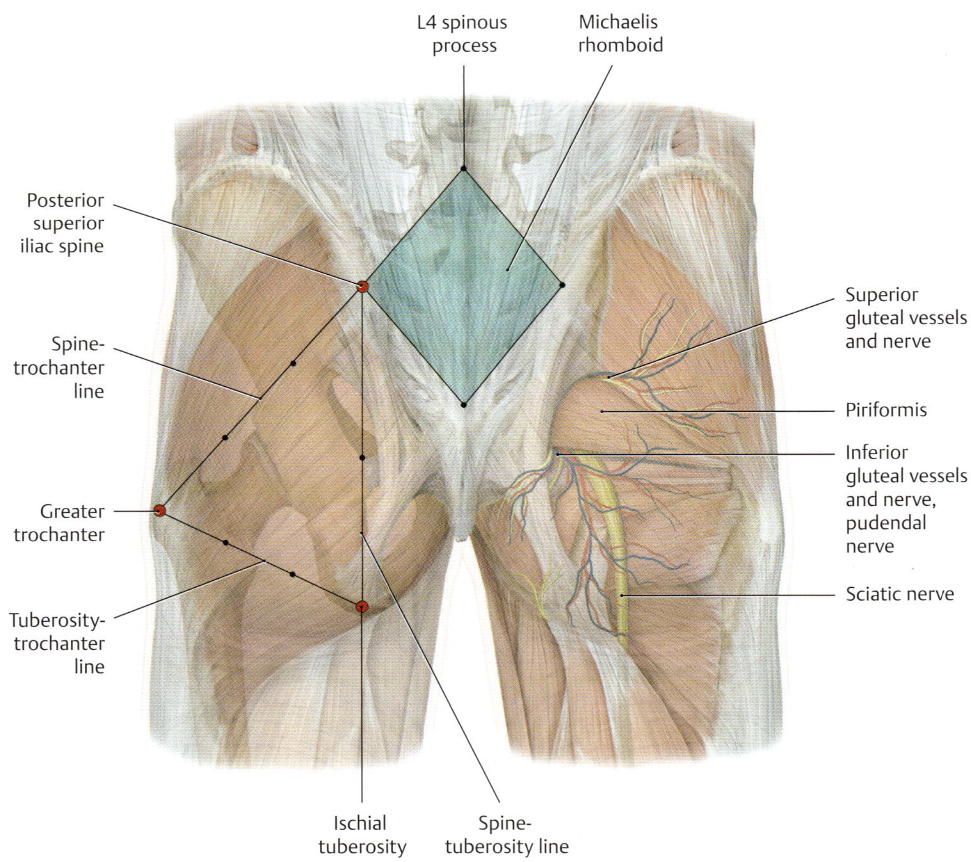

C Reference lines used for locating neurovascular structures in the gluteal region

Right and left gluteal regions, posterior view. The reference lines are drawn between the following points: the posterior superior iliac spine (lateral point of the Michaelis rhomboid), the ischial tuberosity, and the greater trochanter.

- **Spine-trochanter line:** The superior gluteal vessels emerge from the suprapiriform foramen between the middle and upper thirds of this line.
- **Tuberosity-trochanter line:** The sciatic nerve runs downward between the middle and medial thirds of this line.
- **Spine-tuberosity line:** The sciatic, inferior gluteal, and pudendal nerves and the inferior gluteal vessels emerge from the infrapiriform foramen at the midpoint of this line.

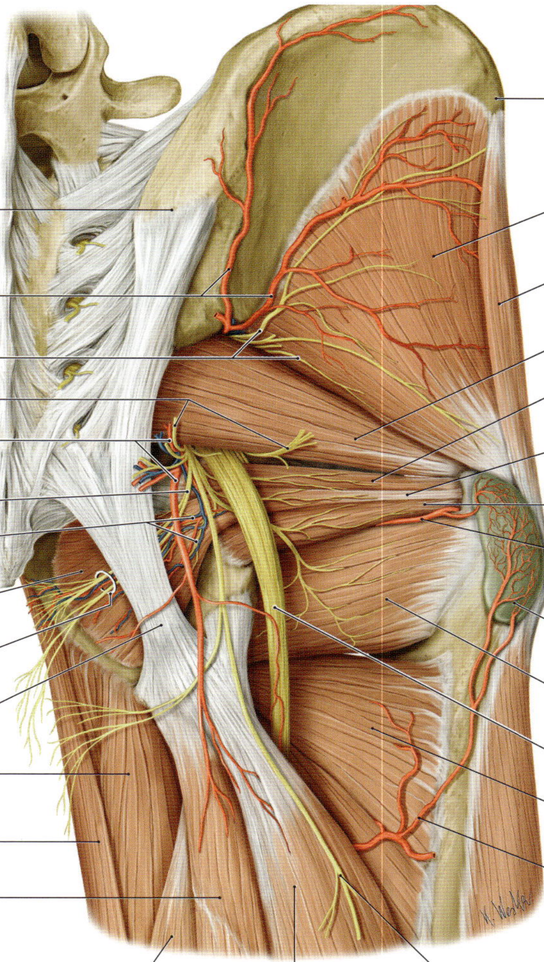

D The vessels and nerves of the gluteal region and ischioanal fossa

Right gluteal region, posterior view, with the gluteus maximus and medius removed.

Note the course of the pudendal vessels and pudendal nerve in the lateral wall of the ischioanal fossa. They run in the pudendal canal (Alcock's canal), which is formed by the obturator internus fascia (see p. 496).

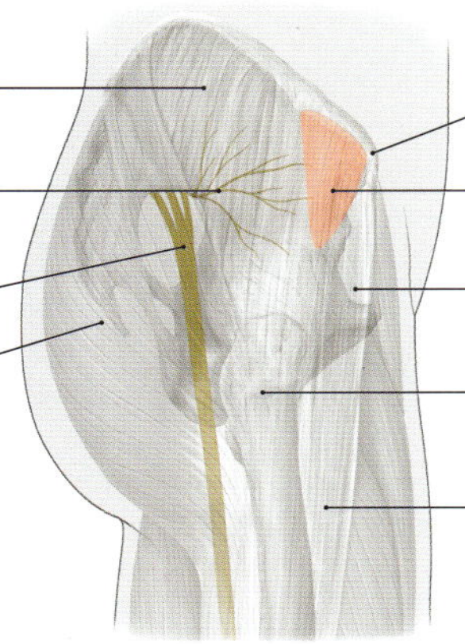

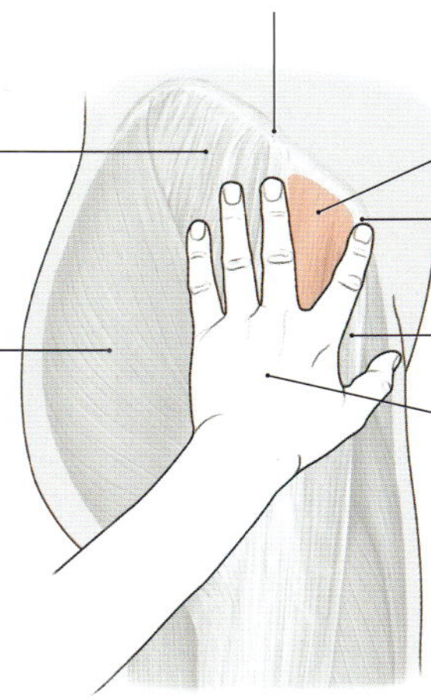

E Location of the sciatic nerve and superior gluteal nerves and their protection during intragluteal injections

Right gluteal region, lateral view.

a Two very important nerves are found in the gluteal region: the sciatic nerve and the superior gluteal nerve. To avoid jeopardizing these nerves during intramuscular injections, the needle should be inserted with the greatest possible safety margin with respect to these structures. Making the injection within the "von Hochstetter triangle" ensures that this safety margin is maintained.

b Locating the von Hochstetter triangle: The target site is located in the anterolateral gluteal region (accounting for the term "ventrogluteal injection"). To give an intramuscular injection on the right side, for example, place the palm of the left hand on the greater trochanter and the tip of the index finger on the anterior superior iliac spine. Keeping the hand in place, abduct the middle finger away from the index finger and introduce the needle perpendicular to the skin surface within the triangular zone between the two fingers and the iliac crest.

5.7 The Ischioanal Fossa

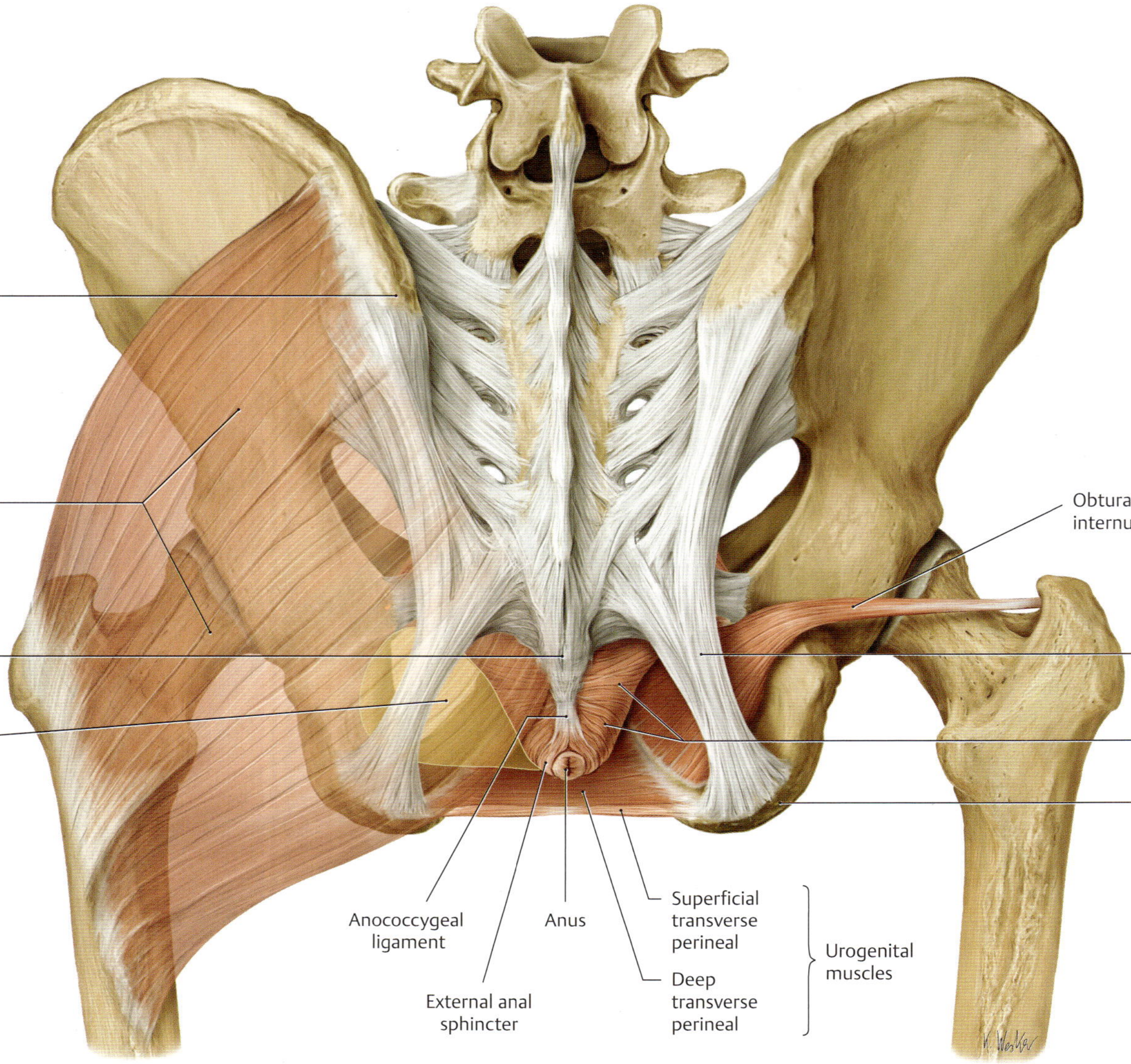

A The muscular boundaries of the ischioanal fossa
Left and right gluteal region, posterior view. The ischioanal fossa is a pyramid-shaped space located lateral to the levator ani muscle on each side. The tip of the three-sided pyramid points toward the symphysis, and the base of the pyramid faces posteriorly. The ischioanal fossa is bounded by the following muscles:

- Superamedially by the levator ani.
- Laterally by the obturator internus.
- Inferiorly by the deep transverse perineal.
- The entrance to the ischioanal fossa is bounded posteriorly by the gluteus maximus and sacrotuberous ligament.

The fatty tissue that occupies most of the ischioanal fossa (the fat pad of the ischioanal fossa) functions as a mobile pad that can slide downward and backward, for example, during bowel evacuation or during labor. It is traversed by the branches of the internal pudendal vessels and pudendal nerve (see **B**), whose trunks run in the pudendal, or Alcock's, canal; see **A**, p. 576.

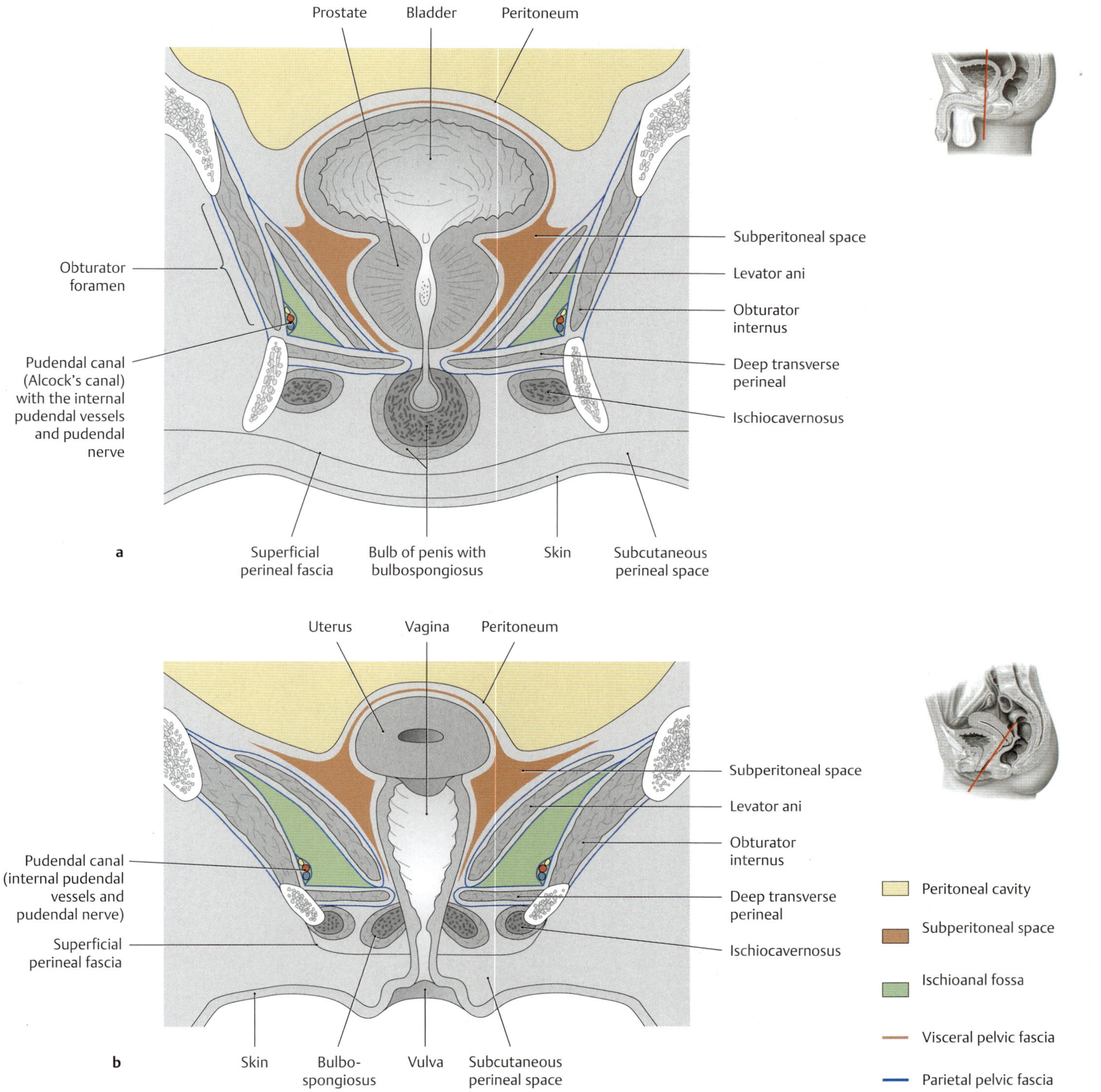

B The ischioanal fossa

a Coronal section through the male pelvis at the level of the prostate.

b Oblique coronal section through the female pelvis at the level of the vagina.

The pelvic cavity is defined by the peritoneum and the pelvic floor is divided into three levels:

- Upper level: peritoneal cavity of the lesser pelvis
- Middle level: subperitoneal (supralevator) space
- Lower level: infralevator space (ischioanal fossa)

The pelvic organs make varying contributions to the shape of the peritoneal cavity and subperitoneal space, but they are not represented in the ischioanal fossa. While the peritoneal cavity is lined by the parietal peritoneum and the visceral peritoneum (on intraperitoneal organs such as the ovary), the subperitoneal space is lined by the pelvic fasciae (composed of parietal and visceral layers, see p. 191).

5.8 The Pudendal Canal and Perineal Region (Urogenital and Anal Region)

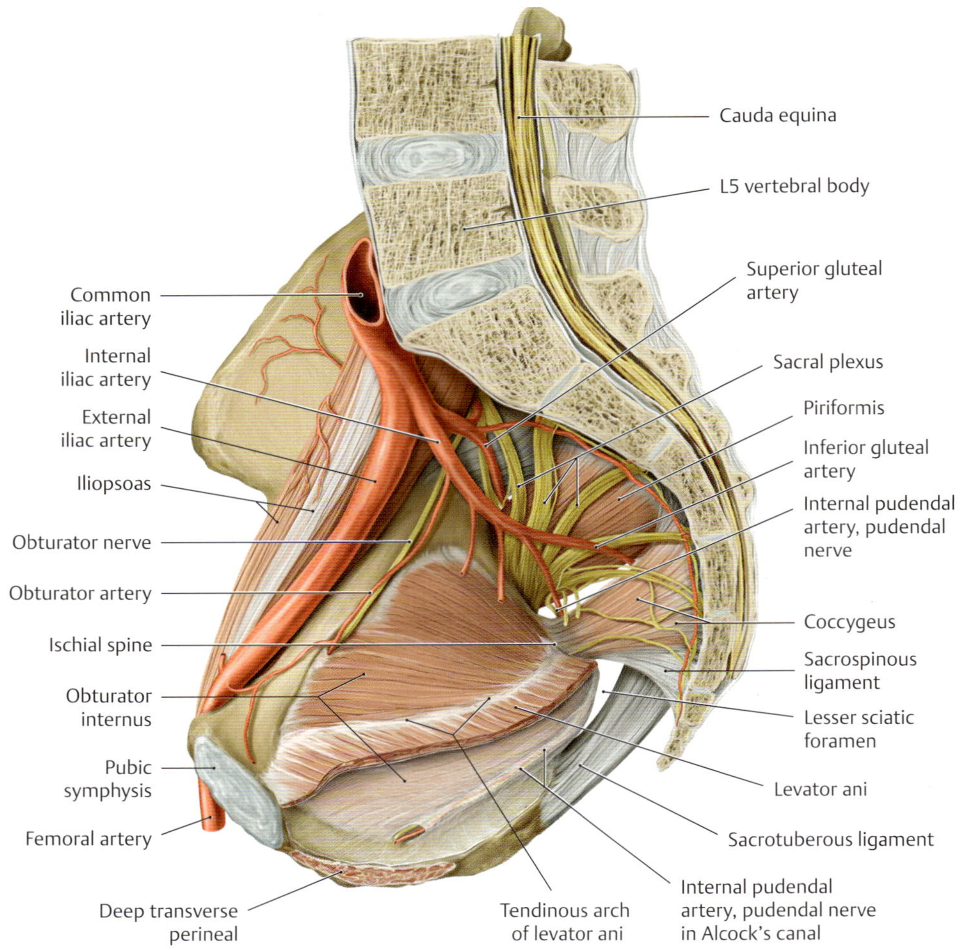

A Location of the pudendal canal (Alcock's canal) and the neurovascular structures it contains

Right half of the pelvis, medial view. All of the muscles have been removed except for the psoas major, piriformis, and obturator internus. For clarity, the individual veins are not shown. The pudendal canal is formed by the obturator internus fascia. It begins just below the ischial spine and courses in the lateral wall of the ischioanal fossa below the tendinous arch of levator ani, passing toward the pubic symphysis and the posterior border of the urogenital muscles (see p. 170). The neurovascular structures that are transmitted by the canal (the internal pudendal vessels, of which only the artery is shown, and the pudendal nerve, see **B**) exit the lesser pelvis through the greater sciatic foramen and enter the pudendal canal through the lesser sciatic foramen. They pass through the canal toward the pubic symphysis and the posterior border of the urogenital muscles.

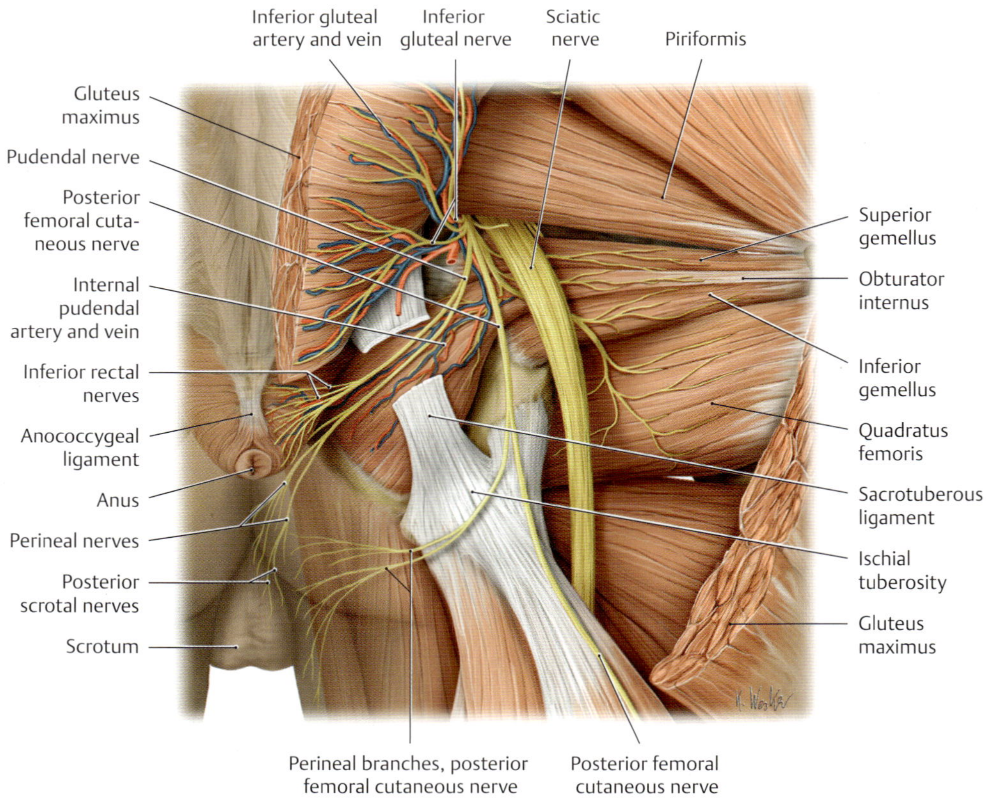

B Distribution of the pudendal nerve and internal pudendal vessels to the anus, perineum, and external genitalia

Gluteal region and ischioanal fossa on the right side, posterior view. The gluteus maximus and sacrotuberous ligament have been partially removed, and all fatty tissue has been removed from the ischioanal fossa to demonstrate the course of the pudendal nerve and the internal pudendal vessels. On their way through the pudendal canal (not shown here in order to display the course of the nerve and vessels below the sacrotuberous ligament), the various neural and vascular branches are successively distributed in a fan-shaped pattern to the anus, perineum, and external genitalia. It is very common in obstetrics to perform a pudendal nerve block by anesthetizing the pudendal nerve at the level of the ischial spine (i. e., before it branches into the inferior rectal, perineal, dorsal clitoral, and posterior labial nerves; see p. 560).

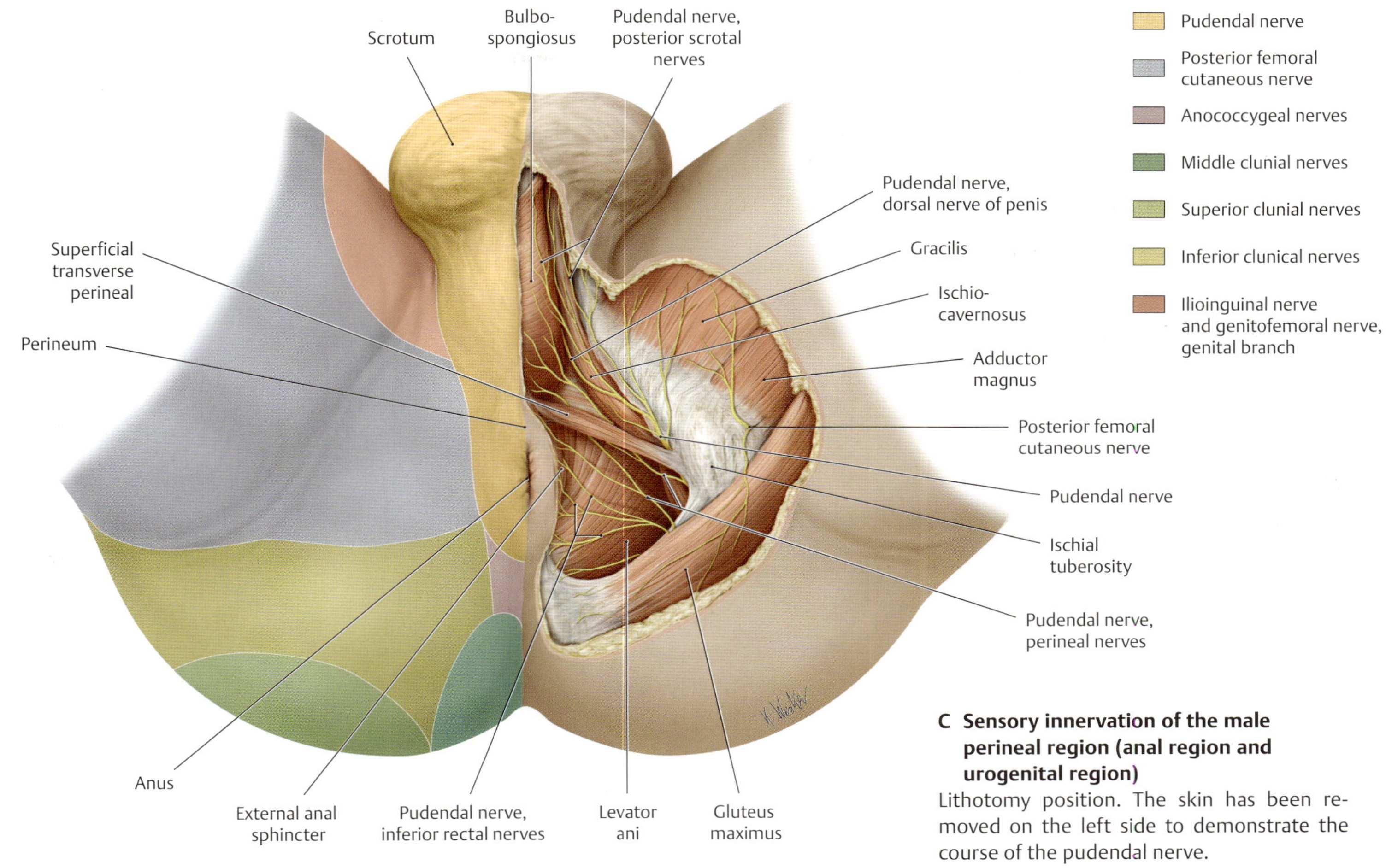

C Sensory innervation of the male perineal region (anal region and urogenital region)

Lithotomy position. The skin has been removed on the left side to demonstrate the course of the pudendal nerve.

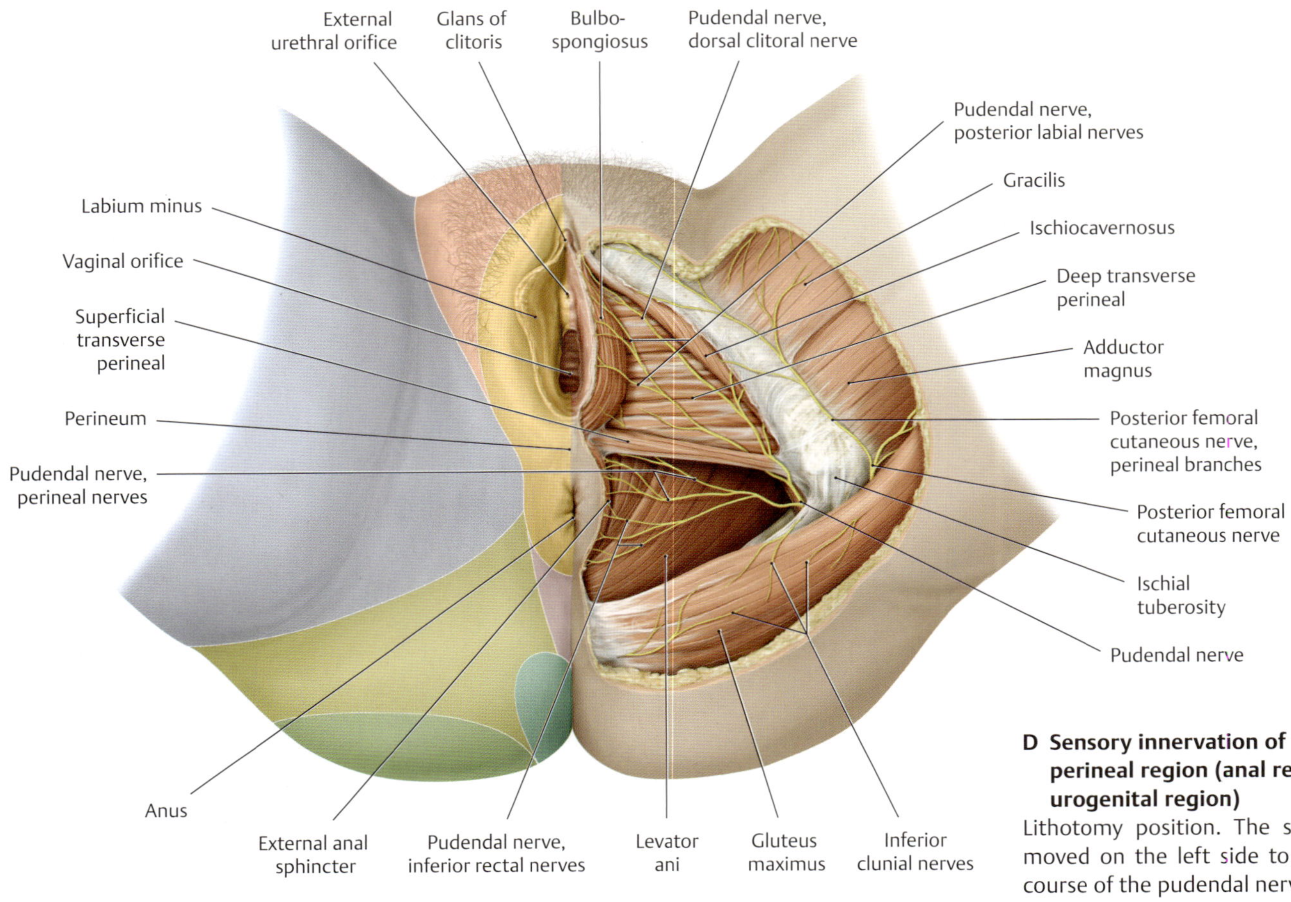

D Sensory innervation of the female perineal region (anal region and urogenital region)

Lithotomy position. The skin has been removed on the left side to demonstrate the course of the pudendal nerve.

5.9 The Posterior Thigh Region and Popliteal Region

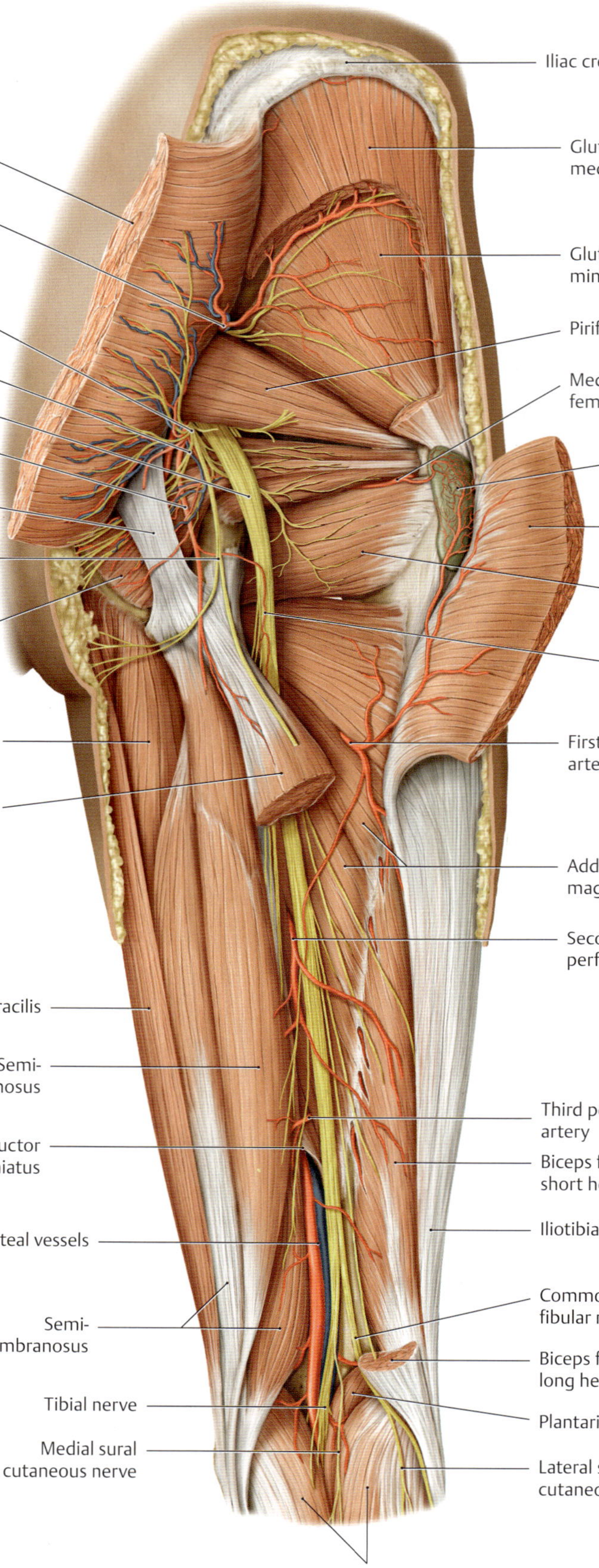

A The vessels and nerves of the posterior thigh

Right thigh, posterior view. To demonstrate the vessels and nerves in their course from the gluteal region down the back of the thigh to the popliteal fossa (see **C**), the skin and muscle fasciae have been removed, and the following muscles have been partially removed: gluteus maximus, gluteus medius, and biceps femoris. The semimembranosus has been retracted slightly medially to display the adductor hiatus (transmits the femoral artery and vein) (deeper neurovascular structures in the popliteal fossa are shown in **F**). The posterior femoral region receives most of its blood supply from branches of the deep femoral artery (the first through third perforating arteries) and the deep branch (not shown) of the medial circumflex femoral artery. The proximal part of the sciatic nerve is supplied by the sciatic artery, which is a branch of the inferior gluteal artery, and its distal part is supplied by branches of the first through third perforating arteries.

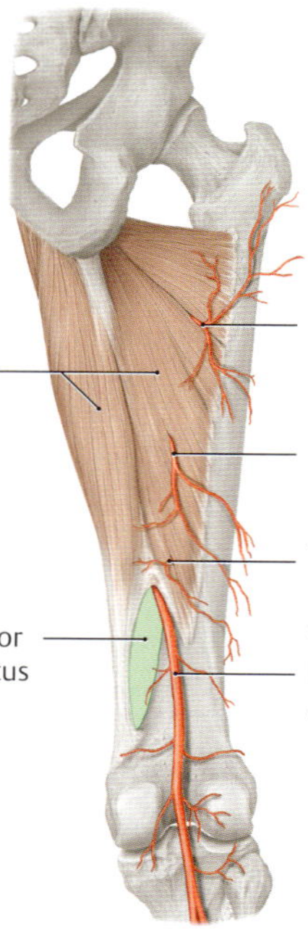

B Sites of emergence of the perforating arteries on the back of the thigh

Right thigh, posterior view. All of the muscles have been removed except for the adductor magnus.

Note: The femoral artery enters the popliteal fossa through the adductor hiatus, thereafter becoming the popliteal artery.

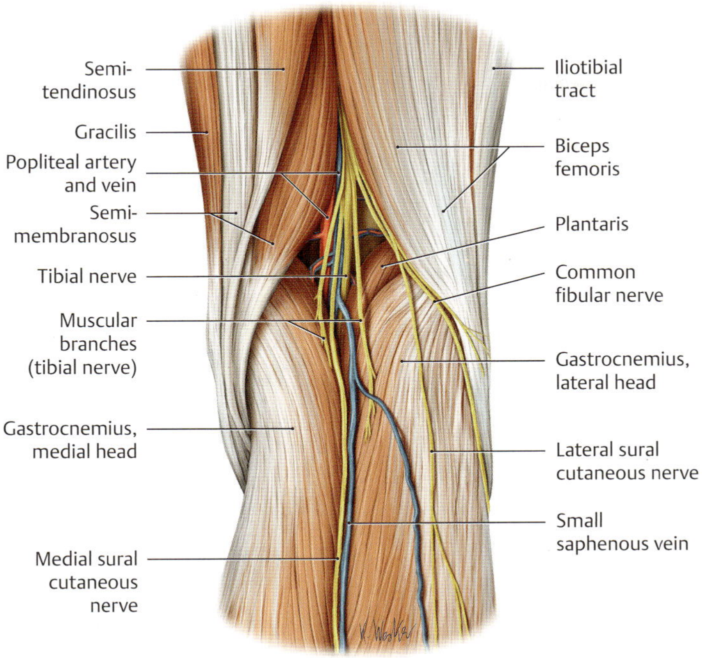

C The muscular boundaries of the popliteal fossa

Right popliteal fossa, posterior view. For clarity the skin, fasciae, and fat pads have been removed.

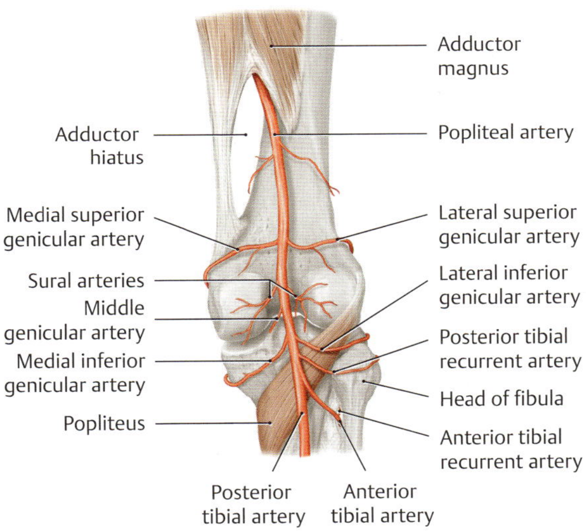

D Branches of the popliteal artery that course in the popliteal fossa

Right knee joint, posterior view. The popliteal artery begins at the outlet of the adductor canal and ends at the level of the popliteus muscle, where it divides into the anterior tibial artery and posterior tibial artery.

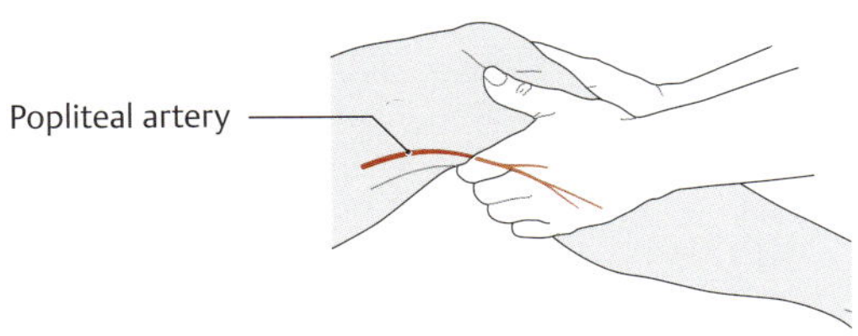

E Palpation of the popliteal artery in the popliteal fossa

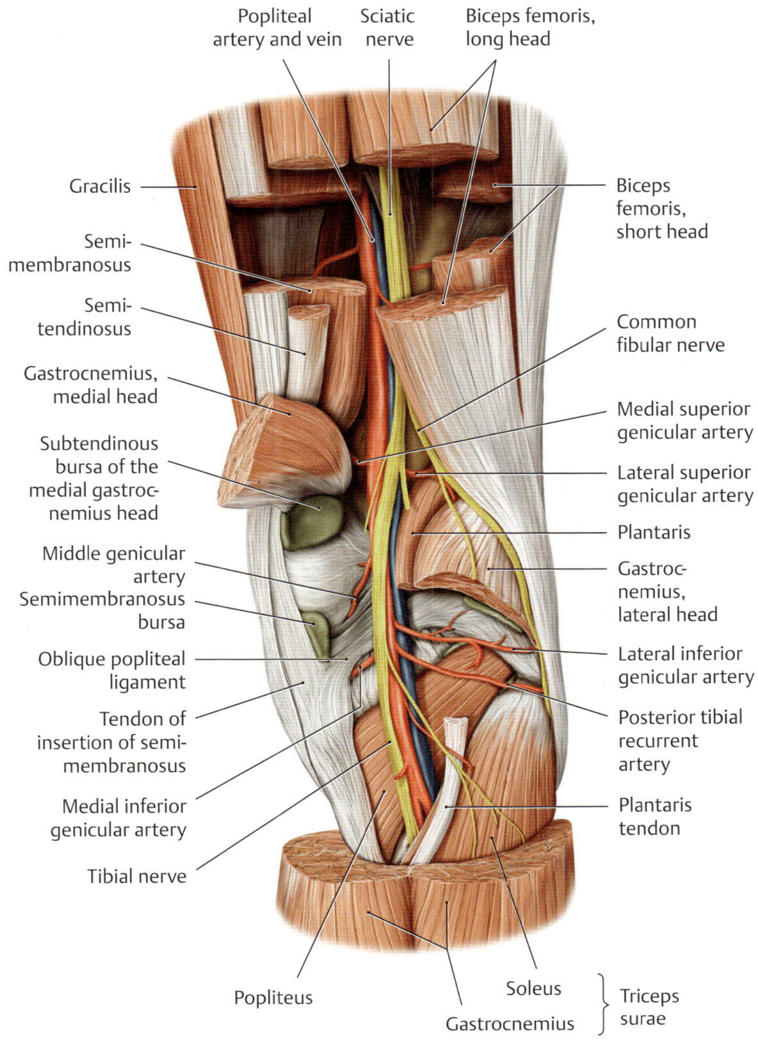

F Deep neurovascular structures in the popliteal fossa

Right knee joint, posterior view. Portions of both heads of the gastrocnemius and portions of the hamstrings have been removed to demonstrate the course of the deep neurovascular structures in the popliteal fossa. Five vessels, some paired, branch from the middle portion of the popliteal artery to supply the knee joint (see **D**):

- Lateral and medial superior genicular arteries
- Middle genicular artery
- Lateral and medial inferior genicular arteries

One of these vessels, the middle genicular artery, pierces the joint capsule of the knee in the area of the oblique popliteal ligament and supplies the cruciform ligaments. The other vessels run forward on the medial and lateral sides to form the arterial network (articular rete) of the knee. The posterior and anterior tibial recurrent arteries contribute to the articular rete. The paired sural arteries supply the two heads of the gastrocnemius (see **D**, removed in **F**).

Note on the medial side the subtendinous bursa of the gastrocnemius, which consistently communicates with the joint cavity of the knee, and the bursa of the semimembranosus, which occasionally communicates with the bursa of the gastrocnemius head (this forms an extensive recess in the joint cavity of the knee, which may become abnormally enlarged to form a Baker cyst; see p. 456).

5.10 The Posterior Leg Region and the Tarsal Tunnel

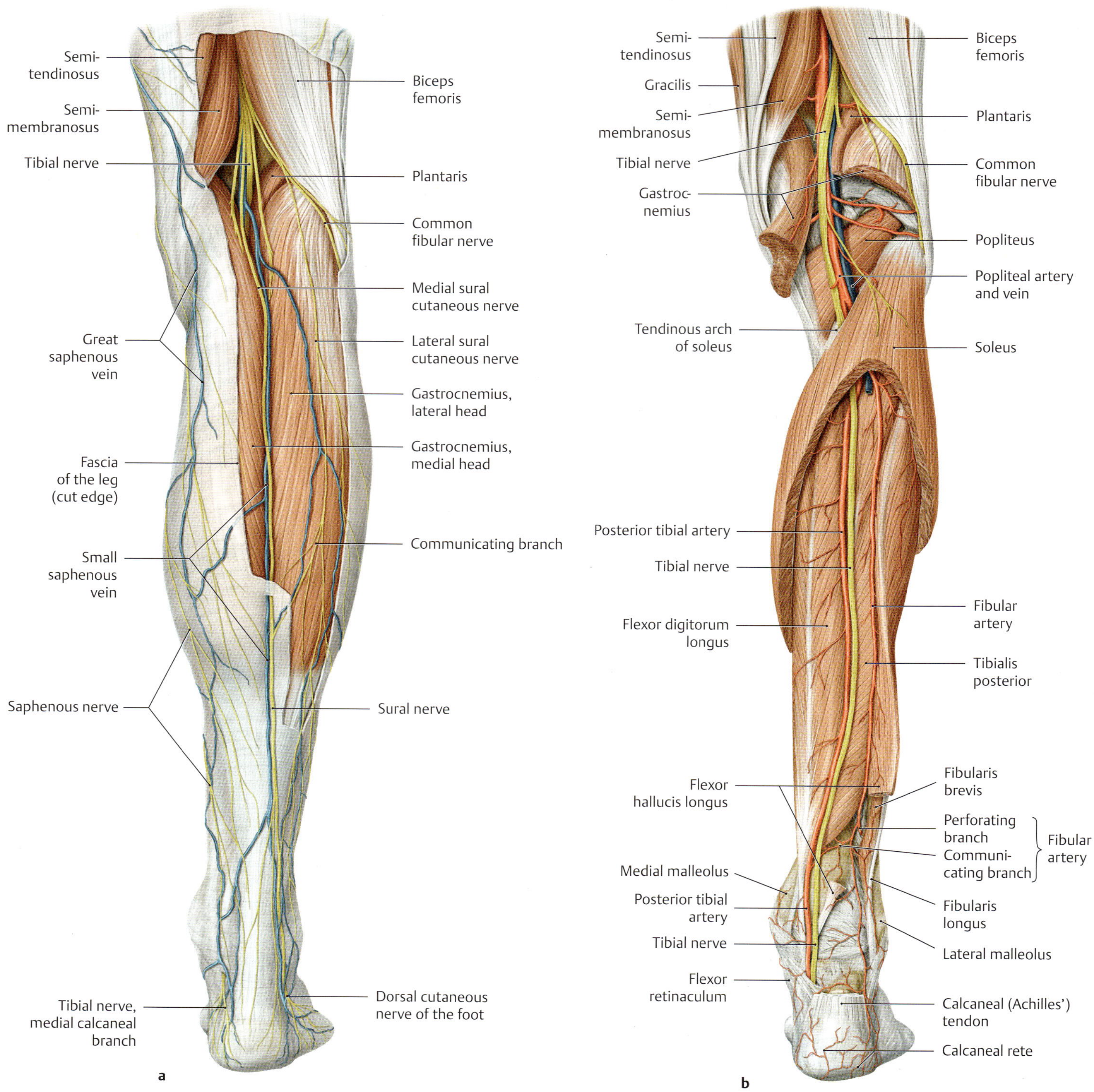

A The neurovascular structures in the superficial and deep posterior compartments
Right leg, posterior view.

a Neurovascular structures in the *superficial posterior compartment*: The superficial layer of the fascia of the leg ensheaths the triceps surae and has been partially removed proximally.

b Neurovascular structures in the *deep posterior compartment* after partial removal of the triceps surae and the deep layer of the fascia of the leg. The popliteal artery divides into the anterior and posterior tibial arteries at the distal border of the popliteus. The anterior tibial artery pierces the interosseous membrane (not shown here, see **B**) and passes to the anterior side of the leg, entering the anterior compartment. The posterior tibial artery, accompanied by the tibial nerve, passes below the tendinous arch of the soleus into the deep posterior compartment, almost immediately gives off the fibular artery, and then continues distally behind the medial malleolus to the plantar side of the foot. The deep posterior compartment is one of four poorly distensible muscle compartments in the leg ("fibro-osseous canals"), which are potential sites for the development of a compartment syndrome following a vascular injury (see p. 585).

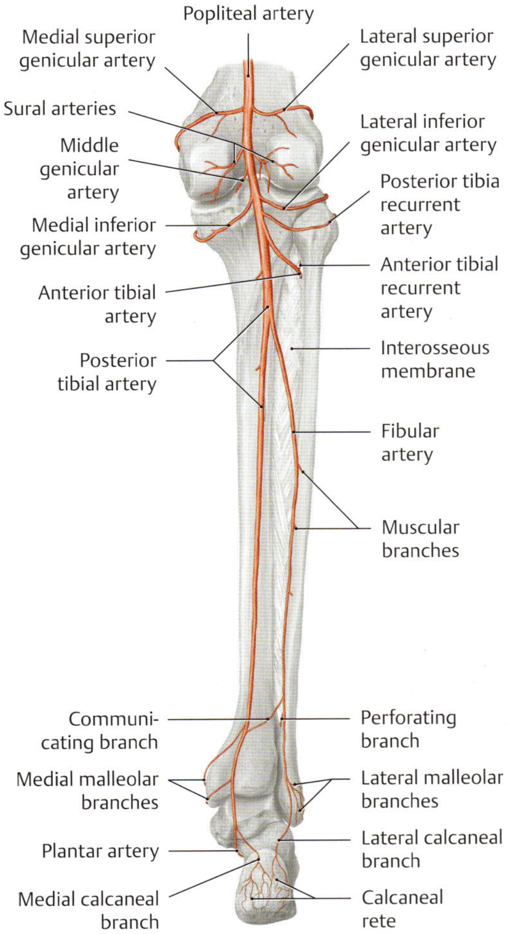

B The arteries of the leg
Posterior view.

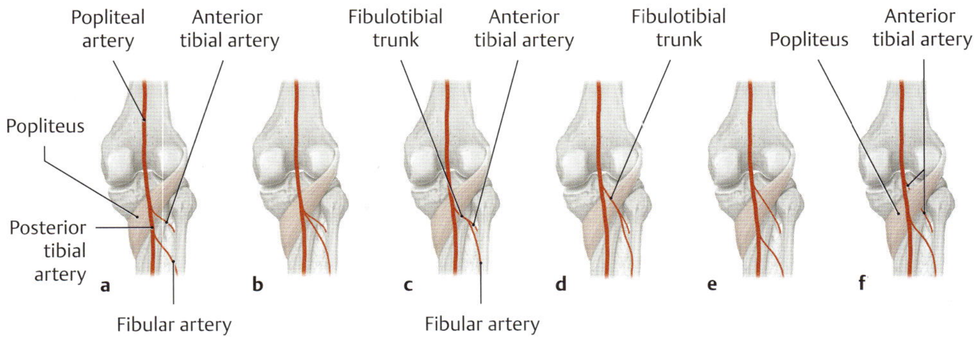

C Typical branching pattern of the popliteal artery and variants (after Lippert and Pabst)

- **a** Typical pattern: The anterior tibial artery arises from the popliteal artery behind the popliteus muscle, and the popliteal artery continues as the posterior tibial artery (i. e., the popliteal artery divides into an anterior and posterior tibial artery). Farther distally, the fibular artery branches from the posterior tibial artery.
- **b** The anterior tibial artery and fibular artery arise *by a common trunk* from the popliteal artery behind the popliteus (4% of cases).
- **c** The fibular artery arises from the anterior tibial artery (= fibulotibial trunk, 1% of cases).
- **d** The fibulotibial trunk arises proximal to the popliteus (1% of cases).
- **e** The anterior tibial artery arises from the popliteal artery proximal to the popliteus (1% of cases).
- **f** The anterior tibial artery runs between the popliteus and the tibia (1% of cases).

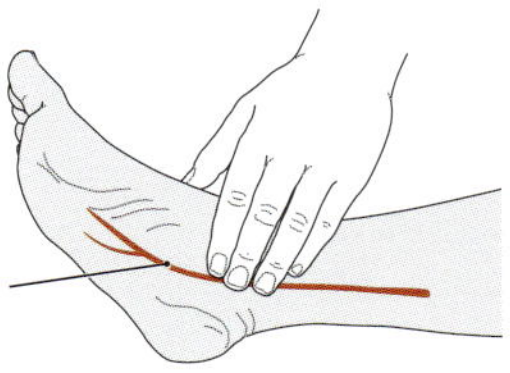

D Palpation of the posterior tibial artery below the medial malleolus

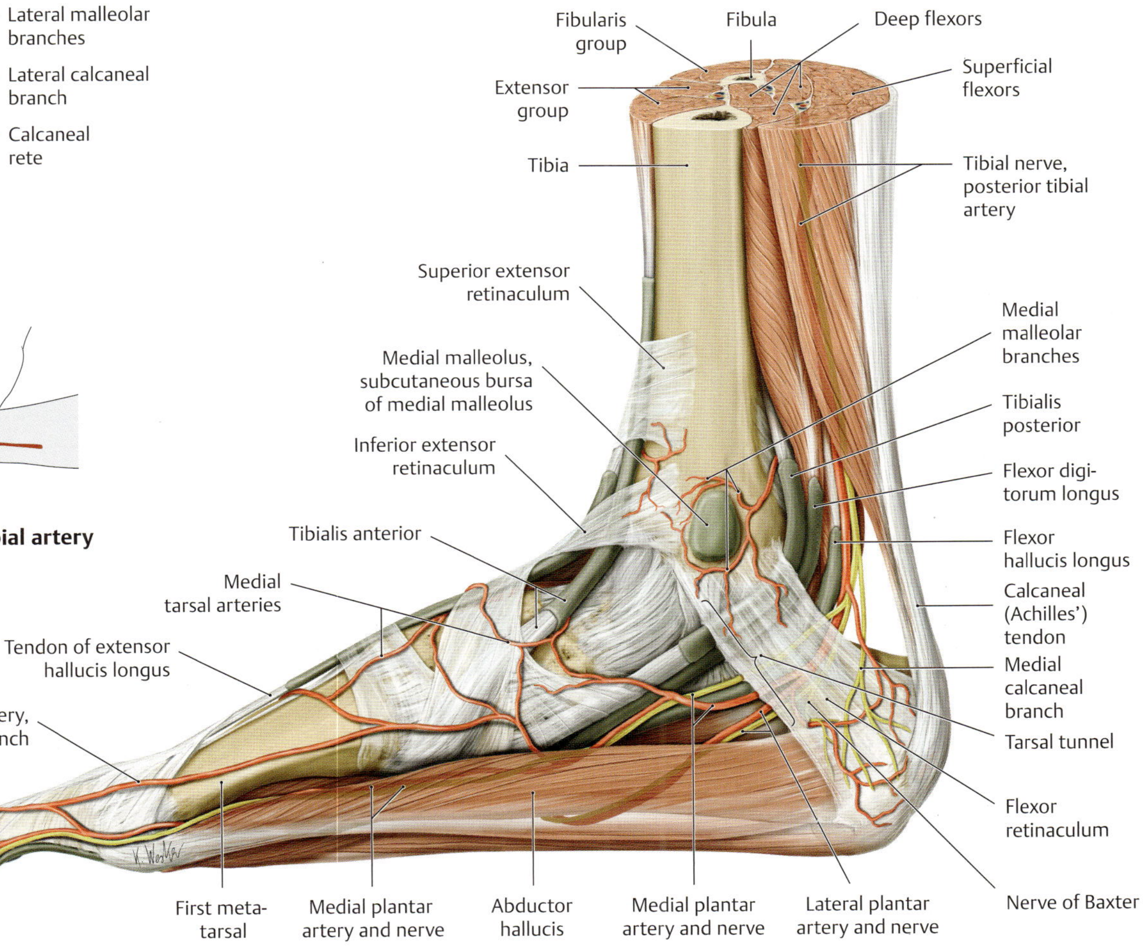

E The neurovascular structures of the medial malleolar region
Right foot, medial view. The neurovascular structures pass from the deep flexor compartment to the plantar side of the foot through the tarsal tunnel, between the flexor retinaculum and medial malleolus. They are accompanied by the tendons of insertion of the long flexors (tibialis posterior, flexor digitorum longus, flexor hallucis longus) in their synovial sheaths.

Note the division of the tibial nerve into the medial and lateral plantar nerves and the division of the posterior tibial artery into the medial and lateral plantar arteries within the malleolar canal. Compression of the nerves at this site can cause a medial or posterior tarsal tunnel syndrome (see p. 559). Refer to p. 556 for the nerve of Baxter.

5.11 The Sole of the Foot

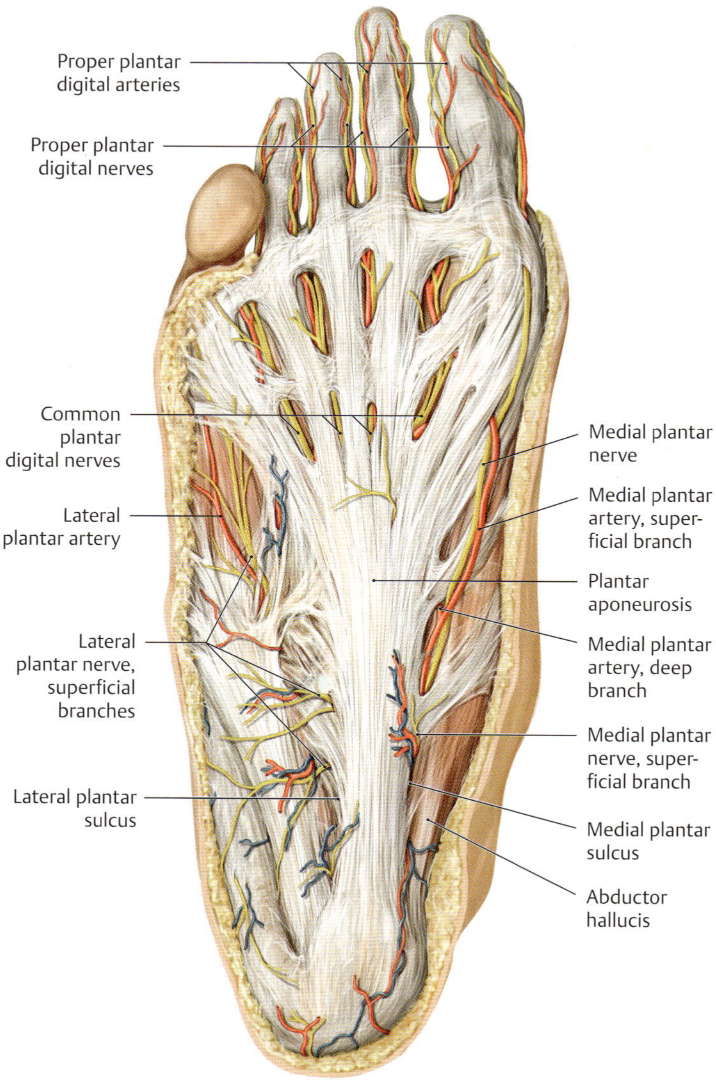

A The plantar arteries and nerves of the foot (superficial layer)
Right foot, plantar view. The skin and subcutaneous tissue have been removed to demonstrate the plantar aponeurosis and superficial neurovascular structures.

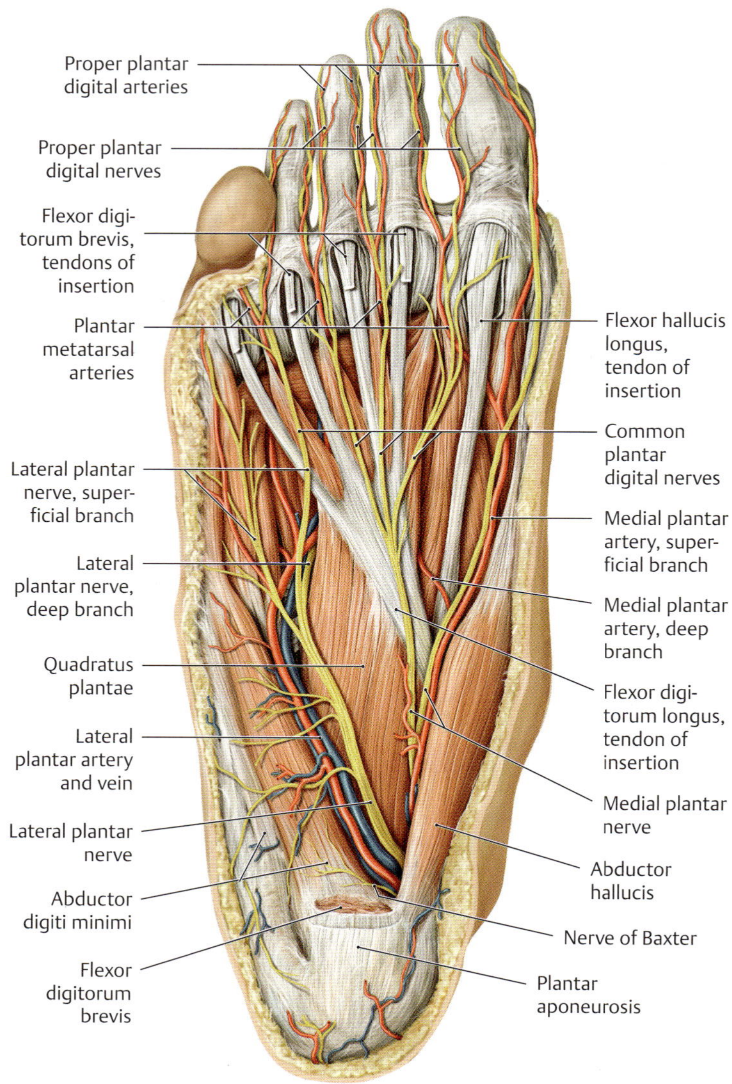

B The plantar arteries and nerves of the foot (middle layer)
Right foot, plantar view, with the plantar aponeurosis and flexor digitorum brevis removed. Refer to p. 556 for the nerve of Baxter.

C The plantar arteries of the foot: possible variants
Right foot, plantar view. Any of four basic anatomic variants may be seen (after Lippert and Pabst):

- **a** The deep plantar arch and the plantar metatarsal arteries arising from it are supplied entirely by the deep plantar branch of the dorsal pedal artery (53% of cases).
- **b** The first through third plantar metatarsal arteries are supplied by the deep plantar branch of the dorsal pedal artery, the fourth plantar metatarsal artery by the deep branch of the lateral plantar artery (19% of cases).
- **c** The first and second plantar metatarsal arteries are supplied by the deep plantar branch of the dorsal pedal artery, the third and fourth metatarsal arteries by the deep branch of the lateral plantar artery (13% of cases).
- **d** The deep plantar arch and the first through fourth plantar metatarsal arteries are supplied entirely by the deep branch of the lateral plantar artery (7% of cases).

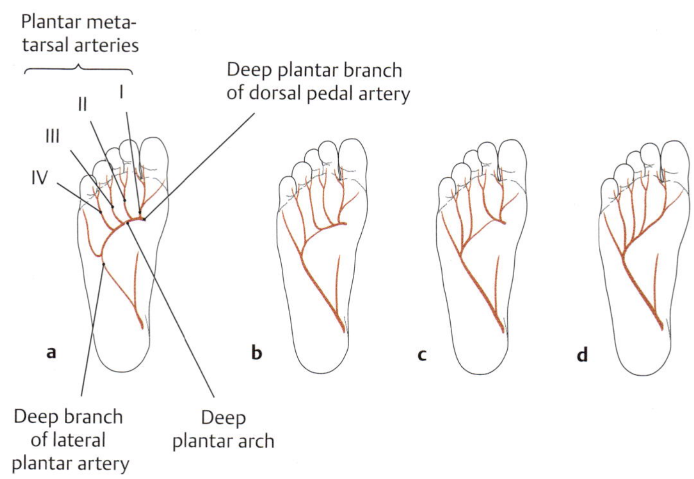

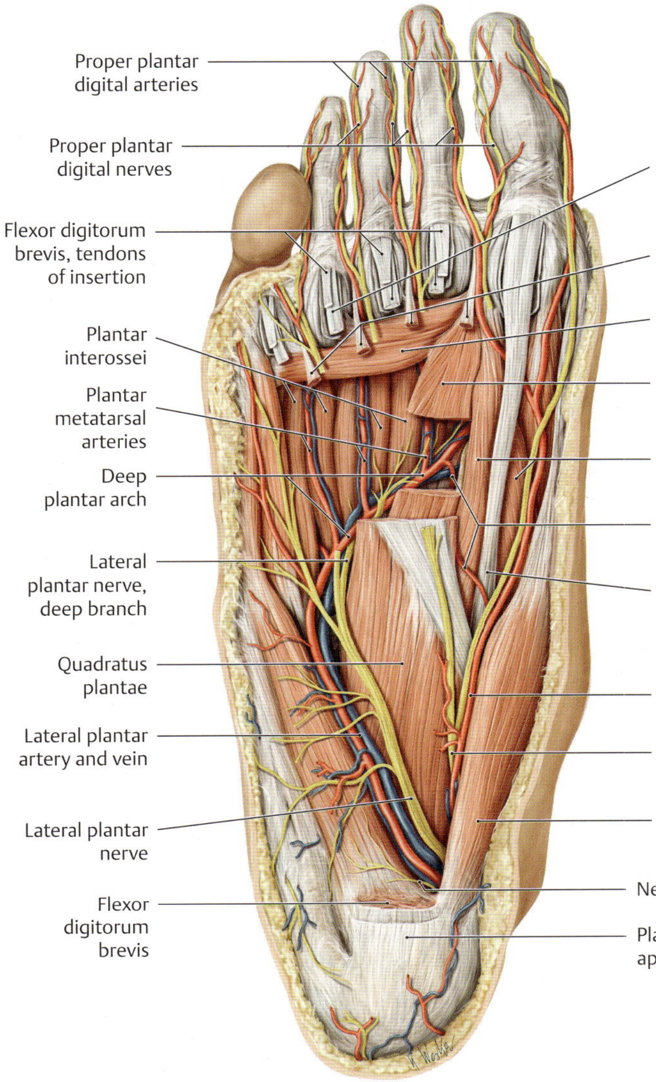

D The plantar arteries and nerves of the foot (deep layer)

Right foot, plantar view. The plantar aponeurosis, flexor digitorum brevis, the tendons of flexor digitorum longus, and the oblique head of adductor hallucis have been removed to demonstrate the deep plantar arch and the deep branch of the lateral plantar nerve.

Note: The superficial branch of the lateral plantar nerve and the lateral plantar artery course in the *lateral plantar sulcus*, while the superficial branches of the medial plantar nerve and medial plantar artery course in the *medial plantar sulcus* (see **A**). The superficial branches of the medial and lateral plantar arteries help to supply the critically important pressure-chamber system in the sole of the foot (see p. 490).

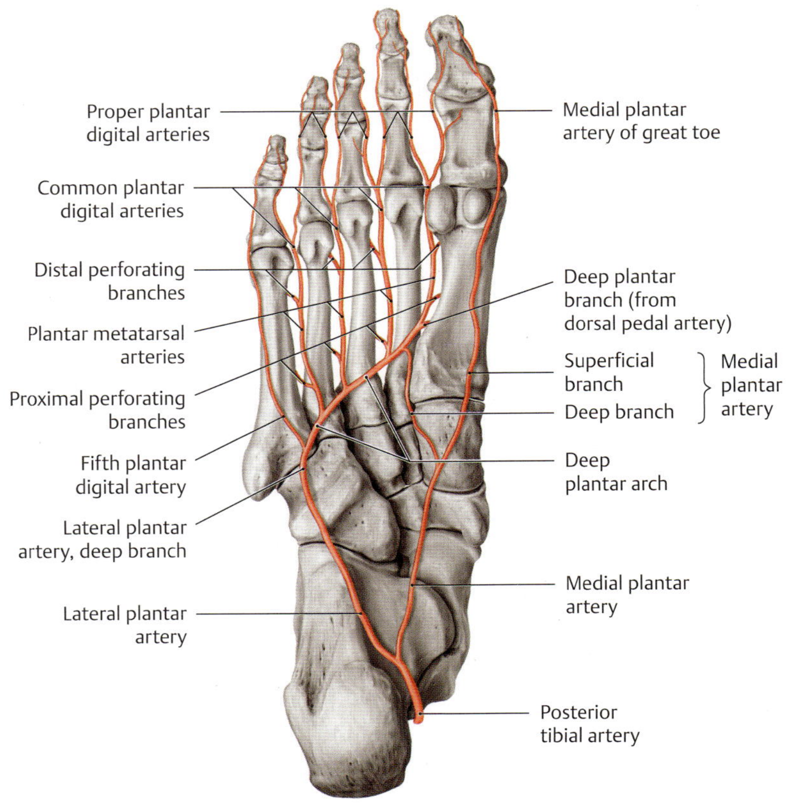

E Overview of the plantar arteries of the foot

Right foot, plantar view.

The *deep plantar arch* is an arterial arcade in the sole of the foot that is formed by the deep plantar branch of the *dorsal pedal artery* and also by the deep branch of the *lateral plantar artery*. Often these two arteries that supply the deep plantar arch differ in size and thus make different contributions to the first through fourth plantar metatarsal arteries that consistently branch from the deep plantar arch (see **C**).

5.12 The Anterior Leg Region and Dorsum of the Foot: Cutaneous Innervation

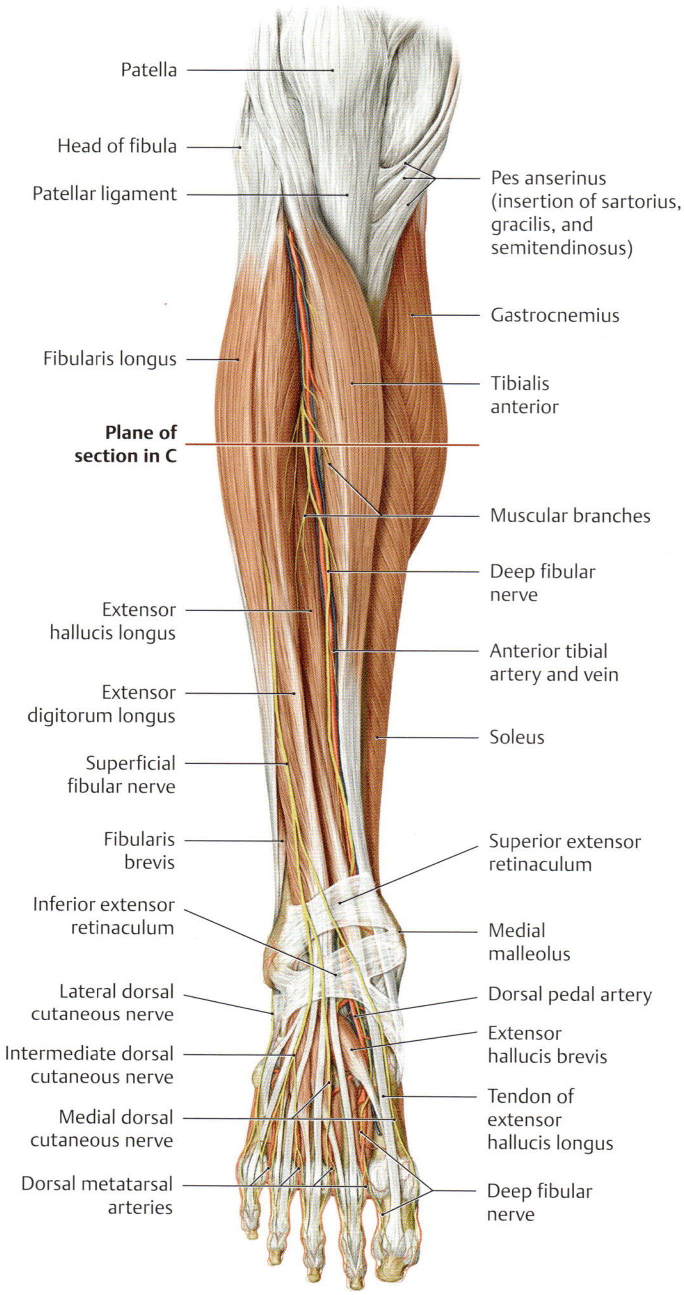

A The neurovascular structures of the anterior compartment and dorsum of the foot

Right leg with the foot in plantar flexion, anterior view. The skin, subcutaneous tissue, and fasciae have been removed and the tibialis anterior and extensor hallucis longus muscles have been retracted to demonstrate the anterior tibial vessels (= anterior tibial artery and vein). The *anterior tibial artery* crosses beneath the tendon of the extensor hallucis longus at the junction of the leg with the dorsum of the foot. Below the extensor retinaculum it becomes the *dorsal pedal artery*, which runs lateral to the hallucis longus tendon on the dorsum of the foot, accompanied by the terminal branch of the *deep fibular nerve* (the site for taking the pedal pulse is shown in **E**). The *deep fibular nerve* may be compressed in its passage beneath the inferior extensor retinaculum (with sensory disturbances affecting the first and second toes).

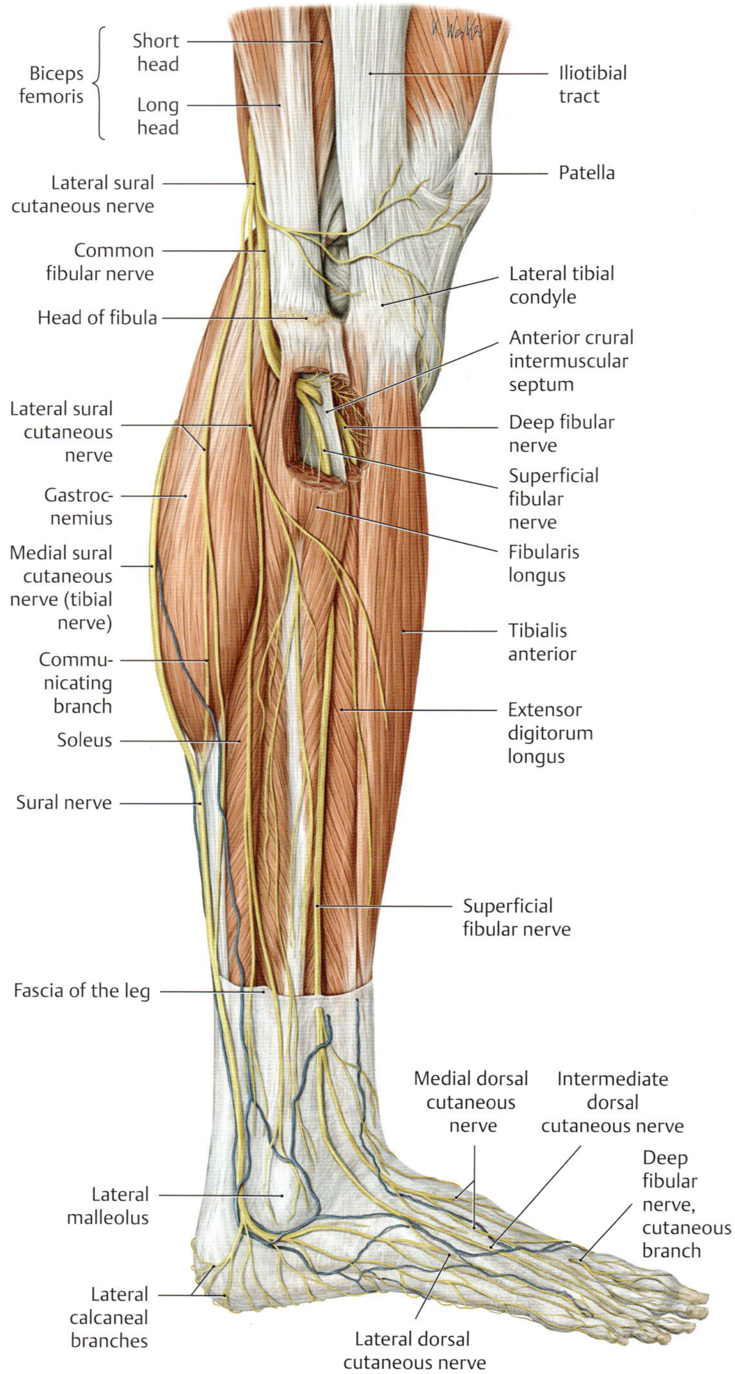

B Division of the common fibular nerve into the deep and superficial fibular nerves

Right leg, lateral view. The origins of the fibularis longus and extensor digitorum longus have been excised below the head of the fibula and the lateral tibial condyle. After the *common fibular nerve* bifurcates in the proximal part of the lateral compartment, the *superficial fibular nerve* remains in the lateral compartment. The *deep fibular nerve* pierces the anterior intermuscular septum and descends with the anterior tibial vessels in the extensor compartment (**C** gives a sectional view of the compartments in the leg).

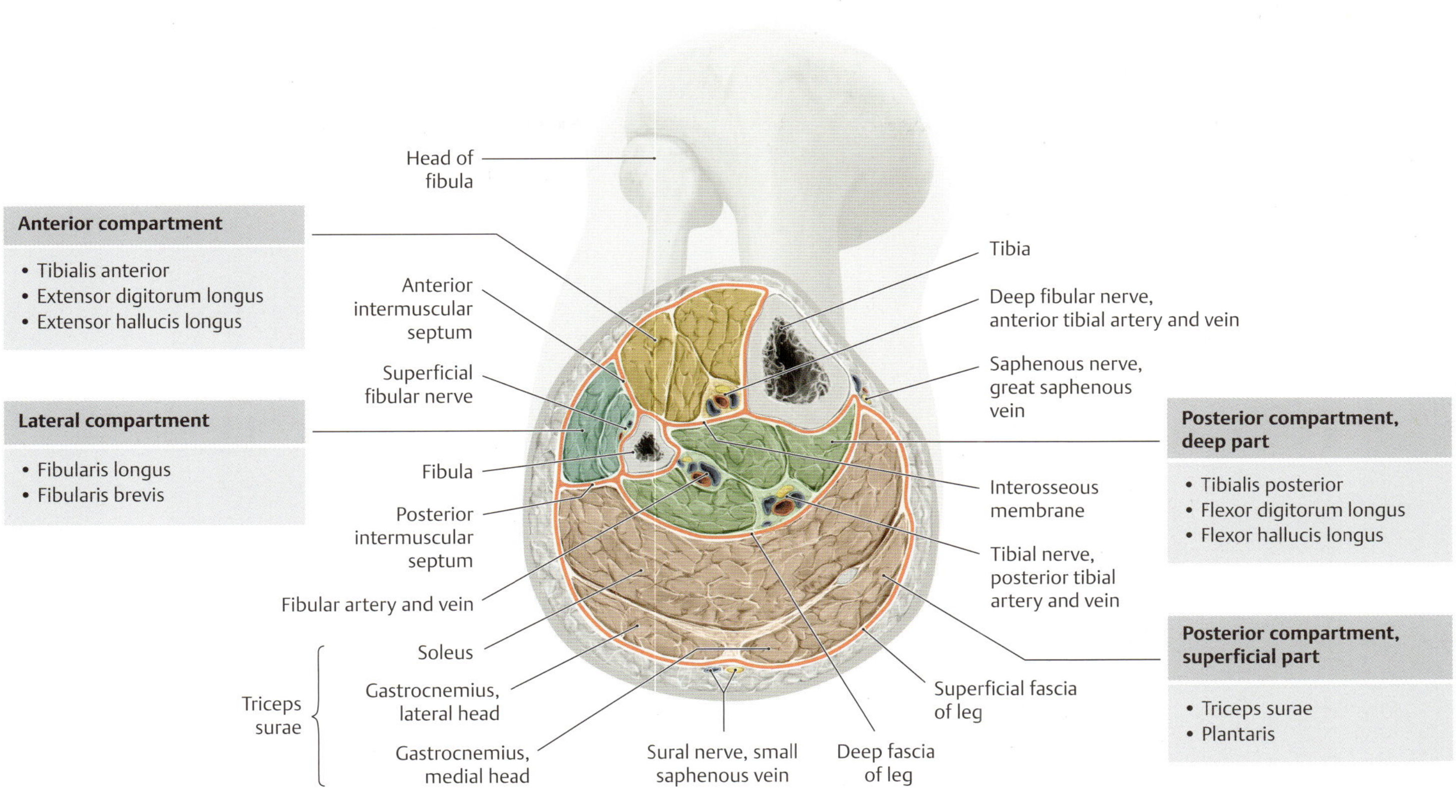

C The compartments and neurovascular structures in the leg
Cross section through a right leg one handwidth below the neck of the fibula, distal view (the level of the section is shown in **A**). The intermuscular septa and interosseous membrane, together with the superficial and deep layers of the fascia of the leg, define the boundaries of four distinct, poorly distensible fibro-osseous compartments in which the neurovascular structures descend through the leg. A rise in tissue pressure, which may result from conditions such as muscular edema or a fracture hematoma, can lead to neurovascular compression, inducing a local ischemia that can cause irreversible neuromuscular damage within a few hours (compartment syndromes such as the tibialis anterior syndrome). At greatest risk are the neurovascular structures of the deep posterior compartment (the posterior tibial artery and veins and the tibial nerve) and the anterior compartment (the anterior tibial artery and veins and the deep fibular nerve). Tibialis anterior syndrome is characterized in its acute stage by severe pain and an inability to dorsiflex the toes due to the unopposed action of the plantar flexors. This causes the toes to "claw up." Generally, the only effective treatment option at this stage is emergency incision of the fascia of the leg. This immediately decompresses the compartment and relieves the pressure on the vessels that supply the muscles.

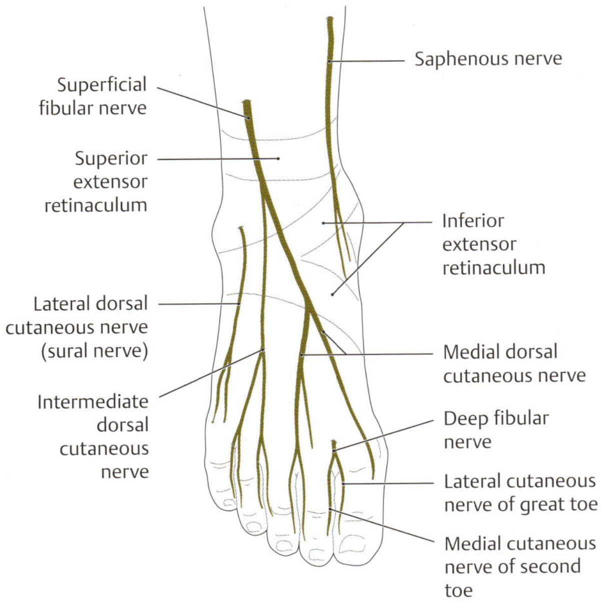

D The cutaneous nerves on the dorsum of the foot
Right foot, dorsal view.

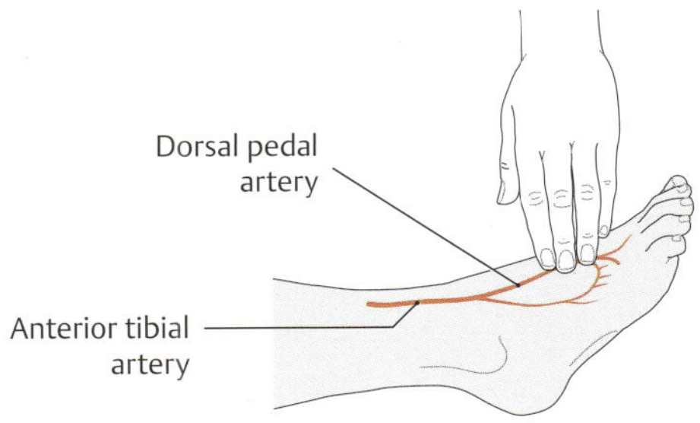

E Palpation of the pedal pulse
The dorsal pedal artery is palpable at the level of the navicular tuberosity, just lateral to the extensor hallucis longus tendon. In addition to determining regional skin temperature, checking the pedal pulse is an important step in the examination of patients with suspected lower limb arterial disease (one foot is markedly colder or paler than the other due to diminished blood flow). It is generally best to begin by palpating the femoral artery at the groin crease, then proceed distally to the popliteal fossa (popliteal artery), the medial malleolus (posterior tibial artery), and finally to the dorsum of the foot (dorsal pedal artery, which is the terminal branch of the anterior tibial artery). The palpable pulses should always be compared between the right and left sides. It should be noted that the pedal pulses may be difficult or impossible to palpate when peripheral edema is present, so it is best to examine the supine patient.

5.13 The Arteries of the Dorsum of the Foot

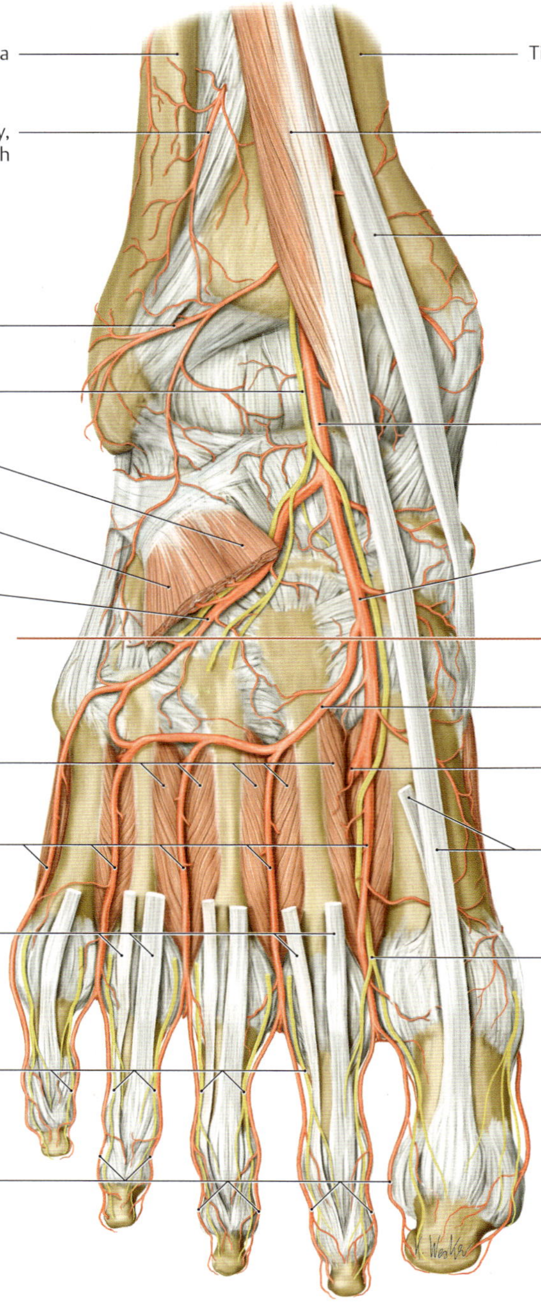

A The dorsal arteries and nerves of the foot
Right foot in plantar flexion, dorsal view.
The skin, subcutaneous tissue, and superficial and deep layers of the dorsal pedal fascia have been removed for clarity, along with the extensor digitorum longus tendons and the extensor digitorum brevis and extensor hallucis brevis muscles. Possible variants of the arteries are shown in **D**.

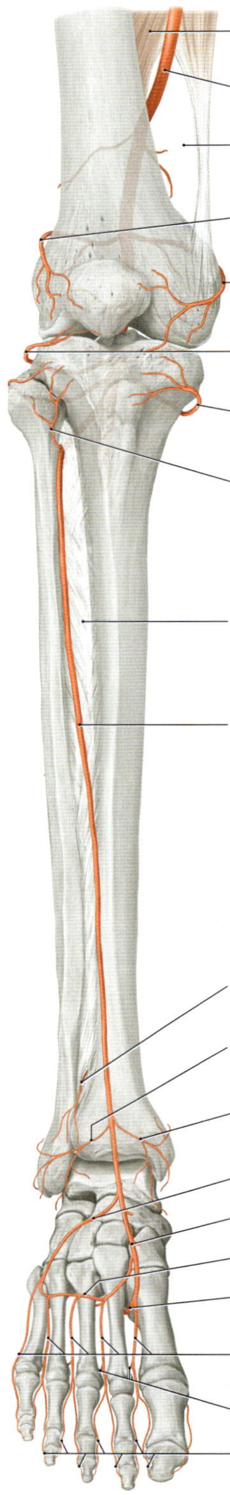

B The arteries of the leg and foot
Right lower limb, anterior view.
Note: The dorsum of the foot is supplied mainly by branches of the anterior tibial artery.

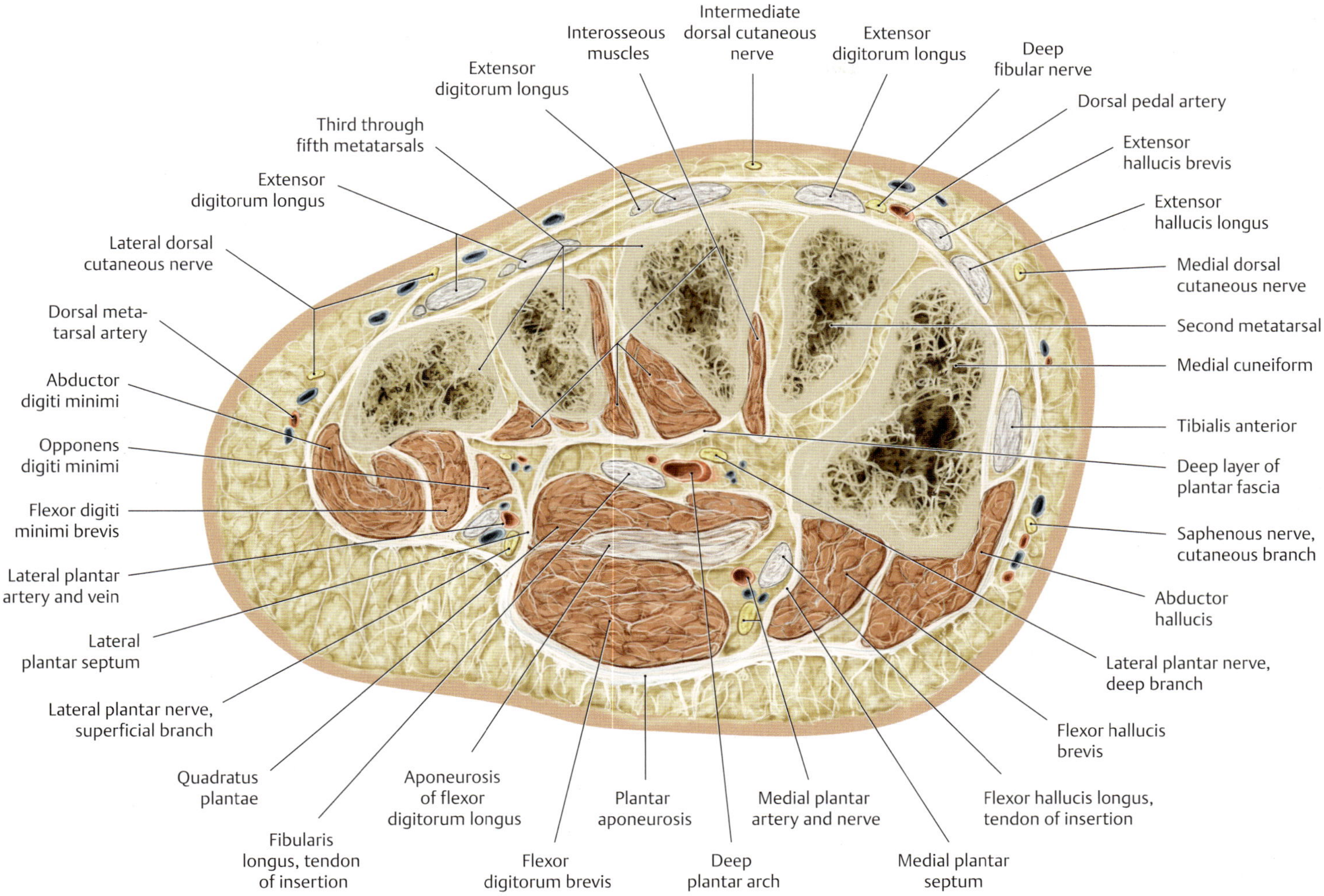

C The neurovascular structures in the sole of the foot
Cross section through the right foot at the level of the medial cuneiform bone (the location of the section is shown in **A**), distal view.

Note the deep layer of the plantar fascia, in which the deep neurovascular structures of the sole (the deep plantar arch and deep branch of the lateral plantar nerve) are embedded in connective tissue that cushions and protects them (for the arrangement of the pedal compartments, see p. 541).

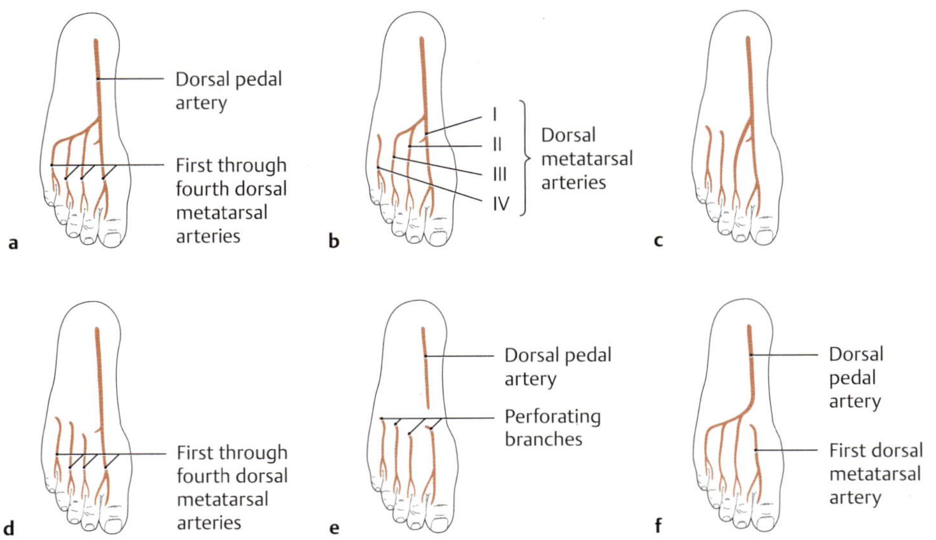

D Variants in the dorsal arterial supply of the foot (after Lippert and Pabst)

- **a** All of the dorsal metatarsal arteries arise from the dorsal pedal artery (20% of cases).
- **b** The fourth dorsal metatarsal artery is supplied by a perforating branch from the plantar side of the foot (6% of cases).
- **c** The third and fourth dorsal metatarsal arteries are supplied by perforating branches from the plantar metatarsal arteries (5% of cases).
- **d** The first dorsal metatarsal artery is the only branch of the dorsal pedal artery (40% of cases).
- **e** All of the dorsal metatarsal arteries are supplied by perforating branches from the plantar metatarsal arteries (10%).
- **f** Only the first dorsal metatarsal artery is supplied by a perforating branch (5%).

Appendix

References

Agur AMR. Grants Anatomie. Lehrbuch und Atlas. Stuttgart: Enke; 1999

Bähr M, Frotscher M. Neurologisch-topische Diagnostik. 10th ed. Stuttgart: Thieme; 2014
Baskin L, Shena J, Sinclair A et al. Development of the Human Penis and Clitoris. Differentiation 2018; 103: 74–85
Battermann N et al. Int J Sports Med 2011; 32: 211–215
Baumgartl F. Das Kniegelenk. Berlin: Springer; 1969
Bohndorf K, Imhof H, Fischer W, Hrsg. Radiologische Diagnostik der Knochen und Gelenke. 2nd ed. Stuttgart: Thieme; 2006
Bohndorf K, Imhof H, Wörtler K, Hrsg. Radiologische Diagnostik der Knochen und Gelenke. 4th ed. Stuttgart: Thieme; 2017
Böhni U, Lauper M, Locher H, Hrsg. Manuelle Medizin 2. 2nd ed. Stuttgart: Thieme; 2020
Buckup K. Klinische Tests an Knochen, Gelenken und Muskeln. 5th ed. Stuttgart: Thieme; 2012

Chassard J, Lapiné C. Étude radiologique de l'arcade pubienne chez la femme enceinte. J Radiol Electrol 7 (1923), 113
Christ B, Wachtler F. Medizinische Embryologie. Wiesbaden: Ullstein Medical; 1998

Dauber W. Bild-Lexikon der Anatomie. 10th ed. Stuttgart: Thieme; 2008
Debrunner AM. Orthopädie. Die Störungen des Bewegungsapparates in Klinik und Praxis. 2nd ed. Stuttgart: Hans Huber; 1985
Debrunner HU. Gelenkmessung (Neutral-0-Methode), Längenmessung, Umfangmessung. Bern: AO-Bulletin; 1971
Di Marino V, Lepidi H. Anatomic Study of the Clitoris and Bulbo-Clitoral Organ. Schweiz: Springer International Publishing; 2014
Drews U. Taschenatlas der Embryologie. Stuttgart: Thieme; 1993

Echtermeyer V, Bartsch S. Praxisbuch Schulter. 2nd revised ed. Stuttgart: Thieme; 2004
Efinger K, Kildal D, Hrsg. Bildgebende Diagnostik beim Polytrauma. Stuttgart: Thieme; 2019

Faller A. Anatomie in Stichworten. Stuttgart: Enke; 1980
Ficat P. Pathologie Fémoro-Patellaire. Paris: Masson; 1970
Földi M, Földi E, Kubik S. Lehrbuch Lymphologie. 7th ed. Stuttgart: Urban & Fischer, Elsevier; 2010
Frick H, Leonhardt H, Starck D. Allgemeine und spezielle Anatomie. Taschenlehrbuch der gesamten Anatomie, Bd. 1 u. 2. 4th ed. Stuttgart: Thieme; 1992
Fritsch H, Kühnel W. Taschenatlas der Anatomie. Bd. 2. 11th ed. Stuttgart: Thieme; 2013

Gertz SD, Liebman M. Basiswissen Neuroanatomie. 4th ed. Stuttgart: Thieme; 2003
Goerke K. Taschenatlas der Geburtshilfe. 2nd ed. Stuttgart: Thieme; 2006
Graf, R. Fundamentals of sonografic diagnosis in infant hip dysplasia. J Pediatr Orthop 1984, 4, 735–740
Graumann W, Sasse D. CompactLehrbuch Anatomie. Bd. 4. Stuttgart: Schattauer; 2005

Haag-Wackernagel D. 2021a. Die Klitoris – das zentrale Organ der weiblichen Lust. Teil 1: Entdeckt, ignoriert und verleugnet – die erstaunliche Geschichte des Bulbo-Klitoralorgans. FRAUENARZT 62 (6), 402–407.
Haag-Wackernagel D. 2021b. Die Klitoris – das zentrale Organ der weiblichen Lust. Teil 2: Bau und Funktion der äußeren weiblichen Genitalien. FRAUENARZT 63 (7), 484–489
Haag-Wackernagel D. 2022. Sensorische Nervenendigungen – der Schlüssel zur weiblichen Lust. Sexuologie 29 (1-2), 5–20
Hansen K, Schliack K. Segmentale Innervation. 2nd ed. Stuttgart: Thieme; 1962
Hees H. Grundriss und Atlas der Mikroskopischen Anatomie des Menschen. Bd. 1. Zytologie und Allgemeine Histologie. 12th ed. Stuttgart: Gustav Fischer; 1996
Henne-Bruns D. Duale Reihe Chirurgie. 4th ed. Stuttgart: Thieme; 2012
Hepp WR. Radiologie des Femoro-Patellargelenkes. Bücherei des Orthopäden. Bd. 37. Stuttgart: Enke; 1983
Hilgenreiner H. Zur Frühdiagnose der angeborenen Hüftgelenksverrenkung. Med Klein 21 1925. Stuttgart: Hippokrates; 1981
Hochschild J. Strukturen und Funktionen begreifen. Bd.1. 4th ed. and Bd 2. 3rd ed. Stuttgart: Thieme; 2014 and 2012
von Hochstetter A, von Rechenberg HK, Schmidt R. Die intragluteale Injektion. Stuttgart: Thieme; 1958
Hüter-Becker A, Schewe H, Heipertz W. Physiotherapie. Bd. 1. Biomechanik, Arbeitsmedizin, Ergonomie. Stuttgart: Thieme; 1999

Junghanns H. Die funktionelle Pathologie der Zwischenwirbelscheibe als Grundlage für klinische Betrachtungen. Langenbecks. Arch Klin Chir 1951; 267: 393–417

Kahle W, Frotscher M. Taschenatlas der Anatomie. Bd. 1. 11th ed. Stuttgart: Thieme; 2013
Kapandji AI. Funktionelle Anatomie der Gelenke. 6th ed. Stuttgart: Thieme; 2016
Kaufmann P. Reife Plazenta. In: Becker V, Schiebler TH, Kubli F, Hrsg. Die Plazenta des Menschen. Stuttgart: Thieme; 1981
Kilka HG, Geiger P, Mehrkens HH. Die vertikale infraklavikuläre Blockade des Plexus brachialis. Anästhesist 1995; 44: 339–344
Kobelt GL. Die männlichen und weiblichen Wollustorgane des Menschen und einiger Säugethiere in anatomisch-pyhsiologischer Beziehung. Freiburg im Breisgau: Druck und Verlag von Adolph Emmerling; 1844
Koebke J. Anatomie des Handgelenkes und der Handwurzel. Unfallchirurgie 1988; 14: 74–79
Konermann W, Gruber G. Ultraschalldiagnostik der Bewegungsorgane. Kursbuch nach den Richtlinien der DEGUM und der DGOOC, Buch und DVD. 3rd revised ed. Stuttgart: Thieme; 2011
Kristic RV. General Histology of the Mammals. Berlin: Springer; 1985
Kubik S. Lymphsystem der oberen Extremität. In: Földi M, Kubik S, Hrsg. Lehrbuch der Lymphologie für Mediziner und Physiotherapeuten. Stuttgart: Gustav Fischer; 1989
Kummer B. Biomechanik der Wirbelgelenke. In: Meinicke FW. Die Wirbelbogengelenke. Stuttgart: Hippokrates; 1983

von Lanz T, Wachsmuth W. Praktische Anatomie. Bd. I/3 Arm. 2nd ed. Berlin: Springer; 1959
von Lanz T, Wachsmuth W. Praktische Anatomie. Bd. I/4 Bein und Statik. Berlin: Springer; 1972
Lehnert G. Dopplersonographische Diagnostik der erektilen Dysfunktion unter Anwendung des Papaverintests [Dissertation] Kiel: Universität Kiel, Medizinische Fakultät; 1995
Lelièvre J. Pathologie du Pied. 2nd ed. Paris: Masson; 1961
Lippert H, Pabst R. Arterial Variations in Man. München: Bergmann; 1985

Loeweneck H. Diagnostische Anatomie. Berlin: Springer; 1981

Lüllmann-Rauch R. Taschenlehrbuch Histologie. 5th ed. Stuttgart: Thieme; 2015

Lundborg G, Myrhage R, Rydevik B. The vascularization of human flexor tendons within the digital synovial sheath region – structural and functional aspects. J Hand Surg 1977; 2: 417–427

Luschka H. Die Halbgelenke des menschlichen Körpers. Berlin: Reiner; 1858

Masuhr KF, Neumann M. Neurologie. Duale Reihe. 7th ed. Stuttgart: Thieme; 2013

Matzen P. Praktische Orthopädie. 3rd completely revised and updated ed. Stuttgart: J. A. Barth Verlag in Thieme Verlag; 2002

Meier G, Bauereis C, Maurer H, Meier Th. Interscalenäre Plexus-blockade. Anästhesist 2001; 50: 333–341

Mense S. Muskeln, Faszien und Schmerz. Stuttgart: Thieme; 2021

Merk H, Jerosch J, Hrsg. Arthroskopie des Schultergelenks. Stuttgart: Thieme; 2000

Möller TB, Reif E. Taschenatlas der Röntgenanatomie. 6th ed. Stuttgart: Thieme; 2016

Möller TB, Reif E. Taschenatlas der Schnittbildanatomie. Bd. 3: Extremitäten, Gelenke, Wirbelsäule. Stuttgart: Thieme; 2007

Mow VC, Hou JS, Owens JM. Biphasic and quasilinear viscoelastic theories for hydrated soft tissue. In: Mow JC, Ratcliffe A, Woo SL. Biomechanics of Diarthrodial Joints. Springer: New York; 1990. Vol. I: 215–260

Mubarak SJ, Hargens AR. Compartment Syndromes and Volkamn's Contracture. Philadelphia: W. B. Saunders; 1981

Müller L, Hollinger B, Burkhart K, Hrsg. Expertise Ellenbogen. Stuttgart: Thieme; 2016

Müller-Vahl H, Mumenthaler M, Stöhr M. Läsionen peripherer Nerven und radikuläre Syndrome. 10th ed. Stuttgart: Thieme; 2014

Netter FH. Farbatlanten der Medizin. Stuttgart: Thieme; 2000

Niethard FU, Pfeil J. Orthopädie. Duale Reihe. 8th ed. Stuttgart: Thieme; 2017

Niethard FU. Kinderorthopädie. 2nd ed. Stuttgart: Thieme; 2009

Noback CR, Strominger NL, Demarest RJ. The Human Nervous System. 4th ed. Philadelphia: Lea & Febiger; 1991

O'Dey DM, Bozkurt A, Pallua N. The anterior Obturator Artery Perforator (aOAP) flap: surgical anatomy and application of a method for vulvar reconstruction. Gynecol Oncol. 2010. 119 (3): 526–30

O'Dey DM. Anatomisch-funktionelle Rekonstruktion des weiblichen Genitales nach ritueller Beschneidung. Journal für Ästhetische Chirurgie 2015; 8(4)

O'Dey DM. Complex reconstruction of the vulva following female genital mutilation/cutting. Der Urologe 2017; 56(10): 1298–1301

O'Dey DM. Vulvar Reconstruction following female genital Mutilation/Cutting (FGM/C) and other acquired Deformaties. Cham: Springer Nature Switzerland; 2019

O'Rahilly, Müller RF. Developmental Stages in Human Embryos. Carnegie Institution of Washington: Publication 637; 1987

Pape HC, Kurtz A, Silbernagl S. Physiologie. 7th ed. Stuttgart: Thieme; 2014

Pauwels F. Eine neue Theorie über den Einfluss mechanischer Reize auf die Differenzierung der Stützgewebe (X). Beitrag zur funktionellen Anatomie und kausalen Morphologie des Stützapparates. Z Anat Entwickl Gesch 1968; 121: 478–515

Pauwels F. Atlas zur Biomechanik der gesunden und kranken Hüfte: Prinzipien, Technik und Resultate einer kausalen Therapie. Heidelberg: Springer; 1973

Petersen W, Tillmann B. Structure and vascularization of the cruciate ligaments of the human knee joint. Z Orthop 1999; 137: 31–37

Petersen W, Zantop T. Das vordere Kreuzband – Grundlagen und aktuelle Praxis der operativen Therapie. Köln: Deutscher Ärzte-Verlag; 2009

Pette D. Das adaptive Potential des Skelettmuskels. Dtsch Z Sportmed 1999; 50: 262–271

Pette D, Staron RS. Transitions of muscle fiber phenotypic profiles. Histochem Cell Biol 2001; 115 (5): 359–379

Platzer W. Taschenatlas der Anatomie. Bd. 1. 11th ed. Stuttgart: Thieme; 2013

Platzer W. Atlas der topographischen Anatomie. Stuttgart: Thieme; 1982

Rauber A, Kopsch F. Anatomie des Menschen. Bd. 1–4. Stuttgart: Thieme; Bd. 1. 2nd ed. 1997; Bde. 2 and 3 1987; Bd. 4 1988

Reiser M, Kuhn FP, Debus J. Radiologie. Duale Reihe. 2nd ed. Stuttgart: Thieme; 2006

Rockwood CA. Subluxations and dislocations about the shoulder. In: Rockwood CA, Green DP (eds). Fractures. Philadelphia: Lippincott; 1984: 722–985

Rohen, JW. Topographische Anatomie. 10th ed. Stuttgart: Schattauer; 2007

Rohen JW, Yokochi C, Lütjen-Drecoll E. Anatomie. 8th ed. Stuttgart: Schattauer; 2015

Romer AS, Parson TS. Vergleichende Anatomie der Wirbeltiere. 5th ed. Hamburg und Berlin: Paul Parey; 1983

Rudigier J. Kurzgefasste Handchirurgie. 6th revised ed. Stuttgart: Thieme; 2014

Scheldrup, EW. Tendon sheath patterns in the hand. Surg Gynec-Obestetr 1951; 93: 16–22

Schmidt HM, Lanz U. Chirurgische Anatomie der Hand. 2nd ed. Stuttgart: Thieme; 2003

Schünke M. Funktionelle Anatomie – Topographie und Funktion des Bewegungssystems. 2nd ed. Stuttgart: Thieme; 2014

Schumpelick V. Hernien. 5th ed. Stuttgart: Thieme; 2015

Silbernagl S, Despopoulos A. Taschenatlas der Physiologie. 8th ed. Stuttgart: Thieme; 2012

Sökeland J, Rübben H. Taschenlehrbuch Urologie. 14th ed. Stuttgart: Thieme; 2007

Starck D. Embryologie. 3rd ed. Stuttgart: Thieme; 1975

Stäbler A, Ertl-Wagner B, Hrsg. Radiologie-Trainer: Bewegungsapparat. 4th ed. Stuttgart: Thieme; 2022

Streater GL. Developmental horizons in human embryos: age group XI, 13–20 somites and age group XII, 21–29 somites. Contrib Embryol 1942; 30: 211–245

Tossy JD, Newton CM, Sigmond HM. Acromioclavicular separations: useful and practical classification for treatment. Clin Orthop 1963; 28: 111–119

Uhthoff HK. The Embryology of the Human Locomotor System. Berlin: Springer; 1990

Vahlensieck M, Reiser M. MRT des Bewegungsapparates. 4th ed. Stuttgart: Thieme; 2014

Vega J, Francesc M et al. The lateral fibulotalocalcaneal ligament complex: an ankle stabilizing isometric structure. Knee Surg Sports Traumatol Arthrosc. 2020; 28(1): 8–17

Weber U, Greulich M, Sparmann M. Orthopädische Mikrochirurgie. Stuttgart: Thieme; 1993

Wiberg G. Roentgenographic and anatomic studies on the femoropatellar joint. With special reference to chondromalacia patellae. Acta Orthop Scand 1941; 12: 319–410

Wiberg G. Studies on dysplastic acetabulum and congenital subluxation of the hip joint with special reference of the complication of osteoarthritis. Acta Chir Scand 1939; 83 (Suppl. 58)

Wolpert L, Beddington R, Brockes J et al. Entwicklungsbiologie. Weinheim: Spektrum; 1999

Index

Note: Imaging, regional anatomy, and surface anatomy are found under these main headings.

C

I

J